Applied Regression Analysis and Other Multivariable Methods

Duxbury Titles of Related Interest

To order copies, contact your local bookstore or call 1-800-354-9706. For more information, contact Duxbury Press at 10 Davis Drive, Belmont, CA 94002, or go to www.duxbury.com.

DUXBURY

Applied Regression Analysis and Other Multivariable Methods

Fourth Edition

David G. Kleinbaum
Emory University

Lawrence L. Kupper
University of North Carolina, Chapel Hill

Azhar Nizam
Emory University

Keith E. Muller
University of Florida

BROOKS/COLE
CENGAGE Learning‍

Australia • Brazil • Japan • Korea • Mexico • Singapore • Spain • United Kingdom • United States

BROOKS/COLE
CENGAGE Learning™

Applied Regression Analysis and Other Multivariable Methods, Fourth Edition

David G. Kleinbaum, Lawrence L. Kupper, Azhar Nizam and Keith E. Muller

Acquisitions Editor: Carolyn Crockett

Assistant Editor: Natasha Coats

Editorial Assistant: Ashley Summers

Managing Marketing Manager: Mandy Jellerichs

Marketing Assistant: Ashley Pickering

Marketing Communications Manager: Jessica Perry

Production Project Manager, Editorial Production: Janet Hill

Creative Director: Rob Hugel

Art Director: Vernon Boes

Print Buyer: Doreen Suruki

Production Service: Matrix Productions

Art Editor: Merrill Peterson

Copy Editor: Ellen Brownstein

Illustrator: Lori Heckelman

Cover Designer: Denise Davidson

Cover Image: Ellen Cary, Getty Images/ Phototonica

Compositor: International Typesetting & Composition

For product information and technology assistance, contact us at **Cengage Learning Customer & Sales Support, 1-800-354-9706**

For permission to use material from this text or product, submit all requests online at **www.cengage.com/permissions**
Further permissions questions can be e-mailed to **permissionrequest@cengage.com**

Library of Congress Control Number: 2006940618

Student Edition:
ISBN-13: 978-0-495-38496-0
ISBN-10: 0-495-38496-8

Brooks/Cole Cengage Learning
20 Davis Drive
Belmont, CA 94002-3098
USA

Cengage Learning is a leading provider of customized learning solutions with office locations around the globe, including Singapore, the United Kingdom, Australia, Mexico, Brazil, and Japan. Locate your local office at **www.cengage.com/global**

Cengage Learning products are represented in Canada by Nelson Education, Ltd.

To learn more about Brooks/Cole, visit **www.cengage.com/brookscole**

Purchase any of our products at your local college store or at our preferred online store **www.cengagebrain.com**

Printed in the United States of America
2 3 4 5 14 13 12 11

David Kleinbaum:
To my wife Edna, my daughter Toby, my son-in-law Jeff, and my granddaughters Emma and Anna.

Larry Kupper:
To Dr. Bill Mendenhall (my favorite role model), to Mark and Chieko (my wonderful son and daughter-in-law), and to Sandy (the light of my life).

Azhar Nizam:
To the family that inspires me: my loving and wonderful wife, Janet, and children, Zainab and Sohail.

All authors:
To all teachers who have never received deserved recognition for the long hours of course preparation, development of innovative course materials and textbooks, mentoring of students, and devotion to successful and creative teaching, all of which is often done in lieu of more lucrative salaries and related recognition for obtaining large grants and publishing research articles.

Preface

This is the third revision of our second-level statistics text, originally published in 1978 and revised in 1987 and 1998. As in previous versions, this text is intended primarily for advanced undergraduates, graduate students, and working professionals in the health, social, biological, and behavioral sciences who engage in applied research in their fields. The text may also provide professional statisticians with some new insights into the application of advanced statistical techniques to realistic research problems.

We have attempted in this revision to retain the basic structure and flavor of the earlier editions, while at the same time making changes to keep pace with current analytic practices and computer usage in applied research. Notable changes in this fourth edition, discussed in more detail later, include:

1. Reorganization of some topics (Chapters 12–15, 21–23); note that Chapters 1–11, 16–20 and 24 are unchanged from the previous ordering;

2. Clarification of content and/or terminology as suggested by reviewers and readers (e.g., sums of squares notation in Chapters 7–9, interaction versus effect modification in Chapter 11, multiple-comparison methods in Chapter 17);

3. Expanded and updated coverage of some content areas (regression diagnostics in newly renumbered Chapter 14, logistic regression in newly renumbered Chapter 22, Poisson regression in Chapter 24);

4. Entire revision of some chapters (newly renumbered Chapter 21 on maximum likelihood [ML] estimation, and newly numbered Chapters 25 and 26 on analysis of correlated data that replace previously numbered Chapter 21 on repeated measures analysis);

5. New chapters on Polytomous and Ordinal Logistic Regression (Chapter 23) and Sample Size Determination (Chapter 27);

6. Some new problems for the reader;

7. Updated SAS computer output;

8. A publisher's website containing links to freely downloadable datasets; a Computer Appendix on the use of SAS, STATA, and SPSS packages to carry out regression modeling; updates on errata; ordering information; and other information about the material contained in the fourth edition.

In this fourth edition, as in our previous editions, we emphasize the intuitive logic and assumptions that underlie the techniques covered, the purposes for which these techniques are designed, the advantages and disadvantages of the techniques, and valid interpretations based on the techniques. Although we describe the statistical calculations required for the techniques we cover, we rely on computer output to provide the results of such calculations, so the reader can concentrate on how to apply a given technique rather than how to carry out the calculations. The mathematical formulas that we do present require no more than simple algebraic manipulations. Proofs are of secondary importance and are generally omitted. Calculus is not explicitly used anywhere in the main text. We introduce matrix notation to a limited extent in Chapters 25 and 26 because we believe that the use of matrices provides a more convenient way to understand some of the complicated mathematical aspects of the analysis of correlated data. We also have continued to include an appendix on matrices for the interested reader.

This edition, as with the previous editions, is *not* intended to be a general reference work dealing with all the statistical techniques available for analyzing data involving several variables. Instead we focus on the techniques we consider most essential for use in applied research. We want the reader to understand the concepts and assumptions involved in these techniques and how the techniques can be applied in practice, including how computer packages can help make it easier to perform data analyses.

The most notable features of this fourth edition, including the material that has not been modified from the previous edition, are the following:

1. Regression analysis (Chapters 1–16) and analysis of variance (Chapters 17–20) are discussed in considerable detail and with pedagogical care that reflects the authors' extensive experience and insight as teachers of such material.

2. The relationship between regression analysis and analysis of variance is highlighted.

3. The connection between multiple regression analysis and multiple and partial correlation analysis is discussed in detail.

4. Several advanced topics are presented in a unique, non-mathematical manner, including a revised ML methods chapter (21), an updated logistic regression chapter (22), a new chapter on polytomous and ordinal logistic regression (23), a Poisson regression chapter (24) with an expanded introduction on the Poisson distribution, and two new chapters (25–26) on the analysis of correlated data (described further below). The expansion of the material on ML methods from three chapters in the previous edition to six chapters in the fourth edition provides a strong foundation for understanding why ML estimation is the most widely used method for fitting mathematical models involving several variables.

5. An up-to-date discussion of the issues and procedures involved in fine-tuning a regression analysis is presented on confounding and interaction in regression (Chapter 11), selecting the best regression model (Chapter 16), and regression diagnostics (new Chapter 14). Chapters 12 through 15 in this fourth edition have been reordered from the third edition to provide a more logical presentation of the content of these

chapters. The new ordering is as follows: dummy variables (12), analysis of covariance (13), regression diagnostics (14), and polynomial regression (15).

6. A new chapter (23) has been added on polytomous and ordinal logistic regression methods. This chapter extends the standard (binary) logistic model to outcome variables that have more than two categories. Polytomous logistic regression is used when the outcome categories do not have any natural order, whereas ordinal logistic regression is appropriate when the outcome categories have a natural order.

7. Two new chapters (25–26) on the analysis of correlated data replace the chapter (21) in the third edition on repeated measures analysis. These new chapters describe the ML/REML linear mixed model approach incorporated into SAS's MIXED procedure. Since ML estimation is assumed, these chapters are logically ordered after the current chapter (21) on ML estimation. In Chapter 25 we describe the general form of the linear mixed model, introduce the terms "correlation structure" and "robust/empirical standard errors," and illustrate how to model correlated data when only fixed effects are considered. In Chapter 26, which serves as part 2 of this topic, we focus on linear mixed models that contain random effects. Chapter 26 also provides a link to ANOVA Chapters 17–20, alternatively formulating the linear mixed model approach in terms of an ANOVA that partitions sources of variation from various predictors into sums of squares and corresponding mean square terms of a summary ANOVA table.

8. A new chapter (27) on sample size determination for linear and logistic regression models has also been added as the final chapter in the text. This chapter begins with a review of basic concepts and formulae relevant to sample size planning. Two approaches for sample size calculation are then described. The first is an approximate, but simple, approach that has been shown to yield fairly accurate sample sizes and for which even manual computation is feasible. The second is based on more traditional theory for sample size determination; it is best implemented using computer software and is illustrated using popular programs from the SAS and PASS packages.

9. Representative computer results from packaged programs (primarily using the SAS package) are used to illustrate concepts in the body of the text, as well as to provide a basis for problems for the reader. In this edition, we have revised the computer output to reflect the most recent version of SAS, and, in many instances, have annotated comments on the output so that it is easier to read.

10. Numerous examples and problems illustrate applications to real studies in a wide variety of disciplines. Problems are included in each new chapter (23, 25, 26, 27) and new problems have been added to several chapters.

11. As in previous editions, answers to selected problems are provided in Appendix C. There also is an *Instructor's Solutions Manual*, containing solutions to all problems, that will be completed within a few months after publication of this fourth edition. A course instructor or individual learner not enrolled in a course can obtain an electronic copy of the *Instructor's Solutions Manual* by contacting the publisher. A *Student's Solutions Manual* containing complete solutions to selected problems will also be available on the publisher's website.

12. A publisher's website for the fourth edition of this text has been developed to provide a variety of information for both instructors and readers of the text. The website contains links to freely downloadable datasets; a Computer Appendix on the use of

SAS, STATA, and SPSS packages to carry out linear regression modeling; updates on errata; ordering information; and other information. In the previous edition, the datasets for most problems were provided on a computer disk that was bound into a copy of each book. Such a disk is not provided for the fourth edition because the datasets can be obtained from website links instead.

13. The Computer Appendix mentioned in item 12 will be a freely downloadable electronic document providing computer guidelines for multiple linear regression models. (Other textbooks by Kleinbaum and Klein have computer appendices for SAS, STATA, and SPSS use with logistic models and Cox proportional hazards and extended Cox models for survival data.) The Computer Appendix will provide a quick and easy reference guide to help readers avoid spending considerable time finding information from more complicated help guides in packages like SAS.

Suggestions for Instructors or Individual Learners

For formal classroom instruction or individual/distance learning, the chapters fall naturally into four clusters:

- Course 1: Chapters 4–16 on multiple linear regression analysis;
- Course 2: Chapter 17–20 on the analysis of variance;
- Course 3: Chapters 21–24 on maximum likelihood methods and important applications involving logistic and Poisson regression modeling;
- Course 4: Chapters 25–26 on the analysis of correlated data involving linear mixed models.

Portions of Chapter 27 on sample size determination could be added, as appropriate, to any of Courses 1–3 above. Courses 1 and 2 have often been combined into one course on regression and ANOVA methods. For a first course in regression analysis, some of Chapters 11–16 may be considered too specialized. For example, Chapter 15 on selecting the best regression model and Chapter 16 on regression diagnostics might be used in a continuation course on regression modeling, which might also include some of the advanced topics covered in Chapters 21–27.

Acknowledgments

We wish to acknowledge several people who contributed to the development of this text, including early editions as well as this fourth edition. Drs. Kleinbaum and Kupper continue to be indebted to John Cassel and Bernard Greenberg, two mentors who have provided us with inspiration and the professional and administrative guidance that enabled us at the beginning of our careers to gain the broad experience necessary to write this text.

Dr. Kleinbaum also wishes to thank John Boring, former Chair of the Epidemiology Department at Emory University, for his strong support and encouragement during the writing of the third and fourth editions and for his deep commitment to teaching excellence. Dr. Kleinbaum also wishes to thank Dr. Mitch Klein of Emory's Department of Epidemiology for his colleagueship, including thoughtful suggestions and review of chapter drafts for this fourth edition.

Dr. Kupper will be forever grateful to Dr. William Mendenhall, who gave him the opportunity, and inspired him, to study the discipline of statistics.

Mr. Nizam wishes to thank Michael Kutner, Chair of the Biostatistics Department at Emory University, for his strong support during the writing of this revision.

We all wish to thank Edna Kleinbaum, Sandy Martin, Janet Nizam, and Sally Muller for their encouragement and support during the writing of various revisions.

We thank our reviewers of the third and fourth editions for their helpful suggestions:

Donald Hedeker, Univeristy of Illinois at Chicago
Vinaya Kelkar, University of North Carolina, Greensboro
Brian Marx, Louisiana State University
E. John Orav, Harvard School of Public Health
Kathy Prewitt, Arizona State University
Pali Sen, University of North Florida
Donald J. Slymen, San Diego State University
Jessica Thomson, Louisiana State University Health Sciences Center

We thank the Thomson Duxbury Brooks/Cole Statistics and Applied Mathematics book team, especially acquisitions editor Carolyn J. Crockett for guiding us through the publication process for the fourth edition, as well as assistant editor Natasha Cole, editorial assistant Ashley Summers, and production project manager Janet Hill. Finally, thanks to our production project editor, Merrill Peterson at Matrix Productions, whose extensive expertise in book production saw us through to publication.

David G. Kleinbaum
Lawrence L. Kupper
Azhar Nizam
Keith E. Muller

Contents

A   APPENDIX A—TABLES   819

B   APPENDIX B—MATRICES AND THEIR RELATIONSHIP TO REGRESSION ANALYSIS   841

C   APPENDIX C—ANSWERS TO SELECTED PROBLEMS   853

1

Concepts and Examples of Research

1.1 Concepts

The purpose of most research is to assess relationships among a set of variables. *Multivariable*[1] *techniques* are concerned with the statistical analysis of such relationships, particularly when at least three variables are involved. Regression analysis, our primary focus, is one type of multivariable technique. Other techniques will also be described in this text. Choosing an appropriate technique depends on the purpose of the research and on the types of variables under investigation (a subject discussed in Chapter 2).

Research may be classified broadly into three types: *experimental, quasi-experimental,* or *observational.* Multivariable techniques are applicable to all such types, yet the confidence one may reasonably have in the results of a study can vary with the research type. In most types, one variable is usually taken to be a *response* or *dependent variable*—that is, a variable to be predicted from other variables. The other variables are called *predictor* or *independent variables.*

If observational units (subjects) are randomly assigned to levels of important predictors, the study is usually classified as an *experiment.* Experiments are the most controlled type of study; they maximize the investigator's ability to isolate the observed effect of the predictors from the distorting effects of other (independent) variables that might also be related to the response.

If subjects are assigned to treatment conditions without randomization, the study is called *quasi-experimental* (Campbell and Stanley 1963). Such studies are often more feasible

---

[1] The term *multivariable* is preferable to *multivariate.* Statisticians generally use the term *multivariate analysis* to describe a method in which several dependent variables can be considered simultaneously. Researchers in the biomedical and health sciences who are not statisticians, however, use this term to describe any statistical technique involving several variables, even if only one dependent variable is considered at a time. In this text we prefer to avoid the confusion by using the term *multivariable analysis* to denote the latter, more general description.

and less expensive than experimental studies, but they provide less control over the study situation.

Finally, if all observations are obtained without either randomization or artificial manipulation (i.e., allocation) of the predictor variables, the study is said to be *observational*. Experiments offer the greatest potential for drawing definitive conclusions, and observational studies the least; however, experiments are the most difficult studies to implement, and observational studies the easiest. A researcher must consider this trade-off between interpretive potential and complexity of design when choosing among types of studies (Kleinbaum, Kupper, and Morgenstern 1982, chap. 3).

To assess a relationship between two variables, one must measure both of them in some manner. Measurement inherently and unavoidably involves error. The need for statistical design and analysis emanates from the presence of such error. Traditionally, statistical inference has been divided into two kinds: estimation and hypothesis testing. *Estimation* refers to describing (i.e., quantifying) characteristics and strengths of relationships. *Testing* refers to specifying hypotheses about relationships, making statements of probability about the reasonableness of such hypotheses, and then providing practical conclusions based on such statements.

This text focuses on regression and correlation methods involving one response variable and one or more predictor variables. In these methods, a mathematical model is specified that describes how the variables of interest are related to one another. The model must somehow be developed from study data, after which inference-making procedures (e.g., testing hypotheses and constructing confidence intervals) are conducted about important parameters of interest. Although other multivariable methods will be discussed, regression techniques are emphasized for three reasons: they have wide applicability; they can be the most straightforward to implement; and many other, more complex statistical procedures can be better appreciated once regression methods are understood.

1.2 Examples

The examples that follow concern *real* problems from a variety of disciplines and involve variables to which the methods described in this book can be applied. We shall return to these examples later when illustrating various methods of multivariable analysis.

■ **Example 1.1** *Study of the associations among the physician–patient relationship, perception of pregnancy, and the outcome of pregnancy, illustrating the use of regression analysis and logistic regression analysis.*

Thompson (1972) and Hulka and others (1971) looked at both the process and the outcomes of medical care in a cohort of 107 pregnant married women in North Carolina. The data were obtained through patient interviews, questionnaires completed by physicians, and a review of medical records. Several variables were recorded for each patient.

One research goal of primary interest was to determine what association, if any, existed between SATISfaction[2] with medical care and a number of variables meant to describe patient perception of pregnancy and the physician–patient relationship. Three perception-of-pregnancy variables measured the patient's WORRY during pregnancy, her desire (WANT) for the baby,

---

[2] Capital letters denote the abbreviated variable name.

and her concern about childBIRTH. Two other variables measured the physician–patient relationship in terms of informational communication (INFCOM) concerning prescriptions and affective communication (AFFCOM) concerning perceptions. Other variables considered were AGE, social class (SOCLS), EDUCation, and PARity.

Regression analysis was used to describe the relationship between scores measuring patient satisfaction with medical care and the preceding variables. From this analysis, variables found not to be related to SATIS could be eliminated, while those found to be associated with SATIS could be ranked in order of importance. Also, the effects of confounding variables such as AGE and SOCLS could be considered, to three ends: any associations found could not be attributed solely to such variables; measures of the strength of the relationship between SATIS and other variables could be obtained; and a functional equation predicting level of patient satisfaction in terms of the other variables found to be important in describing satisfaction could be developed.

Another question of interest in this study was whether patient perception of pregnancy and/or the physician–patient relationship was associated with COMPlications of pregnancy. A variable describing complications was defined so that the value 1 could be assigned if the patient experienced one or more complications of pregnancy and 0 if she experienced no complications. *Logistic regression analysis* was used to evaluate the relationship between complications of pregnancy and other variables. This method, like regression analysis, allows the researcher to determine and rank important variables that can distinguish between patients who have complications and patients who do not.   ∎

■ **Example 1.2**   *Study of the relationship between water hardness and sudden death, illustrating the use of multiple regression analysis.*

Hamilton's (1971) study of the effects of environmental factors on mortality used 88 North Carolina counties as the observational units—in contrast to an earlier study, which used individual patients as the observational units. Hamilton's primary goal was to determine the relationship, if any, between the sudden death rate of residents of a county and the measure of water hardness for that county. However, the researcher also wanted to compare the mortality–water hardness relationship among four regions of the state and to determine whether this relationship was affected by other variables, such as the habits of the county coroner in recording deaths, the distance from the county seat to the main hospital, the per capita income, and the population per physician.

Regression analysis was used in this study. The four regions could have been compared by doing a separate analysis for each region and then comparing the results. Alternatively, though, the comparative analysis could have been performed in a single step by defining a number of artificial, or dummy, variables to represent the regions.   ∎

■ **Example 1.3**   *Comparative study of the effects of two instructional designs for teaching statistics, illustrating the use of analysis of covariance.*

Kleinbaum and Kleinbaum (1976) used a classroom experiment to compare two approaches for teaching probability to graduate students taking an introductory course in biostatistics. A class of 52 students was randomly split into two groups stratified on the basis of the students' fields of study. Both groups were taught in a lecture format by the same instructor. One group, the control, was taught with the standard lecture method, in which an instructor (using chalk and a blackboard) lectured after a handout had been distributed. For the experimental group, 16 transparencies, systematically designed to fit a carefully defined set of objectives, were used along with a set of practice problems that were done and reviewed in class. Both groups were given the same pretests and posttests, as well as questionnaires to measure attitudes. Both received a copy of the objectives, the handout, and a homework assignment. The primary question of interest in this study was whether the instructional design for the experimental group was more effective than that for the control group, as measured by cognitive learning and attitudes.

In comparing the experimental and control groups, the researchers had to take into account pretest scores so that any differences found in posttest scores could not be attributed solely to differing levels of ability or knowledge between the two groups at the beginning of the study. One appropriate method of analysis for this situation was *analysis of covariance,* which showed that posttest scores, when adjusted for pretest scores, were significantly higher for the experimental group.   ■

■ **Example 1.4**   *Study of race and social influence in cooperative problem-solving dyads, illustrating the use of analysis of variance and analysis of covariance.*

James (1973) conducted an experiment on 140 seventh- and eighth-grade males to investigate the effects of two factors—race of the experimenter (E) and race of the comparison norm (N)—on social influence behaviors in three types of dyads: white–white; black–black; and white–black. Subjects played a game of strategy called Kill the Bull, in which 14 separate decisions must be made for proceeding toward a defined goal on a game board. In the game, each pair of players (dyad) must reach a consensus on a direction at each decision step, after which they signal the E, who then rolls a die to determine how far they can advance along their chosen path of six squares. Photographs of the current champion players (N) (either two black youths [black norm] or two white youths [white norm]) were placed above the game board.

Four measures of social influence activity were used as the outcome variables of interest. One of these, called performance output, was a measure of the number of times a given subject attempted to influence his dyad to move in a particular direction.

The major research question focused on the outcomes for biracial dyads. Previous research of this type had used only white investigators and implicit white comparison norms, and the results indicated that the white partner tended to dominate the decision making. James's study sought to determine whether such an "interaction disability," previously attributed to blacks, would be maintained, removed, or reversed when the comparison norm, the experimenter, or both were black. One approach to analyzing this problem was to perform a *two-way analysis of variance* on social-influence-activity difference scores between black and white partners, to assess whether such differences were affected by either the race of E or the race of N. No such significant effects were found, however, implying that neither E nor N influenced biracial dyad interaction. Nevertheless, through use of *analysis of covariance,* it was shown that, controlling for factors such as age, height, grade, and verbal and mathematical test scores, there was no statistical evidence of white dominance in any of the experimental conditions.

Furthermore, when combined output scores for both subjects in same-race dyads (white–white or black–black) were analyzed using a *three-way analysis of variance* (the three factors being race of dyad, race of E, and race of N), subjects in all-black dyads were found to be more verbally active (i.e., exhibited a greater tendency to influence decisions) under a black E than under a white E; the same result was found for white dyads under a white E. This property is generally referred to in statistical jargon as a "race of dyad" by "race of E" interaction. The property continued to hold up after *analysis of covariance* was used to control for the effects of age, height, and verbal and mathematical test scores.   ■

■ **Example 1.5**   *Study of the relationship of cultural change to health, illustrating the use of factor analysis and analysis of variance.*

Patrick and others (1974) studied the effects of cultural change on health in the U.S. Trust Territory island of Ponape. Medical and sociological data were obtained on a sample of about 2,000 people by means of physical exams and a sociological questionnaire. This Micronesian island has experienced rapid Westernization and modernization since American occupation in 1945. The question of primary interest was whether rapid social and cultural change caused a rise in blood pressure and in the incidence of coronary heart disease. A specific hypothesis guiding the research was that persons with high levels of cultural ambiguity and incongruity and low levels of supportive affiliations with others have high levels of blood pressure and are at high risk for coronary heart disease.

A preliminary step in the evaluation of this hypothesis involved measuring three variables: attitude toward modern life; preparation for modern life; and involvement in modern life. Each of these variables was created by isolating specific questions from a sociological questionnaire. Then a *factor analysis*[3] determined how best to combine the scores on specific questions into a single overall score that defined the variable under consideration. Two cultural incongruity variables were then defined. One involved the discrepancy between attitude toward modern life and involvement in modern life; the other was defined as the discrepancy between preparation for modern life and involvement in modern life.

These variables were then analyzed to determine their relationship, if any, to blood pressure and coronary heart disease. Individuals with large positive or negative scores on either of the two incongruity variables were hypothesized to have high blood pressure and to be at high risk for coronary heart disease.

One approach to analysis involved categorizing both discrepancy scores into high and low groups. Then a *two-way analysis of variance* could be performed using blood pressure as the outcome variable. We will see later that this problem can also be described as a regression problem.   ■

1.3 Concluding Remarks

The five examples described in Section 1.2 indicate the variety of research questions to which multivariable statistical methods are applicable. In Chapter 2, we will provide a broad overview of such techniques; in the remaining chapters, we will discuss each technique in detail.

References

Campbell, D. T., and Stanley, J. C. 1963. *Experimental and Quasi-experimental Designs for Research.* Chicago: Rand McNally.

Hamilton, M. 1971. "Sudden Death and Water Hardness in North Carolina Counties in 1956–1964." Master's thesis, Department of Epidemiology, University of North Carolina, Chapel Hill.

---

[3] Factor analysis was described in Chapter 24 of the second edition of this text, but this topic is not included as a topic in this (fourth) edition.

Hulka, B. S.; Kupper, L. L.; Cassel, J. C.; and Thompson, S. J. 1971. "A Method for Measuring Physicians' Awareness of Patients' Concerns." *HSMHA Health Reports* 86: 741–51.

James, S. A. 1973. "The Effects of the Race of Experimenter and Race of Comparison Norm on Social Influence in Same Race and Biracial Problem-Solving Dyads." Ph.D. dissertation, Department of Clinical Psychology, Washington University, St. Louis, Mo.

Kleinbaum, D. G., and Kleinbaum, A. 1976. "A Team Approach for Systematic Design and Evaluation of Visually Oriented Modules." In J. R. O'Fallon and J. Service, eds., *Modular Instruction in Statistics—Report of ASA Study,* pp. 115–21. Washington, D.C.: American Statistical Association.

Kleinbaum, D. G.; Kupper, L. L.; and Morgenstern, H. 1982. *Epidemiologic Research.* Belmont, Calif.: Lifetime Learning Publications.

Patrick, R.; Cassel, J. C.; Tyroler, H. A.; Stanley, L.; and Wild, J. 1974. "The Ponape Study of Health Effects of Cultural Change." Paper presented at the annual meeting of the Society for Epidemiologic Research, Berkeley, Calif.

Thompson, S. J. 1972. "The Doctor–Patient Relationship and Outcomes of Pregnancy." Ph.D. dissertation, Department of Epidemiology, University of North Carolina, Chapel Hill.

2

Classification of Variables
and the Choice of Analysis

2.1 Classification of Variables

Variables can be classified in a number of ways. Such classifications are useful for determining which method of data analysis to use. In this section we describe three methods of classification: by gappiness, by descriptive orientation, and by level of measurement.

2.1.1 Gappiness

In the classification scheme we call *gappiness,* we determine whether gaps exist between successively observed values of a variable (Figure 2.1). If gaps exist between observations, the variable is said to be *discrete;* if no gaps exist, the variable is said to be *continuous.* To speak more precisely, a variable is discrete if, between any two potentially observable values, a value exists that is not possibly observable. A variable is continuous if, between any two potentially observable values, another potentially observable value exists.

Examples of continuous variables are age, blood pressure, cholesterol level, height, and weight. Examples of discrete variables are sex (e.g., 0 if male and 1 if female), number of deaths, group identification (e.g., 1 if group A and 2 if group B), and state of disease (e.g., 1 if a coronary heart disease case and 0 if not a coronary heart disease case).

In analyses of actual data, the sampling frequency distributions for continuous variables are represented differently from those for discrete variables. Data on a continuous variable are usually *grouped* into class intervals, and a relative frequency distribution is determined by counting

FIGURE 2.1   Discrete versus continuous variables

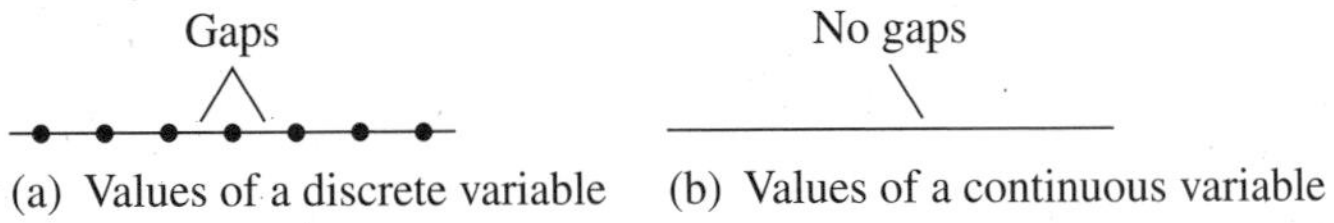

(a) Values of a discrete variable   (b) Values of a continuous variable

FIGURE 2.2   **Sample frequency distributions of a continuous and a discrete variable**

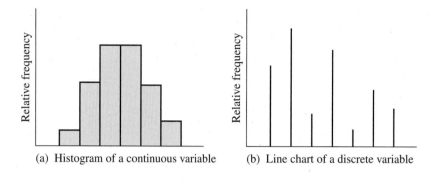

(a) Histogram of a continuous variable   (b) Line chart of a discrete variable

FIGURE 2.3   **Discrete variable that may be treated as continuous**

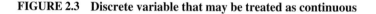

the proportion of observations in each interval. Such a distribution is usually represented by a histogram, as shown in Figure 2.2(a). Data on a discrete variable, on the other hand, are usually not grouped but are represented instead by a line chart, as shown in Figure 2.2(b).

Discrete variables can sometimes be treated for analysis purposes as continuous variables. This is possible when the values of such a variable, even though discrete, are not far apart and cover a wide range of numbers. In such a case, the possible values, although technically gappy, show such small gaps between values that a visual representation would approximate an interval (Figure 2.3).

Furthermore, a line chart, like the one in Figure 2.2(b), representing the frequency distribution of data on such a variable would probably show few frequencies greater than 1 and thus would be uninformative. As an example, the variable "social class" is usually measured as discrete; one popular measure of social class[1] takes on integer values between 11 and 77. When data on this variable are grouped into classes (e.g., 11–15, 16–20, etc.), the resulting frequency histogram gives a clearer picture of the characteristics of the variable than a line chart does. Thus, in this case, treating social class as a continuous variable is sometimes more useful than treating it as discrete.

Just as it is often useful to treat a discrete variable as continuous, some fundamentally continuous variables may be grouped into categories and treated as discrete variables in a given analysis. For example, the variable "age" can be made discrete by grouping its values into two categories, "young" and "old." Similarly, "blood pressure" becomes a discrete variable if it is categorized into "low," "medium," and "high" groups, or into deciles.

2.1.2 Descriptive Orientation

A second scheme for classifying variables is based on whether a variable is intended to *describe* or *be described* by other variables. Such a classification depends on the study

---

[1] Hollingshead's "Two-Factor Index of Social Position," a description of which can be found in Green (1970).

objectives rather than on the inherent mathematical structure of the variable itself. If the variable under investigation is to be described in terms of other variables, we call it a *response* or *dependent variable.* If we might be using the variable in conjunction with other variables to describe a given response variable, we call it a *predictor* or *independent variable.* Other variables may affect relationships but be of no intrinsic interest in a particular study. Such variables may be referred to as *control* or *nuisance variables* or, in some contexts, as *covariates* or *confounders.*

For example, in Thompson's (1972) study of the relationship between patient perception of pregnancy and patient satisfaction with medical care, the perception variables are independent and the satisfaction variable is dependent (Figure 2.4). Similarly, in studying the relationship of water hardness to sudden death rate in North Carolina counties, the water hardness index measured in each county is an independent variable, and the sudden death rate for that county is the dependent variable (Figure 2.5).

Usually, the distinction between independent and dependent variables is clear, as it is in the examples we have given. Nevertheless, a variable considered as dependent for purposes of evaluating one study objective may be considered as independent for purposes of evaluating a different

FIGURE 2.4   Descriptive orientation for Thompson's (1972) study of satisfaction with medical care

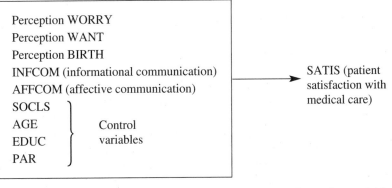

FIGURE 2.5   Descriptive orientation for Hamilton's (1971) study of water hardness and sudden death

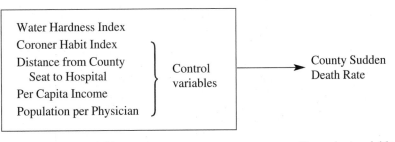

objective. For example, in Thompson's study, in addition to determining the relationship of perceptions as independent variables to patient satisfaction, the researcher sought to determine the relationships of social class, age, and education to perceptions treated as dependent variables.

2.1.3 Level of Measurement

A third classification scheme deals with the preciseness of measurement of the variable. There are three such levels: nominal, ordinal, and interval.

The numerically weakest level of measurement is the *nominal*. At this level, the values assumed by a variable simply indicate different categories. The variable "sex," for example, is nominal: by assigning the numbers 1 and 0 to denote male and female, respectively, we distinguish the two sex categories. A variable that describes treatment group is also nominal, provided that the treatments involved cannot be ranked according to some criterion (e.g., dosage level).

A somewhat higher level of measurement allows not only *grouping* into separate categories but also *ordering* of categories. This level is called *ordinal*. The treatment group may be considered ordinal if, for example, different treatments differ by dosage. In this case, we could tell not only which treatment group an individual falls into but also who received a heavier dose of the treatment. Social class is another ordinal variable, since an ordering can be made among its different categories. For example, all members of the upper middle class are higher in some sense than all members of the lower middle class.

A limitation—perhaps debatable—in the preciseness of a measurement such as social class is the amount of information supplied by the magnitude of the differences between different categories. Thus, although upper middle class is higher than lower middle class, it is debatable *how much* higher.

A variable that can give not only an ordering but also a meaningful measure of the distance between categories is called an *interval* variable. To be interval, a variable must be expressed in terms of some standard or well-accepted physical unit of measurement. Height, weight, blood pressure, and number of deaths all satisfy this requirement, whereas subjective measures such as perception of pregnancy, personality type, prestige, and social stress do not.

An interval variable that has a scale with a true zero is occasionally designated as a *ratio* or *ratio-scale variable*. An example of a ratio-scale variable is the height of a person. Temperature is commonly measured in degrees Celsius, an interval scale. Measurement of temperature in degrees Kelvin is based on a scale that begins at absolute zero, and so is a ratio variable. An example of a ratio variable common in health studies is the concentration of a substance (e.g., cholesterol) in the blood.

Ratio-scale variables often involve measurement errors that follow a nonnormal distribution and are proportional to the size of the measurement. We will see in Chapter 5 that such proportional errors violate an important assumption of linear regression—namely, equality of error variance for all observations. Hence, the presence of a ratio variable is a signal to be on guard for a possible violation of this assumption. In Chapter 14 (on regression diagnostics), we will describe methods for detecting and dealing with this problem.

As with variables in other classification schemes, the same variable may be considered at one level of measurement in one analysis and at a different level in another analysis. Thus, "age" may be considered as interval in a regression analysis or, by being grouped into categories, as nominal in an analysis of variance.

The various levels of mathematical preciseness are cumulative. An ordinal scale possesses all the properties of a nominal scale plus ordinality. An interval scale is also nominal and ordinal. The cumulativeness of these levels allows the researcher to drop back one or more levels of measurement in analyzing the data. Thus, an interval variable may be treated as nominal or ordinal for a particular analysis, and an ordinal variable may be analyzed as nominal.

2.2 Overlapping of Classification Schemes

The three classification schemes described in Section 2-1 overlap in the sense that any variable can be labeled according to each scheme. "Social class," for example, may be considered as ordinal, discrete, and independent in a given study; "blood pressure" may be considered interval, continuous, and dependent in the same or another study.

    The overlap between the level-of-measurement classification and the gappiness classification is shown in Figure 2.6. The diagram does not include classification into dependent or independent variables, because that dimension is entirely a function of the study objectives and not of the variable itself. In reading the diagram, one should consider any variable as being representable by some point within the triangle. If the point falls below the dashed line within the triangle, it is classified as discrete; if it falls above that line, it is continuous. Also, a point that falls into the area marked "interval" is classified as an interval variable; and similarly for the other two levels of measurement.

    As Figure 2.6 indicates, any nominal variable must be discrete, but a discrete variable may be nominal, ordinal, or interval. Also, a continuous variable must be either ordinal or interval, although ordinal or interval variables may exist that are not continuous. For example, "sex" is nominal and discrete; "age" may be considered interval and continuous or, if grouped into categories, nominal and discrete; and "social class," depending on how it is measured and on the viewpoint of the researcher, may be considered ordinal and continuous, ordinal and discrete, or nominal and discrete.

FIGURE 2.6   Overlap of variable classification

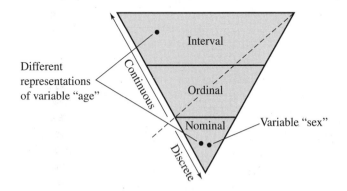

2.3 Choice of Analysis

Any researcher faced with the need to analyze data requires a rationale for choosing a particular method of analysis. Four considerations should enter into such a choice: the purpose of the investigation; the mathematical characteristics of the variables involved; the statistical assumptions made about these variables; and how the data are collected (e.g., the sampling procedure). The first two considerations are generally sufficient to determine an appropriate analysis. However, the researcher must consider the latter two items before finalizing initial recommendations.

Here we focus on the use of variable classification, as it relates to the first two considerations noted at the beginning of this section, in choosing an appropriate method of analysis. Table 2.1 provides a rough guide to help the researcher in this choice when several variables are involved. The guide distinguishes among various multivariable methods. It considers the types of variable sets usually associated with each method and gives a general description of the purposes of each method. In addition to using the table, however, one must carefully check the statistical assumptions being made. These assumptions will be described fully later in the text. Table 2.2 shows how these guidelines can be applied to the examples given in Chapter 1.

Several methods for dealing with multivariable problems are *not* included in Table 2.1 or in this text—among them, nonparametric methods of analysis of variance, multivariate multiple regression, and multivariate analysis of variance (which are extensions of the corresponding methods given here that allow for *several* dependent variables), as well as methods of cluster analysis. In this book, we will cover only the multivariable techniques used most often by health and social researchers.

TABLE 2.1   Rough guide to multivariable methods

	Classification of Variables		
Method	Dependent	Independent	General Purpose
Multiple regression analysis	Continuous	Classically all continuous, but in practice any type(s) can be used	To describe the extent, direction, and strength of the relationship between several independent variables and a continuous dependent variable
Analysis of variance	Continuous	All nominal	To describe the relationship between a continuous dependent variable and one or more nominal independent variables
Analysis of covariance	Continuous	Mixture of nominal variables and continuous variables (the latter used as control variables)*	To describe the relationship between a continuous dependent variable and one or more nominal independent variables, controlling for the effect of one or more continuous independent variables
Logistic regression analysis	Dichotomous	A mixture of various types can be used	To determine how one or more independent variables are related to the probability of the occurrence of one of two possible outcomes
Poisson regression analysis	Discrete	A mixture of various types can be used	To determine how one or more independent variables are related to the rate of occurrence of some outcome

* Generally, a *control variable* is a variable that must be considered before any relationships of interest can be quantified; this is because a control variable may be related to the variables of primary interest and must be taken into account in studying the relationships among the primary variables. For example, in describing the relationship between blood pressure and physical activity, we would probably consider "age" and "sex" as control variables because they are related to blood pressure and physical activity, and unless taken into account, could confound any conclusions regarding the primary relationship of interest.

TABLE 2.2   Application of Table 2.1 to examples in Chapter 1

Study	Multivariable Method	Dependent Variable	Independent Variables	Purpose
Example 1.1	Multiple regression analysis	Patient satisfaction with medical care (SATIS), a continuous variable	WANT, WORRY, BIRTH, INFCOM, AFFCOM, AGE, EDUC, SOCLS, PAR	To describe the relationship between SATIS and the variables WANT, WORRY, etc.
Example 1.1	Logistic regression analysis	Complications of pregnancy (0 = no, 1 = yes), a nominal variable	WANT, WORRY, BIRTH, INFCOM, AFFCOM, AGE, EDUC, SOCLS, PAR	To determine whether and to what extent the independent variables are related to the probability of having pregnancy complications
Example 1.2	Multiple regression analysis	Sudden death rate for a county in North Carolina during a 5-year period, a continuous variable	Water hardness index for county, distance between hospital and county seat, per capita income, population per physician	To describe the relationship between sudden death rate and the independent variables
Example 1.3	Analysis of covariance	Posttest score, a continuous variable	Pretest score, group designation (e.g., 1 = experimental, 0 = control)	To compare posttest scores for experimental and control groups, after adjusting for the possible effect of pretest scores
Example 1.4	Analysis of covariance	Social influence activity score, a continuous variable	Race of subject (e.g., 1 = white, 2 = black), age, height, etc.	To determine whether one racial group dominates the other in biracial dyads, after controlling for age, height, etc.
Example 1.4	Two-way analysis of variance	Social influence activity difference score between black and white partners in biracial dyads, a continuous variable	Race of experimenter (e.g., 1 = white, 2 = black), race of comparison norm (e.g., 1 = white, 2 = black)	To determine whether the experimenter's race and the comparison norm's race have any effect on the difference score
Example 1.5	Two-way analysis of variance	Systolic blood pressure (SBP), a continuous variable	Discrepancy between attitude toward and involvement in modern life, categorized as "high" or "low"; discrepancy between preparation for and involvement in modern life, categorized as "high" or "low"	To describe the relationship between nominal discrepancy scores and SBP

References

Green, L. W. 1970. "Manual for Scoring Socioeconomic Status for Research on Health Behaviors." *Public Health Reports* 85: 815–27.

Hamilton, M. 1971. "Sudden Death and Water Hardness in North Carolina Counties in 1956–1964." Master's thesis, Department of Epidemiology, University of North Carolina, Chapel Hill.

Thompson, S. J. 1972. "The Doctor–Patient Relationship and Outcomes of Pregnancy." Ph.D. dissertation, Department of Epidemiology, University of North Carolina, Chapel Hill.

3

Basic Statistics:
A Review

3.1 Preview

This chapter reviews the fundamental statistical concepts and methods that are needed to understand the more sophisticated multivariable techniques discussed in this text. Through this review, we shall introduce the statistical notation (using conventional symbols whenever possible) employed throughout the text.

The broad area associated with the word *statistics* involves the methods and procedures for collecting, classifying, summarizing, and analyzing data. We shall focus on the latter two activities here. The primary goal of most statistical analysis is to make *statistical inferences*—that is, to draw valid conclusions about a *population* of items or measurements based on information contained in a *sample* from that population.

A *population* is any set of items or measurements of interest, and a *sample* is any subset of items selected from that population. Any characteristic of that population is called a *parameter,* and any characteristic of the sample is termed a *statistic.* A statistic may be considered an estimate of some population parameter, and its accuracy of estimation may be good or bad.

Once sample data have been collected, it is useful, prior to analysis, to examine the data using tables, graphs, and *descriptive statistics,* such as the sample mean and the sample variance. Such descriptive efforts are important for representing the essential features of the data in easily interpretable terms.

Following such examination, statistical inferences are made through two related activities: *estimation* and *hypothesis testing.* The techniques involved here are based on certain assumptions about the probability pattern (or *distribution*) of the (*random*) variables being studied.

Each of the preceding key terms—*descriptive statistics, random variables, probability distribution, estimation,* and *hypothesis testing*—will be reviewed in the sections that follow.

3.2 Descriptive Statistics

A *descriptive statistic* may be defined as any single numerical measure computed from a set of data that is designed to describe a particular aspect or characteristic of the data set. The most common types of descriptive statistics are measures of *central tendency* and of *variability* (or *dispersion*).

The central tendency in a sample of data is the "average value" of the variable being observed. Of the several measures of central tendency, the most commonly used is the sample mean, which we denote by $\overline{X}$ whenever our underlying variable is called X. The formula for the sample mean is given by

$$\overline{X} = \frac{\sum\limits_{i=1}^{n} X_i}{n}$$

where n denotes the sample size; $X_1, X_2, \ldots, X_n$ denote the n independent measurements on X; and $\sum$ denotes summation. The sample mean $\overline{X}$—in contrast to other measures of central tendency, such as the median or mode—uses in its computation all the observations in the sample. This property means that $\overline{X}$ is necessarily affected by the presence of extreme X-values, so it may be preferable to use the median instead of the mean. A remarkable property of the sample mean, which makes it particularly useful in making statistical inferences, follows from the *Central Limit Theorem,* which states that *whenever n is moderately large, $\overline{X}$ has approximately a normal distribution, regardless of the distribution of the underlying variable X.*

Measures of central tendency (such as $\overline{X}$) do not, however, completely summarize all features of the data. Obviously, two sets of data with the same mean can differ widely in appearance (e.g., an $\overline{X}$ of 4 results both from the values 4, 4, and 4 and from the values 0, 4, and 8). Thus, we customarily consider, in addition to $\overline{X}$, measures of variability, which tell us the extent to which the values of the measurements in the sample differ from one another.

The two measures of variability most often considered are the *sample variance* and the *sample standard deviation.* These are given by the following formulas when considering observations $X_1, X_2, \ldots, X_n$ on a single variable X:

$$\text{Sample variance} = S^2 = \frac{1}{n-1} \sum_{i=1}^{n} (X_i - \overline{X})^2 \tag{3.1}$$

$$\text{Sample standard deviation} = S = \sqrt{\frac{1}{n-1} \sum_{i=1}^{n} (X_i - \overline{X})^2} \tag{3.2}$$

The formula for S^2 describes variability in terms of an average of squared deviations from the sample mean—although $n - 1$ is used as the divisor instead of n, due to considerations that make S^2 a good estimator of the variability in the entire population.

A drawback to the use of S^2 is that it is expressed in squared units of the underlying variable X. To obtain a measure of dispersion that is expressed in the same units as X, we simply take the square root of S^2 and call it the sample standard deviation S. Using S in combination with $\overline{X}$ thus gives a fairly succinct picture of both the amount of spread and the center of the data, respectively.

When more than one variable is being considered in the same analysis (as will be the case throughout this text), we will use different letters and/or different subscripts to differentiate among the variables, and we will modify the notations for mean and variance accordingly. For example, if we are using X to stand for age and Y to stand for systolic blood pressure, we will denote the sample mean and the sample standard deviation for each variable as $(\overline{X}, S_X)$ and $(\overline{Y}, S_Y)$, respectively.

All statistical analyses begin with an examination of descriptive statistics computed from the data set at hand. However, often the most direct and revealing way to examine the data is to make a series of plots. We describe three types of simple but useful plots: histograms (especially stem-and-leaf versions), schematic plots, and normal probability plots.

Suppose that we have collected data on the amount of error that occurs in measurements taken with a particular type of instrument. We think the error may be related to the age of the instrument; therefore, readings are taken with 17 instruments of varying ages; the age of each instrument and the error in its measurement are recorded.

In our descriptive analysis of these data, first we examine a frequency histogram of the measurement errors, shown in Figure 3.1 (a). We observe that the errors appear to be quite symmetrically distributed around 0 (i.e., the mean and the median error are roughly 0) and that the picture approximates a bell-shaped curve. (See Section 3.3.2 for more information on data that follow this pattern.) No *outliers* (data points that are extreme in value and that may represent data errors) or other anomalies appear to be present.

The frequency histogram conveys even more information if it is converted into a stem-and-leaf diagram, as in Figure 3.1 (b), since it shows the actual data values while maintaining the shape of the histogram. In the stem-and-leaf diagram, the top-most value has a *stem* of 2 and a *leaf* of 7, indicating that the original data value is 2.7. Beneath that is a value of 1.4 (stem 1, leaf 4); after that are two values that both share a stem of 0 and have leaves equal to 1 (i.e., both values are 0.1), followed by a value of 0.2, etc. The last value shown in the plot is -2.8 (stem -2, leaf 8).

The second kind of useful plot is a schematic plot. Figure 3.2 presents a schematic plot of the measurement error data. A schematic plot is based entirely on the order of the values in the sample. *Quartiles* are the most important order-based statistics for the schematic plot. The *first quartile*, or 25th percentile, is the value at or below which 25% of the data values lie; the *second quartile*, or 50th percentile (or median), is the value at or below which 50% of the data values lie; the *third quartile*, or 75th percentile, is the value at or below which 75% of the data values lie. The *interquartile range* (IQR), calculated as the value of the third quartile minus the value of the first quartile, is a measure of the spread of a distribution, like the variance. One important difference between the IQR and the variance, however, is illustrated by the fact that, whereas

FIGURE 3.1   Frequency histogram and stem-and-leaf diagram of instrument error data ($n = 17$)

```
 3 │  *            2 │ 7
 2 │  *            1 │ 4
 1 │  *******      0 │ 1124557
 0 │  ******      -0 │ 865542
-1 │  *           -1 │ 3
-2 │  *           -2 │ 8
-3 │
```

(a) Histogram   (b) Stem-and-leaf diagram

FIGURE 3.2   Schematic plot of instrument error data ($n = 17$)

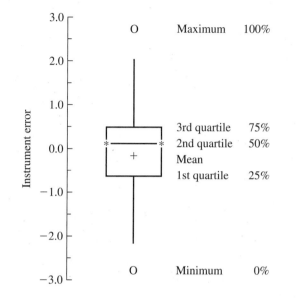

doubling the largest value in the sample would, in general, increase variance dramatically, it would not change the IQR. For the error measurements, the first, second, and third quartiles are approximately −0.6, 0.1, and 0.5 respectively, with an interquartile range of 1.1.

A schematic plot is sometimes called a *box-and-whisker plot*, or simply a *boxplot*, due to its appearance. The box is outlined by three horizontal lines, which mark the values of the first quartile, the second quartile (the median, indicated by asterisks), and the third quartile (see Figure 3.2). The scale is determined by the units and range of the data. The mean is indicated by a + on the backbone of the plot. If the data are symmetric, the mean and median will be similar (i.e., the + will be marked on or close to the middle horizontal line) and the distances between the first and second quartile, and second and third quartile will be similar in size. The whiskers (vertical lines) extend from the box as far as the data extend up or down, to a limit of 1.5 IQRs (in the vertical direction). An O at the end of a whisker indicates a moderate outlier; a value that is more than 3 IQRs away from the nearest quartile is indicated with an ∗ and is a severe outlier. Referring to Figure 3.2, we see one positive outlier and one negative moderate outlier.

3.3 Random Variables and Distributions

The term *random variable* is used to denote a variable whose observed values may be considered outcomes of a stochastic or random experiment (e.g., the drawing of a random sample). The values of such a variable in a particular sample, then, cannot be anticipated with certainty before the sample is gathered. Thus, if we select a random sample of persons from some community and determine the systolic blood pressure (W), cholesterol level (X), race (Y), and sex (Z) of each person, then $W, X, Y,$ and Z are four random variables whose particular realizations (or observed values) for a given person in the sample cannot be known for sure beforehand. In this text, we shall denote random variables by capital italic letters.

FIGURE 3.3   **Discrete and continuous distributions.** $P(X = a)$ **is read: "The probability that** X **takes the value** a**."**

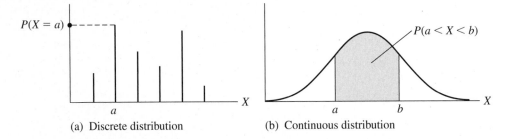

(a) Discrete distribution   (b) Continuous distribution

The probability pattern that gives the relative frequencies associated with all the possible values of a random variable in a population is generally called the *probability distribution* of the random variable. We represent such a distribution by a table, graph, or mathematical expression that provides the probabilities corresponding to the different values or ranges of values taken on by a random variable.

Discrete random variables (such as the number of deaths in a sample of patients, or the number of arrivals at a clinic), whose possible values are countable, have (gappy) distributions that are graphed as a series of lines; the heights of these lines represent the probabilities associated with the various possible discrete outcomes (Figure 3.3(a)). *Continuous* random variables (such as blood pressure and weight), whose possible values are uncountable, have (nongappy) distributions that are graphed as smooth curves; an *area* under such a curve represents the probability associated with a *range of values* of the continuous variable (Figure 3.3(b)). We note in passing that the probability of a continuous random variable taking on one particular value is 0, because there can be no area above a single point.

In the next two subsections, we will discuss two particular distributions of enormous practical importance: the binomial (which is discrete) and the normal (which is continuous).

3.3.1 The Binomial Distribution

A *binomial* random variable describes the number of occurrences of a particular event in a series of n trials, under the following four conditions:

1.   The n trials are identical.

2.   The outcome of any one trial is independent of (i.e., is not affected by) the outcome of any other trial.

3.   There are two possible outcomes of each trial: "success" (i.e., the event of interest occurs) or "failure" (i.e., the event of interest does not occur), with probabilities π and $1 - \pi$, respectively.

4.   The probability of success, π, remains the same for all trials.

For example, the distribution of the number of lung cancer deaths in a random sample of $n = 400$ persons would be considered binomial only if the four conditions were all satisfied, as would the distribution of the number of persons in a sample of $n = 70$ who favor a certain form of legislation.

The two elements of the binomial distribution that one must specify to determine the precise shape of the probability distribution and to compute binomial probabilities are the sample size n and the parameter π. The usual notation for this distribution is, therefore, $B(n, \pi)$. If X has a binomial distribution, it is customary to write

$$X \frown B(n, \pi)$$

where $\frown$ stands for "is distributed as." The probability formula for this discrete random variable X is given by the expression

$$P(X = j) = {_nC_j} \pi^j (1 - \pi)^{n-j} \quad j = 0, 1, \ldots, n$$

where ${_nC_j} = n!/[j! (n - j)!]$ denotes the number of combinations of n distinct objects selected j at a time.

3.3.2 The Normal Distribution

The *normal distribution,* denoted as $N(\mu, \sigma)$, where μ and σ are the two parameters, is described by the well-known bell-shaped curve (Figure 3.4). The parameters μ (the mean) and σ (the standard deviation) characterize the center and the spread, respectively, of the distribution. We generally attach a subscript to the parameters μ and σ to distinguish among variables; that is, we often write

$$X \frown N(\mu_X, \sigma_X)$$

to denote a normally distributed X.

An important property of any normal curve is its *symmetry,* which distinguishes it from some other continuous distributions that we will discuss later. This symmetry property is quite helpful when using tables to determine probabilities or percentiles of the normal distribution.

Probability statements about a normally distributed random variable X that are of the form $P(a \le X \le b)$ require for computation the use of a single table (Table A-1 in Appendix A). This table gives the probabilities (or areas) associated with the *standard normal distribution,* which is a normal distribution with $\mu = 0$ and $\sigma = 1$. It is customary to denote a standard normal random variable by the letter Z, so we write

$$Z \frown N(0, 1)$$

To compute the probability $P(a \le X \le b)$ for an X that is $N(\mu, \sigma)$, we must transform (i.e., *standardize*) X to Z by applying the conversion formula

$$Z = \frac{X - \mu}{\sigma} \tag{3.3}$$

FIGURE 3.4   The normal distribution

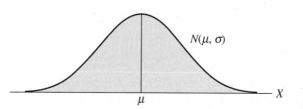

FIGURE 3.5 The (100p)th percentiles of X and Z

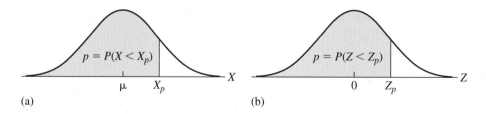

(a)   (b)

to each of the elements in the probability statement about X, as follows:

$$P(a \leq X \leq b) = P\left(\frac{a - \mu}{\sigma} \leq Z \leq \frac{b - \mu}{\sigma}\right)$$

We then look up the equivalent probability statement about Z in the N(0, 1) tables.

This rule also applies to the sample mean $\overline{X}$ whenever the underlying variable X is normally distributed or whenever the sample size is moderately large (by the Central Limit Theorem). But because the standard deviation of $\overline{X}$ is $\sigma/\sqrt{n}$, the conversion formula has the form

$$Z = \frac{\overline{X} - \mu}{\sigma/\sqrt{n}}$$

An inverse procedure for computing a probability for a range of values of X is to find a percentile of the distribution of X. A *percentile* is a value of the variable X *below which* the area under the probability distribution has a certain specified value. We denote the (100p)th percentile of X by X_p and picture it as in Figure 3.5, where p is the amount of area under the curve to the left of X_p. In determining X_p for a given p, we must again use the conversion formula (3.3). Since the procedure requires that we first determine Z_p and then convert back to X_p, however, we generally rewrite the conversion formula as

$$X_p = \mu + \sigma Z_p \tag{3.4}$$

For example, if $\mu = 140$ and $\sigma = 40$, and we want to find $X_{.95}$, the N(0, 1) table first gives us $Z_{.95} = 1.645$, which we convert back to $X_{.95}$ as follows:

$$X_{.95} = 140 + (40)Z_{.95} = 140 + 40(1.645) = 205.8$$

Formulas (3.3) and/or (3.4) can also be used to approximate probabilities and percentiles for the binomial distribution $B(n, \pi)$ whenever n is moderately large (e.g., $n > 20$). Two conditions are usually required for this approximation to be accurate: $n\pi > 5$ and $n(1 - \pi) > 5$. Under such conditions, the mean and the standard deviation of the approximating normal distribution are

$$\mu = n\pi \quad \text{and} \quad \sigma = \sqrt{n\pi(1 - \pi)}$$

A *normal probability plot* can be used to assess whether data are normally distributed. The ordered data values are plotted versus the cumulative relative frequencies (i.e., i/n where i is the ith ordered data value) up to each value. Plots that are linear in appearance are consistent with normality, since the cumulative relative frequencies for a normal distribution plot as a straight line.

The *skewness* and *kurtosis* statistics can also be helpful in assessing normality. *Skewness* indicates the degree of asymmetry of a distribution. Just as variance is the average squared deviation of observations about the mean, skewness is the average of cubed deviations about the mean. To simplify comparisons between samples and to help account for estimation in small samples, skewness is usually computed as

$$\text{sk}(X) = \left(\frac{n}{n-2}\right)\left(\frac{1}{n-1}\right)\sum_{i=1}^{n}\left(\frac{X_i - \overline{X}}{S_X}\right)^3$$

For large n, sk(X) should be approximately equal to 0 for a random sample size of n from any symmetric probability distribution (such as a normal distribution). Positive values of sk(X) indicate that relatively more values are above the mean than below it; the sample values are thus said to be "positively skewed." A negative value for sk(X) indicates that relatively more values are below the mean than above it.

Kurtosis indicates the heaviness of the tails relative to the middle of the distribution. Because kurtosis is the average of the fourth power of the deviations about the mean, it is always nonnegative. Standardized kurtosis may be computed as

$$\text{Kur}(X) = \left[\frac{n(n+1)}{(n-2)(n-3)}\right]\left(\frac{1}{n-1}\right)\sum_{i=1}^{n}\left(\frac{X_i - \overline{X}}{S_X}\right)^4$$

The term in brackets, which approaches 1.00 as n increases, helps to account for estimation based on a small sample. Since standardized kurtosis for a standard normal distribution is 3.0, this value is often subtracted from Kur(X). The resulting statistic can be as small as -3 for flat distributions with short tails; it is approximately zero for moderate to large random samples from a normal distribution, and it is positive for heavy-tailed distributions. Thus, the positive kurtosis value in our example (the reader is encouraged to do the required calculations) suggests a distribution with tails heavier than for a normal distribution. Skewness and kurtosis statistics are highly variable in small samples and hence are often difficult to interpret.

3.4 Sampling Distributions of t, χ^2, and F

The Student's t, chi-square (χ^2), and Fisher's F distributions are particularly important in statistical inference making.

The (*Student's*) *t distribution* (Figure 3.6(a)), which like the standard normal distribution is symmetric about 0, was originally developed to describe the behavior of the random variable

$$T = \frac{\overline{X} - \mu}{S/\sqrt{n}} \tag{3.5}$$

which represents an alternative to

$$Z = \frac{\overline{X} - \mu}{\sigma/\sqrt{n}}$$

whenever the population variance σ^2 is unknown and is estimated by S^2. The denominator of (3.5), $S/\sqrt{n}$, is the *estimated standard error of* $\overline{X}$. When the underlying distribution of X is normal, and when $\overline{X}$ and S^2 are calculated from a random sample from that distribution, then (3.5) has the t *distribution with* $n-1$ *degrees of freedom,* where $n-1$ is the quantity that must be specified in order to look up tabulated percentiles of this distribution. We denote all this by writing

$$T = \frac{\overline{X} - \mu}{S/\sqrt{n}} \frown t_{n-1}$$

FIGURE 3.6   The t, χ^2, and F distributions

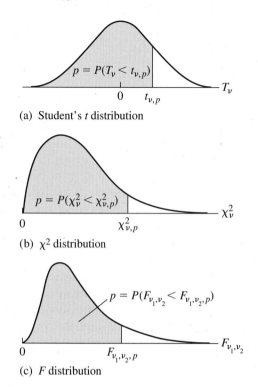

(a) Student's t distribution

(b) χ^2 distribution

(c) F distribution

It has generally been shown by statisticians that the t distribution is sometimes appropriate for describing the behavior of a random variable of the general form

$$T = \frac{\hat{\theta} - \mu_{\hat{\theta}}}{S_{\hat{\theta}}} \tag{3.6}$$

where $\hat{\theta}$ is any random variable that is normally distributed with mean $\mu_{\hat{\theta}}$ and standard deviation $\sigma_{\hat{\theta}}$, where $S_{\hat{\theta}}$ is the estimated standard error of $\hat{\theta}$, and where $\hat{\theta}$ and $S_{\hat{\theta}}$ are statistically independent. For example, when random samples are taken from two normally distributed populations with the same standard deviation (e.g., from $N(\mu_1, \sigma)$ and $N(\mu_2, \sigma)$), and we consider $\hat{\theta} = \bar{X}_1 - \bar{X}_2$ in (3.6), we can write

$$T = \frac{(\bar{X}_1 - \bar{X}_2) - (\mu_1 - \mu_2)}{S_p\sqrt{\dfrac{1}{n_1} + \dfrac{1}{n_2}}} \frown t_{n_1+n_2-2}$$

where

$$S_p^2 = \frac{(n_1 - 1)S_1^2 + (n_2 - 1)S_2^2}{n_1 + n_2 - 2} \tag{3.7}$$

estimates the common variance σ^2 in the two populations. The quantity S_p^2 is called a *pooled sample variance,* since it is calculated by pooling the data from both samples in order to estimate the common variance σ^2.

The *chi-square* (or χ^2) *distribution* (Figure 3.6(b)) is a nonsymmetric distribution and describes, for example, the behavior of the nonnegative random variable

$$\frac{(n-1)S^2}{\sigma^2} \tag{3.8}$$

where S^2 is the sample variance based on a random sample of size n from a normal distribution. The variable given by (3.8) has the chi-square distribution with $n-1$ degrees of freedom:

$$\frac{(n-1)S^2}{\sigma^2} \frown \chi^2_{n-1}$$

Because of the nonsymmetry of the chi-square distribution, both upper and lower percentage points of the distribution need to be tabulated, and such tabulations are solely a function of the degrees of freedom associated with the particular χ^2 distribution of interest. The chi-square distribution has widespread application in analyses of categorical data.

The *F distribution* (Figure 3.6(c)), which like the chi-square distribution is skewed to the right, is often appropriate for modeling the probability distribution of the ratio of independent estimators of two population variances. For example, given random samples of sizes n_1 and n_2 from $N(\mu_1, \sigma_1)$ and $N(\mu_2, \sigma_2)$, respectively, so that estimates S_1^2 and S_2^2 of σ_1^2 and σ_2^2 can be calculated, it can be shown that

$$\frac{S_1^2/\sigma_1^2}{S_2^2/\sigma_2^2} \tag{3.9}$$

has the F distribution with $n_1 - 1$ and $n_2 - 1$ degrees of freedom, which are called the *numerator* and *denominator* degrees of freedom, respectively. We write this as

$$\frac{S_1^2\sigma_2^2}{S_2^2\sigma_1^2} \frown F_{n_1-1, n_2-1}$$

The F distribution can also be related to the t distribution, when the numerator degrees of freedom equal 1; that is, the square of a variable distributed as Student's t with v degrees of freedom has the F distribution with 1 and v degrees of freedom. In other words,

$$T^2 \frown F_{1, v} \quad \text{if and only if} \quad T \frown t_v$$

Percentiles of the t, χ^2, and F distributions may be obtained from Tables A-2, A-3, and A-4 in Appendix A. The shapes of the curves that describe these probability distributions, together with the notation we will use to denote their percentile points are given in Figure 3.6.

3.5 Statistical Inference: Estimation

Two general categories ical inference—estimation and hypothesis testing—can be distinguished by their differing purposes: estimation is concerned with quantifying the specific value of an unknown population parameter; hypothesis testing is concerned with making a decision about a hypothesized value of an unknown population parameter.

In estimation, which we focus on in this section, we want to estimate an unknown parameter θ by using a random variable $\hat{\theta}$ ("theta hat," called a *point estimator* of θ). This point estimator takes the form of a formula or rule. For example,

$$\bar{X} = \frac{1}{n} \sum_{i=1}^{n} X_i \quad \text{or} \quad S^2 = \frac{1}{n-1} \sum_{i=1}^{n} (X_i - \bar{X})^2$$

tells us how to calculate a specific point estimate, given a particular set of data.

To estimate a parameter of interest (e.g., a population mean μ, a binomial proportion π, a difference between two population means $\mu_1 - \mu_2$, or a ratio of two population standard deviations σ_1/σ_2), the usual procedure is to select a random sample from the population or populations of interest, calculate the point estimate of the parameter, and then associate with this estimate a measure of its variability, which usually takes the form of a confidence interval for the parameter of interest.

As its name implies, a *confidence interval* (often abbreviated CI) consists of two random boundary points between which we have a certain specified *level of confidence* that the population parameter lies. More specifically, a 95% confidence interval for a parameter θ consists of lower and upper limits determined so that, in many repeated sets of samples of the same size, about 95% of all such intervals would be expected to contain the parameter θ. Care must be taken when interpreting such a confidence interval not to consider θ a random variable that either falls or does not fall in the calculated interval; rather, θ is a fixed (unknown) constant, and the random quantities are the lower and upper limits of the confidence interval, which vary from sample to sample.

We illustrate the procedure for computing a confidence interval with two examples—one involving estimation of a single population mean μ, and one involving estimation of the difference between two population means $\mu_1 - \mu_2$. In each case, the appropriate confidence interval has the following general form:

$$\left(\begin{array}{c} \text{Point estimate of} \\ \text{the parameter} \end{array} \right) \pm \left[\left(\begin{array}{c} \text{Percentile of} \\ \text{the } t \text{ distribution} \end{array} \right) \cdot \left(\begin{array}{c} \text{Estimated standard} \\ \text{error of the estimate} \end{array} \right) \right] \qquad (3.10)$$

This general form also applies to confidence intervals for other parameters considered in the remainder of the text (e.g., those considered in multiple regression analysis).

■ **Example 3.1**   Suppose that we have determined the Quantitative Graduate Record Examination (QGRE) scores for a random sample of nine student applicants to a certain graduate department in a university, and we have found that $\bar{X} = 520$ and $S = 50$. If we want to estimate with 95% confidence the population mean QGRE score (μ) for all such applicants to the department, and we are willing to assume that the population of such scores from which our random sample was selected is approximately normally distributed, the confidence interval for μ is given by the general formula

$$\bar{X} \pm t_{n-1, 1-\alpha/2} \left(\frac{S}{\sqrt{n}} \right) \qquad (3.11)$$

which gives the $100(1 - \alpha)\%$ (small-sample) confidence interval for μ when σ is unknown. In our problem, $\alpha = 1 - .95 = .05$ and $n = 9$; therefore, by substituting the given information into (3.11), we obtain

$$520 \pm t_{8, 0.975} \left(\frac{50}{\sqrt{9}} \right)$$

Since $t_{8,0.975} = 2.3060$, this formula becomes

$$520 \pm 2.3060\left(\frac{50}{\sqrt{9}}\right)$$

or

$$520 \pm 38.43.$$

Our 95% confidence interval for μ is thus given by

(481.57, 558.43)

If we wanted to use this confidence interval to help determine whether 600 is a likely value for μ (i.e., if we were interested in making a decision about a specific value for μ), we would conclude that 600 is not a likely value, since it is not contained in the 95% confidence interval for μ just developed. This helps clarify the connection between estimation and hypothesis testing.   ■

■ **Example 3.2**   Suppose that we want to compare the change in health status of two groups of mental patients who are undergoing different forms of treatment for the same disorder. Suppose that we have a measure of change in health status based on a questionnaire given to each patient at two different times, and we are willing to assume that this measure of change in health status is approximately normally distributed and has the same variance in the populations of patients from which we selected our independent random samples. The data obtained are summarized as follows:

Group 1: $n_1 = 15, \overline{X}_1 = 15.1, S_1 = 2.5$

Group 2: $n_2 = 15, \overline{X}_2 = 12.3, S_2 = 3.0$

where the underlying variable X denotes the change in health status between time 1 and time 2.

A 99% confidence interval for the true mean difference ($\mu_1 - \mu_2$) in health status change between these two groups is given by the following formula, which assumes equal population variances (i.e., $\sigma_1^2 = \sigma_2^2$):

$$(\overline{X}_1 - \overline{X}_2) \pm t_{n_1+n_2-2, 1-\alpha/2} S_p \sqrt{\frac{1}{n_1} + \frac{1}{n_2}} \tag{3.12}$$

where S_p is the pooled standard deviation derived from S_p^2, the pooled sample variance given by (3.7). Here we have

$$S_p^2 = \frac{(15-1)(2.5)^2 + (15-1)(3.0)^2}{15+15-2} = 7.625$$

so

$$S_p = \sqrt{7.625} = 2.76$$

Since $\alpha = .01$, our percentile in (3.12) is given by $t_{28,0.995} = 2.7633$. So the 99% confidence interval for $\mu_1 - \mu_2$ is given by

$$(15.1 - 12.3) \pm 2.7633(2.76)\sqrt{\frac{1}{15} + \frac{1}{15}}$$

which reduces to

$$2.80 \pm 2.78$$

yielding the following 99% confidence interval for $\mu_1 - \mu_2$:

(0.02, 5.58)

Since the value 0 is not contained in this interval, we conclude that there is evidence of a difference in health status change between the two groups.   ■

3.6 Statistical Inference: Hypothesis Testing

Although closely related to confidence interval estimation, hypothesis testing has a slightly different orientation. When developing a confidence interval, we use our sample data to estimate what we think is a *likely* set of values for the parameter of interest. When performing a statistical test of a null hypothesis concerning a certain parameter, we use our sample data to *test* whether our estimated value for the parameter is *different enough* from the hypothesized value to support the conclusion that the null hypothesis is *unlikely* to be true.

The general procedure used in testing a statistical null hypothesis remains basically the same, regardless of the parameter being considered. This procedure (which we will illustrate by example) consists of the following seven steps:

1.   Check the assumptions regarding the properties of the underlying variable(s) being measured that are needed to justify use of the testing procedure under consideration.

2.   State the null hypothesis H_0 and the alternative hypothesis H_A.

3.   Specify the significance level α.

4.   Specify the test statistic to be used and its distribution under H_0.

5.   Form the decision rule for rejecting or not rejecting H_0 (i.e., specify the rejection and nonrejection regions for the test).

6.   Compute the value of the test statistic from the observed data.

7.   Draw conclusions regarding rejection or nonrejection of H_0.

■ **Example 3.3**   Let us again consider the random sample of nine student applicants with mean QGRE score $\overline{X} = 520$ and standard deviation $S = 50$. The department chairperson suspects that, because of the declining reputation of the department, this year's applicants are not quite as good quantitatively as those from the previous five years for whom the average QGRE score was 600. If we assume that the population of QGRE scores from which our random sample has been selected is normally distributed, we can test the null hypothesis that the population mean score associated with this year's applicants is 600 versus the alternative hypothesis that it is less than 600. The *null hypothesis,* in mathematical terms, is H_0: $\mu = 600$, which asserts that the population mean μ for this year's applicants does not differ from what it has generally been in the past. The *alternative hypothesis* is stated as H_A: $\mu < 600$ which asserts that the QGRE scores, on average, have gotten worse.

We have thus far considered the first two steps of our testing procedure:

1.   Assumptions: The variable QGRE score has a normal distribution, from which a random sample has been selected.

2.   Hypotheses: H_0: $\mu = 600$; H_A: $\mu < 600$.

Our next step is to decide what error or probability we are willing to tolerate for incorrectly rejecting H_0 (i.e., making a Type I error, as discussed later in this chapter). We call this probability of making a Type I error the *significance level* α.[1]

We usually assign a value such as .1, .05, .025, or .01 to α. Suppose, for now, that we choose $\alpha = .025$. Then Step 3 is

3. Use $\alpha = .025$.

Step 4 requires us to specify the test statistic that will be used to test H_0. In this case, with H_0: $\mu = 600$, we have

4. $T = \dfrac{\overline{X} - 600}{S/\sqrt{9}} \frown t_8$ under H_0: $\mu = 600$.

Step 5 requires us to specify the decision rule that we will use to reject or not reject H_0. In determining this rule, we divide the possible values of T into two sets: the *rejection region* (or *critical region*), which consists of values of T for which we reject H_0; and the *nonrejection region,* which consists of those T-values for which we do not reject H_0. If our computed value of T falls in the rejection region, we conclude that the observed results deviate far enough from H_0 to cast considerable doubt on the validity of the null hypothesis.

In our example, we determine the critical region by choosing from t tables a point called the *critical point,* which defines the boundary between the nonrejection and rejection regions. The value we choose is

$$-t_{8,\,0.975} = -2.306$$

in which case the probability that the test statistic takes a value of less than -2.306 under H_0 is exactly $\alpha = .025$, the significance level (Figure 3.7). We thus have the following decision rule:

5. Reject H_0 if $T = \dfrac{\overline{X} - 600}{S/\sqrt{9}} < -2.306$; do not reject H_0 otherwise.

FIGURE 3.7   The critical point for Example 3.3

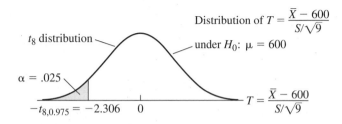

----

[1] Two types of errors can be made when performing a statistical test. A Type II error occurs if we fail to reject H_0 when H_0 is actually false. We denote the probability of a Type II error as β and call $(1 - \beta)$ the *power* of the test. For a fixed sample size, α and β for a given test are inversely related; that is, lowering one has the effect of increasing the other. In general, the power of any statistical test can be raised by increasing the sample size.

Now we simply apply the decision rule to our data by computing the observed value of T. In our example, since $\overline{X} = 520$ and $S = 50$, our computed T is

6. $T = \dfrac{\overline{X} - 600}{S/\sqrt{9}} = \dfrac{520 - 600}{50/3} = -4.8.$

The last step is to make the decision about H_0 based on the rule given in Step 5:

7. Since $T = -4.8$, which lies below -2.306, we reject H_0 at significance level .025 and conclude that there is evidence that students currently applying to the department have QGRE scores *significantly lower* than 600.

In addition to performing the procedure just described, we often want to compute a *P-value*, which quantifies *exactly how unusual the observed results would be if H_0 were true.* An equivalent way of describing the P-value is as follows: *The P-value gives the probability of obtaining a value of the test statistic that is at least as unfavorable to H_0 as the observed value* (Figure 3.8).

To get an idea of the approximate size of the P-value in this example, our approach is to determine from the table of the distribution of T under H_0 the two percentiles that bracket the observed value of T. In this case, the two percentiles are

$$-t_{8,\,0.995} = -3.355 \quad \text{and} \quad -t_{8,\,0.9995} = -5.041$$

Since the observed value of T lies between these two values, we conclude that the area we seek lies between the two areas corresponding to these two percentiles:

$.0005 < P < .005$

In interpreting this inequality, we observe that the P-value is *quite small,* indicating that we have observed a highly unusual result if H_0 is true. In fact, this P-value is so small as to lead us to reject H_0. Furthermore, the size of this P-value means that we would reject H_0 even for an α as small as .005.

For the general computation of a P-value, the appropriate P-value for a two-tailed test is twice that for the corresponding one-tailed test. If an investigator wants to draw conclusions about a test on the basis of the P-value (e.g., in lieu of specifying α a priori), the following guidelines are recommended:

1. If P is small (less than .01), reject H_0.

2. If P is large (greater than .1), do not reject H_0.

3. If $.01 < P < .1$, the significance is borderline, since we reject H_0 for $\alpha = .1$ but not for $\alpha = .01$.

Notice that, if we actually do specify α a priori, we reject H_0 when $P < \alpha$. ∎

FIGURE 3.8 The P-value

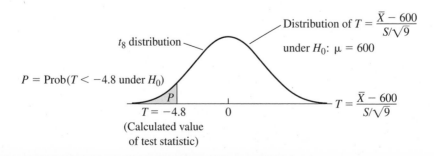

■ **Example 3.4**   We now look at one more worked example about hypothesis testing—this time involving a comparison of two means, μ_1 and μ_2. Consider the following data on health status change, which were discussed earlier:

$$\text{Group 1:} \quad n_1 = 15, \bar{X}_1 = 15.1, S_1 = 2.5$$
$$\text{Group 2:} \quad n_2 = 15, \bar{X}_2 = 12.3, S_2 = 3.0$$
$$(S_p = 2.76)$$

Suppose that we want to test at significance level .01 whether the true average change in health status differs between the two groups. The steps required to perform this test are as follows:

1. Assumptions: We have independent random samples from two normally distributed populations. The population variances are assumed to be equal.

2. Hypotheses: $H_0: \mu_1 = \mu_2;\ H_A: \mu_1 \neq \mu_2$.

3. Use $\alpha = .01$.

4. $T = \dfrac{(\bar{X}_1 - \bar{X}_2) - 0}{S_p\sqrt{\dfrac{1}{n_1} + \dfrac{1}{n_2}}} \frown t_{28}$ under H_0.

5. Reject H_0 if $|T| \geq t_{28,\,0.995} = 2.763$; do not reject H_0 otherwise (Figure 3.9).

6. $T = \dfrac{(\bar{X}_1 - \bar{X}_2) - 0}{S_p\sqrt{\dfrac{1}{n_1} + \dfrac{1}{n_2}}} = \dfrac{15.1 - 12.3}{2.76\sqrt{\dfrac{1}{15} + \dfrac{1}{15}}} = 2.78.$

7. Since $T = 2.78$ exceeds $t_{28,\,0.995} = 2.763$, we reject H_0 at $\alpha = .01$ and conclude that there is evidence that the true average change in health status differs between the two groups.

The P-value for this test is given by the shaded area in Figure 3.10. For the t distribution with 28 degrees of freedom, we find that $t_{28,\,0.995} = 2.763$ and $t_{28,\,0.9995} = 3.674$. Thus, $P/2$ is given by the inequality

$$1 - .9995 < \frac{P}{2} < 1 - .995$$

so

$$.001 < P < .01$$ ■

FIGURE 3.9   Critical region for the health status change example

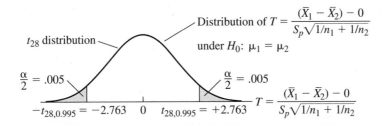

FIGURE 3.10   *P*-value for the health status change example

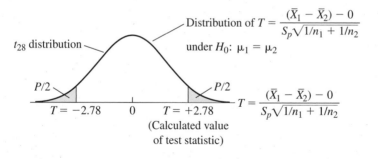

Distribution of $T = \dfrac{(\bar{X}_1 - \bar{X}_2) - 0}{S_p\sqrt{1/n_1 + 1/n_2}}$

under H_0: $\mu_1 = \mu_2$

t_{28} distribution

$P/2$

$P/2$

$T = -2.78$ 0 $T = +2.78$ $T = \dfrac{(\bar{X}_1 - \bar{X}_2) - 0}{S_p\sqrt{1/n_1 + 1/n_2}}$

(Calculated value
of test statistic)

3.7 Error Rates, Power, and Sample Size

Table 3.1 summarizes the decisions that result from hypothesis testing. If the true state of nature is that the null hypothesis is true and if the decision is made that the null hypothesis is true, then a correct decision has been made. Similarly, if the true state of nature is that the alternative hypothesis is true and if the decision is made that the alternative is true, then a correct decision has been made. On the other hand, if the true state of nature is that the null hypothesis is true but the decision is made to choose the alternative, then a false positive error (commonly referred to as a *Type I error*) has been made. And if the true state of nature supports the alternative hypothesis but the decision is made that the null hypothesis is true, then a false negative error (commonly referred to as a *Type II error*) has been made.

Table 3.2 summarizes the probabilities associated with the outcomes of hypothesis testing just described. If the true state of nature corresponds to the null hypothesis but the alternative hypothesis is chosen, then a Type I error has been made, with probability denoted by the symbol α. Hence, the probability of making a correct choice of H_0 when H_0 is true must be $1 - \alpha$. In turn, if the actual state of nature is that the alternative hypothesis is true but the null hypothesis is chosen, then a Type II error has occurred, with probability denoted by β. In turn, $1 - \beta$ is the probability of choosing the alternative hypothesis when it is true, and this probability is often called the *power of the test.*

When we design a research study, we would like to use statistical tests for which both α and β are small (i.e., for which there is a small chance of making either a Type I or a Type II error). For a given α, we can sometimes determine the sample size required in the study to ensure that β is no larger than some desired value for a particular alternative hypothesis of interest. Such a design consideration generally involves the use of a *sample size formula* pertinent to the

TABLE 3.1   Outcomes of hypothesis testing

Hypothesis Chosen	True State of Nature	
	H_0	H_A
H_0	Correct decision	False negative decision (Type II error)
H_A	False positive decision (Type I error)	Correct decision

TABLE 3.2   Probabilities of outcomes of hypothesis testing

Hypothesis Chosen	True State of Nature	
	H_0	H_A
H_0	$1 - \alpha$	β
H_A	α	$1 - \beta$

research question(s). This formula usually requires the researcher to guess values for some of the unknown parameters to be estimated in the study (see Cohen 1977; Muller and Peterson 1984; Kupper and Hafner 1989).

For example, the classical sample size formula used for a one-sided test of $H_0: \mu_1 = \mu_2$ versus $H_0: \mu_2 > \mu_1$, when a random sample of size n is selected from each of two normal populations with common variance σ^2, is as follows:

$$n \geq \frac{2(Z_{1-\alpha} + Z_{1-\beta})^2 \sigma^2}{\Delta^2}$$

For chosen values of α, β, and σ^2, this formula provides the minimum sample size n required to detect a specified difference $\Delta = \mu_2 - \mu_1$ between μ_1 and μ_2 (i.e., to reject $H_0: \mu_2 - \mu_1 = 0$ in favor of $H_A: \mu_2 - \mu_1 = \Delta > 0$ with power $1 - \beta$). Thus, in addition to picking α and β, the researcher must guess the size of the population variance σ^2 and specify the difference Δ to be detected. An educated guess about the value of the unknown parameter σ^2 can sometimes be made by using information obtained from related research studies. To specify Δ intelligently, the researcher has to decide on the smallest population mean difference $(\mu_2 - \mu_1)$ that is practically (as opposed to statistically) meaningful for the study.

For a fixed sample size, α and β are inversely related in the following sense. If one tries to guard against making a Type I error by choosing a small rejection region, the nonrejection region (and hence β) will be large. Conversely, protecting against a Type II error necessitates using a large rejection region, leading to a large value for α. Increasing the sample size generally decreases β; of course, α remains unaffected. A detailed discussion about power and sample size determination is provided in Chapter 27.

It is common practice to conduct several statistical tests using the same data set. If such a data-set-specific series of tests is performed and each test is based on a size α rejection region, the probability of making at least one Type I error will be much larger than α. This multiple-testing problem is pervasive and bothersome. One simple—but not optimal—method for addressing this problem is to employ the so-called *Bonferroni correction*. For example, if k tests are to be conducted and the overall Type I error rate (i.e., the probability of making at least one Type I error in k tests) is to be no more than α, then a rule of thumb is to conduct *each individual* test at a Type I error rate of α/k.

This simple adjustment ensures that the overall Type I error rate will (at least approximately) be no larger than α. In many situations, however, this correction leads to such a small rejection region for each individual test that the power of each test may be too low to detect important deviations from the null hypotheses being tested. Resolving this antagonism between Type I and Type II error rates requires a conscientious study design and carefully considered error rates for planned analyses.

Problems

1. **a.** Give two examples of discrete random variables.
   b. Give two examples of continuous random variables.

2. Name the four levels of measurement, and give an example of a variable at each level.

3. Assume that Z is a normal random variable with mean 0 and variance 1.
   a. $P(Z \geq -1) = ?$
   b. $P(Z \leq ?) = .20$

4. **a.** $P(\chi_7^2 \geq ?) = .01$
   b. $P(\chi_{12}^2 \leq 14) = ?$

5. **a.** $P(T_{13} \leq ?) = .10$
   b. $P(|T_{28}| \geq 2.05) = ?$

6. **a.** $P(F_{6, 24} \geq ?) = .05$
   b. $P(F_{5, 40} \geq 2.9) = ?$

7. What are the **(a)** mean, **(b)** median, and **(c)** mode of the standard normal distribution?

8. An $F_{1, \nu}$ random variable can be thought of as the square of what kind of random variable?

9. Find the **(a)** mean, **(b)** median, and **(c)** variance for the following set of scores:

   $$\{0, 2, 5, 6, 3, 3, 3, 1, 4, 3\}$$

   d. Find the set of Z scores for the data.

10. Which of the following statements about descriptive statistics is correct?
    a. *All* of the data are used to compute the median.
    b. The mean should be preferred to the median as a measure of central tendency if the data are noticeably skewed.
    c. The variance has the same units of measurement as the original observations.
    d. The variance can never be 0.
    e. The variance is like an average of squared deviations from the mean.

11. Suppose that the weight W of male patients registered at a diet clinic has the normal distribution with mean 190 and variance 100.
    a. For a random sample of patients of size $n = 25$, the expression $P(\overline{W} < 180)$, in which $\overline{W}$ denotes the mean weight, is equivalent to saying $P(Z > ?)$. [*Note:* Z is a standard normal random variable.]
    b. Find the interval (a, b) such that $P(a < \overline{W} < b) = .80$ for the same random sample in part (a).

12. The limits of a 95% confidence interval for the mean μ of a normal population with unknown variance are found by adding to and subtracting from the sample mean a certain multiple of the estimated standard error of the sample mean. If the sample size on which this confidence interval is based is 28, the *multiple* referred to in the previous sentence is the number _____ .

13. A random sample of 32 persons attending a certain diet clinic was found to have lost (over a three-week period) an average of 30 pounds, with a sample standard deviation of 11. For these data, a 99% confidence interval for the true mean weight loss by all patients attending the clinic would have the limits (?, ?).

14. From two normal populations assumed to have the same variance, independent random samples of sizes 15 and 19 were drawn. The first sample (with $n_1 = 15$) yielded mean and standard deviation 111.6 and 9.5, respectively, while the second sample ($n_2 = 19$) gave mean and standard deviation 100.9 and 11.5, respectively. The estimated standard error of the difference in sample means is _____ .

15. For the data of Problem 14, suppose that a test of H_0: $\mu_1 = \mu_2$ versus H_A: $\mu_1 > \mu_2$ yielded a computed value of the appropriate test statistic equal to 2.55.
    a. What conclusions should be drawn for $\alpha = .05$?
    b. What conclusions should be drawn for $\alpha = .01$?

16. Test the hypothesis that average body weight is the same for two independent diagnosis groups from one hospital:

    Diagnosis group 1 data: {132, 145, 124, 122, 165, 144, 151}
    Diagnosis group 2 data: {141, 139, 172, 131, 150, 125}

    You may assume that the populations from which the data come are each normally distributed, with equal population variances. What conclusion should be drawn, with $\alpha = .05$?

17. Independent random samples are drawn from two normal populations, which are assumed to have the same variance. One sample (of size 5) yields mean 86.4 and standard deviation 8.0, and the other sample (of size 7) has mean 78.6 and standard deviation 10. The limits of a 99% confidence interval for the difference in population means are found by adding to and subtracting from the difference in sample means a certain multiple of the estimated standard error of this difference. This *multiple* is the number _____ .

18. If a 99% confidence interval for $\mu_1 - \mu_2$ is 4.8 to 9.2, which of the following conclusions can be drawn *based on this interval*?
    a. Do not reject H_0: $\mu_1 = \mu_2$ at $\alpha = .05$ if the alternative is H_A: $\mu_1 \neq \mu_2$.
    b. Reject H_0: $\mu_1 = \mu_2$ at $\alpha = .01$ if the alternative is H_A: $\mu_1 \neq \mu_2$.
    c. Reject H_0: $\mu_1 = \mu_2$ at $\alpha = .01$ if the alternative is H_A: $\mu_1 < \mu_2$.
    d. Do not reject H_0: $\mu_1 = \mu_2$ at $\alpha = .01$ if the alternative is H_A: $\mu_1 \neq \mu_2$.
    e. Do not reject H_0: $\mu_1 = \mu_2 + 3$ at $\alpha = .01$ if the alternative is H_A: $\mu_1 \neq \mu_2 + 3$.

19. Assume that we gather data, compute a T, and reject the null hypothesis. If, in fact, the null hypothesis is true, we have made **(a)** _____ . If the null hypothesis is false, we have made **(b)** _____ . Assume instead that our data lead us to not reject the null hypothesis. If, in fact, the null hypothesis is true, we have made **(c)** _____ . If the null hypothesis is false, we have made **(d)** _____ .

20. Suppose that the critical region for a certain test of hypothesis is of the form $|T| \geq 2.5$ and the computed value of T from the data is -2.75. Which, if any, of the following statements is correct?
    a. H_0 should be rejected.
    b. The significance level α is the probability that, under H_0, T is either greater than 2.75 or less than -2.75.
    c. The nonrejection region is given by $-3.5 < T < 3.5$.
    d. The nonrejection region consists of values of T above 3.5 or below -3.5.
    e. The P-value of this test is given by the area to the right of $T = 3.5$ for the distribution of T under H_0.

21. Suppose that $\bar{X}_1 = 125.2$ and $\bar{X}_2 = 125.4$ are the mean systolic blood pressures for two samples of workers from different plants in the same industry. Suppose, further, that a test of $H_0: \mu_1 = \mu_2$ using these samples is rejected for $\alpha = .001$. Which of the following conclusions is most reasonable?
    a. There is a meaningful difference (clinically speaking) in population means but not a statistically significant difference.
    b. The difference in population means is both statistically and meaningfully significant.
    c. There is a statistically significant difference but not a meaningfully significant difference in population means.
    d. There is neither a statistically significant nor a meaningfully significant difference in population means.
    e. The sample sizes used must have been quite small.

22. The choice of an alternative hypothesis (H_A) should depend primarily on (choose all that apply):
    a. the data obtained from the study.
    b. what the investigator is interested in determining.
    c. the critical region.
    d. the significance level.
    e. the power of the test.

23. For each of the areas in the accompanying figure, labeled a, b, c, and d, select an answer from the following: α, $1 - \alpha$, β, $1 - \beta$.

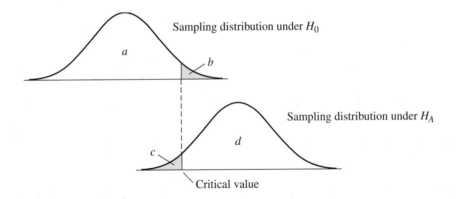

24. Suppose that $H_0: \mu_1 = \mu_2$ is the null hypothesis and that $.10 < P < .25$. What is the most appropriate conclusion?

25. Suppose that $H_0: \mu_1 = \mu_2$ is the null hypothesis and that $.005 < P < .01$. Which of the following conclusions is most appropriate?
    a. Do not reject H_0 because P is small.
    b. Reject H_0 because P is small.
    c. Do not reject H_0 because P is large.
    d. Reject H_0 because P is large.
    e. Do not reject H_0 at $\alpha = .01$.

References

Cohen, J. 1977. *Statistical Power Analysis for the Behavioral Sciences,* 2d ed. New York: Academic Press.

Kupper, L. L., and Hafner, K. B. 1989. "How Appropriate Are Popular Sample Size Formulas?" *The American Statistician* 43(2): 101–5.

Muller, K. E., and Peterson, B. L. 1984. "Practical Methods for Computing Power in Testing the Multivariate General Linear Hypothesis." *Computational Statistics and Data Analysis* 2: 143–58.

4

Introduction to Regression Analysis

4.1 Preview

Regression analysis is a statistical tool for evaluating the relationship of one or more independent variables $X_1, X_2, \ldots, X_k$ to a single, continuous dependent variable Y. It is most often used when the independent variables cannot be controlled, as when they are collected in a sample survey or other observational study. Nevertheless, it is equally applicable to more-controlled experimental situations.

In practice, a regression analysis is appropriate for several possibly overlapping situations, including the following.

Application 1   You want to *characterize the relationship* between the dependent and independent variables by determining the extent, direction, and strength of the association. For example ($k = 2$), in Thompson's (1972) study described in Chapter 1, one of the primary questions was to describe the extent, direction, and strength of the association between "patient satisfaction with medical care" (Y) and the variables "affective communication between patient and physician" (X_1) and "informational communication between patient and physician" (X_2).

Application 2   You *seek a quantitative formula* or equation to describe (e.g., predict) the dependent variable Y as a function of the independent variables $X_1, X_2, \ldots, X_k$. For example ($k = 1$), a quantitative formula may be desired for a study of the effect of dosage of a blood-pressure-reducing treatment (X_1) on blood pressure change (Y).

Application 3   You want to describe quantitatively or qualitatively the relationship between $X_1, X_2, \ldots, X_k$ and Y but *control for the effects of still other variables* $C_1, C_2, \ldots, C_p$, which you believe have an important relationship with the dependent variable. For example ($k = 2, p = 2$), a study of the epidemiology of chronic diseases might describe the relationship of blood pressure (Y) to smoking habits (X_1) and social class (X_2), controlling for age (C_1) and weight (C_2).

Application 4   You want to *determine which of several independent variables are important and which are not* for describing or predicting a dependent variable. You may want to control for other variables. You may also want to *rank* independent variables in their order of importance. In Thompson's (1972) study, for example ($k = 4$, $p = 2$), the researcher sought to determine for the dependent variable "satisfaction with medical care" (Y) which of the following independent variables were important descriptors: WORRY (X_1), WANT (X_2), INFCOM (X_3), and AFFCOM (X_4). It was also considered necessary to control for AGE (C_1) and EDUC (C_2).

Application 5   You want to *determine the best mathematical model* for describing the relationship between a dependent variable and one or more independent variables. Any of the previous examples can be used to illustrate this.

Application 6   You want to *compare several derived regression relationships*. An example would be a study to determine whether smoking (X_1) is related to blood pressure (Y) in the same way for males as for females, controlling for age (C_1).

Application 7   You want to *assess the interactive effects of two or more independent variables* with regard to a dependent variable. For example, you may want to determine whether the relationship of alcohol consumption (X_1) to blood pressure level (Y) is different depending on smoking habits (X_2). In particular, the relationship between alcohol and blood pressure might be quite strong for heavy smokers but very weak for nonsmokers. If so, we would say that there is *interaction* between alcohol and smoking. Then, any conclusions about the relationship between alcohol and blood pressure must take into account whether—and possibly how much—a person smokes. More generally, if X_1 and X_2 interact in their joint effect on Y, then the relationship of either X variable to Y depends on the value of the other X variable.

Application 8   You want to *obtain a valid and precise estimate of one or more regression coefficients* from a larger set of regression coefficients in a given model. For example, you may want to obtain an accurate estimate of the coefficient of a variable measuring alcohol consumption (X_1) in a regression model that relates hypertension status (Y), a dichotomous response variable, to X_1 and several other control variables (e.g., age and smoking status). Such an estimate may be used to quantify the effect of alcohol consumption on hypertension status after adjustment for the effects of certain control variables also in the model.

4.2 Association versus Causality

A researcher must be cautious about interpreting the results obtained from a regression analysis or, more generally, from any form of analysis seeking to quantify an association (e.g., via a correlation coefficient) among two or more variables. Although the statistical computations used to produce an estimated measure of association may be correct, the estimate itself may be biased. Such bias may result from the method used to select subjects for the study, from errors in the information used in the statistical analyses, or even from other variables that can account for the observed association but that have not been measured or appropriately considered in the analysis. (See Kleinbaum, Kupper, and Morgenstern 1982 or Kleinbaum 2002, for a discussion of validity in epidemiologic research.)

For example, if diastolic blood pressure and physical activity level were measured on a sample of individuals at a particular time, a regression analysis might suggest that, on the average, blood pressure decreases with increased physical activity; further, such an analysis may provide evidence (e.g., based on a confidence interval) that this association is of moderate strength and is statistically significant. If the study involved only healthy adults, however, or if physical activity level was measured inappropriately, or if such other factors as age, race, and sex were not correctly taken into account, the above conclusions might be rendered invalid or at least questionable.

Continuing with the preceding example, if the investigators were satisfied that the findings were basically valid (i.e., the observed association was not spurious), could they then conclude that a low level of physical activity is a cause of high blood pressure? The answer is an unequivocal *no!*

The finding of a "statistically significant" association in a particular study (no matter how well done) does not establish a causal relationship. To evaluate claims of causality, the investigator must consider criteria that are external to the specific characteristics and results of any single study.

It is beyond the scope of this text to discuss causal inference making. Nevertheless, we will briefly review some key ideas on this subject. Most strict definitions of causality (e.g., Blalock 1971; Susser 1973) require that a *change* in one variable (X) always *produce* a change in another variable (Y).[1] This suggests that, to demonstrate a cause–effect relationship between X and Y, *experimental proof* is required that a change in Y results from a change in X. Although it is needed, such experimental evidence is often impractical, infeasible, or even unethical to obtain, especially when considering risk factors (e.g., cigarette smoking or exposure to chemicals) that are potentially harmful to human subjects. Consequently, alternative criteria based on information *not* involving direct experimental evidence are typically employed when attempting to make causal inferences regarding variable relationships in human populations.

One school of thought regarding causal inference has produced a collection of procedures commonly referred to as *path analysis* (Blalock 1971) or *structural equations analysis* (Bollen 1989). To date, such procedures have been applied primarily to sociological and political science studies. Essentially, these methods attempt to assess causality indirectly, by eliminating competing causal explanations via data analysis and finally arriving at an acceptable causal model that is not obviously contradicted by the data at hand. Thus, these methods, rather than attempting to establish a particular causal theory directly, arrive at a final causal model through a process of elimination. In this procedure, literature relevant to the research question must be considered in order to postulate causal models; in addition, various estimated correlation ("path") coefficients must be compared by means of data analysis.

A second, more widely used approach for making causal conjectures, particularly in the health and medical sciences, employs a judgmental (and more qualitative than quantitative)

---

[1] An imperfect approximation to this ideal for real-world phenomena might be that, on the average, a change in Y is produced by a change in X.

evaluation of the combined results from several studies, using a set of operational criteria generally agreed on as necessary (but not sufficient) for supporting a given causal theory. Efforts to define such a set of criteria were made in the late 1950s and early 1960s by investigators reviewing research on the health hazards of smoking. A list of general criteria for assessing the extent to which available evidence supports a causal relationship was formalized by Bradford Hill (1971), and this list has subsequently been adopted by many epidemiologic researchers. The list contains seven criteria:

1. *Strength of association.* The stronger an observed association appears over a series of different studies, the less likely it is that this association is spurious because of bias.

2. *Dose–response effect.* The value of the dependent variable (e.g., the rate of disease development) changes in a meaningful pattern (e.g., increases) with the dose (or level) of the suspected causal agent under study.

3. *Lack of temporal ambiguity.* The hypothesized cause precedes the occurrence of the effect. (The ability to establish this time pattern depends on the study design used.)

4. *Consistency of the findings.* Most or all studies concerned with a given causal hypothesis produce similar results. Of course, studies dealing with a given question may all have serious bias problems that can diminish the importance of observed associations.

5. *Biological and theoretical plausibility of the hypothesis.* The hypothesized causal relationship is consistent with current biological and theoretical knowledge. The current state of knowledge may nonetheless be insufficient to explain certain findings.

6. *Coherence of the evidence.* The findings do not seriously conflict with accepted facts about the outcome variable being studied (e.g., knowledge about the natural history of some disease).

7. *Specificity of the association.* The study factor (i.e., the suspected cause) is associated with only one effect (e.g., a specific disease). Many study factors have multiple effects, however, and most diseases have multiple causes.

Clearly, applying the above criteria to a given causal hypothesis is hardly a straightforward matter. Even if these criteria are all satisfied, a causal relationship cannot be claimed with complete certainty. Nevertheless, in the absence of solid experimental evidence, the use of such criteria may be a logical and practical way to address the issue of causality, especially with regard to studies on human populations.

4.3 Statistical versus Deterministic Models

Although *causality* cannot be established by statistical analyses, associations among variables can be well quantified in a *statistical* sense. With proper statistical design and analysis, an investigator can model the extent to which changes in independent variables are related to changes in dependent variables. However, *statistical models* developed by using regression or other multivariable methods must be distinguished from *deterministic models.*

The law of falling bodies in physics, for example, is a deterministic model that assumes an ideal setting: the dependent variable varies in a completely prescribed way according to a perfect (error-free) mathematical function of the independent variables.

Statistical models, on the other hand, allow for the possibility of error in describing a relationship. For example, in a study relating blood pressure to age, persons of the same age are unlikely to have exactly the same observed blood pressure. Nevertheless, with proper statistical methods, we might be able to conclude that, on the average, blood pressure increases with age. Further, appropriate statistical modeling can permit us to predict the expected blood pressure for a given age and to associate a measure of variability with that prediction. Through the use of probability and statistical theory, such statements take into account the uncertainty of the real world by means of measurement error and individual variability. Of course, because such statements are necessarily nondeterministic, they require careful interpretation. Unfortunately, such interpretation is often quite difficult to make.

4.4 Concluding Remarks

In this short chapter, we have informally introduced the general regression problem and indicated a variety of situations to which regression modeling can be applied. We have also cautioned the reader about the types of conclusions that can be drawn from such modeling efforts.

We now turn to the actual quantitative details involved in fitting a regression model to a set of data and then in estimating and testing hypotheses about important parameters in the model. In the next chapter, we will discuss the simplest form of regression model—a straight line. In subsequent chapters, we will consider more complex forms.

References

Blalock, H. M., Jr., ed. 1971. *Causal Models in the Social Sciences.* Chicago: Aldine Publishing.

Bollen, K. A. 1989. *Structural Equations with Latent Variables.* New York: John Wiley & Sons.

Hill, A. B. 1971. *Principles of Medical Statistics,* 9th ed. New York: Oxford University Press.

Kleinbaum, D. G.; Kupper, L. L.; and Morgenstern, H. 1982. *Epidemiologic Research: Principles and Quantitative Methods.* Belmont, Calif.: Lifetime Learning Publications.

Kleinbaum, D. G. 2002. *ActivEpi.* New York and Berlin: Springer Publishers.

Susser, M. 1973. *Causal Thinking in the Health Sciences.* New York: Oxford University Press.

Thompson, S. J. 1972. "The Doctor–Patient Relationship and Outcomes of Pregnancy." Ph.D. dissertation, Department of Epidemiology, University of North Carolina, Chapel Hill.

Straight-line Regression Analysis

5.1 Preview

The simplest (but by no means trivial) form of the general regression problem deals with one dependent variable Y and one independent variable X. We have previously described the general problem in terms of k independent variables $X_1, X_2, \ldots, X_k$. Let us now restrict our attention to the special case $k = 1$ but denote X_1 as X to keep our notation as simple as possible. To clarify the basic concepts and assumptions of regression analysis, we find it useful to begin with a single independent variable. Furthermore, researchers often begin by looking at one independent variable at a time even when several independent variables are eventually jointly considered.

5.2 Regression with a Single Independent Variable

We begin this section by describing the statistical problem of finding the *curve* (straight line, parabola, etc.) that *best fits* the data, closely approximating the true (but unknown) relationship between X and Y.

5.2.1 The Problem

Given a sample of n individuals (or other study units, such as geographical locations, time points, or pieces of physical material), we observe for each a value of X and a value of Y. We thus have n pairs of observations that can be denoted by $(X_1, Y_1), (X_2, Y_2), \ldots, (X_n, Y_n)$, where the subscripts now refer to different individuals rather than different variables. Because these pairs may be considered as points in two-dimensional space, we can plot them on a graph. Such a graph is called a *scatter diagram*. For example, measurements of age and systolic blood pressure for 30 individuals might yield the scatter diagram given in Figure 5.1.

FIGURE 5.1   Scatter diagram of age and systolic blood pressure

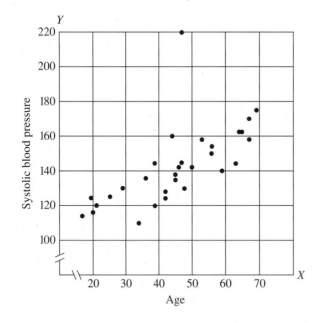

5.2.2 Basic Questions to Be Answered

Two basic questions must be dealt with in any regression analysis:

1. What is the most appropriate mathematical model to use—a straight line, a parabola, a log function, or what?

2. Given a specific model, what do we mean by and how do we determine the best-fitting model for the data? In other words, if our model is a straight line, how do we find the best-fitting line?

5.2.3 General Strategy

Several general strategies can be used to study the relationship between two variables by means of regression analysis. The most common of these is called the *forward method*. This strategy begins with a simply structured model—usually a straight line—and adds more complexity to the model in successive steps, if necessary. Another strategy, called the *backward method*, begins with a complicated model—such as a high-degree polynomial—and successively simplifies it, if possible, by eliminating unnecessary terms. A third approach uses *a model suggested from experience or theory*, which is revised either toward or away from complexity, as dictated by the data.

The strategy chosen depends on the type of problem and on the data; there are no hard-and-fast rules. The quality of the results often depends more on the skill with which a strategy is applied than on the particular strategy chosen. It is often tempting to try many strategies and then to use the results that provide the most "reasonable" interpretation of the relationship between the response and predictor variables. This exploratory approach demands particular care to ensure the reliability of any conclusions.

In Chapter 16, we will discuss in detail the issue of choosing a model-building strategy. For reasons discussed there, we often prefer the backward strategy. The forward method, however, corresponds more naturally to the usual development of theory from simple to complex. In some simple situations, forward and backward strategies lead to the same final model. In general, however, this is not the case!

Since it is the simplest method to understand and can therefore be used as a basis for understanding other methods, we begin by offering a step-by-step description of the forward strategy:

1. Assume that a straight line is the appropriate model. Later the validity of this assumption can be investigated.

2. Find the best-fitting straight line, which is the line among all possible straight lines that best agrees (as will be defined later) with the data.

3. Determine whether the straight line found in Step 2 significantly helps to describe the dependent variable Y. Here it is necessary to check that certain basic statistical assumptions (e.g., normality) are met. These assumptions will be discussed in detail subsequently.

4. Examine whether the assumption of a straight-line model is correct. One approach for doing this is called *testing for lack of fit,* although other approaches can be used instead.

5. If the assumption of a straight line is found to be invalid in Step 4, fit a new model (e.g., a parabola) to the data, determine how well it describes Y (i.e., repeat Step 3), and then decide whether the new model is appropriate (i.e., repeat Step 4).

6. Continue to try new models until an appropriate one is found.

A flow diagram for this strategy is given in Figure 5.2.

Since the usual (forward) approach to regression analysis with a single independent variable begins with the assumption of a straight-line model, we will consider this model first. Before describing the *statistical* methodology for this special case, let us review some basic straight-line mathematics. You may want to skip the next section if you are already familiar with its contents.

FIGURE 5.2   Flow diagram of the forward method

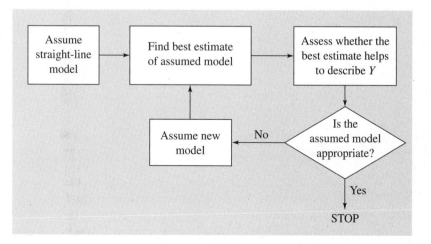

5.3 Mathematical Properties of a Straight Line

Mathematically, a straight line can be described by an equation of the form

$$y = \beta_0 + \beta_1 x \tag{5.1}$$

We have used lowercase letters y and x, instead of capital letters, in this equation to emphasize that we are treating these variables in a purely mathematical, rather than statistical, context. The symbols β_0 and β_1 have constant values for a given line and are therefore not considered variables; β_0 is called the *y-intercept* of the line, and β_1 is called the *slope*. Thus, $y = 5 - 2x$ describes a straight line with intercept 5 and slope -2, whereas $y = -4 + 1x$ describes a different line with intercept -4 and slope 1. These two lines are shown in Figure 5.3.

The intercept β_0 is the value of y when $x = 0$. For the line $y = 5 - 2x$, $y = 5$ when $x = 0$. For the line $y = -4 + 1x$, $y = -4$ when $x = 0$. The slope β_1 is the amount of change in y for each 1-unit increase in x. For any given straight line, this rate of change is always constant. Thus, for the line $y = 5 - 2x$, when x changes 1 unit from 3 to 4, y changes -2 units (the value of the slope) from $5 - 2(3) = -1$ to $5 - 2(4) = -3$; and when x changes from 1 to 2, also 1 unit, y changes from $5 - 2(1) = 3$ to $5 - 2(2) = 1$, also -2 units.

The properties of any straight line can be viewed graphically as in Figure 5.3. To graph a given line, plot any two points on the line and then connect them with a ruler. One of the two points often used is the y-intercept. This point is given by $(x = 0, y = 5)$ for the line $y = 5 - 2x$ and by $(x = 0, y = -4)$ for $y = -4 + 1x$. The other point for each line may be determined by arbitrarily selecting an x and finding the corresponding y. An x of 3 was used in our two examples. Thus, for $y = 5 - 2x$, an x of 3 yields a y of $5 - 2(3) = -1$; and for $y = -4 + 1x$, an x of 3 yields a y of $-4 + 1(3) = -1$. The line $y = 5 - 2x$ can then be drawn by connecting the points $(x = 0, y = 5)$ and $(x = 3, y = -1)$, and the line $y = -4 + 1x$ can be drawn from the points $(x = 0, y = -4)$ and $(x = 3, y = -1)$.

As Figure 5.3 illustrates, in the equation $y = 5 - 2x$, y decreases as x increases. Such a line is said to have *negative* slope. Indeed, this definition agrees with the sign of the slope -2 in

FIGURE 5.3   Straight-line plots

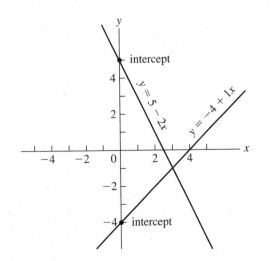

the equation. Conversely, the line $y = -4 + 1x$ is said to have *positive* slope, since y increases as x increases.

5.4 Statistical Assumptions for a Straight-line Model

Suppose that we have tentatively assumed a straight-line model as the first step in the forward method for determining the best model to describe the relationship between X and Y. We now want to determine the best-fitting line. Certainly, we will have no trouble deciding what is meant by "best fitting" if the data allow us to draw a single straight line through every point in the scatter diagram. Unfortunately, this will never happen with real-life data. For example, persons of the same age are unlikely to have the same blood pressure, height, or weight.

Thus, the straight line we seek can only approximate the true state of affairs and cannot be expected to predict precisely each individual's Y from that individual's X. In fact, this need to approximate would exist even if we measured X and Y on the whole population of interest instead of on just a sample from that population. In addition, the fact that the line is to be determined from the sample data and not from the population requires us to consider the problem of how to estimate unknown population parameters.

What are these parameters? The ones of primary concern at this point are the intercept β_0 and the slope β_1 of the straight line of the general mathematical form of (5.1) that best fits the X-Y data for the entire population. To make inferences from the sample about this population line, we need to make five statistical assumptions covering existence, independence, linearity, homoscedasticity, and normality.

5.4.1 Statement of Assumptions

Assumption 1: Existence   *For any fixed value of the variable X, Y is a random variable with a certain probability distribution having finite mean and variance.* The (population) mean of this distribution will be denoted as $\mu_{Y|X}$ and the (population) variance as $\sigma^2_{Y|X}$. The notation "$Y|X$" indicates that the mean and the variance of the random variable Y depend on the value of X.

This assumption applies to any regression model, whether a straight line or not. Figure 5.4 illustrates the assumption. The different distributions are drawn vertically to correspond to different values of X. The dots denoting the mean values $\mu_{Y|X}$ at different X's have been connected to form the *regression equation,* which is the population model to be estimated from the data.

Assumption 2: Independence   *The Y-values are statistically independent of one another.* This assumption is appropriate in many, but not all, situations. In particular, Assumption 2 is usually violated when different observations are made on the same individual at different times. For example, if weight is measured on an individual at different times (i.e., longitudinally over time), we can expect that the weight at one time is related to the weight at a later time. As another example, if blood pressure is measured on a given individual longitudinally over time, we can expect the blood pressure value at one time to be in the same range as the blood pressure value at the previous or following time. When Assumption 2 is not satisfied, ignoring the dependency among the Y-values can often lead to invalid statistical conclusions.

FIGURE 5.4 **General regression equation**

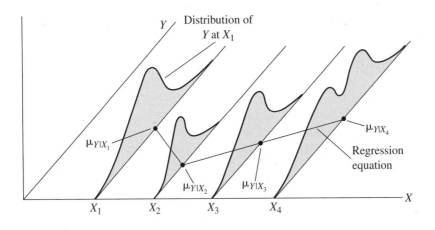

When the Y-values are not independent, special methods can be used to find the best-fitting model and to make valid statistical inferences. The method chosen depends on the characteristics of the response variable, the type of dependence, and the complexity of the problem. In Chapter 26, we describe a "mixed model" analysis of variance approach for designs involving repeated measurements on study subjects. In some cases, multivariate linear models are appropriate. See Morrison (1976) or Timm (1975) for a general introduction to multivariate linear models. More recently, Zeger and Liang (1986) introduced the "generalized estimating equations" (GEE) approach for analyzing correlated response data, and an excellent book on this very general and useful methodology is now available (Diggle, Heagerty, Liang, and Zeger 2002).

Assumption 3: Linearity *The mean value of Y, $\mu_{Y|X}$, is a straight-line function of X.* In other words, if the dots denoting the different mean values $\mu_{Y|X}$ are connected, a straight line is obtained. This assumption is illustrated in Figure 5.5.

Using mathematical symbols, we can describe Assumption 3 by the equation

$$\mu_{Y|X} = \beta_0 + \beta_1 X \qquad (5.2)$$

where β_0 and β_1 are the intercept and the slope of this (population) straight line, respectively. Equivalently, we can express (5.2) in the form

$$Y = \beta_0 + \beta_1 X + E \qquad (5.3)$$

where E denotes a random variable that has mean 0 at fixed X (i.e, $\mu_{E|X} = 0$ for any X). More specifically, since X is fixed and not random, (5.3) represents the dependent variable Y as the sum of a constant term ($\beta_0 + \beta_1 X$) and a random variable (E).[1] Thus, the probability distributions of Y and E differ only in the value of this constant term; that is, since E has mean 0, Y must have mean $\beta_0 + \beta_1 X$.

---

[1] For our purposes, the practical implication of the statement "X is fixed and not random" for making statistical inferences from sample to population is that the *only* random component on the right-hand side of (5.3) is E when X is fixed.

FIGURE 5.5   Straight-line assumption

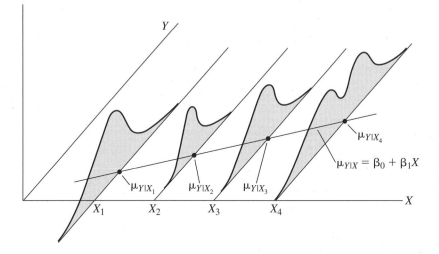

Equations (5.2) and (5.3) describe a *statistical* model. These equations should be distinguished from the *mathematical* model for a straight line described by (5.1), which does not consider Y as a random variable.

The variable E describes how distant an individual's response can be from the population regression line (Figure 5.6). In other words, what we observe at a given X (namely, Y) is in *error* from that expected on the average (namely, $\mu_{Y|X}$) by an amount E, which is random and varies from individual to individual. For this reason, E is commonly referred to as the *error component* in the model (5.3). Mathematically, E is given by the formula

$$E = Y - (\beta_0 + \beta_1 X)$$

or by

$$E = Y - \mu_{Y|X}$$

This concept of an error component is particularly important for defining a good-fitting line, since, as we will see in the next section, a line that fits data well ought to have small deviations (or errors) between what is observed and what is predicted by the fitted model.

FIGURE 5.6   Error component E

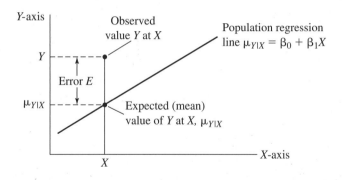

Assumption 4: Homoscedasticity   *The variance of Y is the same for any X. (Homo-* means "same," and *-scedastic* means "scattered.") An example of the violation of this assumption (called *heteroscedasticity*) is shown in Figure 5.5, where the distribution of Y at X_1 has considerably more spread than the distribution of Y at X_2. This means that $\sigma^2_{Y|X_1}$, the variance of Y at X_1, is greater than $\sigma^2_{Y|X_2}$, the variance of Y at X_2.

In mathematical terms, the homoscedastic assumption can be written as

$$\sigma^2_{Y|X} \equiv \sigma^2$$

for all X. This formula is a short-hand way of saying that, since $\sigma^2_{Y|X_i} = \sigma^2_{Y|X_j}$ for *any* two different values of X, we might as well simplify our notation by giving the common variance a single name—say, σ^2—that does not involve X at all.

A number of techniques of varying statistical sophistication can be used to determine whether the homoscedastic assumption is satisfied. Some of these procedures will be discussed in Chapter 14.

Assumption 5: Normal Distribution   *For any fixed value of X, Y has a normal distribution.* This assumption makes it possible to evaluate the statistical significance (e.g., by means of confidence intervals and tests of hypotheses) of the relationship between X and Y, as reflected by the fitted line.

Figure 5.5 provides an example in which this assumption is violated. In addition to the variances not being all equal in this figure, the distributions of Y at X_3 and at X_4 are not normal. The distribution at X_3 is skewed, whereas the normal distribution is symmetric. The distribution at X_4 is bimodal (two humps), whereas the normal distribution is unimodal (one hump). Methods for determining whether the normality assumption is tenable are described in Chapter 14.

If the normality assumption is not *badly* violated, the conclusions reached by a regression analysis in which normality is assumed will generally be reliable and accurate. This stability property with respect to deviations from normality is a type of *robustness*. Consequently, we recommend giving considerable leeway before deciding that the normality assumption is so badly violated as to require alternative inference-making procedures.

If the normality assumption is deemed unsatisfactory, the Y-values may be transformed by using a log, square root, or other function to see whether the new set of observations is approximately normal. Care must be taken when using such transformations to ensure that other assumptions, such as variance homogeneity, are not violated for the transformed variable. Fortunately, in practice such transformations usually help satisfy both the normality and variance homogeneity assumptions.

5.4.2 Summary and Comments

The assumptions of homoscedasticity and normality apply to the distribution of Y when X is fixed (i.e., Y given X), and not to the distributions of Y associated with different X-values. Many people find it more convenient to describe these two assumptions in terms of the error E. It is sufficient to say that the random variable E has a normal distribution with mean 0 and variance σ^2 for all observations. Of course, the linearity, existence, and independence assumptions must also be specified.

It is helpful to maintain distinctions among such concepts as *random variables, parameters,* and *point estimates.* The variable Y is a random variable, and an observation of it yields a particular value or "realization"; the variable X is assumed to be measured without error. The constants β_0 and β_1 are parameters with unknown but specific values for a particular population. The variable E is a random, unobservable variable. Using some estimation procedure (e.g., least squares), one constructs point estimates $\hat{\beta}_0$ and $\hat{\beta}_1$ of β_0 and β_1, respectively. Once $\hat{\beta}_0$ and $\hat{\beta}_1$ are obtained, a point estimate of E at the value X is calculated as

$$\hat{E} = Y - \hat{Y} = Y - (\hat{\beta}_0 + \hat{\beta}_1 X)$$

The estimated error $\hat{E}$ is typically called a *residual.* If there are n (X, Y) pairs $(X_1, Y_1), (X_2, Y_2), \ldots,$ (X_n, Y_n), then there are n residuals $\hat{E}_i = Y_i - \hat{Y}_i = Y_i - (\hat{\beta}_0 + \hat{\beta}_1 X_i)$, $i = 1, 2, \ldots, n$.

Some statisticians refer to a normally distributed random variable as having a Gaussian distribution. This terminology avoids confusing *normal* with its other meaning of "customary" or "usual"; it emphasizes the fact that the term *Gaussian* refers to a *particular* bell-shaped function; and it appropriately honors the mathematician Carl Gauss (1777–1855).

5.5   Determining the Best-fitting Straight Line

By far the simplest and quickest method for determining a straight line is to choose the line that can best be drawn by eye. Although this method often paints a reasonably good picture, it is extremely subjective and imprecise and is worthless for statistical inference. We now consider two analytical approaches for finding the best-fitting straight line.

5.5.1   The Least-squares Method

The *least-squares method* determines the best-fitting straight line as the line that *minimizes* the sum of squares of the lengths of the vertical-line segments (Figure 5.7) drawn from the observed data points on the scatter diagram to the fitted line. The idea here is that the smaller the deviations of observed values from this line (and consequently the smaller the sum of squares of these deviations), the closer or "snugger" the best-fitting line will be to the data.

In mathematical notation, the least-squares method is described as follows. Let $\hat{Y}_i$ denote the estimated response at X_i based on the fitted regression line; in other words, $\hat{Y}_i = \hat{\beta}_0 + \hat{\beta}_1 X_i$, where $\hat{\beta}_0$ and $\hat{\beta}_1$ are the intercept and the slope of the fitted line, respectively. The vertical

FIGURE 5.7   Deviations of observed points from the fitted regression line

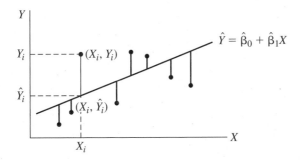

distance between the observed point (X_i, Y_i) and the corresponding point $(X_i, \hat{Y}_i)$ on the fitted line is given by the absolute value $|Y_i - \hat{Y}_i|$ or $|Y_i - (\hat{\beta}_0 + \hat{\beta}_1 X_i)|$. The sum of the squares of all such distances is given by

$$\sum_{i=1}^{n} (Y_i - \hat{Y}_i)^2 = \sum_{i=1}^{n} (Y_i - \hat{\beta}_0 - \hat{\beta}_1 X_i)^2$$

The least-squares solution is defined to be the choice of $\hat{\beta}_0$ and $\hat{\beta}_1$ for which the sum of squares just described is a minimum. In standard jargon, $\hat{\beta}_0$ and $\hat{\beta}_1$ are termed the *least-squares esti-mates* of the parameters β_0 and β_1, respectively, in the statistical model given by (5.3).

The *minimum sum of squares* corresponding to the least-squares estimates $\hat{\beta}_0$ and $\hat{\beta}_1$ is usually called the *sum of squares about the regression line,* the *residual sum of squares,* or the *sum of squares due to error* (SSE). The measure SSE is of great importance in assessing the quality of the straight-line fit, and its interpretation will be discussed in Section 5.6.

Mathematically, the essential property of the measure SSE can be stated in the following way. If β_0^* and β_1^* denote any other possible estimators of β_0 and β_1, we must have

$$\text{SSE} = \sum_{i=1}^{n} (Y_i - \hat{\beta}_0 - \hat{\beta}_1 X_i)^2 \leq \sum_{i=1}^{n} (Y_i - \beta_0^* - \beta_1^* X_i)^2$$

5.5.2 The Minimum-variance Method

The *minimum-variance method* is more classically statistical than the method of least squares, which can be viewed as a purely mathematical algorithm. In this second approach, determining the best fit becomes a statistical estimation problem. The goal is to find point estimators of β_0 and β_1 with good statistical properties. In this regard, under the previous assumptions, the best line is determined by the estimators $\hat{\beta}_0$ and $\hat{\beta}_1$ that are *unbiased* for their unknown population counterparts β_0 and β_1, respectively, and have minimum variance among all unbiased (linear) estimators of β_0 and β_1.

5.5.3 Solution to the Best-fit Problem

Fortunately, both the least-squares method and the minimum-variance method yield exactly the same solution, which we will state without proof.[2]

Let $\overline{Y}$ denote the sample mean of the observations on Y, and let $\overline{X}$ denote the sample mean of the values of X. Then the best-fitting straight line is determined by the formulas

$$\hat{\beta}_1 = \frac{\sum_{i=1}^{n} (X_i - \overline{X})(Y_i - \overline{Y})}{\sum_{i=1}^{n} (X_i - \overline{X})^2} \tag{5.4}$$

---

[2] Another general method of parameter estimation is called *maximum likelihood.* Under the assumptions of a Gaussian distribution for Y_i given X_i fixed and mutual independence of the Y_i's, the maximum-likelihood estimates of β_0 and β_1 are exactly the same as the least-squares and minimum-variance estimates. A general discussion of maximum-likelihood methods is given in Chapter 21.

$$\hat{\beta}_0 = \bar{Y} - \hat{\beta}_1 \bar{X} \qquad (5.5)$$

In calculating $\hat{\beta}_0$ and $\hat{\beta}_1$, we recommend using a computer program for regression modeling from a convenient computer package. Many computer packages with regression programs are now available, the most popular of which are SAS, SPSS, STATA, and MINITAB. In this text, we will use SAS exclusively to present computer output, although we recognize that other packages may be preferred by particular users.

The least-squares line may generally be represented by

$$\hat{Y} = \hat{\beta}_0 + \hat{\beta}_1 X \qquad (5.6)$$

or equivalently by

$$\hat{Y} = \bar{Y} + \hat{\beta}_1(X - \bar{X}) \qquad (5.7)$$

Either (5.6) or (5.7) may be used to determine predicted Y's that correspond to X's actually observed or to other X-values in the region of experimentation. Simple algebra can be used to demonstrate the equivalence of (5.6) and (5.7). The right-hand side of (5.7), $\bar{Y} + \hat{\beta}_1(X - \bar{X})$, can be written as $\bar{Y} + \hat{\beta}_1 X - \hat{\beta}_1 \bar{X}$, which in turn equals $\bar{Y} - \hat{\beta}_1 \bar{X} + \hat{\beta}_1 X$, which is equivalent to (5.6) since $\hat{\beta}_0 = \bar{Y} - \hat{\beta}_1 \bar{X}$.

Table 5.1 lists observations on systolic blood pressure and age for a sample of 30 individuals. The scatter diagram for this sample was presented in Figure 5.1. For this data set, the output from the SAS package's PROC REG routine is also shown.

**TABLE 5.1   Observations on systolic blood pressure (SBP)
and age for a sample of 30 individuals**

Individual (i)	SBP (Y)	Age (X)	Individual (i)	SBP (Y)	Age (X)
1	144	39	16	130	48
2	220	47	17	135	45
3	138	45	18	114	17
4	145	47	19	116	20
5	162	65	20	124	19
6	142	46	21	136	36
7	170	67	22	142	50
8	124	42	23	120	39
9	158	67	24	120	21
10	154	56	25	160	44
11	162	64	26	158	53
12	150	56	27	144	63
13	140	59	28	130	29
14	110	34	29	125	25
15	128	42	30	175	69

Edited SAS Output for Data of Table 5.1

```
                        REGRESSION OF SBP(Y) ON AGE (X)
                           Descriptive Statistics
```

Variable	Sum	Mean	Uncorrected SS	Variance	Standard Deviation
Intercept	30.00000	1.00000	30.00000	0	0
X	1354.00000	$\overline{X} \to$ 45.13333	67894	$S_X^2 \to$ 233.91264	15.29420
Y	4276.00000	$\overline{Y} \to$ 142.53333	624260	$S_Y^2 \to$ 509.91264	22.58125

```
                           Analysis of Variance
```

Source	DF	Sum of Squares	Mean Square	F Value	Pr > F
Model	1	6394.02269	6394.02269	21.33	<.0001
Error	28	8393.44398	299.76586		
Corrected Total	29	14787			

$\nwarrow S_{Y|X}^2$

r^2 (see Chapter 6) $\downarrow$

Root MSE	$S_{Y|X} \to$ 17.31375	R-square	0.4324	
Dependent Mean	142.53333	Adj R-sq	0.4121	
Coeff Var	12.14716			

t-test statistics and P-values for tests on the intercept and slope ↙ ↘

```
                           Parameter Estimates
```

Variable	DF	Parameter Estimate	Standard Error	t Value	Pr > |t|
Intercept	1	$\hat{\beta}_0 \longrightarrow$ 98.71472	10.00047	9.87	<.0001
X	1	$\hat{\beta}_1 \longrightarrow$ 0.97087	0.21022	4.62	<.0001

The estimated line (5.6) is computed to be $\hat{Y} = 98.71 + 0.97X$ and is graphed as the solid line in Figure 5.8. This line reflects the clear trend that systolic blood pressure increases as age increases. Notice that one point, (47, 220), seems quite out of place with the other data; such an observation is often called an *outlier*. Because an outlier can affect the least-squares estimates, the determination of whether an outlier should be removed from the data is important. Usually this decision can be made only after thorough evaluation of the experimental conditions, the data collection process, and the data themselves. (See Chapter 14 for further discussion of the treatment of outliers.) If the decision is difficult, one can always determine the effect of removing the outlier by refitting the model to the remaining data. In this case, the resulting least-squares line is

$$\hat{Y} = 97.08 + 0.95X$$

and is shown on the graph in Figure 5.8 as the dashed line. As might be expected, this line is slightly below the one obtained by using all the data.

FIGURE 5.8   Best-fitting straight line to age–systolic blood pressure data of Table 5.1

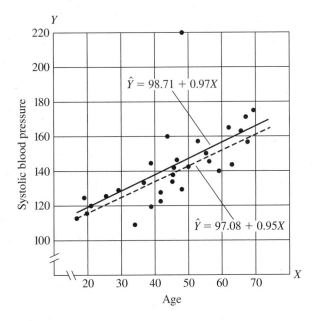

5.6 Measure of the Quality of the Straight-line Fit and Estimate of σ^2

Once the least-squares line is determined, we want to evaluate whether the fitted line actually aids in predicting Y—and if so, to what extent. A measure that helps to answer these questions is provided by

$$\text{SSE} = \sum_{i=1}^{n}(Y_i - \hat{Y}_i)^2$$

where $\hat{Y}_i = \hat{\beta}_0 + \hat{\beta}_1 X_i$. Clearly, if SSE = 0 the straight line fits perfectly; that is, $Y_i = \hat{Y}_i$ for each i, and every observed point lies on the fitted line. As the fit gets worse, SSE gets larger, since the deviations of points from the regression line become larger.

   Two possible factors contribute to the inflation of SSE. First, there may be a lot of variation in the data; that is, σ^2 may be large. Second, the assumption of a straight-line model may not be appropriate. It is important, therefore, to determine the separate effects of each of these components, since they address decidedly different issues with regard to the fit of the model. For the time being, we will assume that the second factor is not at issue. Thus, assuming that the straight-line model is appropriate, we can obtain an estimate of σ^2 by using SSE. Such an estimate is needed for making statistical inferences about the true (i.e., population) straight-line relationship between X and Y. This estimate of σ^2 is given by the formula

$$S^2_{Y|X} = \frac{1}{n-2}\sum_{i=1}^{n}(Y_i - \hat{Y}_i)^2 = \frac{1}{n-2}\text{SSE} \tag{5.8}$$

■ **Example 5.1**   From the computer output provided on p. 52, we find

$$S^2_{Y|X} = 299.76586$$

Readers may wonder why $S^2_{Y|X}$ estimates σ^2—especially since, at first glance, (5.8) looks different from the formula usually used for the sample variance, $\sum_{i=1}^{n}(Y_i - \overline{Y})^2/(n-1)$. The latter formula is appropriate when the Y's are independent, with the same mean μ and variance σ^2. Since μ is unknown in this case (its estimate being, of course, $\overline{Y}$), we must divide by $n-1$ instead of n to make the sample variance an unbiased estimator of σ^2. To put it another way, we subtract 1 from n because to estimate σ^2 we had to estimate *one* other parameter first, μ.

If a straight-line model is appropriate, the population mean response $\mu_{Y|X}$ changes with X. For example, using the least-squares line (5.6) as an approximation to the population line for the age–systolic blood pressure data, the estimated mean of the Y's at $X = 40$ is approximately 138, whereas the estimated mean of the Y's at $X = 70$ is close to 167. Therefore, instead of subtracting $\overline{Y}$ from each Y_i when estimating σ^2, we should subtract $\hat{Y}_i$ from Y_i, because $\hat{Y}_i = \hat{\beta}_0 + \hat{\beta}_1 X_i$ is the estimate of $\mu_{Y|X_i}$. Furthermore, we subtract 2 from n in the denominator of our estimate, since the determination of $\hat{Y}_i$ requires the estimation of two parameters, β_0 and β_1.

When we discuss testing for lack of fit of the assumed model, we will show how it is possible in some circumstances to obtain a reliable estimate of σ^2 that does not assume the correctness of the straight-line model.   ■

5.7 Inferences About the Slope and Intercept

To assess whether the fitted line helps to predict Y from X, and to take into account the uncertainties of using a sample, it is standard practice to compute confidence intervals and/or test statistical hypotheses about the unknown parameters in the assumed straight-line model. Such confidence intervals and tests require, as described in Section 5.4, the assumptions that the random variable Y has a normal distribution at each fixed value of X and that $Y_1, Y_2, \ldots, Y_n$ are mutually independent given $X_1, X_2, \ldots, X_n$. Working from these assumptions, one can deduce that the estimators $\hat{\beta}_0$ and $\hat{\beta}_1$ are each normally distributed, with respective means β_0 and β_1 when (5.2) holds, and with easily derivable variances.[3] These estimators, together with estimators of their variances, can then be used to form confidence intervals and test statistics based on the *t distribution*.

---

[3] An important property that allows the normality assumption on Y to carry over to $\hat{\beta}_0$ and $\hat{\beta}_1$ is that these estimators are *linear functions* of the mutually independent Y's. Such a function is defined by a formula of the structure

$$L = \sum_{i=1}^{n} c_i Y_i$$

or equivalently, $L = c_1 Y_1 + c_2 Y_2 + \cdots + c_n Y_n$, where the c_i's are constants not involving the Y's. A simple example of a linear function is $\overline{Y}$, which can be written as

$$\sum_{i=1}^{n} \frac{1}{n} Y_i$$

Here the c_i's equal $1/n$ for each i. The normality of $\hat{\beta}_0$ and $\hat{\beta}_1$ derives from a statistical theorem stating that linear functions of mutually independent normally distributed observations are themselves normally distributed.

More specifically, to test the hypothesis H_0: $\beta_1 = \beta_1^{(0)}$, where $\beta_1^{(0)}$ is some hypothesized value for β_1, the test statistic used is

$$T = \frac{\hat{\beta}_1 - \beta_1^{(0)}}{S_{\hat{\beta}_1}} \qquad (5.9)$$

where

$$S_{\hat{\beta}_1} = \frac{S_{Y|X}}{S_X \sqrt{n-1}}$$

This test statistic has the t distribution with $n-2$ degrees of freedom when H_0 is true. Here, $S_{Y|X}^2$ denotes the sample estimate of σ^2 defined by (5.8), and S_X is the sample standard deviation of the X's defined by (3.2) on p. 15. The denominator $S_{\hat{\beta}_1}$ in the test statistic is an estimate of the unknown standard error of the estimator $\hat{\beta}_1$, given by

$$\sigma_{\hat{\beta}_1} = \frac{\sigma}{S_X \sqrt{n-1}}$$

Thus, the test statistic (5.9) is the ratio of a normally distributed random variable minus its mean divided by an independent estimator of its standard error. Such a statistic has a t distribution for the kinds of situations encountered in this text.

Similarly, to test the hypothesis H_0: $\beta_0 = \beta_0^{(0)}$, we use the test statistic

$$T = \frac{\hat{\beta}_0 - \beta_0^{(0)}}{S_{Y|X}\sqrt{\dfrac{1}{n} + \dfrac{\overline{X}^2}{(n-1)S_X^2}}} \qquad (5.10)$$

which also has the t distribution with $n-2$ degrees of freedom when H_0: $\beta_0 = \beta_0^{(0)}$ is true. The denominator here estimates the standard error of $\hat{\beta}_0$, given by

$$\sigma_{\hat{\beta}_0} = \sigma \sqrt{\frac{1}{n} + \frac{\overline{X}^2}{(n-1)S_X^2}}$$

The reason why both test statistics (5.9) and (5.10) have $n-2$ degrees of freedom is that both involve $S_{Y|X}^2$, which itself has $n-2$ degrees of freedom and is the only random component in the denominator of both statistics.

In testing either of the preceding hypotheses at a significance level α, we should reject H_0 whenever any of the following occur:

$$\begin{cases} T \geq t_{n-2,\,1-\alpha} \text{ for an upper one-tailed test} & (\text{i.e., } H_A: \beta_1 > \beta_1^{(0)} \text{ or } H_A: \beta_0 > \beta_0^{(0)}) \\ T \leq -t_{n-2,\,1-\alpha} \text{ for a lower one-tailed test} & (\text{i.e., } H_A: \beta_1 < \beta_1^{(0)} \text{ or } H_A: \beta_0 < \beta_0^{(0)}) \\ |T| \geq t_{n-2,\,1-\alpha/2} \text{ for a two-tailed test} & (\text{i.e., } H_A: \beta_1 \neq \beta_1^{(0)} \text{ or } H_A: \beta_0 \neq \beta_0^{(0)}) \end{cases}$$

where $t_{n-2,\,1-\alpha}$ denotes the $100(1-\alpha)\%$ point of the t distribution with $n-2$ degrees of freedom. As an alternative to using a specified significance level, we may compute P-values based on the calculated value of the test statistic T.

Table 5.2 summarizes the formulas needed for performing statistical tests and computing confidence intervals for β_0 and β_1. Also given in this table are formulas for inference-making procedures concerned with prediction using the fitted line; these formulas are described in Sections 5.9 and 5.10. Table 5.3 gives examples illustrating the use of each formula in Table 5.2, using the age–systolic blood pressure data previously considered.

TABLE 5.2   Confidence intervals, tests of hypotheses, and prediction intervals for straight-line regression analysis

Parameter	$100(1 - \alpha)\%$ Confidence Interval	H_0	Test Statistic (T)	Distribution Under H_0
β_1	$\hat{\beta}_1 \pm t_{n-2,\,1-\alpha/2} S_{\hat{\beta}_1}$	$\beta_1 = \beta_1^{(0)}$	$T = \dfrac{(\hat{\beta}_1 - \beta_1^{(0)})}{S_{\hat{\beta}_1}}$	t_{n-2}
β_0	$\hat{\beta}_0 \pm t_{n-2,\,1-\alpha/2} S_{\hat{\beta}_0}$	$\beta_0 = \beta_0^{(0)}$	$T = \dfrac{(\hat{\beta}_0 - \beta_0^{(0)})}{S_{\hat{\beta}_0}}$	t_{n-2}
$\mu_{Y\mid X_0}$	$\bar{Y} + \hat{\beta}_1(X_0 - \bar{X}) \pm t_{n-2,\,1-\alpha/2} S_{\hat{Y}_{X_0}}$	$\mu_{Y\mid X_0} = \mu_{Y\mid X_0}^{(0)}$	$T = \dfrac{\bar{Y} + \hat{\beta}_1(X_0 - \bar{X}) - \mu_{Y\mid X_0}^{(0)}}{S_{\hat{Y}_{X_0}}}$	t_{n-2}
Y_{X_0} [a]	$\bar{Y} + \hat{\beta}_1(X_0 - \bar{X})$ $\pm t_{n-2,\,1-\alpha/2} S_{Y\mid X} \sqrt{1 + \dfrac{1}{n} + \dfrac{(X_0 - \bar{X})^2}{(n-1)S_X^2}}$ [b]			

Note: $\mu_{Y\mid X} = \beta_0 + \beta_1 X$ is the assumed true regression model.

$$\hat{\beta}_1 = \frac{\sum_{i=1}^{n}(X_i - \bar{X})(Y_i - \bar{Y})}{\sum_{i=1}^{n}(X_i - \bar{X})^2}$$

$$\hat{\beta}_0 = \bar{Y} - \hat{\beta}_1 \bar{X}$$

$$\hat{Y} = \hat{\beta}_0 + \hat{\beta}_1 X = \bar{Y} + \hat{\beta}_1(X - \bar{X})$$

$t_{n-2,\,1-\alpha/2}$ is the $100(1 - \alpha/2)\%$ point of the t distribution with $n - 2$ degrees of freedom.

$$S_{Y\mid X}^2 = \frac{1}{n-2}\sum_{i=1}^{n}(Y_i - \hat{Y}_i)^2 \qquad S_{\hat{\beta}_1} = \frac{S_{Y\mid X}}{S_X \sqrt{n-1}}$$

$$S_Y^2 = \frac{1}{n-1}\sum_{i=1}^{n}(Y_i - \bar{Y})^2 \qquad S_{\hat{\beta}_0} = S_{Y\mid X}\sqrt{\frac{1}{n} + \frac{\bar{X}^2}{(n-1)S_X^2}}$$

$$S_X^2 = \frac{1}{n-1}\sum_{i=1}^{n}(X_i - \bar{X})^2 \qquad S_{\hat{Y}_{X_0}} = S_{Y\mid X}\sqrt{\frac{1}{n} + \frac{(X_0 - \bar{X})^2}{(n-1)S_X^2}}$$

[a]Value of Y at $X = X_0$, a random variable, not a parameter.
[b]Prediction interval for a new individual's Y when $X = X_0$.

5.8 Interpretations of Tests for Slope and Intercept

Researchers often make errors when interpreting the results of tests regarding the slope and the intercept. In this section we discuss conclusions that can be drawn based on nonrejection or rejection of the most common null hypotheses involving the slope and the intercept.[4] In our discussion we assume that the usual assumptions about normality, independence, and variance homogeneity are not violated. If these assumptions do not hold, any conclusions based on testing procedures developed under the assumptions are suspect.

---

[4] Statistically speaking, "nonrejection of H_0" really means "determination that there is insufficient evidence to reject H_0."

TABLE 5.3 **Sample calculations of confidence intervals, tests of hypotheses, and prediction intervals for the age–systolic blood pressure data of Table 5.1.**

Parameter	$100(1 - \alpha)\%$ Confidence Interval	H_0	Test Statistic (T)
β_1	For $\alpha = .05$: $0.97 \pm 2.0484(0.21)$ or $(0.54, 1.40)$	$\beta_1 = 0$	$T = \dfrac{(0.97 - 0)}{0.21} = 4.62$ Reject H_0 at $\alpha = .05$ (two-tailed test), since $t_{28,\,0.975} = 2.0484$ ($P < .001$).
β_0	For $\alpha = .05$: $98.71 \pm (2.0484)(10.00)$ or $(78.23, 119.20)$	$\beta_0 = 75$	$T = \dfrac{(98.71 - 75)}{10.00} = 2.37$ Reject H_0 at $\alpha = .05$ (two-tailed test), since $t_{28,\,0.975} = 2.0484$ ($.02 < P < .05$).
$\mu_{Y\mid65}$[a]	For $\alpha = .10$: $142.53 + (0.97)(65 - 45.13) \pm (1.7011)(5.24)$ or $(152.89, 170.72)$	$\mu_{Y\mid65} = 147$	$T = \dfrac{142.53 + (0.97)(65 - 45.13) - 147}{5.24}$ $= 2.82$ Reject H_0 at $\alpha = .10$ (two-tailed test), since $t_{28,\,0.95} = 1.7011$ $(.001 < P < .01)$.
Y_{65}[b]	For $\alpha = .10$: $142.53 + (0.97)(65 - 45.13)$ $\pm (1.7011)(17.31)\sqrt{1 + \dfrac{1}{30} + \dfrac{(65 - 45.13)^2}{(30 - 1)(15.29)^2}}$ or $(131.04, 192.5)$		

Note: $n = 30$, $\hat{\beta}_0 = 98.71$, $\hat{\beta}_1 = 0.97$, $\overline{Y} = 142.53$, $\overline{X} = 45.13$, $S_{Y\mid X} = 17.31$, $S_X = 15.29$,

$$S_{\hat{\beta}_1} = \frac{17.31}{15.29\sqrt{30 - 1}} = 0.21, \quad S_{\hat{\beta}_0} = 17.31\sqrt{\frac{1}{30} + \frac{(45.13)^2}{(30 - 1)(15.29)^2}} = 10.00, \quad S_{\hat{Y}_{X_0}} = 17.31\sqrt{\frac{1}{30} + \frac{(65 - 45.13)^2}{(30 - 1)(15.29)^2}} = 5.24$$

[a]Mean of Y at $X_0 = 65$.

[b]Value of Y at $X_0 = 65$.

5.8.1 Test for Zero Slope

The most important test of hypothesis dealing with the parameters of the straight-line model relates to whether the slope of the regression line differs significantly from zero or, equivalently, whether X helps to predict Y using a straight-line model. The appropriate null hypothesis for this test is $H_0\colon \beta_1 = 0$. Care must be taken in interpreting the result of the test of this hypothesis.

If we ignore for now the ever-present possibilities of making a Type I error (i.e., rejecting a true H_0) or a Type II error (i.e., not rejecting a false H_0), we can make the following interpretations:

1. If $H_0\colon \beta_1 = 0$ is not rejected, one of the following is true.
   a. For a true underlying straight-line model, X provides little or no help in predicting Y; that is, $\overline{Y}$ is essentially as good as $\overline{Y} + \hat{\beta}_1(X - \overline{X})$ for predicting Y (Figure 5.9(a)).
   b. The true underlying relationship between X and Y is *not* linear; that is, the true model may involve quadratic, cubic, or other more complex functions of X (Figure 5.9(b)).

FIGURE 5.9   **Interpreting the test for zero slope**

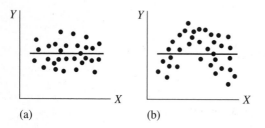

(a)   (b)

Examples when H_0: $\beta_1 = 0$ is rejected

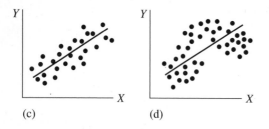

(c)   (d)

Combining (a) and (b), we can say that *not rejecting H_0: $\beta_1 = 0$ implies that a straight-line model in X is not the best model to use and does not provide much help for predicting Y.*

2. If H_0: $\beta_1 = 0$ is rejected, one of the following is true.
    a. X provides significant information for predicting Y; that is, the model $\overline{Y} + \hat{\beta}_1(X - \overline{X})$ is far better than the naive model $\overline{Y}$ for predicting Y (Figure 5.9(c)).
    b. A better model might have, for example, a curvilinear term (e.g., Figure 5.9(d)), although there is statistical evidence of a linear component.

Combining (a) and (b), we can say that *rejecting H_0: $\beta_1 = 0$ implies that a straight-line model in X is better than a model that does not include X at all, although it may well represent only a linear approximation to a truly nonlinear relationship.*

An important point implied by these interpretations is that *whether or not the hypothesis H_0: $\beta_1 = 0$ is rejected, a straight-line model may not be appropriate; instead, some other curve may describe the relationship between X and Y better.*

5.8.2 Test for Zero Intercept

Another hypothesis sometimes tested involves whether the population straight line goes through the origin—that is, whether its Y-intercept β_0 is zero. The null hypothesis here is H_0: $\beta_0 = 0$. If this null hypothesis is not rejected, it may be appropriate to remove the constant from the model, provided that previous experience or a relevant theory suggests that the line may go through the origin and provided that observations are taken around the origin to improve the estimate of β_0. Forcing the fitted line through the origin merely because H_0: $\beta_0 = 0$ cannot be rejected may give a spurious appearance to the regression line. In any case this hypothesis is rarely of interest in most studies, because data are not usually gathered near the origin. For example, when dealing with age (X) and blood pressure (Y), we are not interested in knowing what happens at $X = 0$, and we rarely choose values of X near 0.

5.9 Inferences About the Regression Line $\mu_{Y|X} = \beta_0 + \beta_1 X$

In addition to making inferences about the slope and the intercept, we may also want to perform tests and/or compute confidence intervals concerning the regression line itself. More specifically, for a given $X = X_0$, we may want to find a confidence interval for $\mu_{Y|X_0}$, the mean value of Y at X_0.[5] We may also be interested in testing the hypothesis H_0: $\mu_{Y|X_0} = \mu_{Y|X_0}^{(0)}$ where $\mu_{Y|X_0}^{(0)}$ is some hypothesized value of interest.

The test statistic to use for the hypothesis H_0: $\mu_{Y|X_0} = \mu_{Y|X_0}^{(0)}$ is given by the formula

$$T = \frac{\hat{Y}_{X_0} - \mu_{Y|X_0}^{(0)}}{S_{\hat{Y}_{X_0}}} \tag{5.11}$$

where $\hat{Y}_{X_0} = \hat{\beta}_0 + \hat{\beta}_1 X_0 = \overline{Y} + \hat{\beta}_1(X_0 - \overline{X})$ is the predicted value of Y at X_0 and

$$S_{\hat{Y}_{X_0}} = S_{Y|X}\sqrt{\frac{1}{n} + \frac{(X_0 - \overline{X})^2}{(n-1)S_X^2}} \tag{5.12}$$

This test statistic, like those for slope and intercept, has the t distribution with $n - 2$ degrees of freedom when H_0 is true. The denominator $S_{\hat{Y}_{X_0}}$ is an estimate of the standard error of $\hat{Y}_{X_0}$, which is given by

$$\sigma_{\hat{Y}_{X_0}} = \sigma\sqrt{\frac{1}{n} + \frac{(X_0 - \overline{X})^2}{(n-1)S_X^2}}$$

The corresponding confidence interval for $\mu_{Y|X_0}$ at a given $X = X_0$ is given by the formula

$$\hat{Y}_{X_0} \pm t_{n-2,\, 1-\alpha/2} S_{\hat{Y}_{X_0}} \tag{5.13}$$

In addition to drawing inferences about specific points on the regression line, researchers often find it useful to construct a confidence interval for the regression line over the entire range of X-values. The most convenient way to do this is to plot the upper and lower confidence limits obtained for several specified values of X and then to sketch the two curves that connect these points. Such curves are called *confidence bands* for the regression line. The confidence bands for the data of Table 5.1 are indicated in Figure 5.10.

Sketching confidence bands by hand calculator can be a painful job. Instead, we generally recommend using a computer program for regression analysis to compute confidence intervals

---

[5] The point $(X_0, \mu_{Y|X_0})$, of course, lies on the population regression line.

FIGURE 5.10 **90% confidence and prediction bands for age–systolic blood pressure data of Table 5.1.**

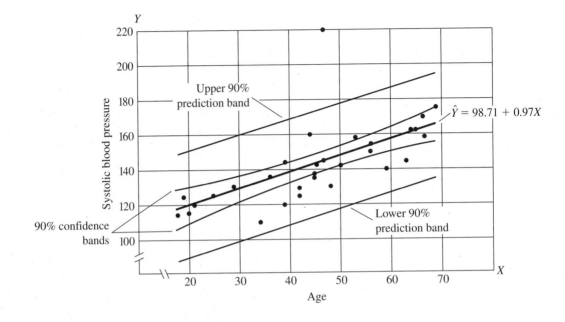

for a range of X_0 values and then to plot these intervals on the same graph that contains the fitted regression line. A convenient way to choose X_0 values is to use $X_0 = \bar{X}$, $X_0 = \bar{X} \pm k$, $X_0 = \bar{X} \pm 2k$, $X_0 = \bar{X} \pm 3k$, etc., where k is chosen so that the range of X-values in the data is uniformly covered.

■ **Example 5.2** For 90% confidence bands for our age–systolic blood pressure data, confidence interval formula (5.13) simplifies to:

$$142.53 + (0.97)(X_0 - 45.13) \pm 29.45\sqrt{0.033 + \frac{(X_0 - 45.13)^2}{6{,}783.48}} \tag{5.14}$$

At $X_0 = \bar{X} = 45.13$, the formula simplifies to $142.53 \pm 29.45\sqrt{0.033}$, which yields a lower limit of 137.18 and an upper limit of 147.90. Notice that the minimum-width confidence interval is always obtained at $X_0 = \bar{X}$, since the second term under the square root sign in (5.14) is zero.
At $X_0 = \bar{X} \pm k$, the confidence interval formula becomes

$$142.53 \pm (0.97)k \pm 29.45\sqrt{0.033 + \frac{k^2}{6{,}783.48}}$$

Thus, when $k = 10$, the confidence limits are 145.78 and 158.70 for $X_0 = \bar{X} + 10$, and 126.37 and 139.28 for $X_0 = \bar{X} - 10$. Figure 5.10 shows the 90% confidence bands for these data, together with the fitted model.
In SAS, PROC REG will compute 95% confidence intervals for $\mu_{Y|X}$, using the sample independent variable values of X_0. In the following output, the lower and upper 95% confidence limits for the systolic blood pressure–age example appear in the columns labeled "95% CL Mean." ■

Edited SAS Output Showing Confidence and Prediction Intervals

REGRESSION OF SBP(Y) ON AGE(X)

$\hat{Y}_{X_0}$ $S_{\hat{Y}_0}$ 95% confidence and prediction intervals

Obs	Dependent Variable	Predicted Value	Std Error Mean Predict	95% CL Mean		95% CL Predict	
1	144.0000	136.5787	3.4139	129.5857	143.5717	100.4302	172.7271
2	220.0000	144.3456	3.1853	137.8208	150.8704	108.2848	180.4064
3	138.0000	142.4039	3.1612	135.9285	148.8792	106.3520	178.4558
4	145.0000	144.3456	3.1853	137.8208	150.8704	108.2848	180.4064
5	162.0000	161.8213	5.2377	151.0923	172.5502	124.7684	198.8742
6	142.0000	143.3748	3.1663	136.8889	149.8606	107.3210	179.4285
7	170.0000	163.7630	5.5787	152.3356	175.1905	126.5018	201.0242
8	124.0000	139.4913	3.2289	132.8771	146.1055	103.4142	175.5684
9	158.0000	163.7630	5.5787	152.3356	175.1905	126.5018	201.0242
10	154.0000	153.0835	3.9001	145.0946	161.0724	116.7292	189.4377
11	162.0000	160.8504	5.0717	150.4616	171.2393	123.8945	197.8063
12	150.0000	153.0835	3.9001	145.0946	161.0724	116.7292	189.4377
13	140.0000	155.9961	4.2999	147.1881	164.8041	119.4531	192.5391
14	110.0000	131.7243	3.9332	123.6676	139.7810	95.3551	168.0935
15	128.0000	139.4913	3.2289	132.8771	146.1055	103.4142	175.5684
16	130.0000	145.3165	3.2180	138.7248	151.9082	109.2435	181.3895
17	135.0000	142.4039	3.1612	135.9285	148.8792	106.3520	178.4558
18	114.0000	115.2195	6.7058	101.4832	128.9558	77.1867	153.2523
19	116.0000	118.1321	6.1568	105.5204	130.7439	80.4909	155.7734
20	124.0000	117.1613	6.3382	104.1781	130.1444	79.3939	154.9286
21	136.0000	133.6661	3.6984	126.0901	141.2420	97.4003	169.9318
22	142.0000	147.2582	3.3225	140.4525	154.0640	111.1455	183.3709
23	120.0000	136.5787	3.4139	129.5857	143.5717	100.4302	172.7271
24	120.0000	119.1030	5.9774	106.8588	131.3472	81.5833	156.6227
25	160.0000	141.4330	3.1700	134.9395	147.9265	105.3779	177.4882
26	158.0000	150.1708	3.5675	142.8632	157.4785	113.9602	186.3815
27	144.0000	159.8796	4.9090	149.8238	169.9353	123.0159	196.7432
28	130.0000	126.8700	4.6362	117.3731	136.3668	90.1549	163.5851
29	125.0000	122.9865	5.2825	112.1657	133.8072	85.9069	160.0661
30	175.0000	165.7048	5.9299	153.5579	177.8517	128.2167	203.192

5.10 Prediction of a New Value of Y at X_0

We have just dealt with estimating the mean $\mu_{Y|X_0}$ at $X = X_0$. In practice, we may want instead to estimate the response Y of a single individual, based on the fitted regression line; that is, we may want to predict an individual's Y given his or her $X = X_0$. The obvious point estimate to use in this case is $\hat{Y}_{X_0} = \hat{\beta}_0 + \hat{\beta}_1 X_0$. Thus, $\hat{Y}_{X_0}$ is used to estimate both the mean $\mu_{Y|X_0}$ and an individual's response Y at X_0.

Of course, some bounds (i.e., limits) must be placed on this estimate to take its variability into account. Here, however, we cannot say that we are constructing a confidence interval for Y, since Y is a random variable and not a parameter; neither can we perform a test of hypothesis, for the same reason. The term used to describe the "hybrid limits" we require is the *prediction interval* (PI), which is given by

$$\bar{Y} + \hat{\beta}_1(X_0 - \bar{X}) \pm t_{n-2,\, 1-\alpha/2} S_{Y|X} \sqrt{1 + \frac{1}{n} + \frac{(X_0 - \bar{X})^2}{(n-1)S_X^2}} \qquad (5.15)$$

We first note that an estimator of an individual's response should naturally have more variability than an estimator of a group's mean response. This is reflected by the extra term 1 under the square root sign in (5.15), which is not found in the square root part of the confidence interval formula for $\mu_{Y|X}$ (see (5.12) and (5.13)). To be more specific, in predicting an actual observed Y for a given individual, there are two sources of error operating: individual error as measured by σ^2 and the error in estimating $\mu_{Y|X_0}$ using $\hat{Y}_{X_0}$. More precisely, this can be expressed by the equation

$$\underbrace{Y - \hat{Y}_{X_0}}_{\substack{\text{Error in} \\ \text{predicting an} \\ \text{individual's} \\ Y \text{ at } X_0}} = \underbrace{(Y - \mu_{Y|X_0})}_{\substack{\text{Deviation of} \\ \text{individual's} \\ Y \text{ from true} \\ \text{mean at } X_0}} + \underbrace{(\mu_{Y|X_0} - \hat{Y}_{X_0})}_{\substack{\text{Deviation of} \\ \hat{Y}_{X_0} \text{ from true} \\ \text{mean at } X_0}}$$

This representation allows us to write the variance of an individual's predicted response at X_0 as

$$\text{Var } Y + \text{Var } \hat{Y}_{X_0} = \sigma^2 \left[1 + \frac{1}{n} + \frac{(X_0 - \overline{X})^2}{(n-1)S_X^2} \right]$$

This variance expression is estimated by replacing σ^2 by its estimate $S_{Y|X}^2$. This accounts for the expression on the right-hand side of the prediction interval in (5.15).

Prediction bands, used to describe individual predictions over the entire range of X-values, may be determined in a manner analogous to that by which confidence bands are computed. Figure 5.10 gives 90% prediction bands for the age–systolic blood pressure data. As expected, the 90% prediction bands in this figure are wider than the corresponding 90% confidence bands.

Once again, SAS can be used to compute 95% prediction intervals. In the previous computer output, the lower and upper limits for these intervals are given in the two columns labeled "95% CL Predict." SAS uses the sample values of the independent variable X_0.

5.11 Assessing the Appropriateness of the Straight-line Model

In Section 5.1, we noted that the usual strategy for regression with a single independent variable is to assume that the straight-line model is appropriate. This assumption is then rejected if the data indicate that a more complex model is warranted.

Many methods may be used to assess whether the straight-line assumption is reasonable. These will be discussed separately later. The basic techniques include tests for lack of fit and are understood most easily in terms of polynomial regression models (Chapter 15). Many regression diagnostics (Chapter 14) also help in evaluating the straight-line assumption, either explicitly or implicitly. With the linear model, the assumptions of linearity, homoscedasticity, and normality are so intertwined that they often are met or violated as a set.

Problems

SAS computer output is provided for most of the problems in this chapter. The same is true for many of the subsequent chapters. In some problems, using the output can significantly reduce the computational and programming effort required to answer the questions. Since the actual data are also often provided, however, instructors and students may choose to do their own computations and/or programming to perform the necessary analyses.

1. The accompanying table gives the dry weights (Y) of 11 chick embryos ranging in age from 6 to 16 days (X). Also given in the table are the values of the common logarithms of the weights (Z).

Age (X) (days)	6	7	8	9	10	11
Dry Weight (Y)	0.029	0.052	0.079	0.125	0.181	0.261
Log_{10} Dry Weight (Z)	−1.538	−1.284	−1.102	−0.903	−0.742	−0.583

Age (X) (days)	12	13	14	15	16
Dry Weight (Y)	0.425	0.738	1.130	1.882	2.812
Log_{10} Dry Weight (Z)	−0.372	−0.132	0.053	0.275	0.449

a. Observe the following two scatter diagrams. Describe the relationships between age (X) and dry weight (Y), and between age and log_{10} dry weight (Z).

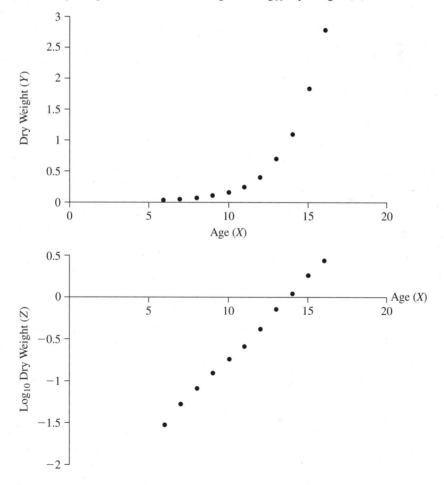

b. State the simple linear regression models for these two regressions: Y regressed on X, and Z regressed on X.

    c. Determine the least-squares estimates of each of the regression lines in part (b).

    d. Sketch each estimated line on the appropriate scatter diagram. Which of the two regression lines has the better fit? Based on your answers to parts (a)–(c), is it more appropriate to run a linear regression of Y on X, or of Z on X? Explain.

    e. For the regression that you chose as being more appropriate in part (d), find 95% confidence intervals for the true slope and intercept. Interpret each interval with regard to the null hypothesis that the true parameter is 0.

    f. For the regression that you chose as being more appropriate in part (d), find and sketch 95% confidence and prediction bands. Using your sketch, find and interpret an approximate 95% confidence interval on the mean response for an 8-day-old chick.

Edited SAS Output (PROC REG) for Problem 1

```
            OUTPUT FOR REGRESSION OF DRY WEIGHT (Y) ON AGE (X)
                            .
                        .[Portion of output omitted]
                            .
        Root MSE            0.48185   R-Square       0.7442

                        Parameter Estimates
                        Parameter        Standard
        Variable    DF    Estimate         Error     t Value    Pr > |t|
        Intercept    1    -1.88453        0.52584     -3.58     <0.0059
        X            1     0.23507        0.04594      5.12     <0.0006

                            Output Statistics
        Dependent   Predicted     Std Error
Obs     Variable      Value     Mean Predict      95% CL Mean       95% CL Predict
 1       0.0290       -0.4741       0.2718     -1.0889    0.1408   -1.7256    0.7774
 2       0.0520       -0.2390       0.2343     -0.7690    0.2909   -1.4510    0.9730
 3       0.0790       -0.003945     0.2003     -0.4570    0.4491   -1.1844    1.1765
 4       0.1250        0.2311       0.1719     -0.1577    0.6200   -0.9262    1.3884
 5       0.1810        0.4662       0.1524      0.1215    0.8109   -0.6770    1.6094
 6       0.2610        0.7013       0.1453      0.3726    1.0299   -0.4372    1.8398
 7       0.4250        0.9363       0.1524      0.5917    1.2810   -0.2069    2.0796
 8       0.7380        1.1714       0.1719      0.7826    1.5603    0.0141    2.3287
 9       1.1300        1.4065       0.2003      0.9535    1.8595    0.2261    2.5869
10       1.8820        1.6416       0.2343      1.1116    2.1715    0.4296    2.8536
11       2.8120        1.8766       0.2718      1.2618    2.4915    0.6252    3.1281

            OUTPUT FOR REGRESSION OF LOG10 DRY WEIGHT (Z) ON AGE (X)
                            .
                        .[Portion of output omitted]
                            .
        Root MSE            0.02807   R-Square       0.9983

                        Parameter Estimates
                        Parameter        Standard
        Variable    DF    Estimate         Error     t Value    Pr > |t|
        Intercept    1    -2.68925        0.03064     -87.78    <.0001
        X            1     0.19589        0.00268      73.18    <.0001

                            Output Statistics
        Dependent   Predicted     Std Error
Obs     Variable      Value     Mean Predict      95% CL Mean       95% CL Predict
 1      -1.5380       -1.5139       0.0158     -1.5497   -1.4781   -1.5868   -1.4410
 2      -1.2840       -1.3180       0.0136     -1.3489   -1.2871   -1.3886   -1.2474
```

```
       OUTPUT FOR REGRESSION OF LOG10 DRY WEIGHT (Z) ON AGE (X) (continued)
                        Output Statistics (continued)
        Dependent     Predicted        Std Error
Obs     Variable        Value        Mean Predict      95% CL Mean        95% CL Predict
 3      -1.1020        -1.1221          0.0117      -1.1485   -1.0957    -1.1909   -1.0534
 4      -0.9030        -0.9262          0.0100      -0.9489   -0.9036    -0.9937   -0.8588
 5      -0.7420        -0.7303         0.008878     -0.7504   -0.7103    -0.7970   -0.6637
 6      -0.5830        -0.5345         0.008465     -0.5536   -0.5153    -0.6008   -0.4681
 7      -0.3720        -0.3386         0.008878     -0.3586   -0.3185    -0.4052   -0.2720
 8      -0.1320        -0.1427          0.0100      -0.1653   -0.1200    -0.2101   -0.0752
 9       0.0530         0.0532          0.0117       0.0268    0.0796    -0.0156    0.1220
10       0.2750         0.2491          0.0136       0.2182    0.2800     0.1785    0.3197
11       0.4490         0.4450          0.0158       0.4092    0.4808     0.3721    0.5179
```

2. The following table gives the systolic blood pressure (SBP), body size (QUET),[6] age (AGE), and smoking history (SMK = 0 if nonsmoker, SMK = 1 if a current or previous smoker) for a hypothetical sample of 32 white males over 40 years old from the town of Angina.

Person	SBP	QUET	AGE	SMK	Person	SBP	QUET	AGE	SMK	Person	SBP	QUET	AGE	SMK
1	135	2.876	45	0	12	138	4.032	51	1	23	137	3.296	53	0
2	122	3.251	41	0	13	152	4.116	64	0	24	132	3.210	50	0
3	130	3.100	49	0	14	138	3.673	56	0	25	149	3.301	54	1
4	148	3.768	52	0	15	140	3.562	54	1	26	132	3.017	48	1
5	146	2.979	54	1	16	134	2.998	50	1	27	120	2.789	43	0
6	129	2.790	47	1	17	145	3.360	49	1	28	126	2.956	43	1
7	162	3.668	60	1	18	142	3.024	46	1	29	161	3.800	63	0
8	160	3.612	48	1	19	135	3.171	57	0	30	170	4.132	63	1
9	144	2.368	44	1	20	142	3.401	56	0	31	152	3.962	62	0
10	180	4.637	64	1	21	150	3.628	56	1	32	164	4.010	65	0
11	166	3.877	59	1	22	144	3.751	58	0					

a. On each of the accompanying scatter diagrams, sketch by eye a line that fits the data reasonably well. Comment on the relationships described.

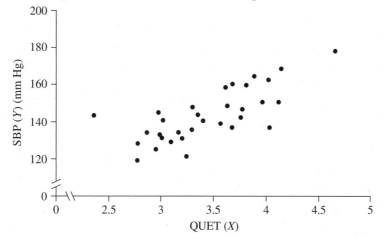

---

[6] QUET stands for "quetelet index," a measure of size defined by QUET = 100 (weight/height^2).

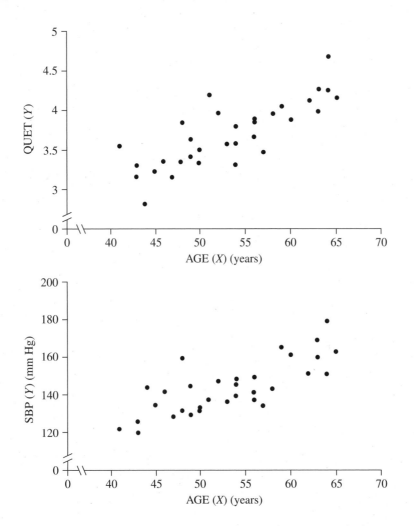

b. **(1)** Determine the least-squares estimates of the slope (β_1) and intercept (β_0) for the straight-line regression of SBP (Y) on QUET (X).

   (2) Sketch the estimated regression line on the scatter diagram involving SBP and QUET. Compare this new line with the line you drew in part (a).

   (3) Test the null hypothesis of zero slope; be sure to interpret the result.

   (4) Based on your test in (3), would you conclude that blood pressure increases as body size increases?

   (5) Find and sketch 95% prediction bands on the appropriate scatter diagram.

   (6) Using your answer to part (b)(5), find an approximate 95% prediction interval for an individual with QUET = 3.4 (the sample mean value of QUET). Interpret your answer.

   (7) Are any assumptions for straight-line regression clearly not satisfied in this example?

c. Repeat questions (1) through (4) in part (b) for the regression of QUET on AGE.

d. Repeat questions (1) through (4) in part (b) for the regression of SBP on AGE.

e. **(1)** Determine the least-squares estimates of the slope and intercept for the straight-line regression of SBP (Y) on SMK (X).

(2) Compare the value of $\hat{\beta}_0$ with the mean SBP for nonsmokers. Compare the values of $\hat{\beta}_0 + \hat{\beta}_1$ with the mean SBP for smokers. Explain the results of these comparisons.

(3) Test the null hypothesis that the true slope is 0; be sure to interpret the result.

(4) Is the test in part (e)(3) equivalent to the usual two-sample t test for the equality of two population means, assuming equal but unknown variances? Demonstrate your answer numerically.

Edited SAS Output (PROC REG) for Problem 2

```
                       OUTPUT FOR REGRESSION OF SBP ON QUET
                                      .
                               .[Portion of output omitted]
                                      .

                 Root MSE          9.81160   R-Square      0.5506

                          Parameter Estimates
                       Parameter        Standard
  Variable      DF     Estimate          Error        t Value     Pr > |t|
  Intercept      1     70.57640         12.32187         5.73      <.0001
  QUET           1     21.49167          3.54515         6.06      <.0001

                             Output Statistics
            Dependent    Predicted    Std Error
  Obs       Variable      Value     Mean Predict      95% CL       Predict     Residual
   1        135.0000     132.3864      2.6499      111.6306      153.1423       2.6136
   2        122.0000     140.4458      1.8608      120.0507      160.8409     -18.4458
   3        130.0000     137.2006      2.1144      116.7026      157.6985      -7.2006
   4        148.0000     151.5570      2.0860      131.0712      172.0428      -3.5570
   5        146.0000     134.6001      2.3858      113.9782      155.2219      11.3999
   6        129.0000     130.5382      2.8873      109.6506      151.4257      -1.5382
   7        162.0000     149.4078      1.9119      128.9930      169.8227      12.5922
   8        160.0000     148.2043      1.8372      127.8181      168.5905      11.7957
   9        144.0000     121.4687      4.1810       99.6873      143.2501      22.5313
  10        180.0000     170.2333      4.5807      148.1191      192.3475       9.7667
  11        166.0000     153.8996      2.3230      133.3077      174.4915      12.1004
  12        138.0000     157.2308      2.7197      136.4373      178.0243     -19.2308
  13        152.0000     159.0361      2.9552      138.1090      179.9632      -7.0361
  14        138.0000     149.5153      1.9194      129.0975      169.9331     -11.5153
  15        140.0000     147.1297      1.7886      126.7623      167.4972      -7.1297
  16        134.0000     135.0084      2.3401      114.4085      155.6084      -1.0084
  17        145.0000     142.7884      1.7581      122.4313      163.1455       2.2116
  18        142.0000     135.5672      2.2792      114.9957      156.1387       6.4328
  19        135.0000     138.7265      1.9812      118.2841      159.1689      -3.7265
  20        142.0000     143.6696      1.7403      123.3189      164.0203      -1.6696
  21        150.0000     148.5482      1.8567      128.1546      168.9418       1.4518
  22        144.0000     151.1917      2.0531      130.7197      171.6636      -7.1917
  23        137.0000     141.4129      1.8091      121.0372      161.7887      -4.4129
  24        132.0000     139.5647      1.9182      119.1473      159.9820      -7.5647
  25        149.0000     141.5204      1.8042      121.1465      161.8943       7.4796
  26        132.0000     135.4168      2.2954      114.8378      155.9958      -3.4168
  27        120.0000     130.5167      2.8901      109.6275      151.4058     -10.5167
  28        126.0000     134.1058      2.4425      113.4563      154.7553      -8.1058
  29        161.0000     152.2447      2.1511      131.7309      172.7586       8.7553
  30        170.0000     159.3800      3.0013      138.4255      180.3344      10.6200
  31        152.0000     155.7264      2.5335      135.0312      176.4216      -3.7264
  32        164.0000     156.7580      2.6601      135.9967      177.5193       7.2420
```

(continued)

```
                    OUTPUT FOR REGRESSION OF QUET ON AGE
                              .
                             .[Portion of output omitted]
                              .
          Root MSE              0.30131   R-Square        0.6444

                          Parameter Estimates
                        Parameter        Standard
Variable       DF        Estimate          Error        t Value    Pr > |t|
Intercept      1          0.38645         0.41769         0.93      0.3622
AGE            1          0.05736         0.00778         7.37      <.0001

                    OUTPUT FOR REGRESSION OF SBP ON AGE
                              .
                             .[Portion of output omitted]
                              .
          Root MSE              9.24543   R-Square        0.6009

                          Parameter Estimates
                        Parameter        Standard
Variable       DF        Estimate          Error        t Value    Pr > |t|
Intercept      1         59.09163        12.81626         4.61      <.0001
AGE            1          1.60450         0.23872         6.72      <.0001

                    OUTPUT FOR REGRESSION OF SBP ON SMK
                              .
                             .[Portion of output omitted]
                              .
          Root MSE             14.18082   R-Square        0.0612

                          Parameter Estimates
                        Parameter        Standard
Variable       DF        Estimate          Error        t Value    Pr > |t|
Intercept      1        140.80000         3.66147        38.45      <.0001
SMK            1          7.02353         5.02350         1.40      0.1723
```

3. For married couples with one or more offspring, a demographic study was conducted to determine the effect of the husband's annual income (at marriage) on the time (in months) between marriage and the birth of the first child. The following table gives the husband's annual income (INC) and the time between marriage and the birth of the first child (TIME) for a hypothetical sample of 20 couples.

INC	TIME	INC	TIME
5,775	16.20	4,608	9.70
9,800	35.00	24,210	20.00
13,795	37.20	19,625	38.20
4,120	9.00	18,000	41.25
25,015	24.40	13,000	44.00
12,200	36.75	5,400	9.20
7,400	31.75	6,440	20.00
9,340	30.00	9,000	40.20
20,170	36.00	18,180	32.00
22,400	30.80	15,385	39.20

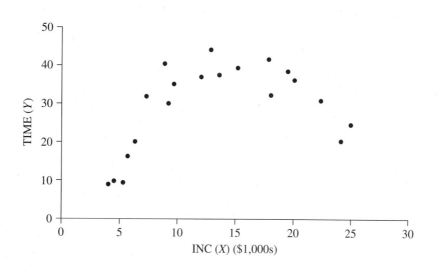

a. On the scatter diagram above, sketch by eye a line that fits the data reasonably well. Comment on the relationship between TIME (Y) and INC (X).

b. Determine the least-squares estimates of the slope (β_1) and intercept (β_0) for the straight-line regression of TIME (Y) on INC (X).

c. Draw the estimated regression line on the accompanying scatter diagram. Comment on how well this line fits the data.

d. Are any of the assumptions for straight-line regression clearly not satisfied in this example?

e. Test the null hypothesis that the true slope is 0. Interpret the results of this test.

f. Can you suggest a model that would describe the TIME–INC relationship better than a straight line does?

Edited SAS Output (PROC REG) for Problem 3

```
                OUTPUT FOR REGRESSION OF TIME (Y) ON INC (X)

                         .[Portion of output omitted]

            Root MSE            10.49580    R-Square        0.1853

                         Parameter Estimates
                         Parameter          Standard
Variable      DF          Estimate            Error        t Value     Pr > |t|
Intercept      1          19.62575          5.21291         3.76        0.0014
X              1          0.00071376        0.00035281       2.02        0.0582
```

4. A sociologist assigned to a correctional institution was interested in studying the relationship between intelligence and delinquency. A delinquency index (ranging from 0 to 50) was formulated to account for both the severity and the frequency of crimes committed, while intelligence was measured by IQ. The following table displays the delinquency index (DI) and IQ of a sample of 18 convicted minors.

DI (Y)	IQ (X)	DI (Y)	IQ (X)
26.20	110	22.10	92
33.00	89	18.60	116
17.50	102	35.50	85
25.25	98	38.00	73
20.30	110	30.00	90
31.90	98	19.70	104
21.10	122	41.10	82
22.70	119	39.60	134
10.70	120	25.15	114

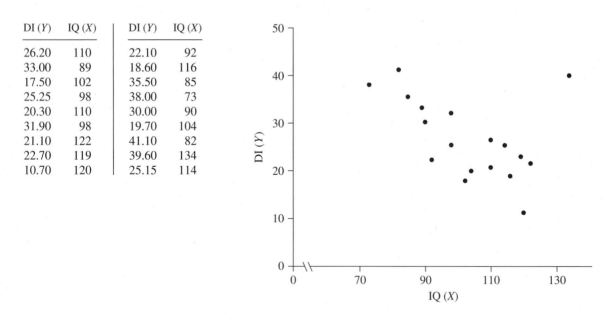

a. Given that $\hat{\beta}_1 = -0.249$ and $\hat{\beta}_0 = 52.273$, draw the estimated regression line on the accompanying scatter diagram.

b. How do you account for the fact that $\hat{Y} = 52.273$ when IQ = 0, even though the delinquency index goes no higher than 50?

c. Find a 95% confidence interval for the true slope β_1, using the fact that $S_{Y|X} = 7.704$ and $S_X = 16.192$.

d. Interpret this confidence interval with regard to testing the null hypothesis of zero slope at the $\alpha = .05$ level.

e. Notice that the convicted minor with IQ = 134 and DI = 39.6 appears to be quite out of place in the data. Decide whether this outlier has any effect on your estimate of the IQ–DI relationship, by looking at the graph for the fitted line obtained when the outlier is omitted. (Note that $\hat{\beta}_0 = 70.846$ and $\hat{\beta}_1 = -0.444$ when the outlier is removed.)

f. Test the null hypothesis of zero slope when the outlier is removed, given that $S_{Y|X} = 4.933$, $S_X = 14.693$, and $n = 17$. (Use $\alpha = .05$.)

g. For these data, would you conclude that the delinquency index decreases as IQ increases?

5. Following a recent congressional election, a political scientist attempted to investigate the relationship between campaign expenditures on television advertisements and subsequent voter turnout. The following table presents the percentage of total campaign expenditures delegated to television advertisements (TVEXP) and the percentage of registered voter turnout (VOTE) for a hypothetical sample of 20 congressional districts.

a. Determine the least-squares line of VOTE on TVEXP, and draw the estimated line on the accompanying scatter diagram.

b. Are any of the assumptions for straight-line regression clearly *not* satisfied in this example?

c. Test the null hypothesis that the true slope is 0; be sure to interpret your result.

d. Test the hypothesis $\mu_{Y|X_0} = 45$ for $X_0 = \overline{X} = 36.99$. Interpret your result.

e. Calculate the corresponding 95% confidence interval for $\mu_{Y|36.99}$, and interpret your result.

VOTE (Y)	TVEXP (X)	VOTE (Y)	TVEXP (X)
35.4	28.5	40.8	31.3
58.2	48.3	61.9	50.1
46.1	40.2	36.5	31.3
45.5	34.8	32.7	24.8
64.8	50.1	53.8	42.2
52.0	44.0	24.6	23.0
37.9	27.2	31.2	30.1
48.2	37.8	42.6	36.5
41.8	27.2	49.6	40.2
54.0	46.1	56.6	46.1

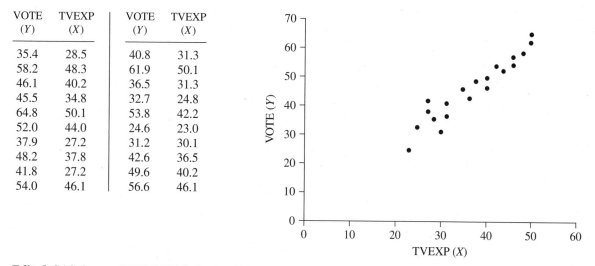

Edited SAS Output (PROC REG) for Problem 5

```
              OUTPUT FOR REGRESSION OF VOTE (Y) ON TVEXP (X)

                           .[Portion of output omitted]
                           .

              Root MSE         3.33177     R-Square        0.9101

                            Parameter Estimates
                         Parameter        Standard
Variable      DF         Estimate          Error         t Value      Pr > |t|
Intercept      1          2.17407         3.30974          0.66        0.5196
X              1          1.17696         0.08718         13.50        <.0001
```

6. A group of 13 children and adolescents (considered healthy) participated in a psychological study designed to analyze the relationship between age and average total sleep time (ATST). To obtain a measure for ATST (in minutes), recordings were taken on each subject on three consecutive nights and then averaged. The results obtained are displayed in the following table.

ATST (min/24 h)	AGE	ATST (min/24 h)	AGE
586.00	4.40	515.20	8.90
461.75	14.00	493.00	11.10
491.10	10.10	528.30	7.75
565.00	6.70	575.90	5.50
462.00	11.50	532.50	8.60
532.10	9.60	530.50	7.20
477.60	12.40		

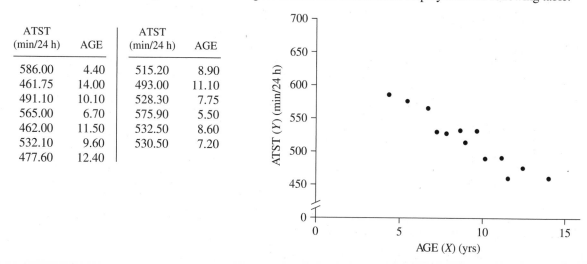

 a. Determine the least-squares estimates of the slope and intercept for the straight-line regression of ATST (Y) on AGE (X). Draw the estimated line on the accompanying scatter diagram, and comment on the fit.

 b. Are any of the assumptions for straight-line regression clearly *not* satisfied in this example?

 c. Test the null hypothesis that the true slope is 0; be sure to interpret your result.

 d. Obtain a 95% confidence interval for β_1. Interpret your result.

 e. Would you reject the null hypothesis H_0: $\beta_1 = 0$ based on the confidence interval you calculated in part (d)? Explain.

 f. Determine and sketch 95% confidence bands on the accompanying scatter diagram. Use your diagram to estimate the mean ATST when AGE = 10. Interpret your result.

Edited SAS Output (PROC REG) for Problem 6

```
            OUTPUT FOR REGRESSION OF ATST (Y) ON AGE (X)

                       .[Portion of output omitted]
                       .

         Root MSE            13.15238    R-Square        0.9054

                      Parameter Estimates
                      Parameter        Standard
 Variable     DF      Estimate          Error        t Value      Pr > |t|
 Intercept    1       646.48334        12.91773        50.05       <.0001
 X            1       -14.04105         1.36812       -10.26       <.0001

                          Output Statistics
         Dependent    Predicted      Std Error
 Obs     Variable       Value       Mean Predict        95% CL Mean         Residual
  1      586.0000      584.7027       7.3425       568.5420   600.8635       1.2973
  2      461.7500      449.9087       7.6829       432.9988   466.8185      11.8413
  3      491.1000      504.6688       3.9166       496.0483   513.2892     -13.5688
  4      565.0000      552.4083       4.8694       541.6908   563.1258      12.5917
  5      462.0000      485.0113       4.9468       474.1234   495.8992     -23.0113
  6      532.1000      511.6893       3.7225       503.4961   519.8824      20.4107
  7      477.6000      472.3743       5.8494       459.4998   485.2488       5.2257
  8      515.2000      521.5180       3.6542       513.4752   529.5608      -6.3180
  9      493.0000      490.6277       4.5950       480.5143   500.7411       2.3723
 10      528.3000      537.6652       4.0629       528.7228   546.6076      -9.3652
 11      575.9000      569.2576       6.0826       555.8700   582.6452       6.6424
 12      532.5000      525.7303       3.7012       517.5841   533.8765       6.7697
 13      530.5000      545.3878       4.4459       535.6024   555.1731     -14.8878
```

 7. Several research workers associated with the Office of Highway Safety were evaluating the relationship between driving speed (MPH) and the distance a vehicle travels once brakes are applied (DIST). The results of 19 experimental tests are displayed in the following table.

 a. Determine the least-squares estimates of the slope and intercept for each of the following straight-line regressions: Y_1 (DIST) on X, and $Y_2(\sqrt{\text{DIST}})$ on X. Draw the estimated lines on the appropriate scatter diagrams.

 b. Which of the two variable pairs mentioned in (a) seems to be better suited for straight-line regression?

 c. For the variable pair Y_2 and X, test the hypothesis that the true slope is equal to 1 (use $\alpha = 0.01$). Be sure to interpret your result.

MPH (X)	DIST (Y_1)	$\sqrt{\text{DIST}}$ (Y_2)	MPH (X)	DIST (Y_1)	$\sqrt{\text{DIST}}$ (Y_2)
25.0	37.4	6.12	50.0	170.0	13.04
35.0	57.7	7.60	20.0	20.0	4.47
60.0	337.6	18.37	15.0	13.5	3.67
45.0	142.5	11.94	27.5	40.8	6.39
50.0	182.4	13.51	55.0	207.8	14.42
37.5	67.5	8.22	40.0	105.0	10.25
30.0	37.5	6.12	45.0	132.6	11.52
55.0	225.0	15.00	17.5	19.1	4.37
60.0	258.1	16.07	22.5	25.0	5.00
65.0	297.4	17.25			

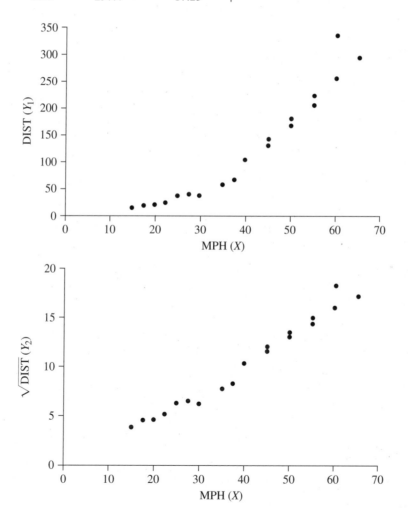

d. Construct a 99% confidence interval for the true slope in part (c). Interpret your result.

e. For the same variable pair considered in parts (c) and (d), calculate and sketch 95% confidence bands on the appropriate scatter diagram. Using the confidence bands, estimate the mean value of Y_2 when $X = 45$. Interpret your result.

Edited SAS Output (PROC REG) for Problem 7

```
                    OUTPUT FOR REGRESSION OF DIST (Y1) ON MPH (X)

                              .[Portion of output omitted]
                              .

              Root MSE              31.64950    R-Square        0.9106

                            Parameter Estimates
                         Parameter          Standard
    Variable      DF      Estimate             Error        t Value      Pr > |t|
    Intercept     1      -122.34459          20.15624         -6.07       <.0001
    X             1         6.22708           0.47319         13.16       <.0001

                 OUTPUT FOR REGRESSION OF SQRT (DIST) (Y2) ON MPH (X)

                              .[Portion of output omitted]
                              .

              Root MSE               0.81097    R-Square        0.9728

                            Parameter Estimates
                         Parameter          Standard
    Variable      DF      Estimate             Error        t Value      Pr > |t|
    Intercept     1        -1.69712           0.51647         -3.29       0.0044
    X             1         0.29878           0.01212         24.64       <.0001

                              Output Statistics
                  Dependent      Predicted        Std Error
    Obs           Variable         Value        Mean Predict        95% CL Mean
     1             6.1200          5.7723          0.2580       5.2280      6.3165
     2             7.6000          8.7600          0.1947       8.3492      9.1708
     3            18.3700         16.2294          0.3082      15.5792     16.8796
     4            11.9400         11.7478          0.1967      11.3328     12.1627
     5            13.5100         13.2416          0.2238      12.7694     13.7139
     6             8.2200          9.5070          0.1880       9.1103      9.9036
     7             6.1200          7.2661          0.2203       6.8013      7.7310
     8            15.0000         14.7355          0.2624      14.1819     15.2892
     9            16.0700         16.2294          0.3082      15.5792     16.8796
    10            17.2500         17.7233          0.3584      16.9671     18.4794
    11            13.0400         13.2416          0.2238      12.7694     13.7139
    12             4.4700          4.2784          0.3031       3.6389      4.9179
    13             3.6700          2.7845          0.3529       2.0399      3.5292
    14             6.3900          6.5192          0.2380       6.0171      7.0213
    15            14.4200         14.7355          0.2624      14.1819     15.2892
    16            10.2500         10.2539          0.1861       9.8613     10.6465
    17            11.5200         11.7478          0.1967      11.3328     12.1627
    18             4.3700          3.5314          0.3276       2.8403      4.2226
    19             5.0000          5.0253          0.2798       4.4350      5.6157
```

8. The following table presents the starting annual salaries (SAL) of a group of 30 college graduates who have recently entered the job market, along with their cumulative grade-point averages (CGPA).

   a. Determine the least-squares estimates of the slope and intercept for the straight-line regression of SAL (Y) on CGPA (X). Draw the estimated line on the accompanying scatter diagram, and comment on the fit.

   b. Are any of the assumptions for straight-line regression clearly *not* satisfied in this example?

SAL (Y)	CGPA (X)	SAL (Y)	CGPA (X)
10455	2.58	8000	2.30
9680	2.31	12548	2.83
7300	2.47	7700	2.37
9388	2.52	10028	2.52
12496	3.22	13176	3.22
11812	3.37	13255	3.55
9224	2.43	13004	3.55
11725	3.08	8000	2.47
11320	2.78	8224	2.47
12000	2.98	10750	2.78
12500	3.55	11669	2.78
13310	3.64	12322	2.98
12105	3.72	11002	2.58
6200	2.24	10666	2.58
11522	2.70	10839	2.58

c. Obtain a 95% confidence interval for β_1.
d. Would you reject the null hypothesis H_0: $\beta_1 = 4000$ at the $\alpha = 0.05$ level?
e. Find and graph 95% confidence and prediction bands.
f. Would you reject the hypothesis H_0: $\mu_{Y|X} = 11{,}500$ at $X_0 = 2.75$?

Edited SAS Output (PROC REG) for Problem 8

```
                    OUTPUT FOR REGRESSION OF SAL (Y) ON CGPA (X)

                           .[Portion of output omitted]
                           .

            Root MSE              1124.71499    R-Square        0.6845

                           Parameter Estimates
                           Parameter          Standard
            Variable      DF     Estimate        Error     t Value    Pr > |t|
            Intercept      1    435.92357    1337.85967       0.33      0.7470
            X              1   3630.56128     465.76874       7.79      <.0001
```

Output Statistics

Obs	Dependent Variable	Predicted Value	Std Error Mean Predict	95% CL Mean		95% CL Predict	
1	10455	9803	237.9998	9315	10290	7448	12158
2	9680	8823	320.5028	8166	9479	6427	11218
3	7300	9403	267.5786	8855	9952	7035	11772
4	9388	9585	253.2786	9066	10104	7223	11947
5	12496	12126	271.6022	11570	12683	9756	14496
6	11812	12671	321.6964	12012	13330	10275	15067
7	9224	9258	279.8892	8685	9832	6884	11632
8	11725	11618	234.1710	11138	12098	9265	13971
9	11320	10529	207.1336	10105	10953	8186	12872
10	12000	11255	215.6850	10813	11697	8909	13601
11	12500	13324	389.9229	12526	14123	10886	15763
12	13310	13651	426.1305	12778	14524	11187	16115
13	12105	13942	459.1316	13001	14882	11453	16430
14	6200	8568	346.1668	7859	9277	6158	10979
15	11522	10238	215.2151	9798	10679	7893	12584

(continued)

```
           OUTPUT FOR REGRESSION OF SAL (Y) ON CGPA (X) (continued)
                         Output Statistics (continued)
         Dependent    Predicted       Std Error
Obs      Variable       Value      Mean Predict      95% CL Mean        95% CL Predict
 16        8000          8786        324.0927      8122     9450      6389     11184
 17       12548         10710        205.3806     10290    11131      8368     13052
 18        7700          9040        299.5814      8427     9654      6656     11425
 19       10028          9585        253.2786      9066    10104      7223     11947
 20       13176         12126        271.6022     11570    12683      9756     14496
 21       13255         13324        389.9229     12526    14123     10886     15763
 22       13004         13324        389.9229     12526    14123     10886     15763
 23        8000          9403        267.5786      8855     9952      7035     11772
 24        8224          9403        267.5786      8855     9952      7035     11772
 25       10750         10529        207.1336     10105    10953      8186     12872
 26       11669         10529        207.1336     10105    10953      8186     12872
 27       12322         11255        215.6850     10813    11697      8909     13601
 28       11002          9803        237.9998      9315    10290      7448     12158
 29       10666          9803        237.9998      9315    10290      7448     12158
 30       10839          9803        237.9998      9315    10290      7448     12158
```

9. In an experiment designed to describe the dose–response curve for vitamin K, individual rats were depleted of their vitamin K reserves and then fed dried liver for 4 days at different dosage levels.[7] The response of each rat was measured as the concentration of a clotting agent needed to clot a sample of its blood in 3 minutes. The results of the experiment on 12 rats are given in the following table; values are expressed in common logarithms for both dose and response.

   a. Determine the least-squares estimates of the slope (β_1) and the intercept (β_0) for the straight-line regression of Y on X.

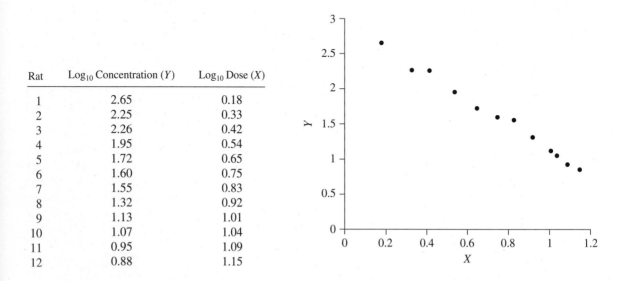

Rat	Log$_{10}$ Concentration (Y)	Log$_{10}$ Dose (X)
1	2.65	0.18
2	2.25	0.33
3	2.26	0.42
4	1.95	0.54
5	1.72	0.65
6	1.60	0.75
7	1.55	0.83
8	1.32	0.92
9	1.13	1.01
10	1.07	1.04
11	0.95	1.09
12	0.88	1.15

---

[7] Adapted from a study by Schønheyder (1936).

b. Draw the estimated regression line on the accompanying scatter diagram. How well does this line fit the data?

c. Determine and sketch 95% confidence bands based on the estimated regression line.

d. Convert the fitted straight line into an equation in the original units $Y' = 10^Y$ and $X' = 10^X$.

e. For the converted equation obtained in part (e), determine 99% confidence intervals for the true mean responses at the maximum and minimum doses used in the experiment.

f. If the values for X and Y on each rat are converted to their original units X' and Y', the following fitted straight line is obtained: $Y' = 237.16095 - 21.32117X'$. How would you evaluate whether using the variables X' and Y' is better or worse than using X and Y for the regression analysis?

Edited SAS Output (PROC REG) for Problem 9

```
        OUTPUT FOR REGRESSION OF LOG10 (CONC) (Y) ON LOG10(DOSE) (X)
                                    .
                            .[Portion of output omitted]
                                    .

            Root MSE              0.05589   R-Square        0.9914

                            Parameter Estimates
                        Parameter            Standard
 Variable      DF        Estimate              Error        t Value      Pr > |t|
 Intercept     1          2.93620            0.04230          69.41       <.0001
 X             1         -1.78501            0.05267         -33.89       <.0001

                            Output Statistics
            Dependent      Predicted       Std Error
  Obs       Variable         Value       Mean Predict        95% CL Mean        Residual
   1         2.6500         2.6149          0.0337        2.5397    2.6901        0.0351
   2         2.2500         2.3472          0.0271        2.2869    2.4074       -0.0972
   3         2.2600         2.1865          0.0234        2.1343    2.2387        0.0735
   4         1.9500         1.9723          0.0193        1.9292    2.0154       -0.0223
   5         1.7200         1.7759          0.0169        1.7384    1.8135       -0.0559
   6         1.6000         1.5974          0.0161        1.5615    1.6334        0.002554
   7         1.5500         1.4546          0.0168        1.4173    1.4920        0.0954
   8         1.3200         1.2940          0.0186        1.2524    1.3355        0.0260
   9         1.1300         1.1333          0.0214        1.0856    1.1811       -0.003343
  10         1.0700         1.0798          0.0225        1.0297    1.1299       -0.009792
  11         0.9500         0.9905          0.0244        0.9362    1.0449       -0.0405
  12         0.8800         0.8834          0.0268        0.8236    0.9433       -0.003441
```

10. The susceptibility of catfish to a certain chemical pollutant was determined by immersing individual fish in 2 liters of an emulsion containing the pollutant and measuring the survival time in minutes.[8] The data in the following table give the common log of survival time (Y) and the common log of concentration (X) of the pollutant in parts per million for 18 fish.

a. Determine and draw the estimated straight line of Y regressed on X on the accompanying scatter diagram. Comment on the fit.

b. Test for the significance of the straight-line regression. Interpret your result.

c. Determine 95% confidence intervals for the true mean survival time $\mu_{Y|X}$ (where $Y = 10^Y$) at values of $X = 5, 4.5$, and 4. (Note $\overline{X} = 4.5$.) Interpret these intervals.

---

[8] Adapted from a study by Nagasawa, Osano, and Kondo (1964).

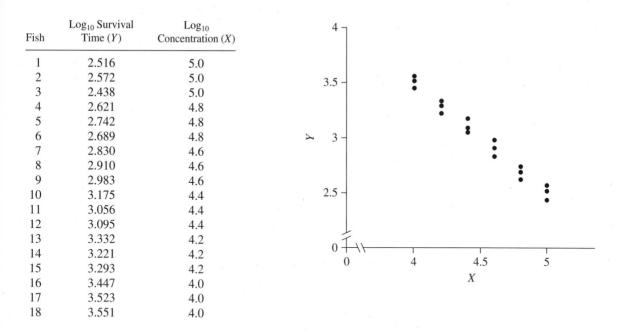

Fish	Log$_{10}$ Survival Time (Y)	Log$_{10}$ Concentration (X)
1	2.516	5.0
2	2.572	5.0
3	2.438	5.0
4	2.621	4.8
5	2.742	4.8
6	2.689	4.8
7	2.830	4.6
8	2.910	4.6
9	2.983	4.6
10	3.175	4.4
11	3.056	4.4
12	3.095	4.4
13	3.332	4.2
14	3.221	4.2
15	3.293	4.2
16	3.447	4.0
17	3.523	4.0
18	3.551	4.0

Edited SAS Output (PROC REG) for Problem 10

```
              OUTPUT FOR REGRESSION OF LOG10 (SURV TIME) (Y) ON LOG10(CONC)

                                   .[Portion of output omitted]
                                   .

                   Root MSE              0.05597    R-Square         0.9766

                             Parameter Estimates
                             Parameter          Standard
    Variable      DF         Estimate            Error          t Value        Pr > |t|
    Intercept     1           7.49110           0.17431          42.97          <.0001
    X             1          -0.99810           0.03863         -25.84          <.0001

                                   Output Statistics
                 Dependent      Predicted       Std Error
    Obs          Variable         Value        Mean Predict      95% CL Mean         Residual
    1             2.5160         2.5006           0.0234      2.4510    2.5502         0.0154

                                   .[Portion of output omitted]
                                   .

    9             2.9830         2.8999           0.0137      2.8707    2.9290         0.0831
    10            3.1750         3.0995           0.0137      3.0703    3.1286         0.0755

                                   .[Portion of output omitted]
                                   .

    18            3.5510         3.4987           0.0234      3.4491    3.5483         0.0523
```

11. An experiment was conducted to determine the extent to which the growth rate of a certain fungus could be affected by filling test tubes containing the same medium at the same temperature with different inert gases.[9] Three such experiments were performed for each of six gases, and the average growth rate over these three tests was used as the response. The following table gives the molecular weight (X) of each gas used and the average growth rate (Y) in milliliters per hour for the three tests.

Gas	Average Growth Rate (Y)	Molecular Weight (X)
A	3.85	4.0
B	3.48	20.2
C	3.27	28.2
D	3.08	39.9
E	2.56	83.8
F	2.21	131.3

a. Find the least-squares estimates of slope and intercept for the straight-line regression of Y on X, and draw the estimated straight line on a scatter diagram for this data set.
b. Test for significant slope of the fitted straight line.
c. What information has not been used that might improve the sensitivity of the analysis?
d. What is the 90% confidence interval for the true average growth rate if the gas used has a molecular weight of 100?
e. Why would it be inappropriate to use the fitted line to estimate the growth rate of a gas whose molecular weight is 200?
f. Based on the choice of X-values used in this study, how would you criticize the accuracy of prediction obtained in this experiment by using the fitted straight line?

Edited SAS Output (PROC REG) for Problem 11

```
           OUTPUT FOR REGRESSION OF AVG. GROWTH RATE (Y) on MOLECULAR WEI

                              Descriptive Statistics
                                   Uncorrected                              Standard
Variable        Sum          Mean          SS          Variance           Deviation
Intercept    6.00000      1.00000      6.00000            0                    0
X          307.40000     51.23333        27073       2264.85867           47.59053
Y           18.45000      3.07500     58.54990          0.36323            0.60269

                            .[Portion of output omitted]
                            .

            Root MSE              0.15123    R-Square      0.9496

                          Parameter Estimates
                        Parameter           Standard
Variable      DF        Estimate              Error        t Value       Pr > |t|
Intercept      1         3.70727            0.09546         38.83         <.0001
X              1        -0.01234            0.00142         -8.68         0.0010
```

---

[9] Adapted from a study by Schreiner, Gregoine, and Lawrie (1962).

12. Consider the data in the following table.[10]

Age		Vocabulary Size	Age		Vocabulary Size
Years	Months		Years	Months	
0	8	0	3	0	896
0	10	1	3	6	1,222
1	0	3	4	0	1,540
1	3	19	4	6	1,870
1	6	22	5	0	2,072
1	9	118	5	6	2,289
2	0	272	6	0	2,562
2	6	446			

Note: Data are from M. E. Smith, "An Investigation of the Development of the Sentence and the Extent of Vocabulary in Young Children," *Studies in Child Welfare* (University of Iowa) 3(5) (1926).

a. First convert ages to decimal years (e.g., 1 year 6 months gives 1.5 years). Draw a scatter diagram for the variable pair vocabulary size (Y) and age (X).

b. Then calculate the least-squares estimates of the parameters of the regression line. Sketch this regression line on the scatter diagram.

c. Add a new observation to the vocabulary data, with values 0.00 years and 0 words. Plot this new point. Logically this value should be on the line of vocabulary growth. Is it near the fitted line from part (b)?

d. Recompute the least-squares estimates of the regression line to include the (0, 0) observation ($n = 16$). Sketch the new line distinctly on the scatter diagram.

e. Is either line fitted acceptably? If not, sketch your idea of the true relationship. If the data included observations through age 30 years, what would the extrapolated curve look like?

f. The data appear to be from one child. If this is true, what assumption of the least-squares approach is most likely violated, and why?

Edited SAS Output (PROC REG) for Problem 12

```
                      REGRESSION OF VOCAB (Y) ON AGE (X)

                          .[Portion of output omitted]

               Root MSE           148.07090    R-Square      0.9776
                            Parameter Estimates
                          Parameter          Standard
Variable      DF          Estimate            Error          t Value      Pr > |t|
Intercept     1          -621.12595          74.08216        -8.38        <.0001
X             1           526.71836          22.13536        23.80        <.0001
                     Y ON X, INCLUDING 16TH OBS.

                          .[Portion of output omitted]

               Root MSE           205.90999    R-Square      0.9558
                            Parameter Estimates
                          Parameter          Standard
Variable      DF          Estimate            Error          t Value      Pr > |t|
Intercept     1          -496.77535          92.13222        -5.39        <.0001
X             1           494.89316          28.43144        17.41        <.0001
```

[10] Taken from Bourne, Ekstrand, and Dominowski (1971, table 14.3).

```
                    OUTPUT FOR REGRESSION OF LATENCY ON WEIGHT
                                         .
                                  .[Portion of output omitted]
                                         .
            Root MSE                 0.72779    R-Square      0.0519

                            Parameter Estimates
                         Parameter          Standard
   Variable      DF       Estimate            Error        t Value      Pr > |t|
   Intercept     1         0.11598           2.50749         0.05        0.9639
   Weight        1         0.00498           0.00615         0.81        0.4333
```

13. The following table gives rat body weights (in grams) and latency to seizure (in minutes), following injection of 40 mg/kg of body weight of metrazol.

Latency	Weight	Latency	Weight
2.30	348	2.00	409
1.95	372	1.70	413
2.90	378	2.00	415
2.30	390	2.95	423
1.10	392	1.25	428
2.50	395	2.05	464
1.30	400	3.70	468

a. Draw a scatter diagram on graph paper, with latency plotted as a function of weight.
b. Determine the least-squares estimates of slope and intercept for the straight-line regression of latency on weight.
c. Test whether the slope is equal to 0. Use $\alpha = .10$.
d. Test whether the intercept is equal to 0. Use $\alpha = .10$.
e. Sketch the estimated regression line.
f. Distinctly sketch your choice for a regression line if it differs from that of part (e). Explain why you agree or disagree with part (e), noting whether any assumptions appear to be violated.

Edited SAS Output (PROC REG) for Problem 13

```
                    REGRESSION OF LATENCY ON WEIGHT
                                     .
                              .[Portion of output omitted]
                                     .
            Root MSE             0.72779    R-Square      0.0519

                        Parameter Estimates
                     Parameter          Standard
   Variable    DF     Estimate            Error        t Value      Pr > |t|
   Intercept   1     0.115983          2.50749283       0.046        0.9639
   Weight      1     0.004983          0.00614559       0.811        0.4333
```

14. Stevens (1966), citing Dimmick and Hubbard (1939), reported data from 20 studies of the color perception of unitary hues. The wavelength (in millimeters) of light called green by subjects in each experiment is given in the following table, along with the year the study was conducted.

Study	Wavelength	Date	Study	Wavelength	Date	Study	Wavelength	Date
1	532	1874	8	506	1909	15	500	1928
2	535	1884	9	509	1911	16	506	1931
3	495	1888	10	520	1912	17	528	1934
4	527	1890	11	514	1920	18	530	1935
5	505	1898	12	504	1922	19	512	1935
6	505	1898	13	515	1926	20	515	1939
7	503	1907	14	498	1927			

a. These data stimulate the question "Is there any linear trend over time in the wavelength of light called green?" Evaluate this question by finding the least-squares estimates of the straight-line regression functions for predicting wavelength from year. [*Hint:* Subtract 500 from the wavelengths and 1850 from the year.]
b. Find a 95% confidence interval for the true slope.
c. Draw a scatter diagram on graph paper.

Edited SAS Output (PROC REG) for Problem 14

```
                    OUTPUT FOR REGRESSION OF WAVELENTH ON DATE
                                 .
                            .[Portion of output omitted]
                                 .
             Root MSE              12.17807    R-Square     0.0310

                          Parameter Estimates
                          Parameter        Standard
    Variable      DF      Estimate          Error       t Value    Pr > |t|
    Intercept     1       19.88171         9.52906        2.09      0.0514
    Date          1       -0.10933         0.14403       -0.76      0.4576
```

15. The data listed in the following table are from a study by Benignus and others (1981). Blood and brain levels of toluene (a commonly used solvent) were measured following a 3-hour inhalation exposure to 50, 100, 500, or 1,000 parts per million (ppm) toluene (PPM_TOLU). Blood toluene (BLOODTOL) and brain toluene (BRAINTOL) are expressed in parts per million, weight in grams, and age in days.

Rat	BLOODTOL	BRAINTOL	PPM_TOLU	WEIGHT	AGE	LN_BLDTL	LN_BRNTL	LN_PPMTL
1	0.553	0.481	50	393	95	−0.593	−0.732	3.912
2	0.494	0.584	50	378	95	−0.706	−0.538	3.912
3	0.609	0.585	50	450	95	−0.495	−0.536	3.912
4	0.763	0.628	50	439	95	−0.270	−0.465	3.912
5	0.420	0.533	50	397	95	−0.868	−0.629	3.912
6	0.397	0.490	50	301	84	−0.923	−0.713	3.912
7	0.503	0.719	50	406	84	−0.687	−0.330	3.912
8	0.534	0.585	50	302	84	−0.628	−0.536	3.912
9	0.531	0.675	50	382	84	−0.633	−0.393	3.912

Rat	BLOODTOL	BRAINTOL	PPM_TOLU	WEIGHT	AGE	LN_BLDTL	LN_BRNTL	LN_PPMTL
10	0.384	0.442	50	355	84	−0.957	−0.816	3.912
11	0.215	0.492	50	405	85	−1.536	−0.709	3.912
12	0.552	0.859	50	405	85	−0.595	−0.152	3.912
13	0.420	0.650	50	387	85	−0.868	−0.431	3.912
14	0.324	0.528	50	358	85	−1.127	−0.639	3.912
15	0.387	0.546	50	311	85	−0.949	−0.605	3.912
16	1.036	1.262	100	355	86	0.035	0.233	4.605
17	1.065	1.584	100	440	86	0.063	0.460	4.605
18	1.084	1.773	100	421	86	0.081	0.573	4.605
19	0.944	1.307	100	370	86	−0.058	0.268	4.605
20	0.994	1.338	100	375	86	−0.006	0.291	4.605
21	1.146	1.180	100	368	83	0.136	0.166	4.605
22	1.167	1.108	100	321	83	0.154	0.103	4.605
23	0.833	0.939	100	359	83	−0.183	−0.063	4.605
24	0.630	0.909	100	367	83	−0.462	−0.095	4.605
25	0.955	1.078	100	363	83	−0.046	0.075	4.605
26	0.687	1.152	100	388	86	−0.376	0.141	4.605
27	0.723	1.796	100	404	86	−0.324	0.586	4.605
28	0.705	1.262	100	454	86	−0.349	0.233	4.605
29	0.696	1.865	100	389	86	−0.363	0.623	4.605
30	0.868	1.892	100	352	86	−0.142	0.638	4.605
31	8.223	19.843	500	367	83	2.107	2.988	6.215
32	10.604	24.450	500	406	83	2.361	3.197	6.215
33	12.085	29.297	500	371	83	2.492	3.377	6.215
34	7.936	18.098	500	408	83	2.071	2.896	6.215
35	11.164	25.196	500	305	83	2.413	3.227	6.215
36	10.289	18.266	500	391	84	2.331	2.905	6.215
37	11.140	19.486	500	396	84	2.411	2.970	6.215
38	9.647	18.479	500	347	84	2.267	2.917	6.215
39	13.343	21.920	500	372	84	2.591	3.087	6.215
40	11.292	20.861	500	331	84	2.424	3.038	6.215
41	7.524	22.130	500	365	85	2.018	3.097	6.215
42	10.783	18.301	500	348	85	2.378	2.907	6.215
43	8.595	17.038	500	416	85	2.151	2.835	6.215
44	9.616	22.423	500	344	85	2.263	3.110	6.215
45	11.956	15.452	500	398	85	2.481	2.738	6.215
46	30.274	44.900	1000	417	93	3.410	3.804	6.908
47	32.923	35.500	1000	351	93	3.494	3.570	6.908
48	28.619	30.800	1000	378	93	3.354	3.428	6.908
49	28.761	38.500	1000	338	93	3.359	3.651	6.908
50	25.402	31.500	1000	433	93	3.235	3.450	6.908
51	35.464	42.330	1000	342	85	3.569	3.745	6.908
52	32.706	34.030	1000	319	85	3.488	3.527	6.908
53	29.347	30.760	1000	440	85	3.379	3.426	6.908
54	26.481	32.360	1000	363	85	3.276	3.477	6.908
55	33.401	41.830	1000	336	85	3.509	3.734	6.908
56	39.541	54.930	1000	378	86	3.677	4.006	6.908
57	28.155	39.780	1000	420	86	3.338	3.683	6.908
58	25.629	49.290	1000	346	86	3.244	3.898	6.908
59	33.188	47.490	1000	413	86	3.502	3.861	6.908
60	33.505	42.660	1000	432	86	3.512	3.753	6.908

a. Provide a scatter diagram, with BLOODTOL as the response and PPM_TOLU as the predictor.
b. Compute least-squares estimates of the straight-line regression coefficients. Plot the line on the scatter diagram.
c. Repeat part (a) for the natural logarithms LN_BLDTL and LN_PPMTL.
d. Repeat part (b) for the natural logarithms.
e. Which transformation leads to the best representation of the data? Note in your comments the validity of the regression assumptions.

Edited SAS Output (PROC REG) for Problem 15

```
                    OUTPUT FOR BLOODTOL REGRESSED ON PPM_TOLU

                            .[Portion of output omitted]

            Root MSE                2.85301    R-Square        0.9497

                            Parameter Estimates
                        Parameter          Standard
Variable      DF         Estimate             Error       t Value      Pr > |t|
Intercept      1         -2.54610           0.54254         -4.69       <.0001
PPM_TOLU       1          0.03196        0.00096570         33.09       <.0001


            SAS PROC REG OUTPUT FOR LN_BLDTL REGRESSED ON LN_PPMTL

                            .[Portion of output omitted]

            Root MSE                0.24144    R-Square        0.9813

                            Parameter Estimates
                        Parameter          Standard
Variable      DF         Estimate             Error       t Value      Pr > |t|
Intercept      1         -6.53158           0.14365        -45.47       <.0001
LN_PPMTL       1          1.43045           0.02592         55.19       <.0001
```

16. Real estate prices depend, in part, on property size. The house size X (in hundreds of square feet) and house price Y (in thousands of dollars) of a random sample of houses in a certain county were recorded as in the following table.

House	1	2	3
X	18	20	25
Y	80	95	104

House	4	5	6	7
X	22	33	19	17
Y	110	175	85	89

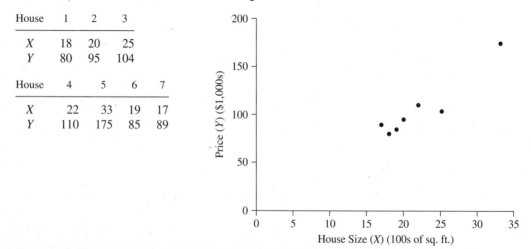

a. On the accompanying scatter diagram, sketch by eye a line that fits the data reasonably well. Comment on the relationship between house size and house price.

b. Determine the least-squares estimates of the slope (β_1) and the intercept (β_0) for the straight-line regression of Y on X.

c. Draw the estimated regression line on the scatter diagram. Comment on how well the line fits the data.

d. Test the null hypothesis that the true slope is 0. Interpret the results of this test.

Edited SAS Output (PROC REG) for Problem 16

```
                OUTPUT FOR THE REGRESSION OF PRICE ON HOUSE SIZE

                           .[Portion of output omitted]
                           .
           Root MSE                 10.70973    R-Square        0.9091

                           Parameter Estimates
                        Parameter           Standard
  Variable     DF        Estimate             Error         t Value     Pr > |t|
  Intercept     1       -17.36491           17.83513         -0.97       0.3750
  X             1         5.58152            0.78953          7.07        0.0009

                              Output Statistics
            Dependent      Predicted      Std Error
  Obs        Variable        Value       Mean Predict      95% CL Mean        Residual
   1         80.0000        83.1025         5.1341      69.9048    96.3002     -3.1025
   2         95.0000        94.2655         4.3450      83.0964   105.4347      0.7345
   3        104.0000       122.1731         4.6900     110.1172   134.2291    -18.1731
   4        110.0000       105.4286         4.0479      95.0231   115.8340      4.5714
   5        175.0000       166.8253         9.5819     142.1944   191.4563      8.1747
   6         85.0000        88.6840         4.6900      76.6281   100.7399     -3.6840
   7         89.0000        77.5210         5.6542      62.9865    92.0554     11.4790
```

17. Sales revenue (Y) and advertising expenditure (X) data for a large retailer for the period 1988–1993 are given in the following table.

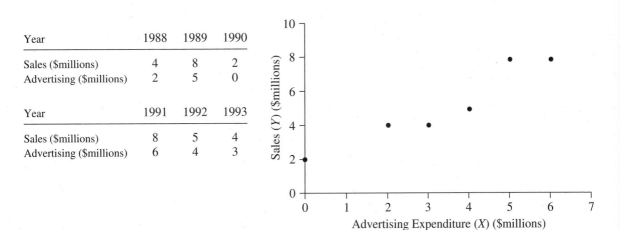

Year	1988	1989	1990
Sales ($millions)	4	8	2
Advertising ($millions)	2	5	0

Year	1991	1992	1993
Sales ($millions)	8	5	4
Advertising ($millions)	6	4	3

    a. Does the plot of Y versus X suggest that a linear relationship exists between X and Y?

    b. Calculate the least-squares estimates of the parameters of the regression line, and draw the estimated line on the accompanying scatter diagram. Does the line appear to fit the data well?

    c. Find a 95% confidence interval for the slope parameter. Based on your interval, is sales revenue linearly related to advertising expenditure? Explain.

    d. Would it be appropriate to use the estimated regression line in part (b) to estimate the sales for a new year in which an advertising expenditure of $10 million is planned? Why or why not?

Edited SAS Output (PROC REG) for Problem 17

```
              OUTPUT FOR REGRESSION OF SALES (Y) ON ADV (X)

                        .[Portion of output omitted]
                        .
        Root MSE              0.83023    R-Square      0.9044

                         Parameter Estimates
                       Parameter          Standard
Variable     DF        Estimate            Error        t Value       Pr > |t|
Intercept     1         1.64286           0.66567        2.47          0.0691
ADV           1         1.05714           0.17187        6.15          0.0035
```

18. The production manager of a plant that manufactures syringes records the marginal cost at various levels of output for 14 randomly selected months. The data are shown below:

Marginal cost Y ($ per 100 units)	Output X (thousands of units)
31.00	3.0
30.00	3.0
28.00	3.5
46.00	1.0
43.00	1.5
35.00	2.0
37.50	2.0
27.00	3.5
25.00	5.5
24.00	7.0
25.00	7.0
29.50	4.5
26.00	4.5
28.00	4.5

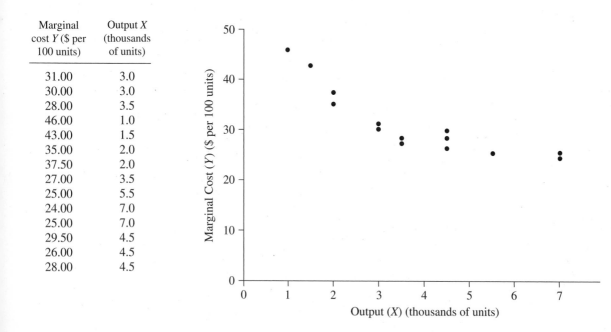

a. On the accompanying scatter diagram of marginal cost (Y) versus output (X), sketch by eye a line that fits the data reasonably well.

b. Find the estimated least-squares equation for the regression of marginal cost on output.

c. Sketch the estimated line on the scatter diagram. Does it seem to fit the data well?

d. Test the null hypothesis that the true slope is zero, at the $\alpha = .05$ significance level. Interpret your result.

e. Can you suggest a model that would describe the marginal cost–output relationship for this manufacturer better than a straight line does?

Edited SAS Output (PROC REG) for Problem 18

```
            OUTPUT FOR REGRESSION OF MARGINAL COST ON OUTPUT
                                      .
                             .[Portion of output omitted]
                                      .
            Root MSE                3.63338    R-Square      0.7405

                          Parameter Estimates
                          Parameter          Standard
Variable     DF           Estimate             Error       t Value     Pr > |t|
Intercept    1            42.84255           2.23377         19.18      <.0001
Output       1            -3.13896           0.53644         -5.85      <.0001
```

19. The data shown in the following table were obtained from the 1990 Census.[11] Included is information on 26 randomly selected Metropolitan Statistical Areas (MSAs). Of interest are factors that potentially are associated with the rate of owner occupancy of housing units. Two variables are included in the data set: OWNEROCC = percentage of housing units that are owner-occupied (as opposed to renter-occupied); OWNCOST = median selected monthly ownership costs (in $).

a. Based on the accompanying scatter diagram of OWNEROCC versus OWNCOST, does there appear to be a linear relationship between these two variables?

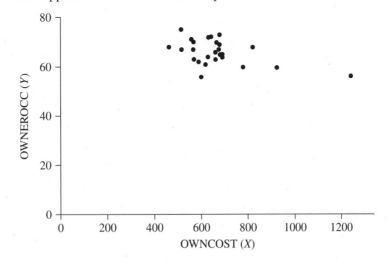

---

[11] The data were randomly sampled on June 20, 1996, from the 1990 U.S. Census.

MSA	OWNEROCC (Y)	OWNCOST(X)
Abilene, TX	62	583
Burlington, NC	72	627
Daytona Beach, FL	72	636
Grand Rapids, MI	73	677
Laredo, TX	61	614
Louisville, KY–IN	67	561
Oklahoma City, OK	64	627
Pine Bluff, AR	67	513
San Francisco–Oakland– San Jose, CA	57	1234
Wichita Falls, TX	63	565
Albany, GA	56	597
Canton, OH	71	555
Des Moines, IA	67	673
Jacksonville, FL	65	687
Johnstown, PA	75	510
Medford, OR	66	660
Omaha, NE–IA	64	687
Provo–Orem, UT	63	659
Williamsburg, PA	70	564
Appleton–Oshkosh– Neenah, WI	70	663
Melbourne–Titusville– Palm Bay, FL	69	675
Redding, CA	65	682
Worcester, MA	60	918
Milwaukee–Racine, WI	60	777
Rochester, NY	68	818
St. Joseph, MO	68	457

b. State the model for the straight-line regression of OWNEROCC (Y) on OWNCOST (X). Determine the least-squares estimates for this regression line. Interpret the estimated values of the slope and the intercept in the context of the problem.

c. Sketch the estimated line on the scatter diagram and assess the fit.

d. Test for the significance of the slope parameter of the model in part (b). Interpret your result.

e. Determine a 95% confidence interval for the true slope in part (d). Interpret your result with regard to the test mentioned in part (d).

Edited SAS Output (PROC REG) for Problem 19

```
                         OWNEROCC REGRESSED ON OWNCOST

                              .[Portion of output omitted]
                              .

             Root MSE              4.41749    R-Square       0.2207

                          Parameter Estimates
                          Parameter            Standard
Variable     DF           Estimate               Error        t Value      Pr > |t|
Intercept    1            76.00764             3.94979         19.24        <.0001
Owncost      1            -0.01517             0.00582         -2.61        0.0155
```

20. Researchers have studied the ecology of ponds in rural Bangladesh. Of particular interest in such studies are the zooplankton, phytoplankton and copepod counts (per liter of water) in these ponds. Copepods are a particular type of zooplankton that are thought to be a natural reservoir of cholera bacteria, but copepod counts can be difficult to obtain in laboratory analyses; total zooplankton counts are easier to obtain. It is also believed that copepod counts will be related to phytoplankton counts since copepods feed on phytoplankton.

The data below show zooplankton, phytoplankton and copepod counts for 100 water samples from ponds in rural Bangladesh.

Obs	Copepods	Zooplankton	Phytoplankton	Obs	Copepods	Zooplankton	Phytoplankton
1	585	1560	475.54	51	150	228	493.92
2	111	1191	96.96	52	60	108	63.54
3	48	537	521.31	53	279	804	24.59
4	33	597	26.30	54	84	252	34.16
5	372	1710	49.02	55	36	279	49.53
6	3	993	86.77	56	240	1317	7069.31
7	7	86	225.45	57	26	100	839.62
8	117	468	209.75	58	42	57	701.06
9	123	714	101.46	59	258	285	48.51
10	153	836	89.52	60	3	597	1031.35
11	18	684	1133.49	61	186	3780	1195.33
12	15	36	69.18	62	573	4932	42.70
13	21	489	84.04	63	521	715	252.26
14	129	888	684.61	64	39	234	28.01
15	60	369	827.06	65	3	84	80.79
16	144	258	360.92	66	12	660	302.68
17	30	45	26.99	67	81	225	17.25
18	1080	1308	42.02	68	36	126	69.69
19	18	321	33.14	69	570	900	1316.26
20	108	210	99.41	70	78	4119	50.22
21	146	1052	65.01	71	420	486	117.52
22	33	264	125.37	72	297	744	44.06
23	258	1527	213.63	73	15	207	13.66
24	39	2151	4554.82	74	27	93	28.69
25	264	696	548.64	75	171	258	235.72
26	120	430	971.28	76	561	1473	141.77
27	269	827	63.04	77	405	465	393.38
28	226	285	419.37	78	54	66	72.77
29	90	174	328.98	79	87	204	5.29
30	237	351	1361.87	80	81	468	449.74
31	708	936	80.62	81	162	189	287.64
32	315	351	62.17	82	402	466	139.86
33	12	336	49.19	83	429	468	43.73
34	74	117	28.02	84	42	340	9.19
35	66	156	81.99	85	252	324	20.84
36	141	156	403.45	86	408	671	227.18
37	192	3483	110.00	87	171	408	104.71
38	12	284	117.52	88	30	432	332.39
39	120	1182	63.19	89	162	261	156.28
40	129	2436	7710.88	90	150	1431	287.64
41	18	504	1075.25	91	66	69	36.33
42	222	264	481.86	92	54	287	233.00
43	18	24	37.58	93	276	360	131.52
44	378	732	35.19	94	48	345	89.50
45	66	924	312.75	95	150	336	99.92
46	3	46	174.00	96	96	873	521.65
47	879	1221	44.07	97	90	450	1122.39
48	60	324	53.63	98	237	276	279.79
49	228	1641	2646.53	99	143	657	21.45
50	729	1809	87.11	100	243	426	39.29

a. Find the estimated least squares equation for the regression of copepod count on zooplankton count.

b. Test the null hypothesis that the true slope is zero, at the $\alpha = 0.05$ significance level. Interpret your result.

c. Find a 95% confidence interval for the true slope in part (a). Interpret your result.

d. Find the estimated least squares equation for the regression of copepod count on phytoplankton count.

e. Test the null hypothesis that the true slope in part (d) is zero, at the $\alpha = 0.05$ significance level. Interpret your result.

References

Benignus, V. A.; Muller, K. E.; Barton, C. N.; and Bittekofer, J. A. 1981. "Toluene Levels in Blood and Brain of Rats during and after Respiratory Exposure." *Toxicology and Applied Pharmacology* 61: 326–34.

Bourne, L. E.; Ekstrand, B. E.; and Dominowski, R. L. 1971. *The Psychology of Thinking.* Englewood Cliffs, N.J.: Prentice-Hall.

Diggle, P. J.; Heagerty, P; Liang, K. Y.; and Zeger, S. L. 2002. *Analysis of Longitudinal Data.* New York: Oxford University Press.

Dimmick, F. L., and Hubbard, M. R. 1939. "The Spectral Location of Psychologically Unique Yellow, Green, and Blue." *American Journal of Psychology* 52: 242.

Morrison, D. F. 1976. *Multivariate Statistical Methods.* New York: McGraw-Hill.

Nagasawa, S.; Osana, S.; and Kondo, K. 1964. "An Analytical Method for Evaluating the Susceptibility of Fish Species to an Agricultural Chemical." *Japanese Journal of Applied Enterological Zoology* 8: 118–22.

Schønheyder, F. 1936. "The Quantitative Determination of Vitamin K." *Biochemistry Journal* 30: 890–96.

Schreiner, H. R.; Gregoine, R. C.; and Lawrie, J. A. 1962. "New Biological Effects of the Gases on the Helium Group." *Science* 136: 653–54.

Siegel, S. 1956. *Nonparametric Statistics for the Behavioral Sciences.* New York: McGraw-Hill.

Smith, M. E. 1926. "An Investigation of the Development of the Sentence and the Extent of Vocabulary in Young Children." *Studies in Child Welfare* 3:5.

Stevens, S. S. 1966. *Handbook of Experimental Psychology.* New York: John Wiley & Sons.

Timm, N. H. 1975. *Multivariate Analysis with Applications in Education and Psychology.* Monterey, Calif.: Brooks/Cole.

Zeger, S. L., and Liang, K. Y. 1986. "Longitudinal Data Analysis for Discrete and Continuous Outcomes," *Biometrics* 42: 121–30.

The Correlation Coefficient and Straight-line Regression Analysis

6.1 Definition of *r*

The correlation coefficient is an often-used statistic that provides a measure of how two random variables are linearly associated in a sample and has properties closely related to those of straight-line regression. We define the *sample correlation coefficient r* for two variables X and Y by the formula

$$r = \frac{\sum_{i=1}^{n}(X_i - \overline{X})(Y_i - \overline{Y})}{\left[\sum_{i=1}^{n}(X_i - \overline{X})^2 \sum_{i=1}^{n}(Y_i - \overline{Y})^2\right]^{1/2}} = \frac{\text{SSXY}}{\sqrt{\text{SSX} \cdot \text{SSY}}} \tag{6.1}$$

where $\text{SSXY} = \sum_{i=1}^{n}(X_i - \overline{X})(Y_i - \overline{Y})$, $\text{SSX} = \sum_{i=1}^{n}(X_i - \overline{X})^2$, and $\text{SSY} = \sum_{i=1}^{n}(Y_i - \overline{Y})^2$.

An equivalent formula for r that illustrates its mathematical relationship to the least-squares estimate of the slope of a fitted regression line is[1]

$$r = \frac{S_X}{S_Y}\hat{\beta}_1 \tag{6.2}$$

■ **Example 6.1**   For the age–systolic blood pressure data in Table 5.1, r is 0.66. This value can be obtained from the SAS output on page 52 by taking the square root of "*R*–square = 0.4324."

---

[1] $S_X^2 = \dfrac{1}{n-1}\text{SSX}$ and $S_Y^2 = \dfrac{1}{n-1}\text{SSY}$ are the estimated sample variances of the X and Y variables, respectively.

Alternatively, using (6.2), we have

$$r = \frac{15.29}{22.58}(0.97) = 0.66$$

Three important mathematical properties are associated with r:

1. The possible values of r range from -1 to 1.

2. r is a dimensionless quantity; that is, r is independent of the units of measurement of X and Y.

3. r is positive, negative, or zero as $\hat{\beta}_1$ is positive, negative, or zero; and vice versa. This property follows directly, of course, from (6.2). ■

6.2 r as a Measure of Association

In the statistical assumptions for straight-line regression analysis discussed earlier, we did not specifically consider the variable X to be a random variable. Nevertheless, it often makes sense to view the regression problem as one in which both X and Y are random variables. The measure r can then be interpreted as an *index of linear association* between X and Y, in the following sense:

1. The more positive r is, the more positive the association is. This means that, when r is close to 1, an individual with a high value for one variable will likely have a high value for the other, and an individual with a low value for one variable will likely have a low value for the other (Figure 6.1(a)).

2. The more negative r is, the more negative the association is; that is, an individual with a high value for one variable will likely have a low value for the other when r is close to -1, and conversely (Figure 6.1(b)).

3. If r is close to 0, there is little, if any, *linear* association between X and Y (Figure 6.1(c) or 6.1(d)).[2]

By *association*, we mean the lack of statistical independence between X and Y. More loosely, the lack of an association means that the value of one variable cannot be reasonably anticipated from knowing the value of the other variable.

Since r is an index obtained from a *sample* of n observations, it can be considered as an estimate of an unknown population parameter. This unknown parameter, called the *population correlation coefficient,* is generally denoted by the symbol ρ_{XY} or more simply ρ (if it is clearly understood which two variables are being considered). We shall agree to use ρ unless confusion is possible. The parameter ρ_{XY} is defined as $\rho_{XY} = \sigma_{XY}/\sigma_X\sigma_Y$, where σ_X and σ_Y denote the population standard deviations of the random variables X and Y and where σ_{XY} is called the population *covariance* between X and Y. The covariance σ_{XY} is a population parameter describing the average amount by which two variables covary. In actuality, it is the population mean of the random variable $SSXY/(n-1)$.

---

[2] Later we will see that a value of r close to 0 does not rule out a possible *nonlinear* association.

FIGURE 6.1   Correlation coefficient as a measure of association

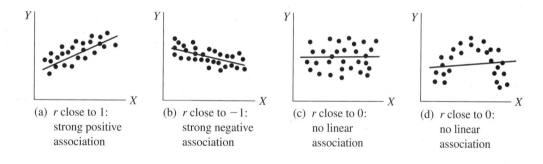

(a) *r* close to 1:   (b) *r* close to −1:   (c) *r* close to 0:   (d) *r* close to 0:
strong positive   strong negative   no linear   no linear
association   association   association   association

Figure 6.2 on the next page provides informative examples of scatter diagrams. Data were generated (via computer simulation) to have means and variances similar to those of the age–systolic blood pressure data of Chapter 5. The six samples observed here were produced by selecting 30 paired observations at random from each of six populations for which the population correlation coefficient ρ varied in value.

In Figure 6.2, the sample correlations range in value from .037 to .894. It should be clear that an eyeball analysis of the relative strengths of association is difficult, even though $n = 30$. For example, the difference between $r = .037$ in Figure 6.2(a) and $r = .220$ in Figure 6.2(b) is apparently due to the influence of just a few points. The study of so-called influential data points will be described in Chapter 14 on regression diagnostics.

In evaluating a scatter diagram, we find it helpful to include reference lines at the X and Y means, as in Figure 6.2(f). Roughly speaking, the proportions of observations in each quadrant reflect the strength of association. Notice that most of the observations in this figure are located in quadrants B and C, which are often referred to as the *positive quadrants.* Quadrants A and D are called the *negative quadrants.* When more observations are in positive quadrants than in negative quadrants, the sample correlation coefficient r is usually positive. On the other hand, if more observations are in negative quadrants, r is usually negative.

To understand why this is so, we need to examine the numerator part of equation (6.1)—namely,

$$\sum_{i=1}^{n}(X_i - \overline{X})(Y_i - \overline{Y})$$

(Notice that the denominator in (6.1) is simply a positive scaling factor ensuring that r is both dimensionless and satisfies the inequality $-1 \le r \le 1$.) The numerator describes how X and Y covary in terms of the n cross-products $(X_i - \overline{X})(Y_i - \overline{Y})$, where $i = 1, 2, \ldots, n$. For a given i, such a cross-product term is either positive or negative (or zero), depending on how X_i compares with $\overline{X}$ and how Y_i compares with $\overline{Y}$. In particular, if the ith observation (X_i, Y_i) is in quadrant B, then $X_i > \overline{X}$ and $Y_i > \overline{Y}$; hence, the product of $(X_i - \overline{X})$ and $(Y_i - \overline{Y})$ must be positive. Similarly, if (X_i, Y_i) is in quadrant C, $X_i < \overline{X}$ and $Y_i < \overline{Y}$, so $(X_i - \overline{X})(Y_i - \overline{Y})$ is again positive. Thus, observations in the positive quadrants B and C contribute positive values to the numerator of (6.1). Conversely, observations in the negative quadrants A and D contribute negative values to this numerator. So, roughly speaking, the sign of the correlation coefficient reflects the distribution of observations in these positive and negative quadrants.

FIGURE 6.2 Examples of a range of observed correlations between age and systolic blood pressure (SBP) for simulated data

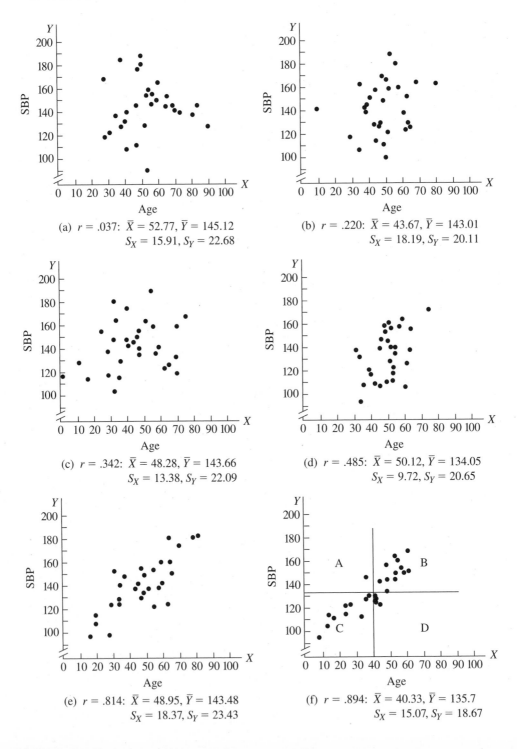

(a) $r = .037$: $\bar{X} = 52.77$, $\bar{Y} = 145.12$
$S_X = 15.91$, $S_Y = 22.68$

(b) $r = .220$: $\bar{X} = 43.67$, $\bar{Y} = 143.01$
$S_X = 18.19$, $S_Y = 20.11$

(c) $r = .342$: $\bar{X} = 48.28$, $\bar{Y} = 143.66$
$S_X = 13.38$, $S_Y = 22.09$

(d) $r = .485$: $\bar{X} = 50.12$, $\bar{Y} = 134.05$
$S_X = 9.72$, $S_Y = 20.65$

(e) $r = .814$: $\bar{X} = 48.95$, $\bar{Y} = 143.48$
$S_X = 18.37$, $S_Y = 23.43$

(f) $r = .894$: $\bar{X} = 40.33$, $\bar{Y} = 135.7$
$S_X = 15.07$, $S_Y = 18.67$

6.3 The Bivariate Normal Distribution[3]

Another way of looking at straight-line regression is to consider X and Y as random variables having the *bivariate normal distribution,* which is a generalization of the *univariate normal distribution.* Just as the univariate normal distribution is described by a density function that appears as a bell-shaped curve when plotted in two dimensions, the bivariate normal distribution is described by a *joint density function* whose plot looks like a bell-shaped surface in three dimensions (Figure 6.3).

One property of the bivariate normal distribution that has implications for straight-line regression analysis is the following: If the bell-shaped surface is cut by a plane *parallel* to the YZ-plane and passing through a specific X-value, the curve, or *trace,* that results is a normal distribution. In other words, the distribution of Y for fixed X is univariate-normal. We call such a distribution the *conditional distribution* of Y at X, and we denote the corresponding random variable as Y_X. Let us denote the mean of this distribution as $\mu_{Y|X}$ and the variance as $\sigma^2_{Y|X}$. Then it follows from statistical theory that the mean and the variance, respectively, of Y_X can be written in terms of μ_X, μ_Y, σ^2_X, σ^2_Y, and ρ_{XY} as follows:

$$\mu_{Y|X} = \mu_Y + \rho_{XY}\frac{\sigma_Y}{\sigma_X}(X - \mu_X) \tag{6.3}$$

and

$$\sigma^2_{Y|X} = \sigma^2_Y(1 - \rho^2_{XY}) \tag{6.4}$$

Now suppose that we let $\beta_1 = \rho_{XY}(\sigma_Y/\sigma_X)$ and $\beta_0 = \mu_Y - \beta_1\mu_X$. Then (6.3) has been transformed into the familiar expression for a straight-line model; that is, $\mu_{Y|X} = \beta_0 + \beta_1X$.

FIGURE 6.3   The bivariate normal distribution

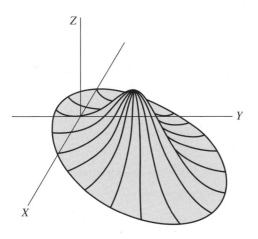

---

[3] This section is not essential for understanding the correlation coefficient as it relates to regression analysis.

Furthermore, if we substitute the estimators $\bar{X}$, $\bar{Y}$, S_X, S_Y, and r for their respective parameters μ_X, μ_Y, σ_X, σ_Y, and ρ_{XY} in (6.3), we obtain the formula

$$\hat{\mu}_{Y|X} = \bar{Y} + r\frac{S_Y}{S_X}(X - \bar{X})$$

The right-hand side of this equation is exactly equivalent to the expression for the least-squares straight line given in (5.7), since

$$\hat{\beta}_1 = r\frac{S_Y}{S_X}$$

Thus, *the least-squares formulas for $\hat{\beta}_0$ and $\hat{\beta}_1$ can be developed by assuming that X and Y are random variables having the bivariate normal distribution and by substituting the usual estimates of μ_X, μ_Y, σ_X, σ_Y, and ρ_{XY} into the expression for $\mu_{Y|X}$, the conditional mean of Y given X.*

Our estimate $S_{Y|X}^2$ of $\sigma_{Y|X}^2$ can also be obtained by substituting the estimates S_Y^2 and r for σ_Y^2 and ρ_{XY} in (6.4). Thus, we obtain

$$S_{Y|X}^2 = S_Y^2(1 - r^2)$$

Finally, (6.4) can be algebraically manipulated into the form

$$\rho_{XY}^2 = \frac{\sigma_Y^2 - \sigma_{Y|X}^2}{\sigma_Y^2} \tag{6.5}$$

This equation describes the square of the population correlation coefficient as the proportionate reduction in the variance of Y due to conditioning on X. The importance of (6.5) in describing the strength of the straight-line relationship will be discussed in the next section.

6.4 *r* and the Strength of the Straight-line Relationship

To quantify what we mean by the *strength* of the linear relationship between X and Y, we should first consider what our predictor of Y would be if we did not use X at all. The best predictor in this case would simply be $\bar{Y}$, the sample mean of the Y's. The sum of the squares of deviations associated with the naive predictor $\bar{Y}$ would then be given by the formula

$$\text{SSY} = \sum_{i=1}^{n}(Y_i - \bar{Y})^2$$

Now, if the variable X is of any value in predicting the variable Y, the residual sum of squares given by

$$\text{SSE} = \sum_{i=1}^{n}(Y_i - \hat{Y}_i)^2$$

should be considerably less than SSY. If so, the least-squares model $\hat{Y} = \hat{\beta}_0 + \hat{\beta}_1 X$ fits the data better than does the horizontal line $\hat{Y} = \bar{Y}$ (Figure 6.4). A quantitative measure of the improvement

FIGURE 6.4 **Predictions of Y using and not using X**

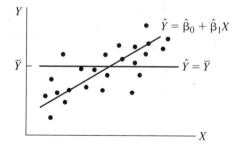

in the fit obtained by using X is given by the *square of the sample correlation coefficient r,* which can be written in the suggestive form

$$r^2 = \frac{\text{SSY} - \text{SSE}}{\text{SSY}} \qquad (6.6)$$

This quantity naturally varies between 0 and 1, since r itself varies between -1 and 1.

What interpretation can be given to the quantity r^2? To answer this question, we first note that the difference, or *reduction,* in SSY due to using X to predict Y may be measured by (SSY $-$ SSE), which is always nonnegative. Furthermore, the *proportionate reduction* in SSY due to using X to predict Y is this difference divided by SSY. Thus, r^2 measures the strength of the linear relationship between X and Y in the sense that it gives the proportionate reduction in the sum of the squares of vertical deviations obtained by using the least-squares line $\hat{Y} = \hat{\beta}_0 + \hat{\beta}_1 X$ instead of the naive model $\hat{Y} = \overline{Y}$ (the predictor of Y if X is ignored). The larger the value of r^2, the greater the reduction in SSE relative to $\sum_{i=1}^{n}(Y_i - \overline{Y})^2$, and the stronger the linear relationship between X and Y.

The largest value that r^2 can attain is 1, which occurs when $\hat{\beta}_1$ is nonzero and when SSE $= 0$ (i.e., when a perfect positive or negative straight-line relationship exists between X and Y). By "perfect" we mean that *all* the data points lie on the fitted straight line. In other words, when $Y_i = \hat{Y}_i$ for all i, we must have

$$\text{SSE} = \sum_{i=1}^{n}(Y_i - \hat{Y}_i)^2 = 0$$

so

$$r^2 = \frac{\text{SSY} - \text{SSE}}{\text{SSY}} = \frac{\text{SSY}}{\text{SSY}} = 1$$

Figure 6.5 illustrates examples of perfect positive and perfect negative linear association.

The smallest value that r^2 may take, of course, is 0. This value means that using X offers no improvement in predictive power; that is, SSE $=$ SSY. Furthermore, appealing to (6.2), we see that a correlation coefficient of 0 implies an estimated slope of 0 and

FIGURE 6.5   Examples of perfect linear association

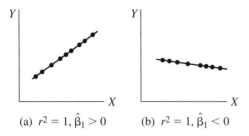

(a) $r^2 = 1, \hat{\beta}_1 > 0$ (b) $r^2 = 1, \hat{\beta}_1 < 0$

consequently the absence of any linear relationship (although a nonlinear relationship is still possible).

Finally, one should *not* be led to a false sense of security by considering the magnitude of r, rather than of r^2, when assessing the strength of the linear association between X and Y. For example, when r is 0.5, r^2 is only 0.25, and it takes $r > 0.7$ to make $r^2 > 0.5$. Also, when r is 0.3, r^2 is 0.09, which indicates that only 9% of the variation in Y is explained with the help of X. For the age–systolic blood pressure data, r^2 is 0.43, compared with an r of 0.66. The r^2 value also appears on the SAS output on p. 52.

6.5 What r Does Not Measure

Two common misconceptions about r (or, equivalently, about r^2) occasionally lead a researcher to make spurious interpretations of the relationship between X and Y. The correct notions are as follows:

1. r^2 *is not a measure of the magnitude of the slope of the regression line.* Even when the value of r^2 is high (i.e., close to 1), the magnitude of the slope $\hat{\beta}_1$ is not necessarily large. This phenomenon is illustrated in Figure 6.5. Notice that r^2 equals 1 in both parts, despite the fact that the slopes are different. Another way to understand this, using (6.2), is

$$\hat{\beta}_1^2 = \frac{S_Y^2}{S_X^2} \quad \text{when } r^2 = 1$$

   Thus, if two different sets of data have the same amount of X variation, but the first set has less Y variation than the second set, the magnitude of the slope for the first set is smaller than that for the second.

2. r^2 *is not a measure of the appropriateness of the straight-line model.* Note that $r^2 = 0$ in parts (a) and (b) of Figure 6.6, even though no evidence of association between X and Y exists in (a) and strong evidence of a nonlinear association exists in (b). Conversely, r^2 is high in parts (c) and (d), even though a straight-line model is quite appropriate in (c) but not entirely appropriate in (d).

FIGURE 6.6   Examples showing that r^2 is not a measure of the appropriateness of the straight-line model

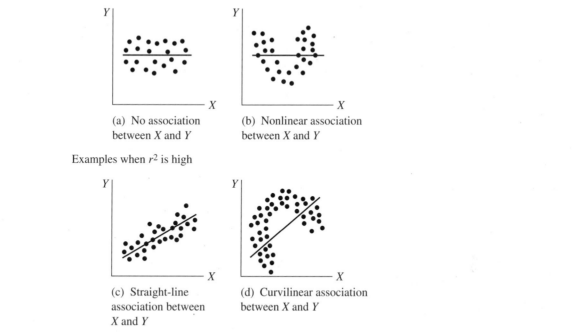

(a) No association
between X and Y

(b) Nonlinear association
between X and Y

Examples when r^2 is high

(c) Straight-line
association between
X and Y

(d) Curvilinear association
between X and Y

6.6   Tests of Hypotheses and Confidence Intervals for the Correlation Coefficient

Researchers interested in the association between two interval variables X and Y often want to test the null hypothesis $H_0: \rho = 0$.

6.6.1   Test of $H_0: \rho = 0$

A test of $H_0: \rho = 0$ turns out to be mathematically equivalent to the test of the null hypothesis $H_0: \beta_1 = 0$ described in Section 5.8. This equivalence is suggested by the formulas $\beta_1 = \rho \sigma_Y / \sigma_X$ and $\hat{\beta}_1 = r S_Y / S_X$, which tell us, for example, that β_1 is positive, negative, or zero as ρ is positive, negative, or zero, and that an analogous relationship exists between $\hat{\beta}_1$ and r. The test statistic for the hypothesis $H_0: \rho = 0$ can be written entirely in terms of r and n, so we can perform the test without having to fit the straight line. This test statistic is given by the formula

$$T = \frac{r\sqrt{n-2}}{\sqrt{1-r^2}} \tag{6.7}$$

which has the t distribution with $n - 2$ degrees of freedom when the null hypothesis $H_0: \rho = 0$ (or equivalently, $H_0: \beta_1 = 0$) is true. Formula (6.7) yields exactly the same numerical answer as does (5.9), given by

$$T = \frac{\hat{\beta}_1 - \beta_1^{(0)}}{S_{\hat{\beta}_1}} \quad \text{when} \quad \beta_1^{(0)} = 0$$

■ **Example 6.2** For the age–systolic blood pressure data of Table 5.1, for which $r = 0.66$, the statistic in (6.7) is calculated as follows:

$$T = \frac{.66\sqrt{30 - 2}}{\sqrt{1 - (.66)^2}} = 4.62$$

which is the same value as obtained for the test for slope in Table 5.3. ■

6.6.2 Test of $H_0: \rho = \rho_0, \rho_0 \neq 0$

A test concerning the null hypothesis $H_0: \rho = \rho_0 \ (\rho_0 \neq 0)$ cannot be directly related to a test concerning β_1; moreover, the hypothesis $H_0: \rho = \rho_0 \ (\rho_0 \neq 0)$ is not equivalent to the hypothesis $H_0: \beta_1 = \beta_1^{(0)}$ for some value $\beta_1^{(0)}$. Nevertheless, a test of $H_0: \rho = \rho_0 \ (\rho_0 \neq 0)$ is meaningful when previous experience or theory suggests a particular value to use for ρ_0.

The test statistic in this case can be obtained by considering the distribution of the sample correlation coefficient r. This distribution happens to be symmetric, like the normal distribution, *only* when ρ is 0. When ρ is nonzero, the distribution of r is skewed. This lack of normality prevents us from using a test statistic of the usual form, which has a normally distributed estimator in the numerator and an independent estimator of its standard deviation in the denominator. But through an appropriate transformation, r can be changed into a statistic that is often approximately normal. This transformation is called *Fisher's Z transformation.*[4] The formula for this transformation is

$$\frac{1}{2} \ln \frac{1 + r}{1 - r} \tag{6.8}$$

This quantity has approximately the normal distribution, with mean $\frac{1}{2} \ln [(1 + \rho)/(1 - \rho)]$ and variance $1/(n - 3)$ when n is not too small (e.g., $n \geq 20$). In testing the null hypothesis $H_0: \rho = \rho_0 \ (\rho_0 \neq 0)$, we can then use the test statistic

$$Z = \frac{\frac{1}{2} \ln [(1 + r)/(1 - r)] - \frac{1}{2} \ln[(1 + \rho_0)/(1 - \rho_0)]}{1/\sqrt{n - 3}} \tag{6.9}$$

This test statistic has approximately the standard normal distribution (i.e., $Z \frown N(0, 1)$) under H_0. To test $H_0: \rho = \rho_0 \ (\rho_0 \neq 0)$, therefore, we use one of the following critical regions for significance level α:

$Z \geq z_{1-\alpha}$	(upper one-tailed alternative $H_A: \rho > \rho_0$)
$Z \leq -z_{1-\alpha}$	(lower one-tailed alternative $H_A: \rho < \rho_0$)
$\|Z\| \geq z_{1-\alpha/2}$	(two-tailed alternative $H_A: \rho \neq \rho_0$)

where $z_{1-\alpha}$ denotes the $100(1 - \alpha)\%$ point of the standard normal distribution. Computation of Z can be aided by using Appendix Table A-5, which gives values of $\frac{1}{2} \ln [(1 + r)/(1 - r)]$ for given values of r.

---

[4] Named after R. A. Fisher, who introduced it in 1925.

■ **Example 6.3**   Suppose that from previous experience we can hypothesize that the true correlation between age and systolic blood pressure is $\rho_0 = 0.85$. To test the null hypothesis $H_0: \rho = 0.85$ against the two-sided alternative $H_A: \rho \neq 0.85$, we perform the following calculations using $r = 0.66$, $\rho_0 = 0.85$, and $n = 30$:

$$\frac{1}{2} \ln \frac{1 + \rho_0}{1 - \rho_0} = \frac{1}{2} \ln \frac{1 + 0.85}{1 - 0.85} = 1.2561 \qquad \text{(from Table A-5)}$$

$$\frac{1}{2} \ln \frac{1 + r}{1 - r} = \frac{1}{2} \ln \frac{1 + 0.66}{1 - 0.66} = .7928 \qquad \text{(from Table A-5)}$$

$$Z = \frac{0.7928 - 1.2561}{1/\sqrt{30 - 3}} = -2.41$$

For $\alpha = .05$, the critical region is

$$|Z| \geq z_{.975} = 1.96$$

Since $|Z| = 2.41$ exceeds 1.96, the hypothesis $H_0: \rho_0 = 0.85$ is rejected at the .05 significance level. Further calculations show that the P-value for this test is $P = 0.0151$, which tells us that the result is not significant at $\alpha = .01$.   ■

6.6.3 Confidence Interval for ρ

A $100(1 - \alpha)\%$ confidence interval for ρ can be obtained by using Fisher's Z transformation (6.8) as follows. First, compute a $100(1 - \alpha)\%$ confidence interval for the parameter $\frac{1}{2} \ln [(1 + \rho)/(1 - \rho)]$ using the formula

$$\frac{1}{2} \ln \frac{1 + r}{1 - r} \pm \frac{z_{1-\alpha/2}}{\sqrt{n - 3}} \tag{6.10}$$

where $z_{1-\alpha/2}$ is as defined previously.

Denote the lower limit of the confidence interval (6.10) by L_Z, and the upper limit by U_Z; then use Appendix Table A-5 (in reverse) to determine the lower and upper confidence limits L_ρ and U_ρ for the confidence interval for ρ. In other words, determine L_ρ and U_ρ from the following formulas[5]:

$$L_Z = \frac{1}{2} \ln \frac{1 + L_\rho}{1 - L_\rho} \quad \text{and} \quad U_Z = \frac{1}{2} \ln \frac{1 + U_\rho}{1 - U_\rho}$$

■ **Example 6.4**   Suppose that we seek a 95% confidence interval for ρ based on the age–systolic blood pressure data for which $r = 0.66$ and $n = 30$. A 95% confidence interval for $\frac{1}{2} \ln [(1 + \rho)/(1 - \rho)]$ is given by

$$\frac{1}{2} \ln \frac{1 + .66}{1 - .66} \pm \frac{1.96}{\sqrt{30 - 3}}$$

---

[5] L_ρ and U_ρ can also be calculated directly using the conversion formulas:

$$L_\rho = \frac{e^{2L_Z} - 1}{e^{2L_Z} + 1} \quad \text{and} \quad U_\rho = \frac{e^{2U_Z} - 1}{e^{2U_Z} + 1}$$

which is equal to

$$0.793 \pm 0.377$$

providing a lower limit of $L_Z = 0.416$ and an upper limit of $U_Z = 1.170$.

To transform these L_Z and U_Z values into lower and upper confidence limits for ρ, we determine the values of L_ρ and U_ρ that satisfy

$$0.416 = \frac{1}{2} \ln \frac{1 + L_\rho}{1 - L_\rho} \quad \text{and} \quad 1.170 = \frac{1}{2} \ln \frac{1 + U_\rho}{1 - U_\rho}$$

Using Table A-5, we see that a value of 0.416 corresponds to an r of about 0.394, so $L_\rho = 0.394$. Similarly, a value of 1.170 corresponds to an r of about 0.824, so $U_\rho = 0.824$. The 95% confidence interval for ρ thus has a lower limit of 0.394 and an upper limit of 0.824.

Notice that the interval (0.394, 0.824) does not contain the value 0.85, which agrees with the conclusion of the previous section that H_0: $\rho = 0.85$ is to be rejected at the 5% level (two-tailed test). ∎

6.7 Testing for the Equality of Two Correlations

Suppose that independent random samples of sizes n_1 and n_2 are selected from two populations. Further, suppose that we want to test H_0: $\rho_1 = \rho_2$ versus, say, H_A: $\rho_1 \neq \rho_2$. An appropriate test statistic can be developed based on the results given in Section 6.6. In this section, we will also consider the situation in which the sample correlations to be compared are calculated by using the same data set; in this case, the sample correlations are themselves "correlated."

6.7.1 Test of H_0: $\rho_1 = \rho_2$ Using Independent Random Samples

Let us assume that independent random samples of sizes n_1 and n_2 have been selected from two populations. For each population, the straight-line regression analysis assumptions given in Chapter 5, including that of normality, will hold.

An approximate test of H_0: $\rho_1 = \rho_2$ can be based on the use of Fisher's Z transformation. Let r_1 be the sample correlation calculated by using the n_1 observations from the first population, and let r_2 be defined similarly. Using (6.8), let

$$Z_1 = \frac{1}{2} \ln \frac{1 + r_1}{1 - r_1} \tag{6.11}$$

and

$$Z_2 = \frac{1}{2} \ln \frac{1 + r_2}{1 - r_2} \tag{6.12}$$

Appendix Table A-5 can be used to determine Z_1 and Z_2.

To test H_0: $\rho_1 = \rho_2$, we can compute the test statistic

$$Z = \frac{Z_1 - Z_2}{\sqrt{1/(n_1 - 3) + 1/(n_2 - 3)}} \tag{6.13}$$

For large n_1 and n_2, this test statistic has (approximately) the standard normal distribution when H_0 is true. Hence, the following critical regions for significance level α should be used:

$$Z \geq z_{1-\alpha} \quad \text{(upper one-tailed alternative } H_A: \rho_1 > \rho_2)$$

$$Z \leq -z_{1-\alpha} \quad \text{(lower one-tailed alternative } H_A: \rho_1 < \rho_2)$$

$$|Z| \geq z_{1-\alpha/2} \quad \text{(two-tailed alternative } H_A: \rho_1 \neq \rho_2)$$

To illustrate this procedure, let us test whether the data sets plotted in Figures 6.2(b) and 6.2(c) reflect populations with different correlations. In other words, we want to test $H_0: \rho_1 = \rho_2$ versus the two-sided alternative $H_A: \rho_1 \neq \rho_2$.

For the data in Figure 6.2(b), $r_1 = 0.220$; for the Figure 6.2(c) data, $r_2 = 0.342$. Using Fisher's Z transformation and Table A-5, we can calculate Z_1 and Z_2 as

$$Z_1 = \frac{1}{2} \ln \frac{1 + r_1}{1 - r_1} = \frac{1}{2} \ln \frac{1 + .220}{1 - .220} = 0.2237$$

and

$$Z_2 = \frac{1}{2} \ln \frac{1 + r_2}{1 - r_2} = \frac{1}{2} \ln \frac{1 + .342}{1 - .342} = 0.3564$$

Then the test statistic (6.13) takes the value

$$Z = \frac{0.2237 - 0.3564}{\sqrt{1/(30 - 3) + 1/(30 - 3)}} = \frac{-0.1327}{0.2722} = -0.488$$

For $\alpha = .01$, the critical region is

$$|Z| \geq z_{.005} = 2.576$$

Since $|Z| = 0.488$ is less than 2.576, we cannot reject $H_0: \rho_1 = \rho_2$ at $\alpha = .01$.

6.7.2 Single Sample Test of $H_0: \rho_{12} = \rho_{13}$

Consider testing the null hypothesis that the correlation ρ_{12} of variable 1 with variable 2 is the same as the correlation ρ_{13} of variable 1 with variable 3. Let us assume that a single random sample of n subjects is selected and that the three sample correlations—r_{12}, r_{13}, and r_{23}—are calculated. Clearly these sample correlations are not independent, since they are computed using the same data set. Under the usual straight-line regression analysis assumptions, it can be shown (we omit the details) that an appropriate large-sample test statistic for testing $H_0: \rho_{12} = \rho_{13}$ is

$$Z = \frac{(r_{12} - r_{13})\sqrt{n}}{\sqrt{(1 - r_{12}^2)^2 + (1 - r_{13}^2)^2 - 2r_{23}^3 - (2r_{23} - r_{12}r_{13})(1 - r_{12}^2 - r_{13}^2 - r_{23}^2)}} \tag{6.14}$$

For large n, this test statistic has approximately the standard normal distribution under $H_0: \rho_{12} = \rho_{13}$ (Olkin and Siotani 1964; Olkin 1967).

■ **Example 6.5**   Assume that the weight, height, and age have been measured for each member of a sample of 12 nutritionally deficient children. Such a small sample brings into question the normal approximation involved in the use of (6.14). The data to be analyzed appear in Table 8.1 in Chapter 8. For these data, the three sample correlations are:

$$r_{12} = r_{\text{(weight, height)}} = 0.814$$

$$r_{13} = r_{\text{(weight, age)}} = 0.770$$

$$r_{23} = r_{\text{(height, age)}} = 0.614$$

We want to test whether height and age are equally correlated with weight (i.e., $H_0: \rho_{12} = \rho_{13}$) versus the two-tailed alternative that they are not (i.e., $H_A: \rho_{12} \neq \rho_{13}$). Using (6.14), the test statistic takes the value

$$Z = \frac{(.814 - .770)\sqrt{12}}{\sqrt{\begin{array}{c} [1 - (.814)^2]^2 + [1 - (.770)^2]^2 - 2(.614)^3 \\ -[2(.614) - (.814)(.770)][1 - (.814)^2 - (.770)^2 - (.614)^2] \end{array}}}$$

$$= \frac{0.1524}{\sqrt{0.1968}} = 0.3435$$

It is clear that, for these data, we cannot reject the null hypothesis of equal correlation of weight with height and age. ∎

Problems

1. Using the data set of Problem 1 in Chapter 5, perform the following operations.
   a. Determine the sample correlation coefficients of (1) age with dry weight, and (2) age with $\log_{10}$ dry weight. Interpret your results.
   b. Using Fisher's Z transformation, obtain a 95% confidence interval for ρ based on each of the correlations obtained in part (a).
   c. For each straight-line regression, determine r^2 directly by squaring the r obtained in part (a); also determine r^2 from the computer output or from the formula $r^2 = (SSY - SSE)/SSY$. Interpret your results.
   d. Based on the preceding results, which of the two regression lines provides the better fit? Explain. Does this agree with your earlier conclusion in Problem 1(d) of Chapter 5?

2. Examine the five pairs of data points given in the following table.

i	1	2	3	4	5
X_i	-2	-1	0	1	2
Y_i	4	1	0	1	4

   a. What is the mathematical relationship between X and Y?
   b. Show by computation that, for the straight-line regression of Y on X, $\hat{\beta}_1 = 0$.
   c. Show by computation that $r = 0$.
   d. Why is there apparently no relationship between X and Y, as indicated by the estimates of β_1 and ρ?

3. Consider the data in the following table.

i	1	2	3	4	5	6	7	8	9	10
X_i	1	1	1	2	2	2	3	3	3	20
Y_i	1	2	3	1	2	3	1	2	3	20

    a. Find the sample correlation coefficient r. Interpret your result.

    b. Show that the test statistic $T = \hat{\beta}_1/(S_{Y|X}/S_X\sqrt{n-1})$ for testing $H_0: \beta_1 = 0$ (based on a straight-line regression relationship between Y and X) is exactly equivalent to the test statistic $T' = r\sqrt{n-2}/\sqrt{1-r^2}$ for testing $H_0: \rho = 0$. [*Hint:* Use $\hat{\beta}_1 = rS_Y/S_X$ and $S_{Y|X}^2 = [(n-1)/(n-2)](S_Y^2 - \hat{\beta}_1^2 S_X^2).$]

    c. Using T', test $H_0: \rho = 0$ versus $H_A: \rho \neq 0$.

    d. Despite the conclusion you obtained in part (c), why should you be reluctant to conclude that the two variables are linearly related? (*Hint:* "A graph is worth a thousand words.")

4.–6. Answer the following questions concerning the straight-line regressions of Y on X referred to in parts (c), (d), and (e) of Problem 2 in Chapter 5.

    a. Determine r and r^2, and interpret your results.

    b. Find a 99% confidence interval for ρ, and interpret your result with regard to the test of $H_0: \rho = 0$ versus $H_A: \rho \neq 0$ at $\alpha = .01$.

7.–12. Answer the following questions for each of the data sets of Problems 3–8 in Chapter 5.

    a. Determine r and r^2 for each variable pair, and interpret your results.

    b. Test $H_0: \rho = 0$ versus $H_A: \rho \neq 0$, and interpret your findings.

    c. Find a 95% confidence interval for ρ. Interpret your result with regard to the test in part (b).

13. Suppose that, in a study on geographic variation in a certain species of beetle,[6] the mean tibia length (U) and the mean tarsus length (V) were obtained for samples of size 50 from each of 10 different regions spanning five southern states. Suppose further that the results were as given in the following table.

Region	1	2	3	4	5	6	7	8	9	10
U	7.500	7.164	7.512	8.544	7.380	7.860	7.836	8.100	7.584	7.344
V	1.680	1.596	1.680	1.908	1.632	1.752	1.776	1.860	1.692	1.680

    a. Determine the sample correlation coefficient between mean tarsus length and mean tibia length.

    b. Find a 99% confidence interval for ρ. Interpret your results with respect to the hypotheses: $H_0: \rho = 0$ versus $H_A: \rho \neq 0$.

14. In a sample of 23 young adult men, the correlation between total hemoglobin (THb) measured from venipuncture and measured from a finger needle puncture was 0.82. For a sample of 32 women of similar age, the correlation was 0.74. The two samples from each person were collected within 1 hour of each other. Assume that the straight-line regression assumptions hold.

---

[6] Adapted from a study by Sokal and Thomas (1965).

    a. Test the hypothesis that the two population correlations are equal. Use a two-tailed test. What do you conclude?

    b. If the experimenter had planned to do so before collecting the data, a valid one-tailed test could have been conducted. With this assumption, repeat part (a) but use a one-tailed test to assess whether the correlation for women is lower than that for men. What do you conclude?

    c. Assume that the researcher had planned to conduct a one-tailed test of the hypothesis that the correlation for women is higher than that for men. What test should be conducted? What do you conclude?

15. A university admissions officer regularly administers a test to all entering freshmen. A new version of the test is marketed by the testing company. To evaluate the new form, the admissions officer has 121 freshmen take both the old and the new versions. After the end of the school year, the admissions officer correlates the two scores with each other and with the students' freshman grade point averages (GPA). With 1 indicating the old test version, 2 the new test version, and G the GPA,

$$r_{12} = .6969, \quad r_{1G} = .5514, \quad r_{2G} = .4188$$

Test the hypothesis that the two forms of the test are equally correlated with GPA. Use a two-tailed test with $\alpha = .05$. Assume that the straight-line regression assumptions hold.

16.–26. Answer the following questions for each of the data sets of Problems 12(a), 12(c), 13, 14, 15(a), 15(c), 16, 17, 18, 19, and 20 in Chapter 5.

    a. Determine r and r^2 for each variable pair, and interpret your results.

    b. Test $H_0: \rho = 0$ versus $H_A: \rho \neq 0$, and interpret your findings.

    c. Find a 95% confidence interval for ρ. Interpret your result with respect to the test in part (b).

References

Olkin, I. 1967. "Correlation Revisited." In Julian C. Stanley, ed., *Improving Experimental Design and Statistical Analysis.* Chicago: Rand McNally.

Olkin, I., and Siotani, M. 1964. "Asymptotic Distribution Functions of a Correlation Matrix." Stanford University Laboratory for Quantitative Research in Education, Report No. 6, Stanford, Calif.

Sokal, R. R., and Thomas, P. A. 1965. "Geographic Variation of *Pemphigus populitransversus* in Eastern North America: Stem Mothers and New Data on Alates." *University of Kansas Scientific Bulletin* 46: 201–52.

The Analysis-of-Variance Table

7.1 Preview

An overall summary of the results of any regression analysis, whether straight-line or not, can be provided by a table called an *analysis-of-variance (ANOVA) table*. This name derives primarily from the fact that the basic information in an ANOVA table consists of several estimates of variance. These estimates, in turn, can be used to answer the principal inferential questions of regression analysis. In the straight-line case, there are three such questions: (1) Is the true slope β_1 zero? (2) What is the strength of the straight-line relationship? (3) Is the straight-line model appropriate?

Historically, the name "analysis-of-variance table" was coined to describe the overall summary table for the statistical procedure known as *analysis of variance*. As we observed in Chapter 2 and will see later when discussing the ANOVA method, regression analysis and analysis of variance are closely related. More precisely, analysis-of-variance problems can be expressed in a regression framework. Thus, such a table can be used to summarize the results obtained from either method.

7.2 The ANOVA Table for Straight-line Regression

Various textbooks, researchers, and computer program printouts have slightly different ways of presenting the ANOVA table associated with straight-line regression analysis. This section describes the most common form.

The simplest version of the ANOVA table for straight-line regression is given in the accompanying SAS computer output, as applied to the age–systolic blood pressure data of Table 5.1. The mean-square term is obtained by dividing the sum of squares by its degrees of freedom. The

F statistic (i.e., *F* value in the output) is obtained by dividing the regression (i.e., model) mean square by the residual (i.e., error) mean square.

SAS Output for an ANOVA Table Based on Table 5.1 Data

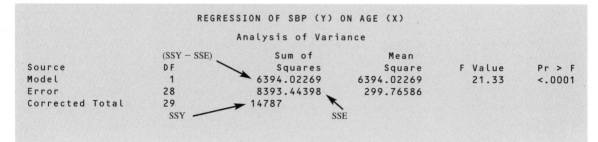

In Chapter 6, when describing the correlation coefficient, we observed in (6.6) that

$$r^2 = \frac{\text{SSY} - \text{SSE}}{\text{SSY}}$$

where $\text{SSY} = \sum_{i=1}^{n}(Y_i - \overline{Y})^2$ is the sum of the squares of deviations of the observed Y's from the mean $\overline{Y}$, and $\text{SSE} = \sum_{i=1}^{n}(Y_i - \hat{Y}_i)^2$ is the sum of squares of deviations of observed Y's from the fitted regression line. Since SSY represents the total variation of Y before accounting for the linear effect of the variable X, we usually call SSY the *total unexplained variation* or the *total sum of squares about* (or *corrected for*) *the mean*. Because SSE measures the amount of variation in the observed Y's that remains after accounting for the linear effect of the variable X, we usually call $(\text{SSY} - \text{SSE})$ the *sum of squares due to* (or *explained by*) *regression*. It turns out that $(\text{SSY} - \text{SSE})$ is mathematically equivalent to the expression

$$\sum_{i=1}^{n}(\hat{Y}_i - \overline{Y})^2$$

which represents the sum of squares of deviations of the predicted values from the mean $\overline{Y}$. We thus have the following mathematical result:

> Total unexplained variation = Variation due to regression
> + Unexplained residual variation

or

$$\sum_{i=1}^{n}(Y_i - \overline{Y})^2 = \sum_{i=1}^{n}(\hat{Y}_i - \overline{Y})^2 + \sum_{i=1}^{n}(Y_i - \hat{Y}_i)^2 \tag{7.1}$$

FIGURE 7.1   Variation explained and unexplained by straight-line regression

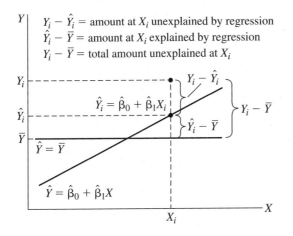

Equation (7.1), which is often called the *fundamental equation of regression analysis,* holds for any general regression situation. Figure 7.1 illustrates this equation.

The mean-square residual term is simply the estimate $S^2_{Y|X}$ presented earlier. If the true regression model is a straight line, then, as mentioned in Section 5.6, $S^2_{Y|X}$ is an estimate of σ^2. On the other hand, the mean-square regression term (SSY − SSE) provides an estimate of σ^2 only if the variable X does not help to predict the dependent variable Y—that is, only if the hypothesis $H_0: \beta_1 = 0$ is true. If in fact $\beta_1 \neq 0$, the mean-square regression term will be inflated in proportion to the magnitude of β_1 and will correspondingly overestimate σ^2.

It can be shown that the mean-square residual and mean-square regression terms are *statistically* independent of one another. Thus, if $H_0: \beta_1 = 0$ is true, the ratio of these terms represents the ratio of two independent estimates of the same variance σ^2. Under the normality and independence assumptions about the Y's, such a ratio has the F distribution, and this F statistic (with the value 21.330 in the accompanying SAS computer output) can be used to test the hypothesis H_0: "No significant straight-line relationship of Y on X" (i.e., $H_0: \beta_1 = 0$ or $H_0: \rho = 0$).

Fortunately, this way of testing H_0 is *equivalent* to using the two-sided t test previously discussed. This is so because, for v degrees of freedom,

$$F_{1,\,v} = T^2_v \tag{7.2}$$

so

$$F_{1,\,v,\,1-\alpha} = t^2_{v,\,1-\alpha/2} \tag{7.3}$$

The expression in (7.3) states that the $100(1 - \alpha)\%$ point of the F distribution with 1 and v degrees of freedom is exactly the same as the square of the $100(1 - \alpha/2)\%$ point of the t distribution with v degrees of freedom.

To illustrate the equivalence of the F and t tests, we can see from our age–systolic blood pressure example that $F = 21.33$ and $T^2 = 4.62^2 = 21.33$, where 4.62 is the figure obtained for T at the end of Section 6.6.1. Also, it can be seen that $F_{1,\,28,\,0.95} = 4.20$ and that $t^2_{28,\,0.975} = (2.05)^2 = 4.20$.

As these equalities establish, the critical region

$$|T| > t_{28, 0.975} = 2.05$$

for testing $H_0: \beta_1 = 0$ against the two-sided alternative $H_A: \beta_1 \neq 0$ is exactly the same as the critical region

$$F > F_{1, 28, 0.95} = 4.20$$

Hence, if $|T|$ exceeds 2.05, then F will exceed 4.20. Similarly, if F exceeds 4.20, then $|T|$ will exceed 2.05. Thus, the null hypothesis $H_0: \beta_1 = 0$ (or equivalently, H_0: "No significant straight-line relationship of Y on X") is rejected at the $\alpha = .05$ level of significance.

An alternative but less common representation of the ANOVA table is given in Table 7.1. This table differs from the SAS output table only in that it splits up the total sum of squares corrected for the mean, SSY, into its two components: the *total uncorrected sum of squares,* $\sum_{i=1}^{n} Y_i^2$; and the *correction factor,* $(\sum_{i=1}^{n} Y_i)^2 / n$. The relationship between these components is given by the equation

$$\sum_{i=1}^{n} (Y_i - \overline{Y})^2 = \sum_{i=1}^{n} Y_i^2 - \frac{\left(\sum_{i=1}^{n} Y_i \right)^2}{n}$$

In the total (uncorrected) sum of squares $\sum_{i=1}^{n} Y_i^2$, the n observations on Y are considered before any estimation of the population mean of Y. The "Regression $\overline{Y}$" listed in Table 7.1 refers to the variability explained by using a model involving only β_0 (which is estimated by $\overline{Y}$). This is necessarily the same amount of variability as is explained by using only $\overline{Y}$ to predict Y, without attempting to account for the linear contribution of X to the prediction of Y. The "Regression $X|\overline{Y}$" describes the contribution of the variable X to predicting Y *over and above* that contributed by $\overline{Y}$ alone. Usually "Regression $X|\overline{Y}$" is written simply as "Regression X," the "given $\overline{Y}$ " part being suppressed for notational simplicity. We will see more of this notation when we discuss multiple regression in subsequent chapters.

TABLE 7.1   Alternative ANOVA table for age–systolic blood pressure data of Table 5.1

Source	Degrees of Freedom (df)	Sum of Squares (SS)	Mean Square (MS)	Variance Ratio (F)
Regression $\begin{cases} \overline{Y} \\ X|\overline{Y} \end{cases}$	1	$\dfrac{\left(\sum_{i=1}^{n} Y_i \right)^2}{n} = 609{,}472.53$		
	1	6,394.02	6,394.02	21.33 $(P < .001)$
Residual	28	8,393.44	299.77	
Total	30	$\sum_{i=1}^{n} Y_i^2 = 624{,}260.00$		

Problems

1. Use the data set of Problem 1 in Chapter 5 to answer the following questions.
   a. Determine the ANOVA tables for the following regressions: (1) dry weight (Y) on age (X), and (2) $\log_{10}$ dry weight (Z) on age (X). The following results will be helpful in reducing computation time:

   $$S_{Y|X}^2 = 0.23218 \quad SSY = 8.16811 \quad S_{Z|X}^2 = 0.0007838 \quad SSZ = 4.2276839$$

   b. Use the tables in part (a) to perform the F test for the significance of each straight-line regression. Interpret your results.

2.–4. Answer the same questions as in parts (a) and (b) of Problem 1 for each regression of Y on X, using the data in parts (b), (c), and (d) of Problem 2 in Chapter 5. The following results will be helpful in reducing computation time:

SBP (Y) regressed on QUET (X):	$S_{Y|X}^2 = 96.26743$	$SSY = 6{,}425.96875$
QUET (Y) regressed on AGE (X):	$S_{Y|X}^2 = 0.09079$	$SSY = 7.65968$
SBP (Y) regressed on AGE (X):	$S_{Y|X}^2 = 85.47795$	

5. Use the data of Problem 3 in Chapter 5 to answer the following questions.
   a.–b. Answer the same questions as in parts (a) and (b) of Problem 1 for the regression of TIME (Y) on INC (X). The following results will be helpful:

   $$S_{Y|X}^2 = 110.16190 \quad SSY = 2433.78137$$

   c. Compare the value of the test statistic F obtained in part (b) with the value of T^2, the square of the test statistic for testing H_0: $\beta_1 = 0$ that was required in part (g) of Problem 3 in Chapter 5.
   d. The p-values for the F test in part (b) and for the t test in part (g) of Problem 3 in Chapter 5 are the same. Intuitively, why does this make sense? (*Hint:* Compare the hypotheses for each of the tests.)

6.–10. Answer the same questions as in parts (a) and (b) of Problem 1 for each of the regressions of Y on X in Problems 5 through 8 and Problem 10 of Chapter 5. The following results will be helpful:

	$S_{Y|X}^2$	SSY
Chapter 5, Problem 5:	11.101	2,223.018
Chapter 5, Problem 6:	172.985	20,123.382
Chapter 5, Problem 7:	1,001.691	190,502.800
	0.658	410.531
Chapter 5, Problem 8:	1,264,983.805	112,278,032.670
Chapter 5, Problem 10:	0.003	2.142

11. A biologist wanted to study the effects of the temperature of a certain medium on the growth of human amniotic cells in a tissue culture. Using the same parent batch, she conducted an experiment in which five cell lines were cultured at each of four temperatures. The procedure involved initially inoculating a fixed number (0.25 million) of cells into a fresh culture flask and then, after 7 days, removing a small sample from the growing surface to use in estimating the total number of cells in the flask. The results are given in the following table, together with a computer printout for straight-line regression.

Number of Cells ($\times 10^{-6}$) after 7 days (Y)	Temperature (X)	Number of Cells ($\times 10^{-6}$) after 7 days (Y)	Temperature (X)
1.13	40	2.30	80
1.20	40	2.15	80
1.00	40	2.25	80
0.91	40	2.40	80
1.05	40	2.49	80
1.75	60	3.18	100
1.45	60	3.10	100
1.55	60	3.28	100
1.64	60	3.35	100
1.60	60	3.12	100

a. Complete the ANOVA table shown in the accompanying computer output.
b. Perform the F test for the significance of the straight-line regression of Y on X, and interpret your results.

Edited SAS Output (PROC REG) for Problem 11

```
                    # Of Cells (Y) Regressed on Temperature (X)

                                Analysis of Variance

                            Sum of          Mean
Source              DF      Squares        Square       F Value        Pr > F
Model                1      _____       _____       _____       <.0001
Error               18      0.36618       _____
Corrected Total     19     13.19690


              Root MSE            0.14263       R-Square       0.9723
              Dependent Mean      2.04500       Adj R-Sq       0.9707
              Coeff Var           6.97454


                            Parameter Estimates

                          Parameter      Standard
Variable            DF     Estimate        Error      t Value       Pr > |t|
Intercept            1     -0.46240       0.10481      -4.41         0.0003
X                    1      0.03582       0.00143      25.11         <.0001
```

12.–21. Answer the same questions as in parts (a) and (b) of Problem 1 for each of the data sets in Problems 12(a), 12(c), 13, 14, 15(a), 15(c), 16, 17, 18, and 19 in Chapter 5. The following results will be useful:

	$S^2_{Y\mid X}$	SSY
Chapter 5, Problem 12(a):	21,924.992	12,699,346.400
Chapter 5, Problem 12(c):	42,398.926	13,439,939.000
Chapter 5, Problem 13:	0.530	6.704
Chapter 5, Problem 14:	384.701	3,770.900
Chapter 5, Problem 15(a):	8.140	9,386.654
Chapter 5, Problem 15(c):	0.058	180.911
Chapter 5, Problem 16:	114.698	6,305.714
Chapter 5, Problem 17:	0.689	28.833
Chapter 5, Problem 18:	13.201	610.429
Chapter 5, Problem 19:	19.514	600.962

8

Multiple Regression Analysis: General Considerations

8.1 Preview

Multiple regression analysis can be looked upon as an extension of straight-line regression analysis (which involves only one independent variable) to the situation in which more than one independent variable must be considered. Several general applications of multiple regression analysis[1] were described in Chapter 4, and specific examples were given in Chapter 1. In this chapter we will describe the multiple regression method in detail, stating the required assumptions, describing the procedures for estimating important parameters, explaining how to make and interpret inferences about these parameters, and providing examples that illustrate how to use the techniques of multiple regression analysis. Dealing with several independent variables simultaneously in a regression analysis is considerably more difficult than dealing with a single independent variable for the following reasons:

1. It is more difficult to choose the best model, since several reasonable candidates may exist.

2. It is more difficult to visualize what the fitted model looks like (especially if there are more than two independent variables), since it is not possible to plot either the data or the fitted model directly in more than three dimensions.

3. It is sometimes more difficult to interpret what the best-fitting model means in real-life terms.

4. Computations are virtually impossible without access to a high-speed computer and a reliable packaged computer program.

---

[1] We shall generally refer to multiple regression analysis simply as "regression analysis" throughout the remainder of the text.

8.2 Multiple Regression Models

One example of a multiple regression model is given by any second- or higher-order polynomial. Adding higher-order terms (e.g., an X^2 or X^3 term) to a model can be considered as equivalent to adding new independent variables. Thus, if we rename X as X_1 and X^2 as X_2, the second-order model

$$Y = \beta_0 + \beta_1 X + \beta_2 X^2 + E$$

can be rewritten as

$$Y = \beta_0 + \beta_1 X_1 + \beta_2 X_2 + E$$

Of course, in polynomial regression we have only one basic independent variable, the others being simple mathematical functions of this basic variable. In more general multiple regression problems, however, the number of basic independent variables may be greater than one. The general form of a regression model for k independent variables is given by

$$Y = \beta_0 + \beta_1 X_1 + \beta_2 X_2 + \cdots + \beta_k X_k + E$$

where $\beta_0, \beta_1, \beta_2, \ldots, \beta_k$ are the *regression coefficients* that need to be estimated. The *independent variables* $X_1, X_2, \ldots, X_k$ may all be separate basic variables, or some may be functions of a few basic variables.

■ **Example 8.1** Suppose that we want to investigate how weight (WGT) varies with height (HGT) and age (AGE) for children with a particular kind of nutritional deficiency.[2] The dependent variable here is Y = WGT, and our two basic independent variables are X_1 = HGT and X_2 = AGE.

Suppose that, as outlined in Example 6.5, a random sample consists of 12 children who attend a certain clinic. The WGT, HGT, and AGE data obtained for each child are given in Table 8.1.

In describing the relationship of WGT to HGT and AGE, we may want to consider the model

$$Y = \beta_0 + \beta_1 X_1 + \beta_2 X_2 + E$$

TABLE 8.1 WGT, HGT, and AGE of a random sample of 12 nutritionally deficient children

Child	1	2	3	4	5	6	7	8	9	10	11	12
WGT (Y)	64	71	53	67	55	58	77	57	56	51	76	68
HGT (X_1)	57	59	49	62	51	50	55	48	42	42	61	57
AGE (X_2)	8	10	6	11	8	7	10	9	10	6	12	9

---

[2] Perhaps the main question associated with this type of study is whether the relationship for nutritionally deficient children is the same as that for "normal" children. To answer this question would require additional data on normal children and some kind of comparison of the models obtained for each group. Although we will learn how to deal with this kind of question in Chapter 12, we focus here on the methods needed to describe the relationship of weight to height and age for this single group of nutritionally deficient children.

if we are interested only in first-order terms. If we want to consider, in addition, the higher-order term X_1^2, our model is given by

$$Y = \beta_0 + \beta_1 X_1 + \beta_2 X_2 + \beta_3 X_3 + E$$

where $X_3 = X_1^2$. To consider all possible first- and second-order terms, we look at the model

$$Y = \beta_0 + \beta_1 X_1 + \beta_2 X_2 + \beta_3 X_3 + \beta_4 X_4 + \beta_5 X_5 + E$$

where $X_3 = X_1^2$, $X_4 = X_2^2$, and $X_5 = X_1 X_2$, or equivalently

$$Y = \beta_0 + \beta_1 X_1 + \beta_2 X_2 + \beta_3 X_1^2 + \beta_4 X_2^2 + \beta_5 X_1 X_2 + E$$

If we want to find the best predictive model, we might consider all of the preceding models (as well as some others) and then choose the best model according to some reasonable criterion.

We discuss the question of model selection in Chapter 16; the interpretation of product terms such as $X_1 X_2$ as interaction effects is explained in Chapter 11. For now our focus is on the methods used and the interpretations that can be made when the choice of independent variables to use in the model is not at issue. ∎

8.3 Graphical Look at the Problem

When we are dealing with only one independent variable, our problem can easily be described graphically as that of finding the curve that best fits the scatter of points $(X_1, Y_1), (X_2, Y_2), \ldots, (X_n, Y_n)$ obtained on n individuals. Thus, we have a *two-dimensional* representation involving a plot of the form shown in Figure 8.1. Furthermore, the *regression equation* for this problem is defined as the path described by the mean values of the distribution of Y when X is allowed to vary.

When the number k of (basic) independent variables is two or more, the (graphical) dimension of the problem increases. The regression equation ceases to be a curve in two-dimensional space and becomes instead a *hypersurface in $(k + 1)$-dimensional space*. Obviously, we will not be able to represent in a single plot either the scatter of data points or the regression equation if more than two basic independent variables are involved. In the special case $k = 2$, as in the example just given where $X_1 = \text{HGT}$, $X_2 = \text{AGE}$, and $Y = \text{WGT}$, the problem is to find the *surface* in three-dimensional space that best fits the scatter of points (X_{11}, X_{12}, Y_1),

FIGURE 8.1   Scatter plot for a single independent variable

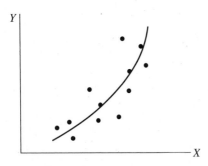

(X_{21}, X_{22}, Y_2), . . . , (X_{n1}, X_{n2}, Y_n), where (X_{i1}, X_{i2}, Y_i) denotes the X_1, X_2, and Y-values for the ith individual in the sample. The *regression equation* in this case, therefore, is the surface described by the mean values of Y at various combinations of values of X_1 and X_2; that is, corresponding to *each* distinct pair of values of X_1 and X_2 is a distribution of Y values with mean $\mu_{Y|X_1, X_2}$ and variance $\sigma^2_{Y|X_1, X_2}$.

Just as the simplest curve in two-dimensional space is a straight line, the simplest surface in three-dimensional space is a *plane,* which has the statistical model form $Y = \beta_0 + \beta_1 X_1 + \beta_2 X_2 + E$. Thus, finding the best-fitting plane is frequently the first step in determining the best-fitting surface in three-dimensional space when two independent variables are relevant, just as fitting the best straight line is the first step when one independent variable is involved. A graphical representation of a planar fit to data in the three-dimensional situation is given in Figure 8.2.

For the three-dimensional case, the least-squares solution that gives the best-fitting plane is determined by minimizing the sum of squares of the distances between the observed values Y_i and the corresponding predicted values $\hat{Y}_i = \hat{\beta}_0 + \hat{\beta}_1 X_{i1} + \hat{\beta}_2 X_{i2}$, based on the fitted plane. In other words, the quantity

$$\sum_{i=1}^{n} (Y_i - \hat{Y}_i)^2 = \sum_{i=1}^{n} (Y_i - \hat{\beta}_0 - \hat{\beta}_1 X_{i1} - \hat{\beta}_2 X_{i2})^2$$

is minimized to find the least-squares estimates $\hat{\beta}_0$ of β_0, $\hat{\beta}_1$ of β_1, and $\hat{\beta}_2$ of β_2.

How much can one learn by considering the independent variables in the multivariable problem separately? Probably the best answer is that we can learn something about what is going on, but there are too many separate (univariable) pieces of information to permit us to complete the (multivariable) puzzle. For example, consider the data previously given for Y = WGT, X_1 = HGT, and X_2 = AGE. If we plot separate scatter diagrams of WGT on HGT, WGT on AGE, and AGE on HGT, we get the results shown in Figure 8.3.

FIGURE 8.2   Best-fitting plane for three-dimensional data

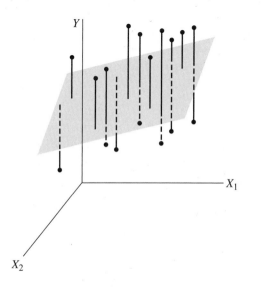

FIGURE 8.3   Separate scatter diagrams of WGT versus HGT, WGT versus AGE, and AGE versus HGT

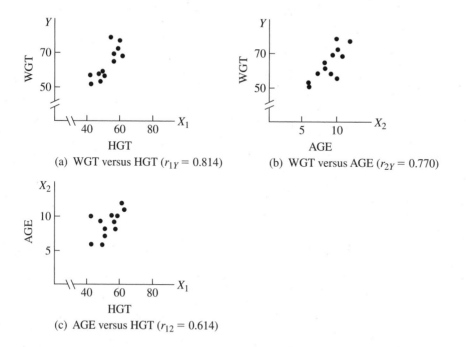

(a) WGT versus HGT ($r_{1Y} = 0.814$)   (b) WGT versus AGE ($r_{2Y} = 0.770$)

(c) AGE versus HGT ($r_{12} = 0.614$)

HGT is highly positively correlated with WGT ($r_{1Y} = 0.814$), as is AGE ($r_{2Y} = 0.770$). Thus, if we used each of these independent variables separately, we would likely find two separate, significant straight-line regressions. Does this mean that the best-fitting plane with both variables in the model together will also have significant predictive ability? The answer is probably yes. But what will the plane look like? This is difficult to say. We can get some idea of the difficulty if we consider the plot of HGT versus AGE in part (c), which reflects a positive correlation ($r_{12} = 0.614$). If, instead, these two variables were negatively correlated, we would expect a different orientation of the plane, although we could not clearly quantify either orientation. Thus, treating each independent variable separately does not help very much because the relationships between the independent variables themselves are not taken directly into account. The techniques of multiple regression, however, account for all these intercorrelations with regard to both estimation and inference making.

8.4 Assumptions of Multiple Regression

In the previous section we described the multiple regression problem in some generality and also hinted at some of the assumptions involved. We now state these assumptions somewhat more formally.

8.4.1 Statement of Assumptions

Assumption 1: Existence   *For each specific combination of values of the (basic) independent variables $X_1, X_2, \ldots, X_k$ (e.g., $X_1 = 57, X_2 = 8$ for the first child in Example 8.1),*

Y is a (univariate) random variable with a certain probability distribution having finite mean and variance.

Assumption 2: Independence   *The Y observations are statistically independent of one another.* As with straight-line regression, this assumption is usually violated when several *Y* observations are made on the same subject. Methods for dealing with regression modeling of correlated data include *generalized estimating equations (GEE)* techniques (Zeger and Liang 1986; Diggle, Heagerty, Liang, and Zeger 2002), and *mixed model* techniques (described in Chapters 25 and 26) using SAS's MIXED procedure, (SAS release 9.1, SAS Institute 2002–2003).

Assumption 3: Linearity   *The mean value of Y for each specific combination of X_1, $X_2, \ldots, X_k$ is a linear function*[3] *of $\beta_0, \beta_1, \ldots, \beta_k$.* That is,

$$\mu_{Y|X_1, X_2, \ldots, X_k} = \beta_0 + \beta_1 X_1 + \beta_2 X_2 + \cdots + \beta_k X_k \qquad (8.1)$$

or

$$Y = \beta_0 + \beta_1 X_1 + \beta_2 X_2 + \cdots + \beta_k X_k + E \qquad (8.2)$$

where E is the error component reflecting the difference between an individual's observed response Y and the true average response $\mu_{Y|X_1, X_2, \ldots, X_k}$. Some comments are in order regarding Assumption 3:

1. The surface described by (8.1) is called the *regression equation* (or *response surface* or *regression surface*).

2. If some of the independent variables are higher-order functions of a few basic independent variables (e.g., $X_3 = X_1^2$, $X_5 = X_1 X_2$), the expression $\beta_0 + \beta_1 X_1 + \beta_2 X_2 + \cdots + \beta_k X_k$ is nonlinear in the basic variables (hence the use of the word *surface* rather than *plane*).

3. Consonant with its meaning in straight-line regression, E is the amount by which any individual's observed response deviates from the response surface. Thus, E is the *error component* in the model.

Assumption 4: Homoscedasticity   *The variance of Y is the same for any fixed combination of $X_1, X_2, \ldots, X_k$.* That is,

$$\sigma^2_{Y|X_1, X_2, \ldots, X_k} = \text{Var}(Y|X_1, X_2, \ldots, X_k) \equiv \sigma^2 \qquad (8.3)$$

---

[3] The techniques of multiple regression that we will be describing are applicable as long as the model under consideration is *inherently linear* in the regression coefficients (regardless of how the independent variables are defined). For example, a model of the form $\mu_{Y|X} = \beta_0 e^{(\beta_1 X + \beta_2 X^2)}$ is inherently linear because it can be transformed into the equivalent form $\mu^*_{Y|X} = \beta_0^* + \beta_1 X + \beta_2 X^2$ where $\mu^*_{Y|X} = \ln \mu_{Y|X}$ and $\beta_0^* = \ln \beta_0$. However, the model $\mu_{Y|X_1, X_2} = e^{\beta_1 X_1} + e^{\beta_2 X_2}$ cannot be transformed directly into a form that is linear in β_1 and β_2; so estimating β_1 and β_2 requires the use of *nonlinear regression* procedures (see, e.g., Gallant 1975). A discussion of these procedures is beyond the scope of this text.

As before, this is called the assumption of homoscedasticity. An alternative (but equivalent) definition of homoscedasticity, based on (8.2), is that

$$\sigma^2_{E \mid X_1, X_2, \ldots, X_k} \equiv \sigma^2$$

This assumption may seem very restrictive. But variance heteroscedasticity needs to be considered only when the data show very obvious and significant departures from homogeneity. In general, mild departures do not have significant adverse effects on the results.

Assumption 5: Normality   *For any fixed combination of* $X_1, X_2, \ldots, X_k$, *the variable* Y *is normally distributed.* In other words,

$$Y \frown N(\mu_{Y \mid X_1, X_2, \ldots, X_k}, \sigma^2)$$

or equivalently,

$$E \frown N(0, \sigma^2) \tag{8.4}$$

This assumption is not necessary for the least-squares fitting of the regression model, but it is required in general for inference making. The usual parametric tests of hypotheses and confidence intervals used in a regression analysis are robust in the sense that only extreme departures of the distribution of Y from normality yield spurious results. (This statement is based on both theoretical and experimental evidence.) If the normality assumption does not hold, one typically seeks a transformation of Y—say, log Y or $\sqrt{Y}$—to produce a transformed set of Y observations that are approximately normal (see Section 14.4.1). If the Y variable is either categorical or ordinal, however, alternative regression methods such as logistic regression (for binary Y's) or Poisson regression (for discrete Y's) are typically required (see Chapters 22 and 24).

8.4.2 Summary and Comments

Our assumptions for simple linear (i.e., straight-line) regression analysis can be generalized to multiple linear regression analysis. Here, homoscedasticity and normality apply to $Y \mid X_1, X_2, \ldots, X_k$, rather than to Y (i.e., to the conditional distribution of Y given $X_1, X_2, \ldots, X_k$, rather than to the so-called unconditional or marginal distribution of Y).

The assumptions for multiple linear regression analysis dictate that the random error component E have a normal distribution with mean 0 and variance σ^2. Of course, the linearity, existence, and independence assumptions must also hold.

Again, Y is an observable random variable, while $X_1, X_2, \ldots, X_k$ are assumed to be measured without error. The constants $\beta_0, \beta_1, \ldots, \beta_k$ are unknown population parameters, and E is an unobservable random variable. If one estimates $\beta_0, \beta_1, \ldots, \beta_k$ with $\hat{\beta}_0, \hat{\beta}_1, \ldots, \hat{\beta}_k$, then an acceptable estimate of E_i for the i-th subject is

$$\hat{E}_i = Y_i - \hat{Y}_i = Y_i - (\hat{\beta}_0 + \hat{\beta}_1 X_{i1} + \cdots + \hat{\beta}_k X_{ik})$$

The estimated error $\hat{E}_i$ is usually called a *residual*.

The assumption of a Gaussian distribution is needed to justify the use of procedures of statistical inference involving the t and F distributions.

8.5 Determining the Best Estimate of the Multiple Regression Equation

As with straight-line regression, there are two basic approaches to estimating a multiple regression equation: the least-squares approach and the minimum-variance approach. In the straight-line case, both approaches yield the same solution. (We are assuming, as previously noted, that we already know the best form of regression model to use; that is, we have already settled on a fixed set of k independent variables $X_1, X_2, \ldots, X_k$. The problem of determining the best model form via algorithms for choosing the most important independent variables will be discussed in detail in Chapter 16.) The multiple regression model may also be fitted by using other statistical methodology, such as maximum likelihood (see Chapter 21). Given fixed X's and assuming Y_i is Gaussian with the Y_i's mutually independent, then the least-squares estimates of the regression coefficients are identical to the maximum-likelihood estimates.

8.5.1 Least-squares Approach

In general, the least-squares method chooses as the best-fitting model the one that minimizes the sum of squares of the distances between the observed responses and those predicted by the fitted model. Again, the better the fit, the smaller the deviations of observed from predicted values. Thus, if we let

$$\hat{Y} = \hat{\beta}_0 + \hat{\beta}_1 X_1 + \hat{\beta}_2 X_2 + \cdots + \hat{\beta}_k X_k$$

denote the fitted regression model, the sum of squares of deviations of observed Y-values from corresponding values predicted by using the fitted regression model is given by

$$\sum_{i=1}^{n}(Y_i - \hat{Y}_i)^2 = \sum_{i=1}^{n}(Y_i - \hat{\beta}_0 - \hat{\beta}_1 X_{i1} - \cdots - \hat{\beta}_k X_{ik})^2 \tag{8.5}$$

The least-squares solution then consists of the values $\hat{\beta}_0, \hat{\beta}_1, \ldots, \hat{\beta}_k$ (called the "least-squares estimates") for which the sum in (8.5) is a minimum. This minimum sum of squares is generally called the *residual sum of squares* (or equivalently, the *error sum of squares* or the *sum of squares about regression*); as in the case of straight line regression, it is referred to as the SSE.

8.5.2 Minimum-variance Approach

As in the straight-line case, the minimum-variance approach to estimating the multiple regression equation identifies as the best-fitting surface the one utilizing the minimum-variance (linear) unbiased estimates $\hat{\beta}_0, \hat{\beta}_1, \ldots, \hat{\beta}_k$ of $\beta_0, \beta_1, \ldots, \beta_k$, respectively.

8.5.3 Comments on the Least-squares Solutions

In this text we do not present matrix formulas for calculating the least-squares estimates $\hat{\beta}_0, \hat{\beta}_1, \ldots, \hat{\beta}_k$, since computer programs are readily available to perform the necessary calculations. Even so, we provide in Appendix B a discussion of matrices and their use in regression analysis; by using matrix mathematics, one can represent the general regression model and the associated least-squares methodology in compact form. Also, an understanding of the matrix formulation for regression analysis carries over to more complex modeling problems, such as those involving multivariate data (i.e., data relating to two or more dependent variables).

The least-squares solutions have several important properties:

1. Each of the estimators $\hat{\beta}_0, \hat{\beta}_1, \ldots, \hat{\beta}_k$ is a linear function of the Y-values. This linearity property makes determining the statistical properties of these estimators fairly straightforward. In particular, since the Y-values are assumed to be normally distributed and to be statistically independent of one another, each of the estimators $\hat{\beta}_0, \hat{\beta}_1, \ldots, \hat{\beta}_k$ will be normally distributed, with easily computable standard errors.

2. The least-squares regression equation $\hat{Y} = \hat{\beta}_0 + \hat{\beta}_1 X_1 + \hat{\beta}_2 X_2 + \cdots + \hat{\beta}_k X_k$ is the unique linear combination of the independent variables $X_1, X_2, \ldots, X_k$ that has maximum possible correlation with the dependent variable. In other words, of all possible linear combinations of the form $b_0 + b_1 X_1 + b_2 X_2 + \cdots + b_k X_k$, the linear combination $\hat{Y}$ is such that the correlation

$$r_{Y,\hat{Y}} = \frac{\sum\limits_{i=1}^{n}(Y_i - \overline{Y})(\hat{Y}_i - \overline{\hat{Y}})}{\sqrt{\sum\limits_{i=1}^{n}(Y_i - \overline{Y})^2 \sum\limits_{i=1}^{n}(\hat{Y}_i - \overline{\hat{Y}})^2}} \qquad (8.6)$$

is a maximum, where $\hat{Y}_i$ is the predicted value of Y for the ith individual and $\overline{\hat{Y}}$ is the mean of the $\hat{Y}_i$'s. Incidentally, it is always true that $\overline{\hat{Y}} = \overline{Y}$; that is, the mean of the predicted values is equal to the mean of the observed values. The quantity $r_{Y,\hat{Y}}$ is called the *multiple correlation coefficient*.

3. Just as straight-line regression is related to the bivariate normal distribution, multiple regression can be related to the multivariate normal distribution. We will return to this point in Section 10.4 of Chapter 10.

■ **Example 8.2** For the data given in Table 8.1 on the variables $Y = $ WGT, $X_1 = $ HGT, and $X_2 = $ AGE, the least-squares algorithm applied to the model

$$\text{WGT} = \beta_0 + \beta_1\text{HGT} + \beta_2\text{AGE} + \beta_3(\text{AGE})^2 + E$$

produces the estimated equation

$$\text{WGT} = 3.438 + 0.724\text{HGT} + 2.777\text{AGE} - 0.042(\text{AGE})^2$$

so

$$\hat{\beta}_0 = 3.438, \quad \hat{\beta}_1 = 0.724, \quad \hat{\beta}_2 = 2.777, \quad \hat{\beta}_3 = -0.042 \qquad ■$$

8.6 The ANOVA Table for Multiple Regression

As with straight-line regression, an ANOVA table can be used to provide an overall summary of a multiple regression analysis. The particular form of an ANOVA table may vary, depending on how the contributions of the independent variables are to be considered (e.g., individually or collectively in some fashion). A simple form reflects the contribution that all independent variables considered collectively make to prediction. For example, consider Table 8.2, an ANOVA table based on the use of HGT, AGE, and $(AGE)^2$ as independent variables for the data of Table 8.1.

As before, the term $SSY = \sum_{i=1}^{n}(Y_i - \overline{Y})^2 = 888.25$ is called the *total sum of squares,* and this figure represents the total variability in the Y observations before accounting for the joint effect of using the independent variables HGT, AGE, and $(AGE)^2$. The term $SSE = \sum_{i=1}^{n}(Y_i - \hat{Y}_i)^2 = 195.19$. is the *residual sum of squares* (or the *sum of squares due to error*), which represents the amount of Y variation left unexplained after the independent variables have been used in the regression equation to predict Y. Finally, $SSY - SSE = \sum_{i=1}^{n}(\hat{Y}_i - \overline{Y})^2 = 693.06$ is called the *regression sum of squares* and measures the reduction in variation (or the variation explained) due to the independent variables in the regression equation. We thus have the familiar partition

Total sum of squares = Regression sum of squares + Residual sum of squares

or

$$\sum_{i=1}^{n}(Y_i - \overline{Y})^2 \quad = \quad \sum_{i=1}^{n}(\hat{Y}_i - \overline{Y})^2 \quad + \quad \sum_{i=1}^{n}(Y_i - \hat{Y}_i)^2$$

In Table 8.2, as in ANOVA tables for straight-line regression, the SS column identifies the various sums of squares. The d.f. column gives the corresponding degrees of freedom: the regression degrees of freedom is k (the number of independent variables in the model); the residual degrees of freedom is $n - k - 1$; and the total degrees of freedom is $n - 1$. The MS column contains the mean-square terms, obtained by dividing the sum-of-squares terms by their corresponding degrees-of-freedom values. The F ratio is obtained by dividing the mean-square regression by the mean-square residual; the interpretation of this F ratio will be discussed in Chapter 9 on hypothesis testing.

The R^2 in Table 8.2 (with the value 0.7802) provides a quantitative measure of how well the fitted model containing the variables HGT, AGE, and $(AGE)^2$ predicts the dependent variable WGT. The computational formula for R^2 is

$$R^2 = \frac{SSY - SSE}{SSY} \tag{8.7}$$

The quantity R^2 lies between 0 and 1. If the value is 1, we say that the fit of the model is perfect. R^2 always increases as more variables are added to the model, but a very small increase in R^2 may be neither practically nor statistically important. Additional properties of R^2 are discussed in Chapter 10.

TABLE 8.2 ANOVA table for WGT regressed on HGT, AGE, and $(AGE)^2$

Source	d.f.	SS	MS	F	R^2
Regression	$k = 3$	SSY − SSE = 693.06	231.02	9.47**	0.7802
Residual	$n - k - 1 = 8$	SSE = 195.19	24.40		
Total	$n - 1 = 11$	SSY = 888.25			

Note: The ** next to the computed F denotes significance at the 0.01 level; i.e., $P < 0.01$.

8.7 Numerical Examples

We conclude this chapter with some examples of the type of computer output to be expected from a typical regression program. This output generally consists of the values of the estimated regression coefficients, their estimated standard errors, the associated partial F (or T^2) statistics,[4] and an ANOVA table. For the data of Table 8.1, the six models that follow are by no means the only ones possible; for instance, no interaction terms were included.

Model 1   WGT $= \beta_0 + \beta_1$HGT $+ E$

Edited SAS Output for Regression of WGT on HGT

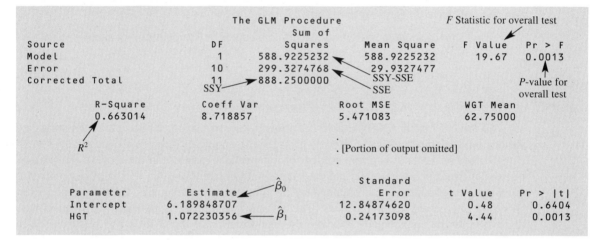

The GLM Procedure

Source	DF	Sum of Squares	Mean Square	F Value	Pr > F
Model	1	588.9225232	588.9225232	19.67	0.0013
Error	10	299.3274768	29.9327477		
Corrected Total	11	888.2500000			

F Statistic for overall test → F Value

$\hat{\beta}_0 \leftarrow$ SSY (888.2500000), SSY−SSE (588.9225232), SSE (29.9327477)

P-value for overall test

R-Square	Coeff Var	Root MSE	WGT Mean
0.663014	8.718857	5.471083	62.75000

R^2

. [Portion of output omitted]

Parameter	Estimate	Standard Error	t Value	Pr > |t|
Intercept	6.189848707	12.84874620	0.48	0.6404
HGT	1.072230356	0.24173098	4.44	0.0013

$\hat{\beta}_0$ ← Estimate; $\hat{\beta}_1$ ← 1.072230356

Model 2   WGT $= \beta_0 + \beta_2$AGE $+ E$

Edited SAS Output for Regression of WGT on AGE

The GLM Procedure

Source	DF	Sum of Squares	Mean Square	F Value	Pr > F
Model	1	526.3928571	526.3928571	14.55	0.0034
Error	10	361.8571429	36.1857143		
Corrected Total	11	888.2500000			

R-Square	Coeff Var	Root MSE	WGT Mean
0.592618	9.586385	6.015456	62.75000

Parameter	Estimate	Standard Error	t Value	Pr > |t|
Intercept	30.57142857	8.61370526	3.55	0.0053
AGE	3.64285714	0.95511512	3.81	0.0034

---

[4] Partial F statistics will be discussed in Chapter 9 on hypothesis testing.

Model 3   WGT $= \beta_0 + \beta_3(AGE)^2 + E$

Edited SAS Output for Regression of WGT on $(AGE)^2$

Source	DF	Sum of Squares	Mean Square	F Value	Pr > F
		The GLM Procedure			
Model	1	521.9320473	521.9320473	14.25	0.0036
Error	10	366.3179527	36.6317953		
Corrected Total	11	888.2500000			

R-Square	Coeff Var	Root MSE	WGT Mean
0.587596	9.645292	6.052421	62.75000

Parameter	Estimate	Standard Error	t Value	Pr > |t|
Intercept	45.99764279	4.76964028	9.64	<.0001
AGESQ	0.20597161	0.05456692	3.77	0.0036

Model 4   WGT $= \beta_0 + \beta_1 HGT + \beta_2 AGE + E$

Edited SAS Output for Regression of WGT on HGT and AGE

Source	DF	Sum of Squares	Mean Square	F Value	Pr > F
		The GLM Procedure			
Model	2	692.8226065	346.4113033	15.95	0.0011
Error	9	195.4273935	21.7141548		
Corrected Total	11	888.2500000			

R-Square	Coeff Var	Root MSE	WGT Mean
0.779986	7.426048	4.659845	62.75000

Parameter	Estimate	Standard Error	t Value	Pr > |t|
Intercept	6.553048251	10.94482708	0.60	0.5641
HGT	0.722037958	0.26080506	2.77	0.0218
AGE	2.050126352	0.93722561	2.19	0.0565

Test statistics and *P*-values for partial tests on model parameters (see Section 9.3)

Model 5   WGT $= \beta_0 + \beta_1 HGT + \beta_3(AGE)^2 + E$

Edited SAS Output for Regression of WGT on HGT and $(AGE)^2$

Source	DF	Sum of Squares	Mean Square	F Value	Pr > F
		The GLM Procedure			
Model	2	689.6499511	344.8249755	15.63	0.0012
Error	9	198.6000489	22.0666721		
Corrected Total	11	888.2500000			

R-Square	Coeff Var	Root MSE	WGT Mean
0.776414	7.486084	4.697518	62.75000

(continued)

```
                           The GLM Procedure (continued)
                                     Standard
         Parameter           Estimate            Error      t value      Pr > |t|
         Intercept        15.11753900      11.79690059         1.28        0.2321
         HGT               0.72597651       0.26333057         2.76        0.0222
         AGESQ             0.11480164       0.05373319         2.14        0.0614
```

Model 6 $WGT = \beta_0 + \beta_1 HGT + \beta_2 AGE + \beta_3 (AGE)^2 + E$

Edited SAS Output for Regression of WGT on HGT, AGE, and (AGE)^2

```
                              The GLM Procedure
                                   Sum of
Source                        DF       Squares    Mean Square    F Value    Pr > F
Model                          3    693.0604634    231.0201545       9.47    0.0052
Error                          8    195.1895366     24.3986921
Corrected Total               11    888.2500000
            R-Square      Coeff Var        Root MSE       WGT Mean
            0.780254       7.871718        4.939503       62.75000
                                   Standard
    Parameter           Estimate            Error     t value     Pr > |t|
    Intercept         3.438426001      33.61081984        0.10       0.9210
    HGT               0.723690241       0.27696316        2.61       0.0310
    AGE               2.776874563       7.42727877        0.37       0.7182
    AGESQ            -0.041706699       0.42240715       -0.10       0.9238
```

Although we will discuss model selection more fully in Chapter 16, it may already be clear from these results that model 4, involving HGT and AGE, is the best of the lot if we use R^2 and model simplicity as our criteria for selecting a model. The R^2-value of 0.7800 achieved by using this model is, for all practical purposes, the same as the maximum R^2-value of 0.7803 obtained by using all three variables.

Problems

1. The multiple regression relationships of SBP (Y) to AGE (X_1), SMK (X_2), and QUET (X_3) are studied using the data in Problem 2 of Chapter 5. Three regression models are considered, yielding least-squares estimates and ANOVA tables as shown in the accompanying SAS output.

   a. Use the model that includes all three independent variables (X_1, X_2, and X_3) to answer the following questions. (1) What is the predicted SBP for a 50-year-old smoker with a quetelet index of 3.5? (2) What is the predicted SBP for a 50-year-old nonsmoker with a quetelet index of 3.5? (3) For 50-year-old smokers, estimate the change in SBP corresponding to an increase in quetelet index from 3.0 to 3.5.

   b. Using the computer output, determine and compare the R^2-values for the three models. If you use R^2 and model simplicity as the criteria for selecting a model, which of the three models appears to be the best?

Edited SAS Output (GLM PROC) for Problem 1

```
                    SAS PROC GLM OUTPUT FOR REGRESSION OF SBP ON AGE
                                     Sum of
Source                   DF          Squares       Mean Square     F Value     Pr > F
Model                     1       3861.630375      3861.630375       45.18     <.0001
Error                    30       2564.338375        85.477946
Corrected Total          31       6425.968750

                              . [Portion of output omitted]

                                   Standard
       Parameter        Estimate      Error      t value    Pr > |t|
       Intercept      59.09162500  12.81626145      4.61     <.0001
       AGE             1.60450000   0.23871593      6.72     <.0001

              SAS PROC GLM OUTPUT FOR REGRESSION OF SBP ON AGE AND SMK
                                     Sum of
Source                   DF          Squares       Mean Square     F Value     Pr > F
Model                     2       4689.684229      2344.842114       39.16     <.0001
Error                    29       1736.284521        59.871880
Corrected Total          31       6425.968750

                              . [Portion of output omitted]

                                   Standard
       Parameter        Estimate      Error      t value    Pr > |t|
       Intercept      48.04960299  11.12955962      4.32     0.0002
       AGE             1.70915965   0.20175872      8.47     <.0001
       SMK            10.29439180   2.76810685      3.72     0.0009

           SAS PROC GLM OUTPUT FOR REGRESSION OF SBP ON AGE, SMK AND QUET
                                     Sum of
Source                   DF          Squares       Mean Square     F Value     Pr > F
Model                     3       4889.825697      1629.941899       29.71     <.0001
Error                    28       1536.143053        54.862252
Corrected Total          31       6425.968750

                              . [Portion of output omitted]

                                   Standard
       Parameter        Estimate      Error      t value    Pr > |t|
       Intercept      45.10319242  10.76487511      4.19     0.0003
       AGE             1.21271462   0.32381922      3.75     0.0008
       SMK             9.94556782   2.65605655      3.74     0.0008
       QUET            8.59244866   4.49868122      1.91     0.0664
```

2. A psychiatrist wants to know whether the level of pathology (Y) in psychotic patients 6 months after treatment can be predicted with reasonable accuracy from knowledge of pre-treatment symptom ratings of thinking disturbance (X_1) and hostile suspiciousness (X_2). The following table lists data collected on 53 patients.

Patient	Y	X_1	X_2	Patient	Y	X_1	X_2	Patient	Y	X_1	X_2
1	44	2.80	6.1	19	21	2.81	6.0	37	50	2.90	6.7
2	25	3.10	5.1	20	22	2.80	6.4	38	9	2.74	5.5
3	10	2.59	6.0	21	60	3.62	6.8	39	13	2.70	6.9
4	28	3.36	6.9	22	10	2.74	8.4	40	22	3.08	6.3
5	25	2.80	7.0	23	60	3.27	6.7	41	23	2.18	6.1
6	72	3.35	5.6	24	12	3.78	8.3	42	31	2.88	5.8
7	45	2.99	6.3	25	28	2.90	5.6	43	20	3.04	6.8
8	25	2.99	7.2	26	39	3.70	7.3	44	65	3.32	7.3
9	12	2.92	6.9	27	14	3.40	7.0	45	9	2.80	5.9
10	24	3.23	6.5	28	8	2.63	6.9	46	12	3.29	6.8
11	46	3.37	6.8	29	11	2.65	5.8	47	21	3.56	8.8
12	8	2.72	6.6	30	7	3.26	7.2	48	13	2.74	7.1
13	15	3.47	8.4	31	23	3.15	6.5	49	10	3.06	6.9
14	28	2.70	5.9	32	16	2.60	6.3	50	4	2.54	6.7
15	26	3.24	6.0	33	26	2.74	6.8	51	18	2.78	7.2
16	27	2.65	6.0	34	8	2.72	5.9	52	10	2.81	5.2
17	4	3.41	7.6	35	11	3.11	6.8	53	7	3.26	6.6
18	14	2.58	6.2	36	12	2.79	6.7				

a. The least-squares equation involving both independent variables is given by $\hat{Y} = -.0635 + 23.451X_1 - 7.073X_2$. Using this equation, determine the predicted level of pathology for a patient with pretreatment scores of 2.80 on thinking disturbance and 7.0 on hostile suspiciousness. How does this predicted value compare with the value actually obtained for patient 5?

b. Sums of squares are shown next for three regression models. Determine the R^2 values for each of these models. If you use R^2 and model simplicity as selection criteria, which model appears to be the best?

Y regressed on X_1:   SSY = 13,791.1698,   SSE = 12,255.3128

Y regressed on X_2:   SSE = 13,633.3225

Y regressed on X_1 and X_2:   SSE = 11,037.2985

3. The following table presents the weight (X_1), age (X_2), and plasma lipid levels of total cholesterol (Y) for a hypothetical sample of 25 patients suffering from hyperlipoproteinemia, before drug therapy.

a. Three estimated regression models, along with their sums-of-squares results, are as follows:

$\hat{Y} = 77.983 + 0.417X_1 + 5.217X_2$ SSY = 145377.0400,   SSE = 42806.2254

$\hat{Y} = 199.2975 + 1.622X_1$ SSE = 135145.3138

$\hat{Y} = 102.5751 + 5.321X_2$ SSE = 43444.3743

For each of these models, determine the predicted cholesterol level (Y) for patient 4, and compare these predicted cholesterol levels with the observed value. Comment on your findings.

Patient	Total Cholesterol (Y) (mg/100 ml)	Weight (X₁) (kg)	Age (X₂) (yr)	Patient	Total Cholesterol (Y) (mg/100 ml)	Weight (X₁) (kg)	Age (X₂) (yr)
1	354	84	46	14	254	57	23
2	190	73	20	15	395	59	60
3	405	65	52	16	434	69	48
4	263	70	30	17	220	60	34
5	451	76	57	18	374	79	51
6	302	69	25	19	308	75	50
7	288	63	28	20	220	82	34
8	385	72	36	21	311	59	46
9	402	79	57	22	181	67	23
10	365	75	44	23	274	85	37
11	209	27	24	24	303	55	40
12	290	89	31	25	244	63	30
13	346	65	52				

b. Determine R^2-values for each of the three models considered in part (a). If you use R^2 and model simplicity as selection criteria, which model appears to be the best predictive model?

4. A sociologist investigating the recent increase in the incidence of homicide throughout the United States studied the extent to which the homicide rate per 100,000 population (Y) is associated with the city's population size (X_1), the percentage of families with yearly income less than \$5,000 (X_2), and the rate of unemployment (X_3). Data are provided in the following table for a hypothetical sample of 20 cities.

City	Y	X₁ (thousands)	X₂	X₃	City	Y	X₁ (thousands)	X₂	X₃
1	11.2	587	16.5	6.2	11	14.5	7,895	18.1	6.0
2	13.4	643	20.5	6.4	12	26.9	762	23.1	7.4
3	40.7	635	26.3	9.3	13	15.7	2,793	19.1	5.8
4	5.3	692	16.5	5.3	14	36.2	741	24.7	8.6
5	24.8	1,248	19.2	7.3	15	18.1	625	18.6	6.5
6	12.7	643	16.5	5.9	16	28.9	854	24.9	8.3
7	20.9	1,964	20.2	6.4	17	14.9	716	17.9	6.7
8	35.7	1,531	21.3	7.6	18	25.8	921	22.4	8.6
9	8.7	713	17.2	4.9	19	21.7	595	20.2	8.4
10	9.6	749	14.3	6.4	20	25.7	3,353	16.9	6.7

a. Using the regression results presented next, determine the R^2-values for each two-variable model, and comment on which appears to be the best.

Y regressed on X_1 and X_2:   SSY = 1855.2020,   SSE = 537.4036

Y regressed on X_1 and X_3:   SSE = 431.9700

Y regressed on X_2 and X_3:   SSE = 367.3426

b. When Y is regressed on X_1, X_2, and X_3, we get SSE = 337.0571. Determine and comment on the increase in R^2 in going from a model with just X_2 and X_3 to a model that includes all three independent variables.

c. Consider the model with independent variables X_2, X_3, and X_4, where $X_4 = X_2X_3$.[5] The ANOVA table for this model, from SAS's GLM procedure, is given next. Does the addition of X_4 lead to a large improvement in fit over the model with X_2 and X_3 as the only independent variables? Explain.

Edited SAS Output (PROC GLM) for Problem 4

```
              SAS PROC GLM OUTPUT FOR Q.4: Y REGRESSED ON X2, X3 AND X4
                                  Sum of
Source                   DF      Squares      Mean Square     F Value      Pr > F
Model                    3     1490.571279     496.857093      21.80      <.0001
Error                   16      364.630721      22.789420
Corrected Total         19     1855.202000
```

5. A panel of educators in a large urban community wanted to evaluate the effects of educational resources on student performance. They examined the relationship between twelfth-grade mean verbal SAT scores (Y) and the following independent variables for a random sample of 25 high schools: $X_1 = $ Per pupil expenditure (in dollars); $X_2 = $ Percentage of teachers with a master's degree or higher; and $X_3 = $ Pupil–teacher ratio. The sums of squares shown next can be used to summarize the key results from the regression of Y on X_1, X_2 and X_3:

$$SSY = 28{,}222.23 \quad SSE = 2{,}248.23$$

a. Determine the ANOVA table for the regression of Y on X_1, X_2, and X_3.
b. Determine the R^2-value for the model in part (a). Based on this value, comment on whether the three educational resource variables appear to be associated with student performance.

6. A team of environmental epidemiologists used data from 23 counties to investigate the relationship between respiratory cancer mortality rates (Y) for a given year and the following three independent variables: $X_1 = $ Air pollution index for the county; $X_2 = $ Mean age (over 21) for the county; and $X_3 = $ Percentage of workforce in the county employed in a certain industry. The sums of squares shown next can be used to summarize the key results from the regression of Y on X_1, X_2, and X_3:

$$SSY = 2{,}387.653 \quad SSE = 551.723$$

a. Determine the ANOVA table for the regression of Y on X_1, X_2, and X_3.
b. Determine the R^2-value for the model in part (a). Based on this value, comment on whether the pollution level, mean age, and percentage of people working in the particular industry appear to be related (as a group) to respiratory cancer mortality rates.

---

[5] The coefficient of the product term X_4 measures an *interaction effect* associated with the variables X_2 and X_3, which concerns whether the relationship between Y and one of these two variables depends on the levels of the other variable. A more detailed discussion of the concept of interaction is given in Chapter 11.

7. In an experiment to describe the toxic action of a certain chemical on silkworm larvae, the relationship of $\log_{10}$ dose and $\log_{10}$ larva weight to $\log_{10}$ survival time was sought.[6] The data, obtained by feeding each larva a precisely measured dose of the chemical in an aqueous solution and then recording the survival time (i.e., time until death), are given in the following table. Relevant computer results are also provided.

Larva	1	2	3	4	5	6	7	8
$\log_{10}$ survival time (Y)	2.836	2.966	2.687	2.679	2.827	2.442	2.421	2.602
$\log_{10}$ dose (X_1)	0.150	0.214	0.487	0.509	0.570	0.593	0.640	0.781
$\log_{10}$ weight (X_2)	0.425	0.439	0.301	0.325	0.371	0.093	0.140	0.406

Larva	9	10	11	12	13	14	15
$\log_{10}$ survival time (Y)	2.556	2.441	2.420	2.439	2.385	2.452	2.351
$\log_{10}$ dose (X_1)	0.739	0.832	0.865	0.904	0.942	1.090	1.194
$\log_{10}$ weight (X_2)	0.364	0.156	0.247	0.278	0.141	0.289	0.193

a. Compute R^2 for each of the three models. Which independent variable appears to be the single best predictor of survival time?

b. Which model involving one or both of the independent variables do you prefer? Why?

Edited SAS Output (PROC GLM) for Problem 7

```
                    SAS PROC GLM OUTPUT FOR Y REGRESSED ON X1
                              Sum of
Source               DF       Squares      Mean Square    F Value    Pr > F
Model                1      0.36327405     0.36327405      31.91    <.0001
Error               13      0.14799288     0.01138407
Corrected Total     14      0.51126693

                        . [Portion of output omitted]
                        .

                                    Standard
   Parameter        Estimate          Error        t value    Pr > |t|
   Intercept       2.952199058      0.07355501       40.14     <.0001
   X1             -0.549855934      0.09733756       -5.65     <.0001


                    SAS PROC GLM OUTPUT FOR Y REGRESSED ON X2
                            The GLM Procedure
                              Sum of
Source               DF       Squares      Mean Square    F Value    Pr > F
Model                1      0.33667410     0.33667410      25.07    0.0002
Error               13      0.17459283     0.01343022
Corrected Total     14      0.51126693

                        . [Portion of output omitted]
                        .

                                          Standard
   Parameter        Estimate          Error        t value    Pr > |t|
   Intercept       2.184705838      0.08199583       26.64     <0.0001
   X2              1.375578800      0.27474016        5.01      0.0002
```

(continued)

---

[6] Adapted from a study by Bliss (1936).

```
            SAS PROC GLM OUTPUT FOR Y REGRESSED ON X1 AND X2 (continued)
                                Sum of
Source              DF          Squares        Mean Square     F Value      Pr > F
Model               2          0.46420205       0.23210102      59.18       <.0001
Error               12         0.04706489       0.00392207
Corrected Total     14         0.51126693

                         . [Portion of output omitted]
                         .

                                        Standard
        Parameter          Estimate       Error          t value     Pr > |t|
        Intercept         2.588995674    0.08360793        30.97      <.0001
        X1               -0.378480493    0.06637411        -5.70      <.0001
        X2                0.874974778    0.17248352         5.07       0.0003
```

8. An experiment to evaluate the effects of certain variables on soil erosion was performed on 10-foot-square plots of sloped farmland subjected to 2 inches of artificial rain applied over a 20-minute period.[7] The data and related ANOVA table are as follows:

Plot	1	2	3	4	5	6	7	8	9	10	11
SL (Y)	27.1	35.6	31.4	37.8	40.2	39.8	55.5	43.6	52.1	43.8	35.7
SG (X_1)	0.43	0.47	0.44	0.48	0.48	0.49	0.53	0.50	0.55	0.51	0.48
LOBS (X_2)	1.95	5.13	3.98	6.25	7.12	6.50	10.67	7.08	9.88	8.72	4.96
PGC (X_3)	0.34	0.32	0.29	0.30	0.25	0.26	0.10	0.16	0.19	0.18	0.28

Note: SL denotes soil lost (in pounds/acre), SG denotes slope gradient of the plot, LOBS denotes length (in inches) of the largest opening of bare soil on any boundary, and PGC denotes percentage of ground cover.

Source	d.f.	SS
Regression	3	680.4912
Residual	7	16.0942
Total	10	696.5855

a. Compute R^2, and comment on the fit of the model.

b. The fitted model involving all three independent variables is given by $\hat{Y} = -1.879 + 77.326X_1 + 1.559X_2 - 23.904X_3$. Compute and compare observed and predicted values of Y for plots 1, 5, and 7.

9. In a study by Yoshida (1961), the oxygen consumption of wireworm larva groups was measured at five temperatures. The rate of oxygen consumption per larva group (in milliliters per hour)—the dependent variable—was transformed to 0.5 less than the common logarithm. Another independent variable (other than temperature) of importance was larva group weight, which was also transformed to common logarithms. The data are given in the following table.

---

[7] Adapted from a study by Packer (1951).

Oxygen Consumption (Y) (log ml/hr −0.5)	Larva Group Weight (X₁) (log cg)	Temperature (X₂) (°C)	Oxygen Consumption (Y) (log ml/hr −0.5)	Larva Group Weight (X₁) (log cg)	Temperature (X₂) (°C)
0.054	0.130	15.5	0.482	0.053	30.0
0.154	0.215	15.5	0.477	0.114	30.0
0.073	0.250	15.5	0.551	0.137	30.0
0.182	0.267	15.5	0.516	0.190	30.0
0.241	0.389	15.5	0.561	0.210	30.0
0.316	0.490	15.5	0.588	0.230	30.0
0.290	0.491	15.5	0.561	0.240	30.0
0.061	0.004	20.0	0.580	0.260	30.0
0.143	0.164	20.0	0.674	0.389	30.0
0.188	0.225	20.0	0.718	0.470	30.0
0.176	0.314	20.0	0.754	0.521	30.0
0.248	0.447	20.0	0.800	0.544	30.0
0.357	0.477	20.0	0.654	−0.004	35.0
0.403	0.505	20.0	0.744	0.033	35.0
0.342	0.537	20.0	0.711	0.049	35.0
0.335	−0.046	25.0	0.855	0.140	35.0
0.408	0.176	25.0	0.932	0.204	35.0
0.366	0.199	25.0	0.927	0.210	35.0
0.482	0.292	25.0	0.914	0.215	35.0
0.545	0.380	25.0	0.914	0.265	35.0
0.596	0.483	25.0	0.973	0.346	35.0
0.590	0.491	25.0	1.000	0.462	35.0
0.631	0.491	25.0	0.998	0.468	35.0
0.610	0.519	25.0			

a. The fitted multiple regression model containing both X_1 and X_2 is given by

$$\hat{Y} = -0.6838 + 0.5921X_1 + 0.0394X_2$$

On the basis of this fitted model, how much of a change in oxygen consumption would be predicted for a larva group with fixed weight X_1 if the temperature were increased from $X_2 = 20$ to $X_2 = 25$?

b. For a temperature of 20°C, compute and compare the predicted values of $\hat{Y}$ for weights of 0.250 and 0.500.

c. What is R^2 for each of the three models?

Edited SAS Output (PROC GLM) for Problem 9

```
                 SAS PROC GLM OUTPUT FOR Q.9: Y REGRESSED ON X1
                              Sum of
Source              DF       Squares      Mean Square    F Value    Pr > F
Model                1      0.06607190     0.06607190       0.89    0.3505
Error               45      3.33995010     0.07422111
Corrected Total     46      3.40602200
```

(continued)

```
            SAS PROC GLM OUTPUT FOR Q.9: Y REGRESSED ON X2 (continued)
                                Sum of
Source                  DF      Squares      Mean Square    F Value    Pr > F
Model                    1    2.77418353      2.77418353     197.58    <.0001
Error                   45    0.63183847      0.01404085
Corrected Total         46    3.40602200


            SAS PROC GLM OUTPUT FOR Q.9: Y REGRESSED ON X1 AND X2
                                Sum of
Source                  DF      Squares      Mean Square    F Value    Pr > F
Model                    2    3.21119764      1.60559882     362.62    <.0001
Error                   44    0.19482436      0.00442783
Corrected Total         46    3.40602200
```

10. Residential real estate prices depend, in part, on property size and number of bedrooms. The house size X_1 (in hundreds of square feet), number of bedrooms X_2, and house price Y (in thousands of dollars) of a random sample of houses in a certain county were observed. The data are listed in the following table.

House	1	2	3	4	5	6	7
House size (X_1)	18	20	25	22	33	19	17
Number of bedrooms (X_2)	3	3	4	4	5	4	3
House price (Y)	80	95	104	110	175	85	89

Use the accompanying output to answer the following questions.
a. Determine the least-squares estimates for the model in which house price is regressed on both house size and number of bedrooms. Find the predicted average price of a 2,500-square-foot home that has four bedrooms.
b. Compare the R^2-value for the regression in part (a) with the r^2 value in Problem 22 of Chapter 6. Does adding X_2 to a model that already contains X_1 appear to be useful in predicting the house price?

Edited SAS Output (PROC GLM) for Problem 10

```
            SAS PROC GLM OUTPUT FOR Q.10: Y REGRESSED ON X1 AND X2
                                Sum of
Source                  DF      Squares      Mean Square    F Value    Pr > F
Model                    2    5733.321290     2866.660645     20.03    0.0082
Error                    4     572.392996      143.098249
Corrected Total          6    6305.714286


                                         Standard
        Parameter       Estimate          Error        t Value    Pr > |t|
        Intercept     -16.09338521      24.64693805      -0.65      0.5494
        X1              5.72178988       1.82778969       3.13      0.0352
        X2             -1.17315175      13.38993701      -0.09      0.9344
```

11. Data on sales revenues Y, television advertising expenditures X_1, and print media advertising expenditures X_2 for a large retailer for the period 1988–1993 are given in the following table.

Year	1988	1989	1990	1991	1992	1993
Sales ($millions)	4.0	8.0	2.0	8.0	5.0	4.0
TV advertising ($millions)	1.5	4.5	0.0	5.0	3.0	1.5
Print advertising ($millions)	0.5	0.5	0.0	1.0	1.0	1.5

a. State the model for the regression of sales revenue on television advertising expenditures and print advertising expenditures.

Use the accompanying computer output to answer the following questions.

b. State the estimate of the model in part (a). What is the estimated change in sales revenue for every $1,000 increase in television advertising expenditures? What is the estimated change in sales revenue for every $1,000 increase in print advertising expenditures?

c. Find the R^2-value for the regression of Y on X_1 and X_2. Interpret your result.

d. Predict the sales for a year in which $5 million was spent on TV advertising, and $1 million was spent on print advertising.

Edited SAS Output (PROC GLM) for Problem 11

```
                 SAS PROC GLM OUTPUT FOR Q.11: Y REGRESSED ON X1 AND X2
Source               DF          Squares      Mean Square      F Value      Pr > F
Model                2         28.11853189     14.05926594       59.01       0.0039
Error                3          0.71480144      0.23826715
Corrected Total      5         28.83333333
```

. [Portion of output omitted]

```
                                         Standard
        Parameter        Estimate           Error         t value     Pr > |t|
        Intercept       2.104693141       0.42196591         4.99        0.0155
        X1              1.241877256       0.11913357        10.42        0.0019
        X2             -0.194945848       0.43944079        -0.44        0.687
```

12. Radial keratotomy is a type of refractive surgery in which radial incisions are made in the cornea of myopic (nearsighted) patients in an effort to reduce their myopia. Theoretically, the incisions allow the curvature of the cornea to become less steep, thereby reducing the refractive error of the patient. This and other vision correction surgery grew in popularity in the 1980s and 1990s, both among the public and among ophthalmologists.

The Prospective Evaluation of Radial Keratotomy (PERK) study began in 1983 to investigate the effects of radial keratotomy. Lynn et al. (1987) examined the factors associated with the five-year postsurgical change in refractive error (Y, measured in diopters, D). Two independent variables under consideration were baseline refractive error (X_1, in diopters) and baseline curvature of the cornea (X_2, in diopters). [*Note:* Myopic patients have negative refractive errors. Patients who are far-sighted have positive refractive errors. Patients who are neither near nor far-sighted have zero refractive error.]

The accompanying computer output is based on data adapted from the PERK study. Use it to answer the following questions.

a. State the estimated least-squares equation for the regression of change in refractive error (Y) on baseline refractive error (X_1) and baseline curvature (X_2).

b. Using your answer to (a), give a point estimate for the change in refractive error for a patient who, at baseline, has a refractive error of $-8.00D$, and a corneal curvature of 44D. [*Note:* Myopic patients have negative refractive errors.]

c. Find the R^2-value for the regression in (a), and comment on the fit of the model.

Edited SAS Output (PROC GLM) for Problem 12

```
                    SAS PROC GLM OUTPUT FOR Q.12: Y REGRESSED ON X1 AND X2
                                        Sum of
Source                      DF          Squares       Mean Square      F Value      Pr > F
Model                        2        17.62277191      8.81138596        7.18       0.0018
Error                       51        62.63016929      1.22804254
Corrected Total             53        80.25294120
```

. [Portion of output omitted]

```
                                        Standard
        Parameter         Estimate        Error       t value     Pr > |t|
        Intercept       12.36001523     5.08621177       2.43       0.0187
        X1              -0.29160125     0.09165295      -3.18       0.0025
        X2              -0.22039615     0.11482391      -1.92       0.0605
```

13. In 1990, *Business Week* magazine compiled financial data on the 1,000 companies that had the biggest impact on the U.S. economy.[8] Data sampled from the top 500 companies in *Business Week*'s report are presented in the following table. In addition to the company name, four variables are shown:

1990 rank:	Based on company's market value (share price on March 16, 1990, multiplied by available common shares outstanding)
1989 rank:	Rank in 1989 compilation
P-E ratio:	Price-to-earnings ratio, based on 1989 earnings and March 16, 1990 share price
Yield:	Annual dividend rate as a percentage of March 16, 1990 share price

Company	1990 Rank	1989 Rank	P-E Ratio	Yield	Company	1990 Rank	1989 Rank	P-E Ratio	Yield
AT&T	4	4	17	2.87	ITT	81	57	8	2.98
Merck	7	7	19	2.54	Humana	162	209	15	2.62
Boeing	27	41	24	1.72	Salomon	236	172	7	2.91
American Home Products	32	37	14	4.26	Walgreen	242	262	17	1.87
Walt Disney	33	42	23	0.41	Lincoln National	273	274	9	4.73
Pfizer	46	46	14	4.10	Citizens Utilities	348	302	21	0.00
MCI Communications	52	72	19	0.00	MNC Financial	345	398	6	5.46
Dunn & Bradstreet	55	48	15	4.27	Bausch & Lomb	354	391	15	1.99
United Telecommunications	63	93	22	2.61	Medtronic	356	471	16	1.10
Warner Lambert	77	91	17	2.94	Circuit City	497	514	14	0.33

[8] "The Business Week 1000," a special issue of *Business Week* magazine, April 13, 1990

Answer the following questions about these data.

a. What is the estimated least-squares equation for the regression of yield (Y) on 1989 rank (X_2) and P-E ratio (X_3)?

b. Using your answer to part (a), give a point estimate for the yield for a company that had a 1989 ranking of 200 and a P-E ratio of 10.

c. Find the R^2-value for the regression in part (a), and comment on the fit of the model.

Edited SAS Output (PROC GLM) for Problem 13

```
          SAS PROC GLM OUTPUT FOR Q.13: Y REGRESSED ON X2 AND X3
                                Sum of
Source                  DF      Squares      Mean Square    F Value    Pr > F
Model                    2    26.58272135    13.29136068     10.51     0.0011
Error                   17    21.49757365     1.26456316
Corrected Total         19    48.08029500


            R-Square          Coeff Var        Root MSE      Y Mean
            0.552882          45.24353         1.124528      2.485500

                               . [Portion of output omitted]

                                       Standard
        Parameter        Estimate        Error       t value    Pr > |t|
        Intercept      6.873525074     0.99001534      6.94      <.0001
        X2            -0.004138085     0.00164749     -2.51      0.0224
        X3            -0.234451675     0.05292277     -4.43      0.0004
```

14. This problem refers to the 1990 Census data presented in Problem 19 of Chapter 5. In addition to median selected monthly ownership costs (OWNCOST), another independent variable studied was the proportion of the total metropolitan statistical area (MSA) population living in urban areas (URBAN).

Use the accompanying computer output to answer the following questions about the regression of OWNEROCC on OWNCOST and URBAN.

a. What is the estimated least-squares equation for the regression of OWNEROCC on OWNCOST and URBAN?

b. Using your answer to part (a), give a point estimate for the rate of owner occupancy for an MSA for which the urban proportion is 0.73 and median selected monthly ownership cost is $513.

c. Find the R^2-value for the regression in part (a), and comment on the fit of the model.

Edited SAS Output (PROC REG) for Problem 14

```
                 EDITED SAS PROC REG OUTPUT FOR PROBLEM 19:
                 OWNEROCC REGRESSED ON OWNCOST AND URBAN
                        Analysis of Variance
                            Sum of          Mean
Source                DF    Squares        Square      F Value    Pr > F
Model                  2   255.77851     127.88925      8.52      0.0017
Error                 23   345.18303      15.00796
Corrected Total       25   600.96154
```

(continued)

```
              EDITED SAS PROC REG OUTPUT FOR PROBLEM 19:
           OWNEROCC REGRESSED ON OWNCOST AND URBAN (continued)
           Root MSE             3.87401       R-Square         0.4256
           Dependent Mean      65.96154       Adj R-Sq         0.3757
           Coeff Var            5.87314

                           Parameter Estimates
                        Parameter       Standard
        Variable     DF   Estimate         Error      t Value     Pr > |t|
        Intercept     1   86.75877        5.10721       16.99      <.0001
        OWNCOST       1   -0.01113        0.00529       -2.10       0.0467
        URBAN         1  -16.87557        5.89098       -2.86       0.0088
```

15. This problem refers to the pond ecology data of Chapter 5, Problem 20.
  a. What is the estimated least-squares equation for the multiple regression of copepod count on zooplankton and phytoplankton counts.
  b. Find the R^2 value for the model in (a) and comment on the fit of the model.

References

Bliss, C. I. 1936. "The Size Factor in Action of Arsenic upon Silkworms' Larvae." *Journal of Experimental Biology* 13: 95–110.

Business Week. 1990. "The Business Week 1000, America's Most Valuable Companies." Special issue of *Business Week,* April 13, 1990.

Diggle, P. J.; Heagerty, P.; Liang, K. Y.; and Zeger, S. L. 2002. *Analysis of Longitudinal Data.* New York: Oxford University Press.

Gallant, A. R. 1975. "Non-linear Regression." *American Statistician* 29: 73–81.

Lynn, M. J.; Waring, G. O.; Sperduto, R. D.; et al. 1987. "Factors Affecting Outcome and Predictability of Radial Keratotomy in the PERK Study." *Archives of Ophthalmology* 105: 42–51.

Packer, P. E. 1951. "An Approach to Watershed Protection Criteria." *Journal of Forestry* 49: 638–44.

Yoshida, M. 1961. "Ecological and Physiological Researches on the Wireworm, *Melanotus caudex* Lewis. Iwata." *Shizuoka Pref.,* Japan.

Zeger, S. L., and Liang, K. Y. 1986. "Longitudinal Data Analysis for Discrete and Continuous Outcomes." *Biometrics* 42: 121–30.

9

Testing Hypotheses in Multiple Regression

9.1 Preview

Once we have fit a multiple regression model and obtained estimates for the various parameters of interest, we want to answer questions about the contributions of various independent variables to the prediction of Y. Such questions raise the need for three basic types of tests:

1. *Overall test.* Taken collectively, does the *entire set* of independent variables (or equivalently, the fitted model itself) contribute significantly to the prediction of Y?

2. *Test for addition of a single variable.* Does the addition of *one* particular independent variable of interest add significantly to the prediction of Y achieved by other independent variables already present in the model?

3. *Test for addition of a group of variables.* Does the addition of some *group* of independent variables of interest add significantly to the prediction of Y obtained through other independent variables already present in the model?

These questions are typically answered by performing statistical tests of hypotheses. The null hypotheses for the tests can be stated in terms of the unknown parameters (the regression coefficients) in the model. The form of these hypotheses differs depending on the question being asked. (In Chapter 10 we will look at alternative but equivalent ways to state such null hypotheses in terms of population correlation coefficients.)

In the sections that follow, we will describe the statistical test appropriate for each of the preceding questions. Each of these tests can be expressed as an F test; that is, the test statistic will have an F distribution when the stated null hypothesis is true. In some cases, the test may be equivalently expressed as a t test. (For a review of material concerning the F and t distributions, refer to Chapter 3.)

All F tests used in regression analyses involve a ratio of two independent estimates of variance—say, $F = \hat{\sigma}_0^2 / \hat{\sigma}^2$. Under the assumptions for the standard multiple linear regression analysis given earlier, the term $\hat{\sigma}_0^2$ estimates σ^2 if H_0 is true; the term $\hat{\sigma}^2$ estimates σ^2 whether H_0

139

is true or not. The specific forms that these variance estimates take will be described in subsequent sections. In general, each is a mean-square term that can be found in an appropriate ANOVA table. If H_0 is not true, then $\hat{\sigma}_0^2$ estimates some quantity larger than σ^2. Thus, we would expect a value of F that is close to $1(= \sigma^2/\sigma^2)$ if H_0 is true, but larger than 1 if H_0 is not true. The larger the value of F, then, the likelier H_0 is to be untrue.

Another general characteristic of the tests to be discussed in this chapter is that *each test can be interpreted as a comparison of two models*. One of these models will be referred to as the *full* or *complete* model; the other will be called the *reduced* model (i.e., the model to which the complete model reduces under the null hypothesis).

As a simple example, consider the following two models:

$$Y = \beta_0 + \beta_1 X_1 + \beta_2 X_2 + E$$

and

$$Y = \beta_0 + \beta_1 X_1 + E$$

Under H_0: $\beta_2 = 0$, the larger (full) model reduces to the smaller (reduced) model. A test of H_0: $\beta_2 = 0$ is thus essentially equivalent to determining which of these two models is more appropriate.

As this example suggests, the set of independent variables in the reduced model (namely, X_1) is a subset of the independent variables in the full model (namely, X_1 and X_2). This is a characteristic common to all the basic types of tests to be described in this chapter. (More generally, this subset characteristic need not always be present. Suppose, for example, that we have H_0: $\beta_1 = \beta_2$ ($= \beta$, say). Then, the reduced model may be written as $Y = \beta_0 + \beta X + E$, with $\beta = \beta_1 = \beta_2$ and $X = X_1 + X_2$.)

9.2 Test for Significant Overall Regression

We now reconsider our first question, regarding an overall test for a model containing k independent variables—say,

$$Y = \beta_0 + \beta_1 X_1 + \beta_2 X_2 + \cdots + \beta_k X_k + E$$

The null hypothesis for this test may be generally stated as H_0: "All k independent variables considered together do not explain a significant amount of the variation in Y." Equivalently, we may state the null hypothesis as H_0: "There is no significant overall regression using all k independent variables in the model", or as H_0: $\beta_1 = \beta_2 = \cdots = \beta_k = 0$. Under this last version of H_0, the full model is reduced to a model that contains only the intercept term β_0.

To perform the test, we use the mean-square quantities provided in our ANOVA table (see Table 8.2 of Chapter 8). We calculate the F statistic

$$F = \frac{\text{Regression MS}}{\text{Residual MS}} = \frac{(\text{SSY} - \text{SSE})/k}{\text{SSE}/(n - k - 1)} \tag{9.1}$$

where $\text{SSY} = \sum_{i=1}^{n} (Y_i - \bar{Y})^2$ and $\text{SSE} = \sum_{i=1}^{n} (Y_i - \hat{Y}_i)^2$ are the total and error sums of squares, respectively. The computed value of F can then be compared with the critical point $F_{k,n-k-1, 1-\alpha}$, where α is the preselected significance level. We would reject H_0 if the computed F exceeded the critical point. Alternatively, we could compute the P-value for this test as the area under the curve of the $F_{k,n-k-1}$ distribution to the right of the computed F statistic. It can be shown that an equivalent expression for (9.1) in terms of R^2 is

$$F = \frac{R^2/k}{(1 - R^2)/(n - k - 1)} \tag{9.2}$$

For the example summarized in Table 8.2, which concerns the regression of WGT on HGT, AGE, and $(AGE)^2$ for a sample of $n = 12$ children, we have $k = 3$, Regression MS $= 231.02$, Residual MS $= 24.40$, and $R^2 = 0.7802$, so that

$$F = \frac{231.02}{24.40} = \frac{0.7802/3}{(1 - 0.7802)/(12 - 3 - 1)} = 9.47$$

The critical point for $\alpha = .01$ is $F_{3,8,0.99} = 7.59$. Thus, we would reject H_0 at $\alpha = .01$, because the P-value is less than .01. (We usually denote $P < .01$ by putting a double ** next to the computed F, as in Table 8.2. When $.01 < P < .05$, we usually use only one *.)

In interpreting the results of this test, we can conclude that, based on the observed data, the set of variables HGT, AGE, and $(AGE)^2$ significantly help to predict WGT. This conclusion does not mean that *all three* variables are needed for significant prediction of Y; perhaps only one or two of them are sufficient. In other words, a more parsimonious model than the one involving all three variables may be adequate. To determine this requires further tests, to be described in the next section.

The mean-square residual term in the overall F test, which is the denominator of the F in (9.1), is given by the formula

$$\frac{1}{n - k - 1} \text{SSE} = \frac{1}{n - k - 1} \sum_{i=1}^{n} (Y_i - \hat{Y}_i)^2$$

This quantity provides an estimate of σ^2 under the assumed model. The mean-square regression term $\sum_{i=1}^{n} (\hat{Y}_i - \bar{Y})^2/k$, which is the numerator of the F in (9.1), provides an independent estimate of σ^2 only if the null hypothesis of no significant overall regression is true. However, if H_0 is not true (i.e., one or more of the $\beta_1, \beta_2, \ldots, \beta_k$ are not equal to zero), then the numerator overestimates σ^2; this is why an F-value that is "too large" favors rejection of H_0. Thus, the F statistic (9.1) is the ratio of two independent estimates of the same variance only if the null hypothesis $H_0: \beta_1 = \beta_2 = \cdots = \beta_k = 0$ is true.

9.3 Partial F Test

Some important additional information regarding the fitted regression model can be obtained by presenting the ANOVA table as shown in Table 9.1. In this representation, we have partitioned the regression sum of squares into three components:

1. $\text{SS}(X_1)$: the sum of squares explained by using only $X_1 = \text{HGT}$ to predict Y

2. $\text{SS}(X_2 | X_1)$: the extra sum of squares explained by using $X_2 = \text{AGE}$ in addition to X_1 to predict Y

3. $\text{SS}(X_3 | X_1, X_2)$: the extra sum of squares explained by using $X_3 = (AGE)^2$ in addition to X_1 and X_2 to predict Y

TABLE 9.1   **ANOVA table for WGT regressed on HGT, AGE, and $(AGE)^2$, containing components of the regression sum of squares**

Source		d.f.	SS	MS	F	R^2
	$\begin{cases} X_1 \end{cases}$	1	588.92	588.92	19.67^{**}	0.7802
Regression	$\begin{cases} X_2\|X_1 \end{cases}$	1	103.90	103.90	$4.78 \ (.05 < P < .10)$	
	$\begin{cases} X_3\|X_1, X_2 \end{cases}$	1	0.24	0.24	0.01	
Residual		8	195.19	24.40		
Total		11	888.25			

We can use the extra information in Table 9.1 to answer the following questions:

1. Does $X_1 = $ HGT alone significantly aid in predicting Y?

2. Does the addition of $X_2 = $ AGE significantly contribute to the prediction of Y after we account (or control) for the contribution of X_1?

3. Does the addition of $X_3 = (AGE)^2$ significantly contribute to the prediction of Y after we account for the contribution of X_1 and X_2?

To answer question 1, we simply fit a straight-line regression model, using $X_1 = $ HGT as the single independent variable. The value 588.92, therefore, is the regression sum of squares for this straight-line regression model. The SSE for this model can be obtained from Table 9.1 by adding 195.19, 0.24, and 103.90 together, which yields the sum of squares 299.33, having 10 degrees of freedom (i.e., $10 = 8 + 1 + 1$). The F statistic for testing whether there is significant straight-line regression when we use only $X_1 = $ HGT is then given by $F = (588.92/1)/(299.33/10) = 19.67$, which has a P-value of less than .01 (i.e., X_1 contributes significantly to the linear prediction of Y).

To answer questions 2 and 3, we must use what is called a *partial F test*. This test assesses whether the addition of any specific independent variable, given others already in the model, significantly contributes to the prediction of Y. The test, therefore, allows for the deletion of variables that do not help in predicting Y and thus enables one to reduce the set of possible independent variables to an economical set of "important" predictors.

9.3.1 The Null Hypothesis

Suppose that we want to test whether adding a variable X^* significantly improves the prediction of Y, given that variables $X_1, X_2, \ldots, X_p$ are already in the model. The null hypothesis may then be stated as H_0: "X^* does not significantly add to the prediction of Y, given that $X_1, X_2, \ldots, X_p$ are already in the model," or equivalently, as H_0: $\beta^* = 0$ in the model $Y = \beta_0 + \beta_1 X_1 + \beta_2 X_2 + \cdots + \beta_p X_p + \beta^* X^* + E$.

As can be inferred from the second statement, the test procedure essentially compares two models: the *full* model contains $X_1, X_2, \ldots, X_p$ and X^* as independent variables; the *reduced* model contains $X_1, X_2, \ldots, X_p$, but not X^* (since $\beta^* = 0$ under the null hypothesis). The goal is to determine which model is more appropriate based on how much additional information X^* provides about Y over that already provided by $X_1, X_2, \ldots, X_p$. In the next chapter, we shall see that an equivalent statement of H_0 can be given in terms of a partial correlation coefficient.

9.3.2 The Procedure

To perform a partial F test involving a variable X^*, given that variables $X_1, X_2, \ldots, X_p$ are already in the model, we must first compute the extra sum of squares from adding X^* given $X_1, X_2, \ldots, X_p$, which we place in our ANOVA table under the source heading "Regression $X^* \mid X_1, X_2, \ldots, X_p$." This sum of squares is computed by the formula

$$
\begin{bmatrix}
\text{Extra sum of} \\
\text{squares from} \\
\text{adding } X^*, \text{ given} \\
X_1, X_2, \ldots, X_p
\end{bmatrix}
=
\begin{bmatrix}
\text{Regression sum} \\
\text{of squares when} \\
X_1, X_2, \ldots, X_p \\
\text{and } X^* \text{ are } all \\
\text{in the model}
\end{bmatrix}
-
\begin{bmatrix}
\text{Regression sum} \\
\text{of squares when} \\
X_1, X_2, \ldots, X_p \\
(\text{and } not \ X^*) \text{ are} \\
\text{in the model}
\end{bmatrix}
\tag{9.3}
$$

or, more compactly

$$
\begin{aligned}
\text{SS}(X^* \mid X_1, X_2, \ldots, X_p) &= \text{Regression SS}(X_1, X_2, \ldots, X_p, X^*) \\
&\quad - \text{Regression SS}(X_1, X_2, \ldots, X_p)
\end{aligned}
$$

[For any model, $\sum_{i=1}^{n}(Y_i - \bar{Y})^2$ can be split into two components—the regression sum of squares and the residual sum of squares. Therefore,

$$
\text{SS}(X^* \mid X_1, X_2, \ldots, X_p) = \text{Residual SS}(X_1, X_2, \ldots, X_p) - \text{Residual SS}(X_1, X_2, \ldots, X_p, X^*)
$$

is an equivalent expression.]

Thus, for our example,

$$
\begin{aligned}
\text{SS}(X_2 \mid X_1) &= \text{Regression SS}(X_1, X_2) - \text{Regression SS}(X_1) \\
&= 692.82 - 588.92 \\
&= 103.90
\end{aligned}
$$

and

$$
\begin{aligned}
\text{SS}(X_3 \mid X_1, X_2) &= \text{Regression SS}(X_1, X_2, X_3) - \text{Regression SS}(X_1, X_2) \\
&= 693.06 - 692.82 \\
&= 0.24
\end{aligned}
$$

To test the null hypothesis H_0: "The addition of X^* to a model already containing $X_1, X_2, \ldots, X_p$ does not significantly improve the prediction of Y," we compute

$$
F(X^* \mid X_1, X_2, \ldots, X_p) = \frac{\text{Extra sum of squares from adding } X^*, \text{ given } X_1, X_2, \ldots, X_p}{\text{Residual MS for the model containing all the variables } X_1, X_2, \ldots, X_p, X^*}
$$

or more compactly,

$$
F(X^* \mid X_1, X_2, \ldots, X_p) = \frac{\text{SS}(X^* \mid X_1, X_2 \ldots, X_p)}{\text{Residual MS}(X_1, X_2, \ldots, X_p, X^*)}
\tag{9.4}
$$

This F statistic has an F distribution with 1 and $n - p - 2$ degrees of freedom under H_0, so we should reject H_0 if the computed F exceeds $F_{1, n-p-2, 1-\alpha}$. For our example, the partial F statistics are (from Table 9.1).

$$F(X_2 \mid X_1) = \frac{SS(X_2 \mid X_1)}{\text{Residual MS } (X_1, X_2)} = \frac{103.90}{(195.19 + 0.24)/9} = 4.78$$

and

$$F(X_3 \mid X_1, X_2) = \frac{SS(X_3 \mid X_1, X_2)}{\text{Residual MS } (X_1, X_2, X_3)} = \frac{0.24}{24.40} = 0.01$$

The quantity Residual MS (X_1, X_2) can be obtained directly from the ANOVA table for only X_1 and X_2, or indirectly from the partitioned ANOVA table for X_1, X_2, and X_3, by using the formula

$$\text{Residual MS } (X_1, X_2) = \frac{\text{Residual SS}(X_1, X_2, X_3) + SS(X_3 \mid X_1, X_2)}{8 + 1}$$

The statistic $F(X_2 \mid X_1) = 4.78$ has a P-value satisfying $.05 < P < .10$, since $F_{1, 9, 0.90} = 3.36$ and $F_{1, 9, 0.95} = 5.12$. Thus, we should reject H_0 at $\alpha = .10$ and conclude that the addition of X_2 after accounting for X_1 significantly adds to the prediction of Y at the $\alpha = .10$ level. At $\alpha = .05$, however, we would not reject H_0.

The statistic $F(X_3 \mid X_1, X_2)$ equals 0.01, so obviously H_0 should not be rejected regardless of the significance level; we therefore conclude that, once $X_1 = $ HGT and $X_2 = $ AGE are in the model, the addition of $X_3 = (\text{AGE})^2$ is superfluous.

9.3.3 The t Test Alternative

An equivalent way to perform the partial F test for the variable added last is to use a t test. (You may recall that an F statistic with 1 and $n - k - 1$ degrees of freedom is the square of a t statistic with $n - k - 1$ degrees of freedom.) The t test alternative focuses on a test of the null hypothesis H_0: $\beta^* = 0$, where β^* is the coefficient of X^* in the regression equation $Y = \beta_0 + \beta_1 X_1 + \beta_2 X_2 + \cdots + \beta_p X_p + \beta^* X^* + E$. The equivalent statistic for testing this null hypothesis is

$$T = \frac{\hat{\beta}^*}{S_{\hat{\beta}^*}} \tag{9.5}$$

where $\hat{\beta}^*$ is the corresponding estimated coefficient and $S_{\hat{\beta}^*}$ is the estimate of the standard error of $\hat{\beta}^*$, both of which are produced by standard regression programs.

In performing this test, we reject H_0: $\beta^* = 0$ if

$$\begin{cases} |T| > t_{n-p-2, 1-\alpha/2} & \text{(two-sided test; } H_A: \beta^* \neq 0) \\ T > t_{n-p-2, 1-\alpha} & \text{(upper one-sided test; } H_A: \beta^* > 0) \\ T < -t_{n-p-2, 1-\alpha} & \text{(lower one-sided test; } H_A: \beta^* < 0) \end{cases}$$

It can be shown that a two-sided t test is equivalent to the partial F test described earlier. For example, in testing H_0: $\beta_3 = 0$ in the model $Y = \beta_0 + \beta_1 X_1 + \beta_2 X_2 + \beta_3 X_3 + E$ fit to the data in Table 8.1, we compute

$$T = \frac{\hat{\beta}_3}{S_{\hat{\beta}_3}} = \frac{-0.0417}{0.4224} = -0.10$$

Squaring, we get

$$T^2 = 0.01 = \text{partial } F(X_3 | X_1, X_2)$$

from Table 9.1.

9.3.4 Comments

An important general application of the partial F test concerns the control of extraneous variables (e.g., confounders, which will be discussed in Chapter 11). Consider, for example, a situation with one main study variable of interest, S, and p control variables $C_1, C_2, \ldots, C_p$. The effect of S on the outcome variable Y, controlling for $C_1, C_2, \ldots, C_p$, may be assessed by considering the model

$$Y = \beta_0 + \beta_1 C_1 + \beta_2 C_2 + \cdots + \beta_p C_p + \beta_{p+1} S + E$$

The appropriate null hypothesis is $H_0: \beta_{p+1} = 0$. The partial F statistic in this situation is given by $F(S \mid C_1, C_2, \ldots, C_p)$, using (9.4) with $X^* = S$ and $C_i = X_i$, for $i = 1, 2, \ldots, p$.

When several study variables (i.e., several S's) are involved, the task includes determining which of the S's are important and perhaps even rank-ordering them by their relative importance. Such a task amounts to finding a best model, a topic we will address in Chapter 16 (where the term *best* will be carefully defined). For now, we note that one strategy (detailed in Chapter 16) is to work backward by *deleting* study (S) variables, one at a time, until a best model is obtained. This requires performing several partial F tests (as described in Chapter 16). If the starting model of interest is

$$Y = \beta_0 + \beta_1 C_1 + \cdots + \beta_p C_p + \beta_{p+1} S_1 + \cdots + \beta_{p+k} S_k + E$$

then the first backward step involves considering k partial F tests, $F(S_i | C_1, C_2, \ldots, C_p, S_1, S_2, \ldots, S_k$ except $S_i)$, where $i = 1, 2, \ldots, k$. The corresponding (separate) null hypotheses are $H_0: \beta_i = 0$, where $i = p + 1, p + 2, \ldots, p + k$. The usual backward procedure identifies the variable S_l associated with the smallest partial F value. This variable becomes the first to be deleted from the model, provided that its partial F is not significant. Then the elimination process starts all over again for the reduced model with S_l removed. Of course, if the smallest partial F value is significant, no S variables are deleted.

Each partial F test made at the first backward step weighs the contribution of a specific S variable, given that it is the last S variable to enter the model. It is therefore inappropriate to delete more than one S variable at this first step. For example, it is inappropriate to delete simultaneously *all* S variables from the model if all partial F's are nonsignificant at this first step. This is because, given that one particular S variable (say, S_l) is deleted, the remaining S variables may become important (based on consideration of their partial F's under the reduced model).

For example, suppose that we fit the model

$$Y = \beta_0 + \beta_1 C_1 + \beta_2 S_1 + \beta_3 S_2 + E$$

and obtain the following partial F results:

$$F(S_1 | C_1, S_2) = 0.01 \ (P = .90)$$
$$F(S_2 | C_1, S_1) = 0.85 \ (P = .25)$$

Then, S_1 is "less significant" than S_2, controlling for C_1 and the other S variable. Under the strategy of backward elimination, S_1 should be deleted before the elimination of S_2 is considered; however, to delete both S_1 and S_2 at this point would be incorrect. In fact, when considering the reduced model $Y = \beta_0 + \beta_1C_1 + \beta_2S_2 + E$, the partial F statistic $F(S_2|C_1)$ may be highly significant. In other words, if S_1 is not significant given S_2 and C_1, and if S_2 is not significant given S_1 and C_1, it does not necessarily follow that S_2 is unimportant in a reduced model containing S_2 and C_1 but not S_1.

9.4 Multiple Partial F Test

This testing procedure addresses the more general problem of assessing the additional contribution of two or more independent variables over and above the contribution made by other variables already in the model. For the example involving $Y =$ WGT, $X_1 =$ HGT, $X_2 =$ AGE, and $X_3 = (\text{AGE})^2$, we may be interested in testing whether the AGE variables, taken collectively, significantly improve the prediction of WGT given that HGT is already in the model. In contrast to the partial F test discussed in Section 9.3, the multiple partial F test addresses the simultaneous addition of two or more variables to a model. Nevertheless, the test procedure is a straightforward extension of the partial F test.

9.4.1 The Null Hypothesis

We want to test whether the addition of the k variables $X_1^*, X_2^*, \ldots, X_k^*$ significantly improves the prediction of Y, given that the p variables $X_1, X_2, \ldots, X_p$ are already in the model. The (full) model of interest is thus

$$Y = \beta_0 + \beta_1X_1 + \cdots + \beta_pX_p + \beta_1^*X_1^* + \cdots + \beta_k^*X_k^* + E$$

Then, the null hypothesis of interest may be stated as H_0: "$X_1^*, X_2^*, \ldots, X_k^*$ do not significantly add to the prediction of Y given that $X_1, X_2, \ldots, X_p$ are already in the model," or equivalently, H_0: $\beta_1^* = \beta_2^* = \cdots = \beta_k^* = 0$ in the (full) model.[1]
From the second version of H_0, it follows that the *reduced* model is of the form

$$Y = \beta_0 + \beta_1X_1 + \beta_2X_2 + \cdots + \beta_pX_p + E$$

(i.e., terms involving $X_1^*, \ldots, X_k^*$ are dropped from the full model).
For the preceding example, the (full) model is

$$\text{WGT} = \beta_0 + \beta_1\text{HGT} + \beta_1^*\text{AGE} + \beta_2^*(\text{AGE})^2 + E$$

The null hypothesis here is H_0: $\beta_1^* = \beta_2^* = 0$.

9.4.2 The Procedure

As in the case of the partial F test, we must compute the extra sum of squares due to the addition of terms involving $X_1^*, \ldots, X_k^*$ to the model. In particular, we have

$$SS(X_1^*, X_2^*, \ldots, X_k^* \mid X_1, X_2, \ldots, X_p)$$

$$= \text{Regression } SS(X_1, X_2, \ldots, X_p, X_1^*, X_2^*, \ldots, X_k^*) - \text{Regression } SS(X_1, X_2, \ldots, X_p)$$

$$= \text{Residual } SS(X_1, X_2, \ldots, X_p) - \text{Residual } SS(X_1, X_2, \ldots, X_p, X_1^*, X_2^*, \ldots, X_k^*)$$

---

[1] In Chapter 10, an equivalent expression for this null hypothesis will be given in terms of a multiple partial correlation coefficient.

Using this extra sum of squares, we obtain the following F statistic:

$$F(X_1^*, X_2^*, \ldots, X_k^* | X_1, X_2, \ldots, X_p) = \frac{\text{SS}(X_1^*, X_2^*, \ldots, X_k^* | X_1, X_2, \ldots, X_p)/k}{\text{Residual MS }(X_1, X_2, \ldots, X_p, X_1^*, X_2^*, \ldots, X_k^*)} \quad (9.6)$$

This F statistic has an F distribution with k and $n - p - k - 1$ degrees of freedom under H_0: $\beta_1^* = \beta_2^* = \cdots = \beta_k^* = 0$.

In (9.6), we must divide the extra sum of squares by k, the number of regression coefficients specified to be zero under the null hypothesis of interest. This number k is also the numerator degrees of freedom for the F statistic. The denominator of the F statistic is the Residual MS for the full model; its degrees of freedom is $n - (p + k + 1)$, which is $(n - 1)$ minus the number of predictor variables in this model (namely, $p + k$).

An alternative way to write this F statistic is

$$F(X_1^*, X_2^*, \ldots, X_k^* | X_1, X_2, \ldots, X_p) = \frac{[\text{Regression SS(full)} - \text{Regression SS(reduced)}]/k}{\text{Residual MS(full)}}$$

$$= \frac{[\text{Residual SS(reduced)} - \text{Residual SS(full)}]/k}{\text{Residual MS(full)}}$$

Using the information in Table 9.1 involving WGT, HGT, AGE, and $(\text{AGE})^2$, we can test $H_0: \beta_1^* = \beta_2^* = 0$ in the model WGT $= \beta_0 + \beta_1 \text{HGT} + \beta_1^* \text{AGE} + \beta_2^* (\text{AGE})^2 + E$, as follows:

$$F(\text{AGE}, (\text{AGE})^2 | \text{HGT})$$

$$= \frac{\{\text{Regression SS}[\text{HGT, AGE, }(\text{AGE})^2] - \text{Regression SS(HGT)}\}/2}{\text{Residual MS}[\text{HGT, AGE, }(\text{AGE})^2]}$$

$$= \frac{[(588.92 + 103.90 + 0.24) - 588.92]/2}{24.40}$$

$$= 2.13$$

For $\alpha = .05$, the critical point is

$$F_{k, n-p-k-1, 0.95} = F_{2, 12-1-2-1, 0.95} = 4.46$$

so H_0 would not be rejected at $\alpha = .05$.

In the preceding calculation, we used the relationship

Regression SS$[\text{HGT, AGE, }(\text{AGE})^2]$

$= $ Regression SS(HGT) $+$ Regression SS(AGE|HGT)

$\quad +$ Regression SS$[(\text{AGE})^2 | \text{HGT, AGE}]$

$= 588.92 + 103.90 + 0.24$

Alternatively, we could form two ANOVA tables (Table 9.2)—one for the full and one for the reduced model—and then extract the appropriate regression and/or residual sum-of-squares terms from these tables. More examples of partial F calculations are given at the end of this chapter.

9.4.3 Comments

The multiple partial F test is useful for assessing the simultaneous importance of particular subsets of a set of predictor variables. In particular, it is often used to test whether a "chunk"

TABLE 9.2 **ANOVA tables for WGT regressed on HGT, AGE, and $(AGE)^2$**

Full Model					Reduced Model				
Source	d.f.	SS	MS		Source	d.f.	SS	MS
Regression (X_1, X_2, X_3)	3	693.06	231.02		Regression (X_1)	1	588.92	588.92
Residual	8	195.19	24.40		Residual	10	299.33	29.93
Total	11	888.25			Total	11	888.25	

(i.e., a group) of variables having some trait in common is important when considered together. An example of a chunk is a collection of variables that are all of a certain order (e.g., $(AGE)^2$, HGT × AGE, and $(HGT)^2$ are all of order 2).

Another example is a collection of two-way product terms (e.g., X_1X_2, X_1X_3, X_2X_3); this latter group is sometimes referred to as a set of interaction variables (see Chapter 11). It is often of interest to assess the importance of interaction effects collectively before trying to consider individual interaction terms in a model. In fact, the initial use of such a chunk test can reduce the total number of tests to be performed, since variables may be dropped from the model as a group. This, in turn, helps provide better control of overall Type I error rates, which may be inflated due to multiple testing (Abt 1981; Kupper, Stewart, and Williams 1976).

9.5 Strategies for Using Partial F Tests

In applying the ideas presented in this chapter, readers will typically use a computer program to carry out the numerical calculations required. Therefore, we will briefly describe the computer output for typical regression programs. To help readers understand and use such output, we discuss two strategies for using partial F tests: *variables-added-in-order tests* and *variables-added-last tests*.

The accompanying computer output is from a typical regression computer program[2] for the model

$$WGT = \beta_0 + \beta_1 HGT + \beta_2 AGE + \beta_3 (AGE)^2 + E$$

The results here were computed with *centered* predictors[3] (see Section 14.5.2), so (HGT − 52.75), (AGE − 8.833) and $(AGE - 8.833)^2$ were used, with mean HGT = 52.75 and mean AGE = 8.833. The computer output consists of five sections, labeled A through E. Section A provides the overall ANOVA table for the regression model. Computer output typically presents numbers with far more significant digits than can be justified. Section B provides the multiple R^2 value, the mean $(\bar{Y})$ of the dependent variable (WGT), the WGT residual standard deviation or "Root MSE" (S), and the coefficient of variation $(100S/\bar{Y})$.

---

[2] This particular output was produced by the SAS program GLM.

[3] *Centering* is the process of transforming a variable, say AGE, by subtracting from AGE its sample mean, e.g., 8.833 in the preceding data. The newly defined centered variable (AGE − 8.833) will then have zero as its overall mean. Centering is frequently done to gain computational accuracy when estimating higher-order (e.g., polynomial) terms in a regression model. See Section 14.5.2 for further discussion.

Edited SAS Output for Data from Table 8.1 (HGT and AGE Centered)

```
                              The GLM Procedure
Dependent Variable: Weight
                                    Sum of
Source                      DF      Squares      Mean Square     F Value     Pr > F
Model                        3    693.0604634    231.0201545        9.47     0.0052  } A
Error                        8    195.1895366     24.3986921
Corrected Total             11    888.2500000

            R-Square            Coeff Var             Root MSE      Weight Mean  } B
            0.780254             7.871718             4.939503         62.75000

Source           DF      Type I SS       Mean Square     F Value     Pr > F
Height            1    588.9225232      588.9225232       24.14      0.0012  }
Age               1    103.9000834      103.9000834        4.26      0.0730  } C
Age*Age           1      0.2378569        0.2378569        0.01      0.9238  }

Source           DF      Type III SS     Mean Square     F Value     Pr > F
Height            1    166.5819549      166.5819549        6.83      0.0310  }
Age               1    101.8175237      101.8175237        4.17      0.0753  } D
Age*Age           1      0.2378569        0.2378569        0.01      0.9238  }

                                     Standard
Parameter          Estimate          Error        t Value     Pr > |t|
Intercept        62.88718379       1.99573190       31.51      <.0001  }
Height            0.72369024        0.27696316        2.61      0.0310  }
Age              2.04008401        0.99866553        2.04      0.0753  } E
Age*Age          -0.04170670       0.42240715       -0.10      0.9238  }
```

Section C provides certain (Type I) tests for assessing the importance of each predictor in the model; section D provides a different set of (Type III) tests regarding these predictors; and section E provides yet a third set of (*t*) tests.

9.5.1 Basic Principles

Two methods (or strategies) are widely used for evaluating whether a variable should be included in a model: partial (Type I) *F* tests for variables added in order, and partial (Type III) *F* tests for variables added last.[4] For the first (variables-added-in-order) method, the following procedure is employed: (1) an order for adding variables one at a time is specified; (2) the significance of the (straight-line) model involving only the variable ordered first is assessed; (3) the significance of adding the second variable to the model involving only the first variable is assessed; (4) the significance of adding the third variable to the model containing the first and second variables is assessed; and so on.

---

[4] Searle (1971), among others, refers to these methods as "ignoring" and "eliminating" tests, respectively.

Instead of considering variables added in order, it may be of interest to consider *variables deleted in order*. The latter strategy would apply, for example, in polynomial regression, where a backward selection algorithm is used to determine the proper degree of the polynomial. As can be seen from the accompanying computer output, the same set of sums of squares is produced whether the variables are considered to be added in one order or deleted in the reverse order.

For the second (variables-added-last) method, the following procedure is used: (1) an initial model containing two or more variables is specified; (2) the significance of each variable in the initial model is assessed separately, as if it were the last variable to enter the model (i.e., if k variables are in the initial model, then k variables-added-last tests are conducted). In either method, each test is conducted using a partial F test for the addition of a single variable.

Variables-added-in-order tests can be illustrated with the weight example. One possible ordering is HGT first, followed by AGE, and then $(AGE)^2$. For this ordering, the smallest model considered is

$$WGT = \beta_0 + \beta_1 HGT + E$$

The overall regression F test of $H_0: \beta_1 = 0$ is used to assess the contribution of HGT. Next, the model

$$WGT = \beta_0 + \beta_1 HGT + \beta_2 AGE + E$$

is fit. The significance of adding AGE to a model already containing HGT is then assessed by using the partial $F(AGE \mid HGT)$. Finally, the full model is fit by using HGT, AGE, and $(AGE)^2$. The importance of the last variable is tested with the partial $F[(AGE)^2 \mid HGT, AGE]$. The tests used are those discussed in this chapter and summarized in Table 9.1. These are also the tests provided in section C of the earlier computer output (using Type I sums of squares). However, each test in Table 9.1 involves a different residual sum of squares, while those in the computer output use a common residual sum of squares. More will be said about this issue shortly.

To describe variables-added-last tests, consider again the full model

$$WGT = \beta_0 + \beta_1 HGT + \beta_2 AGE + \beta_3 (AGE)^2 + E$$

The contribution of HGT, when added last, is assessed by comparing the full model to the model with HGT deleted—namely,

$$WGT = \beta_0 + \beta_2 AGE + \beta_3 (AGE)^2 + E$$

The partial F statistic, based on (9.4), has the form $F[(HGT \mid AGE, (AGE)^2]$. The sum of squares for HGT added last is then the appropriate difference in the error sum of squares (or the regression sum of squares) for the two preceding models. Similarly, the reduced model with AGE deleted is

$$WGT = \beta_0 + \beta_1 HGT + \beta_3 (AGE)^2 + E$$

for which the corresponding partial F statistic is $F[AGE \mid HGE, (AGE)^2]$; and the reduced model with $(AGE)^2$ omitted is

$$WGT = \beta_0 + \beta_1 HGT + \beta_2 AGE + E$$

for which the partial F statistic is $F[(AGE)^2 \mid HGT, AGE]$. The three F statistics just described are provided in section D of the earlier computer output (using Type III sums of squares).[5]

---

[5] It is important to remember that the preceding computer output was based on the use of the *centered* predictor variables (HGT $-$ 52.75), (AGE $-$ 8.833), and (AGE $-$ 8.833)^2, and not on the use of the original (uncentered) predictor variables HGT, AGE, and $(AGE)^2$. As a result, certain numerical results (specifically, some of those in sections D and E) for the centered predictors differ from those for the uncentered predictors (see, e.g., the computer output for Model 6 on page 126 in Chapter 8).

An important characteristic of variables-added-in-order sums of squares is that they decompose the regression sum of squares into a set of mutually exclusive and exhaustive pieces. For example, the sums of squares provided in section C of the computer output (588.922, 103.900, and 0.238) add to 693.060, which is the regression sum of squares given in section A. The variables-added-last sums of squares do not generally have this property (e.g., the sums of squares given in section D of the computer output do not add to 693.060).

Each of these two testing strategies has its own advantages, and the situation being considered determines which is preferable. For example, if all variables are considered to be of equal importance, the variables-added-last tests are usually preferred. Such tests treat all variables equally in the sense that each variable is assessed as if it were the last variable to enter the model.

In contrast, if the order in which the predictors enter the model is an important consideration, the variables-added-in-order testing approach may be better. An example where the entry order is important is one where main effects (e.g., X_1, X_2, and X_3) are forced into the model, followed by their cross-products (X_1X_2, X_1X_3, and X_2X_3) or so-called interaction terms (see Chapter 11). Such tests evaluate the contribution of one or more variables and adjust *only* for the variables just preceding them into the model.

9.5.2 Commentary

As discussed in the preceding subsection, section C of the earlier computer output provides variables-added-in-order tests, which are also given for the same data in Table 9.1. Section D of the computer output provides variables-added-last tests. Finally, section E provides t tests (which are equivalent to the variables-added-last F tests in section D), as well as regression coefficient estimates and their standard errors, for the *centered* predictor variables.

Table 9.3 gives an ANOVA table for the variables-added-last tests for the weight example. (We recommend that readers consider how this table was extracted from the earlier computer output.) The variables-added-last tests usually give a different ANOVA table from one based on the variables-added-in-order tests. A different residual sum of squares is used for each variables-added-in-order test in Table 9.1, while the same residual sum of squares (based on the three-variable model involving the *centered* versions of HGT, AGE, and $(AGE)^2$) is used for all the variables-added-last tests in Table 9.3.

An argument can be made that it is preferable to use the residual sum of squares for the three-variable model (i.e., the largest model containing all candidate predictors) for all tests. This is because the error variance σ^2 will *not* be correctly estimated by a model that ignores important predictors, but it will be correctly estimated (under the usual regression assumptions) by a model that contains all candidate predictors (even if some are not important). In other

TABLE 9.3   ANOVA table for WGT regressed on HGT, AGE, and $(AGE)^2$, using variables-added-last tests

Source	d.f.	SS	MS	F	R^2
$X_1\mid X_2, X_3$	1	166.58	166.58	6.83[*]	.7802
$X_2\mid X_1, X_3$	1	101.81	101.82	4.17	
$X_3\mid X_1, X_2$	1	0.24	0.24	0.01	
Residual	8	195.19	24.40		
Total	11	888.25			

[*] Exceeds .05 critical value of 5.32 for F with 1 and 8 degrees of freedom.

words, overfitting a model in estimating σ^2 is safer than underfitting it. Of course, extreme overfitting results in lost precision, but it still provides a valid estimate of residual variation. We generally prefer using the residual sum of squares based on fitting the "largest" model, although some statisticians disagree.

9.5.3 Models Underlying the Source Tables

Tables 9.4 and 9.5 present the models being compared based on the earlier computer output and the associated ANOVA tables. Table 9.4 summarizes the models and residual sums of squares needed to conduct variables-added-last tests for the full model containing the *centered* versions of HGT, AGE, and $(AGE)^2$. Table 9.5 lists the models that must be fitted to provide variables-added-in-order tests for the order of entry HGT, then AGE, and then $(AGE)^2$, again using *centered* versions of these predictors.

Table 9.6 details computations of regression sums of squares for both types of tests. For example, the first line, where $24{,}226.2637 = 24{,}421.4532 - 195.1895$, is the difference in the error sums of squares for models 1 and 5 given in Table 9.4. These results can then be used to produce any of the F tests given in Tables 9.1 and 9.2 and in the computer output.

TABLE 9.4   **Variables-added-last regression models and residual sums of squares for data from Table 8.1**

Model No.	Model	SSE
1	$WGT = \beta_1 HGT + \beta_2 AGE + \beta_3 (AGE)^2 + E$	24,421.4532
2	$WGT = \beta_0 + \beta_2 AGE + \beta_3 (AGE)^2 + E$	361.7715
3	$WGT = \beta_0 + \beta_1 HGT + \beta_3 (AGE)^2 + E$	297.0071
4	$WGT = \beta_0 + \beta_1 HGT + \beta_2 AGE + E$	195.4274
5	$WGT = \beta_0 + \beta_1 HGT + \beta_2 AGE + \beta_3 (AGE)^2 + E$	195.1895

TABLE 9.5   **Variables-added-in-order regression models and residual sums of squares for data from Table 8.1**

Model No.	Model	SSE
6	$WGT = + E$	48,139.0000
7	$WGT = \beta_0 + E$	888.2500
8	$WGT = \beta_0 + \beta_1 HGT + E$	299.3275
4	$WGT = \beta_0 + \beta_1 HGT + \beta_2 AGE + E$	195.4274
5	$WGT = \beta_0 + \beta_1 HGT + \beta_2 AGE + \beta_3 (AGE)^2 + E$	195.1895

TABLE 9.6   **Computations for regression sum of squares for data from Table 8.1**

		Regression SS		
Parameter	Variable	Added Last		In Order
β_0	Intercept	$SSE(1) - SSE(5) = 24{,}226.2637$		$SSE(6) - SSE(7) = 47{,}250.7500$
β_1	HGT	$SSE(2) - SSE(5) = \phantom{24{,}2}166.5820$		$SSE(7) - SSE(8) = \phantom{47{,}2}588.9225$
β_2	AGE	$SSE(3) - SSE(5) = \phantom{24{,}2}101.8176$		$SSE(8) - SSE(4) = \phantom{47{,}2}103.9001$
β_3	$(AGE)^2$	$SSE(4) - SSE(5) = \phantom{24{,}226.26}0.2379$		$SSE(4) - SSE(5) = \phantom{47{,}250.75}0.2379$

9.6 Tests Involving the Intercept

Inferences about the intercept β_0 are occasionally of interest in multiple regression analysis. A test of $H_0: \beta_0 = 0$ is usually carried out with an intercept-added-last test, although an intercept-added-in-order test is also feasible (where the intercept is the first term added to the model). Many computer programs provide only a t test involving the intercept. The t-test statistic for the intercept in the earlier computer output corresponds exactly to a partial F test for adding the intercept last. The two models being compared are

$$Y = \beta_0 + \beta_1 X_1 + \beta_2 X_2 + \cdots + \beta_k X_k + E$$

and

$$Y = \beta_1 X_1 + \beta_2 X_2 + \cdots + \beta_k X_k + E$$

The null hypothesis of interest is $H_0: \beta_0 = 0$ versus $H_A: \beta_0 \neq 0$. The test is computed as

$$F = \frac{(\text{SSE without } \beta_0 - \text{SSE with } \beta_0)/1}{\text{SSE with } \beta_0/(n - k - 1)}$$

This F statistic has 1 and $n - k - 1$ degrees of freedom and is equal to the square of the t statistic used for testing $H_0: \beta_0 = 0$. For the weight example involving the *centered* predictors, an intercept-added-last test is reported in the output on page 149 as a t test. The corresponding partial F equals $(31.51)^2 = 992.88$ and has 1 and 8 degrees of freedom.

An intercept-added-in-order test can also be conducted. In this case, the two models being compared are

$$Y = E$$

and

$$Y = \beta_0 + E$$

Again, the null hypothesis is $H_0: \beta_0 = 0$ versus the alternative $H_A: \beta_0 \neq 0$. The special nature of this test leads to the simple expression

$$F = \frac{n\bar{Y}^2}{\text{SSY}/(n - 1)}$$

which represents an F statistic with 1 and $n - 1$ degrees of freedom. This statistic involves SSY, the residual sum of squares from a model with just an intercept (such as model 7 in Table 9.5). Alternatively, the residual sum of squares from the "largest" model may be used. When we use the latter approach, the F statistic for the weight data becomes (see Table 9.5)

$$F = \frac{[\text{SSE}(6) - \text{SSE}(7)]/1}{\text{SSE}(5)/8}$$
$$= \frac{(48{,}139.00 - 888.25)/1}{195.1895/8}$$
$$= 1936.61$$

where SSE(5) denotes the residual (i.e., error) sum of squares for model 5 (which is the largest of the models in Table 9.5); this F statistic has 1 and 8 degrees of freedom. In general, using the residual from the largest model (with k predictors) gives $n - k - 1$ error degrees of freedom, so the F statistic is compared to a critical value with 1 and $n - k - 1$ degrees of freedom.

Problems

1. Use the data of Problem 1 in Chapter 8 to answer the following questions.
   a. Conduct the overall F tests for significant regression for the three models in Problem 1 of Chapter 8. Be sure to state the null and alternative hypotheses for each test, and interpret each result.
   b. Based on your results in part (a), which of the three models is best? Compare your answer here with your answer to Problem 1(b) in Chapter 8.

2. Use the information given in Problem 2 of Chapter 8 as well as the computer output given here to answer the following questions about the data from that problem.
   a. Conduct overall regression F tests for these three models: Y regressed on X_1 and X_2; Y regressed on X_1 alone; and Y regressed on X_2 alone. Based on your answers, how would you rate the importance of the two variables in predicting Y? Comment on how your answer compares with the answer to Problem 2(b) in Chapter 8.
   b. Provide variables-added-in-order tests for both variables, with thinking disturbances (X_1) added first. Use $\alpha = .05$.
   c. Provide variables-added-in-order tests for both variables, with hostile suspiciousness (X_2) added first. Use $\alpha = .05$.
   d. Provide a table of variables-added-last tests.
   e. What, if any, differences are present in the approaches in parts (a) through (d)?
   f. Which predictors appear to be necessary? Why?

Edited SAS Output (PROC GLM) for Problem 2

```
                      Y REGRESSED ON X1 AND X2
                         THE GLM Procedure
Dependent Variable: Y
                                Sum of
Source                  DF      Squares     Mean Square   F Value   Pr > F
Model                    2    2753.87136    1376.93568      6.24    0.0038
Error                   50   11037.29845     220.74597
Corrected Total         52   13791.16981

            R-Square      Coeff Var     Root MSE     Y Mean
            0.199684      65.45708      14.85752     22.69811

Source                  DF     Type I SS    Mean Square   F Value   Pr > F
X1                       1   1535.856965    1535.856965     6.96    0.0111
X2                       1   1218.014394    1218.014394     5.52    0.0228

Source                  DF    Type III SS   Mean Square   F Value   Pr > F
X1                       1   2596.024025    2596.024025    11.76    0.0012
X2                       1   1218.014394    1218.014394     5.52    0.0228

                                            Standard
         Parameter        Estimate           Error      t Value   Pr > |t|
         Intercept      -0.63535020       20.96833367    -0.03     0.9759
         X1             23.45143521        6.83850964     3.43     0.0012
         X2             -7.07260920        3.01092434    -2.35     0.0228
```

3. A psychologist examined the regression relationship between anxiety level (Y)—measured on a scale ranging from 1 to 50, as the average of an index determined at three points in a 2-week period—and the following three independent variables: X_1 = Systolic blood pressure; X_2 = IQ; and X_3 = Job satisfaction (measured on a scale ranging from 1 to 25). The

following ANOVA table summarizes results obtained from a variables-added-in-order regression analysis on data involving 22 outpatients who were undergoing therapy at a certain clinic.

Source		d.f.	SS
Regression	(X_1)	1	981.326
	$(X_2 \mid X_1)$	1	190.232
	$(X_3 \mid X_1, X_2)$	1	129.431
Residual		18	442.292

a. Test for the significance of each independent variable as it enters the model. State the null hypothesis for each test in terms of regression coefficient parameters.
b. Test for the significance of adding both X_2 and X_3 to a model already containing X_1. State the null hypothesis in terms of regression coefficient parameters.
c. In terms of regression sums of squares, identify the test that corresponds to comparing the two models

$$Y = \beta_0 + \beta_1 X_1 + \beta_2 X_2 + \beta_3 X_3 + E$$

and

$$Y = \beta_0 + \beta_3 X_3 + E$$

Why can't this test be done by using the ANOVA table? Describe the appropriate test procedure.
d. Based on the tests made, what would you recommend as the most appropriate statistical model? Use $\alpha = .05$.

4. An educator examined the relationship between the number of hours devoted to reading each week (Y) and the independent variables social class (X_1), number of years of school completed (X_2), and reading speed (X_3), in pages read per hour. The following ANOVA table was obtained from a stepwise regression analysis on data for a sample of 19 women over 60.

Source		d.f.	SS
Regression	(X_3)	1	1,058.628
	$(X_2 \mid X_3)$	1	183.743
	$(X_1 \mid X_2, X_3)$	1	37.982
Residual		15	363.300

a. Test for the significance of each variable as it enters the model.
b. Test $H_0: \beta_1 = \beta_2 = 0$ in the model $Y = \beta_0 + \beta_1 X_1 + \beta_2 X_2 + \beta_3 X_3 + E$.
c. Why can't we test $H_0: \beta_1 = \beta_3 = 0$ by using the ANOVA table given? What formula would you use for this test?
d. Based on your results in parts (a) and (b), what is the most appropriate model to use?

5. An experiment was conducted regarding a quantitative analysis of factors found in high-density lipoprotein (HDL) in a sample of human blood serum. Three variables thought to be predictive of, or associated with, HDL measurement (Y) were the total cholesterol (X_1) and total triglyceride (X_2) concentrations in the sample, plus the presence or absence of a certain sticky component of the serum called sinking pre-beta, or SPB (X_3) coded as 0 if absent and 1 if present. The data obtained are shown in the following table.
a. Test whether X_1, X_2, or X_3 alone significantly helps to predict Y.
b. Test whether X_1, X_2, and X_3 taken together significantly helps to predict Y.

Y	X_1	X_2	X_3	Y	X_1	X_2	X_3	Y	X_1	X_2	X_3
47	287	111	0	63	339	168	1	36	318	180	0
38	236	135	0	40	161	68	1	42	270	134	0
47	255	98	0	59	324	92	1	41	262	154	0
39	135	63	0	56	171	56	1	42	264	86	0
44	121	46	0	76	265	240	1	39	325	148	0
64	171	103	0	67	280	306	1	27	388	191	0
58	260	227	0	57	248	93	1	31	260	123	0
49	237	157	0	57	192	115	1	39	284	135	0
55	261	266	0	42	349	408	1	56	326	236	1
52	397	167	0	54	263	103	1	40	248	92	1
49	295	164	0	60	223	102	1	58	285	153	1
47	261	119	1	33	316	274	0	43	361	126	1
40	258	145	1	55	288	130	0	40	248	226	1
42	280	247	1	36	256	149	0	46	280	176	1

c. Test whether the true coefficients of the product terms X_1X_3 and X_2X_3 are simultaneously zero in the model containing X_1, X_2, and X_3 plus these product terms. Specify the two models being compared, and state the null hypothesis in terms of regression coefficients. If this test is not rejected, what can you conclude about the relationship of Y to X_1 and X_2 when X_3 equals 1, compared to the same relationship when X_3 equals 0?

d. Test (at $\alpha = .05$) whether X_3 is associated with Y, after taking into account the combined contribution of X_1 and X_2. What does your result, together with your answer to part (c), tell you about the relationship of Y with X_1 and X_2 when SPB is present compared to the same relationship when it is absent?

Edited SAS Output (PROC GLM) for Problem 5

```
                          Y REGRESSED ON X2
                          THE GLM Procedure

Dependent Variable: Y

                                    Sum of
Source                     DF       Squares       Mean Square     F Value    Pr > F
Model                       1     21.339709        21.339709        0.19     0.6687
Error                      40   4592.279339       114.806983
Corrected Total            41   4613.619048


              R-Square      Coeff Var      Root MSE       Y Mean
              0.004625      22.43378       10.71480      47.76190


Source                     DF     Type I SS       Mean Square     F Value    Pr > F
X2                          1    21.33970871       21.33970871      0.19     0.6687

                          Y REGRESSED ON X3

                           . [Portion of output omitted]


Source                     DF     Type I SS       Mean Square     F Value    Pr > F
X3                          1    735.2054113      735.2054113       7.58     0.0088
```

(continued)

```
           Y REGRESSED ON X1, X2, X3, X1X3 AND X2X3 (continued)

                               . [Portion of output omitted]
                               .
                               .
```

Source	DF	Type I SS	Mean Square	F Value	Pr > F
X1	1	46.2355575	46.2355575	0.45	0.5078
X2	1	89.1465681	89.1465681	0.86	0.3591
X3	1	684.3651973	684.3651973	6.62	0.0143
X1X3	1	48.6904324	48.6904324	0.47	0.4968
X2X3	1	26.1296111	26.1296111	0.25	0.6181

   e. What overall conclusion can you draw about the association of Y with the three independent variables for this data set? Specify the two models being compared, and state the
   appropriate null hypothesis in terms of regression coefficients.

6. Use the results from Problem 3 in Chapter 8, as well as the computer output given here, to
   answer the following questions about the data from that problem.

   a. Conduct overall regression F tests for these three models: Y regressed on weight and age;
   Y regressed on weight alone; and Y regressed on age alone. Based on your answers, how
   would you rate the importance of the two variables in predicting Y? Comment on how
   your answer here compares with your answer to Problem 3(b) in Chapter 8.

   b. Provide variables-added-in-order tests for both variables, with weight (X_1) added first.
   Use $\alpha = .05$.

   c. Provide variables-added-in-order tests for both variables, with age (X_2) added first. Use
   $\alpha = .05$.

   d. Provide a table of variables-added-last tests for both weight and age.

   e. What, if any, differences are discernible in the results of the approaches in parts (a)
   through (d)?

Edited SAS Output (PROC GLM) for Problem 6

```
                    TOT. CHOL (Y) REGRESSED ON WEIGHT (X1)
                             Sum of
```

Source	DF	Squares	Mean Square	F Value	Pr > F
Model	1	10231.7262	10231.7262	1.74	0.2000
Error	23	135145.3138	5875.8832		
Corrected Total	24	145377.0400			

```
                    TOT. CHOL (Y) REGRESSED ON AGE (X2)
                             Sum of
```

Source	DF	Squares	Mean Square	F Value	Pr > F
Model	1	101932.6657	101932.6657	53.96	<.0001
Error	23	43444.3743	1888.8858		
Corrected Total	24	145377.0400			

```
                    Y REGRESSED ON X1 AND X2
                             Sum of
```

Source	DF	Squares	Mean Square	F Value	Pr > F
Model	2	102570.8147	51285.4073	26.36	<.0001
Error	22	42806.2253	1945.7375		
Corrected Total	24	145377.0400			

7. Use the results from Problem 4 in Chapter 8, as well as the computer output given here, to answer the following questions about the data from that problem.

a. Conduct the overall regression F test for the model where Y is regressed on X_1, X_2, and X_3. Use $\alpha = .05$. Interpret your result.

b. Provide variables-added-in-order tests for the order X_2, X_1, and X_3.

c. Provide variables-added-in-order tests for the order X_3, X_1, and X_2.

d. List all orders that can be tested, using the computer results below and in Problem 4 of Chapter 8. List all orders that *cannot* be so computed.

Edited SAS Output (PROC GLM) for Problem 7

Y REGRESSED ON X2 AND X1

. [Portion of output omitted]

Source	DF	Type I SS	Mean Square	F Value	Pr > F
X2	1	1308.339479	1308.339479	41.39	<.0001
X1	1	9.458932	9.458932	0.30	0.5915

Source	DF	Type III SS	Mean Square	F Value	Pr > F
X2	1	1309.445931	1309.445931	41.42	<.0001
X1	1	9.458932	9.458932	0.30	0.5915

Y REGRESSED ON X3 AND X1

. [Portion of output omitted]

Source	DF	Type I SS	Mean Square	F Value	Pr > F
X3	1	1387.599718	1387.599718	54.61	<.0001
X1	1	35.632307	35.632307	1.40	0.2526

Source	DF	Type III SS	Mean Square	F Value	Pr > F
X3	1	1414.879544	1414.879544	55.68	<.0001
X1	1	35.632307	35.632307	1.40	0.2526

Y REGRESSED ON X3 AND X2

. [Portion of output omitted]

Source	DF	Type I SS	Mean Square	F Value	Pr > F
X3	1	1387.599718	1387.599718	64.22	<.0001
X2	1	100.259687	100.259687	4.64	0.0459

Source	DF	Type III SS	Mean Square	F Value	Pr > F
X3	1	179.5199260	179.5199260	8.31	0.0103
X2	1	100.2596869	100.2596869	4.64	0.0459

Y REGRESSED ON X1 X2 AND X3

. [Portion of output omitted]

Source	DF	Type I SS	Mean Square	F Value	Pr > F
X1	1	8.352481	8.352481	0.40	0.5378
X2	1	1309.445931	1309.445931	62.16	<.0001
X3	1	200.346530	200.346530	9.51	0.0071

Source	DF	Type III SS	Mean Square	F Value	Pr > F
X1	1	30.2855361	30.2855361	1.44	0.2480
X2	1	94.9129162	94.9129162	4.51	0.0497
X3	1	200.3465296	200.3465296	9.51	0.0071

e. Provide variables-added-last tests for X_1, X_2, and X_3.

f. Provide the variables-added-last test for $X_4 = X_2X_3$, given X_2 and X_3 already in the model. Does X_4 significantly improve the prediction of Y, given that X_2 and X_3 are already in the model?

8. The following ANOVA table is based on the data discussed in Problem 5 of Chapter 8. Use $\alpha = .05$.

Source		d.f.	SS
	(X_1)	1	18,953.04
Regression	$(X_3\|X_1)$	1	7,010.03
	$(X_2\|X_1, X_3)$	1	10.93
Residual		21	2,248.23
Total		24	28,222.23

a. Provide a test to compare the following two models:

$$Y = \beta_0 + \beta_1 X_1 + \beta_2 X_2 + \beta_3 X_3 + E$$

and

$$Y = \beta_0 + \beta_1 X_1 + E$$

b. Provide a test to compare the following two models:

$$Y = \beta_0 + \beta_1 X_1 + \beta_3 X_3 + E$$

and

$$Y = \beta_0 + E$$

c. State which two models are being compared in the computation

$$F = \frac{(18,953.04 + 7,010.03 + 10.93)/3}{2,248.23/21}$$

9. Residential real estate prices are thought to depend, in part, on property size and number of bedrooms. The house size X_1 (in hundreds of square feet), number of bedrooms X_2, and house price Y (in thousands of dollars) of a random sample of houses in a certain county were observed. The resulting data and some associated computer output were presented in Problem 10 of Chapter 8. Additional portions of the output are shown here. Use all of the output to answer the following questions.

a. Perform the overall F test for the regression of Y on both independent variables. Interpret your result.

b. Perform variables-added-in-order tests for both independent variables, with X_1 added first.

c. Perform variables-added-in-order tests for both independent variables, with X_2 added first.

d. Provide a table of variables-added-last tests.

e. Which predictors appear to be necessary? Why?

Edited SAS Output (PROC GLM) for Problem 9

```
                        Y REGRESSED ON X1 AND X2
                                   .
                                   . [Portion of output omitted]
                                   .
Source      DF        Type I SS       Mean Square     F Value      Pr > F
X1          1       5732.222826       5732.222826      40.06       0.0032
X2          1          1.098464          1.098464       0.01       0.9344

Source      DF      Type III SS       Mean Square     F Value      Pr > F
X1          1       1402.315337       1402.315337       9.80       0.0352
X2          1          1.098464          1.098464       0.01       0.9344
```

10. Data on sales revenue Y, television advertising expenditures X_1, and print media advertising expenditures X_2 for a large retailer for the period 1988–1993 were given in Problem 11 of Chapter 8. Use the computer output for that problem, along with the additional portions of the output shown here, to answer the following questions.

   a. Perform the overall F test for the regression of Y on both independent variables. Interpret your result.

   b. Perform variables-added-in-order tests for both independent variables, with X_1 added first.

   c. Perform variables-added-in-order tests for both independent variables, with X_2 added first.

   d. Provide a table of variables-added-last tests.

   e. Which predictors appear to be necessary? Why?

Edited SAS Output (PROC GLM) for Problem 10

```
                        Y REGRESSED ON X1 AND X2
                                   .
                                   . [Portion of output omitted]
                                   .
Source      DF        Type I SS       Mean Square     F Value      Pr > F
X1          1        28.07164068       28.07164068     117.82      0.0017
X2          1         0.04689121        0.04689121       0.20      0.6874

Source      DF      Type III SS       Mean Square     F Value      Pr > F
X1          1        25.89125916       25.89125916     108.66      0.0019
X2          1         0.04689121        0.04689121       0.20      0.6874
```

11. Use the computer output for the radial keratotomy data of Problem 12 in Chapter 8, along with the additional output here, to answer the following questions.

   a. Perform the overall F test for the regression of Y on both independent variables. Interpret your result.

   b. Perform variables-added-in-order tests for both independent variables, with X_1 added first.

   c. Perform variables-added-in-order tests for both independent variables, with X_2 added first.

   d. Provide a table of variables-added-last tests.

   e. Which predictors appear to be necessary? Why?

Edited SAS Output (PROC GLM) for Problem 11

```
                          Y REGRESSED ON X1 AND X2
                                        .
                                        . [Portion of output omitted]
                                        .
```

Source	DF	Type I SS	Mean Square	F Value	Pr > F
X1	1	13.09841552	13.09841552	10.67	0.0020
X2	1	4.52435640	4.52435640	3.68	0.0605

Source	DF	Type III SS	Mean Square	F Value	Pr > F
X1	1	12.43080596	12.43080596	10.12	0.0025
X2	1	4.52435640	4.52435640	3.68	0.0605

12. Use the computer output for the *Business Week* magazine data of Problem 13 in Chapter 8, as well as the additional output here, to answer the following questions.
    a. Perform the overall F test for the regression of Y on both independent variables, X_2 and X_3. Interpret your result.
    b. Perform variables-added-in-order tests for both independent variables, with X_2 added first.
    c. Perform variables-added-in-order tests for both independent variables, with X_3 added first.
    d. Provide a table of variables-added-last tests.
    e. Which predictors appear to be necessary? Why?

Edited SAS Output (PROC GLM) for Problem 12

```
                          Y REGRESSED ON X2 AND X3
                                        .
                                        . [Portion of output omitted]
                                        .
```

Source	DF	Type I SS	Mean Square	F Value	Pr > F
X2	1	1.76498949	1.76498949	1.40	0.2537
X3	1	24.81773186	24.81773186	19.63	0.0004

Source	DF	Type III SS	Mean Square	F Value	Pr > F
X2	1	7.97795163	7.97795163	6.31	0.0224
X3	1	24.81773186	24.81773186	19.63	0.0004

13. This problem refers to the 1990 Census data presented in Problem 19 of Chapter 5 and in Problem 14 of Chapter 8.

    Use the computer output from Problem 14 of Chapter 8, along with the additional output shown here, to answer the following questions about the regression of OWNEROCC on OWNCOST and URBAN.
    a. Perform the overall F test for the regression of OWNEROCC on OWNCOST and URBAN. Interpret your result.
    b. Perform variables-added-in-order tests for both independent variables, with OWNCOST added first.
    c. Perform variables-added-in-order tests for both independent variables, with URBAN added first.
    d. Provide a table of variables-added-last tests.
    e. Which predictors appear to be necessary? Why?

Edited SAS Output (PROC GLM) for Problem 13

```
                    OWNEROCC REGRESSED ON OWNCOST AND URBAN
                                       .
                                     . [Portion of output omitted]
                                       .
Source          DF        Type I SS      Mean Square     F Value    Pr > F
OWNCOST          1       132.6203242     132.6203242       8.84     0.0068
URBAN            1       123.1581842     123.1581842       8.21     0.0088

Source          DF       Type III SS     Mean Square     F Value    Pr > F
OWNCOST          1        66.3318457      66.3318457       4.42     0.0467
URBAN            1       123.1581842     123.1581842       8.21     0.0088
```

14. This problem refers to the pond ecology data of Chapter 5, Problem 20.
   a. Perform the overall F test for the multiple regression of copepod count on zooplankton and phytoplankton counts.
   b. Perform variables-added-in-order tests for both independent variables, with zooplankton count added first.
   c. Perform variables-added-last tests for the regression in (a).
   d. Which variables appear to be necessary? Why?

References

Abt, K. 1981. "Problems of Repeated Significance Testing." *Controlled Clinical Trials* 1: 377–81.

Kupper, L. L.; Stewart, J. R.; and Williams, K. A. 1976. "A Note on Controlling Significance Levels in Stepwise Regression." *American Journal of Epidemiology* 103(1): 13–15.

Searle, S. R. 1971. *Linear Models.* New York: John Wiley & Sons.

Correlations: Multiple, Partial, and Multiple Partial

10.1 Preview

We saw in Chapter 5 that the following essential features of straight-line regression (excluding the quantitative prediction formula provided by the fitted regression equation) can also be described in terms of the correlation coefficient r. These features are summarized as follows:

1. The squared correlation coefficient r^2 measures the strength of the linear relationship between the dependent variable Y and the independent variable X. The closer r^2 is to 1, the stronger the linear relationship; the closer r^2 is to 0, the weaker the linear relationship.

2. $r^2 = (\text{SSY} - \text{SSE})/\text{SSY}$ is the proportionate reduction in the total sum of squares achieved by using a straight-line model in X to predict Y.

3. $r = \hat{\beta}_1(S_X/S_Y)$, where $\hat{\beta}_1$ is the estimated slope of the regression line.

4. r (or r_{XY}) is an estimate of the population parameter ρ (or ρ_{XY}), which describes the correlation between X and Y, both considered as random variables.

5. Assuming that X and Y have a bivariate normal distribution with parameters μ_X, μ_Y, σ_X^2, σ_Y^2, and ρ_{XY}, the conditional distribution of Y given X is $N\left(\mu_{Y|X}, \sigma_{Y|X}^2\right)$, where

$$\mu_{Y|X} = \mu_Y + \rho \frac{\sigma_Y}{\sigma_X}(X - \mu_X) \quad \text{and} \quad \sigma_{Y|X}^2 = \sigma_Y^2(1 - \rho^2)$$

Here, r^2 estimates ρ^2, which can be expressed as

$$\rho^2 = \frac{\sigma_Y^2 - \sigma_{Y|X}^2}{\sigma_Y^2}$$

6. r can be used as a general index of linear association between two random variables, in the following sense:

a. The more highly positive r is, the more positive is the linear association; that is, an individual with a high value of one variable will likely have a high value of the other, and an individual with a low value of one variable will probably have a low value of the other.

b. The more highly negative r is, the more negative is the linear association; that is, an individual with a high value of one variable will likely have a low value of the other, and conversely.

c. If r is close to 0, there is little evidence of linear association, which indicates that there is a nonlinear association or no association at all.

This connection between regression and correlation can be extended to the multiple regression case, as we will discuss in this chapter. When several independent variables are involved, however, the essential features of regression are described not by a single correlation coefficient, as in the straight-line case, but by several correlations. These include a set of zero-order correlations such as r, plus a whole group of higher-order indices called *multiple correlations, partial correlations,* and *multiple partial correlations.*[1] These higher-order correlations allow us to answer many of the same questions that can be answered by fitting a multiple regression model. In addition, the correlation analog has been found to be particularly useful in uncovering spurious relationships among variables, identifying intervening variables, and making certain types of causal inferences.[2]

10.2 Correlation Matrix

When dealing with more than one independent variable, we can represent the collection of all zero-order correlation coefficients (i.e., the r's between all possible pairs of variables) most compactly in *correlation matrix form*. For example, given $k = 3$ independent variables X_1, X_2, and X_3, and one dependent variable Y, there are $C_2^4 = 6$ zero-order correlations, and the correlation matrix has the general form

$$
\begin{array}{c c}
 & \begin{array}{cccc} Y & X_1 & X_2 & X_3 \end{array} \\
\begin{array}{c} Y \\ X_1 \\ X_2 \\ X_3 \end{array} &
\left[\begin{array}{cccc}
1 & r_{Y1} & r_{Y2} & r_{Y3} \\
 & 1 & r_{12} & r_{13} \\
 & & 1 & r_{23} \\
 & & & 1
\end{array} \right]
\end{array}
$$

Here, r_{Yj} ($j = 1, 2, 3$) is the correlation between Y and X_j, and r_{ij} ($i, j = 1, 2, 3$) is the correlation between X_i and X_j.

---

[1] The *order* of a correlation coefficient, as the term is used here, is the number of variables being controlled or adjusted for (see Section 10.5).

[2] See Blalock (1971) and Bollen (1989) for discussions of techniques for causal inference using regression modeling.

For the data in Table 8.1, this matrix takes the form

$$
\begin{array}{c} \\ \text{WGT} \\ \text{HGT} \\ \text{AGE} \\ (\text{AGE})^2 \end{array}
\begin{array}{cccc}
\text{WGT} & \text{HGT} & \text{AGE} & (\text{AGE})^2 \\
\left[\begin{array}{cccc}
1 & .814 & .770 & .767 \\
 & 1 & .614 & .615 \\
 & & 1 & .994 \\
 & & & 1
\end{array}\right]
\end{array}
$$

Taken separately, each of these correlations describes the strength of the linear relationship between the two variables involved. In particular, the correlations $r_{Y1} = .814$, $r_{Y2} = .770$, and $r_{Y3} = .767$ measure the strength of the linear association with the dependent variable WGT for each of the independent variables taken separately. As we can see, HGT ($r_{Y1} = .814$) is the independent variable with the strongest linear relationship to WGT, followed by AGE and then $(\text{AGE})^2$.

Nevertheless, these zero-order correlations do not describe (1) the overall relationship of the dependent variable WGT to the independent variables HGT, AGE, and $(\text{AGE})^2$ considered together; or (2) the relationship between WGT and AGE after controlling for[3] HGT; or (3) the relationship between WGT and the combined effects of AGE and $(\text{AGE})^2$ after controlling for HGT. The measure that describes relationship (1) is called the *multiple correlation coefficient* of WGT on HGT, AGE, and $(\text{AGE})^2$. The measure that describes relationship (2) is called the *partial correlation coefficient* between WGT and AGE controlling for HGT. Finally, the measure that describes relationship (3) is called the *multiple partial correlation coefficient* between WGT and the combined effects of AGE and $(\text{AGE})^2$ controlling for HGT.

Even though AGE and $(\text{AGE})^2$ are very highly correlated in our example, it is possible—if the general relationship of AGE to WGT is nonlinear—that $(\text{AGE})^2$ will be significantly correlated with WGT even after AGE has been controlled for. In fact, this is what happens in general when the addition of a second-order term in polynomial regression significantly improves the prediction of the dependent variable.

10.3 Multiple Correlation Coefficient

The *multiple correlation coefficient*, denoted as $R_{Y|X_1, X_2, \ldots, X_k}$, is a measure of the overall *linear association* of one (dependent) variable Y with several other (independent) variables $X_1, X_2, \ldots, X_k$. By "linear association" we mean that $R_{Y|X_1, X_2, \ldots, X_k}$ measures the strength of the association between Y and the best-fitting linear combination of the X's, which is the least-squares solution $\hat{Y} = \hat{\beta}_0 + \hat{\beta}_1 X_1 + \hat{\beta}_2 X_2 + \cdots + \hat{\beta}_k X_k$. In fact, no other linear combination of the X's will have as great a correlation with Y. Also, $R_{Y|X_1, X_2, \ldots, X_k}$ is always nonnegative.

The multiple correlation coefficient is thus a direct generalization of the simple correlation coefficient r to the case of several independent variables. We have dealt with this measure up to now under the name R^2, which is the square of the multiple correlation coefficient.

---

[3] In this case the phrase "controlling for" pertains to determining the extent to which the variables WGT and AGE are related after removing the effect of HGT on WGT and AGE.

Two computational formulas provide useful interpretations of the multiple correlation coefficient $R_{Y|X_1, X_2, \ldots, X_k}$ and its square:

$$R_{Y|X_1, X_2, \ldots, X_k} = \frac{\sum_{i=1}^{n}(Y_i - \bar{Y})(\hat{Y}_i - \bar{\hat{Y}})}{\sqrt{\sum_{i=1}^{n}(Y_i - \bar{Y})^2 \sum_{i=1}^{n}(\hat{Y}_i - \bar{\hat{Y}})^2}}$$

and   (10.1)

$$R_{Y|X_1, X_2, \ldots, X_k}^2 = \frac{\sum_{i=1}^{n}(Y_i - \bar{Y})^2 - \sum_{i=1}^{n}(Y_i - \hat{Y}_i)^2}{\sum_{i=1}^{n}(Y_i - \bar{Y})^2} = \frac{\text{SSY} - \text{SSE}}{\text{SSY}}$$

where $\hat{Y}_i = \hat{\beta}_0 + \hat{\beta}_1 X_{i1} + \hat{\beta}_2 X_{i2} + \cdots + \hat{\beta}_k X_{ik}$ (the predicted value for the ith individual) and $\bar{\hat{Y}} = \sum_{i=1}^{n}\hat{Y}_i/n$. The second formula in (10.1), which we have seen several times before (as R^2), is most useful for assessing the fit of the regression model. The first formula in (10.1) indicates that $R_{Y|X_1, X_2, \ldots, X_k} = r_{Y, \hat{y}}$, the simple linear correlation between the observed values Y and the predicted values $\hat{Y}$.

As a numerical example, let us again consider the data of Table 8.1, where $X_1 = $ HGT, $X_2 = $ AGE, and $Y = $ WGT. Using only X_1 and X_2 in the model, the fitted regression equation is $\hat{Y} = 6.553 + 0.722X_1 + 2.050X_2$, and the observed and predicted values are as given in Table 10.1.

One can check that $\bar{Y} = \bar{\hat{Y}} = 62.75$. As we mentioned in Chapter 8, this is no coincidence: it is a mathematical fact that $\bar{Y} = \bar{\hat{Y}}$.

The following SAS output shows that the computed value of R^2 is 0.7800 for the model containing X_1 and X_2. This tells us that 78% of the variation in Y is explained by the regression model. The corresponding multiple correlation coefficient $R_{Y|X_1, X_2} = R$ is 0.8832 since R is defined as the *positive* square root of R^2.

TABLE 10.1 **Observed and predicted values for the regression of WGT on HGT and AGE**

Child	1	2	3	4	5	6
observed	64	71	53	67	55	58
predicted	64.11	69.65	54.23	73.87	59.78	57.01

Child	7	8	9	10	11	12
observed	77	57	56	51	76	68
predicted	66.77	59.66	57.38	49.18	75.20	66.16

Edited SAS Output for Regression of WGT on HGT and AGE

```
Dependent Variable: WGT

                              Analysis of Variance

                          Sum of          Mean
Source              DF    Squares        Square      F Value      Pr > F
Model                2   692.82261     346.41130      15.953      0.0011
Error                9   195.42739      21.71415
Corrected Total     11   888.25000
              Root MSE        4.65984         R-square        0.7800
              Dep Mean       62.75000         Adj R-sq        0.7311
              Coeff Var       7.42605
```

10.4 Relationship of $R_{Y|X_1, X_2, \ldots, X_k}$ to the Multivariate Normal Distribution[4]

An informative way of looking at the sample multiple correlation coefficient $R_{Y|X_1, X_2, \ldots, X_k}$ is to consider it as an estimator of a population parameter characterizing the joint distribution of all the variables $Y, X_1, X_2, \ldots, X_k$ taken together. When we had two variables X and Y and assumed that their joint distribution was bivariate normal $N_2(\mu_Y, \mu_X, \sigma_Y^2, \sigma_X^2, \rho_{XY})$, we saw that r_{XY} estimated ρ_{XY}, which satisfied the formula $\rho_{XY}^2 = (\sigma_Y^2 - \sigma_{Y|X}^2)/\sigma_Y^2$, where $\sigma_{Y|X}^2$ was the variance of the conditional distribution of Y given X. Now, when we have k independent variables and one dependent variable, we get an analogous result if we assume that their joint distribution is *multivariate normal*. Let us now consider what happens with just two independent variables. In this case the *trivariate normal distribution* of Y, X_1, and X_2 can be described as

$$N_3(\mu_Y, \mu_{X_1}, \mu_{X_2}, \sigma_Y^2, \sigma_{X_1}^2, \sigma_{X_2}^2, \rho_{Y1}, \rho_{Y2}, \rho_{12})$$

where μ_Y, μ_{X_1}, and μ_{X_2} are the three (unconditional) means; $\sigma_Y^2, \sigma_{X_1}^2$, and $\sigma_{X_2}^2$ are the three (unconditional) variances; and ρ_{Y1}, ρ_{Y2}, and ρ_{12} are the three correlation coefficients. The *conditional distribution of Y given X_1 and X_2* is then a univariate normal distribution with a (conditional) mean denoted by $\mu_{Y|X_1, X_2}$ and a (conditional) variance denoted by $\sigma_{Y|X_1, X_2}^2$; we usually write this compactly as

$$Y|X_1, X_2 \frown N(\mu_{Y|X_1, X_2}, \sigma_{Y|X_1, X_2}^2)$$

In fact, it turns out that

$$\mu_{Y|X_1, X_2} = \mu_Y + \rho_{Y1|2}\frac{\sigma_{Y|X_2}}{\sigma_{X_1|X_2}}(X_1 - \mu_{X_1}) + \rho_{Y2|1}\frac{\sigma_{Y|X_1}}{\sigma_{X_2|X_1}}(X_2 - \mu_{X_2})$$

and

$$\sigma_{Y|X_1, X_2}^2 = (1 - \rho_{Y, \mu_{Y|X_1, X_2}}^2)\sigma_Y^2$$

where $\rho_{Y, \mu_{Y|X_1, X_2}}$ is the population correlation coefficient between the random variables Y and $\mu_{Y|X_1, X_2} = \beta_0 + \beta_1 X_1 + \beta_2 X_2$ (where we are considering X_1 and X_2 as random variables) and where $\rho_{Y1|2}$ and $\rho_{Y2|1}$ are partial correlations (to be discussed in Section 10.5). Also, $\sigma_{Y|X_2}^2, \sigma_{Y|X_1}^2,$

---

[4] This section is not essential for the application-oriented reader.

$\sigma^2_{X_1|X_2}$, and $\sigma^2_{X_2|X_1}$ are the conditional variances, respectively, of Y given X_2, Y given X_1, X_1 given X_2, and X_2 given X_1.

The parameter $\rho_{Y,\,\mu_{Y|X_1,X_2}}$ is the *population analog of the sample multiple correlation coefficient* $R_{Y|X_1,X_2}$, and we write $\rho_{Y,\,\mu_{Y|X_1,X_2}}$ simply as $\rho_{Y|X_1,X_2}$. Furthermore, from the formula for $\sigma^2_{Y|X_1,X_2}$, we can confirm (with a little algebra) that

$$\rho^2_{Y|X_1,X_2} = \frac{\sigma^2_Y - \sigma^2_{Y|X_1,X_2}}{\sigma^2_Y}$$

which is the *proportionate reduction in the unconditional variance of Y due to conditioning on X_1 and X_2.*

Generalizing these findings to the case of k independent variables, we may summarize the characteristics of the multiple correlation coefficient $R_{Y|X_1,X_2,\ldots,X_k}$ as follows:

1. $R^2_{Y|X_1,X_2,\ldots,X_k}$ measures the proportionate reduction in the total sum of squares $\sum_{i=1}^{n}(Y_i - \bar{Y})^2$ to $\sum_{i=1}^{n}(Y_i - \hat{Y}_i)^2$ due to the multiple linear regression of Y on $X_1, X_2, \ldots, X_k$.

2. $R_{Y|X_1,X_2,\ldots,X_k}$ is the correlation $r_{Y\hat{Y}}$ of the observed values (Y) with the predicted values ($\hat{Y}$), and this correlation is always nonnegative.

3. $R_{Y|X_1,X_2,\ldots,X_k}$ is an estimate of $\rho_{Y|X_1,X_2,\ldots,X_k}$, the correlation of Y with the true regression equation $\beta_0 + \beta_1 X_1 + \beta_2 X_2 + \cdots + \beta_k X_k$, where the X's are considered to be random.

4. $R^2_{Y|X_1,X_2,\ldots,X_k}$ is an estimate of the proportionate reduction in the unconditional variance of Y due to conditioning on $X_1, X_2, \ldots, X_k$, that is, it estimates

$$\rho^2_{Y|X_1,X_2,\ldots,X_k} = \frac{\sigma^2_Y - \sigma^2_{Y|X_1,X_2,\ldots,X_k}}{\sigma^2_Y}$$

10.5 Partial Correlation Coefficient

The *partial correlation coefficient* is a measure of the strength of the linear relationship between two variables after we control for the effects of other variables. If the two variables of interest are Y and X, and the control variables are $Z_1, Z_2, \ldots, Z_p$, then we denote the corresponding partial correlation coefficient by $r_{YX|Z_1,Z_2,\ldots,Z_p}$. The order of the partial correlation depends on the number of variables that are being controlled for. Thus, *first-order* partials have the form $r_{YX|Z}$, *second-order* partials have the form $r_{YX|Z_1,Z_2}$, and in general, *p*th-order partials have the form $r_{YX|Z_1,Z_2,\ldots,Z_p}$.

For the three independent variables HGT, AGE, and (AGE)^2 in our example, the highest-order partial possible is second order. The values of most of the partial correlations that can be computed from this data set are given in Table 10.2.

The easiest way to obtain a partial correlation is to use a standard computer program. A computing formula that helps highlight the structure of the partial correlation coefficient will be given a little later. First, however, let us see how we can use the information in Table 10.2 to describe our data.

Looking back at our (zero-order) correlation matrix, we see that the variable most highly correlated with WGT is HGT ($r_{Y1} = .814$). Thus, of the three independent variables we

10.5.5 Semipartial Correlations

An alternative form of partial correlation is sometimes considered. The partial correlation $r_{YX|Z}$ was just characterized as a correlation between Y adjusted for Z and X adjusted for Z. Some statisticians refer to this as a *full partial,* since both variables being correlated have been adjusted for Z.

The *semipartial correlation* (or "part" correlation) may be characterized as the correlation between two variables when only one of the two has been adjusted for a third variable. For example, one may consider the semipartial correlation between Y and X with only X adjusted for Z or with only Y adjusted for Z. The first will be denoted by $r_{Y(X|Z)}$, and the second by $r_{X(Y|Z)}$. Thus, we have $r_{Y(X|Z)} = r_{Y,X-\hat{X}}$ and $r_{X(Y|Z)} = r_{Y-\hat{Y},X}$, where $\hat{X}$ and $\hat{Y}$ are obtained from straight-line regressions on Z.

Another way of describing these semipartials is in terms of zero-order correlations, as follows:

$$r_{Y(X|Z)} = \frac{r_{YX} - r_{YZ}r_{XZ}}{\sqrt{1 - r_{XZ}^2}} \tag{10.5}$$

and

$$r_{X(Y|Z)} = \frac{r_{YX} - r_{YZ}r_{XZ}}{\sqrt{1 - r_{YZ}^2}} \tag{10.6}$$

It is instructive to compare these formulas with formula (10.3) for the full partial. The numerator is the same in all three expressions: the partial covariance between Y and X adjusted for Z (with all three variables standardized to have variance 1). Clearly, then, these correlations all have the same sign; and if any one equals 0, they all do. For significance testing, it is appropriate to use the extra-sum-of-squares test described earlier.

These three correlations have different interpretations. They each describe the relationship between Y and X, but with adjustment for different quantities. The semipartial $r_{Y(X|Z)}$ is the correlation between Y and X with X adjusted for Z; the semipartial $r_{X(Y|Z)}$ is the correlation between Y and X with Y adjusted for Z. Finally, the full partial $r_{YX|Z}$ is the correlation between Y and X with both Y and X adjusted for Z.

Choosing the proper correlation coefficient depends on the relationship among the three variables X, Y, and Z (the nuisance variable). Table 10.3 shows the four possible types of relationships. Case 1 involves assessing the relationship between X and Y without a nuisance variable present; here, the simple correlation r_{XY} should be used. Case 2 illustrates the situation in which the nuisance variable Z is related to both X and Y, so that the use of $r_{YX|Z}$ is appropriate. In cases 3 and 4, the nuisance variable affects just one of the two variables X and Y. In these cases, semipartial correlations permit just one of two primary variables to be adjusted for the effects of a nuisance variable.

10.5.6 Summary of the Features of the Partial Correlation Coefficient

1. The partial correlation $r_{YX|Z_1, Z_2, \ldots, Z_p}$ measures the strength of the linear relationship between two variables X and Y while controlling for variables $Z_1, Z_2, \ldots, Z_p$.

TABLE 10.3 **Possible relationships among variables X and Y and nuisance variable Z**

Case	Nuisance Relationship	Diagram	Preferred Correlation
1	Neither X nor Y affected by Z	$X \leftrightarrow Y$	r_{XY}
2	Both X and Y affected by Z	$X \leftrightarrow Y$ Z	$r_{YX\|Z}$
3	Only X affected by Z	$X \leftrightarrow Y$ Z	$r_{Y(X\|Z)}$
4	Only Y affected by Z	$X \leftrightarrow Y$ Z	$r_{X(Y\|Z)}$

2. The square of the partial correlation $r_{YX|Z_1, Z_2, \ldots, Z_p}$ measures the proportion of the residual sum of squares that is accounted for by the addition of X to a regression model already involving $Z_1, Z_2, \ldots, Z_p$; that is,

$$r^2_{YX|Z_1, Z_2, \ldots, Z_p} = \frac{\left[\begin{array}{c} \text{Extra sum of squares due to adding} \\ X \text{ to a model already containing } Z_1, Z_2, \ldots, Z_p \end{array}\right]}{\text{Residual SS (using only } Z_1, Z_2, \ldots, Z_p \text{ in the model)}}$$

3. The partial correlation coefficient $r_{YX|Z_1, Z_2, \ldots, Z_p}$ is an estimate of the population parameter $\rho_{YX|Z_1, Z_2, \ldots, Z_p}$, which is the correlation between Y and X in the conditional joint distribution of Y and X given $Z_1, Z_2, \ldots, Z_p$. The square of this population partial correlation coefficient is given by the equivalent formula

$$\rho^2_{YX|Z_1, Z_2, \ldots, Z_p} = \frac{\sigma^2_{Y|Z_1, Z_2, \ldots, Z_p} - \sigma^2_{Y|X, Z_1, Z_2, \ldots, Z_p}}{\sigma^2_{Y|Z_1, Z_2, \ldots, Z_p}}$$

where $\sigma^2_{Y|Z_1, Z_2, \ldots, Z_p}$ is the variance of the conditional distribution of Y given $Z_1, Z_2, \ldots, Z_p$ (and where $\sigma^2_{Y|X, Z_1, Z_2, \ldots, Z_p}$ is similarly defined).

4. The partial F statistic $F(X|Z_1, Z_2, \ldots, Z_p)$ is used to test $H_0: \rho_{YX|Z_1, Z_2, \ldots, Z_p} = 0$.

5. The (first-order) partial correlation coefficient $r_{YX|Z}$ is an adjustment of the (zero-order) correlation r_{YX} that takes into account the effect of the control variable Z. This can be seen from the formula

$$r_{YX|Z} = \frac{r_{YX} - r_{YZ} r_{XZ}}{\sqrt{(1 - r^2_{YZ})(1 - r^2_{XZ})}}$$

Higher-order partial correlations are computed by reapplying this formula, using the next-lower-order partials.

6. The partial correlation $r_{YX|Z}$ can be defined as the correlation of the residuals of the straight-line regressions of Y on Z and of X on Z; that is, $r_{YX|Z} = r_{Y - \hat{Y}, X - \hat{X}}$.

10.6 Alternative Representation of the Regression Model

With the correlation analog to multiple regression, we can express the regression model $\mu_{Y|X_1, X_2, \ldots, X_k} = \beta_0 + \beta_1 X_1 + \beta_2 X_2 + \cdots + \beta_k X_k$ in terms of partial correlation coefficients and conditional variances. When $k = 3$, this representation takes the form

$$\mu_{Y|X_1, X_2, X_3} = \mu_Y + \rho_{Y1|23}\frac{\sigma_{Y|23}}{\sigma_{1|23}}(X_1 - \mu_{X_1}) \qquad (10.7)$$
$$+ \rho_{Y2|13}\frac{\sigma_{Y|13}}{\sigma_{2|13}}(X_2 - \mu_{X_2})$$
$$+ \rho_{Y3|12}\frac{\sigma_{Y|12}}{\sigma_{3|12}}(X_3 - \mu_{X_3})$$

where, for example, $\rho_{Y1|23} = \rho_{YX_1|X_2, X_3}$, and where we define

$$\beta_1 = \rho_{Y1|23}\frac{\sigma_{Y|23}}{\sigma_{1|23}}, \quad \beta_2 = \rho_{Y2|13}\frac{\sigma_{Y|13}}{\sigma_{2|13}}, \quad \beta_3 = \rho_{Y3|12}\frac{\sigma_{Y|12}}{\sigma_{3|12}}$$

Notice the similarity between this representation and the one for the straight-line case, where β_1 is equal to $\rho(\sigma_Y/\sigma_X)$. Also, here

$$\beta_0 = \mu_Y - \beta_1\mu_{X_1} - \beta_2\mu_{X_2} - \beta_3\mu_{X_3}$$

An equivalent method to that of least squares for estimating the coefficients $\beta_0, \beta_1, \beta_2,$ and β_3 is to substitute for the population parameters in the preceding formulas the corresponding estimates

$$\hat{\mu}_Y = \overline{Y}, \quad \hat{\mu}_{X_1} = \overline{X}_1, \quad \hat{\mu}_{X_2} = \overline{X}_2, \quad \hat{\mu}_{X_3} = \overline{X}_3$$

$$\hat{\beta}_1 = r_{Y1|23}\frac{S_{Y|23}}{S_{1|23}}, \quad \hat{\beta}_2 = r_{Y2|13}\frac{S_{Y|13}}{S_{2|13}}, \quad \hat{\beta}_3 = r_{Y3|12}\frac{S_{Y|12}}{S_{3|12}}$$

10.7 Multiple Partial Correlation

10.7.1 The Coefficient and Its Associated *F* Test

The *multiple partial correlation coefficient* is used to describe the overall relationship between a dependent variable and two or more independent variables while controlling for still other variables. For example, suppose that we consider the variable $X_1^2 = (\text{HGT})^2$ and the product term $X_1 X_2 = \text{HGT} \times \text{AGE}$, in addition to the independent variables $X_1 = \text{HGT}, X_2 = \text{AGE},$ and $X_2^2 = (\text{AGE})^2$. Our complete regression model is then of the form

$$Y = \beta_0 + \beta_1 X_1 + \beta_2 X_2 + \beta_{11} X_1^2 + \beta_{22} X_2^2 + \beta_{12} X_1 X_2 + E$$

We call such a model a *complete second-order model,* since it includes all possible variables up through second-order terms. For such a complete model, we frequently want to know whether any of the second-order terms are important—in other words, whether a first-order model involving only X_1 and X_2 (i.e., $Y = \beta_0 + \beta_1 X_1 + \beta_2 X_2 + E$) is adequate. There are two equivalent ways to represent this question as a hypothesis-testing problem: one is to test $H_0: \beta_{11} = \beta_{22} = \beta_{12} = 0$ (i.e., all second-order coefficients are zero); the other is to test the hypothesis H_0: $\rho_{Y(X_1^2, X_2^2, X_1X_2)|X_1, X_2} = 0$, where $\rho_{Y(X_1^2, X_2^2, X_1X_2)|X_1, X_2}$ is the population multiple partial correlation of Y with the second-order variables, controlling for the effects of the first-order variables. (In general, we write the multiple partial as $\rho_{Y(X_1, X_2, \ldots, X_k)|Z_1, Z_2, \ldots, Z_p}$.) This parameter is estimated by the sample multiple partial correlation $r_{Y(X_1^2, X_2^2, X_1X_2)|X_1, X_2}$, which describes the overall multiple contribution of adding the second-order terms to the model after the effects of the first-order terms are partialed out or controlled for (hence the term *multiple partial*). Two equivalent formulas for $r_{Y(X_1^2, X_2^2, X_1X_2)|X_1, X_2}^2$ are:

$$r_{Y(X_1^2, X_2^2, X_1X_2)|X_1, X_2}^2 \tag{10.8}$$

$$= \frac{\begin{bmatrix} \text{Residual SS (only } X_1 \text{ and } X_2 \text{ in the model)} \\ - \text{ Residual SS (all first- and second-order terms in the model)} \end{bmatrix}}{\text{Residual SS (only } X_1 \text{ and } X_2 \text{ in the model)}}$$

$$= \frac{\begin{bmatrix} \text{Extra sum of squares due to the addition of the second-order terms } X_1^2, X_2^2, \text{ and} \\ X_1X_2 \text{ to a model containing only the first-order terms } X_1 \text{ and } X_2 \end{bmatrix}}{\text{Residual SS (only } X_1 \text{ and } X_2 \text{ in the model)}}$$

and

$$r_{Y(X_1^2, X_2^2, X_1X_2)|X_1, X_2}^2 = \frac{R_{Y|X_1, X_2, X_1^2, X_2^2, X_1X_2}^2 - R_{Y|X_1, X_2}^2}{1 - R_{Y|X_1, X_2}^2}$$

In most applications, estimating a multiple partial correlation is rarely of interest, so the preceding formulas are infrequently used. Nevertheless, there often is interest in testing hypotheses about a collection (or "chunk") of higher-order terms, and such tests involve the multiple partial correlation. For instance, to test either H_0: $\rho_{Y(X_1^2, X_2^2, X_1X_2)|X_1, X_2} = 0$ (or equivalently, H_0: $\beta_{11} = \beta_{22} = \beta_{12} = 0$), we calculate the multiple partial F statistic given in general form by expression (9.6) in Chapter 9. For this example, the formula becomes

$$F(X_1^2, X_2^2, X_1X_2|X_1, X_2)$$

$$= \frac{[\text{Regression SS}(X_1, X_2, X_1^2, X_2^2, X_1X_2) - \text{Regression SS}(X_1, X_2)]/3}{\text{Residual MS}(X_1, X_2, X_1^2, X_2^2, X_1X_2)}$$

$$= \frac{[\text{Residual SS}(X_1, X_2) - \text{Residual SS}(X_1, X_2, X_1^2, X_2^2, X_1X_2)]/3}{\text{Residual MS}(X_1, X_2, X_1^2, X_2^2, X_1X_2)}$$

We would reject H_0 at the α level of significance if the calculated value of $F(X_1^2, X_2^2, X_1X_2|X_1, X_2)$ exceeded the critical value $F_{3, n-6, 1-\alpha}$.

In general, the null hypothesis H_0: $\rho_{Y(X_1, X_2, \ldots, X_k)|Z_1, Z_2, \ldots, Z_p} = 0$ is equivalent to the hypothesis H_0: $\beta_1^* = \beta_2^* = \cdots = \beta_k^* = 0$ in the model $Y = \beta_0 + \beta_1 Z_1 + \beta_2 Z_2 + \cdots + \beta_p Z_p + \beta_1^* X_1 + \beta_2^* X_2 + \cdots + \beta_k^* X_k + E$. The appropriate test statistic is the multiple partial F given by

$$F(X_1, X_2, \ldots, X_k|Z_1, Z_2, \ldots, Z_p)$$

which has the F distribution with k and $n - p - k - 1$ degrees of freedom under H_0.

The general F statistic given by (9.6) may be expressed in terms of squared multiple correlations (R^2 terms) involving the two models being compared. The general form for this alternative expression is

$$F = \frac{[R^2(\text{large model}) - R^2 (\text{smaller model})]/[\text{regression df (larger model)} - \text{regression df (smaller model)}]}{[1 - R^2(\text{larger model})]/\text{residual df (larger model)}} \qquad (10.9)$$

10.8 Concluding Remarks

As we have seen throughout this chapter, a regression F test is associated with many equivalent null hypotheses. As an example, the following null hypotheses all make the same statement but in different forms:

1. H_0: "adding variables to the smaller model to form the larger model does not significantly improve the prediction of Y"

2. H_0: "the population regression coefficients for the variables in the larger model but not in the smaller model are all equal to 0"

3. H_0: "the population multiple-partial correlation between Y and variables added to produce the larger model, controlling for the variables in the smaller model, is 0"

4. H_0: "the value of the population squared multiple correlation coefficient for the larger model is not greater than the value of that parameter for the smaller model"

The first two null hypotheses involve prediction, while the latter two involve association. The investigator, for interpretive purposes, can choose either group depending on whether his or her focus is on the predictive ability of the regression model or on a particular association of interest.

Problems _____

1. The correlation matrix obtained for the variables SBP (Y), AGE (X_1), SMK (X_2), and QUET (X_3), using the data from Problem 2 in Chapter 5, is given by

	SBP	AGE	SMK	QUET
SBP	1	.7752	.2473	.7420
AGE	.7752	1	−.1395	.8028
SMK	.2473	−.1395	1	−.0714
QUET	.7420	.8028	−.0714	1

a. Based on this matrix, which of the independent variables AGE, SMK, and QUET explains the largest proportion of the total variation in the dependent variable SBP?

b. Using the available computer outputs, determine the partial correlations $r_{SBP, SMK|AGE}$ and $r_{SBP, QUET|AGE}$.

c. Test for the significance of $r_{SBP, SMK|AGE}$, using the ANOVA results given in Problem 1 of Chapter 8. Express the appropriate null hypothesis in terms of a population partial correlation coefficient.

d. Determine the second-order partial $r_{SBP, QUET|AGE, SMK}$, and test for the significance of this partial correlation (again, using the computer output here and in Chapter 8, Problem 1).

e. Based on the results you obtained in parts (a) through (d), how would you rank the independent variables in terms of their importance in predicting Y? Which of these variables are relatively unimportant?

f. Compute the squared multiple partial correlation $r^2_{SBP(QUET, SMK)|AGE}$ using the output here and in Problem 1 of Chapter 8. Test for the significance of this correlation. Does this test result alter your decision in part (e) about which variables to include in the regression model?

Edited SAS Output (PROC REG) for Problem 1

```
                        SBP  Regressed  on  AGE  and  SMK

                                  .
                                  .  [Portion of output omitted]
                                  .

                          Parameter  Estimates
```

Variable	DF	Parameter Estimate	Standard Error	t Value	Pr > |t|
INTERCEPT	1	48.049603	11.12955962	4.317	0.0002
AGE	1	1.709160	0.20175872	8.471	0.0001
SMK	1	10.294392	2.76810685	3.719	0.0009

Variable	DF	Squared Partial Corr Type I	Squared Partial Corr Type II
INTERCEPT	1	.	.
AGE	1	0.60094136	0.71219596
SMK	1	0.32291131	0.32291131

```
                        SBP  Regressed  on  AGE  and  QUET

                                  .
                                  .  [Portion of output omitted]
                                  .

                          Parameter  Estimates
```

Variable	DF	Parameter Estimate	Standard Error	t value	Pr > |t|
INTERCEPT	1	55.323436	12.53474644	4.414	0.0001
AGE	1	1.045157	0.38605667	2.707	0.0113
QUET	1	9.750732	5.40245598	1.805	0.0815

Variable	DF	Squared Partial Corr Type I	Squared Partial Corr Type II
INTERCEPT	1	.	.
AGE	1	0.60094136	0.20174580
QUET	1	0.10098584	0.10098584

(continued)

```
            SBP  Regressed  or  AGE,  SMK  and  QUET  (continued)

                                       .
                                       . [Portion of output omitted]
                                       .

                              Parameter  Estimates

                          Parameter              Standard
Variable          DF        Estimate                 Error      t Value      Pr > |t|
INTERCEPT          1       45.103192             10.76487511       4.190        0.0003
AGE                1        1.212715              0.32381922       3.745        0.0008
SMK                1        9.945568              2.65605655       3.744        0.0008
QUET               1        8.592449              4.49868122       1.910        0.0664

                              Squared               Squared
                              Partial               Partial
Variable          DF        Corr  Type  I         Corr  Type  II
INTERCEPT          1            .
AGE                1        0.60094136            0.33373458
SMK                1        0.32291131            0.33366934
QUET               1        0.11526997            0.11526997
```

2. An (equivalent) alternative to performing a partial F test for the significance of adding a new variable to a model while controlling for variables already in the model is to perform a t test using the appropriate partial correlation coefficient. If the dependent variable is Y, the independent variable of interest is X, and the controlling variables are $Z_1, Z_2, \ldots, Z_p$, then the t test for H_0: $\rho_{YX|Z_1, Z_2, \ldots, Z_p} = 0$ versus H_A: $\rho_{YX|Z_1, Z_2, \ldots, Z_p} \neq 0$ is given by the test statistic

$$T = r_{YX|Z_1, Z_2, \ldots, Z_p} \frac{\sqrt{n - p - 2}}{\sqrt{1 - r_{YX|Z_1, Z_2, \ldots, Z_p}^2}}$$

which has a t distribution under H_0 with $n - p - 2$ degrees of freedom. The critical region for this test is therefore given by

$$|T| \geq t_{n-p-2, \, 1-\alpha/2}$$

a. In a study of the relationship between water hardness and sudden death rates in $n = 88$ North Carolina counties, the following partial correlations were obtained:

$$r_{YX_3|X_1} = .124 \quad \text{and} \quad r_{YX_3|X_1, X_2} = .121$$

where

Y = Sudden death rate in county

X_1 = Distance from county seat to main hospital center

X_2 = Population per physician

X_3 = Water hardness index for county

Test separately the following null hypotheses:

$$\rho_{YX_3|X_1} = 0 \quad \text{and} \quad \rho_{YX_3|X_1, X_2} = 0$$

b. An alternative way to form the ANOVA table associated with a regression analysis is to use partial correlation coefficients. For example, if three independent variables X_1, X_2, and X_3 are used, the ANOVA table is as shown next.

Source		d.f.	SS
Regression	$\begin{cases} X_1 \\ X_2 \mid X_1 \\ X_3 \mid X_1, X_2 \end{cases}$	1	$r^2_{YX_1} SSY$
		1	$r^2_{YX_2 \mid X_1}(1 - r^2_{YX_1})SSY$
		1	$r^2_{YX_3 \mid X_1, X_2}(1 - r^2_{YX_2 \mid X_1})(1 - r^2_{YX_1})SSY$
Residual		$n - 4$	$(1 - r^2_{YX_3 \mid X_1, X_2})(1 - r^2_{YX_2 \mid X_1})(1 - r^2_{YX_1})SSY$
Total		$n - 1$	$\sum_{i=1}^{n}(Y_i - \bar{Y})^2 = SSY$

Determine the ANOVA table for the three independent variables in the water hardness study, using the following information:

$$r_{YX_1} = -.196, \quad r_{YX_2} = .033, \quad r_{X_1 X_2} = .038,$$

$$r_{YX_3 \mid X_1, X_2} = .121, \quad SSY = 21.05, \quad n = 88$$

c. In addition to the independent variables X_1, X_2, and X_3 already mentioned, the predictive abilities of the following independent variables were studied:

$X_4 = $ Per capita income

$X_5 = $ Coroner habit

$Z_1 = 1$ if county is in Piedmont area, and 0 otherwise

$Z_2 = 1$ if county is in Coastal Plains area, and 0 otherwise

$Z_3 = 1$ if county is in Tidewater area, and 0 otherwise

Furthermore, 25 first-order product (i.e., interaction) terms of the form $X_1 X_2$, $X_1 Z_1$, and so on, were also considered (excluding $Z_1 Z_2$, $Z_1 Z_3$, and $Z_2 Z_3$). The following three ANOVA tables were obtained.

Only X_1 Used

Source	d.f.	SS	MS
Regression	1	0.3846	0.3846
Residual	86	20.6703	0.2404

Only "Main Effects" Used

Source	d.f.	SS	MS
Regression	8	2.6853	0.3357
Residual	79	18.3696	0.2325

Main Effects Plus First-Order Interactions Used

Source	d.f.	SS	MS
Regression	33	7.4143	0.2247
Residual	54	13.6406	0.2258

Test whether adding all the interaction terms to the model significantly aids in predicting the dependent variable, after controlling for the main effects. State the null hypothesis for this test in terms of the appropriate multiple partial correlation coefficient.

d. Test whether there is significant overall prediction based on each of the three ANOVA tables. Determine the multiple R^2-values for each of the three tables.

e. What can you conclude from these results about the relationship of water hardness to sudden death?

3. Two variables X and Y are said to have a *spurious correlation* if their correlation solely reflects each variable's relationship to a third (antecedent) variable Z (and possibly to other variables). For example, the correlation between the total annual income (from all sources) of members of the U.S. Congress (Y) and the number of persons owning color television sets (X) is quite high. Simultaneously, however, a general upward trend has occurred in buying power (Z_1) and in wages of all types (Z_2), which would naturally be reflected in increased purchases of color TVs as well as in increased income of members of Congress. Thus, the high correlation between X and Y probably only reflects the influence of inflation on each of these two variables. Therefore, this correlation is spurious because it misleadingly suggests a relationship between color TV sales and the income of members of Congress.

a. How would you attempt to detect statistically whether a correlation between X and Y like the one described is spurious?

b. In a hypothetical study investigating socioecological determinants of respiratory morbidity for a sample of 25 communities, the following correlation matrix was obtained for four variables.

	Unemployment Level (X_1)	Average Temperature (X_2)	Air Pollution Level (X_3)	Respiratory Morbidity Rate (Y)
Unemployment level (X_1)	1	.51	.41	.35
Average temperature (X_2)	—	1	.29	.65
Air pollution level (X_3)	—	—	1	.50
Respiratory morbidity rate (Y)	—	—	—	1

(1) Determine the partial correlations $r_{YX_1|X_2}$, $r_{YX_1|X_3}$, and $r_{YX_1|X_2, X_3}$.

(2) Use the results in part (1) to determine whether the correlation of .35 between unemployment level (X_1) and respiratory morbidity rate (Y) is spurious. (Use the testing formula given in Problem 2 to make the appropriate tests.)

c. Describe a relevant example of spurious correlation in your field of interest. (Use only interval variables, and define them carefully.)

4. a. Using the information provided in Problem 2 of Chapter 9, determine the proportion of residual variation that is explained by the addition of X_2 to a model already containing X_1; that is, compute

$$Q = \frac{\text{Regression SS}(X_1, X_2) - \text{Regression SS}(X_1)}{\text{Residual SS}(X_1)}$$

b. How is the formula given in part (a) related to the partial correlation $r_{YX_2|X_1}$?

c. Test the hypothesis $H_0: \rho_{YX_2|X_1} = 0$, using both an F test and a two-sided t test. Check to confirm that these tests are equivalent.

5. Refer to Problem 7 of Chapter 9 to answer the following questions about the relationship of homicide rate (Y) to city population size (X_1), percentage of families with yearly incomes less than \$5,000 (X_2), and unemployment rate (X_3).

a. Determine the squared partial correlations $r^2_{YX_1|X_3}$ and $r^2_{YX_2|X_3}$ using the computer output here. Check the computation of $r^2_{YX_2|X_3}$ by means of an alternative formula, using the information that $r_{YX_2} = .8398$, $r_{YX_3} = .8648$, and $r_{X_2X_3} = .8154$.

b. Based on the results you obtained in part (a), which variable (if any) should next be considered for entry into the model, given that X_3 is already in the model?

c. Test H_0: $\rho_{YX_2|X_3} = 0$, using the t test described in Problem 2.

d. Determine the squared partial correlation $r^2_{YX_1|X_2, X_3}$ from the output here and/or in Problem 7 of Chapter 9, and then test H_0: $\rho_{YX_1|X_2, X_3} = 0$.

e. Determine the squared multiple partial correlation $r^2_{Y(X_1, X_2)|X_3}$, and test H_0: $\rho_{Y(X_1, X_2)|X_3} = 0$.

f. Based on the results you obtained in parts (a) through (e), which variables would you include in your final regression model? Use $\alpha = .05$.

Edited SAS Output (PROC REG) for Problem 5

Y Regressed on X3, X2, and X1

Analysis of Variance

Source	DF	Sum of Squares	Mean Square	F Value	Pr > F
Model	3	1518.14494	506.04831	24.022	0.0001
Error	16	337.05706	21.06607		
Corrected Total	19	1855.20200			

Root MSE	4.58978	R-square	0.8183
Dep Mean	20.57000	Adj R-sq	0.7843
Coeff Var	22.31297		

Parameter Estimates

Variable	DF	Parameter Estimate	Standard Error	t Value	Pr > |t|
INTERCEPT	1	-36.764925	7.01092577	-5.244	0.0001
X3	1	4.719821	1.53047547	3.084	0.0071
X2	1	1.192174	0.56165391	2.123	0.0497
X1	1	0.000763	0.00063630	1.199	0.2480

Variable	DF	Squared Partial Corr Type I	Squared Partial Corr Type II
INTERCEPT	1	.	.
X3	1	0.74795075	0.37280460
X2	1	0.21441231	0.21972110
X1	1	0.08244493	0.08244493

6. Using the ANOVA table given in Problem 8 of Chapter 9, which deals with the regression relationship of twelfth-grade mean verbal SAT scores (Y) to per pupil expenditures (X_1), percentage of teachers with advanced degrees (X_2), and pupil–teacher ratio (X_3), test the following null hypotheses

a. H_0: $\rho_{YX_3|X_1} = 0$

b. H_0: $\rho_{YX_2|X_1, X_3} = 0$

c. H_0: $\rho_{Y(X_2, X_3)|X_1} = 0$

d. Based on these results, and assuming that X_1 is an important predictor of Y, what additional variables would you include in your regression model?

7. Using the following ANOVA table based on data in Problem 6 of Chapter 8 about the regression relationship of respiratory cancer mortality rates (Y) to air pollution index (X_1), mean age (X_2), and percentage of workforce employed in a certain industry (X_3), test the following hypotheses:

a. H_0: $\rho_{YX_2|X_1} = 0$

b. H_0: $\rho_{YX_3|X_1, X_2} = 0$

Source	d.f.	SS
X_1	1	1,523.658
$X_2\|X_1$	1	181.743
$X_3\|X_1, X_2$	1	130.529
Residual	19	551.723
Total	22	2,387.653

c. H_0: "The addition of X_2 and X_3 to a model already containing X_1 does not significantly improve the prediction of Y."

d. Based on these results, which variables are important predictors of Y? Use $\alpha = .05$.

e. State the results in part (a) in terms of equivalent tests about population semipartial correlations.

8. Refer to the following ANOVA tables and to SAS output given here (from data in Problem 8 of Chapter 8) to answer the following questions dealing with factors related to soil erosion.

a. Using the accompanying SAS output, compute $r_{YX_2|X_1}$ and $r_{YX_3|X_1}$.

b. Based on your results in part (a), which variable (if any) should next be entered into a regression model that already contains X_1?

c. Test H_0: $\rho_{YX_2|X_1} = 0$, using the t test described in Problem 2.

d. Determine the squared multiple partial correlation $r^2_{Y(X_2, X_3)|X_1}$, and test H_0: $\rho_{Y(X_2, X_3)|X_1} = 0$.

Fitting X_1 First, Then Letting X_2 and X_3 Enter Stepwise

Source	d.f.	SS	Source	d.f.	SS
X_2	1	667.7279	X_1	1	640.4249
$X_3\|X_2$	1	5.8228	$X_2\|X_1$	1	32.7819
$X_1\|X_3, X_2$	1	6.9406	$X_3\|X_1, X_2$	1	7.2844
Residual	7	16.0942	Residual	7	16.0942

Edited SAS Output (PROC REG) for Problem 8

```
                        Y Regressed on X1 and X2
                        Analysis of Variance

                              Sum of            Mean
Source              DF        Squares          Square      F Value    Pr > F
Model               2        673.20680       336.60340     115.183    0.0001
Error               8         23.37865         2.92233
Corrected Total    10        696.58545

                    Root MSE        1.70948      R-square      0.9664
                    Dep Mean       40.23636      Adj R-sq      0.9580
                    Coeff Var       4.24860

                        Parameter Estimates

                            Parameter        Standard
Variable            DF       Estimate           Error       t Value    Pr >|t|
INTERCEPT           1       -8.084810     20.06314836        -0.403     0.6975
X1                  1       68.250681     49.84520939         1.369     0.2081
X2                  1        2.293871      0.68488343         3.349     0.0101

                            Squared         Squared
                            Partial         Partial
Variable            DF    Corr Type I    Corr Type II
INTERCEPT           1         .               .
X1                  1     0.91937735      0.18986133
X2                  1     0.58371760      0.58371760
```

(continued)

```
                      Y Regressed on X1 and X3 (continued)
                           Analysis of Variance

                              Sum of        Mean
Source                  DF    Squares      Square     F Value    Pr > F
Model                    2   670.13094   335.06547    101.326    0.0001
Error                    8    26.45452     3.30681
Corrected Total         10   696.58545

                 Root MSE        1.81846      R-square       0.9620
                 Dep Mean       40.23636      Adj R-sq       0.9525
                 Coeff Var       4.51946

                          Parameter Estimates

                          Parameter        Standard
Variable        DF        Estimate          Error      t value    Pr > |t|
INTERCEPT        1       -26.704505      16.62113471    -1.607     0.1468
X1               1       157.267135      28.44374038     5.529     0.0006
X3               1       -39.925950      13.32102838    -2.997     0.0171

                          Squared          Squared
                          Partial          Partial
Variable        DF       Corr Type I      Corr Type II
INTERCEPT        1            .                .
X1               1        0.91937735       0.79258760
X3               1        0.52894851       0.52894851

                      Y Regressed on X1, X2 and X3
                              Correlation

Corr            X1              X2              X3               Y
X1            1.0000          0.9515         -0.8191          0.9588
X2            0.9515          1.0000         -0.8785          0.9791
X3           -0.8191         -0.8785          1.0000         -0.9038
Y             0.9588          0.9791         -0.9038          1.0000

                          Analysis of Variance

                              Sum of        Mean
Source                  DF    Squares      Square     F Value    Pr > F
Model                    3   680.49122   226.83041     98.657    0.0001
Error                    7    16.09423     2.29918
Corrected Total         10   696.58545

                 Root MSE        1.51630      R-Square       0.9769
                 Dep Mean       40.23636      Adj R-Sq       0.9670
                 Coeff Var       3.76849

                          Parameter Estimates

                          Parameter        Standard
Variable        DF        Estimate          Error      t value    Pr > |t|
INTERCEPT        1        -1.879316      18.13419563    -0.104     0.9204
X1               1        77.325781      44.50547040     1.737     0.1259
X2               1         1.559100       0.73447037     2.123     0.0714
X3               1       -23.903783      13.42935668    -1.780     0.1183

                          Squared          Squared
                          Partial          Partial
Variable        DF       Corr. Type I     Corr Type II
INTERCEPT        1            .                .
X1               1        0.91937735       0.30130747
X2               1        0.58371760       0.39162631
X3               1        0.31158432       0.31158432
```

9. Use the computer results from Problem 9 of Chapter 8 to answer the following questions.

   a. Test $H_0: \rho_{YX_1} = 0$ and $H_0: \rho_{YX_2} = 0$.

b. Test H_0: $\rho_{YX_1|X_2} = 0$ and H_0: $\rho_{YX_2|X_1} = 0$.

c. Based on your results in parts (a) and (b), which variables (if any) should be included in the regression model, and what is their order of importance?

10. Use the correlation matrix from Problem 1 to answer the following questions.

   a. Compute the semipartial $r_{SBP(SMK|AGE)}$.

   b. Compute the semipartial $r_{SMK(SBP|AGE)}$.

   c. Compare these correlations to the (full) partial $r_{SBP, SMK|AGE}$, computed in Problem 1.

11. For the data discussed in Problem 7, provide numerical values for the following quantities:

   a. $r^2_{YX_1}$

   b. $R^2_{Y|X_1,X_2}$

   c. $R^2_{Y|X_1, X_2, X_3}$

   d. $r^2_{YX_3|X_1, X_2}$

   e. $r^2_{YX_2|X_1}$

12. Use the results in Problem 8 to answer the following questions.

   a. Provide a numerical value for $r^2_{YX_1|X_3, X_2}$.

   b. Which two models should you compare in testing whether the correlation in part (a) is zero in the population?

   c. Provide a numerical value for $r^2_{YX_1}$.

   d. What does the difference between parts (a) and (c) say about the relationships among the three predictor variables?

   e. Which two models should you compare in testing whether the correlations in parts (a) and (c) differ in the population?

13. The accompanying SAS computer output relates to the house price data of Problem 10 in Chapter 8. Use this output and, if necessary, the output associated with Problem 9 in Chapter 9 to answer the following questions.

   a. Determine $r^2_{Y|X_1, X_2}$, the squared multiple correlation between house price (Y) and the independent variables house size (X_1) and number of bedrooms (X_2).

   b. Determine $r_{YX_2|X_1}$, the partial correlation of Y with X_2 given that X_1 is in the model.

   c. Determine $r_{YX_1|X_2}$, the partial correlation of Y with X_1 given that X_2 is in the model.

   d. Using the t test technique of Problem 2, test H_0: $\rho_{YX_2|X_1} = 0$. Compare your test statistic value with the partial t test statistic for X_2 shown on the SAS output here.

   e. Using the t test technique of Problem 2, test H_0: $\rho_{YX_1|X_2} = 0$. Compare your test statistic value with the partial t test statistic for X_1 shown on the SAS output here.

   f. Based on your answers to parts (a) through (e), which variables (if any) should be included in the regression model, and what is their order of importance?

Edited SAS Output (PROC REG) for Problem 13

Y Regressed on X1 and X2
Analysis of Variance

Source	DF	Sum of Squares	Mean Square	F Value	Pr > F
Model	2	5733.32129	2866.66064	20.033	0.0082
Error	4	572.39300	143.09825		
Corrected Total	6	6305.71429			

(continued)

```
                         Y Regressed on X1 and X2 (continued)
                         Analysis of Variance (continued)
                  Root MSE              11.96237        R-square       0.9092
                  Dep Mean             105.42857        Adj R-sq       0.8638
                  Coeff Var             11.34642

                              Parameter Estimates

                              Parameter              Standard
Variable              DF       Estimate                Error       t value    Pr >|t|
INTERCEPT             1      -16.093385             24.64693805     -0.653     0.5494
X1                   1        5.721790              1.82778969      3.130     0.0352
X2                   1       -1.173152             13.38993701     -0.088     0.9344

                              Squared               Squared
                              Partial               Partial
Variable              DF     Corr Type I           Corr Type II
INTERCEPT            1          .                      .
X1                   1       0.90905210             0.71013795
X2                   1       0.00191540             0.00191540
```

14. The accompanying SAS computer output relates to the sales revenue data from Problem 11 in Chapter 8. Use this output and, if necessary, the output for Problem 10 in Chapter 9 to answer the following questions.

   a. Determine $r^2_{Y|X_1, X_2}$, the squared multiple correlation between sales revenue (Y) and the independent variables TV advertising expenditures (X_1) and print advertising expenditures (X_2).

   b.–f. For the sales revenue data, answer the questions posed in Problem 13, parts (b) through (f).

Edited SAS Output (PROC REG) for Problem 14

```
                              Y Regressed on X1 and X2
                              Analysis of Variance

                              Sum of          Mean
Source              DF        Squares         Square      F Value    Pr > F
Model               2         28.11853        14.05927    59.006     0.0039
Error               3         0.71480         0.23827
Corrected Total     5         28.83333
                  Root MSE     0.48813        R-square       0.9752
                  Dep Mean     5.16667        Adj R-sq       0.9587
                  Coeff Var    9.44760

                              Parameter Estimates
                              Parameter              Standard
Variable              DF       Estimate                Error       t value    Pr > |t|
INTERCEPT            1        2.104693              0.42196591      4.988      0.0155
X1                   1        1.241877              0.11913357     10.424      0.0019
X2                   1       -0.194946              0.43944079     -0.444      0.6874
                              Squared               Squared
                              Partial               Partial
Variable              DF     Corr Type I           Corr Type II
INTERCEPT            1          .                      .
X1                   1       0.97358291             0.97313389
X2                   1       0.06156185             0.06156185
```

15. The accompanying SAS computer output relates to the radial keratotomy data from Problem 12 in Chapter 8. Use this output and, if necessary, the output from Problem 11 in Chapter 9 to answer the following questions.

   a. Determine $r^2_{Y|X_1, X_2}$, the squared multiple correlation between change in refractive error (Y) and the independent variables baseline refractive error (X_1) and baseline curvature (X_2).

   b–f. For the radial keratotomy data, answer the questions posed in Problem 13, parts (b) through (f).

Edited SAS Output (PROC REG) for Problem 15

```
                         Y Regressed on X1 and X2
                           Analysis of Variance
                                  Sum of         Mean
Source                   DF      Squares        Square      F Value      Pr > F
Model                     2     17.62277       8.81139        7.175      0.0018
Error                    51     62.63017       1.22804
Corrected Total          53     80.25294

                  Root MSE            1.10817     R-square          0.2196
                  Dep Mean            3.83343     Adj R-sq          0.1890
                  Coeff Var          28.90811

                           Parameter Estimates
                         Parameter              Standard
Variable         DF       Estimate                 Error     t value      Pr >|t|
INTERCEPT         1      12.360015            5.08621177       2.430       0.0187
X1                1      -0.291601            0.09165295      -3.182       0.0025
X2                1      -0.220396            0.11482391      -1.919       0.0605

                         Squared               Squared
                         Partial               Partial
Variable         DF    Corr Type I          Corr Type II
INTERCEPT         1        .                     .
X1                1     0.16321415            0.16560944
X2                1     0.06737232            0.06737232
```

16. The accompanying SAS computer output relates to the *Business Week* data from Problem 13 in Chapter 8. Use this output and, if necessary, the output from Problem 12 in Chapter 9 to answer the following questions.

   a. Determine $r^2_{Y|X_2, X_3}$, the squared multiple correlation between the yield (Y) and the independent variables 1989 rank (X_2) and P-E ratio (X_3).

   b.–f. For the *Business Week* data, answer the questions posed in Problem 13, parts (b) through (f).

Edited SAS Output (PROC REG) for Problem 16

```
          Yield (Y) Regressed on 1989 Rank (X2) and P-E Ratio (X3)
                                    .
                                    .   [Portion of output omitted]
                                    .
                                  Squared                       Squared
                                  Partial                       Partial
Variable             DF        Corr Type I                   Corr Type II
INTERCEPT             1            .                               .
X2                    1        0.03670921                    0.27066359
X3                    1        0.53584299                    0.53584299
```

17. The accompanying SAS computer output relates to the 1990 Census data from Problem 14 in Chapter 8. Use this output and, if necessary, the output from Problem 13 in Chapter 9 to answer the following questions.

   a. Determine $r^2_{Y|X_1, X_2}$, the squared multiple correlation between the rate of owner occupancy (Y = OWNEROCC) and the independent variables for monthly ownership costs (X_1 = OWNCOST) and proportion of population living in urban areas (X_2 = URBAN).

   b.–f. For the Census data, answer the questions posed in Problem 13, parts (b) through (f).

Edited SAS Output (PROC REG) for Problem 17

```
                    OWNEROCC Regressed on OWNCOST and URBAN
                                        .
                                        .  [Portion of output omitted]
                                        .
                                           Squared                      Squared
                                           Partial                      Partial
Variable              DF            Corr. Type I                 Corr Type II
INTERCEPT             1                   .                            .
OWNCOST              1             0.22068022                   0.16119308
URBAN                1             0.26296676                   0.26296676
```

18. This problem refers to the pond ecology data of Chapter 5, Problem 20.

   a. Determine the squared multiple correlation for the regression of copepod count on zooplankton and phytoplankton counts.

   b. Determine the partial correlation between each independent variable in (a) and copepod count, controlling for the other independent variable. Using the technique of Problem 2, perform tests of significance of the partial correlations.

   c. Based on your answers in (b), which independent variables should be included in the regression model, and in what order of importance?

References

Blalock, H. M., Jr., ed. 1971. *Causal Models in the Social Sciences.* Chicago: Aldine Publishing.

Bollen, K. A. 1989. *Structural Equations with Latent Variables.* New York: John Wiley & Sons.

Confounding and Interaction
in Regression

11.1 Preview

A regression analysis may have two different goals: to predict the dependent variable by using a set of independent variables; and to quantify the relationship of one or more independent variables to a dependent variable. The first of these goals focuses on finding a model that fits the observed data and predicts future data as well as possible, whereas the second pertains to producing valid and precise statistical inferences about one or more regression coefficients in the model. The second goal is of particular interest when the research question concerns disease etiology, such as trying to identify one or more determinants of a disease or other health-related outcome.

  Confounding and *interaction* are two methodological concepts relevant to attaining the second goal. In this chapter, we use regression terminology to describe these concepts. More general discussion of this subject can be found elsewhere (e.g., Kleinbaum, Kupper, and Morgenstern 1982) within the context of epidemiological research, which typically addresses etiologic questions involving the second goal. We begin here with a general overview of these concepts, after which we discuss the regression formulation of each concept separately. In Chapter 13, we describe a popular regression procedure, analysis of covariance (ANACOVA), which may be used to adjust or correct for problems of confounding. Subsequently, in Chapter 16, we briefly describe a strategy for obtaining a "best" regression model that incorporates the assessment of both confounding and interaction.

11.2 Overview

Both confounding and interaction involve the assessment of an association between two or more variables so that additional variables that may affect this association are accounted for. The measure of association chosen usually depends on the characteristics of the variables of interest. For example, if both variables are continuous, as in the classic regression context, the measure of association is typically a regression coefficient. The additional variables to be considered are synonymously referred to as *extraneous variables, control variables,* or *covariates.* The essential questions about these variables are whether and how they should be incorporated into a model that can be used to estimate the association of interest.

Suppose that we are conducting a study to assess whether physical activity level (PAL) is associated with systolic blood pressure (SBP), accounting (i.e., controlling) for AGE. The extraneous variable here is AGE. We need to determine whether we can ignore AGE in our analysis and still correctly assess the PAL–SBP association. In particular, we need to address the following two questions: (1) Is the estimate of the association between PAL and SBP meaningfully different depending on whether or not we ignore AGE? (2) Is the estimate of the association between PAL and SBP meaningfully different for different values of AGE? The first question is concerned with confounding; the second question, with interaction.

In general, *confounding exists if meaningfully different interpretations of the relationship of interest result when an extraneous variable is ignored or included in the data analysis.* In practice, the assessment of confounding requires a comparison between a *crude* estimate of an association (which ignores the extraneous variable(s)) and an *adjusted* estimate of the association (which accounts in some way for the extraneous variables). If the crude and adjusted estimates are meaningfully different, confounding is present and one or more extraneous variables must be included in our data analysis. This definition does not require a statistical test but rather a comparison of estimates obtained from the data (see Kleinbaum, Kupper, and Morgenstern 1982, chap. 13, or Kleinbaum 2002, Lesson 10 for further discussion of this point).

For example, a crude estimate of the relationship between PAL and SBP (ignoring AGE) is given by the regression coefficient—say, $\hat{\beta}_1$—of the variable PAL in the straight-line model that predicts SBP using just PAL. In contrast, an adjusted estimate is given by the regression coefficient $\hat{\beta}_1^*$ of the same variable PAL in the multiple regression model that uses both PAL and AGE to predict SBP. In particular, if PAL is defined dichotomously (e.g., PAL = 1 or 0 for high or low physical activity, respectively), then the crude estimate is simply the crude difference between the mean systolic blood pressures in each physical activity group, and the adjusted estimate represents an adjusted difference in these two mean systolic blood pressures that controls for AGE. In general, confounding is present if any meaningful difference exists between the crude and adjusted estimates.

Interaction is the condition in which the relationship of interest is different at different levels (i.e., values) of the extraneous variable(s). In contrast to confounding, the assessment of interaction does not consider either a crude estimate or an (overall) adjusted estimate; instead, it focuses on describing the relationship of interest at different values of the extraneous variables. For example, in assessing interaction due to AGE in describing the PAL–SBP relationship, we must determine whether some description (i.e., estimate) of the relationship varies with different values of AGE (e.g., whether the relationship is strong at older ages and weak at younger ages). If the PAL–SBP relationship does vary with AGE, then we say that an AGE × (read "by") PAL interaction exists. To assess interaction, we may employ a statistical test in addition to subjective evaluation of the meaningfulness (e.g., clinical importance) of an estimated interaction effect. Again, for further discussion, see Kleinbaum, Kupper, and Morgenstern (1982) or Kleinbaum (2002).

When both confounding and interaction are considered for the same data set, using an overall (adjusted) estimate as a summary index of the relationship of interest tends to mask any (strong) interaction effects that may be present. For example, if the PAL–SBP association differs meaningfully at different values of AGE, using a single overall estimate, such as the regression coefficient of PAL in a multiple regression model containing both AGE and PAL, would hide this interaction finding. This illustrates the following important principle: *Interaction should be assessed before confounding is assessed; the use of a summary (adjusted) estimate that controls for confounding is recommended only when there is no meaningful interaction* (Kleinbaum, Kupper, and Morgenstern 1982, chap. 13).

A variable may manifest both confounding and interaction, neither, or only one of the two. But if strong interaction is found, an adjustment for confounding is inappropriate.

We are now ready to use regression terminology to address how these concepts can be employed, assuming a linear model and a continuous dependent variable. A regression analog for a dichotomous outcome variable might, for example, involve a logistic rather than a linear model. Logistic regression analysis is discussed in detail in Chapter 22. A more detailed discussion in which confounding and interaction are considered can be found in Kleinbaum, Kupper, and Morgenstern (1982, chaps. 20–24) or Kleinbaum and Klein (2002, chaps. 6–7).

11.3 Interaction in Regression

In this section, we describe how two independent variables can interact to affect a dependent variable and how such an interaction can be represented by an appropriate regression model.

11.3.1 First Example

To illustrate the concept of interaction, let us consider the following simple example. Suppose that we wish to determine how two independent variables—temperature (T) and catalyst concentration (C)—jointly affect the growth rate (Y) of organisms in a certain biological system. Further, suppose that two particular temperature levels (T_0 and T_1) and two particular levels of catalyst concentration (C_0 and C_1) are to be examined, and that an experiment is performed in which an observation on Y is obtained for each of the four combinations of temperature–catalyst concentration level: (T_0, C_0), (T_0, C_1), (T_1, C_0), and (T_1, C_1). In statistical parlance, this experiment is called a *complete factorial experiment,* because observations on Y are obtained for all combinations of settings for the independent variables (or factors). The advantage of a factorial experiment is that any existing interaction effects can be detected and estimated efficiently.

Now, let us consider two graphs based on two hypothetical data sets for the experimental scheme just described. Figure 11.1(a) suggests that the rate of change[1] in the growth rate as a function of temperature remains the same regardless of the level of catalyst concentration; in other words, the relationship between Y and T does not in any way depend on C.

We are not saying that Y and C are unrelated, but that the relationship between Y and T does not vary as a function of C. When this is the case, we say that T and C do not interact or, equivalently, that there is no $T \times C$ interaction effect. Practically speaking, this means that we can investigate the effects of T and C on Y independently of one another and that we can legitimately talk about the separate effects (sometimes called the *main effects*) of T and C on Y.

---

[1] For readers familiar with calculus, the phrase "rate of change" is related to the notion of a derivative of a function. In particular, Figure 11.1(a) portrays a situation in which the partial derivative with respect to T of the response function relating the mean of Y to T and C is independent of C.

FIGURE 11.1 **Graphs of noninteracting and interacting independent variables**

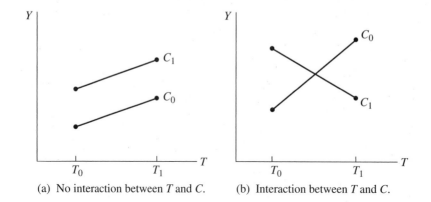

(a) No interaction between T and C.   (b) Interaction between T and C.

One way to quantify the relationship depicted in Figure 11.1(a) is with a regression model of the form

$$\mu_{Y|T,\,C} = \beta_0 + \beta_1 T + \beta_2 C \qquad\qquad (11.1)$$

Here, the change in the mean of Y for a 1-unit change in T is equal to β_1, regardless of the level of C. In fact, changing the level of C in (11.1) has only the effect of shifting the straight line relating $\mu_{Y|T,\,C}$ and T either up or down, without affecting the value of the slope β_1, as seen in Figure 11.1(a). In particular, $\mu_{Y|T,\,C_0} = (\beta_0 + \beta_2 C_0) + \beta_1 T$ and $\mu_{Y|T,\,C_1} = (\beta_0 + \beta_2 C_1) + \beta_1 T$.

In general, then, we might say that no interaction is synonymous with parallelism, in the sense that the response curves of Y versus T for fixed values of C are parallel; in other words, these response curves (which may be linear or nonlinear) all have the same general shape, differing from one another only by additive constants independent of T (see, e.g., Figure 11.2).

In contrast, Figure 11.1(b) depicts a situation in which the relationship between Y and T depends on C; in particular, Y appears to increase with increasing T when $C = C_0$ but to decrease with increasing T when $C = C_1$. In other words, the behavior of Y as a function of temperature cannot be considered independently of catalyst concentration. When this is the case, we say that T and C interact or, equivalently, that there is a $T \times C$ interaction effect. Practically speaking, this

FIGURE 11.2 **Response curves illustrating no interaction between T and C**

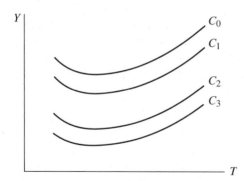

means that it does not make much sense to talk about the separate (or main) effects of T and C on Y, since T and C do not operate independently of one another in their effects on Y.

One way to represent such interaction effects mathematically is to use a regression model of the form

$$\mu_{Y|T, C} = \beta_0 + \beta_1 T + \beta_2 C + \beta_{12} TC \tag{11.2}$$

Here the change in the mean value of Y for a 1-unit change in T is equal to $\beta_1 + \beta_{12}C$, which clearly depends on the level of C. In other words, introducing a product term such as $\beta_{12}TC$ in a regression model of the type given in (11.2) is one way to account for the fact that two such factors as T and C do not operate independently of one another. For our particular example, when $C = C_0$, model (11.2) can be written as

$$\mu_{Y|T, C_0} = (\beta_0 + \beta_2 C_0) + (\beta_1 + \beta_{12} C_0)T$$

and when $C = C_1$, model (11.2) becomes

$$\mu_{Y|T, C_1} = (\beta_0 + \beta_2 C_1) + (\beta_1 + \beta_{12} C_1)T$$

In particular, Figure 11.1(b) suggests that the interaction effect β_{12} is negative, with the linear effect $(\beta_1 + \beta_{12} C_0)$ of T at C_0 being positive and the linear effect $(\beta_1 + \beta_{12} C_1)$ of T at C_1 being negative. A negative interaction effect is to be expected here, since Figure 11.1(b) suggests that the slope of the linear relationship between Y and T decreases (goes from positive to negative in sign) as C changes from C_0 to C_1. Of course, it is possible for β_{12} to be positive, in which case the interaction effect manifests itself as a larger positive value for the slope when $C = C_1$ than when $C = C_0$.

11.3.2 Interaction Modeling in General

As the preceding illustration suggests, interaction among independent variables can generally be described in terms of a regression model that involves product terms. Unfortunately, no precise rules exist for specifying such terms. For example, if interaction involving three variables X_1, X_2, and X_3 is of interest, one model to consider is

$$Y = \beta_0 + \beta_1 X_1 + \beta_2 X_2 + \beta_3 X_3 + \beta_4 X_1 X_2 + \beta_5 X_1 X_3 + \beta_6 X_2 X_3 \\ + \beta_7 X_1 X_2 X_3 + E \tag{11.3}$$

Here, the two-factor products of the form $X_i X_j$ are often referred to as *first-order interactions*; the three-factor products, like $X_1 X_2 X_3$, are called *second-order interactions*; and so on for higher-order products. The higher the order of interaction, the more difficult it becomes to interpret its meaning.

Model (11.3) is not the most general model possible for considering the three variables X_1, X_2, and X_3. Additional product terms such as $X_i X_j^2$, $X_i X_j^3$, $X_i^2 X_j^2$, and so on can be included. Nevertheless, there is a limit on the total number of such terms: a model with an intercept (β_0) term cannot contain more than $n - 1$ independent variables when n is the total number of observations in the data. Moreover, it may not even be possible to fit a model with fewer than $n - 1$ variables reliably if some of the variables (e.g., higher-order products) are highly correlated with other variables in the model, as would be the case when the model contains several interaction terms. This problem, called *collinearity*, is discussed in Chapter 14.

On the other hand, model (11.3) may be considered too general if one is focusing on particular interactions of interest. For example, if the purpose of one's study is to describe the relationship between X_1 and Y, controlling for the possible confounding and/or interaction effects of X_2 and X_3, the following simpler model may be of more interest than (11.3):

$$Y = \beta_0 + \beta_1 X_1 + \beta_2 X_2 + \beta_3 X_3 + \beta_4 X_1 X_2 + \beta_5 X_1 X_3 + E \qquad (11.4)$$

The terms $X_1 X_2$ and $X_1 X_3$ describe the interactions of X_2 and X_3, respectively, with X_1. In contrast, the term $X_2 X_3$, which is not contained in model (11.4), has no relevance to confounding or interaction involving X_1.

In using statistical testing to evaluate interaction for a given regression model, we have a number of available options. (A more detailed discussion of how to select variables is given in Chapter 16.) One approach is to test globally for the presence of any kind of interaction and then, if significant interaction is found, to identify particular interaction terms of importance by using other tests. For example, in considering model (11.3), we could first test $H_0: \beta_4 = \beta_5 = \beta_6 = \beta_7 = 0$, using the multiple partial F statistic

$$F(X_1 X_2, X_1 X_3, X_2 X_3, X_1 X_2 X_3 | X_1, X_2, X_3)$$

which has an $F_{4, n-8}$ distribution when H_0 is true. If this F statistic is found to be significant, individually important interaction terms may then be identified by using selected partial F tests.

A second way to assess interaction is to test for interaction in a hierarchical sequence, beginning with the highest-order terms and then proceeding sequentially through lower-order terms if higher-order terms are not significant. Using model (11.3), for example, we might first test $H_0: \beta_7 = 0$, which considers the second-order interaction, and then test $H_0: \beta_4 = \beta_5 = \beta_6 = 0$ in a reduced model (excluding the three-way product term $X_1 X_2 X_3$) if the result of the first test is nonsignificant.

11.3.3 Second Example

We present here an example of how to assess interaction by using SAS computer output. In Chapters 5 through 7, the age–systolic blood pressure example was used to illustrate the main principles and methods of straight-line regression analysis. These hypothetical data were shown, upon analysis, to support the commonly found epidemiological observation that blood pressure increases with age.

Another question that can be answered by such data is whether an interaction exists between age and gender: Does the slope of the straight-line relating systolic blood pressure to age significantly differ for males and for females? We will continue with our age–systolic blood pressure example to investigate this possible interaction.

In Chapter 5, we observed that the data point (47, 220) is an outlier quite distinct from the rest of the data. We will discard this data point in all further analyses; henceforth we will assume that all 29 remaining observations on age and systolic blood pressure considered previously were made on females and that a second sample of observations on age and systolic blood pressure was collected on 40 males. The data set for the 40 males is given in Table 11.1 and the data set for females appears in Table 5.1 in Chapter 5.

The accompanying SAS computer output is given for the following regression model:

$$Y = \beta_0 + \beta_1 X + \beta_2 Z + \beta_3 XZ + E$$

where Z represents SEX ($Z = 0$ if male, $Z = 1$ if female) and XZ is the interaction between AGE and SEX.

TABLE 11.1   **Data on systolic blood pressure (SBP) and age for 40 males and 29 females, together with associated data for comparing two straight-line regression equations**

Male (i)	SBP (Y)	Age (X)	Male (i)	SBP (Y)	Age (X)	Male (i)	SBP (Y)	Age (X)
1	158	41	15	142	44	28	144	33
2	185	60	16	144	50	29	139	23
3	152	41	17	149	47	30	180	70
4	159	47	18	128	19	31	165	56
5	176	66	19	130	22	32	172	62
6	156	47	20	138	21	33	160	51
7	184	68	21	150	38	34	157	48
8	138	43	22	156	52	35	170	59
9	172	68	23	134	41	36	153	40
10	168	57	24	134	18	37	148	35
11	176	65	25	174	51	38	140	33
12	164	57	26	174	55	39	132	26
13	154	61	27	158	65	40	169	61
14	124	36						

Edited SAS OUTPUT: SBP Regressed on Age (X), Sex (Z), and Interaction (XZ)

```
                          The GLM Procedure
Source                  DF            SS       Mean Square      F Value      Pr > F
Model                    3     18010.329       6003.443         75.02       <.0001
Error                   65      5201.439         80.022
Corrected Total         68     23211.768

            R-Square      Coeff Var          Root MSE          Y Mean
            0.7759         6.0148             8.9455          148.7246

Source       DF          Type I SS        Mean Square       F Value      Pr > F
X             1         14951.2546        14951.2546         186.84      <.0001
Z             1          3058.5248         3058.5248          38.22      <.0001
XZ            1             0.5494            0.5494           0.01       0.9342

Source       DF        Type III SS        Mean Square       F Value      Pr > F
X             1          7971.0071         7971.0071         99.61       <.0001
Z             1           273.4433          273.4433          3.42       0.0691
XZ            1             0.5494            0.5494           0.01       0.9342

Parameter        Estimate          StdErr          t Value        Pr > |t|
Intercept        110.0386          4.7361           23.23          <.0001
X                  0.9614          0.0963            9.98          <.0001
Z                -12.9614          7.0117           -1.85          0.0691
XZ                -0.0120          0.1452           -0.08          0.9342
```

From the output, we see that when $Z = 0$ (i.e., when SEX = male), the estimated model reduces to

$$Y = 110.04 + 0.96X + (-12.96)(0) + (-0.01)X(0)$$

$$= 110.04 + 0.96X$$

This is the estimated regression line for males. When $Z = 1$, the estimated model gives the estimated regression line for females:

$$Y = 110.04 + 0.96X + (-12.96)(1) + (-0.01)X(1)$$

$$= 97.08 + 0.95X$$

Plotting these two lines (Figure 11.3), we see that they appear to be almost parallel, indicating that there is probably no statistically significant interaction. We can confirm this lack of significance by inspecting the output: the partial F test for the significance of β_3, given that X and Z are in the model, has a p-value of 0.9342. Therefore, the slopes of the straight lines relating systolic blood pressure and age do not statistically significantly differ for males and for females: there is no interaction between age and sex in this situation.

11.3.4 Third Example

We now consider a study to assess physical activity level (PAL) as a predictor of systolic blood pressure (SBP), controlling for AGE and SEX. A model that allows for possible interactions of both AGE with PAL and SEX with PAL is given by

$$SBP = \beta_0 + \beta_1(PAL) + \beta_2(AGE) + \beta_3(SEX) + \beta_4(PAL \times AGE)$$

$$+ \beta_5(PAL \times SEX) + E$$

Notice the absence of any term involving AGE $\times$ SEX; such a term does not indicate interaction associated with the study variable of interest (PAL).

To assess interaction for this model, we might first perform a multiple partial F test of H_0: $\beta_4 = \beta_5 = 0$; if this test was found to be significant, we could conduct partial F tests to determine whether one or more of these product terms should be kept in the model. If the first test was found to be nonsignificant, we could then simplify the full model by removing these two product terms entirely, leaving the reduced model SBP $= \beta_0 + \beta_1(PAL) + \beta_2(AGE) + \beta_3(SEX) + E$. At this point the interaction phase of model building would be

FIGURE 11.3   Comparison by sex of straight-line regressions of systolic blood pressure on age

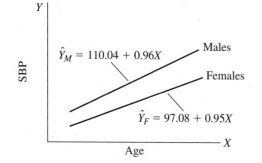

complete, and the next step would involve the assessment of confounding, as discussed in the next section.

11.3.5 Interaction versus Effect Modification

The term **effect modification** is often used interchangeably with the term **interaction,** particularly for health-related research situations considered in the field of epidemiology.[2] From an epidemiological perspective, "effect modification" is used to describe an association between a predictor (i.e., exposure) of interest and a health outcome that is "modified" (i.e., is different) depending on the value of one or more "control" variables. Such control variables are called **effect modifiers** of the relationship between exposure and outcome. In contrast, the term "interaction" is used to describe the statistical property of a mathematical model containing a predictor variable defined as the product of two or more component predictor variables (e.g., $X_k = X_i \times X_j$). Such a product term represents a combination (i.e., *interaction*) of its component terms to form a predictor that explains additional statistical variation in the outcome variable over and above the independent contributions of each component predictor.

To illustrate effect modification, consider the regression model provided in the previous section to describe the association of the predictor, physical activity level (PAL), with the health outcome, systolic blood pressure (SBP), controlling for AGE and SEX:

$$SBP = \beta_0 + \beta_1(PAL) + \beta_2(AGE) + \beta_3(SEX) + \beta_4(PAL \times AGE) + \beta_5(PAL \times SEX) + E$$

This model contains two interaction terms, PAL $\times$ SEX and PAL $\times$ AGE. As previously described, the assessment of interaction is typically supported using the results of statistical testing, where, in this case, the null hypothesis is H_0: $\beta_4 = \beta_5 = 0$. Assuming that such a statistical test is significant and that both interaction terms need to remain in the model, the fitted model would be written as:

$$\widehat{SBP} = \hat{\beta}_0 + \hat{\beta}_1(PAL) + \hat{\beta}_2(AGE) + \hat{\beta}_3(SEX) + \hat{\beta}_4(PAL \times AGE) + \hat{\beta}_5(PAL \times SEX)$$

We would then conclude that there is *effect modification* of the association of PAL with SBP, and that AGE and SEX are *effect modifiers* of this association. In other words, the association of PAL with SBP differs (i.e., is *modified*) depending on a person's AGE and/or SEX.

For example, suppose we compare the association between PAL and SBP for a 40-year-old female with the corresponding association for a 30-year-old male. Suppose, also, that AGE

---

[2] Although the terms *effect modification* and *interaction* are often used interchangeably, there is some controversy in the epidemiologic literature about the precise definitions of effect modification and interaction (see Kleinbaum, Kupper, and Morgenstern, 1982, chapter 19; and Kleinbaum, 2002, Lesson 10). One distinction frequently made is that effect modification describes a nonquantitative clinical or biological attribute of a population, whereas interaction is typically quantitative and data-specific and, in particular, depends on the scale on which the "interacting" variables are measured. Nevertheless, this conceptual distinction is often overlooked in applied research studies.

is a continuous variable and SEX is coded as 1 if female and 0 if male. Then the predicted SBP for a 40-year-old female (i.e., AGE = 40, SEX = 1) is

$$\widehat{SBP}_{40F} = \hat{\beta}_0 + \hat{\beta}_1(PAL) + 40\hat{\beta}_2 + \hat{\beta}_3 + 40\hat{\beta}_4(PAL) + \hat{\beta}_5(PAL)$$

$$= (\hat{\beta}_0 + 40\hat{\beta}_2 + \hat{\beta}_3) + (\hat{\beta}_1 + 40\hat{\beta}_4 + \hat{\beta}_5)(PAL)$$

$$= \hat{\beta}_{0(40F)} + \hat{\beta}_{1(40F)}PAL$$

where $\hat{\beta}_{0(40F)} = (\hat{\beta}_0 + 40\hat{\beta}_2 + \hat{\beta}_3)$ and $\hat{\beta}_{1(40F)} = (\hat{\beta}_1 + 40\hat{\beta}_4 + \hat{\beta}_5)$. Note that once AGE and SEX are specified, the model simplifies to a straight line model for predicting SBP from PAL. The estimated slope for this model, $\hat{\beta}_{1(40F)}$, represents the measure of association relating PAL to SBP.

Similarly, the predicted SBP for a 30-year-old male (AGE = 30, SEX = 0) is

$$\widehat{SBP}_{30M} = \hat{\beta}_0 + \hat{\beta}_1(PAL) + 30\hat{\beta}_2 + 0 + 30\hat{\beta}_4(PAL) + 0$$

$$= (\hat{\beta}_0 + 30\hat{\beta}_2) + (\hat{\beta}_1 + 30\hat{\beta}_4)(PAL)$$

$$= \hat{\beta}_{0(30M)} + \hat{\beta}_{1(30M)}PAL$$

where $\hat{\beta}_{0(30M)} = (\hat{\beta}_0 + 30\hat{\beta}_2)$ and $\hat{\beta}_{1(30M)} = (\hat{\beta}_1 + 30\hat{\beta}_4)$. This model is also a straight line model for predicting SBP from PAL, but here the slope of the model that represents the measure of association is $\hat{\beta}_{1(30M)}$. It follows that the two estimated slopes will differ, i.e., $\hat{\beta}_{(40F)} \neq \hat{\beta}_{(30M)}$, whenever $\hat{\beta}_4 \neq 0$ and/or $\hat{\beta}_5 \neq 0$. Thus, the association between PAL and SBP varies with the values of AGE and/or SEX, i.e., the association is **modified** by AGE and SEX.

11.4 Confounding in Regression

We emphasized in Section 11.1 that the assessment of confounding is questionable in the presence of interaction. Thus, our discussion of confounding here assumes throughout that no interaction is present.[3]

11.4.1 Controlling for One Extraneous Variable

Suppose that we are interested in describing the relationship between an independent variable T and a continuous dependent variable Y, taking into account the possible confounding effect of a third variable C. As described in Section 11.2, the assessment of confounding requires us to

---

[3] It is possible, however, to assess confounding for variables that are not components of interaction terms. For example, if we consider the model $Y = \beta_0 + \beta_1 X_1 + \beta_2 X_2 + \beta_3 X_3 + \beta_4 X_1 X_3 + E$, where X_1 is the study variable of interest, we might want to consider whether X_2 is a confounder, since it is not a component of $X_1 X_3$, the only interaction term in the model. For more realistic examples, see Kleinbaum, Kupper, and Morgenstern (1982, chap. 23).

compare a crude estimate of the *T–Y* relationship—which ignores the effect of the control variable (*C*)—with an estimate of the relationship that accounts (or controls) for this variable. This comparison can be expressed in terms of the following two regression models:

$$Y = \beta_0 + \beta_1 T + \beta_2 C + E \qquad (11.5)$$

and

$$Y = \beta_0 + \beta_1 T + E \qquad (11.6)$$

The assumption of no $T \times C$ interaction precludes the need to consider a product term of the form *TC* in these models.

From model (11.5), we can express the relationship between *T* and *Y*, adjusted for the variable *C*, in terms of the (partial) regression coefficient (β_1) of the *T* variable. The estimate of β_1 (which we will denote by $\hat{\beta}_{1|C}$), obtained from least-squares fitting of model (11.5), is an adjusted-effect measure in the sense that it gives the estimated change in *Y* per unit change in *T* after accounting for *C* (i.e., with *C* in the model).

A crude estimate of the *T–Y* relationship is the estimated coefficient of *T* (namely, $\hat{\beta}_1$) based on model (11.6)—a model that does not involve the variable *C*.

Thus, we have the following general rule for assessing the presence of confounding when only one independent variable is to be controlled: confounding is present if the estimate of the coefficient (β_1) of the study variable *T* meaningfully changes when the variable *C* is removed from model (11.5)—that is, if

$$\hat{\beta}_{1|C} \neq \hat{\beta}_1 \qquad (11.7)$$

where $\hat{\beta}_{1|C}$ denotes the (adjusted for *C*) estimate of β_1 using model (11.5), and $\hat{\beta}_1$ denotes the (crude) estimate of β_1 using model (11.6).

The $\neq$ sign in expression (11.7) indicates that a subjective decision is required as to whether the two estimates are meaningfully different; that is, we must determine subjectively whether the two estimates each describe a different interpretation of the *T–Y* association in question. A statistical test is neither required nor appropriate (Kleinbaum, Kupper, and Morgenstern 1982, chap. 13, or Kleinbaum 2002, Lesson 10).

As an example, suppose that *Y* denotes SBP, *T* denotes PAL, and *C* denotes AGE. For some set of data, suppose we found that

$$\hat{\beta}_{1|\text{AGE}} = 4.1 \quad \text{and} \quad \hat{\beta}_1 = 15.9$$

Then, we could conclude that a 1-unit change in PAL yields about a 16-unit change in SBP when AGE is ignored, and that a 1-unit change in PAL yields only a 4.1-unit change in SBP when AGE is controlled; that is, the association between PAL and SBP is much weaker after controlling for AGE. (As a special case, if PAL is a 0–1 variable, then $\hat{\beta}_1$ gives the crude difference in mean systolic blood pressures between the two PAL groups, and $\hat{\beta}_{1|\text{AGE}}$ gives an adjusted [for AGE] difference in mean blood pressures.) Thus, we would treat AGE as a confounder and control for it in the analysis.

As another example, suppose that

$$\hat{\beta}_{1|\text{AGE}} = 6.2 \quad \text{and} \quad \hat{\beta}_1 = 6.1$$

Here, we would be inclined to say that AGE is not a confounder because there is no meaningful difference between the estimates 6.2 and 6.1. Unfortunately, an investigator may have to deal with much more problematic comparisons, such as $\hat{\beta}_{1|AGE} = 4.1$ versus $\hat{\beta}_1 = 5.5$. In comparing such estimates numerically, one must consider the clinical importance of the numerical difference between estimates, based on (a priori) knowledge of the variable(s) involved. For instance, since the coefficients 4.1 and 5.5 estimate, respectively, adjusted and crude differences in mean blood pressures between high and low PAL groups, it is important to decide whether a mean difference of 5.5 is clinically more important than a mean difference of 4.1. One approach to this problem is to control for any variable (as a confounder) that changes the crude effect estimate by some prespecified amount determined by clinical judgment.

One approach sometimes used (incorrectly) to assess confounding is, for example, to conduct a statistical test of H_0: $\beta_2 = 0$ in model (11.5). Such a test does not address confounding, but rather *precision;* that is, such a test evaluates whether significant additional variation in Y is explained by adding C to a model already containing T. An almost equivalent approach is to determine whether a confidence interval for β_1, the coefficient of T, is considerably narrower when C is in the model than when it is not. Precision is often an important issue to assess when considering extraneous factors, but it differs fundamentally from confounding. For etiologic questions, confounding, which concerns validity (i.e., do you have the right answer?), usually takes precedence over precision. Another reason for not focusing on β_2 is that, if $\hat{\beta}_2 \neq 0$, it does not follow that $\hat{\beta}_{1|C} \neq \hat{\beta}_1$. In other words, $\hat{\beta}_2 \neq 0$ is not a sufficient condition for confounding.[4]

What type of variables (i.e., covariates) should be considered for control as potential confounders? Although our answer is somewhat debatable, we consider that a list of eligible variables should be constructed based on prior knowledge and/or research about the relationship of the dependent variable to each covariate under consideration. In particular, we recommend that only variables known to be reasonably predictive of (i.e., associated with) the dependent variable should be considered as potential confounders and/or effect modifiers. In epidemiological terms, such variables are generally referred to as *risk factors* (Kleinbaum 2002). The idea here is to restrict attention to controlling only the (previously studied) extraneous variables that the investigator anticipates may account for the hypothesized relationship between T and Y currently being studied. To develop such a list, the investigators have to make a subjective decision.[5]

11.4.2 Controlling for Several Extraneous Variables

Suppose that we want to describe the association between T and Y, taking into account several covariates $C_1, C_2, \ldots, C_p$. Similarly to our approach for handling one covariate, we can assess confounding by comparing a crude estimate of the T–Y relationship to some

---

[4] Suppose that $n = 6$ and that we have the following data for (T, C, Y): $(1, 0, 4)$, $(1, 1, 5)$, $(1, 2, 6)$, $(0, 0, 1)$, $(0, 1, 2)$, and $(0, 2, 3)$. Then unweighted least-squares fitting gives $\hat{Y} = 1 + 3T + C$ when T and C are predictors, whereas $\hat{Y} = 2 + 3T$ when C is ignored. Thus, $\hat{\beta}_2 = 1 (\neq 0)$, yet there is no confounding since $\hat{\beta}_1 = 3 = \hat{\beta}_{1|C}$.

[5] As a caveat to these recommendations, we note that certain variables usually referred to as *intervening variables* should not be considered as potential confounders (Kleinbaum, Kupper, and Morgenstern 1982). A variable C is said to intervene between T and Y if T causes C and then C causes Y. Controlling intervening variables may spuriously reduce or eliminate any manifestation in the data of a true association between T and Y.

adjusted estimate. As before, the crude estimate can be defined in terms of a regression model like (11.6), which describes the relationship between T and Y while ignoring all covariates. To obtain the adjusted estimate, however, we must now consider an extended model defined as

$$Y = \beta_0 + \beta_1 T + \beta_2 C_1 + \beta_3 C_2 + \cdots + \beta_{p+1} C_p + E \qquad (11.8)$$

(Like model (11.5), model (11.8) assumes that there is no interaction involving T, since no product terms of the form TC_i are included.)

Using this model, we can define confounding that involves several variables as follows: Confounding is present if the estimate of the regression coefficient (β_1) of T in a regression model like (11.6), which ignores the variables $C_1, C_2, \ldots, C_p$, is meaningfully different from the corresponding estimate of β_1 based on a model like (11.8), which controls for $C_1, C_2, \ldots, C_p$—that is, if

$$\hat{\beta}_{1|C_1, C_2, \ldots, C_p} \neq \hat{\beta}_1 \qquad (11.9)$$

where $\hat{\beta}_{1|C_1, C_2, \ldots, C_p}$ denotes the (adjusted) estimate of β_1 using (11.8) and $\hat{\beta}_1$ is the (crude) estimate of β_1 using (11.6).

One problem with applying this definition is that it addresses the question of whether confounding is present without directly identifying specific variables to be controlled.[6] In other words, when confounding is deemed to be present based on (11.9), it may still be the case that only a subset of $C_1, C_2, \ldots, C_p$ is required for adequate control. How does one identify such a subset? More specifically, why bother to identify such a subset rather than simply to control for all variables $C_1, C_2, \ldots, C_p$?

One answer to the latter question is that, when addressing the control of covariates, we should consider the possible gains in *precision*[7] in addition to the control of confounding. In particular, we may prefer a subset of C_j variables to the entire set because the subset may provide equivalent control of confounding (i.e., may give essentially the same adjusted estimate) while providing greater precision in estimating the adjusted association of interest. However, there is no guarantee that precision will be increased by using a subset; in fact, precision may frequently be reduced. In any case, confounding should take precedence over precision in the sense that no subset should be considered unless it gives almost the same adjusted-effect estimate as is obtained when the researcher controls for all C_j's.

---

[6] Another problem concerns the assessment of confounding when there are two or more study variables—say, T_1 and T_2—of interest. For this general situation, confounding may be defined to be present if (11.9) is satisfied for the coefficient of *any* study variable of interest, given a model containing all such study variables and all control variables. Unfortunately, this definition has the practical drawback of requiring the researcher to make a subjective decision for each study variable of interest.

[7] The term *precision* refers to the size of an estimator's variance or, equivalently, the narrowness of a confidence interval for the parameter being estimated. The smaller the variance of the estimator, the higher is the precision of the estimator. Equivalently, since the width of a confidence interval depends on the variance estimate, the narrower is the width of the confidence interval, the higher is the precision of the estimator.

To illustrate, suppose that $p = 5$; that is, we consider controlling for $C_1, C_2, \ldots, C_5$ using model (11.8). Suppose, too, that the estimate of β_1 takes on the following values, depending on which sets of $C_1, C_2, \ldots, C_5$ are controlled:

$$\hat{\beta}_{1|C_1, C_2, \ldots, C_5} = 4.0, \quad \hat{\beta}_{1|C_1, C_2} = 4.3, \quad \hat{\beta}_1 = 16.0$$

Then, because 16.0 is much different from 4.0, one can argue that confounding is present. Yet since 4.0 is not meaningfully different from 4.3, it can also be argued that C_3, C_4, and C_5 do not need to be controlled, since essentially the same (adjusted) estimate is obtained when we control only for C_1 and C_2 as when we adjust for all C_j's.

Thus, for this example, we have identified two sets of C_j variables that we can use for control of confounding. Which set do we choose? The answer depends on an evaluation of precision. One approach is to compare interval estimates for some parameter of interest—one interval being derived from a model that controls for C_1 and C_2 only, and the other interval coming from a model that controls for C_1 through C_5. The logical parameter for this example is the population regression coefficient β_1 of the variable T, when controlling for a particular set of C_j's. In other words, we may compare an interval estimate for β_1 when only C_1 and C_2 are controlled to a corresponding interval estimate for β_1 when C_1 through C_5 are controlled. The narrower interval of the two is the interval reflecting the greater precision. For example, if the two 95% interval estimates are $(2.6, 7.4)$ for $\beta_{1|C_1, C_2}$ and $(1.7, 7.6)$ for $\beta_{1|C_1, C_2, \ldots, C_5}$, then the former interval is narrower; in this case, some precision is gained by dropping C_3, C_4, and C_5 from the model.

An alternative, but not exactly equivalent, approach to evaluating precision is to perform a statistical test for the significance of the addition of C_3, C_4, and C_5 to a model containing T, C_1, and C_2. The null hypothesis for this test may be stated as H_0: $\beta_4 = \beta_5 = \beta_6 = 0$ in model (11.8), with $p = 5$. If this test is not significant, we could argue that retaining C_3, C_4, and C_5 does not provide additional precision (i.e., explanation of variance). This would indicate that only C_1 and C_2 should be controlled for greater precision.

Because this testing approach does not always lead to the same conclusion as the internal estimation approach, the investigator may need to choose between them. In most situations, however, both approaches lead to similar results.

How do we identify which set of C_j variables to control? We have seen, by example, that we must first identify a baseline-adjusted estimate (i.e., a "gold standard") against which to make comparisons. The ideal gold standard is the regression coefficient estimate that controls for all C_j's that are being considered. Then, any subset of C_j's that gives essentially the same adjusted estimate (i.e., an estimate that is not meaningfully different from the gold standard when only the C_j's in that subset are controlled) is a candidate set for control. Several such candidates may be possible (Kleinbaum, Kupper, and Morgenstern 1982, chap. 14, or Kleinbaum and Klein 2002, chap. 7).

Which set should we finally use? The answer, again, is based on precision: we should use the set that gives the greatest precision (e.g., the tightest confidence interval for the adjusted effect under study). (For "political" reasons—that is, to convince people that all variables have been controlled—it might be better to control for $C_1, C_2, \ldots, C_p$, unless some subset of C_j's leads to a large increase in precision.)

To illustrate, suppose that the candidate sets in Table 11.2 can be identified when $p = 5$ in model (11.8). All three proper subsets of C_1, C_2, C_3, C_4, and C_5 may be considered candidates for control, since they all give adjusted estimates that are roughly equal to the gold standard $\hat{\beta}_{1|C_1, C_2, \ldots, C_5} = 4.8$. Of these candidates, the subset involving C_1, C_2, and C_4 gives the best precision (narrowest confidence interval); therefore, this subset can be used both to control confounding and to enhance precision.

TABLE 11.2   An example of candidate sets for control

Candidate Set	$\hat{\beta}_{1\,|\,\text{Candidate Set}}$	95% Confidence Interval for $\beta_{1\,|\,\text{Candidate Set}}$
C_1, C_2, C_3, C_4, C_5 (baseline)	4.8	(2.3, 7.2)
C_1, C_2	5.1	(2.6, 7.6)
C_1, C_4	4.6	(2.1, 7.0)
C_1, C_2, C_4	4.7	(2.6, 6.8)

11.4.3 An Example Revisited

In Section 11.3.3 we considered a hypothetical study to assess the relationship between physical activity level (PAL) and systolic blood pressure (SBP) while controlling for both AGE and SEX. We considered a model that allows for possible interactions of AGE and SEX with PAL, and we described methods for testing for such interactions. Assuming that no significant interaction effects are found, the resulting reduced model is

$$\text{SBP} = \beta_0 + \beta_1(\text{PAL}) + \beta_2(\text{AGE}) + \beta_3(\text{SEX}) + E$$

Given this no-interaction model, we next assess confounding: Does the coefficient of PAL change when AGE and/or SEX is dropped from the model? To answer this, we can examine the estimate of the coefficient of PAL in four models—namely, one including both AGE and SEX, one involving either AGE or SEX but not both, and one involving neither. The gold standard model for comparison contains both control variables and PAL. Then, for example, if the estimate of β_1 changes considerably when at least one control variable is dropped from this gold standard model, we need to control for both AGE and SEX. However, if we obtain essentially the same estimate of β_1 (as obtained using the gold standard model) when only AGE is in the model, we do not need to retain SEX in the model to control for confounding. Even so, including the SEX variable in addition to AGE may increase or decrease precision. Thus, the decision as to whether to control for just AGE or for both AGE and SEX depends, for example, on a comparison of confidence intervals for β_1. If the confidence interval is considerably narrower when only AGE is controlled, we should not retain SEX in the model.

Finally, once we make a decision about which variables to control (i.e., which model is the best for providing a valid and precise estimate of the coefficient of PAL), we make statistical inferences about the true PAL–SBP relationship. Given the no-interaction model, this involves testing $H_0: \beta_1 = 0$ in the best model and then obtaining an interval estimate of β_1.

11.5 Summary and Conclusions

Confounding and interaction are two methodological concepts that pertain to assessing a relationship between independent and dependent variables.

Interaction, which takes precedence over confounding, exists when the relationship of interest differs at different levels of extraneous (control) variables. In multiple linear regression, interaction is evaluated by using statistical tests about product terms involving basic independent variables in the model.

Confounding, which is not evaluated with statistical testing, is present when the effect of interest differs depending on whether an extraneous variable is ignored or retained in the analysis. In regression terms, confounding is assessed by comparing crude versus adjusted regression coefficients from different models.

When several potential confounders are being considered, it may be worthwhile to identify nonconfounders that can be dropped from the model to gain precision; this may not be possible (i.e., precision may be lost by dropping variables) in some situations.

When there is strong interaction involving a certain extraneous variable, the assessment of confounding for that extraneous variable becomes irrelevant. Moreover, in such a situation, the assessment of confounding involving other extraneous variables, though possible, is quite complex and extremely subjective. Consequently, the assessment of confounding is not usually recommended when important interaction effects have been identified.

Problems

1. Consider the numerical examples given in Section 8.7 of Chapter 8, involving assessment of the relationship of the independent variables HGT, AGE, and $(AGE)^2$ to the dependent variable WGT. Suppose that HGT is the independent variable of primary concern, so interest lies in evaluating the relationship of HGT to WGT, controlling for the possible confounding effects of AGE and $(AGE)^2$.

   a. Assuming that no interaction of any kind exists, state an appropriate regression model to use as the baseline (i.e., standard) for decisions about confounding.

   b. Using an appropriate regression coefficient given in part (a) as your measure of association, determine whether confounding exists due to AGE and/or $(AGE)^2$.

   c. Can $(AGE)^2$ be dropped from your initial model in part (a) because it is not needed to control adequately for confounding? Explain your answer (using a regression coefficient as your measure of association).

   d. Should $(AGE)^2$ be retained in the final model for the sake of precision? Explain.

   e. In light of both confounding and precision, what should be your final model? Why?

   f. How would you modify your initial model in part (a) to allow for assessing interactions?

   g. Regarding your answer to part (f), how would you test for interaction?

2. Consider the following computer results, which describe regression analyses involving two independent variables X_1 and X_2 and a dependent variable Y. Assume that your goal is to assess the relationship of X_1 with Y, controlling for the possible confounding effects of X_2.

Edited SAS Output (PROC REG) for Problem 2

Y Regressed on X1 and X2

Correlation

CORR	X1	X2	Y
X1	1.0000	0.5000	0.6527
X2	0.5000	1.0000	0.9494
Y	0.6527	0.9494	1.0000

(continued)

SAS Output for Problem 2 (Continued)

Y Regressed on X1 and X2

Analysis of Variance

Source	DF	Sum of Squares	Mean Square	F Value	Pr > F
Model	2	268.00000	134.00000	108.875	0.0001
Error	13	16.00000	1.23077		
Corrected Total	15	284.00000			

. [Portion of output omitted]

Parameter Estimates

Variable	DF	Parameter Estimate	Standard Error	t value	Pr > |t|
INTERCEP	1	5.000000	0.42365927	11.802	0.0001
X1	1	2.000000	0.64051262	3.122	0.0081
X2	1	7.000000	0.64051262	10.929	0.0001

Variable	DF	Squared Partial Corr Type I	Squared Partial Corr Type II
X1	1	0.42605634	0.42857143
X2	1	0.90184049	0.90184049

Y Regressed on X1

Analysis of Variance

Source	DF	Sum of Squares	Mean Square	F Value	Pr > F
Model	1	121.00000	121.00000	10.393	0.0061
Error	14	163.00000	11.64286		
Corrected Total	15	284.00000			

. [Portion of output omitted]

Parameter Estimates

Variable	DF	Parameter Estimate	Standard Error	t value	Pr > |t|
INTERCEP	1	6.750000	1.20638184	5.595	0.0001
X1	1	5.500000	1.70608156	3.224	0.0061

a. Using an appropriate regression coefficient as your measure of association, determine whether confounding exists. Explain.

b. Suppose that confounding was defined to require a comparison of crude versus adjusted (partial) correlation coefficients. What conclusion would you draw? Explain.

c. What is the moral of this example?

3. a–c. Consider the accompanying computer results, which describe regression analyses involving two independent variables X_1 and X_2 and a dependent variable Y (using a different data set from the one used in Problem 2). Answer the same questions as in Problem 2 for this new printout.

d. What does this example illustrate about using a test of the hypothesis H_0: $\beta_2 = 0$ to assess confounding?

Edited SAS Output (PROC REG) for Problem 3

Y Regressed on X1 and X2

Correlation

CORR	X1	X2	Y
X1	1.0000	0.0000	0.2649
X2	0.0000	1.0000	0.9272
Y	0.2649	0.9272	1.0000

Analysis of Variance

Source	DF	Sum of Squares	Mean Square	F Value	Pr > F
Model	2	106.00000	53.00000	33.125	0.0013
Error	5	8.00000	1.60000		
Corrected Total	7	114.00000	.		

. [Portion of output omitted]

Parameter Estimates

Variable	DF	Parameter Estimate	Standard Error	t value	Pr > |t|
INTERCEP	1	5.000000	0.77459667	6.455	0.0013
X1	1	2.000000	0.89442719	2.236	0.0756
X2	1	7.000000	0.89442719	7.826	0.0005

Variable	DF	Squared Partial Corr Type I	Squared Partial Corr Type II
X1	1	0.07017544	0.50000000
X2	1	0.92452830	0.92452830

Y Regressed on X1

Analysis of Variance

Source	DF	Sum of Squares	Mean Square	F Value	Pr > F
Model	1	8.00000	8.00000	0.453	0.5260
Error	6	106.00000	17.66667		
Corrected Total	7	114.00000			

. [Portion of output omitted]

Parameter Estimates

Variable	DF	Parameter Estimate	Standard Error	t value	Pr > |t|
INTERCEP	1	8.500000	2.10158670	4.045	0.0068
X1	1	2.000000	2.97209242	0.673	0.5260

4. A regression analysis of data on $n = 53$ males considered the following variables:

$Y =$ SBPSL (estimated slope based on the straight-line regression of an individual's blood pressure over time)

$X_1 =$ SBP1 (initial blood pressure)

$X_2 =$ RW (relative weight)

$X_3 = X_1 X_2 =$ SR (product of SBP1 and RW)

The accompanying computer printout was obtained by using a standard stepwise regression program (SPSS). Using this output, complete the following exercises.

a. Fill in the following ANOVA table for the fit of the model $Y = \beta_0 + \beta_1 X_1 + \beta_2 X_2 + \beta_3 X_3 + E$.

Source		d.f.	SS	MS
Regression $\begin{cases} X_1 \\ X_2\vert X_1 \\ X_3\vert X_1, X_2 \end{cases}$				
Residual				
Total		52		

b. Test $H_0: \rho_{YX_2\vert X_1} = 0$.

c. Test H_0: "The addition of X_3 to the model, given that X_1 and X_2 are already in the model, is not significant."

d. Test $H_0: \rho_{Y(X_2, X_3)\vert X_1} = 0$.

e. Based on the tests in parts (b) through (d), what is the most appropriate regression model? Use $\alpha = .05$.

f. Based on the information provided, can you assess whether X_1 is a confounder of the X_2–Y relationship? Explain.

5. An experiment involved a quantitative analysis of factors found in high-density lipoprotein (HDL) in a sample of human blood serum. Three variables thought to be predictive of or associated with HDL measurement (Y) were the total cholesterol (X_1) and total triglyceride (X_2) concentrations in the sample, plus the presence or absence of a certain sticky component called sinking pre-beta, or SPB (X_3), which was coded as 0 if absent and 1 if present. The data obtained are shown in the table on page 209 and the accompanying computer results.

Edited SPSS Output for Problem 4

```
DEPENDENT VARIABLE.  . SBPSL
VARIABLE(S) ENTERED ON STEP NUMBER 1.. SBP 1
                         ANALYSIS OF VARIANCE   DF   SUM OF SQUARES   MEAN SQUARE          F
MULTIPLE R       0.45834   REGRESSION           1.       14.79083      14.79083     13.56308
R SQUARE         0.21007   RESIDUAL            51.       55.61661       1.09052
STANDARD ERROR   1.04428

             VARIABLES IN THE EQUATION-----               -----VARIABLES NOT IN THE EQUATION-----
VARIABLE        B       BETA   STD ERROR B      F      VARIABLE    BETA   IN PARTIAL   TOLERANCE      F
SBP 1      -0.04660  -0.45834    0.01265      13.563     RW      0.23166    0.26007     0.99553     3.627
(CONSTANT)  5.10797                                      SR      0.23074    0.25953     0.99933     3.611

VARIABLE(S) ENTERED ON STEP NUMBER 2.. RW
                         ANALYSIS OF VARIANCE   DF   SUM OF SQUARES   MEAN SQUARE          F
MULTIPLE R       0.51332   REGRESSION           2.       18.55240       9.27620      8.94435
R SQUARE         0.26350   RESIDUAL            50.       51.85504       1.03710
STANDARD ERROR   1.01838

             VARIABLES IN THE EQUATION-----               -----VARIABLES NOT IN THE EQUATION-----
VARIABLE        B       BETA   STD ERROR B      F      VARIABLE    BETA   IN PARTIAL   TOLERANCE      F
SBP 1      -0.04817  -0.47382    0.01237      15.174     SR      0.04646    0.00450     0.00690     0.001
RW          0.02252   0.23166    0.01182       3.627
(CONSTANT)  5.38484

VARIABLE(S) ENTERED ON STEP NUMBER 3.. SR
                         ANALYSIS OF VARIANCE   DF   SUM OF SQUARES   MEAN SQUARE          F
MULTIPLE R       0.51334   REGRESSION           3.       18.55345       6.18448      5.84409
R SQUARE         0.26352   RESIDUAL            49.       51.85399       1.05824
STANDARD ERROR   1.02871

             VARIABLES IN THE EQUATION-----               -----VARIABLES NOT IN THE EQUATION-----
VARIABLE        B       BETA   STD ERROR B      F      VARIABLE    BETA   IN PARTIAL   TOLERANCE      F
SBP 1      -0.04798  -0.47193    0.01391      11.899
RW          0.01801   0.18527    0.14372       0.016
SR          0.00004   0.04646    0.00122       0.001
(CONSTANT)  5.36183
```

From Nie et al., *Statistical Package for the Social Sciences*. Copyright© 1975 by McGraw-Hill Book Company and Dr. Norman Nie, President, SPSS, Inc.

Dataset for Problem 5

Y	X_1	X_2	X_3	Y	X_1	X_2	X_3
47	287	111	0	57	192	115	1
38	236	135	0	42	349	408	1
47	255	98	0	54	263	103	1
39	135	63	0	60	223	102	1
44	121	46	0	33	316	274	0
64	171	103	0	55	288	130	0
58	260	227	0	36	256	149	0
49	237	157	0	36	318	180	0
55	261	266	0	42	270	134	0
52	397	167	0	41	262	154	0
49	295	164	0	42	264	86	0
47	261	119	1	39	325	148	0
40	258	145	1	27	388	191	0
42	280	247	1	31	260	123	0
63	339	168	1	39	284	135	0
40	161	68	1	56	326	236	1
59	324	92	1	40	248	92	1
56	171	56	1	58	285	153	1
76	265	240	1	43	361	126	1
67	280	306	1	40	248	226	1
57	248	93	1	46	280	176	1

Edited SAS Output (PROC GLM) for Problem 5

```
                        General Linear Models Procedure
                            Y Regressed on X1
                                         Sum of
Source                    DF             Squares           F Value        Pr > F
Model                      1         46.23555746              0.40        0.5282
Error                     40       4567.38349016
Corrected Total           41       4613.61904762

                     R-Square          Coeff Var          Y Mean
                     0.010022          22.37289        47.7619048

                                        Type I
Source             DF                      SS           F Value        Pr > F
X1                  1              46.23555746              0.40        0.5282

                                       Type III
Source             DF                      SS           F Value        Pr > F
X1                  1              46.23555746              0.40        0.5282

Parameter            Estimate        Standard Error     t value        Pr > |t|
INTERCEPT         52.47018272           7.58057266         6.92        0.0001
X1                -0.01758070           0.02762815        -0.64        0.5282

                            Y Regressed on X2
                                         Sum of
Source                    DF             Squares           F Value        Pr > F
Model                      1         21.33970871              0.19        0.6687
Error                     40       4592.27933891
Corrected Total           41       4613.61904762
```

(continued)

SAS Output for Problem 5 (Continued)

Y Regressed on X2 (Continued)

	R-Square	Coeff Var	Y Mean
	0.004625	22.43378	47.7619048

Source	DF	Type I SS	F Value	Pr > F
X2	1	21.33970871	0.19	0.6687

Source	DF	Type III SS	F Value	Pr > F
X2	1	21.33970871	0.19	0.6687

Parameter	Estimate	Standard Error of Estimate	t value	Pr > |t|
INTERCEPT	46.24519280	3.88711502	11.90	0.0001
X2	0.00978223	0.02268966	0.43	0.6687

Y Regressed on X3

Source	DF	Sum of Squares	F Value	Pr > F
Model	1	735.20541126	7.58	0.0088
Error	40	3878.41363636		
Corrected Total	41	4613.61904762		

	R-Square	Coeff Var	Y Mean
	0.159355	20.61652	47.7619048

Source	DF	Type I SS	F Value	Pr > F
X3	1	735.20541126	7.58	0.0088

Source	DF	Type III SS	F Value	Pr > F
X3	1	735.20541126	7.58	0.0088

Parameter	Estimate	Standard Error	t value	Pr > |t|
INTERCEPT	43.77272727	2.09935424	20.85	0.0001
X3	8.37727273	3.04225332	2.75	0.0088

Y Regressed on X1 and X2

Source	DF	Sum of Squares	F Value	Pr > F
Model	2	135.38212555	0.59	0.5595
Error	39	4478.23692207		
Corrected Total	41	4613.61904762		

	R-Square	Coeff Var	Y Mean
	0.029344	22.43570	47.7619048

Source	DF	Type I SS	F Value	Pr > F
X1	1	46.23555746	0.40	0.5294
X2	1	89.14656809	0.78	0.3837

Source	DF	Type III SS	F Value	Pr > F
X1	1	114.04241683	0.99	0.3251
X2	1	89.14656809	0.78	0.3837

Parameter	Estimate	Standard Error	t value	Pr > |t|
INTERCEPT	52.76400680	7.60916422	6.93	0.0001
X1	-0.03216055	0.03227093	-1.00	0.3251
X2	0.02328833	0.02643061	0.88	0.3837

(continued)

SAS Output for Problem 5 (Continued)

```
                    General Linear Models Procedure
                  Y Regressed on X1 and X3 (Continued)
                                 Sum of
Source                  DF       Squares        F Value      Pr > F
Model                    2    783.16906883         3.99      0.0266
Error                   39   3830.44997878
Corrected Total         41   4613.61904762

             R-Square        Coeff Var         Y Mean
             0.169752        20.74966       47.7619048

                              Type I
Source        DF                  SS        F Value      Pr > F
X1             1         46.23555746           0.47      0.4967
X3             1        736.93351138           7.50      0.0092

                             Type III
Source        DF                  SS        F Value      Pr > F
X1             1         47.96365758           0.49      0.4888
X3             1        736.93351138           7.50      0.0092

Parameter         Estimate      Standard Error      t value      Pr > |t|
INTERCEPT      48.56350948          7.17377750         6.77        0.0001
X1             -0.01790642          0.02562390        -0.70        0.4888
X3              8.38720265          3.06193225         2.74        0.0092

                      Y Regressed on X2 and X3
                                 Sum of
Source                  DF       Squares        F Value      Pr > F
Model                    2    737.80689080         3.71      0.0334
Error                   39   3875.81215682
Corrected Total         41   4613.61904762

             R-Square        Coeff Var         Y Mean
             0.159919        20.87216       47.7619048
                              Type I
Source        DF                  SS        F Value      Pr > F
X2             1         21.33970871           0.21      0.6457
X3             1        716.46718209           7.21      0.0106
                             Type III
Source        DF                  SS        F Value      Pr > F
X2             1          2.60147955           0.03      0.8723
X3             1        716.46718209           7.21      0.0106

Parameter         Estimate      Standard Error      t value      Pr > |t|
INTERCEPT      48.26641927          3.78286563        11.44        0.0001
X2              0.00343683          0.02124209         0.16        0.8723
X3              8.32148668          3.09921587         2.69        0.0106

                 Y Regressed on X1, X2, X3, X1X3, and X2X3
                                 Sum of
Source                  DF       Squares        F Value      Pr > F
Model                    5    894.56736631         1.73      0.1523
Error                   36   3719.05168130
Corrected Total         41   4613.61904762

             R-Square        Coeff Var         Y Mean
             0.193897        21.28057       47.7619048
```

(continued)

SAS Output for Problem 5 (Continued)

```
                        General Linear Models Procedure
               Y Regressed on X1, X2, X3, X1X3, and X2X3 (Continued)
```

Type I

Source	DF	SS	F Value	Pr > F
X1	1	46.23555746	0.45	0.5078
X2	1	89.14656809	0.86	0.3591
X3	1	684.36519731	6.62	0.0143
X1X3	1	48.69043235	0.47	0.4968
X2X3	1	26.12961110	0.25	0.6181

Type III

Source	DF	SS	F Value	Pr > F
X1	1	154.23113774	1.49	0.2297
X2	1	47.48756478	0.46	0.5021
X3	1	1.99625065	0.02	0.8902
X1X3	1	74.26755996	0.72	0.4021
X2X3	1	26.12961110	0.25	0.6181

Parameter	Estimate	Standard Error	t value	Pr > |t|
INTERCEPT	52.33465853	9.25548949	5.65	0.0001
X1	-0.04955819	0.04055966	-1.22	0.2297
X2	0.03188430	0.04702749	0.68	0.5021
X3	-2.07933755	14.95830269	-0.14	0.8902
X1X3	0.05438698	0.06414462	0.85	0.4021
X2X3	-0.02821521	0.05610244	-0.50	0.6181

5. a. Test whether X_1, X_2, or X_3 alone significantly helps in predicting Y.

b. Test whether X_1, X_2, and X_3 taken together significantly help to predict Y.

c. Test whether the true coefficients of the product terms X_1X_3 and X_2X_3 are simultaneously zero in the model containing X_1, X_2, and X_3 plus these product terms. State the null hypothesis in terms of a multiple partial correlation coefficient. If this test is not rejected, what can you conclude about the relationship of Y to X_1 and X_2 when X_3 equals 1, as compared with when X_3 equals 0?

d. Using $\alpha = .05$, test whether X_3 is associated with Y, after the combined contribution of X_1 and X_2 is taken into account. State the appropriate null hypothesis in terms of a partial correlation coefficient. What does your result, together with your answer to part (c), tell you about the relationship of Y with X_1 and X_2 when SPB is present, as compared with when it is absent?

e. How would you determine whether X_1, X_2, or both X_1 and X_2 need to be retained in the model to control for confounding and possibly to enhance precision? Assume that no interaction occurs and that the study variable of interest is X_3.

f. Based on the information provided, can confounding of X_1 and/or X_2 be assessed in evaluating the relationship of X_3 to Y? Explain.

6. Use the computer output from Problem 10 in Chapter 8 and from Problem 16 in Chapter 5 to answer the following questions. (Assume that no interaction occurs between house size (X_1) and number of rooms (X_2).)

a. Does the number of rooms (X_2) confound the relationship between house price (Y) and house size (X_1)?

b. Should the number of rooms be included in a model that already contains house size, based on considerations of precision?

c. In light of your answers to parts (a) and (b), should the final model include both predictors? Explain.

7. Use the output from Problem 11 in Chapter 8 and the output given here to answer the following questions. (Assume that no interaction occurs between TV advertising expenditure (X_1) and print advertising expenditure (X_2).)

  a. Does the print advertising expenditure confound the relationship between sales (Y) and TV advertising expenditure (X_1)?

  b. Should print advertising expenditure be included in a model that already contains TV advertising expenditure, based on considerations of precision?

  c. In light of your answers to parts (a) and (b), should the final model include both types of advertising expenditures as predictors? Explain.

Edited SAS Output (PROC GLM) for Problem 7

Y Regressed on X1

. [Portion of output omitted]

Parameter	Estimate	Standard Error	t value	Pr > |t|
INTERCEPT	2.002227171	0.31569714	6.34	0.0032
X1	1.224944321	0.10088866	12.14	0.0003

8. Use the computer output from Problem 12 in Chapter 8 and the output given here to answer the following questions.

  a. State the model that relates change in refraction (Y) to baseline refraction (X_1), baseline curvature (X_2), and the interaction of X_1 and X_2. Is the partial F test for the interaction significant?

  b. Is it appropriate to assess confounding, given your answer to part (a)? Explain.

  c. If your answer to part (b) is yes, does X_2 confound the relationship between Y and X_1?

  d. If your answer to part (b) is yes, does X_1 confound the relationship between Y and X_2?

  e. In light of your answers to parts (a) through (d), and on considerations of precision, which predictor(s) should be included in the model?

Edited SAS Output (PROC GLM) for Problem 8

Y Regressed on X1, X2, and X1X2

. [Portion of output omitted]

Parameter	Estimate	Standard Error	t value	Pr > |t|
INTERCEPT	25.85215663	16.33395716	1.58	0.1199
X1	2.69854313	3.41991591	0.79	0.4339
X2	-0.52636869	0.37006537	-1.42	0.1613
X1X2	-0.06778472	0.07750534	-0.87	0.3861

(continued)

SAS Output for Problem 8 (Continued)

```
                              Y Regressed on X1
                                      .
                                      . [Portion of output omitted]
                                      .
Parameter                 Estimate         Standard Error      t value      Pr > |t|
INTERCEPT               2.619102900            0.41260537          6.35        0.0001
X1                     -0.298729328            0.09463300         -3.16        0.0027

                              Y Regressed on X2
                                      .
                                      . [Portion of output omitted]
                                      .
Parameter                 Estimate         Standard Error      t value      Pr > |t|
INTERCEPT              14.14611679            5.52569396          2.56        0.0135
X2                     -0.23446227            0.12544713         -1.87        0.0674

                           Y Regressed on X1 and X2

Dependent Variable: Y                      Sum of
Source                     DF              Squares          F Value        Pr > F
Model                       2          17.53454762             7.02        0.0021
Error                      50          62.41955709
Corrected Total            52          79.95410472

                        R-Square              Coeff Var        Y mean
                        0.219308              29.22453     3.82320755
                                      .
                                      . [Portion of output omitted]
                                      .
Parameter                 Estimate         Standard Error      t value      Pr > |t|
INTERCEPT              12.29334748            5.13074580          2.40        0.0204
X1                     -0.29135396            0.09241113         -3.15        0.0027
X2                     -0.21905406            0.11581741         -1.89        0.0644
```

9. Refer to the data in Problem 19 of Chapter 5. The relationship between the rate of owner occupancy of housing units (OWNEROCC) and the median monthly ownership costs (OWNCOST) was studied in that problem in connection with a random sample of data from 26 Metropolitan Statistical Areas (MSAs). The results of the regression of OWNEROCC on OWNCOST and a third variable, the median household income (INCOME), are presented next.

     Use the output given here and in Problem 19 of Chapter 5 to answer the following questions.

a. Test whether OWNCOST and INCOME, taken together, significantly help to predict the rate of owner occupancy.

b. Test whether OWNCOST is associated with the rate of owner occupancy, after taking into account the contribution of INCOME.

c. Test whether INCOME is associated with the rate of owner occupancy, after taking into account the contribution of OWNCOST.

d. Should INCOME be included in the model to control for confounding? Explain your answer, including discussion of any assumptions you made in reaching your answer.

Edited SAS Output (PROC GLM) for Problem 9

```
                    OWNEROCC Regressed on OWNCOST and INCOME
                         General Linear Models Procedure
Dependent Variable: OWNEROCC
                                    Sum of              Mean
Source                   DF        Squares             Square      F Value      Pr > F
Model                     2    214.4833906        107.2416953         6.38      0.0062
Error                    23    386.4781479         16.8033977
Corrected Total          25    600.9615385

            R-Square            Coeff Var        Root MSE       OWNEROCC Mean
            0.356900             6.214523        4.099195           69.96154

                                   Type I               Mean
Source                   DF           SS             Square      F Value      Pr > F
OWNCOST                   1    132.6203242        132.6203242        7.89      0.0100
INCOME                    1     81.8630664         81.8630664        4.87      0.0375

                                  Type III              Mean
Source                   DF           SS             Square      F Value      Pr > F
OWNCOST                   1    192.6978217        192.6978217       11.47      0.0025
INCOME                    1     81.8630664         81.8630664        4.87      0.0375

Parameter            Estimate        Standard Error      t value      Pr > |t|
INTERCEPT          70.03201034          4.55666725         15.37        0.0001
OWNCOST            -0.03334031          0.00984532         -3.39        0.0025
INCOME             0.00064489          0.00029217          2.21        0.0375
```

10. Refer to the *Business Week* magazine data in Problem 13 of Chapter 8. The relationship between the yield (Y), 1989 rank (X_2), and P-E ratio (X_3) was studied in that problem, using a random sample of data from the magazine's compilation of information on the top 1,000 companies. The results of the regression of yield on 1989 rank, P-E ratio, and 1990 rank (X_1) are presented next. Use this output and the output from Problem 13 in Chapter 8 to answer the following questions.

Edited SAS Output (PROC GLM) for Problem 10

```
                 YIELD (Y) Regressed on 1990 RANK (X1), 1989 RANK (X2),
                                  and P-E RATIO (X3)

                         General Linear Models Procedure
Dependent Variable: Y
                                    Sum of              Mean
Source                   DF        Squares             Square      F Value      Pr > F
Model                     3     29.97725832          9.99241944        8.83      0.0011
Error                    16     18.10303668          1.13143979
Corrected Total          19     48.08029500

            R-Square            Coeff Var        Root MSE        Y Mean
            0.623483            42.79588         1.063692       .485500

                                   Type I               Mean
Source                   DF           SS             Square      F Value      Pr > F
X1                        1      1.82894871          1.82894871        1.62      0.2218
X2                        1      0.00159198          0.00159198        0.00      0.9705
X3                        1     28.14671763         28.14671763       24.88      0.0001
```

(continued)

SAS Output for Problem 10 (Continued)

```
         YIELD (Y) Regressed on 1990 RANK (X1), 1989 RANK (X2),
                    and P-E RATIO (X3) (continued)
```

Source	DF	Type III SS	Mean Square	F Value	Pr > F
X1	1	3.39453697	3.39453697	3.00	0.1025
X2	1	1.35441258	1.35441258	1.20	0.2901
X3	1	28.14671763	28.14671763	24.88	0.0001

Parameter	Estimate	Standard Error	t value	Pr > |t|
INTERCEPT	7.421187723	0.98839344	7.51	0.0001
X1	-0.013351684	0.00770835	-1.73	0.1025
X2	0.007620880	0.00696539	1.09	0.2901
X3	-0.261846279	0.05249866	-4.99	0.0001

a. Test whether 1989 rank, P-E ratio, and 1990 rank, taken together, significantly help to predict the yield.

b. Test whether 1990 rank is associated with the yield, after taking into account the contribution of 1989 rank and P-E ratio.

c. Should 1990 rank be included in the model to control for confounding? Explain your answer, including discussion of any assumptions you made in reaching your answer.

References

Kleinbaum, D. G. 2002. *ActivEpi*, New York and Berlin: Springer.

Kleinbaum, D. G.; and Klein M. 2002, *Logistic Regression: A Self-Learning Text*, Second Edition. New York and Berlin: Springer.

Kleinbaum, D. G.; Kupper, L. L.; and Morgenstern, H. 1982. *Epidemiologic Research*. Belmont, Calif.: Lifetime Learning Publications.

Nie, N., et al. 1975. *Statistical Package for the Social Sciences*. New York: McGraw-Hill.

Dummy Variables in Regression

12.1 Preview

To this point we have mainly considered only continuous variables as predictors, but the methods of regression analysis can be generalized to treat multi-category categorical predictors as well. The generalization is based entirely on the use of dummy variables, the central idea of this chapter.

By using dummy variables, we can broaden the application of regression analysis. In particular, dummy variables allow us to employ regression analysis to produce the same information obtained by such seemingly distinct analytical procedures as analysis of covariance (Chapter 13) and analysis of variance (Chapters 17 through 20).

In this chapter we focus on one important application of dummy variables: the comparison of several regression equations via the use of a single multiple regression model. We also describe an alternative method that can be used if only two equations are being compared.

12.2 Definitions

A *dummy,* or *indicator, variable* is any variable in a regression equation that takes on a finite number of values so that different categories of a nominal variable can be identified. The term *dummy* reflects the fact that the values taken on by such variables (usually values like 0, 1, and -1) do not indicate meaningful measurements but rather the categories of interest.

Examples of dummy variables include the following:

$$X_1 = \begin{cases} 1 & \text{if treatment A is used} \\ 0 & \text{otherwise} \end{cases}$$

$$X_2 = \begin{cases} 1 & \text{if subject is female} \\ -1 & \text{if subject is male} \end{cases}$$

$$Z_1 = \begin{cases} 1 & \text{if residence is in western United States} \\ 0 & \text{if residence is in central United States} \\ -1 & \text{if residence is in eastern United States} \end{cases}$$

$$Z_2 = \begin{cases} 0 & \text{if residence is in western United States} \\ 1 & \text{if residence is in central United States} \\ -1 & \text{if residence is in eastern United States} \end{cases}$$

The variable X_1 indicates a nominal variable describing the category "treatment group" (either treatment A or not treatment A); the variable X_2 indexes the levels of the nominal variable "sex"; and variables Z_1 and Z_2 work in tandem to describe the nominal variable "geographical residence." In the last case, the three categories of geographical residence are described by the following combination of the two variables Z_1 and Z_2:

Residence in western United States: $Z_1 = 1, Z_2 = 0$
Residence in central United States: $Z_1 = 0, Z_2 = 1$
Residence in eastern United States: $Z_1 = -1, Z_2 = -1$

12.3 Rule for Defining Dummy Variables

The following simple rule should always be applied to avoid collinearity (see Chapter 14) in defining a dummy variable for regression analysis: *If the nominal independent variable of interest has k categories, then exactly k − 1 dummy variables must be defined to index these categories, provided that the regression model contains a constant term (i.e., an intercept β_0). If the regression model does not contain an intercept, then k dummy variables are needed to index the k categories of interest.* For example, given $k = 3$ categories, the number of dummy variables should be $k - 1 = 2$ for a model containing an intercept. If an intercept is not included in an overall regression model designed to compare several regression equations, however, the dummy variables can be defined so that each of the regression equations derived from the overall model has its own intercept. Thus, using an intercept in the overall model generally depends on how the investigator prefers to code the dummy variables.

Applying this rule raises several significant points:

1. If an intercept is used in the regression equation, proper definition of the $k - 1$ dummy variables automatically indexes all k categories.

2. If k dummy variables are used to describe a nominal variable with k categories in a model containing an intercept, all the coefficients in the model cannot be uniquely estimated (i. e., collinearity is present).

3. The $k - 1$ dummy variables for indexing the k categories of a given nominal variable can be properly defined in many different ways. For example, two equivalent ways to describe the nominal variable "geographical residence" (represented earlier by Z_1 and Z_2) are

$$Z_1^* = \begin{cases} 1 & \text{if residence is in western United States} \\ 0 & \text{otherwise} \end{cases}$$

$$Z_2^* = \begin{cases} 1 & \text{if residence is in central United States} \\ 0 & \text{otherwise} \end{cases}$$

and

$$Z_1' = \begin{cases} 1 & \text{if residence is in western United States} \\ 0 & \text{otherwise} \end{cases}$$

$$Z_2' = \begin{cases} 1 & \text{if residence is in eastern United States} \\ 0 & \text{otherwise} \end{cases}$$

Among the coding schemes available for regression, we recommend the method often referred to as *reference cell coding,* which uses $k - 1$ dummy variables as suggested earlier. Each variable takes on only values of 1 and 0, and each variable indicates group membership (1 for a specific group, 0 otherwise). Some computer programs use 1 and -1 for coding. Since the choice of coding scheme affects analysis and interpretation, it is important to specify which coding method is being used.

We now illustrate how to use dummy variables to compare two or more regression models. We begin by considering two straight-line models, and then we extend the discussion to comparisons of more than two multiple regression models.

12.4 Comparing Two Straight-line Regression Equations: An Example

In Chapter 11, the age–systolic blood pressure example was used to assess whether there is a significant relationship between blood pressure and age. Here, we use the data to investigate the common observation that males tend to have higher blood pressure than females of similar age. To do this, we must compare the straight-line regression of systolic blood pressure versus age for females against the corresponding regression for males. The entire data set for this example was presented in Table 11.1. Table 12.1 also provides the information needed to compare the two fitted straight lines. For each data set, this information consists of the sample size (n), the intercept ($\hat{\beta}_0$), the slope ($\hat{\beta}_1$), the sample mean $\bar{X}$ of the X's, the sample mean $\bar{Y}$ of the Y's, the sample variance S_X^2 of the X's, and the residual mean-square error ($S_{Y|X}^2$). To distinguish between male and female data, we have used the subscripts M and F, respectively. Thus, n_M, $\hat{\beta}_{0M}$, $\hat{\beta}_{1M}$, and $S_{Y|X_M}^2$ denote the sample size, intercept, slope, and mean-square error for the male data, whereas n_F, $\hat{\beta}_{0F}$, $\hat{\beta}_{1F}$, and $S_{Y|X_F}^2$ denote the corresponding information for the female data.

The least-squares lines are then given as follows:[1]

Males: $\hat{Y}_M = 110.04 + 0.96X$

Females: $\hat{Y}_F = 97.08 + 0.95X$

TABLE 12.1

Group	n	$\hat{\beta}_0$	$\hat{\beta}_1$	$\bar{X}$	$\bar{Y}$	S_X^2	$S_{Y|X}^2$
Males	40	110.04	0.96	46.93	155.15	221.15	71.90
Females	29	97.08	0.95	45.07	139.86	242.14	91.46

[1] It can be shown that the straight-line model for males, like that for females, is appropriate based on a lack-of-fit test (see Chapter 15).

FIGURE 12.1   **Comparison by sex of straight-line regressions of
systolic blood pressure on age**

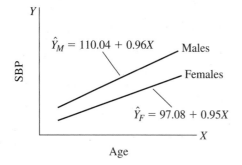

These lines are sketched in Figure 12.1. It can be seen from the figure that the male line lies completely above the female line. This fact alone supports the contention that males have higher blood pressure than females over the age range being considered. Nevertheless, it is necessary to explore statistically whether the observed differences between the regression lines could have occurred by chance. In other words, to be statistically precise when comparing two regression lines, we must consider the sampling variability of the data by using statistical test(s) and/or confidence interval(s). The sections that follow describe a number of statistical procedures for dealing with this comparison problem.

12.5 Questions for Comparing Two Straight Lines

There are three basic questions to consider when comparing two straight-line regression equations:

1. Are the two slopes the same or different (regardless of whether the intercepts are different)?[2]

2. Are the two intercepts the same or different (regardless of whether the slopes are different)?

3. Are the two lines coincident (that is, the same), or do they differ in slope and/or intercept?

Situations pertaining to these three questions are illustrated in Figure 12.2.

For our particular age–systolic blood pressure example, concluding that the lines are parallel (Figure 12.2(a)) is equivalent to finding that one sex has a consistently higher systolic blood pressure than the other at all ages, but that the rate of change with respect to age is the same for both sexes. If we conclude that the two lines have a common intercept but different slopes (Figure 12.2(b)), we find that the two sexes begin at an early age with the same average SBP but that average SBP changes with respect to age at different rates for each sex. If the two lines have

---

[2] If the two slopes are not different, we say that the two lines are *parallel*.

FIGURE 12.2   Possible conclusions from comparing two straight-line regressions

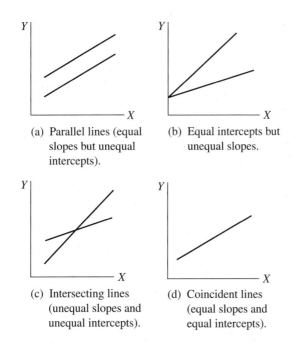

(a) Parallel lines (equal slopes but unequal intercepts).

(b) Equal intercepts but unequal slopes.

(c) Intersecting lines (unequal slopes and unequal intercepts).

(d) Coincident lines (equal slopes and equal intercepts).

different slopes and different intercepts[3] (Figure 12.2(c)), it means that the relationship between age and mean systolic blood pressure differs for the two sexes with regard to both the origins and the rates of change. Furthermore, if the lines intersect in the range of X-values of interest, this indicates that at early ages one sex has a higher average systolic blood pressure than the other, but at later ages the other sex does.

12.6 Methods of Comparing Two Straight Lines

There are two general approaches to answering the earlier three questions related to comparing two straight lines:

Method I   Treat the male and female data separately by fitting the two separate regression equations

$$Y_M = \beta_{0M} + \beta_{1M}X + E_M \tag{12.1}$$

and

$$Y_F = \beta_{0F} + \beta_{1F}X + E_F \tag{12.2}$$

and then conduct appropriate two-sample t tests.

---

[3] It is possible for two lines to have unequal slopes and unequal intercepts and yet not intersect within the range of X-values of interest. This is illustrated by our example in Figure 12.1.

Method II   Define the dummy variable Z to be 0 if the subject is male and 1 if female. Thus, for the n_M observations on males, $Z = 0$; and for the n_F observations on females, $Z = 1$. Our data will then be of the form:

Males: $(X_{1M}, Y_{1M}, 0), (X_{2M}, Y_{2M}, 0), \ldots, (X_{n_M M}, Y_{n_M M}, 0)$

Females: $(X_{1F}, Y_{1F}, 1), (X_{2F}, Y_{2F}, 1), \ldots, (X_{n_F F}, Y_{n_F F}, 1)$

Then, for the combined data above, the single multiple regression model

$$Y = \beta_0 + \beta_1 X + \beta_2 Z + \beta_3 XZ + E \tag{12.3}$$

yields the following two models for the two values of Z:

$$\begin{cases} Z = 0: & Y_M = \beta_0 + \beta_1 X + E \\ Z = 1: & Y_F = (\beta_0 + \beta_2) + (\beta_1 + \beta_3)X + E \end{cases}$$

This allows us to write the regression coefficients for the separate models for method I in terms of the coefficients of model (12.3) as follows:

$$\beta_{0M} = \beta_0, \quad \beta_{0F} = \beta_0 + \beta_2, \quad \beta_{1M} = \beta_1, \quad \beta_{1F} = \beta_1 + \beta_3$$

Thus, *model (12.3) incorporates the two separate regression equations within a single model* and allows for different slopes (β_1 for males and $\beta_1 + \beta_3$ for females) and different intercepts (β_0 for males and $\beta_0 + \beta_2$ for females). We now consider the details involved in making statistical inferences for these two methods.

12.7 Method I: Using Separate Regression Fits to Compare Two Straight Lines

12.7.1 Testing for Parallelism

From (12.1) and (12.2), we can conclude that the appropriate null hypothesis for comparing the slopes (i.e., for conducting a test of parallelism) is given by

$$\boxed{H_0: \beta_{1M} = \beta_{1F}}$$

When the null hypothesis $H_0: \beta_{1M} = \beta_{1F}$ is true, the two regression lines simplify to $Y_M = \beta_{0M} + \beta_1 X + E_M$ for males and $Y_F = \beta_{0F} + \beta_1 X + E_F$ for females, where $\beta_1(= \beta_{1M} = \beta_{1F})$ is the common slope. An estimate of this common slope β_1 is given by the following formula, which is a weighted average of the two separate slope estimates:

$$\hat{\beta}_1 = \frac{(n_M - 1)S_{X_M}^2 \hat{\beta}_{1M} + (n_F - 1)S_{X_F}^2 \hat{\beta}_{1F}}{(n_M - 1)S_{X_M}^2 + (n_F - 1)S_{X_F}^2}$$

Notice that $\hat{\beta}_1$ equals the slope computed by fitting a straight line to the pooled data.

Any of the following three alternative hypotheses can be used:

$$H_A: \begin{cases} \beta_{1M} > \beta_{1F} & \text{(one sided)} \\ \beta_{1M} < \beta_{1F} & \text{(one sided)} \\ \beta_{1M} \neq \beta_{1F} & \text{(two sided)} \end{cases}$$

The test statistic for evaluating parallelism is then given by

$$T = \frac{\hat{\beta}_{1M} - \hat{\beta}_{1F}}{S_{(\hat{\beta}_{1M} - \hat{\beta}_{1F})}} \qquad (12.4)$$

where

$$\hat{\beta}_{1M} = \text{Least-squares estimate of the slope } \beta_{1M}, \text{ using the } n_M \text{ observations (on males)}$$

$$\hat{\beta}_{1F} = \text{Least-squares estimate of the slope } \beta_{1F}, \text{ using the } n_F \text{ observations (on females)}$$

$$S_{(\hat{\beta}_{1M} - \hat{\beta}_{1F})} = \text{Estimate of the standard error of the estimated difference between slopes } (\hat{\beta}_{1M} - \hat{\beta}_{1F})$$

This standard error involves pooling and summing the estimated variances of the slopes of the fitted regression lines.[4] It is equal to the square root of the following variance:

$$S^2_{(\hat{\beta}_{1M} - \hat{\beta}_{1F})} = S^2_{P,\,Y|X} \left[\frac{1}{(n_M - 1)S^2_{X_M}} + \frac{1}{(n_F - 1)S^2_{X_F}} \right] \qquad (12.5)$$

where

$$S^2_{P,\,Y|X} = \frac{(n_M - 2)S^2_{Y|X_M} + (n_F - 2)S^2_{Y|X_F}}{n_M + n_F - 4} \qquad (12.6)$$

is a pooled estimate of σ^2 based on combining residual mean-square errors for males and females, and where

$$S^2_{Y|X_M} = \text{Residual mean-square error for the male data}$$

$$S^2_{Y|X_F} = \text{Residual mean-square error for the female data}$$

$$S^2_{X_M} = \text{Variance of the } X\text{'s for the male data}$$

$$S^2_{X_F} = \text{Variance of the } X\text{'s for the female data}$$

The test statistic given by (12.4) will, under the usual regression assumptions, be distributed as a Student's t with $n_M + n_F - 4$ degrees of freedom when H_0 is true. We then have the following critical regions for different hypotheses and significance level α:

$$\begin{cases} T \geq t_{n_M+n_F-4,\,1-\alpha} & \text{for } H_A: \beta_{1M} > \beta_{1F} \\ T \leq -t_{n_M+n_F-4,\,1-\alpha} & \text{for } H_A: \beta_{1M} < \beta_{1F} \\ |T| > t_{n_M+n_F-4,\,1-\alpha/2} & \text{for } H_A: \beta_{1_M} \neq \beta_{1_F} \end{cases}$$

---

[4] This pooling is valid only if the variance of E_M—say, σ^2_M—in (12.1) is equal to the variance of E_F—say, σ^2_F—in (12.1); that is, only if the assumption of homogeneity of error variance holds (i.e., $\sigma^2_M = \sigma^2_F = \sigma^2$); we make this assumption for all of the developments to follow.

The associated $100(1 - \alpha)\%$ confidence interval for $(\beta_{1M} - \beta_{1F})$ is of the form

$$(\hat{\beta}_{1M} - \hat{\beta}_{1F}) \pm t_{n_M + n_F - 4, \, 1 - \alpha/2} S_{(\hat{\beta}_{1M} - \hat{\beta}_{1F})}$$

■ **Example 12.1**   Using the data given in Table 12.1, we can compute the estimates $S^2_{P, Y|X}$, $S^2_{(\hat{\beta}_{1M} - \hat{\beta}_{1F})}$, $S^2_{\hat{\beta}_{1M}}$, and $S^2_{\hat{\beta}_{1F}}$ as follows:

$$S^2_{P, Y|X} = \frac{(n_M - 2)S^2_{Y|X_M} + (n_F - 2)S^2_{Y|X_F}}{n_M + n_F - 4} = \frac{38(71.90) + 27(91.46)}{40 + 29 - 4}$$

$$= \frac{5{,}201.62}{65} = 80.02$$

$$S^2_{(\hat{\beta}_{1M} - \hat{\beta}_{1F})} = S^2_{P, Y|X}\left[\frac{1}{(n_M - 1)S^2_{X_M}} + \frac{1}{(n_F - 1)S^2_{X_F}}\right]$$

$$= 80.02\left[\frac{1}{39(221.15)} + \frac{1}{28(242.14)}\right]$$

$$= 0.021$$

$$S^2_{\hat{\beta}_{1M}} = \frac{S^2_{Y|X_M}}{(n_M - 1)S^2_{X_M}} = \frac{71.90}{39(221.15)} = 0.0083$$

$$S^2_{\hat{\beta}_{1F}} = \frac{S^2_{Y|X_F}}{(n_F - 1)S^2_{X_F}} = \frac{91.46}{28(242.14)} = 0.0135$$

The test statistic (12.4) is then computed as

$$T = \frac{\hat{\beta}_{1M} - \hat{\beta}_{1F}}{S_{(\hat{\beta}_{1M} - \hat{\beta}_{1F})}} = \frac{0.96 - 0.95}{\sqrt{0.021}} = \frac{0.01}{0.145} = 0.069$$

For this test statistic, the critical value for a two-sided test (i.e., $H_A: \beta_{1M} \neq \beta_{1F}$) with $\alpha = .05$ is given by

$$t_{65, \, 0.975} = 1.9964$$

Since $|T| = 0.069$ does not exceed 1.9964 (i.e., $p > 0.9$), we do not reject H_0. Thus, we conclude that there is insufficient evidence to permit us to reject the hypothesis of parallelism (namely, that the lines for males and females have the same slope).   ■

12.7.2 Comparing Two Intercepts

We now describe how to use separate regression fits to determine whether both straight lines have the same intercept, regardless of the two slopes. The null hypothesis in this case is given by

$$H_0: \beta_{0M} = \beta_{0F}$$

If $H_0: \beta_{0M} = \beta_{0F}$ is true, the two regression lines simplify to $Y_M = \beta_0 + \beta_{1M}X + E_M$ and $Y_F = \beta_0 + \beta_{1F}X + E_F$, where $\beta_0(= \beta_{0M} = \beta_{0F})$ is the common intercept. An estimate of this common intercept β_0 is given by the equation

$$\hat{\beta}_0 = \frac{n_M\hat{\beta}_{0M} + n_F\hat{\beta}_{0F}}{n_M + n_F}$$

which is a weighted average of the two separate intercept estimates.

The test statistic in this case is given by:

$$T = \frac{\hat{\beta}_{0M} - \hat{\beta}_{0F}}{S_{(\hat{\beta}_{0M} - \hat{\beta}_{0F})}} \qquad (12.7)$$

where $\hat{\beta}_{0M}$ and $\hat{\beta}_{0F}$ are the intercept estimates for males and females, respectively, and where $S^2_{(\hat{\beta}_{0M} - \hat{\beta}_{0F})}$ estimates the variance of the estimated difference between the intercepts by means of the formula

$$S^2_{(\hat{\beta}_{0M} - \hat{\beta}_{0F})} = S^2_{P,Y|X}\left[\frac{1}{n_M} + \frac{1}{n_F} + \frac{\overline{X}^2_M}{(n_M - 1)S^2_{X_M}} + \frac{\overline{X}^2_F}{(n_F - 1)S^2_{X_F}}\right] \qquad (12.8)$$

The statistic T given in (12.7) will have the t distribution with $n_M + n_F - 4$ degrees of freedom when $H_0: \beta_{0M} = \beta_{0F}$ is true and when $\sigma^2_M = \sigma^2_F$. We therefore have the following critical regions for different hypotheses and significance level α:

$$\begin{cases} T \geq t_{n_M+n_F-4,\, 1-\alpha} & \text{for } H_A: \beta_{0M} > \beta_{0F} \\ T \leq -t_{n_M+n_F-4,\, 1-\alpha} & \text{for } H_A: \beta_{0M} < \beta_{0F} \\ |T| \geq t_{n_M+n_F-4,\, 1-\alpha/2} & \text{for } H_A: \beta_{0M} \neq \beta_{0F} \end{cases}$$

The associated $100(1 - \alpha)\%$ confidence interval for $\beta_{0M} - \beta_{0F}$ is

$$(\hat{\beta}_{0M} - \hat{\beta}_{0F}) \pm t_{n_M+n_F-4,\, 1-\alpha/2}\, S_{(\hat{\beta}_{0M} - \hat{\beta}_{0F})}$$

■ **Example 12.2**   For the data in Table 12.1 and the value of $S^2_{P,Y|X}$ obtained in Example 12.1, the estimates $S^2_{(\hat{\beta}_{0M} - \hat{\beta}_{0F})}, S^2_{\hat{\beta}_{0M}}$, and $S^2_{\hat{\beta}_{0F}}$ are computed as follows:

$$\begin{aligned} S^2_{(\hat{\beta}_{0M} - \hat{\beta}_{0F})} &= S^2_{P,Y|X}\left[\frac{1}{n_M} + \frac{1}{n_F} + \frac{\overline{X}^2_M}{(n_M - 1)S^2_{X_M}} + \frac{\overline{X}^2_F}{(n_F - 1)S^2_{X_F}}\right] \\ &= 80.02\left[\frac{1}{40} + \frac{1}{29} + \frac{(46.93)^2}{39(221.15)} + \frac{(45.07)^2}{28(242.14)}\right] \\ &= 80.02(0.0250 + 0.0345 + 0.2554 + 0.2996) \\ &= 49.17 \end{aligned}$$

$$S^2_{\hat{\beta}_{0M}} = S^2_{Y|X_M}\left[\frac{1}{n_M} + \frac{\overline{X}^2_M}{(n_M - 1)S^2_{X_M}}\right] = 71.90\left[\frac{1}{40} + \frac{(46.93)^2}{39(221.15)}\right]$$

$$= 71.90(0.0250 + 0.2554) = 20.16$$

$$S^2_{\hat{\beta}_{0F}} = S^2_{Y|X_F}\left[\frac{1}{n_F} + \frac{\overline{X}^2_F}{(n_F - 1)S^2_{X_F}}\right] = 91.46\left[\frac{1}{29} + \frac{(45.07)^2}{28(242.14)}\right]$$

$$= 91.46(0.0345 + 0.2996) = 30.56$$

From these results, we compute the T statistic of (12.7) as

$$T = \frac{\hat{\beta}_{0M} - \hat{\beta}_{0F}}{S_{(\hat{\beta}_{0M} - \hat{\beta}_{0F})}} = \frac{110.04 - 97.08}{\sqrt{49.17}} = \frac{12.96}{7.01} = 1.85$$

For a two-sided test ($H_A: \beta_{0M} \neq \beta_{0F}$) with $\alpha = .05$, we find that $|T| = 1.85$ does not exceed $t_{65,\,0.975} = 1.9964$ (i.e., $.05 < P < .1$). Thus, the null hypothesis of common intercepts is not rejected at $\alpha = .05$ but it is rejected at $\alpha = .1$. ∎

12.7.3 Testing for Coincidence from Separate Straight-line Fits

Two straight lines are coincident if their slopes and their intercepts are equal. In considering the male–female regression equations given by (12.1) and (12.2), we may conclude that the null hypothesis of coincidence is therefore equivalent to testing $H_0: \beta_{0M} = \beta_{0F}$ and $\beta_{1M} = \beta_{1F}$ *simultaneously.* If so, the two regression models both reduce to the general form

$$Y = \beta_0 + \beta_1 X + E$$

where $\beta_0(= \beta_{0M} = \beta_{0F})$ and $\beta_1(= \beta_{1M} = \beta_{1F})$ are the common intercept and slope, respectively. The estimates of the common slope β_1 and common intercept β_0 are obtained by pooling all observations on males and females together and determining the usual least-squares slope and intercept estimates using the pooled data set.

A preferred way to test the null hypothesis of coincident lines is to employ a multiple regression model involving dummy variables. Another, generally less efficient (e.g., not as powerful) procedure is often convenient when separate models are fit. In practice, this procedure frequently yields the same conclusion as is obtained from using dummy variables.

Using separate regression fits, we perform both the test of $H_0: \beta_{0M} = \beta_{0F}$ of equal intercepts and the test of $H_0: \beta_{1M} = \beta_{1F}$ of equal slopes. If either one or both of these null hypotheses are rejected, we can conclude that statistical evidence indicates that the two lines do not coincide. If neither is rejected, we must conclude that no evidence of noncoincidence exists in the data.

A valid criticism of this testing procedure, which calls into question its power, is that it involves two separate tests rather than a single test. This fact raises two difficulties:

1. The procedure does not precisely test for coincidence.

2. If α is the significance level of each separate test, the overall significance level for the two tests combined is greater than α; that is, there is more chance of rejecting a true H_0 (i.e., of making a Type I error).

One reasonable (but fairly conservative) way to get around the second difficulty is to use $\alpha/2$ for each separate test to guarantee an overall significance level of no more than α (the Bonferroni correction). Nevertheless, using $\alpha/2$ for each test is conservative (i.e., makes it harder to reject either H_0), thus making it difficult to detect a real difference between the two lines.

With regard to the first difficulty, even if both tests are not rejected, it is still possible (although unlikely) that the two lines do not coincide. This is so because each separate test (e.g., the test of H_0: $\beta_{1M} = \beta_{1F}$) allows the remaining parameters (β_{0M} and β_{0F}) to be unequal. In other words, the test for equal slopes does not assume equal intercepts; nor does the test for equal intercepts assume equal slopes. The multiple regression procedure, which involves using a single model containing a dummy variable for group status, avoids this drawback and permits testing for common slope and common intercept *simultaneously*.

■ **Example 12.3**   We saw in Examples 12.1 and 12.2 that the null hypothesis of equal slopes (regardless of the intercepts) was not rejected ($P > .90$), and that the null hypothesis of equal intercepts (regardless of the slopes) was associated with a P-value of between .05 and .1. Putting these two facts together, we would be inclined to support the conclusion that there is no *strong* evidence for noncoincidence. As we shall see shortly, the more appropriate test procedure involving a single model yields a different conclusion.   ■

12.8 Method II: Using a Single Regression Equation to Compare Two Straight Lines

Another approach for comparing regression equations uses a single multiple regression model that contains one or more dummy variables to distinguish the groups being compared. The model for comparing two straight lines is given by (12.3), which we restate here:

$$Y = \beta_0 + \beta_1 X + \beta_2 Z + \beta_3 XZ + E$$

where $Y = $ SBP, $X = $ AGE, and Z is a dummy variable indicating gender (1 if female, 0 if male). For the data in Table 12.1 ($n_M = 40$, $n_F = 29$), the fitted model is

$$\hat{Y} = 110.04 + 0.96X - 12.96Z - 0.012XZ$$

which yields the following separate straight-line equations:

$$Z = 0: \quad \hat{Y}_M = 110.04 + 0.96X$$
$$Z = 1: \quad \hat{Y}_F = 97.08 + 0.95X$$

These two straight-line equations are identical to those obtained in Section 12.4 by fitting separate regressions.

Table 12.2 provides ANOVA results needed to answer statistical inference questions about this model. This table provides variables-added-in-order tests for the fitted regression equation and allows us to perform appropriate tests for parallelism, for equal intercepts, and for coincidence.

12.8.1 Test of Parallelism: Single-model Approach

Referring again to the dummy variable model of (12.3), we know the null hypothesis that the two regression lines are parallel is equivalent to H_0: $\beta_3 = 0$. If $\beta_3 = 0$, then the slope for

TABLE 12.2 **Three models for the age–systolic blood pressure example**

Source	d.f.	SS	MS	F
Regression (X)	1	14,951.25	14,951.25	121.27
Residual	67	8,260.51	123.29	
Regression (X, Z)	2	18,009.78	9,004.89	114.25
Residual	66	5,201.99	78.82	
Regression (X, Z, XZ)	3	18,010.33	6,003.44	75.02
Residual	65	5,201.44	80.02	

females, $\beta_{1F} = \beta_1 + \beta_3$, simplifies to β_1, which is the slope for males (i.e., the two lines are parallel). The test statistic for testing $H_0: \beta_3 = 0$ is the partial F statistic (or equivalent t test) for the significance of the addition of the variable XZ to a model already containing X and Z.[5]

In our example, this test statistic is computed as follows:

$$F(XZ \mid X, Z) = \frac{\text{Regression SS}(X, Z, XZ) - \text{Regression SS}(X, Z)}{\text{Residual MS}(X, Z, XZ)}$$

$$= \frac{18,010.33 - 18,009.78}{80.02}$$

$$= 0.007 \qquad (P = .9342)$$

This F statistic, with 1 and 65 degrees of freedom, is extremely small (P is very large); so we do not reject H_0 and, therefore, have no statistical basis for believing that the two lines are not parallel. This was the same decision we reached on the basis of separate regression fits. In fact, the F computed here is (theoretically) the square of the corresponding T computed when using separate straight-line fits in Section 12.7.1, although the numerical answers may not exactly agree due to round-off errors.

12.8.2 Test of Equal Intercepts: Single-model Approach

The hypothesis that the two intercepts are equal, allowing for unequal slopes, is equivalent to $H_0: \beta_2 = 0$ for the overall model (12.3). The test compares the overall model

$$Y = \beta_0 + \beta_1 X + \beta_2 Z + \beta_3 XZ + E$$

to the reduced model

$$Y = \beta_0 + \beta_1 X + \beta_3 XZ + E$$

This is a variables-added-last test considering Z, the sex group dummy variable.[6] Another approach involves a variables-added-in-order test comparing

$$Y = \beta_0 + \beta_1 X + \beta_2 Z + E$$

---

[5] If H_0 is not rejected by the test of $H_0: \beta_3 = 0$, model (12.3) can be revised to eliminate the β_3 term. This revised (or reduced) model becomes $Y = \beta_0 + \beta_1 X + \beta_2 Z + E$, which has the form of an analysis-of-covariance model (see Chapter 13).

[6] The partial F statistic for the variables-added-last test for equal intercepts, $F(Z \mid X, XZ)$, is the square of the statistic given by (12.7) when fitting separate straight-line models.

to the reduced model

$$Y = \beta_0 + \beta_1 X + E$$

The latter test presumes equal slopes, so it is essentially a test for coincidence, assuming parallelism. Not surprisingly, neither test is uniformly preferred (see Section 12.10). As discussed in Chapters 9 and 15, we recommend using the residual from the full model, (12.3), for either test.

For the example under consideration, we opt for the second approach, since the slope test was not significant. The variables-added-in-order test for H_0: $\beta_2 = 0$ is then computed as follows:

$$F(Z|X) = \frac{\text{Regression SS}(Z, X) - \text{Regression SS}(X)}{\text{Residual MS}(X, Z, XZ)}$$

$$= \frac{18{,}009.78 - 14{,}951.25}{80.02}$$

$$= 38.22$$

The preceding test statistic is modified from the usual partial F statistic in that we are now using for the denominator the mean-square residual for the full model, which contains X, Z, and XZ; the usual partial F statistic would use the mean-square residual for the model containing only X and Z.

Table 12.3 indicates that, with 1 and 65 degrees of freedom, $P < .0001$. Hence the intercepts are judged to be different for the male and female straight-line models.

12.8.3 Test of Coincidence: Single-model Approach

The hypothesis that the two regression lines coincide is H_0: $\beta_2 = \beta_3 = 0$. When both β_2 and β_3 are 0, the model for females, $Y_F = (\beta_0 + \beta_2) + (\beta_1 + \beta_3)X + E$, reduces to $Y_M = \beta_0 + \beta_1 X + E$, the model for males (i.e., the two lines coincide). The test of H_0: $\beta_2 = \beta_3 = 0$ is thus a multiple-partial F test, since it involves a subset of regression coefficients.[7] The two models being compared are therefore

$$Y = \beta_0 + \beta_1 X + \beta_2 Z + \beta_3 XZ + E$$

and

$$Y = \beta_0 + \beta_1 X + E$$

TABLE 12.3   ANOVA table for method II for the age–systolic blood pressure example

Source	d.f.	SS	MS	F	P
X (AGE)	1	14,951.25	14,951.25	186.84	<.0001
Z (SEX)$\|X$	1	3,058.52	3,058.52	38.22	<.0001
$XZ\|X, Z$	1	0.55	0.55	0.01	.9342
Residual	65	5,201.44	80.02		
Total (corrected)	68	23,211.76			

[7] If the test for coincidence is not rejected, model (12.3) can be reduced to the form $Y = \beta_0 + \beta_1 X + E$.

For our example, the information in either Table 12.2 or Table 12.3 leads to the following computation:

$$F(XZ, Z|X) = \frac{[\text{Regression SS}(X, Z, XZ) - \text{Regression SS}(X)]/2}{\text{Residual MS}(X, Z, XZ)}$$

$$= \frac{(18{,}010.33 - 14{,}951.25)/2}{80.02}$$

$$= 19.1$$

Comparing this F with $F_{2,65,0.999} = 7.72$, we reject H_0 with $P < .001$ and conclude that very strong evidence exists that the two lines are *not* coincident. This conclusion contradicts our earlier conclusion (Section 12.7.3) based on the results from separate tests for equal slopes and equal intercepts.

12.9 Comparison of Methods I and II

Does the method that uses dummy variables differ from the method that fits two separate regression equations? And if so, is one of the methods preferable to the other?

In deciding whether one method is preferable, we first observe that the two methods yield exactly the same estimated regression coefficients for the two straight-line models. That is, if we fit the model $Y = \beta_0 + \beta_1 X + \beta_2 Z + \beta_3 XZ + E$ by the least-squares method to obtain estimated coefficients $\hat{\beta}_0, \hat{\beta}_1, \hat{\beta}_2,$ and $\hat{\beta}_3$, the straight-line equations obtained by setting Z equal to 0 and to 1 in this estimated model will be the same as those obtained by fitting the two straight lines separately. In particular, if $\hat{\beta}_{0M}, \hat{\beta}_{0F}, \hat{\beta}_{1M},$ and $\hat{\beta}_{1F}$ denote the estimated regression coefficients based on separate regression fits, then $\hat{\beta}_{0M} = \hat{\beta}_0, \hat{\beta}_{0F} = \hat{\beta}_0 + \hat{\beta}_2, \hat{\beta}_{1M} = \hat{\beta}_1,$ and $\hat{\beta}_{1F} = \hat{\beta}_1 + \hat{\beta}_3$. As for statistical tests involving regression coefficients estimated by the two methods, the following two points are valid:

1. *The tests for parallel lines are exactly equivalent;* that is, the T statistic with $n_1 + n_2 - 4$ degrees of freedom computed for testing H_0: $\beta_3 = 0$ in the dummy variable model is exactly the same as the T statistic given by (12.4) for testing H_0: $\beta_{1M} = \beta_{1F}$ based on fitting two separate models.

2. *The tests for coincident lines differ,* and the one using the dummy variable model is generally preferable. The approach using separate regressions tests H_0: $\beta_{1M} = \beta_{1F}$ and H_0: $\beta_{0M} = \beta_{0F}$ separately, and then rejects the null hypothesis of coincident lines if either or both null hypotheses are rejected. This is exactly equivalent to performing two separate tests of H_0: $\beta_2 = 0$ and H_0: $\beta_3 = 0$ and using the same decision rule for the dummy variable approach; but it is not equivalent to testing the single null hypothesis H_0: $\beta_2 = \beta_3 = 0$ (i.e., testing whether β_2 and β_3 are both simultaneously 0).

12.10 Testing Strategies and Interpretation: Comparing Two Straight Lines

Several strategies can be used to identify a best model for comparing two straight lines. Strategies for more general situations are described in Chapter 16. We prefer a backward strategy for most situations—that is, starting with the largest model of interest and then trying to reduce the model through a sequence of hypothesis tests. A flow diagram of this strategy for comparing two straight lines is given in Figure 12.3. In this case, the largest model to be considered is model (12.3), which contains X, Z, and XZ as independent variables. To reduce the

FIGURE 12.3   Comparing two straight lines: Backward testing strategy

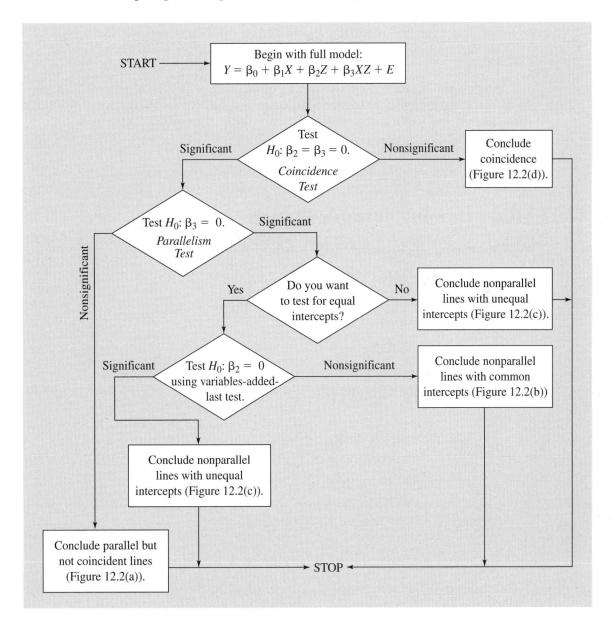

model, we perform tests for coincidence, then for parallelism, and then for equal intercepts, as follows:

1. If the test for coincidence is nonsignificant, we stop further testing and conclude that the best model is $Y = \beta_0 + \beta_1 X + E$ (i.e., coincident lines, Figure 12.2(d)).

2. If the test for coincidence is significant and the test for parallelism is nonsignificant, the data support a finding of parallel but noncoincident lines (Figure 12.2(a)).

3. If the test for coincidence is significant and the test for parallelism is significant, we might not even be interested in a test for equal intercepts; if we are, the appropriate test procedure involves the variables-added-last statistic $F(Z|X, XZ)$, which does not assume parallel lines. If this test produces a significant result, we would argue for Figure 12.2(c); if it produces a nonsignificant result, we would tend to support Figure 12.2(b).

Applying this strategy to the age–systolic blood pressure data, we would conclude, based on the preceding tests, that the test for coincidence is significant and the test for parallelism is nonsignificant. Our overall conclusion, therefore, is that the best model has the form

$$Y = \beta_0 + \beta_1 X + \beta_2 Z + E$$

In words, we assume parallel (and noncoincident) lines, as shown in Figure 12.2(a).

12.11 Other Dummy Variable Models

Two other dummy variable models could have been used instead of (12.3):

$$Y = \beta_0^* + \beta_1^* X + \beta_2^* Z^* + \beta_3^* XZ^* + E \tag{12.9}$$

for which

$$Z^* = \begin{cases} 1 & \text{if subject is male} \\ -1 & \text{if subject is female} \end{cases}$$

and

$$Y = \beta_{0M} Z_1' + \beta_{0F} Z_2' + \beta_{1M} XZ_1' + \beta_{1F} XZ_2' + E \tag{12.10}$$

for which

$$Z_1' = \begin{cases} 1 & \text{if subject is male} \\ 0 & \text{if subject is female} \end{cases}$$

$$Z_2' = \begin{cases} 0 & \text{if subject is male} \\ 1 & \text{if subject is female} \end{cases}$$

For model (12.9), the separate regression equations are

$$Z^* = 1: \quad Y_M = (\beta_0^* + \beta_2^*) + (\beta_1^* + \beta_3^*)X + E$$

and

$$Z^* = -1: \quad Y_F = (\beta_0^* - \beta_2^*) + (\beta_1^* - \beta_3^*)X + E$$

The test for parallel lines is equivalent to testing $H_0: \beta_3^* = 0$; the test for equal intercepts is equivalent to testing $H_0: \beta_2^* = 0$; and the test for coincident lines is equivalent to testing $H_0: \beta_2^* = \beta_3^* = 0.$[8]

For model (12.10), the separate regression equations are

$$Z_1' = 1, Z_2' = 0: \qquad Y_M = \beta_{0M} + \beta_{1M}X + E$$

and

$$Z_1' = 0, Z_2' = 1: \qquad Y_F = \beta_{0F} + \beta_{1F}X + E$$

The test for parallel lines here is equivalent to testing $H_0: \beta_{1M} = \beta_{1F}$ (not necessarily equal to 0); the test for equal intercepts is equivalent to testing $H_0: \beta_{0M} = \beta_{0F}$ (not necessarily equal to 0); and the test for coincident lines is equivalent to testing $H_0: \beta_{1M} = \beta_{1F}$ and $\beta_{0M} = \beta_{0F}$ simultaneously.[9] The test of $H_0: \beta_{1M} = \beta_{1F}$ and $\beta_{0M} = \beta_{0F}$ differs from previously discussed multiple-partial F tests because the coefficients under this H_0 are not equal to 0. The testing procedure in such a case is given as follows:

1. Reduce the model according to the specifications under the null hypothesis; for the test of coincidence, for example, the full model (12.10) becomes

$$Y = \beta_{0M}(Z_1' + Z_2') + \beta_{1M}(XZ_1' + XZ_2') + E$$
$$= \beta_{0M} + \beta_{1M}X + E \text{ since } Z_1' + Z_2' = 1$$

2. Find the residual sum of squares for this reduced model.

3. Compute the following F statistic:

$$F = \frac{[\text{Residual SS(reduced model)} - \text{Residual SS(full model)}]/v^*}{\text{Residual MS(full model)}}$$

where v^* is the number of linearly independent parametric functions specified to be 0 under H_0 (in our case $v^* = 2$, since the null hypothesis of coincidence specifies that $\beta_{1M} - \beta_{1F} = 0$ and that $\beta_{0M} - \beta_{0F} = 0$).

4. Test H_0 using F tables with v^* and $n - 4$ degrees of freedom, where 4 is the number of parameters in the full model.

Model (12.10) does not include an overall intercept. Care must be taken in using this particular model with most computer programs (which generally include an overall intercept by default). Collinearity (often labeled "LESS THAN FULL RANK MODEL") may occur.

---

[8] We can express the coefficients of model (12.9) in terms of the regression coefficients for the separate male and female models (12.1) and (12.2) as follows:

$$\beta_0^* = \frac{\beta_{0M} + \beta_{0F}}{2}, \qquad \beta_1^* = \frac{\beta_{1M} + \beta_{1F}}{2}, \qquad \beta_2^* = \frac{\beta_{0M} - \beta_{0F}}{2}, \qquad \beta_3^* = \frac{\beta_{1M} - \beta_{1F}}{2}$$

[9] The coefficients of model (12.10) are identical to the correspondingly labeled coefficients for the separate male and female models (12.1) and (12.2).

12.12 Comparing Four Regression Equations

Suppose that we want to compare the separate multiple regressions of systolic blood pressure on age and weight for four social class groups. For each individual in each social class group, we observe values of the variables $Y = $ SBP, $X_1 = $ AGE, and $X_2 = $ WEIGHT. Further, let us suppose that there are n_i individuals in the ith social class (SC) group, $i = 1, 2, 3, 4$. We begin by defining three dummy variables Z_1, Z_2, and Z_3:

$$Z_1 = \begin{cases} 1 & \text{if SC2 member} \\ 0 & \text{otherwise} \end{cases} \qquad Z_2 = \begin{cases} 1 & \text{if SC3 member} \\ 0 & \text{otherwise} \end{cases}$$

$$Z_3 = \begin{cases} 1 & \text{if SC4 member} \\ 0 & \text{otherwise} \end{cases}$$

The complete model to be used (if no interaction between AGE and WEIGHT is considered) is given as follows:

$$\begin{aligned} Y = {} & \beta_0 + \beta_1 X_1 + \beta_2 X_2 + \beta_3 Z_1 + \beta_4 Z_2 + \beta_5 Z_3 + \beta_6 X_1 Z_1 + \beta_7 X_2 Z_1 \\ & + \beta_8 X_1 Z_2 + \beta_9 X_2 Z_2 + \beta_{10} X_1 Z_3 + \beta_{11} X_2 Z_3 + E \end{aligned} \qquad (12.11)$$

For each particular social class, model (12.11) specializes as follows:

$$\begin{aligned} \text{SC1}(Z_1 = Z_2 = Z_3 = 0)\colon \quad & Y = \beta_0 + \beta_1 X_1 + \beta_2 X_2 + E \\ \text{SC2}(Z_1 = 1, Z_2 = Z_3 = 0)\colon \quad & Y = (\beta_0 + \beta_3) + (\beta_1 + \beta_6) X_1 + (\beta_2 + \beta_7) X_2 + E \\ \text{SC3}(Z_1 = Z_3 = 0, Z_2 = 1)\colon \quad & Y = (\beta_0 + \beta_4) + (\beta_1 + \beta_8) X_1 + (\beta_2 + \beta_9) X_2 + E \\ \text{SC4}(Z_1 = Z_2 = 0, Z_3 = 1)\colon \quad & Y = (\beta_0 + \beta_5) + (\beta_1 + \beta_{10}) X_1 + (\beta_2 + \beta_{11}) X_2 + E \end{aligned}$$

12.12.1 Tests of Hypotheses

The following hypotheses involving the parameters in model (12.11) are of interest.

1. *All four regression equations are coincident* (i.e., test H_0: $\beta_3 = \beta_4 = \beta_5 = \beta_6 = \beta_7 = \beta_8 = \beta_9 = \beta_{10} = \beta_{11} = 0$). When H_0 is true, all four social class models reduce to the form

$$Y = \beta_0 + \beta_1 X_1 + \beta_2 X_2 + E$$

The test statistic is the multiple partial

$$F(Z_1, Z_2, Z_3, X_1 Z_1, X_1 Z_2, X_2 Z_1, X_2 Z_2, X_1 Z_3, X_2 Z_3 \mid X_1, X_2)$$

which has 9 and $n_1 + n_2 + n_3 + n_4 - 12$ degrees of freedom.

2. *All four regression equations are parallel* (i.e., test H_0: $\beta_6 = \beta_7 = \beta_8 = \beta_9 = \beta_{10} = \beta_{11} = 0$). When H_0 is true, the models for the four social classes reduce to

$$\begin{aligned} \text{SC1:} \quad & Y = \beta_0 + \beta_1 X_1 + \beta_2 X_2 + E \\ \text{SC2:} \quad & Y = (\beta_0 + \beta_3) + \beta_1 X_1 + \beta_2 X_2 + E \\ \text{SC3:} \quad & Y = (\beta_0 + \beta_4) + \beta_1 X_1 + \beta_2 X_2 + E \\ \text{SC4:} \quad & Y = (\beta_0 + \beta_5) + \beta_1 X_1 + \beta_2 X_2 + E \end{aligned}$$

Thus, the coefficients of X_1 and X_2 are the same for each social class when H_0 is true (i.e., the four regression equations are said to be parallel). The test statistic for testing H_0 is given by the multiple partial

$$F(X_1Z_1, X_2Z_1, X_1Z_2, X_2Z_2, X_1Z_3, X_2Z_3 \mid X_1, X_2, Z_1, Z_2, Z_3)$$

which has 6 and $n_1 + n_2 + n_3 + n_4 - 12$ degrees of freedom.

12.12.2 An Alternate Dummy Variable Model

Another dummy variable coding scheme for comparing the four social class groups begins by defining

$$Z_1^* = \begin{cases} -1 & \text{if SC1 member} \\ 1 & \text{if SC2 member} \\ 0 & \text{if SC3 member} \\ 0 & \text{if SC4 member} \end{cases} \qquad Z_2^* = \begin{cases} -1 & \text{if SC1 member} \\ 0 & \text{if SC2 member} \\ 1 & \text{if SC3 member} \\ 0 & \text{if SC4 member} \end{cases}$$

$$Z_3^* = \begin{cases} -1 & \text{if SC1 member} \\ 0 & \text{if SC2 member} \\ 0 & \text{if SC3 member} \\ 1 & \text{if SC4 member} \end{cases}$$

Then we use the model

$$\begin{aligned} Y = {}& \beta_0^* + \beta_1^* X_1 + \beta_2^* X_2 + \beta_3^* Z_1^* + \beta_4^* Z_2^* + \beta_5^* Z_3^* + \beta_6^* X_1 Z_1^* + \beta_7^* X_2 Z_1^* \\ & + \beta_8^* X_1 Z_2^* + \beta_9^* X_2 Z_2^* + \beta_{10}^* X_1 Z_3^* + \beta_{11}^* X_2 Z_3^* + E \end{aligned} \qquad (12.12)$$

For the preceding dummy variable coding, the four regression equations for the four social classes based on model (12.12) are

SC1 $(Z_1^* = Z_2^* = Z_3^* = -1)$:
$$Y = (\beta_0^* - \beta_3^* - \beta_4^* - \beta_5^*) + (\beta_1^* - \beta_6^* - \beta_8^* - \beta_{10}^*)X_1 + (\beta_2^* - \beta_7^* - \beta_9^* - \beta_{11}^*)X_2 + E$$

SC2 $(Z_1^* = 1, Z_2^* = Z_3^* = 0)$:
$$Y = (\beta_0^* + \beta_3^*) + (\beta_1^* + \beta_6^*)X_1 + (\beta_2^* + \beta_7^*)X_2 + E$$

SC3 $(Z_1^* = 0, Z_2^* = 1, Z_3^* = 0)$:
$$Y = (\beta_0^* + \beta_4^*) + (\beta_1^* + \beta_8^*)X_1 + (\beta_2^* + \beta_9^*)X_2 + E$$

SC4 $(Z_1^* = Z_2^* = 0, Z_3^* = 1)$:
$$Y = (\beta_0^* + \beta_5^*) + (\beta_1^* + \beta_{10}^*)X_1 + (\beta_2^* + \beta_{11}^*)X_2 + E$$

So the null hypotheses to be tested in connection with parallelism and coincidence (using appropriate multiple partial F tests) are

Parallelism: $H_0: \beta_6^* = \beta_7^* = \beta_8^* = \beta_9^* = \beta_{10}^* = \beta_{11}^* = 0$

Coincidence: $H_0: \beta_3^* = \beta_4^* = \beta_5^* = \beta_6^* = \beta_7^* = \beta_8^* = \beta_9^* = \beta_{10}^* = \beta_{11}^* = 0$

12.13 Comparing Several Regression Equations Involving Two Nominal Variables

Suppose that we want to compare eight regression equations of SBP (Y) on AGE (X_1) and WEIGHT (X_2), corresponding to the eight combinations of SEX (Q) and social class (SC) groups. Then the following regression model can be used:

$$
\begin{aligned}
Y = {} & \beta_0 + \beta_1 X_1 + \beta_2 X_2 + \beta_3 Z_1 + \beta_4 Z_2 + \beta_5 Z_3 + \beta_6 Q + \beta_7 Z_1 Q + \beta_8 Z_2 Q \\
& + \beta_9 Z_3 Q + \beta_{10} X_1 Z_1 + \beta_{11} X_2 Z_1 + \beta_{12} X_1 Z_2 + \beta_{13} X_2 Z_2 + \beta_{14} X_1 Z_3 \\
& + \beta_{15} X_2 Z_3 + \beta_{16} X_1 Q + \beta_{17} X_2 Q + \beta_{18} X_1 Z_1 Q + \beta_{19} X_2 Z_1 Q \\
& + \beta_{20} X_1 Z_2 Q + \beta_{21} X_2 Z_2 Q + \beta_{22} X_1 Z_3 Q + \beta_{23} X_2 Z_3 Q + E
\end{aligned}
\tag{12.13}
$$

in which the dummy variables are defined as

$$
Z_1 = \begin{cases} 1 & \text{if SC2 member} \\ 0 & \text{otherwise} \end{cases}
\qquad
Z_2 = \begin{cases} 1 & \text{if SC3 member} \\ 0 & \text{otherwise} \end{cases}
$$

$$
Z_3 = \begin{cases} 1 & \text{if SC4 member} \\ 0 & \text{otherwise} \end{cases}
\qquad
Q = \begin{cases} 1 & \text{if subject is male} \\ 0 & \text{if subject is female} \end{cases}
$$

For each SEX–SC combination, we have

SC1–male: $Y = (\beta_0 + \beta_6) + (\beta_1 + \beta_{16})X_1 + (\beta_2 + \beta_{17})X_2 + E$

SC2–male: $Y = (\beta_0 + \beta_3 + \beta_6 + \beta_7) + (\beta_1 + \beta_{10} + \beta_{16} + \beta_{18})X_1$
          $+ (\beta_2 + \beta_{11} + \beta_{17} + \beta_{19})X_2 + E$

SC3–male: $Y = (\beta_0 + \beta_4 + \beta_6 + \beta_8) + (\beta_1 + \beta_{12} + \beta_{16} + \beta_{20})X_1$
          $+ (\beta_2 + \beta_{13} + \beta_{17} + \beta_{21})X_2 + E$

SC4–male: $Y = (\beta_0 + \beta_5 + \beta_6 + \beta_9) + (\beta_1 + \beta_{14} + \beta_{16} + \beta_{22})X_1$
          $+ (\beta_2 + \beta_{15} + \beta_{17} + \beta_{23})X_2 + E$

SC1–female: $Y = \beta_0 + \beta_1 X_1 + \beta_2 X_2 + E$

SC2–female: $Y = (\beta_0 + \beta_3) + (\beta_1 + \beta_{10})X_1 + (\beta_2 + \beta_{11})X_2 + E$

SC3–female: $Y = (\beta_0 + \beta_4) + (\beta_1 + \beta_{12})X_1 + (\beta_2 + \beta_{13})X_2 + E$

SC4–female: $Y = (\beta_0 + \beta_5) + (\beta_1 + \beta_{14})X_1 + (\beta_2 + \beta_{15})X_2 + E$

The model therefore includes SEX $\times$ SC interaction, but not WEIGHT $\times$ AGE interaction. It is best to make decisions about which interactions to include based on subject matter knowledge. However, if necessary, we could test whether there are no interaction effects involving SEX and SC. This is a test of

$$
H_0: \beta_7 = \beta_8 = \beta_9 = \beta_{18} = \beta_{19} = \beta_{20} = \beta_{21} = \beta_{22} = \beta_{23} = 0
$$

given that all the other terms in (12.13) are included in the model.

Once the decision to include the interaction effect between SEX and SC has been made, several hypotheses concerning model (12.13) are of interest.

1. *All eight regression equations are coincident.* This is a test of

    $$
    H_0: \beta_3 = \beta_4 = \cdots = \beta_{23} = 0,
    $$

    given that X_1 and X_2 are in the model. If this null hypothesis is rejected, we can proceed with further testing.

2. *All eight regression equations are parallel.* This is a test of

$$H_0: \beta_{10} = \beta_{11} = \cdots = \beta_{23} = 0$$

given that X_1, X_2, Z_1, Z_2, Z_3, Q, Z_1Q, Z_2Q, and Z_3Q are in the model. If this null hypothesis is true, the eight equations are all of the form

$$Y = \beta_{0(j)} + \beta_1X_1 + \beta_2X_2 + E \qquad \text{for } j = 1, 2, \ldots, 8$$

(i.e., the eight models differ only in intercept).

Also, the specific effects of SEX and SC can be investigated via the following hypotheses.

3. *Male and female regression equations are coincident (controlling for SC).* This is a test of

$$H_0: \beta_6 = \beta_7 = \beta_8 = \beta_9 = \beta_{16} = \beta_{17} = \beta_{18} = \beta_{19} = \beta_{20} = \beta_{21} = \beta_{22} = \beta_{23} = 0$$

4. *Male and female regression equations are parallel (controlling for SC).* This is a test of

$$H_0: \beta_{16} = \beta_{17} = \beta_{18} = \beta_{19} = \beta_{20} = \beta_{21} = \beta_{22} = \beta_{23} = 0$$

When this null hypothesis is true, the eight equations above reduce to

$$\begin{cases} \text{SC1–male:} & Y = (\beta_0 + \beta_6) + \beta_1X_1 + \beta_2X_2 + E \\ \text{SC1–female:} & Y = \beta_0 + \beta_1X_1 + \beta_2X_2 + E \end{cases}$$

$$\begin{cases} \text{SC2–male:} & Y = (\beta_0 + \beta_3 + \beta_6 + \beta_7) + (\beta_1 + \beta_{10})X_1 + (\beta_2 + \beta_{11})X_2 + E \\ \text{SC2–female:} & Y = (\beta_0 + \beta_3) + (\beta_1 + \beta_{10})X_1 + (\beta_2 + \beta_{11})X_2 + E \end{cases}$$

$$\begin{cases} \text{SC3–male:} & Y = (\beta_0 + \beta_4 + \beta_6 + \beta_8) + (\beta_1 + \beta_{12})X_1 + (\beta_2 + \beta_{13})X_2 + E \\ \text{SC3–female:} & Y = (\beta_0 + \beta_4) + (\beta_1 + \beta_{12})X_1 + (\beta_2 + \beta_{13})X_2 + E \end{cases}$$

$$\begin{cases} \text{SC4–male:} & Y = (\beta_0 + \beta_5 + \beta_6 + \beta_9) + (\beta_1 + \beta_{14})X_1 + (\beta_2 + \beta_{15})X_2 + E \\ \text{SC4–female:} & Y = (\beta_0 + \beta_5) + (\beta_1 + \beta_{14})X_1 + (\beta_2 + \beta_{15})X_2 + E \end{cases}$$

Thus, within any specific social class group, the male and female regression equations are parallel (since they have the same X_1 and X_2 coefficients).

5. *All four social class equations are coincident (controlling for SEX).* This is a test of

$$H_0: \beta_3 = \beta_4 = \beta_5 = \beta_7 = \beta_8 = \beta_9 = \beta_{10} = \beta_{11} = \beta_{12} = \beta_{13} = \beta_{14} = \beta_{15}$$
$$= \beta_{18} = \beta_{19} = \beta_{20} = \beta_{21} = \beta_{22} = \beta_{23} = 0$$

6. *All four social class equations are parallel (controlling for SEX).* This is a test of

$$H_0: \beta_{10} = \beta_{11} = \beta_{12} = \beta_{13} = \beta_{14} = \beta_{15} = \beta_{18} = \beta_{19} = \beta_{20} = \beta_{21} = \beta_{22} = \beta_{23} = 0$$

When this hypothesis is true, the eight equations reduce to

$$\begin{cases} \text{SC1–male:} & Y = (\beta_0 + \beta_6) + (\beta_1 + \beta_{16})X_1 + (\beta_2 + \beta_{17})X_2 + E \\ \text{SC2–male:} & Y = (\beta_0 + \beta_3 + \beta_6 + \beta_7) + (\beta_1 + \beta_{16})X_1 + (\beta_2 + \beta_{17})X_2 + E \\ \text{SC3–male:} & Y = (\beta_0 + \beta_4 + \beta_6 + \beta_8) + (\beta_1 + \beta_{16})X_1 + (\beta_2 + \beta_{17})X_2 + E \\ \text{SC4–male:} & Y = (\beta_0 + \beta_5 + \beta_6 + \beta_9) + (\beta_1 + \beta_{16})X_1 + (\beta_2 + \beta_{17})X_2 + E \end{cases}$$

$$\begin{cases} \text{SC1–female:} & Y = \beta_0 + \beta_1X_1 + \beta_2X_2 + E \\ \text{SC2–female:} & Y = (\beta_0 + \beta_3) + \beta_1X_1 + \beta_2X_2 + E \\ \text{SC3–female:} & Y = (\beta_0 + \beta_4) + \beta_1X_1 + \beta_2X_2 + E \\ \text{SC4–female:} & Y = (\beta_0 + \beta_5) + \beta_1X_1 + \beta_2X_2 + E \end{cases}$$

Thus, within any given sex group, all four regression equations have the same coefficients for X_1 and X_2.

■ **Example 12.4**   Suppose that, in a study similar to that of Chapter 5, problem 2, we are interested in comparing the multiple regressions of Systolic Blood Pressure on Quetelet Index for four age groups and two smoking status categories. For each individual in each age group, we observe the values of the variables $Y = $ SBP, $X = $ QUET and SMK $= $ smoking status. The relevant dummy variables (Z_1, Z_2, Z_3 and Q) can be defined as:

$$Z_1 = \begin{cases} 1 & \text{if age 40–49 years} \\ 0 & \text{otherwise} \end{cases} \qquad Z_2 = \begin{cases} 1 & \text{if age 50–59 years} \\ 0 & \text{otherwise} \end{cases}$$

$$Z_3 = \begin{cases} 1 & \text{if age 60 or older} \\ 0 & \text{otherwise} \end{cases} \qquad Q = \begin{cases} 0 & \text{if never a smoker (SMK} = 0) \\ 1 & \text{if ever a smoker (SMK} = 1) \end{cases}$$

Note that $Z_1 = Z_2 = Z_3 = 0$ for ages 30–39.

The complete multiple regression model to be used is as follows:

$$\begin{aligned} Y = {} & \beta_0 + \beta_1 X + \beta_2 Z_1 + \beta_3 Z_2 + \beta_4 Z_3 + \beta_5 Q \\ & + \beta_6 Z_1 Q + \beta_7 Z_2 Q + \beta_8 Z_3 Q + \beta_9 X Z_1 + \beta_{10} X Z_2 + \beta_{11} X Z_3 \\ & + \beta_{12} XQ + \beta_{13} X Z_1 Q + \beta_{14} X Z_2 Q + \beta_{15} X Z_3 Q + E \end{aligned}$$

The data for this hypothetical example are shown in Table 12.4.

For each age-smoking status combination, the regression model above specializes as follows:

30–39 yrs, nonsmoker:	$Y = \beta_0 + \beta_1 X + E$
30–39 yrs, smoker:	$Y = (\beta_0 + \beta_5) + (\beta_1 + \beta_{12})X + E$
40–49 yrs, nonsmoker:	$Y = (\beta_0 + \beta_2) + (\beta_1 + \beta_9)X + E$
40–49 yrs, smoker:	$Y = (\beta_0 + \beta_2 + \beta_5 + \beta_6) + (\beta_1 + \beta_9 + \beta_{12} + \beta_{13})X + E$
50–59 yrs, nonsmoker:	$Y = (\beta_0 + \beta_3) + (\beta_1 + \beta_{10})X + E$
50–59 yrs, smoker:	$Y = (\beta_0 + \beta_3 + \beta_5 + \beta_7) + (\beta_1 + \beta_{10} + \beta_{12} + \beta_{14})X + E$
60+ yrs, nonsmoker:	$Y = (\beta_0 + \beta_4) + (\beta_1 + \beta_{11})X + E$
60+ yrs, smoker:	$Y = (\beta_0 + \beta_4 + \beta_5 + \beta_8) + (\beta_1 + \beta_{11} + \beta_{12} + \beta_{15})X + E$

Several hypothesis tests will be performed in order to compare the multiple regressions of SBP on QUET for different age groups and smoking status categories.

1.   First, we will determine whether the SBP–QUET relationship is coincident for non-smokers and smokers, within each age group, by testing the null hypothesis

$$H_0\text{: } \beta_5 = \beta_6 = \beta_7 = \beta_8 = \beta_{12} = \beta_{13} = \beta_{14} = \beta_{15} = 0$$

given that X, Z_1, Z_2, Z_3, XZ_1, XZ_2, and XZ_3 are in the model. The test statistic for this test of coincidence is

$$F(Q, Z_1 Q, Z_2 Q, Z_3 Q, XQ, XZ_1 Q, XZ_2 Q, XZ_3 Q, | X, Z_1, Z_2, Z_3, XZ_1, XZ_2, XZ_3)$$

$$= \frac{\begin{bmatrix} \text{Regression SS}(X, Z_1, Z_2, Z_3, Q, Z_1 Q, Z_2 Q, Z_3 Q, XZ_1, XZ_2, XZ_3, XQ, XZ_1 Q, XZ_2 Q, XZ_3 Q) \\ - \text{Regression SS}(X, Z_1, Z_2, Z_3, XZ_1, XZ_2, XZ_3) \end{bmatrix}}{\text{Residual MS}(X, Z_1, Z_2, Z_3, Q, Z_1 Q, Z_2 Q, Z_3 Q, XZ_1, XZ_2, XZ_3, XQ, XZ_1 Q, XZ_2 Q, XZ_3 Q)}$$

TABLE 12.4   Data for Example 12.1

Person	SBP (Y)	QUET (X)	SMK	Age Group	Person	SBP (Y)	QUET (X)	SMK	Age Group
1	120	2.789	0	1	36	160	3.612	1	2
2	152	4.116	0	4	37	142	3.401	0	3
3	135	3.171	0	3	38	140	3.562	1	3
4	180	4.637	1	4	39	144	2.368	1	1
5	137	3.296	0	2	40	142	3.024	1	1
6	162	3.668	1	4	41	142	3.401	0	3
7	130	3.100	0	2	42	161	3.800	0	4
8	162	3.668	1	4	43	130	3.100	0	2
9	122	3.251	0	1	44	144	3.751	0	3
10	132	3.210	0	2	45	146	2.979	1	3
11	138	3.673	0	3	46	166	3.877	1	4
12	180	4.637	1	4	47	129	2.790	1	1
13	135	2.876	0	1	48	138	4.032	1	2
14	137	3.296	0	2	49	148	3.768	0	2
15	120	2.789	0	1	50	132	3.017	1	2
16	149	3.301	1	3	51	150	3.628	1	3
17	145	3.360	1	2	52	129	2.790	1	1
18	161	3.800	0	4	53	166	3.877	1	4
19	122	3.251	0	1	54	170	4.132	1	4
20	145	3.360	1	2	55	134	2.998	1	2
21	144	3.751	0	3	56	138	3.673	0	3
22	140	3.562	1	3	57	134	2.998	1	2
23	132	3.017	1	2	58	164	4.010	0	4
24	164	4.010	0	4	59	126	2.956	1	1
25	152	3.962	0	4	60	170	4.132	1	4
26	135	2.876	0	1	61	149	3.301	1	3
27	126	2.956	1	1	62	160	3.612	1	2
28	150	3.628	1	3	63	142	3.024	1	1
29	148	3.768	0	2	64	132	3.210	0	2
30	135	3.171	0	3					
31	152	4.116	0	4					
32	152	3.962	0	4	*Note*: Age Group = 1	if age is 30–39 years			
33	146	2.979	1	3	2	if age is 40–49 years			
34	138	4.032	1	2	3	if age is 50–59 years			
35	144	2.368	1	1	4	if age is 60 + years			

For this example, this multiple-partial F-test has 8 numerator d.f. and 48 denominator d.f. and value 3.69 ($P = 0.002$). At the $\alpha = 0.05$ level of significance, we reject the null hypothesis that the regression lines for nonsmokers and smokers, controlling for age-group, are coincident.

2. Next, since the regression lines for nonsmokers and smokers have been found to differ by age group, we will perform a test of parallelism. The null hypothesis of interest is

$$H_0: \beta_{12} = \beta_{13} = \beta_{14} = \beta_{15} = 0$$

given that X, Z_1, Z_2, Z_3, Q, Z_1Q, Z_2Q, Z_3Q, XZ_1, XZ_2, and XZ_3 are in the model. The test statistic is

$$F(XQ, XZ_1Q, XZ_2Q, XZ_3Q | X, Z_1, Z_2, Z_3, Q, Z_1Q, Z_2Q, Z_3Q, XZ_1, XZ_2, XZ_3)$$

$$= \frac{[\text{Regression SS}(X, Z_1, Z_2, Z_3, Q, Z_1Q, Z_2Q, Z_3Q, XZ_1, XZ_2, XZ_3, XQ, XZ_1Q, XZ_2Q, XZ_3Q) - \text{Regression SS}(X, Z_1, Z_2, Z_3, Q, Z_1Q, Z_2Q, Z_3Q, XZ_1, XZ_2, XZ_3)]/4}{\text{Residual MS}(X, Z_1, Z_2, Z_3, Q, Z_1Q, Z_2Q, Z_3Q, XZ_1, XZ_2, XZ_3, XQ, XZ_1Q, XZ_2Q, XZ_3Q)}$$

which, for this example, has 4 and 48 degrees of freedom and value 1.85 ($P = 0.14$). At the $\alpha = 0.05$ level of significance, we fail to reject the null hypothesis that the regression lines for nonsmokers and smokers, controlling for age group, are parallel.

Therefore, we conclude that the age group-specific straight lines for nonsmokers and smokers are parallel but not coincident.

3. We will determine whether the SBP–QUET relationship is coincident for the four age groups, controlling for smoking status, by testing the following null hypothesis:

$$H_0: \beta_2 = \beta_3 = \beta_4 = \beta_6 = \beta_7 = \beta_8 = \beta_9 = \beta_{10} = \beta_{11} = \beta_{13} = \beta_{14} = \beta_{15} = 0$$

given that X, Q, and XQ are in the model. The test statistic for the relevant multiple-partial F-test is

$$F(Z_1, Z_2, Z_3, Z_1Q, Z_2Q, Z_3Q, XZ_1, XZ_2, XZ_3, XZ_1Q, XZ_2Q, XZ_3Q | X, Q, XQ,)$$

$$= \frac{[\text{Regression SS}(X, Z_1, Z_2, Z_3, Q, Z_1Q, Z_2Q, Z_3Q, XZ_1, XZ_2, XZ_3, XQ, XZ_1Q, XZ_2Q, XZ_3Q) - \text{Regression SS}(X, Q, XQ)]/12}{\text{Residual MS}(X, Z_1, Z_2, Z_3, Q, Z_1Q, Z_2Q, Z_3Q, XZ_1, XZ_2, XZ_3, XQ, XZ_1Q, XZ_2Q, XZ_3Q)}$$

which, for this example, has 12 and 48 degrees of freedom and value 5.65 ($P < 0.0001$). At the $\alpha = 0.05$ level of significance, we reject the null hypothesis of coincident lines for the four different age groups, controlling for smoking status.

4. Finally, since the regression lines for the four age groups have been found to differ (controlling for smoking status), we will perform a test of parallelism. The null hypothesis is

$$H_0: \beta_9 = \beta_{10} = \beta_{11} = \beta_{13} = \beta_{14} = \beta_{15} = 0$$

given that X, Z_1, Z_2, Z_3, Q, Z_1Q, Z_2Q, Z_3Q and XQ are in the model. The test statistic is

$$F(XZ_1, XZ_2, XZ_3, XZ_1Q, XZ_2Q, XZ_3Q | X, Z_1, Z_2, Z_3, Q, Z_1Q, Z_2Q, Z_3Q, XQ)$$

$$= \frac{[\text{Regression SS}(X, Z_1, Z_2, Z_3, Q, Z_1Q, Z_2Q, Z_3Q, XZ_1, XZ_2, XZ_3, XQ, XZ_1Q, XZ_2Q, XZ_3Q) - \text{Regression SS}(X, Z_1, Z_2, Z_3, Q, Z_1Q, Z_2Q, Z_3Q, XQ)]/6}{\text{Residual MS}(X, Z_1, Z_2, Z_3, Q, Z_1Q, Z_2Q, Z_3Q, XZ_1, XZ_2, XZ_3, XQ, XZ_1Q, XZ_2Q, XZ_3Q)}$$

which, for this example, has 6 and 48 degrees of freedom and value 3.22 ($P = 0.01$). At the $\alpha = 0.05$ level of significance, we reject the null hypothesis that the regression lines for the four age groups, within each of the two categories of smoking status, are parallel.

It is not of interest to test for differences in intercepts in this example; therefore, we simply conclude that the sets of straight lines, controlling for either SC or SMK, are not parallel. From the form of the complete model (12.13), each of the coefficients in the parallelism null hypotheses considered corresponds (as expected) to a product term of the general form Z_jQ or XZ_jQ. ∎

Problems

1. Using the data from Problem 2 in Chapter 5 and/or the SAS output given here, answer the following questions about the separate straight-line regressions of SBP on QUET for smokers (SMK = 1) and nonsmokers (SMK = 0).

   a. Determine the least-squares line of SBP (Y) on QUET (X) separately for smokers and nonsmokers.

   b. Test H_0: "The slopes are the same for the populations of smokers and nonsmokers being sampled" versus H_A: "Nonsmokers have a more positive slope."

   c. Test H_0: "The intercepts are the same for the populations of smokers and nonsmokers being sampled" versus H_A: "The intercepts are different."

   d. Test H_0: "The straight lines coincide for the populations of smokers and nonsmokers being sampled" versus H_A: "The straight lines do not coincide."

Edited SAS Output (PROC REG) for Problem 1

```
                        SBP Regressed on QUET for Nonsmokers
                              Descriptive Statistics

Variables                   Sum                    Mean              Uncorrected SS
QUET                     52.174             3.4782666667                 183.93641
SBP                        2112                    140.8                    299700

              Variables              Variance        Std Deviation
              QUET                0.1758089238       0.4192957474
              SBP                 166.45714286      12.901827113

                            Analysis of Variance

                                 Sum of          Mean
Source              DF          Squares         Square      F Value    Pr > F
Model                1       1702.83966     1702.83966       35.275    <.0001
Error               13        627.56034       48.27387
Corrected Total     14       2330.40000

              Root MSE            6.94794      R-square       0.7307
              Dep Mean          140.80000      Adj R-sq       0.7100
              Coeff Var           4.93462

                            Parameter Estimates

                            Parameter       Standard
Variable       DF           Estimate           Error      t value    Pr > |t|
INTERCEP        1          49.311759      15.50814374        3.180      0.0072
QUET            1          26.302825       4.42865240        5.939      <.0001
```

(continued)

Edited SAS Output (PROC REG) for Problem 1 (Continued)

```
                    SBP Regressed on QUET for Smokers (continued)
                              Descriptive Statistics

Variables                   Sum                      Mean              Uncorrected SS
QUET                      57.941               3.4082941176              202.639313
SBP                        2513                147.82352941                  375183

            Variables              Variance          Standard Deviation
            QUET                0.3224589706              0.567854709
            SBP                 231.40441176             15.211982506

                              Analysis of Variance

                               Sum of              Mean
Source                DF       Squares            Square        F Value        Pr > F
Model                  1     2088.16977         2088.16977       19.403        0.0005
Error                 15     1614.30082          107.62005
Corrected Total       16     3702.47059

            Root MSE              10.37401          R-square           0.5640
            Dep Mean             147.82353          Adj R-sq           0.5349
            Coeff Var              7.01783

                              Parameter Estimates

                         Parameter           Standard
Variable       DF         Estimate             Error        t value       Pr > |t|
INTERCEP        1        79.255330          15.76836893       5.026         0.0002
QUET            1        20.118041           4.56719317       4.405         0.0005
```

2. A topic of major concern to demographers and economists is the effect of a high fertility rate on per capita income. The first two accompanying tables display values of per capita income (PCI) and population percentage under age 15 (YNG) for a hypothetical sample of developing countries in Latin America and Africa, respectively. The third table summarizes the results of straight-line regressions of PCI (Y) on YNG (X) for each group of countries.

   a.–d. Repeat parts (a) through (d) of Problem 1 for the straight-line regressions of PCI (Y) on YNG (X) for Latin American and African countries.

Latin American Countries

YNG (X)	PCI (Y)	YNG (X)	PCI (Y)	YNG (X)	PCI (Y)
32.2	788	44.0	292	35.0	685
47.0	202	44.0	321	47.4	220
34.0	825	43.0	300	48.0	195
36.0	675	43.0	323	37.0	605
38.7	590	40.0	484	38.4	530
40.9	408	37.0	625	40.6	480
45.0	324	39.0	525	35.8	690
45.4	235	44.6	340	36.0	685
42.2	338	33.0	765		

African Countries

YNG (X)	PCI (Y)	YNG (X)	PCI (Y)	YNG (X)	PCI (Y)
34.0	317	41.0	188	39.0	225
36.0	270	42.0	166	39.0	232
38.2	208	45.0	132	37.0	260
43.0	150	36.0	290	37.0	250
44.0	105	42.6	160	46.0	92
44.0	128	33.0	300	45.6	110
45.0	85	33.0	320	42.0	180
48.0	75	47.0	85	38.8	235
40.0	210	47.0	75		

Summary of Separate Straight-Line Fits

Location	n	$\hat{\beta}_0$	$\hat{\beta}_1$	$\overline{X}$	$\overline{Y}$	S_X^2	$S_{Y|X}^2$	r
Latin America	26	2170.67	−42.0	40.277	478.846	21.633	1391.756	−.983
Africa	26	897.519	−17.39	40.892	186.462	20.244	188.919	−.985

e. Test H_0: "The population correlation coefficients are equal for the two groups of countries under study." Use $\alpha = .05$. Does your conclusion here clash with your findings regarding the equality of slopes? [*Hint:* See Section 6.7.]

For each of the preceding tests, assume that the alternative hypothesis is two-sided.

3. A team of anthropologists and nutrition experts investigated the influence of protein content in diet on the relationship between AGE and height (HT) for New Guinean children. The first two accompanying tables display values of HT (in centimeters) and AGE for a hypothetical sample of children with protein-rich and protein-poor diets, respectively.

Protein-Rich Diet

AGE (X)	0.2	0.5	0.8	1.0	1.0	1.4	1.8	2.0	2.0	2.5	2.5	3.0	2.7
HT (Y)	54	54.3	63	66	69	73	82	83	80.3	91	93.2	94	94

Protein-Poor Diet

AGE (X)	0.4	0.7	1.0	1.0	1.5	2.0	2.0	2.4	2.8	3.0	1.3	1.8	0.2	3.0
HT (Y)	52	55	61	63.4	66	68.5	67.9	72	76	74	65	69	51	77

Summary of Separate Straight-line Fits

Diet	n	$\hat{\beta}_0$	$\hat{\beta}_1$	$\overline{X}$	$\overline{Y}$	S_X^2	$S_{Y|X}^2$	r
Protein-Rich	13	50.324	16.009	1.646	76.677	0.808	5.841	.937
Protein-Poor	14	51.225	8.686	1.650	65.557	0.873	4.598	.969

a.–d. Repeat parts (a) through (d) of Problem 1 for the straight-line regressions of HT (Y) on AGE (X) for the two diets. (Consider a two-sided alternative in each case.)

e. Test whether the population correlation coefficient for children with a protein-rich diet differs significantly from that for children with a protein-poor diet. (Consider a two-sided alternative.)

4. For the data involving regression of DI (Y) on IQ (X) in Problem 4 in Chapter 5, assume that the sample of 17 observations (with the outlier removed) consists of males only. Now suppose that another sample of observations on DI (Y) and IQ (X) has been obtained for 14 females. The information needed to compare the straight-line regression equations for males and females is given in the following table.

Sex Group	n	$\hat{\beta}_0$	$\hat{\beta}_1$	$\overline{X}$	$\overline{Y}$	S_X^2	$S_{Y\|X}^2$	r
Males	17	70.846	−0.444	101.411	25.812	215.882	24.335	−.807
Females	19	61.871	−0.438	101.053	17.579	175.497	16.692	−.825

a.–d. Repeat parts (a) through (d) of Problem 1 for the straight-line regressions of DI (Y) on IQ (X) for the two sexes. (Consider two-sided alternatives.)

e. Perform a two-sided test of whether the population correlation coefficients for males and females are equal.

5. Assume that the data involving the regression of VOTE (Y) on TVEXP (X) in Problem 5 in Chapter 5 came from congressional districts in New York. Now, suppose that researchers selected a second sample of 17 congressional districts in California and recorded the same information for it. The following table provides the information needed to compare the straight-line regression equations for New York and California.

Location	n	$\hat{\beta}_0$	$\hat{\beta}_1$	$\overline{X}$	$\overline{Y}$	S_X^2	$S_{Y\|X}^2$	r
New York	20	2.174	1.177	36.99	45.71	76.870	11.101	.954
California	17	8.030	1.036	36.371	45.706	97.335	13.492	.945

a.–d. Repeat parts (a) through (d) of Problem 1 for the straight-line regression of VOTE (Y) on TVEXP (X) for the two states. Consider the one-sided alternative H_A: $\beta_{1(CAL)} < \beta_{1(NY)}$ for the test for slope, and consider the one-sided alternative H_A: $\beta_{0(CAL)} < \beta_{0(NY)}$ for the test for intercept.

e. Perform a two-sided test of whether the correlation coefficients for New York and California are equal.

6. The data in the following table represent four-week growth rates for depleted chicks at different dosage levels of vitamin B, by sex.[10]

---

[10] Adapted from a study by Clark, Lechyeka, and Cook (1940).

Males		Females	
Growth Rate (Y)	Log_{10} Dose (X)	Growth Rate (Y)	Log_{10} Dose (X)
17.1	0.301	18.5	0.301
14.3	0.301	22.1	0.301
21.6	0.301	15.3	0.301
24.5	0.602	23.6	0.602
20.6	0.602	26.9	0.602
23.8	0.602	20.2	0.602
27.7	0.903	24.3	0.903
31.0	0.903	27.1	0.903
29.4	0.903	30.1	0.903
30.1	1.204	28.1	0.903
28.6	1.204	30.3	1.204
34.2	1.204	33.0	1.204
37.3	1.204	35.8	1.204
33.3	1.505	32.6	1.505
31.8	1.505	36.1	1.505
40.2	1.505	30.5	1.505

Use the information provided in the next table to answer the following questions.
a. Determine the dose–response straight lines separately for each sex, and plot them on the same graph.
b. Test whether the slopes for males and females differ.
c. Find a 99% confidence interval for the true difference between the male and female slopes.
d. Test for coincidence of the two straight lines.

Sex Group	n	$\hat{\beta}_0$	$\hat{\beta}_1$	$\bar{X}$	$\bar{Y}$	S_X^2	$S_{Y|X}^2$	r
Males	16	14.178	14.825	0.922	27.84	0.187	10.5553	.807
Females	16	15.656	12.735	0.903	27.16	0.1812	8.4719	.788

7. The results in the first of the following tables were obtained in a study of the amount of energy metabolized by two similar species of birds under constant temperature.[11] Information based on separate straight-line fits to each data set is summarized in the second table.
a. Plot the least-squares straight lines for each species on the same graph.
b. Test whether the two lines are parallel.
c. Test whether the two lines have the same intercept.
d. Give a 95% confidence interval for the true difference between the mean amounts of energy metabolized by each species at 15°C. [*Hint:* Use a confidence interval of the form

$$\left(\hat{Y}_{15}^A - \hat{Y}_{15}^B\right) \pm t_{n_A+n_B-4,\, 1-\alpha/2} \sqrt{S_{\hat{Y}_{15}^A}^2 + S_{\hat{Y}_{15}^B}^2}$$

[11] Adapted from a study by Davis (1955).

Species A		Species B	
Calories (Y)	Temperature (X) (°C)	Calories (Y)	Temperature (X) (°C)
36.9	0	41.1	0
35.8	2	40.6	2
34.6	4	38.9	4
34.3	6	37.9	6
32.8	8	37.0	8
31.7	10	36.1	10
31.0	12	36.3	12
29.8	14	34.2	14
29.1	16	33.4	16
28.2	18	32.8	18
27.4	20	32.0	20
27.8	22	31.9	22
25.5	24	30.7	24
24.9	26	29.5	26
23.7	28	28.5	28
23.1	30	27.7	30

Species	n	$\hat{\beta}_0$	$\hat{\beta}_1$	$\overline{X}$	$\overline{Y}$	S_X^2	$S_{Y \mid X}^2$	r
A	16	36.579	−0.4528	15.00	29.79	90.6667	0.1662	−.9959
B	16	40.839	−0.4368	15.00	34.29	90.6667	0.1757	−.9953

where $\hat{Y}_{15}^i$ is the predicted value at 15°C for species i (i = A, B), and $S_{\hat{Y}_{15}^i}^2$ is the estimated variance of $\hat{Y}_{15}^i$ given by the general formula (for $X = X_0$)

$$S_{\hat{Y}_{X_0}^i}^2 = S_{P,\, Y \mid X}^2 \left[\frac{1}{n_i} + \frac{(X_0 - \overline{X})^2}{(n_i - 1) S_{X_i}^2} \right]$$

in your calculations.]

8. In Problem 1, separate straight-line regressions of SBP on QUET were compared for smokers (SMK = 1) and nonsmokers (SMK = 0).
   a. Define a single multiple regression model that uses the data for both smokers and non-smokers and that defines straight-line models for each group with possibly differing intercepts and slopes. Define the intercept and slope for each straight-line model in terms of the regression coefficients of the single regression model.
   b. Using the accompanying computer output, determine and plot on graph paper the two fitted straight lines obtained from the fit of the regression model.
   c. Test the following null hypotheses:

   H_0: "The two lines are parallel."

   H_0: "The two lines are coincident."

   For each of these tests, state the appropriate null hypothesis in terms of the regression coefficients of the regression model.
   d. Compare your answers in parts (b) and (c) with those you obtained by fitting separate regressions in Problem 1.

Edited SAS Output (PROC REG) for Problem 8

```
                        Regression of SBP on QUET, SMK, and QUET x SMK

                              Analysis of Variance

                                  Sum of          Mean
   Source·           DF         Squares          Square       F Value      Pr > F
   Model             3        4184.10759       1394.70253      17.419       <.0001
   Error            28        2241.86116         80.06647
   Corrected Total  31        6425.96875

                    Root MSE         8.94799      R-square       0.6511
                    Dep Mean       144.53125      Adj R-sq       0.6137
                    Coeff Var        6.19104

                              Parameter Estimates

                          Parameter        Standard
   Variable        DF      Estimate           Error       t value      Pr > |t|
   INTERCEP         1     49.311759       19.97234635       2.469        0.0199
   QUET             1     26.302825        5.70349238       4.612        <.0001
   SMK              1     29.943571       24.16355268       1.239        0.2256
   QUETSMK          1     -6.184785        6.93170667      -0.892        0.3799

   Variable        DF     Type I SS
   INTERCEP         1        668457
   QUET             1   3537.945739
   SMK              1    582.420754
   QUETSMK          1     63.741095
```

9. Use the computer output shown next to compare the separate regressions of SBP on AGE and QUET for smokers and nonsmokers (based on the data from Problem 2 in Chapter 5), as follows.
   a. State the appropriate regression model, simultaneously incorporating equations for both smokers and nonsmokers.
   b. Determine the fitted regression equations for smokers and nonsmokers separately, using the fitted regression model.
   c. Test for parallelism of the two models, stating the null hypothesis in terms of appropriate regression coefficients.
   d. Test for coincidence of the two models, stating the null hypothesis in terms of regression coefficients.

Edited SAS Output (PROC REG) for Problem 9

```
                 SBP Regressed on AGE, QUET, SMK, AGESMK, and QUETSMK

                              Analysis of Variance

                                  Sum of          Mean
   Source           DF         Squares          Square       F Value      Pr > F
   Model             5        4915.63040        983.12608      16.924       <.0001
   Error            26        1510.33835         58.08994
   Corrected Total  31        6425.96875
```

(continued)

Edited SAS Output (PROC REG) for Problem 9 (Continued)

```
        SBP Regressed on AGE, QUET, SMK, AGESMK, and QUETSMK (continued)

                        Analysis of Variance (continued)

              Root MSE            7.62168          R-square         0.7650
              Dep Mean          144.53125          Adj R-sq         0.7198
              Coeff Var           5.27338

                            Parameter Estimates

                          Parameter            Standard
      Variable     DF     Estimate              Error          t value      Pr > |t|
      INTERCEP     1      48.612699          17.01536941         2.857        0.0083
      AGE          1       1.028915           0.50176589         2.051        0.0505
      QUET         1      10.451041           9.13014561         1.145        0.2628
      SMK          1      -0.537436          23.23003664        -0.023        0.9817
      AGESMK       1       0.437325           0.71279040         0.614        0.5449
      QUETSMK      1      -3.706822          10.76763124        -0.344        0.7334

      Variable     DF     Type I SS
      INTERCEP     1        668457
      AGE          1   3861.630375
      QUET         1    258.961870
      SMK          1    769.233452
      AGESMK       1     18.920333
      QUETSMK      1      6.884366
```

10. The results presented in the following tables are based on data from a study by Gruber (1970) to determine how and to what extent changes in blood pressure over time depend on initial blood pressure (at the beginning of the study), the sex of the individual, and the relative weight of the individual. Data were collected on $n = 104$ persons, for which the $k = 7$ independent variables were examined by multiple regression. The following variables were used in the study:

$Y =$ SBPSL (estimated slope based on the straight-line regression of an individual's blood pressure over time)[12]

$X_1 =$ SBP1 (initial systolic blood pressure)

$X_2 =$ SEX (male $= 1$, female $= -1$)

$X_3 =$ RW (relative weight)

$X_4 = X_1 X_2, \quad X_5 = X_1 X_3$

$X_6 = X_2 X_3, \quad X_7 = X_1 X_2 X_3$

Use only the ANOVA table to answer the following questions.

a. Determine the form of the separate fitted regression models of SBPSL (Y) on SBP1 (X_1), RW (X_3), and SBP1 $\times$ RW (X_5) for each sex in terms of the estimated regression coefficients for the fitted regression model.

---

[12] This choice of dependent variable can be criticized because observations of an individual's blood pressure taken over time are *not* independent, thus violating Assumption 2 in Chapter 8. A statistic preferable to the one Gruber used could be obtained by weighted regression (see Chapter 14) or by growth curve analysis (see Allen and Grizzle 1969), although these are not the only analysis options.

Variable	$\hat{\beta}$	$S_{\hat{\beta}}$	Partial $F(\hat{\beta}^2/S_{\hat{\beta}}^2)$
X_1(SBP1)	−0.045	0.00762	34.987
X_2(SEX)	0.695	0.86644	0.643
X_3(RW)	0.027	0.07049	0.149
$X_4(X_1X_2)$	−0.0029	0.00762	0.145
$X_5(X_1X_3)$	−0.00018	0.00062	0.084
$X_6(X_2X_3)$	−0.0092	0.07049	0.017
$X_7(X_1X_2X_3)$	0.00022	0.00062	0.125
(Intercept)	4.667		

Source		d.f.	SS	MS	F
Overall regression		7	37.148	5.307	$\dfrac{5.307}{76.246/96} = 6.68^{**}$
Regression	X_1	1	24.988	24.988	$\dfrac{24.988}{88.406/102} = 28.83^{**}$
	$X_2\vert X_1$	1	7.886	7.886	$\dfrac{7.886}{80.520/101} = 9.89^{**}$
	$X_3\vert X_1, X_2$	1	1.057	1.057	$\dfrac{1.057}{79.463/100} = 1.33$
	$X_4\vert X_1, X_2, X_3$	1	0.020	0.020	$\dfrac{0.020}{79.443/99} = 0.025$
	$X_5\vert X_1, X_2, X_3, X_4$	1	0.254	0.254	$\dfrac{0.254}{79.189/98} = 0.314$
	$X_6\vert X_1, X_2, X_3, X_4, X_5$	1	2.844	2.844	$\dfrac{2.844}{76.345/97} = 3.613$
	$X_7\vert X_1, X_2, X_3, X_4, X_5, X_6$	1	0.099	0.099	$\dfrac{0.099}{76.246/96} = 0.125$
Residual		96	76.246	0.794	
Total (corrected)		103	113.394		

Note: $F_{7,96,0.95} = 2.11$, $F_{1,96,0.95} = 3.95$, $F_{1,102,0.95} = 3.94$.

    b. Why can't you use this ANOVA table to test whether the two regression equations are either parallel or coincident? Describe the appropriate testing procedure in each case.

11. For the data from Problem 2 in this chapter, address the following questions, using the information provided in the accompanying SAS output.

    a. State the regression model that incorporates the straight-line models for each group of countries.

    b. Determine and plot the separate fitted straight lines, based on the fitted regression model given in part (a).

    c. Test for parallelism of the two straight lines.

    d. Test for coincidence of the two straight lines.

    e. Compare your results in parts (b) through (d) to those obtained in Problem 2.

Edited SAS Output (PROC REG) for Problem 11

```
                    Regression of PCI on YNG, Z, and YNG x Z
                            Analysis of Variance

                                Sum of              Mean
Source                  DF      Squares            Square      F Value      Pr > F
Model                    3   2218613.5581      739537.85268    935.724      <.0001
Error                   48     37936.21118        790.33773
Corrected Total         51   2256549.7692

              Root MSE            28.11295       R-square         0.9832
              Dep Mean           332.65385       Adj R-sq         0.9821
              Coeff Var            8.45111

                            Parameter Estimates

                         Parameter          Standard
Variable        DF        Estimate            Error       t value      Pr > |t|
INTERCEP         1       897.518610       51.39769142      17.462        <.0001
YNG              1       -17.388529        1.24965131     -13.915        <.0001
Z                1      1273.151485       71.01247197      17.929        <.0001
YNGZ             1       -24.616267        1.73867221     -14.158        <.0001

Variable        DF      Type I SS
INTERCEP         1       5754246
YNG              1       1089772
Z                1        970417
YNGZ             1        158424
```

12. Using the information given here, answer the same questions as in Problem 11 about the regression of height (Y) on age (X) for children in one of two diet categories. (This problem is based on the data in Problem 3.)

Edited SAS Output (PROC REG) for Problem 12

```
                    Height Regressed on Age, Z, and Age x Z
                    Where Z = 1 if Protein Rich, 0 Otherwise
                            Analysis of Variance

                                Sum of              Mean
Source                  DF      Squares            Square      F Value      Pr > F
Model                    3    4174.20665        1391.40222    267.981      <.0001
Error                   23     119.42001           5.19217
Corrected Total         26    4293.62667

              Root MSE             2.27863       R-square         0.9722
              Dep Mean            70.91111       Adj R-sq         0.9686
              Coeff Var            3.21337

                            Parameter Estimates

                         Parameter          Standard
Variable        DF        Estimate            Error       t value      Pr > |t|
INTERCEP         1        51.225175        1.27112426      40.299        <.0001
AGE              1         8.686041        0.67620919      12.845        <.0001
Z                1        -0.901476        1.86193670      -0.484        <.6329
AGEZ             1         7.322927        0.99647349       7.349        <.0001
```

(continued)

Edited SAS Output (PROC REG) for Problem 12 (Continued)

```
                    Height Regressed on Age, Z, and Age x Z
              Where Z = 1 if Protein Rich, 0 Otherwise (continued)
Variable            DF              Type I SS
INTERCEP            1                 135766
AGE                 1              3053.348351
Z                   1               840.452385
AGEZ                1               280.405918
```

13. In Gruber's (1970) study of $n = 104$ individuals (discussed in Problem 10), the relationship between blood pressure change (SBPSL) and relative weight (RW), controlling for initial blood pressure (SBP1), was compared for three different geographical backgrounds and for three different psychosocial orientations, using the following 15 variables:

$Y = $ SBPSL

$X_1 = $ SBP1 (initial blood pressure)

$X_2 = $ R (1 if rural background, 0 if town, -1 if urban)

$X_3 = $ T (1 if town background, 0 if rural, -1 if urban)

$X_4 = $ TD (1 if traditional orientation, 0 if transitional, -1 if modern)

$X_5 = $ TN (1 if transitional orientation, 0 if traditional, -1 if modern)

$X_6 = $ RW (relative weight)

$X_7 = $ T $\times$ TD

$X_8 = $ T $\times$ TN

$X_9 = $ R $\times$ TD

$X_{10} = $ R $\times$ TN

$X_{11} = $ R $\times$ TD $\times$ RW

$X_{12} = $ R $\times$ TN $\times$ RW

$X_{13} = $ T $\times$ TD $\times$ RW

$X_{14} = $ T $\times$ TN $\times$ RW

A standard stepwise-regression program was run using these data, yielding the following ANOVA table (variables were forced to enter in the order presented) based on the model

$$Y = \beta_0 + \beta_1 X_1 + \beta_2 X_2 + \cdots + \beta_{14} X_{14} + E$$

a. Using this regression model, determine the form of the nine fitted regression equations corresponding to the nine possible combinations of background with orientation (i.e., R = 1 and TD = 1, R = 0 and TD = 1, R = -1 and TD = 1, etc.). [*Note:* Each of the nine equations will be of the form $\hat{Y} = \hat{\beta}_0^* + \hat{\beta}_1^*(\text{SBP1}) + \hat{\beta}_2^*(\text{RW})$.]

b. Test the null hypothesis that the nine regression equations determined in part (a) are parallel. State the null hypothesis in terms of the regression coefficients of the original 14-variable regression model.

c. Test the hypothesis H_0: "The three regression equations corresponding to the three backgrounds (rural, town, and urban) are parallel (but not necessarily coincident)," against the alternative H_A: "They are not parallel."

d. Set up the multiple-partial F-test formula for testing H_0: "The nine regression equations dealt with in part (a) are coincident," against H_A: "They are not coincident." State the null hypothesis in terms of the coefficients in the regression equation.

Source		d.f.	SS	F
Regression	X_1	1	24.9878	28.830
	$X_2\|X_1$	1	0.5218	0.600
	$X_3\|X_1, X_2$	1	0.0057	0.006
	$X_4\|X_1, X_2, X_3$	1	1.0520	1.199
	$X_5\|X_1, X_2, X_3, X_4$	1	1.1116	1.271
	$X_6\|X_1-X_5$	1	0.8321	0.951
	$X_7\|X_1-X_6$	1	0.2919	0.331
	$X_8\|X_1-X_7$	1	1.6601	1.902
	$X_9\|X_1-X_8$	1	0.5843	0.667
	$X_{10}\|X_1-X_9$	1	0.2266	0.257
	$X_{11}\|X_1-X_{10}$	1	1.1916	1.355
	$X_{12}\|X_1-X_{11}$	1	2.0853	2.407
	$X_{13}\|X_1-X_{12}$	1	1.5915	1.854
	$X_{14}\|X_1-X_{13}$	1	0.0208	0.024
Residual		89	77.2303	
Total		103	113.3934	

14. The Environmental Protection Agency conducted an experiment to assess the characteristics of sampling procedures designed to measure the suspended particulate concentration (X) in a particular city. At each of two distinct locations (designated as location 1 and location 2), two identical sampling units were set up side by side, and readings were taken on each of 10 days. The data are given here in tabular form, where X_{ij1} and X_{ij2} are, respectively, the measured concentration for samplers 1 and 2 at location i on day j ($i = 1, 2$; $j = 1, 2,\ldots, 10$). Researchers hypothesized that the inherent variation in the observations depends on the level of concentration being measured. To quantify this hypothesis, they proposed to fit a model of the form

$$|d_{ij}| = \beta_0 + \beta_1 Z + \beta_2(X_{ij1} + X_{ij2}) + \beta_3(X_{ij1} + X_{ij2})Z + E$$

where Z is 1 if the observation pertains to location 1 and is 0 otherwise, and where $|d_{ij}| = |X_{ij1} - X_{ij2}|$.

a. Using the results provided in the following table, determine and interpret the fitted straight-line relationship between $|d_{ij}|$ and $(X_{ij1} + X_{ij2})$ at each location.

	Location 1			**Location 2**		
Day	X_{1j1}	X_{1j2}	$d_{1j} = X_{1j1} - X_{1j2}$	X_{2j1}	X_{2j2}	$d_{2j} = X_{2j1} - X_{2j2}$
1	4	3	1	6	5	1
2	8	6	2	3	1	2
3	12	16	−4	1	2	−1
4	1	1	0	10	12	−2
5	7	6	1	17	17	0
6	11	8	3	4	7	−3
7	14	10	4	8	6	2
8	10	12	−2	12	12	0
9	2	2	0	10	9	1
10	15	20	−5	20	19	1

b. Test for parallelism of the two lines.

c. Test for coincidence of the two lines.

d. How would you test whether level of concentration is significantly related to inherent variation in the observations for at least one of the two locations?

Edited SAS Output (PROC REG) for Problem 14

```
          ABS(d) Regressed on X (=X1+X2), Z, and X x Z Where Z = 1 if
                            Location 1, 0 Otherwise

                          Analysis of Variance

                                 Sum of          Mean
Source                    DF      Squares        Square      F Value       Pr > F
Model                      3     31.33952       10.44651      19.873       <.0001
Error                     16      8.41048        0.52565
Corrected Total           19     39.75000

             Root MSE          0.72502          R-square        0.7884
             Dep Mean          1.75000          Adj R-sq        0.7487
             Coeff Var        41.42974

                          Parameter Estimates

                         Parameter        Standard
Variable        DF        Estimate          Error       t value      Pr > |t|
INTERCEP         1        2.008561       0.43196210       4.650        0.0003
X                1       -0.039147       0.02022626      -1.935        0.0708
Z                1       -2.427916       0.61773149      -3.930        0.0012
XZ               1        0.195061       0.03022850       6.453        <.0001

Variable        DF       Type I SS
INTERCEP         1      61.250000
X                1       4.834820
Z                1       4.616603
XZ               1      21.888102
```

15. A biologist compared the effect of temperature for each of two media on the growth of human amniotic cells in a tissue culture. The data shown in the following table were obtained.

Medium A				Medium B			
No. of Cells $\times 10^{-6}$ (Y)	Temperature (°F) (X)	No. of Cells $\times 10^{-6}$ (Y)	Temperature (°F) (X)	No. of Cells $\times 10^{-6}$ (Y)	Temperature (°F) (X)	No. of Cells $\times 10^{-6}$ (Y)	Temperature (°F) (X)
1.13	40	2.30	80	0.98	40	2.20	80
1.20	40	2.15	80	1.05	40	2.10	80
1.00	40	2.25	80	0.92	40	2.20	80
0.91	40	2.40	80	0.90	40	2.30	80
1.05	40	2.49	80	0.89	40	2.26	80
1.75	60	3.18	100	1.60	60	3.10	100
1.45	60	3.10	100	1.45	60	3.00	100
1.55	60	3.28	100	1.40	60	3.13	100
1.64	60	3.35	100	1.50	60	3.20	100
1.60	60	3.12	100	1.56	60	3.07	100

a. Assuming that a parabolic model is appropriate for describing the relationship between Y and X for each medium, provide a single regression model that incorporates two separate parabolic models, one corresponding to each medium.

b. Use the SAS output provided here to determine and plot the separate fitted parabolas for each medium. [*Note:* $Z = 0$ for medium A and $Z = 1$ for medium B.]

c. Test for "parallelism" of the two parabolas.

d. Test for coincidence of the two parabolas.

e. Is it possible to test whether a quadratic term should be included in the model for each medium, using only the provided SAS output? Explain.

Edited SAS Output (PROC REG) for Problem 15

```
              Y Regressed on X, X2, Z, ZX, and ZX2 Where Z = 1
                        if Medium B, 0 Otherwise

                          Analysis of Variance

                              Sum of        Mean
Source                 DF     Squares      Square     F Value    Pr > F
Model                   5    26.06864     5.21373     583.965    <.0001
Error                  34     0.30356     0.00893
Corrected Total        39    26.37220

                 Root MSE        0.09449     R-square    0.9885
                 Dep Mean        1.99275     Adj R-sq    0.9868
                 Coeff Var       4.74163

                          Parameter Estimates

                       Parameter          Standard
Variable      DF        Estimate            Error       t value   Pr > |t|
INTERCEP       1        0.494600         0.24256234      2.039     0.0493
X              1        0.005370         0.00745505      0.720     0.4763
X2             1        0.000217         0.00005282      4.118     0.0002
Z              1       -0.143700         0.34303495     -0.419     0.6779
ZX             1        0.001235         0.01054303      0.117     0.9074
ZX2            1       -0.000008750      0.00007470     -0.117     0.9074

Variable      DF        Type I SS
INTERCEP       1       158.842103
X              1        25.668613
X2             1         0.290702
Z              1         0.109202
ZX             1         0.000000500
ZX2            1         0.000122
```

16. Answer the following questions, using the accompanying SPSS output, which is based on data on the growth rates (Y) of depleted chicks at different (log) dosage levels (X) of vitamin B for males and females.

a. Define a single multiple regression model that incorporates different straight-line models for males and females.

b. Plot the fitted straight lines for each sex on graph paper.

c. Test for parallelism.

d. Test for coincidence.

Edited SPSS Output for Problem 16

VARIABLE(S) ENTERED ON STEP NUMBER 1.. X

		ANALYSIS OF VARIANCE	DF	SUM OF SQUARES	MEAN SQUARE	F
MULTIPLE R	0.88583					
R SQUARE	0.78470	REGRESSION	1.	1041.33728	1041.33728	109.33735
ADJUSTED R SQUARE	0.77752	RESIDUAL	30.	285.72230	9.52408	
STANDARD ERROR	3.08611					

------VARIABLES IN THE EQUATION------

VARIABLE	B	BETA	STD ERROR B	F
X	13.85501	0.88583	1.32502	109.337
(CONSTANT)	14.72828			

------VARIABLES NOT IN THE EQUATION------

VARIABLE	BETA IN	PARTIAL	TOLERANCE	F
Z	0.01293	0.02784	0.99791	0.022
XZ	0.04475	0.08801	0.83267	0.226

VARIABLE(S) ENTERED ON STEP NUMBER 2.. Z (= 1 if MALE, = 0 if FEMALE)

		ANALYSIS OF VARIANCE	DF	SUM OF SQUARES	MEAN SQUARE	F
MULTIPLE R	0.88592					
R SQUARE	0.78486	REGRESSION	2.	1041.55874	520.77937	52.89862
ADJUSTED R SQUARE	0.77002	RESIDUAL	29.	285.50084	9.84486	
STANDARD ERROR	3.13765					

------VARIABLES IN THE EQUATION------

VARIABLE	B	BETA	STD ERROR B	F
X	13.84577	0.88524	1.34856	109.413
Z	0.16655	0.01293	1.11049	0.022
(CONSTANT)	14.65352			

------VARIABLES NOT IN THE EQUATION------

VARIABLE	BETA IN	PARTIAL	TOLERANCE	F
XZ	0.19139	0.15355	0.13847	0.676

VARIABLE(S) ENTERED ON STEP NUMBER 3.. XZ

		ANALYSIS OF VARIANCE	DF	SUM OF SQUARES	MEAN SQUARE	F
MULTIPLE R	0.88878					
R SQUARE	0.78993	REGRESSION	3.	1048.28988	349.42996	35.09721
ADJUSTED R SQUARE	0.76743	RESIDUAL	28.	278.76970	9.95606	
STANDARD ERROR	3.15532					

------VARIABLES IN THE EQUATION------

VARIABLE	B	BETA	STD ERROR B	F
X	12.73533	0.81424	1.91389	44.278
Z	-1.88944	-0.14670	2.73852	0.476
XZ	2.23020	0.19139	2.71233	0.676
(CONSTANT)	15.65624			

------VARIABLES NOT IN THE EQUATION------

VARIABLE	BETA IN	PARTIAL	TOLERANCE	F

From Nie et al., *Statistical Package for the Social Sciences.* Copyright © 1975 by McGraw-Hill Book Company and Dr. Norman Nie, President, SPSS Inc.

17. Market research was conducted for a national retail company to compare the relationship between sales and advertising during the warm Spring and Summer seasons as compared with the cool Fall and Winter seasons. The data shown in the following table were collected over a period of several years.

Season (Warm = 0, Cool = 1)	Advertising Expenditure ($millions)	Sales Revenue ($millions)	Season (Warm = 0, Cool = 1)	Advertising Expenditure ($millions)	Sales Revenue ($millions)
0	17.0	156.1	1	10.0	131.0
0	12.5	142.6	1	13.8	136.8
0	20.5	166.8	1	15.0	141.5
0	16.0	155.4	1	19.5	151.8
0	15.0	150.5	1	17.0	148.3
0	14.5	147.5	1	12.5	133.3
0	17.5	156.9	1	14.5	138.0
0	12.5	138.8	1	12.5	135.9
0	11.5	134.3	1	12.0	132.0

a. Identify a single regression model that uses the data for both warm and cool seasons and that defines straight-line models relating sales revenue (Y) to advertising expenditure (X) for each season.

b. Using the computer output given next, determine and plot the fitted straight lines for each season.

c. Test whether the straight lines for cool and warm seasons coincide.

d. Test H_0: "The lines are parallel" versus H_A: "The lines are not parallel."

e. In light of your answers to parts (c) and (d), comment on differences and similarities in the sales–advertising expenditure relationship between cooler and warmer seasons.

Edited SAS Output (PROC REG) for Problem 17

```
              Sales Revenue (Y) Regressed on Adv. Expenditure (X),
                         Season (Z), and XZ
                      Descriptive Statistics
Variables          Sum                    Mean              Uncorrected SS
INTERCEP            18                      1                       18
X                263.8             14.655555556                4002.94
Z                  9                     0.5                        9
XZ               126.8              7.0444444444               1851.44
Y               2597.5             144.30555556              376638.73
```

(continued)

Edited SAS Output (PROC REG) for Problem 17 (Continued)

```
                Sales Revenue (Y) Regressed on Adv. Expenditure (X),
                        Season (Z), and XZ (continued)

                       Descriptive Statistics (continued)

          Variables                  Variance              Standard Deviation
          INTERCEP                          0                               0
          X                      8.0473202614                    2.8367799106
          Z                      0.2647058824                    0.5144957554
          XZ                     56.36496732                     7.5076605757
          Y                      106.17937908                    10.304337877

                             Analysis of Variance

                                  Sum of          Mean
Source                    DF      Squares         Square      F Value     Pr > F
Model                      3    1755.87993      585.29331     166.650     <.0001
Error                     14      49.16951        3.51211
Corrected Total           17    1805.04944

          Root MSE              1.87406       R-square        0.9728
          Dep Mean            144.30556       Adj R-sq        0.9669
          Coeff Var             1.29868

                             Parameter Estimates

                          Parameter       Standard
Variable       DF         Estimate         Error       t Value     Pr > |t|
INTERCEP        1         96.830446      3.56515723     27.160      <.0001
X               1          3.484861      0.23058405     15.113      <.0001
Z               1          7.172687      4.88169856      1.469       0.1639
XZ              1         -1.019784      0.32745582     -3.114       0.0076

Variable       DF         Type I SS
INTERCEP        1           374834
X               1         1461.750224
Z               1          260.067028
XZ              1           34.062678
```

18. A testing laboratory studies and compares the relationship between tire tread wear per 1,000 miles (Y) and average driving speed (X) for two competing tire types (denoted A and B). The data shown in the following table were collected for a random sample of 20 tires, 10 of A and 10 of B.

Tire	Tread Wear Per 100 Miles of Travel (% of tread thickness)	Average Speed (mph)	Tire	Tread Wear Per 100 Miles of Travel (% of tread thickness)	Average Speed (mph)
A	0.5	65	B	0.7	65
A	0.4	55	B	0.7	64
A	0.6	70	B	0.8	69
A	0.6	68	B	0.7	66
A	0.6	70	B	0.9	70
A	0.5	66	B	0.9	69
A	0.4	55	B	0.5	58
A	0.3	50	B	0.6	62
A	0.5	64	B	0.6	64
A	0.6	68	B	0.8	68

a. Identify a single regression model that uses the data for both tire types and that defines straight-line models relating tread wear (Y) to average speed (X) for each tire.

b. Using the computer output given next, determine and plot the fitted straight lines for each tire type.

c. Test the null hypothesis that the straight lines for the two tire types coincide.

d. Test H_0: "The lines are parallel" versus H_A: "The lines are not parallel."

e. In light of your answers to parts (c) and (d), comment on differences and similarities in the tread wear–average speed relationship for the two tire types.

Edited SAS Output (PROC REG) for Problem 18

```
          Tread Wear (Y) Regressed on Avg. Speed (X), Tire (Z), and XZ
                    (Z = 0 for Tire A, Z = 1 for Tire B)
                          Descriptive Statistics
```

Variables	Sum	Mean	Uncorrected SS
INTERCEP	20	1	20
X	1286	64.3	83302
Z	10	0.5	10
XZ	655	32.75	43027
Y	12.2	0.61	7.94

Variables	Variance	Standard Deviation
INTERCEP	0	0
X	32.221052632	5.6763590999
Z	0.2631578947	0.512989176
XZ	1135.5657895	33.698157063
Y	0.0262105263	0.1618966532

```
                         Analysis of Variance
```

Source	DF	Sum of Squares	Mean Square	F Value	Pr > F
Model	3	0.47861	0.15954	131.640	<.0001
Error	16	0.01939	0.00121		
Corrected Total	19	0.49800			

Root MSE	0.03481	R-square	0.9611	
Dep Mean	0.61000	Adj R-sq	0.9538	
Coeff Var	5.70697			

```
                         Parameter Estimates
```

Variable	DF	Parameter Estimate	Standard Error	t Value	Pr > |t|
INTERCEP	1	-0.407518	0.10313225	-3.951	0.0011
X	1	0.014382	0.00162509	8.850	<.0001
Z	1	-1.082121	0.22917196	-4.722	0.0002
XZ	1	0.019353	0.00351783	5.501	<.0001

Variable	DF	Type I SS
INTERCEP	1	7.442000
X	1	0.295057
Z	1	0.146875
XZ	1	0.036678

19. A random sample of data were collected on residential sales in a large city. The following table shows the sales price Y (in $1,000s), area X_1 (in hundreds of square feet), number of bedrooms X_2, total number of rooms X_3, age X_4 (in years), and location (dummy variables Z_1 and Z_2, defined as follows: $Z_1 = Z_2 = 0$ for intown; $Z_1 = 1$, $Z_2 = 0$ for inner suburbs; $Z_1 = 0$, $Z_2 = 1$ for outer suburbs) of each house.

House	Y	X_1	X_2	X_3	X_4	Z_1	Z_2
1	84.0	13.8	3	7	10	1	0
2	93.0	19.0	2	7	22	0	1
3	83.1	10.0	2	7	15	0	1
4	85.2	15.0	3	7	12	0	1
5	85.2	12.0	3	7	8	0	1
6	85.2	15.0	3	7	12	0	1
7	85.2	12.0	3	7	8	0	1
8	63.3	9.1	3	6	2	0	1
9	84.3	12.5	3	7	11	0	1
10	84.3	12.5	3	7	11	0	1
11	77.4	12.0	3	7	5	1	0
12	92.4	17.9	3	7	18	0	0
13	92.4	17.9	3	7	18	0	0
14	61.5	9.5	2	5	8	0	0
15	88.5	16.0	3	7	11	0	0
16	88.5	16.0	3	7	11	0	0
17	40.6	8.0	2	5	5	0	0
18	81.6	11.8	3	7	8	0	1
19	86.7	16.0	3	7	9	1	0
20	89.7	16.8	2	7	12	0	0
21	86.7	16.0	3	7	9	1	0
22	89.7	16.8	2	7	12	0	0
23	75.9	9.5	3	6	6	0	1
24	78.9	10.0	3	6	11	1	0
25	87.9	16.5	3	7	15	1	0
26	91.0	15.1	3	7	8	0	1
27	92.0	17.9	3	8	13	0	1
28	87.9	16.5	3	7	15	1	0
29	90.9	15.0	3	7	8	0	1
30	91.9	17.8	3	8	13	0	1

a. Identify a single regression model that uses the data for all three locations and that defines straight-line models relating sales price (Y) to area (X_1) for each location.
b. Using the computer output given next, determine and plot the fitted straight lines for each location.
c. Test the null hypothesis that the straight lines for the three locations coincide.
d. Test H_0: "The lines are parallel" versus H_A: "The lines are not parallel."
e. In light of your answers to parts (c) and (d), comment on differences and similarities in the sales price–area relationship for the three locations.

Edited SAS Output (PROC REG) for Problem 19

```
                Sales Price (Y) Regressed on Area (X1), Location (Z), and XZ
                            Descriptive Statistics
Variables                   Sum                        Mean                Uncorrected SS
INTERCEPT                    30                           1                          30
X1                       423.9                       14.13                     6276.55
Z1                           7                0.2333333333                           7
Z2                          15                         0.5                          15
X1Z1                     100.8                        3.36                     1490.94
X1Z2                     204.2                6.8066666667                     2914.06
Y                       2504.9                83.496666667                   212704.81

                    Variables               Variance          Standard Deviation
                    INTERCEP                       0                           0
                    X1                   9.891137931                 3.145017954
                    Z1                 0.1850574713                 0.4301830672
                    Z2                 0.2586206897                 0.5085476277
                    X1Z1                39.732827586                 6.3033980983
                    X1Z2                52.556505747                 7.2495865915
                    Y                   122.55205747                 11.070323278

                            Analysis of Variance

                            Sum of          Mean
Source              DF      Squares        Square     F Value      Pr > F
Model                5    3158.41441     631.68288      38.323      <.0001
Error               24     395.59526      16.48314
Corrected Total     29    3554.00967

            Root MSE            4.05994       R-square         0.8887
            Dep Mean           83.49667       Adj R-sq         0.8655
            Coeff Var           4.86240

                            Parameter Estimates
                            Parameter      Standard
Variable        DF          Estimate          Error     t Value     Pr > |t|
INTERCEP         1          8.968611     6.07754231       1.476       0.1530
X1               1          4.806990     0.39734913      12.098       <.0001
Z1               1         52.122387    11.22484137       4.643       <.0001
Z2               1         48.558241     7.79709969       6.228       <.0001
X1Z1             1         -3.201206     0.75896494      -4.218       0.0003
X1Z2             1         -2.803086     0.52980716      -5.291       <.0001

Variable        DF          Type I SS
INTERCEP         1             209151
X1               1       2271.713503
Z1               1          0.016309
Z2               1        336.002950
X1Z1             1         89.282266
X1Z2             1        461.399382
```

20. In Problem 19 in Chapter 5 and Problem 14 in Chapter 8, data from the 1990 Census for 26 randomly selected Metropolitan Statistical Areas (MSAs) were discussed. Of interest were factors potentially associated with the rate of owner occupancy of housing units. The following three variables were included in the data set:

OWNEROCC: Proportion of housing units that are owner-occupied (as opposed to renter-occupied)

OWNCOST: Median selected monthly ownership costs, in $

URBAN: Proportion of population living in urban areas

It is also of interest to see whether the owner occupancy rate–ownership cost relationship is different in metropolitan areas where 75% or more of the population lives in urban areas than in metropolitan areas where less than 75% of the population lives in urban areas. For this purpose, the following additional variable is defined:

$$X_1 = \begin{cases} 1 \text{ if proportion of population living in urban areas} \geq 0.75 \\ 0 \text{ otherwise} \end{cases}$$

a. State a single regression model that defines straight-line models relating OWNEROCC to OWNCOST both for MSAs with high percentages ($\geq 75\%$) of urban populations and for MSAs with lower percentages of urban populations ($< 75\%$).

b. Using the computer output given next, test whether the straight lines for the two MSA types described in part (a) coincide. [*Note:* In the accompanying SAS output, the variable OWNCOST has been centered to avoid collinearity problems.]

c. Test H_0: "The lines are parallel" versus H_A: "The lines are not parallel."

d. In light of your answers to parts (b) and (c), comment on differences and similarities in the owner occupancy rate–ownership cost relationship for the two types of MSA.

Edited SAS Output (PROC REG) for Problem 20

OWNEROCC Regressed on OWNCOST, X1, and X1*OWNCOST (X1OWNCOS)

Analysis of Variance

Source	DF	Sum of Squares	Mean Square	F Value	Pr > F
Model	3	214.19496	71.39832	4.061	0.0194
Error	22	386.76658	17.58030		
Corrected Total	25	600.96154			

. [Portion of output omitted]

OWNEROCC Regressed on OWNCOST and X1

Analysis of Variance

Source	DF	Sum of Squares	Mean Square	F Value	Pr > F
Model	2	213.49860	106.74930	6.337	0.0064
Error	23	387.46294	16.84621		
Corrected Total	25	600.96154			

. [Portion of output omitted]

Parameter Estimates

Variable	DF	Parameter Estimate	Standard Error	t Value	Pr > |t|
INTERCEP	1	68.665453	1.47336083	46.605	<.0001
OWNCOST	1	-0.012668	0.00552565	-2.293	0.0314
X1	1	-3.905669	1.78250451	-2.191	0.0388

21. This problem involves the PERK study data discussed in Problem 12 in Chapter 8. Suppose that we want to compare the relationship between change in refraction five years after surgery and baseline refractive error, for males and females. To this end, we define the following dummy variable:

$$Z = \begin{matrix} 1 & \text{if patient is male} \\ 0 & \text{otherwise} \end{matrix}$$

a. State a single regression model that defines straight-line models relating change in refraction five years after surgery and baseline refractive error for both males and females.

b. Using the computer output given next, test whether the straight lines for males and females coincide.

c. Test H_0: "The lines are parallel" versus H_A: "The lines are not parallel."

d. In light of your answers to parts (b) and (c), comment on differences and similarities in the change in refraction–baseline refractive error relationship for males and females.

Edited SAS Output (PROC REG) for Problem 21

Y Regressed on X1, Z, and X1Z

Analysis of Variance

Source	DF	Sum of Squares	Mean Square	F Value	Pr > F
Model	3	19.65170	6.55057	5.405	0.0027
Error	50	60.60124	1.21202		
Corrected Total	53	80.25294			

Root MSE	1.10092	R-square	0.2449	
Dep Mean	3.83343	Adj R-sq	0.1996	
Coeff Var	28.71896			

Parameter Estimates

Variable	DF	Parameter Estimate	Standard Error	t value	Pr > |t|
INTERCEP	1	3.178210	0.46488413	6.837	<.0001
X1	1	-0.201008	0.10785972	-1.864	0.0683
Z	1	-1.995126	0.88972206	-2.242	0.0294
X1Z	1	-0.383826	0.20265951	-1.894	0.0640

Y Regressed on X1, Z

Analysis of Variance

Source	DF	Sum of Squares	Mean Square	F Value	Pr > F
Model	2	15.30414	7.65207	6.009	0.0045
Error	51	64.94880	1.27351		
Corrected Total	53	80.25294			

Root MSE	1.12850	R-square	0.1907	
Dep Mean	3.83343	Adj R-sq	0.1590	
Coeff Var	29.43835			

Parameter Estimates

Variable	DF	Parameter Estimate	Standard Error	t value	Pr > |t|
INTERCEP	1	2.752647	0.41716942	6.598	<.0001
X1	1	-0.309731	0.09360197	-3.309	0.0017
Z	1	-0.412878	0.31372310	-1.316	0.1940

References

Allen, D. M., and Grizzle, J. E. 1969. "Analysis of Growth and Dose Response Curves." *Biometrics* 25: 357–82.

Armitage, P. 1971. *Statistical Methods in Medical Research.* Oxford: Blackwell Scientific Publications.

Clark, M. F.; Lechyeka, M.; and Cook, C. A. 1940. "The Biological Assay of Riboflavia." *Journal of Nutrition* 20: 133–44.

Davis, E. A., Jr. 1955. "Seasonal Changes in the Energy Balance of the English Sparrow." *Auk.,* 72(4): 385–411.

Gruber, F. J. 1970. "Industrialization and Health." Ph.D. dissertation, Department of Epidemiology, University of North Carolina, Chapel Hill, N.C.

Nie, N., et al. 1975. *Statistical Package for the Social Sciences.* New York: McGraw-Hill.

Analysis of Covariance and Other Methods for Adjusting Continuous Data

13.1 Preview

In Chapter 11, we discussed the issue of controlling for extraneous variables when assessing an association of interest. Three reasons for considering control are to assess *interaction,* to correct for *confounding,* and to increase the *precision* in estimating the association of interest. In regression, the usual approach for carrying out such control is to fit a regression model that contains as independent variables not only the study factors (exposure variables) of interest, but also extraneous variables considered to be important (and perhaps even product terms involving these variables). The focus then becomes one of determining the effects of the study factors on the response variable *adjusted* for the presence of the control variables in the model.

In this chapter, we describe how to carry out this process of adjustment by using a popular procedure for regression modeling called analysis of covariance (ANACOVA). This technique involves a multiple regression model in which the study factors of interest are all treated as nominal variables, whereas the variables being controlled—that is, the *covariates*—may be measurements on any measurement scale. As discussed in Chapter 12, nominal variables are incorporated into regression models by means of dummy variables. Thus, the general ANACOVA model usually contains a mixture of dummy variables and other types of variables, and the dependent variable is considered continuous. In using the ANACOVA model, we also assume that there is no interaction of covariates with study variables, although (as discussed later) this assumption should be assessed in the analysis.

In addition to considering ANACOVA, we briefly review other regression-type methods for controlling for extraneous factors.

13.2 Adjustment Problem

Suppose, as in the example in Section 12.4, that we are considering a sample of observations made on the dependent variable systolic blood pressure and the independent variable age for $n_F = 29$ women and $n_M = 40$ men.

Two questions are often of interest in analyses of such data:

1. Is the true straight-line relationship between blood pressure and age (given that a straight-line model is adequate) the same for male and female populations?

2. Do the mean blood pressure levels for the male and female groups differ significantly after one takes into account (i.e., after one adjusts for or controls for) the possible confounding effect of there being differing age distributions in the two groups?

Although the statistical techniques required to answer these questions are related, the questions nevertheless differ in emphasis: The first focuses on a comparison of straight-line regression equations, whereas the second focuses on a comparison of the mean blood pressure levels in the two groups.

We have already considered question 1 (in Chapter 12) through use of the model

$$Y = \beta_0 + \beta_1 X + \beta_2 Z + \beta_3 XZ + E \tag{13.1}$$

where Z is a dummy variable identifying the sex group ($Z = 1$ if female and $Z = 0$ if male). By using appropriate tests of hypotheses about the parameters in this model, we may reach one of three important conclusions regarding question 1:

1. The lines are coincident (i.e., $\beta_2 = \beta_3 = 0$).

2. The lines are parallel but not coincident (i.e., $\beta_3 = 0$, but $\beta_2 \neq 0$).

3. The lines are not parallel ($\beta_3 \neq 0$).

These conclusions greatly influence the answer to question 2. If conclusion 1 is appropriate, we say that the two sex groups do not differ in mean blood pressure level after the effect of age is controlled. If conclusion 2 holds, we say that the sex group associated with the higher straight line has a higher mean blood pressure level at all ages. If conclusion 3 is valid, we must look closer at the orientation of the two straight lines: if they do not intersect in the age range of interest, we say that the sex group associated with the higher curve has a higher mean blood pressure level at each age; if they do intersect in the age range of interest, we say that one sex group has a higher mean blood pressure level than the other group at lower ages and a lower mean blood pressure level at higher ages (i.e., there is an age–sex group interaction effect).

Thus, by considering question 1 as described, we may draw reasonable inferences about the relationship between the true average blood pressure levels in the two groups as a function of age. Nevertheless, we must take additional statistical considerations into account to estimate the true adjusted mean difference and the adjusted means for each group. In this regard, question 2 raises the problem of determining an appropriate method for adjusting the sample mean blood pressure levels to take into account the effect of age, as well as the problem of providing a statistical test to compare these adjusted mean scores.

In the example we are considering, age is a factor known to be strongly associated with blood pressure, and the two groups, as sampled, may have widely different age distributions. Without adjusting the sample mean values to reflect any difference in the age distributions in the two groups, we could not determine (e.g., through the use of a two-sample t test based on the unadjusted sample means) whether a significant difference was attributable solely to age or to other factors. With adjustment, however, we could determine whether any findings were solely attributable, for example, to the fact that the females in the sample were older than the males (or vice versa).

13.3 Analysis of Covariance

The usual statistical technique for handling the adjustment problem described in Section 13.2 is called the *analysis of covariance.* In this approach, we fit a regression model of the form

$$Y = \beta_0 + \beta_1 X + \beta_2 Z + E \qquad (13.2)$$

where X (age) is referred to as the *covariate* and Z is a dummy variable that indexes the two groups to be compared ($Z = 1$ if female, $Z = 0$ if male). This model assumes—in contrast to model (13.1), which contains a $\beta_3 XZ$ term—that the regression lines for males and females are parallel. Under this model, *the adjusted mean scores for males and females are defined to be the predicted values obtained by evaluating the model at $Z = 0$ and $Z = 1$ when X is set equal to the overall mean age for the two groups. A partial F test of the hypothesis H_0: $\beta_2 = 0$ is then used to determine whether these adjusted mean scores are significantly different.*

In computing these adjusted scores, we need to consider the two straight lines obtained by fitting model (13.2):

$$
\begin{aligned}
\text{Males } (Z = 0): \quad & \hat{Y}_M = \hat{\beta}_0 + \hat{\beta}_1 X \\
\text{Females } (Z = 1): \quad & \hat{Y}_F = (\hat{\beta}_0 + \hat{\beta}_2) + \hat{\beta}_1 X
\end{aligned}
\qquad (13.3)
$$

Explicit formulas for the estimated coefficients in (13.3) are

$$\hat{\beta}_1 = \frac{(n_M - 1)S_{X_M}^2 \hat{\beta}_{1M} + (n_F - 1)S_{X_F}^2 \hat{\beta}_{1F}}{(n_M - 1)S_{X_M}^2 + (n_F - 1)S_{X_F}^2} \qquad (13.4)$$

$$\hat{\beta}_0 = \overline{Y}_M - \hat{\beta}_1 \overline{X}_M$$

$$\hat{\beta}_0 + \hat{\beta}_2 = \overline{Y}_F - \hat{\beta}_1 \overline{X}_F$$

where $\hat{\beta}_{1M}$ and $\hat{\beta}_{1F}$ are the estimated slopes based on separate straight-line fits for males and females, $S_{X_M}^2$ and $S_{X_F}^2$ are the sample variances (of X) for males and females, $\overline{Y}_M$ and $\overline{Y}_F$ are the mean blood pressures for the male and female samples, and $\overline{X}_M$ and $\overline{X}_F$ are the mean ages for the male and female samples, respectively. Notice that $\hat{\beta}_1$ is a weighted average of the slopes $\hat{\beta}_{1M}$ and $\hat{\beta}_{1F}$, which are estimated separately from the male and female data sets.

Based on (13.3) and (13.4), two alternative formulas for computing adjusted mean scores can be used:

Sex	Z	Adjusted Score	Formula 1	Formula 2	
Male	0	$\overline{Y}_M$ (adj)	$\hat{\beta}_0 + \hat{\beta}_1 \overline{X}$	$\overline{Y}_M - \hat{\beta}_1(\overline{X}_M - \overline{X})$	(13.5)
Female	1	$\overline{Y}_F$ (adj)	$(\hat{\beta}_0 + \hat{\beta}_2) + \hat{\beta}_1 \overline{X}$	$\overline{Y}_F - \hat{\beta}_1(\overline{X}_F - \overline{X})$	

In this table, $\overline{X}$ is the overall mean age for the combined data on males and females:

$$\overline{X} = \frac{n_M \overline{X}_M + n_F \overline{X}_F}{n_M + n_F}$$

Formula 1 is useful when model (13.2) has been estimated directly using a standard multiple regression program; formula 2, on the other hand, can be used without resorting to multiple regression procedures, although separate straight lines must be fitted to the male and female data sets.

Given that the parallel straight-line assumption of model (13.2) is appropriate (we discuss this assumption in Sections 13.4 and 13.6.2), the two formulas provide a comparison of the mean blood pressure levels for the two sex groups as if they both had the same age distribution. In this regard, the covariance approach just described attempts artificially to equate the age distributions by treating both sex groups as if they had the same mean age, the best estimate of which is $\overline{X}$. The adjusted scores, then, represent the predicted $\hat{Y}$-values for each fitted line at $\overline{X}$, the assumed common mean age. This is depicted graphically in Figure 13.1.

That the partial F test of H_0: $\beta_2 = 0$ addresses the question of whether a significant difference exists between the adjusted means follows because, from formula 1 in (13.5), the difference in the two adjusted mean scores is exactly equal to $\hat{\beta}_2$; that is,

$$\hat{\beta}_2 = \overline{Y}_F(\text{adj}) - \overline{Y}_M(\text{adj}) = [(\hat{\beta}_0 + \hat{\beta}_2) + \hat{\beta}_1\overline{X}] - (\hat{\beta}_0 + \hat{\beta}_1\overline{X})$$

■ **Example 13.1** The least-squares fitting of the model (13.2) for the male–female data we have been discussing yields the following estimated model:

$$\hat{Y} = 110.29 + 0.96X - 13.51Z$$

The separate fitted equations for males and females, respectively, are

Males $(Z = 0)$: $\qquad \hat{Y}_M = 110.29 + 0.96X$

Females $(Z = 1)$: $\qquad \hat{Y}_F = 96.78 + 0.96X$

The adjusted mean scores obtained from these equations, using formula 1 of (13.5) with $\overline{X} = 46.14$, are

$$\overline{Y}_M(\text{adj}) = 110.29 + 0.96(46.14) = 154.58$$
$$\overline{Y}_F(\text{adj}) = 96.78 + 0.96(46.14) = 141.07$$

FIGURE 13.1 **Adjusted mean systolic blood pressure (SBP) scores for males and females, controlling for age by using analysis of covariance**

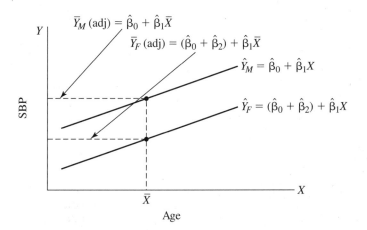

A comparison of these adjusted mean scores with the unadjusted means yields the following statistics:

Sex	Unadjusted $\bar{Y}$	Adjusted
Male	155.15	154.58
Female	139.86	141.07

Notice that the adjusted mean for males is slightly lower than the unadjusted mean for males, whereas the female adjusted mean is slightly higher than its unadjusted counterpart. The direction of these changes accurately reflects the fact that, in this sample, the males are somewhat older on the average ($\bar{X}_M = 46.93$) than the females ($\bar{X}_F = 45.07$). Using the adjusted mean scores, in effect, removes the influence of age on the comparison of mean blood pressures by considering what the mean blood pressures in the two groups would be if both groups had the same mean age ($\bar{X} = 46.14$).

But whether adjusted or not, the mean blood pressure for males in this example appears to be considerably higher than that for females. In fact, the covariance adjustment has done little to change this impression: the discrepancy between the male and female groups is 15.29 using unadjusted mean scores and 13.51 using adjusted mean scores. To test whether this difference in adjusted mean scores is significant, we use the partial F test of the hypothesis H_0: $\beta_2 = 0$ based on model (13.2), which can be computed from the following analysis-of-variance presentation:

Source		SS	d.f.	MS
Reduced model ($\beta_2 = 0$)	Regression (X)	14,951.25	1	14,951.25
	Residual	8,260.51	67	123.29
Complete model (13.2)	Regression (X, Z)	18,009.78	2	9,004.89
	Residual	5,201.99	66	78.82

From this presentation, we can obtain the appropriate partial F statistic as follows:

$$F(Z|X) = \frac{\text{Regression SS}(X, Z) - \text{Regression SS}(X)}{\text{Residual MS}(X, Z)}$$

$$= \frac{18,009.78 - 14,951.25}{78.82}$$

$$= 38.80$$

which has 1 and 66 degrees of freedom. The P-value for this test satisfies $P < .001$, so we reject H_0 and conclude that the two adjusted scores differ significantly. ∎

13.4 Assumption of Parallelism: A Potential Drawback

A potential problem raised by using ANACOVA involves the assumption of parallelism of the regression lines. In certain applications, these regression lines may have different slopes. In such cases, the parallelism assumption is invalid and the covariance method of adjustment just described should be avoided. *To guard against applying the covariance method of adjustment incorrectly, we recommend conducting a test for parallelism before proceeding with ANACOVA.*

This amounts to testing $H_0: \beta_3 = 0$ for the complete model (13.1). We saw in Chapter 12 that the parallelism hypothesis ($H_0: \beta_3 = 0$) is not rejected for the age–systolic blood pressure data. This result supports the use of the ANACOVA model for these data (given, of course, that the assumption of variance homogeneity also holds).

If the test for parallelism supports the conclusion that the regression lines are not truly parallel, what, if anything, should be done about adjustment? Usually no adjustment at all should be made to the sample means, since any such adjustment would be misleading; that is, a direct comparison of means is not appropriate when the true difference between the mean blood pressure levels in the two groups varies with age (i.e., when there is an age–sex interaction). In this case, since the main feature of the data is that the two regression lines describe very different relationships between age and blood pressure, an analysis that allows two separate regression lines to be fitted (without assuming parallelism) and that quantifies how the lines differ is sufficient.

13.5 Analysis of Covariance: Several Groups and Several Covariates

In the example discussed in previous sections, we compared two groups and adjusted for the single covariate, age. In general, ANACOVA may be used to provide adjusted scores when there are several (say, k) groups and when it is necessary to adjust simultaneously for several (say, p) covariates. The regression model describing this general situation is written

$$Y = \beta_0 + \beta_1 X_1 + \beta_2 X_2 + \cdots + \beta_p X_p + \beta_{p+1} Z_1 + \beta_{p+2} Z_2$$
$$+ \cdots + \beta_{p+k-1} Z_{k-1} + E$$

$$(13.6)$$

This model includes p covariates $X_1, X_2, \ldots, X_p$ and k groups, which are represented by the $k - 1$ dummy variables $Z_1, Z_2, \ldots, Z_{k-1}$. As discussed in Chapter 12, we have some leeway in defining these dummy variables; but for our purposes we assume that the Z's are defined as follows:

$$Z_j = \begin{cases} 1 & \text{if group } j \\ 0 & \text{otherwise} \end{cases} \quad j = 1, 2, \ldots, k - 1$$

The fitted regression equations for the k different groups are then determined by specifying the appropriate combinations of values for the Z's. These are

Group 1 ($Z_1 = 1$, other $Z_j = 0$): $\hat{Y}_1 = (\hat{\beta}_0 + \hat{\beta}_{p+1}) + \hat{\beta}_1 X_1$
$$+ \hat{\beta}_2 X_2 + \cdots + \hat{\beta}_p X_p$$

Group 2 ($Z_2 = 1$, other $Z_j = 0$): $\hat{Y}_2 = (\hat{\beta}_0 + \hat{\beta}_{p+2}) + \hat{\beta}_1 X_1$
$$+ \hat{\beta}_2 X_2 + \cdots + \hat{\beta}_p X_p$$

$\vdots$ $\vdots$ (13.7)

Group $k - 1$ ($Z_{k-1} = 1$, other $Z_j = 0$): $\hat{Y}_{k-1} = (\hat{\beta}_0 + \hat{\beta}_{p+k-1}) + \hat{\beta}_1 X_1$
$$+ \hat{\beta}_2 X_2 + \cdots + \hat{\beta}_p X_p$$

Group k (all $Z_j = 0$): $\hat{Y}_k = \hat{\beta}_0 + \hat{\beta}_1 X_1 + \hat{\beta}_2 X_2$
$$+ \cdots + \hat{\beta}_p X_p$$

From (13.7) we see that the corresponding coefficients of the covariates $X_1, X_2, \ldots, X_p$ in each of the k equations are identical. Thus, these regression equations represent "parallel" hypersurfaces in $(p + 1)$-space, which is a natural generalization of the situation for the single covariate case. The adjusted mean score for a particular group is then computed as the predicted Y-value obtained by evaluating the fitted equation for that group at the mean values $\bar{X}_1, \bar{X}_2, \ldots, \bar{X}_p$ of the p covariates, based on the combined data for all k groups:

$$\bar{Y}_j(\text{adj}) = (\hat{\beta}_0 + \hat{\beta}_{p+j}) + \hat{\beta}_1\bar{X}_1 + \cdots + \hat{\beta}_p\bar{X}_p \qquad j = 1, 2, \ldots, k - 1$$
$$\bar{Y}_k(\text{adj}) = \hat{\beta}_0 + \hat{\beta}_1\bar{X}_1 + \cdots + \hat{\beta}_p\bar{X}_p$$

(13.8)

To determine whether the k adjusted mean scores differ significantly from one another, we test the null hypothesis

$$H_0: \beta_{p+1} = \beta_{p+2} = \cdots = \beta_{p+k-1} = 0$$

using a multiple partial F test with $k - 1$ and $n - p - k$ degrees of freedom, based on model (13.6). If H_0 is rejected, we conclude that significant differences exist among the adjusted means (although we cannot, without further inspection, determine where the major differences are).

■ **Example 13.2**   In the Ponape study (Patrick et al. 1974) of the effect of rapid cultural change on health status, one research goal was to determine whether blood pressure was associated with a measure of the strength of a Ponapean male's prestige in the modern (i.e., western) part of his culture relative to the traditional culture. A measure of prestige discrepancy (PD) was developed and then measured by questionnaire for each of 550 Ponapean males. Of particular interest was whether a higher prestige discrepancy score corresponded to higher blood pressure. The effects of the covariates age and body size were adjusted or controlled for when considering these questions.

To perform this analysis, the researchers categorized prestige discrepancy into three groups:

Group 1: Modern prestige much higher than traditional prestige

Group 2: Modern prestige not much different from traditional prestige

Group 3: Traditional prestige much higher than modern prestige

Then an ANACOVA was carried out with diastolic blood pressure (DBP) as the dependent variable and with AGE and quetelet index (QUET, a measure of body size) as the two covariates. An analysis of covariance model in this case is given by

$$\text{DBP} = \beta_0 + \beta_1\text{AGE} + \beta_2\text{QUET} + \beta_3Z_1 + \beta_4Z_2 + E \qquad (13.9)$$

where

$$Z_1 = \begin{cases} 1 & \text{if group 1} \\ 0 & \text{otherwise} \end{cases} \qquad \text{and} \qquad Z_2 = \begin{cases} 1 & \text{if group 2} \\ 0 & \text{otherwise} \end{cases}$$

A test of the parallelism assumption implicit in model (13.9) considers the null hypothesis $H_0: \beta_5 = \beta_6 = \beta_7 = \beta_8 = 0$ in the full model

$$\text{DBP} = \beta_0 + \beta_1\text{AGE} + \beta_2\text{QUET} + \beta_3Z_1 + \beta_4Z_2 + \beta_5Z_1\text{AGE} + \beta_6Z_2\text{AGE}$$
$$+ \beta_7Z_1\text{QUET} + \beta_8Z_2\text{QUET} + E$$

This multiple-partial F test (with 4 and 541 degrees of freedom) was not rejected using the Ponape data, thus supporting the use of model (13.9).

TABLE 13.1   ANACOVA example using Ponape study data

$\overline{DBP}$	PD Group 1 ($n_1 = 87$)	PD Group 2 ($n_2 = 383$)	PD Group 3 ($n_3 = 80$)	P-Value for F Test of $H_0: \beta_3 = \beta_4 = 0^*$
Unadjusted	71.68	68.16	68.55	$.001 < P < .005$
Adjusted	72.07	68.02	68.80	

*Using model (13.9) with 2 and 545 d.f.

The researchers then determined adjusted DBP mean scores for the three groups by substituting the values of the overall means $\overline{AGE}$ and $\overline{QUET}$ into the fitted equations for the three groups, as follows:

$$\overline{DBP}_1(\text{adj}) = (\hat{\beta}_0 + \hat{\beta}_3) + \hat{\beta}_1\overline{AGE} + \hat{\beta}_2\overline{QUET}$$

$$\overline{DBP}_2(\text{adj}) = (\hat{\beta}_0 + \hat{\beta}_4) + \hat{\beta}_1\overline{AGE} + \hat{\beta}_2\overline{QUET}$$

$$\overline{DBP}_3(\text{adj}) = \hat{\beta}_0 + \hat{\beta}_1\overline{AGE} + \hat{\beta}_2\overline{QUET}$$

The test for equality of these adjusted means was based on use of a multiple partial F test of the null hypothesis $H_0: \beta_3 = \beta_4 = 0$ under model (13.9). Table 13.1 summarizes these calculations.

The results indicate the presence of highly significant differences among the adjusted mean blood pressure scores, with group 1 having a somewhat higher adjusted mean blood pressure than the other two groups. Despite the statistical significance found, however, the adjusted mean blood pressure of 72.07 for group 1 is close enough (clinically speaking) to the adjusted means for the other groups to cast doubt on the clinical significance of these results.   ■

13.6 Comments and Cautions

13.6.1 Rationale for Adjustment

ANACOVA adjusts for disparities in covariate distributions over groups by artificially assuming that all groups have the same set of mean covariate values. For example, if age and weight are the covariates and two groups are being compared, the ANACOVA adjustment procedure treats both groups as if they had the same mean age and the same mean weight.

The ANACOVA adjustment procedure is equivalent to assuming a common covariate *distribution* based on the combined sample over all groups. That is, not only are the means assumed to be equal, but the entire distribution of the covariates in the combined sample is assumed to be the same as the distribution of the covariates in each group. The adjusted score for any group can be expressed as the average over the combined sample of the predicted scores for that group; that is,

$$\hat{Y}_j(\text{adj}) = \frac{1}{n}\sum_{i=1}^{n}\hat{Y}_{ij} \qquad j = 1, 2, \ldots, k$$

where $\hat{Y}_j(\text{adj})$ is the adjusted score for group j defined by (13.8), n is the number of subjects in the combined sample, and $\hat{Y}_{ij}$ is the predicted response in the jth group, based on the set of equations (13.7) and the covariate values of the ith individual in the combined sample. For example,

$$\hat{Y}_{i1} = (\hat{\beta}_0 + \hat{\beta}_{p+1}) + \hat{\beta}_1 X_{i1} + \hat{\beta}_2 X_{i2} + \cdots + \hat{\beta}_p X_{ip}$$

is the predicted group 1 response for the ith individual in the combined sample, where $(X_{i1}, X_{i2}, \ldots, X_{ip})$ are the covariate values for that individual. Thus, the adjusted score for a given group can be obtained by artificially assuming that all n persons in the combined sample constitute the given group; then the covariate distribution of the group is assumed to be that of the combined sample.

Thus the method of adjustment using ANACOVA corrects the disparity in covariate distributions over groups by assuming a common distribution (not just a common set of means).

13.6.2 The Parallelism Assumption

As stated earlier, the ANACOVA method is inappropriate when the relationship between the covariates and the response is not the same in each group. Such nonparallelism or interaction might be reflected, for example, in a finding that males have higher blood pressures than females at older ages and that females have higher blood pressures than males at younger ages. Consequently, using a standard ANACOVA could lead to adjusted (for age) mean scores for each (gender) group that are roughly equal; this would create the misleading impression that there was little difference between male and female blood pressures, when, in fact, there were large differences in certain age categories. This example illustrates why we recommend that no method of adjustment be used in the presence of interaction; instead, the nature of the interaction itself should be quantified. The method for assessing interaction in this context, as previously indicated, requires first that the ANACOVA model (13.6) be expanded to include product terms between covariates and group variables and second that the coefficients of these product terms be tested for significance. The extended ANACOVA model that allows for such interaction terms has the form

$$Y = \beta_0 + \sum_{i=1}^{p} \beta_i X_i + \sum_{j=1}^{k-1} \beta_{p+j} Z_j + \sum_{i=1}^{p}\sum_{j=1}^{k-1} \gamma_{ij} X_i Z_j + E \tag{13.10}$$

In this model, a chunk test for parallelism would test H_0: All $\gamma_{ij} = 0$, and would involve a multiple partial F statistic of the form

$F \, (\text{All } X_i Z_j \text{ product terms} \mid \text{All } X_i \text{ and } Z_j \text{ terms})$

which would have an F distribution under H_0 with $p(k-1)$ and $n-1-p-(k-1)-p(k-1)$ degrees of freedom. The use of ANACOVA-adjusted scores is appropriate only when the preceding test for interaction yields the conclusion that interaction effects are not present.

13.6.3 Validity and Precision

As discussed in Chapter 11, validity and precision are two reasons to consider controlling covariates. In using ANACOVA, validity is achieved by adjusting for confounding, thereby obtaining an estimate of association that would have been distorted (biased) if the covariate(s) of interest had been ignored in the analysis.

Although validity should be the first consideration, it is possible to find no confounding in one's data and still control for one or more covariates to gain precision. To assess precision, we can consider either the variances of the estimators of the association(s) of interest (the smaller these variances, the greater the precision) or confidence intervals for the association(s) of interest (the narrower the confidence intervals, the greater the precision). For example, consider an ANACOVA model involving two groups—say, males and females—with systolic blood pressure as the dependent variable and age as the only covariate of interest. If the age distribution for males were identical to that for females in the data, then age would not be a confounder. Nevertheless, because age is strongly positively associated with systolic blood pressure, precision will probably be increased by adjusting for age, even though confounding (i.e., validity) is not at issue.

13.6.4 Alternatives to ANACOVA

When adjusting for covariates, we can fit a model to contain the covariates and the study variables of primary interest (and perhaps even product terms, so interaction can be assessed) without having to make the study variables categorical. A best model would then have to be derived by means of criteria for selecting variables (see Chapter 16). If, for such a model, significant interaction is found, it is inappropriate (as noted earlier with regard to categorical study variables) to derive adjusted scores. Moreover, even in the absence of interaction, it is impossible to obtain adjusted scores for groups unless the study variables are defined categorically.

Nevertheless, predicted values based on the best regression model can be treated as adjusted values, since the covariates are being taken into account in the modeling process. Furthermore, adjusted scores for distinct values of continuous study variables can be obtained by computing predicted values using the overall mean covariate values in the best model, as was done for ANACOVA using categorical study variables. For example, if we determine that the best model relating systolic blood pressure to age and weight is

$$\widehat{\text{SBP}} = \hat{\beta}_0 + \hat{\beta}_1 \text{AGE} + \hat{\beta}_2 \text{WEIGHT}$$

then we can compute adjusted blood pressure scores for persons weighing 150 and 175 pounds (controlling for age) as

$$\widehat{\text{SBP}}_{150}(\text{adj}) = \hat{\beta}_0 + \hat{\beta}_1 \overline{\text{AGE}} + 150\hat{\beta}_2$$

and

$$\widehat{\text{SBP}}_{175}(\text{adj}) = \hat{\beta}_0 + \hat{\beta}_1 \overline{\text{AGE}} + 175\hat{\beta}_2$$

These adjusted values are similar to group adjusted scores: the value of 150 can be thought of as representing a group of people whose weights are all close to 150; and a similar interpretation can be given to the 175-pound value.

In a more restrictive alternative to ANACOVA, *all* variables—even covariates—are treated as categorical. If we distinguish the set of dummy variables defining the covariates from the set of dummy variables defining the study variables, we can treat the regression model under consideration as a two-way analysis-of-variance model with unequal cell numbers (see Chapter 20). This alternative, however, might be inappropriate if the underlying scales of measurement of some of the covariates are actually noncategorical (e.g., are continuous). If, for example, the inherently continuous variable age is categorized into three age groups, a completely categorical model based on that arbitrary categorization might lead to different results from those obtained by using a model that treats age continuously.

13.7 Summary

In this chapter, we have examined a long-established approach to controlling for covariates, ANACOVA. To use ANACOVA, the study variables of interest must be treated as categorical variables, whereas the covariates are not so restricted. ANACOVA also requires the assumption that there is no interaction between covariates and study variables. This assumption can be checked by testing for the significance of appropriate product terms in an extended ANACOVA model. If the test for interaction yields a finding of nonsignificance, adjusted scores for different groups can be obtained by substituting the mean covariate values for the combined sample into the group-specific ANACOVA model. If the test for interaction is significant, adjusted scores should not be used; instead, the nature of the interaction should be characterized.

Problems

1. Problem 8 in Chapter 12 involved comparing straight-line regression fits of SBP on QUET for smokers and nonsmokers, and it was found that these straight lines could be considered parallel. Use results based on that problem and the computer output given next (and the fact that the overall sample mean value $\overline{\text{QUET}} = 3.441$) to address the following issues.

   a. State the appropriate ANACOVA regression model to use for comparing the mean blood pressures in the two smoking categories, controlling for QUET.

   b. Determine the adjusted SBP means for smokers and nonsmokers. Compare these values to the unadjusted mean values

   $$\overline{\text{SBP}}(\text{smokers}) = 147.823 \qquad \text{and} \qquad \overline{\text{SBP}}(\text{nonsmokers}) = 140.800$$

   c. Test whether the true adjusted mean blood pressures in the two groups are equal. State the null hypothesis in terms of the regression coefficients in the ANACOVA model given in part (a).

   d. Obtain a 95% confidence interval for the true difference in adjusted SBP means.

Edited SAS Output (PROC REG) for Problem 1

```
                     REGRESSION OF SBP ON QUET AND SMK

                          Analysis of Variance

                              Sum of          Mean
Source              DF       Squares         Square      F Value    Pr > F
Model                2      4120.36649     2060.18325      25.91    <.0001
Error               29      2305.60226       79.50353
Corrected Total     31      6425.96875

            Root MSE              8.91647      R-Square      0.6412
            Dependent Mean      144.53125      Adj R-Sq      0.6165
            Coeff Var             6.16924

                          Parameter Estimates
                      Parameter      Standard
Variable      DF      Estimate         Error     t Value    Pr > |t|     Type I SS
Intercept      1      63.87603       11.46811       5.57     <.0001        668457
QUET           1      22.11560        3.22996       6.85     <.0001       3537.94574
SMK            1       8.57101        3.16670       2.71      0.0113        582.42075
```

2. a.–d. Answer the same questions as in parts (a) through (d) in Problem 1 regarding an analysis of covariance designed to control for *both* AGE and QUET. [*Note:* $\overline{AGE} = 53.250$.] Use the results from Problem 9 in Chapter 12.

   e. Is it necessary to control for *both* AGE and QUET as opposed to controlling just one of the two covariates?

Edited SAS Output (PROC REG) for Problem 2

```
        PROC REG OUTPUT FOR REGRESSION OF SBP ON SMK, AGE AND QUET
                          Analysis of Variance

                              Sum of            Mean
Source              DF        Squares          Square       F Value      Pr > F
Model                3      4889.82570      1629.94190        29.71      <.0001
Error               28      1536.14305        54.86225
Corrected Total     31      6425.96875
             Root MSE              7.40691      R-Square       0.7609
             Dependent Mean      144.53125      Adj R-Sq       0.7353
             Coeff Var             5.12478

                            Parameter Estimates
                          Parameter     Standard
Variable            DF     Estimate        Error     t Value    Pr > |t|    Type I SS
Intercept            1     45.10319     10.76488        4.19      0.0003       668457
SMK                  1      9.94557      2.65606        3.74      0.0008     393.09816
AGE                  1      1.21271      0.32382        3.75      0.0008    4296.58607
QUET                 1      8.59245      4.49868        1.91      0.0664     200.14147
```

3. In an experiment conducted at the National Institute of Environmental Health Sciences, the absorption (or uptake) of a chemical by a rat on one of two different diets, I or II, was known to be affected by the weight (or size) of the rat. A completely randomized design utilizing four rats on each diet was employed in the experiment, and the initial weight of each rat was recorded so that the diets could be compared after the researchers adjusted for the effect of initial weight. The data for the experiment are given in the following table.

Initial Weight (X)	3	1	4	4	5	2	3	2
Diet (Z)	I	I	I	I	II	II	II	II
Response (Y)	14	13	14	15	16	15	15	14

a. Using the initial weight as a covariate, state the ANACOVA regression model for comparing the two diets (set $Z = -1$ if diet I is used, and $Z = 1$ if diet II is used).

b. Use the computer results given next to determine the adjusted mean responses for each diet, controlling for initial weight.

c. Use the ANACOVA regression model defined in part (a) to test whether the two diets differ significantly.

d. Test whether the two diets differ significantly, completely ignoring the covariate. How do the two testing procedures compare?

e. Determine a 95% confidence interval for the true difference in the adjusted mean responses.

Edited SAS Output (PROC REG) for Problem 3

```
PROC REG OUTPUT FOR REGRESSION OF RESPONSE (Y) ON INITIAL WEIGHT (X) AND DIET (Z)
                            Analysis of Variance

                                    Sum of          Mean
Source                      DF      Squares         Square      F Value      Pr > F
Model                        2      5.00000        2.50000       12.50       0.0113
Error                        5      1.00000        0.20000
Corrected Total              7      6.00000
                 Root MSE           0.44721     R-Square          0.8333
                 Dependent Mean    14.50000     Adj R-Sq          0.7667
                 Coeff Var          3.08423

                            Parameter Estimates
                        Parameter     Standard
Variable            DF   Estimate       Error      t Value    Pr > |t|     Type I SS
Intercept            1   13.00000      0.41833      31.08      <.0001      1682.00000
X                    1    0.50000      0.12910       3.87      0.0117         3.00000
Z                    1    0.50000      0.15811       3.16      0.0250         2.0000
```

4. A political scientist developed a questionnaire to determine political tolerance scores (Y) for a random sample of faculty members at her university. She wanted to compare mean scores adjusted for age (X) in each of three categories: full professors, associate professors, and assistant professors. The data results are given in the accompanying tables (the higher the score, the more tolerant the individual).

Group 1: Full Professors (Z1 = 1, Z2 = 0)

Age (X)	65	61	47	52	49	45	41	41	40	39
Tolerance (Y)	3.03	2.70	4.31	2.70	5.09	4.02	3.71	5.52	5.29	4.62

Group 2: Associate Professors (Z1 = 0, Z2 = 1)

Age (X)	34	31	30	35	49	31	42	43	39	49
Tolerance (Y)	4.62	5.22	4.85	4.51	5.12	4.47	4.50	4.88	5.17	5.21

Group 3: Assistant Professors (Z1 = Z2 = 0)

Age (X)	26	33	48	32	25	33	42	30	31	27
Tolerance (Y)	5.20	5.86	4.61	4.55	4.47	5.71	4.77	5.82	3.67	5.29

a. State an ANACOVA regression model that can be used to compare the three groups, controlling for age.

b. What model should be used to check whether the ANACOVA model in part (a) is appropriate? Carry out the appropriate test. Use $\alpha = .01$.

c. Using ANACOVA, determine adjusted mean tolerance scores for each group, and test whether these adjusted scores differ significantly from one another. Also, compare the adjusted means with the unadjusted means. [*Note:* $\bar{X}$(overall) $= 39.667, \bar{Y}$(group 1) $= 4.10$, $\bar{Y}$(group 2) $= 4.86$, and $\bar{Y}$(group 3) $= 5.00$.]

Edited SAS Output (PROC REG) for Problem 4

```
     PROC REG OUTPUT FOR REGRESSION OF TOLERANCE (Y) ON AGE (X), Z1 AND Z2
                                       .
                                       . [Portion of output omitted]
                                       .
                              Parameter Estimates
                          Parameter        Standard
       Variable      DF     Estimate         Error        t Value      Pr > |t|
       Intercept      1      6.18372        0.61189         10.11        <.0001
       X              1     -0.03635        0.01741         -2.09        0.0467
       Z1             1     -0.33981        0.41457         -0.82        0.4199
       Z2             1      0.06357        0.33234          0.19        0.8498

           PROC REG OUTPUT FOR REGRESSION OF Y ON X, Z1, Z2, XZ1 AND XZ2
                              Analysis of Variance
                                      Sum of          Mean
  Source                      DF      Squares         Square      F Value     Pr > F
  Model                        5     10.20934        2.04187        5.02      0.0027
  Error                       24      9.76275        0.40678
  Corrected Total             29     19.97210
                      Root MSE            0.63779      R-Square        0.5112
                      Dependent Mean      4.64967      Adj R-Sq        0.4093
                      Coeff Var          13.71699

                              Parameter Estimates
                          Parameter        Standard
  Variable      DF         Estimate         Error      t Value     Pr > |t|     Type I SS
  Intercept      1          5.42706        0.98483       5.51       <.0001       648.58200
  X              1         -0.01321        0.02948      -0.45       0.6580         6.20495
  Z1             1          2.78490        1.51591       1.84       0.0786         0.62494
  Z2             1         -1.22343        1.50993      -0.81       0.4258         0.01847
  XZ1            1         -0.07247        0.03779      -1.92       0.0671         3.14678
  XZ2            1          0.03022        0.04165       0.73       0.4751         0.21420
```

5. A psychological experiment was performed to determine whether in problem-solving dyads containing one male and one female, "influencing" behavior depended on the sex of the experimenter. The problem for each dyad was a strategy game called "Rope a Steer," which required 20 separate decisions about which way to proceed toward a defined goal on a game board. For each subject in the dyad, a verbal-influence activity score was derived as a function of the number of statements made by the subject to influence the dyad to move in a particular direction. The difference in verbal-influence activity scores within a dyad was denoted as the variable VIAD, which was then used as the dependent variable in an ANACOVA designed to control for the effects of differing IQ scores of the male and female in each dyad. The relevant data are given in the following tables.

a. State an ANACOVA model appropriate for these data.

b. Determine adjusted mean VIAD scores for each group and compare these to the unadjusted means for each group.

c. Test whether the adjusted mean scores differ significantly.

d. Find a 95% confidence interval for the true difference in adjusted mean scores.

Group 1: Male Experimenter (Z = 0)

VIAD	−10	−4	9	−15	−15	5	−8	−4	−1	13
IQ_M	115	112	106	123	125	105	115	122	138	110
IQ_F	100	110	108	135	115	112	121	132	135	126

Group 2: Female Experimenter (Z = 1)

VIAD	8	−5	2	−7	15	−10	−3	10	2	4
IQ_M	120	130	110	113	102	141	120	113	114	102
IQ_F	141	128	104	98	106	130	128	105	107	111

Edited SAS Output (PROC REG) for Problem 5

```
             PROC REG OUTPUT FOR REGRESSION OF VIAD ON IQM, IQF AND Z
                           Descriptive Statistics

                                                    Uncorrected                    Standard
Variable              Sum           Mean                 SS         Variance       Deviation
Intercept        20.00000        1.00000           20.00000              0               0
IQM            2336.00000      116.80000             275060      116.58947        10.79766
IQF            2352.00000      117.60000             279924      175.20000        13.23631
Z                10.00000        0.50000           10.00000        0.26316         0.51299
VIAD            -14.00000       -0.70000         1518.00000       79.37895         8.90949
                                              .
                                              . [Portion of output omitted]
                                              .

                                Parameter Estimates
                                Parameter       Standard
Variable          DF            Estimate          Error        t Value      Pr > |t|
Intercept          1            44.59001        17.85879          2.50        0.0238
IQM                1            -0.69301         0.19389         -3.57        0.0025
IQF                1             0.28109         0.15967          1.76        0.0974
Z                  1             5.19610         3.13292          1.66        0.1167
                                              .
                                              . [Portion of output omitted]
                                              .

          PROC REG OUTPUT FOR REGRESSION OF VIAD ON IQM, IQF, Z, IQM × Z, AND IQF × Z
                              Analysis of Variance
                                      Sum of          Mean
Source                      DF       Squares         Square      F Value      Pr > F
Model                        5     746.73852       149.34770       2.75       0.0623
Error                       14     761.46148        54.39011
Corrected Total             19    1508.20000
                    Root MSE            7.37496      R-Square     0.4951
                    Dependent Mean     -0.70000      Adj R-Sq     0.3148
                    Coeff Var       -1053.56640
```

(continued)

Edited SAS Output for Problem 5 (Continued)

```
PROC REG OUTPUT FOR REGRESSION OF VIAD ON IQM, IQF, Z, IQM × Z, AND IQF × Z (continued)
                                Parameter Estimates
                        Parameter        Standard
Variable      DF        Estimate          Error      t Value    Pr > |t|     Type I SS
Intercept      1         43.99054        29.67269      1.48       0.1604        9.80000
IQM            1         -0.73495         0.31358     -2.34       0.0344      502.25850
IQF            1          0.32723         0.25688      1.27       0.2234      109.79250
Z              1          6.03875        38.43473      0.16       0.8774      131.46715
IQMZ           1          0.07613         0.41737      0.18       0.8579        0.06709
IQFZ           1         -0.08265         0.34325     -0.24       0.8132        3.15327
```

6. An experiment was conducted to compare the effects of four different drugs (A, B, C, and D) in delaying atrophy of denervated muscles. A certain leg muscle in each of 48 rats was deprived of its nerve supply by surgical severing of the appropriate nerves. The rats were then put randomly into four groups, and each group was assigned treatment with one of the drugs. After 12 days, four rats from each group were sacrificed and the weight W (in grams) of the denervated muscle was obtained for each rat, as listed in the following table.

Theoretically, atrophy should be measured as the loss in weight of the muscle, but the initial weight of the muscle could not have been obtained without sacrificing the rat. Consequently, the initial total body weight X (in grams) of the rat was measured. It was assumed that this figure is closely related to the initial weight of the leg muscle.

Drugs A and C were large and small dosages, respectively, of atropine sulfate. Drug B was quinidine sulfate. Drug D acted as a control; it was simply a saline solution and could not have had any effect on atrophy. Use ANACOVA to compare the effects of the four drugs, controlling for initial total body weight (X). Use the results given in the SAS output shown next to perform your analysis. [*Note:* $\overline{X} = 226.125$.]

Drug A $(Z1 = 1, Z2 = Z3 = 0)$		Drug B $(Z2 = 1, Z1 = Z3 = 0)$		Drug C $(Z3 = 1, Z1 = Z2 = 0)$		Drug D $(Z1 = Z2 = Z3 = 0)$	
X	W	X	W	X	W	X	W
198	0.34	233	0.41	204	0.57	186	0.81
175	0.43	250	0.87	234	0.80	286	1.01
199	0.41	289	0.91	211	0.69	245	0.97
224	0.48	255	0.87	214	0.84	215	0.87

Edited SAS Output (PROC REG) for Problem 6

```
                PROC REG OUTPUT FOR REGRESSION OF W ON X, Z1, Z2 AND Z3
                               Analysis of Variance
                                    Sum of         Mean
Source                   DF        Squares        Square      F Value     Pr > F
Model                     4        0.61847       0.15462       10.71      0.0009
Error                    11        0.15873       0.01443
Corrected Total          15        0.77720
                 Root MSE           0.12013      R-Square      0.7958
                 Dependent Mean     0.70500      Adj R-Sq      0.7215
                 Coeff Var         17.03922
```

(continued)

Edited SAS Output for Problem 6 (Continued)

```
          PROC REG OUTPUT FOR REGRESSION OF W ON X, Z1, Z2 AND Z3 (continued)
                                 Parameter Estimates
                        Parameter       Standard
Variable      DF        Estimate         Error        t Value    Pr > |t|    Type I SS
Intercept      1        0.17283         0.30373         0.57      0.5808      7.95240
X              1        0.00319         0.00128         2.49      0.0299      0.32015
Z1             1       -0.39170         0.09541        -4.11      0.0017      0.20177
Z2             1       -0.22565         0.09020        -2.50      0.0294      0.06236
Z3             1       -0.13505         0.08776        -1.54      0.1521      0.03418
```

7. Trough urine samples were analyzed for sodium content for each of two collection periods, one before and one after administration of Mercuhydrin, for each of 30 dogs. The experimenter used 7 dogs as a control group for the study; these dogs were not administered the drug, but their urine samples were collected for two similar time periods.

	Experimental Group (Z = 1)			**Experimental Group (Z = 1)**	
Animal	First Collection Period (X) ([Na], mM/l)	Second Collection Period (Y) ([Na], mM/l)	Animal	First Collection Period (X) ([Na], mM/l)	Second Collection Period (Y) ([Na], mM/l)
1	17.5	22.1	18	6.3	12.7
2	9.4	12.0	19	9.7	17.1
3	10.0	15.2	20	7.1	9.5
4	7.4	23.1	21	7.2	11.0
5	8.8	9.8	22	5.3	8.2
6	18.9	26.9	23	14.3	15.8
7	10.8	11.1	24	7.9	9.7
8	8.8	13.6	25	14.1	14.7
9	8.8	12.8	26	12.8	17.0
10	9.2	7.5	27	12.8	20.2
11	8.1	8.1	28	10.7	13.9
12	10.3	27.5	29	5.9	11.8
13	10.1	11.2	30	3.8	9.0
14	7.3	11.0			
15	11.1	15.3	Mean	9.7	13.9
16	9.4	11.5	S.D.	3.3	5.3
17	8.2	8.4			

	Control Group (Z = 0)	
Animal	First Collection Period (X) ([Na], mM/l)	Second Collection Period (Y) ([Na], mM/l)
1	11.1	9.4
2	5.1	5.9
3	6.5	14.8
4	17.2	15.5
5	11.8	23.4
6	6.6	7.3
7	4.1	8.2
Mean	8.9	12.1
S.D.	4.7	6.4

7. a. Use ANACOVA (computer results are given next), with the before measure of sodium content as the covariate, to find the adjusted mean sodium contents for the experimental and control groups.

b. Test whether the two adjusted means differ significantly.

c. What alternative testing approach (involving a t test) could be used for these data? Carry out this test. [*Hint:* Variances of before–after differences for each group are 16.785 (experimental) and 25.915 (control).]

d. Are the two testing methods (t test versus covariance analysis) equivalent in this problem? Explain by comparing regression models appropriate for each method.

e. What do the results of a lack-of-fit test, based on using the computer output given, indicate about the appropriateness of the covariance model used in parts (a) and (b)? [*Note:* Compare the value $6.3568(= 21.0913/3.3179)$ to $F_{30, 4, 0.95}$; see Chapter 15 for further discussion of lack-of-fit tests.]

Dummy of Computer Results for Problem 7

	Multiple $R^2 = .406$				ANOVA		
Variable	$\hat{\beta}$	S.D. of $\hat{\beta}$	$\hat{\beta}$/S.D.		Source	d.f.	MS
X	0.96155	0.20418	4.709		X	1	434.4861
Z	1.06435	1.83729	0.5793		Added by Z	1	6.3764
(intercept)	3.49992				Lack of fit	30	21.0913
					Pure error	4	3.3179

8. Consider again the data in Problem 4.

a. How would you compute appropriate cross-product variables to allow testing of whether an interaction exists between age and faculty rank?

b. State the associated regression model.

c. Using a computer, fit the model. Provide estimates of the regression coefficients.

d. Provide a multiple-partial F test for interaction, controlling for age and faculty rank. Use $\alpha = .05$.

e. Does the F test in part (d) indicate that ANACOVA is valid?

9. This problem involves data from Problem 15 in Chapter 5. Treat LN_BRNTL as the dependent variable, and dosage level of toluene (PPM_TOLU) as a categorical predictor (four levels). The experimenter wanted to explore the possibility of using WEIGHT as a control variable.

a. State the appropriate ANACOVA model. Use the 50-ppm exposure group as the reference group when coding dummy variables.

b. Use a computer program to fit the model and provide estimated regression coefficients.

c. Provide adjusted mean estimates. Compare these with the unadjusted means. (Do not perform any tests.)

d. Test whether the adjusted means are all equal. Use $\alpha = .05$. State the null hypothesis in terms of the regression coefficients.

10. Repeat Problem 9, using LN_BLDTL as the dependent variable.

11. a. How would you compute appropriate cross-product terms for testing the interaction of WEIGHT and dosage level of toluene for Problem 9?

b. State the appropriate regression model.

c. State the null hypothesis (of no interaction) in terms of regression coefficients.

12. **a.** For the model fit in Problem 9, compute adjusted predicted scores and residuals.
    b. Plot residuals against predicted scores.
    c. Compute estimates of the residual variances separately for each dosage group. Are they approximately equal?
    d. Do these diagnostics suggest any problems?

13. Refer to the residential sales data in Problem 19 in Chapter 12. Use the computer output given below to answer the following questions.
    a. State an ANACOVA regression model that can be used to compare outer suburbs with other locations (intown and inner suburbs), controlling for age. [*Hint:* Let $Z = 0$ if intown or inner suburbs, 1 if outer suburbs.]
    b. Identify the model that should be used to check whether the ANACOVA model in part (a) is appropriate. Carry out the appropriate test.
    c. Using ANACOVA, determine adjusted mean sales prices for the two locations, and test whether they differ significantly from one another. [*Note:* Mean house age = 10.8667; unadjusted mean sales price for intown and inner suburbs = 82.1867; unadjusted mean sales price for outer location = 84.8067.]

Edited SAS Output (PROC REG) for Problem 13

```
                SALES PRICE (Y) REGRESSED ON AGE (X4), AND LOCATION (Z)
                           Analysis of Variance
                                     Sum of          Mean
Source                      DF       Squares        Square      F Value      Pr > F
Model                        3     1646.14368     548.71456        7.48      0.0009
Error                       26     1907.86599      73.37946
Corrected Total             29     3554.00967
                Root MSE                   8.56618     R-Square      0.4632
                Dependent Mean            83.49667     Adj R-Sq      0.4012
                Coeff Var                 10.25931

                           Parameter Estimates
                    Parameter      Standard
Variable      DF    Estimate        Error      t Value    Pr > |t|      Type I SS
Intercept      1    55.45706       6.85947        8.08      <.0001        209151
X4             1     2.37245       0.57631        4.12      0.0003     1324.84735
Z              1    17.90588       8.90573        2.01      0.0548      114.99710
X4Z            1    -1.27910       0.76286       -1.68      0.1056      206.29922

                SALES PRICE (Y) REGRESSED ON AGE (X4), AND LOCATION (Z)
                           Analysis of Variance
                                     Sum of          Mean
Source                      DF       Squares        Square      F Value      Pr > F
Model                        2     1439.84445     719.92223        9.19      0.0009
Error                       27     2114.16521      78.30242
Corrected Total             29     3554.00967
                Root MSE                   8.84887     R-Square      0.4051
                Dependent Mean            83.49667     Adj R-Sq      0.3611
                Coeff Var                 10.59787

                           Parameter Estimates
                    Parameter      Standard
Variable      DF    Estimate        Error      t Value    Pr > |t|
Intercept      1    63.68190       4.95305       12.86      <.0001
X4             1     1.64244       0.39005        4.21      0.0003
Z              1     3.93395       3.24618        1.21      0.2361
```

14. A company wants to compare three different point-of-sale promotions for its snack foods. The three promotions are:

Promotion 1: Buy two items, get a third free.

Promotion 2: Mail in a rebate for $1.00 with any $2.00 purchase.

Promotion 3: Buy reduced-price multi-packs of each snack food.

The company is interested in the average increase in sales volume due to the promotions. Fifteen grocery stores were selected in a targeted market, and each store was randomly assigned one of the promotion types. During the month-long run of the promotions, the company collected data on increase in sales volume (Y, in hundreds of units) at each store, to be gauged against average monthly sales volume (X, in hundreds of units) prior to the promotions. Let $Z_1 = 1$ if promotion type 1, or 0 otherwise. Let $Z_2 = 1$ if promotion type 2, or 0 otherwise. The sample data are shown in the following table.

Store	Promotion	Y	X
1	1	12	39
2	1	23	42
3	2	11	23
4	3	17	39
5	3	15	37
6	3	18	31
7	1	12	36
8	2	19	38
9	3	21	33
10	1	13	44
11	1	7	26
12	2	5	20
13	2	8	32
14	3	17	36
15	2	19	29

a. State an ANACOVA regression model for comparing the three promotion types, controlling for average pre-promotion monthly sales.

b. Identify the model that should be used to check whether the ANACOVA model in part (a) is appropriate. Carry out the appropriate test.

c. Using ANACOVA, determine adjusted mean increases in sales volume for the three promotions, and test whether they differ significantly from one another. [*Note:* Mean pre-promotional average sales volume $= 33.6667$; unadjusted mean increases in sales volume were 13.4 for promotion 1, 12.4 for promotion 2, and 17.6 for promotion 3.]

Edited SAS Output (PROC REG) for Problem 14

```
                        Y Regressed on X, Z1, Z2, XZ1, XZ2
                            Analysis of Variance
                               Sum of              Mean
Source              DF        Squares             Square       F Value      Pr > F
Model                5       219.94330           43.98866        2.25       0.1369
Error                9       175.79003           19.53223
Corrected Total     14       395.73333
```

(continued)

Edited SAS Output for Problem 14 (Continued)

```
                    Y Regressed on X, Z1, Z2, XZ1, XZ2
                      Analysis of Variance (Continued)
              Root MSE              4.41953     R-Square      0.5558
              Dependent Mean       14.46667     Adj R-Sq      0.3090
              Coeff Var            30.54973

                              Parameter Estimates
                         Parameter        Standard
Variable        DF        Estimate         Error       t Value   Pr > |t|     Type I SS
Intercept        1        31.92157        24.43508        1.31     0.2238     3139.26667
x                1        -0.40686         0.69190       -0.59     0.5710      115.60187
z1               1       -39.96273        27.16860       -1.47     0.1754       61.21220
z2               1       -36.29584        26.03369       -1.39     0.1967        6.85169
xz1              1         0.98016         0.75946        1.29     0.2290        2.41391
xz2              1         0.99751         0.75757        1.32     0.2205       33.86363

                          Y Regressed on X, Z1, Z2
                           Analysis of Variance
                                     Sum of           Mean
Source                    DF         Squares          Square      F Value    Pr > F
Model                      3        183.66577        61.22192        3.18     0.0674
Error                     11        212.06757        19.27887
Corrected Total           14        395.73333           .
                  Root MSE              4.39077     R-Square      0.4641
                  Dependent Mean       14.46667     Adj R-Sq      0.3180
                  Coeff Var            30.35095

                              Parameter Estimates
                         Parameter        Standard
Variable        DF        Estimate         Error       t Value   Pr > |t|     Type I SS
Intercept        1         0.30045         7.58360        0.04     0.9691     3139.26667
x                1         0.49146         0.20810        2.36     0.0377      115.60187
z1               1        -5.28122         2.81445       -1.88     0.0874       61.21220
z2               1        -1.85804         3.11672       -0.60     0.5631        6.85169
```

15. In Problem 19 in Chapter 5 and Problem 14 in Chapter 8, data from the 1990 Census for 26 randomly selected Metropolitan Statistical Areas (MSAs) were discussed. Of interest were factors potentially associated with the rate of owner occupancy of housing units. The following three variables were included in the data set:

OWNEROCC: Proportion of housing units that are owner-occupied (as opposed to renter-occupied)

OWNCOST: Median selected monthly ownership costs, in dollars

URBAN: Proportion of population living in urban areas

It is also of interest to see whether the average owner occupancy rate differs between metropolitan areas where 75% or more of the population lives in urban areas and metropolitan areas where less than 75% of the population lives in urban areas, while controlling for ownership costs. For this purpose, the following additional variable is defined:

$$Z = \begin{matrix} 1 \\ 0 \end{matrix} \quad \begin{matrix} \text{if proportion of population living in urban areas} \geq 0.75 \\ \text{otherwise} \end{matrix}$$

a. State an ANACOVA regression model that can be used to compare the two MSA types, controlling for average monthly ownership costs.

b. State the model that should be used to check whether the ANACOVA model in part (a) is appropriate. Using the computer output given next, carry out the appropriate test. [*Note:* In the SAS output, the variable OWNCOST has been centered to avoid collinearity problems.]

c. Using ANACOVA methods and the accompanying output, determine adjusted mean owner occupancy rates for the two types of MSAs, and test whether they significantly differ from one another. [*Note:* The unadjusted average owner occupancy rate for MSAs where at least 75% of the population lives in urban areas = 64.5%; the unadjusted average owner occupancy rate for MSAs where less than 75% of the population lives in urban areas = 69.25%.]

Edited SAS Output for Problem 15

```
            OWNEROCC Regressed on OWNCOST, Z and OWNCOST*Z
                       Analysis of Variance
                          Sum of          Mean
Source              DF    Squares         Square      F Value    Pr > F
Model               3     214.19496       71.39832     4.06      0.0194
Error               22    386.76658       17.58030
Corrected Total     25    600.96154
                              .
                              . [Portion of output omitted]
                              .
             OWNEROCC Regressed on OWNCOST and Z
                     Analysis of Variance
                        Sum of          Mean
Source              DF    Squares         Square      F Value    Pr > F
Model               2     213.49860       106.74930    6.34      0.0064
Error               23    387.46294       16.84621
Corrected Total     25    600.96154
                              .
                              . [Portion of output omitted]
                              .
                      Parameter Estimates
                      Parameter      Standard
Variable        DF    Estimate       Error        t Value    Pr > |t|
Intercept       1     68.66545       1.47336       46.60     <.0001
owncost         1     -0.01267       0.00553       -2.29      0.0314
z               1     -3.90567       1.78250       -2.19      0.0388
```

16. This problem refers to the radial keratotomy study data from Problem 12 in Chapter 8. Suppose that we want to compare the average change in refraction for males and females, controlling for baseline refractive error and baseline curvature. To this end, we define the following dummy variable:

$$Z = \begin{array}{ll} 1 & \text{if patient is male} \\ 0 & \text{otherwise} \end{array}$$

a. State an ANACOVA regression model that can be used to compare the mean change in refraction (Y) for males and females, controlling for baseline refractive error (X_1) and baseline curvature (X_2).

b. State the model that should be used to check whether the ANACOVA model in part (a) is appropriate. Using the computer output given next, carry out the appropriate test.

c. Using ANACOVA methods and the SAS output, determine adjusted mean changes in refractive error for males and females, and test whether they differ significantly from one another. [*Note:* Unadjusted average change in refraction for males = 3.64 diopters; unadjusted average change in refraction for females = 3.965 diopters; overall mean baseline refraction = −4.03 diopters; overall mean baseline curvature = 44.02 diopters.]

Edited SAS Output for Problem 16

Y REGRESSED ON X1, X2, Z, X1Z, X2Z
Analysis of Variance

Source	DF	Sum of Squares	Mean Square	F Value	Pr > F
Model	5	24.71516	4.94303	4.27	0.0027
Error	48	55.53778	1.15704		
Corrected Total	53	80.25294			

Root MSE	1.07566	R-Square	0.3080
Dependent Mean	3.83343	Adj R-Sq	0.2359
Coeff Var	28.05993		

Parameter Estimates

Variable	DF	Parameter Estimate	Standard Error	t Value	Pr > |t|
Intercept	1	13.06563	6.97300	1.87	0.0671
X1	1	-0.20220	0.10539	-1.92	0.0610
X2	1	-0.22389	0.15756	-1.42	0.1618
Z	1	-0.91427	10.01057	-0.09	0.9276
X1Z	1	-0.35838	0.19864	-1.80	0.0775
X2Z	1	-0.02426	0.22572	-0.11	0.9149

Y REGRESSED ON X1, X2 and Z
Analysis of Variance

Source	DF	Sum of Squares	Mean Square	F Value	Pr > F
Model	3	20.90031	6.96677	5.87	0.0016
Error	50	59.35263	1.18705		
Corrected Total	53	80.25294			

Root MSE	1.08952	R-Square	0.2604
Dependent Mean	3.83343	Adj R-Sq	0.2161
Coeff Var	28.42156		

Parameter Estimates

Variable	DF	Parameter Estimate	Standard Error	t Value	Pr > |t|
Intercept	1	13.71986	5.06713	2.71	0.0093
X1	1	-0.30382	0.09041	-3.36	0.0015
X2	1	-0.24770	0.11408	-2.17	0.0347
Z	1	-0.50859	0.30608	-1.66	0.1028

Reference

Patrick, R.; Cassel, J. C.; Tyroler, H. A.; Stanley, L.; and Wild, J. 1974. "The Ponape Study of the Health Effects of Cultural Change." Paper presented at the annual meeting of the Society for Epidemiologic Research, Berkeley, California.

14

Regression Diagnostics

14.1 Preview

This chapter provides an introduction to *regression diagnostics*, namely, statistical techniques for detecting conditions which can lead to inaccurate or invalid regression results. Such techniques include methods for detecting outliers, checking regression assumptions, and detecting the presence of collinearity.

Outliers (individual data values that are much larger or smaller than the rest of the values in the data set) may represent recording errors. Clearly, such values must be detected and corrected before the analysis proceeds. Sometimes, however, outliers are not recording errors. Even so, they may influence the fit of the regression model and, therefore, may affect our understanding of the relationship between the dependent and independent variables. Therefore, outliers need to be identified, their plausibility carefully and scientifically judged, and a decision made as to whether the analysis should be revised in light of their presence.

If regression assumptions are violated, then, naturally, regression results may be invalid. Assumption checking can be a subjective, difficult, and time-consuming process, and remedies for assumption violations can themselves be difficult to implement. Nevertheless, it is essential that a thorough check of assumptions be performed so that potential limitations of the analysis are identified and attempts to address them can be made.

Collinearity is a fairly complex mathematical problem that exists when there are strong linear relationships among two or more *predictor variables*. When severe collinearity exists, regression results are unstable in the sense that small, practically unimportant changes in data values can result in meaningfully large changes in the estimated model. Such instability can lead to inaccurate estimates of regression coefficients, variability, and P-values, and is clearly undesirable.

We present a strategy for regression diagnostics, including simple methods that will help the analyst avoid and diagnose such problems. Suggestions are also made for corrective action when problems exist.

14.2 Simple Approaches to Diagnosing Problems in Data

In statistical analyses, it is important to be thoroughly familiar with the basic characteristics of the data, and regression analyses are no exception. Such familiarity helps avoid many errors. For example, it is essential to know the following:

1. The type of subject or experimental unit (e.g., small pine tree needles, elderly male humans)

2. The procedure for data collection

3. The unit of measurement for each variable (e.g., kilograms, meters, cubic centimeters)

4. A plausible range of values and a typical value for each variable

This knowledge, combined with a thorough descriptive statistical analysis of each variable, can be very useful in detecting errors in the data and pinpointing potential violations of regression assumptions. With this in mind, we recommend, as a first step in a regression diagnostics strategy, that the following simple descriptive analyses be performed:[1]

1. Examine the five largest and five smallest values for every numeric variable. This is a simple but extremely powerful technique for examining data. Combined with knowledge of the basic characteristics of the data described above, examination of these extreme values may help pinpoint the most extreme recording errors and other outliers. For large data sets, it may be necessary to examine more than just these ten values for some variables.

When an outlier is detected, the data value in question should be compared against data collection forms or other original records to determine whether a recording error has occurred. If it has, every attempt should be made to recover the correct data value before proceeding with the analysis. If the correct value cannot be recovered, the data value in question is usually set to missing in the computerized data set so that it is not inadvertently used in the fitting of the model.[2] Note that, since subsequent regression models may not require the use of the variable with the outlying value, the entire observation is usually not deleted—only the value of the variable in question is set to missing.

If it is determined that the data value has not been recorded in error, it is important to judge the plausibility of the value in question. For example, suppose we have collected data on body temperature in human subjects. The value 38.1 is plausible if the units are degrees Celsius. But, if the units are degrees Fahrenheit, 38.1 is implausible. More generally, one may classify any observation as being impossible, highly implausible, or plausible. Impossible values should be set to missing. Plausible values are not themselves changed, but it may be necessary to consider alternative forms of the model (i.e., a model with interaction or other higher-order terms)

---

[1] Indeed, every statistical analysis should begin with steps 1 and 2, not just regression analyses.

[2] Modern statistical techniques for data imputation may also be employed to estimate missing data. These techniques will be useful if there are more than just a few missing values. See Rubin (1987) and Shafer (1997) for more information.

that might fit the data better. For highly implausible values, a decision should be made as to whether to remove the value in question from the analysis. When values are removed, this action, and the scientific justification for it, should be documented and presented along with any regression results.

This simple approach may not be sufficient for detecting all important outliers. We discuss additional methods focused on outlier detection in 14.3.1.

2. Examine the appropriate descriptive statistics for each variable. For categorical data, frequency tables should be produced. For continuous variables, the mean, median, standard deviation, interquartile range, and range should be examined, at a minimum. Graphical approaches such as bar charts, histograms, and box-and-whisker plots are also useful, especially when the number of variables is large. The information should be compared with what is expected from the study design and from scientific knowledge about the variables.

3. For a simple linear regression with a continuous predictor, create a scatterplot of the dependent variable versus the independent variable. Since a linear relationship is assumed, the plot should depict, at least roughly, a linear scatter of the plotted points. The Pearson correlation between the pair of variables should also be calculated to help further describe the strength and direction of the linear relationship observed. A distinct but non-linear pattern may indicate a potential violation of the linearity assumption, and a need for revision of the model. A fairly non-descript or random scatter of points may indicate that there will be no significant linear association between the independent and dependent variables in the subsequent regression analysis.

If these plots indicate significant problems, the regression model may need to be reviewed and reformulated before proceeding further. Usually, however, model reformulation is postponed until further diagnostics analyses (described below) have been performed.

Scatterplots are also useful for identifying strong outliers in the data.

4. For multiple regression models, the simple plots of the dependent variable versus individual predictor variables described in (3) above may be misleading since they do not account for other predictor variables. In general, independent variables will not be completely uncorrelated with each other, and these associations may impact relationships with the dependent variable.

Partial regression plots do take the other predictor variables into account, and should be used for multiple regression instead of, or in addition to, the simple scatterplots described in (3) above. In a partial regression plot for a particular predictor, we plot the dependent variable against the predictor of interest, adjusting for the remaining $(k - 1)$ predictors. In particular, assume that the overall model of primary interest is

$$Y = \beta_0 + \beta_1 X_1 + \beta_2 X_2 + \cdots + \beta_k X_k + E$$

To produce the partial regression plot for the kth predictor, we first fit two models:

$$Y = \beta_0 + \beta_1 X_1 + \beta_2 X_2 + \cdots + \beta_{k-1} X_{k-1} + E$$

and

$$X_k = \alpha_0 + \alpha_1 X_1 + \alpha_2 X_2 + \cdots + \alpha_{k-1} X_{k-1} + E$$

Then we plot the residuals from the two models against each other, i.e., we plot $(Y - \hat{Y})$ versus $(X_k - \hat{X}_k)$. The residuals represent the original Y and X_k variables, adjusted for the remaining $(k - 1)$ predictors. The partial correlation, $r_{YX_k|X_1,X_2,\ldots,X_{k-1}}$, between Y and X_k should also be calculated to help further describe the strength and direction of the linear relationship observed.

5. Calculate pairwise correlations between independent variables. Strong linear associations between independent variables may signal collinearity problems, which we will discuss in detail later.

■ **Example 14.1** We will illustrate many of the techniques described in this chapter using a data set taken from Lewis and Taylor (1967). The variables WEIGHT (in kilograms), HEIGHT (in meters), and AGE (in years) were all recorded for a sample of boys and girls. We consider only the data on 127 boys. The goal of the regression analysis is to model WEIGHT, the dependent variable, as a linear function of HEIGHT and AGE.

Output for the relevant descriptive statistical analyses can be easily produced using SAS and is shown below, in edited form. From the PROC UNIVARIATE output, we see that the summary statistics (mean, median, standard deviation, range, and interquartile range, shown in shaded boxes on the output) for AGE seem consistent with what might be expected for such data on boys; the same is true for HEIGHT. Furthermore, the five largest and five smallest data values

Edited SAS Output for Example 14.1 (PROC UNIVARIATE Output)

```
                              Variable: AGE

          Location                                 Variability
Mean              13.68110          Std Deviation              1.45449
Median            13.50000          Variance                   2.11553
Mode              12.00000          Range                      5.58333
                                    Interquartile Range        2.16667

                        Extreme Observations

          ---------Lowest--------          ---------Highest--------
              Value         Obs                Value         Obs
            11.5833          68              16.6667          59
            11.5833           3              16.9167          58
            11.6667         120              17.0833          32
            11.6667          55              17.1667          19
            11.6667          34              17.1667         103

                           Variable: HEIGHT

          Location                                 Variability
Mean             1.575760          Std Deviation              0.10980
Median           1.569720          Variance                   0.01206
Mode             1.536700          Range                      0.54610
                                   Interquartile Range        0.17780

                        Extreme Observations

          ---------Lowest--------          ---------Highest--------
              Value         Obs                Value         Obs
            1.28270          42              1.76530         103
            1.33350          94              1.77292         114
            1.35382          18              1.80340          59
            1.36652         127              1.80340          76
            1.39700         112              1.82880          67
```

(continued)

Edited SAS Output for Example 14.1 (continued)

Variable: WEIGHT

Location			Variability	
Mean	47.25396		Std Deviation	9.76021
Median	47.62719		Variance	95.26163
Mode	50.80234		Range	57.15263
			Interquartile Range	12.47379

Extreme Observations

---------Lowest--------		---------Highest--------	
Value	Obs	Value	Obs
31.7515	112	68.0388	67
32.4318	86	68.2656	77
32.6586	85	77.7911	32
33.3390	68	77.7911	103
34.0194	123	88.9041	127

for these variables do not seem to be outliers. For WEIGHT, the statistics also look reasonable for the most part; however, the standard deviation appears somewhat large and the largest value (88.9 kg) seems to be quite a bit larger than the next nearest value. It would be prudent to verify that no recording errors have occurred before proceeding. If not, then the plausibility of this largest value of WEIGHT should be examined. Suppose that it is decided that the value, although unusual, is scientifically plausible; it will, therefore, remain unchanged.

The plots of WEIGHT versus HEIGHT and AGE (Figures 14.1 and 14.2) support the idea that there are linear associations between WEIGHT and these predictors.[3] Note that the plotting symbols represent the number of points plotted at each location. The plots provide a clear indication that the observation in which WEIGHT is approximately 90 kg (circled in the figures) is an outlier. This is, of course, observation 127 identified above. The partial regression plots, (Figures 14.3 and 14.4), which are appropriate to examine for multiple regression models, roughly suggest linear relationships. However, the relationships are different than in Figures 14.1 and 14.2, because they have been adjusted for the effects of the second predictor. After the adjustment, the relationships look less obvious, especially for AGE. This is an indication that the apparent relationship between WEIGHT and HEIGHT in the simple scatterplot in Figure 14.1 is partly attributable to the children's ages. Similarly, the apparent relationship between WEIGHT and AGE in the simple scatterplot in Figure 14.2 is partly attributable to the children's heights. In this case, when both predictors will be included in a multiple regression model, neither will appear to have as strong a relationship with WEIGHT compared with being used separately in a simple linear regression model. (*Note*: The inclusion of the outlier, observation 127, in Figure 14.4 forces the vertical axis to be extended in such a way that it is difficult to see the linear relationship

---

[3] The temptation to overanalyze these plots should be avoided. For example, some analysts might perceive a very subtle curvilinear pattern in Figure 14.1 and decide that the regression model should be revised to reflect that curvature. However, while these very simple plots are useful for detecting gross departures from ideal conditions, other, more subtle, departures that are apparent may simply be artifacts of sample size, scale, and style. Therefore, further diagnostics analyses should be undertaken before decisions to revise the model are made. Likewise, even if no anomalies are apparent in the plots, that does not necessarily mean that there are no problems with the data or the regression assumptions; further diagnostics analyses may reveal such problems.

FIGURE 14.1   Children's body weight as a function of height; Lewis and Taylor data (1967) ($n = 127$)

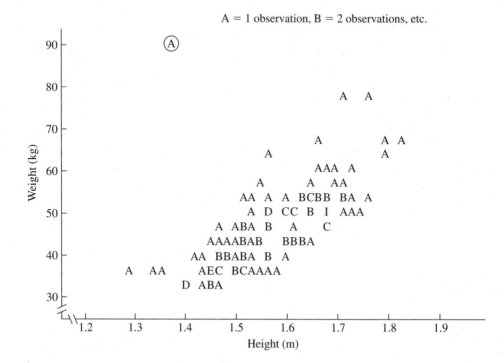

FIGURE 14.2   Children's body weight as a function of age; Lewis and Taylor data (1967) ($n = 127$)

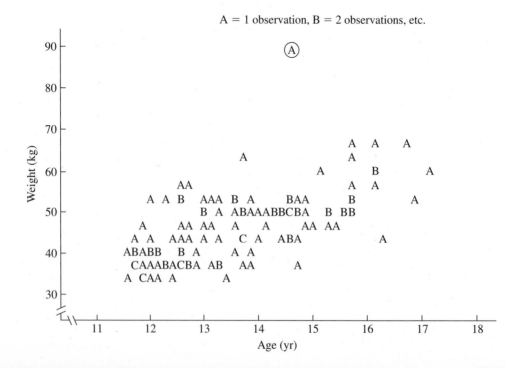

FIGURE 14.3   **Partial regression plot (adjusting for age) of children's body weight versus height; data from Lewis and Taylor (1967) ($n = 127$)**

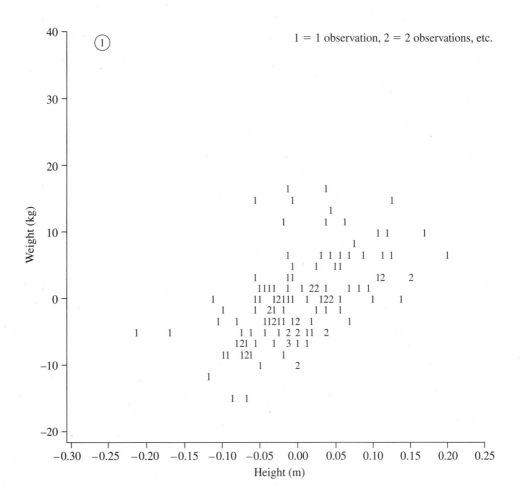

Edited SAS Output for Example 14.1 (PROC CORR Output)

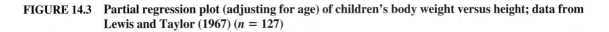

	The CORR Procedure		
	3 Variables: WEIGHT HEIGHT AGE		
	Pearson Correlation Coefficients, N = 127		
	WEIGHT	HEIGHT	AGE
WEIGHT	1.00000	0.65470	0.66497
HEIGHT	0.65470	1.00000	0.75328
AGE	0.66497	0.75328	1.00000

FIGURE 14.4 **Partial regression plot (adjusting for height) of children's body weight versus age; data from Lewis and Taylor (1967) (*n* = 127)**

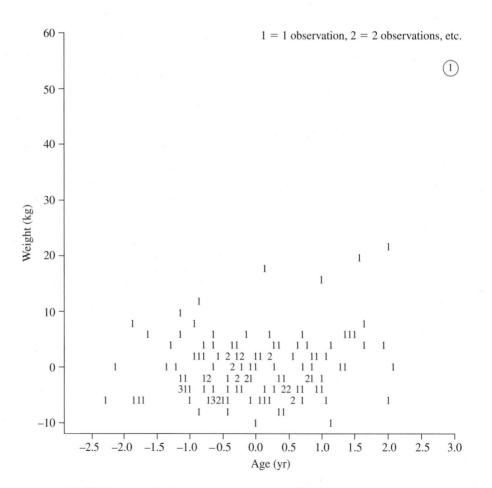

between WEIGHT and AGE for the rest of the data. Plotting without the outlier reveals the linear association slightly more clearly, though it still remains weak.)

The SAS PROC CORR output shows that the simple Pearson correlations between WEIGHT and HEIGHT and WEIGHT and AGE are moderately large (0.65470 and 0.66497 respectively), supporting the idea that the assumption of linearity may be reasonable. The partial correlations (output not shown) between WEIGHT and each of the predictors are $r_{\text{WEIGHT, HEIGHT}|\text{AGE}} = 0.31$, and $r_{\text{WEIGHT, AGE}|\text{HEIGHT}} = 0.35$, confirming the information contained in the partial regression plots: the linear associations of HEIGHT with WEIGHT and AGE with WEIGHT are not as strong once the other predictor is controlled for. (*Note:* Once again the outlier affects these partial correlations; without the outlier, we would find that $r_{\text{WEIGHT, HEIGHT}|\text{AGE}} = 0.55$ and $r_{\text{WEIGHT, AGE}|\text{HEIGHT}} = 0.23$.)

Both Figure 14.5 and the SAS PROC CORR output suggest that there is a linear association between the two independent variables ($r_{\text{HEIGHT, AGE}} = 0.75$). This is a preliminary indication of potential collinearity problems that could cause invalid regression results. However, further examination of the problem is necessary to determine whether the collinearity

FIGURE 14.5   **Children's height as a function of age; data from Lewis and Taylor (1967) ($n = 127$)**

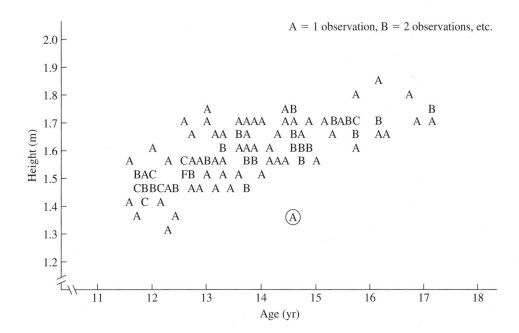

problem is severe enough to warrant corrective action; relevant methods will be described in Section 14.5.   ■

14.3 Residual Analysis: Detecting Outliers and Violations of Model Assumptions

Although the simple techniques described in Section 14.2 are useful for detecting the largest outliers in a data set, they often fail to detect more subtle outliers that may still be influential in the analysis. The methods of Section 14.2 also are not sufficient, by themselves, for detecting regression assumption violations. The second stage of a diagnostics analysis usually involves examination of regression residuals to detect additional outliers and assumption violations.

In Section 8.4.2, we defined the regression residual, $\hat{E}_i$, for the ith observation as the difference between the observed and predicted response:

$$\hat{E}_i = Y_i - \hat{Y}_i \qquad i = 1, 2, \ldots, n$$

It is reasonable to expect that the residuals will be large for observations that are outliers, since the outliers lie "far" from the bulk of the data and therefore will also be relatively far from the best-fitting regression surface. We will describe a few methods for examining the sizes of residuals. Also, the $\hat{E}_i$ are estimates of the unobserved model error terms, E_i, $i = 1$, $2, \ldots, n$. The regression assumptions, described in Chapter 8, are that the E_i are independent, have a mean of zero, have a common variance σ^2, and follow a normal distribution. It is reasonable to expect that the residuals, $\hat{E}_i$, should roughly exhibit properties consistent with these assumptions. We will describe methods for assessing whether the residuals follow a normal distribution, have constant variance, and are consistent with the linearity and independence assumptions.[4]

Readers should note that, instead of examining the ordinary residuals, $\hat{E}_i$, we often study the following functions of the residuals: the *standardized* residuals, $z_i = \frac{\hat{E}_i}{S}$, where S is the square root of the mean squared error (MSE); the *studentized* residuals, $r_i = \frac{\hat{E}_i}{S\sqrt{1 - h_i}}$, where h_i is the "leverage" of the ith observation (described in Section 14.3.1); or the *jackknife* residuals, $r_{(-i)} = \frac{\hat{E}_i}{S_{(-i)}\sqrt{1 - h_i}}$, where $S_{(-i)}$ is the square root of the MSE calculated excluding observation i. The z_i have unit variance in the sense that

$$\frac{1}{n - k - 1}\sum_{i=1}^{n} z_i^2 = \frac{1}{n - k - 1}\sum_{i=1}^{n}\left(\frac{\hat{E}_i}{S}\right)^2 = \frac{1}{S^2}\left(\frac{1}{n - k - 1}\sum_{i=1}^{n}\hat{E}_i^2\right) = 1$$

Also, if regression assumptions are satisfied, each r_i has a Student's t distribution with $(n - k - 1)$ degrees of freedom; and each $r_{(-i)}$ has a Student's t distribution with $(n - k - 2)$ degrees of freedom. The latter jackknife residuals are particularly useful, and will be discussed in more detail below.

14.3.1 Detecting Outliers

Substantial differences exist among possible types of extreme values. An *outlier* is any rare or unusual observation that appears at one of the extremes of the data range. All regression observations—and, hence, outliers in particular—may be evaluated with respect to three criteria: reasonableness, given knowledge of the variable; response extremeness; and predictor extremeness. The goals in a diagnostics analysis are to identify recording errors and to identify *influential* observations that significantly affect either the choice of variables in the model or the accuracy of estimates of the regression coefficients and associated standard errors.

Methods for detecting outliers have received a great deal of attention in the last few decades (see, e.g., Belsey, Kuh, and Welsch, 1980; Cook and Weisberg 1982; and Stevens 1984). In this section, we describe three regression diagnostic statistics for evaluating outliers: leverages, jackknife residuals, and Cook's distance. Many other closely related statistics also exist. Most of these statistics are easily produced using SAS and other standard statistical software.

Leverages: The leverage, h_i, is a measure of the extremeness of an observation with respect to the independent variables. For a data set containing predictor variables $X_1, X_2, \ldots,$ X_k, it is the geometric distance of the ith predictor point $(X_{i1}, X_{i2}, \ldots, X_{ik})$ from the center

---

[4] The residuals are not independent random variables—this is obvious from the fact that they sum to zero. However, in general, if the number of residuals is large relative to the number of independent variables, the dependency effect can, for all practical purposes, be ignored in any analysis of the residuals (see Anscombe and Tukey, 1963, for a discussion of the effect of this dependency on graphical procedures involving residuals).

point $(\overline{X}_1, \overline{X}_2, \ldots, \overline{X}_k)$ of the predictor space. Therefore, the larger the value, the farther is the outlier from the center of the data. It can be shown that the average leverage value is $(k + 1)/n$, and that $0 \leq h_i \leq 1$. Hoaglin and Welsch (1978) recommend scrutinizing all observations for which $h_i > 2(k + 1)/n$.

Jackknife Residuals: Generally, large outliers will result in large $\hat{E}_i$ values. Therefore, the sizes of the $\hat{E}_i$ (or the standardized or studentized residuals) can be studied in order to detect outliers. Sometimes, however, outliers can mask their own effects. That is, the outlier exerts influence on the fitted regression surface, pulling the fitted regression surface away from the main body of the data and toward itself, thereby reducing the size of the associated residual. The jackknife residual avoids this problem and therefore unmasks outliers. For the ith observation, the jackknife residual, $r_{(-i)}$, is calculated as:

$$r_{(-i)} = \frac{\hat{E}_i}{S_{(-i)} \sqrt{1 - h_i}}$$

The quantity $S_{(-i)}^2$ is the mean square error computed with the ith observation deleted. The unmasking is achieved by this deletion of the ith observation during the computation of the mean square error. The idea is that the ordinary mean square error, S^2, will be larger than $S_{(-i)}^2$ if the outlier is masking its effect, since the outlier will pull the surface toward itself and away from the main body of the data. Therefore, for the ith observation, $r_{(-i)}$ will be larger than the studentized residual, r_i in the case of an outlier.

From the formula for $r_{(-i)}$, we see that the jackknife residuals may be large if an observation is an outlier in the response variable Y (resulting in a large numerator), or in the predictor space of $X_1, X_2, \ldots, X_k$ (resulting in a large h_i), or if it strongly affects the fit of the model (as reflected in the difference between S^2 and $S_{(-i)}^2$). It can be shown that each jackknife residual has a t distribution with $n - k - 2$ degrees of freedom if the usual regression assumptions are met. Therefore, if the absolute value of $r_{(-i)}$ is greater than the 95th percentile of the relevant t distribution (i.e., $\alpha = 0.10$), then it would be reasonable to say that observation i is an outlier and should be scrutinized further.[5]

Cook's Distance: Cook's distance measures the extent to which the estimates of the regression coefficients change when an observation is deleted from the analysis. When the error terms have mean 0 and equal variance, and when the model predictors are uncorrelated, Cook's distance, d_i, for the ith observation, is proportional to:

$$\sum_{j=0}^{k} [\hat{\beta}_j - \hat{\beta}_{j(-i)}]^2 = [\hat{\beta}_0 - \hat{\beta}_{0(-i)}]^2 + [\hat{\beta}_1 - \hat{\beta}_{1(-i)}]^2 + \cdots + [\hat{\beta}_k - \hat{\beta}_{k(-i)}]^2$$

In general, for the ith observation, Cook's distance is given by:

$$d_i = \left(\frac{1}{k + 1}\right)\left(\frac{h_i}{1 - h_i}\right) r_i^2$$

Clearly, d_i may be large either because the observation is extreme in the predictor space (i.e., resulting in a large h_i) or because it has a large studentized residual r_i (indicating an outlier

---

[5] For large datasets consisting of several thousand observations, this may be an impractical criterion, as a large number of observations will need to be scrutinized. Smaller α levels are advisable in this case, for example, 0.05, 0.01, or even 0.005. Another approach would be to divide α by the number of observations, resulting in a Bonferroni type of correction—this is described in more detail in Section 17.7.1.

with respect to either the predictors or the response). Cook and Weisberg (1982) suggested that any observations with $d_i > 1$ may deserve closer scrutiny, although more recent research indicates that this may not be the optimal critical value in all situations.[6] In the absence of other hard and fast rules, we suggest that analysts should use the $d_i > 1$ rule, but should also use Cook's distance in conjunction with other outlier statistics, rather than in isolation.

As described in Section 14.2, once identified, outliers that are not easily identifiable as recording errors should be judged as to plausibility.

■ **Example 14.2**   For the regression of WEIGHT on HEIGHT and AGE, using the data of Example 14.1, outlier detection statistics were computed using SAS. The observations for which at least one of the diagnostic statistics exceeded the critical value (i.e., $d_i > 1$, $h_i > 2(2 + 1)/127 = 0.047$, and $|r_{(-i)}| > t_{127-2-2,0.95} \approx 1.66$) are shown in Table 14.1. For each of these observations, we would first determine whether any data values were recorded in error. If not, we then examine each data value for each observation in more detail.

For observation 5, only the leverage value is large ($h_i = 0.057$), indicating that either the value of HEIGHT or AGE (or both) should be examined. For this observation, HEIGHT = 1.68 m, which is one of the taller heights (but not one of the extremes) and so is not highly implausible; and AGE = 12.58 years, which is also not an extreme value and thus is not highly implausible for this study. Since both values are plausible, and assuming that they have been recorded correctly, no corrective action is necessary. This observation simply appears as an outlier (albeit a moderate one) because it represents one of the younger but taller children in the study.

Let us also consider observation 127. From Table 14.1, we see that all three outlier diagnostic statistics are large. Since the leverage is large, we should examine the HEIGHT and AGE

TABLE 14.1   Outliers with either large Cook's distance (d_i), leverage (h_i) or jackknife residual ($r_{(-i)}$) values for regression of WEIGHT on HEIGHT and AGE using Lewis and Taylor data (1967) ($n = 127$)

Observation	d_i	h_i	$r_{(-i)}$
5	0.01235	0.05673	0.78378
19	0.00027	0.05578	−0.11583
32	0.12519	0.05586	2.57621
41	0.02372	0.01505	2.19094
42	0.00187	0.08374	0.24686
52	0.02261	0.02140	1.77651
58	0.01712	0.05475	−0.94113
67	0.01552	0.05015	0.93846
69	0.04869	0.04612	−1.75257
75	0.02172	0.01891	−1.85646
77	0.02675	0.02312	1.85939
93	0.00067	0.07225	0.16055
94	0.00012	0.05804	0.07542
103	0.09529	0.05357	2.28530
126	0.01580	0.00813	2.45374
127	2.01754	0.10997	8.96226

[6] Muller and Chen Mok (1997) performed simulations that suggested that the $d_i > 1$ rule can be inaccurate. They suggested an alternate approach, which is presented in Table A. 10.

values. Indeed, the data reveal that this child was 14.5 years old (close to the average for the study) but was only 1.37 m tall (one of the shortest heights in the study). The plausibility of this HEIGHT and AGE combination needs to be judged. Since Cook's distance and the jackknife residual are also quite large, it is worth examining the dependent variable WEIGHT as well (although it is possible that the independent variables alone are responsible for all the outlier statistics being large in magnitude). It turns out that this child's WEIGHT, 88.9 kg, is the largest in the data set. For this observation, the combined effect of all three variables causes it to stand out as an extreme outlier. Although we would probably decide that the observation is unusual but plausible and that nothing should be changed (indeed, that is what happens in most cases), it is possible that, on a rare occasion, the decision to remove an observation due to implausibility will be made or a decision to revise the model (by adding interaction or other higher-order terms) will be made in order to improve fit. We would examine each of the other observations in Table 14.1 in similar fashion.

Note that, for the WEIGHT/HEIGHT/AGE data, while the simple plots and univariate statistics in Section 14.2 did reveal observation 127 to be an outlier, they did not readily identify the remaining outliers shown in Table 14.1. This is precisely why the simple approaches of Section 14.2 are not, by themselves, sufficient for a reliable outlier detection analysis.   ∎

14.3.2 Assessing the Linearity, Homoscedasticity, and Independence Assumptions Using Graphical Analyses of Residuals

Plots of the residuals (ordinary, studentized, or jackknife) versus predicted values are very useful for checking regression assumptions. These plots are commonly referred to simply as *residual plots*. Some of the general patterns that may emerge in the plots are shown in Figure 14.6.

FIGURE 14.6   Typical jackknife residual plots as a function of predicted value, $\hat{Y}$, or time of data collection for hypothetical data

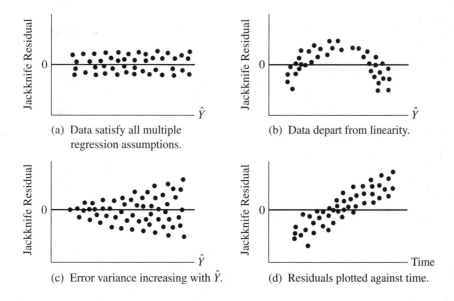

(a) Data satisfy all multiple regression assumptions.

(b) Data depart from linearity.

(c) Error variance increasing with $\hat{Y}$.

(d) Residuals plotted against time.

Figure 14.6(a) illustrates the type of pattern to be expected when all basic assumptions hold: Since the assumption is that model errors are independent with mean 0 and homogeneous variance, a random scatter around the horizontal line at residual = 0 should be obtained with no hint of any systematic trends.

Figure 14.6(b) illustrates a systematic pattern that could occur when the data depart from linearity. In this case, the plot indicates that current variables in the model may need to be transformed or that additional curvilinear terms need to be added to introduce curvature. Naturally, different types of model inappropriateness result in different residual patterns and different corrective actions.

The pattern in Figure 14.6(c) shows that the variance of the residuals is not constant, suggesting a violation of the homoscedasticity assumption. Transformations of the data, or advanced techniques such as weighted least-squares analysis, are often used to address heteroscedasticity (see Section 14.4).

Figure 14.6(d) is a plot of the residuals versus time. A time-related effect is clearly present. This may reflect the omission of an omitted covariate, namely, TIME. But it may instead be suggestive of a violation of the independence assumption. There are few simple remedies for violations of the independence assumption. If corrective action does become necessary, advanced methods (e.g., time-series analysis or correlated data analysis methods [see Chapters 25 and 26]) may be helpful in completing the analysis.

When a residual plot indicates possible assumption violations, it is useful to create further plots of the residuals versus each predictor in order to identify the specific predictors involved in the violations. Any model transformations undertaken to correct violations would then focus on the dependent variable and/or on the identified predictors. Note that the residual plots also provide one more opportunity to detect outliers.

It is important to note that, sometimes, patterns apparent in these plots are not due to assumption violations, but due to small sample size or sparse data for certain combinations of predictor values. The best way to avoid such a situation is, of course, to have a large enough sample size and to design the study in such a way that a sufficient number of replicate observations is taken at each predictor value. Unfortunately, researchers usually do not have this level of control over the sampled predictor variable values in observational studies. We suggest that analysts look only for very distinct patterns or *gross* assumption violations in these plots, rather than focusing on more subtle patterns and violations.

■ **Example 14.3** Figure 14.7 shows the jackknife residual plot for the WEIGHT/HEIGHT/AGE example. The plot appears to reflect random scattering around the value zero, so that no assumption violations are readily apparent. One outlier is apparent (circled); further inspection will reveal that it is observation 127 (already identified in Sections 14.2 and 14.3).[7]

---

[7] Once again, these plots should not be overanalyzed (see footnote 3, earlier in this chapter). Some investigators might suggest that the variance of the residuals is not constant based on Figure 14.7. However, this impression is likely the result of the sparseness of the data in some areas on the plot.

FIGURE 14.7   **Jackknife residual plot for regression of children's body weight as a function of height and age; data from Lewis and Taylor (1967) ($n = 127$)**

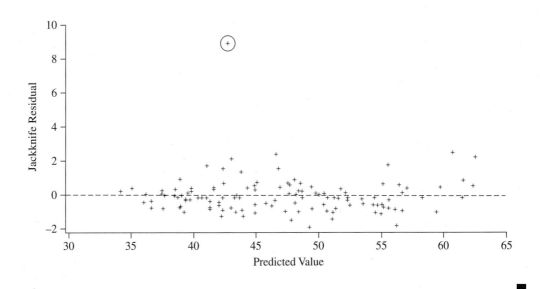

14.3.3 Assessing the Normality Assumption

The normality assumption can be assessed by examining statistics such as skewness and kurtosis, as well as figures such as normal probability plots of ordinary, studentized, or jackknife residuals (see Chapter 3 for a review of these statistics and plots). Furthermore, goodness-of-fit tests such as the Kolmogorov-Smirnov test (Stephens, 1974) [and, in the case of small sample sizes (e.g., $n < 50$), the Shapiro-Wilks (1965) test] can be performed. These methods are covered in many introductory statistics courses and textbooks, and can be easily implemented using SAS and other standard statistical software.

■ **Example 14.4**   For the regression of WEIGHT on HEIGHT and AGE using the data of Example 14.1, the estimated residuals were computed and, using the UNIVARIATE procedure in SAS, descriptive statistics produced (shown below). The output shows that the skewness and kurtosis statistics are quite different from the ideal value of 0, and the stem-and-leaf and box-and-whisker plots also reflect strongly skewed data. The Kolmogorov-Smirnov test for normality has a small P-value (0.0264), indicating that the null hypothesis of normality is to be rejected at a significance level of $\alpha = 0.05$. However, upon further inspection, this lack of normality is due to the one extreme outlier in the data set (observation 127). Without this outlier, the skewness statistic value ($= 0.87$) is reasonably close to 0 and the kurtosis statistic value ($= 1.37$) is consistent with values for t distributions (which have slightly heavier tails than normal distributions). The normal probability plot in Figure 14.8 also reveals a fairly linear pattern consistent with normality, the slight curvature at the extremes being, in part, an indication that the jackknife residuals follow a t distribution rather than a normal distribution. No gross violation of the normality assumption seems to be present for these data after eliminating the extreme outlier.

FIGURE 14.8   **SAS normal probability plot of jackknife residuals for regression of children's body weight as a function of height and age using 126 observations (excluding the outlier); Lewis and Taylor data (1967) (*n* = 127)**

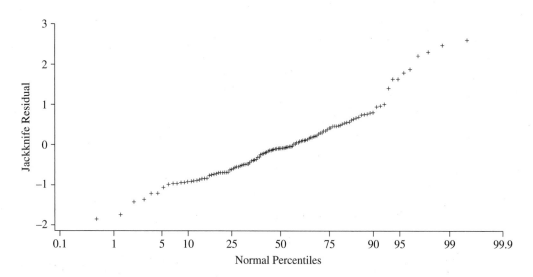

Edited SAS Output for Example 14.4 (PROC UNIVARIATE Output)

```
                           The SAS System
                       The UNIVARIATE Procedure
                      Variable: RESIDUAL (Residual)
                               Moments
N                            127          Sum Weights                      127
Mean                           0          Sum Observations                   0
Std Deviation          6.92312568         Variance                   47.9296692
Skewness               2.70909774         Kurtosis                   14.986965
Uncorrected SS         6039.13832         Corrected SS               6039.13832
Coeff Variation               .           Std Error Mean             0.61432806
                       Tests for Normality
        Test                   --Statistic----        ------p Value-----
        Shapiro-Wilk           W      0.709149         Pr < W     0.0011
        Kolmogorov-Smirnov     D      0.278162         Pr > D     0.0264
 Stem Leaf                                             #            Boxplot
   4 6                                                 1              *
   4
   3
   3
   2
   2
   1 5577                                              4              0
   1 01123                                             5              |
   0 5555556677                                        10             |
   0 00000111111111111222222333333344444              35         +--+--+
  -0 44444444433333332222221111111111111111000        42         *-----*
  -0 9877777766666666555555555                         26             |
  -1 3200                                              4              |
   ---+----+----+----+----+----+----+----+----
 Multiply Stem.Leaf by 10**+1
```

14.4 Alternate Strategies of Analysis

When one or more basic assumptions of regression are clearly not satisfied and/or when numerical problems are identified, the analyst may want to turn to alternate strategies of analysis. In the subsections that follow, we briefly describe transformations of variables that may remedy certain assumption violations, we list some alternatives to classical linear regression methods, and we briefly mention generalizations of multiple linear regression that may be appropriate in some applications.

14.4.1 Transformations

There are three primary reasons for using data transformations: (1) *to stabilize* the variance of the dependent variable, if the homoscedasticity assumption is violated; (2) *to normalize* (i.e., to transform to the normal distribution) the dependent variable, if the normality assumption is noticeably violated; (3) *to linearize* the regression model, if the original data suggest a model that is nonlinear in either the regression coefficients or the original variables (dependent or independent). Fortunately, the same transformation often helps to accomplish the first two goals and sometimes even the third, rather than achieving one goal at the expense of one or both of the other two.

   A more complete discussion of the properties of various transformations can be found in Armitage (1971), Draper and Smith (1981), and Neter, Wasserman, and Kutner (1983). In addition, Box and Cox (1964) describe an approach to making an exploratory search for one of a family of transformations (see also Box and Cox 1984, and Carroll and Ruppert 1984). Nevertheless, we consider it useful to describe a few of the more commonly used transformations:

1. The *log transformation* ($Y' = \log Y$) is used (provided Y takes on only positive values) to stabilize the variance of Y, if the variance increases markedly with increasing Y; to normalize the dependent variable, if the distribution of the residuals for Y is positively skewed; and to linearize the regression model, if the relationship of Y to some independent variable suggests a model with consistently increasing slope.

2. The *square root transformation* ($Y' = \sqrt{Y}$) is used to stabilize the variance, if the variance is proportional to the mean of Y. This is particularly appropriate if the dependent variable has the Poisson distribution.

3. The *reciprocal transformation* ($Y' = 1/Y$) is used to stabilize the variance, if the variance is proportional to the fourth power of the mean of Y (which indicates that a huge increase in variance occurs above some threshold value of Y). This transformation minimizes the effect of large values of Y, since the transformed Y'-values for these values will be close to 0, and large increases in Y will cause only trivial decreases in Y'.

4. The *square transformation* ($Y' = Y^2$) is used to stabilize the variance, if the variance decreases with the mean of Y; to normalize the dependent variable, if the distribution of the residuals for Y is negatively skewed; and to linearize the model, if the original relationship with some independent variable is curvilinear downward (i.e., if the slope consistently decreases as the independent variable increases).

5. The *arcsin transformation* ($Y' = \arcsin \sqrt{Y} = \sin^{-1} \sqrt{Y}$) is used to stabilize the variance, if Y is a proportion of rate.

14.4.2 Weighted Least-Squares Analysis

The *weighted least-squares* method of analysis is a modification of standard regression analysis procedures that is used when a regression model is to be fit to a set of data for which the assumptions of variance homogeneity and/or independence do not hold. We shall briefly describe here the weighted least-squares approach for dealing with variance heterogeneity. For discussions of the general method of weighted regression (which incorporate a discussion of treating noninde-pendence), see Draper and Smith (1981) and Neter, Wasserman, and Kutner (1983).

Weighted least-squares analysis can be used when the variance of Y varies for different values of the independent variable(s), provided that these variances (i.e., σ_i^2 for the ith obser-vation on Y) are known or can be assumed to be of the form $\sigma_i^2 = \sigma^2/W_i$, where the weights $\{W_i\}$ are known. The methodology then involves determining the regression coefficients $\hat{\beta}_0', \hat{\beta}_1', \ldots, \hat{\beta}_k'$ that minimize the expression

$$\sum_{i=1}^{n} W_i (Y_i - \hat{\beta}_0' - \hat{\beta}_1' X_{i1} - \hat{\beta}_2' X_{i2} - \cdots - \hat{\beta}_k' X_{ik})^2$$

where the weight W_i is given by $1/\sigma_i^2$ (when the $\{\sigma_i^2\}$ are known) or is exactly the W_i in the expression σ^2/W_i (when this form applies).

The specific weighted least-squares solution for the straight-line regression case (i.e., $Y = \beta_0 + \beta_1 X + E$) is given by the formulas

$$\hat{\beta}_1' = \frac{\sum_{i=1}^{n} W_i (X_i - \bar{X}')(Y_i - \bar{Y}')}{\sum_{i=1}^{n} W_i (X_i - \bar{X}')^2}$$

and

$$\hat{\beta}_0' = \bar{Y}' - \hat{\beta}_1' \bar{X}'$$

in which

$$\bar{Y}' = \frac{\sum_{i=1}^{n} W_i Y_i}{\sum_{i=1}^{n} W_i} \quad \text{and} \quad \bar{X}' = \frac{\sum_{i=1}^{n} W_i X_i}{\sum_{i=1}^{n} W_i}$$

Under the usual normality and mutual independence assumptions for the Y_i's, the same general procedures are applicable as are used in the unweighted least squares case regarding t tests, F tests, and confidence intervals about the various regression parameters. For example, to test $H_0 : \beta_1 = 0$, we may use the following test statistic:

$$T = \frac{\hat{\beta}_1' - 0}{S_{Y|X}'/S_X' \sqrt{n - 1}} \frown t_{n-2} \quad \text{under } H_0$$

in which

$$S_{Y|X}'^2 = \frac{1}{n - 2} \sum_{i=1}^{n} W_i (Y_i - \hat{\beta}_0' - \hat{\beta}_1' X_i)^2$$

and

$$S_X'^2 = \frac{1}{n - 1} \sum_{i=1}^{n} W_i (X_i - \bar{X}')^2$$

14.4.3 Alternate Approaches

If the analyst decides that the regression model used does not fit the data and cannot be made to fit via simple generalizations of standard multiple linear regression (such as weighted least squares), other methods must be used. For situations where the response variable cannot be modeled as a linear function of the regression coefficients, analysis methods do exist for nonlinear functions (Gallant 1975). If the problem concerns nonnormal distributions, methods of rank analysis (Conover and Iman 1981) or categorical data analysis (Agresti 1990) may be appropriate.

Some rank analysis methods may be thought of as being essentially two-step processes. The first step is to replace the original data with appropriate ranks. The second is to conduct multiple linear regression analysis on the ranks.

14.4.4 Generalizations of Multiple Linear Regression

Survey of Methods

Generalizations of multiple linear regression may be loosely grouped into exact and approximate methods. Exact methods have procedures of estimation and hypothesis testing with known properties for finite samples; approximate methods have only asymptotic results available and, therefore, must be used with caution in small samples. The generalized approximate methods include regression on principal components, ridge regression, and robust regression. Exact generalizations include multivariate techniques and exact weighted least squares (in which weights are known without error).

Robust regression involves weighting or transforming the data so as to minimize the effects of extreme observations. The goal is to make the analysis more robust (i.e., less sensitive) to any particular observation and also less sensitive to the basic assumptions of regression analysis. (See Huber 1981 for a discussion of such procedures.)

Multivariate methods for the analysis of continuous responses include multivariate multiple regression, multivariate analysis of variance, canonical correlation analysis, and growth curve analysis, among others. In all these procedures, one can account for nonindependence among observations by explicitly modeling such nonindependence. The types of predictors and response variables that may be used in these methods differ, as do the hypotheses that are testable and methods by which the nonindependence is modeled. Many multivariate books are available; see, e.g., Timm (1975) and Morrison (1976). Also, Chapters 25 and 26 in this book discuss and illustrate the use of *mixed linear models* for analysing data where response variable dependencies and variance heterogeneities are present.

14.5 Collinearity

Detection of collinearity is the last component of the regression diagnostic methods that we will discuss, but it is by no means the least important. As mentioned earlier, collinearity exists when there are strong linear relationships among *independent variables*. We begin with simple examples illustrating the symptoms and effects of collinearity, after which we present in more detail the mathematics behind the collinearity problem. We then present three approaches for diagnosing the presence of collinearity and some simple remedies for the problem.

14.5.1 Simple Examples Illustrating Collinearity

For the data of Lewis and Taylor (1967), consider the following three competing models:

$$\text{WEIGHT} = \beta_0 + \beta_1 \text{ HEIGHT} + \beta_2 \text{ AGE} + E \qquad \text{(Model 1)}$$
$$\text{WEIGHT} = \beta_0' + \beta_1' \text{ HEIGHT} + \beta_2' \text{ AGE} + \beta_3' \text{ AGE}^2 + E \qquad \text{(Model 2)}$$
$$\text{WEIGHT} = \beta_0'' + \beta_1'' \text{ HEIGHT} + \beta_2'' \text{ AGE}_\text{p} + \beta_3'' \text{ (AGE}_\text{p})^2 + E \quad \text{(Model 3)}$$

We have already conducted several diagnostics analyses for the first model. Users who perceived a slight curvilinear relationship in Figure 14.1 may be interested in considering Model 2. The third model is similar to the second, except that we will fit it using data for AGE that have been perturbed slightly by adding a randomly generated value between -0.25 and 0.25 to each AGE value. Note that this perturbation changes individual AGE values to AGE_p values by no more than 2%, an amount that is unimportant, practically speaking, and that causes virtually no change in summary statistics for AGE and AGE_p. Therefore, such a perturbation should not yield regression results that are very different from the original data if the regression model is "good" (i.e., is valid and precise). The parameter estimates and their standard errors for each model are shown in Table 14.2 below.

Comparing Models 1 and 2, we see that the estimates of the regression coefficient for AGE differ in several important ways: the signs of the estimates are different, the magnitudes are considerably different, and the standard error of the estimate is very large in Model 2 [as a result, the regression coefficient is insignificant (i.e., is not significantly different from 0) in Model 2, even though the corresponding coefficient in Model 1 is significant]. Indeed, a multiple partial test for the joint significance of the AGE and AGE^2 coefficients, controlling for HEIGHT, in model 2 turns out to be non-significant, indicating that child's age is not important, even though it was important in Model 1. It is impossible to discern the true nature of the association between AGE and WEIGHT because of these inconsistencies. Such inconsistencies often occur in the presence of collinearity and are an indication of the instability caused by this undesirable condition.

In this example, it is easily concluded that collinearity exists because of the obvious association between the predictors AGE and AGE^2 (which can be regarded as having a linear component). However, it is not always this easy to diagnose the presence of collinearity or to identify predictors involved in collinearity problems; in Section 14.5.3 we present techniques that can be used for this purpose.

TABLE 14.2   **Parameter estimates for three regression models based on Lewis and Taylor data (1967) ($n = 127$)**

	Parameter Estimates (Std. Errors)					
Parameter	Model 1 (Without AGE^2)		Model 2 (With AGE^2)		Model 3 (AGE Values Perturbed)	
Intercept	$-39.0^\dagger$	(9.0)	52.9	(54.6)	66.5	(54.0)
Regression Coefficient						
For HEIGHT	$31.6^\dagger$	(8.6)	$34.0^\dagger$	(8.7)	$34.6^\dagger$	(8.4)
For AGE	$2.7^\dagger$	(0.6)	-11.0	(8.0)	$-13.1^\dagger$	(7.9)
For AGE^2	--		$0.5^\dagger$	(0.3)	$0.6^\dagger$	(0.3)

†Statistically significant at the 10% significance level (two-tailed test).

A comparison of Models 2 and 3 further highlights the instability that collinearity can cause. The models are the same and are fitted with data that are similar except for a slight, practically meaningless, perturbation of the children's ages. However, the coefficient esti-mate for AGE in Model 3 is almost 20% more negative than in Model 2 (a considerable difference) and is now statistically significant. The fact that the regression results for AGE can be so different for data that are not very different should be of concern. In actual analy-ses, models are usually not re-fitted with perturbed data; therefore, this aspect of the serious instability due to collinearity may go undetected.

14.5.2 Mathematical Concepts in Collinearity

In this section we explore some of the mathematical concepts related to the collinearity prob-lem. This will be helpful in understanding the statistics used to diagnose the presence of collinearity; these statistics are presented in Section 14.5.3. We recognize that some applied regression analysts may not require the detail presented here, and that others may be very inter-ested in delving much more deeply into the mathematical roots of the collinearity problem. In the former case, readers may simply skip this subsection; in the latter, we refer readers to the detailed discussion of the topic in the book by Belsley, Kuh, and Welsch (1980).

Collinearity with Two Predictors   The problems emanating from collinearity can be illustrated with simple two-variable regression examples. Consider fitting the model

$$Y_i = \beta_0 + \beta_1 X_{i1} + \beta_2 X_{i2} + E_i$$

to produce $\hat{\beta}_0$, $\hat{\beta}_1$, and $\hat{\beta}_2$. In general, we can show that

$$\hat{\beta}_j = c_j \left[\frac{1}{1 - r_{12}^2} \right]$$

for $j = 1$ or $j = 2$. Here c_j is a value that depends on the data, and r_{12}^2 is the squared correlation between X_1 and X_2. In turn,

$$\hat{\beta}_0 = \bar{Y} - \hat{\beta}_1 \bar{X}_1 - \hat{\beta}_2 \bar{X}_2$$

so

$$\bar{Y} - \hat{\beta}_0 = \hat{\beta}_1 \bar{X}_1 + \hat{\beta}_2 \bar{X}_2 = \left[\frac{1}{1 - r_{12}^2} \right] (c_1 \bar{X}_1 + c_2 \bar{X}_2)$$

Here, $\bar{X}_1$ and $\bar{X}_2$ are the means of X_1 and X_2, respectively. These expressions tell us that $\hat{\beta}_1$, $\hat{\beta}_2$, and $(\bar{Y} - \hat{\beta}_0)$ are all proportional to $1/[1 - r_{12}^2]$. More specifically, it is informative to consider fitting the model

$$Y_i = \theta_0 + \theta_1 X_{i1} + \theta_2 X_{i1} + E_i$$

In this case, a single variable, X_1, is included in the model twice. What are the estimates of the regression coefficients? Since $r_{12}^2 = r_{11}^2 = 1$, it follows that $1 - r_{12}^2 = 0$, and

$$\hat{\theta}_j = c_j \left(\frac{1}{0} \right) = ?$$

From this we conclude that the estimates of the regression coefficients are indeterminate. Since the estimates of the variances of the regression coefficients are proportional to the "inflation factor" $1/(1 - r_{12}^2)$, they are indeterminate as well. In turn, the P-values for tests about the coefficients are also indeterminate, since they involve the estimates just discussed. The preceding model can be rewritten in the form

$$Y_i = \theta_0 + (\theta_1 + \theta_2)X_{i1} + E_i$$

which establishes that an infinite number of pairs of θ_1 and θ_2 values add up to the same coefficient value; thus an estimate of the coefficient of X_1—say, $\widehat{\theta_1 + \theta_2}$—does not permit a unique determination of the individual estimates $\hat{\theta}_1$ and $\hat{\theta}_2$. Consequently, θ_1 and θ_2 cannot be estimated separately. For example, if the actual estimate for the slope of X_1 is 12.5, then $\hat{\theta}_1 = 12.5, \hat{\theta}_2 = 0$ are possible components, as are $\hat{\theta}_1 = 0, \hat{\theta}_2 = 12.5$ and $\hat{\theta}_1 = 6, \hat{\theta}_2 = 6.5$, and so on. In this extreme example, one of the predictor variables is a *perfect linear combination* of the other—namely,

$$X_1 = \alpha + \beta X_2 = 0 + (1)X_2 = X_2$$

Geometrically the points (X_{i1}, X_{i2}) all fall on the straight line $X_1 = X_2$; hence, the term *collinear* is applicable.

The following data provide a slightly more general example:

X_1	X_2
0	3
1	5
3	9
4	11
7	17

Plotting X_2 against X_1 yields a very simple picture, as shown in Figure 14.9. The plot illustrates that all of the data points (X_{i1}, X_{i2}) fall exactly on the straight line $X_2 = 3 + 2X_1$, which demonstrates that X_1 and X_2 are exactly collinear. Moreover, $r_{12}^2 = 1$ directly measures this perfect collinearity. The collinearity issue involves only the predictor variables and does not depend on the relationship between the response and any of the predictors. As r_{12}^2 decreases, the collinearity problem between X_1 and X_2 becomes less severe, with the ideal situation occurring when X_1 and X_2 are uncorrelated.

FIGURE 14.9 **Perfectly collinear set of pairs of variable values: $X_2 = 3 + 2X_1$**

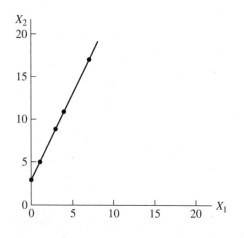

In general, for a two-variable model, if r_{12}^2 is nearly 1.0, then near collinearity is present. Although the regression coefficient estimates can be computed, they are highly unstable. In particular, this instability is reflected in large estimates of the coefficient variances, since such variance estimates are proportional to

$$\frac{1}{1 - r_{12}^2}$$

As r_{12}^2 gets closer to 1.0, this factor becomes large, thereby inflating the estimated variances of the regression coefficients. Later we will see that this factor is a special case of a variance inflation factor.

Collinearity Concepts   We now generalize the discussion to treat any number of predictors. In addition, we describe methods for quantifying the degree of collinearity in fitting a regression model to a particular set of data.

As we observed in the examples just discussed, collinearity involves relationships among the predictor variables and does not directly involve the response variable. As such, one informative way to examine collinearity is to consider what happens if each predictor variable is treated as the response variable in a multiple regression model in which the independent variables are all of the remaining predictors. For k predictors, then, we have k such models. For example, with the four-predictor model

$$Y_i = \beta_0 + \beta_1 X_{i1} + \beta_2 X_{i2} + \beta_3 X_{i3} + \beta_4 X_{i4} + E_i$$

four models would be fitted as follows:

$$X_{i1} = \alpha_{01} \qquad\qquad + \alpha_{21} X_{i2} + \alpha_{31} X_{i3} + \alpha_{41} X_{i4} + E_i$$
$$X_{i2} = \alpha_{02} + \alpha_{12} X_{i1} \qquad\qquad + \alpha_{32} X_{i3} + \alpha_{42} X_{i4} + E_i$$
$$X_{i3} = \alpha_{03} + \alpha_{13} X_{i1} + \alpha_{23} X_{i2} \qquad\qquad + \alpha_{43} X_{i4} + E_i$$
$$X_{i4} = \alpha_{04} + \alpha_{14} X_{i1} + \alpha_{24} X_{i2} + \alpha_{34} X_{i3} \qquad\qquad + E_i$$

To assess collinearity, we need to know the associated R^2-values based on fitting these four models—namely, $R^2_{X_1|X_2, X_3, X_4}$, $R^2_{X_2|X_1, X_3, X_4}$, $R^2_{X_3|X_1, X_2, X_4}$, and $R^2_{X_4|X_1, X_2, X_3}$. If any of these multiple R^2-values equals 1.0, a perfect collinearity is said to exist among the set of predictors. The term *collinearity* is used to indicate that one of the predictors is an *exact* linear combination of the others. As described in earlier examples, in the special case $k = 2$, perfect collinearity means that X_1 is a straight-line function of X_2—say, $X_1 = \alpha + \beta X_2$—so that the points (X_{i1}, X_{i2}) lie on a straight line.

Consider, for example, predicting college grade point average (CGPA) on the basis of high school grade point average (HGPA) and College Board scores on the mathematics (MATH) and verbal (VERB) tests. The model is

$$CGPA_i = \beta_0 + \beta_1 HGPA_i + \beta_2 MATH_i + \beta_3 VERB_i + E_i$$

Now imagine trying to improve prediction by adding the total (combined) board scores (TOT = MATH + VERB) to create a new model:

$$CGPA_i = \alpha_0 + \alpha_1 HGPA_i + \alpha_2 MATH_i + \alpha_3 VERB_i + \alpha_4 TOT_i + E_i$$

This model has a perfect collinearity, which means that the parameters in the model cannot be estimated uniquely. To see this, begin by rewriting the model as

$$CGPA_i = \alpha_0 + \alpha_1 HGPA_i + \alpha_2 MATH_i + \alpha_3 VERB_i + \alpha_4(MATH_i + VERB_i) + E_i$$

It follows that

$$CGPA_i = \alpha_0 + \alpha_1 HGPA_i + (\alpha_2 + \alpha_4)MATH_i + (\alpha_3 + \alpha_4)VERB_i + E_i$$

With this version of the model, consider choosing $\alpha_4 = 0$. Then $\alpha_2 = \beta_2$ and $\alpha_3 = \beta_3$ give the correct original model. Next, choose $\alpha_4 = 3$. Then $\alpha_2 = \beta_2 - 3$ and $\alpha_3 = \beta_3 - 3$ also give the correct original model. In fact, for any choice of α_4, we can choose values for α_2 and α_3 that provide the correct model. Since the α_4 parameter is irrelevant under the circumstances (it could best be set equal to 0), a model containing a perfect collinearity is sometimes said to be *overparameterized*.

Near collinearity arises if the multiple R^2 relating one predictor with the remaining predictors is nearly 1. For a general model involving k predictors such as

$$Y_i = \beta_0 + \beta_1 X_{i1} + \beta_2 X_{i2} + \cdots + \beta_k X_{ik} + E_i$$

the multiple R^2-value of interest for the first predictor is $R^2_{X_1|X_2, X_3, \ldots, X_k}$, the multiple R^2-value of interest for the second predictor is $R^2_{X_2|X_1, X_3, \ldots, X_k}$, and so on. These quantities are generalizations of the statistic r^2_{12} for a $k = 2$ variable model. For convenience, we denote by R^2_j the squared multiple correlation based on regressing X_j on the remaining $(k - 1)$ predictors.

The *variance inflation factor* (VIF) is often used to measure collinearity in a multiple regression analysis. It may be computed as

$$VIF_j = \frac{1}{1 - R^2_j}, \quad j = 1, 2, \ldots, k$$

The quantity VIF_j generalizes the variance inflation factor for a two-predictor model, $1/(1 - r^2_{12})$. Clearly, $VIF_j \geq 0$. As for the two-variable case, the estimates of the variances for the regression coefficients are proportional to the VIFs—namely,

$$S^2_{\hat{\beta}_j} = c^*_j(VIF_j), \quad j = 1, 2, \ldots, k$$

This expression suggests that the larger the value of VIF_j, the more troublesome is the variable X_j. A rule of thumb for evaluating VIFs is to be concerned with any value larger than 10.0. For VIF_j, this corresponds to $R^2_j > .90$ or, equivalently, $R_j > .95$. Some people prefer to consider

$$Tolerance_j = \frac{1}{VIF_j} = 1 - R^2_j$$

The choice among R^2_j, $1 - R^2_j$, and VIF_j is a matter of personal preference since they all contain exactly the same information. As R^2_j goes to 1.0, the tolerance $(1 - R^2_j)$ goes to 0, and VIF_j goes to infinity.

Because of its special nature, the intercept requires separate treatment in evaluating collinearity. For the general model involving k predictors,

$$Y_i = \beta_0 + \beta_1 X_{i1} + \beta_2 X_{i2} + \cdots + \beta_k X_{ik} + E_i$$

we find regression coefficient estimates $\hat{\beta}_0, \hat{\beta}_1, \ldots, \hat{\beta}_k$. The intercept estimate can be expressed simply as

$$\hat{\beta}_0 = \bar{Y} - (\hat{\beta}_1 \bar{X}_1 + \hat{\beta}_2 \bar{X}_2 + \cdots + \hat{\beta}_k \bar{X}_k) = \bar{Y} - \sum_{j=1}^{k} \hat{\beta}_j \bar{X}_j$$

Here, $\bar{Y} = \sum_{i=1}^{n} Y_i/n$ is the mean of the response values, and $\bar{X}_j = \sum_{i=1}^{n} X_{ij}/n$ is the mean of the values for predictor X_j. From this, we can deduce that the estimated intercept is affected by the VIF_j's, $j = 1, 2, \ldots, k$, since it is a function of the $\hat{\beta}_j$'s. The problem disappears if the means of all X_j's are 0 (e.g., if the predictor data are centered). In that case, $\bar{Y}$ is the estimated intercept.

In general, even if the predictor data are not centered, a variance inflation factor VIF_0 for $\hat{\beta}_0$ can be defined and interpreted in the same way as for VIF_j. First, define

$$\text{VIF}_0 = \frac{1}{1 - R_0^2}$$

Here R_0^2 is the generalized squared multiple correlation for the regression model

$$I_i = \alpha_1 X_{i1} + \alpha_2 X_{i2} + \cdots + \alpha_k X_{ik} + E_i$$

in which I_i is identically 1 (which may be thought of as the score for the intercept variable). No intercept is included in this model (as a predictor), which is why R_0^2 is called a *generalized* squared correlation. As with the VIF_j's, $\text{VIF}_0 \geq 0$, and

$$S_{\hat{\beta}_0}^2 = c_0^*(\text{VIF}_0)$$

Hence, the interpretation for VIF_0 is the same as for VIF_j, $j = 1, 2, \ldots, k$.

Controversy surrounds the treatment of the intercept in regression diagnostics. For some, it is simply another predictor; for others, it should be eliminated from discussion. We take a middle position, arguing that the model and the data at hand determine the role of the intercept. (See, for example, the discussions following Belsley 1984.) This leads us to discuss diagnostics both with and without the intercept included in the regression model, corresponding to the cases with and without centering of the predictors and response variables.

The presence of collinearity or (more typically) near collinearity presents computational difficulties in calculating numerically reliable estimates of the R_j^2-values, tolerances, and the VIF_j's on the basis of standard regression procedures. This apparent impasse can be solved in at least three ways. The first way is to use computational algorithms that detect collinearity problems as they arise in the midst of the calculations. A discussion of such algorithms is beyond the scope of this text.

A second way to avoid the impasse is to scale the data appropriately. By *scaling*, we mean the choice of measurement unit (e.g., degrees Celsius versus degrees Fahrenheit) and the choice of measurement origin (e.g., degrees Celsius versus degrees Kelvin). We shall consider only linear changes in scale, such as

$$X_1 = \alpha + \beta X_2$$

$$C = \frac{5}{9}(F - 32) = -32\left(\frac{5}{9}\right) + \left(\frac{5}{9}\right)(F)$$

or

$$K = (-273) + (1)(C)$$

where C, F, and K are temperatures in degrees Celsius, degrees Fahrenheit, and degrees Kelvin, respectively.

Often scaling refers just to multiplying by a constant rather than also to adding or subtracting a constant. An example of scaling is the conversion from feet to inches. One important case of subtracting a constant, a form of scaling, is *centering*. A set of values $\{X_{ij}\}$ for predictor X_j is centered by subtracting the mean $\bar{X}_j$ of the values for predictor X_j from each individual value for that predictor, giving

$$X_{ij}^* = X_{ij} - \bar{X}_j$$

in which $\bar{X}_j = \sum_{i=1}^{n} X_{ij}/n$ for predictor X_j.

Computing standardized scores (z scores) is a closely related method of scaling. In particular, the standardized score corresponding to X_{ij} is

$$z_{ij} = \frac{X_{ij} - \bar{X}_j}{S_j}$$

in which $S_j^2 = \sum_{i=1}^{n}(X_{ij} - \bar{X}_j)^2/(n-1)$. Centered and standardized scores have mean 0, since $\sum_{i=1}^{n}X_{ij}^* = \sum_{i=1}^{n}z_{ij} = 0$. Also, $\sum_{i=1}^{n}z_{ij}^2/(n-1) = 1$, so the set $\{z_{ij}\}$ of standardized scores has a variance equal to 1. We shall delay the discussion of detecting and fixing scaling problems until later in this chapter. For now, we need the concept of scaling in order to understand the following discussion of other diagnostic statistics.

A third way to avoid the impasse created by collinearity and near collinearity is to use alternate computational methods to diagnose collinearity. One especially popular method for characterizing near and/or exact collinearities among the predictors involves computing the *eigenvalues* of the predictor variable *correlation matrix*. The eigenvalues are connected with the *principal component analysis* of the predictors. The principal components of the predictors are a set of new variables that are linear combinations of the original predictors. These components have two special properties: they are not correlated with each other; and each, in turn, has maximum variance, given that all components are mutually uncorrelated. The principal components provide idealized predictor variables that still retain all of the same information as the original variables. The variances of these components (the new variables) are called *eigenvalues*. The larger the eigenvalue, the more important is the associated principal component in representing the information in the predictors. *As an eigenvalue approaches zero, the presence of a near collinearity among the original predictors is indicated.* The presence of an eigenvalue of exactly 0 means that a perfect linear dependency (i.e., an exact collinearity) exists among the predictors.

If a set of k predictor variables does *not* involve an *exact* collinearity, then k principal components are needed to reproduce exactly all of the information contained in the original variables. If one of the predictors is a perfect linear combination of the others, then only $(k-1)$ principal components are needed to provide all of the information in the original variables. *The number of zero (or near-zero) eigenvalues is the number of collinearities (or near collinearities) among the predictors.* Even a single eigenvalue near 0 presents a serious problem that must be resolved.

In using the eigenvalues to determine the presence of near collinearity, researchers usually employ three kinds of statistics: the *condition index* (CI), the *condition number* (CN), and the *variance proportions*.

Consider again the general linear model

$$Y_i = \beta_0 + \beta_1 X_{i1} + \beta_2 X_{i2} + \cdots + \beta_k X_{ik} + E_i$$

Often collinearity is assessed by considering only the k parameters β_1 through β_k and ignoring the intercept β_0. This is accomplished by centering the response and predictor variables, which corresponds to fitting the model

$$Y_i - \bar{Y} = \beta_1(X_{i1} - \bar{X}_1) + \beta_2(X_{i2} - \bar{X}_2) + \cdots + \beta_k(X_{ik} - \bar{X}_k) + E_i$$

Noting that $\hat{\beta}_0 = \bar{Y} - \sum_{j=1}^{k}\hat{\beta}_j\bar{X}_j$ in the original model, we observe that centering the predictors and the response forces the estimated intercept in that centered model to be 0. Hence, it can be dropped from that model.

It is also common to assess collinearity after centering and standardizing both the predictors and the response. This leads to the standardized model

$$\frac{Y_i - \overline{Y}}{S_Y} = \beta_1^* \frac{(X_{i1} - \overline{X}_1)}{S_1} + \beta_2^* \frac{(X_{i2} - \overline{X}_2)}{S_2} + \cdots + \beta_k^* \frac{(X_{ik} - \overline{X}_k)}{S_k} + E_i^*$$

The coefficients for this standardized model are often called the *standardized regression coefficients;* in particular,

$$\beta_j^* = \beta_j \left(\frac{S_j}{S_Y} \right), \quad j = 1, 2, \ldots, k$$

Table 14.3 summarizes an eigenanalysis of the predictor correlation matrix R_{xx} for a hypothetical four-predictor ($k = 4$) standardized regression model. With $k = 4$ predictors, four eigenvalues (denoted by λ's) exist. Later we will include the intercept in the analysis and have ($k + 1$) eigenvalues. In the present case, $\lambda_1 = 2.0$, $\lambda_2 = 1.0$, $\lambda_3 = 0.6$, and $\lambda_4 = 0.4$. (The sum of the eigenvalues for a correlation matrix involving k predictors is *always* equal to k.) It is customary to list the eigenvalues from largest (λ_1) to smallest (λ_k). A condition index can be computed for each eigenvalue as

$$CI_j = \sqrt{\lambda_1/\lambda_j}$$

In particular, CI_3 for this example is $\sqrt{2.0/0.6} = 1.87$. The first (largest) eigenvalue always has an associated condition index (CI_1) of 1.0. The largest CI_j, called the *condition number*, always involves the largest (λ_1) and smallest (λ_k) eigenvalues. It is given by the formula

$$CN = \sqrt{\lambda_1/\lambda_k}$$

For this example, $CN = \sqrt{2.0/0.4} = 2.24$. Since eigenvalues are variances, CI_j and CN are ratios of standard deviations (of principal components, the idealized predictors). Like VIFs, values of CI_j and CN are nonnegative, and larger values suggest potential near collinearity. Belsley, Kuh, and Welsch (1980) recommended interpreting a CN of 30 or more as reflecting moderate to severe collinearity, worthy of further investigation. Of course, such a CN may be associated with two or more CI_j's that are greater than or equal to 30.

A variance proportion indicates, for each predictor, the proportion of the estimated variance of its estimated regression coefficient that is associated with a particular principal component.

TABLE 14.3 Eigenanalysis of the predictor correlation matrix for the hypothetical four-variable model

Variable	Eigenvalue	Condition Index	Variance Proportions			
			X_1	X_2	X_3	X_4
1	2.0	1.00	.09	.11	.08	.15
2	1.0	1.41	.32	.10	.25	.07
3	0.6	1.87	.40	.52	.12	.13
4	0.4	2.24	.19	.27	.55	.65

The variance proportions suggest collinearity problems if more than one predictor has a high variance proportion (loads highly) on a principal component having a high condition index. The presence of two or more proportions of at least .5 for such a component suggests a problem. One should definitely be concerned when two or more loadings greater than .9 appear on a component with a large condition index.

Table 14.3 includes variance proportions for the hypothetical four-predictor model under consideration. The entries represent a typical pattern of condition indices and proportions for a regression analysis with no major collinearity problems. Each column sums to a total proportion of 1.00 because each estimated regression coefficient has its own estimated variance partitioned among the four components. The last row involves two loadings greater than .5, corresponding to the smallest eigenvalue; but since the smallest eigenvalue has a CI of only 2.24 (far from the suggested warning level of 30.0 for moderate to severe collinearity), the proportion pattern does not indicate a major problem.

The intercept can play an important role in collinearity. In defining regression models, researchers follow the standard practice of including the intercept term β_0. The intercept is a regression coefficient for a variable that has a constant value of 1. Consequently, any variable with near-zero variance will be nearly a constant multiple of the intercept and hence will be nearly collinear with it. This problem can arise spuriously when variables are improperly scaled. To evaluate this possibility, the eigenanalysis discussed earlier can be modified to include the intercept by basing it on the scaled cross-products matrix.[5] The eigenanalysis including the intercept may suggest that this constant is nearly collinear with one or more predictors.

Centering may help decrease collinearity. From a purely theoretical perspective, statisticians disagree as to when centering helps regression calculations (for example, see Belsley 1991; Belsley 1984; Smith and Campbell 1980; and related comments in the same issues). For actual computations, centering can increase numerical accuracy in many situations. Polynomial regression (Chapter 15) is one situation where centering is recommended. In the example that follows, we numerically illustrate the effects of centering on procedures for diagnosing collinearity.

14.5.3 Collinearity Diagnostics

There are three steps that should be taken to diagnose the presence of collinearity. These steps will also identify the particular predictor variables involved in the collinearity problem—information that is needed before the appropriate corrective action can be taken.

The first step is taken during the simple descriptive analyses mentioned in Section 14.1, and serves to bring the most obvious possible collinearity problems to our attention. Scatter plots of each continuous predictor versus each of the other continuous predictors, and correlations between predictors, can quickly reveal potentially strong and problematic linear associations. For example, Figure 14.5 and the SAS PROC CORR output for Example 14.1 suggested that there may be a collinearity problem involving HEIGHT and AGE.

Next, the *variance inflation factor* (VIF) values should be examined for each predictor in the model. We demonstrated, in Example 14.5.1, that collinearity can result in inflated standard

---

[5] The eigenanalysis would then be based on the $(k + 1) \times (k + 1)$ cross-products matrix $\mathbf{X}'\mathbf{X}$ suitably scaled to have 1's on the diagonal, rather than on the $k \times k$ correlation matrix $\mathbf{R}_{xx}$. The sum of the eigenvalues for this scaled cross-products matrix is equal to $(k + 1)$.

errors for the fitted model. As discussed in Section 14.5.2, the VIF values provide an indication of the degree to which the estimated standard errors of regression parameter estimates are affected by the linear relationships among predictor variables. Generally, VIF values greater than 10 indicate collinearities among predictor variables that are strong enough to warrant corrective action.

In the case of Model 1, described in Section 14.5.1, the VIF values for HEIGHT and AGE are each equal to 2.3. These are not large enough to conclude that there is a collinearity problem needing correction in the model with two predictors HEIGHT and AGE.

It should be noted that, in models with just two predictors, the two VIF values will be identical (see Section 14.5.2 for details on VIF calculations). In larger models, the VIF values will, generally, all be different, and identifying which variables are strongly associated with other predictors may not be easy. Indeed, it may be the case that three or more predictors are simultaneously collinear with each other. In such a situation, the analyst must use good scientific judgment to determine which predictors are collinear with each other, and should also proceed to the third step of the collinearity investigation.

In the third step of the collinearity investigation, the condition index and variance proportion statistics should be examined. As described in Section 14.5.2, the condition indices for a data set describe the degree to which the data are "ill-conditioned," i.e., the degree to which small changes in data values result in large changes in the regression parameter estimates. There are as many condition indices as there are parameters in the regression model. Condition indices greater than 30 reflect moderate collinearity or worse. For the largest condition index greater than 30, the variance proportion statistics should be examined in order to determine which predictor variables are primarily responsible for the large condition index. Model terms with variance proportions higher than 0.5 can be considered to be involved in the collinearity problem. Additionally, if the intercept is involved in the collinearity (i.e., it has a high variance proportion), it is recommended that intercept-adjusted collinearity diagnostic statistics (i.e., diagnostic statistics computed after centering all predictor and response variables) be examined rather than unadjusted statistics using uncentered variables. Corrective action is taken (see Section 14.5.4) and the statistics are then re-examined for additional collinearity problems.

Variance inflation factors, condition indices, and variance proportions are all easily produced using SAS.

■ **Example 14.5**   Consider model 2 from Section 14.5.1:

$$\text{WEIGHT} = \beta_0 + \beta_1 \, \text{HEIGHT} + \beta_2 \, \text{AGE} + \beta_3 \, \text{AGE}^2 + E$$

The relevant SAS output is shown below.

Edited SAS Output for Example 14.5 (PROC CORR Output)

```
                          The CORR Procedure

                 Pearson Correlation Coefficients, N = 127

                                                                     AGE_
                         HEIGHT              AGE                    SQUARED
HEIGHT                  1.00000           0.75328                  0.74648
AGE                     0.75328           1.00000                  0.99853
AGE_SQUARED             0.74648           0.99853                  1.00000
```

Edited SAS Output for Example 14.5 (PROC REG Output)

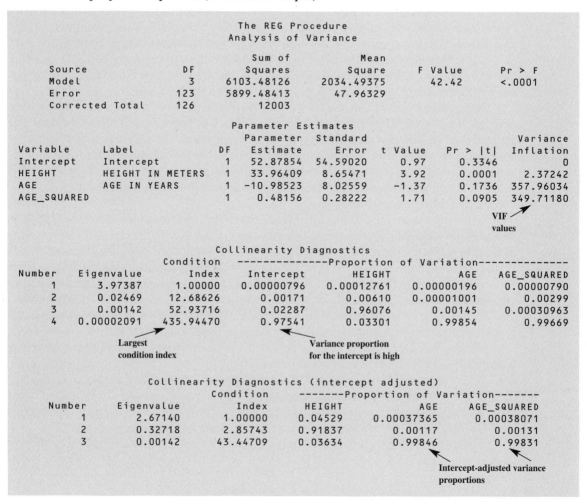

The REG Procedure
Analysis of Variance

Source	DF	Sum of Squares	Mean Square	F Value	Pr > F
Model	3	6103.48126	2034.49375	42.42	<.0001
Error	123	5899.48413	47.96329		
Corrected Total	126	12003			

Parameter Estimates

Variable	Label	DF	Parameter Estimate	Standard Error	t Value	Pr > |t|	Variance Inflation
Intercept	Intercept	1	52.87854	54.59020	0.97	0.3346	0
HEIGHT	HEIGHT IN METERS	1	33.96409	8.65471	3.92	0.0001	2.37242
AGE	AGE IN YEARS	1	-10.98523	8.02559	-1.37	0.1736	357.96034
AGE_SQUARED		1	0.48156	0.28222	1.71	0.0905	349.71180

VIF values

Collinearity Diagnostics

Number	Eigenvalue	Condition Index	Proportion of Variation — Intercept	HEIGHT	AGE	AGE_SQUARED
1	3.97387	1.00000	0.00000796	0.00012761	0.00000196	0.00000790
2	0.02469	12.68626	0.00171	0.00610	0.00001001	0.00299
3	0.00142	52.93716	0.02287	0.96076	0.00145	0.00030963
4	0.00002091	435.94470	0.97541	0.03301	0.99854	0.99669

Largest condition index

Variance proportion for the intercept is high

Collinearity Diagnostics (intercept adjusted)

Number	Eigenvalue	Condition Index	Proportion of Variation — HEIGHT	AGE	AGE_SQUARED
1	2.67140	1.00000	0.04529	0.00037365	0.00038071
2	0.32718	2.85743	0.91837	0.00117	0.00131
3	0.00142	43.44709	0.03634	0.99846	0.99831

Intercept-adjusted variance proportions

From the scatterplot in Figure 14.10, we see that there is an approximate linear relationship between HEIGHT and AGE^2. This is not surprising, since there was such a relationship between HEIGHT and AGE. A scatterplot of AGE versus AGE^2 is not necessary; clearly there is a relationship, and all that remains to be determined is whether that relationship is the source of moderate or strong collinearity. The PROC CORR output confirms the strong associations.

The PROC REG output shows VIF values of 358 and 350 for AGE and AGE^2, respectively. Clearly, these terms are involved in a severe collinearity problem. The VIF for HEIGHT is small, so it is not involved (even though early indications were that the HEIGHT–AGE relationship might be a problem).

For illustration purposes, we discuss the use of the condition indices, although, in this example, it is clear that the collinearity is due to the relationship between AGE and AGE^2. The largest condition index in the SAS output is 435.9, indicating the presence of strong collinearity. The variance proportions in the same row indicate that the intercept is involved in the collinearity (variance

FIGURE 14.10   SAS scatterplot of HEIGHT vs AGE² for Example 14.5

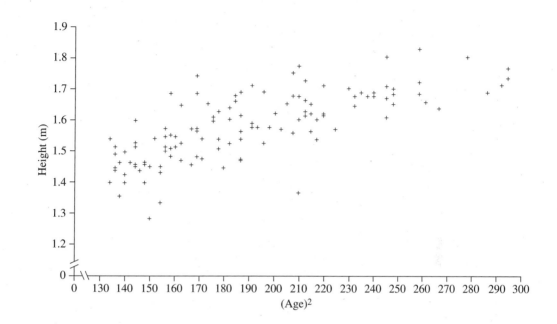

proportion for the intercept $= 0.97541$). As a result, we consider the intercept-adjusted condition indices. Here, the variance proportions corresponding to the largest condition index are examined (the condition indices are ordered, so the largest index appears in the last row). The variance proportions for AGE and AGE² are high (0.99846 and 0.99831, respectively), indicating that they are involved in the collinearity problem. Corrective action should be taken as described in Section 14.5.4.

14.5.4 Treating Collinearity Problems

Collinearity problems are often easily remedied by eliminating one or more of the predictors in the collinear set. At first, this may seem to be a drastic step; however, the strategy does make sense from the point of view that, since strongly collinear variables are closely related, they are very similar and can be regarded as measuring much the same thing. In that case, there may not be a great loss of information if one variable is dropped. In Example 14.5, AGE² would probably be removed. The collinearity diagnostics should then be re-examined to ensure that the problem has been adequately addressed. If not (i.e., if VIF values or condition indices are still high), another variable may need to be removed. The process repeats until collinearity is no longer judged to be a problem.[8]

---

[8] Note that it is usually best to remove just one variable from a collinear set, re-examine the collinearity diagnostics, and then remove another variable if necessary, and so on. Often, removing just one variable will greatly reduce or eliminate collinearity problems.

Decisions about which variables to remove from a collinear set can be difficult to make. Judgments should not be made on the basis of *P*-values, since hypothesis test results may be misleading in the presence of collinearity. Instead, researchers should use their scientific expertise and previous experience to identify the variables that, from a scientific point of view, would be acceptable to drop. Practical concerns, such as cost and difficulty of collecting data for each variable, may also be important to take into account. Throughout, a conservative approach should be taken that ensures that only the least scientifically interesting predictors are dropped and that the original objectives of the study can still be met.

In models in which *natural polynomials* or powers of a continuous predictor are included in a regression model (e.g., AGE^2 is included along with AGE), an alternative to dropping higher-order terms to reduce the naturally occurring collinearity is to use the method of orthogonal polynomials. This topic is discussed in detail in Chapter 15.

The improper use of dummy variables may inadvertently introduce severe collinearity problems. Discussion of how to avoid this problem is found in Section 12.3.

Interaction terms generally create the atmosphere for collinearity problems, especially if such terms are overused. For example, if the predictors are AGE, HEIGHT and AGE $\times$ HEIGHT, then collinearity may be a problem due to the close functional relationship between the product term and the two basic predictors. In general, fitting a model that contains several interaction terms almost guarantees that some collinearity problems will arise. Analysts are therefore advised to proceed conservatively, and to limit interaction terms in a model to just those few terms believed, a priori, to have the most scientific relevance.

Researchers designing studies can attempt to head off potential collinearity problems by collecting data that break the pattern of collinearity. For example, suppose that family income and years of education are to be used as predictors in a regression analysis. Since these variables are likely to be highly positively correlated, researchers can use stratified sampling schemes that deliberately sample from sub-populations of highly educated subjects who earn low incomes, and higher-income subjects who have few years of education. If enough such subjects are sampled, collinearity problems between the two predictors might be lessened or eliminated. However, this approach may not be possible (for example, in observational studies with simple random sampling) and may not be successful in reducing collinearity problems regardless of the researcher's efforts.

There are some other practical techniques that may be helpful in eliminating collinearity. In some cases, using centered data (data in which the predictor variables have been transformed by subtracting the mean value of the variable from each value) can alleviate collinearity problems. A discussion about centering was presented in Section 14.5.2. However, analysts should proceed with caution. Belsley et al. (1980) have shown that centering can, in some situations, render the usual collinearity diagnostics, VIF values and condition indices, ineffective. That is, the instability characteristic of collinearity may remain even though the VIF values and condition indices appear normal when data are centered.

Other advanced techniques, such as regression on *principal components* and the technique known as *ridge regression*, may also be considered in the treatment of collinearity, although we do not favor the use of these methods. In the analysis of principal components, the original predictor variables are replaced by a set of mutually uncorrelated variables, the principal components. If necessary, components associated with eigenvalues near zero are dropped from the analysis, thus eliminating the attendant collinearity problem. Ridge regression involves perturbing the eigenvalues of the original predictor variable crossproducts matrix to push them away from zero, thus reducing the amount of collinearity. A detailed discussion of these methods is beyond the scope of this book (e.g., see Gunst and Mason, 1980). Both methods lead to biased estimators of the regression coefficients under the assumed model. In addition, *P*-values for statistical tests may be optimistically small when based on such biased estimation methods.

14.6 Scaling Problems

A general class of problems in regression analysis may arise due to improper scaling of the predictors and/or the response variable. Specifically, such problems involve the loss of computational accuracy. The resulting inaccuracy can sometimes be so great as to produce regression coefficient estimates with incorrect signs. A scaling problem may occur if a predictor has too wide a range of values. For example, recording adult human body weight in grams might be problematic. Another scaling problem might occur with the use of data for people who consume vastly greater quantities of, say, vitamin C than the average person. Properly measured values of such vitamin C intake would still yield outliers, and their use could lead to modeling problems. Similarly, if the mean of a variable is large and has little variability, computational problems may arise. A simple example of this would involve recording human body temperature in degrees Kelvin (degrees Celsius minus 273). Most scaling problems can be avoided by using proper data validation and rescaling methods before performing regression analyses.

14.7 Diagnostics Example

We now present a detailed example illustrating the implementation of the regression diagnostics methods discussed and recommended in this chapter.

■ **Example 14.6**   The data for this example arise from a hypothetical experiment concerned with environmental monitoring of surface water pollutants. The concentrations (X_1 and X_2) of two pollutants, A and B, and the instrument reading (Y) for a third pollutant, C, were recorded (see Table 14.4). Instrument readings for pollutant C are difficult and expensive to obtain; therefore, it would be useful to develop a predictive model based on the more easily measured concentrations of pollutants A and B. The regression model under consideration is:

$$Y = \beta_0 + \beta_1 X_1 + \beta_2 X_2 + E$$

The experiment was designed so that several specific levels of pollutant A, between 0.1 and 5.0 parts-per-million (ppm) were studied. Based on past experience, it is thought that a plausible range of concentrations for pollutant B would be 0.1 to 3.0 ppm, and would be 0 to 60 ppm for pollutant C. It is believed that there will be a positive association between the pollutant levels of pollutants A and B.

As a first step in the diagnostics analysis, we examine simple descriptive statistics for each variable. The statistics (mean, median, standard deviation, range, interquartile range) in the SAS PROC UNIVARIATE output below do not, for the most part, look unusual given the prior

TABLE 14.4   Data for Hypothetical Pollutant Concentration Example

Observation	Pollutant A Concentration (X_1)	Pollutant B Concentration (X_2)	Instrument Reading (Y)
1	0.1	0.5	10.7
2	0.1	1.1	14.2
3	0.1	2.3	16.7
4	0.5	0.4	19.1
5	0.5	1.2	24.9
6	0.5	2.0	25.4
7	1.0	0.1	32.3
8	1.0	1.2	33.8
9	1.0	2.1	39.6
10	1.5	0.8	33.3
11	1.5	1.5	37.2
12	1.5	2.9	37.8
13	2.0	0.2	37.5
14	2.0	1.5	38.6
15	2.0	2.5	42.6
16	2.5	0.7	44.3
17	2.5	1.6	45.2
18	2.5	2.3	47.2
19	3.0	0.3	45.2
20	3.0	1.3	46.4
21	3.0	2.3	65.3
22	3.5	0.9	47.5
23	3.5	1.7	48.2
24	3.5	2.6	48.2
25	4.0	0.5	47.4
26	4.0	1.9	48.6
27	4.0	2.2	48.9
28	4.5	0.4	48.0
29	4.5	1.7	48.9
30	4.5	2.4	50.1
31	5.0	0.5	48.5
32	5.0	1.7	48.9
33	5.0	2.7	50.2

knowledge about plausible values for X_1, X_2, and Y. However, the largest value of Y is 65.3 (see the Extreme Observations section of the output for Y), which is quite a bit larger than the next largest value and is larger than expected based on the plausible range for Y. Efforts should be made to ensure that this value was not recorded in error; if it was, the error should now be corrected. If not recorded in error, this potentially influential observation should be noted, and further attention paid to it during the formal check for outliers in the next step.

Edited SAS Output for Example 14.6 (PROC UNIVARIATE Output)

```
                        The UNIVARIATE Procedure
                             Variable: Y
                               Moments

N                              33      Sum Weights                    33
Mean                    40.0212121     Sum Observations           1320.7
Std Deviation           12.2732625     Variance               150.632973
Skewness                -0.8440522     Kurtosis               0.45212782

                     Basic Statistical Measures

           Location                            Variability

    Mean           40.02121     Std Deviation              12.27326
    Median         45.20000     Variance                  150.63297
    Mode           48.90000     Range                      54.60000
                                Interquartile Range        14.40000

                      Extreme Observations

    ··········Lowest··········          ··········Highest··········
    Value      OBS     Obs             Value      OBS      Obs
    10.7         1       1              48.9       29       29
    14.2         2       2              48.9       32       32
    16.7         3       3              50.1       30       30
    19.1         4       4              50.2       33       33
    24.9         5       5              65.3       21       21

    Stem Leaf                            #          Boxplot
       6 5                               1             |
       6                                               |
       5                                               |
       5 00                              2             |
       4 55677888889999                 14         +-----+
       4 034                            3          |  +  |
       3 7889                           4          |     |
       3 234                            3          +-----+
       2 55                             2             |
       2                                               |
       1 79                             2             |
       1 14                             2             0
         ····+····+····+····+
    Multiply Stem.Leaf by 10**+1

                               Variable: X1
                               Moments

N                              33      Sum Weights                    33
Mean                    2.50909091     Sum Observations             82.8
Std Deviation           1.59125808     Variance               2.53210227
Skewness                0.02636322     Kurtosis               -1.2491643
Uncorrected SS              288.78     Corrected SS           81.0272727
Coeff Variation         63.419706      Std Error Mean         0.27700248

                     Basic Statistical Measures

           Location                            Variability

    Mean           2.509091     Std Deviation               1.59126
    Median         2.500000     Variance                    2.53210
    Mode           0.100000     Range                       4.90000
                                Interquartile Range         3.00000

NOTE: The mode displayed is the smallest of 11 modes with a count of 3.
```

(continued)

Edited SAS Output for Example 14.6 (continued)

```
                          Variable: X1 (Continued)
                          Extreme Observations

          .........Lowest..........              .........Highest.........
          Value        OBS        Obs            Value        OBS        Obs
           0.1          3          3              4.5         29         29
           0.1          2          2              4.5         30         30
           0.1          1          1              5.0         31         31
           0.5          6          6              5.0         32         32
           0.5          5          5              5.0         33         33


      Stem Leaf                          #                       Boxplot
         5 000                           3                          |
         4 555                           3                          |
         4 000                           3                       +-----+
         3 555                           3                       |     |
         3 000                           3                       |     |
         2 555                           3                       *--+--*
         2 000                           3                       |     |
         1 555                           3                       |     |
         1 000                           3                       +-----+
         0 555                           3                          |
         0 111                           3                          |
           ----+----+----+----+
```

```
                          Variable: X2
                            Moments

N                              33       Sum Weights                    33
Mean                    1.45454545      Sum Observations               48
Std Deviation           0.82994386      Variance               0.68880682
Skewness               -0.0394651       Kurtosis              -1.2489157
Uncorrected SS               91.86      Corrected SS           22.0418182
Coeff Variation         57.0586407      Std Error Mean         0.14447468

                    Basic Statistical Measures

            Location                          Variability
       Mean       1.454545          Std Deviation          0.82994
       Median     1.500000          Variance               0.68881
       Mode       0.500000          Range                  2.80000
                                    Interquartile Range    1.50000

NOTE: The mode displayed is the smallest of 3 modes with a count of 3.

                          Extreme Observations

          .........Lowest..........              .........Highest.........
          Value        OBS        Obs            Value        OBS        Obs
           0.1          7          7              2.4         30         30
           0.2         13         13              2.5         15         15
           0.3         19         19              2.6         24         24
           0.4         28         28              2.7         33         33
           0.4          4          4              2.9         12         12
```

(continued)

Edited SAS Output for Example 14.6 (continued)

```
                    Extreme Observations (continued)

        Stem Leaf                        #              Boxplot
         28 0                            1                 |
         26 00                           2                 |
         24 00                           2                 |
         22 0000                         4              +-----+
         20 00                           2              |     |
         18 0                            1              |     |
         16 0000                         4              |     |
         14 00                           2              *--+--*
         12 000                          3              |     |
         10 0                            1              |     |
          8 00                           2              |     |
          6 0                            1              +-----+
          4 00000                        5                 |
          2 00                           2                 |
          0 0                            1                 |
            ....+....+....+....+
        Multiply Stem.Leaf by 10**-1
```

Edited SAS Output for Example 14.6 (PROC CORR Output)

```
                          The CORR Procedure
               Pearson Correlation Coefficients, N = 33

                        Y                  X1                  X2
    Y             1.00000             0.84142             0.28431
    X1            0.84142             1.00000             0.13969
    X2            0.28431             0.13969             1.00000
```

The scatterplot of Y versus X_1 (Figure 14.11) shows an obvious relationship, although perhaps a curvilinear one. The scatterplot of Y versus X_2 (Figure 14.12) shows no obvious relationship. The SAS PROC CORR output shows that the association between Y and X_1 has a strong linear component ($r_{YX_1} = 0.84$), although, as we have seen, the association may be better described as curvilinear. The correlation between Y and X_2 is weak ($r_{YX_2} = 0.28$). Ultimately, the regression model may need to be adjusted to account for a possible curvilinear pattern relating X_1 to Y. No adjustment of X_2 is necessary, however; X_2 may simply not be important in the final model. It is prudent to perform other checks for assumption violations before adjusting the model.

An outlier is circled in each figure. This is likely to be the same value identified in the PROC UNIVARIATE output for Y. No other outliers are immediately apparent. However, a more formal check using outlier diagnostics may reveal others.

Since a multiple regression is to be performed, partial regression plots are also inspected (Figures 14.13 and 14.14). Since there are no strong linear relationships between Y and X_2 or between X_1 and X_2 ($r_{X_1X_2} = 0.14$ from the PROC CORR output), the partial regression plot of Y versus X_1 adjusted for X_2 is very similar in appearance to the unadjusted plot in Figure 14.11. The partial regression plot of Y versus X_2 adjusted for X_1 shows a slightly more distinct (though still fairly weak) positive linear association than does the unadjusted scatterplot in Figure 14.12. Partial correlation calculations confirm these findings ($r_{YX_1|X_2} = 0.84$, $r_{YX_2|X_1} = 0.31$; SAS output not shown).

FIGURE 14.11   Plot of pollutant C concentration level (Y) versus pollutant A concentration level (X_1) for hypothetical environmental monitoring data

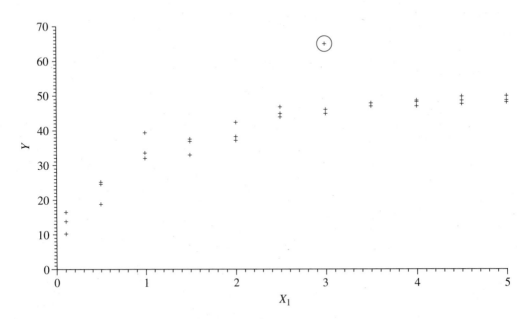

FIGURE 14.12   Plot of pollutant C concentration level (Y) versus pollutant B concentration level (X_2) for hypothetical environmental monitoring data

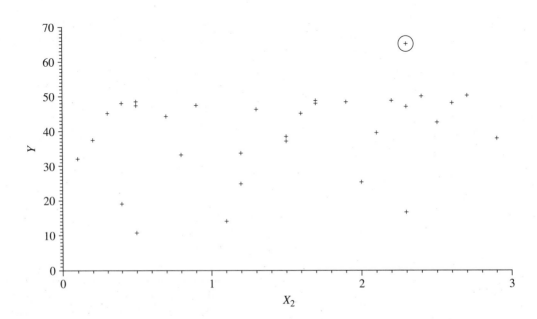

FIGURE 14.13   **Partial regression plot (adjusted for pollutant B) of pollutant C concentration level (Y) versus pollutant A concentration level (X_1) for hypothetical environmental monitoring data**

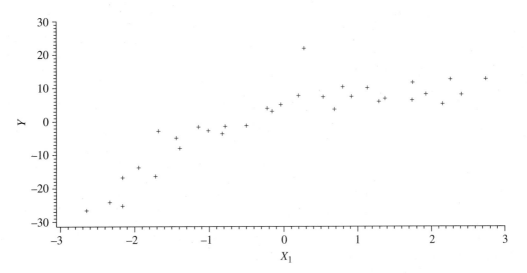

In the next stage of the diagnostics analysis, we will inspect outlier diagnostic statistics. The observations for which at least one of the diagnostic statistics exceeded the corresponding critical value (i.e., $d_i > 1$, $h_i > 2(2 + 1)/33 = 0.18$, and $|r_{(-i)}| > t_{33-2-2, 0.95} \approx 1.7$) are shown in Table 14.5; only the jackknife residual cutoff value is exceeded (highlighted in table). For each of these observations, we would first determine whether any data values were recorded in error. If they were, then every effort should be made to recover the correct value, and the analysis repeated. If not, we examine each variable value for each observation and proceed accordingly.

FIGURE 14.14   **Partial regression plot (adjusted for pollutant A) of pollutant C concentration level (Y) versus pollutant B concentration level (X_2) for hypothetical environmental monitoring data**

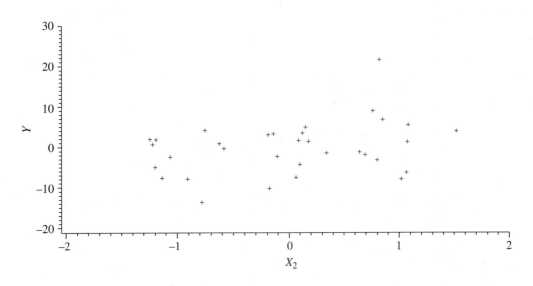

TABLE 14.5 **Outliers in Environmental Monitoring Example**

Observation	d_i	h_i	$r_{(-i)}$
1	0.186	0.130	2.030
3	0.172	0.150	1.769
21	0.230	0.064	3.848

Suppose we judge observations 1 and 3 to be plausible, since they are within the ranges identified prior to the study. Why then do they have large jackknife residual values and appear to be outliers? It is probably because they are at the end of the X_1 scale, where the curvilinear relationship between X_1 and Y departs to the greatest extent from the fitted regression function $\hat{Y} = \hat{\beta}_0 + \hat{\beta}_1 X_1 + \hat{\beta}_2 X_2$. Once the curvature issue is dealt with, it is likely that the residuals for these observations will not be overly large in magnitude.

From Table 14.4, we see that the value of Y for observation 21 is clearly beyond the plausible range for Y identified prior to the study. If the correct value of Y cannot be recovered for this observation, then the value of Y may be set to missing (effectively excluding observation 21 from the analysis). This is the course of action we recommend.

Next we examine the validity of regression assumptions. Figure 14.15 presents a plot of jackknife residuals versus predicted values. Instead of the desired random scatter of points, a systematic pattern is apparent. Examination of Figures 14.16 and 14.17 reveals that the same pattern is apparent in the plot of jackknife residuals versus X_1 (but not in the analogous plot involving X_2). A modification of the model should be attempted in order to eliminate the systematic

FIGURE 14.15 **SAS plot of jackknife residuals versus predicted values for regression of pollutant C concentration level (Y) on pollutant A concentration level (X_1) and pollutant B concentration level (X_2)**

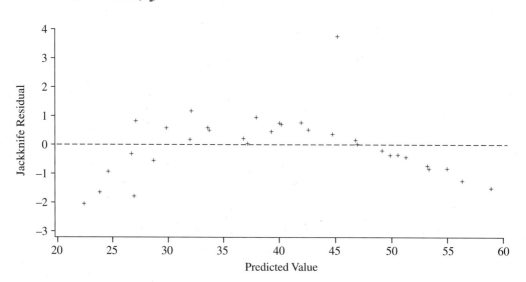

FIGURE 14.16   **SAS plot of jackknife residuals versus pollutant A concentration level (X_1) for regression of pollutant C concentration level (Y) on pollutant A concentration level (X_1) and pollutant B concentration level (X_2)**

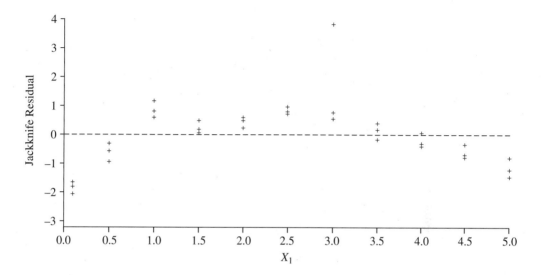

FIGURE 14.17   **SAS plot of jackknife residuals versus pollutant B concentration level (X_2) for regression of pollutant C concentration level (Y) on pollutant A concentration level (X_1) and pollutant B concentration level (X_2)**

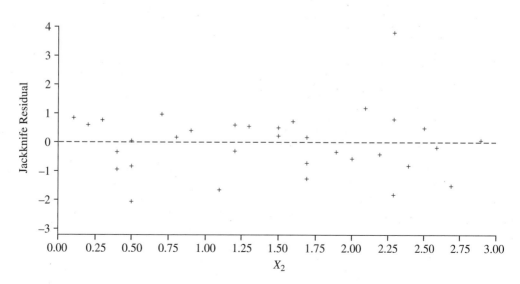

FIGURE 14.18   **SAS normal quantile-quantile plot of jackknife residuals from the regression of pollutant C concentration level (Y) on pollutant A concentration level (X_1) and pollutant B concentration level (X_2)**

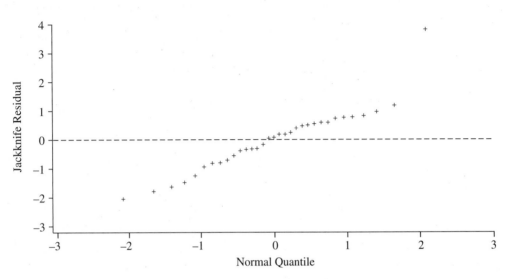

pattern in the jackknife residuals reflected in Figures 14.15 and 14.16. Based on these residual plots, there does not appear to be any gross violation of the homoscedasticity assumption.

The normal quantile-quantile plot of jackknife residuals shown in Figure 14.18 exhibits a fairly linear pattern, indicating that the normality assumption is not badly violated.

Finally we conduct a check for collinearity problems. The correlation between X_1 and X_2 shown in the SAS PROC CORR output was not high, indicating that there is no moderate or strong relationship between the two predictors and, therefore, probably no collinearity problem. The variance inflation factors confirm this; they are small (both equal to 1.02; SAS output not shown). Hence, there is no real need to inspect any more diagnostic statistics such as condition indices and variance proportions.

In summary, in this example, it appears that there are two issues that need to be addressed: the effect of the outlier (observation 21) and the non-linear relationship between Y and X_1. A reasonable strategy would be to eliminate observation 21 from the data set and then to fit the model $Y = \beta_0 + \beta_1 X_1 + \beta_2 X_2 + \beta_3 X_1^2 + E$ to the revised data set. The reader is encouraged to carry out this analysis, including appropriate regression diagnostic methods. ∎

14.8 An Important Caution

All of the techniques described in this chapter involve checking the validity of the assumptions required for a valid multiple linear regression analysis. To that end, outlier analysis and collinearity diagnosis may suggest deleting observations and/or predictor variables in order to improve the quality of the model. Naturally, the observations and variables that are most at odds with the observed data patterns and fitted models are the ones to be considered for deletion. All of these techniques thus lead to possible underestimation of true variability and may lead to *P*-values that are optimistically small. The methods discussed in this chapter are safest to use with large samples and are least reliable with small samples. Unfortunately, of course, the potential influence of a single observation is largest in small samples. This is a strong argument against using small samples in regression analysis.

In some cases, a problem can be resolved only by evaluating a second sample of data. Picard and Cook (1984) discuss the issue of optimism in selecting a regression model and recommend considering split samples. In a split-sample design, some of the data are used for exploratory data analysis, and the rest of the data are used for confirming the validity and reliability of the exploratory results. This technique is described in Chapter 16 in the context of selecting the best regression model.

Problems

The reader may use the provided computer output to answer the questions posed below; alternatively, the reader may use his or her own statistical software to create the necessary output. For some of the questions, be aware that the data sets have hidden problems: outliers, collinearities, and variables in need of transformation. In response to certain requests, some computer programs may either balk or produce invalid results. This state of affairs is realistic. Other problems use "nice" data that, while not realistic, do nevertheless allow us to focus on the particular statistical concepts discussed in this chapter.

1. Consider the data of Problem 1, Chapter 5. For the model with dry weight as the response and age as the predictor,
  a. Examine a plot of the studentized or jackknife residuals versus the predicted values. Are any regression assumption violations apparent? If so, suggest possible remedies.
  b. Examine numerical descriptive statistics, histograms, box-and-whisker plots, and normal probability plots of jackknife residuals. Is the normality assumption violated? If so, suggest possible remedies.
  c. Examine outlier diagnostics, including Cook's distance, leverage statistics, and jackknife residuals, and identify any potential outliers. What course of action, if any, should be taken when outliers are identified?

2. a.–c. Repeat Problem 1 using $\log_{10}$(dry weight) as the response.

3.–14. a.–c. Repeat parts (a) through (c) of Problem 1 using the data of Problems 3 through 14 of Chapter 5.

15. a.–c. Consider the data of Problem 15, Chapter 5. For the model with BLOODTOL as the response and PPM_TOLU as the predictor, repeat (a) through (c) of Problem 1.

16. a.–c. Repeat Problem 15 using LN_BLOODTOL as the response and LN_PPMTL as the predictor.
  d. Which approach—using the original variables or using the logged variables—leads to fewer diagnostic problems?

17. a.–c. Repeat Problem 15 using BLOODTOL as the response and BRAINTOL as the predictor.

18. a.–d. Repeat Problem 16 using LN_BLDTL as the response and LN_BRNTL as the predictor.

19. The data in the following table come from an article by Bethel et al. (1985). All subjects are asthmatics. For the model with FEV_1 as the response and HEIGHT, WEIGHT, and AGE as the predictors,

Subject	AGE (yr)	Sex	Height (cm)	Weight (kg)	FEV$_1^*$ (L)	Subject	AGE (yr)	Sex	Height (cm)	Weight (kg)	FEV$_1^*$ (L)
1	24	M	175	78.0	4.7	11	26	M	180	70.5	3.5
2	36	M	172	67.6	4.3	12	29	M	163	75.0	3.2
3	28	F	171	98.0	3.5	13	33	F	180	68.0	2.6
4	25	M	166	65.5	4.0	14	31	M	180	65.0	2.0
5	26	F	166	65.0	3.2	15	30	M	180	70.4	4.0
6	22	M	176	65.5	4.7	16	22	M	168	63.0	3.9
7	27	M	185	85.5	4.3	17	27	M	168	91.2	3.0
8	27	M	171	76.3	4.7	18	46	M	178	67.0	4.5
9	36	M	185	79.0	5.2	19	36	M	173	62.0	2.4
10	24	M	182	88.2	4.2						

*Forced expiratory volume in 1 second.

a.–c. Repeat (a) through (c) in Problem 1.

d. Examine variance inflation factors, condition indices (unadjusted and adjusted for the intercept), and variance proportions. Are there any important collinearity problems? If so, suggest possible remedies.

20. Repeat Problem 19(d), this time including SEX as a predictor (coded SEX = 1 if female, SEX = 0 if male).

21. Repeat Problem 20, this time including three interactions: SEX and AGE, SEX and HEIGHT, SEX and WEIGHT.

22. In an analysis of daily soil evaporation (EVAP), Freund (1979) identified the following predictor variables:

MAXAT = Maximum daily air temperature

MINAT = Minimum daily air temperature

AVAT = Integrated area under the daily air temperature curve (i.e., a measure of average air temperature)

MAXST = Maximum daily soil temperature

MINST = Minimum daily soil temperature

AVST = Integrated area under the soil temperature curve

MAXH = Maximum daily relative humidity

MINH = Minimum daily relative humidity

AVH = Integrated area under the daily humidity curve

WIND = Total wind, measured in miles per day

In addition, Freund provided the following overall ANOVA table and significance tests about the regression coefficients.

Source	d.f.	SS	MS	F
Regression	10	8159.35	815.94	19.27
Residual	35	1482.76	42.36	
Total	45	9642.11		

Variable	$\hat{\beta}$	t	VIF
MAXAT	0.5011	0.88	8.828
MINAT	0.3041	0.39	8.887
AVAT	0.09219	0.42	22.21
MAXST	2.232	2.22	39.29
MINST	0.2049	0.19	14.08
AVST	0.7426	−2.12	52.36
MAXH	1.110	0.98	1.981
MINH	0.7514	1.54	25.38
AVH	−0.5563	−3.44	24.12
WIND	0.00892	0.97	1.985

a. Compute R^2, and provide a test of significance. Use $\alpha = .01$.
b. Compute R_j^2 for each predictor.
c. Which, if any, variables are implicated as possibly inducing collinearity?
d. Freund (1979) noted that "some of the coefficients have 'wrong' signs." Using your knowledge of evaporation, explain which coefficients have wrong signs, paying particular attention to those with extreme t-values.

23. The raw data for Problem 22, from Freund (1979) appear below. For the model of Problem 22,
   a. Fit the model.
   b. What discrepancies do you note between the results of (a) and the data summary presented in Problems 22?
   c. Examine the correlation matrix for all predictor variables in this problem. Are any collinearity problems apparent?
   d. Determine the variance inflation factors, condition indices (unadjusted and adjusted for the intercept), and variance proportions. Are any collinearity problems apparent? If so, suggest a remedy.
   e. Examine outlier diagnostics. Are any outliers apparent? If so, suggest a course of action.
   f. Determine eigenvalues, condition indexes, and condition numbers for the correlation matrix (excluding the intercept).
   g. Determine eigenvalues, condition indexes, and condition numbers for the scaled cross-products matrix (including the intercept).
   h. Determine residuals (preferably studentized) and leverage values. Do any observations seem bothersome? Explain.
   i. Does there appear to be any problem with collinearity? Explain.

Observation	Month	Day	MAXST	MINST	AVST	MAXAT	MINAT	AVAT	MAXH	MINH	AVH	WIND	EVAP
1	6	6	84	65	147	85	59	151	95	40	398	273	30
2	6	7	84	65	149	86	61	159	94	28	345	140	34
3	6	8	79	66	142	83	64	152	94	41	388	318	33
4	6	9	81	67	147	83	65	158	94	50	406	282	26
5	6	10	84	68	167	88	69	180	93	46	379	311	41
6	6	11	74	66	131	77	67	147	96	73	478	446	4
7	6	12	73	66	131	78	69	159	96	72	462	294	5
8	6	13	75	67	134	84	68	159	95	70	464	313	20
9	6	14	84	68	161	89	71	195	95	63	430	455	31
10	6	15	86	72	169	91	76	206	93	56	406	604	38
11	6	16	88	73	178	91	76	208	94	55	393	610	43
12	6	17	90	74	187	94	76	211	94	51	385	520	47
13	6	18	88	72	171	94	75	211	96	54	405	663	45
14	6	19	88	72	171	92	70	201	95	51	392	467	45
15	6	20	81	69	154	87	68	167	95	61	448	184	11
16	6	21	79	68	149	83	68	162	95	59	436	177	10
17	6	22	84	69	160	87	66	173	95	42	392	173	30
18	6	23	84	70	160	87	68	177	94	44	392	76	29
19	6	24	84	70	168	88	70	169	95	48	398	72	23
20	6	25	77	67	147	83	66	170	97	60	431	183	16
21	6	26	87	67	166	92	67	196	96	44	379	76	37
22	6	27	89	69	171	92	72	199	94	48	393	230	50
23	6	28	89	72	180	94	72	204	95	48	394	193	36
24	6	29	93	72	186	92	73	201	94	47	386	400	54
25	6	30	93	74	188	93	72	206	95	47	385	339	44
26	7	1	94	75	199	94	72	208	96	45	370	172	41
27	7	2	93	74	193	95	73	214	95	50	396	238	45
28	7	3	93	74	196	95	70	210	96	45	380	118	42
29	7	4	96	75	198	95	71	207	93	40	365	93	50
30	7	5	95	76	202	95	69	202	93	39	357	269	48
31	7	6	84	73	173	96	69	173	94	58	418	128	17
32	7	7	91	71	170	91	69	168	94	44	420	423	20
33	7	8	88	72	179	89	70	189	93	50	399	415	15
34	7	9	89	72	179	95	71	210	98	46	389	300	42
35	7	10	91	72	182	96	73	208	95	43	384	193	44
36	7	11	92	74	196	97	75	215	96	46	389	195	41
37	7	12	94	75	192	96	69	198	95	36	380	215	49
38	7	13	96	75	195	95	67	196	97	24	354	185	53
39	7	14	93	76	198	94	75	211	93	43	364	466	53
40	7	15	88	74	188	92	73	198	95	52	405	399	21
41	7	16	88	74	178	90	74	197	95	61	447	232	1
42	7	17	91	72	175	94	70	205	94	42	380	275	44
43	7	18	92	72	190	95	71	209	96	44	379	166	44
44	7	19	92	73	189	96	72	208	93	42	372	189	46
45	7	20	94	75	194	95	71	208	93	43	373	164	47
46	7	21	96	76	202	96	71	208	94	40	368	139	50

24. Real estate prices depend, in part, on property size. The house size X (in hundreds of square feet) and house price Y (in $1,000s) of a random sample of houses in a certain county were observed. The data (which first appeared in Problem 16 of Chapter 5) are as follows:

X	18	20	25	22	33	19	17
Y	80	95	104	110	175	85	89

Edited SAS Output for Problem 24

Plot of JACKKNIFE by X. Legend: A = 1 obs, B = 2 obs, etc.

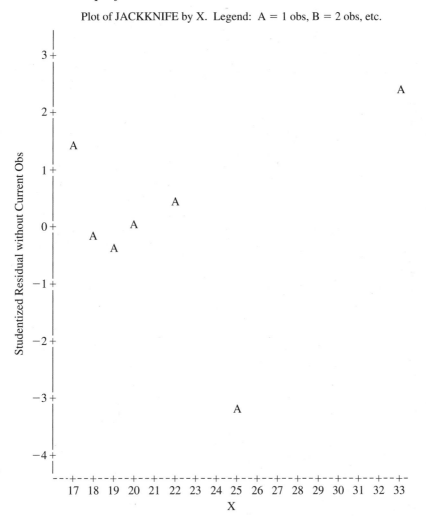

Univariate Procedure

.

. [Portion of output omitted]

.

Variable-JACKKNIFE			Studentized Residual without Current Obs	
Stem	Leaf		#	Boxplot
2	4		1	1
1	4		1	+------------------+
0	14		2	*----------+--------*
-0	33		2	+------------------+
-1				
-2				
-3	1		1	0
	------+------+------+------+			

a. Fit a straight line model, with house price Y as the response and house size X as the predictor.

b. Provide a plot of studentized or jackknife residuals versus the predictor.

c. Provide a frequency histogram and a schematic plot of the residuals.

d. Report and interpret a test of whether any residuals have extreme values.

e. Do these analyses highlight any potentially troublesome observations? Why or why not?

25. Data on sales revenues (Y) and advertising expenditures (X) for a large retailer for the period 1988–1993 are given in the following table.

Year	1988	1989	1990	1991	1992	1993
Sales Y ($millions)	4	8	2	8	5	4
Advertising X ($millions)	2	5	0	6	4	3

These data first appeared in Problem 17 of Chapter 5.

a.–e. Repeat parts (a) through (e) of Problem 24 for the current data set, using sales revenues Y as the response and advertising expenditures X as the predictor.

26. The production manager of a plant that manufactures syringes has recorded the marginal cost Y at various levels of output X for 14 randomly selected months. The data are shown in the following table.

Marginal Cost Y (per 100 units)	Output X (thousands of units)	Marginal Cost Y (per 100 units)	Output X (thousands of units)
31.00	3.0	27.00	3.5
30.00	3.0	25.00	5.5
28.00	3.5	24.00	7.0
46.00	1.0	25.00	7.0
43.00	1.5	29.50	4.5
35.00	2.0	26.00	4.5
37.50	2.0	28.00	4.5

These data first appeared in Problem 18 of Chapter 5.

Edited SAS Output for Problem 25

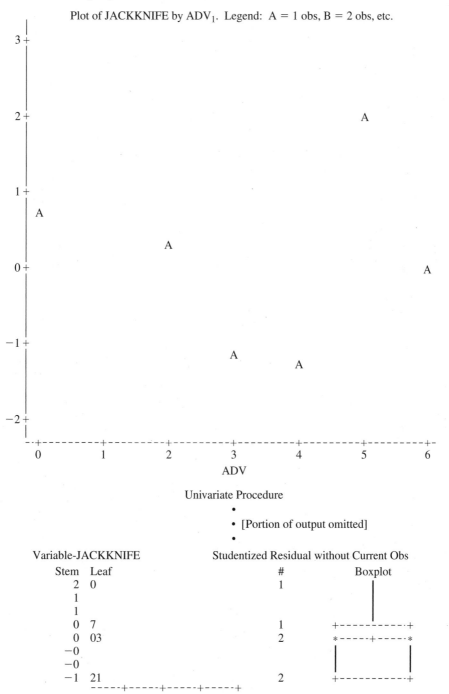

Plot of JACKKNIFE by ADV_1. Legend: A = 1 obs, B = 2 obs, etc.

Univariate Procedure

• [Portion of output omitted]

Variable-JACKKNIFE		Studentized Residual without Current Obs	
Stem	Leaf	#	Boxplot
2	0	1	
1			
1			
0	7	1	+----------+
0	03	2	*-----+-----*
-0			
-0			
-1	21	2	+----------+

a.–e. Repeat parts (a) through (e) of Problem 24 for the current data set, using marginal cost *Y* as the response and output *X* as the predictor.

Edited SAS Output for Problem 26

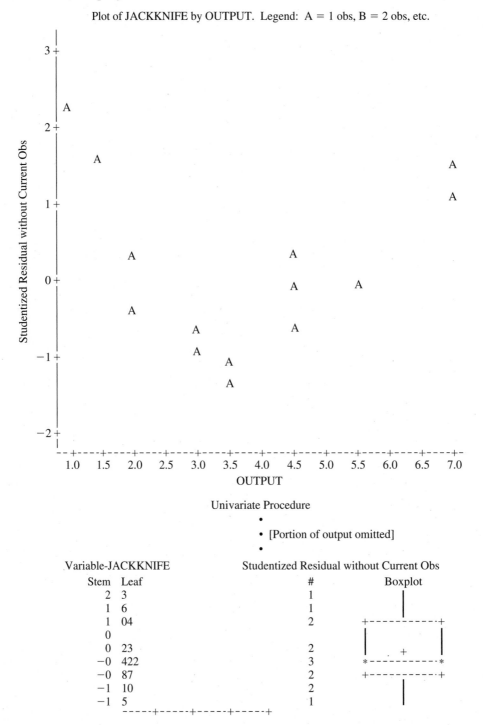

Plot of JACKKNIFE by OUTPUT. Legend: A = 1 obs, B = 2 obs, etc.

Univariate Procedure

•

• [Portion of output omitted]

•

Variable-JACKKNIFE

Studentized Residual without Current Obs

Stem	Leaf	#	Boxplot
2	3	1	
1	6	1	
1	04	2	+----------+
0			
0	23	2	
−0	422	3	*----------*
−0	87	2	+----------+
−1	10	2	
−1	5	1	

Parameter Estimates (Continued)

Obs	Dep Var OWNEROCC	Predict Value	Standard Err Predict	Residual	Standard Err Residual	Student Residual
4	73.0000	66.1494	0.777	6.8506	3.795	1.805
5	61.0000	64.2598	1.166	-3.2598	3.694	-0.882
6	67.0000	66.6296	0.968	0.3704	3.751	0.099
7	64.0000	64.6950	1.003	-0.6950	3.742	-0.186
8	67.0000	68.8092	1.095	-1.8092	3.716	-0.487
9	57.0000	56.8102	3.019	0.1898	2.427	0.078
10	63.0000	64.5294	1.362	-1.5294	3.627	-0.422
11	56.0000	66.5931	0.839	-10.5931	3.782	-2.801
12	71.0000	68.0833	0.952	2.9167	3.755	0.777
13	67.0000	64.5928	0.870	2.4072	3.775	0.638
14	65.0000	64.1798	0.913	0.8202	3.765	0.218
15	75.0000	74.1348	2.317	0.8652	3.104	0.279
16	66.0000	68.4097	1.135	-2.4097	3.704	-0.651
17	64.0000	63.9992	0.949	0.000769	3.756	0.000
18	63.0000	63.7493	1.096	-0.7493	3.716	-0.202
19	70.0000	71.3159	1.627	-1.3159	3.516	-0.374
20	70.0000	66.7189	0.806	3.2811	3.789	0.866
21	69.0000	63.7457	1.039	5.2543	3.732	1.408
22	65.0000	68.8408	1.349	-3.8408	3.632	-1.058
23	60.0000	63.7172	1.614	-3.7172	3.522	-1.056
24	60.0000	62.9960	1.050	-2.9960	3.729	-0.803
25	68.0000	65.7428	1.330	2.2572	3.639	0.620
26	68.0000	66.9432	1.493	1.0568	3.575	0.296

Obs	-2-1-0	1 2	Cook's D	Rstudent	Hat Diag H	Cov Ratio
1	*		0.022	-0.8492	0.0816	1.1294
2		*	0.023	0.8935	0.0787	1.1145
3		***	0.046	1.7925	0.0454	0.7952
4		***	0.046	1.9054	0.0402	0.7529
5	*		0.026	-0.8780	0.0906	1.1332
6			0.000	0.0966	0.0624	1.2171
7			0.001	-0.1818	0.0670	1.2192
8			0.007	-0.4786	0.0799	1.2039
9			0.003	0.0765	0.6075	2.9087
10			0.008	-0.4140	0.1235	1.2737
11	*****		0.129	-3.3746	0.0469	0.3430
12		*	0.013	0.7698	0.0604	1.1229
13		*	0.007	0.6292	0.0505	1.1407
14			0.001	0.2133	0.0556	1.2024
15			0.014	0.2730	0.3578	1.7614
16	*		0.013	-0.6422	0.0858	1.1821
17			0.000	0.0002	0.0600	1.2156
18			0.001	-0.1974	0.0800	1.2355
19			0.010	-0.3672	0.1764	1.3622
20		*	0.011	0.8610	0.0433	1.0813
21		**	0.051	1.4404	0.0720	0.9396
22	**		0.051	-1.0604	0.1212	1.1196
23	**		0.078	-1.0583	0.1736	1.1914
24	*		0.017	-0.7971	0.0735	1.1323
25		*	0.017	0.6119	0.1178	1.2314
26			0.005	0.2897	0.1485	1.3267

.
. [Portion of output omitted]
.

Sum of Residuals	0
Sum of Squared Residuals	345.1830
Predicted Resid SS (Press)	396.8002

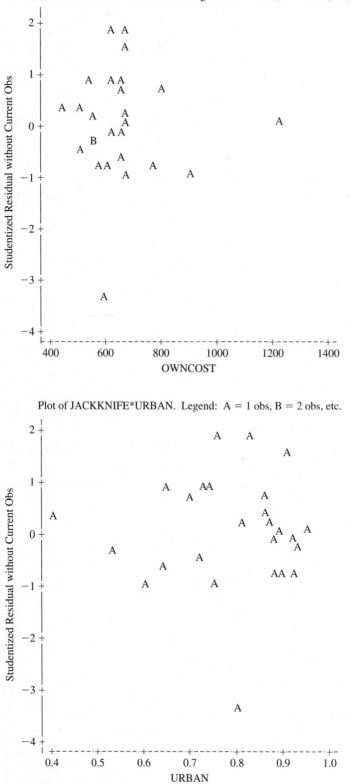

Plot of JACKKNIFE*OWNCOST. Legend: A = 1 obs, B = 2 obs, etc.

Plot of JACKKNIFE*URBAN. Legend: A = 1 obs, B = 2 obs, etc.

References

The following list includes sources not cited in Chapter 12 that nonetheless offer helpful discussions of topics covered here.

Agresti, A. 1990. *Categorical Data Analysis*. New York: John Wiley & Sons.

Anscombe, F. J. 1960. "Rejection of Outliers." *Technometrics* 2: 123–47.

Anscombe, F. J., and Tukey, J. W. 1963. "The Examination and Analysis of Residuals." *Technometrics* 5: 141–60.

Armitage, P. 1971. *Statistical Methods in Medical Research*. Oxford: Blackwell Scientific.

Barnett, V., and Lewis, T. 1978. *Outliers in Statistical Data*. New York: John Wiley & Sons.

Bartlett, M. S. 1947. "The Use of Transformations." *Biometrics* 3: 39–52.

Belsley, D. A. 1984. "Demeaning Conditioning Diagnostics Through Centering." *The American Statistician* 38: 73–77.

———.1991. *Conditioning Diagnostics, Collinearity and Weak Data in Regression*. New York: John Wiley & Sons.

Belsley, D. A.; Kuh, E.; and Welsch, R. E. 1980. *Regression Diagnostics: Identifying Influential Data and Sources of Collinearity*. New York: John Wiley & Sons.

Bethel, R. A.; Sheppard, D.; Geffroy, B.; Tam, E.; Nadel, J. A.; and Boushey, J. A. 1985. "Effect of 0.25 ppm Sulfur Dioxide on Airway Resistance in Freely Breathing, Heavily Exercising, Asthmatic Subjects." *American Review of Respiratory Diseases* 131: 659–61.

Box, G. E. P., and Cox, D. R. 1964. "An Analysis of Transformations." *Journal of the Royal Statistical Society* B26: 211–43 (with discussion, 244–52).

———.1984. "An Analysis of Transformations Revisited, Rebuttal." *Journal of the American Statistical Association* 17: 209–10.

Carroll, R. J., and Ruppert, D. 1984. "Power Transformations When Fitting Theoretical Models to Data." *Journal of the American Statistical Association* 79: 321–28.

Conover, W. J., and Iman, R. L. 1981. "Rank Transformations as a Bridge Between Parametric and Nonparametric Statistics." *American Statistician* 35: 124–28.

Cook, R. D., and Weisberg, S. 1982. *Residuals and Influence in Regression*. New York: Chapman & Hall.

Draper, N. R., and Smith, H. 1981. *Applied Regression Analysis*. New York: John Wiley & Sons.

Durbin, J., and Watson, G. S. 1951. "Testing for Serial Correlation in Least Squares Regression." *Biometrika* 37: 409–28.

Freund, R. J. 1979. "Multicollinearity etc., Some 'New' Examples." *Proceedings of the Statistical Computing Section, American Statistical Association*, pp. 111–12.

Gallant, A. R. 1975. "Nonlinear Regression." *American Statistician* 29: 73–81.

Gunst, R. F., and Mason, R. L. 1980. *Regression Analysis and Its Application*. New York: Marcel Dekker.

Hackney, O. J., and Hocking, R. R. 1979. "Diagnostic Techniques for Identifying Data Problems in Multiple Linear Regression." *Proceedings of the Statistical Computing Section, American Statistical Association*, pp. 94–98.

Hinkley, D. V., and Runger, G. 1984. "The Analysis of Transformed Data." *Journal of the American Statistical Association*, 79: 302–20.

Hoaglin, D. C., and Welsch, R. E. 1978. "The Hat Matrix in Regression and ANOVA." *American Statistician* 32: 17–22.

Hocking, R. R. 1983. "Developments in Linear Regression Methodology: 1959–1982." *Technometrics* 25: 219–48.

Huber, P. J. 1981. *Robust Statistics*. New York: John Wiley & Sons.

Jensen, D. R., and Ramirez, D. E. 1996. " Computing the CDF of Cook's D_I Statistic." In A. Prat and E. Ripoll, eds., *Proceedings of the 12th Symposium in Computational Statistics*, pp. 65–66. Barcelona, Spain: Institut d'Estadística de Catalunya.

———.1998. "Some Exact Properties of Cook's D_I." In C. R. Rao and N. Balakrishnan, eds., *Handbook of Statistics-16: Order Statistics and Their Applications*. Amsterdam: North Holland.

Lewis, T., and Taylor, L. R. 1967. *Introduction to Experimental Ecology*. New York: Academic Press.

McCabe, G. P. 1984. "Principal Variables." *Technometrics* 26: 137–44.

Morrison, D. F. 1976. *Multivariate Statistical Methods*. New York: McGraw-Hill.

Mosteller, F., and Tukey, J. W. 1977. *Data Analysis and Regression*. Reading, Mass.: Addison-Wesley.

Muller, K. E., and Chen Mok, M. 1997. "The Distribution of Cook's Statistic." *Communications in Statistics: Theory and Methods* 26(3) pp. 525–546.

Neter, J.; Wasserman, W.; and Kutner, M. H. 1983. *Applied Linear Regression Models*. Homewood, Ill.: Richard D. Irwin.

Obenchain, R. L. 1977. Letter to the Editor. *Technometrics* 19: 348–49.

Picard, R. R., and Cook, R. D. 1984. " Cross-validation of Regression Models." *Journal of the American Statistical Association* 79: 575–83.

Rubin, D. B. 1987. *Multiple Imputation for Nonresponse in Surveys*. New York: John Wiley & Sons, Inc.

Schafer, J. L. 1997. *Analysis of Incomplete Multivariate Data*. New York: Chapman and Hall.

Shapiro, S. S., and Wilks, M. B. 1965. "An Analysis of Variance Test for Normality (Complete Samples)." *Biometrika* 52: 591–611.

Siegel, S. 1956. *Nonparametric Statistics for the Behavioral Sciences*. New York: McGraw-Hill.

Smith, G., and Campbell, F. 1980. "A Critique of Some Ridge Regression Methods." *Journal of the American Statistical Association* 75: 74–81.

Stephens, M. A. 1974. "EDF Statistics for Goodness of Fit and Some Comparisons." *Journal of the American Statistical Association* 69: 730–37.

Stevens, J. P. 1984. "Outliers and Influential Data Points in Regression Analysis." *Psychology Bulletin* 95: 334–44.

Timm, N. H. 1975. *Multivariate Analysis with Applications in Education and Psychology*. Belmont, Calif.: Wadsworth.

Tukey, J. W. 1977. *Exploratory Data Analysis*. Reading, Mass.: Addison-Wesley.

Weisberg, S. 1980. *Applied Linear Regressions*. New York: John Wiley & Sons.

15

Polynomial Regression

15.1 Preview

In this chapter, we focus on a special case of the multiple regression model, the *polynomial model,* which is often of interest when only *one basic* independent variable—say, X—is to be considered. We initially considered a straight-line model (Chapter 5) for this situation; however, we may want to determine whether prediction can be improved significantly by increasing the complexity of the fitted straight-line model. The simplest extension of the straight-line model is the second-order polynomial, or *parabola,* which involves a second term, X^2, in addition to X. Adding high-order terms like X^2 and X^3, which are simple functions of a single basic variable, can be considered equivalent to adding new independent variables. Thus, if we rename X as X_1, and X^2 as X_2, the second-order model

$$Y = \beta_0 + \beta_1 X + \beta_2 X^2 + E$$

becomes

$$Y = \beta_0 + \beta_1 X_1 + \beta_2 X_2 + E$$

In general, polynomial models are special cases of the general multiple regression model. Since only one basic independent variable is being considered, however, any polynomial model can be represented by a curvilinear plot on a two-dimensional graph (rather than as a surface in higher-dimensional space). As mentioned in Chapter 5, when only one basic independent variable X is being considered, the fundamental goal is to find the curve that best fits the data so that the relationship between X and Y is appropriately described. Because a higher-order curve may be more appropriate than a straight line, a discussion of how to fit and evaluate such (polynomial) curves is important.

We first consider methods for fitting and evaluating the second-order (parabolic) model, after which we consider higher-order polynomial models. Because these models are special cases of the general multiple regression model, the procedure for fitting these models and the methods for inference are essentially the same as those described more generally in Chapters 8 and 9. Since the independent variables in a polynomial model are functions of the same basic variable (X), they are inherently correlated. This, in turn, can lead to computational difficulties due to collinearity (Chapter 14). Fortunately, techniques such as centering and the use of orthogonal polynomials are available to help remedy such problems; these procedures are discussed later in this chapter. We shall also see that the use of orthogonal polynomials helps simplify hypothesis testing.

15.2 Polynomial Models

The most general kind of curve usually considered for describing the relationship between a single independent variable X and a response Y is called a *polynomial*. Mathematically, a polynomial of order k in x is an expression of the form

$$y = c_0 + c_1 x + c_2 x^2 + \cdots + c_k x^k$$

in which the c's and k (which must be a nonnegative integer) are constants. We have already considered the simple polynomial corresponding to $k = 1$ (namely, the straight line having the form $y = c_0 + c_1 x$). The second-order polynomial corresponding to $k = 2$ (namely, the parabola) has the general form $y = c_0 + c_1 x + c_2 x^2$.

In going from a *mathematical* model to a *statistical* model, as we did in the straight-line case, we may write a parabolic model in either of the following forms:

$$\mu_{Y|X} = \beta_0 + \beta_1 X + \beta_2 X^2 \tag{15.1}$$

or

$$Y = \beta_0 + \beta_1 X + \beta_2 X^2 + E \tag{15.2}$$

In these equations, capital Y's and X's denote statistical variables; β_0, β_1, and β_2 denote the unknown parameters called *regression coefficients*; $\mu_{Y|X}$ denotes the mean of Y at a given X; and E denotes the error component, which represents the difference between the observed response Y at X and the true average response $\mu_{Y|X}$ at X.

If we tentatively assume that a parabolic model—as given by either (15.1) or (15.2)—is appropriate for describing the relationship between X and Y, we must then determine a specific estimated parabola that best fits the data. As in the straight-line case, this best-fitting parabola may be determined by employing the least-squares method.

15.3 Least-squares Procedure for Fitting a Parabola

The least-squares estimates of the parameters β_0, β_1, and β_2 in a parabolic model are chosen so as to minimize the sum of squares of deviations of observed points from corresponding points on the fitted parabola (Figure 15.1). Letting $\hat{\beta}_0$, $\hat{\beta}_1$, and $\hat{\beta}_2$ denote the least-squares estimates of

FIGURE 15.1   **Deviations of observed points from the least-squares parabola**

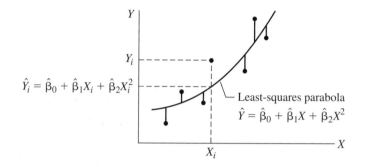

the unknown regression coefficients in the parabolic model (15.1), and letting $\hat{Y}$ denote the value of the predicted response at X, we can write the estimated parabola as

$$\hat{Y} = \hat{\beta}_0 + \hat{\beta}_1 X + \hat{\beta}_2 X^2 \qquad (15.3)$$

The minimum sum of squares obtained by using this least-squares parabola is

$$\text{SSE} = \sum_{i=1}^{n}(Y_i - \hat{Y}_i)^2 = \sum_{i=1}^{n}(Y_i - \hat{\beta}_0 - \hat{\beta}_1 X_i - \hat{\beta}_2 X_i^2)^2 \qquad (15.4)$$

As with the general regression model, we do not find it necessary to present the precise formulas for calculating the least-squares estimates $\hat{\beta}_0$, $\hat{\beta}_1$, and $\hat{\beta}_2$. These formulas are quite complex and become even more so for polynomials of orders higher than two. The researcher is not likely to employ such polynomial regression methods without using a packaged computer program, which can perform the necessary calculations and print the numerical results. (Appendix B contains a discussion of matrices and their relationship to regression analysis; by using matrix mathematics, we can compactly represent the general regression model and the associated least-squares methodology.)

■ **Example 15.1**   For the age–systolic blood pressure data of Table 5.1, with the outlier removed,[1] the least-squares estimates for the parabolic regression coefficients are computed to be

$$\hat{\beta}_0 = 113.41, \qquad \hat{\beta}_1 = 0.088, \qquad \hat{\beta}_2 = 0.010$$

The fitted model given by (15.3) then becomes

$$\hat{Y} = 113.41 + 0.088X + 0.010X^2 \qquad (15.5)$$

This equation can be compared with the straight-line equation obtained in Section 5.5 for these data with the outlier removed—namely,

$$\hat{Y} = 97.08 + 0.95X \qquad (15.6)$$

---

[1] As described in Section 5.5.3, the outlier corresponds to the data point ($X = 47$, $Y = 220$), for the second individual listed in Table 5.1.

A comparison between (15.5) and (15.6) reveals that the estimates of β_0 and β_1 differ in the two models, indicating that the estimation of β_2 affects the estimation of β_0 and β_1 in the quadratic model. ∎

15.4 ANOVA Table for Second-order Polynomial Regression

As in the straight-line case, the essential results gathered from a second- or higher-order polynomial model can be summarized in an ANOVA table. The ANOVA table for a parabolic fit to the age–systolic blood pressure data of Table 5.1 (with the outlier removed) is given in Table 15.1.

The contents of Table 15.1 deserve comment. First, only variables-added-in-order tests are described. Natural variable orderings suggest themselves, either from the largest to the smallest power of the predictor or vice versa. Consequently, a variables-added-last test for each term should be avoided with polynomial models. Using variables-added-in-order tests aids in choosing the most parsimonious yet relevant model possible. Such tests should also utilize the residual mean square from the largest model considered; this notion will be discussed more fully in section 15.10.

TABLE 15.1   **ANOVA table for a parabola fit to the age–systolic blood pressure data of Table 5.1, with the outlier removed**

Source	d.f.	SS	MS	F
Regression $\begin{cases} X \\ X^2\vert X \end{cases}$	1	6,110.10	6,110.10	68.89
	1	163.30	163.30	1.84
Residual	26	2,306.05	88.69	
Total (corrected)[*]	28	8,579.45		

Note: The residual from the largest model was used for all tests.

[*]$R^2 = .731$

15.5 Inferences Associated with Second-order Polynomial Regression

Three basic inferential questions are associated with second-order polynomial regression:

1. Is the overall regression significant? That is, is more of the variation in Y explained by the second-order model than by ignoring X completely (and just using $\overline{Y}$)?

2. Does the second-order model provide significantly more predictive power than the straight-line model does?

3. Given that a second-order model is more appropriate than a straight-line model, should we add higher-order terms (X^3, X^4, etc.) to the second-order model?

15.5.1 Test for Overall Regression and Strength of the Overall Parabolic Relationship

Determining whether the overall regression is significant involves testing the null hypothesis H_0: "There is no significant overall regression using X and X^2" (i.e., $\beta_1 = \beta_2 = 0$). The testing procedure used for this null hypothesis involves the overall F test described in Chapter 9—namely, computing

$$F = \frac{\text{Regression MS}}{\text{Residual MS}}$$

and then comparing the value of this F statistic with an appropriate critical point of the F distribution, which (in our example) has 2 and 26 degrees of freedom in the numerator and denominator, respectively. For $\alpha = .001$, we find that $F = 35.37 > F_{2,26,0.999} = 9.12$, so we reject the null hypothesis of nonsignificant overall regression ($P < .001$).

To obtain a quantitative measure of how well the second-order model predicts the dependent variable, we can use the squared multiple correlation coefficient (the multiple R^2). As with r^2 in straight-line regression, R^2 represents the proportionate reduction in the error sum of squares obtained by using X and X^2 instead of the naive predictor $\overline{Y}$. The formula for calculating R^2 is given by

$$R^2(\text{second-order model}) = \frac{\text{SSY} - \text{SSE (second-order model)}}{\text{SSY}} \qquad (15.7)$$

For this example, $R^2 = .731$. The preceding F test (with $P < .001$) tells us that this R^2 is significantly different from 0. (It is possible, although not likely, that the overall F test for the second-order model will not lead to the rejection of H_0 even if the t test—or the equivalent F test—for significant regression of the straight-line model leads to rejection. This possibility arises because the loss of 1 degree of freedom in SSE in going from the straight-line model to the second-order model may result in a smaller computed F, coupled with an altered critical point of the F distribution. In our example, the computed F is reduced from 66.81 (68.89 for the variables-added-in-order test in Table 15.1) for the straight-line model to 35.37 for the second-order model, and the critical point of the F distribution for $\alpha = .001$ is reduced from $F_{1,27,0.999} = 13.6$ to $F_{2,26,0.999} = 9.12$.

15.5.2 Test for the Addition of the X^2 Term to the Model

To answer the second question, about increased predictive power, we must perform a partial F test of the null hypothesis H_0: "The addition of the X^2 term to the straight-line model does not significantly improve the prediction of Y over and above that achieved by the straight-line model itself" (i.e., $\beta_2 = 0$). To test this null hypothesis, we compute the partial F statistic

$$F(X^2|X) = \frac{(\text{Extra SS due to adding } X^2)/1}{\text{Residual MS (second-order model)}} \qquad (15.8)$$

and then compare this F value to an appropriate F percentage point (which is an $F_{1,26}$ value in our example). Since X^2 is the last variable added, this is a variables-added-last test. Alternatively, we could divide the estimated coefficient $\hat{\beta}_2$ by its estimated standard error to form a statistic that has a t distribution under H_0 with 26 degrees of freedom.

The ANOVA information needed to compute the F test for our example is given in Table 15.1. The extra sum of squares for $X^2|X$ is 163.30 and is computed as the difference between the sum-of-squares regression values for the first- and second-order models. The partial F statistic (15.8) is then computed to be

$$F = \frac{163.30}{88.69} = 1.84$$

Since $F_{1,26,0.90} = 2.91$, we would not reject H_0 at the $\alpha = .10$ level. Furthermore, the P-value for this test satisfies the inequality $.10 < P < .25$. Thus, for this example, we conclude that adding a quadratic term to the straight-line model does not significantly improve prediction. Corroboration of the results of this partial F test is provided by a scatter diagram of the data (Figure 5.8 in Chapter 5), which offers no evidence of a parabolic relationship between X and Y. Finally, the conclusion is also supported by the small increase in R^2 when the X^2 term is added to the straight-line model. Since $r^2 = R^2 = .712$ for the straight-line model, and $R^2 = .731$ for the parabolic model, the increase in R^2 is $.731 - .712 = .019$.

15.5.3 Testing for Adequacy of the Second-order Model

The preceding analysis of the Table 5.1 data has shown that a straight-line model fits the data adequately, is significantly predictive of the response, and is preferable to a parabolic model. Consequently, it would be superfluous in this case to evaluate whether a model of an order higher than two would be significantly better than a straight-line model. Nevertheless, in general, any question of model adequacy can be addressed (with a *lack-of-fit* test) for any model (of any order) that is being considered at a given stage of an analysis. Any such lack-of-fit test can be characterized by a partial or multiple-partial F test for the addition of one or more terms to the model under study. More detailed discussion of lack-of-fit tests is provided in the subsequent sections.

15.6 Example Requiring a Second-order Model

We now turn to another hypothetical example to illustrate the methods of polynomial regression. This example will lead us to a different conclusion regarding the appropriateness of a second-order model.

Suppose that a laboratory study is undertaken to determine the relationship between the dosage (X) of a certain drug and weight gain (Y). Eight laboratory animals of the same sex, age, and size are selected and randomly assigned to one of eight dosage levels.[2]

The gain in weight (in dekagrams) is measured for each animal after a two-week time period during which all animals were subject to the same dietary regimen and general laboratory

---

[2] The study design here can certainly be criticized for not having more than one animal receive a particular dosage, as well as for involving such a small total sample size. Replication at each dosage would provide a reliable estimate of animal-to-animal variation in the data. However, for some laboratory studies, sufficient numbers of animals are not easily obtainable; cost and time are often limiting factors, as well. Finally, the data for this example have been contrived to simplify the analysis and to present a relationship that is clearly second-order in nature.

conditions. The data are given in Table 15.2, and a scatter diagram for these data in Figure 15.2. By simply eyeballing this diagram, one can see that a parabolic curve is a more appropriate model than a straight line. We shall now proceed to quantify this visual impression.

The complete ANOVA table, based on fitting a parabola to the data in Table 15.2, is given in Table 15.3. The equation for the least-squares parabola on which the ANOVA table is based has the form

$$Y = 1.35 - 0.41X + 0.17X^2$$

TABLE 15.2   **Weight gain after two weeks as a function of dosage level**

Dosage level (X)	1	2	3	4	5	6	7	8
Weight gain (Y) (dag)	1	1.2	1.8	2.5	3.6	4.7	6.6	9.1

FIGURE 15.2   **Scatter diagram of hypothetical data for the animal weight gain study**

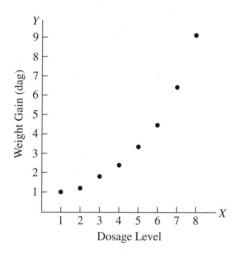

TABLE 15.3   **Regression ANOVA table for the quadratic model fit to the weight gain data**

Source		d.f.	SS	MS	F
Regression	X	1	52.04	52.04	61.95
	$X^2\|X$	1	4.83	4.83	120.75
Residual		5	0.20	0.04	
Total (corrected)[*]		7	57.07		

[*]$R^2 = .997$.

Let us investigate the information contained in this ANOVA table. First, by combining the regression $X^2|X$ sum of squares with the residual sum of squares, we can test whether there is a significant straight-line regression effect before we add the X^2 term to the model; in particular, the ANOVA table for straight-line regression derived from Table 15.3 is given in Table 15.4. The least-squares line is $\hat{Y} = -1.20 + 1.11X$. The null hypothesis of no significant linear regression is clearly rejected, since an F of 62.053 exceeds $F_{1,6,0.999} = 35.51$ ($P < .001$). Our next step is to examine the complete ANOVA table and decide whether adding the X^2 term significantly improves the prediction of Y over and above what is obtained from a simple straight-line model. In doing so, we are asking whether the increase in R^2 of .085 (.997 − .912), obtained by including the X^2 term in the model, significantly improves the fit. The appropriate test statistic to use in answering this question is the partial F statistic

$$F = \frac{(\text{Extra sum of squares due to adding } X^2)/1}{\text{Residual MS (second-order model)}}$$

$$= \frac{4.83}{0.04} = 120.75$$

which exceeds $F_{1,5,0.999} = 47.18$ ($P < .001$). Therefore, adding the X^2 term to the model significantly improves prediction. As might be expected, a test for overall significant regression of the second-order model yields a highly significant F—namely,

$$F = \frac{\text{Regression MS (second-order model)}}{\text{Residual MS (second-order model)}}$$

$$= \frac{(52.04 + 4.83)/2}{0.04} = 710.88$$

Up to this point, we have concluded that a first-order (straight-line) model is not as good as a second-order model. We now need to determine whether adding higher-order terms to the second-order model is warranted. For example, we can add an X^3 term (with associated regression coefficient β_3) to the second-order model and then test whether prediction is significantly improved. Fitting this third-order model by least squares results in the ANOVA table given in Table 15.5. To test whether adding the third-order term significantly improves the fit, we calculate the following statistic:

$$F = \frac{(\text{Extra sum of squares due to adding } X^3)/1}{\text{Residual MS (third-order model)}}$$

$$= \frac{0.14}{0.014} = 10.00$$

This F statistic has an F distribution with 1 and 4 degrees of freedom under H_0: "The addition of the X^3 term is not worthwhile" (i.e., $\beta_3 = 0$). Since $F_{1,4,0.95} = 7.71$ and $F_{1,4,0.975} = 12.22$, we have

TABLE 15.4   Regression ANOVA table for the linear model fit to the weight gain data

Source	d.f.	SS	MS	F
Regression (X)	1	52.04	52.04	62.05
Residual	6	5.03	0.84	
Total (corrected)[*]	7	57.07		

[*]$R^2 = .9118$.

TABLE 15.5 **Regression ANOVA table for the third-order model fit to the weight gain data**

Source	d.f.	SS	MS	F
X	1	52.037	52.04	
Regression $X^2|X$	1	4.835	4.84	
$X^3|X, X^2$	1	0.141	0.14	10.00[a]
Residual	4	0.056	0.014	
Total (corrected)[b]	7	57.066		

[a] $.025 < P < .05$
[b] $R^2 = .999$

$.025 < P < .05$. This P-value would thus reject H_0 for $\alpha = .05$ but not for $\alpha = .025$. This makes the decision of whether to include the X^3 term in the model somewhat difficult. However, several other factors should be taken into consideration: (1) the R^2-value for the parabolic fit is very high (namely, .997); (2) the R^2-value only increases from .997 to .999 in going from a second-order model to a third-order model; (3) the scatter diagram clearly suggests a second-order curve; (4) when in doubt, the simpler model is preferable because it is easier to interpret. All things considered, then, it is most sensible to conclude that the second-order model is most appropriate.

In summary, for the data in Table 15.2, the best-fitting model is

$$\hat{Y} = 1.35 - 0.41X + 0.17X^2$$

with an R^2 of .997.

Finally, it is also valuable to have the standard deviations (or standard errors) of the estimated regression coefficients. These are difficult to compute by hand for models involving two or more predictors. However, all commonly used computer regression programs print the numerical values of the estimated coefficients and their estimated standard errors. For the second-order model fit to the data in Table 15.2, we obtain $S_{\hat{\beta}_1} = 0.141$ and $S_{\hat{\beta}_2} = 0.015$. For example, then, a $100(1 - \alpha)\%$ confidence interval for β_2 would be

$$\hat{\beta}_2 \pm t_{5, 1-\alpha/2}S_{\hat{\beta}_2}$$

where the degrees of freedom for the appropriate critical t value are the degrees of freedom associated with the residual sum of squares in Table 15.3. In particular, a 95% confidence interval for β_2 in our example is

$$0.17 \pm (2.571)(0.015)$$

or $(0.13, 0.21)$. This interval does not include 0, which agrees with our ANOVA table conclusion concerning the importance of the X^2 term in the quadratic model.

15.7 Fitting and Testing Higher-order Models

So far we have seen how the basic ideas of multiple regression may be applied to fitting and testing quadratic and cubic polynomial models. These same methods generalize to all higher-order polynomial models. Nevertheless, several related issues need to be discussed: the use of orthogonal polynomials, and strategies for choosing a polynomial model.

How large an order of polynomial model to consider depends on the problem being studied and the amount and type of data being collected. For studies in the biological and social sciences, one important consideration is whether the regression relationship can be described by a monotonic function (i.e., one that is always increasing or decreasing). If only monotonic functions are of interest, a second- or third-order model usually suffices (although monotonicity is not guaranteed, since, for example, some parabolas increase and then decrease). A large number of well-placed predictor values and a small error variance are needed to obtain reliable fits for models of higher order than cubic.

A more general consideration is the number of *bends* (more technically, *relative extrema*) in the polynomial curve that one wishes to fit. For example, a first-order model has no bends; a second-order model has no more than one bend, and each higher-order term adds another potential bend. In practice, fitting polynomial models of orders higher than three usually leads to models that are neither always decreasing nor always increasing. Substantial theoretical and/or empirical evidence should exist to support the employment of such complicated nonmonotonic models.

The quantity of data directly limits the maximum order of a polynomial that may be fit. Consider the weight gain data (Table 15.2). Given those eight distinct values, a polynomial of order seven would fit the eight points perfectly, giving an SSE value of 0 and an R^2-value of 1. (Nevertheless, because the fitted equation would have eight estimated parameters, no gain in parsimony is made over simply listing the eight data points.) *Generally, the maximum-order polynomial that may be fit is one less than the number of distinct X-values.* For example, consider the age–systolic blood pressure data in Table 5.1 (with the outlier removed). Of the 29 observations, 5 are replicates, which implies that 24 distinct X-values exist. Hence, a polynomial of order 23 could be fit to these data, although it would be absurd to consider fitting such a model.

15.8 Lack-of-fit Tests

Given that a polynomial model has been fitted and the estimated regression coefficients have been tested for their significance, how can one be confident that a model of order higher than the highest order tested is probably not needed? A lack-of-fit (LOF) test can be used to address this question. Conceptually, a lack-of-fit test evaluates a model more complex than the one under primary consideration.

The classical lack-of-fit test can be applied only if there are replicate observations. The term *replicate* means that an experimental unit (subject) has the same X-value as another experimental unit. With n total observations, if d X-values are distinct, the number of replicates is $r = n - d$. Recall that a polynomial curve of order $d - 1$ can pass through exactly d distinct points. A classical lack-of-fit test compares the fit of a polynomial of order $d - 1$ with the fit of the polynomial model currently under consideration.

For the age–systolic blood pressure example (with the outlier removed), 5 X-values out of the total of 29 involve replicates (i.e., $r = 5$). These are listed in Table 15.6. (Notice that these replicate data in Table 15.6 call into question the validity of the variance homogeneity assumption.) In a classical lack-of-fit test of these data, we consider a polynomial of order $d - 1 = n - r - 1 = 29 - 5 - 1 = 23$, which is the higher order of the two polynomial models being compared. The lower-order polynomial is the model of primary interest, such as the second-order model fit to the age–systolic blood pressure data. For our example, the two models would be

$$Y = \beta_0 + \beta_1 X + \beta_2 X^2 + E$$

TABLE 15.6   Replicates and pure-error estimates from repeated observations for the age–systolic blood pressure data of Table 5.1

X	Y	SS*	d.f.
39	144, 120	288.0	1
42	124, 128	8.0	1
45	138, 135	4.5	1
56	154, 150	8.0	1
67	170, 158	72.0	1
		380.5 (SS$_{PE}$)	5

*The sum of squares for a given X is calculated by using the formula $\sum_{m}(Y_{mx} - \bar{Y}_x)^2$, where Y_{mx} is the mth observation of Y at $X = x$ and $\bar{Y}_x$ is the mean of all replicates at $X = x$.

and

$$Y = \beta_0 + \beta_1 X + \beta_2 X^2 + \beta_3 X^3 + \cdots + \beta_{23} X^{23} + E$$

Table 15.7 contains the ANOVA information needed to compare these two models. The F test for such a comparison is a multiple partial F test of the null hypothesis $H_0: \beta_j = 0$, $j = 3, 4, \ldots, 23$. (Trying to fit the larger 23rd-degree polynomial directly would lead to serious collinearity problems; a possible alternative method of fitting, described in Section 15.9, involves the use of orthogonal polynomials.)

The ANOVA framework for the classical lack-of-fit test (see Table 15.7) partitions the residual sum of squares (SSE) for the model whose fit is being questioned into two components:

TABLE 15.7   Regression ANOVA table for the classical lack-of-fit test of the second-order polynomial fit to the age–systolic blood pressure data of Table 5.1

Source		d.f.
X		1
$X^2 \vert X$		1
Residual $\begin{cases} \text{Lack of fit } (X^3, X^4, \ldots, X^{23} \vert X, X^2) \\ \text{Pure error} \end{cases}$		21
		5
Total		28

Note: The lack-of-fit test statistic is given by

$$F = F(X^3, X^4, \ldots, X^{23} \vert X, X^2)$$

$$= \frac{[\text{Regression SS}(X, X^2, \ldots, X^{23}) - \text{Regression SS}(X, X^2)]/21}{\text{Residual SS}(X, X^2, \ldots, X^{23})/5}$$

$$= \frac{MS_{LOF}}{MS_{PE}}$$

Under H_0: "No lack of fit of second-order model," this F statistic should have an F distribution with 21 and 5 degrees of freedom.

a pure-error sum of squares SS_{PE} (with degrees of freedom df_{PE}), and a lack-of-fit sum of squares SS_{LOF}. The test statistic, which is equivalent to the multiple partial F test just described, can be written as $F = MS_{LOF}/MS_{PE}$. In actuality, SS_{PE} is the error sum of squares for the higher-degree polynomial model being compared, of order $d - 1 = n - r - 1$. The quantity SS_{LOF} is the extra sum of squares due to the addition to the lower-order model of all higher-order terms needed to construct the higher-order model. In the example under consideration,

$$SS_{LOF} = \text{Regression } SS(X, X^2, \ldots, X^{23}) - \text{Regression } SS(X, X^2)$$

and

$$SS_{PE} = \text{Residual } SS(X, X^2, \ldots, X^{23})$$

With the availability of standard computer regression packages, using a multiple partial F test for lack of fit can be less computationally cumbersome than identifying replicate observations in order to compute SS_{PE} directly. Orthogonal polynomials *must* be used to avoid serious inaccuracy in the multiple regression computations leading to SS_{LOF}.

15.9 Orthogonal Polynomials

In Chapter 14, we illustrated collinearity problems that can arise in work with polynomial models, and we demonstrated that centering the predictor helped remedy such problems for a second-order polynomial model. A more sophisticated approach is needed for higher-order models. The polynomials we have discussed so far have all been *natural polynomials*. This terminology derives from the fact that each of the independent variables (X, X^2, X^3, etc.) in a polynomial of the form

$$Y = \beta_0 + \beta_1 X + \beta_2 X^2 + \cdots + \beta_j X^j + E$$

is a simple polynomial by itself. Another method of fitting polynomial models involves *orthogonal polynomials*, which constitute a new set of independent variables that are defined in terms of the simple polynomials but have more complicated structures. In this section we explain the method and describe its advantages and disadvantages. The basic motivation for using orthogonal polynomials is to avoid the serious collinearity inherent in using natural polynomials.

In Section 15.2, we alluded to the natural polynomial model

$$Y = \beta_0 + \beta_1 X + \beta_2 X^2 + \cdots + \beta_k X^k + E$$

The simple polynomials $X, X^2, \ldots, X^k$ are the predictors. Orthogonal polynomial variables are new predictor variables that consist of linear combinations of these simple polynomials. Denoting orthogonal polynomials as $X_1^*, X_2^*, \ldots, X_k^*$, we can write them as linear combinations of the form

$$X_1^* = a_{01} + a_{11}X$$
$$X_2^* = a_{02} + a_{12}X + a_{22}X^2$$
$$\vdots$$
$$X_k^* = a_{0k} + a_{1k}X + a_{2k}X^2 + \cdots + a_{kk}X^k$$

where the a's are constants that relate the X^*'s to the original predictors. Each linear combination has the form of a polynomial. It can be shown that each simple polynomial can be written as a linear combination of the X^*'s as follows:

$$X = b_{01} + b_{11}X_1^*$$
$$X^2 = b_{02} + b_{12}X_1^* + b_{22}X_2^*$$
$$\vdots$$
$$X^k = b_{0k} + b_{1k}X_1^* + b_{2k}X_2^* + \cdots + b_{kk}X_k^*$$

where the b's are constants. With no loss of information, we can write either

$$Y = \beta_0 + \beta_1 X + \beta_2 X^2 + \cdots + \beta_k X^k + E$$

or

$$Y = \beta_0^* + \beta_1^* X_1^* + \beta_2^* X_2^* + \cdots + \beta_k^* X_k^* + E$$

The parameters $\{\beta_j^*\}$ in the latter model differ numerically from those in the former. Moreover, whereas the simple polynomials are highly correlated with one another, the orthogonal polynomials (via a judicious choice of the a's) are pairwise uncorrelated.[3] However, although the parameters and predictors in the two models have very different interpretations it can be shown that the multiple R^2-values and the overall regression F tests obtained by fitting these two models are exactly the same.

In summary, the orthogonal polynomial values represent a recoding of the original predictors with two desirable basic properties: *the orthogonal polynomial variables contain exactly the same information as the simple polynomial variables,* and *the orthogonal polynomial variables are uncorrelated with each other.* The first property means that all questions about the natural polynomial model can be answered by using the orthogonal polynomial model. For example, one can assess the overall strength of the regression relationship, conduct the overall regression F test, and even compute partial F tests while using the orthogonal polynomial model. The second property, zero pairwise correlation, completely eliminates any collinearity.

With regard to partial F tests, it can be shown that the partial F test of H_0: $\beta_j^* = 0$ for the orthogonal polynomial model of order k is equivalent to the partial F test of H_0: $\beta_j = 0$ in the reduced natural polynomial model

$$Y = \beta_0 + \beta_1 X + \beta_2 X^2 + \cdots + \beta_j X^j + E$$

for any $j \leq k$. Thus, only one orthogonal polynomial model has to be fit (of the highest order— say, k—of interest) in order to test sequentially for the significance of each simple polynomial term in the original natural polynomial model. This may be summarized by saying that the k tests of H_0: $\beta_j^* = 0$, for $j = 1, 2, \ldots, k$, using the orthogonal polynomial model are equivalent to the k variables-added-in-order tests of H_0: $\beta_j = 0$, for $j = 1, 2, \ldots, k$, using an appropriate sequence of k natural polynomial models.

As an illustration, consider a third-order natural polynomial model

$$Y = \beta_0 + \beta_1 X + \beta_2 X^2 + \beta_3 X^3 + E$$

Suppose that it is of interest to determine a best model by proceeding backward, starting with a test of H_0: $\beta_3 = 0$ in the preceding cubic model. If this test is nonsignificant, one would proceed to a test of H_0: $\beta_2 = 0$ in a (reduced) parabolic model; in turn, if H_0: $\beta_2 = 0$ is not rejected, one

---

[3] $r_{X_j^*, X_{j'}^*} = 0$ for all $j \neq j'$.

would perform a test of H_0: $\beta_1 = 0$ in a (reduced) straight-line model. One way to conduct these three tests would be to fit three separate natural polynomial models (cubic, parabolic, and straight-line models) and then to perform partial F (or t) tests—starting with the cubic model—to assess the significance of the highest-order term in each model, stopping at any point if significance is obtained. Unfortunately, collinearity will generally compromise reliable fitting of the cubic and (without centering) quadratic models. To avoid this problem, one can fit a *single* third-order orthogonal polynomial model, and then test sequentially H_0: $\beta_3^* = 0$, H_0: $\beta_2^* = 0$, and finally H_0: $\beta_1^* = 0$, stopping at any point if significance is obtained. This approach generally ensures computational accuracy.

15.9.1 Transforming from Natural to Orthogonal Polynomials

In this section we describe a method for expressing orthogonal polynomials in terms of the original simple polynomials. This procedure involves specifying the a's in the earlier equations relating the X^*'s to powers of X. To illustrate the method, assume that the scientist who collected the data in Table 15.8 decided to try to confirm the conclusions made in the first study by collecting new data. The same eight dosage values, from 1 to 8, were used. Three new observations

TABLE 15.8   **Observed weight gains and simple polynomial values**

Observation	WGTGAIN (Y)	DOSE1 (X)	DOSE2 (X^2)	DOSE3 (X^3)	DOSE4 (X^4)	DOSE5 (X^5)	DOSE6 (X^6)	DOSE7 (X^7)
1	0.9	1	1	1	1	1	1	1
2	1.1	2	4	8	16	32	64	128
3	1.6	3	9	27	81	243	729	2187
4	2.3	4	16	64	256	1024	4096	16384
5	3.5	5	25	125	625	3125	15625	78125
6	5.0	6	36	216	1296	7776	46656	279936
7	6.6	7	49	343	2401	16807	117649	823543
8	8.7	8	64	512	4096	32768	262144	2097152
9	0.9	1	1	1	1	1	1	1
10	1.1	2	4	8	16	32	64	128
11	1.6	3	9	27	81	243	729	2187
12	2.1	4	16	64	256	1024	4096	16384
13	3.4	5	25	125	625	3125	15625	78125
14	4.5	6	36	216	1296	7776	46656	279936
15	6.7	7	49	343	2401	16807	117649	823543
16	8.6	8	64	512	4096	32768	262144	2097152
17	0.8	1	1	1	1	1	1	1
18	1.2	2	4	8	16	32	64	128
19	1.4	3	9	27	81	243	729	2187
20	2.2	4	16	64	256	1024	4096	16384
21	3.2	5	25	125	625	3125	15625	78125
22	4.8	6	36	216	1296	7776	46656	279936
23	6.7	7	49	343	2401	16807	117649	823543
24	8.8	8	64	512	4096	32768	262144	2097152

Note: The linear predictor variable X is called DOSE1. The quadratic predictor variable X^2 is called DOSE2, etc.; in general, the variable X^j is called DOSEj.

TABLE 15.9   Observed weight gains and orthogonal polynomial values

Observation	WGTGAIN	ODOSE1	ODOSE2	ODOSE3	ODOSE4	ODOSE5	ODOSE6	ODOSE7
1	0.9	−7	7	−7	7	−7	1	−1
2	1.1	−5	1	5	−13	23	−5	7
3	1.6	−3	−3	7	−3	−17	9	−21
4	2.3	−1	−5	3	9	−15	−5	35
5	3.5	1	−5	−3	9	15	−5	−35
6	5.0	3	−3	−7	−3	17	9	21
7	6.6	5	1	−5	−13	−23	−5	−7
8	8.7	7	7	7	7	7	1	1
9	0.9	−7	7	−7	7	−7	1	−1
10	1.1	−5	1	5	−13	23	−5	7
11	1.6	−3	−3	7	−3	−17	9	−21
12	2.1	−1	−5	3	9	−15	−5	35
13	3.4	1	−5	−3	9	15	−5	−35
14	4.5	3	−3	−7	−3	17	9	21
15	6.7	5	1	−5	−13	−23	−5	−7
16	8.6	7	7	7	7	7	1	1
17	0.8	−7	7	−7	7	−7	1	−1
18	1.2	−5	1	5	−13	23	−5	7
19	1.4	−3	−3	7	−3	−17	9	−21
20	2.2	−1	−5	3	9	−15	−5	35
21	3.2	1	−5	−3	9	15	−5	−35
22	4.8	3	−3	−7	−3	17	9	21
23	6.7	5	1	−5	−13	−23	−5	−7
24	8.8	7	7	7	7	7	1	1

Note: The linear orthogonal polynomial predictor variable is called ODOSE1, the quadratic orthogonal polynomial variable is called ODOSE2, etc.; in general, the *j*th-order variable is called ODOSE*j*.

were taken at each dosage, giving a total of 24 observations. Since eight distinct *X*-values are available, the highest-order polynomial that can be fitted is of the seventh order. To demonstrate the use of orthogonal polynomials and an associated lack-of-fit test, let us consider fitting third- and seventh-order polynomials to these data.

Table 15.8 provides the data from the new study and also gives the values of the $X^2, X^3, \ldots, X^7$ terms (which are labeled DOSE2, DOSE3, ..., DOSE7 in the table). These are the natural polynomial values. Table 15.9 includes the same response (*Y*) values, but the predictor (*X*) values have been replaced by orthogonal polynomial values (labeled ODOSE1, ODOSE2, ..., ODOSE7). These values were obtained by using Table A.7 in Appendix A as follows.

Table A.7 can be used only if the predictor (X) values are equally spaced and if the same number of observations (i.e., replicates) occur at each value. If either of these conditions is not satisfied, Table A.7 cannot be used. In that case, a reasonable alternative is to use a computer program (such as the ORPOL function in SAS PROC IML) to calculate the appropriate orthogonal polynomial values.

In describing the transition from Table 15.8 to Table 15.9, we begin by noting that the variable "dosage" has eight distinct, equally spaced values and that three replicates are taken at each of these values. Hence, Table A.7 can be used. In Table A.7, k indicates the number of distinct values of X—in our case, 8. The row in Table A.7 labeled "Linear" for $k = 8$ consists of the entries $(-7, -5, -3, -1, 1, 3, 5, 7)$. These eight entries will be used to replace the eight original values for the linear term X. Thus, a subject who received a dosage of 1 (as did the first subject) is given a linear orthogonal polynomial score of -7; a subject who received a dosage of 2 is given a linear orthogonal polynomial score of -5; and so on, with a dosage of 8 corresponding to a score of 7. This pattern is repeated in Table 15.9 twice more for the replicate observations. Analogous operations yield the remaining columns appearing in Table 15.9. For example, the eight quadratic entries for $k = 8$ $(7, 1, -3, -5, -5, -3, 1, 7)$ give the orthogonal polynomial values for the variable ODOSE2 in Table 15.9, which replace the original eight values for the quadratic term X^2 (given in the corresponding column of Table 15.8).

Finally, the rightmost column of Table A.7 gives the sum of squared values for each row in the table. For $k = 8$, dividing each linear orthogonal polynomial score by $\sqrt{168}$, each quadratic score by $\sqrt{168}$, each cubic score by $\sqrt{264}$, and so on, makes the variance of each set of orthogonal polynomial scores (i.e., each set of row entries) equal to 1. This helps in two ways. First, numerical accuracy is improved by the avoidance of scaling problems. Second, the estimated standard errors of the resulting estimated regression coefficients are all equal, which simplifies the task of comparing and interpreting such regression coefficients.

TABLE 15.10 **Third-order natural polynomial model for the weight gain data** $(n = 24)$

Source	d.f.	SS	MS
Model[a]	3	169.7122	56.5707
Error	20	0.3274	0.0164
Total (corrected)[b]	23	170.0396	

Source	d.f.	Variables-Added-in-Order SS	F
DOSE1 (X)	1	155.6667	9508.98 $(P < .0001)$
DOSE2 $(X^2\|X)$	1	14.0334	857.23 $(P < .0001)$
DOSE3 $(X^3\|X, X^2)$	1	0.0121	0.74 $(P = .3995)$

Source	d.f.	Variables-Added-Last SS	F
DOSE1 $(X\|X^2, X^3)$	1	0.0395	2.41 $(P = 0.1359)$
DOSE2 $(X^2\|X, X^3)$	1	0.1662	10.15 $(P = 0.0046)$
DOSE3 $(X^3\|X, X^2)$	1	0.0121	0.74 $(P = 0.3995)$

Parameter	Estimate	$S_{\hat{\beta}_j}$	T for H_0: Parameter = 0
Intercept	1.0261	0.182	5.64 $(P = .0001)$
DOSE1	−0.2559	0.165	−1.55 $(P = .1359)$
DOSE2	0.1316	0.041	3.19 $(P = .0046)$
DOSE3	0.0026	0.003	0.86 $(P = .3995)$

[a] $F = 3{,}455.65$ $(P < .0001)$.

[b] $R^2 = .998075$.

15.9.2   Regression Analysis with Orthogonal Polynomials

Let us now consider the regression ANOVA tables based on fitting third- and seventh-order polynomial models to the weight gain data in Tables 15.8 and 15.9. Table 15.10 summarizes the results for the third-order natural polynomial model. Clearly, the model fits extremely well ($R^2 = .998$). Furthermore, the variable-added-in-order tests give P-values of .0001, .0001, and .3995 for linear, quadratic, and cubic tests, respectively. These results argue persuasively for a second-order model. Since the simple polynomials X, X^2, and X^3 are correlated, the variable-added-last tests are more difficult to interpret. Nevertheless, the cubic test again gives $P = .3995$, since it is the last variable to enter; the linear effect is not significant ($P = .1359$) but the quadratic effect is ($P = .0046$). When using variable-added-last tests, analysts should proceed backward, starting with the cubic model and including in the final model all lower-order terms below the highest term deemed significant. Then, the series of variable-added-last tests also supports the choice of the second-order model.

The preference for a second-order model is further supported by comparing R^2-values for the second- and third-order models. It can be shown, using the variable-added-in-order SS entries in Table 15.10, that the R^2-value for the second-order model is

$$R^2 = \frac{155.6667 + 14.0334}{170.040} = .998$$

which is the same (within round-off error) as the R^2-value for the third-order model.

The P-values for the t tests are the same as the P-values for the variable-added-last F tests (.1359, .0046, and .3995, respectively, for linear, quadratic, and cubic); this is because these t tests are equivalent to the variable-added-last F tests (indeed, $T_{20}^2 = F_{1,20}$). All such tests use the residual mean square from the largest model, which has 20 degrees of freedom.

For comparison, consider the results of fitting the third-order orthogonal polynomial model, as summarized in Table 15.11. The most important difference between Tables 15.10 and 15.11 is that the two types of sum of squares (namely, variable-added-in-order and variable-added-last) are equal in the latter. Consequently, the corresponding partial F test results are the same and coincide with the t test results. Furthermore, the standard errors of the estimated linear, quadratic, and cubic regression coefficients are identical. As a result, the estimated regression coefficients can be compared directly.

The preceding results, whether from analysis using natural or orthogonal polynomials, indicate that we need not consider any polynomial model higher than second order. Nevertheless, it is instructive to use these data to illustrate particular problems that arise if the model order does exceed two. Specifically, we address the (hidden) problem of collinearity. The existence of this collinearity problem can be demonstrated by considering the predictor correlations given in Table 15.12 for both the simple polynomials and their centered counterparts—that is, for $(X - \overline{X})$, $(X - \overline{X})^2$, . . . , as well as for X, X^2. . . . We first focus on the correlations among the linear, quadratic, and cubic terms in a third-order model; this information is contained in the upper left corner of each array (above and to the left of the dashed lines). The existence of a collinearity problem is suggested by the presence of three very high correlations (all above .93) for the noncentered predictor data. Centering X helps reduce collinearity: Two of the three correlations become 0 after centering. Turning to the complete array, we again see, for the noncentered polynomials, that the smallest off-diagonal correlation is .779 and that four are greater than .990. As before, centering X helps reduce collinearity substantially, since correlations between odd and even powers are then all 0. Nevertheless, the remaining correlations are high, with two correlations greater than .99. The analogous correlation array using orthogonal polynomials would have all zeros off the diagonal.

TABLE 15.11   Third-order orthogonal polynomial model for the weight gain data ($n = 24$)

Source	d.f.	SS	MS
Model[a]	3	169.7122	56.5707
Error	20	0.3274	0.0164
Total (corrected)[b]	23	170.0396	

Source	d.f.	Variables-Added-in-Order SS	F
ODOSE1	1	155.6667	9508.98 ($P < .0001$)
ODOSE2	1	14.0334	857.23 ($P < .0001$)
ODOSE3	1	0.0121	0.74 ($P = .3995$)

Source	d.f.	Variables-Added-Last SS	F
ODOSE1	1	155.6667	9508.98 ($P < .0001$)
ODOSE2	1	14.0334	857.23 ($P < .0001$)
ODOSE3	1	0.0121	0.74 ($P = .3995$)

Parameter	Estimate	$S_{\hat{\beta}_j^*}$	T for H_0: Parameter $= 0$
Intercept	3.6542	0.0261	139.91 ($P < .0001$)
ODOSE1	0.5558	0.0057	97.51 ($P < .0001$)
ODOSE2	0.1669	0.0057	29.27 ($P < .0001$)
ODOSE3	0.0039	0.0045	0.86 ($P = .3995$)

[a]$F = 3,455.65$ ($P < .0001$).

[b]$R^2 = .998075$.

In Table 15.13, the collinearity problem suggested by Table 15.12 can be examined further. Predictor correlation matrix eigenvalues (see Chapter 14) are reported here for third- and seventh-order natural, centered, and orthogonal polynomial models. For the third-order natural polynomial model, the condition number (CN; see Chapter 14) is 70.8, suggesting a collinearity problem. If the X-variable is centered, the condition number is reduced to 5.1, indicating no severe collinearity problem. Unfortunately, as shown by the correlation arrays in Table 15.12, such centering does not solve the collinearity problem for higher-order models. For a seventh-order model based on centered dosage data, the condition number is a disturbing 430.0. And the condition number for an uncentered, seventh-order natural polynomial model is an extremely alarming 569,664. In contrast, the condition number for both the third-order and the seventh-order orthogonal polynomial models is 1.0. This will be true for any fitted orthogonal polynomial model.

The preceding condition numbers lead us to recommend using only orthogonal polynomials (i.e., we rule out using natural polynomials) to conduct lack-of-fit tests. Table 15.14 summarizes the ANOVA results based on fitting a seventh-order orthogonal polynomial model to

TABLE 15.12 **Predictor correlations for the weight gain data ($n = 24$)**

Simple Polynomials

	X	X^2	X^3	X^4	X^5	X^6	X^7
X	1	.976	.932	.887	.846	.810	.779
X^2		1	.988	.963	.935	.908	.882
X^3			1	.993	.978	.960	.941
X^4				1	.996	.986	.973
X^5					1	.997	.990
X^6						1	.998
X^7							1

Centered Simple Polynomials

	$(X - \bar{X})$	$(X - \bar{X})^2$	$(X - \bar{X})^3$	$(X - \bar{X})^4$	$(X - \bar{X})^5$	$(X - \bar{X})^6$	$(X - \bar{X})^7$
$(X - \bar{X})$	1	0	.926	0	.855	0	.813
$(X - \bar{X})^2$		1	0	.969	0	.932	0
$(X - \bar{X})^3$			1	0	.985	0	.965
$(X - \bar{X})^4$				1	0	.992	0
$(X - \bar{X})^5$					1	0	.996
$(X - \bar{X})^6$						1	0
$(X - \bar{X})^7$							1

TABLE 15.13 **Eigenvalues of predictor correlation matrices for polynomial models for the weight gain data ($n = 24$)**

Third-order Model

	Natural Polynomial		Orthogonal Polynomial
Eigenvalue	Uncentered	Centered	
1	2.931	1.926	1.000
2	0.069	1.000	1.000
3	6×10^{-4}	0.074	1.000
CN	70.8	5.1	1.0

Seventh-order Model

	Natural Polynomial		Orthogonal Polynomial
Eigenvalue	Uncentered	Centered	
1	6.638	3.773	1.000
2	0.348	2.929	1.000
3	0.013	0.221	1.000
4	3×10^{-4}	0.071	1.000
5	4×10^{-6}	0.006	1.000
6	2×10^{-8}	5×10^{-4}	1.000
7	2×10^{-11}	2×10^{-5}	1.000
CN	569,664.0	430.0	1.0

TABLE 15.14   **Seventh-order orthogonal polynomial model for the weight gain data ($n = 24$)**

Source	d.f.	SS	MS
Model[a]	7	169.7796	24.2542
Error	16	0.2600	0.0163
Total (corrected)[b]	23	170.0396	

Source	d.f.	Variables-Added-in-Order SS	F
ODOSE1	1	155.6667	9579.49 ($P < .0001$)
ODOSE2	1	14.0334	863.59 ($P < .0001$)
ODOSE3	1	0.0121	0.75 ($P = .4003$)
ODOSE4	1	0.0509	3.13 ($P = .0958$)
ODOSE5	1	0.0000	0.00 ($P = .9924$)
ODOSE6	1	0.0036	0.22 ($P = .6420$)
ODOSE7	1	0.0128	0.79 ($P = .3871$)

Source	d.f.	Variables-Added-Last SS	F
ODOSE1	1	155.6667	9579.49 ($P < .0001$)
ODOSE2	1	14.0334	863.59 ($P < .0001$)
ODOSE3	1	0.0121	0.75 ($P = .4003$)
ODOSE4	1	0.0509	3.13 ($P = .0958$)
ODOSE5	1	0.0000	0.00 ($P = .9924$)
ODOSE6	1	0.0036	0.22 ($P = .6420$)
ODOSE7	1	0.0128	0.79 ($P = .3871$)

Parameter	Estimate	$S_{\hat{\beta}_j^*}$	T for H_0: Parameter $= 0$
Intercept	3.6541	0.0260	140.54 ($P = .0001$)
ODOSE1	0.5558	0.0057	98.00 ($P = .0001$)
ODOSE2	0.1669	0.0057	29.43 ($P = .0001$)
ODOSE3	0.0039	0.0045	0.86 ($P = .4003$)
ODOSE4	−0.0052	0.0030	−1.77($P = .0958$)
ODOSE5	0.0000	0.0016	0.01 ($P = .9924$)
ODOSE6	−0.0021	0.0045	0.47 ($P = .6420$)
ODOSE7	−0.0011	0.0013	−0.89 ($P = .3871$)

[a]$F = 1{,}492.57$ ($P = .0001$).

[b]$R^2 = .998075$.

the $n = 24$ dosage–weight gain observations. The lower-order polynomial results (up to, say, order three) remain virtually unchanged. This follows from the fact that the quadratic model fits so well and from the properties of orthogonal polynomials. After the .0001 P-values for the linear and quadratic terms, the next smallest P-value is .0958 for the fourth-order term. As before, a second-order polynomial model seems most appropriate.

6. For the data given in Problem 11 in Chapter 7, which concerns the relationship between the temperature (X) of a certain medium and the growth (Y) of human amniotic cells in a tissue culture, researchers wish to evaluate whether a parabolic model is more appropriate than a straight-line model. Use the accompanying computer output to answer the following questions:

a. Plot the fitted straight-line model and the fitted quadratic model on the same scatter diagram.

b. Test for the significance of adding the X^2 term to the model.

c. Determine the change in R^2 in going from a straight-line to a parabolic model.

d. How do your results in parts (b) and (c) compare with the results in Problem 11 in Chapter 7 for the test of adequacy of fit of the straight-line model?

e. Which model is more appropriate—straight line or parabolic?

7. The skin response in rats to different concentrations of a newly developed vaccine was measured in an experiment, resulting in the data, models, and computer output that follow.

Concentration (X) (ml/l)	0.5	0.5	1.0	1.0	1.5	1.5	2.0	2.0	2.5	2.5	3.0	3.0
Skin response (Y) (mm)	13.90	13.81	14.08	13.99	13.75	13.60	13.32	13.39	13.45	13.53	13.59	13.64

a. Plot the straight-line, quadratic, and cubic equations on the scatter diagram for this data set.

b. Test sequentially for significant straight-line fit, for significant addition of X^2, and for significant addition of X^3 to the model.

c. Which of the three models do you recommend, and why? (*Note*: You might also want to consider R^2 for each model.)

Edited SAS Output for Problem 7

Linear Regression of Y on X
Analysis of Variance

Source	DF	Sum of Squares	Mean Square	F Value	Pr > F
Model	1	0.28440	0.28440	8.218	0.0168
Error	10	0.34609	0.03461		
Corrected Total	11	0.63049			

Parameter Estimates

Variable	DF	Parameter Estimate	Standard Error	t value	Pr > |t|
INTERCEP	1	13.986333	0.12246336	114.208	<0.0001
X	1	-0.180286	0.06289138	-2.867	0.0168

Quadratic Regression of Y on X
Analysis of Variance

Source	DF	Sum of Squares	Mean Square	F Value	Pr > F
Model	2	0.35362	0.17681	5.747	0.0246
Error	9	0.27688	0.03076		
Corrected Total	11	0.63049			

Parameter Estimates

Variable	DF	Parameter Estimate	Standard Error	t value	Pr > |t|
INTERCEP	1	14.270500	0.22186125	64.322	<0.0001
X	1	-0.606536	0.29029538	-2.089	0.0663
X2	1	0.121786	0.08119290	1.500	0.1679

(continued)

Edited SAS Output for Problem 7 (continued)

Cubic Regression of Y on X
Analysis of Variance

Source	DF	Sum of Squares	Mean Square	F Value	Prob > F
Model	3	0.52218	0.17406	12.857	0.0020
Error	8	0.10831	0.01354		
Corrected Total	11	0.63049			

Parameter Estimates

Variable	DF	Parameter Estimate	Standard Error	t value	Pr > |t|
INTERCEP	1	13.361667	0.29664990	45.042	<0.0001
X	1	1.679974	0.67600948	2.485	0.0378
X2	1	-1.392937	0.43263967	-3.220	0.0122
X3	1	0.288519	0.08176643	3.529	0.0077

8. This problem uses the data presented in the table for Problem 12 in Chapter 5. Use $\alpha = .05$.
   a. Use a computer program to fit a natural polynomial cubic model for predicting vocabulary size as a function of age in years. Provide estimated regression coefficients.
   b. Using variables-added-in-order tests, determine the best model.
   c. Report variables-added-last tests. Explain any differences between your results here and those you obtained in part (b).
   d. Report appropriate collinearity diagnostics for the model, and evaluate them. Include predictor correlations.
   e. Plot jackknife (or studentized) residuals against predicted values for the best model based on your results in part (b). Provide a frequency histogram or schematic plot of the residuals, and comment on it.
   f. Compare your results here with those you obtained in Problem 12 in Chapter 5.

9. a.–e. Repeat Problem 8, parts (a) through (e), after centering the predictor (age).
   f. Compare the results obtained here to those obtained in Problem 8.

10. a.–e. Repeat Problem 8, (a) through (e), but use orthogonal polynomials.
   f. Compare the results obtained here to those obtained in Problems 8 and 9. (*Hint:* Table A.7 cannot be used.)

11. This problem uses the data from Problem 13 in Chapter 5. Use $\alpha = .10$.
   a. Use a computer program to fit a quadratic natural polynomial model for predicting latency as a function of weight minus average weight.
   b.–e. Repeat parts (b) through (e) from Problem 8 for this analysis.
   f. Compare the results obtained here to those obtained in Chapter 5.

12. This problem uses the data from Problem 15 in Chapter 5.
   a. Using a computer program, fit a cubic polynomial model with BLOODTOL as the response and PPM_TOLU as the predictor. Center PPM_TOLU. Provide the prediction equation.
   b.–e. Repeat parts (b) through (e) from Problem 8 for this analysis.
   f. For each predictor value, compute an estimate of variance of the response variable. Are these estimates approximately equal?
   g. Compare the results obtained here to those obtained in Chapter 5.

13. a. For the data from Problem 15 in Chapter 5, specify the orthogonal polynomial codings needed for coding PPM_TOLU linear, quadratic, and cubic terms. Repeat part (a) of Problem 12, but use the orthogonal coding. (*Hint:* Table A.7 cannot be used.)
   b.–e. Repeat parts (b) through (e) from Problem 8 for this analysis.

14. To promote safe driving habits and to better protect its customers, an insurance company offers a discount of between 5% and 20% on renewal insurance premiums to customers who have completed a defensive driving course. The following table of data shows the number of customers who have applied for the discount at various discount levels, over a period of 12 months.

Month	Discount (X, %)	Number of Renewing Customers Applying for Discount
1	5	485
2	5	1,025
3	5	1,056
4	10	1,020
5	10	1,149
6	10	1,100
7	15	1,800
8	15	1,805
9	15	1,725
10	20	2,225
11	20	2,325
12	20	2,650

Use the accompanying computer output to answer the following questions.
a. Determine the estimated equation of the straight-line regression of the number of customers applying for the discount (Y) on the discount level (X).
b. Determine the estimated equation of the quadratic regression of the number of customers applying for the discount (Y) on the discount level (X).
c. Plot both estimated models, along with a scatterplot of the data. Which model appears to fit the data better?
d. Conduct variables-added-in-order tests for the model in part (b).
e. Carry out tests for the significance of the straight-line regression in part (a) and for the adequacy of fit of the estimated regression line.
f. Carry out tests for the significance of the quadratic regression in part (b) and for the adequacy of fit of the estimated regression line.
g. Based on the results from parts (a) through (f), which of the two regressions appears to be more appropriate for predicting the number of customers that apply for the discount?

Edited SAS Output (PROC REG and PROC RSREG) for Problem 14

```
                        Straight-Line Regression of Y on X
                              Analysis of Variance
                                    Sum of            Mean
Source                    DF        Squares           Square       F Value    Pr > F
Model                      1      4246956.15        4246956.15     90.183     0.0001
Error                     10      470928.76667      47092.87667
Corrected Total           11      4717884.9167

                    Root MSE       217.00893        R-square       0.9002
                    Dep Mean       1530.41667       Adj R-sq       0.8902
                    Coeff Var      14.17973
```

(continued)

Edited SAS Output for Problem 14 (continued)

```
              Straight-line Regression of Y on X (Continued)
                          Parameter Estimates
                      Parameter          Standard
Variable      DF      Estimate            Error           t value      Pr > |t|
INTERCEP      1      200.166667        153.44848756         1.304        0.2213
X             1      106.420000         11.20629307         9.496       <0.0001
                       Quadratic Regression of Y on X
                          Analysis of Variance
                                      Sum of           Mean
Source                     DF        Squares          Square       F Value      Pr > F
Model                      2       4360446.9        2180223.45      54.896       0.0001
Error                      9      357438.01667     39715.33519
Corrected Total           11      4717884.9167

                   Root MSE          199.28707        R-square        0.9242
                   Dep Mean         1530.41667        Adj R-sq        0.9074
                   Coeff Var          13.02175

                          Parameter Estimates
                      Parameter          Standard
Variable      DF      Estimate            Error           t value      Pr > |t|
INTERCEP      1      686.416667        320.30914634         2.143        0.0607
X             1        9.170000         58.44244028         0.157        0.8788
X2            1        3.890000          2.30116884         1.690        0.1252

                              Variance
Variable      DF          Inflation
INTERCEP      1          0.00000000
X             1         32.25000000
X2            1         32.25000000

                    Quadratic Regression of Y on X
                                  .
                                  .  [Portion of output omitted]
                                  .

                Degrees         Type I
                  of            Sum of
Regression      Freedom         Squares        R-Square      F-Ratio       Pr > F
Linear             1           4246956         0.9002        106.9        <0.0001
Quadratic          1            113491         0.0241         2.858        0.1252
Crossproduct       0                 0         0.0000           .            .
Total Regress      2           4360447         0.9242        54.896       <0.0001

                Degrees
                  of            Sum of          Mean
Residual        Freedom         Squares        Square        F-Ratio       Pr > F
Lack of Fit        1             39990          39990         1.008        0.3448
Pure Error         8            317448          39681
Total Error        9            357438          39715
```

15. Refer to Problem 14. Using the computer output for that problem, and the accompanying output here, answer the following questions.

   a. Determine the variance inflation factors for the estimated model in part (b) of Problem 14. Does collinearity appear to be a problem?

   b. Determine the estimated equation of the quadratic regression of the number of customers applying for the discount (Y) on the centered discount level (Z).

   c. Determine the variance inflation factors for the estimated model in part (b) of this problem. Does collinearity appear to be a problem?

   d. Conduct variables-added-in-order tests for the model in part (b).

   e. Carry out tests for the significance of the quadratic regression in part (b) and for the adequacy of fit of the estimated regression line.

f. Based on the results from Problem 14 and parts (a) through (e) of this problem, which of the two regressions appears to be more appropriate for predicting the number of customers that apply for the discount?

Edited SAS Output (PROC REG and PROC RSREG) for Problem 15

Quadratic Regression of Y on Z
Analysis of Variance

Source	DF	Sum of Squares	Mean Square	F Value	Pr > F
Model	2	4360446.9	2180223.45	54.896	<0.0001
Error	9	357438.01667	39715.33519		
Corrected Total	11	4717884.9167			

Root MSE	199.28707	R-square	0.9242	
Dep Mean	1530.41667	Adj R-sq	0.9074	
Coeff Var	13.02175			

Parameter Estimates

Variable	DF	Parameter Estimate	Standard Error	t value	Pr > |t|
INTERCEP	1	1408.854167	92.09168729	15.298	<0.0001
Z	1	106.420000	10.29113990	10.341	<0.0001
Z2	1	3.890000	2.30116884	1.690	0.1252

Variable	DF	Variance Inflation
INTERCEP	1	0.00000000
Z	1	1.00000000
Z2	1	1.00000000

Quadratic Regression of Y on X

Regression	Degrees of Freedom	Type I Sum of Squares	R-Square	F-Ratio	Pr > F
Linear	1	4246956	0.9002	106.9	0.0000
Quadratic	1	113491	0.0241	2.858	0.1252
Crossproduct	0	0	0.0000	.	.
Total Regress	2	4360447	0.9242	54.896	0.0000

Residual	Degrees of Freedom	Sum of Squares	Mean Square	F-Ratio	Pr > F
Lack of Fit	1	39990	39990	1.008	0.3448
Pure Error	8	317448	39681		
Total Error	9	357438	39715		

16. Columbus Airlines introduced a new, specially discounted line of air fares in 1995. Annual ticket revenues Y (in $1,000s) are shown in the following table, along with the time period X (in months, with $X = 1$ for January 1995).

Month	Ticket Revenues
1	34.9
2	38.8
3	41.5
4	45.1
5	48.3
6	51.2
7	56.6
8	59.9
9	65.4

Use the accompanying computer output to answer the following questions.

16. a. Determine the estimated equation for the straight-line regression of ticket revenues (Y) on month (X).
   b. Determine the estimated equation for the quadratic regression of ticket revenues (Y) on month (X).
   c. Plot both estimated models, along with a scatterplot of the data. Which model appears to fit the data better?
   d. Conduct variables-added-in-order tests for the model in part (b).
   e. Carry out tests for the significance of the straight-line regression in part (a).
   f. Carry out tests for the significance of the quadratic regression in part (b).
   g. Examine the variance inflation factors for the estimated model in part (b). Does there appear to be a problem with collinearity?
   h. Based on the results from parts (a) through (g), which of the two regression models appears to be more appropriate for predicting ticket revenues?

Edited SAS Output (PROC REG) for Problem 16

```
                         Straight-Line Regression of Y on X
                                Analysis of Variance
                                    Sum of          Mean
Source                    DF       Squares         Square      F Value     Pr > F
Model                      1      818.44267      818.44267     858.563     0.0001
Error                      7        6.67289        0.95327
Corrected Total            8      825.11556

                    Root MSE          0.97636      R-square     0.9919
                    Dep Mean         49.07778      Adj R-sq     0.9908
                    Coeff Var         1.98940

                              Parameter Estimates
                         Parameter        Standard
Variable          DF      Estimate          Error          t value     Pr > |t|
INTERCEP           1     30.611111       0.70930574         43.156      <0.0001
X                  1      3.693333       0.12604694         29.301      <0.0001

                          Quadratic Regression of Y on X
                                Analysis of Variance
                                    Sum of          Mean
Source                    DF       Squares         Square      F Value     Pr > F
Model                      2      822.91956      411.45978    1124.211     0.0001
Error                      6        2.19599        0.36600
Corrected Total            8      825.11556

                    Root MSE          0.60498      R-square     0.9973
                    Dep Mean         49.07778      Adj R-sq     0.9965
                    Coeff Var         1.23269

                              Parameter Estimates
                         Parameter        Standard
Variable          DF      Estimate          Error          t value     Pr > |t|
INTERCEP           1     32.821429       0.76978509         42.637      <0.0001
X                  1      2.487706       0.35345534          7.038       0.0004
X2                 1      0.120563       0.03447183          3.497       0.0129

                                        Variance
Variable          DF                   Inflation
INTERCEP           1                  0.00000000
X                  1                 20.48051948
X2                 1                 20.48051948
```

Selecting the Best Regression Equation

16.1 Preview

The general problem to be discussed in this chapter is as follows: we have one response variable Y and a set of k predictor variables $X_1, X_2, \ldots, X_k$; and we want to determine the *best* (*most important* or *most valid*) subset of the k predictors and the corresponding *best-fitting* regression model for describing the relationship between Y and the X's.[1] What exactly we mean by "best" depends in part on our overall goal for modeling. Two such goals were described in Chapter 11 and are briefly reviewed now.

One goal is to find a model that provides the best prediction of Y, given $X_1, X_2, \ldots, X_k$, for some new observation or for a batch of new observations. In practice, we emphasize estimating the regression of Y on the X's (see Chapter 8)—$E(Y \mid X_1, X_2, \ldots, X_k)$—which expresses the mean of Y as a function of the predictors. Using this goal, we may say that our best model is *reliable* if it predicts well in a new sample. The details of the model may be of little or no consequence, such as whether we include any particular variable or what magnitude or sign we use for its regression coefficient. For example, in considering a sample of systolic blood pressures, we may simply wish to predict systolic blood pressure (SBP) as a function of demographic variables like AGE, RACE, and GENDER. We may not care which variables are in the model or how they are defined, as long as the final model obtained gives the best prediction possible.

Alongside the question of prediction is the question of validity—that is, of obtaining accurate estimates for one or more regression coefficient parameters in a model and then making inferences about these parameters of interest. The goal here is to quantify the relationship between one or more independent variables of interest and the dependent variable, controlling for the other variables. For example, we might wish to describe the relationship of SBP to AGE, controlling for RACE and GENDER. In this case, we are focusing on regression coefficients

---

[1] Hocking (1976) provided an exceptionally thorough and well-written review of this topic. His presentation is at a technical level somewhat higher than that of this text.

involving AGE (including functions of AGE); although RACE and GENDER may remain in the model for control purposes, their regression coefficients are not of primary interest.

In this chapter we focus on strategies for selecting the best model when the primary goal of analysis is prediction. In the last section we briefly mention a strategy for modeling in situations where the validity of the estimates of one or more regression coefficients is of primary importance (see also Chapter 11).

16.2 Steps in Selecting the Best Regression Equation

To select the best regression equation, carry out the following steps:

1. Specify the maximum model (defined in Section 16.3) to be considered.

2. Specify a criterion for selecting a model.

3. Specify a strategy for selecting variables.

4. Conduct the specified analysis.

5. Evaluate the reliability of the model chosen.

By following these steps, you can convert the fuzzy idea of finding the best predictors of Y into simple, concrete actions. Each step helps to ensure reliability and to reduce the work required. Specifying the maximum model forces the analyst to state the analysis goal clearly, recognize the limitations of the data at hand, and describe the range of plausible models explicitly. All computer programs for selecting models require that a maximum model be specified. In doing so, the analyst should consider all available scientific knowledge. In turn, specifying the criterion for selecting a model and the strategy for applying that criterion (i.e., for selecting variables) simplifies and speeds the analysis process. Finally, whether the primary goal is prediction or validity, the reliability of the chosen model must be demonstrated.

We illustrate the recommended process of analysis via two examples. The first of these was introduced in Chapter 8. The data (given in Table 8.1) are the hypothetical results of measuring a response variable weight (WGT) and predictor variables height (HGT) and age (AGE) on 12 children. The second example uses real data from a study involving more than 200 subjects (Lewis and Taylor 1967, discussed in Chapter 14).

16.3 Step 1: Specifying the Maximum Model

The *maximum model* is defined to be the largest model (the one having the most predictor variables) considered at any point in the process of model selection. All other possible models can be created by deleting predictor variables from the maximum model. A model created by deleting predictors from the maximum model is called a *restriction* of the maximum model.

Throughout this chapter we assume that the maximum model with k variables or some restriction of it with $p \leq k$ variables is the *correct model* for the population. An important implication of this assumption is that the *population* squared multiple correlation for the maximum model—namely, $\rho^2_{Y|X_1, X_2, \ldots, X_k}$—is no larger than that for the correct model (which may have fewer variables); as a result, adding more predictors to the correct model does not increase the *population* squared multiple correlation for the correct model. In turn, for the sample at hand, $R^2_{Y|X_1, X_2, \ldots, X_k}$ for the maximum model is always at least as large as the corresponding R^2 for any subset model. However, other sample-based criteria may not necessarily suggest that the largest model is best.

To illustrate, consider the data from Table 8.1, reproduced in the following table.

Child	1	2	3	4	5	6	7	8	9	10	11	12
Y (WGT)	64	71	53	67	55	58	77	57	56	51	76	68
X_1 (HGT)	57	59	49	62	51	50	55	48	42	42	61	57
X_2 (AGE)	8	10	6	11	8	7	10	9	10	6	12	9

One possible model (not necessarily the maximum one) to consider is given by

$$Y = \beta_0 + \beta_1 X_1 + \beta_2 X_2 + E$$

This model allows for only a linear (planar) relationship between the response (WGT) and the two predictors (HGT and AGE). Nevertheless, the nature of growth suggests that the relationship, although monotonic in both HGT and AGE, may well be nonlinear. This implies that at least one quadratic (squared) term may have to be included in the model. Since HGT and AGE are highly correlated, and since the sample size is very small,[2] only $(AGE)^2$ will be considered. Should the interaction between AGE and HGT be considered? Should transformations of the predictors and/or the response be considered (e.g., replacing AGE and $(AGE)^2$ with log(AGE))? The limitations of the data lead us to define the maximum model as

$$Y = \beta_0 + \beta_1 X_1 + \beta_2 X_2 + \beta_3 X_3 + E$$

with $X_1 = $ HGT, $X_2 = $ AGE, and $X_3 = (AGE)^2$.

The most important reason for choosing a large maximum model is to minimize the chance of making a Type II (false negative) error. In a regression analysis, a Type II error corresponds to *omitting* a predictor that has a truly nonzero regression coefficient in the population. Other reasons for considering a large maximum model are the wishes to include the following elements:

1. All conceivable basic predictors

2. High-order powers of basic predictors ($(AGE)^2$, $(HGT)^2$)

3. Other transformations of predictors (log(AGE), $1/HGT$)

4. Interactions among predictors (e.g., AGE $\times$ HGT), including two-way and higher-order interactions

5. All possible "control" variables, as well as their powers and interactions

Recall that overfitting a model (including variables in the model with truly zero regression coefficients in the population) will not introduce bias when population regression coefficients are estimated, if the usual regression assumptions are met. We must be careful, however, to

---

[2] Because the sample size here is so small, it almost precludes making reliable conclusions from any regression analysis. This small sample size is used to simplify our discussion.

ensure that overfitting does not introduce harmful collinearity (Chapter 14). Underfitting (i.e., leaving important predictors out of the final model), however, will introduce bias in the estimated regression coefficients.

There are also important reasons for working with a small maximum model. With a prediction goal, the need for reliability strongly argues for a small maximum model; and with a validity goal, we want to focus on a few important variables. In either case, we wish to avoid a Type I (false positive) error. In a regression analysis, a Type I error corresponds to *including* a predictor that has a zero population regression coefficient. The desire for parsimony is another important reason for choosing a small maximum model: practically unimportant but statistically significant predictors can greatly confuse the interpretation of regression results, and complex interaction terms are particularly troublesome in this regard.

The particular sample of data to be analyzed imposes certain constraints on the choice of the maximum model. In general, the smaller the sample size, the smaller the maximum model should be. The idea here is that a larger number of independent observations are needed to estimate reliably a larger number of regression coefficients. This notion has led to various guidelines about the size of a maximum model. The most basic constraint is that the error degrees of freedom must be positive. Symbolically, we require

$$\text{d.f. error} = n - k - 1 > 0$$

which is equivalent to the constraint

$$n > k + 1$$

As always, n is the number of observations and k is the number of predictors, giving $k + 1$ regression coefficients (including the intercept). With negative error degrees of freedom, the model has at least one perfect collinearity; consequently, unique estimates of regression coefficients and variances cannot be computed. With zero error degrees of freedom ($n = k + 1$), unique estimates of the regression coefficients can be computed, but unique estimates of variances cannot (recall that $\hat{\sigma}^2 = \text{SSE/d.f. error}$). Furthermore, even if the population squared multiple correlation is 0, $R^2 = 1.00$ when d.f. error $= 0$, which reflects the fact that the model exactly fits the observed data (i.e., SSE $= 0$). In such a situation, we have exchanged n values of Y for n estimated regression coefficients. Hence we have gained nothing, since the dimensionality of the problem remains the same.

The question then arises as to how many error degrees of freedom are needed. The weakest requirement is for a minimum of approximately 10 error degrees of freedom—namely,

$$n - k - 1 \geq 10$$

or

$$n \geq 10 + k + 1$$

Another suggested rule of thumb for regression is to have at least 5 (or 10) observations per predictor—namely, to require that

$$n \geq 5k \qquad (\text{or } n \geq 10k)$$

Assume, for example, that we wish to consider a maximum model involving 30 predictors. To have 10 error degrees of freedom, we need a sample of size 41, and $n \geq 5k$ demands a sample of size 150. Split-sample approaches, discussed later in this chapter, may reduce the required sample size substantially.

Another constraint on the maximum model involves the amount of variability present in the predictor values, considered either individually or jointly. If a predictor has the same value for all subjects, it obviously cannot be used in any model. For example, consider the variable SEX of child (male = 1, female = 0) as a candidate predictor for the weight example. If all of the subjects are male, then SEX = 1 for all subjects, the sample variance for the variable is 0.00, and it is perfectly collinear with the intercept variable. Clearly, if all subjects are of one gender, comparisons of gender cannot be made.

Similarly, consider the variables SEX, RACE (white = 0, black = 1), and their interaction (SEXRACE = SEX $\times$ RACE). If no black females are represented in the data, the 1 degree of freedom interaction effect cannot be estimated. If a race–sex cell is nearly empty, the estimated interaction coefficient may be very unstable (i.e., have a large estimated variance).

Polynomial terms (e.g., X^2 and X^3) and other transformations merit particular consideration when the maximum model is being specified. For the weight example, we might wish to consider AGE, $(AGE)^2$, and exp(AGE) as possible predictors. Near collinearities (Chapter 14) can lead to very unstable and often uninterpretable results when we try to find the best model. (See Marquardt and Snee 1975, for further discussion of this topic.) Consequently, we should attempt to reduce such collinearity if possible. Centering, if applicable, almost always helps increase numerical accuracy, as does multiplying variables by various constants (a special form of scaling; see Chapter 14) to produce nearly equal variances for all predictors (say, variances around 1).

If the collinearity problem cannot be overcome, then we have three recourses: conduct separate analyses for each form of X (for example, one analysis using X and X^2 and then a separate analysis using exp (X)); eliminate some variables; or impose structure (e.g., a fixed order of tests) on the testing procedure. If we do nothing about severe collinearity, the estimated regression coefficients in the best model may be highly unstable (i.e., have high variances) and may be quite far from the true parameter values.

16.4 Step 2: Specifying a Criterion for Selecting a Model

The second step in selecting the best model is to specify the selection criterion. A *selection criterion* is an index that can be computed for each candidate model and used to compare models. Thus, given one particular selection criterion, candidate models can be ordered from best to worst. This helps automate the process of choosing the best model. As we shall see, this selection-criterion-specific process may not find the *best* model in a global sense. Nonetheless, using a specific selection criterion can substantially reduce the work involved in finding a *good* model.

Obviously, the selection criterion should be related to the goal of the analysis. For example, if the goal is reliable prediction of future observations, the selection criterion should be somewhat liberal, to avoid missing useful predictors. Distinctions can be drawn among *numerical* differences, *statistically significant* differences, and *scientifically important* differences. Numerical differences may or may not correspond to significant or important differences. With sensitive tests (typical of analyses of large samples), significant differences may or may not correspond to important differences. And, with insensitive tests (typical of analyses of very small samples), important differences may not be significant.

Many selection criteria for choosing the best model have been suggested. Hocking (1976), for example, reviewed eight candidates. We consider four criteria: R_p^2, F_p, MSE(p), and C_p. But before discussing these, we must define the notation needed to understand them. All four criteria attempt to compare two model equations: the maximum model with k predictors and a restricted model with p predictors ($p \leq k$). The maximum model is

$$Y = \beta_0 + \beta_1 X_1 + \beta_2 X_2 + \cdots + \beta_p X_p + \beta_{p+1} X_{p+1} + \cdots + \beta_k X_k + E \tag{16.1}$$

and the reduced model (a restriction of the maximum model) is

$$Y = \beta_0 + \beta_1 X_1 + \beta_2 X_2 + \cdots + \beta_p X_p + E \tag{16.2}$$

Let SSE(k) be the error sum of squares for the k-variable model, let SSE(p) be the error sum of squares for the p-variable model, and so on. Also, SSY $= \sum_{i=1}^{n} (Y_i - \overline{Y})^2$ is the total (corrected) sum of squares for the response Y. Often $p = k - 1$, which is the case when we evaluate the addition or deletion of a single variable. We assume that the $(k - p)$ variables under consideration for addition or deletion are denoted $X_{p+1}, X_{p+2}, \ldots, X_k$, for notational convenience.

The sample squared multiple correlation R^2 is a natural candidate for deciding which model is best, so we will discuss this criterion first. The multiple R^2 for the p-variable model is

$$R_p^2 = R_{Y|X_1, X_2, \ldots, X_p}^2 = 1 - \frac{\text{SSE}(p)}{\text{SSY}} \tag{16.3}$$

Unfortunately, R_p^2 has three potentially misleading characteristics. First, it tends to overestimate ρ_p^2, the corresponding population value. Second, adding predictors, even useless ones, can never decrease R_p^2. In fact, adding variables invariably increases R_p^2, at least slightly. Finally, R_p^2 is always largest for the maximum model, even though a better model may be obtained by deleting some (or even many) variables. The reduced (or restricted) model may be better because it may sacrifice only a negligible amount of predictive strength while substantially simplifying the model.

Another reasonable criterion for selecting the best model is the F test statistic for comparing the full and restricted models. This statistic, F_p, can be expressed in terms of sums of squares for error (SSEs) as

$$F_p = \frac{[\text{SSE}(p) - \text{SSE}(k)]/(k - p)}{\text{SSE}(k)/(n - k - 1)} = \frac{[\text{SSE}(p) - \text{SSE}(k)]/(k - p)}{\text{MSE}(k)} \tag{16.4}$$

This statistic may be compared to an F distribution with $k - p$ and $n - k - 1$ degrees of freedom. The criterion F_p tests whether SSE(p) $-$ SSE(k), the difference between the residual sum of squares for the p-variable model and the residual sum of squares for the maximum model, differs significantly from 0. If F_p is *not* significant, we can use the smaller (p-variable) model and achieve roughly the same predictive ability as that yielded by the full model. Hence, a reasonable rule for selecting variables is to retain p variables if F_p is not significant and if p is as small as possible. An often-used special case of F_p occurs when $p = k - 1$, in which case F_p is a test of $H_0: \beta_k = 0$ in the full model.

A third criterion to consider in selecting the best model is the estimated error variance for the p-variable model—namely,

$$\text{MSE}(p) = \frac{\text{SSE}(p)}{n - p - 1} \tag{16.5}$$

The quantity MSE(p) is an inviting choice for a selection criterion, since we wish to find a model with small residual variance.

A less obvious candidate for a selection criterion involving SSE(p) is Mallow's C_p:

$$C_p = \frac{\text{SSE}(p)}{\text{MSE}(k)} - [n - 2(p + 1)] \qquad (16.6)$$

The C_p criterion helps us to decide how many variables to put in the best model, since it achieves a value of approximately $p + 1$ if MSE(p) is roughly equal to MSE(k) (i.e., if the correct model is of size p). Knowing the correct model size greatly aids in choosing the best model.

The criteria F_p, R_p^2, MSE(p), and C_p are intimately related. For example, the F_p test can be expressed in terms of multiple squared correlations as follows:

$$F_p = \frac{(R_k^2 - R_p^2)/(k - p)}{(1 - R_k^2)/(n - k - 1)} \qquad (16.7)$$

and the C_p statistic is the following simple function of the F_p statistic:

$$C_p = (k - p)F_p + (2p - k + 1) \qquad (16.8)$$

The reason to consider more than one criterion is that no single criterion is always best. In practice, the alternatives can lead to different model choices. In the remainder of the chapter we shall consider all of the criteria mentioned, at least to some extent. An important aspect of our discussion will be a demonstration of the limitations of R_p^2 as the sole criterion for selecting a model. We favor C_p because it tends to simplify the decision about how many variables to retain in the final model.

16.5 Step 3: Specifying a Strategy for Selecting Variables

The third step in choosing the best model is to specify the strategy for selecting variables. Such a strategy is concerned with determining how many variables and also which particular variables should be in the final model. Traditionally, such strategies have focused on deciding whether a single variable should be added to a model (a forward selection method) or whether a single variable should be deleted from a model (a backward elimination method). As computers became more powerful, methods for considering more than one variable per step became practical (by generalizing single-variable methods to deal with sets, or *chunks,* of variables). Before discussing these strategies in detail, we consider an algorithm for evaluating models that, if feasible to conduct, should be the method of choice.

16.5.1 All Possible Regressions Procedure

Whenever practical, the *all possible regressions procedure* is to be preferred over any other variable selection strategy. It is the only method guaranteed to find the model having the largest R_p^2, the smallest MSE(p), and so on. This strategy is not always used because the amount of calculation necessary becomes impractical when the number of variables k in the maximum model is large. The all possible regressions procedure requires that we fit each possible regression equation associated with all possible sets of the k independent variables. For our example, we must fit the seven models corresponding to the following seven sets of independent variables: HGT; AGE; $(AGE)^2$; HGT and AGE; HGT and $(AGE)^2$; AGE and $(AGE)^2$; and HGT, AGE, and $(AGE)^2$. For k independent variables, the number of models to be fitted would be $(2^k - 1)$; for example, if $k = 10$, then $(2^{10} - 1) = 1,023$.

Once all $(2^k - 1)$ models have been fitted, we assemble the fitted models into sets involving from 1 to k variables and then order the models within each set according to some criterion (e.g., R_p^2, F_p, MSE(p), or C_p).

For our data, a summary of the results of the all possible regressions procedure appears in Table 16.1. From the table, the leaders (in terms of R_p^2-values) in each of the sets involving one, two, and three variables are

One-variable set: HGT with $R_1^2 = .6630$

Two-variable set: HGT, AGE with $R_2^2 = .7800$

Three-variable set: HGT, AGE, $(AGE)^2$ with $R_3^2 = .7802$

Of the three models (models 1, 4, and 7, respectively, in Table 16.1), model 4, involving HGT and AGE, should clearly be our choice, since its R^2-value is essentially the same as that for model 7 and is much higher than the value for model 1. Thus, our choice of the best regression equation based on the all possible regressions procedure with R_p^2 as the criterion is

$$\text{WGT} = 6.553 + 0.722\text{HGT} + 2.050\text{AGE}$$

Now consider the partial F statistics in Table 16.1. For a given variable in a given model, the associated partial F statistic assesses the contribution made by that variable to the prediction of Y(WGT) over and above the contributions made by other variables already in the given model. For example, for model 4, which involves only X_1 and X_2, the partial F for X_2 is $F(X_2 \mid X_1) = 4.785$; but for model 7, which includes X_1, X_2, and X_3, the partial F for X_2 is $F(X_2 \mid X_1, X_3) = 0.140$.

TABLE 16.1   Summary of results of all possible regressions procedure

Model	No. of Variables	Variables Used	Estimated Coefficients $\hat{\beta}_0$	$\hat{\beta}_1$	$\hat{\beta}_2$	$\hat{\beta}_3$	Partial F Statistics X_1	X_2	X_3	Overall F Statistic	R_p^2	MSE(p)	C_p
1	1	HGT (X_1)	6.190	1.073			19.67**			19.67**	.6630	29.93	4.27
2	1	AGE (X_2)	30.571		3.643			14.55**		14.55**	.5926	36.18	6.83
3	1	$(AGE)^2$ (X_3)	45.998			0.206			14.25**	14.25**	.5876	36.63	7.01
4	2	HGT, AGE	6.553	0.722	2.050		7.665*	4.785		15.95**	.7800	21.71	2.01
5	2	HGT, $(AGE)^2$	15.118	0.726		0.115	7.601*		4.565	15.63**	.7764	22.07	2.14
6	2	AGE, $(AGE)^2$	32.404		3.205	0.025		0.113	0.002	6.55*	.5927	40.20	8.83
7	3	HGT, AGE, $(AGE)^2$	3.438	0.724	2.777	−0.042	6.827*	0.140	0.010	9.47**	.7802	24.40	4.00

These partial F's are variables-added-last tests, each based on the MSE for the corresponding model being fit for that row. Such tests must be treated with some caution. Any test based on a model with fewer terms than the correct model will be biased, perhaps substantially, because the test involves using biased error terms. Furthermore, so many tests are computed that the Type I error rate should be higher than the nominal rate α.

Overall F tests are also affected by the estimate of the error variance used. In variable selection, this estimate is important because biased tests conducted early in a stepwise algorithm may stop the procedure prematurely and miss important predictors. As calculated in Table 16.1, each overall F statistic is based on the corresponding MSE listed in the same row. If the MSE for the largest model, number 7, had been used in the denominator of each overall F test instead, each resulting test statistic would have been an F_p statistic. Such a statistic involves a comparison of the R_p^2-value for a reduced model (models 1 through 6) and the R_k^2-value for the largest model (model 7). For example, for model 4,

$$F_p = \frac{(R_k^2 - R_p^2)/(k - p)}{(1 - R_k^2)/(n - k - 1)} = \frac{(.7802 - .7800)/(3 - 2)}{(1 - .7802)/(12 - 3 - 1)} = 0.007$$

In general, such a test is a multiple-partial F test. The small F-value here indicates that the predictive abilities of the maximum model and of model 4 do not differ significantly.

Table 16.1 also provides C_p-values. The value of C_p is expected to be close to $(p + 1)$ if the correct model—or a larger model that contains the correct model—is considered. If important predictors are omitted, C_p should be larger than $(p + 1)$. Also, if $F_p < 1$, then $C_p < (p + 1)$; this can occur when R_p^2 is close enough to R_k^2 in value. Users prefer models with C_p not too far from $(p + 1)$. Therefore, for a model with one variable, C_p is compared with the value 2.0; for two variables, with the value 3.0; and so on. For this example, no one-variable model has a C_p-value near 2.0. For the only three-variable model, C_p is exactly 4.00. The full model with k predictors is guaranteed to have C_p exactly equal to $(k + 1)$; to see this, examine equation (16.6), the formula for C_p. The two-variable model with AGE and $(AGE)^2$ has a C_p-value much greater than 3, while models 4 and 5 have C_p-values near the minimum possible C_p-value of

$$\frac{(n - k - 1)MSE(k)}{MSE(k)} - [n - 2(p + 1)] = (2p - k + 1)$$

which equals 2 when $k = 3$ and $p = 2$. (We can see from (16.8) that this lower bound for C_p is attained when $F_p = 0$, and that it can be negative.) Such a value is better than the value of $(p + 1) = 3$.

The all possible regressions procedure was presented first, since it is preferred whenever practical. It alone is guaranteed to find the best model, in the sense that any selection criterion will be numerically optimized for the particular sample under study. Naturally, using it does not guarantee finding the correct (or population) model. In fact, in many situations, several reasonable candidates for the best model can be found, with different selection criteria suggesting different best models. Furthermore, such findings may vary from sample to sample, even though all the samples are chosen from the same population. Consequently, the choice of the best model may vary from sample to sample. These considerations motivate the discussion in Section 16.7 of evaluating the reliability of the chosen regression equation.

As mentioned earlier, the all possible regressions algorithm is often impractical, since $(2^k - 1)$ models must be fitted when k candidate predictors are being evaluated. Many methods have been suggested as computationally feasible alternatives for approximating the all possible regressions procedure. Although these methods are not guaranteed to find the best model, they can (with careful use) glean essentially all the information from the data needed to choose the best model.

16.5.2 Backward Elimination Procedure

In the *backward elimination procedure,* we proceed as follows:

1. *Determine the fitted regression equation containing all independent variables* (i.e., the estimated maximum model). From the accompanying SAS computer output, we obtain

$$\widehat{\text{WGT}} = 3.438 + 0.724\text{HGT} + 2.777\text{AGE} - 0.042(\text{AGE})^2$$

2. *Determine the partial F statistic for every variable in the model as though it were the last variable to enter, and determine the P-values associated with the test statistics.* The partial F statistics and P-values are indicated in the Step 0 portion of the SAS output. (Recall that the partial F statistics test whether adding the last variable to the model significantly helps predict the dependent variable, given that the other variables are already in the model.)

3. *Focus on the lowest observed partial F statistic (or, equivalently, on the highest P-value).* From the Step 0 portion of the SAS output, we see that we should focus on $(\text{AGE})^2$.

4. *Compare the P-value with a preselected significance level (say, 10%) and decide whether to remove the variable under consideration. If the variable is dropped, recompute the regression equation for the remaining variables, and repeat backward elimination procedure Steps 2, 3, and 4. If the variable is not dropped, the backward elimination procedure ends, and the selected model consists of variables remaining in the model.* From the Step 1 portion of the SAS output, we see that $(\text{AGE})^2$ is dropped and the regression is recomputed. With $(\text{AGE})^2$ out of the picture, the smallest partial F statistic is 4.78, with an associated P-value of .0565. The P-value is less than .10, so we stop here with this model, which is the same model that we arrived at using the all possible regressions procedure:

$$\widehat{\text{WGT}} = 6.553 + 0.722\text{HGT} + 2.050\text{AGE}$$

Edited SAS Output (PROC REG) for Backward Elimination Procedure

```
            Backward Elimination Procedure for Dependent Variable WGT
Step 0 All Variables Entered R-square = 0.78025383
       C(p) = 4.00000000

                                Sum of              Mean
                      DF        Squares             Square        F Value    Pr > F
Regression            3      693.06046340        231.02015447       9.47     0.005
Error                 8      195.18953660         24.39869208
Corrected Total      11      888.25000000
```

(continued)

Edited SAS Output for Backward Elimination Procedure (continued)

```
                    Backward Elimination Procedure for Dependent
                          Variable WGT (Continued)              Partial F
Step 0 (Continued)                                             statistics      P-value
                                                                   ↓              ↓
                    Parameter         Standard         Type II
Variable            Estimate          Error            SS          F Value      Pr > F
INTERCEP            3.43842600        33.61081984      0.25534520   0.01         0.9210
HGT                 0.72369024        0.27696316       166.58195495 6.83         0.0310
AGE                 2.77687456        7.42727877       3.41051231   0.14         0.7182
AGE2                -0.04170670       0.42240715       0.23785686  (0.01)       (0.9238)

Bounds on condition number:      89.96948,              543.7942
```

```
Step 1 Variable AGE2 Removed R-square = 0.77998605
       C(p) = 2.00974875

                              Sum of           Mean
                    DF        Squares          Square        F Value      Pr > F
Regression          2         692.82260654     346.41130327  15.95        0.0011
Error               9         195.42739346     21.71415483
Corrected Total     11        888.25000000

                    Parameter         Standard         Type II
Variable            Estimate          Error            SS          F Value      Pr > F
INTERCEP            6.55304825        10.94482708      7.78416178   0.36         0.5641
HGT                 0.72203796        0.26080506       166.42974940 7.66         0.0218
AGE                 2.05012635        0.93722561       103.90008336 (4.78)      (0.0565)

Bounds on condition number:      1.604616,              6.418463
```

```
All variables left in the model are significant at the 0.1000 level.
Summary of Backward Elimination Procedure for Dependent Variable WGT

            Variable    Number    Partial    Model
Step        Removed     In        R**2       R**2       C(p)      F Value      Pr > F
1           AGE2        2         0.0003     0.7800     2.0097    0.0097       0.923
```

16.5.3 Forward Selection Procedure

In the *forward selection procedure,* we proceed as follows:

1. *Select as the first variable to enter the model the variable most highly correlated with the dependent variable, and then fit the associated straight-line regression equation. If the overall F test for this regression is not significant, stop and conclude that no independent variables are important predictors. If the test is significant, include this variable in the model, and proceed to Step 2 of the procedure.* From the accompanying SAS output, we see that the highest squared correlation is for X_1, HGT. The straight-line regression equation relating WGT and HGT is

$$\widehat{WGT} = 6.190 + 1.072HGT$$

2. *Determine the partial F statistic and P-value associated with each remaining variable, based on a regression equation containing that variable and the variable*

initially selected. For our data, the statistics are shown in the Step 2 portion of the SAS output.

3. *Focus on the variable with the largest partial F statistic (i.e., the variable with the smallest P-value).* From the Step 2 portion of the SAS output, we see that AGE has the largest partial F statistic, with a P-value of .0565.

4. *Test for the significance of the partial F statistic associated with the variable from Step 3 of the forward selection procedure. If this test is significant, add the new variable to the regression equation. If it is not significant, use in the model only the variable added in step 1. above.* For our data, since the P-value for AGE is less than .10, we add AGE to get the following two-variable model:

$$\widehat{WGT} = 6.553 + 0.722HGT + 2.050AGE$$

5. *At each subsequent step, determine the partial F statistics for the variables not yet in the model, and then add to the model the variable with the largest partial F value (if it is statistically significant). At any step, if the largest partial F is not significant, no more variables are included in the model and the process is terminated.* For our example we have already added HGT and AGE to the model. We now check to see whether we should add $(AGE)^2$. The partial F for $(AGE)^2$, controlling for HGT and AGE, is .01, with a P-value of .9238. This value is not statistically significant at $\alpha = .10$. Again, we have arrived at the same two-variable model chosen via the previously discussed methods.

Edited SAS Output (PROC REG) for Forward Selection Procedure

```
              Forward Selection Procedure for Dependent Variable WGT
                       Statistics for Entry: Step 0
                              DF = 1,10

                                      Model
Variable          Tolerance           R**2            F Value        Pr > F
HGT               1.000000           (0.6630)         19.6749        (0.0013)
AGE               1.000000            0.5926          14.5470         0.0034
AGE2              1.000000            0.5876          14.2481         0.0036

Step 1  Variable HGT Entered R-square = 0.66301438
        C(p) = 4.26817716

                              Sum of            Mean
                   DF         Squares          Square      F Value    Pr > F
Regression          1      588.92252318    588.92252318     19.67     0.0013
Error              10      299.32747682     29.93274768
Corrected Total    11      888.25000000

                 Parameter        Standard         Type II
Variable         Estimate          Error             SS        F Value    Pr > F
INTERCEP        6.18984871       12.84874620      6.94680569      0.23     0.6404
HGT             1.07223036        0.24173098    588.92252318     19.67     0.0013

Bounds on condition number:               1,             1
```

(continued)

Edited SAS Output for Forward Selection Procedure (continued)

```
              Forward Selection Procedure for Dependent Variable WGT (Continued)
                            Statistics for Entry: Step 2
                                     DF = 1, 9

                                                 Model
              Variable        Tolerance          R**2         F Value        Pr > F
              AGE             0.623202           0.7800        4.7849         0.0565
              AGE2            0.621230           0.7764        4.5647         0.0614

       Step 2 Variable AGE Entered R-square = 0.77998605
          C(p) = 2.00974875

                                         Sum of              Mean
                             DF          Squares             Square     F Value     Pr > F
       Regression             2       692.82260654        346.41130327    15.95     0.0011
       Error                  9       195.42739346         21.71415483
       Corrected Total       11       888.25000000

                        Parameter        Standard          Type II
       Variable         Estimate         Error             SS          F Value     Pr > F
       INTERCEP         6.55304825      10.94482708        7.78416178     0.36      0.5641
       HGT              0.72203796       0.26080506      166.42974940     7.66      0.0218
       AGE              2.05012635       0.93722561      103.90008336     4.78      0.0565

       Bounds on condition number:          1.604616,          6.418463

       ------------------------------------------------------------------------------------

                              Statistics for Entry: Step 3
                                     DF = 1, 8
                                         Model
              Variable        Tolerance          R**2         F Value        Pr > F
              AGE2            0.011115           0.7803        0.0097         0.9238
                   No other variable met the 0.1000 significance level for
                                entry into the model.
       Summary of Forward Selection Procedure for Dependent Variable WGT
                    Variable    Number    Partial     Model
       Step         Entered     In        R**2        R**2        C(p)     F Value     Pr > F
         1          HGT          1        0.6630      0.6630      4.2682   19.6749     0.0013
         2          AGE          2        0.1170      0.7800      2.0097    4.7849     0.0565
```

16.5.4 Stepwise Regression Procedure

Stepwise regression is a modified version of forward regression that permits reexamination, at every step, of the variables incorporated in the model in previous steps. A variable that entered at an early stage may become superfluous at a later stage because of its relationship with other variables subsequently added to the model. To check this possibility, at each step we make a partial F test for each variable currently in the model, as though it were the most recent variable entered, irrespective of its actual entry point into the model. The variable with the smallest non-significant partial F statistic (if there is such a variable) is removed; the model is refitted with the remaining variables; the partial F's are obtained and similarly examined; and so on. The whole process continues until no more variables can be entered or removed.

For our example, the first step, as in the forward selection procedure, is to add the variable HGT to the model, since it has the highest significant correlation with WGT. Next, as before, we add AGE to the model, since it has a higher significant partial correlation with WGT than does

$(AGE)^2$, controlling for HGT. Now, before testing to see whether $(AGE)^2$ should also be added to the model, we look at the partial F of HGT, given that AGE is already in the model, to see whether HGT should now be removed. This partial is given by $F(X_1 | X_2) = 7.665$ (see Model 4, Table 16.1), which exceeds $F_{1,9,0.90} = 3.36$. Thus, we do not remove HGT from the model. Next we check to see whether we should add $(AGE)^2$; of course, the answer is no, since we have dealt with this situation before.

The analysis-of-variance table that best summarizes the results obtained for our example is as follows:

Source		d.f.	SS	MS	F	R^2
Regression	X_1	1	588.92	588.92	19.67**	.7800
	$X_2 \| X_1$	1	103.90	103.90	4.79 $(P < .10)$	
Residual		9	195.43	21.71		
Total		11	888.25			

The ANOVA table that considers all variables is:

Source		d.f.	SS	MS	F	R^2
	X_1	1	588.92	588.92	19.67**	.7802
Regression	$X_2 \| X_1$	1	103.90	103.90	4.79 $(P < .10)$	
	$X_3 \| X_1, X_2$	1	0.24	0.24	0.01	
Error		8	195.19	24.40		
Total		11	888.25			

16.5.5   Using Computer Programs

So far we have discussed decisions in stepwise variable selection in terms of F_p and its P-value, R_p^2, C_p, and $MSE(p)$. In backward elimination, the F statistic is often called an "F-to-leave"; while in forward selection, the F statistic is often called an "F-to-enter." Stepwise computer programs require specifying comparison or critical values for these F's or specifying their associated significance levels. These values must not be interpreted as they are in a single-hypothesis test. Their limitation derives from the fact that the probability of finding at least one significant independent variable when there are actually none increases rapidly as the number k of candidate independent variables increases—an approximate upper bound on this overall significance level being $1 - (1 - \alpha)^k$, where α is the significance level of any one test.

To prevent this overall significance level from exceeding α in value, a conservative but easily implemented approach is to conduct any one test at level α/k (see Pope and Webster 1972, and Kupper, Stewart, and Williams 1976, for further discussion). In Section 16.7 we discuss other techniques for helping to ensure the reliability of our conclusions.

Since almost all model selection methods involve using the data to generate hypotheses for further study (often called "exploratory data analysis"), P-values and other variable selection criteria must be utilized cleverly. For example, model selection programs often allow you to specify a P-value to help decide whether a variable is to be considered significant enough to be included in a model. It is very helpful to specify a "P-to-enter" of 1.00 for forward selection.

This guarantees that the process will go as far as possible, thereby providing the maximum information for choosing the best model. Similarly, specifying a value of 0.00 for the *P*-to-leave guarantees that a backward elimination process will remove all variables.

16.5.6 Chunkwise Methods

The methods just described for selecting single variables can be generalized in a very useful way. The basic idea is that any selection method in which a single variable is added or deleted can be generalized to apply to adding or deleting a group of variables. Consider, for example, using backward elimination to build a model for which the response variable is blood cholesterol level. Assume that three groups of predictors are available: demographic (gender, race, age, and their pairwise interactions), anthropometric (height, weight, and their interaction), and diet recall (amounts of five food types); hence, a total of $6 + 3 + 5 = 14$ predictors exists. The three groups of variables constitute *chunks*—sets of predictors that are logically related and equally important (within a chunk) as candidate predictors.

Several possible chunkwise testing methods are available. Choosing among them depends on two factors: the analyst's preference for backward, forward, or other selection strategies; and the extent to which the analyst can logically group variables (i.e., form chunks) and then order the groups in importance. In many applications, an a priori order exists among the chunks. For this example, the researcher may wish to consider diet variables *only* after controlling for important demographic and anthropometric variables. Imposing order in this fashion simplifies the analysis, typically increases reliability, and increases the chance of finding a scientifically plausible model. Hence, whenever possible, we recommend imposing an order on chunk selection.

We illustrate the use of chunkwise testing methods by describing a backward elimination strategy. Other strategies may be preferred, depending on the situation. One approach to a backward elimination method for chunkwise testing involves requiring all but one specified chunk of variables to stay in the model, the variables in that specified chunk being candidates for deletion.[3] For our example, assume that the set of diet variables is the first chunk considered for deletion. (If an a priori order among chunks exists, that order determines which chunk is considered first for deletion. Otherwise, we choose first the chunk that makes the least important predictive contribution (e.g., because it has the smallest multiple partial *F* statistic).)

Thus, in our example, all of the demographic and anthropometric variables are forced to remain in the model. If, for example, the chunkwise multiple-partial *F* test for the set of diet variables is not significant, the entire chunk can be deleted. If this test shows that the chunk is significant, at least one of the variables in the diet chunk should be retained.

Of course, the simplest chunkwise method adds or deletes all variables in a chunk together. A more sensitive approach is to manipulate single variables within a significant chunk while keeping the other chunks in place. If we assume that the diet chunk is important, we must decide which of the diet variables to retain as important predictors. A reasonable second step here is then to require all demographic variables and the important individual diet variables to be retained, while considering the (second) chunk of anthropometric variables for deletion. The final step in this three-chunk example requires that the individual variables selected from the first two chunks remain in the model, while variables in the third (demographic) chunk become

---

[3] Stepwise regression computer packages permit this approach by letting the user specify variables to be forced into the model. The same feature can be used for subsequent chunkwise steps.

candidates for deletion. Forward and stepwise single-variable selection methods can also be generalized for use in chunkwise testing.

Chunkwise methods for selecting variables can have substantial advantages over single-variable selection methods. First, chunkwise methods effectively incorporate into the analysis prior scientific knowledge and preferences about sets of variables. Second, the number of possible models to be evaluated is reduced. If a chunk test is not significant and the entire chunk of variables is deleted, then no tests about individual variables in that chunk are carried out. In many situations, such testing for group (or chunk) significance is more effective and reliable than testing variables one at a time.

16.6 Step 4: Conducting the Analysis

Having specified the maximum model, the criterion for selecting a model, and the strategy for applying the criterion, we must conduct the analysis as planned. Obviously this requires some type of computer program. The goodness of fit of the model chosen should certainly be examined. Also, the regression diagnostic methods of Chapter 14, such as residual analysis, are needed to demonstrate that the model chosen is reasonable for the data at hand. In the next section, we discuss methods for evaluating whether the model chosen has a good chance of fitting new samples of data from the same (or similar) populations.

16.7 Step 5: Evaluating Reliability with Split Samples

Having chosen a model that is best for a particular sample of data, we have no assurance that the model can reliably be applied to other samples. In effect, we are now asking the question "Will our conclusions generalize?" If the chosen model predicts well for subsequent samples from the population of interest, we say that the model is *reliable.* In this section, we discuss methods for evaluating the reliability of a model. Most generally accepted methods for assessing model reliability involve some form of a *split-sample* approach. We discuss three approaches to assessing model reliability here: the follow-up study, the split-sample analysis, and the holdout sample.

The most compelling way to assess the reliability of a chosen model is to conduct a new study and test the fit of the chosen model to the new data. However, this approach is usually expensive and sometimes intractable. Can a single study achieve the two goals of finding the best model and assessing its reliability?

A split-sample analysis attempts to do so. For the cholesterol example in 16.5.6, the simplest split-sample analysis would proceed as follows. First, we randomly assign each observation to one or the other of two groups, the training group or the holdout group. This must be done before any analysis is conducted. Subjects may be grouped in strata based on one or more important categorical variables. For the cholesterol example, appropriate strata might be defined by various gender–race combinations. If such strata are important, we may use a stratum-specific split-sample scheme. With this method, we randomly assign each subject within a stratum to either the training group or the holdout group; such random assignment is done separately for each stratum. The goal of stratum-specific random assignment is to ensure that the two groups of observations (training and holdout) are equally representative of the parent population.

An alternative to stratified random splitting is a pair-matching assignment scheme. With pair-matching assignment, we finds pairs of subjects that are as similar as possible and then randomly assign one member of the pair to the training sample and the other to the holdout sample. Unfortunately, the differences between the resulting training and holdout groups tend to be fewer than corresponding differences among subsequent randomly chosen samples. This tends to create an unrealistically optimistic opinion about model reliability, which we wish to avoid.

Either of two alternatives is usually recommended as the second step in a split-sample analysis. The first is to conduct model selection separately for each of the two groups of data. Typically, any difference in predictor variables chosen by the two selection processes is taken as an indication of unreliability. In practice, the two models obtained *almost always* differ somewhat, which is the primary reason model selection methods have a reputation for being somewhat unreliable. This approach for assessing reliability is far too stringent when prediction is the goal. More realistically, we suggest that a *very good* predictive model should accomplish two things: predict as well in any new sample as it does in the sample at hand; and pass all regression diagnostic tests for model adequacy applied to any new sample. These comments lead us to consider a second approach to split-sample analysis for assessing reliability.

The second approach attempts to answer the question "Will the chosen model predict well in a new sample?" This approach thus addresses a relatively modest, yet very desirable goal. With this second approach, we begin by conducting a model-building process using the data for the training group. Suppose that the fitted prediction equation obtained for the training group data is denoted as

$$\hat{Y}_1 = \hat{\beta}_0 + \hat{\beta}_1 X_1 + \hat{\beta}_2 X_2 + \cdots + \hat{\beta}_p X_p$$

For the children's weight example, this equation takes the form

$$\widehat{\text{WGT}} = 6.553 + 0.722\text{HGT} + 2.050\text{AGE}$$

Next, we use the estimated prediction equation, which is based only on the training group data, to compute predicted values for this group. We denote this set of predicted values as $\{\hat{Y}_{i1}\}$, $i = 1, 2, \ldots, n_1$, with the subscript i indexing the subject and the subscript 1 denoting the training group. For this training sample, we let

$$R^2(1) = R^2_{Y_1 | X_1, X_2, \ldots, X_p} = r^2_{Y_1, \hat{Y}_1}$$

denote the sample squared *multiple* correlation, which equals the sample squared *univariate* correlation between the observed and predicted response values.

Next, we use the prediction equation for the training group to compute predicted values for the holdout (or "validation") sample. We denote this set of predicted values as

$$\{\hat{Y}^*_{i2}\}, \qquad i = n_1 + 1, n_1 + 2, \ldots, n_1 + n_2$$

where n_2 is the number of observations in the holdout group and $n = n_1 + n_2$.

Finally, we compute the univariate correlation between these predicted values and the observed responses in the holdout sample (group 2)—namely,

$$R^2_*(2) = r^2_{Y_2, \hat{Y}^*_2}$$

The quantity $R^2_*(2)$ is called the *cross-validation correlation,* and the quantity

$$R^2(1) - R^2_*(2)$$

is called the *shrinkage on cross-validation*. Hence, the shrinkage statistic is almost always positive. How large must shrinkage be to cast doubt on model reliability? No firm rules can be given. Certainly the fitted model is unreliable if shrinkage is .90 or more. In contrast, shrinkage values less than .10 indicate a reliable model.

Using only half the data to estimate the prediction equation appears rather wasteful. If the shrinkage is small enough to indicate reliability of the model, it seems reasonable to pool the data and calculate pooled estimates of the regression coefficients.

Depending on the situation, using only half of the data for the training sample analysis may be inadvisable. A useful rule of thumb is to increase the relative size of the training sample as the total sample size decreases and to decrease the relative size of the training sample as the total sample size increases. To illustrate this, consider two situations. In the first situation, data from over 3,000 subjects were available for regression analysis. The maximum model included fewer than 20 variables, consisting of a chunk of demographic control variables and a chunk of pollution exposure variables. The primary goal of the study was to test for the presence of a relationship between the response variable (a measure of pulmonary function) and the pollution variables, controlling for demographic effects. As a framework for choosing the best set of demographic control variables, a 10% stratified random sample ($n \approx 300$) was used as a training sample.

The second situation involved a study of the relationship between measurements of amount of body fat by X-ray methods (CAT scans) and by traditional parameters such as skinfold thickness. The primary goal was to provide an equation to predict the X-ray-measured body fat from demographic information (gender and age) and simple anthropometric measures (body weight, height, and three readings of skinfold thickness). The maximum model contained fewer than 20 predictors. Data from approximately 200 subjects were available. These data constituted nearly one year's collection from the hospital X-ray laboratory and demanded substantial human effort and computer processing to extract. Consequently, approximately 75% of the data were used as a training sample. These two examples illustrate the general concept that the splitting proportion should be tailored to the problem under study.

16.8 Example Analysis of Actual Data

We are now ready to apply the methods discussed in this chapter to a set of actual data. In Chapter 14 (on regression diagnostics), we introduced data on more than 200 children, reported by Lewis and Taylor (1967). The categories reported for all subjects include body weight (WGT), standing height (HGT), AGE, and SEX. Our goal here is to provide a reliable equation for predicting weight.

The first step is to choose a maximum model. The number of subjects is large enough that we can consider a fairly wide range of models. It is natural to want to include the *linear* effects of AGE, HGT, and SEX in the model. Quadratic and cubic terms for both AGE and HGT will also be included, since WGT is expected to increase in a nonlinear way as a function of AGE and HGT. Because the same model is unlikely to hold for both males and females, the interaction of the dichotomous variable SEX with each power of AGE and HGT will be included. Terms involving cross-products between AGE and HGT powers will not be included, since no biological fact suggests that such interactions are important. Hence, $k = 13$ variables will be included in the maximum model as predictors of WGT: AGE; AGE2 $= (\text{AGE})^2$; AGE3 $= (\text{AGE})^3$; HGT;

HT2 = (HGT)^2; HT3 = (HGT)^3; SEX = 1 for a boy, and 0 for a girl; SEX × AGE; SEX × AGE2; SEX × AGE3; SEX × HGT; SEX × HT2; and SEX × HT3. (In the analysis to follow, AGE and HGT denote centered variables.)

The second step is to choose a selection criterion. We choose to emphasize C_p, while also looking at R_p^2. The former criterion is helpful in deciding the size of the best model, while the latter provides an easily interpretable measure of predictive ability.

The third step is to choose a strategy for selecting variables. We prefer backward elimination methods, with as much structure imposed on the process as possible. As mentioned earlier, one procedure is to group the variables into chunks and then to order the chunks by degree of importance. However, we shall not consider a chunkwise strategy here.

After the (seemingly) best model is chosen, the fourth step is to evaluate the reliability of the model. Since the goal here is to produce a reliable prediction equation, this step is especially important. To implement a split-sample approach to assessing reliability, we first randomly assigned each subject to either data set ONE (the training group) or data set TWO (the holdout group). Random splitting was sex-specific. The following table summarizes the observed split-sample subject distribution.

Sample	Female	Male	Total
ONE	55	63	118
TWO	56	63	119
Total	111	126	237

Data set ONE will be used to build the best model. Then data set TWO will be used to compute cross-validation correlation and shrinkage statistics.

16.8.1   Analysis of Data Set ONE

Table 16.2 provides descriptive statistics for data set ONE, and Table 16.3 provides the matrix of zero-order correlations. To avoid computational problems, the predictor variables were centered (see Table 16.2 and Chapter 14).

Table 16.4 summarizes the backward elimination analysis based on data set ONE. For data set ONE, SSY = 35,700.70 and MSE = 120.00 for the full model. Table 16.4 specifies the order of variable elimination, the variables in the best model for each possible model size, and C_p and R_p^2 for each such model. Since the maximum model includes $k = 13$ variables, the top row corresponds to $p = 13$, for which $R_p^2 = .65042$. Since the maximum model always has $C_p = k + 1$,

TABLE 16.2   Descriptive statistics for data set ONE ($n = 118$)

Variable	Mean	S.D.	Minimum	Maximum
WGT (lb)	100.05	17.47	63.5	150.0
AGE[a] (yr)	0	1.50	−2.00	3.91
HGT[b] (in.)	0	3.7	−8.47	10.73

[a]Equals (Age − $\overline{\text{AGE}}_1$), where $\overline{\text{AGE}}_1 \doteq 13.59$
[b]Equals (Height − $\overline{\text{HGT}}_1$), where $\overline{\text{HGT}}_1 \doteq 61.27$

TABLE 16.3   **Correlations ($\times 100$) for data set ONE with centered age and height variables ($n = 118$)**

WGT													
AGE	59												
AGE2	25	44											
AGE3	48	78	80										
HGT	77	67	16	43									
HT2	05	10	23	11	15								
HT3	63	46	14	32	79	35							
SEX	03	−01	−04	−03	18	18	24						
SEX × AGE	46	72	28	50	62	29	46	00					
SEX × AGE2	18	26	55	38	27	35	32	51	37				
SEX × AGE3	38	58	49	60	46	30	40	12	81	70			
SEX × HGT	57	55	16	34	81	41	68	10	77	27	57		
SEX × HT2	26	22	19	17	43	78	70	50	31	53	36	47	
SEX × HT3	48	38	17	27	64	65	88	16	54	32	46	78	75

TABLE 16.4   **Backward elimination based on data set ONE ($n = 118$)**

p	C_p	R_p^2	Variables Used
13	14.00	.65042	HT3
12	12.00	.65042	SEX × AGE
11	10.00	.65042	AGE2
10	8.00	.65040	SEX × AGE2
9	6.00	.65040	SEX × AGE3
8	4.03	.65032	HT2
7	2.08	.65015	AGE
6	0.38	.64914	SEX
5	−0.82	.64644	SEX × HT3
4	1.15	.63312	SEX × HT2
3	−0.12	.63066	SEX × HGT
2	0.52	.62177	AGE3
1	6.72	.59422	HGT

the C_p-value of 14 provides no information regarding the best model size. The second row corresponds to a $p = 12$ variable model. The variable HT3 (listed in the preceding row) has been deleted, leaving $C_p = 12$. Similarly, the bottom row tells us that the best single-variable model includes HGT ($C_p = 6.72, R_p^2 = .59422$), the best two-variable model includes HGT and AGE3 ($C_p = 0.52, R_p^2 = .62177$), and so on. In general, the variable identified in row p in the table and all variables listed in rows below it are included in the best p-variable model, while all variables listed in rows above row p are excluded.

Table 16.4 does not suggest any obvious choice for a best model. The maximum R^2 is about .650, and the minimum is around .594, which is a small operating range, indicating that many different-size models predict about equally well. This is not unusual with moderate to strong predictor intercorrelations—say, above .50 in absolute value (see Table 16.3). Since

$R_7^2 \doteq .650$ and $R_{13}^2 \doteq .650$, it seems unreasonable to use more than seven predictors. However, on the basis of R_p^2 alone, it is difficult to decide whether a model containing fewer than seven predictors is appropriate.

In contrast, the C_p statistic suggests that only two variables are needed. (Recall that C_p should be compared with the value $p + 1$.) C_p is greater than $p + 1$ only when $p = 1$, which calls into question only the one-variable model in Table 16.4. Unfortunately, the $p = 2$ model with HGT and AGE3 as predictors is not appealing, since the linear and quadratic terms AGE and AGE2 are not included. As discussed in detail in Chapter 15, we strongly recommend including such lower-order terms in polynomial-type regression models.

An apparently nonlinear relationship may indicate the need to transform the response and/or predictor variables. Since quadratic and cubic terms are included as candidate predictors, $\log(\text{WGT})$ and $(\text{WGT})^{1/2}$ can be tried as alternative response variables. Such transformations can sometimes linearize a relationship (see Chapter 14). Based on all 13 predictors, $R^2 = .668$ for the log transformation and .659 for the square root transformation. Because neither transformation produces a substantial gain in R^2, they will not be considered further.

In additional exploratory analyses of the data in data set ONE, models were fitted in two fixed orders: an interaction ordering (Table 16.5), and a power ordering (Table 16.6). We can interpret Tables 16.5 and 16.6 similarly to Table 16.4; each row gives p, the number of variables in the model, and the associated C_p- and R_p^2-values. All variables on or below line p are included in the fitted model, and all those above line p are excluded.

In both tables, certain values of p are natural break points for model evaluation. For example, in Table 16.5, $p = 3$ corresponds to including just the linear terms of the predictors HGT, AGE, and SEX. The fact that $4.98 > (3 + 1)$ leads us to consider larger models. Including the pairwise interactions SEX $\times$ HGT and SEX $\times$ AGE gives $p = 5$ and $C_5 = 5.36 < (5 + 1)$, which suggests that this five-variable model is reasonable. Notice that $R_5^2 = .626$ and that each subsequent variable addition increases R^2 very little. In fact, the best model might be of size $p = 4$, since $C_4 = 3.60$ and R^2 is not appreciably reduced. Taken together, these comments suggest choosing a four-variable model containing HGT, AGE, SEX, and SEX $\times$ HGT, with $R_4^2 = .625$ and $C_4 = 3.60$.

TABLE 16.5   **Interaction-ordered fitting based on data set ONE ($n = 118$)**

p	C_p	R_p^2	Variables Used
13	14.00	.65042	SEX $\times$ AGE3
12	12.01	.65040	SEX $\times$ HT3
11	10.39	.64909	SEX $\times$ AGE2
10	8.39	.64909	SEX $\times$ HT2
9	6.94	.64727	AGE3
8	6.97	.64045	HT3
7	6.37	.63575	AGE2
6	7.35	.62571	HT2
5	5.36	.62567	SEX $\times$ AGE
4	3.60	.62486	SEX $\times$ HGT
3	4.98	.61352	SEX
2	5.86	.60383	AGE
1	6.72	.59422	HGT

TABLE 16.6 **Power-ordered fitting based on data set ONE ($n = 118$)**

p	C_p	R_p^2	Variables Used
13	14.00	.65042	SEX × AGE3
12	12.01	.65040	SEX × HT3
11	10.39	.64909	SEX × AGE2
10	8.39	.64909	SEX × HT2
9	6.94	.64727	SEX × AGE
8	8.00	.64698	SEX × HGT
7	4.77	.64112	AGE3
6	5.05	.63343	HT3
5	4.83	.62747	AGE2
4	6.27	.61589	HT2
3	4.98	.61352	SEX
2	5.86	.60383	AGE
1	6.72	.59422	HGT

Table 16.6 summarizes a similar analysis in which the powers of the continuous predictors are entered into the model in a logical ordering. Here, as in Table 16.5, $p = 3$ gives $C_3 = 4.98$, encouraging us to consider larger models. The model with all three original variables and the two quadratic terms gives $p = 5$, $C_p = 4.83$, and $R_p^2 = .62747$. Adding higher-order powers or any interactions does not improve R^2 appreciably, nor does it lead to C_p-values noticeably greater than $p + 1$. All models smaller than $p = 5$, on the other hand, do have C_p-values noticeably greater than $p + 1$. The results in Table 16.6 lead to choosing a five-variable model containing HGT, AGE, SEX, HT2, and AGE2, with $R_5^2 = .627$ and $C_5 = 4.83$.

In this example, then, two possible models have been suggested, both with roughly the same R^2-value. Personal preference may be exercised within the constraints of parsimony and scientific knowledge. In this case, we prefer the five-variable model with HGT, AGE, SEX, HT2, and AGE2 as predictors. We choose this model as best because we expect growth to be nonlinear and because we prefer to avoid using interaction terms. The chosen model is thus of the general form

$$\widehat{WGT} = \hat{\beta}_0 + \hat{\beta}_1 SEX + \hat{\beta}_2 HGT + \hat{\beta}_3 AGE + \hat{\beta}_4 HT2 + \hat{\beta}_5 AGE2$$

A predicted weight for the ith subject may be computed with the formula

$$\widehat{WGT}_i = 100.9 - 3.153(SEX_i) + 3.53(HGT_i - 61.27) + .4199(AGE_i - 13.59)$$
$$- .0731(HGT_i - 61.27)^2 + .8067(AGE_i - 13.59)^2$$

Recall that SEX has a value of 1 for boys and 0 for girls. For males, the estimated intercept is $100.9 - 3.153(1) = 97.747$; while for females, it is $100.9 - 3.153(0) = 100.9$. For these particular data, then, the fitted model predicts that, on average, girls outweigh boys of the same height and age by about 3 pounds. After selecting a model, we must consider residual analysis and other regression diagnostic procedures (see Chapter 14). Some large residuals are present, but these do not justify a more complex model. Since they turn out not to be influential, they need not be deleted to improve estimation accuracy.

16.8.2 Analysis of Sample TWO

Since we used exploratory analysis of data set ONE to choose the best (most appealing) model, we will use analysis of the holdout data set TWO to evaluate the reliability of the chosen prediction equation. Table 16.7 provides descriptive statistics for data set TWO. These appear very similar to those of data set ONE, as one would hope. In particular, the AGE and HGT variables are centered using the sample ONE means. This continuity is necessary to maintain the definitions of these variables as used in the fitted model. Cross-validation analysis would look spuriously bad if this were not done.

In comparison to Table 16.4, Table 16.8 illustrates the instability of stepwise methods for selecting variables. Despite using the same backward elimination strategy, we find that almost no variable appears in the same place as it did in the analysis for data set ONE.

Our recommended cross-validation analysis begins by using the regression equation estimated from data set ONE (with predictors HGT, AGE, SEX, HT2, and AGE2) to predict WGT values for data set TWO. The squared multiple correlation between these predicted values and the observed WGT values in data set TWO is $R_*^2(2) = .621$, and this is the cross-validation squared correlation. Since $R^2(1) = .627$ for sample ONE, shrinkage is $.627 - .621 = .006$, which is quite small and indicates excellent reliability of estimation.

TABLE 16.7   Descriptive statistics for data set TWO ($n = 119$)

Variable	Mean	S.D.	Minimum	Maximum
WGT (lb)	102.6	21.22	50.5	171.5
AGE[a] (yr)	0.20	1.46	−1.92	3.58
HGT[b] (in.)	0.18	4.16	−10.77	9.73

[a]Equals (Age $-\ \overline{\text{AGE}}_1$), where $\overline{\text{AGE}}_1 \doteq 13.59$
[b]Equals (Height $-\ \overline{\text{HGT}}_1$), where $\overline{\text{HGT}}_1 \doteq 61.27$

TABLE 16.8   Backward elimination based on data set TWO ($n = 119$)

p	C_p	R_p^2	Variables Used
13	14.00	.72328	AGE2
12	12.00	.72328	SEX
11	10.02	.72323	AGE3
10	8.31	.72245	SEX × AGE2
9	7.22	.72006	HT2
8	6.06	.71781	SEX × HT2
7	4.15	.71763	SEX × HGT
6	3.16	.71496	HT3
5	1.52	.71402	SEX × HT3
4	2.58	.70595	SEX × AGE
3	7.99	.68642	AGE
2	9.60	.67690	SEX × AGE3
1	33.90	.60762	HGT

An important aspect of assessing the reliability of a model involves considering difference scores of the form

$$Y_i - \hat{Y}_i$$

or (in our example)

$$\text{WGT}_i - \widehat{\text{WGT}}_i$$

where only data set TWO subjects are used and where the data set ONE equation is used to compute the predicted values. These "unstandardized residuals" can be subjected to various residual analyses. The most helpful such analysis entails univariate descriptive statistics, a box-and-whisker plot, and a plot of the difference scores $(Y_i - \hat{Y}_i)$ versus the predicted values $(\hat{Y}_i)$ (such plots are described in Chapter 14). In our example, a few large positive residuals are present, but they are not sufficiently implausible or influential to require further investigation. Their presence does hint at why cubic terms and interactions keep trying to creep into the model. These residuals could be reduced in size, but only by adding many more predictors.

For prediction purposes, if the model is finally deemed acceptable, we should pool all the data and report the coefficient estimates based on the combined data. If the model is not acceptable, we must review the model-building process, paying particular attention to variable selection and large differences between training and holdout groups of data.

16.9 Issues in Selecting the Most Valid Model

In this chapter, we have considered variable selection strategies for situations where the primary study goal is prediction. In contrast, validity-oriented strategies are directed toward providing valid estimates of one or more regression coefficients in a model. Essentially, a valid estimate accurately reflects the true relationship(s) in the target population being studied. As described in Chapter 11, in a validity-based strategy for selecting variables, both confounding and interaction must be considered. Moreover, the sample of data under consideration must accurately reflect the population of interest; and the assumptions attending the model and analysis selected must be reasonably well satisfied.

More specifically, a variable selection strategy with a validity goal first involves determining important interaction effects, followed by evaluating the effect of confounding, which is contingent on the results of assessing interaction in the analysis. For examples of this type of strategy, see the epidemiologic-research-oriented text of Kleinbaum, Kupper, and Morgenstern (1982, Chapters 21–24), and Kleinbaum and Klein (2002, chapters 6–7).

Problems _____

1. Add the variables $(\text{HGT})^2$ and $(\text{AGE} \times \text{HGT})$ to the data set for WGT, HGT, AGE, and $(\text{AGE})^2$ given in Section 16.3. Then, using an available regression program and the accompanying computer output, determine the best regression model relating WGT to the five independent variables as follows.
   a. Use the forward approach.
   b. Use the backward approach.

5. The data set listed in the following table contains information on AGE, SEX (1 = male, 2 = female), work problems index (WP), marital conflict index (MC), and depression index (DEP) for a sample of 39 new admissions to a psychiatric clinic at a large university hospital. For each sex *separately,* determine (using α = .10) the best regression model relating DEP to MC and WP, controlling for AGE, using the following sequential procedure: (1) force AGE into the model first; (2) use the all possible regressions approach on the remaining two independent variables WP and MC; (3) determine whether the interaction

ID No.	AGE	SEX	WP	MC	DEP
1	45	2	90	70	69
2	35	1	90	75	75
3	32	2	70	32	35
4	32	2	80	30	73
5	39	2	85	55	86
6	25	2	85	6	161
7	22	1	75	20	202
8	30	2	70	63	91
9	49	2	75	4	113
10	47	1	84	12	68
11	48	1	64	11	109
12	49	2	85	7	92
13	45	2	80	8	80
14	41	2	80	15	82
15	45	2	82	6	156
16	59	2	72	5	198
17	42	2	70	17	170
18	35	1	70	29	188
19	31	2	70	80	82
20	45	1	70	126	37
21	28	1	85	30	194
22	37	1	90	9	294
23	29	1	80	14	94
24	29	1	70	24	126
25	31	1	80	21	192
26	29	1	60	11	232
27	29	1	70	10	184
28	23	2	80	10	238
29	44	2	78	19	112
30	28	1	70	22	141
31	32	2	70	21	108
32	36	2	74	77	87
33	22	2	78	67	33
34	46	2	70	25	73
35	21	1	70	14	168
36	34	1	80	17	218
37	27	2	80	18	175
38	31	2	80	42	126
39	19	2	75	36	135

term (MC × WP) should be added to the model. Compare and discuss the results obtained for each sex.

6. For the data in Problem 15 in Chapter 5, use LN_BRNTL as the response variable and LN_PPMTL, LN_BLDTL, AGE, and WEIGHT as predictors. (Use $\alpha = .05$.)

   a. Indicate a plausible fixed order for testing predictors, based on the nature of the data. Briefly defend your choice.

   b. Use a computer program to fit the full model. Provide a test of whether or not the population multiple correlation is 0.

   c. Using the *fixed* order, choose a best model, adding variables in a forward fashion. Do not adhere to a particular α, but use your judgment.

   d. Repeat part (c) but delete the variables in the fixed order (a backward method).

   e. Using a computer program employing a stepwise procedure, use a backward algorithm to find the best model.

   f. Repeat part (e), using a forward algorithm.

   g. Compare and contrast your conclusions for parts (c), (d), (e), and (f). Indicate your preferred model.

   h. Using the ideas presented in Chapter 14, indicate how you could evaluate the adequacy of the best model and the validity of the underlying assumptions.

   i. Suggest a practical split-sample approach for this particular set of data. Include recommended sample sizes, any stratification variables, and a variable selection strategy.

7. Consider the data of Bethel and others (1985), discussed in Problem 19 in Chapter 14. Delete the three female subjects, leaving 16 observations. Use FEV_1 as the response, and AGE, WEIGHT, and HEIGHT as predictors.

   a. Use the all possible regressions procedure to suggest a best model.

   b. Consider a model with centered AGE, WEIGHT, HEIGHT, and their squares as predictors. Suggest a plausible forward chunkwise strategy for choosing a model, and implement it.

   c. Use the all possible regressions procedure for the expanded model to choose a best model.

   d. Compare results from parts (a), (b), and (c). What model seems most plausible? How do the data limit your conclusions?

8. Use the data from Freund (1979), presented in Problem 22 in Chapter 14. Taking the model discussed there as the maximum model, repeat parts (a) through (h) of Problem 6. In part (h), note the possible role of collinearity.

9. A random sample of data were collected on residential sales in a large city. The accompanying table shows the selling price (Y, in \$1,000s), area (X_1, in hundreds of square feet), number of bedrooms (X_2), total number of rooms (X_3), house age (X_4, in years), and location ($Z = 0$ for in-town and inner suburbs, $Z = 1$ for outer suburbs).

   In parts (a) through (c), use variables X_1, X_2, X_3, X_4, and Z as the predictor variables. Use the accompanying computer output to answer parts (a)–(d).

   a. Use the all possible regressions procedure to suggest a best model.

   b. Use the stepwise regression algorithm to suggest a best model.

   c. Use the backward elimination algorithm to suggest a best model.

   d. Which of the models selected in parts (a), (b), and (c) seems to be the best model, and why?

House	Y	X_1	X_2	X_3	X_4	Z
1	84.0	13.8	3	7	10	0
2	93.0	19.0	2	7	22	1
3	83.1	10.0	2	7	15	1
4	85.2	15.0	3	7	12	1
5	85.2	12.0	3	7	8	1
6	85.2	15.0	3	7	12	1
7	85.2	12.0	3	7	8	1
8	63.3	9.1	3	6	2	1
9	84.3	12.5	3	7	11	1
10	84.3	12.5	3	7	11	1
11	77.4	12.0	3	7	5	0
12	92.4	17.9	3	7	18	0
13	92.4	17.9	3	7	18	0
14	61.5	9.5	2	5	8	0
15	88.5	16.0	3	7	11	0
16	88.5	16.0	3	7	11	0
17	40.6	8.0	2	5	5	0
18	81.6	11.8	3	7	8	1
19	86.7	16.0	3	7	9	0
20	89.7	16.8	2	7	12	0
21	86.7	16.0	3	7	9	0
22	89.7	16.8	2	7	12	0
23	75.9	9.5	3	6	6	1
24	78.9	10.0	3	6	11	0
25	87.9	16.5	3	7	15	0
26	91.0	15.1	3	7	8	1
27	92.0	17.9	3	8	13	1
28	87.9	16.5	3	7	15	0
29	90.9	15.0	3	7	8	1
30	91.9	17.8	3	8	13	1

Edited SAS Output (PROC REG) for Problem 9

N = 30	Regression models for Dependent Variable: Y			
Number in Model	R-square	C(p)	MSE	Variables in Model
1	0.74168607	11.59122	32.78751	X3
1	0.63919733	26.50593	45.79629	X1
1	0.37277539	65.27708	79.61294	X4
1	0.11031111	103.47228	112.92725	X2
1	0.01448589	117.41728	125.09024	Z
2	0.80692546	4.09724	25.41440	X1 X3
2	0.80634720	4.18139	25.49052	X3 X4
2	0.75352898	11.86778	32.44298	X3 Z
2	0.74172797	13.58513	33.99634	X2 X3

(continued)

Edited SAS Output for Problem 9 (continued)

```
N = 30   Regression models for Dependent Variable: Y (Continued)
Number in     R-square         C(p)            MSE          Variables
Model                                                       in Model

      2       0.70553756       18.85175        38.76009     X1  Z
      2       0.69355902       20.59493        40.33682     X1  X2
      2       0.64446398       27.73950        46.79920     X1  X4
      2       0.57233554       38.23602        56.29347     X2  X4
      2       0.40513240       62.56831        78.30242     X4  Z
      2       0.11465711      104.83983       116.53767     X2  Z
------------------------------------------------------------------------------
      3       0.82237293        3.84924        24.28032     X1  X3  X4
      3       0.81899611        4.34065        24.74191     X2  X3  X4
      3       0.81039947        5.59168        25.91700     X1  X2  X3
      3       0.80863284        5.84877        26.15849     X3  X4  Z
      3       0.80801358        5.93889        26.24314     X1  X3  Z
      3       0.75353085       13.86751        33.69053     X2  X3  Z
      3       0.74058026       15.75215        35.46078     X1  X2  Z
      3       0.72307049       18.30026        37.85424     X1  X2  X4
      3       0.70977138       20.23562        39.67213     X1  X4  Z
      3       0.58509132       38.37973        56.71498     X2  X4  Z
------------------------------------------------------------------------------
      4       0.83459939        4.06998        23.51341     X1  X2  X3  X4
      4       0.82270177        5.80139        25.20478     X1  X3  X4  Z
      4       0.82090242        6.06324        25.46058     X2  X3  X4  Z
      4       0.81180550        7.38707        26.75380     X1  X2  X3  Z
      4       0.76269650       14.53367        33.73516     X1  X2  X4  Z
------------------------------------------------------------------------------
      5       0.83508027        6.00000        24.42193     X1  X2  X3  X4  Z
------------------------------------------------------------------------------

              Stepwise Procedure for Dependent Variable Y

Step 1 Variable X3 Entered R-square = 0.74168607
       C(p) = 11.59122300
                                    Sum of          Mean
                      DF           Squares          Square      F Value     Pr > F
Regression             1      2635.95947489    2635.95947489      80.40    <0.0001
Error                 28       918.05019178      32.78750685
Corrected Total       29      3554.00966667

                  Parameter        Standard        Type II
Variable          Estimate           Error              SS      F Value     Pr > F
INTERCEP       -17.08438356      11.26623604       75.39617765     2.30     0.1406
X3              14.71917808       1.64160396     2635.95947489     80.40    <0.0001

Bounds on condition number:                      1,           1
------------------------------------------------------------------------------

Step 2 Variable X1 Entered   R-square = 0.80692546
       C(p) = 4.09723891
                                    Sum of          Mean
                      DF           Squares          Square      F Value     Pr > F
Regression             2      2867.82088345    1433.91044172      56.42    <0.0001
Error                 27       686.18878322      25.41439938
Corrected Total       29      3554.00966667
```

(continued)

Edited SAS Output for Problem 9 (continued)

```
Step 2 (Continued)
              Parameter        Standard          Type II
Variable      Estimate          Error                 SS    F Value    Pr > F
INTERCEP    -4.21355813      10.79550361         3.87161368     0.15    0.6994
X1           1.30267900       0.43128376       231.86140856     9.12    0.0055
X3          10.14195666       2.09410969       596.10738063    23.46   <0.0001

Bounds on condition number:          2.099378,          8.397511
```

```
Step 3 Variable X4 Entered R-square = 0.82237293
         C(p) = 3.84924057

                            Sum of            Mean
                   DF       Squares          Square    F Value    Pr > F
Regression          3   2922.72134013    974.24044671     40.12   <0.0001
Error              26    631.28832653     24.28032025
Corrected Total    29   3554.00966667

              Parameter        Standard          Type II
Variable      Estimate          Error                 SS    F Value    Pr > F
INTERCEP    -4.92269366      10.56242154         5.27391395     0.22    0.6451
X1           0.81475833       0.53197084        56.95559271     2.35    0.1377
X3          10.52638005       2.06275692       632.29012250    26.04   <0.0001
X4           0.45796577       0.30455957        54.90045668     2.26    0.1447

Bounds on condition number:          3.343224,          22.37951
```

```
All variables left in the model are significant at the 0.1500 level.
No other variable met the 0.1500 significance for entry into
the model.
                    Summary of Stepwise Procedure for
                         Dependent Variable Y

         Variable
         Entered    Number    Partial    Model
Step     Removed      In        R**2      R**2      C(p)    F Value    Pr > F
 1        X3           1       0.7417    0.7417   11.5912    80.3952   <0.0001
 2        X1           2       0.0652    0.8069    4.0972     9.1232    0.0055
 3        X4           3       0.0154    0.8224    3.8492     2.2611    0.1447
```

10. In Problem 9, the first-order interactions between Z and the predictor variables X_1, X_3, and X_4 were not included in the model selection process. Investigate whether this was appropriate, as follows (using the accompanying computer output).

    a. Conduct a test to investigate the importance of the interactions, given that X_1, X_2, X_3, X_4, and Z are in the model.

    b. In view of your answer in part (a), and in light of the model selected in Problem 9, was it appropriate to exclude the interactions?

    c. Why is it necessary to ignore the interaction term $X_2 \times Z$?

11. In 1990, *Business Week* magazine compiled financial data on the 1,000 companies that had the biggest impact on the U.S. economy.[4] Data from a sample of the top 500 companies in

---

[4] "The 1990 Business Week 1000," in *Business Week* magazine, April 13, 1990.

Edited SAS Output (PROC REG) for Problem 10

```
                              Analysis of Variance

                                 Sum of            Mean
Source                   DF      Squares          Square       F Value      Pr > F
Model                     8    3100.73976       387.59247      17.957      <0.0001
Error                    21     453.26991        21.58428
Corrected Total          29    3554.00967

                 Root MSE           4.64589       R-square        0.8725
                 Dep Mean          83.49667       Adj R-sq        0.8239
                 Coeff Var          5.56416

                              Parameter Estimates

                         Parameter        Standard           t Value
Variable      DF         Estimate           Error        Parameter = 0     Pr > |T|
INTERCEP       1        -24.707976       16.10563434        -1.534          0.1399
X1             1          0.277195        1.30211784         0.213          0.8335
X2             1          0.283568        3.48718098         0.081          0.9360
X3             1         13.607861        4.98736294         2.728          0.0126
X4             1          1.006553        0.57142756         1.761          0.0927
Z              1         53.467221       26.89803697         1.988          0.0600
X1Z            1          0.706832        1.55572552         0.454          0.6542
X3Z            1         -8.124224        5.76509590        -1.409          0.1734
X4Z            1         -0.676632        0.83800926        -0.807          0.4285

Variable      DF         Type I SS
INTERCEP       1           209151
X1             1        2271.713503
X2             1         193.201955
X3             1         415.252079
X4             1          86.006770
Z              1           1.709055
X1Z            1          89.289861
X3Z            1          29.494884
X4Z            1          14.071648
```

Business Week's report were presented in Problem 13 in Chapter 8. In addition to the company name, the following variables were shown:

1990 Rank (X_1):   Based on company's market value (share price on March 16, 1990, multiplied by available common shares outstanding).

1989 Rank (X_2):   Rank in 1989 compilation.

P-E Ratio (X_3):   Price-to-earning ratio based on 1989 earnings and March 16, 1990 share price.

Yield (Y):   Annual dividend rate as a percentage of March 16, 1990 share price.

In parts (a) and (b), use variables X_1, X_2, and X_3 as the predictor variables.

11. a. Use the all possible regressions procedure to suggest a best model.
  b. Use the stepwise regression algorithm to suggest a best model.
  c. Which model seems to be the better one? Why?

Edited SAS Output (PROC REG) for Problem 11

```
                    ALL POSSIBLE REGRESSIONS OUTPUT
N = 20  Regression Models for Dependent Variable: Y

Number in        R-square          C(p)            MSE        Variables in Model
Model
       1        0.38695207      10.05134       1.6375292           X3
       1        0.03803947      24.87831       2.5695192           X1
       1        0.03670921      24.93484       2.5730725           X2
-------------------------------------------------------------------------------
       2        0.59531344       3.19707       1.1445558           X1 X3
       2        0.55288183       5.00019       1.2645632           X2 X3
       2        0.03807258      26.87690       2.7205738           X1 X2
-------------------------------------------------------------------------------
       3        0.62348324       4.00000       1.1314398           X1 X2 X3
-------------------------------------------------------------------------------

                       STEPWISE REGRESSION OUTPUT

            Stepwise Procedure for Dependent Variable Y

Step 1 Variable X3 Entered R-square = 0.38695207
       C(p) = 10.05134226
                                Sum of          Mean
                    DF          Squares         Square      F Value     Pr > F
Regression           1       18.60476972     18.60476972    11.36       0.0034
Error               18       29.47552528      1.63752918
Corrected Total     19       48.08029500
                  Parameter       Standard        Type II
Variable          Estimate         Error             SS       F Value     Pr > F
INTERCEP          5.49230032      0.93681453     56.28470078    34.37     <0.0001
X3               -0.19274361      0.05718240     18.60476972    11.36      0.0034
Bounds on condition number:            1,              1
-------------------------------------------------------------------------------
Step 2 Variable X1 Entered R-square = 0.59531344
       C(p) = 3.19706996
                                Sum of          Mean
                    DF          Squares         Square      F Value     Pr > F
Regression           2       28.62284574     14.31142287    12.50       0.0005
Error               17       19.45744926      1.14455584
Corrected Total     19       48.08029500
                  Parameter       Standard        Type II
Variable          Estimate         Error             SS       F Value     Pr > F
INTERCEP          7.21134513      0.97521036     62.58538586    54.68     <0.0001
X1               -0.00513171      0.00173456     10.01807602     8.75      0.0088
X3               -0.24882557      0.05142752     26.79389702    23.41      0.0002
Bounds on condition number:  1.157227,      4.628907
-------------------------------------------------------------------------------

All variables left in the model are significant at the 0.0500 level.
No other variables met the 0.0500 significance level for entry into the model.
       Summary of Stepwise Procedure for Dependent Variable Y

            Variable
            Entered      Number     Partial    Model
Step        Removed        In        R**2      R**2        C(p)      F Value     Pr > F
   1        X3             1        0.3870     0.3870    10.0513    11.3615     0.0034
   2        X1             2        0.2084     0.5953     3.1971     8.7528     0.0088
```

12. Radial keratotomy is a type of refractive surgery in which radial incisions are made in a myopic (nearsighted) patient's cornea to reduce the person's myopia. Theoretically, the incisions allow the curvature of the cornea to become less steep, thereby reducing the patient's refractive error. This and other vision-correction surgery techniques have been growing in popularity in the 1980s and 1990s, both among the public and among ophthalmologists.

The Prospective Evaluation of Radial Keratotomy (PERK) clinical trial was begun in 1983 to evaluate the effects of radial keratotomy. As part of the study, Lynn et al. (1987) examined the variables associated with the sample patients' five-year postsurgical change in refractive error (Y, measured in diopters, D). Several independent variables were under consideration:

Baseline refractive error (X_1, diopters)

Patient age (X_2, in years)

Patient's sex (X_3)

Baseline average central keratometric power (X_4, a measure of corneal curvature, in diopters)

Depth of incision scars (X_5, in mm)

Baseline horizontal corneal diameter (X_6, in mm)

Baseline intraocular pressure (X_7, in mm Hg)

Baseline central corneal thickness (X_8, in mm)

Diameter of clear zone (X_9, in mm). (The clear zone is the circular central portion of the cornea that is left uncut during the surgery; the surgical incisions are made radially from the periphery of the cornea to the edge of the clear zone. Smaller clear zones are used for more myopic patients, the thinking being that "more" surgery, in the form of longer incisions, is probably needed for such patients.)

Some of the PERK study results from the all possible subsets analysis that was performed are shown in the following table.

Variable Added to Model	Model R^2 as Variables Are Added
Diameter of clear zone (X_9)	0.28
Patient age (X_2)	0.40
Depth of incision scars (X_5)	0.44
Baseline refractive error (X_1)	0.45
Baseline horizontal corneal diameter (X_6)	0.47
Baseline avg. central keratometric power (X_4)	0.48
Baseline intraocular pressure (X_7)	0.49
Patient's sex (X_3)	0.49
Baseline central corneal thickness (X_8)	0.49

a. Using this table of R^2 values, perform an all possible regressions analysis to suggest a best model.

b. The PERK study researchers concluded that "The regression analysis of the factors affecting the outcome of radial keratotomy [i.e., the change in refractive error] showed that the diameter of the clear zone, patient age, and the average depth of the incision scars were the most important factors." Do you agree with this assessment? Explain.

References

Bethel, R. A.; Sheppard, D.; Geffroy, B.; Tam, E.; Nadel, J. A.; and Boushey, J. A. 1985. "Effect of 0.25 ppm Sulfur Dioxide on Airway Resistance in Freely Breathing, Heavily Exercising, Asthmatic Subjects." *American Review of Respiratory Diseases* 131: 659–61.

Freund, R. J. 1979. "Multicollinearity etc., Some 'New' Examples." *Proceedings of the Statistical Computing Section,* American Statistical Association, pp. 111–12.

Hocking, R. R. 1976. "The Analysis and Selection of Variables in Linear Regression." *Biometrics* 32: 1–49.

Kleinbaum, D. G., and Klein, M. 2002. *Logistic Regression: A Self-Learning Text, Second Edition.* New York: Springer.

Kleinbaum, D. G.; Kupper, L. L.; and Morgenstern, H. 1982. *Epidemiologic Research: Principles and Quantitative Methods.* Belmont, Calif.: Lifetime Learning Publications.

Kupper, L. L.; Stewart, J. R.; and Williams, K. A. 1976. "A Note on Controlling Significance Levels in Stepwise Regression." *American Journal of Epidemiology* 103(1): 13–15.

Lewis, T., and Taylor, L. R. 1967. *Introduction to Experimental Ecology.* New York: Academic Press.

Lynn, M. J.; Waring, G. O., III; Sperduto, R. D.; et al. 1987. "Factors Affecting Outcome and Predictability of Radial Keratotomy in the PERK Study." *Archives of Ophthalmology* 105: 42–51.

Marquardt, D. W., and Snee, R. D. 1975. "Ridge Regression in Practice." *The American Statistician* 29(1): 3–20.

Pope, P. T., and Webster, J. T. 1972. "The Use of an F-statistic in Stepwise Regression Procedures." *Technometrics* 14(2): 327–40.

One-way Analysis of Variance

17.1 Preview

This chapter is the first of four that focus on *analysis of variance* (ANOVA). Earlier we described ANOVA as a technique for assessing how several *nominal* independent variables affect a *continuous* dependent variable. An example given in Table 2.2 (the Ponape study) involved describing the effects on blood pressure of two cultural incongruity indices, each dichotomized into "high" and "low" categories (see Example 1.5). In this case, the dependent variable (blood pressure) was continuous, and the two independent variables (the cultural incongruity indices) were both nominal.

The fact that ANOVA is generally restricted to use with nominal independent variables suggests an interesting interpretation of the purpose of the technique. Loosely speaking, *ANOVA is usually employed in comparisons involving several population means.* In fact, in the simplest special case (involving a comparison of two population means), the ANOVA comparison procedure is equivalent to the usual two-sample *t* test, which requires the assumption of equal population variances.

The population means to be compared can generally be easily specified by cross-classifying the nominal independent variables under consideration to form different combinations of categories.[1] In the example dealing with the Ponape study, we need only cross-classify the HI and LO categories of incongruity index 1 with the HI and LO categories of index 2. This yields the four population means corresponding to the four combinations HI–HI, HI–LO, LO–HI, and LO–LO, as indicated in the following configuration:

---

[1] Such specification is not possible if the categories of any nominal variable are viewed as being only a sample from a much larger population of categories of interest. We consider such situations later when discussing *random-effects models.*

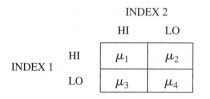

Assessing whether the two indices have some effect on the dependent variable "blood pressure" is equivalent to determining what kind of differences, if any, exist among the four population means.

17.1.1 Why the Name ANOVA?

If the ANOVA technique usually involves comparing means, it seems somewhat inappropriate to call it analysis of *variance*. Why not instead use the acronym ANOME, where *ME* stands for *means*? Actually, the designation *ANOVA* is quite justifiable: although typically means are compared, the comparisons are made using estimates of variance. As with regression analysis, the ANOVA test statistics are F statistics and are actually ratios of estimates of variance. In fact, it is even possible, and in some cases appropriate, to specify the null hypotheses of interest in terms of population variances.

17.1.2 ANOVA versus Regression

Another general distinction relates to the difference between an "ANOVA problem" and a "regression problem." For ANOVA, *all* independent variables must be treated as nominal; for regression analysis, any mixture of measurement scales (nominal, ordinal, or interval) is permitted for the independent variables. In fact, ANOVA is often viewed as a special case of regression analysis, and almost any ANOVA model can be represented by a regression model whose parameters can be estimated and inferred about in the usual manner. The same may be said for certain other multivariable techniques, such as analysis of covariance. Hence, we may view the various names given to these techniques as indicators of different (linear) models having the same general form

$$Y = \beta_0 + \beta_1 X_1 + \beta_2 X_2 + \cdots + \beta_p X_p + E$$

yet involving different types of variables and perhaps different assumptions about these variables. The choice of method can thus be regarded as equivalent to the choice of an appropriate linear model.

17.1.3 Factors and Levels

Some additional terms must be introduced at this point. In Chapter 12, in connection with using dummy variables in regression, we saw that a nominal variable with k categories can generally be incorporated into a regression model if we define $(k - 1)$ dummy variables. These $(k - 1)$ variables collectively describe the *basic* nominal variable under consideration. To refer to a basic variable without having to identify the specific dummy variables used to define it in the regression model, we can follow the approach of calling the basic nominal variable a *factor*. The different categories of the factor are often referred to as its *levels*.

For example, if we wanted to compare the effects of several drugs on some human health response, we would consider the nominal variable "drugs" as a single factor and the specific drug categories as the levels. If we were comparing k drugs, we would incorporate them into a regression model by defining $(k - 1)$ dummy variables. If, in addition to comparing the drugs, we wanted to consider whether males and females responded differently, we would consider the nominal variable "sex" as a second factor and the specific categories (male and female) as the levels of this dichotomous factor.

17.1.4 Fixed versus Random Factors

A *random factor* is a factor whose levels may be regarded as a sample from some large population of levels.[2] A *fixed factor* is a factor whose levels are the only ones of interest. The distinction is important in any ANOVA, since different tests of significance are required for different configurations of random and fixed factors. We will see this more specifically when considering two-way ANOVA. For now, we can simply look at some examples of random and fixed factors (summarized in Table 17.1):

1. "Subjects" or "litters" is usually considered a random factor, since we ordinarily want to make inferences about a large population of potential subjects on the basis of the subjects sampled.

2. "Observers" is a random factor we often consider when examining the effect of different observers on the response variable of interest.

3. "Days," "weeks," and so on are usually considered random factors in investigations of the effect of time on a response variable observed during different time periods. We normally use many levels for such temporal factors, to represent a large number of time periods.

4. "Sex" is always a fixed factor, since its two levels include all possible levels of interest.

5. "Locations" (for example, cities, plants, or states) may be fixed or random, depending on whether a set of specific sites or a larger geographical universe is to be considered.

6. "Age" is usually treated as a fixed factor regardless of how the different age groups are defined.

TABLE 17.1   Examples of random and fixed factors

Random	Fixed	Random or Fixed
Subjects	Sex	Locations
Litters	Age	Treatments
Observers	Marital status	Drugs
Days	Education	Exposures
Weeks		

---

[2] In practice, the experimental levels of a random factor need not be selected at random as long as they are reasonably representative of the larger population of levels of interest.

7. "Treatments," "drugs," "exposures," and so on are usually considered fixed factors, but they may be considered random if their levels represent a much larger group of possible levels.

8. "Marital status" is treated as a fixed factor.

9. "Education" is treated as a fixed factor.

17.2 One-way ANOVA: The Problem, Assumptions, and Data Configuration

One-way ANOVA deals with the effect of a single factor on a single response variable. When that one factor is a fixed factor, one-way ANOVA (often referred to as *fixed-effects one-way ANOVA*) involves a comparison of several (two or more) population means.[3] The different populations effectively correspond to the different levels of the single-factor "populations."

17.2.1 The Problem

The main analysis problem in fixed-effects one-way ANOVA is to determine whether the population means are all equal or not. Thus, given k means (denoted as $\mu_1, \mu_2, \ldots, \mu_k$), the basic null hypothesis of interest is

$$H_0: \mu_1 = \mu_2 = \cdots = \mu_k \qquad (17.1)$$

The alternative hypothesis is given by

H_A: "The k population means are not all equal."

If the null hypothesis (17.1) is rejected, the next problem is to find out where the differences are. For example, if $k = 3$ and $H_0: \mu_1 = \mu_2 = \mu_3$ is rejected, we might wish to determine whether the main differences are between μ_1 and μ_2, between μ_1 and μ_3, between μ_1 and the average of the other two means, etc. Such questions fall under the general statistical subject of multiple-comparison procedures, which are discussed in Section 17.7.

17.2.2 The Assumptions

Four assumptions must be made for fixed-effects one-way ANOVA:

1. Random samples (individuals, animals, etc.) have been selected from each of k populations or groups.

2. A value of a specified dependent variable has been recorded for each experimental unit (individual, animal, etc.) sampled.

---

[3] In this section, we focus on situations involving fixed factors. Random factors are discussed in Section 17.6.

3.   The dependent variable is normally distributed in each population.

4.   The variance of the dependent variable is the same in each population (this common variance is denoted as σ^2).

Although these assumptions provide the theoretical justification for applying this method, it is sometimes necessary to use fixed-effects one-way ANOVA to compare several means even when the necessary assumptions are not clearly satisfied. Indeed, these assumptions rarely hold exactly. It is therefore important to consider the consequences of applying fixed-effects one-way ANOVA when the assumptions are in question.

In general, fixed-effects one-way ANOVA can be applied as long as none of the assumptions is badly violated. This is true for more complex ANOVA situations as well as for fixed-effects one-way ANOVA. The term generally used to denote this property of broad applicability is *robustness*: a procedure is robust if moderate departures from the basic assumptions do not adversely affect its performance in any meaningful way.

We must nevertheless avoid asserting robustness as an automatic justification for carelessly applying the ANOVA method. Certain facts should be kept in mind when considering the use of ANOVA in a given situation. It is true that the normality assumption does not have to be exactly satisfied as long as we are dealing with relatively large samples (e.g., 20 or more observations from each population), although the consequences of large deviations from normality are somewhat more severe for random factors than for fixed factors. Similarly, the assumption of variance homogeneity can be mildly violated without serious risk, provided that the numbers of observations selected from each population are more or less the same (again, the consequences are more severe for random factors).

On the other hand, an inappropriate assumption of independence of the observations can lead to serious errors in inference for both fixed and random cases. In general, great care should be taken to ensure that the observations are independent. This concern arises primarily in studies where repeated observations are recorded on the same experimental subjects: the level of response of a subject on one occasion commonly has a decided effect on subsequent responses.

What should we do when one or more of these assumptions are in serious question? One option is to transform the data (e.g., by means of a log, square root, or other transformation) so that they more closely satisfy the assumptions. Another strategy is to select a more appropriate method of analysis (e.g., nonparametric ANOVA methods or growth curve analysis or longitudinal data analysis procedures).[4] A detailed discussion concerning the analysis of longitudinal data is provided in Chapters 25 and 26.

17.2.3 Data Configuration for One-way ANOVA

Computations necessary for one-way ANOVA can easily be performed even with an ordinary calculator when the data are conveniently arranged. Table 17.2 illustrates a useful way of presenting the data for the general one-way situation. Clearly, the number of observations selected from each population does *not* have to be the same; that is, there are n_i observations

---

[4] Descriptions of nonparametric methods that can be used when these assumptions are clearly and strongly violated can be found in Siegel (1956), Lehmann (1975), and Hollander and Wolfe (1973). A discussion of growth curve analysis can be found in Allen and Grizzle (1969). An excellent discussion of longitudinal data analysis methods can be found in Diggle, Heagerty, Liang, and Zeger (2002).

TABLE 17.2   General data configuration for one-way ANOVA

Population	Sample Size	Observations	Total	Sample Mean
1	n_1	$Y_{11}, Y_{12}, Y_{13}, \ldots, Y_{1n_1}$	$T_1 = Y_1.$	$\overline{Y}_1. = T_1/n_1$
2	n_2	$Y_{21}, Y_{22}, Y_{23}, \ldots, Y_{2n_2}$	$T_2 = Y_2.$	$\overline{Y}_2. = T_2/n_2$
3	n_3	$Y_{31}, Y_{32}, Y_{33}, \ldots, Y_{3n_3}$	$T_3 = Y_3.$	$\overline{Y}_3. = T_3/n_3$
$\vdots$	$\vdots$	$\vdots$	$\vdots$	$\vdots$
k	n_k	$Y_{k1}, Y_{k2}, Y_{k3}, \ldots, Y_{kn_k}$	$T_k = Y_k.$	$\overline{Y}_k. = T_k/n_k$

$$n = \sum_{i=1}^{k} n_i \qquad\qquad G = Y.. \qquad\qquad \overline{Y} = G/n$$

from the ith population, and n_i need not equal n_j if $i \neq j$. Double-subscript notation (Y_{ij}) is used to distinguish one observation from another. The first subscript for a given observation denotes the population number, and the second distinguishes that observation from the others in the sample from that particular population. Thus, Y_{23} denotes the third observation from the second population, Y_{62} denotes the second observation from the sixth population, and Y_{kn_k} denotes the last observation from the kth population. The totals for each sample (from each population) are denoted alternatively by T_i or $Y_i.$ for the ith sample, where the $\cdot$ denotes that we are summing over all values of j (i.e., we are adding together all observations making up the given sample). The grand total over all samples is denoted as $Y.. = G$. The sample means are alternatively denoted by $\overline{Y}_i.$ or T_i/n_i for the ith sample; these statistics are particularly important because they represent the estimates of the population means of interest. Finally, the grand mean over all samples is $\overline{Y} = G/n$.

■ **Example 17.1**   In a study by Daly (1973) of the effects of neighborhood characteristics on health, a stratified random sample of 100 households was selected—25 from each of four turnkey neighborhoods included in the study. The data configuration of Cornell Medical Index (CMI) scores for female heads of household is given in Table 17.3. Such scores are measures (derived from questionnaires) of the overall (self-perceived) health of individuals; the higher the score, the poorer the health. Each of the turnkey neighborhoods differed in total number of households and in percentage of blacks in the surrounding neighborhoods. The racial composition of the turnkey neighborhoods themselves was over 95% black. Daly's main thesis was that the health of persons living in similar federal or state housing projects varied according to the racial composition of the surrounding neighborhoods: the "friendlier" (in terms of similar racial composition) the surrounding neighborhood was, the better would be the health of the residents in the project. According to Daly, federal housing planners had never considered information about overall neighborhood racial composition and its relationship to health as criteria for selecting areas for such projects. This study, it was hoped, might provide some concrete recommendations for improved federal planning.

The sample means of the data in Table 17.3 vary. To determine whether the observed differences in these sample means are attributable solely to chance, we can perform a one-way fixed-effects ANOVA. The possibility of violations of the assumptions underlying this methodology are not of great concern for this data set, since the sample sizes are equal and reasonably large and since observations on women from different households may be treated as independent.

TABLE 17.3   **Cornell Medical Index scores for a sample of women from different households in four turnkey housing neighborhoods**

Neighborhood	No. of Households	% Blacks in Surrounding Neighborhoods	Sample Size (n_i)	Observations (Y_{ij})	Total (T_i)	Sample Mean $(\bar{Y}_{i\cdot})$
Cherryview	98	17	25	49, 12, 28, 24, 16, 28, 21, 48, 30, 18, 10, 10, 15, 7, 6, 11, 13, 17, 43, 18, 6, 10, 9, 12, 12	$T_1 = 473$	$\bar{Y}_{1\cdot} = 18.92$
Morningside	211	100	25	5, 1, 44, 11, 4, 3, 14, 2, 13, 68, 34, 40, 36, 40, 22, 25, 14, 23, 26, 11, 20, 4, 16, 25, 17	$T_2 = 518$	$\bar{Y}_{2\cdot} = 20.72$
Northhills	212	36	25	20, 31, 19, 9, 7, 16, 11, 17, 9, 14, 10, 5, 15, 19, 29, 23, 70, 25, 6, 62, 2, 14, 26, 7, 55	$T_3 = 521$	$\bar{Y}_{3\cdot} = 20.84$
Easton	40	65	25	13, 10, 20, 20, 22, 14, 10, 8, 21, 35, 17, 23, 17, 23, 83, 21, 17, 41, 20, 25, 49, 41, 27, 37, 57	$T_4 = 671$	$\bar{Y}_{4\cdot} = 26.84$

$$\sum_{i=1}^{4} n_i = 100 \qquad\qquad G = 2{,}183 \qquad \bar{Y} = 21.83$$

17.3 Methodology for One-way Fixed-effects ANOVA

The null hypothesis of equal population means (H_0: $\mu_1 = \mu_2 = \cdots = \mu_k$) is tested by using an F test. The test statistic is calculated as follows:

$$F = \frac{\text{MST}}{\text{MSE}} \tag{17.2}$$

where

$$\text{MST} = \frac{\sum_{i=1}^{k}(T_i^2/n_i) - G^2/n}{k - 1} \tag{17.3}$$

and

$$\text{MSE} = \frac{\sum_{i=1}^{k}\sum_{j=1}^{n_i} Y_{ij}^2 - \sum_{i=1}^{k}(T_i^2/n_i)}{n-k} \qquad (17.4)$$

When H_0 is true (i.e., when the population means are all equal), the F statistic of (17.2) has an F distribution with $(k-1)$ numerator and $(n-k)$ denominator degrees of freedom. Thus, for a given α, we would reject H_0 and conclude that some (i.e., at least two) of the population means differ from one another if

$$F \geq F_{k-1,\,n-k,\,1-\alpha}$$

where $F_{k-1,\,n-k,\,1-\alpha}$ is the $100(1-\alpha)\%$ point of the F distribution with $(k-1)$ and $(n-k)$ degrees of freedom. The critical region for this test involves only upper percentage points of the F distribution, since only large values of the F statistic (usually values much greater than 1) will provide significant evidence for rejecting H_0.

17.3.1 Numerical Illustration

For the data given in Table 17.3, the calculations needed to perform the F test proceed as follows:

$$\underbrace{\sum_{i=1}^{4}\sum_{j=1}^{25} Y_{ij}^2 = (49)^2 + (12)^2 + \cdots + (37)^2 + (57)^2}_{\text{Sum of 100 squared observations}} = 72{,}851.00$$

$$\sum_{i=1}^{4}\frac{T_i^2}{n_i} = \frac{(473)^2}{25} + \frac{(518)^2}{25} + \frac{(521)^2}{25} + \frac{(671)^2}{25} = 48{,}549.40$$

$$\frac{G^2}{n} = \frac{(2183)^2}{100} = 47{,}654.89$$

$$\begin{aligned}
\text{MST} &= \frac{\sum_{i=1}^{4}(T_i^2/n_i) - G^2/n}{4-1} \\[2mm]
&= \frac{48{,}549.40 - 47{,}654.89}{3} \\[2mm]
&= 298.17
\end{aligned}$$

$$\begin{aligned}
\text{MSE} &= \frac{\sum_{i=1}^{4}\sum_{j=1}^{25} Y_{ij}^2 - \sum_{i=1}^{4}(T_i^2/n_i)}{100-4} \\[2mm]
&= \frac{72{,}851.00 - 48{,}549.40}{96} \\[2mm]
&= 253.14
\end{aligned}$$

$$F = \frac{MST}{MSE}$$
$$= \frac{298.17}{253.14}$$
$$= 1.178$$

The preceding calculations can be conveniently performed by a computer program. An example of computer output for these data (using SAS's GLM procedure) is shown next.

SAS Output for ANOVA of CMI Scores

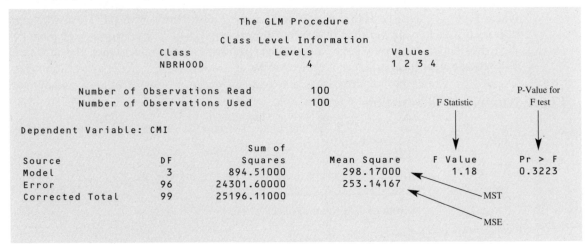

Using these calculations, we may test H_0: $\mu_1 = \mu_2 = \mu_3 = \mu_4$ (i.e., the hypothesis that there are *no* differences among the true mean CMI scores for the four neighborhoods) against H_A: "There are differences among the true mean CMI scores." For example, if $\alpha = .10$, we would find from the F tables that $F_{3,96,0.90} = 2.15$, which is greater than the computed F of 1.178. Thus, we would not reject H_0 at $\alpha = .10$.

To find the P-value for this test, we first note that $F_{3,96,0.75} = 1.41$, which also exceeds the computed F. Thus, we know that $P > .25$. From the preceding SAS output, we see that the P-value is in fact .3223. Therefore, we conclude (as did Daly) that the observed mean CMI scores for the four neighborhoods do not significantly differ.

If a significant difference among the sample means had been found, it would still have been up to the investigator to determine whether the actual magnitude of the difference(s) was meaningful in a practical sense and whether the pattern of the difference(s) was as hypothesized. In our example, the distribution of percentages of blacks in the surrounding neighborhoods (see Table 17.3) indicates that the observed differences among the sample means clearly does not match the pattern hypothesized. Under Daly's conjecture, Cherryview (with 17% black in the surrounding neighborhood) would have been expected to register the highest observed mean CMI score, followed by Northhills (36%), Easton (65%), and finally Morningside (100%). This was not the order actually obtained. Daly also examined whether her conjecture was supported when she controlled for other possibly relevant factors, such as "months lived in the neighborhood," "number of children," and "marital status." However, no significant results were obtained from these analyses either.

17.3.2 Rationale for the *F* Test in One-way ANOVA

The use of the *F* test described in the preceding subsection may be motivated by various considerations. We therefore offer an intuitive theoretical appreciation of its purpose and also provide some insight into the rationale behind more complex ANOVA testing procedures.

1. The F test in one-way fixed-effects ANOVA is a generalization of the two-sample t test. We can easily show with a little algebra that the numerator and denominator components in the *F* statistic (17.2) for one-way ANOVA are simple generalizations of the corresponding components in the square of the ordinary two-sample *t* test statistic. In fact, when $k = 2$, the *F* statistic for one-way ANOVA is exactly equal to the square of the corresponding *t* statistic. Such a result is intuitively reasonable, since the numerator degrees of freedom of *F* when $k = 2$ is 1, and we have previously noted that the square of a *t* statistic with v degrees of freedom has the *F* distribution with 1 and v degrees of freedom in numerator and denominator, respectively.

In particular, recall that the two-sample *t* test statistic is given by the formula

$$T = \frac{(\bar{Y}_{1.} - \bar{Y}_{2.})/\sqrt{1/n_1 + 1/n_2}}{S_p}$$

where the pooled sample variance S_p^2 is given by

$$S_p^2 = \frac{1}{n_1 + n_2 - 2} \sum_{i=1}^{2} \sum_{j=1}^{n_i} (Y_{ij} - \bar{Y}_{i.})^2$$

or, equivalently, by

$$S_p^2 = \frac{(n_1 - 1)S_1^2 + (n_2 - 1)S_2^2}{n_1 + n_2 - 2}$$

where S_1^2 and S_2^2 are the sample variances for groups 1 and 2, respectively. Focusing first on the denominator (MSE) of the *F* statistic (17.2), we can show with some algebra that

$$\text{MSE} = \frac{\sum_{i=1}^{k} \sum_{j=1}^{n_i} Y_{ij}^2 - \sum_{i=1}^{k} (T_i^2/n_i)}{n - k}$$

$$= \frac{(n_1 - 1)S_1^2 + (n_2 - 1)S_2^2 + \cdots + (n_k - 1)S_k^2}{n_1 + n_2 + \cdots + n_k - k}$$

Thus, MSE is a pooled estimate of the common population variance σ^2, since it is a weighted sum of the *k* estimates of σ^2 obtained by using the *k* different sets of observations. Furthermore, when $k = 2$, MSE is equal to S_p^2.

Looking at the numerator (MST) of the F statistic, we can show that

$$\text{MST} = \frac{\sum\limits_{i=1}^{k}(T_i^2/n_i) - G^2/n}{k-1}$$

$$= \frac{1}{k-1}\sum\limits_{i=1}^{k} n_i(\bar{Y}_{i\cdot} - \bar{Y})^2$$

which simplifies to $(\bar{Y}_{1\cdot} - \bar{Y}_{2\cdot})^2/(1/n_1 + 1/n_2)$ when $k = 2$. Thus, the equivalence is established.

2. The F statistic is the ratio of two variance estimates.   We have already seen that MSE is a pooled estimate of the common population variance σ^2; that is, the true average (or mean) value (μ_{MSE}, say) of MSE is σ^2. It turns out, however, that MST estimates σ^2 *only* when H_0 is true—that is, *only* when the population means $\mu_1, \mu_2, \ldots, \mu_k$ are all equal. In fact, the true mean value (μ_{MST}, say) of MST has the general form

$$\mu_{\text{MST}} = \sigma^2 + \frac{1}{k-1}\sum\limits_{i=1}^{k} n_i(\mu_i - \bar{\mu})^2 \tag{17.5}$$

where $\bar{\mu} = \sum_{i=1}^{k} n_i\mu_i/n$. By inspection of expression (17.5), we can see that MST estimates σ^2 *only* when all the μ_i are equal, in which case $\mu_i = \bar{\mu}$ for every i, and so $\sum_{i=1}^{k} n_i(\mu_i - \bar{\mu})^2 = 0$. Otherwise, both terms on the right-hand side of (17.5) are positive and MST estimates something greater in value than σ^2. In other words,

$$\mu_{\text{MST}} = \sigma^2 \quad \text{when } H_0 \text{ is true}$$

and

$$\mu_{\text{MST}} > \sigma^2 \quad \text{when } H_0 \text{ is not true.}$$

Loosely speaking, then, the F statistic MST/MSE may be viewed as approximating in some sense the ratio of population means

$$\frac{\mu_{\text{MST}}}{\mu_{\text{MSE}}} = \frac{\sigma^2 + \dfrac{1}{(k-1)}\sum\limits_{i=1}^{k} n_i(\mu_i - \bar{\mu})^2}{\sigma^2} \tag{17.6}$$

When H_0 is true, the numerator and denominator of (17.6) both equal σ^2, and the F statistic is the ratio of two estimates of the same variance. Furthermore, F can be expected to give different values depending on whether H_0 is true or not; that is, F should take a value close to 1 if H_0 is true (since in that case it approximates $\sigma^2/\sigma^2 = 1$), whereas F should be larger than 1 if H_0 is false (since the numerator of (17.6) is greater than the denominator).

3. The F statistic compares the variability between groups to the variability within groups.   As with regression analysis, the total variability in the observations in a one-way ANOVA situation is measured by a total sum of squares:

$$\text{SSY} = \sum\limits_{i=1}^{k}\sum\limits_{j=1}^{n_i}(Y_{ij} - \bar{Y})^2 \tag{17.7}$$

Furthermore, it can be shown that

$$SSY = SST + SSE \qquad (17.8)$$

where

$$SST = (k - 1)MST = \sum_{i=1}^{k} n_i (\overline{Y}_{i\cdot} - \overline{Y})^2$$

and

$$SSE = (n - k)MSE = \sum_{i=1}^{k} \sum_{j=1}^{n_i} (Y_{ij} - \overline{Y}_{i\cdot})^2$$

SST can be considered to be a measure of the variability *between* (or *across*) populations. (The designation SST is read "sum of squares due to treatments," since the populations often represent treatment groups.) It involves components of the general form $(\overline{Y}_{i\cdot} - \overline{Y})$, which is the difference between the ith group mean and the overall mean.

SSE is a measure of the variability *within* populations and gives no information about variability between populations. It involves components of the general form $(Y_{ij} - \overline{Y}_{i\cdot})$, which is the difference between the jth observation in the ith group and the mean for the ith group.

If SST is quite large in comparison to SSE, we know that most of the total variability is due to differences *between* populations rather than to differences *within* populations. Thus, it is natural in such a case to suspect that the population means are not all equal.

By writing the F statistic (17.2) in the form

$$F = \frac{SST}{SSE} \left(\frac{n - k}{k - 1} \right)$$

we can see that F will be large whenever SST accounts for a much larger proportion of the total sum of squares than does SSE.

17.3.3 ANOVA Table for One-way ANOVA

As in regression analysis, the results of any ANOVA procedure can be summarized in an ANOVA table. The ANOVA table for one-way ANOVA is given in general form in Table 17.4. Table 17.5 is the ANOVA table for our example involving the CMI data; this ANOVA table was also previously shown in the SAS output on page 428.

TABLE 17.4   General ANOVA table for one-way ANOVA (k populations)

Source	d.f.	SS	MS	F
Between	$k - 1$	SST	$MST = \dfrac{SST}{k - 1}$	$\dfrac{MST}{MSE}$
Within	$n - k$	SSE	$MSE = \dfrac{SSE}{n - k}$	
Total	$n - 1$	SSY		

TABLE 17.5 **ANOVA table for CMI data ($k = 4$)**

Source	d.f.	SS	MS	F
Between (neighborhoods)	3	894.51	298.17	1.18
Within (error)	96	24,301.60	253.14	
Total	99	25,196.11		

The "Source" and "SS" columns in Table 17.4 display the components of the fundamental equation of one-way ANOVA:

$$SSY = SST + SSE$$

The "MS" column contains the sums of squares divided by their corresponding degrees of freedom. The two mean squares are then used to form the numerator and denominator for the F test.

17.4 Regression Model for Fixed-effects One-way ANOVA

We observed earlier that most ANOVA procedures can also be considered in a regression analysis setting; this can be done by defining appropriate dummy variables in a regression model.[5] The ANOVA F tests are then formulated in terms of hypotheses concerning the coefficients of the dummy variables in the regression model.[6]

■ **Example 17.2**   For the example involving the CMI data of Daly's (1973) study (see Table 17.3), a number of alternative regression models could be used to describe the situation, depending on the coding schemes used for the dummy variables. One such model is

$$Y = \mu + \alpha_1 X_1 + \alpha_2 X_2 + \alpha_3 X_3 + E \tag{17.9}$$

where the regression coefficients are denoted as μ, α_1, α_2, and α_3, and the independent variables are defined as

$$X_1 = \begin{cases} 1 & \text{if neighborhood 1} \\ -1 & \text{if neighborhood 4} \\ 0 & \text{if otherwise} \end{cases} \qquad X_2 = \begin{cases} 1 & \text{if neighborhood 2} \\ -1 & \text{if neighborhood 4} \\ 0 & \text{if otherwise} \end{cases}$$

$$X_3 = \begin{cases} 1 & \text{if neighborhood 3} \\ -1 & \text{if neighborhood 4} \\ 0 & \text{if otherwise} \end{cases}$$

---

[5] As mentioned earlier, we are restricting our attention here entirely to models with fixed factors. Models involving random factors will be treated in Section 17.6.

[6] We will see later that a regression formulation is often desirable, if not mandatory, for dealing with certain nonorthogonal ANOVA problems involving two or more factors. We will discuss such problems in Chapter 20.

Although we previously used the Greek letter β with subscripts to denote regression coefficients, we have changed the notation for our ANOVA regression model so that these coefficients correspond to the parameters in the classical fixed-effects ANOVA model described in Section 17.5.

The coding scheme used here to define the dummy variables is called an *effect* coding scheme. The coefficients μ, α_1, α_2, and α_3 for this (dummy variable) model can each be expressed in terms of the underlying population means (μ_1, μ_2, μ_3, and μ_4), as follows:

$$\mu = \frac{\mu_1 + \mu_2 + \mu_3 + \mu_4}{4} \qquad (=\overline{\mu}^*, \text{ say})$$

$$\alpha_1 = \mu_1 - \overline{\mu}^* \qquad\qquad\qquad (17.10)$$

$$\alpha_2 = \mu_2 - \overline{\mu}^*$$

$$\alpha_3 = \mu_3 - \overline{\mu}^*$$

The coefficients in (17.10) are related as follows: $\mu_{Y|X_1, X_2, X_3} = \mu + \alpha_1 X_1 + \alpha_2 X_2 + \alpha_3 X_3$. Thus,

$$\mu_1 = \mu_{Y|1,0,0} = \mu + \alpha_1 \qquad\qquad \text{since } X_1 = 1, X_2 = 0, X_3 = 0 \text{ for neighborhood 1}$$

$$\mu_2 = \mu_{Y|0,1,0} = \mu + \alpha_2 \qquad\qquad \text{since } X_1 = 0, X_2 = 1, X_3 = 0 \text{ for neighborhood 2}$$

$$\mu_3 = \mu_{Y|0,0,1} = \mu + \alpha_3 \qquad\qquad \text{since } X_1 = 0, X_2 = 0, X_3 = 1 \text{ for neighborhood 3}$$

$$\mu_4 = \mu_{Y|-1,-1,-1}$$

$$\quad = \mu - \alpha_1 - \alpha_2 - \alpha_3 \qquad \text{since } X_1 = X_2 = X_3 = -1 \text{ for neighborhood 4}$$

Adding the left-hand sides and right-hand sides of these equations yields

$$\mu_1 + \mu_2 + \mu_3 + \mu_4 = 4\mu$$

or

$$\mu = \frac{1}{4}\sum_{i=1}^{4} \mu_i \qquad (= \overline{\mu}^*)$$

Solution (17.10) is obtained by replacing μ with $\overline{\mu}^*$ in the preceding equations and then solving for the regression coefficients α_i in terms of μ_1, μ_2, μ_3, and μ_4.

Model (17.9) involves coefficients that describe separate comparisons of the first three population means with the overall unweighted mean $\overline{\mu}^*$. In this model, $(\mu_4 - \overline{\mu}^*)$ can be expressed as the negative sum of α_1, α_2, and α_3. Moreover, model (17.9) can be fitted to provide *exactly* the same F statistic as is required in one-way ANOVA for the test of $H_0: \mu_1 = \mu_2 = \mu_3 = \mu_4$. The equivalent regression null hypothesis is $H_0: \alpha_1 = \alpha_2 = \alpha_3 = 0$,[7] the regression F statistic will

---

[7] When $\alpha_1 = \alpha_2 = \alpha_3 = 0$, it follows from simple algebra based on (17.10) that $\mu_1 = \mu_2 = \mu_3 = \mu_4$ (e.g., $\alpha_1 = \mu_1 - \overline{\mu}^* = 0$ implies that $\mu_1 = \overline{\mu}^*$; $\alpha_2 = \mu_2 - \overline{\mu}^* = 0$ implies that $\mu_2 = \overline{\mu}^* = \mu_1$; etc.).

have the same degrees of freedom (i.e., $k - 1 = 3$ and $n - k = 96$ as given previously, and the ANOVA table will be exactly the same as the one given in the last section (where we pooled the dummy variable effects into one source of variation with 3 degrees of freedom).

Other coding schemes for the independent variables yield exactly the same ANOVA table and F test as model (17.9), although the regression coefficients themselves represent different parameters and have different least-squares estimators. One frequently used coding scheme defines the independent variables as

$$X_i = \begin{cases} 1 & \text{if neighborhood } i \\ 0 & \text{otherwise} \end{cases} \qquad i = 1, 2, 3$$

This coding scheme is an example of *reference cell* coding. The referent group in this case is group 4, and the regression coefficients describe separate comparisons of the first three population means with μ_4:

$$\mu = \mu_4$$

$$\alpha_1 = \mu_1 - \mu_4$$

$$\alpha_2 = \mu_2 - \mu_4$$

$$\alpha_3 = \mu_3 - \mu_4 \qquad \blacksquare$$

17.4.1 Effect Coding Model

For the general situation involving k populations, the following model using effect coding is analogous to that of the previous section:

$$Y = \mu + \alpha_1 X_1 + \alpha_2 X_2 + \cdots + \alpha_{k-1} X_{k-1} + E \qquad (17.11)$$

in which

$$X_i = \begin{cases} 1 & \text{for population } i \\ -1 & \text{for population } k \\ 0 & \text{otherwise} \end{cases} \qquad i = 1, 2, \ldots, k - 1$$

The coefficients of this model can be expressed in terms of the k population means $\mu_1, \mu_2, \ldots, \mu_k$ as

$$\mu = \frac{\mu_1 + \mu_2 + \cdots + \mu_k}{k} = \overline{\mu}^*$$

$$\alpha_1 = \mu_1 - \overline{\mu}^* \qquad (17.12)$$

$$\alpha_2 = \mu_2 - \overline{\mu}^*$$

$$\vdots$$

$$\alpha_{k-1} = \mu_{k-1} - \overline{\mu}^*$$

$$-(\alpha_1 + \alpha_2 + \cdots + \alpha_{k-1}) = \mu_k - \overline{\mu}^*$$

for model (17.11). The F statistic for one-way ANOVA with k populations can be obtained equivalently by testing the null hypothesis H_0: $\alpha_1 = \alpha_2 = \cdots = \alpha_{k-1} = 0$ in model (17.11).

17.4.2 Reference Cell Coding Model

Another coding scheme for one-way ANOVA of k populations uses the following reference cell coding:

$$X_i = \begin{cases} 1 & \text{for population } i \\ 0 & \text{otherwise} \end{cases} \qquad i = 1, 2, \ldots, k - 1$$

Again, only $k - 1$ such dummy variables are needed, and the population "left out" becomes the reference population (group or cell). Thus, given X_i as defined here, group k is the reference group. (If X_1 had been left out instead of X_k, then group 1 would have been the reference group.)

With the specified reference cell coding, the responses in each group under the model $Y = \mu + \alpha_1 X_1 + \alpha_2 X_2 + \cdots + \alpha_{k-1} X_{k-1} + E$ are as follows:

Group 1: $\qquad Y = \mu + \alpha_1 \ + E$

Group 2: $\qquad Y = \mu + \alpha_2 \ + E$

$\qquad\qquad\quad \vdots$

Group $k - 1$: $\quad Y = \mu + \alpha_{k-1} + E$

Group k: $\qquad Y = \mu \qquad + E$

In turn, the regression coefficients can be written in terms of group means as follows:

$\alpha_1 = \mu_1 - \mu_k$

$\alpha_2 = \mu_2 - \mu_k$

$\qquad \vdots$

$\alpha_{k-1} = \mu_{k-1} - \mu_k$

$\mu = \mu_k$

Thus, different coding schemes (e.g., an effect coding or reference cell coding) yield regression coefficients representing different parameters (for example, $\alpha_1 = \mu_1 - \overline{\mu}^*$ for the effect coding, but $\alpha_1 = \mu_1 - \mu_k$ for the reference cell coding). Regardless of the coding scheme used, the test of the hypothesis H_0: $\mu_1 = \mu_2 = \cdots = \mu_k$ can be obtained equivalently by testing H_0: $\alpha_1 = \alpha_2 = \cdots = \alpha_{k-1} = 0$ in the regression model (17.11). In other words, the correct SST, SSE, and $F_{k-1, n-k}$ values are obtained from the regression analysis regardless of the (legitimate) coding scheme chosen.

17.5 Fixed-effects Model for One-way ANOVA

Many textbooks and articles that deal strictly with ANOVA procedures use a more classical type of model than the regression model given earlier to describe the fixed-effects one-way ANOVA situation. The more classical type of model is often referred to as a *fixed-effects ANOVA model;*

in it, all factors under consideration are fixed (i.e., the levels of each factor are the only levels of interest). The *effects* referred to in this type of model represent measures of the influence (i.e., the effect) that different levels of the factor have on the dependent variable.[8] Such measures are often expressed in the form of differences between a given mean and an overall mean; that is, the effect of the ith population is often measured as the amount by which the ith population mean differs from an overall mean.

■ **Example 17.3** For the CMI data ($k = 4$), the fixed-effects ANOVA model is

$$Y_{ij} = \mu + \alpha_i + E_{ij} \qquad i = 1, 2, 3, 4; \qquad j = 1, 2, \ldots, 25 \qquad (17.13)$$

where

Y_{ij} = jth observation from the ith population

$$\mu = \frac{\mu_1 + \mu_2 + \mu_3 + \mu_4}{4} = \overline{\mu}^* \text{ (the overall unweighted mean), since } n_i = 25 \text{ for all } i$$

$\alpha_1 = \mu_1 - \mu$ = Differential effect of neighborhood 1

$\alpha_2 = \mu_2 - \mu$ = Differential effect of neighborhood 2

$\alpha_3 = \mu_3 - \mu$ = Differential effect of neighborhood 3

$\alpha_4 = \mu_4 - \mu$ = Differential effect of neighborhood 4

$E_{ij} = Y_{ij} - \mu - \alpha_i$ = Error component associated with the jth observation from the ith population

One important property of this model is that the sum of the four α effects is 0; that is, $\alpha_1 + \alpha_2 + \alpha_3 + \alpha_4 = 0$. Thus, these effects represent differentials from the overall population mean μ that average out to 0. Nevertheless, the effect of one level (i.e., a neighborhood) may differ considerably from the effect of another. If this proved to be the case, we would probably find that our F test leads to rejection of the null hypothesis of equal population mean CMI scores for the four neighborhoods.

Another important property of this model is that the effects α_1, α_2, α_3, and α_4, which are population parameters defined in terms of population means, can each be estimated from the data by appropriately substituting the usual estimates of the means into the expressions for the effects. For our example, the estimated effects are given by

$\hat{\alpha}_1 = \overline{Y}_1. - \overline{Y}$ = Sample mean CMI score for neighborhood 1 − Overall sample mean CMI score for all neighborhoods

$\hat{\alpha}_2 = \overline{Y}_2. - \overline{Y}$ = Sample mean CMI score for neighborhood 2 − Overall sample mean CMI score for all neighborhoods

$\hat{\alpha}_3 = \overline{Y}_3. - \overline{Y}$ = Sample mean CMI score for neighborhood 3 − Overall sample mean CMI score for all neighborhoods

$\hat{\alpha}_4 = \overline{Y}_4. - \overline{Y}$ = Sample mean CMI score for neighborhood 4 − Overall sample mean CMI score for all neighborhoods

---

[8] In situations involving models with two or more factors, *effects* can also refer to measures of the influence of combinations of levels of the different factors on the dependent variable.

The actual numerical values obtained from these formulas are as follows:

$$\hat{\alpha}_1 = 18.92 - 21.83 = -2.91$$

$$\hat{\alpha}_2 = 20.72 - 21.83 = -1.11$$

$$\hat{\alpha}_3 = 20.84 - 21.83 = -0.99$$

$$\hat{\alpha}_4 = 26.84 - 21.83 = 5.01$$

Like the population effects, the estimated effects sum to 0; that is, $\sum_{i=1}^{4} \hat{\alpha}_i = 0$.

If we consider the general one-way ANOVA situation (with k populations and n_i observations from the ith population), the fixed-effects one-way ANOVA model may be written as follows:

$$Y_{ij} = \mu + \alpha_i + E_{ij} \qquad i = 1, 2, \ldots, k; \qquad j = 1, 2, \ldots, n_i \qquad (17.14)$$

where

$Y_{ij} = j$th observation from the ith population

$$\mu = \frac{\mu_1 + \mu_2 + \cdots + \mu_k}{k} \ (= \bar{\mu}^*)$$

$\alpha_i = \mu_i - \mu =$ Differential effect of population i

$E_{ij} = Y_{ij} - \mu - \alpha_i =$ Error component associated with the jth observation from the ith population

Here, it is easy to show that the sum of the α effects is 0; that is, $\sum_{i=1}^{k} \alpha_i = 0$. Similarly, the estimated effects, $\hat{\alpha}_i^* = (\bar{Y}_{i\cdot} - \bar{Y}^*)$, where $\bar{Y}^* = \sum_{i=1}^{k} \bar{Y}_{i\cdot}/k$, satisfy the constraint $\sum_{i=1}^{k} \hat{\alpha}_i^* = 0$.

An alternative definition of μ is $\bar{\mu} = \sum_{i=1}^{k} n_i \mu_i / n$, the overall weighted mean of the means. In this case, the weighted sum $\sum_{i=1}^{k} n_i \alpha_i = 0$, and the weighted sum of the estimated effects $\hat{\alpha}_i = (\bar{Y}_{i\cdot} - \bar{Y})$, where $\bar{Y} = \sum_{i=1}^{k} n_i \bar{Y}_{i\cdot}/n$, satisfies $\sum_{i=1}^{k} n_i \hat{\alpha}_i = 0$.

Model (17.14) corresponds in structure to the regression model given by (17.9): the regression coefficients $\alpha_1, \alpha_2, \ldots, \alpha_{k-1}$ are precisely the effects $\alpha_1 = \mu_1 - \bar{\mu}^*$, $\alpha_2 = \mu_2 - \bar{\mu}^*, \ldots$, $\alpha_{k-1} = \mu_{k-1} - \bar{\mu}^*$; the regression constant μ represents the overall (unweighted) mean $\bar{\mu}^*$; and the negative sum of the regression coefficients $(-\sum_{i=1}^{k-1} \alpha_i)$ represents the effect $\alpha_k = \mu_k - \bar{\mu}^*$. This is why we have defined each of these models using the same notation for the unknown parameters:

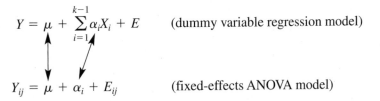

$$Y = \mu + \sum_{i=1}^{k-1} \alpha_i X_i + E \qquad \text{(dummy variable regression model)}$$

$$Y_{ij} = \mu + \alpha_i + E_{ij} \qquad \text{(fixed-effects ANOVA model)}$$

Notice that μ represents the unweighted average of the k population means, $\bar{\mu}^*$, rather than the weighted average $\bar{\mu}$, even though the sample sizes can be different in the different populations. Correspondingly, the least-squares estimate of μ is $\sum_{i=1}^{k} \bar{Y}_{i\cdot}/k$, the unweighted average of the

k sample means, rather than $\overline{Y}$. Nevertheless, the dummy variables in the regression model can be redefined to obtain a least-squares solution yielding $\overline{Y}$ as the estimate of μ. The following dummy variable definitions

$$X_i = \begin{cases} -n_i & \text{if population } k \\ n_k & \text{if population } i \\ 0 & \text{otherwise} \end{cases} \qquad i = 1, 2, \ldots, k - 1$$

are required.   ■

17.6 Random-effects Model for One-way ANOVA

In Section 17.1 we distinguished between fixed and random factors. We have also observed that, in ANOVA situations involving two or more factors, the F tests required for making inferences differ depending on whether all factors are fixed, all factors are random, or some of both are present. The null hypotheses to be tested must be stated in different terms when random factors are involved than when only fixed factors are involved.

■ **Example 17.4**   To get some insight into the structure of random-effects models, reconsider Daly's (1973) study (see Table 17.3). It might be argued that the four different turnkey neighborhoods form a representative sample of a larger population of similar types of neighborhoods (some of which might even be predominantly white with differing percentages of blacks in the surrounding neighborhoods). If so, the neighborhood factor would have to be considered random, and the appropriate ANOVA model would be a *random-effects* one-way ANOVA model.[9] Its form would be essentially the same as that given in (17.13), except that the α components would be treated differently; that is, the random-effects model would be of the form[10]

$$Y_{ij} = \mu + A_i + E_{ij} \qquad i = 1, 2, 3, 4; \qquad j = 1, 2, \ldots, 25 \qquad (17.15)$$

In this model, the A_i's can be viewed as constituting a random sample of random variables that have a common distribution—one that represents the entire population of possible effects (in our example, neighborhoods).

To perform the appropriate analysis, we must assume that the distribution of A_i is normal with zero mean:

$$A_i \frown N(0, \sigma_A^2) \qquad i = 1, 2, 3, 4 \qquad (17.16)$$

----

[9] This type of model is also referred to as a *variance-components model.*

[10] Our usual convention has been to use Latin letters (X, Y, Z, etc.) to denote random variables and to use Greek letters (β, μ, σ, τ) to denote parameters. This requires using A_i's rather than α_i's to denote random effects.

where σ_A^2 denotes the variance of A_i. We must also assume that the A_i's are independent of the E_{ij}'s and of each other.[11]

The requirement of zero mean in (17.16) is similar in philosophy to the requirement that $\sum_{i=1}^{k} \alpha_i = 0$ for the fixed-effects model. When the random-effects model (17.15) applies, we assume that the average (i.e., mean) effect of neighborhoods is 0 over the entire population of neighborhoods; that is, we assume that $\mu_{A_i} = 0$, $i = 1, 2, 3, 4$.

How do we state our null hypothesis? Because we have specified that the neighborhood effects average out to 0 over the entire population of possible effects, the only way to assess whether any significant neighborhood effects are present at all involves considering σ_A^2. If there is no variability (i.e., $\sigma_A^2 = 0$), all neighborhood effects must be 0. If there is variability (i.e., $\sigma_A^2 > 0$), some nonzero effects must exist in the population of neighborhood effects.

Thus, our null hypothesis of no neighborhood effects should be stated as follows:

$$H_0: \sigma_A^2 = 0 \qquad\qquad (17.17)$$

This hypothesis is analogous to the null hypothesis (17.1) used in the fixed-effects case, although it happens to be stated in terms of a population variance rather than in terms of population means.

We must still explain why the F test given by (17.2) for the fixed-effects model is exactly the same as that used for the random-effects model.[12] Such an explanation is best made by considering the properties of the mean squares MST and MSE. In Section 17.3.2, in connection with the fixed-effects model, we saw that the F statistic, MST/MSE, could be considered a rough approximation to the ratio of the means of these mean squares,

$$\frac{\mu_{\text{MST}}}{\mu_{\text{MSE}}} = \frac{\sigma^2 + \dfrac{1}{(k-1)} \sum_{i=1}^{k} n_i (\mu_i - \bar{\mu})^2}{\sigma^2}$$

The parameters μ_{MST} and μ_{MSE} are often called *expected mean squares*.

A similar argument can be made with regard to the F statistic for the random-effects model. In particular, for the random-effects model as well as for the fixed-effects model, the denominator MSE estimates σ^2; that is,

$$\mu_{\text{MSE}} = \sigma^2$$

Furthermore, for the random-effects model applied to our example ($k = 4$, $n_i = 25$), it can be shown that MST estimates

$$\mu_{\text{MST(random)}} = \sigma^2 + 25\sigma_A^2 \qquad\qquad (17.18)$$

Thus, for the random-effects model, F approximates the ratio

$$\frac{\mu_{\text{MST(random)}}}{\mu_{\text{MSE}}} = \frac{\sigma^2 + 25\sigma_A^2}{\sigma^2} \qquad\qquad (17.19)$$

---

[11] Since the *same* random variable A_i defined by (17.16) appears in (17.15) for each of the observations from population i, it follows that corr $(Y_{ij}, Y_{ij'}) = \sigma_A^2 / (\sigma_A^2 + \sigma^2)$, the so-called intraclass (or intrapopulation) correlation. Thus, in contrast to the one-way *fixed-effects* ANOVA model, the one-way *random-effects* ANOVA model introduces a dependency among the set of observations from the same population.

[12] Again, the F tests are computationally equivalent for fixed-effects and random-effects models only in one-way ANOVA. When dealing with two-way or higher-way ANOVA, the testing procedures may be different.

Since the null hypothesis in this case is $H_0: \sigma_A^2 = 0$, the ratio (17.19) simplifies to $\sigma^2/\sigma^2 = 1$ when H_0 is true. Thus, the F statistic under H_0 again consists of the ratio of two estimates of the same variance σ^2. Furthermore, because $\sigma_A^2 > 0$ when H_0 is not true, the greater the variability among neighborhood effects is, the larger should be the observed value of F.

In general, the random-effects model for one-way ANOVA is given by

$$Y_{ij} = \mu + A_i + E_{ij} \qquad\qquad i = 1, 2, \ldots, k; \qquad j = 1, 2, \ldots, n_i \qquad\qquad (17.20)$$

where A_i and E_{ij} are independent random variables satisfying $A_i \frown N(0, \sigma_A^2)$ and $E_{ij} \frown N(0, \sigma^2)$[13] For this model, F approximates the following ratio of expected mean squares:

$$\frac{\mu_{\text{MST(random)}}}{\mu_{\text{MSE}}} = \frac{\sigma^2 + n_0\sigma_A^2}{\sigma^2}$$

where

$$n_0 = \frac{\sum_{i=1}^{k} n_i - \left(\sum_{i=1}^{k} n_i^2 \Big/ \sum_{i=1}^{k} n_i \right)}{k-1}$$

functions as an average of the n_i observations selected from each population.[14] The F statistic for the random-effects model is therefore the ratio of two estimates of σ^2 when $H_0: \sigma_A^2 = 0$ is true.

Table 17.6 summarizes the similarities and the differences between the fixed- and random-effects models. Tables with similar formats will be used in subsequent chapters to highlight distinctions for ANOVA situations with more than two factors.

TABLE 17.6   **Combined one-way ANOVA table for fixed- and random-effects models**

Source	d.f.	MS	F	Expected Mean Square (EMS)	
				Fixed Effects	Random Effects
Between	$k-1$	MST	$\dfrac{\text{MST}}{\text{MSE}}$	$\sigma^2 + \dfrac{1}{k-1}\sum_{i=1}^{k} n_i(\mu_i - \bar{\mu})^2$	$\sigma^2 + n_0\sigma_A^2$
Within	$n-k$	MSE		σ^2	σ^2
Total	$n-1$				
				$H_0: \mu_1 = \mu_2 = \cdots = \mu_k$	$H_0: \sigma_A^2 = 0$

■

---

[13] Under these assumptions, $Y_{ij} \sim N(\mu, \sigma_A^2 + \sigma^2)$ for all (i, j). And, corr$(Y_{ij}, Y_{ij'}) = \sigma_A^2/(\sigma_A^2 + \sigma^2)$ for all $j \neq j'$ with fixed i. In other words, observations from *different* populations are independent, but observations from the *same* population are correlated.

[14] When all the n_i are equal, as in the Daly (1973) example (i.e., $n_i = n^*$), then n_0 is equal to n^*, since

$$n_0 = \frac{kn^* - (kn^{*2}/kn^*)}{k-1} = n^*$$

In the Daly (1973) example, $n^* = 25$.

17.7 Multiple-comparison Procedures for Fixed-effects One-way ANOVA

When we find that an ANOVA F test for simultaneously comparing several population means is statistically significant, our next step customarily is to determine which *specific* differences exist among the population means. For example, if we are comparing four means (fixed-effects case) and the null hypothesis $H_0: \mu_1 = \mu_2 = \mu_3 = \mu_4$ is rejected,[15] we next try to determine which subgroups of means are different, by considering more specific hypotheses such as $H_{01}: \mu_1 = \mu_2$, $H_{02}: \mu_2 = \mu_3$, $H_{03}: \mu_3 = \mu_4$, or even $H_{04} \, (\mu_1 + \mu_2)/2 = (\mu_3 + \mu_4)/2$, which compares the average effect of populations 1 and 2 with the average effect of populations 3 and 4. Such specific comparisons may have been of interest to us before (a priori) the data were collected, or they may arise in completely exploratory studies only after (a posteriori) the data have been examined. In either event, a seemingly reasonable first approach to drawing inferences about differences among the population means would be to conduct several t tests and to focus on all the tests found significant. For example, if all pairwise comparisons among the means are desired, then $_4C_2 = 6$ such tests must be performed with regard to 4 means (or in general, $_kC_2 = k(k - 1)/2$ tests with regard to k means). Thus, in testing $H_0: \mu_i = \mu_j$ at the α level of significance, we could reject this H_0 when

$$|T| \geq t_{n-k, \, 1-\alpha/2}$$

where

$$T = \frac{(\bar{Y}_i - \bar{Y}_j) - 0}{\sqrt{\mathrm{MSE}(1/n_i + 1/n_j)}}$$

and where n is the total number of observations; k is the number of means under consideration; n_i and n_j are the sizes of the samples selected from the ith and jth populations, respectively; $\bar{Y}_i$ and $\bar{Y}_j$ are the corresponding sample means; and MSE is the mean-square-error term, with $(n - k)$ degrees of freedom, that estimates the (homoscedastic) variance σ^2. MSE is used instead of a simple two-sample estimate of σ^2 based entirely on data from groups i and j; this is because MSE is a better estimate of σ^2 (in terms of degrees of freedom) under the assumption of variance homogeneity over all k populations.

Equivalently, one could reject H_0 if the $100(1 - \alpha)\%$ confidence interval

$$(\bar{Y}_i - \bar{Y}_j) \pm t_{n-k, \, 1-\alpha/2}\sqrt{\mathrm{MSE}\left(\frac{1}{n_i} - \frac{1}{n_j}\right)}$$

does not include 0.

Unfortunately, performing several such t tests has a serious drawback: the more null hypotheses there are to be tested, the more likely it is that one of them will be rejected even if

---

[15] This section deals only with fixed-effects ANOVA problems. The random-effects model treats the observed factor levels as a sample from a larger population of levels of interest and therefore is not directed exclusively at comparisons of the sampled levels.

all the null hypotheses are actually true. In other words, if several such tests are made, each at the α level, the probability of incorrectly rejecting *at least one* H_0 is much larger than α and continues to increase with each additional test made. Moreover, if in an exploratory study the investigator decides to compare only the sample means that are most discrepant (e.g., the largest versus the smallest), the testing procedure becomes biased in favor of rejecting H_0 because only the comparisons most likely to be significant are made. This bias will be reflected in the fact that the actual probability of incorrectly rejecting a given true null hypothesis exceeds the α level specified for the test.

Many different hypothesis testing procedures have been devised to provide an overall significance level of α even when several tests are performed. All of these procedures are grouped under the heading "multiple-comparison procedures." We shall focus here on three such approaches—the first due to Bonferroni, the second due to Tukey and Kramer, and the third due to Scheffé. Detailed mathematical discussions of these and other multiple-comparison methods can be found in Miller (1966), Guenther (1964), Lindman (1974) and Neter and Wasserman (1974).

17.7.1 The Bonferroni Approach

An approximate way to circumvent the problem of distorted significance levels when performing several tests involves reducing the significance levels used for each individual test sufficiently to fix the *overall significance level* (i.e.,the probability of incorrectly rejecting at least one of the null hypotheses being tested) at some desired level (say α). If we perform l such tests, the maximum possible value for this overall significance level is $l\alpha$. Thus, one simple way to ensure an overall significance level of at most α is to use α/l as the significance level for each separate test. This approach is often referred to as the Bonferroni method. For example, if five contrasts have been identified, a priori, as being of interest, then performing each of the five relevant hypothesis tests at the $0.05/5 = 0.01$ level of significance will ensure an overall significance level of no more than $\alpha = 0.05$.

One disadvantage of the Bonferroni method is that the *true* overall significance level may be considerably lower than α, and, in extreme situations, it may be so low that none of the individual tests will be rejected (i.e., the overall power of the method will be low). However, the true overall significance level using the Bonferroni approach will generally not be too low when the number of tests performed is close to the number of levels of the factor in question.

■ **Example 17.5**   Consider the set of data given in Table 17.7, which was collected from an experiment designed to compare the relative potencies of four cardiac substances. In the experiment, a suitable dilution of one of the substances was slowly infused into an anesthetized guinea pig, and the dosage at which the pig died was recorded. Ten guinea pigs were used for each substance, and the laboratory environment and the measurement procedures were assumed to be identical for each guinea pig. The main research goal was to determine whether any differences existed among the potencies of the four substances and, if so, to quantify those differences. The overall ANOVA table for comparing the mean potencies of the four cardiac substances is given in Table 17.8.

The global F test strongly rejects ($P < .001$) the null hypothesis of equality of the four population means. Therefore, the multiple-comparison question arises: What is the best way to account for the differences found? As a crude first step, we can examine the nature of the

TABLE 17.7 Potencies (dosages at death) of four cardiac substances

Substance	Sample Size (n_i)	Dosage at Death (Y_{ij})	Total	Sample Mean ($\bar{Y}_i$)	Sample Variance (S_i^2)
1	10	29, 28, 23, 26, 26, 19, 25, 29, 26, 28	259	25.9	9.4333
2	10	17, 25, 24, 19, 28 21, 20, 25, 19, 24	222	22.2	12.1778
3	10	17, 16, 21, 22, 23 18, 20, 17, 25, 21	200	20.0	8.6667
4	10	18, 20, 25, 24, 16 20, 20, 17, 19, 17	196	19.6	8.7111

TABLE 17.8 ANOVA table for data of Table 17.7

Source	d.f.	SS	MS	F
Substances	3	249.875	83.292	8.545 ($P < .001$)
Error	36	350.900	9.747	
Total	39	600.775		

differences with the help of a schematic diagram of ordered sample means (Figure 17.1). In the diagram, an overbar has been drawn over the labels for substances 3 and 4 to indicate that the sample mean potencies for these two substances are quite similar. On the other hand, no overbar has been drawn connecting 1 and 2 with each other or with 3 and 4, suggesting that substances 1 and 2 differ from each other as well as from both 3 and 4.

Such an overall quantification of the differences among the population means is desired from a multiple-comparison analysis. Nevertheless, the purely descriptive approach taken does not account for the sampling variability associated with any estimated comparison of interest. As a result, two sample means that seem practically different may not, in fact, be statistically different. Since the only multiple-comparison method we have discussed so far is the Bonferroni method, let us consider how to apply this method to the data of Table 17.7, using an overall significance level of $\alpha = .05$ for all pairwise comparisons of the mean potencies of the four cardiac substances. This approach requires computing $_4C_2 = 6$ confidence intervals, each associated with a significance level of $\alpha/6 = .05/6 = .0083$, utilizing the formula

$$(\bar{Y}_i - \bar{Y}_j) \pm t_{36,1-0.0083/2}\sqrt{MSE\left(\frac{1}{10} + \frac{1}{10}\right)}$$

FIGURE 17.1 Crude comparison of sample means for potency data

The right-hand side of this expression is calculated as

$$t_{36,0.99585}\sqrt{9.747\left(\frac{1}{5}\right)} = 2.79(1.396) = 3.895$$

Thus, the confidence intervals for the six population mean differences are given as:

$\mu_1 - \mu_4$: 6.3 ± 3.895; i.e., $(2.405, 10.195)^*$

$\mu_1 - \mu_3$: 5.9 ± 3.895; i.e., $(2.005, 9.795)^*$

$\mu_1 - \mu_2$: 3.7 ± 3.895; i.e., $(-0.195, 7.595)$

$\mu_2 - \mu_4$: 2.6 ± 3.895; i.e., $(-1.295, 6.495)$

$\mu_2 - \mu_3$: 2.2 ± 3.895; i.e., $(-1.695, 6.095)$

$\mu_3 - \mu_4$: 0.4 ± 3.895; i.e., $(-3.495, 4.295)$

The results are ordered by the value of the sample mean difference (largest to smallest). The preceding intervals reveal only two significant comparisons (the ones starred) and translate into the diagrammatic overall ranking shown in Figure 17.2. These results are somewhat ambiguous due to overlapping "sets of similarities," which indicate that substances 2, 3, and 4 have essentially the same potency; that 1 and 2 have about the same potency; but also that 1 differs from both 3 and 4. In other words, one possible conclusion is that 2, 3, and 4 are to be grouped together and that 1 and 2 are to be grouped together—which is difficult to reconcile because substance 2 is common to both groups. Having to confront this ambiguity is quite fortuitous from a pedagogical standpoint, since such ambiguous results are not infrequently encountered in connection with multiple-comparison procedures. In our case the results indicate that the procedure used was not sensitive enough to permit an adequate evaluation of substance 2. Repeating the analysis with a larger data set would help clear up the ambiguity. Alternatively, since the Bonferroni approach tends to be conservative (i.e., the confidence intervals tend to be wider than necessary to achieve the overall significance level desired), other multiple-comparison methods—such as that of Tukey and Kramer—may provide more precise results (i.e., narrower confidence intervals) and so may reduce or eliminate any ambiguity.

The Bonferroni approach can also be implemented by using SAS's GLM procedure. In the SAS output shown below, a graphic similar to that of Figure 17.2 is displayed. In the SAS output, population means that are similar (i.e., not significantly different) at $\alpha = 0.05$ are labeled with the same letter.

FIGURE 17.2 **Bonferroni comparison of sample means for potency data**

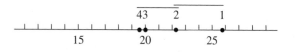

Edited SAS Output for Data of Table 17.7 Using the Bonferroni Approach

```
                    The GLM Procedure

             Bonferroni (Dunn) t Tests for DOSAGE

        Alpha                                    0.05
        Error Degrees of Freedom                   36
        Error Mean Square                    9.747222
        Critical Value of t                   2.79197
        Minimum Significant Difference         3.8982


    Means with the same letter are not significantly different.
        Bon Grouping            Mean        N        SUBSTANCE
                    A         25.900        10            1
                    A
                B   A         22.200        10            2
                B
                B             20.000        10            3
                B
                B             19.600        10            4
```

17.7.2 The Tukey-Kramer Method

The Tukey-Kramer method[16] is applicable when pairwise comparisons of population means are of primary interest; that is, null hypotheses of the form H_0: $\mu_i = \mu_j$ are to be considered. To use the Tukey-Kramer method, we compute the following confidence interval for the population mean difference ($\mu_i - \mu_j$):

$$(\bar{Y}_i - \bar{Y}_j) \pm \frac{q_{k,n-k,1-\alpha}}{\sqrt{2}} \sqrt{MSE\left(\frac{1}{n_i} + \frac{1}{n_j}\right)} \tag{17.21}$$

where $q_{k,\,n-k,\,1-\alpha}$ is the $100(1-\alpha)\%$ point of the studentized range distribution,[17] with k and $(n-k)$ degrees of freedom (see Table A.6 in Appendix A); k is the number of populations or groups, and n is the total number of observations.

In the set of all ${}_kC_2$ Tukey-Kramer pairwise confidence intervals of the form (17.21), when all of the group sample sizes are equal (that is, $n_i \equiv n^*$ for all $i = 1, 2, \ldots, k$), the probability is $(1-\alpha)$ that these intervals simultaneously contain the associated population mean differences

---

[16] The Tukey-Kramer method allows for unequal group sample sizes; it is the more general version of the method proposed by Tukey (1953), in which equal group sample sizes are assumed, and later extended by Kramer (1956).

[17] The studentized range distribution with k and r degrees of freedom is the distribution of a statistic of the form R/S, where $R = \{\max_i(Y_i) - \min_i(Y_i)\}$ is the range of a set of k independent observations $Y_1, Y_2, \ldots, Y_k$ from a normal distribution with mean μ and variance σ^2, and where S^2 is an estimate of σ^2 based on r degrees of freedom (which is independent of the Y's). In particular, when k means are being compared in fixed-effects one-way ANOVA, the statistic $\{\max_i(\bar{Y}_i) - \min_i(\bar{Y}_i)\}/\sqrt{MSE/n^*}$ (where n^* is the common sample size for each population group), has the studentized range distribution with k and $(n-k)$ degrees of freedom under H_0: $\mu_1 = \mu_2 = \cdots = \mu_k$, where $n_i = n^*$ for each i and where $n = kn^*$.

that are being estimated; that is, $100(1-\alpha)$ is the *overall confidence level* for all pairwise confidence intervals taken together. In particular, if each confidence interval is used to evaluate the corresponding pairwise null hypothesis of the general form H_0: $\mu_i = \mu_j$ by determining whether the value 0 is contained in the calculated interval, the probability of falsely rejecting the null hypothesis for *at least one* of the $_kC_2$ comparisons is equal to α. When the group sample sizes are unequal, or when fewer than all $_kC_2$ pairwise comparisons are considered, the Tukey-Kramer method is conservative; that is, the overall significance level for the pairwise comparisons performed is less than α and the overall confidence level is greater than $(1 - \alpha)$.

Let us now perform all pairwise comparisons of the group means for the potency data of Table 17.7, using the Tukey-Kramer method with an overall significance level of $\alpha = 0.05$. Since the value of $\dfrac{q_{k,n-k,1-\alpha}}{\sqrt{2}}\sqrt{MSE(\frac{1}{n_i} + \frac{1}{n_j})}$ is needed for all of the six possible confidence intervals, we compute it first. We have $n_i = 10$ for all i ($= 1, 2, 3, 4$), $k = 4$, $n = 40$, MSE $= 9.747$, and $q_{k,\,n-k,\,1-\alpha} = 3.81$. (This was obtained from Table A.6 by interpolation.) Therefore,

$$\frac{q_{k,\,n-k,1-\alpha}}{\sqrt{2}}\sqrt{MSE\left(\frac{1}{n_i} + \frac{1}{n_j}\right)} = \frac{3.81}{\sqrt{2}}\sqrt{9.747\left(\frac{1}{10} + \frac{1}{10}\right)} = 3.762$$

Thus, the pairwise Tukey-Kramer confidence intervals are:

$\mu_1 - \mu_4$: 6.3 ± 3.762; i.e., $(2.538, 10.062)^*$

$\mu_1 - \mu_3$: 5.9 ± 3.762; i.e., $(2.138,\ 9.662)^*$

$\mu_1 - \mu_2$: 3.7 ± 3.762; i.e., $(-0.062, 7.462)$

$\mu_2 - \mu_4$: 2.6 ± 3.762; i.e., $(-1.162, 6.362)$

$\mu_2 - \mu_3$: 2.2 ± 3.762; i.e., $(-1.562, 5.962)$

$\mu_3 - \mu_4$: 0.4 ± 3.762; i.e., $(-3.362, 4.162)$

Because the first confidence interval, comparing population means for groups 1 and 4, does not contain 0, we have statistical evidence (based on an overall significance level of $\alpha = 0.05$) that $\mu_1 \neq \mu_4$. Similarly, based on the second confidence interval, we conclude $\mu_1 \neq \mu_3$. All of the other intervals contain 0; therefore, all other pairwise mean difference comparisons are non-significant. (Note that, once again, the intervals were sorted in descending order based on the values of the sample mean differences; therefore, once the third pairwise mean difference in the list was determined to be nonsignificant, all remaining differences were also nonsignificant.)

Like the Bonferroni method, the Tukey-Kramer method leaves some ambiguity as to results: substance 2 has again been associated with substance 1 and also with substances 3 and 4. Such ambiguity is not uncommon. In this instance, it suggests that the amount of data being collected was insufficient to permit a clear characterization of substance 2.

Note that the Tukey-Kramer confidence intervals are narrower than the Bonferroni intervals for the potency data example. Indeed, when only all possible pairwise comparisons are being investigated and group sample sizes are equal, the Tukey-Kramer approach guarantees an overall significance level of α, whereas the Bonferroni approach guarantees an overall significance level $\leq \alpha$; in other words, for this situation, the Bonferroni approach usually will be less powerful than the Tukey-Kramer approach, and will, at best, be equally as powerful. This usually will also be true when the group sample sizes are unequal. Therefore, it is recommended that, when only all possible pairwise comparisons are being made, the Tukey-Kramer approach should be used rather than the Bonferroni approach.

The Tukey-Kramer approach also can be implemented by using SAS's GLM procedure. In the graphic shown in the SAS output below, sample means that are similar (i.e., not significantly different) at $\alpha = 0.05$ are labeled with the same letter.

Edited SAS Output for Data of Table 17.7 Using the Tukey-Kramer Approach

```
                        The GLM Procedure

            Tukey's Studentized Range (HSD) Test for DOSAGE

            Alpha                                        0.05
            Error Degrees of Freedom                       36
            Error Mean Square                        9.747222
            Critical Value of Studentized Range       3.80880
            Minimum Significant Difference             3.7604

      Means with the same letter are not significantly different.

            Tukey Grouping            Mean        N       SUBSTANCE
                          A         25.900        10           1
                          A
                      B   A         22.200        10           2
                      B
                      B             20.000        10           3
                      B
                      B             19.600        10           4
```

17.7.3 Scheffé's Method

Scheffé's method is generally recommended when comparisons other than simple pairwise differences between means are of interest, and when these more general comparisons are not planned a priori (i.e., are, instead, suggested by observed data patterns during the course of the analysis).

These more general comparisons are referred to as *contrasts*. To illustrate the meaning of the word *contrast*, suppose that the investigator who collected the potency data of Table 17.7 suspected that substances 1 and 3 had similar potencies, that substances 2 and 4 had similar potencies, and that the potencies of 1 and 3, on average, differed significantly from those of 2 and 4. Then it would be of interest to compare the average results obtained for 1 and 3 with the average results for 2 and 4, namely, to assess whether $(\mu_1 + \mu_3)/2$ really differs from $(\mu_2 + \mu_4)/2$. In other words, we could consider the contrast

$$L_1 = \frac{\mu_1 + \mu_3}{2} - \frac{\mu_2 + \mu_4}{2}$$

which would be 0 if the null hypothesis H_0: $(\mu_1 + \mu_3)/2 = (\mu_2 + \mu_4)/2$ were true. We can rewrite L_1 as follows:

$$L_1 = \frac{\mu_1 + \mu_3}{2} - \frac{\mu_2 + \mu_4}{2} = \frac{1}{2}\mu_1 - \frac{1}{2}\mu_2 + \frac{1}{2}\mu_3 - \frac{1}{2}\mu_4$$

or

$$L_1 = \sum_{i=1}^{4} c_{1i}\mu_i$$

so that L_1 is a *linear* function of the population means, with $c_{11} = \frac{1}{2}$, $c_{12} = -\frac{1}{2}$, $c_{13} = \frac{1}{2}$, and $c_{14} = -\frac{1}{2}$. Further,

$$c_{11} + c_{12} + c_{13} + c_{14} = \frac{1}{2} - \frac{1}{2} + \frac{1}{2} - \frac{1}{2} = 0$$

In general, a *contrast* is defined as any linear function of the population means, say,

$$L = \sum_{i=1}^{k} c_i\mu_i$$

such that

$$\sum_{i=1}^{k} c_i = 0$$

The associated null hypothesis is

$$H_0: \sum_{i=1}^{k} c_i\mu_i = 0$$

and the two-sided alternative hypothesis is

$$H_A: \sum_{i=1}^{k} c_i\mu_i \neq 0$$

The data in Table 17.7 suggest that the mean potency of substance 1 is definitely higher than the mean potencies of the other three substances. Such an observation invites a comparison of the mean potency of substance 1 with the average potency of substances 2, 3, and 4. In this case, the appropriate contrast to consider is

$$L_2 = \mu_1 - \frac{\mu_2 + \mu_3 + \mu_4}{3}$$

or

$$L_2 = \sum_{i=1}^{4} c_{2i}\mu_i$$

where $c_{21} = 1$, $c_{22} = c_{23} = c_{24} = -\frac{1}{3}$. (Again, $\sum_{i=1}^{4} c_{2i} = 0$.)

FIGURE 17.4   Recommended procedure for choosing a multiple-comparison technique in ANOVA

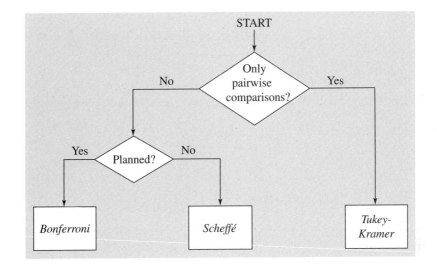

tend to be fairly robust; the Tukey-Kramer method is less so. Scheffé's method is always the least powerful, since all possible comparisons are considered. The Bonferroni method is generally presumed to be insensitive (i.e., to have low power), but this presumption can be somewhat discounted if the sets of comparisons are well planned.

Although many acceptable multiple-comparison procedures are available, most are limited to a rather narrow range of application (e.g., the Tukey-Kramer method is limited to pairwise contrasts). Both the Bonferroni and Scheffé procedures are completely general methods—the former for planned (a priori) and the latter for unplanned (a posteriori) multiple comparisons. The tasks of discussing all multiple-comparison methods completely is beyond the scope of this book.

17.9 Orthogonal Contrasts and Partitioning an ANOVA Sum of Squares

Our previous discussions of multiple regression have touched on the notion of a partitioned sum of squares in regression analysis, where the sum of squares due to regression (SSR) is broken down into various components reflecting the relative contributions of various terms in the fitted model. In an ANOVA framework, it is possible (via the use of orthogonal contrasts) to partition the treatment sum of squares SST into meaningful components associated with certain specific comparisons of particular interest. To illustrate how such a partitioning can be accomplished, we must discuss two new concepts: *orthogonal contrasts* and the *sum of squares associated with a contrast.*

In the notation of Section 17.7, two estimated contrasts

$$\hat{L}_A = \sum_{i=1}^{k} c_{Ai} \overline{Y}_i \qquad \text{and} \qquad \hat{L}_B = \sum_{i=1}^{k} c_{Bi} \overline{Y}_i$$

are *orthogonal* to each other (i.e., are orthogonal contrasts) if

$$\sum_{i=1}^{k} \frac{c_{Ai}c_{Bi}}{n_i} = 0 \qquad (17.24)$$

In the special case where the n_i's are equal, then equation (17.24) reduces to the condition

$$\sum_{i=1}^{k} c_{Ai}c_{Bi} = 0 \qquad (17.25)$$

Consider the three contrasts discussed earlier with regard to the potency data of Table 17.7:

$$\hat{L}_1 = \frac{\overline{Y}_1 + \overline{Y}_3}{2} - \frac{\overline{Y}_2 + \overline{Y}_4}{2} = \frac{1}{2}\overline{Y}_1 - \frac{1}{2}\overline{Y}_2 + \frac{1}{2}\overline{Y}_3 - \frac{1}{2}\overline{Y}_4$$

where $c_{11} = c_{13} = \frac{1}{2}$ and $c_{12} = c_{14} = -\frac{1}{2}$;

$$\hat{L}_2 = \overline{Y}_1 - \frac{1}{3}(\overline{Y}_2 + \overline{Y}_3 + \overline{Y}_4) = \overline{Y}_1 - \frac{1}{3}\overline{Y}_2 - \frac{1}{3}\overline{Y}_3 - \frac{1}{3}\overline{Y}_4$$

where $c_{21} = 1$ and $c_{22} = c_{23} = c_{24} = -\frac{1}{3}$; and

$$\hat{L}_3 = \frac{\overline{Y}_1 + \overline{Y}_2}{2} - \frac{\overline{Y}_3 + \overline{Y}_4}{2} = \frac{1}{2}\overline{Y}_1 + \frac{1}{2}\overline{Y}_2 - \frac{1}{2}\overline{Y}_3 - \frac{1}{2}\overline{Y}_4$$

where $c_{31} = c_{32} = \frac{1}{2}$ and $c_{33} = c_{34} = -\frac{1}{2}$. Since $n_i = 10$ for every i, we need only verify that condition (17.25) holds to demonstrate orthogonality. In particular, for $\hat{L}_1$ and $\hat{L}_2$, we have

$$\sum_{i=1}^{4} c_{1i}c_{2i} = \left(\frac{1}{2}\right)(1) + \left(-\frac{1}{2}\right)\left(-\frac{1}{3}\right) + \left(\frac{1}{2}\right)\left(-\frac{1}{3}\right) + \left(-\frac{1}{2}\right)\left(-\frac{1}{3}\right) = \frac{2}{3} \neq 0$$

For $\hat{L}_1$ and $\hat{L}_3$, we have

$$\sum_{i=1}^{4} c_{1i}c_{3i} = \left(\frac{1}{2}\right)\left(\frac{1}{2}\right) + \left(-\frac{1}{2}\right)\left(\frac{1}{2}\right) + \left(\frac{1}{2}\right)\left(-\frac{1}{2}\right) + \left(-\frac{1}{2}\right)\left(-\frac{1}{2}\right) = 0$$

And, for $\hat{L}_2$ and $\hat{L}_3$, we have

$$\sum_{i=1}^{4} c_{2i}c_{3i} = (1)\left(\frac{1}{2}\right) + \left(-\frac{1}{3}\right)\left(\frac{1}{2}\right) + \left(-\frac{1}{3}\right)\left(-\frac{1}{2}\right) + \left(-\frac{1}{3}\right)\left(-\frac{1}{2}\right) = \frac{2}{3} \neq 0$$

Thus, we conclude that $\hat{L}_1$ and $\hat{L}_3$ are orthogonal to one another but that neither is orthogonal to $\hat{L}_2$.

Orthogonality is a desirable property for several reasons. Suppose that SST denotes the treatment sum of squares with $(k - 1)$ degrees of freedom for a fixed-effects one-way ANOVA, and suppose that $\hat{L}_1, \hat{L}_2, \ldots, \hat{L}_t$ are a set of $t(\leq k - 1)$ mutually orthogonal contrasts of the k sample means (*mutually orthogonal* means that any two contrasts selected from the set of t contrasts are orthogonal to one another). Then SST can be partitioned into $(t + 1)$ statistically independent sums of squares, with t of these sums of squares having 1 degree of freedom each and being associated with the t orthogonal contrasts, and with the remaining sum of squares having $(k - t - 1)$ degrees of freedom and being associated with what remains after the t orthogonal contrast sums of squares have been accounted for. In other words, we can write

$$\text{SST} = \text{SS}(\hat{L}_1) + \text{SS}(\hat{L}_2) + \cdots + \text{SS}(\hat{L}_t) + \text{SS(remainder)}$$

where $SS(\hat{L})$ is the notation for the sum of squares (with 1 degree of freedom) associated with the contrast $\hat{L}$. In particular, it can be shown that

$$SS(\hat{L}) = \frac{(\hat{L})^2}{\sum\limits_{i=1}^{k} (c_i^2/n_i)} \qquad \text{when} \qquad \hat{L} = \sum_{i=1}^{k} c_i \bar{Y}_i, \qquad (17.26)$$

that

$$\frac{SS(\hat{L})}{MSE} \frown F_{1,\, n-k}$$

under $H_0\colon L = \sum_{i=1}^{k} c_i \mu_i = 0$, and, in general, that

$$\frac{[SS(\hat{L}_1) + SS(\hat{L}_2) + \cdots + SS(\hat{L}_t)]/t}{MSE} \frown F_{t,\, n-k}$$

under $H_0\colon L_1 = L_2 = \cdots = L_t = 0$ when $\hat{L}_1, \hat{L}_2, \ldots, \hat{L}_t$ are mutually orthogonal. Thus, by partitioning SST as described, we can test hypotheses concerning sets of orthogonal contrasts that are of more specific interest than the global hypothesis $H_0\colon \mu_1 = \mu_2 = \cdots = \mu_k$.

For example, to test $H_0\colon L_2 = \mu_1 - \frac{1}{3}(\mu_2 + \mu_3 + \mu_4) = 0$ for the potency data in Table 17.7, we first use (17.26) to calculate

$$SS(\hat{L}_2) = \frac{(5.30)^2}{[(1)^2 + (-1/3)^2 + (-1/3)^2 + (-1/3)^2]/10} = 210.675$$

and then we form the ratio

$$\frac{SS(\hat{L}_2)}{MSE} = \frac{210.675}{9.747} = 21.614$$

which is highly significant ($P < .001$ based on the $F_{1,36}$ distribution). Modifying Table 17.8 to reflect this partitioning of SST (the sum of squares for "substances") yields

Source		d.f.	SS	MS	F
Substances $\begin{cases} 1 \text{ vs. } (2,3,4) \\ \text{Remainder} \end{cases}$	1 vs. (2,3,4)	1	210.675	210.675	21.614 ($P < .001$)
	Remainder	2	39.200	irrelevant	—
Error		36	350.900	9.747	
Total		39	600.775		

Similarly, the partitioned ANOVA table for testing

$$H_0\colon L_3 = \frac{\mu_1 + \mu_2}{2} - \frac{\mu_3 + \mu_4}{2} = 0$$

is as follows:

Source		d.f.	SS	MS	F
Substances $\begin{cases} (1, 2) \text{ vs. } (3, 4) \\ \text{Remainder} \end{cases}$		1	180.625	180.625	18.531 ($P < .001$)
		2	69.25	irrelevant	—
Error		36	350.900	9.747	
Total		39	600.775		

This partition follows from the fact that

$$SS(\hat{L}_3) = \frac{(4.25)^2}{[(1/2)^2 + (1/2)^2 + (-1/2)^2 + (-1/2)^2]/10} = 180.625$$

Finally, since $\hat{L}_1$ and $\hat{L}_3$ are orthogonal, we can represent the independent contributions of these two contrasts to SST in one ANOVA table:

Source		d.f.	SS	MS	F
Substances $\begin{cases} (1, 3) \text{ vs. } (2, 4) \\ (1, 2) \text{ vs. } (3, 4) \\ \text{Remainder} \end{cases}$		1	42.025	42.025	4.312 (n.s.)
		1	180.625	180.625	18.531 ($P < .001$)
		1	27.225	irrelevant	—
Error		36	350.900	9.747	
Total		39	600.775		

Here,

$$SS(\hat{L}_1) = \frac{\left(\dfrac{25.9 + 20.0}{2} - \dfrac{22.2 + 19.6}{2} \right)^2}{[(1/2)^2 + (-1/2)^2 + (1/2)^2 + (-1/2)^2]/10} = 42.025$$

It would not be valid to present a partitioned ANOVA table that simultaneously included partitions due to $\hat{L}_2$ as well as $\hat{L}_1$ and/or $\hat{L}_3$. This is because $\hat{L}_2$ is not orthogonal to these contrasts, and so its sum of squares does not represent a separate and independent contribution to SST; this can easily be confirmed by observing from the preceding tables that

$$SST = 249.875 \neq SS(\hat{L}_1) + SS(\hat{L}_2) + SS(\hat{L}_3) = 42.025 + 210.675 + 180.625$$

A final example of using orthogonal contrasts involves the need to assess whether the sample means exhibit a trend of some sort; this need often arises when the treatments (or populations) being studied represent, for example, different levels of the same factor (e.g., different concentrations of the same material, or different temperature or pressure settings). In such situations it is of interest to quantify how the sample means vary with changes in the level of the factor—that is, to clarify whether the change in mean response takes place in a linear, quadratic, or other way as the level of the factor increases or decreases.

A qualitative first step in assessing such a trend is to plot the observed treatment means as a function of the factor levels. This may yield some general idea of the pattern (if any) present. Standard regression techniques can then be used to quantify any trends suggested by such plots, but our goal here is not to fit a regression model; rather, it is to evaluate a possible general trend in the sample means statistically, instead of forming an opinion on the basis of a simple examination of a plot of these means.

In a standard regression approach, the independent variable ("treatments") could possibly be considered as an interval variable, and the actual value of the variable at each treatment setting could be used. To test for a linear trend by using regression, we would apply the model $Y = \beta_0 + \beta_1 X + E$, where X denotes the (interval) treatment variable. With this model, the test for linear trend is the usual test for zero slope. In general, this test is not equivalent to the test for a linear trend in *mean* response using the orthogonal polynomials described in this section, except when the pure error mean square is used in place of the residual mean square in the denominator of the F statistic to test for zero slope (because then the regression pure error mean square and the one-way ANOVA error mean square are identical). The usual test for lack of fit of a straight-line regression model is equivalent to the test for a nonlinear trend in mean response discussed in this section. (See Problem 10 at the end of this chapter for more discussion about this issue.)

A statistical trend analysis may be carried out by determining how much of the sum of squares due to treatments (SST) is associated with each of the terms (linear, quadratic, cubic, etc.) in a polynomial regression. If the various levels of the treatment or factor being studied are equally spaced, this determination is best carried out by using the method of orthogonal polynomials. (For a discussion, see Armitage (1971) and also Chapter 15.)

To illustrate the use of orthogonal polynomials, let us again turn to the potency data of Table 17.7. Further, let us suppose that the four substances actually represent four equally spaced concentrations of some toxic material, with substance 1 the least concentrated solution and substance 4 the most concentrated. For example, substance 1 might represent a 10% solution of the toxic material, substance 2 a 20% solution, substance 3 a 30% solution, and substance 4 a 40% solution. In this case, a plot of the four sample means versus concentration takes the form shown in Figure 17.5. This plot suggests at least a linear and possibly a quadratic relationship between concentration and response, and this general impression can be quantified via the use of orthogonal polynomials. In particular, given $k = 4$ sample means, it is possible to fit up to a third-order ($k - 1 = 3$) polynomial to these means; the cubic model (with four terms) would pass through all four points on the Figure 17.5 graph, thus explaining all the variation in the four sample means (or, equivalently, in SST). Because the concentrations are equally spaced, it is possible via the use of orthogonal polynomials to define three orthogonal contrasts of the four sample means—one (say, $\hat{L}_l$) measuring the strength of the linear component of the third-degree polynomial, one (say, $\hat{L}_q$) the quadratic component contribution, and one (say, $\hat{L}_c$) the cubic component effect. The sums of squares associated with these three orthogonal contrasts each have 1 degree of freedom, are statistically independent, and satisfy the relationship

$$\text{SST} = \text{SS}(\hat{L}_l) + \text{SS}(\hat{L}_q) + \text{SS}(\hat{L}_c)$$

FIGURE 17.5 **Plot of the sample means versus concentration**

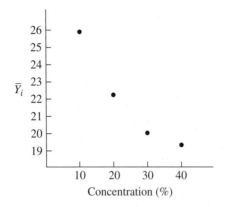

It can be shown that these three orthogonal contrasts have the following coefficients:

Contrast	Coefficient of				Calculated Value of Contrast
	$\bar{Y}_1 = 25.9$	$\bar{Y}_2 = 22.2$	$\bar{Y}_3 = 20.0$	$\bar{Y}_4 = 19.6$	
$\hat{L}_l$	-3	-1	1	3	-21.1
$\hat{L}_q$	1	-1	-1	1	3.3
$\hat{L}_c$	-1	3	-3	1	0.3

Condition (17.25) clearly holds for these three sets of coefficients, which is sufficient to establish mutual orthogonality here, since the n_i's are all equal. These coefficients were taken from Table A.7 in Appendix A. The table may be used only for equally spaced treatment values and equal cell-specific sample sizes. Kirk (1969) summarizes an algorithm for the general case that does not require these two restrictions. More conveniently, computer software can be used to obtain orthogonal contrast coefficients.

Now, from (17.26), the sums of squares for these three particular contrasts are

$$SS(\hat{L}_l) = \frac{(-21.1)^2}{[(-3)^2 + (-1)^2 + (1)^2 + (3)^2]/10} = 222.605$$

$$SS(\hat{L}_q) = \frac{(3.3)^2}{[(1)^2 + (-1)^2 + (-1)^2 + (1)^2]/10} = 27.225$$

$$SS(\hat{L}_c) = \frac{(0.3)^2}{[(-1)^2 + (3)^2 + (-3)^2 + (1)^2]/10} = 0.045$$

Notice that

$$SST = 249.875 = 222.605 + 27.225 + 0.045$$

Finally, to assess the significance of these sums of squares, we form the following partitioned ANOVA table:

Source		d.f.	SS	MS	F
Substances	$\hat{L}_l$	1	222.605	222.605	22.838 $(P < .001)$
	$\hat{L}_q$	1	27.225	27.225	2.793 $(.1 < P < .25)$
	$\hat{L}_c$	1	0.045	0.045	< 1 (n.s.)
Error		36	350.900	9.747	
Total		39	600.775		

It is clear from the preceding F tests that the relationship between potency (as measured by the amount of injected material needed to cause death) and concentration is strongly linear, with no evidence of higher-order effects.

Problems

1. Five treatments for fever blisters, including a placebo, were randomly assigned to 30 patients. For each of the five treatments, the data in the accompanying table identify the number of days from initial appearance of the blisters until healing is complete.

Treatment	No. of Days
Placebo (1)	5, 8, 7, 7, 10, 8
2	4, 6, 6, 3, 5, 6
3	6, 4, 4, 5, 4, 3
4	7, 4, 6, 6, 3, 5
5	9, 3, 5, 7, 7, 6

   a. Compute the sample means and the sample standard deviations for each treatment.
   b. Complete the ANOVA table in the accompanying computer output for the data given.
   c. Do the effects of the five treatments differ significantly with regard to healing fever blisters? In other words, test H_0: $\mu_1 = \mu_2 = \mu_3 = \mu_4 = \mu_5$ against H_A: "At least two treatments have different population means."
   d. What are the estimates of the true effects $(\mu_i - \mu)$ of the treatments? Verify that the sum of these estimated effects is 0. [*Note*: μ_i is the population mean for the ith treatment, and $\mu = \frac{1}{5}\sum_{i=1}^{5} \mu_i$ is the overall population mean.]
   e. Using dummy variables, create an appropriate regression model that describes this experiment. Give two possible ways of defining these dummy variables (one using 0's and 1's, and the other using 1's and -1's), and describe for each coding scheme how the regression coefficients are related to the population means $\mu_1, \mu_2, \mu_3, \mu_4, \mu_5$, and μ.
   f. For the data of this problem, carry out the Scheffé, Tukey-Kramer, and Bonferroni multiple-comparison procedures for examining pairwise differences between means, as described in Section 17.7. Also, compare the widths of the confidence intervals obtained by the three procedures.

Edited SAS Output (PROC GLM) for Problem 1

```
                              The GLM Procedure
Dependent Variable: days

                                 Sum of
Source                DF        Squares        Mean Square     F Value      Pr > F
Model                 __     _____      _____    _____     0.0136
Error                 25     58.50000000        2.34000000
Corrected Total       29     _____

          R-Square        Coeff Var        Root MSE       days Mean
          0.383994        27.15454        1.529706        5.633333

Source                DF       Type I SS       Mean Square     F Value      Pr > F
treat                  4     36.46666667       9.11666667        3.90       0.0136

Source                DF      Type III SS      Mean Square     F Value      Pr > F
treat                  4     36.46666667       9.11666667        3.90       0.0136

               Tukey's Studentized Range (HSD) Test for days
               Error Degrees of Freedom                   25
               Error Mean Square                         2.34
               Critical Value of Studentized Range     4.15337
               Minimum Significant Difference           2.5938

        Means with the same letter are not significantly different.
           Tukey Grouping           Mean          N        treat
                          A        7.5000          6          1
                          A
                    B     A        6.1667          6          5
                    B     A
                    B     A        5.1667          6          4
                    B     A
                    B     A        5.0000          6          2
                    B
                    B              4.3333          6          3

                     Bonferroni (Dunn) t Tests for days
               Error Degrees of Freedom                   25
               Error Mean Square                         2.34
               Critical Value of t                    3.07820
               Minimum Significant Difference           2.7186

        Means with the same letter are not significantly different.
            Bon Grouping            Mean          N        treat
                          A        7.5000          6          1
                          A
                    B     A        6.1667          6          5
                    B     A
                    B     A        5.1667          6          4
                    B     A
                    B     A        5.0000          6          2
                    B
                    B              4.3333          6          3

                        Scheffe's Test for days
               Error Degrees of Freedom                   25
               Error Mean Square                         2.34
               Critical Value of F                    2.75871
               Minimum Significant Difference           2.9338
```

(continued)

Edited SAS Output (PROC GLM) for Problem 1 (continued)

```
                    The GLM Procedure (continued)

        Means with the same letter are not significantly different.
              Scheffe Grouping           Mean        N        treat
                          A             7.5000        6          1
                          A
                    B     A             6.1667        6          5
                    B     A
                    B     A             5.1667        6          4
                    B     A
                    B     A             5.0000        6          2
                    B
                    B                   4.3333        6          3
```

2. The following data are replicate measurements of the sulfur dioxide concentration in each of three cities:

   City I: 2, 1, 3

   City II: 4, 6, 8

   City III: 2, 5, 2

   a. Complete the ANOVA table in the accompanying computer output for simultaneously comparing the mean sulfur dioxide concentrations in the three cities.
   b. Test whether the three cities differ significantly in mean sulfur dioxide concentration levels.
   c. What is the estimated effect associated with each city?
   d. State precisely the appropriate ANOVA fixed-effects model for these data.
   e. Using dummy variables, state precisely the regression model that corresponds to the fixed-effects model in part (d). What is the relationship between the coefficients in the regression model and the effects in the ANOVA model?
   f. Using the t distribution, find a 90% confidence interval for the true difference between the effects of cities I and II (making sure to use the best estimate of σ^2 provided by the data).

Edited SAS Output (PROC GLM) for Problem 2

```
                           The GLM Procedure
   Dependent Variable: SO2

                                    Sum of
   Source                 DF        Squares     Mean Square    F Value    Pr > F
   Model                   2     26.00000000    13.00000000    _____    0.0553
   Error                  __     _____    _____
   Corrected Total         8     _____
```

	R-Square	Coeff Var	Root MSE	SO2 Mean
	0.619048	44.53618	1.632993	3.666667

```
   Source                 DF      Type I SS     Mean Square    F Value    Pr > F
   city                    2     26.00000000    13.00000000      4.88      0.0553

   Source                 DF      Type III SS    Mean Square    F Value    Pr > F
   city                    2     26.00000000    13.00000000      4.88      0.0553
```

3. Each of three chemical laboratories performed four replicate determinations of the concentration of suspended particulate matter in a certain area using the "Hi-Vol" method of analysis. The resulting data are presented next.

Lab 1	Lab II	Lab III
4	2	5
4	2	2
6	5	5
10	3	8

a. Identify the appropriate ANOVA table for these data.
b. Test the null hypothesis of no differences among the laboratories.
c. With large-scale interlaboratory studies, analysts usually make inferences about a large population of laboratories of which only a random sample (e.g., laboratories I, II, and III) can be investigated. In such a case, describe the appropriate random-effects model for the data.
d. Two quantities of particular interest in a large-scale interlaboratory study are repeatability (i.e., a measure of the variability among replicate measurements within a single laboratory) and reproducibility (i.e., a measure of the variability between results from different laboratories). Using the random-effects model defined in part (c), define what you think are reasonable measures of repeatability and reproducibility, and obtain estimates of the quantities you have defined using the data in the accompanying computer output.

Edited SAS Output (PROC GLM) for Problem 3

The GLM Procedure

Dependent Variable: conc

Source	DF	Sum of Squares	Mean Square	F Value	Pr > F
Model	2	18.66666667	9.33333333	1.75	0.2280
Error	9	48.00000000	5.33333333		
Corrected Total	11	66.66666667			

R-Square	Coeff Var	Root MSE	conc Mean
0.280000	49.48717	2.309401	4.666667

Source	DF	Type I SS	Mean Square	F Value	Pr > F
Lab	2	18.66666667	9.33333333	1.75	0.2280

Source	DF	Type III SS	Mean Square	F Value	Pr > F
Lab	2	18.66666667	9.33333333	1.75	0.2280

4. Ten randomly selected mental institutions were examined to determine the effects of three different antipsychotic drugs on patients with the same types of symptoms. Each institution used one and only one of the three drugs exclusively for a one-year period. The proportion of treated patients in each institution who were discharged after one year of treatment is as follows for each drug used:

Drug 1: 0.10, 0.12, 0.08, 0.14 $(\bar{Y}_1 = 0.11, S_1 = 0.0192)$
Drug 2: 0.12, 0.14, 0.19 $(\bar{Y}_2 = 0.15, S_2 = 0.0361)$
Drug 3: 0.20, 0.25, 0.15 $(\bar{Y}_3 = 0.20, S_3 = 0.0500)$

a. Determine the appropriate ANOVA table for this data set, using the accompanying computer output.
b. Test to see whether significant differences exist among drugs with regard to the average proportion of patients discharged.
c. What other factors should be considered in comparing the effects of the three drugs?
d. What basic ANOVA assumptions might be violated here?

Edited SAS Output (PROC GLM) for Problem 4

```
                              The GLM Procedure

Dependent Variable: disch

                              Sum of
Source                 DF     Squares        Mean Square     F Value     Pr > F
Model                  2      0.01389000     0.00694500      5.06        0.0436
Error                  7      0.00960000     0.00137143
Corrected Total        9      0.02349000


           R-Square        Coeff Var        Root MSE        disch Mean
           0.591315        24.85423         0.037033        0.149000


Source                 DF     Type I SS      Mean Square     F Value     Pr > F
drug                   2      0.01389000     0.00694500      5.06        0.0436


Source                 DF     Type III SS    Mean Square     F Value     Pr > F
drug                   2      0.01389000     0.00694500      5.06        0.0436
```

5. Suppose that a random sample of five active members in each of four political parties in a certain western European country was given a questionnaire purported to measure (on a 100-point scale) the extent of "general authoritarian attitude toward interpersonal relationships." The means and standard deviations of the authoritarianism scores for each party are given in the following table.

	Party 1	Party 2	Party 3	Party 4
$\bar{Y}_i$	85	80	95	50
S_i	6	7	4	10
n_i	5	5	5	5

a. Determine the appropriate ANOVA table for this data set.
b. Test to see whether significant differences exist among parties with respect to mean authoritarianism scores.
c. Using dummy variables, state an appropriate regression model for this experimental situation.
d. Apply the Tukey-Kramer method of multiple comparisons to identify the pairs in which the means significantly differ from one another. (Use $\alpha = .05$.)

6. A psychosociological questionnaire was administered to a random sample of 200 persons on an island in the South Pacific that has become increasingly westernized over the past 30 years. From the questionnaire data, each of the 200 persons was classified into one of three groups—HI-POS, NO-DIF, and HI-NEG—according to the discrepancy between the amount of prestige in that person's traditional culture and the amount of prestige in the modern (westernized) culture. On the basis of the questionnaire data, a measure of "anomie" (i.e., social disorientation), denoted as Y, was determined on a 100-point scale, with the results summarized in the following table.

Group	n_i	$\bar{Y}_i$	S_i
HI-POS	50	65	9
NO-DIF	75	50	11
HI-NEG	75	55	10

a. Determine the appropriate ANOVA table.
b. Test whether the three different categories of prestige discrepancy have significantly different sample mean anomie scores.
c. How would you test whether a significant difference exists between the NO-DIF category and the other two categories combined?

7. To determine whether reading skills of the average high school graduate have declined over the past 10 years, the average verbal college aptitude scores (VSAT) were compared for a random sample of five big-city high schools for the years 1965, 1970, and 1975. The results are shown in the following table and accompanying computer output.

Year	School 1	School 2	School 3	School 4	School 5
1965	550	560	535	545	555
1970	545	560	528	532	541
1975	536	552	526	527	530

a. Determine the sample means for each year.
b. Comment on the independence assumption required for using one-way ANOVA on the data given.
c. Determine the one-way ANOVA table for these data.
d. Test by means of one-way ANOVA whether any significant differences exist among the three VSAT average scores for the years 1965, 1970, and 1975.
e. Use Scheffé's method to locate any significant differences between pairs of means. (Use $\alpha = .05$.)

Edited SAS Output (PROC GLM) for Problem 7

The GLM Procedure

Dependent Variable: vsat

Source	DF	Sum of Squares	Mean Square	F Value	Pr > F
Model	2	548.133333	274.066667	2.26	0.1466
Error	12	1453.600000	121.133333		
Corrected Total	14	2001.733333			

(continued)

Edited SAS Output (PROC GLM) for Problem 19 (continued)

```
                      The GLM Procedure (Continued)

Dependent Variable: OPI

                                Sum of
Source                DF        Squares      Mean Square      F Value      Pr > F
Model                 __      282.0545833    94.0181944       _____       _____
Error                 20      255.5050000    12.7752500
Corrected Total       23      537.5595833


         R-Square         Coeff Var        Root MSE        OPI Mean
         0.524695         3.966062         3.574248        90.12083


Source                DF      Type I SS      Mean Square      F Value      Pr > F
CAT                   3      282.0545833    94.0181944         7.36       0.0016


Source                DF      Type III SS    Mean Square      F Value      Pr > F
CAT                   3      282.0545833    94.0181944         7.36       0.0016


              Tukey's Studentized Range (HSD) Test for OPI
              Alpha                                         0.05
              Error Degrees of Freedom                        20
              Error Mean Square                         12.77525
              Critical Value of Studentized Range        3.95829
              Minimum Significant Difference              5.7759


        Means with the same letter are not significantly different.
              Tukey Grouping          Mean          N         CAT
                            A         93.617         6          2
                            A
                            A         92.383         6          3
                            A
                        B   A         89.800         6          1
                        B
                        B             84.683         6          4
```

20. This problem refers to the data of Problem 19.
   a. Suppose that an ANOVA is to be performed to compare the average OPI values for funds with different volatilities. State precisely the ANOVA model. Is fund volatility a fixed- or random-effects factor?
   b. In the SAS output that follows, complete the ANOVA table.
   c. Test whether the average overall performance indices differ significantly by volatility rating. Interpret your results.

Edited SAS Output (PROC GLM) for Problem 20

```
                           The GLM Procedure

Dependent Variable: OPI

                                Sum of
Source                DF        Squares      Mean Square      F Value      Pr > F
Model                 __      291.4399405    41.6342772       _____       _____
Error                 16      246.1196429    15.3824777
Corrected Total       23      537.5595833
```

(continued)

Edited SAS Output (PROC GLM) for Problem 20 (continued)

	R-Square	Coeff Var	Root MSE	OPI Mean		
	0.542154	4.351991	3.922050	90.12083		
Source	DF	Type I SS	Mean Square	F Value		Pr > F
VOL	7	291.4399405	41.6342772	2.71		0.0470
Source	DF	Type III SS	Mean Square	F Value		Pr > F
VOL	7	291.4399405	41.6342772	2.71		0.0470

21. Radial keratotomy is a type of refractive surgery in which radial incisions are made in a myopic (nearsighted) patient's cornea to reduce the person's myopia. Theoretically, the incisions allow the curvature of the cornea to become less steep, thereby reducing the patient's refractive error. [*Note*: Myopic patients have negative refractive errors. Patients who are farsighted have positive refractive errors. Patients who are neither near- nor farsighted have zero refractive error.]

The incisions extend radially from the periphery toward the center of the cornea. A circular central portion of the cornea, known as the clear zone, remains uncut. The diameter of the clear zone is determined by the baseline refraction of the patient. Patients with a greater degree of myopia may receive longer incisions, leaving them with smaller clear zones. The thinking here is that "more surgery" is needed to correct the worse initial vision of these patients.

Radial keratotomy and other vision-correction surgery techniques have been growing in popularity in the 1980s and 1990s, both among the public and among ophthalmologists. The Prospective Evaluation of Radial Keratotomy (PERK) study was begun in 1983 to evaluate the effects of radial keratotomy. Lynn et al. (1987) examined the variables associated with the five-year postsurgical change in refractive error (Y, measured in diopters, D). One of the independent variables under consideration was diameter of the clear zone (X). In the PERK study, three clear zone sizes were used: 3.0 mm, 3.5 mm, and 4.0 mm.

The following computer output is based on data adapted from the PERK study.

a. Suppose that an ANOVA is to be performed to compare the average five-year change in refraction for patients with different clear zones. State precisely the ANOVA model. Is clear zone size a fixed- or random-effects factor?

b. In the SAS output that follows, complete the ANOVA table.

c. Test whether the average five-year change in refraction differs significantly by clear zone size. Interpret your results.

d. Use the Tukey-Kramer method to locate any significant pairwise differences between clear zones. (Use $\alpha = .05$.) Interpret your results.

Edited SAS Output (PROC GLM) for Problem 21

```
                        The GLM Procedure

                    Class Level Information

              Class           Levels        Values
              CLRZONE            3           3.0 3.5 4.0

          Number of Observations Read         54
          Number of Observations Used         51
```

(continued)

Edited SAS Output (PROC GLM) for Problem 21 (continued)

```
                          The GLM Procedure (continued)

Dependent Variable: Y

                                      Sum of
Source                  DF            Squares      Mean Square      F Value      Pr > F
Model                   __        14.70441590
Error                   48        64.59163802       1.34565913
Corrected Total         50        79.29605392

                  R-Square        Coeff Var        Root MSE        Y Mean
                  0.185437        30.43590         1.160025        3.811373

Source                  DF          Type I SS      Mean Square      F Value      Pr > F
CLRZONE                  2        14.70441590       7.35220795         5.46      0.0073

Source                  DF         Type III SS     Mean Square      F Value      Pr > F
CLRZONE                  2        14.70441590       7.35220795         5.46      0.0073

              Tukey's Studentized Range (HSD) Test for Y
              Alpha                                      0.05
              Error Degrees of Freedom                   48
              Error Mean Square                     1.345659
              Critical Value of Studentized Range   3.42026

      Comparisons significant at the 0.05 level are indicated by ***
                              Difference
              CLRZONE         Between                Simultaneous 95%
           Comparison          Means         Confidence           Limits
            3.0 - 3.5          0.7021          -0.2562            1.6603
            3.0 - 4.0          1.2778           0.3368            2.2188  ***
            3.5 - 3.0         -0.7021          -1.6603            0.2562
            3.5 - 4.0          0.5757          -0.4326            1.5840
            4.0 - 3.0         -1.2778          -2.2188           -0.3368  ***
            4.0 - 3.5         -0.5757          -1.5840            0.4326
```

22. Data on law and business schools were sampled from *U.S. News & World Report*'s 1996 report on the "America's Best Graduate Schools 1996 Annual Guide."[19] The school's reputation rank among academics, and the 1995 median starting salary for graduates, are shown in the accompanying table for 12 randomly sampled law schools and 12 randomly sampled business schools.

---

[19] "America's Best Graduate Schools," *U.S. News & World Report* (March 18, 1996), pp. 79–91.

University	School	Reputation Rank by Academics	1995 Median Starting Salaries (in $1000s)
Vanderbilt University	Law	17	62
University of Chicago	Law	2	70
Brigham Young University	Law	45	45.5
George Washington University	Law	24	60
Cornell University	Law	11	70
Rutgers University	Law	45	62
University of Pennsylvania	Law	6	70
University of Illinois (Urbana-Champaign)	Law	23	53
Villanova University	Law	65	62.2
University of Florida	Law	37	43.4
Tulane University	Law	49	53
Case Western Reserve University	Law	41	47
University of Georgia	Business	45	45
Tulane University	Business	40	50
University of Southern California	Business	24	55
Brigham Young University	Business	57	68
Emory University	Business	33	58
University of Illinois (Urbana-Champaign)	Business	24	45.6
Cornell University	Business	10	60
University of North Carolina (Chapel Hill)	Business	16	59
Massachusetts Institute of Technology	Business	1	75
University of Texas (Austin)	Business	19	55
College of William and Mary	Business	67	48
Penn State University	Business	33	49.7

a. Suppose that an ANOVA is to be performed to compare the average 1995 starting salaries (SAL) for business and law schools (let the variable SCHOOL = Law or Business, depending on the type of school). State precisely the ANOVA model. Is SCHOOL a fixed- or random-effects factor?

b. In the SAS output that follows, complete the ANOVA table.

c. Test whether the average 1995 starting salaries differ significantly for graduates of law schools and of business schools.

d. Suppose that an ANOVA is to be performed to compare the average 1995 starting salaries (SAL) for the top 25 schools and schools not in the top 25 in terms of reputation rank (let the variable REP = 1 if a school's reputation rank is 25 or less, REP = 2 if the rank is 26 or more). State precisely the ANOVA model.

e. In the second part of the SAS output that follows, complete the ANOVA table.

f. Test whether the average 1995 starting salaries differ significantly between top 25 schools and schools not in the top 25 in terms of reputation rank.

Edited SAS Output (PROC GLM) for Problem 22

```
                          The GLM Procedure
                       Class Level Information

                 Class          Levels          Values
                 SCHOOL              2           Bus Law

              Number of Observations Read          24
              Number of Observations Used          24
```

Dependent Variable: SAL

Source	DF	Sum of Squares	Mean Square	F Value	Pr > F
Model	—	37.001667	37.001667		
Error	22	1929.391667	87.699621		
Corrected Total	23	1966.393333			

R-Square	Coeff Var	Root MSE	SAL Mean
0.018817	16.44873	9.364808	56.93333

Source	DF	Type I SS	Mean Square	F Value	Pr > F
SCHOOL	1	37.00166667	37.00166667	0.42	0.5227

Source	DF	Type III SS	Mean Square	F Value	Pr > F
SCHOOL	1	37.00166667	37.00166667	0.42	0.5227

```
                 Class          Levels          Values
                 REP                 2           1 2

              Number of Observations Read          24
              Number of Observations Used          24
```

Dependent Variable: SAL

Source	DF	Sum of Squares	Mean Square	F Value	Pr > F
Model	—	440.326667	440.326667		
Error	22	1526.066667	69.366667		
Corrected Total	23	1966.393333			

R-Square	Coeff Var	Root MSE	SAL Mean
0.223926	14.62880	8.328665	56.93333

Source	DF	Type I SS	Mean Square	F Value	Pr > F
REP	1	440.3266667	440.3266667	6.35	0.0195

Source	DF	Type III SS	Mean Square	F Value	Pr > F
REP	1	440.3266667	440.3266667	6.35	0.0195

23. In September 1996, *U.S. News & World Report* published a report on America's health maintenance organizations (HMOs).[20] The report was intended to serve as a consumer guide to HMO quality. For each HMO included in the report, data were provided on several variables, including the following:

---

[20] "Rating the HMOs," *U.S. News & World Report* (September 2, 1996), pp. 52–63.

PHYSTURN: Physician turnover rate (%).

PHYSCERT: Percentage of doctors who were board certified.

PREV: Prevention score, indicating how well the HMO meets Public Health Service goals in various measures of preventive care (including immunizations and prenatal care). The results can be negative (indicating that the HMO falls short of the goals), or positive (indicating that goals are exceeded).

HMO	PREV	PHYSTURN	PHYSCERT
HMO Kentucky	−114	1	64
Kaiser Foundation (HI Region)	−6	7	92
CIGNA HealthCare of LA	−97	6	80
CIGNA HealthCare of S. California	−82	8	81
CIGNA HealthCare of N. California	−42	21	82
HIP Health Plan of Florida	−8	3	80
CIGNA HealthCare of San Diego	−36	5	76
NYLCare of the Mid-Atlantic	−9	6	80
Personalcare Insurance of Illinois	−26	13	74
Prudential Healthcare Tri-state	−78	9	71
CIGNA HealthCare of S. Florida	−26	4	77
Health New England	11	5	79
Group Health Northwest	4	3	90
Kaiser Foundation—Mid-Atlantic	29	4	92
Health Alliance Medical Plans	−28	6	82
Pilgrim Health Care	22	2	83
Partners National Health, NC	−1	6	84
Healthsource of New Hampshire	11	5	81

a. In the *U.S. News & World Report* story, the average physician turnover rate for HMOs is reported to be 6%. Suppose that an ANOVA is to be performed to compare the average prevention scores for HMOs with low physician turnover rates ($\leq 6\%$) and HMOs with higher rates ($> 6\%$). State precisely the ANOVA model.

b. In the SAS output that follows, complete the ANOVA table. To produce the output, the following coding scheme was followed: **TURN** = 1 if PHYSTURN $\leq 6\%$; 2 otherwise.

c. Test whether the average prevention scores differ significantly for the two types of HMOs mentioned in part (a).

d. Suppose that an ANOVA is to be performed to compare the average prevention scores for HMOs with low percentages of board-certified primary care physicians ($\leq 75\%$) and HMOs with higher percentages of board-certified physicians ($> 75\%$). State precisely the ANOVA model.

e. In the second part of the SAS output that follows, complete the ANOVA table. To produce the output, the following coding scheme was followed: **CERT** = 1 if PHYSCERT $\leq 75\%$; 2 otherwise.

f. Test whether the average prevention scores differ significantly for the two types of HMOs mentioned in part (d).

Edited SAS Output (PROC GLM) for Problem 23

```
                         The GLM Procedure
                     Class Level Information

                  Class        Levels        Values
                  TURN           2             1 2

              Number of Observations Read      18
              Number of Observations Used      18
```

Dependent Variable: PREV

Source	DF	Sum of Squares	Mean Square	F Value	Pr > F
Model	1	2868.56752	2868.56752	1.72	0.2085
Error	16	26717.87692	1669.86731		
Corrected Total	17	29586.44444			

R-Square	Coeff Var	Root MSE	PREV Mean
0.096955	-154.5278	40.86401	-26.44444

Source	DF	Type I SS	Mean Square	F Value	Pr > F
TURN	1	2868.567521	2868.567521	1.72	0.2085

Source	DF	Type III SS	Mean Square	F Value	Pr > F
TURN	1	2868.567521	2868.567521	1.72	0.2085

```
                         The GLM Procedure
                     Class Level Information

                  Class        Levels        Values
                  BORDCERT       2             1 2

              Number of Observations Read      18
              Number of Observations Used      18
```

Dependent Variable: PREV

Source	DF	Sum of Squares	Mean Square	F Value	Pr > F
Model	1	7691.37778	7691.37778	5.62	0.0306
Error	16	21895.06667	1368.44167	.	
Corrected Total	17	29586.44444			

R-Square	Coeff Var	Root MSE	PREV Mean
0.259963	-139.8874	36.99245	-26.44444

Source	DF	Type I SS	Mean Square	F Value	Pr > F
BORDCERT	1	7691.377778	7691.377778	5.62	0.0306

Source	DF	Type III SS	Mean Square	F Value	Pr > F
BORDCERT	1	7691.377778	7691.377778	5.62	0.0306

References

Allen, D. M., and Grizzle, J. E. 1969. "Analysis of Growth and Dose Response Curves." *Biometrics* 25: 357–82.

Armitage, P. 1971. *Statistical Methods in Medical Research.* Oxford: Blackwell Scientific.

Cotton, R. B.; Stahlman, M. T.; Kovar, I.; and Catterton, W. Z. 1979. "Medical Management of Small Preterm Infants with Symptomatic Patent Ductus Arteriosus." *Journal of Pediatrics* 2: 467–73.

Daly, M. B. 1973. "The Effect of Neighborhood Racial Characteristics on the Attitudes, Social Behavior, and Health of Low Income Housing Residents." Ph.D. dissertation, Department of Epidemiology, University of North Carolina, Chapel Hill, N.C.

Diggle, P. J.; Heagerty, P. J.; Liang, K. Y.; and Zeger, S. L. 2002. *Analysis of Longitudinal Data, Second Edition.* Oxford: Oxford University Press.

Guenther, W. C. 1964. *Analysis of Variance.* Englewood Cliffs, N.J.: Prentice-Hall.

Hollander, M., and Wolfe, D. A. 1973. *Nonparametric Statistical Methods.* New York: John Wiley & Sons.

Kirk, R. E. 1969. *Experimental Design: Procedures for the Behavioral Sciences.* Belmont, Calif.: Wadsworth.

Kramer, C. Y.1956. Extension of the Multiple Range Test to Group Means with Unequal Numbers of Replications. *Biometrics* 12: 307–310.

Lehmann, E. L. 1975. *Non-parametrics: Statistical Methods Based on Ranks.* San Francisco: Holden-Day.

Lindman, H. R. 1974. *Analysis of Variance in Complex Experimental Designs.* San Francisco: W. H. Freeman.

Lynn, M. J.; Waring, G. O., III; Sperduto, R. D.; et al. 1987. "Factors Affecting Outcome and Predictability of Radial Keratotomy in the PERK Study." *Archives of Ophthalmology* 105: 42–51.

Miller, R. G., Jr. 1966. *Simultaneous Statistical Inference.* New York: McGraw-Hill.

———. 1981. *Simultaneous Statistical Inference,* 2nd ed. New York: Springer-Verlag.

Neter, J., and Wasserman, W. 1974. *Applied Linear Statistical Models.* Homewood, Ill.: Richard D. Irwin.

Siegel, S. 1956. *Nonparametric Statistics for the Behavioral Sciences.* New York: McGraw-Hill.

Tukey, J. W. 1953. The Problem of Multiple Comparisons. Mimeographed Notes, Princeton University.

U.S. News & World Report. 1996a. "1996 Mutual Funds Guide, Best Mutual Funds," *U.S. News & World Report* (January 29, 1996), pp. 88–100.

———. 1996b. "America's Best Graduate Schools," *U.S. News & World Report* (March 18, 1996), pp. 79–91.

———. 1996c. "Rating the HMOs," *U.S. News & World Report* (September 2, 1996), pp. 52–63.

Varvarigou, A.; Bardin, C. L.; Beharry, K.; Chemtob, S.; Papageorgiou, A.; and Aranda, J. 1996. "Early Ibuprofen Administration to Prevent Patent Ductus Arteriosus in Premature Newborn Infants." *Journal of American Medical Association* 275: 539–44.

For the pattern in Figure 18.1(c), cells in the same column have the same number of observations, whereas cells in the same row are in the ratio 4:2:3. For this table, each of the four cell frequencies in the jth column is equal to the same fraction of the corresponding total column frequency (i.e., $n_{ij} = n_{.j}/4$ in this case). Note, for example, that $n_{.1}/4 = 16/4 = 4$, which is the number of observations in any cell in column 1.

For Figure 18.1(d), the cells in a given column are in the ratio $1 : 2 : 3 : 2$, whereas the cells in a given row are in the ratio $4 : 2 : 3$. This pattern results because n_{ij} is determined as

$$n_{ij} = \frac{n_{i.}.n_{.j}}{n_{..}}$$

which means that any cell frequency can be obtained by multiplying the corresponding row and column marginal frequencies together and then dividing by the total number of observations. Thus, for cell (1, 2) in Figure 18.1(d), we have $n_1.n_{.2}/n_{..} = 9(16)/72 = 2$, which equals n_{12}. Similarly, for cell (4, 3), $n_4.n_{.3}/n_{..} = 18(24)/72 = 6$, which equals n_{43}.

There is no mathematical rule for describing the pattern of cell frequencies in Figure 18.1(e), so we say that such a pattern is nonsystematic. As we will see in Chapter 20, the ANOVA procedures required for the patterns in Figure 18.1(c) and 18.1(d) differ from those required for the irregular pattern in Figure 18.1(e). For the former two patterns, the same computational procedure may be used as when an equal number of observations exists per cell. For the nonsystematic case, a different procedure is required.

18.1.2 The Case of a Single Observation per Cell

A two-way table with a single observation in each cell can arise in a number of different experimental situations. Consider the following three examples:

1. Six hypertensive individuals, matched pairwise by age and sex, are randomly assigned (within each pair) to either a treatment or a control group. For each individual, a measure of change in self-perception of health is determined after 1 year. The main question of interest is whether the true mean change in self-perception for the treatment group differs from that for the control group.

2. Six growth-inducing treatment combinations are randomly assigned to six mice from the same litter. The treatment combinations are defined by the cross-classification of the levels of two factors: factor A (drug A1 or placebo A0) and factor B (drug B2 [high dose], drug B1 [low dose], or placebo B0). The dependent variable of interest is weight gain measured 1 week after treatment is initiated. The questions to be considered include (a) whether the effect of drug A1 differs from that of placebo A0; (b) whether differences exist among the effects of drugs B1 and B2 and placebo B0; and (c) whether the drug A1 effect differs from that of placebo A0 in the same way at each level of factor B.

3. Scores of satisfaction with medical care are recorded for six hypertensive patients assigned to one of six categories depending on whether the nurse practitioner assigned to the patient was measured to have high or low autonomy (factor A) and high, medium, or low knowledge of hypertension (factor B). The main questions of interest are (a) whether mean satisfaction scores differ between patients with high-autonomy

nurses and those with low-autonomy nurses, and (b) whether these differences in mean satisfaction scores differ in the same way at each level of knowledge grouping.

Each of these experiments may be represented by a 2 × 3 two-way table, as shown in Figure 18.2. In the figure, the third example involves two factors whose levels (or categories) were determined after the data were gathered. Such a study is often referred to as an *observational study* rather than as an *experiment,* since the latter term is usually reserved for studies involving factors whose levels are specified beforehand. Epidemiologic, sociological, and psychological studies are usually observational rather than experimental. For such studies, the levels of the various factors of interest are determined after the observed frequency distributions of these factors have been considered.[2] For this example, the autonomy groupings were determined after the frequency distribution of autonomy scores on nurses was considered. Similarly, the knowledge categories were determined using the observed knowledge scores. Thus, one patient was in each of the six groups because of a posteriori (rather than a priori) considerations. In actual practice it is often impossible to arrange things so nicely, especially when large samples are involved. That is why most such observational studies have unequal numbers of observations in various cells.

FIGURE 18.2   Different experimental situations resulting in two-way tables with a single observation per cell

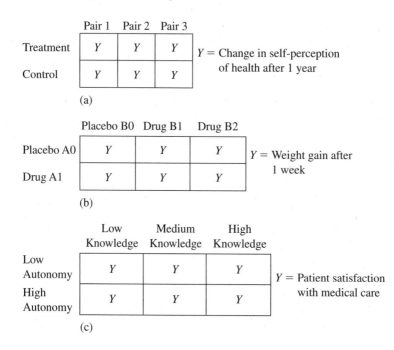

(a)

(b)

(c)

---

[2] In this chapter we focus on the case where the levels of each factor are considered fixed; nevertheless, even if one or both factors were considered random, the tests of hypotheses of interest, as with one-way ANOVA, would be computed in exactly the same way as in the fixed-factor case. This is not so when more than one observation occurs per cell, as discussed in Chapter 19.

The second example, Figure 18.2(b), involves two factors whose levels were determined before the data were collected; the resulting six *treatment combinations* can be viewed, in one sense, as representing the different levels of a single factor that have been randomly assigned to the six individuals. Although it may be necessary because of limited experimental material or prohibitive cost to apply or assign each treatment combination to only a single experimental unit, considerably more information is obtained if each treatment combination has several replications.

This brings us to Figure 18.2(a), which is of the general type to be considered in this chapter. Like Figure 18.2(b), this example represents a designed rather than an observational study. It differs from the other tables in Figure 18.2, however, in its allocation of individuals to cells (i.e., treatment combinations). The allocation here was done by randomization within each pair rather than, say, by randomization across all six individuals, ignoring any pairing. Another feature unique to this table is that the effect of only one of the two factors involved (in this case "treatment versus control") is of primary interest. The other factor, "pair" (with three levels corresponding to the three pairs), is used only to help increase the precision of the comparison between the treatment and control groups. Thus, if pair matching (on age and sex) is used and significant differences are found between the Y values for paired treatment and control subjects, such differences cannot solely be attributed to one group being older, say, or having a different sex composition than the other. The pairing, therefore, serves to eliminate or block out noise that otherwise would affect the comparison between treatment and control groups due to the confounding effects of age and sex. Such pairs are often referred to as *blocks,* and the associated experimental design is called a *randomized-blocks design.*

The analysis required for data arranged as in Figure 18.2(a) is described in most introductory statistics texts. Since two groups are involved (treatment and control) and since matching has been done, the generally recommended method of analysis involves using the paired-difference t test, which focuses on differences in Y values within pairs. Hence, the key test statistic involved is of the form $T = \bar{d}\sqrt{n}/S_d$, where $\bar{d}$ is the difference between the treatment group mean and the control group mean, S_d is the standard deviation of the difference scores for all pairs, and n is the number of pairs (3, in Figure 18.2(a)). This statistic has the t distribution with $(n-1)$ degrees of freedom under H_0; so, the critical region is of the form $|T| \geq t_{n-1,1-\alpha/2}$ for a two-sided test of the null hypothesis of a true average difference score of zero.

The paired-difference t test can be viewed as a special case of the general F test used in a randomized-blocks ANOVA, involving more than two treatments per block. In fact, this randomized-blocks F test represents a generalization of the paired-difference t test just as the one-way ANOVA F test represents a generalization of the two-sample t test.

18.2 Equivalent Analysis of a Matched-pairs Experiment

Let us make the matched-pairs example of the previous section more realistic by considering the data given in Table 18.1, which involves 15 pairs of individuals matched on age and sex. The main inferential question for these data involves whether the mean change in self-perception score for the treatment (T) group significantly differs from that for the control (C) group. Stated in terms of population means, the null hypothesis is

$$H_0\colon \mu_T = \mu_C$$

TABLE 18.1   **Matched-pairs design involving change scores in self-perception of health (Y) among hypertensives**

Group	|	1	2	3	4	5	6	7	8	9	10	11	12	13	14	15	Total	Mean
												Pair						
Treatment		10	12	8	8	13	11	15	16	4	13	2	15	5	6	8	146	9.73
Control		6	5	7	9	10	12	9	8	3	14	6	10	1	2	1	103	6.87
Total		16	17	15	17	23	23	24	24	7	27	8	25	6	8	9	249	8.30
Difference	|	4	7	1	−1	3	−1	6	8	1	−1	−4	5	4	4	7	43	2.86

where μ_T and μ_C denote the population means for the treatment and control groups, respectively. The alternative hypothesis is then

$$H_A: \mu_T \neq \mu_C$$

assuming that the analyst did not theorize in advance that one particular group would have a higher or lower population mean than the other. (If, however, the analyst thought a priori that the treatment group would have a significantly higher mean change score than the control group, the alternative would be one-sided: $H_A: \mu_T > \mu_C$.)

18.2.1 Paired-difference *t* Test

One method for testing the null hypothesis H_0 was described in the previous section—the paired-difference t test. To perform this test, we first determine from Table 18.1 that $\bar{d} = \bar{Y}_T - \bar{Y}_C = 2.86$ and

$$S_d^2 = \frac{1}{14}\left[\sum_{i=1}^{15} d_i^2 - \frac{\left(\sum_{i=1}^{15} d_i\right)^2}{15}\right] = 12.695$$

where d_i is the observed difference between the scores for the treatment and control groups for the ith pair. Then we compute the test statistic as

$$T = \frac{\bar{d}\sqrt{n}}{S_d} = \frac{2.86\sqrt{15}}{\sqrt{12.695}} = 3.109$$

Since $t_{14,0.995} = 2.976$ and $t_{14,0.9995} = 4.140$, the P-value for this two-sided test is given by

$$.001 < P < .01$$

We therefore reject H_0 and conclude that the observed mean change score for the treatment group differs significantly from that for the control group.

18.2.2 Randomized-blocks *F* Test

Another way to test the null hypothesis $H_0: \mu_T = \mu_C$ for the data of Table 18.1 is to use an F test, based on the ANOVA table given in Table 18.2.

This ANOVA table differs in form from the one used for one-way ANOVA in that the total sum of squares is partitioned into three components (treatments, pairs, and errors) instead of just two (between and within). The treatment component in Table 18.2 is similar to the between

TABLE 18.2   ANOVA table for matched-pairs data of Table 18.1

Source	SS	d.f.	MS	F
Treatment	61.63	1	61.63	9.71 (.005 < P < .01)
Pairs (blocks)	391.80	14	27.99	4.41 (.001 < P < .005)
Error	88.87	14	6.35	
Total	542.30	29		

(i.e., treatment) source in the one-way ANOVA case; and the error component here is analogous to the within component in the one-way ANOVA case, because this source is used as an estimate of the population variance σ^2. Finally, we have a pairs (or blocks) component in Table 18.2 that has no corresponding component in the one-way ANOVA case.

To recap, whereas the total sum of squares in one-way ANOVA may be split up into two components, as indicated by the equation

SS(Total) = SS(Between) + SS(Within),

the total sum of squares in a randomized-blocks ANOVA can be partitioned into three meaningful components:

$$SS(Total) = SS(Treatments) + SS(Blocks) + SS(Error) \tag{18.1}$$

The last two components on the right-hand side of (18.1) can be seen as representing a partition of the experimental error (or within) sum of squares associated with one-way ANOVA (i.e., with the blocking ignored). By separating out the blocking effect, we obtain a more precise estimate of experimental error—one not contaminated by any noise due to the effects of the blocking variables (in our case, age and sex).

The computed sums of squares in expression (18.1) are shown in the ANOVA table in Table 18.2. They are

SS(Treatments)	= 61.63
SS(Blocks)	= 391.80
SS(Error)	= 88.87
SS(Total)	= 542.30.

The degress of freedom (d.f.) corresponding to the sums of squares are shown in the second column of Table 18.2. The d.f. for "treatments" is always equal to 1 less than the number of treatments (in our example, 2 − 1 = 1). The d.f. for "blocks" is equal to 1 less than the number of blocks (in our example, 15 − 1 = 14). The d.f. for "error" is obtained as the product of the treatment degrees of freedom and the block degrees of freedom:

Error d.f. = [Treatment d.f.][Block d.f.] = 1(14) = 14

Finally, the d.f. for the total sum of squares is equal to 1 less than the number of observations (in our example, 30 − 1 = 29).

The mean squares, as usual, are obtained by dividing the sums of squares by their corresponding degrees of freedom. Finally, the F statistic for the test of the null hypothesis that no differences exist among the treatments is given by the formula

$$F = \frac{MS(Treatments)}{MS(Error)} \tag{18.2}$$

In particular, to test $H_0: \mu_T = \mu_C$ for the data of Table 18.1, we calculate

$$F = \frac{61.63/1}{88.87/14} = \frac{61.63}{6.35} = 9.71$$

which is the observed value for an F random variable with 1 and 14 degrees of freedom under H_0. Thus, H_0 is rejected at significance level α if the observed value of F is greater than $F_{1,14,1-\alpha}$. Since $F_{1,14,0.99} = 8.86$ and $F_{1,14,0.995} = 11.06$, the P-value for this test satisfies the inequality $.005 < P < .01$. This is quite small, so it is reasonable to reject H_0 and to conclude that the treatment group has a significantly different observed mean change score compared to that of the control group.

Thus, the conclusions reached via the paired-difference t test and the randomized-blocks F test are exactly the same: reject H_0. This agreement occurs because the two tests are completely equivalent. It can be shown mathematically that the square of a paired-difference T statistic with v degrees of freedom is exactly equal to a randomized-blocks F statistic with 1 and v degrees of freedom (i.e., $F_{1,v} = T_v^2$), and that $F \geq F_{1,v,1-\alpha}$ is equivalent to $|T| \geq t_{v,1-\alpha/2}$.

In our example,

$$T_{14}^2 = (3.109)^2 = 9.67 = F_{1,14}$$

and

$$t_{14,0.995}^2 = (2.977)^2 = 8.86 = F_{1,14,0.99}$$

An F test of the null hypothesis H_0: "No significant differences exist among the blocks" may also be performed. This test is not of primary interest, since the very use of a randomized-blocks design is based on the a priori assumption that significant block-to-block variation exists. Nevertheless, an a posteriori F test may be used to check the reasonableness of this assumption. The test statistic to be used in this case is

$$F = \frac{\text{MS(Blocks)}}{\text{MS(Error)}}$$

which, for the example of Table 18.1, has the F distribution under H_0 with 14 degrees of freedom in the numerator and 14 degrees of freedom in the denominator. From Table 18.2, we find that this F statistic is computed to be 4.41, with a P-value satisfying $.001 < P < .005$. The conclusion for this test, as expected, is to reject the null hypothesis that no block differences exist.

18.3 Principle of Blocking

For the matched-pairs example, we indicated that the primary reason for "pairing up" the data was to prevent the confounding factors age and sex from blurring the comparison between the treatment and control groups. In other words, using the matched-pairs design represented an attempt to account for the fact that the experimental units (i.e., the subjects) were not homogeneous with regard to factors (other than experimental group membership status) that were likely to affect the response variable. The key point here is not simply that subjects of different age and sex are different, but more precisely that age and sex are likely to affect the response variable. In another experimental situation, age and sex might not be important covariables and so would not have to be controlled or adjusted for.

In a situation where the experimental units under study are heterogeneous relative to certain concomitant variables that affect the response variable (but are not of primary interest), using a randomized-blocks design requires the following two steps:

1. The experimental units (e.g., people, animals) that are homogeneous are collected together to form a block.

2. The various treatments are assigned at random to the experimental units within each block.

These steps are illustrated in Figure 18.3, where six blocks are formed, each consisting of three homogeneous experimental units. Three treatments (labeled A, B, and C) are then assigned at random to the three units within each block.

Figure 18.4, on the other hand, provides an example of incorrect blocking. In this case, one type of experimental unit might predominantly receive one kind of treatment, and another type might not get that treatment at all. For example, the experimental unit type ☆ was assigned

FIGURE 18.3   Steps used in forming randomized blocks

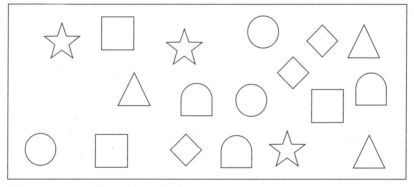

(a) Heterogeneous Experimental Units

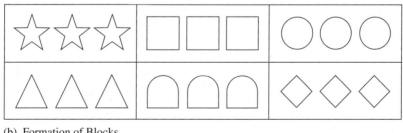

(b) Formation of Blocks

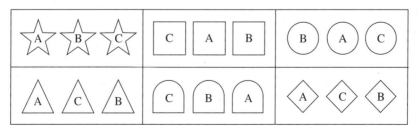

(c) Randomization of Treatments A, B, and C Within Each Block

FIGURE 18.4 **Example of incorrect blocking**

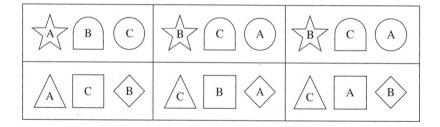

treatment B twice, whereas experimental unit types $\bigcirc$ and $\triangle$ were not assigned treatment B at all. If the blocking had been done correctly, every distinct type of experimental unit would have been assigned each treatment exactly once.

With regard to our matched-pairs design in Table 18.1, if we had not blocked on age, the treatment might have been assigned mostly to older subjects, and the control group might have consisted mostly of younger subjects. Any treatment-control differences found subsequently might very well have been due entirely to age differences between the two groups and not to a real differential effect of the treatment relative to control.

18.4 Analysis of a Randomized-blocks Experiment

In this section we discuss the appropriate analysis for a randomized-blocks experiment, focusing on the case where both factors (treatments and blocks) are considered fixed, although the tests of hypotheses of interest are computed in exactly the same way even if one or both factors are considered random; that is, practically speaking, it does not matter how the factors are defined.[3]

18.4.1 Data Configuration

Table 18.3 gives the general layout of the data for a randomized-blocks design involving k treatments and b blocks. In this table Y_{ij} denotes the value of the observation on the dependent variable Y corresponding to the ith treatment in the jth block (e.g., Y_{23} denotes the value of the observation associated with treatment 2 in block 3). The (row) total for treatment i is denoted by T_i; that is, $T_i = \sum_{j=1}^{b} Y_{ij}$. The (column) total for block j is denoted by B_j; that is, $B_j = \sum_{i=1}^{k} Y_{ij}$. The grand total of all bk observations is $G = \sum_{i=1}^{k}\sum_{j=1}^{b} Y_{ij}$. Finally, the treatment (row) means are denoted by $\overline{Y}_{i.}$ for the ith treatment; the block (column) means are denoted by $\overline{Y}_{.j}$ for the jth block; and $\overline{Y}_{..}$ denotes the grand mean.

A special case of this general format was given in Table 18.1, where $k = 2$ and $b = 15$. Another example (with $k = 4$ and $b = 8$) appears in Table 18.4. Although based on artificial data, this example illustrates the type of information being considered in several ongoing, large-scale

---

[3] In Chapter 19, which discusses the situation of equal replications (at least two) per cell, the required F tests differ depending on how the factors are defined.

TABLE 18.3   Data layout for a randomized-block design

Treatment	Block 1	2	3	...	b	Total	Mean
1	Y_{11}	Y_{12}	Y_{13}	...	Y_{1b}	T_1	$\bar{Y}_{1\cdot}$
2	Y_{21}	Y_{22}	Y_{23}	...	Y_{2b}	T_2	$\bar{Y}_{2\cdot}$
3	Y_{31}	Y_{32}	Y_{33}	...	Y_{3b}	T_3	$\bar{Y}_{3\cdot}$
$\vdots$	$\vdots$	$\vdots$	$\vdots$		$\vdots$	$\vdots$	$\vdots$
k	Y_{k1}	Y_{k2}	Y_{k3}	...	Y_{kb}	T_k	$\bar{Y}_{k\cdot}$
Total	B_1	B_2	B_3	...	B_b	G	—
Mean	$\bar{Y}_{\cdot 1}$	$\bar{Y}_{\cdot 2}$	$\bar{Y}_{\cdot 3}$	...	$\bar{Y}_{\cdot b}$	—	$\bar{Y}_{\cdot\cdot}$

U.S. intervention studies dealing with risk factors associated with heart disease. Table 18.4 presents the data for a small experiment designed to assess the effects of four different cholesterol-reducing diets on persons who have hypercholesterolemia. In such a study it would seem logical to try to take into account (or adjust for) the effects of sex, age, and body size on the dependent variable Y, which represents reduction in cholesterol level after one year. One way to do this is by using a randomized-blocks design, where the blocks are chosen to represent combinations of the various sex–age–body size categories of interest. Thus, in Table 18.4, block 1 consists of males over 50 years of age with a quetelet index ($= 100[weight]/[height]^2$) above 3.5; block 3 consists of males under 50 with a quetelet index above 3.5; and so on. For each block of subjects, the four diets are randomly assigned to the sample of four persons in the block. Each subject is then followed for one year, after which the change in cholesterol level (Y) is recorded.

The primary research question of interest in this study concerns whether any of the observed average reductions in cholesterol level achieved by the four diets significantly differ. By inspecting Table 18.4, we can see that diet 1 appears to be the best, since it is associated with the largest average reduction (12.58 units). Nevertheless, this assessment needs to be evaluated statistically, which involves determining whether the observed differences among the mean

TABLE 18.4   Randomized-blocks experiment for comparing the effects of four cholesterol-reducing diets on persons with hypercholesterolemia (Y = reduction in cholesterol level after one year)

Treatment (Diet)	Block 1 (Male, Age > 50, QUET >3.5)	2 (Male, Age > 50, QUET <3.5)	3 (Male, Age < 50, QUET >3.5)	4 (Male, Age < 50, QUET <3.5)	5 (Female, Age > 50, QUET >3.5)	6 (Female, Age > 50, QUET <3.5)	7 (Female, Age < 50, QUET >3.5)	8 (Female, Age < 50, QUET <3.5)	Total	Mean
1	11.2	6.2	16.5	8.4	14.1	9.5	21.5	13.2	100.6	12.58
2	9.3	4.1	14.2	6.9	14.2	8.9	15.2	10.1	82.9	10.36
3	10.4	5.1	14.0	6.2	11.1	8.4	17.3	11.2	83.7	10.46
4	9.0	4.9	13.7	6.1	11.8	8.4	15.9	9.7	79.5	9.94
Total	39.9	20.3	58.4	27.6	51.2	35.2	69.9	44.2	346.7	—
Mean	9.98	5.08	14.60	6.90	12.80	8.80	17.48	11.05	—	10.83

reductions for the four diets can be attributed solely to chance. The randomized-blocks F test provides a method for making such a statistical evaluation.

Further inspection of Table 18.4 indicates that considerable differences exist among the block means. This is to be expected, since the reason for blocking was that different blocks (or equivalently, different categories of the covariates age, sex, and body size) were expected to have different effects on the response. We can nevertheless perform a statistical test to satisfy ourselves that such block differences are statistically significant. In fact, if this test yields a finding of nonsignificance, we probably should not have used these blocks in the first place, since they cost us degrees of freedom for estimating σ^2 and consequently produce a less sensitive comparison of treatment effects.

18.4.2 Hypotheses Tested in a Randomized-blocks Analysis

The primary null hypothesis of interest in a randomized-blocks analysis is the equality of treatment means:[4]

$$H_0\colon \mu_1. = \mu_2. = \cdots = \mu_k. \tag{18.3}$$

where $\mu_i.$ denotes the population mean response associated with the ith treatment. The alternative hypothesis may be stated as

$$H_A\colon \text{``Not all the } \mu_i.\text{'s are equal.''}$$

If performing this test leads to rejection of H_0, it becomes of interest to determine where the important differences are among the treatment effects. One qualitative way to make such a determination is simply to look at the observed treatment means and to make visual comparisons. Thus, from Table 18.4, rejecting the null hypothesis that no differences exist among the diets leads to the conclusion that diet 1 is best in reducing cholesterol level, that diets 2 and 3 are next in line and are about equally effective, and that diet 4 is the worst of the four. To verify such conclusions statistically, we can use multiple-comparison techniques to conduct appropriate tests, as described in Chapter 17.

As mentioned earlier, another null hypothesis sometimes of interest is that no significant differences exist among the block means. Since use of a randomized-blocks design is based on the a priori opinion that the block means are different, such a hypothesis is usually tested only to check whether the blocking was justified.

18.5 ANOVA Table for a Randomized-blocks Experiment

Table 18.5 summarizes the procedures necessary to test the null hypotheses of equality among of the treatment means and among the block means. Since computation of the ANOVA table sums of

---

[4] Again, we are considering the null hypothesis for the fixed-effects case. In two-way ANOVA, this hypothesis would be stated differently if the treatment effects were considered random, but the test of hypothesis would be computed in exactly the same way.

TABLE 18.5   ANOVA table for a randomized-blocks experiment with k treatments and b blocks

Source	d.f.	SS	MS	F
Treatments	$k-1$	SST	$\text{MST} = \dfrac{\text{SST}}{k-1}$	$\dfrac{\text{MST}}{\text{MSE}}$
Blocks	$b-1$	SSB	$\text{MSB} = \dfrac{\text{SSB}}{b-1}$	$\dfrac{\text{MSB}}{\text{MSE}}$
Error	$(k-1)(b-1)$	SSE	$\text{MSE} = \dfrac{\text{SSE}}{(k-1)(b-1)}$	
Total	$kb-1$	SSY		

squares is routinely carried out by a computer program, (e.g., by SAS's GLM procedure), specific computational formulas are not included in the SS column of Table 18.5. The ANOVA table quantities for the data given in Table 18.4 are shown in the accompanying SAS computer output.

From Table 18.5 and the SAS output, it is clear that the total (corrected) sum of squares, SSY, is split up into the three components SST (treatments), SSB (blocks), and SSE (error) using the fundamental equation

$$\text{SSY} = \text{SST(Treatments)} + \text{SSB(Blocks)} + \text{SSE(Error)} \qquad (18.4)$$

Edited SAS Output (PROC GLM) for Data of Table 18.4

```
                        Class Level Information
                   Class            Levels         Values
                   trt                 4            1 2 3 4
                   blk                 8            1 2 3 4 5 6 7 8
                                      SSE
                        Sum of
Source             DF   Squares              Mean Square    F Value     Pr > F
Model              10   496.4206250     /    49.6420625       52.89     <.0001
Error              21    19.7115625          0.9386458                 ← MSE
Corrected Total    31   516.1321875  ← SSY

          R-Square    Coeff Var    Root MSE           y Mean
          0.961809    8.942254     0.968837           10.83438        F Statistics

Source       DF   SST      Type I SS      Mean Square  MST  F Value      Pr > F
trt           3            33.5609375     11.1869792        11.92        <.0001
blk           7            462.8596875    66.1228125 ←MSB   70.44        <.0001

Source       DF   SSB      Type III SS    Mean Square       F Value      Pr > F
trt           3            33.5609375     11.1869792        11.92        <.0001
blk           7            462.8596875    66.1228125        70.44        <.0001
```

The formula for the total sum of squares is:

$$\text{SSY} = \sum_{i=1}^{k} \sum_{j=1}^{b} (Y_{ij} - \bar{Y}..)^2 \qquad (18.5)$$

As is the case in any ANOVA situation, SSY measures the total unexplained variation in the data, corrected for the grand mean. The degrees of freedom associated with SSY are $(kb - 1)$. From the SAS output, we see that SSY is 516.132.

The treatment sum of squares SST is defined as

$$\text{SST} = b \sum_{i=1}^{k} (\overline{Y}_{i.} - \overline{Y}..)^2 \qquad (18.6)$$

SST reflects the variation among the observed treatment means. Its basic components are of the form $(\overline{Y}_{i.} - \overline{Y}..)$, which is the difference between the ith treatment mean and the grand mean. SST has $(k - 1)$ degrees of freedom. For the SAS output, SST is computed as 33.561.

The block sum of squares SSB is defined as

$$\text{SSB} = k \sum_{j=1}^{b} (\overline{Y}_{.j} - \overline{Y}..)^2 \qquad (18.7)$$

with $(b - 1)$ degrees of freedom. The basic components of SSB take the form $(\overline{Y}_{.j} - \overline{Y}..)$, which is the difference between the jth block mean and the grand mean. From the SAS output, we see that SSB is computed as 462.860.

Finally, the residual sum of squares SSE is given by

$$\text{SSE} = \sum_{i=1}^{k} \sum_{j=1}^{b} (Y_{ij} - \overline{Y}_{i.} - \overline{Y}_{.j} + \overline{Y}..)^2 \qquad (18.8)$$

with degrees of freedom equal to $(k - 1)(b - 1)$. For the SAS output, SSE is computed as 19.712.

The complexity of expression (18.8) indicates that SSE is not easily recognizable as an estimate of σ^2. In fact, its basic component, $(Y_{ij} - \overline{Y}_{i.} - \overline{Y}_{.j} + \overline{Y}..)$, can be written in the form

$$(Y_{ij} - \overline{Y}_{.j}) - (\overline{Y}_{i.} - \overline{Y}..)$$

which is an estimate of the difference between the effect of the ith treatment relative to the average effect of all treatments in the jth block (i.e., $Y_{ij} - \overline{Y}_{.j}$) and the overall effect of the ith treatment relative to the overall mean (i.e., $\overline{Y}_{i.} - \overline{Y}..$). Actually, SSE measures block–treatment interaction here, in the sense that SSE is large when the treatment effects vary from block to block. Although we discuss the concept of interaction more thoroughly in Chapter 19, using a randomized-blocks design requires us to assume that no such block-treatment interaction exists; under this assumption, the residual variation reflected in SSE can then be attributed solely to experimental error.[5] The expected effect of violating the no-interaction assumption is a reduction in the power of the tests of treatment and block effects. This assumption must not be violated; otherwise, we would have no way of obtaining an unbiased estimator of σ^2 to use in the denominator of the F statistics. After all, for each block in a randomized-blocks design, no treatment is applied more than once, which makes it impossible to obtain a pure error estimate of σ^2 associated with a particular treatment in a given block. In the case of two-way ANOVA with

---

[5] A method for testing the validity of the assumption of no block–treatment interaction, developed by Tukey (1949), is described in Problem 3(f) at the end of this chapter.

more than one observation per cell, a pure error estimate of σ^2 can be developed by utilizing the available information on within-cell variability.

Each mean-square term in Table 18.5 is, as usual, obtained by dividing the corresponding sum-of-squares term by its degrees of freedom. The F statistics are formed as the ratios of mean-square terms, with MSE in the denominator in each case (regardless of whether each factor is treated as fixed or random).

18.5.1 *F* Test for the Equality of Treatment Means

The test of H_0: $\mu_1. = \mu_2. = \cdots = \mu_k.$ is performed by using the following F statistic (defined in terms of the notation of Table 18.3):

$$F = \frac{\text{MST}}{\text{MSE}} \qquad (18.9)$$

where

$$\text{MST} = \frac{1}{k-1}(\text{SST}) = \frac{1}{k-1}\left(\frac{1}{b}\sum_{i=1}^{k} T_i^2 - \frac{G^2}{bk}\right)$$

and

$$\text{MSE} = \frac{1}{(k-1)(b-1)}(\text{SSE})$$

The assumptions required for the validity of this test are as follows:

1. The observations are statistically independent of one another.

2. Each observation is selected from a normally distributed population.

3. Each observation is selected from a population with variance σ^2 (i.e., variance homogeneity is assumed).

4. There is no block–treatment interaction effect (i.e., the true extent to which treatments differ is the same, regardless of the block considered).

If H_0 is true, the F statistic in (18.9) has the F distribution with $(k-1)$ degrees of freedom in the numerator and $(k-1)(b-1)$ degrees of freedom in the denominator. Thus, for a given α, we would reject H_0 and conclude that not all treatments have the same effect on the response when

$$F \geq F_{k-1,(k-1)(b-1),1-\alpha}$$

From the SAS output shown in Section 18.5, we see that MST is equal to 11.187 and MSE is equal to 0.939, so

$$F = \frac{11.187}{0.939} = 11.914$$

Since $F_{3,21,0.999} = 7.94$, the P-value for this test satisfies $P < .001$. Thus, we reject the null hypothesis H_0 and conclude that significant differences exist among the four diets.

18.5.2 *F* Test for the Equality of Block Means

As previously mentioned, this test is rarely used except as an a posteriori check to confirm that blocking was effective. To perform this test, we calculate the following *F* statistic:

$$F = \frac{\text{MSB}}{\text{MSE}} \tag{18.10}$$

where

$$\text{MSB} = \frac{1}{b-1}(\text{SSB}) = \frac{1}{b-1}\left(\frac{1}{k}\sum_{j=1}^{b}B_j^2 - \frac{G^2}{bk}\right)$$

and where MSE is calculated as before.

Under the null hypothesis H_0: "There are no differences among the true block means," the *F* statistic (18.10) has the *F* distribution with $(b-1)$ and $(k-1)(b-1)$ degrees of freedom. Thus, H_0 is rejected at significance level α when *F* exceeds $F_{b-1,(k-1)(b-1),1-\alpha}$. As shown in the SAS output in Section 18.5, this test yields the following *F* statistic:

$$F = \frac{66.123}{0.939} = 70.419$$

The *P*-value for this test satisfies $P < .001$, so H_0 is rejected, as expected.

18.6 Regression Models for a Randomized-blocks Experiment

The randomized-blocks experiment, like the one-way ANOVA situation, can be described by either a regression model or a classical ANOVA effects model. The regression formulation, which we consider in this section, is technically equivalent to the fixed-effects ANOVA approach in terms of estimating the unknown parameters in each model. Nevertheless, for testing purposes, mean-square terms in the ANOVA table obtained from the fit of a proper regression model can be used to compute appropriate *F* statistics, regardless of whether the actual ANOVA model is fixed, random, or mixed.

An appropriate regression model for the randomized-blocks experiment should contain $(k-1)$ dummy variables for the *k* treatments and $(b-1)$ dummy variables for the *b* blocks. One such model formulation is

$$Y = \mu + \sum_{i=1}^{k-1}\alpha_i X_i + \sum_{j=1}^{b-1}\beta_j Z_j + E \tag{18.11}$$

where

$$X_i = \begin{cases} -1 & \text{if treatment } k \\ 1 & \text{if treatment } i \\ 0 & \text{otherwise} \end{cases} \quad \text{and} \quad Z_j = \begin{cases} -1 & \text{if block } b \\ 1 & \text{if block } j \\ 0 & \text{otherwise} \end{cases}$$

(for $i = 1, 2, \ldots, k-1$; and $j = 1, 2, \ldots, b-1$).

were applied separately to the three squares on each rat. The squares were then rated from 0 to 10, depending on the degree of irritation. The data are as shown in the following table.

Chemical	Rat								Total
	1	2	3	4	5	6	7	8	
I	6	9	6	5	7	5	6	6	50
II	5	9	9	8	8	7	7	7	60
III	3	4	3	6	8	5	5	6	40
Total	14	22	18	19	23	17	18	19	150

a. What are the blocks and what are the treatments in this randomized-blocks design?
b. Complete the ANOVA table in the accompanying computer output for the data set given.
c. Do the data provide sufficient evidence to indicate a significant difference in the toxic effects of the three chemicals?
d. Using a confidence interval of the form

$$(\bar{Y}_{1.} - \bar{Y}_{2.}) \pm t_{v,1-\alpha/2}\sqrt{MSE\left(\frac{1}{n_1} + \frac{1}{n_2}\right)}$$

where v is the degrees of freedom, find a 98% confidence interval for the true difference in the toxic effects of chemicals I and II.
e. Provide a reasonable measure of the proportion of total variation that is explained by the particular statistical model used in analyzing this data set.
f. State the fixed-effects ANOVA model and the corresponding regression model for this analysis.
g. State the assumptions on which the validity of the analysis depends.

Edited SAS Output (PROC GLM) for Problem 1

```
Dependent Variable: score
                                     Sum of
Source                   DF          Squares        Mean Square      F Value      Pr > F
Model                     9       43.50000000        4.83333333         2.71       0.0463
Error                    14       _____       _____
Corrected Total          23       68.50000000

             R-Square            Coeff Var          Root MSE       score Mean
             0.635036            21.38090           1.336306        6.250000

Source                   DF        Type I SS        Mean Square      F Value      Pr > F
chem                      2        _____      _____       7.00       0.0078
rat                       7       18.50000000        2.64285714         1.48       0.2518

Source                   DF       Type III SS        Mean Square      F Value      Pr > F
chem                      2       25.00000000       12.50000000         7.00       0.0078
rat                       7       18.50000000        2.64285714         1.48       0.251
```

2. In a study of the psychosocial changes in individuals who participated in a community-based intervention program, individuals who were clinically identified as hypertensives were randomly assigned to one of two treatment groups (with n individuals in each group). Group 1 was given the usual care that existing community facilities provide for hypertensives; group 2 was given the special care that the intervention study team provided. Among the variables measured on each individual were: SP1, an index of the individual's self-perception of health immediately after being identified as hypertensive but before being assigned to one of the two groups; SP2, an index of the individual's self-perception of health one year after assignment to one of the two groups; AGE; and SEX. One main research question of interest was whether the change in self-perception of health after one year would be greater for individuals in group 2 than for those in group 1. Several different analytical approaches to answering this question are possible, depending on the dependent variable chosen and on the treatment of the variables SP1, AGE, and SEX in the analysis. Among these approaches are the following six:

- Matching pairwise on AGE and SEX (which we assume is possible) and then performing a paired-difference t test to determine whether the mean of group 1 change scores significantly differs from the mean of group 2 change scores. [*Note:* This is equivalent to performing a randomized-blocks analysis, where the blocks are the pairs of individuals.]
- Matching pairwise on AGE and SEX and then performing a regression analysis, with the change score $Y = (SP2 - SP1)$ as the dependent variable and with SP1 as one of the independent variables.
- Matching pairwise on AGE and SEX and then performing a regression analysis, with SP2 as the dependent variable and with SP1 as one of the independent variables.
- Controlling for AGE and SEX (without any prior matching) via analysis of covariance, with the change score $Y = (SP2 - SP1)$ as the dependent variable.
- Controlling for AGE and SEX via analysis of covariance, with the change score $Y = (SP2 - SP1)$ as the dependent variable and with SP1 as one of the independent variables.
- Controlling for AGE and SEX via analysis of covariance, with SP2 as the dependent variable and with SP1 as one of the independent variables.

   a. What regression model is associated with each of the preceding six approaches? Make sure to define your variables carefully.

   b. For each of the preceding regression models, state the appropriate null hypothesis (in terms of regression coefficients) for testing for group differences with respect to self-perception scores. For each regression model, indicate how to set up the appropriate ANOVA table to carry out the desired test.

   c. Assuming that you have decided to match pairwise on AGE and SEX, which of the preceding regression models would you prefer to use, and why? [*Note:* Actually, you have only two models to choose from, since the second and third models described in the preceding list produce exactly the same test statistic for comparing the two groups.]

3. An experiment was conducted at the University of North Carolina to see whether the BOD test for water pollution is biased by the presence of copper. In this test, the amount of dissolved oxygen in a sample of water is measured at the beginning and at the end of a five-day period; the difference in dissolved oxygen content is ascribed to the action of bacteria on the impurities in the sample and is called the *biochemical oxygen demand* (BOD). The question is whether dissolved copper retards the bacterial action and produces artificially low responses for the test.

The data in the following table are partial results from this experiment. The three samples (which are from different sources) are split into five subsamples, and the concentration of copper ions in each subsample is given. The BOD measurements are given for each subsample–copper ion concentration combination.

Sample	Copper Ion Concentration (ppm)					Mean
	0	0.1	0.3	0.5	0.75	
1	210	195	150	148	140	168.60
2	194	183	135	125	130	153.40
3	138	98	89	90	85	100.00
Mean	180.67	158.67	124.67	121.00	118.33	140.67

a. Using dummy variables and treating the copper ion concentration as a nominal variable, provide an appropriate regression model for this experiment. Is this a randomized-blocks experiment?

b. If copper ion concentration is treated as an interval (continuous) variable, one appropriate regression model would be

$$Y = \beta_0 + \beta_1 Z_1 + \beta_2 Z_2 + \beta_3 X + \beta_4 Z_1 X + \beta_5 Z_2 X + E$$

where

$$Z_1 = \begin{cases} 1 & \text{for sample 1} \\ 0 & \text{for sample 2} \\ -1 & \text{for sample 3} \end{cases} \quad Z_2 = \begin{cases} 0 & \text{for sample 1} \\ 1 & \text{for sample 2} \\ -1 & \text{for sample 3} \end{cases}$$

and X = copper ion concentration. What are the advantages and disadvantages of using the models in parts (a) and (b)? Which model would you prefer to use, and why?

c. Compare (without doing any statistical tests) the average BOD responses at the various levels of copper ion concentration.

d. Use the following table, which is based on a randomized-blocks analysis, to test (at $\alpha = .05$) the null hypothesis that copper ion concentration has no effect on the BOD test.

Source	d.f.	SS	MS	F
Samples	2	12,980.9333	6,490.4667	56.83
Concentrations	4	9,196.6667	2,299.1667	20.13
Error	8	913.7333	114.2167	

e. Judging from the ANOVA table and the observed block means, does blocking appear to be justified?

f. The randomized-blocks analysis assumes that the relative differences in BOD responses at different levels of copper ion concentration are the same regardless of the sample used; in other words, there is no copper ion concentration–sample interaction. One method (see Tukey 1949) for testing whether such an interaction effect actually exists is *Tukey's test for additivity*. It addresses the null hypothesis

H_0: No interaction exists (i.e., the model is additive in the block and treatment effects).

versus the alternative hypothesis

H_A: The model is not additive, and a transformation $f(Y)$ exists that removes the nonadditivity in the model for Y.

Tukey's test statistic is given by

$$F = \frac{SSN}{(SSE - SSN)/[(k-1)(b-1)-1]}$$

(which is distributed as $F_{1,(k-1)(b-1)-1}$ under H_0), where

$$SSN = \frac{\left[\sum\limits_{i=1}^{k} \sum\limits_{j=1}^{b} Y_{ij}(\bar{Y}_{i\cdot} - \bar{Y}_{\cdot\cdot})(\bar{Y}_{\cdot j} - \bar{Y}_{\cdot\cdot}) \right]^2}{\sum\limits_{i=1}^{k}(\bar{Y}_{i\cdot} - \bar{Y}_{\cdot\cdot})^2 \sum\limits_{j=1}^{b}(\bar{Y}_{\cdot j} - \bar{Y}_{\cdot\cdot})^2}$$

using the notation in this chapter.

Given that the computed F statistic equals 4.37 in Tukey's test for additivity, is there significant evidence of nonadditivity? (Use $\alpha = .05$.)

g. The following tables are based on fitting the multiple regression model given in part (b). Use these tables to test whether evidence exists of a significant effect due to copper ion concentration (i.e., test $H_0: \beta_3 = \beta_4 = \beta_5 = 0$). Use the result that $R^2 = .888$.

Variable	Regression Coefficient	Beta Coefficient
(intercept)	167.195	
Z_1	32.513	0.67659
Z_2	16.972	0.35319
X	-80.389	-0.55585
$Z_1 X$	-13.877	-0.12337
$Z_2 X$	-12.844	-0.11419

Source	d.f.	MS
Z_1	1	11,764.90
$Z_2 \mid Z_1$	1	1,216.03
$X \mid Z_1, Z_2$	1	7,134.57
$Z_1 X \mid Z_1, Z_2, X$	1	303.27
$Z_2 X \mid Z_1, Z_2, X, Z_1 X$	1	91.07
Error	9	286.83

4. Using the accompanying computer output based on the data in Problem 7 in Chapter 17, conduct a randomized-blocks analysis, treating the high schools as blocks, to test whether significant differences exist among the mean VSAT scores for the years 1965, 1970, and 1975. Do the results obtained from this randomized-blocks analysis differ from those obtained earlier from the one-way ANOVA?

Edited SAS Output (PROC GLM) for Problem 4

Dependent Variable: VSAT

Source	DF	Sum of Squares	Mean Square	F Value	Pr > F
Model	6	1875.20000	312.53333	19.76	0.0002
Error	8	126.53333	15.81667		
Corrected Total	14	2001.73333			

R-Square	Coeff Var	Root MSE	VSAT Mean
0.936788	0.734490	3.97702	541.467

Source	DF	Type I SS	Mean Square	F Value	Pr > F
YEAR	2	548.13333	274.06667	17.33	0.0012
SCHOOL	4	1327.06667	331.76667	20.98	0.0003

Source	DF	Type III SS	Mean Square	F Value	Pr > F
YEAR	2	548.13333	274.06667	17.33	0.0012
SCHOOL	4	1327.06667	331.76667	20.98	0.0003

5. Using the accompanying computer output based on the data in Problem 8 in Chapter 17, conduct a randomized-blocks analysis, treating the days as blocks, to test whether the three persons have significantly different ESP ability. Does blocking on days seem appropriate?

Edited SAS Output (PROC GLM) for Problem 5

Dependent Variable: SCORE

Source	DF	Sum of Squares	Mean Square	F Value	Pr > F
Model	6	111.466667	18.577778	6.12	0.0113
Error	8	24.266667	3.033333		
Corrected Total	14	135.733333			

R-Square	Coeff Var	Root MSE	SCORE Mean
0.821218	9.231343	1.74165	18.8667

Source	DF	Type I SS	Mean Square	F Value	Pr > F
PERSON	2	91.7333333	45.8666667	15.12	0.0019
DAY	4	19.7333333	4.9333333	1.63	0.2585

Source	DF	Type III SS	Mean Square	F Value	Pr > F
PERSON	2	91.7333333	45.8666667	15.12	0.0019
DAY	4	19.7333333	4.9333333	1.63	0.2585

6. The promotional policies of four companies in a certain industry were compared to determine whether their rates for promoting blacks and whites differed. Data on the variable "rate discrepancy" were obtained for the four companies in each of three different two-year periods. This variable was defined as

$$d = \hat{p}_W - \hat{p}_B$$

where

$$\hat{p}_W = \frac{[\text{Number of whites promoted}](100)}{[\text{Number of whites eligible for promotion}]}$$

and

$$\hat{p}_B = \frac{[\text{Number of blacks promoted}](100)}{[\text{Number of blacks eligible for promotion}]}$$

The resulting data are reproduced in the following table.

Period	Company			
	1	2	3	4
1	3	5	5	4
2	4	4	3	5
3	8	12	10	9

a. Using dummy variables, create an appropriate regression model for this data set. Is this a randomized-blocks design?

b. Use the following table to test whether any significant differences exist among the average rate discrepancies for the four companies. Tukey's F for testing additivity equals 3.583 with 1 and 5 degrees of freedom.

Source	d.f.	SS	MS	F
Periods	2	84.5000	42.2500	33.800
Companies	3	6.0000	2.0000	1.600
Error	6	7.5000	1.2500	

c. Does Tukey's test for these data indicate that a removable interaction effect is present?

d. If no significant differences are found among the rate discrepancies for the four companies, would this support the contention that none of the companies has a discriminatory promotional policy?

e. Comment on the suitability of this analysis in view of the fact that the response variable is a difference in proportions.

7. Suppose that, in a study to compare body sizes of three genotypes of fourth-instar silkworm, the mean lengths (in millimeters) for separately reared cocoons of heterozygous (HET), homozygous (HOM), and wild (WLD) silkworms were determined at five laboratory sites, as detailed in the following table and in the accompanying computer output.[7]

---

[7] Adapted from a study by Sokal and Karlen (1964).

Variable	Site				
	1	2	3	4	5
HOM	29.87	28.16	32.08	30.84	29.44
HET	32.51	30.82	34.17	33.46	32.99
WLD	35.76	33.14	36.29	34.95	35.89

a. Assuming that this is a randomized-blocks experiment, what are the blocks and what are the treatments?

b. Why do you think a randomized-blocks analysis is appropriate (or inappropriate) for this experiment?

c. Carry out an appropriate analysis of the data for this experiment, and state your conclusions. Be sure to state the null hypothesis for each test performed.

Edited SAS Output (PROC GLM) for Problem 7

```
Dependent Variable: LENGTH

                            Sum of            Mean
Source              DF      Squares           Square        F Value      Pr > F
Model                6      84.7616800        14.1269467     46.97        <0.0001
Error                8      2.4062933         0.3007867
Corrected Total     14      87.1679733

          R-Square           Coeff Var         Root MSE        LENGTH Mean
          0.972395           1.677632          0.54844         32.6913

Source             DF       Type I SS         Mean Square     F Value      Pr > F
VARIABLE            2        65.8139733        32.9069867      109.40       <0.0001
SITE               4        18.9477067        4.7369267       15.75        0.0007

Source             DF       Type III SS       Mean Square     F Value      Pr > F
VARIABLE            2        65.8139733        32.9069867      109.40       <0.0001
SITE               4        18.9477067        4.7369267       15.75        0.0007
```

8. In Problem 3, the independent variable of interest, copper ion concentration, is an interval-scaled variable.

a. Treating copper ion concentration as a five-level categorical predictor (as in the randomized-blocks analysis) corresponds to including a collection of polynomial terms for copper ion concentration (e.g., X, X^2, . . .) in the model. What is the highest order power of X allowed?

b. State, in terms of k levels of a predictor, a general principle based on the example in part (a).

c. State the expansion of the regression model given in Problem 3(b) that encompasses the polynomial terms discussed in part (a).

d. Use a computer program to fit the model stated in part (c). Center X by subtracting 0.330. (Why?) Provide estimated regression coefficients.

e. Provide a multiple partial F test comparing the model in parts (b) and (g) of Problem 3 to the one fitted here; the latter is equivalent to the model considered in part (d) of Problem 3.

9. Assume that the following table came from the analysis of a randomized-blocks design ANOVA.

Source	d.f.	SS	MS	F
Treatments	4	b	e	5.00
Blocks	a	c	48.00	6.00
Error	20	d	f	

a.–f. Complete the source table by determining the numerical values for the entries a through f. Show your work.

g. Provide tests concerning treatment and block effects using $\alpha = .05$. What are your conclusions?

10. An educator has obtained class average scores on a standardized test for classes from each grade from first through seventh at each of four schools in a city. These are shown in the following table. To assess differences among schools, the educator decides to conduct a randomized-blocks ANOVA, treating grade as a blocking factor.

Observation	School	Grade	Score
1	1	1	70.4
2	1	2	74.5
3	1	3	49.6
4	1	4	72.9
5	1	5	60.3
6	1	6	59.5
7	1	7	77.0
8	2	1	73.8
9	2	2	71.0
10	2	3	65.0
11	2	4	73.2
12	2	5	79.7
13	2	6	82.5
14	2	7	60.9
15	3	1	71.3
16	3	2	62.7
17	3	3	71.9
18	3	4	99.4
19	3	5	74.4
20	3	6	82.8
21	3	7	74.4
22	4	1	68.5
23	4	2	79.5
24	4	3	72.1
25	4	4	77.0
26	4	5	86.5
27	4	6	101.8
28	4	7	100.7

a. Conduct the appropriate analysis, using the accompanying computer output.

b. Provide F tests about school and grade differences, using $\alpha = .05$. What do you conclude?

c. Why might using the children's individual scores be better than using mean scores?

Edited SAS Output (PROC GLM) for Problem 10

```
Dependent Variable: SCORE

                               Sum of            Mean
Source              DF        Squares          Square      F Value      Pr > F
Model                9     2005.67821       222.85313        2.07      0.0905
Error               18     1939.26857       107.73714
Corrected Total     27     3944.94679

            R-Square        Coeff Var         Root MSE        SCORE Mean
            0.508417         13.88383          10.3797           74.7607

Source              DF       Type I SS     Mean Square      F Value      Pr > F
SCHOOL               3     1131.06393       377.02131        3.50      0.0370
GRADE                6      874.61429       145.76905        1.35      0.2857

Source              DF     Type III SS     Mean Square      F Value      Pr > F
SCHOOL               3     1131.06393       377.02131        3.50      0.0370
GRADE                6      874.61429       145.76905        1.35      0.2857
```

11. A company evaluated three new production-line technologies, one involving high automation, the second involving moderate automation, and the third involving low automation. All three technologies were intended to improve productivity through increased automation, but they continued to require operator intervention. Three groups were identified, each consisting of three line operators with similar years of production-line experience. Within each group, the operators were randomly assigned to one of the three new production-line technologies. The output (in units per hour) was recorded for each operator, and is shown in the accompanying table.

Operator Experience	Technology		
	High Automation (H)	Moderate Automation (M)	Low Automation (L)
<1 Year	10	8	5
1–2 Years	18	12	10
>2 Years	19	13	12

a. What are the blocks, and what are the treatments in this randomized-blocks design?
b. Complete the ANOVA table in the accompanying computer output for the data set given.
c. Do the data provide sufficient evidence to indicate that any significant differences exist among the outputs for the three technologies?
d. Using a confidence interval of the form

$$(\bar{Y}_{1\cdot} - \bar{Y}_{2\cdot}) \pm t_{v,1-\alpha/2}\sqrt{MSE\left(\frac{1}{n_1} + \frac{1}{n_2}\right)}$$

where v represents the degrees of freedom, find a 99% confidence interval for the true difference in average output between the high and moderate automation lines.

Edited SAS Output (PROC GLM) for Problem 11

```
                           The GLM Procedure
                        Class Level Information
                    Class        Levels        Values
                    line           3            H L M
                    exper          3            1-2 <1 >2

                  Number of Observations Read        9
                  Number of Observations Used        9
```

Dependent Variable: output

Source	DF	Sum of Squares	Mean Square	F Value	Pr > F
Model	4	153.1111111	38.2777778		0.0039
Error	4				
Corrected Total	8	158.8888889			

R-Square	Coeff Var	Root MSE	output Mean
0.963636	10.10902	1.201850	11.88889

Source	DF	Type I SS	Mean Square	F Value	Pr > F
line	2	70.22222222	35.11111111		
exper	2	82.88888889	41.44444444		

Source	DF	Type III SS	Mean Square	F Value	Pr > F
line	2	70.22222222	35.11111111		
exper	2	82.88888889	41.44444444		

Level of line	N	--------------output------------ Mean	Std Dev
H	3	15.6666667	4.93288286
L	3	9.0000000	3.60555128
M	3	11.0000000	2.64575131

Level of exper	N	--------------output------------ Mean	Std Dev
1-2	3	13.3333333	4.16333200
<1	3	7.6666667	2.51661148
>2	3	14.6666667	3.7859389

12. An advertising company evaluated three types of television ads for a new low-cost subcompact automobile: visual-appeal ads, budget-appeal ads, and feature-appeal ads. To control for age differences, the company randomly selected viewers from four age groups to evaluate the persuasiveness of the ads (measured on a scale ranging from 1 to 10, with 1 being the lowest level of persuasion, and 10 the highest). Within each age group, subgroups of three viewers were randomly assigned to view one or another of the three types of ads. The sample persuasion scores are reproduced in the following table and processed as shown in the accompanying computer output.

Viewer Age	Type of Ad		
	Visual Appeal (V)	Budget Appeal (B)	Feature Appeal (F)
18–25 Years	7	8	5
26–35 Years	6	9	4
36–45 Years	8	9	4
46 and Older	10	10	5

a. Identify the blocks and treatments in this randomized-blocks design.

b. Complete the ANOVA table for the data set given.

c. Do the data provide sufficient evidence to indicate significant differences among the persuasion scores for the three types of ads?

d. Using a confidence interval of the form

$$(\bar{Y}_{i\cdot} - \bar{Y}_{j\cdot}) \pm t_{v,1-\alpha/2} \sqrt{\text{MSE}\left(\frac{1}{n_i} + \frac{1}{n_j}\right)}$$

where v is the error degrees of freedom, find a 99% confidence interval for the true difference in average output between the visual-appeal and budget-appeal ads.

e. Repeat part (d) for visual-appeal and feature-appeal ads.

f. Repeat part (d) for budget-appeal and feature-appeal ads.

g. Based on your results from parts (b)–(f), what do you conclude about the average persuasion scores for the different types of ads?

Edited SAS Output (PROC GLM) for Problem 12

```
                        General Linear Models Procedure
                           Class Level Information

                    Class         Levels         Values
                    AD               3           B F V
                    AGE              4           18-25 26-35 36-45 >45

                    Number of observations in data set = 12
```

Dependent Variable: SCORE

Source	DF	Sum of Squares	Mean Square	F Value	Pr > F
Model	5	50.0833333	10.0166667	☐	0.0040
Error	6	☐	☐		
Corrected Total	11	54.9166667			

R-Square	C.V.	Root MSE	SCORE Mean
0.911988	12.67098	0.89753	7.08333

Source	DF	Type I SS	Mean Square	F Value	Pr > F
AD	2	43.1666667	21.5833333	☐	☐
AGE	3	6.9166667	2.3055556	☐	☐

Source	DF	Type III SS	Mean Square	F Value	Pr > F
AD	2	43.1666667	21.5833333	☐	☐
AGE	3	6.9166667	2.3055556	☐	☐

Level of AD	N	------------SCORE------------ Mean	SD
B	4	9.00000000	0.81649658
F	4	4.50000000	0.57735027
V	4	7.75000000	1.70782513

Level of AGE	N	------------SCORE------------ Mean	SD
18-25	3	6.66666667	1.52752523
26-35	3	6.33333333	2.51661148
36-45	3	7.00000000	2.64575131
>45	3	8.33333333	2.88675135

References

Armitage, P. 1971. *Statistical Methods in Medical Research.* Oxford: Blackwell Scientific.

Ostle, B. 1963. *Statistics in Research,* 2d ed. Ames, Iowa: Iowa State University Press.

Peng, K. C. 1967. *Design and Analysis of Scientific Experiments.* Reading, Mass.: Addison-Wesley.

Sokal, R. R., and Karlen, I. 1964. "Competition Among Genotypes in *Tibolium castaneum* at Varying Densities and Gene Frequencies (the Black Locus)." *Genetics* 49: 195–211.

Tukey, J. W. 1949. "One Degree of Freedom for Nonadditivity." *Biometrics* 5: 232.

TABLE 19.5   Forced expiratory volume classified by plant and toxic exposure

Plant	Toxic Substance			Row Total
	A	B	C	
1	4.64, 5.92, 5.25, 6.17, 4.20, 5.90, 5.07, 4.13, 4.07, 5.30, 4.37, 3.76 ($T_{11} = 58.78$)	3.21, 3.17, 3.88, 3.50, 2.47, 4.12, 3.51, 3.85, 4.22, 3.07, 3.62, 2.95 ($T_{12} = 41.57$)	3.75, 2.50, 2.65, 2.84, 3.09, 2.90, 2.62, 2.75, 3.10, 1.99, 2.42, 2.37 ($T_{13} = 32.98$)	$R_1 = 133.33$
2	5.12, 6.10, 4.85, 4.72, 5.36, 5.41, 5.31, 4.78, 5.08, 4.97, 5.85, 5.26 ($T_{21} = 62.81$)	3.92, 3.75, 4.01, 4.64, 3.63, 3.46, 4.01, 3.39, 3.78, 3.51, 3.19, 4.04 ($T_{22} = 45.33$)	2.95, 3.21, 3.15, 3.25, 2.30, 2.76, 3.01, 2.31, 2.50, 2.02, 2.64, 2.27 ($T_{23} = 32.37$)	$R_2 = 140.51$
3	4.64, 4.32, 4.13, 5.17, 3.77, 3.85, 4.12, 5.07, 3.25, 3.49, 3.65, 4.10 ($T_{31} = 49.56$)	4.95, 5.22, 5.16, 5.35, 4.35, 4.89, 5.61, 4.98, 5.77, 5.23, 4.86, 5.15 ($T_{32} = 61.52$)	2.95, 2.80, 3.63, 3.85, 2.19, 3.32, 2.68, 3.35, 3.12, 4.11, 2.90, 2.75 ($T_{33} = 37.65$)	$R_3 = 148.73$
Column Total	$C_1 = 171.15$	$C_2 = 148.42$	$C_3 = 103.00$	$G = 422.57$

TABLE 19.6   Cell means for data of Table 19.5

Plant	Toxic Substance			Row Mean
	A	B	C	
1	4.90	3.46	2.75	3.70
2	5.23	3.78	2.70	3.90
3	4.13	5.13	3.14	4.13
Column Mean	4.75	4.12	2.86	3.91

(4.13). This suggests that the workers in plant 1 have poorer respiratory health than those in plant 2, and so on. Nevertheless, if the 3.70 value for plant 1 is considered clinically normal, then—despite the differences observed—all plants will be given a "clean bill of health." Furthermore, these differences among plants might have occurred solely by chance (i.e., might not be statistically significant).

In addition, it can be seen from Table 19.6 that exposure to substance C is associated with the poorest respiratory health (2.86), whereas exposure to substance A (4.75) and exposure to substance B (4.12) are associated with considerably better respiratory health. Again, we must decide whether to view the 2.86 value as meaningfully low and determine whether the differences among substances A, B, and C are statistically significant.

Finally, Table 19.6 indicates that the differences among plants depend somewhat on the toxic exposure being considered. For example, with respect to toxic substance A, plant 3 has the lowest mean FEV (4.13). With respect to toxic substance B, however, plant 1 has the lowest mean (3.46); and with respect to toxic substance C, plant 2 has the lowest mean (2.70). Furthermore, the magnitude of the differences among plants varies with the toxic substance. For toxic substance B, the difference between the highest and lowest plant means is $5.13 - 3.46 = 1.67$, whereas for toxic substances A and C the maximum differences are smaller ($5.23 - 4.13 = 1.10$) and $3.14 - 2.70 = 0.44$, respectively). Such fluctuations suggest that a significant interaction effect may be present, although this must be verified statistically. ■

19.3.2 ANOVA Table for Two-way ANOVA

Table 19.7 gives the general form of the two-way ANOVA table with r levels of the row factor, c levels of the column factor, and n observations per cell. The computer output given right after Table 19.7 shows the corresponding ANOVA results associated with the FEV data of Table 19.5. From Table 19.7 we see that the total (corrected) sum of squares (SSY) has been divided into the four components—SSR (rows), SSC (columns), SSRC (row × column interaction), and SSE (error)—based on the following fundamental equation:

$$SSY = SSR + SSC + SSRC + SSE, \qquad (19.1)$$

or equivalently

$$\sum_{i=1}^{r}\sum_{j=1}^{c}\sum_{k=1}^{n}(Y_{ijk} - \bar{Y}...)^2 = cn\sum_{i=1}^{r}(\bar{Y}_{i..} - \bar{Y}...)^2 + rn\sum_{j=1}^{n}(\bar{Y}_{.j.} - \bar{Y}...)^2$$

$$+ n\sum_{i=1}^{r}\sum_{j=1}^{c}(\bar{Y}_{ij.} - \bar{Y}_{i..} - \bar{Y}_{.j.} + \bar{Y}...)^2 + \sum_{i=1}^{r}\sum_{j=1}^{c}\sum_{k=1}^{n}(Y_{ijk} - \bar{Y}_{ij.})^2$$

TABLE 19.7   General (balanced) two-way ANOVA table[2]

Source	d.f.	SS	MS	F Fixed	F Mixed or Random
Row (main effect)	$r - 1$	SSR	$MSR = \dfrac{SSR}{r - 1}$	$\dfrac{MSR}{MSE}$	$\dfrac{MSR}{MSRC}$
Column (main effect)	$c - 1$	SSC	$MSC = \dfrac{SSC}{c - 1}$	$\dfrac{MSC}{MSE}$	$\dfrac{MSC}{MSRC}$
Row × column (interaction)	$(r - 1)(c - 1)$	SSRC	$MSRC = \dfrac{SSRC}{(r - 1)(c - 1)}$	$\dfrac{MSRC}{MSE}$	$\dfrac{MSRC}{MSE}$
Error	$rc(n - 1)$	SSE	$MSE = \dfrac{SSE}{rc(n - 1)}$		
Total	$rcn - 1$	SSY			

---

[2] See Section 19.7 for a discussion of the rationale for these F tests in terms of "expected mean squares."

Edited SAS Output (PROC GLM) for the Data of Table 19.5

```
                           Class Level Information
                        Class        Levels      Values
                        PLANT          3          1 2 3
                        SUBSTANCE      3          A B C

Dependent Variable: FEV
                                    Sum of
Source                    DF         Squares      Mean Square      F Value    Pr > F
Model                      8       94.6984630     11.8373079        44.10     <.0001
Error                     99       26.5758583      0.2684430
Corrected Total          107      121.2743213
           R-Square              Coeff Var          Root MSE        FEV Mean
           0.780862              13.24193           0.518115        3.912685

Source            DF       Type I SS     Mean Square     F Value              Pr > F
PLANT              2      3.29889630     1.64944815        6.14               0.0031
SUBSTANCE          2     66.88936852    33.44468426      124.59               <.0001
PLANT*SUBSTANCE    4     24.51019815     6.12754954       22.83               <.0001

Source            DF      Type III SS    Mean Square     F Value              Pr > F
PLANT              2      3.29889630     1.64944815        6.14               0.0031
SUBSTANCE          2     66.88936852    33.44468426      124.59               <.0001
PLANT*SUBSTANCE    4     24.51019815     6.12754954       22.83               <.0001


Source                 Type III Expected Mean Square
PLANT                  Var(Error) + 12 Var(PLANT*SUBSTANCE) + 36 Var(PLANT)
SUBSTANCE              Var(Error) + 12 Var(PLANT*SUBSTANCE) + 36 Var(SUBSTANCE)
PLANT*SUBSTANCE        Var(Error) + 12 Var(PLANT*SUBSTANCE)


          Tests of Hypotheses for Random Model Analysis of Variance
Source                  DF       Type III SS    Mean Square     F Value     Pr > F
PLANT                    2        3.298896       1.649448        0.27       0.7768
SUBSTANCE                2       66.889369      33.444684        5.46       0.0719
Error                    4       24.510198       6.127550
Error: MS(PLANT*SUBSTANCE)

Source                  DF       Type III SS    Mean Square     F Value     Pr > F
PLANT*SUBSTANCE          4       24.510198       6.127550       22.83       <.0001
Error: MS(Error)        99       26.575858       0.268443
```

Annotations on the output: "Fixed-Effect-tests" (pointing to Type III F values), "For random-effects model" (pointing to Type III Expected Mean Square), "Random and Mixed-effects tests" (pointing to the Tests of Hypotheses F values).

Since the sums of squares for each of these sources are routinely calculated by computer programs for ANOVA (e.g., SAS's GLM procedure), we do not provide specific computation formulas in the SS column of Table 19.7.

For the FEV data of Table 19.5, the sums of squares obtained (see the above SAS output) are:

$$\text{SSR(Plant)} = 3.299$$
$$\text{SSC(Toxsub)} = 66.889$$
$$\text{SSRC(Plant} \times \text{Toxsub)} = 24.510$$
$$\text{SSE} = 26.576$$
$$\text{SSY} = 121.274$$

The degrees of freedom associated with these sums of squares are

SSR has $(r - 1)$ d.f.

SSC has $(c - 1)$ d.f.

SSRC has $(r - 1)(c - 1)$ d.f.   *(19.2)*

SSE has $rc(n - 1)$ d.f.

SSY has $(rcn - 1)$ d.f.

Each mean-square term is obtained (as usual) by dividing the corresponding sum of squares by its associated degrees of freedom. Again, the appropriate F statistics to use depend on whether the row and column factors are classified as fixed or as random. These F tests are described in Section 19.4.

19.4 *F* Tests for Two-way ANOVA

The null hypotheses of interest for two-way ANOVA—as well as the basic statistical assumptions required for validly testing them—can be stated quite generally to encompass the four possible schemes of factor classification.[3]

Null Hypotheses

1. $H_0(R)$: *There is no row factor (main) effect* (i.e., there are no differences among the effects of the levels of the row factor).

2. $H_0(C)$: *There is no column factor (main) effect* (i.e., there are no differences among the effects of the levels of the column factor).

3. $H_0(RC)$: *There is no interaction effect between rows and columns* (i.e., the row level effects within any one column are the same as those within any other column; equivalently, the column level effects within any one row are the same as those within any other row).

Assumptions

1. All observations are *statistically independent* of one another for fixed-effects models.[4]

2. Each observation comes from a *normally distributed population*.

3. Each observation has the same population variance (i.e., the usual assumption of *variance homogeneity* applies).

---

[3] However, the null hypotheses are quite different when stated more precisely in terms of population cell means and/or variances.

[4] The responses are *not* mutually independent when random effects are included in the regression model for two-way ANOVA with equal cell numbers.

As previously stated, the appropriate *F* statistics depend on whether the row and column factors are classified as fixed or random. The mean-square term associated with a given source of variation will estimate different quantities, depending on whether the row and column factors are fixed or random; this explains why different denominators are used in the various *F* tests for two-way ANOVA.

Regardless of how the factors are classified, however, the *F* statistic used to test $H_0(RC)$ of no row–column interaction always has the form

$$F(RC) = \frac{MSRC}{MSE}$$

with $(r - 1)(c - 1)$ and $rc(n - 1)$ degrees of freedom.[5]

The *F* statistics for evaluating main effects differ as follows with respect to the factor classification scheme:

1. *Rows and columns fixed.* Divide the mean squares for rows and for columns by the mean square for error:

    $$F(R) = \frac{MSR}{MSE}$$

    with $(r - 1)$ and $rc(n - 1)$ degrees of freedom, or

    $$F(C) = \frac{MSC}{MSE}$$

    with $(c - 1)$ and $rc(n - 1)$ degrees of freedom.

2. *Rows and columns random, or one factor fixed and the other factor random.* Divide the mean squares for rows and columns by the mean square for interaction:

    $$F(R) = \frac{MSR}{MSRC}$$

    with $(r - 1)$ and $(r - 1)(c - 1)$ degrees of freedom, or

    $$F(C) = \frac{MSC}{MSRC}$$

    with $(c - 1)$ and $(r - 1)(c - 1)$ degrees of freedom.

For the FEV data, the classification of the factors depends on the point of view of the researcher. If, for example, the plants and toxic substances were the only ones of interest, both factors would be considered fixed. However, if the plants were considered to represent a sample from a large population of plants of interest and if the toxic substances likewise were viewed as representing a population of toxic agents of interest, both factors would be treated as random. Of course, the classification would be mixed if one of these factors were considered fixed and the other random.

We will not argue for any particular classification scheme for the factors in the FEV example, since the example is artificial. The decisions regarding certain null hypotheses differ,

---

[5] $F(RC)$ denotes the *F* test of $H_0(RC)$. Similarly, $F(R)$ and $F(C)$ denote the *F* tests of $H_0(R)$ and $H_0(C)$, respectively.

however, depending on how the factors are classified. This can be seen from the earlier SAS output, as follows:

1. *Both factors fixed.* Both main effects are significant, since $F(\text{Plant}) = 6.14^{**}$ (with 2 and 99 degrees of freedom) and $F(\text{Toxsub}) = 124.59^{**}$ (with 2 and 99 degrees of freedom).

2. *Both factors random, or one factor fixed and the other factor random.* Neither main effect is significant, since $F(\text{Plant}) = 0.27$ (with 2 and 4 degrees of freedom) and $F(\text{Toxsub}) = 5.46$ (with 2 and 4 degrees of freedom).

Despite these differences among the main-effect test results, the most important finding from this analysis is that the interaction effect is significant:

$$F(\text{Plant} \times \text{Toxsub}) = 22.83^{**} \text{ (with 4 and 99 d.f.)}$$

In Section 19.6 we discuss the interpretation of such interaction effects. Certainly, it does not make much sense to talk about the separate or independent effects (i.e., main effects) of Plant and Toxsub on FEV, since there is strong evidence that these factors do not affect FEV independently of one another.

Regardless of whether the factors are fixed or random, eight distinct patterns of significance may result, depending on the significance (or nonsignificance) of the three tests involved (two main-effect tests and the interaction test). Suppose that one factor is labeled A and the second factor B. Then Table 19.8 summarizes the possible outcomes (with significant results indicated by asterisks). In pattern I, the F tests for the A effect, the B effect, and the A $\times$ B interaction are all nonsignificant. In pattern VIII, all three are significant.

Each pattern deserves some comment. Pattern I leads us to conclude that there is no evidence of any relationship between the response variable and the predictors (considered either separately or together). Pattern II implies that only factor A is related to the response variable. Similarly, Pattern III implies that only factor B is related to the response variable. In Pattern IV, we conclude that both A and B affect the response. Furthermore, the nature of the relationship between factor A and the response remains the same at all levels of factor B studied, and conversely. Patterns V, VI, VII, and VIII all involve significant interaction. Most statisticians recommend focusing only on this interaction in such cases. A significant interaction implies that both factors are important and that the level of one of the factors must be known to characterize the effect of the other factor on the response. Section 19.6 discusses interpreting interactions.

TABLE 19.8 **Patterns of significance possible in two-way ANOVA with equal cell numbers**

Source	Pattern							
	I	II	III	IV	V	VI	VII	VIII
A		*		*		*		*
B			*	*			*	*
A $\times$ B					*	*	*	*

Asterisk denotes significance for the effect in that row.

19.5 Regression Model for Fixed-effects Two-way ANOVA

In this section we describe a particular regression model[6] and a related classical fixed-effects ANOVA model for two-way ANOVA when both factors are considered fixed. Random-effects and mixed-effects models are discussed in detail in Section 19.7. As in the cases of one-way ANOVA and randomized-blocks ANOVA, a regression model for two-way ANOVA can be interpreted in terms of the cell, marginal, and overall means associated with the two-way layout (see Table 19.9). In the table,

$$\mu_{i \cdot} = \frac{1}{c} \sum_{j=1}^{c} \mu_{ij} \qquad i = 1, 2, \ldots, r$$

$$\mu_{\cdot j} = \frac{1}{r} \sum_{i=1}^{r} \mu_{ij} \qquad j = 1, 2, \ldots, c$$

$$\mu_{\cdot \cdot} = \frac{1}{rc} \sum_{i=1}^{r} \sum_{j=1}^{c} \mu_{ij}$$

19.5.1 Regression Model

When there are r rows and c columns, a regression model can be formulated involving $(r - 1)$ dummy variables for the row factor, $(c - 1)$ dummy variables for the column factor, and $(r - 1)(c - 1)$ interaction dummy variables, which are constructed by forming products of each of the row dummy variables with each of the column dummy variables. Such a model can be expressed as

$$Y = \mu + \sum_{i=1}^{r-1} \alpha_i X_i + \sum_{j=1}^{c-1} \beta_j Z_j + \sum_{i=1}^{r-1} \sum_{j=1}^{c-1} \gamma_{ij} X_i Z_j + E \qquad (19.3)$$

TABLE 19.9   Table of population cell means for two-way layout

Row	Column				Row Mean
	1	2	. . .	c	
1	μ_{11}	μ_{12}	. . .	μ_{1c}	$\mu_{1 \cdot}$
2	μ_{21}	μ_{22}	. . .	μ_{2c}	$\mu_{2 \cdot}$
$\vdots$	$\vdots$	$\vdots$		$\vdots$	$\vdots$
r	μ_{r1}	μ_{r2}	. . .	μ_{rc}	$\mu_{r \cdot}$
Column Mean	$\mu_{\cdot 1}$	$\mu_{\cdot 2}$	. . .	$\mu_{\cdot c}$	$\mu_{\cdot \cdot}$

---

[6] Several other regression models can be defined, of course, depending on the coding scheme for the dummy variables. The regression model given here is the one most commonly used, due to its natural connection with the classical fixed-effects two-way ANOVA model.

in which

$$X_i = \begin{cases} -1 & \text{for level } r \text{ of the row factor} \\ 1 & \text{for level } i \text{ of the row factor} \\ 0 & \text{otherwise} \end{cases}$$

and

$$Z_j = \begin{cases} -1 & \text{for level } c \text{ of the column factor} \\ 1 & \text{for level } j \text{ of the column factor} \\ 0 & \text{otherwise} \end{cases}$$

(for $i = 1, 2, \ldots, r - 1$; and $j = 1, 2, \ldots, c - 1$).

The formulas relating the coefficients α_i, β_j, and γ_{ij} to the various means of Table 19.9 are

$$\mu = \mu_{..}$$

$$\alpha_i = \mu_{i.} - \mu_{..} \qquad\qquad i = 1, 2, \ldots, r - 1$$

$$\beta_j = \mu_{.j} - \mu_{..} \qquad\qquad j = 1, 2, \ldots, c - 1$$

$$\gamma_{ij} = \mu_{ij} - \mu_{i.} - \mu_{.j} + \mu_{..} \qquad i = 1, 2, \ldots, r - 1; j = 1, 2, \ldots, c - 1$$

$$-\sum_{i=1}^{r-1} \alpha_i = \mu_{r.} - \mu_{..} \tag{19.4}$$

$$-\sum_{j=1}^{c-1} \beta_j = \mu_{.c} - \mu_.$$

$$-\sum_{i=1}^{r-1} \gamma_{ij} = \mu_{rj} - \mu_{r.} - \mu_{.j} + \mu_{..} \quad j = 1, 2, \ldots, c - 1$$

$$-\sum_{j=1}^{c-1} \gamma_{ij} = \mu_{ic} - \mu_{i.} - \mu_{.c} + \mu_{..} \quad i = 1, 2, \ldots, r - 1$$

As with the ANOVA regression analogies made in earlier chapters, the same F tests given in Table 19.7 (for the case where both factors are fixed) can be obtained by using the appropriate multiple partial F tests concerning subsets of the coefficients in the regression model (19.3). Specifically, the multiple partial F test of H_0: $\alpha_1 = \alpha_2 = \cdots = \alpha_{r-1} = 0$ for model (19.3) yields exactly the same F statistic as that used in standard (balanced) two-way fixed-effects ANOVA for testing the significance of the main effect of the row factor (i.e., $F = \text{MSR}/\text{MSE}$). Similarly, the multiple partial F test of H_0: $\beta_1 = \beta_2 = \cdots = \beta_{c-1} = 0$ in model (19.3) yields exactly the same F statistic as is used in standard (balanced) two-way fixed-effects ANOVA to test the significance of the main effect of the column factor (i.e., $F = \text{MSC}/\text{MSE}$). Finally, the multiple partial F test of H_0: $\gamma_{ij} = 0$ (for $i = 1, 2, \ldots, r - 1$; and $j = 1, 2, \ldots, c - 1$) is identical to the (balanced) two-way fixed-effects ANOVA F test for interaction (i.e., $F = \text{MSRC}/\text{MSE}$).

The preceding formulas may also be used to express the cell, row marginal, and column marginal means as functions of the regression coefficients, as follows:

$$\mu_{ij} = \mu + \alpha_i + \beta_j + \gamma_{ij} \qquad\qquad i = 1, 2,\ldots, r - 1; j = 1, 2,\ldots, c - 1$$

$$\mu_{rj} = \mu - \sum_{i=1}^{r-1}\gamma_{ij} - \sum_{i=1}^{r-1}\alpha_i + \beta_j \qquad j = 1, 2,\ldots, c - 1$$

$$\mu_{ic} = \mu - \sum_{j=1}^{c-1}\gamma_{ij} - \sum_{j=1}^{c-1}\beta_j + \alpha_i \qquad i = 1, 2,\ldots, r - 1$$

$$\mu_{rc} = \mu - \sum_{i=1}^{r-1}\alpha_i - \sum_{j=1}^{c-1}\beta_j + \sum_{i=1}^{r-1}\sum_{j=1}^{c-1}\gamma_{ij}$$

The row marginal means are

$$\mu_{i\cdot} = \mu + \alpha_i \qquad i = 1, 2,\ldots, r - 1$$

$$\mu_{r\cdot} = \mu - \sum_{i=1}^{r-1}\alpha_i$$

and the column marginal means are

$$\mu_{\cdot j} = \mu + \beta_j \qquad j = 1, 2,\ldots, c - 1$$

$$\mu_{\cdot c} = \mu - \sum_{j=1}^{c-1}\beta_j$$

If reference cell coding is used, the general form of model (19.3) stays the same, but the definitions of X_i and Z_j change. For example, we can define

$$X_i = \begin{cases} 1 & \text{for level } i \text{ of the row factor} \\ 0 & \text{otherwise} \end{cases} \qquad i = 1, 2,\ldots, r - 1$$

and

$$Z_j = \begin{cases} 1 & \text{for level } j \text{ of the column factor} \\ 0 & \text{otherwise} \end{cases} \qquad j = 1, 2,\ldots, c - 1$$

With these definitions in place, the parameters in model (19.3) have different interpretations from those given earlier. In particular, the cell means (for the above reference cell coding) can be expressed as the following functions of the parameters in model (19.3):

$$\mu_{ij} = \mu + \alpha_i + \beta_j + \gamma_{ij} \qquad i = 1, 2,\ldots, r - 1; j = 1, 2,\ldots, c - 1$$
$$\mu_{rj} = \mu + \beta_j \qquad\qquad\qquad j = 1, 2,\ldots, c - 1$$
$$\mu_{ic} = \mu + \alpha_i \qquad\qquad\qquad i = 1, 2,\ldots, r - 1$$
$$\mu_{rc} = \mu$$

The row marginal means are

$$\mu_{i\cdot} = \mu + \alpha_i + \sum_{j=1}^{c-1} \frac{(\beta_j + \gamma_{ij})}{c} \qquad i = 1, 2, \ldots, r-1$$

$$\mu_{r\cdot} = \mu + \sum_{j=1}^{c-1} \frac{\beta_j}{c}$$

and the column marginal means are

$$\mu_{\cdot j} = \mu + \beta_j + \sum_{i=1}^{r-1} \frac{(\alpha_i + \gamma_{ij})}{r} \qquad j = 1, 2, \ldots, c-1$$

$$\mu_{\cdot c} = \mu + \sum_{i=1}^{r-1} \frac{\alpha_i}{r}$$

These expressions must be modified when cell-specific sample sizes are not all equal. Chapter 20 addresses two-way ANOVA with unequal cell-specific sample sizes.

19.5.2 Classical Two-way Fixed-effects ANOVA Model

When both factors are considered fixed, there are three types of effects to consider:

1. *Main effects of the row factor:* The differences between the various row means and the overall mean (i.e., $\mu_{i\cdot} - \mu_{\cdot\cdot}$, $i = 1, 2, \ldots, r$).

2. *Main effects of the column factor:* The differences between the various column means and the overall mean (i.e., $\mu_{\cdot j} - \mu_{\cdot\cdot}$, $j = 1, 2, \ldots, c$).

3. *Interaction effects:* The differences between differences of the form $(\mu_{ij} - \mu_{i\cdot}) - (\mu_{\cdot j} - \mu_{\cdot\cdot})$ or $(\mu_{ij} - \mu_{\cdot j}) - (\mu_{i\cdot} - \mu_{\cdot\cdot})$, for $i = 1, 2, \ldots, r$ and $j = 1, 2, \ldots, c$.

The classical two-way fixed-effects ANOVA model involving such effects is of the following form:

$$Y_{ijk} = \mu + \alpha_i + \beta_j + \gamma_{ij} + E_{ijk} \qquad (19.5)$$

where

$$\mu = \mu_{\cdot\cdot} = \text{Overall mean}$$
$$\alpha_i = \mu_{i\cdot} - \mu_{\cdot\cdot} = \text{Effect of row } i$$
$$\beta_j = \mu_{\cdot j} - \mu_{\cdot\cdot} = \text{Effect of column } j$$
$$\gamma_{ij} = \mu_{ij} - \mu_{i\cdot} - \mu_{\cdot j} + \mu_{\cdot\cdot} = \text{Interaction effect associated with cell } (i, j)$$
$$E_{ijk} = Y_{ijk} - \mu - \alpha_i - \beta_j - \gamma_{ij} = \text{Error (or residual) associated with the } k\text{th observation in cell } (i, j)$$

(for $i = 1, 2, \ldots, r$; $j = 1, 2, \ldots, c$; and $k = 1, 2, \ldots, n$).

The following relationships are clearly satisfied by the effects in the preceding model:

$$\sum_{i=1}^{r} \alpha_i = 0, \qquad \sum_{j=1}^{c} \beta_j = 0, \qquad \sum_{i=1}^{r} \gamma_{ij} = 0, \qquad \sum_{j=1}^{c} \gamma_{ij} = 0 \qquad (19.6)$$

A comparison of the regression coefficients in model (19.3) with the ANOVA effects in model (19.5) makes clear that the models are completely equivalent.

Finally, each of the effects in model (19.5) can be simply estimated by using sample means, as follows:

$$\hat{\mu} = \overline{Y}...$$

$$\hat{\alpha}_i = \overline{Y}_{i..} - \overline{Y}... \qquad\qquad i = 1, 2, \ldots, r$$

$$\hat{\beta}_j = \overline{Y}_{.j.} - \overline{Y}... \qquad\qquad j = 1, 2, \ldots, c$$

$$\hat{\gamma}_{ij} = \overline{Y}_{ij.} - \overline{Y}_{i..} - \overline{Y}_{.j.} + \overline{Y}... \qquad i = 1, 2, \ldots, r; j = 1, 2, \ldots, c$$

19.6 Interactions in Two-way ANOVA

In this section we investigate several ways to evaluate interaction in two-way ANOVA. For convenience, we focus on the fixed-effects case. Nevertheless, even though the parameters involved and the test statistics used for making inferences in the fixed-effects case differ from those used in the random- and mixed-effects cases, the interpretations of interactions are generally the same.

19.6.1 Concept of Interaction

An interaction exists between two factors if the relationship among the effects associated with the levels of one factor differs according to the levels of the second factor. In other words, an interaction represents an effect due to the joint influence of two factors, over and above the effects of each factor considered separately.

More specifically, if we consider the two-way table of cell means and the various ways of writing the statistical model in the fixed-effects case, there are three equivalent ways of describing or representing an interaction in statistical terms.

Method 1: Interaction as a Difference in Differences of Means   In two-way ANOVA, an interaction exists between the row and column factors if any of the following equivalent statements are true:

1. For some pair of columns, the difference between the means in these columns for a given row is not equal to the difference between the corresponding means for some other row. For example, for rows 1 and 2 and columns 1 and 2, $\mu_{11} - \mu_{12} \neq \mu_{21} - \mu_{22}$.

2. For some pair of rows, the difference between the means in these rows for a given column is not equal to the difference between the corresponding means for some other column. For example, for rows 1 and 2 and columns 1 and 2, $\mu_{11} - \mu_{21} \neq \mu_{12} - \mu_{22}$.

3. For some cell in the table, the difference between that cell's mean and its associated marginal row mean is not equal to the difference between its associated marginal column mean and the overall mean. For example, for the (i, j)th cell, $\mu_{ij} - \mu_{i.} \neq \mu_{.j} - \mu_{..}$, or $\mu_{ij} - \mu_{i.} - \mu_{.j} + \mu_{..} \neq 0$.

4. For some cell in the table, the difference between that cell's mean and its associated marginal column mean is not equal to the difference between its associated marginal row mean and the overall mean. For example, $\mu_{ij} - \mu_{.j} \neq \mu_{i.} - \mu_{..}$, or $\mu_{ij} - \mu_{i.} - \mu_{.j} - \mu_{..} \neq 0$.

Thus, when there is no interaction, the relationship among the column effects (β_j's) is the same regardless of the row being considered, and vice versa. And, since, from statements 3 and 4,

$$\mu_{ij} - \mu_{i.} - \mu_{.j} - \mu_{..} = 0$$

when there is no interaction, it follows that

$$\text{MSRC} = \frac{n}{(r-1)(c-1)} \sum_{i=1}^{r} \sum_{j=1}^{c} (Y_{ij} - \overline{Y}_{i..} - \overline{Y}_{.j.} + \overline{Y}_{...})^2$$

(which estimates

$$\frac{n}{(r-1)(c-1)} \sum_{i=1}^{r} \sum_{j=1}^{c} (\mu_{ij} - \mu_{i.} - \mu_{.j} + \mu_{..})^2$$

for the population) is small when there is no interaction and large when there is interaction.

Method 2: Interaction as an Effect in the Fixed-effects Model   An interaction exists if the appropriate fixed-effects ANOVA model has the form

$$Y_{ijk} = \mu + \alpha_i + \beta_j + \gamma_{ij} + E_{ijk}$$

where $\gamma_{ij} \neq 0$ for at least one (i, j) pair.

Methods 1 and 2 are completely equivalent. For example, if there is no interaction, then $\mu_{ij} = \mu + \alpha_i + \beta_j$, so $\mu_{1j} - \mu_{2j} = (\mu + \alpha_1 + \beta_j) - (\mu + \alpha_2 + \beta_j) = \alpha_1 - \alpha_2$, which is independent of j. In general, $\mu_{11} - \mu_{21} = \mu_{12} - \mu_{22} = \cdots = \mu_{1c} - \mu_{2c} = \alpha_1 - \alpha_2$.

Method 3: Interaction as a Term in a Regression Model   An interaction exists if the appropriate regression model (using dummy variables) contains a term that involves the product (or in general, any function) of variables from different factors, for example, if the appropriate model is of the form

$$Y = \mu + \sum_{i=1}^{r-1} \alpha_i X_i + \sum_{j=1}^{c-1} \beta_j Z_j + \sum_{i=1}^{r-1}\sum_{j=1}^{c-1} \gamma_{ij} X_i Z_j + E$$

where at least one of the γ_{ij} is not 0.

When $r = c = 2$, the model simplifies to

$$Y = \mu + \alpha_1 X_1 + \beta_1 Z_1 + \gamma_{11} X_1 Z_1 + E$$

where

$$X_1 = \begin{cases} -1 & \text{if level 2 of the row factor} \\ 1 & \text{if level 1 of the row factor} \end{cases}$$

$$Z_1 = \begin{cases} -1 & \text{if level 2 of the column factor} \\ 1 & \text{if level 1 of the column factor} \end{cases}$$

Then, $\mu = \mu_{..}$, $\alpha_1 = \mu_{1.} - \mu_{..}$, $\beta_1 = \mu_{.1} - \mu_{..}$, and $\gamma_{11} = \mu_{11} - \mu_{1.} - \mu_{.1} + \mu_{..}$.

Method 3 is equivalent to the other two methods, provided that both independent variables (i.e., factors) are considered nominal and so are represented by dummy variables. If, however, both independent variables are continuous, a regression model with any product term will exhibit a somewhat different interaction effect, not necessarily characterized by a nonzero difference of mean differences.

19.6.2   Some Hypothetical Examples

We now consider some hypothetical two-way tables of population cell means that illustrate different patterns of interaction. These tables pertain to the example in Section 19.2, for which the factors are NHT and PSN and the dependent variable is CMI score. Subsequently, we will examine the table of sample cell means actually obtained (Table 19.2), keeping in mind that the statistical test for interaction may negate any tentative trends suggested by the sample means. We will also examine the example of Section 19.3 in this light.

Row and Column Main Effects but No Interaction Effect

Table 19.10 presents three alternative layouts, each representing the general situation involving *both* a row main effect and a column main effect but no interaction effect. Each of these tables gives *population* (and not sample) mean values, so there is no sampling variation to consider.

The main effects are reflected in the differences between marginal row means and between marginal column means in each table. The lack of an interaction effect can be established by comparing the differences among the cell means, as discussed in Section 19.6.1. From Table 19.10(a), for example, we have $\mu_{11} - \mu_{21} = \mu_{21} - \mu_{22}$, since $26 - 23 = 3 = 20 - 17$. For this same table, $\mu_{11} - \mu_{1.} - \mu_{.1} + \mu_{..} = 26 - 24.5 - 23 + 21.5 = 0$; and, similar terms associated with the other three cells in the table are also 0. Furthermore, for this table, the model (19.5) can be shown to have the following specific structure:

$$\mu_{ij} = 21.5 + \alpha_i + \beta_j$$

where

$$\alpha_i = \begin{cases} 3 & \text{if } i = 1 \\ -3 & \text{if } i = 2 \end{cases} \quad \text{and} \quad \beta_j = \begin{cases} 1.5 & \text{if } j = 1 \\ -1.5 & \text{if } j = 2 \end{cases}$$

TABLE 19.10   Alternative layouts for main effects but no interaction

(a) NHT	PSN Low	PSN High	Row Mean	(b) NHT	PSN Low	PSN High	Row Mean	(c) NHT	PSN Low	PSN High	Row Mean
Low	26	23	24.5	Low	18	26	22	Low	18	26	22
High	20	17	18.5	High	20	28	24	High	16	24	20
Column Mean	23	20	21.5	Column Mean	19	27	23	Column Mean	17	25	21

This model does not involve any γ_{ij} term (i.e., there is no interaction term in the model). Thus, we have

$$\mu_{11} = 21.5 + 3 + 1.5 = 26$$
$$\mu_{12} = 21.5 + 3 - 1.5 = 23$$
$$\mu_{21} = 21.5 - 3 + 1.5 = 20$$
$$\mu_{22} = 21.5 - 3 - 1.5 = 17$$

The layouts for tables (b) and (c) also represent no-interaction models; they have the particular forms, respectively, of

$$\mu_{ij} = 23 + \alpha_i + \beta_j$$

where

$$\alpha_i = \begin{cases} -1 & \text{if } i = 1 \\ 1 & \text{if } i = 2 \end{cases} \quad \text{and} \quad \beta_j = \begin{cases} -4 & \text{if } j = 1 \\ 4 & \text{if } j = 2 \end{cases}$$

and

$$\mu_{ij} = 21 + \alpha_i + \beta_j$$

where

$$\alpha_i = \begin{cases} 1 & \text{if } i = 1 \\ -1 & \text{if } i = 2 \end{cases} \quad \text{and} \quad \beta_j = \begin{cases} -4 & \text{if } j = 1 \\ 4 & \text{if } j = 2 \end{cases}$$

Exactly One Main Effect and No Interaction Effect

This situation is depicted in Table 19.11. Table 19.11(a) contains a main effect due to PSN but no main effect due to NHT. Table 19.11(b) contains a main effect due to NHT but no main effect due to PSN. There is no PSN × NHT interaction.

Same-direction Interaction

Three examples of same-direction interaction are given in Table 19.12. In Table 19.12(a) we see that $\mu_{11} - \mu_{12} = 26 - 23 = 3$, whereas $\mu_{21} - \mu_{22} = 20 - 13 = 7$. Also, $\mu_{11} - \mu_{1.} = 26 - 24.5 = 1.5$ and $\mu_{.1} - \mu_{..} = 23 - 20.5 = 2.5$, so $\mu_{11} - \mu_{1.} - \mu_{.1} + \mu_{..} = -1.0$. The

TABLE 19.11   Layouts for one main effect and no interaction

(a) Main Effect Due to PSN

NHT	PSN Low	PSN High	Row Mean
Low	18	26	22
High	18	26	22
Column Mean	18	26	22

(b) Main Effect Due to NHT

NHT	PSN Low	PSN High	Row Mean
Low	19	19	19
High	24	24	24
Column Mean	21.5	21.5	21.5

TABLE 19.12   Layouts for same-direction interaction

(a) NHT	PSN Low	PSN High	Row Mean
Low	26	23	24.5
High	20	13	16.5
Column Mean	23	18	20.5

(b) NHT	PSN Low	PSN High	Row Mean
Low	18	26	22
High	20	36	28
Column Mean	19	31	25

(c) NHT	PSN Low	PSN High	Row Mean
Low	18	26	22
High	12	24	18
Column Mean	15	25	20

other interactions of the general form $(\mu_{ij} - \mu_{i.} - \mu_{.j} + \mu_{..})$ are similarly determined to be either 1.0 (for $i = 1, j = 2$; or for $i = 2, j = 1$) or -1.0 (for $i = 2, j = 2$).

These hypothetical results in Table 19.12(a) indicate that in *both* low- and high-NHT neighborhoods, persons in friendly surroundings (i.e., high PSN) are healthier (i.e., have a lower CMI score) than persons in unfriendly surroundings (i.e., low PSN), but that the extent of this difference is greater when there is a large number of households (high NHT) than when there is a small number (low NHT). In other words, at each level of NHT, the difference between the PSN-level effects lies in the same direction (i.e., high PSN is associated with a lower mean CMI score than is low PSN), but the magnitude of the difference depends on the NHT level. This is the meaning of *same-direction interaction*.

The model for Table 19.12(a) may be expressed as

$$\mu_{ij} = 20.5 + \alpha_i + \beta_j + \gamma_{ij}$$

where

$$\alpha_i = \begin{cases} 4.0 & \text{if } i = 1 \\ -4.0 & \text{if } i = 2 \end{cases} \qquad \beta_j = \begin{cases} 2.5 & \text{if } j = 1 \\ -2.5 & \text{if } j = 2 \end{cases}$$

$$\gamma_{ij} = \begin{cases} -1.0 & \text{if } i = 1, j = 1 \\ 1.0 & \text{if } i = 1, j = 2 \\ 1.0 & \text{if } i = 2, j = 1 \\ -1.0 & \text{if } i = 2, j = 2 \end{cases}$$

Thus,

$$\mu_{11} = 20.5 + 4.0 + 2.5 - 1.0 = 26$$

$$\mu_{12} = 20.5 + 4.0 - 2.5 + 1.0 = 23$$

$$\mu_{21} = 20.5 - 4.0 + 2.5 + 1.0 = 20$$

$$\mu_{22} = 20.5 - 4.0 - 2.5 - 1.0 = 13$$

The same general type of model holds for tables (b) and (c).

TABLE 19.13   Layouts for reverse interaction

(a)

NHT	PSN Low	PSN High	Row Mean
Low	18	26	22
High	22	20	21
Column Mean	20	23	21.5

(b)

NHT	PSN Low	PSN High	Row Mean
Low	26	22	24
High	18	24	21
Column Mean	22	23	22.5

Reverse Interaction

Two examples of reverse interaction are given in Table 19.13. In these instances, the direction of the difference between two cell means for one row (column) is opposite to, or reversed from, the direction of the difference between the corresponding cell means for some other row (column).

In Table 19.13(a), we see that $\mu_{11} - \mu_{12} = 18 - 26 = -8$, whereas $\mu_{21} - \mu_{22} = 22 - 20 = 2$. Also, $\mu_{21} - \mu_{2.} = 22 - 21 = 1$ and $\mu_{.1} - \mu_{..} = 20 - 21.5 = -1.5$, so $\mu_{21} - \mu_{2.} - \mu_{.1} + \mu_{..} = 2.5$.

These hypothetical results for this table indicate that, for neighborhoods with a small number of households (low NHT), persons in unfriendly surroundings (low PSN) are healthier than are persons in friendly surroundings; but for neighborhoods with a large number of households, the situation is reversed. In other words, the difference between the effects of the high and low PSN levels is positive for low NHT but negative for high NHT (i.e., there is a reversal in sign). This is the meaning of *reverse interaction*.

The model in this case is given as

$$\mu_{ij} = 21.5 + \alpha_i + \beta_j + \gamma_{ij}$$

where

$$\alpha_i = \begin{cases} 0.5 & \text{if } i = 1 \\ -0.5 & \text{if } i = 2 \end{cases} \qquad \beta_j = \begin{cases} -1.5 & \text{if } j = 1 \\ 1.5 & \text{if } j = 2 \end{cases}$$

$$\gamma_{ij} = \begin{cases} -2.5 & \text{if } i = 1, j = 1 \\ 2.5 & \text{if } i = 1, j = 2 \\ 2.5 & \text{if } i = 2, j = 1 \\ -2.5 & \text{if } i = 2, j = 2 \end{cases}$$

Thus,

$$\mu_{11} = 21.5 + 0.5 - 1.5 - 2.5 = 18$$

$$\mu_{12} = 21.5 + 0.5 + 1.5 + 2.5 = 26$$

$$\mu_{21} = 21.5 - 0.5 - 1.5 + 2.5 = 22$$

$$\mu_{22} = 21.5 - 0.5 + 1.5 - 2.5 = 20$$

It can be shown that a reverse interaction is indicated in Table 19.13(b), as well.

19.6.3 Interaction Effects for Data of Table 19.1

The table of sample cell means actually obtained for the CMI example of Section 19.2 is given in Table 19.14. From this table, the following differences of means can be determined:

$$\bar{Y}_{11\cdot} - \bar{Y}_{12\cdot} = 18.92 - 26.84 = -7.92 \text{ whereas } \bar{Y}_{21\cdot} - \bar{Y}_{22\cdot} = 20.84 - 20.72 = 0.12$$

$$\bar{Y}_{11\cdot} - \bar{Y}_{21\cdot} = 18.92 - 20.84 = -1.92 \text{ whereas } \bar{Y}_{12\cdot} - \bar{Y}_{22\cdot} = 26.84 - 20.72 = 6.12$$

$$\bar{Y}_{11\cdot} - \bar{Y}_{1\cdot\cdot} - \bar{Y}_{\cdot1\cdot} + \bar{Y}_{\cdots} = 18.92 - 22.88 - 19.88 + 21.83 = -2.01$$

$$\bar{Y}_{12\cdot} - \bar{Y}_{1\cdot\cdot} - \bar{Y}_{\cdot2\cdot} + \bar{Y}_{\cdots} = 26.84 - 22.88 - 23.78 + 21.83 = 2.01$$

$$\bar{Y}_{21\cdot} - \bar{Y}_{2\cdot\cdot} - \bar{Y}_{\cdot1\cdot} + \bar{Y}_{\cdots} = 20.84 - 20.78 - 19.88 + 21.83 = 2.01$$

$$\bar{Y}_{22\cdot} - \bar{Y}_{2\cdot\cdot} - \bar{Y}_{\cdot2\cdot} + \bar{Y}_{\cdots} = 20.72 - 20.78 - 23.78 + 21.83 = -2.01$$

These comparisons suggest a possible reverse interaction. Specifically, for small turnkey neighborhoods, persons from friendly surroundings (high PSN) appear to have worse health (higher mean CMI scores) than do persons from unfriendly surroundings; but little difference in mean CMI scores is found on the friendliness variable for large turnkey neighborhoods. As was mentioned in Section 19.2, this pattern runs counter to what Daly (1973) expected to find. However, these observed differences are subject to sampling variation (i.e., they are sample values and not population values); and the test for interaction for these data does not reveal significant interaction.

TABLE 19.14   Cell means for data of Table 19.1

NHT	PSN		Row Mean
	Low	High	
Low	18.92	26.84	22.88
High	20.84	20.72	20.78
Column Mean	19.88	23.78	21.83

19.6.4 Interaction Effects for Data of Table 19.5

The table of sample cell means for the FEV example (Table 19.5) is as shown in Table 19.15. This table of means is slightly more difficult to interpret than the one for the CMI data, because it contains three rows and columns instead of two. Nevertheless, we can immediately observe that the relative magnitudes of the means vary from column to column. For example, for Toxsub A, the order of Plants by increasing mean FEV level is 3, 1, 2. For Toxsub B, on the other hand, the order is 1, 2, 3, and for Toxsub C the order is 2, 1, 3. These differences in ordering indicate some interaction, the significance of which was established earlier. The following comparisons of cell means should help in interpreting the nature of this significant interaction effect:

$$\bar{Y}_{11\cdot} - \bar{Y}_{12\cdot} = 4.90 - 3.46 = 1.44 \quad \text{whereas} \quad \bar{Y}_{31\cdot} - \bar{Y}_{32\cdot} = 4.13 - 5.13 = -1.00$$

$$\bar{Y}_{21\cdot} - \bar{Y}_{31\cdot} = 5.23 - 4.13 = 1.10 \quad \text{whereas} \quad \bar{Y}_{22\cdot} - \bar{Y}_{32\cdot} = 3.78 - 5.13 = -1.35$$

$$\bar{Y}_{21\cdot} - \bar{Y}_{\cdot1\cdot} = 5.23 - 4.75 = 0.48 \quad \text{whereas} \quad \bar{Y}_{2\cdot\cdot} - \bar{Y}_{\cdots} = 3.90 - 3.91 = -0.01$$

TABLE 19.15 **Cell means for data of Table 19.5**

Plant	Toxsub			Row Mean
	A	B	C	
1	4.90	3.46	2.75	3.70
2	5.23	3.78	2.70	3.90
3	4.13	5.13	3.14	4.13
Column Mean	4.75	4.12	2.86	3.91

TABLE 19.16 **Interaction effects ($\hat{\gamma}_{ij}$ values) for data of Table 19.5**

Plant	Toxsub			Row Mean
	A	B	C	
1	0.35	−0.45	0.10	0.00
2	0.49	−0.34	−0.15	0.00
3	−0.84	0.78	0.06	0.00
Column Mean	0.00	0.00	0.00	0.00

The set of interaction effects of the form $\hat{\gamma}_{ij} = \bar{Y}_{ij\cdot} - \bar{Y}_{i\cdot\cdot} - \bar{Y}_{\cdot j\cdot} + \bar{Y}_{\cdots}$ is given in Table 19.16. These patterns demonstrate that some plants are associated with better respiratory health than others for one kind of toxic exposure but are worse for other kinds of such exposure. Therefore, we cannot conclude that one plant was better overall than another. Rather, the differences in respiratory health among plants depend on which toxic substance is being considered.

19.7 Random- and Mixed-effects Two-way ANOVA Models

In this section we examine the classical two-way ANOVA statistical models appropriate when both factors are random or when one factor is fixed and the other is random. We will specify the appropriate null hypotheses of interest and the expected mean squares associated with each model. The expected mean square for a particular source is the true average (population) value of the mean-square term in the ANOVA table.

19.7.1 Random-effects Model[7]

When both factors are random, the two-way ANOVA model is given as

$$Y_{ijk} = \mu + A_i + B_j + C_{ij} + E_{ijk} \qquad (19.7)$$

---

[7] As previously footnoted in Chapter 17 for one-way ANOVA models, whenever random effects occur in a two-way ANOVA model, we can no longer assume that the Y_{ijk} are mutually independent.

where A_i, B_j, C_{ij}, and E_{ijk} are mutually independent random variables satisfying

$$A_i \frown N\!\left(0, \sigma_R^2\right)$$

$$B_j \frown N\!\left(0, \sigma_C^2\right)$$

$$C_{ij} \frown N\!\left(0, \sigma_{RC}^2\right)$$

$$E_{ijk} \frown N\!\left(0, \sigma^2\right) \qquad i = 1, 2, \ldots, r; \quad j = 1, 2, \ldots, c; \quad k = 1, 2, \ldots, n$$

19.7.2   Mixed-effects Model with Fixed Row Factor and Random Column Factor

One particular model[8] is

$$Y_{ijk} = \mu + \alpha_i + B_j + C_{ij} + E_{ijk} \tag{19.8}$$

where each α_i is a constant such that $\sum_{i=1}^{r} \alpha_i = 0$, and where B_j, C_{ij}, and E_{ijk} are mutually independent random variables satisfying

$$B_j \frown N\!\left(0, \sigma_C^2\right)$$

$$C_{ij} \frown N\!\left(0, \sigma_{RC}^2\right) \qquad i = 1, 2, \ldots, r; \quad j = 1, 2, \ldots, c; \quad k = 1, 2, \ldots, n$$

$$E_{ijk} \frown N\!\left(0, \sigma^2\right)$$

19.7.3   Mixed-effects Model with Random Row Factor and Fixed Column Factor

One particular model is

$$Y_{ijk} = \mu + A_i + \beta_j + C_{ij} + E_{ijk} \tag{19.9}$$

---

[8] There are several ways to define the assumptions for a two-way mixed model. In particular, one must choose between two alternative forms of summation restrictions for the random components of a two-way mixed model (see Searle et al. 1992). The assumptions we have made (in the body of the text) for mixed models (19.8) and (19.9) include summation restrictions for the fixed effects—e.g., $\sum_{i=1}^{r} \alpha_i = 0$ for model (19.8)—but no such summation restrictions for random effects, particularly the (random) interaction effects C_{ij}. Alternatively, we can choose to assume (for either model 19.8 or model 19.9) that the sum of the random interaction factors C_{ij} over the levels of the fixed factor are zero; i.e., $\sum_{i=1}^{r} C_{ij} = 0$ for each j in model (19.8). The latter summation restriction is what we assumed in previous editions of this text. Moreover, this restriction implied that C_{ij} and $C_{i'j}$, $i \neq i'$, are correlated, which leads to a different formula for the F statistic for the test of hypothesis about the random main effects in the mixed model; specifically, the denominator mean square is MSE instead of MSRC. The question of which form of mixed model to use—one without the restrictions on the C_{ij} or one with them—remains open (Searle et al. 1992). Nevertheless, we have chosen to assume (in the main body of the text) no restrictions on the C_{ij}, because the F statistics based on no restrictions correspond to those calculated by standard computer programs for ANOVA models with mixed effects, (including SAS's GLM, whose output we illustrate in the text). Also, assumptions not involving summation restrictions correspond to assumptions statisticians typically make when considering the analysis of repeated measures data, which we discuss in Chapters 25 and 26.

where each β_j is a constant such that $\sum_{j=1}^c \beta_j = 0$, and where A_j, C_{ij}, and E_{ijk} are mutually independent random variables satisfying

$$A_i \frown N(0, \sigma_R^2)$$

$$C_{ij} \frown N(0, \sigma_{RC}^2) \qquad i = 1, 2, \ldots, r; \quad j = 1, 2, \ldots, c; \quad k = 1, 2, \ldots, n$$

$$E_{ijk} \frown N(0, \sigma^2)$$

19.7.4 Null Hypotheses and Expected Mean Squares for Two-way ANOVA Models

For fixed-, random-, and mixed-effects models, Table 19.17 gives the specific null hypotheses being tested, with regard to row main effects, column main effects, and interaction effects. Table 19.18 gives the expected mean square for each factor in each of the models. These two tables demonstrate why different F statistics are required for testing the various hypotheses of interest. In this regard, the primary consideration is the choice of the appropriate denominator mean squares to use in the various F statistics. The numerator mean square always corresponds to the factor being considered; for example, if the factor is "rows," the numerator mean square is MSR, regardless of the type of model. Similarly, if the factor is "columns" or "interaction," the numerator mean square is MSC or MSRC, respectively. The denominator mean square, however, is chosen to correspond to the expected mean square to which the numerator expected mean square reduces under the null hypothesis of interest. For example, in a test for significant row effects in a random-effects model, the numerator expected mean square, $(\sigma^2 + n\sigma_{RC}^2 + cn\sigma_R^2)$ from Table 19.18, reduces to $(\sigma^2 + n\sigma_{RC}^2)$ under $H_0: \sigma_R^2 = 0$. This requires that the denominator mean square be MSRC, since the expected mean square of MSRC under the random-effects model is exactly $(\sigma^2 + n\sigma_{RC}^2)$.

Thus, the ratio of expected mean squares

$$\frac{\text{EMS(R)}}{\text{EMS(RC)}} = \frac{\sigma^2 + n\sigma_{RC}^2 + cn\sigma_R^2}{\sigma^2 + n\sigma_{RC}^2}$$

reduces to $(\sigma^2 + n\sigma_{RC})/(\sigma^2 + n\sigma_{RC}^2) = 1$ under $H_0: \sigma_R^2 = 0$, so the F statistic MSR/MSRC is the ratio of two estimators of the same variance under H_0.

TABLE 19.17   Null hypotheses for two-way ANOVA

Source	Fixed Effects	Random Effects	Mixed Effects	
			Rows Fixed, Columns Random	Rows Random, Columns Fixed
Rows	$\alpha_1 = \alpha_2 = \cdots = \alpha_r = 0$	$\sigma_R^2 = 0$	$\alpha_1 = \alpha_2 = \cdots = \alpha_r = 0$	$\sigma_R^2 = 0$
Columns	$\beta_1 = \beta_2 = \cdots = \beta_c = 0$	$\sigma_C^2 = 0$	$\sigma_C^2 = 0$	$\beta_1 = \beta_2 = \cdots = \beta_c = 0$
Interactions	$\gamma_{ij} = 0$ for all i, j	$\sigma_{RC}^2 = 0$	$\sigma_{RC}^2 = 0$	$\sigma_{RC}^2 = 0$

TABLE 19.18 **Expected mean squares for two-way ANOVA (r rows, c columns, n observations per cell)**

	Model Type			
			Mixed Effects	
Source	Fixed Effects	Random Effects	Rows Fixed, Columns Random	Rows Random, Columns Fixed
Rows	$\sigma^2 + cn\sum\limits_{i=1}^{r}\dfrac{\alpha_i^2}{r-1}$	$\sigma^2 + n\sigma_{RC}^2 + cn\sigma_R^2$	$\sigma^2 + n\sigma_{RC}^2 + cn\sum\limits_{i=1}^{r}\dfrac{\alpha_i^2}{r-1}$	$\sigma^2 + n\sigma_{RC}^2 + cn\sigma_R^2$
Columns	$\sigma^2 + rn\sum\limits_{j=1}^{c}\dfrac{\beta_j^2}{c-1}$	$\sigma^2 + n\sigma_{RC}^2 + rn\sigma_C^2$	$\sigma^2 + n\sigma_{RC}^2 + rn\sigma_C^2$	$\sigma^2 + n\sigma_{RC}^2 + rn\sum\limits_{j=1}^{c}\dfrac{\beta_j^2}{c-1}$
Interactions	$\sigma^2 + n\sum\limits_{i=1}^{r}\sum\limits_{j=1}^{c}\dfrac{\gamma_{ij}^2}{(r-1)(c-1)}$	$\sigma^2 + n\sigma_{RC}^2$	$\sigma^2 + n\sigma_{RC}^2$	$\sigma^2 + n\sigma_{RC}^2$
Error	σ^2	σ^2	σ^2	σ^2

As another example, let us consider the F test for significant row effects based on the mixed-effects model with the row factor fixed and column factor random. The test statistic in this case, $F = MSR/MSRC$, involves the following ratio of expected mean squares (see Table 19.18):

$$\frac{\text{EMS(R)}}{\text{EMS(RC)}} = \frac{\sigma^2 + n\sigma_{RC}^2 + cn\sum\limits_{i=1}^{r}\alpha_i^2/(r-1)}{\sigma^2 + n\sigma_{RC}^2}$$

Under $H_0: \alpha_1 = \alpha_2 = \cdots = \alpha_r = 0$, this ratio simplifies to $(\sigma^2 + n\sigma_{RC}^2)/(\sigma^2 + n\sigma_{RC}^2) = 1$. Thus, the F statistic is the ratio of two estimators of the same variance under H_0.

As a final example, we consider the F test for significant row effects based on the mixed-effects model with the row factor random and the column factor fixed. The test statistic is $F = MSR/MSRC$, which involves

$$\frac{\text{EMS(R)}}{\text{EMS(RC)}} = \frac{\sigma^2 + n\sigma_{RC}^2 + cn\sigma_R^2}{\sigma^2 + n\sigma_{RC}^2}$$

Under $H_0: \sigma_R^2 = 0$, this ratio simplifies to

$$\frac{\sigma^2 + n\sigma_{RC}^2}{\sigma^2 + n\sigma_{RC}^2} = 1$$

as desired.

Problems

1. The data in the following table and accompanying computer output come from an animal experiment designed to investigate whether levorphanol reduces stress as reflected in the cortical sterone level. The four treatment groups contained five animals each.

Control	Levorphanol Only	Epinephrine Only	Levorphanol and Epinephrine
1.90	0.82	5.33	3.08
1.80	3.36	4.84	1.42
1.54	1.64	5.26	4.54
4.10	1.74	4.92	1.25
1.89	1.21	6.07	2.57

1. **a.** These data may be analyzed by means of two-way ANOVA. What are the two factors?
   b. Classify each factor as either fixed or random.
   c. Rearrange the data into a two-way table appropriate for two-way ANOVA.
   d. Form the table of sample means, and comment on the patterns observed.
   e. Complete the ANOVA table shown below.
   f. Analyze the data to determine whether significant main effects exist due to levorphanol and epinephrine, and whether a significant interaction effect exists between epinephrine and levorphanol.

Edited SAS Output (PROC GLM) for Problem 1

Dependent Variable: LEVEL

Source	DF	Sum of Squares	Mean Square	F Value	Pr > F
Model	3	37.57844000	12.52614667	_____	0.0002
Error	16	_____	_____		
Corrected Total	19	_____			

R-Square	Coeff Var	Root MSE	LEVEL Mean
0.697495	34.05076	1.009265	2.964000

Source	DF	Type I SS	Mean Square	F Value	Pr > F
L	1	12.83202000	12.83202000	12.60	0.0027
E	1	18.58592000	18.58592000	18.25	0.0006
L*E	1	6.16050000	6.16050000	6.05	0.0257

Source	DF	Type III SS	Mean Square	F Value	Pr > F
L	1	12.83202000	12.83202000	12.60	0.0027
E	1	18.58592000	18.58592000	18.25	0.0006
L*E	1	6.16050000	6.16050000	6.05	0.0257

2. The following table gives the performance competency scores for a random sample of family nurse practitioners (FNPs) with different specialties, from hospitals in three cities.

Specialty	City 1	City 2	City 3
Pediatrics	91.7, 74.9, 88.2, 79.5	86.3, 88.1, 92.0, 69.5	82.3, 78.7, 89.8, 84.5
Obstetrics and gynecology	80.1, 76.2, 70.3, 89.5	71.3, 73.4, 76.9, 87.2	90.1, 65.6, 74.6, 79.1
Diabetes and hypertension	71.5, 49.8, 55.1, 75.4	80.2, 76.1, 44.2, 50.5	48.7, 54.4, 60.1, 70.8

2. **a.** Classify each factor as either fixed or random, and justify your classification.
   b. Form the table of sample means (use the accompanying computer output to do so), and then comment on the patterns observed.
   c. Using the computer output, compute the appropriate F statistic for each of the four possible factor classification schemes (i.e., both factors fixed, both random, and one factor of each type).
   d. Analyze the data based on each possible factor classification scheme. How do the results compare?
   e. Using Scheffé's method, as described in Chapter 17, find a 95% confidence interval for the true difference in mean scores between pediatric FNPs and ob-gyn FNPs.
   f. Using the sample means obtained in part (b), and assuming each factor to be fixed, state a regression model appropriate for the two-way ANOVA table, and provide estimates of the regression coefficients associated with the main effects of each factor.

Edited SAS Output (PROC GLM) for Problem 2

```
                          Class Level Information
                 Class             Levels                    Values
                 SPEC                 3                      D&H OBGYN PED
                 CITY                 3                      1 2 3

Dependent Variable: SCORE
                                    Sum of           Mean
Source                 DF           Squares          Square       F Value      Pr > F
Model                   8         3288.950000      411.118750       3.82       0.0040
Error                  27         2902.180000      107.488148
Corrected Total        35         6191.130000
              R-Square          Coeff Var         Root MSE        SCORE Mean
              0.531236          13.94438          10.36765        74.35000

Source         DF            Type I SS         Mean Square      F Value      Pr > F
SPEC            2          3229.871667        1614.935833        15.02       <.0001
CITY            2            24.541667          12.270833         0.11       0.8925
SPEC*CITY       4            34.536667           8.634167         0.08       0.9877

Source         DF            Type III SS       Mean Square      F Value      Pr > F
SPEC            2          3229.871667        1614.935833        15.02       <.0001
CITY            2            24.541667          12.270833         0.11       0.8925
SPEC*CITY       4            34.536667           8.634137         0.08       0.9877

Source                Type III Expected Mean Square
SPEC                  Var(Error) + 4 Var(SPEC*CITY) + 12 Var(SPEC)
CITY                  Var(Error) + 4 Var(SPEC*CITY) + 12 Var(CITY)
SPEC*CITY             Var(Error) + 4 Var(SPEC*CITY)

                Test of Hypotheses for Random Model Analysis of Variance
Dependent Variable: SCORE
Source                DF        Type III SS       Mean Square     F Value     Pr > F
SPEC                   2       3229.871667       1614.935833      187.04      0.0001
CITY                   2         24.541667         12.270833        1.42      0.3417
Error: MS(SPEC*CITY)   4         34.536667          8.634167

Source                DF        Type III SS       Mean Square     F Value     Pr > F
SPEC*CITY              4         34.536667          8.634167        0.08      0.9877
Error: MS(Error)      27       2902.180000        107.488148
```

(continued)

Edited SAS Output (PROC GLM) for Problem 2 (continued)

```
            Proc GLM Output for Chapter 19, Problem 2 (Continued)
     Level of        Level of                       ------------SCORE------------
      SPEC             CITY         N          Mean                Std Dev
      D&H               1           4       62.9500000          12.4184003
      D&H               2           4       62.7500000          18.0452210
      D&H               3           4       58.5000000           9.4286797
      OBGYN             1           4       79.0250000           8.0619993
      OBGYN             2           4       77.2000000           7.0554943
      OBGYN             3           4       77.3500000          10.1857744
      PED               1           4       83.5750000           7.7301897
      PED               2           4       83.9750000           9.9389386
      PED               3           4       83.8250000           4.6456969
```

3. The following table gives the average patient waiting time in minutes for patients from a random sample of 16 physicians, classified by type of practice and type of physician.

Physician Type	Type of Practice	
	Group	Solo
General practitioner	15, 20, 25, 20	20, 25, 30, 25
Specialist	30, 25, 30, 35	25, 20, 30, 30

a. Classify each factor as either fixed or random, and justify your classification scheme.
b. Using the accompanying computer output, compute the F statistic corresponding to each of the four possible factor classification schemes.
c. Discuss the analysis of the data when both factors are considered fixed.
d. What are the estimates of the (fixed) effect due to "general practitioner," the (fixed) effect due to "group practice," and the interaction effect ($\mu_{11} - \mu_{12} - \mu_{21} + \mu_{22}$), where μ_{ij} denotes the cell mean in the ith row and jth column of the table of cell means?
e. Interpret the interaction effect observed.
f. What is an appropriate regression model for this two-way ANOVA?
g. How might you modify the model in part (f) to reflect the conclusions made in part (c)?

Edited SAS Output (PROC GLM) for Problem 3

```
                          Class Level Information
            Class                   Levels                Values
            PHYS                       2                  GEN SPEC
            PRACTICE                   2                  GROUP SOLO

Dependent Variable: TIME
                              Sum of              Mean
Source              DF        Squares            Square      F Value      Pr > F
Model                3      204.6875000        68.2291667      3.74       0.0415
Error               12      218.7500000        18.2291667
Corrected Total     15      423.4375000
            R-Square          Coeff Var         Root MSE      TIME Mean
            0.483395          16.86741          4.269563      25.31250
```

(continued)

Edited SAS Output for Problem 3 (continued)

```
                        Output for Problem 3 (Continued)

Source              DF        Type I SS      Mean Square    F Value      Pr > F
PHYS                1        126.5625000    126.5625000      6.94        0.0218
PRACTICE            1          1.5625000      1.5625000      0.09        0.7747
PHYS*PRACTICE       1         76.5625000     76.5625000      4.20        0.0629

Source              DF       Type III SS     Mean Square    F Value      Pr > F
PHYS                1        126.5625000    126.5625000      6.94        0.0218
PRACTICE            1          1.5625000      1.5625000      0.09        0.7747
PHYS*PRACTICE       1         76.5625000     76.5625000      4.20        0.0629

Source                 Type III Expected Mean Square
PHYS                   Var(Error) + 4 Var(PHYS*PRACTICE) + 8 Var(PHYS)
PRACTICE               Var(Error) + 4 Var(PHYS*PRACTICE) + 8 Var(PRACTICE)
PHYS*PRACTICE          Var(Error) + 4 Var(PHYS*PRACTICE)

              Tests of Hypotheses for Random Model Analysis of Variance
Dependent Variable: TIME
Source              DF       Type III SS     Mean Square    F Value      Pr > F
PHYS                1        126.562500     126.562500       1.65        0.4208
PRACTICE            1          1.562500       1.562500       0.02        0.9097
Error               1         76.562500      76.562500
Error: MS(PHYS*PRACTICE)

Source              DF       Type III SS     Mean Square    F Value      Pr > F
PHYS*PRACTICE       1         76.562500      76.562500       4.20        0.0629
Error: MS(Error)    12       218.750000      18.229167

        Level of       Level of                   ------------TIME------------
         PHYS          PRACTICE         N             Mean           Std Dev
         GEN           GROUP            4         20.0000000       4.08248290
         GEN           SOLO             4         25.0000000       4.08248290
         SPEC          GROUP            4         30.0000000       4.08248290
         SPEC          SOLO             4         26.2500000       4.78713554
```

4. A study was undertaken to measure and compare sexist attitudes of students at various types of colleges. Random samples of 10 undergraduate seniors of each sex were selected from each of three types of colleges. A questionnaire was then administered to each student, from which a score for "degree of sexism"—defined as the extent to which a student considered males and females to have different life roles—was determined (the higher the score, the more sexist the attitude). The resulting data are given in the following table.

College Type	Male	Female
Coed with 75% or more males	50, 35, 37, 32, 46, 38, 36, 40, 38, 41	38, 27, 34, 30, 22, 32, 26, 24, 31, 33
Coed with less than 75% males	30, 29, 31, 27, 22, 20, 31, 22, 25, 30	28, 31, 28, 26, 20, 24, 31, 24, 31, 26
Not coed	45, 40, 32, 31, 26, 28, 39, 27, 37, 35	40, 35, 32, 29, 24, 26, 36, 25, 35, 35

a. Form the table of cell means, and interpret the results obtained (see the accompanying computer printout).

4. b. Using the computer output, calculate the *F* statistics corresponding to a model with both factors fixed.
  c. Discuss the analysis of the data for this fixed-effects model case.

Edited SAS Output (PROC GLM) for Problem 4

.
.   [Portion of output omitted]
.

Dependent Variable: SCORE

Source	DF	Sum of Squares	Mean Square	F Value	Pr > F
Model	5	1144.883333	228.976667	9.06	<.0001
Error	54	1365.300000	25.283333		
Corrected Total	59	2510.183333			

R-Square	Coeff Var	Root MSE	SCORE Mean
0.456096	16.02205	5.028254	31.38333

Source	DF	Type I SS	Mean Square	F Value	Pr > F
COLLEGE	2	657.4333333	328.7166667	13.00	<.0001
SEX	1	228.1500000	228.1500000	9.02	0.0040
COLLEGE*SEX	2	259.3000000	129.6500000	5.13	0.0091

.
.   [Portion of output omitted]
.

Level of COLLEGE	Level of SEX	N	------------SCORE------------	
			Mean	Std Dev
<75%	FEMALE	10	26.9000000	3.63470922
<75%	MALE	10	26.7000000	4.16466619
>75%	FEMALE	10	29.7000000	4.92273637
>75%	MALE	10	39.3000000	5.31350481
NOT	FEMALE	10	31.7000000	5.41705127
NOT	MALE	10	34.0000000	6.27162924

5. Random samples of 100 persons awaiting trial on felony charges were selected from rural, urban, and suburban court locations in each of two states, one (state 1) in the Northeast and the other (state 2) in the South. The following table summarizes the data on the time $\bar{Y}$ (in months) between arrest and beginning of trial for these random samples.

	Court Location		
State	Rural	Suburban	Urban
1	$\bar{Y} = 3.4, S = 1.3$	$\bar{Y} = 5.8, S = 1.2$	$\bar{Y} = 6.8, S = 1.5$
2	$\bar{Y} = 2.4, S = 1.5$	$\bar{Y} = 3.5, S = 1.7$	$\bar{Y} = 4.7, S = 1.7$

a. Do the sample means in the table suggest that the average waiting times for state 1 vary by court location differently from how they vary for state 2? Is there an interaction effect?

b. Analyze these data. Use the following ANOVA table. Assume that both factors are fixed.

Source	d.f.	SS	MS
States	1	486.00	486.00
Court locations	2	826.33	413.17
Interaction	2	49.00	24.50
Error	594	1,327.591	2.235

5. c. Define an appropriate regression model for this two-way ANOVA.

   d. How might one revise the model in part (c) and the associated ANOVA table in order to investigate whether a linear trend exists between waiting time and degree of urbanization (as determined by treating the categories rural, suburban, and urban on an ordinal scale)? What difficulty does one encounter when considering such a model?

6. An experiment was conducted at a large state university to determine whether two different instructional methods for teaching a beginning statistics course would yield different levels of cognitive achievement. One instructional method involved using a self-instructional format, including a sequence of slide-tape presentations; the other method utilized the standard lecture format. The 100 students who registered for the course were randomly assigned to one of four sections, 25 per section, corresponding to the combinations of one of the two methods with one of two instructors. The results obtained from identical final exams given to each section are summarized in the following table.

Instructor	**Method**	
	Lecture	Self-instruction
A	$\bar{Y} = 71.2, S = 13.8$	$\bar{Y} = 80.2, S = 12.1$
B	$\bar{Y} = 73.8, S = 11.7$	$\bar{Y} = 77.5, S = 14.1$

   a. What do the results suggest about the comparative effects of the two instructional methods?

   b. Classify each factor as either fixed or random, and explain your classification.

Source	d.f.	SS	MS
INSTRUC	1	6.2500E − 02	6.2500E − 02
METHOD	1	6.0081E + 03	1.0081E + 03
INSTRUC × METHOD	1	1.7556E + 02	1.7556E + 02
Error	96	1.6141E + 04	1.6814E + 02

   c. Using the ANOVA table, perform the appropriate F tests for each of the four types of factor classification schemes. Compare the conclusions reached under each scheme.

   d. What factors should be controlled for in this experiment?

   e. Given a continuous variable C to be controlled for, write an appropriate regression model for this data set that takes C into account. What general method of analysis is characterized by such a model?

7. The following table and accompanying computer output present data on the uric acid level found in the bloodstreams of persons with Down's syndrome, and in the bloodstreams of non–Down's syndrome subjects. All subjects were between the ages of 21 and 25. Analyze these data, using the ANOVA table to determine whether evidence of a higher uric acid level exists in the Down's syndrome group, making sure to characterize any sex relationships that exist.

Group	Sex	
	Male	Female
Down's syndrome	5.84, 6.30, 6.95, 5.92, 7.94	4.90, 6.95, 6.73, 5.32, 4.81
Others	5.50, 6.08, 5.12, 7.58, 6.78	4.94, 7.20, 5.22, 4.60, 3.88

Edited SAS Output (PROC GLM) for Problem 7

```
                                      ·
                                      · [Portion of output omitted]
                                      ·
Dependent Variable: ACID
                                 Sum of
Source                    DF     Squares     Mean Square    F Value    Pr > F
Model                      3    5.65548000    1.88516000      1.74     0.1987
Error                     16   17.31484000    1.08217750
Corrected Total           19   22.97032000
          R-Square        Coeff Var         Root MSE       ACID Mean
          0.246208        17.54854          1.040278       5.928000

Source            DF    Type I SS     Mean Square    F Value    Pr > F
GROUP              1    1.13288000     1.13288000      1.05     0.3215
SEX                1    4.47458000     4.47458000      4.13     0.0589
GROUP*SEX          1    0.04802000     0.04802000      0.04     0.8358

                                      ·
                                      · [Portion of output omitted]
                                      ·
        Level of          Level of                -----------ACID------------
        GROUP             SEX          N           Mean            Std Dev
        MONGOLOID         FEMALE       5        5.74200000        1.02360637
        MONGOLOID         MALE         5        6.59000000        0.87286883
        OTHERS            FEMALE       5        5.16800000        1.24149909
        OTHERS            MALE         5        6.21200000        0.98879725
```

8. An experiment was conducted to investigate the survival of diplococcus pneumonia bacteria in chick embryos under relative humidities (RH) of 0%, 25%, 50%, and 100% and under temperatures (Temp) of 10°C, 20°C, 30°C, and 40°C, using 10 chicks for each RH–Temp combination.[9] The partially completed ANOVA table is as given next.

Source	d.f.	MS
RH		2.010
Temp		7.816
Interaction		1.642
Error		0.775
Total		

---

[9] Adapted from a study by Price (1954).

8. a. Should the two factors RH and Temp be considered as fixed or random? Explain.

  b. Carry out the analysis of variance for both the fixed-effects case and the random-effects case. Do your conclusions differ in the two cases?

  c. Write the fixed-effects and the random-effects models that could describe this experiment.

  d. Using dummy variables, provide a regression model that can be used to obtain the results in the ANOVA table.

  e. What regression model would be appropriate for describing the relationship of RH and Temp to survival time (Y) if the data for the independent variables are to be treated as interval rather than as nominal?

9. The diameters (Y) of three species of pine trees were compared at each of four locations, using samples of five trees per species at each location. The resulting data are given in the following table.

	Location			
Species	1	2	3	4
	23	25	21	14
	15	20	17	17
A	26	21	16	19
	13	16	24	20
	21	18	27	24
	28	30	19	17
	22	26	24	21
B	25	26	19	18
	19	20	25	26
	26	28	29	23
	18	15	23	18
	10	21	25	12
C	12	22	19	23
	22	14	13	22
	13	12	22	19

  a. Comment on whether each of the two factors should be considered fixed or random.

  b. Use the following partially completed ANOVA table to carry out your analysis, first considering both factors as fixed and then considering a mixed model with "locations" treated as random. Compare your conclusions.

Source	d.f.	SS
Species		344.9333
Locations		46.0500
Interaction		113.6000
Error		875.6000

10. Consider an ANOVA table of the following form.

Source	d.f.	SS	MS	F	P
A					P_1
B					P_2
A × B					P_3
Error					

Use $\alpha = .05$ for all parts of this problem. In each case, decide what effects (if any) are significant, and what conclusions to draw, based on a two-way ANOVA table with the following P-values:

a. $P_1 = .03, P_2 = .51, P_3 = .31$
b. $P_1 = .001, P_2 = .63, P_3 = .007$
c. $P_1 = .093, P_2 = .79, P_3 = .02$
d. $P_1 = .56, P_2 = .38, P_3 = .24$

11. Assume that a total of 75 subjects were tested in a balanced two-way fixed-effects factorial experiment. A plot of the means from the study is shown next. The dependent variable is Y, and the factors are A and B.

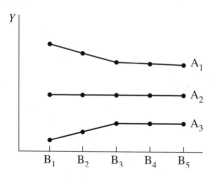

a. Give the left two columns (the source and degrees-of-freedom columns) of the source table for the data plotted.
b. Assume that the plot shows the population means. Indicate which significance tests should hopefully yield significant results in the ANOVA table.

12. Assume that the following ANOVA table came from a balanced two-way fixed-effects ANOVA. Show the formula used (with numbers filled in) and the numerical value of each letter in the table.

Source	d.f.	SS	MS	F
A	a	5.12	g	6.40*
B	b	e	0.76	3.80*
A × B	c	4.32	0.36	i
Error	20	4.00	h	
Total	d	f		

Also, indicate which (if any) family or families of means should be evaluated with multiple comparisons.

13. The data in the following table and accompanying computer output are from a hypothetical study of human body temperature as affected by air temperature and a dietary supplement that is hoped to increase heat tolerance. Body temperatures (in degrees Celsius) were measured for 36 athletes immediately following a standard exercise routine in a room controlled to a fixed air temperature (in degrees Celsius). Each subject had been receiving a steady, fixed dose (in milligrams per kilogram of body weight) of the dietary supplement.

   a. Provide an appropriate two-way ANOVA source table.
   b. Provide F tests of the two main effects and the interaction. Use $\alpha = .05$. What do you conclude?
   c. Define dummy variables, and specify an appropriate multiple regression model corresponding to the analysis done in part (a). [*Hint:* Use dummy variables that have the value -1 for Airtemp $= 21$ and for Dose $= 0$. Why is this a scientifically sensible choice?]
   d. Since both factors are interval-scale variables, specify a corresponding natural-polynomial multiple regression model.

Edited SAS Output (PROC GLM) for Problem 13

```
                                        •  [Portion of output omitted]
                                        •

Dependent Variable: BODYTEMP
                                    Sum of              Mean
Source                  DF          Squares            Square      F Value      Pr > F
Model                   11          0.13888889         0.01262626      0.43      0.9255
Error                   24          0.70000000         0.02916667
Corrected Total         35          0.83888889
            R-Square            Coeff Var          Root MSE       BODYTEMP Mean
            0.165563            0.461644           0.170783           36.99444

Source           DF          Type I SS          Mean Square      F Value      Pr > F
AIRTEMP          3           0.02777778         0.00925926          0.32      0.8126
DOSE             2           0.06055556         0.03027778          1.04      0.3695
AIRTEMP*DOSE     6           0.05055556         0.00842593          0.29      0.9364

Source           DF          Type III SS        Mean Square      F Value      Pr > F
AIRTEMP          3           0.02777778         0.00925926          0.32      0.8126
DOSE             2           0.06055556         0.03027778          1.04      0.3695
AIRTEMP*DOSE     6           0.05055556         0.00842593          0.29      0.9364
```

Observation	Airtemp	Dose	Bodytemp
1	21	0.00	37.2
2	21	0.00	37.2
3	21	0.00	36.8
4	21	0.05	37.1
5	21	0.05	36.9
6	21	0.05	36.8
7	21	0.10	37.1
8	21	0.10	37.1
9	21	0.10	37.1
10	25	0.00	36.9
11	25	0.00	37.0
12	25	0.00	37.1
13	25	0.05	37.1
14	25	0.05	36.7
15	25	0.05	37.0
16	25	0.10	36.9
17	25	0.10	37.0
18	25	0.10	37.3
19	29	0.00	36.9
20	29	0.00	37.0
21	29	0.00	36.8
22	29	0.05	36.9
23	29	0.05	37.0
24	29	0.05	36.9
25	29	0.10	36.9
26	29	0.10	37.0
27	29	0.10	37.2
28	33	0.00	37.1
29	33	0.00	37.3
30	33	0.00	36.7
31	33	0.05	36.9
32	33	0.05	37.0
33	33	0.05	37.0
34	33	0.10	36.9
35	33	0.10	36.8
36	33	0.10	37.2

14. The experiment in Problem 14 in Chapter 18 was repeated—this time using three workers per technology/experience combination rather than just one. For each cell in the following table, the outputs (in number of units per hour) are listed for three randomly chosen workers at the given experience level.

	Technology		
Operator Experience	High Automation (H)	Moderate Automation (M)	Low Automation (L)
<1 Year	16, 14, 12	8, 12, 10	5, 4, 8
1–2 Years	15, 18, 17	10, 10, 12	8, 10, 10
>2 Years	20, 18, 19	13, 13, 14	12, 11, 13

$$\bar{Y}_{i\cdot\cdot} = \frac{1}{n_{i\cdot}} \sum_{j=1}^{c} \sum_{k=1}^{n_{ij}} Y_{ijk} \qquad \text{where} \qquad n_{i\cdot} = \sum_{j=1}^{c} n_{ij}$$

$$\bar{Y}_{\cdot j\cdot} = \frac{1}{n_{\cdot j}} \sum_{i=1}^{r} \sum_{k=1}^{n_{ij}} Y_{ijk} \qquad \text{where} \qquad n_{\cdot j} = \sum_{i=1}^{r} n_{ij}$$

$$\bar{Y}_{\cdot\cdot\cdot} = \frac{1}{n_{\cdot\cdot}} \sum_{i=1}^{r} \sum_{j=1}^{c} \sum_{k=1}^{n_{ij}} Y_{ijk} \qquad \text{where} \qquad n_{\cdot\cdot} = \sum_{i=1}^{r} \sum_{j=1}^{c} n_{ij}$$

TABLE 20.1 Data layout for the unequal-cell-number case (two-way ANOVA)

Row Factor	Column Factor 1	2	...	c	Row Marginals
1	$Y_{111}, Y_{112}, \ldots, Y_{11n_{11}}$ (Sample size $= n_{11}$) (Cell mean $= \bar{Y}_{11\cdot}$)	$Y_{121}, Y_{122}, \ldots, Y_{12n_{12}})$ (Sample size $= n_{12}$) (Cell mean $= \bar{Y}_{12\cdot}$)	...	$Y_{1c1}, Y_{1c2}, \ldots, Y_{1cn_{1c}}$ (Sample size $= n_{1c}$) (Cell mean $= \bar{Y}_{1c\cdot}$)	$n_{1\cdot}, \bar{Y}_{1\cdot\cdot}$
2	$Y_{211}, Y_{212}, \ldots, Y_{21n_{21}}$ (Sample size $= n_{21}$) (Cell mean $= \bar{Y}_{21\cdot}$)	$Y_{221}, Y_{222}, \ldots, Y_{22n_{22}}$ (Sample size $= n_{22}$) (Cell mean $= \bar{Y}_{22\cdot}$)	...	$Y_{2c1}, Y_{2c2}, \ldots, Y_{2cn_{2c}}$ (Sample size $= n_{2c}$) (Cell mean $= \bar{Y}_{2c\cdot}$)	$n_{2\cdot}, \bar{Y}_{2\cdot\cdot}$
$\vdots$	$\vdots$	$\vdots$		$\vdots$	$\vdots$
r	$Y_{r11}, Y_{r12}, \ldots, Y_{r1n_{r1}}$ (Sample size $= n_{r1}$) (Cell mean $= \bar{Y}_{r1\cdot}$)	$Y_{r21}, Y_{r22}, \ldots, Y_{r2n_{r2}}$ (Sample size $= n_{r2}$) (Cell mean $= \bar{Y}_{r2\cdot}$)	...	$Y_{rc1}, Y_{rc2}, \ldots, Y_{rcn_{rc}}$ (Sample size $= n_{rc}$) (Cell mean $= \bar{Y}_{rc\cdot}$)	$n_{r\cdot}, \bar{Y}_{r\cdot\cdot}$
Column Marginals	$n_{\cdot 1}, \bar{Y}_{\cdot 1\cdot}$	$n_{\cdot 2}, \bar{Y}_{\cdot 2\cdot}$	...	$n_{\cdot c}, \bar{Y}_{\cdot c\cdot}$	$n_{\cdot\cdot}, \bar{Y}_{\cdot\cdot\cdot}$

TABLE 20.2 Satisfaction with medical care (Y), classified by patient worry and affective communication between patient and physician

Affective Communication	Worry Negative	Positive	Row Marginals
High	2, 5, 8, 6, 2, 4, 3, 10 ($n_{11} = 8$) ($\bar{Y}_{11\cdot} = 5$)	7, 5, 8, 6, 3, 5, 6, 4, 5, 6, 8, 9 ($n_{12} = 12$) ($\bar{Y}_{12\cdot} = 6$)	$n_{1\cdot} = 20$ $\bar{Y}_{1\cdot\cdot} = 5.60$
Medium	4, 6, 3, 3 ($n_{21} = 4$) ($\bar{Y}_{21\cdot} = 4$)	7, 7, 8, 6, 4, 9, 8, 7 ($n_{22} = 8$) ($\bar{Y}_{22\cdot} = 7$)	$n_{2\cdot} = 12$ $\bar{Y}_{2\cdot\cdot} = 6.00$
Low	8, 7, 5, 9, 9, 10, 8, 6, 8, 10 ($n_{31} = 10$) ($\bar{Y}_{31\cdot} = 8$)	5, 8, 6, 6, 9, 7, 7, 8 ($n_{32} = 8$) ($\bar{Y}_{32\cdot} = 7$)	$n_{3\cdot} = 18$ $\bar{Y}_{3\cdot\cdot} = 7.56$
Column Marginals	$n_{\cdot 1} = 22$ $\bar{Y}_{\cdot 1\cdot} = 6.18$	$n_{\cdot 2} = 28$ $\bar{Y}_{\cdot 2\cdot} = 6.57$	$n_{\cdot\cdot} = 50$ $\bar{Y}_{\cdot\cdot\cdot} = 6.40$

The unequal-cell-number case arises quite frequently in observational studies. In such studies, one or more of the following statements are typically true:

1. Many of the variables of interest are not categorized before the data are collected.

2. New variables are often considered after the data are collected.

3. When all the variables are separately categorized, it is often impractical or even impossible to control in advance how the various categories will combine to form combinations of interest.

The unequal-cell-number case can also arise in experimental studies when a posteriori consideration is given to variables other than those of primary interest, even if the design based on the primary variables calls for equal cell numbers. Furthermore, unequal cell numbers generally result whenever data points are missing, which may occur (for example) because of study dropouts or incomplete records.

The example presented in Table 20.2 is derived from a study by Thompson (1972) of the relationship of two factors—patient perception of pregnancy and physician–patient communication—to patient satisfaction with medical care. Two main variables of interest were the patient's WORRY and a measure of affective communication (AFFCOM). These variables were developed from scales based on questionnaires administered to patients and their physicians. Based on the distribution of scores, the WORRY variable was grouped into the categories "positive" and "negative"; and the AFFCOM variable was grouped into the categories "high," "medium," and "low." Table 20.2 presents artificial data of this type, showing scores for satisfaction with medical care (Y = TOTSAT), classified according to these six combinations of levels of the factors WORRY and AFFCOM.

As the table indicates, the categorization scheme used leads to a two-way table with unequal cell numbers. For WORRY, there are 22 negatives and 28 positives; for AFFCOM, there are 20 high, 12 medium, and 18 low scores. When the separate categories for the two variables are considered together, the resulting six categories have different cell sample sizes, ranging from 4 (for medium AFFCOM, negative WORRY) to 12 (for high AFFCOM, positive WORRY).

20.2 Problem with Unequal Cell Numbers: Nonorthogonality

The key statistical concept associated with the special analytical problems encountered in the unequal-cell-number case pertains to the *nonorthogonality* of the sums of squares usually used to describe the sources of variation in a two-way ANOVA table. To clarify what *orthogonality* means, we first state the general formulas for these sums of squares (given in Section 19.3 for the equal-cell-number case) in terms of unequal cell numbers:

$$\text{SSR} = \sum_{i=1}^{r} \sum_{j=1}^{c} \sum_{k=1}^{n_{ij}} (\bar{Y}_{i..} - \bar{Y}...)^2, \qquad \text{SSC} = \sum_{i=1}^{r} \sum_{j=1}^{c} \sum_{k=1}^{n_{ij}} (\bar{Y}_{.j.} - \bar{Y}...)^2,$$

$$\text{SSRC} = \sum_{i=1}^{r} \sum_{j=1}^{c} \sum_{k=1}^{n_{ij}} (\bar{Y}_{ij.} - \bar{Y}_{i..} - \bar{Y}_{.j.} + \bar{Y}...)^2, \tag{20.1}$$

$$\text{SSE} = \sum_{i=1}^{r} \sum_{j=1}^{c} \sum_{k=1}^{n_{ij}} (Y_{ijk} - \bar{Y}_{ij.})^2, \qquad \text{SSY} = \sum_{i=1}^{r} \sum_{j=1}^{c} \sum_{k=1}^{n_{ij}} (Y_{ijk} - \bar{Y}...)^2$$

Problems

1. Consider hypothetical data based on a study concerning the effects of rapid cultural change on blood pressure levels for native citizens of an island in Micronesia. Blood pressures were taken on a random sample of 30 males over age 40 from a certain province. These persons, who commuted to work in the nearby westernized capital city, were also given a sociological questionnaire from which their social rankings in both their traditional and modern (i.e., westernized) cultures were determined. The results are summarized in the following table.

Modern Rank (Factor A)	Traditional Rank (Factor B)		
	HI	MED	LO
HI	130, 140, 135	150, 145	175, 160, 170, 165, 155
MED	145, 140, 150	150, 160, 155	165, 155, 165, 170, 160
LO	180, 160, 145	155, 140, 135	125, 130, 110

   a. Discuss the table of sample means for this data set.
   b. Give an appropriate regression model for this data set, treating the two factors as nominal variables.
   c. Using the regression ANOVA tables that follow (where X pertains to factor A, and Z to factor B), carry out two different main-effect tests for each factor, and also test for interaction. Compare the results of the two main-effect tests for each factor.

Source	d.f.	MS
X_1	1	469.17985
$X_2 \mid X_1$	1	508.52217
$Z_1 \mid X_1, X_2$	1	187.97673
$Z_2 \mid X_1, X_2, Z_1$	1	7.54570
$X_1 Z_1 \mid X_1, X_2, Z_1, Z_2$	1	3,925.29395
$X_1 Z_2 \mid X_1, X_2, Z_1, Z_2, X_1 Z_1$	1	9.70621
$X_2 Z_1 \mid X_1, X_2, Z_1, Z_2, X_1 Z_1, X_1 Z_2$	1	633.17613
$X_2 Z_2 \mid X_1, X_2, Z_1, Z_2, X_1 Z_1, X_1 Z_2, X_2 Z_1$	1	2.67593
Residual	21	75.83333

Source	d.f.	MS
Z_1	1	278.59213
$Z_2 \mid Z_1$	1	22.71129
$X_1 \mid Z_1, Z_2$	1	391.77041
$X_2 \mid Z_1, Z_2, X_1$	1	480.15062
$X_1 Z_1 \mid Z_1, Z_2, X_1, X_2$	1	3,925.29395
$X_1 Z_2 \mid Z_1, Z_2, X_1, X_2, X_1 Z_1$	1	9.70621
$X_2 Z_1 \mid Z_1, Z_2, X_1, X_2, X_1 Z_1, X_1 Z_2$	1	633.17613
$X_2 Z_2 \mid Z_1, Z_2, X_1, X_2, X_1 Z_1, X_1 Z_2, X_2 Z_1$	1	2.67593
Residual	21	75.83333

1. d. How might you modify the regression model given in part (b) so that any trends in blood pressure levels could be quantified in terms of increasing social rankings for the two factors? (This requires assigning numerical values to the categories of each factor.) What difficulty do you encounter in defining such a model?

2. A study was conducted to assess the combined effects of patient attitude and patient–physician communication on patient satisfaction with medical care during pregnancy. A random sample of 110 pregnant women under the care of private physicians was followed from the first visit with the physician until delivery. On the basis of specially devised questionnaires, the following variables were measured for each patient: Y = Satisfaction score; X_1 = Attitude score; and X_2 = Communication score. Each score was developed as an interval variable, but some question remains as to whether the analysis should treat the attitude and/or communication scores as nominal variables.

a. What would be an appropriate regression model for describing the joint effect of X_1 and X_2 on Y if an interaction between communication and attitude is possible and if all variables are treated as interval variables?

b. What would be an appropriate regression model (using dummy variables) if the analyst wished to allow for an interaction effect but desired only to compare high values versus low values (i.e., to make group comparisons) for both the communication and attitude variables? What kind of ANOVA model would this regression model correspond to?

c. When would the model in part (a) be preferable to that in (b), and vice versa?

d. If both independent variables are treated nominally, as in part (b), would you expect the associated 2×2 table to have equal numbers in each of the four cells?

3. The data listed in the table on the following page are from an unpublished study by Harbin and others (1985). Subjects were young (ages 18 through 29) or old (ages 60 through 86) men. Each subject was exposed to 0 or 100 parts per million (PPM) of carbon monoxide (CO) for a period before and during testing. Median reaction times for 30 trials are reported for two different tasks: (1) simple reaction time (no choice), REACTIM1; (2) two-choice reaction time, REACTIM2. Pilot data analysis from an earlier study established that this dependent variable followed an appropriate Gaussian distribution. For this problem, consider only REACTIM1 as a dependent variable. Use $\alpha = .01$.

a. Observe the cross-tabulation of AGEGROUP and PPM_CO in the accompanying SAS Proc FREQ computer output. Why might this be called a nearly orthogonal two-way design? Any missing data were due to technical problems unrelated to the response variable or treatments.

b. Using dummy variables coded -1 (young, PPM_CO = 0) and 1, define dummy variables and a corresponding multiple regression model for a two-way ANOVA.

c. Use an appropriate computer program to fit the regression model in part (b). Report appropriate tests of the AGEGROUP $\times$ PPM_CO interaction and tests of their main effects in an appropriate summary source table.

d. SAS's Proc GLM procedure automatically codes dummy variables (and properly treats unequal sample sizes). The accompanying Proc GLM computer output shows the two-way ANOVA results for this problem. Complete the F statistics in the ANOVA table, and report the results of the same tests as in part (c). Compare the results to those in part (c), and explain any differences.

Observation	AGEGROUP	PPM_CO	REACTIM1	REACTIM2
1	Young	0	291.5	632.0
2	Young	0	471.0	607.5
3	Young	0	692.0	859.0
4	Young	0	376.0	484.0
5	Young	0	372.5	501.0
6	Young	0	307.0	381.0
7	Young	0	501.0	559.0
8	Young	0	466.0	632.0
9	Young	0	375.0	434.0
10	Young	0	425.0	454.0
11	Young	0	343.0	542.0
12	Young	0	348.0	471.0
13	Young	0	503.0	521.0
14	Young	0	382.5	519.0
15	Young	100	472.5	515.0
16	Young	100	354.0	521.0
17	Young	100	350.0	456.0
18	Young	100	486.0	522.0
19	Young	100	402.0	472.0
20	Young	100	347.0	414.0
21	Young	100	320.0	363.0
22	Young	100	446.0	591.0
23	Young	100	410.0	539.5
24	Young	100	302.0	472.5
25	Young	100	692.5	656.0
26	Young	100	447.5	548.0
27	Young	100	525.5	527.0
28	Young	100	322.5	574.0
29	Young	100	468.5	559.5
30	Young	100	378.0	499.5
31	Young	100	497.5	529.5
32	Old	0	542.0	595.0
33	Old	0	599.0	606.0
34	Old	0	562.0	598.0
35	Old	0	586.0	744.0
36	Old	0	674.0	724.0
37	Old	0	762.0	836.5
38	Old	0	697.0	834.0
39	Old	0	583.0	698.5
40	Old	0	533.5	668.0
41	Old	0	524.5	670.0
42	Old	0	500.0	587.0
43	Old	0	680.0	912.5
44	Old	0	563.5	619.0
45	Old	100	523.5	646.5
46	Old	100	770.0	862.5
47	Old	100	712.0	829.0
48	Old	100	653.0	697.0
49	Old	100	699.5	818.0
50	Old	100	561.0	819.5
51	Old	100	751.0	872.0
52	Old	100	520.5	889.0
53	Old	100	523.0	601.0

Edited SAS Output (PROC FREQ and PROC GLM) for Problem 3

```
                        TABLE OF AGEGROUP BY PPM_CO

              AGEGROUP |     PPM_CO
              Frequency|
              Percent  |
              Row Pct  |
              Col Pct  |      0|     100|   Total
              ---------+--------+--------+
              Old      |     13 |      9 |     22
                       |  24.53 |  16.98 |  41.51
                       |  59.09 |  40.91 |
                       |  48.15 |  34.62 |
              ---------+--------+--------+
              Young    |     14 |     17 |     31
                       |  26.42 |  32.08 |  58.49
                       |  45.16 |  54.84 |
                       |  51.85 |  65.38 |
              ---------+--------+--------+
              Total          27       26       53
                          50.94    49.06   100.00


                     General Linear Models Procedure

                       Class Level Information

                  Class        Levels     Values
                  AGEGROUP        2        Old Young
                  PPM_CO          2        0 100
```

Dependent Variable: REACTIM1

| | Sum of | | Mean | | |
Source	DF	Squares	Square	F Value	Pr > F
Model	3	484794.098	161598.033	17.51	0.0001
Error	49	452176.619	9228.094		
Corrected Total	52	936970.717			

| | R-Square | Coeff Var | Root MSE | REACTIM1 Mean |
| | 0.517406 | 19.14396 | 96.0630 | 501.792 |

SS(AGEGROUP)

Source	DF	Type I SS	Mean Square	F Value	Pr > F
AGEGROUP	1	478181.843	478181.843	☐	☐
PPM_CO	1	4210.892	4210.892	☐	☐
AGEGROUP*PPM_CO	1	2401.364	2401.364	☐	☐

SS(PPM_CO | AGEGROUP)   SS(PPM_CO | AGEGROUP, AGEGROUP*PPM_CO)

Source	DF	Type II SS	Mean Square	F Value	Pr > F
AGEGROUP	1	481452.369	481452.369	☐	☐
PPM_CO	1	4210.892	4210.892	☐	☐
AGEGROUP*PPM_CO	1	2401.364	2401.364	☐	☐

SS(AGEGROUP | PPM_CO), AGEGROUP*PPM_CO)   SS(AGEGROUP*PPM_CO | AGEGROUP, PPM_CO)

4. a.–d. Repeat Problem 3, using REACTIM2 as the dependent variable and referring to the accompanying computer output.

Edited SAS Output (PROC GLM) for Problem 4

```
                    General Linear Models Procedure

                        Class Level Information

                 Class        Levels      Values
                 AGEGROUP        2       Old Young
                 PPM_CO          2        0   100

Dependent Variable: REACTIM2

                               Sum of              Mean
Source             DF          Squares            Square      F Value      Pr > F
Model               3        584722.361        194907.454      20.04       0.0001
Error              49        476633.960          9727.224
Corrected Total    52       1061356.321

           R-Square          Coeff Var        Root MSE      REACTIM2 Mean
           0.550920          16.09215         98.6267          612.887

Source             DF        Type I SS        Mean Square    F Value      Pr > F
AGEGROUP            1       543059.011        543059.011     [     ]      [     ]
PPM_CO              1         3971.106          3971.106     [     ]      [     ]
AGEGROUP*PPM_CO     1        37692.243         37692.243     [     ]      [     ]

Source             DF        Type II SS       Mean Square    F Value      Pr > F
AGEGROUP            1       545527.925        545527.925     [     ]      [     ]
PPM_CO              1         3971.106          3971.106     [     ]      [     ]
AGEGROUP*PPM_CO     1        37692.243         37692.243     [     ]      [     ]
```

5. A crime victimization study was undertaken in a medium-size southern city. The main purpose was to determine the effects of being a crime victim on confidence in law enforcement authority and in the legal system itself. A questionnaire was administered to a stratified random sample of 40 city residents; among the information elicited were data on the number of times victimized, a measure of social class status (SCLS), and a measure of the respondent's confidence in law enforcement and in the legal system. The data are reproduced in the following table.

| No. of Times | Social Class Status | | |
Victimized	LO	MED	HI
0	4, 14, 15, 19, 17, 17, 16	7, 10, 12, 15, 16	8, 19, 10, 17
1	2, 7, 18	6, 19, 12, 12	7, 6, 5, 3, 16
2+	7, 8, 2, 11, 12	1, 2, 4	4, 2, 8, 9

a. Determine the table of sample means, and comment on any patterns noted.
b. Analyze this data set using the following ANOVA table.
c. How would you analyze this data set by using the two tables of regression results that follow?
d. What ANOVA assumption(s) might not hold for these data?

Source	d.f.	SS	MS
VICTIM	2	$4.0000E + 02$	$2.0000E + 02$
SCLS	2	$2.2739E + 01$	$1.1370E + 01$
VICTIM × SCLS	4	$1.0993E + 02$	$2.7483E + 01$
Error	31	$7.0408E + 02$	$2.2712E + 01$

Source	d.f.	MS
Z_1	1	44.03235
$Z_2 \| Z_1$	1	1.03496
$X_1 \| Z_1, Z_2$	1	395.75734
$X_2 \| Z_1, Z_2, X_1$	1	0.06778
$X_1 Z_1 \| Z_1, Z_2, X_1, X_2$	1	1.68985
$X_1 Z_2 \| Z_1, Z_2, X_1, X_2, X_1 Z_1$	1	3.31635
$X_2 Z_1 \| Z_1, Z_2, X_1, X_2, X_1 Z_1, X_1 Z_2$	1	0.40190
$X_2 Z_2 \| Z_1, Z_2, X_1, X_2, X_1 Z_1, X_1 Z_2, X_2 Z_1$	1	94.59353
Residual	31	22.71229

Source	d.f.	MS
X_1	1	407.86993
$X_2 \| X_1$	1	0.52174
$Z_1 \| X_1, X_2$	1	27.98766
$Z_2 \| X_1, X_2, Z_1$	1	4.51309
$X_1 Z_1 \| X_1, X_2, Z_1, Z_2$	1	1.68985
$X_1 Z_2 \| X_1, X_2, Z_1, Z_2, X_1 Z_1$	1	3.31635
$X_2 Z_1 \| X_1, X_2, Z_1, Z_2, X_1 Z_1, X_1 Z_2$	1	0.40190
$X_2 Z_2 \| X_1, X_2, Z_1, Z_2, X_1 Z_1, X_1 Z_2, X_2 Z_1$	1	94.59353
Residual	31	22.71229

Note: X pertains to number of times victimized; *Z* pertains to social class status.

6. The effect of a new antidepressant drug on reducing the severity of depression was studied in manic–depressive patients at two state mental hospitals. In each hospital all such patients were randomly assigned to either a treatment (new drug) or a control (old drug) group. The results of this experiment are summarized in the following table; a high mean score indicates more of a lowering in depression level than does a low mean score.

Hospital	Group	
	Treatment	Control
A	$n = 25, \overline{Y} = 8.5, S = 1.3$	$n = 31, \overline{Y} = 4.6, S = 1.8$
B	$n = 25, \overline{Y} = 2.3, S = 0.9$	$n = 31, \overline{Y} = -1.7, S = 1.1$

a. Without performing any statistical tests, interpret the means in the table.

b. What regression model is appropriate for analyzing the data? For this model, describe how to test whether the new drug has a significant effect.

7. A study was conducted by a television network in a certain state to evaluate the viewing characteristics of adult females. Each individual in a stratified random sample of 480 women was sent a questionnaire; the strata were formed on the basis of the following three factors: season (winter or summer), region (eastern, central, or western), and residence (rural or urban). The averages of the total time reported watching TV (hours per day) are summarized in the accompanying table of sample means and standard deviations.

Residence and Region	Summer			Winter			Marginals	
	n	$\overline{Y}$	S	n	$\overline{Y}$	S	n	$\overline{Y}$
Rural								
East	40	2.75	1.340	40	4.80	0.851	80	3.78
Central	40	2.75	1.380	40	4.85	0.935	80	3.80
West	40	2.65	1.180	40	4.78	0.843	80	3.71
Marginals	120	2.72		120	4.81		240	3.76
Urban								
East	40	3.38	0.958	40	3.65	0.947	80	3.52
Central	40	3.15	1.130	40	4.50	0.743	80	3.83
West	40	3.65	0.779	40	4.05	0.781	80	3.85
Marginals	120	3.39		120	4.07		240	3.73
Marginals	240	3.06		240	4.44		480	3.75

7. a. Suppose that the questionnaire contained items concerning additional factors such as occupation (categorized as housewife, blue-collar worker, white-collar worker, or professional), age (categorized as 20 to 34, 35 to 50, and over 50), and number of children (categorized as 0, 1−2, and 3+). What is the likelihood of obtaining equal cell numbers when carrying out an ANOVA to consider these additional variables?

b. Examine the table of sample means for main effects and interactions. (You may want to form two-factor summary tables for assessing two-factor interactions.)

c. Using the following table of ANOVA results, carry out appropriate F tests, and discuss your results.

d. State a regression model appropriate for obtaining information equivalent to the ANOVA results presented next.

Source	d.f.	SS	MS
RESID	1	1.3333E − 01	1.3333E − 01
REGION	2	2.5527E + 00	1.2763E + 00
SEASON	1	2.2963E + 02	2.2963E + 02
RESID × REGION	2	3.3247E + 00	1.6623E + 00
RESID × SEASON	1	6.0492E + 01	6.0492E + 01
REGION × SEASON	2	7.2247E + 00	3.6123E + 00
RESID × REGION × SEASON	2	6.7460E + 00	3.3730E + 00
Error	468	4.7821E + 02	1.0218E + 00

8. Suppose that the following data were obtained by an investigator studying the influence of estrogen injections on change in the pulse rate of adolescent chimpanzees:

Male $\begin{cases} \text{Control:} & 5.1, -2.3, 4.2, 3.8, 3.2, -1.5, 6.1, -2.5, 1.9, -3.0, -2.8, 1.7 \\ \text{Estrogen:} & 15.0, 6.2, 4.1, 2.3, 7.6, 14.8, 12.3, 13.1, 3.4, 8.5, 11.2, 6.9 \end{cases}$

Female $\begin{cases} \text{Control:} & -2.3, -5.8, -1.5, 3.8, 5.5, 1.6, -2.4, 1.9 \\ \text{Estrogen:} & 7.3, 2.4, 6.5, 8.1, 10.3, 2.2, 12.7, 6.3 \end{cases}$

a. What are the factors in this experiment? Should they be designated as fixed or random?

b. Demonstrate that the cell frequencies for two-way ANOVA in this problem are proportional; that is, $n_{ij} = n_{i.}n_{.j}/n_{..}$ for each of the four cells.

8. c. Use the following table of sample means

Sex	Control	Estrogen
Male	1.158333	8.783333
Female	0.100000	6.975000

and the general formulas

$$SS(\text{rows}) = \sum_{i=1}^{r}\sum_{j=1}^{c}\sum_{k=1}^{n_{ij}} (\overline{Y}_{i..} - \overline{Y}...)^2$$

$$SS(\text{columns}) = \sum_{i=1}^{r}\sum_{j=1}^{c}\sum_{k=1}^{n_{ij}} (\overline{Y}_{.j.} - \overline{Y}...)^2$$

$$SS(\text{cells}) = \sum_{i=1}^{r}\sum_{j=1}^{c}\sum_{k=1}^{n_{ij}} (\overline{Y}_{ij.} - \overline{Y}...)^2$$

to analyze the data for this problem, employing the usual methodology for equal-cell-number two-way ANOVA and the fact that SSE = 530.24078 (i.e., compute the sums of squares for rows, columns, and cells directly, and then obtain the sum of squares for interaction by subtraction).

d. What regression model is appropriate for analyzing this data set?

e. Using the regression analysis results given in the following table, check whether SS(Rows) = Regression SS(SEX) = Regression SS(SEX|TREATMENT) and SS(Columns) = Regression SS(TREATMENT) = Regression SS(TREATMENT|SEX), where SS(Rows) and SS(Columns) are as obtained in part (c). What has been demonstrated here?

Source	d.f.	SS
SEX	1	19.72667
TREATMENT | SEX	1	536.55619
SEX × TREATMENT | SEX, TREATMENT	1	1.35000
Residual	36	530.24078

Source	d.f.	SS
TREATMENT	1	536.55619
SEX | TREATMENT	1	19.72667
SEX × TREATMENT | SEX, TREATMENT	1	1.35000
Residual	36	530.24078

Note: $\text{SEX} = \begin{cases} -1 & \text{if male} \\ 1 & \text{if female} \end{cases}$ and $\text{TREATMENT} = \begin{cases} -1 & \text{if control} \\ 1 & \text{if estrogen} \end{cases}$

9. The data listed in the following table relate to a study by Reiter and others (1981) concerning the effects of injecting triethyl-tin (TET) into rats once at age 5 days. The animals were injected with 0, 3, or 6 mg per kilogram of body weight. The response was the log of the activity count for 1 hour, recorded at 21 days of age. The rat was left to move about freely in a figure 8 maze. Analysis of other studies with this type of activity count confirms that log counts should yield Gaussian errors if the model is correct.

9. **a.** Tabulate the DOSAGE $\times$ SEX cell sample sizes. Explain why this might be called a nearly orthogonal design.

   b. Using the accompanying SAS computer output, conduct a two-way ANOVA with SEX and DOSAGE as factors. (First provide the F statistic values and P-values in the accompanying ANOVA table.)

   c. Using $\alpha = .05$, report your conclusions based on the ANOVA.

   d. Which, if any, families of means should be followed up with multiple-comparison tests? What type of comparisons would you recommend?

Observation	LOGACT21	DOSAGE	SEX	CAGE
1	2.636	0	Male	5
2	2.736	0	Male	6
3	2.775	0	Male	7
4	2.672	0	Male	9
5	2.653	0	Male	11
6	2.569	0	Male	12
7	2.737	0	Male	15
8	2.588	0	Male	16
9	2.735	0	Male	17
10	2.444	3	Male	3
11	2.744	3	Male	5
12	2.207	3	Male	6
13	2.851	3	Male	7
14	2.533	3	Male	9
15	2.630	3	Male	11
16	2.688	3	Male	12
17	2.665	3	Male	15
18	2.517	3	Male	16
19	2.769	3	Male	17
20	2.694	6	Male	3
21	2.845	6	Male	5
22	2.865	6	Male	6
23	3.001	6	Male	7
24	3.043	6	Male	9
25	3.066	6	Male	11
26	2.747	6	Male	12
27	2.894	6	Male	15
28	1.851	6	Male	16
29	2.489	6	Male	17
30	2.494	0	Female	3
31	2.723	0	Female	5
32	2.841	0	Female	6
33	2.620	0	Female	7
34	2.682	0	Female	9
35	2.644	0	Female	11
36	2.684	0	Female	12

Observation	LOGACT21	DOSAGE	SEX	CAGE
37	2.607	0	Female	15
38	2.591	0	Female	16
39	2.737	0	Female	17
40	2.220	3	Female	3
41	2.371	3	Female	5
42	2.679	3	Female	6
43	2.591	3	Female	7
44	2.942	3	Female	9
45	2.473	3	Female	11
46	2.814	3	Female	12
47	2.622	3	Female	15
48	2.730	3	Female	16
49	2.955	3	Female	17
50	2.540	6	Female	3
51	3.113	6	Female	5
52	2.468	6	Female	6
53	2.606	6	Female	7
54	2.764	6	Female	9
55	2.859	6	Female	11
56	2.763	6	Female	12
57	3.000	6	Female	15
58	3.111	6	Female	16
59	2.858	6	Female	17

Edited SAS Output (PROC FREQ and PROC GLM) for Problem 9

```
                           TABLE OF DOSAGE BY SEX

                 DOSAGE              SEX
                 Frequency
                 Percent
                 Row Pct
                 Col Pct   | Female  | Male    |        Total
                 ----------+---------+---------+
                     0     |     10  |      9  |           19
                           |  16.95  |  15.25  |        32.20
                           |  52.63  |  47.37  |
                           |  33.33  |  31.03  |
                 ----------+---------+---------+
                     3     |     10  |     10  |           20
                           |  16.95  |  16.95  |        33.90
                           |  50.00  |  50.00  |
                           |  33.33  |  34.48  |
                 ----------+---------+---------+
                     6     |     10  |     10  |           20
                           |  16.95  |  16.95  |        33.90
                           |  50.00  |  50.00  |
                           |  33.33  |  34.48  |
                 ----------+---------+---------+
                 Total          30        29              59
                             50.85     49.15          100.00
```

Edited SAS Output for Problem 9 (continued)

```
                        General Linear Models Procedure
                          Class Level Information
                   Class          Levels        Values
                   DOSAGE            3           0 3 6
                   SEX              2           Female Male

Dependent Variable: LOGACT21
```

Source	DF	Sum of Squares	Mean Square	F Value	Pr > F
Model	5	0.28197725	0.05639545	1.16	0.3412
Error	53	2.57750079	0.04863209		
Corrected Total	58	2.85947803			

R-Square	Coeff Var	Root MSE	LOGACT21 Mean
0.098611	8.196165	0.22053	2.69061

Source	DF	Type I SS	Mean Square	F Value	Pr > F
DOSAGE	2	0.25750763	0.12875381	☐	☐
SEX	1	0.01054232	0.01054232	☐	☐
DOSAGE*SEX	2	0.01392730	0.00696365	☐	☐

Source	DF	Type II SS	Mean Square	F Value	Pr > F
DOSAGE	2	0.25806594	0.12903297	☐	☐
SEX	1	0.01054232	0.01054232	☐	☐
DOSAGE*SEX	2	0.01392730	0.00696365	☐	☐

10. The experimenters described in Problem 9 hoped that home cage would not affect activity level in any systematic fashion. Explore this question by repeating Problem 9, but replacing SEX with CAGE in your analysis.

Edited SAS Output (PROC FREQ and PROC GLM) for Problem 10

```
                            TABLE OF DOSAGE BY CAGE
```

DOSAGE Frequency Percent Row Pct Col Pct	CAGE 3	5	6	7	9	Total
0	1 1.69 5.26 20.00	2 3.39 10.53 33.33	2 3.39 10.53 33.33	2 3.39 10.53 33.33	2 3.39 10.53 33.33	19 32.20
3	2 3.39 10.00 40.00	2 3.39 10.00 33.33	2 3.39 10.00 33.33	2 3.39 10.00 33.33	2 3.39 10.00 33.33	20 33.90
6	2 3.39 10.00 40.00	2 3.39 10.00 33.33	2 3.39 10.00 33.33	2 3.39 10.00 33.33	2 3.39 10.00 33.33	20 33.90
Total	5 8.47	6 10.17	6 10.17	6 10.17	6 10.17	59 100.00

(continued)

Edited SAS Output for Problem 10 (continued)

```
                        TABLE OF DOSAGE BY CAGE (Continued)

DOSAGE          CAGE
Frequency
Percent
Row Pct
Col Pct             11 |        12 |        15 |        16 |        17 |     Total
          --------+---------+---------+---------+---------+---------+
        0           2          2          2          2          2          19
                  3.39       3.39       3.39       3.39       3.39       32.20
                 10.53      10.53      10.53      10.53      10.53
                 33.33      33.33      33.33      33.33      33.33
          --------+---------+---------+---------+---------+---------+
        3           2          2          2          2          2          20
                  3.39       3.39       3.39       3.39       3.39       33.90
                 10.00      10.00      10.00      10.00      10.00
                 33.33      33.33      33.33      33.33      33.33
          --------+---------+---------+---------+---------+---------+
        6           2          2          2          2          2          20
                  3.39       3.39       3.39       3.39       3.39       33.90
                 10.00      10.00      10.00      10.00      10.00
                 33.33      33.33      33.33      33.33      33.33
          --------+---------+---------+---------+---------+---------+
Total               6          6          6          6          6          59
                 10.17      10.17      10.17      10.17      10.17      100.00

                        General Linear Models Procedure

                          Class Level Information

                    Class          Levels          Values
                    DOSAGE            3             0 3 6
                    CAGE             10             3 5 6 7 9 11 12 15 16 17

Dependent Variable: LOGACT21
                                Sum of            Mean
Source                 DF       Squares          Square       F Value    Pr > F
Model                  29     1.30579603       0.04502745       0.84     0.6786
Error                  29     1.55368200       0.05357524
Corrected Total        58     2.85947803

          R-Square            Coeff Var         Root MSE         LOGACT21 Mean
          0.456655            8.602631          0.23146             2.69061

Source             DF        Type I SS       Mean Square      F Value    Pr > F
DOSAGE              2        0.25750763       0.12875381       [    ]     [    ]
CAGE                9        0.47895137       0.05321682       [    ]     [    ]
DOSAGE*CAGE        18        0.56933703       0.03162984       [    ]     [    ]

Source             DF        Type II SS      Mean Square      F Value    Pr > F
DOSAGE              2        0.26793817       0.13396908       [    ]     [    ]
CAGE                9        0.47895137       0.05321682       [    ]     [    ]
DOSAGE*CAGE        18        0.56933703       0.03162984       [    ]     [    ]
```

11. A manufacturer conducted a pricing experiment to explore the effects of price decreases on sales of one of its breakfast cereals. The two largest supermarket chains in a particular market participated in the experiment. Ten stores from each chain were randomly selected, and each store was assigned a price level for the cereal (either the original price or a 10% reduced price). If the competing chain had a store in the same vicinity, the two stores both were assigned the same price level. Some stores failed to complete the experiment due to competition from other supermarkets chains. Sales volumes (in hundreds of units) over the period of the study were noted for each of the remaining 17 stores, and are shown in the following table.

Supermarket Chain	Price Level	
	Original Price (Price = 0)	10% Reduced Price (Price = R)
1	14, 14, 15	14, 14, 17, 19, 13
2	9, 7, 12, 8	10, 12, 14, 15, 13

11. a. Using the accompanying SAS computer output, conduct a two-way ANOVA with Chain and Price as factors. (Provide the F statistic values and P-values in the accompanying ANOVA table as part of your analysis.)

b. Do the effects of a price decrease on sales volume significantly differ between the different chains? If not, does it appear that a price decrease will significantly increase sales volume? How do the sales volumes compare at the two chains? Report your conclusions using the $\alpha = .05$ level. Also, do the conclusions differ for the different approaches taken in parts (a) and (b)?

Edited SAS Output (PROC GLM) for Problem 11

```
                     General Linear Models Procedure

                         Class Level Information

                    Class        Levels        Values
                    CHAIN           2           1 2
                    PRICE           2           0 R


Dependent Variable: SALES

                              Sum of              Mean
Source              DF        Squares            Square        F Value      Pr > F
Model                3       98.2745098       32.7581699         7.79       0.0031
Error               13       54.6666667        4.2051282
Corrected Total     16      152.9411765


         R-Square            Coeff Var           Root MSE          SALES Mean
         0.642564            15.84586            2.05064            12.9412


Source           DF       Type I SS       Mean Square      F Value       Pr > F
CHAIN             1      64.0522876       64.0522876      [      ]      [      ]
PRICE             1      26.6244821       26.6244821      [      ]      [      ]
CHAIN*PRICE       1       7.5977401        7.5977401      [      ]      [      ]

Source           DF       Type II SS      Mean Square      F Value       Pr > F
CHAIN             1      58.0641646       58.0641646      [      ]      [      ]
PRICE             1      26.6244821       26.6244821      [      ]      [      ]
CHAIN*PRICE       1       7.5977401        7.5977401      [      ]      [      ]
```

12. The manager of a market research company conducted an experiment to investigate the productivity of three employees on each of two computerized data-entry systems. The employees conducted phone surveys, entering the survey data into the computer during the phone call. Productivity was measured as the time (in minutes) taken to complete a call in which the respondent agreed to complete the survey. Each employee used each system for one hour, and the order of use was randomized. The productivity data are recorded in the following table.

Employee	System	
	1	2
1	8, 7, 8, 9, 8	6, 4, 7, 4, 3, 3, 4, 5, 6
2	5, 4, 6, 3, 8, 8, 7	6, 2, 3, 3, 4, 4, 5, 6, 7, 4
3	3, 4, 5, 4, 3, 5, 5, 6	3, 3, 4, 5, 4, 3, 2, 4, 3, 3, 3, 4

12. **a.** Using the accompanying SAS computer output, conduct a two-way ANOVA. (Provide the F statistic values and P-values in the accompanying ANOVA table as part of your analysis.)

 b. Report your conclusions using the $\alpha = .01$ level. Is there an interaction between Employee and System? If not, does it appear that one system is significantly more productive than the other? Do the employees differ in terms of productivity?

Edited SAS Output (PROC GLM) for Problem 12

```
                    General Linear Models Procedure

                        Class Level Information
                  Class           Levels        Values
                  EMPLOYEE          3            1 2 3
                  SYSTEM            2            1 2

Dependent Variable: PRODUCT

                             Sum of              Mean
Source            DF         Squares             Square      F Value      Pr > F
Model              5        85.1276611         17.0255322      9.82       0.0001
Error             45        78.0488095          1.7344180
Corrected Total   50       163.1764706


        R-Square         Coeff Var           Root MSE          PRODUCT Mean
        0.521691         27.64017            1.31697             4.76471


Source            DF        Type I SS        Mean Square      F Value      Pr > F
EMPLOYEE           2       36.2621849        18.1310924      [     ]      [     ]
SYSTEM             1       37.4465293        37.4465293      [     ]      [     ]
EMPLOYEE*SYSTEM    2       11.4189469         5.7094735      [     ]      [     ]

Source            DF        Type II SS       Mean Square      F Value      Pr > F
EMPLOYEE           2       38.4419210        19.2209605      [     ]      [     ]
SYSTEM             1       37.4465293        37.4465293      [     ]      [     ]
EMPLOYEE*SYSTEM    2       11.4189469         5.7094735      [     ]      [     ]
```

13. This question refers to the radial keratotomy data of Problem 21 in Chapter 17.

 a. Suppose that a two-way ANOVA is to be performed, with five-year postoperative change in refractive error as the dependent variable (Y) and diameter of clear zone (CLRZONE) and baseline average corneal curvature (BASECURV $= 1$ if curvature is < 43 diopters; $= 2$ if curvature is between 43 and 44 diopters; and $= 3$ if curvature is > 44 diopters). State precisely the ANOVA model. Are the factors fixed or random?

 b. In the SAS output that follows, complete the ANOVA table.

 c. Analyze the data to determine whether there are significant main effects due to clear zone and baseline curvature, and whether these factors significantly interact. Use $\alpha = .10$.

Edited SAS Output (PROC GLM) for Problem 13

```
                    General Linear Models Procedure
                      Class Level Information
                 Class        Levels      Values
                 CLRZONE         3         3.0 3.5 4.0
                 BASECURV        3         1 2 3

NOTE: Due to missing values, only 51 observations can be used in
      this analysis.

Dependent Variable: Y

Source              DF        Sum of Squares      F Value       Pr > F
Model              [   ]        32.22787390        [    ]       [     ]
Error               42          47.06818002
Corrected Total     50          79.29605392

          R-Square              Coeff Var              Y Mean
          0.406425              27.77523             3.81137255

Source              DF        Type I SS          F Value       Pr > F
CLRZONE              2          14.70441590         6.56        0.0033
BASECURV             2           5.37753783         2.40        0.1031
CLRZONE*BASECURV     4          12.14592017         2.71        0.0428

Source              DF        Type III SS        F Value       Pr > F
CLRZONE              2          12.92686249         5.77        0.0061
BASECURV             2           5.97019625         2.66        0.0814
CLRZONE*BASECURV     4          12.14592017         2.71        0.0428

          Level of                     ----------------Y----------------
          CLRZONE         N                   Mean              SD
          3.0            20              4.41875000        1.15228249
          3.5            15              3.71666667        1.33836476
          4.0            16              3.14093750        0.97595119

          Level of                     ----------------Y----------------
          BASECURV        N                   Mean              SD
          1              11              4.30681818        1.20592759
          2              15              3.60833333        1.03818466
          3              25              3.71520000        1.38615361

Level of      Level of                  ----------------Y----------------
CLRZONE       BASECURV        N                Mean              SD
3.0           1              3            5.00000000        0.99215674
3.0           2              6            3.66666667        0.79320027
3.0           3             11            4.67045455        1.22509276
3.5           1              4            4.65625000        0.71716310
3.5           2              5            4.22500000        1.26676261
3.5           3              6            2.66666667        1.06555932
4.0           1              4            3.43750000        1.42339090
4.0           2              4            2.75000000        0.46770717
4.0           3              8            3.18812500        0.96893881
```

14. This question refers to the *U.S. News & World Report* graduate school data presented in Problem 22 in Chapter 17.

14. a. Suppose that a two-way ANOVA is to be performed, with 1995 starting salary as the dependent variable and with school type (SCHOOL) and reputation rank among academics (REP1 = 1 if reputation rank is in the top 10; = 2 if rank is 11 to 20; = 3 if rank is 21 or worse) as the two factors. State precisely the ANOVA model. Are the factors fixed or random?

b. In the SAS output that follows, complete the ANOVA table.

c. Analyze the data to determine whether there are significant main effects due to school type and reputation rank, and whether these factors significantly interact.

Edited SAS Output (PROC GLM) for Problem 14

```
                    General Linear Models Procedure

                        Class Level Information

                Class         Levels         Values
                SCHOOL           2            1 2
                REP              3            1 2 3

Dependent Variable: SAL

                            Sum of          Mean
Source              DF      Squares        Square      F Value      Pr > F
Model             [    ]    1000.5058      200.1012    [       ]    [      ]
Error               18       965.8875       53.6604
Corrected Total     23      1966.3933

            R-Square          Coeff Var         Root MSE          SAL Mean
            0.508802          12.86650          7.3253             56.933

Source              DF      Type I SS      Mean Square   F Value      Pr > F
SCHOOL               1        37.00167       37.00167      0.69       0.4172
REP                  2       910.36583      455.18292      8.48       0.0025
SCHOOL*REP           2        53.13833       26.56917      0.50       0.6176

Source              DF      Type III SS    Mean Square   F Value      Pr > F
SCHOOL               1        67.78778       67.78778      1.26       0.2758
REP                  2       910.36583      455.18292      8.48       0.0025
SCHOOL*REP           2        53.13833       26.56917      0.50       0.6176

            Level of                    --------------SAL--------------
            SCHOOL          N               Mean                 SD
            1              12           58.1750000          9.65628622
            2              12           55.6916667          9.06396044

            Level of                    --------------SAL--------------
            REP             N               Mean                 SD
            1               4           68.7500000          6.29152870
            2               4           61.5000000          6.35085296
            3              16           52.8375000          7.37688959

    Level of       Level of                 --------------SAL--------------
    SCHOOL         REP           N               Mean                 SD
    1             1             2           70.0000000          0.0000000
    1             2             2           66.0000000          5.6568542
    1             3             8           53.2625000          7.5450906
    2             1             2           67.5000000         10.6066017
    2             2             2           57.0000000          2.8284271
    2             3             8           52.4125000          7.6986896
```

We will assume that the Y_i are normally distributed with variance $\text{Var}(Y_i) = \sigma^2$ not varying with i. We further assume that X_i is measured without error and that the n random variables $Y_1, Y_2, \ldots, Y_n$ are mutually independent.[3]

The expression for the distribution (density function) of the normally distributed random variable Y_i under Model 1 is given by

$$f(Y_i; \beta_0, \beta_1, \sigma^2) = \frac{1}{\sqrt{2\pi\sigma^2}} \exp\left\{-\frac{1}{2\sigma^2}[Y_i - (\beta_0 + \beta_1 X_i)]^2\right\} \qquad (21.6)$$

where $-\infty < Y_i < +\infty$. Note that (21.6) is a function of three parameters—namely, β_0, β_1, and σ^2. Under the assumption that the $\{Y_i, i = 1, 2, \ldots, n\}$ are mutually independent, it can be shown from statistical theory that the joint distribution of $Y_1, Y_2, \ldots, Y_n$ (i. e., the likelihood function) is, from (21.6):

$$L(\mathbf{Y}; \beta_0, \beta_1, \sigma^2) = \prod_{i=1}^{n} f(Y_i; \beta_0, \beta_1, \sigma^2) = \frac{1}{(2\pi\sigma^2)^{n/2}} \exp\left\{-\frac{1}{2\sigma^2} \sum_{i=1}^{n} [Y_i - (\beta_0 + \beta_1 X_i)]^2\right\}$$

$$(21.7)$$

(In our example, $n = 8$.)

A specific algebraic expression for the maximized likelihood $L(\mathbf{Y}; \hat{\beta}_0, \hat{\beta}_1, \hat{\sigma}^2)$ can be written in the form

$$L(Y; \hat{\beta}_0, \hat{\beta}_1, \hat{\sigma}^2) = (2\pi\hat{\sigma}^2 e)^{-n/2} \qquad (21.8)$$

where e is Euler's constant that we approximate as 2.7183. For the particular set of $n = 8$ data points, the ML estimators obtained from maximizing the likelihood function (21.7) are[4]

$$\hat{\beta}_0 = -1.20, \qquad \hat{\beta}_1 = 1.11, \qquad \hat{\sigma}^2 = 0.6288$$

Hence, from (21.8),

$$L(\mathbf{Y}; \hat{\beta}_0, \hat{\beta}_1, \hat{\sigma}^2) = L(\mathbf{Y}; -1.20, 1.11, 0.6288)$$
$$= [(2(3.1416)(0.6288)(2.7183)]^{-8/2} = (10.7397)^{-4}$$

The number that would typically appear in the computer output using an ML estimation program with the likelihood function (21.7) and the dataset under consideration would be $-2 \ln L(\mathbf{Y}; \hat{\beta}_0, \hat{\beta}_1, \hat{\sigma}^2) = -2 \ln(10.7397)^{-4} = 18.9916$.

---

[3] The assumption of *mutual independence* for a set of random variables is the strongest assumption that can be made about the joint behavior of this set. Such an assumption allows precise description of the joint distribution of the variables (i.e., the likelihood function) solely on the basis of knowledge of the separate behavior (i.e., the so-called marginal distribution) of each variable in the set. In particular, the joint distribution under mutual independence is simply the product of its marginal distributions.

[4] The ML estimator $\hat{\sigma}^2$ of σ^2 is actually a biased estimator of σ^2. The unbiased estimator of σ^2 is

$$\left(\frac{n}{n-2}\right)\hat{\sigma}^2 = \frac{\text{SSE}}{n-2}$$

where SSE is the sum of squares of residuals about the fitted straight line using the unweighted least squares estimates for Model 1. (In Table 15.4 of Chapter 15, we found SSE = 5.03). Thus, the ML method does not always produce unbiased estimators of the parameters of interest, although the extent of such bias generally decreases as the sample size decreases when the likelihood function is correctly specified.

The estimated covariance matrix for the ML estimators $\hat{\beta}_0$, $\hat{\beta}_1$, and $\hat{\sigma}^2$ has the following general form:

$$\hat{\mathbf{V}}(\hat{\beta}_0, \hat{\beta}_1, \hat{\sigma}^2) = \begin{bmatrix} \widehat{\text{Var}}(\hat{\beta}_0) & \widehat{\text{Cov}}(\hat{\beta}_0, \hat{\beta}_1) & \widehat{\text{Cov}}(\hat{\beta}_0, \hat{\sigma}^2) \\ \widehat{\text{Cov}}(\hat{\beta}_0, \hat{\beta}_1) & \widehat{\text{Var}}(\hat{\beta}_1) & \widehat{\text{Cov}}(\hat{\beta}_1, \hat{\sigma}^2) \\ \widehat{\text{Cov}}(\hat{\beta}_0, \hat{\sigma}^2) & \widehat{\text{Cov}}(\hat{\beta}_1, \hat{\sigma}^2) & \widehat{\text{Var}}(\hat{\sigma}^2) \end{bmatrix}$$

The estimated variances of the ML estimators appear on the diagonal of this symmetric matrix, and the estimated covariances are given by the off-diagonal elements of the matrix.

For the data under consideration, the following estimated covariance matrix is obtained:

$$\hat{\mathbf{V}}(\hat{\beta}_0, \hat{\beta}_1, \hat{\sigma}^2) = \begin{bmatrix} 0.3818 & -0.0674 & 0 \\ -0.0674 & 0.0150 & 0 \\ 0 & 0 & 0.0988 \end{bmatrix} \tag{21.9}$$

This is the matrix that a computer program would print out based on the use of the likelihood function (21.7) for these data.

Later, we will use the maximized likelihood value of $(10.7397)^{-4}$ to carry out a likelihood ratio test, which involves comparing the ratios of maximized likelihoods for different models. We now focus on the estimated covariance matrix (21.9) and use it to perform certain inference-making exercises.

21.3.1 Hypothesis Testing Using Wald Statistics

For the data we have been considering (Table 21.1), where $n = 8$, we have determined the maximized likelihood value $L(\mathbf{Y}; \hat{\beta}_0, \hat{\beta}_1, \hat{\sigma}^2)$ and the estimated covariance matrix $\hat{\mathbf{V}}(\hat{\beta}_0, \hat{\beta}_1, \hat{\sigma}^2)$ for the straight line model (Model 1)

$$E(Y \mid X) = \beta_0 + \beta_1 X$$

An important question of interest for this model is to determine whether the dosage (X) is a significant predictor of weight gain (Y). This can be formulated as a hypothesis-testing question, and the corresponding null hypothesis can be stated as H_0: $\beta_1 = 0$. We will assume that our alternative hypothesis is two-sided, i.e., H_A: $\beta_1 \neq 0$.

Based on the large-sample properties[5] of ML estimators, it can be shown for Model 1 that the quantity

$$Z = \frac{\hat{\beta}_1 - \beta_1}{\sqrt{\widehat{\text{Var}}(\hat{\beta}_1)}} \tag{21.10}$$

---

[5] The ML method produces estimators whose properties are optimal for *large samples* when the assumed likelihood function is correct. ML estimators are said to be *asymptotically optimal* in the sense that desirable properties such as unbiasedness, minimum variance, and normality hold exactly in the limit only as the amount of data becomes infinitely large. Thus, it is typically reasonable to assume for large data sets that an ML estimator is essentially unbiased, has a small variance, and is approximately normally distributed when the appropriate likelihood function is being used.

The expressions $-2 \ln L_R$ and $-2 \ln L_F$ are called *log-likelihood statistics*. When a computer program is used to carry out the ML estimation procedure, these two log-likelihood statistics are provided in separate outputs for the two models being compared. To carry out the likelihood ratio test, the investigator simply finds $-2 \ln L_R$ and $-2 \ln L_F$ from the output and then subtracts $-2 \ln L_F$ from $-2 \ln L_R$ to obtain the value of the test statistic.

We now provide some numerical examples of likelihood ratio tests. Again, we will consider Models 0, 1, and 2 and the data used earlier. First, we will compare Models 0 and 1 using a likelihood ratio test. Since Model 0 is a special case of Model 1 when $\beta_1 = 0$, this likelihood ratio test addresses $H_0: \beta_1 = 0$ versus $H_A: \beta_1 \neq 0$. The appropriate chi-square distribution for the LR statistic under this H_0 has 1 d.f. (since one parameter is restricted to be equal to 0 under H_0). Using (21.14) with $L_R = L(\mathbf{Y}; \hat{\beta}_0, \hat{\sigma}_0^2)$ and $L_F = L(\mathbf{Y}; \hat{\beta}_0, \hat{\beta}_1, \hat{\sigma}_1^2)$, the LR statistic is given by the following formula:

$$\text{LR} = -2 \ln L_R - (-2 \ln L_F)$$
$$= -2 \ln L(\mathbf{Y}; \hat{\beta}_0, \hat{\sigma}_0^2) - (-2 \ln L(\mathbf{Y}; \hat{\beta}_0, \hat{\beta}_1, \hat{\sigma}_1^2))$$

where the log-likelihood statistics[10] produced by computer output are given by

$$-2 \ln L(\mathbf{Y}; \hat{\beta}_0, \hat{\sigma}_0^2) = 38.4218 \quad \text{and} \quad -2 \ln L(\mathbf{Y}; \hat{\beta}_0, \hat{\beta}_1, \hat{\sigma}_1^2) = 18.9916$$

Substituting the above numerical values into the LR formula, we obtain

$$\text{LR} = 38.4218 - 18.9916 = 19.43$$

which corresponds to a *P*-value of less than 0.0005 and thus provides strong evidence in favor of $H_A: \beta_1 \neq 0$.

The discrepancy between the LR statistic of 19.43 and the Wald statistic of 82.14 is alarming. However, this discrepancy is entirely plausible because of the small sample size ($n = 8$). Because the two test statistics are only asymptotically equivalent, their numerical values are reasonably close only when n is large. Thus, although we have chosen this data set for pedagogical purposes, it is actually an inappropriate one for the application of large-sample statistical procedures. Nevertheless, the LR and Wald tests will not necessarily be identical even when n is considered large. Since the LR statistic is considered to have better statistical properties than the Wald statistic (e.g., the Wald statistic can be more influenced by collinearity problems), we recommend the use of the LR statistic over the Wald statistic when in doubt.

As another illustration, a LR test of $H_0: \beta_2 = 0$ versus $H_A: \beta_2 \neq 0$ involves a comparison of the maximized likelihood values for Models 1 and 2. Using (21.14) again, this time with $L_R = L(\mathbf{Y}; \hat{\beta}_0, \hat{\beta}_1, \hat{\sigma}_1^2)$ and $L_F = L(\mathbf{Y}; \hat{\beta}_0, \hat{\beta}_1, \hat{\beta}_2, \hat{\sigma}_2^2)$, the LR statistic is given by the following formula:

$$\text{LR} = -2 \ln L_R - (-2 \ln L_F)$$
$$= -2 \ln L(\mathbf{Y}; \hat{\beta}_0, \hat{\beta}_1, \hat{\sigma}_1^2) - (-2 \ln L(\mathbf{Y}; \hat{\beta}_0, \hat{\beta}_1, \hat{\beta}_2, \hat{\sigma}_2^2))$$

---

[10] Since we previously stated that $L(\mathbf{Y}; \hat{\beta}_0, \hat{\sigma}_0^2) = (121.8426)^{-4}$, it follows that

$$-2 \ln L(\mathbf{Y}; \hat{\beta}_0, \hat{\sigma}_0^2) = -2(-4) \ln(121.8426) = -2(-4)(4.8027) = 38.4218$$

Also, since we previously stated that $L(\mathbf{Y}; \hat{\beta}_0, \hat{\beta}_1, \hat{\sigma}_1^2) = (10.7397)^{-4}$, it follows that

$$-2 \ln L(\mathbf{Y}; \hat{\beta}_0, \hat{\beta}_1, \hat{\sigma}_1^2) = -2(-4) \ln(10.7397) = -2(-4)(2.3739) = 18.9916$$

where the log-likelihood statistics[11] produced by computer output are given by

$$-2 \ln L(\mathbf{Y}; \hat{\beta}_0, \hat{\beta}_1, \hat{\sigma}_1^2) = 18.9916 \quad \text{and} \quad -2 \ln L(\mathbf{Y}; \hat{\beta}_0, \hat{\beta}_1, \hat{\beta}_2, \hat{\sigma}_2^2) = -6.8078$$

Substituting the above numerical values into the LR formula, we obtain

$$\text{LR} = 18.9916 - (-6.8078) = 25.80$$

which provides strong evidence in favor of H_A: $\beta_2 \neq 0$ (again, assuming that this LR statistic has approximately a chi-square distribution with 1 d.f. under H_0).

As a third illustration, we consider a LR test involving the maximized likelihood values for Models 0 and 2, which tests H_0: $\beta_1 = \beta_2 = 0$ versus H_A: "*At least one of the parameters β_1 and β_2 differs from 0.*" This test is analogous to an overall F test in standard least squares regression analysis because we are testing whether the regression coefficients of all predictors (other than the intercept) are simultaneously equal to 0. The appropriate LR test statistic is

$$\text{LR} = -2 \ln L_R - (-2 \ln L_F)$$
$$= -2 \ln L(\mathbf{Y}; \hat{\beta}_0, \hat{\sigma}_0^2) - (-2 \ln L(\mathbf{Y}; \hat{\beta}_0, \hat{\beta}_1, \hat{\beta}_2, \hat{\sigma}_2^2))$$

which for large samples has approximately a chi-square distribution with 2 d.f. under H_0: $\beta_1 = \beta_2 = 0$. The computed value of this test statistic is

$$\text{LR} = 38.4218 - (-6.8078) = 45.23$$

which is highly significant.

Even though a sample of size $n = 8$ is too small to justify using statistical inference-making procedures whose desirable properties hold only for large samples, the decisions made about the importance of the linear (β_1) and quadratic (β_2) effects in the data agree with the conclusions previously drawn based on the standard regression analysis given in Section 15.6 of Chapter 15.

One way to assess the goodness of fit of the second-degree model in X (i.e., Model 2) is to employ a LR statistic to examine whether adding a cubic term in X (i.e., the term $\beta_3 X^3$) to the second-degree model significantly improves prediction of Y. In other words, we now consider a test of H_0: $\beta_3 = 0$ versus H_A: $\beta_3 \neq 0$. The appropriate test statistic here is computed to be

$$\text{LR} = -2 \ln L(\mathbf{Y}; \hat{\beta}_0, \hat{\beta}_1, \hat{\beta}_2, \hat{\sigma}_2^2) - (-2 \ln L(\mathbf{Y}; \hat{\beta}_0, \hat{\beta}_1, \hat{\beta}_2, \hat{\beta}_3, \hat{\sigma}_3^2))$$
$$= -6.8078 - (-17.0234) = 10.21$$

which corresponds to a P-value between 0.0005 and 0.005 based on a chi-square distribution with 1 d.f. Although this result argues for adding the term $\beta_3 X^3$ to the quadratic model, other information suggests the contrary. Specifically, the change in R^2 in going from a quadratic to a cubic model is negligible ($\Delta R^2 = 0.999 - 0.997 = 0.002$); a plot of the data clearly suggests no more than a second-degree model in X; and the LR test under discussion is based on a data set with only $n = 8$ observations and so is not completely reliable. In light of these considerations, the best conclusion is that a second-degree model in X is appropriate for describing the relationship between X and Y with high precision.

---

[11] Since we previously stated that $L(\mathbf{Y}; \hat{\beta}_0, \hat{\beta}_1, \hat{\beta}_2, \hat{\sigma}_2^2) = (0.4270)^{-4}$, it follows that

$$-2 \ln L(\mathbf{Y}; \hat{\beta}_0, \hat{\beta}_1, \hat{\beta}_2, \hat{\sigma}_2^2) = -2(-4) \ln (0.4270) = -2(-4)(-0.8510) = -6.8078$$

21.3.4 Comparison of Computer Output Using Different SAS Procedures

Table 21.2 below provides a summary of SAS computer output for fitting linear (i.e., straight line) and quadratic multiple linear regression models using the data in Table 21.1 that relates dosage (X) to weight gain (Y). We have used three different SAS computer procedures that fit

TABLE 21.2 **Edited SAS computer output from SAS's REGRESSION, MIXED, and GENMOD procedures applied to the dose (X), weight gain (Y) data of Table 21.1**

SAS Procedure	Model 1: $E(Y\|X) = \beta_0 + \beta_1 X$	Model 2: $E(Y\|X, X^2) = \beta_0 + \beta_1 X + \beta_2 X^2$
REGRESSION (Least Squares)	Variable DF Estimate SE t Pr $>$ |t| Intercept 1 -1.1964 0.7135 -1.68 0.1446 dose 1 1.1131 0.1413 **7.88** 0.0002 Analysis of Variance Source DF SS MS F Pr $>$ F Model 1 52.0372 52.0372 **62.05** 0.0002 Error 6 5.03155 0.8386 Total 7 57.06875 R-Square 0.9118	Variable DF Estimate SE t Pr $>$ |t| Intercept 1 1.3482 0.2767 4.87 0.0046 dose 1 -0.4137 0.1411 -2.93 0.0326 dosesq 1 0.1696 0.0153 **11.09** 0.0001 Analysis of Variance Source DF SS MS F Pr $>$ F Model 2 56.8720 28.4360 **722.73** $<.0001$ Error 5 0.1967 0.0394 Total 7 57.06875 R-Square 0.9966
MIXED-REML	$-2 \ln L_1 = 21.8$ Effect Estimate SE DF t Pr $>$ |t| Intercept 1.1964 0.7135 6 -1.68 0.1446 dose 1.1131 0.1413 6 **7.88** 0.0002 Type 3 Tests of Fixed Effects Effect NDF DDF F Pr $>$ F dose 1 6 **62.05** .0002	$-2 \ln L_1 = 9.0$ Effect Estimate SE DF t Pr $>$ |t| Intercept 1.3482 0.2767 5 4.87 0.0046 dose -0.4137 0.1411 5 -2.93 0.0326 dosesq 0.1696 0.0153 5 **11.09** 0.0001 Type 3 Tests of Fixed Effects Effect NDF DDF F Pr $>$ F dose 1 5 8.60 0.0326 dosesq 1 5 **122.88** 0.0001
MIXED-ML	$-2 \ln L_1 = 19.0$ Effect Estimate SE DF t Pr $>$ |t| Intercept -1.1964 0.6179 6 -1.94 0.1010 dose 1.1131 0.1224 6 **9.10** $<.0001$ Type 3 Tests of Fixed Effects Effect NDF DDF F Pr $>$ F dose 1 6 **82.74** $<.0001$	$-2 \ln L_1 = -6.9$ Effect Estimate SE DF t Pr $>$ |t| Intercept 1.3482 0.2188 5 6.16 0.0016 dose -0.4137 0.1115 5 -3.71 0.0139 dosesq 0.1696 0.0121 5 **14.02** $<.0001$ Type 3 Tests of Fixed Effects Effect NDF DDF F Pr $>$ F dose 1 5 13.76 0.0139 dose sq 1 5 **196.61** $<.0001$
GENMOD-ML	$\ln L_1 = -9.4967$ (i.e., $-2 \ln L_1 = 19.0$) Parameter DF Estimate SE Wald CS Pr $>$ CS Intercept 1 -1.1964 0.6179 3.75 0.0529 dose 1 1.1131 0.1224 **82.74** $<.0001$	$\ln L_1 = 3.4700$ (i.e., $-2 \ln L_1 = -6.94$) Parameter DF Estimate SE Wald CS Pr $>$ CS Intercept 1 1.3482 0.2188 37.98 $<.0001$ dose 1 -0.4137 0.1115 13.76 0.0002 dosesq 1 0.1696 0.0121 **196.61** $<.0001$

multiple linear regression models, namely, SAS's REGRESSION, MIXED, and GENMOD procedures. The method of estimation used by the REGRESSION procedure is unweighted least squares. The MIXED procedure uses ML estimation and a variation of ML estimation called REML (i.e., Restricted ML estimation). The GENMOD procedure fits both linear and nonlinear models and allows for the analysis of repeated measures data in which responses (Y's) on different subjects (or clusters) may be correlated (as discussed later in Chapters 25 and 26). When GENMOD is applied to a linear regression model in which all responses are assumed to be mutually independent, ML estimation is used to fit the model.

From Table 21.2, we highlight the following results:

1. For Models 1 and 2 considered separately, all four SAS procedures (considering the REML and ML approaches using MIXED as different procedures) yield identical estimated regression coefficients; these results demonstrate that unweighted least squares and ML estimates of corresponding regression coefficients are identical for multiple linear regression models (assuming normally distributed and mutually independent responses with homogeneous variance).

2. For Model 1, the REGRESSION procedure and the REML option in the MIXED procedure yield identical standard errors, t statistics (e.g., 7.88 for dose), and F statistics (i.e., 62.05); these results demonstrate that the REML option in MIXED performs unweighted least squares estimation for multiple linear regression models (assuming normally distributed and mutually independent responses with homogeneous variance).

3. For Model 2, the REGRESSION procedure and the REML option in the MIXED procedure yield identical standard errors and t statistics (e.g., 11.09 for dosesq); these results also confirm that the REML option in MIXED performs unweighted least squares estimation for multiple linear regression models.

4. For Model 1, the ML option in the MIXED procedure and the GENMOD procedure yield identical standard errors, Wald chi-square statistics (e.g., 82.74 for dose), and $-2 \ln L_1$ values (i.e., 19.0); these results indicate that the ML option in MIXED and the GENMOD procedure both perform ML estimation for multiple linear regression models (assuming normally distributed and mutually independent responses with homogeneous variance).

5. For Model 2, the ML option in the MIXED procedure and the GENMOD procedure yield identical standard errors, Wald chi-square statistics (e.g., 196.61 for dosesq), and $-2 \ln L_2$ values (i.e., -6.9); these results also indicate that the ML option in MIXED and the GENMOD procedure both perform ML estimation for multiple linear regression models (assuming normally distributed and mutually independent responses with homogeneous variance).

We have previously indicated that, when fitting multiple linear regression models, ML estimators of variance (and corresponding standard errors) are biased when compared to variance estimators based on unweighted least squares estimation (which provides unbiased variance estimators when all assumptions are satisfied). Thus, particularly for small n, inference-making conclusions may differ depending on whether least squares or ML estimation procedures are applied. Consequently, we recommend using unweighted least squares estimation (e.g., as performed by SAS's REGRESSION or MIXED-REML procedures), rather than ML estimation (e.g., SAS's MIXED-ML or GENMOD procedures), when fitting multiple linear regression models assuming normally distributed and mutually independent responses with homogeneous variance. For large n, however, the bias in variance estimators from ML estimation will typically be small, so that inference-making conclusions using least squares and ML estimation approaches typically will not differ very much.

Primarily for pedagogical reasons, we have illustrated the use of ML procedures in this chapter using multiple linear regression models, since we have only considered such models to this point in the text. Nevertheless, we emphasize that ML estimation is the method of choice for estimating the parameters in nonlinear models for non-normal responses such as the logistic regression model (discussed in Chapters 22 and 23) and the Poisson regression model (discussed in Chapter 24).

21.4 Summary

In this chapter, we have discussed the maximum likelihood (ML) method for estimating parameters and for making statistical inferences. In ML estimation, the likelihood function to be maximized must be specified. For multiple linear regression involving mutually independent normally distributed response variables with homogeneous variance, the ML estimates of regression coefficients are identical to the unweighted least-squares estimates. Note, however, that the ML estimators of the variances of estimated regression coefficients are slightly biased when compared to the corresponding variance estimators obtained from unweighted least squares. Nevertheless, ML estimation is particularly useful for estimating regression coefficients in nonlinear models involving non-normal responses (e.g., logistic regression or Poisson regression models discussed in subsequent chapters). The estimation procedure for such models typically requires a computer program that employs an iterative algorithm.

Procedures for testing hypotheses and constructing confidence intervals use maximized likelihood values and related estimated covariance matrices for the various models under study. Two alternative large-sample test procedures for testing hypotheses involve the Wald statistic and the likelihood ratio (LR) statistic. Both statistics frequently yield similar (though not identical) numerical results in small samples; when doubt exists, the likelihood ratio statistic is preferred. Upper and lower limits, respectively, of large-sample confidence intervals for (linear functions) of regression coefficients are obtained by adding to and subtracting from an ML point estimate a percentile of the standard normal distribution multiplied by the standard error of the estimated linear function under consideration.

Problems

1. A certain drug is suspected of lowering blood pressure as a side effect. A clinical trial is conducted to investigate this suspicion. Thirty-two patients are randomized into drug and placebo groups (16 per group). Their initial and posttreatment systolic blood pressures (SBP, measured in mm Hg), and their body sizes as measured by the Quetelet index (QUET), are recorded.

   The researchers conducted a multiple linear regression analysis to investigate whether the mean changes in SBP differ between the placebo and drug groups, controlling for initial SBP and QUET. The dependent variable is posttreatment SBP. One model under consideration involves three predictors: DRUG status, initial SBP, and QUET as main effects only (together with an intercept term). Suppose that a computer program that calculates unweighted least-squares estimates of the regression coefficients (e.g., the REGRESSION procedure in SAS) is used to fit this model.

   a. Are the unweighted least-squares estimates of the regression coefficients identical to the ML estimates of these same coefficients for the above model? Explain briefly.

1. b. Assuming that the estimation procedure is ML, how would one carry out a Wald test for the effect of the DRUG status variable, controlling for initial SBP and QUET? (Specify the null hypothesis being tested, the form of the test statistic, and its large-sample distribution under the null hypothesis.)

   c. Is the Wald test procedure described in part (b) equivalent to a partial F test obtained from unweighted least-squares estimation? Explain briefly.

   d. How would one carry out a likelihood ratio test for the effect of the DRUG status variable, controlling for initial SBP and QUET? (Specify the null hypothesis being tested, the form of the test statistic, and its large-sample distribution under the null hypothesis.)

   e. Should one expect to obtain the same P-value for the likelihood ratio test described in part (d) as for the Wald test described in part (c)? Explain briefly.

   f. Assuming ML estimation, state the formula for a large-sample 95% confidence interval for the effect of the DRUG status variable controlling for initial SBP and QUET.

2. The data for this question consist of a sample of 50 persons from the 1967–1980 Evans County Study (Schoenbach et al. 1986). Two basic independent variables are of interest: AGE and chronic disease status (CHR), where CHR is coded as $0 =$ none and $1 =$ presence of chronic disease. A product term of the form AGE $\times$ CHR is also considered. The dependent variable is time until death, a continuous variable. The primary question of interest is whether CHR, considered as the exposure variable, is related to survival time, controlling for AGE. The computer results, based on ML estimation,[12] are as follows:

Model 1:

Variable	Coeff	S.E.	Chisq	P-value
CHR	0.8595	0.3116	7.61	.0058
	$-2 \ln \hat{L} = 285.74$			

Model 2:

Variable	Coeff	S.E.	Chisq	P-value
CHR	0.8051	0.3252	6.13	.0133
AGE	0.0856	0.0193	19.63	.0000
	$-2 \ln \hat{L} = 264.90$			

Model 3:

Variable	Coeff	S.E.	Chisq	P-value
CHR	1.0009	2.2556	0.20	.6572
AGE	0.0874	0.0276	10.01	.0016
CHR×AGE	-0.0030	0.0345	0.01	.9301
	$-2 \ln \hat{L} = 264.89$			

---

[12] The analysis described in Problem 2 is an example of a *survival analysis,* and the model considered is called the *Cox proportional hazards model.* See Kleinbaum and Klein (2005) for a detailed discussion of survival analysis.

2. a. Assuming a regression model that contains the main effects of CHR and AGE, as well as the interaction effect between CHR and AGE, carry out a Wald test for significant interaction. (Specify the null hypothesis, the form of the test statistic, and its distribution under the null hypothesis.) What are the conclusions about interaction based on this test?

 b. For the same model considered in part (a), how would one carry out a likelihood ratio test for significant interaction? (Specify the null hypothesis, the form of the test statistic, and its distribution under the null hypothesis.) What are the conclusions about interaction based on this test? How do these results compare with those in part (a)?

 c. Assuming no interaction, carry out a Wald test for the significance of the CHR variable, controlling for AGE. (As in the preceding parts of this problem, state the null hypothesis, the form of the test statistic, and its distribution under the null hypothesis.) What are the conclusions based on this test?

 d. For the same no-interaction model considered in part (c), how would one carry out a likelihood ratio test for significance of the CHR variable, controlling for AGE? (Specify the null hypothesis, the form of the test statistic, and its distribution under the null hypothesis.) Why can't one actually carry out this test given the output information provided earlier?

 e. For the no-interaction model considered in part (c), compute a 95% confidence interval for the coefficient of the CHR variable, controlling for AGE. What does the computed confidence interval say about the reliability of the point estimate of the effect of CHR in this model?

 f. What is the overall conclusion about the effect of CHR on survival time based on the computer results from this study?

References

Kleinbaum, D. G., and Klein, M. 2005. *Survival Analysis a Self-Learning Text, Second Edition.* New York: Springer.

Schoenbach, V. J.; Kaplan, B. H.; Fredman, L.; and Kleinbaum, D. G. 1986. "Social Ties and Mortality in Evans County, Georgia." *American Journal of Epidemiology* 123(4): 577–91.

22

Logistic Regression Analysis

22.1 Preview

Logistic regression analysis is the most popular regression technique available for modeling dichotomous dependent variables. This chapter describes the logistic model form and several of its key features, particularly how an odds ratio can be estimated via its use. We also demonstrate how logistic regression may be applied, using a real-life data set.

Maximum likelihood procedures are used to estimate the model parameters of a logistic model. Therefore, the general principles and inference-making procedures described in Chapter 21 on ML estimation directly carry over to the likelihood functions appropriate for logistic regression analysis. We will examine two alternative ML procedures for logistic regression—called unconditional and conditional—which involve different likelihood functions to be maximized. The latter (conditional) method is recommended when the amount of data available for analysis is not large relative to the number of parameters in the model, as is often the case when matching on potential confounders is employed when selecting subjects.

22.2 The Logistic Model

Logistic regression is a mathematical modeling approach that can be used to describe the relationship of several predictor variables $X_1, X_2, \ldots, X_k$ to a *dichotomous* dependent variable Y, where Y is typically coded as 1 or 0 for its two possible categories. The logistic model describes the expected value of Y (i.e., $E(Y)$) in terms of the following "logistic" formula:

$$E(Y) = \frac{1}{1 + \exp\left[-\left(\beta_0 + \sum_{j=1}^{k} \beta_j X_j\right)\right]}$$

For (0, 1) random variables such as Y, it follows from basic statistical principles about expected values[1] that $E(Y)$ is equal to the probability $\text{pr}(Y = 1)$; so, the formula for the logistic model can be written in a form that describes the probability of occurrence of one of the two possible outcomes of Y, as follows:

$$\text{pr}(Y = 1) = \frac{1}{1 + \exp\left[-\left(\beta_0 + \sum_{j=1}^{k}\beta_j X_j\right)\right]} \tag{22.1}$$

The logistic model (22.1) is useful in many important practical situations where the response variable can take one of two possible values. For example, a study of the development of a particular disease in some human population could employ a logistic model to describe in probabilistic terms whether a given individual in the study group will ($Y = 1$) or will not ($Y = 0$) develop the disease in question during a follow-up period of interest.

The first step in logistic regression analysis is to postulate (based on knowledge about, and experience with, the process under study) a mathematical model describing the mean of Y as a function of the X_j and the β_j values. The model is then fitted to the data by maximum likelihood, and eventually appropriate statistical inferences are made (after the model's adequacy of fit is verified, including consideration of relevant regression diagnostic indices).

The mathematical expression given on the right side of the logistic model formula given by (22.1) is of the general mathematical form

$$f(z) = \frac{1}{1 + e^{-z}}$$

where $z = \beta_0 + \sum_{j=1}^{k}\beta_j X_j$. The function $f(z)$ is called the *logistic function*. This function is well-suited to modeling a probability, since the values of $f(z)$ increase from 0 to 1 as z increases from $-\infty$ to $+\infty$. In epidemiologic studies, such a probability can be used to state an individual's risk of developing a disease. The logistic model, therefore, is set up to ensure that, whatever estimated value of risk we obtain, that value always falls between 0 and 1. This is not true for other possible models, which is why the logistic model is often used when a probability must be estimated.

Another reason why the logistic model is popular relates to the general sigmoid shape of the logistic function (see Figure 22.1). A sigmoid shape is particularly appealing to epidemiologists if

FIGURE 22.1   The logistic function $f(z) = \dfrac{1}{1 + e^{-z}}$

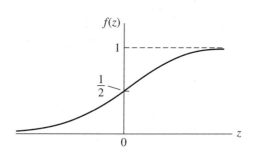

[1] For a (0, 1) random variable Y, $E(Y) = [0 \times \text{pr}(Y = 0)] + [1 \times \text{pr}(Y = 1)] = \text{pr}(Y = 1)$.

the variable z is viewed as representing an index that combines the contributions of several risk factors, so that $f(z)$ represents the risk for a given value of z. In this context, the risk is minimal for low z values, rises over a range of intermediate values of z, and remains close to 1 once z gets large enough. Epidemiologists believe that this sigmoid shape applies to a variety of disease conditions.

22.3 Estimating the Odds Ratio Using Logistic Regression

As in any regression model, the regression coefficients β_j in the logistic model given by (22.1) play an important role in providing information about the relationships of the predictors in the model to the dichotomous dependent variable. For the logistic model, quantification of these relationships involves a parameter called the *odds ratio*.

The odds ratio is a widely used measure of effect in epidemiologic studies. By "measure of effect," we mean a measure that compares two or more groups with regard to the outcome (dependent) variable. To describe an odds ratio, we first define *odds* as the ratio of the probability that some event (e.g., developing lung cancer) will occur divided by the probability that the same event will not occur (e.g., not developing lung cancer), Thus, the odds for some event D is given by the formula

$$\text{odds}(D) = \frac{\text{pr}(D)}{\text{pr}(\text{not } D)} = \frac{\text{pr}(D)}{1 - \text{pr}(D)}$$

For example, if $\text{pr}(D) = .25$, then

$$\text{odds}(D) = \frac{.25}{1 - .25} = \frac{1}{3}$$

An odds of one-third can be interpreted to mean that the probability of event D occurring is $1/3$ the probability of event D not occurring; equivalently, the odds are "3 to 1" that event D will not happen.

Any odds ratio (OR), by definition, is a ratio of two odds; that is,

$$\text{OR}_{A \text{ vs. } B} = \frac{\text{odds}(D_A)}{\text{odds}(D_B)} = \frac{\text{pr}(D_A)}{1 - \text{pr}(D_A)} \bigg/ \frac{\text{pr}(D_B)}{1 - \text{pr}(D_B)}$$

in which the subscripts A and B denote two groups of individuals being compared. For example, suppose that $A = S$ denotes a group of smokers and $B = NS$ denotes a group of nonsmokers; then D_S is the event that a smoker develops lung cancer, and D_{NS} is the event that a nonsmoker develops lung cancer. If $\text{pr}(D_S) = .0025$ and $\text{pr}(D_{NS}) = .0010$, the odds ratio that compares the odds of developing lung cancer for smokers with the odds of developing lung cancer for nonsmokers is given by

$$\text{OR}_{S \text{ vs. } NS} = \frac{.0025}{1 - .0025} \bigg/ \frac{.0010}{1 - .0010} = 2.504$$

In other words, the odds of developing lung cancer for smokers is about 2.5 times the corresponding odds for nonsmokers, suggesting roughly that smokers have roughly a 2.5 times higher risk of developing lung cancer than nonsmokers. An odds ratio of one would mean that the odds for the two groups are the same; that is, it would indicate that there is no effect of smoking on the development of lung cancer.

Where does logistic regression fit in here? To answer this question, we must consider an equivalent way to write the logistic regression model, called the *logit form* of the model. The "logit" is a transformation of the probability $\text{pr}(Y = 1)$, defined as the natural log odds of the event $D = \{Y = 1\}$. In other words,

$$\text{logit}[\text{pr}(Y = 1)] = \log_e[\text{odds}(Y = 1)] = \log_e\left[\frac{\text{pr}(Y = 1)}{1 - \text{pr}(Y = 1)}\right] \tag{22.2}$$

If we then substitute the logistic model formula (22.1) for $\text{pr}(Y = 1)$ into equation (22.2), it follows that

$$\text{logit}[\text{pr}(Y = 1)] = \beta_0 + \sum_{j=1}^{k}\beta_j X_j \tag{22.3}$$

Equation (22.3) is called the *logit form* of the model. The logit form is given by the linear function $(\beta_0 + \sum_{j=1}^{k}\beta_j X_j)$. For convenience, many authors describe the logistic model in its logit form given by (22.3), rather than in its original form defined by (22.1).

For example, if Y denotes lung cancer status ($1 = $ yes, $0 = $ no) and there is only one (i.e., $k = 1$) predictor X_1—say, smoking status ($1 = $ smoker, $0 = $ nonsmoker)—then the logistic model (22.1) can be written equivalently in logit form (22.3) as

$$\text{logit}[\text{pr}(Y = 1)] = \beta_0 + \beta_1 X_1 = \beta_0 + \beta_1(\text{smoking status})$$

To obtain an expression for the odds ratio using a logistic model, we must compare the odds for two groups of individuals. For the preceding example involving one predictor, the two groups are smokers ($X_1 = 1$) and nonsmokers ($X_1 = 0$). Thus, for this example, the log odds for smokers and nonsmokers can be written as

$$\log_e \text{odds}(\text{smokers}) = \beta_0 + (\beta_1)(1) = \beta_0 + \beta_1$$

and

$$\log_e \text{odds}(\text{nonsmokers}) = \beta_0 + (\beta_1)(0) = \beta_0$$

respectively. It follows that the odds ratio comparing smokers to nonsmokers is given by

$$\text{OR}_{S \text{ vs. } NS} = \frac{\text{odds}(\text{smokers})}{\text{odds}(\text{nonsmokers})} = \frac{e^{(\beta_0 + \beta_1)}}{e^{\beta_0}} = e^{\beta_1}$$

In other words, for the simple example involving one (0–1) predictor, the odds ratio comparing the two categories of the predictor is obtained by exponentiating the coefficient of the predictor in the logistic model.

Generally, when computing an odds ratio, we can define the two groups (or individuals) that are to be compared in terms of two different specifications of the set of predictors $X_1, X_2, \ldots, X_k$. We do this by letting $\mathbf{X_A} = (X_{A1}, X_{A2}, \ldots, X_{Ak})$ and $\mathbf{X_B} = (X_{B1}, X_{B2}, \ldots, X_{Bk})$ denote the collection of X's for groups (or individuals) A and B, respectively. For example, if $k = 3$, X_1 is smoking status ($1 = $ yes, $0 = $ no), X_2 is age (continuous), and X_3 is race ($1 = $ black, $0 = $ white), then $\mathbf{X_A}$ and $\mathbf{X_B}$ are two specifications of these three variables—say,

$$\mathbf{X_A} = (1, 45, 1) \text{ and } \mathbf{X_B} = (0, 45, 1).$$

Here, $\mathbf{X_A}$ denotes the group of 45-year-old black smokers, whereas $\mathbf{X_B}$ denotes the group of 45-year-old black nonsmokers.

To obtain a general formula for the odds ratio, we must divide the odds for group (or individual) A by the odds for group (or individual) B and then appropriately employ the logit form of the logistic model given by (22.1) to obtain an expression involving the logistic model parameters. Using some algebra, the following result is obtained:

$$OR_{\mathbf{X_A} \text{ vs. } \mathbf{X_B}} = \frac{\text{odds for } \mathbf{X_A}}{\text{odds for } \mathbf{X_B}} = \frac{e^{(\beta_0 + \sum_{j=1}^{k} \beta_j X_{Aj})}}{e^{(\beta_0 + \sum_{j=1}^{k} \beta_j X_{Bj})}}$$

We can simplify the above expression further[2] to obtain the following general formula for the odds ratio:

$$OR_{\mathbf{X_A} \text{ vs. } \mathbf{X_B}} = e^{\sum_{j=1}^{k}(X_{Aj} - X_{Bj})\beta_j} \tag{22.4}$$

The constant term β_0 in the logistic model (22.1) does not appear in the odds ratio expression (22.4), so the odds ratio depends only on the β_j coefficients and the corresponding differences in values of the X_j's for groups (or individuals) A and B. Expression (22.4) describes a "population" odds ratio parameter because the β_j's in this expression are themselves unknown population parameters. An estimate of this population odds ratio can be obtained by fitting the logistic model using maximum likelihood estimation and substituting the ML estimates $\hat{\beta}_1, \hat{\beta}_2, \ldots, \hat{\beta}_k$, together with values of X_{Aj} and X_{Bj}, into the formula (22.4) to obtain a numerical value for the odds ratio.

For example, if $\mathbf{X_A}$ and $\mathbf{X_B}$ are two specifications of the three variables smoking status, age, and race, so that $\mathbf{X_A} = (1, 45, 1)$ and $\mathbf{X_B} = (0, 45, 1)$ as described earlier, then

$$OR_{\mathbf{X_A} \text{ vs. } \mathbf{X_B}} = e^{(1-0)\beta_1 + (45-45)\beta_2 + (1-1)\beta_3} = e^{\beta_1}$$

If the estimate of the β_1 coefficient from ML estimation turns out to be, say, $\hat{\beta}_1 = 1.32$, then the estimated odds ratio will be $e^{1.32} = 3.74$. Note that the logistic model in this example involves three variables, as follows (in logit form):

$$\text{logit}[\text{pr}(Y = 1)] = \beta_0 + \beta_1(\text{smoking}) + \beta_2(\text{age}) + \beta_3(\text{race})$$

So, the value of the estimate $\hat{\beta}_1$ for this three-variable model will almost surely be numerically different from the value obtained for $\hat{\beta}_1$ in a model containing only smoking status.

In this latest example, the smoking status variable changes from 1 in group A to 0 in group B, whereas the other variables remain the same for each group—namely, age is 45 and race is 1. In general, whenever only one variable (e.g., smoking) changes, while the other variables are fixed, we say that the odds ratio comparing two categories of the changing variable (e.g., smokers versus nonsmokers) is an *adjusted odds ratio* that *controls* for the other variables (i.e., those that are fixed at specific values) in the model. The variable of interest—in this case, smoking

---

[2] For any two values a and b, it follows that $\dfrac{e^a}{e^b} = e^{(a-b)}$.

status—is often referred to as the *exposure* (or *study*) variable; the other variables in the model are often called *control* (or *confounder*) variables. Thus, we have, as an important special case of the odds ratio expression (22.4), the following rule:

> ☐ An adjusted odds ratio can be obtained by exponentiating the coefficient of a (0–1) expo-sure variable in the logistic model (provided that there are no cross-product terms in the model involving the exposure variable).

The examples that we have considered so far have involved only *main effect variables* like smoking, age, and race; we have not considered product terms like "smoking $\times$ age" or "smok-ing $\times$ race," nor have we considered exposure variables other than (0–1) variables. When the model contains product terms (like "smoking $\times$ age") or exposure variables that are not (0–1) variables, the preceding simple rule for obtaining an adjusted odds ratio does not work. In such instances, we must use the general formula given by (22.4).

For example, suppose, as before, that smoking is a (0–1) exposure variable, and that age (continuous) and race (0–1) are control variables; but, suppose now that the logistic model we want to fit is given (in logit form) as

$$\text{logit}[\text{pr}(Y = 1)] = \beta_0 + \beta_1(\text{smoking}) + \beta_2(\text{age}) + \beta_3(\text{race})$$
$$+ \beta_4(\text{smoking} \times \text{age}) + \beta_5(\text{smoking} \times \text{race}) \tag{22.5}$$

Then, to obtain an adjusted odds ratio for the effect of smoking adjusted for age and race, we need to specify two sets of values $\mathbf{X_A}$ and $\mathbf{X_B}$ for the collection of predictor variables defined by

$$\mathbf{X} = (\text{smoking, age, race, smoking} \times \text{age, smoking} \times \text{race})$$

If the two groups being compared are, as before, 45-year-old black smokers (i.e., group A), and 45-year-old black nonsmokers (i.e., group B), then $\mathbf{X_A}$ and $\mathbf{X_B}$ are given as

$$\mathbf{X_A} = (1, 45, 1, 1 \times 45, 1 \times 1) = (1, 45, 1, 45, 1)$$

and

$$\mathbf{X_B} = (0, 45, 1, 0 \times 45, 0 \times 1) = (0, 45, 1, 0, 0)$$

Applying the odds ratio formula (22.4) to model (22.5) with the given values for $\mathbf{X_A}$ and $\mathbf{X_B}$, we obtain the following expression for the adjusted odds ratio:

$$\text{OR}_{\mathbf{X_A} \text{ vs. } \mathbf{X_B}} = e^{(1-0)\beta_1 + (45-45)\beta_2 + (1-1)\beta_3 + (45-0)\beta_4 + (1-0)\beta_5} = e^{\beta_1 + 45\beta_4 + \beta_5}$$

Since this odds ratio describes the effect of smoking adjusted for age $= 45$ and race $= 1$, we can alternatively denote the odds ratio as

$$\text{OR}_{(S \text{ vs. } NS|\text{age}=45, \text{ race}=1)}$$

where the $|$ sign is common mathematical notation for "given." Thus, we have

$$\text{OR}_{(S \text{ vs. } NS|\text{age}=45, \text{ race}=1)} = e^{\beta_1 + 45\beta_4 + \beta_5}$$

Rather than involving only β_1, the preceding expression involves the three coefficients β_1, β_4, and β_5, each of which is a coefficient of a variable in the model that contains the exposure variable (i.e., smoking) in some form. The coefficients β_4 and β_5 are included because they are coeffi-cients of "interaction terms" in the model being fit. The model essentially says that the value of the odds ratio for the effect of smoking should vary depending on the values of the variables age and race (which are components of the two interaction terms in model (22.5)) and of the regression coefficients β_4 and β_5. Since the "fixed" value of age is 45 and the "fixed" value of race is 1, these two values multiply their corresponding regression coefficients in the odds ratio expression. The variables age and race in model (22.5) are examples of *effect modifiers* (see Kleinbaum 2002, Lesson 10, and/or Kleinbaum, Kupper, and Morgenstern 1982, Chapter 13) since the *effect* of the exposure variable smoking status, as quantified by the odds ratio $OR_{(S \text{ vs. } NS|age, race)}$, changes (i.e., *is modified*) depending on the values of age and race.

To illustrate how age and race are effect modifiers of smoking status in model (22.5), suppose that we let the values of age and race in $\mathbf{X_A}$ and $\mathbf{X_B}$ be fixed but unspecified, so we can write $\mathbf{X_A}$ and $\mathbf{X_B}$ as

$$\mathbf{X_A} = (1, \text{age}, \text{race}, 1 \times \text{age}, 1 \times \text{race}) = (1, \text{age}, \text{race}, \text{age}, \text{race})$$

and

$$\mathbf{X_B} = (0, \text{age}, \text{race}, 0 \times \text{age}, 0 \times \text{race}) = (0, \text{age}, \text{race}, 0, 0)$$

Then, substituting $\mathbf{X_A}$ and $\mathbf{X_B}$ into formula (22.4) for the odds ratio, we obtain

$$OR_{(S \text{ vs. } NS|age, race)} = e^{(1-0)\beta_1 + (age-age)\beta_2 + (race-race)\beta_3 + (age-0)\beta_4 + (race-0)\beta_5}$$
$$= e^{\beta_1 + \beta_4(age) + \beta_5(race)}$$

This formula says that the value of the odds ratio for the effect of smoking varies depending on the values of the variables age and race (which are components of the two interaction variables in the model) and of their coefficients β_4 and β_5. So, for example, if we have age = 45 and race = 1, as in our earlier example, the adjusted odds ratio is given by

$$OR_{(S \text{ vs. } NS|age=45, race=1)} = e^{\beta_1 + 45\beta_4 + \beta_5}$$

Similarly, if we have age = 35 and race = 0, the adjusted odds ratio is given by

$$OR_{(S \text{ vs. } NS|age=35, race=0)} = e^{\beta_1 + 35\beta_4}$$

And if age = 20 and race = 1, then

$$OR_{(S \text{ vs. } NS|age=20, race=1)} = e^{\beta_1 + 20\beta_4 + \beta_5}$$

The preceding examples illustrate another general rule that describes an important special case of the general odds ratio formula (22.4):

☐ For a logistic model that contains a (0–1) exposure variable together with interaction terms that are products of the exposure variable with other control variables, the adjusted odds ratio is obtained by exponentiating a linear function of the regression coefficients involving the exposure alone and the product terms in the model involving exposure. Moreover, since the model being fit contains interaction terms, the numerical value of the adjusted odds ratio will vary depending on the values of the control variables (i.e., effect modifiers) that are components of the product terms involving exposure.

We now consider one other important special case of the general odds ratio formula (22.4). Suppose that the exposure variable is a continuous variable, like systolic blood pressure (SBP), so that, for example, our model contains SBP and the control variables age and race; that is, the logit form of the no-interaction model is given by

$$\text{logit}[\text{pr}(Y = 1)] = \beta_0 + \beta_1(\text{SBP}) + \beta_2(\text{age}) + \beta_3(\text{race}) \tag{22.6}$$

To obtain an adjusted odds ratio for the effect of SBP, controlling for age and race, we must specify two values of the exposure variable (SBP) to be compared. Two specified values are needed, even when the exposure variable (like SBP) is continuous, because an odds ratio by definition compares the odds for two groups (or individuals). For example, if the two values of SBP are 160 and 120, and age and race are considered fixed, the general odds ratio expression (22.4) simplifies to

$$\text{OR}_{(\text{SBP}=160 \text{ vs. SBP}=120 \mid \text{age, race})} = e^{(160-120)\beta_1 + (\text{age}-\text{age})\beta_2 + (\text{race}-\text{race})\beta_3} = e^{40\beta_1}$$

More generally, if we specify the two values of SBP to be SBP_1 and SBP_0, then the odds ratio is

$$\text{OR}_{(\text{SBP}_1 \text{ vs. SBP}_0 \mid \text{age, race})} = e^{(\text{SBP}_1-\text{SBP}_0)\beta_1}$$

This odds ratio expression simplifies to e^{β_1} when the difference $(\text{SBP}_1 - \text{SBP}_0)$ equals one. Thus, exponentiating the coefficient of the exposure variable gives an odds ratio for comparing any two groups that differ by one unit of SBP. A one-unit difference in SBP is rarely of interest, however. A more typical choice of SBP values to be compared would be clinically meaningful different values of blood pressure, such as SBP values of 160 and 120. One possible strategy for choosing values of SBP to compare is to categorize the distribution of SBP in a data set into clinically different categories—say, quintiles. Then, using the mean or median SBP in each quintile, we could compute odds ratios for all possible pairs of mean or median SBP values. We would then obtain a table of odds ratios to consider in assessing the relationship of SBP to the outcome variable. (For a more thorough treatment of important special cases for computing the odds ratio, see Chapter 3 of Kleinbaum and Klein 2002).

ML estimates of the regression coefficients are typically obtained by using standard computer packages for logistic regression. These estimates can then be used in appropriate odds ratio formulas based on the general expression (22.4) to obtain numerical values for adjusted odds ratios. Since these are point estimates, researchers typically carry out statistical inferences about the odds ratios that are being estimated. For example, if the adjusted odds ratio is given by the simple expression e^{β_1}, involving a single coefficient, then the null hypothesis that this odds ratio equals 1 can be stated equivalently as $H_0: \beta_1 = 0$, since, under H_0, $e^{\beta_1} = e^0 = 1$. The test for this null hypothesis can be carried out by using either the Wald test or the likelihood ratio test described in Chapter 21 on ML estimation methods. A confidence interval for the adjusted odds ratio can be obtained by first calculating a confidence interval for β_1, as described in Section 21.3, and then exponentiating the lower and upper limits. For this simple situation, an appropriate large-sample $100(1 - \alpha)\%$ confidence interval for β_1 is given by the formula

$$\exp(\hat{\beta}_1 \pm Z_{1-\alpha/2}S_{\hat{\beta}_1})$$

where $\hat{\beta}_1$ is the ML estimate of β_1, $S_{\hat{\beta}_1}$ is the standard error of $\hat{\beta}_1$, and $(1 - \alpha)$ is the confidence level.

More detailed discussion of the properties and applications of logistic regression may be found in several other textbooks, including Kleinbaum and Klein (2002), Hosmer and Lemeshow (1989), Collett (1991), and Kleinbaum, Kupper, and Morgenstern (1982). In particular, a general analysis strategy for selecting the variables to retain in a logistic model is described in Kleinbaum and Klein (2002), Chapters 6 and 7, and in Kleinbaum, Kupper, and Morgenstern (1982), Chapters 21–24. The goal of this strategy is to obtain a *valid* estimate (in the context of the analysis of epidemiologic research data) of the relationship between a specified exposure variable and a particular disease variable, while controlling for other covariates that, if not correctly taken into account, can lead to an incorrect assessment of the strength of the exposure–disease relationship of interest. These covariates can act as confounders and/or effect modifiers, terms we discussed in Chapter 11.

22.4 A Numerical Example of Logistic Regression

Dengue fever is an acute infectious disease caused by a virus transmitted by the *Aedes* mosquito. A retrospective survey of an epidemic of dengue fever[3] was carried out in three Mexican cities in 1984. In this section, we review the analyses of a subset of the data from this study obtained on a two-stage stratified random sample of 196 persons from the city of Puerto Vallarta, 57 of whom were determined to be suffering from dengue fever. The goal of the analyses was to identify risk factors associated with having the disease, especially the effect of the absence of mosquito netting around a subject's bed as a determinant of the disease. The following variables were tracked in a computer file called DENGUE.DAT[4]:

Column 1: Subject ID

Column 2: Dengue fever status (DENGUE): 1 = yes, 2 = no

Column 3: AGE in years

Column 4: Use of mosquito netting (MOSNET): 0 = yes, 1 = no

Column 5: Geographical sector in which the subject lived (SECTOR): 1, 2, 3, 4, or 5.

The variable SECTOR was treated as a categorical variable in the regression analysis, so four dummy variables had to be created to distinguish the five geographical sectors. These variables were defined so that sector 5 was the referent group, as follows:

$$\text{SECTOR1} = \begin{cases} 1 & \text{if sector 1} \\ 0 & \text{if other} \end{cases} \qquad \text{SECTOR2} = \begin{cases} 1 & \text{if sector 2} \\ 0 & \text{if other} \end{cases}$$

$$\text{SECTOR3} = \begin{cases} 1 & \text{if sector 3} \\ 0 & \text{if other} \end{cases} \qquad \text{SECTOR4} = \begin{cases} 1 & \text{if sector 4} \\ 0 & \text{if other} \end{cases}$$

The following block of edited computer output comes from using SAS's LOGISTIC procedure to fit a logistic regression model that regresses the dichotomous outcome variable DENGUE on the predictors MOSNET, AGE, and SECTORj, for $j = 1, 2, 3, 4$. The logit form of the model being fit is given as

$$\begin{aligned} \text{logit}[\text{pr}(Y = 1)] = {} & \beta_0 + \beta_1(\text{AGE}) + \beta_2(\text{MOSNET}) + \beta_3(\text{SECTOR1}) \\ & + \beta_4(\text{SECTOR2}) + \beta_5(\text{SECTOR3}) + \beta_6(\text{SECTOR4}) \end{aligned} \qquad (22.7)$$

where Y denotes the dependent variable DENGUE.

---

[3] Dantes, Koopman, et al. (1988)

[4] The SAS data file dengue.dat can be freely downloaded from the publisher's website http://www.thomsonedu.com/statistics/kleinbaum that contains other data files in examples and problems throughout the text. This website also contains data files for the STATA and SPSS packages in addition to SAS.

```
                           The LOGISTIC Procedure
                              Response Profile
                     Ordered
                      Value          DENGUE           Count
                        1              1                57
                        2              2               139
                 Criteria for Assessing Model Fit
                                     Intercept
                      Intercept         and
   Criterion           only         Covariates      Chi-Square for Covariates
   AIC                238.329         217.706              .
   SC                 241.607         240.653              .
   -2 LOG L           236.329         203.706         32.623 with 6 DF (p = 0.0001)
   Score                 .               .            28.775 with 6 DF (p = 0.0001)

                 Analysis of Maximum Likelihood Estimates
                                       Wald       Pr >
                    Parameter  Standard Chi-      Chi-      Standardized    Odds
   Variable   DF    Estimate    Error   Square    Square     Estimate      Ratio
   INTERCPT    1     -1.9001   1.3254   2.0551    0.1517                   0.150
   AGE         1      0.0243   0.00906  7.1778    0.0074     0.252890      1.025
   MOSNET      1      0.3335   1.2718   0.0688    0.7931     0.034212      1.396
   SECTOR1     1     -2.2200   1.0723   4.2861    0.0384    -0.441811      0.109
   SECTOR2     1     -0.6589   0.5536   1.4164    0.2340    -0.142513      0.517
   SECTOR3     1      0.8121   0.4750   2.9235    0.0873     0.173824      2.253
   SECTOR4     1      0.5310   0.4502   1.3911    0.2382     0.121456      1.701
```

Using the given computer output, we now focus on the information provided under the heading "Analysis of Maximum Likelihood Estimates." From this information, we can see that the ML-based estimated regression coefficients obtained for the fitted model are

$$\hat{\beta}_0 = -1.9001, \hat{\beta}_1 = 0.0243, \hat{\beta}_2 = 0.3335, \hat{\beta}_3 = -2.2200,$$
$$\hat{\beta}_4 = -0.6589, \hat{\beta}_5 = 0.8121, \hat{\beta}_6 = 0.5310$$

so the fitted model is given (in logit form) by

$$\text{logit}[\widehat{\text{pr}(Y = 1)}] = -1.9001 + 0.0243(\text{AGE}) + 0.3335(\text{MOSNET})$$
$$- 2.2200(\text{SECTOR1}) - 0.6589(\text{SECTOR2})$$
$$+ 0.8121(\text{SECTOR3}) + 0.5310(\text{SECTOR4})$$

Based on this fitted model and the information provided in the computer output, we can compute the estimated odds ratio for contracting dengue fever for persons who did not use mosquito netting relative to persons who did use mosquito netting, controlling for age and sector. We do this using the previously stated rule for adjusted odds ratios for (0–1) variables (when there is no interaction) by exponentiating the estimated coefficient ($\hat{\beta}_2$) of the MOSNET variable in the fitted model. We then obtain the following value for the adjusted odds ratio:

$$\widehat{\text{OR}}_{(\text{MOSNET} = 1 \text{ vs.MOSNET} = 0| \text{ age, sector})} = e^{0.3335} = 1.396$$

A 95% confidence interval for e^{β_2} can be obtained by computing

$$\exp [0.3335 \pm 1.96(1.2718)]$$

where 1.2718 is the estimated standard error of the estimator $\hat{\beta}_2$ of β_2 in model (22.7). The resulting lower and upper confidence limits are 0.115 and 16.89, respectively. The Wald (chi-square) statistic for testing the null hypothesis that the adjusted odds ratio e^{β_2} is equal to 1 (or equivalently, that the coefficient β_2 of the MOSNET variable equals 0) is shown in the output to be 0.0688, with a P-value equal to 0.7931.

From these results, it can be concluded that the estimated odds of contracting dengue is about 1.4 times higher for a person who did not use mosquito netting than for a person who did use mosquito netting. The Wald statistic is not statistically significant, however, and the 95% confidence interval is very wide (and includes the null value of 1). Thus, there is no statistical evidence in these data that the nonuse of mosquito netting significantly increases the probability (or risk) of contracting dengue fever.

The preceding computer output contains, under the heading "Criteria for Assessing Model Fit," the log likelihood statistic $(-2 \log \hat{L})$ of 203.706 for the fitted model (in the column labeled "Intercept and Covariates"). To compute the likelihood ratio test for the null hypothesis $H_0: \beta_2 = 0$, we must compare this value of $-2 \log \hat{L}$ to the corresponding value of $-2 \log \hat{L}$ for the "reduced" model, which is obtained under the null hypothesis by dropping the exposure variable (MOSNET) from the "full" model given by (22.7). The reduced model is written in logit form as

$$\text{logit}[\text{pr}(Y = 1)] = \beta_0 + \beta_1(\text{AGE}) + \beta_3(\text{SECTOR1}) + \beta_4(\text{SECTOR2}) + \beta_5(\text{SECTOR3}) + \beta_6(\text{SECTOR4})$$

The following block of edited computer output for the reduced model indicates that the log likelihood statistic for the fitted model has changed slightly to 203.778. The likelihood ratio statistic for comparing these two models is given by the difference:

$$\text{LR} = -2 \log \hat{L}_R - (-2 \log \hat{L}_F) = 203.778 - 203.706 = 0.072$$

Because the null hypothesis $H_0: \beta_2 = 0$ involves only one parameter, the preceding LR statistic is distributed as approximately a chi-square variable with 1 degree of freedom under the null hypothesis. The test is nonsignificant, which agrees with the result from the Wald test described earlier, thereby supporting the previous conclusion that using mosquito netting did not significantly affect a person's probability of contracting dengue fever.

Output for Reduced Model

```
                        The LOGISTIC Procedure
                          Response Profile

                    Ordered
                     Value          DENGUE              Count
                       1              1                  57
                       2              2                 139

                 Criteria for Assessing Model Fit

                                   Intercept
                   Intercept          and
Criterion           Only          Covariates        Chi-Square for Covariates
AIC               238.329           215.778
SC                241.607           235.447
-2 LOG L          236.329           203.778          32.551 with 5 DF (p = 0.0001)
Score                .                 .              28.766 with 5 DF (p = 0.0001)
```

(continued)

where $\hat{L} = \hat{\beta}_2 + \hat{\beta}_7(\text{AGE})$ and $S_{\hat{L}} = \sqrt{\widehat{\text{Var}}(\hat{L})}$, and where $\sqrt{\widehat{\text{Var}}(\hat{L})}$ is computed using the formula for the variance of a linear combination of random variables, which for this example turns out to be

$$\widehat{\text{Var}}(\hat{L}) = \widehat{\text{Var}}(\hat{\beta}_2) + (\text{age})^2\,\widehat{\text{Var}}(\hat{\beta}_7) + 2(\text{age})\widehat{\text{Cov}}(\hat{\beta}_2, \hat{\beta}_7)$$

The numerical values of the variances and the covariances involving the $\hat{\beta}_j$'s are typically printed out as an option by the computer program used. For model (22.8), the numerical values obtained for the preceding variance formula are:

$$\widehat{\text{Var}}(\hat{\beta}_2) = 2.7004, \qquad \widehat{\text{Var}}(\hat{\beta}_7) = 0.001399, \qquad \text{and} \qquad \widehat{\text{Cov}}(\hat{\beta}_2, \hat{\beta}_7) = -0.0435$$

If, for example, we let AGE = 40, then

$$\hat{L} = -0.8043 + 0.0306(40) = 0.4197$$

$$\widehat{\text{Var}}(\hat{L}) = 2.7004 + (40)^2(0.001399) + 2(40)(-0.0435) = 1.4588$$

and

$$S_{\hat{L}} = \sqrt{\widehat{\text{Var}}(\hat{L})} = 1.2078$$

The 95% confidence interval for the adjusted odds ratio $e^{\beta_2 + (40)\beta_7}$ is then given by

$$\exp[0.4197 \pm (1.96)(1.2078)]$$

which yields lower and upper confidence limits of 0.14 and 16.23, respectively. This confidence interval is extremely wide, which implies that the point estimate $e^{0.4197} = 1.52$ when age is 40 is very unreliable. Furthermore, since the confidence interval includes the null value of 1, a test of significance for the adjusted odds ratio is not significant at the 5% level. Computations of 95% intervals for values of age other than 40 also yield very wide intervals and nonsignificant findings.

Finally, we may wish to compare model (22.8), which contains the interaction term MSA, with the no-interaction model (22.7). We can do this by using a likelihood ratio test that takes the following form:

$$\text{LR} = -2\log\hat{L}_R - (-2\log\hat{L}_F) = 203.706 - 202.995 = 0.711$$

The full model (F) in this case is model (22.8), and the reduced model (R) is model (22.7). The null hypothesis is H_0: $\beta_7 = 0$, which involves only one parameter, so that the LR statistic is approximately chi-square with 1 d.f. under H_0. The test statistic value of 0.711 is nonsignificant, indicating that the no-interaction model (22.7) is preferable to the interaction model (22.8). In other words, the estimated odds ratio for the effect of MOSNET adjusted for age and sector status is best expressed by the single value $\exp(0.3335) = 1.396$ based on model (22.7), rather than by the expression $\exp(\hat{L})$, where $\hat{L} = \hat{\beta}_2 + \hat{\beta}_7(\text{AGE})$, based on model (22.8).

22.5 Theoretical Considerations

In this section, we consider the form of the likelihood function to be maximized for a logistic regression model. In particular, we will distinguish between two alternative ML procedures for estimation, called *unconditional* and *conditional*, that involve different likelihood functions.

To describe these two procedures, we must first discuss the distributional properties of the dependent (i.e., outcome) variable underlying the logistic model.

For logistic regression, the basic dependent random variable of interest is a dichotomous variable Y taking the value 1 with probability θ and the value 0 with probability $(1 - \theta)$. Such a random variable is called *Bernoulli* (or point-binomial) and has the simple discrete probability distribution

$$\text{pr}(Y; \theta) = \theta^Y(1 - \theta)^{1-Y}, \quad Y = 0, 1 \tag{22.9}$$

The name *point-binomial* arises because (22.9) is a special case of the binomial distribution $_nC_Y\theta^Y(1 - \theta)^{n-Y}$ when $n = 1$, where $_nC_Y$ denotes the number of combinations of n distinct objects selected Y-at-a-time.

In general, for a study sample of n subjects, suppose (for $i = 1, 2, \ldots, n$) that Y_i denotes the Bernoulli random variable for the ith subject, having the Bernoulli distribution

$$\text{pr}(Y_i; \theta_i) = \theta_i^{Y_i}(1 - \theta_i)^{1-Y_i}, \quad Y_i = 0, 1 \tag{22.10}$$

For example, θ_i could represent the probability that individual i in a random sample of n individuals from some population will develop some particular disease during the follow-up period in question.

22.5.1 Unconditional ML Estimation

Given that $Y_1, Y_2, \ldots, Y_n$ are mutually independent, the likelihood function based on (22.10) is obtained as the product of the marginal distributions for the Y_i's—namely,

$$L(\mathbf{Y}; \boldsymbol{\theta}) = \prod_{i=1}^{n} \text{pr}(Y_i; \theta_i) = \prod_{i=1}^{n} \left[\theta_i^{Y_i}(1 - \theta_i)^{1-Y_i}\right] \tag{22.11}$$

where $\boldsymbol{\theta} = (\theta_1, \theta_2, \ldots, \theta_n)$. Now suppose, without loss of generality, that the first n_1 out of the n individuals in our random sample actually develop the disease in question (so that $Y_1 = Y_2 = \cdots = Y_{n_1} = 1$), and that the remaining $(n - n_1)$ individuals do not (so that $Y_{n_1+1} = Y_{n_1+2} = \cdots = Y_n = 0$). Given this set of observed outcomes, the likelihood expression (22.11) takes the specific form

$$L(\mathbf{Y}; \boldsymbol{\theta}) = \left(\prod_{i=1}^{n_1} \theta_i\right)\left[\prod_{i=n_1+1}^{n} (1 - \theta_i)\right] \tag{22.12}$$

We can work with expression (22.12) to write the likelihood function in terms of the regression coefficients β_j in the logistic model. We let $\mathbf{X}_i = (X_{i1}, X_{i2}, \ldots, X_{ik})$ denote the set of values of the k predictors $X_1, X_2, \ldots, X_k$ specific to individual i. Then the logistic model assumes that the relationship between θ_i and the X_{ij}'s is of the specific form

$$\theta_i = \frac{1}{1 + \exp\left[-\left(\beta_0 + \sum_{j=1}^{k}\beta_j X_{ij}\right)\right]}, \quad i = 1, 2, \ldots, n \tag{22.13}$$

where $\beta_j, j = 0, 1, \ldots, k$, are unknown regression coefficients that must be estimated. (The right side of (22.13) has the same form as the right side of the logistic model form (22.1), where the X_{ij} in (22.13) has been substituted for X_j in (22.1).)

Now, if we replace θ_i in the likelihood (22.12) with the logistic function expression (22.13), we obtain the so-called *unconditional likelihood function* characterizing standard logistic regression analysis—namely,

$$L(\mathbf{Y}; \boldsymbol{\beta}) = \prod_{i=1}^{n_1} \left\{ 1 + \exp\left[-\left(\beta_0 + \sum_{j=1}^{k} \beta_j X_{ij} \right) \right] \right\}^{-1}$$

$$\times \prod_{i=n_1+1}^{n} \left(\exp\left[-\left(\beta_0 + \sum_{j=1}^{k} \beta_j X_{ij} \right) \right] \left\{ 1 + \exp\left[-\left(\beta_0 + \sum_{j=1}^{k} \beta_j X_{ij} \right) \right] \right\}^{-1} \right)$$

where $\boldsymbol{\beta} = (\beta_0, \beta_1, \ldots, \beta_k)$. The term *unconditional likelihood* refers to the unconditional probability of obtaining the particular set of data under consideration. More specifically, the unconditional likelihood function is the joint probability distribution for discrete data or the joint density function for continuous data. A *conditional likelihood,* which we discuss in the next subsection, gives the conditional probability of obtaining the data configuration *actually observed,* given all possible configurations (i.e., permutations) of the observed data values.

With a little algebraic manipulation, we can verify that an equivalent expression for the preceding likelihood function is[6]

$$L(\mathbf{Y}; \boldsymbol{\beta}) = \frac{\prod_{i=1}^{n_1} \exp\left(\beta_0 + \sum_{j=1}^{k} \beta_j X_{ij} \right)}{\prod_{i=1}^{n} \left[1 + \exp\left(\beta_0 + \sum_{j=1}^{k} \beta_j X_{ij} \right) \right]} \qquad (22.14)$$

Because (22.14) is a complex nonlinear function of the elements of $\boldsymbol{\beta}$, maximizing (22.14) to find the ML estimator $\hat{\boldsymbol{\beta}}$ of $\boldsymbol{\beta}$ must involve using appropriate computer algorithms. Such programs (e.g., SAS's LOGISTIC procedure) also produce the maximized likelihood value $L(\mathbf{Y}; \hat{\boldsymbol{\beta}})$ and the estimated (large-sample) covariance matrix $\hat{\mathbf{V}}(\hat{\boldsymbol{\beta}})$ for a given model, which can then be used to make appropriate statistical inferences.

22.5.2 Conditional ML Estimation

An alternative to using the unconditional likelihood function (22.14) for estimating the elements of $\boldsymbol{\beta} = (\beta_0, \beta_1, \beta_2, \ldots, \beta_k)$ is to employ a conditional likelihood function. The primary reason for using conditional likelihood methods is that unconditional methods can lead to

---

[6] The likelihood (22.14) is based on the responses of *individual* subjects, with (22.10) pertaining to the ith subject. In contrast, a categorical data analysis involving the binomial distribution is based on the responses of *groups* of subjects, with, say, Y_i subjects out of n_i in the ith group contracting the disease in question. In this situation, the underlying distribution of Y_i is binomial; that is,

$$\text{pr}(Y_i; \theta_i) = {}_{n_i}C_{Y_i}\theta_i^{Y_i}(1 - \theta_i)^{n_i - Y_i}$$

and the validity of certain categorical data analyses requires that n_i be fairly large in each group. The effects of small samples on the validity of ML analyses based on (22.14) are briefly discussed in the next subsection.

seriously biased estimates of the elements of $\boldsymbol{\beta}$ when the amount of data available for analysis is not large. Although "not large" is admittedly a vague term, it acknowledges the potential problem of using large-sample-based statistical procedures such as maximum likelihood when the number of parameters to be estimated constitutes a fair proportion of the available data. This is often the situation when data involving *matching* must be analyzed.

In some matched case–control studies, for example, each case (i.e., a person with the disease in question) is matched with one or more controls (i.e., persons not having the disease in question) that have the same values (or are in the same categories) as the cases for the covariates involved in the matching. Analyzing such (matched) data requires that the data be cast into strata corresponding to the matched sets (e.g., pairs) with stratum-specific sample sizes that are small (i.e., reflecting sparse data). In particular, for pair-matched case–control data, each stratum contains only two subjects. A logistic model for analyzing matched data requires that the matching be taken into account by including indicator (i.e., dummy) variables in the model to reflect the matching strata. Thus, the model will have as many parameters as there are matched sets (plus parameters for the exposure variable, any other unmatched variables, and possibly even product terms of unmatched variables with exposure); so the total number of parameters in the model is large relative to the number of subjects in the study. See Kleinbaum and Klein (2002), Chapter 8, and Kleinbaum, Kupper, and Morgenstern (1982), Chapter 24, for further discussion of the principles of matching and modeling.

To see why using a conditional likelihood function is appropriate for matched data, let us consider a *pair-matched case–control study* to assess the effect of a (0–1) exposure variable E on a (0–1) health outcome variable D. Suppose that the matching involves the variables age, race, and sex; then, for each case, a control subject is found who has the same age (or age category), race, and sex as the corresponding case. Suppose, too, that the study involves 100 matched pairs, so that the study sample size n is 200. If no predictor variables were considered in the study other than E and the matching variables age, race, and sex, then a (no-interaction) logistic regression model appropriate for analyzing these data is

$$\text{logit}[\text{pr}(D = 1)] = \beta_0 + \beta_1 E + \sum_{i=1}^{99} \gamma_i V_i \tag{22.15}$$

where V_i, $i = 1, \ldots, 99$, denote a set of dummy variables distinguishing the collection of 100 matched pairs; for example, V_i may be defined as

$$V_i = \begin{cases} 1 & \text{for the } i\text{th matched set} \\ 0 & \text{otherwise} \end{cases}$$

Model (22.15) allows us to predict case–control status as a function of exposure status (E) and the matching variables age, race, and sex, where the matching variables are incorporated into the model as dummy variables.[7] The number of parameters in model (22.15) is 101. This is an

---

[7] The analysis of a matched-pair study can (equivalently) be carried out without using logistic regression if no variables are controlled other than those involved in the matching. Such an analysis is a "stratified" analysis, and involves using a Mantel-Haenszel test, odds ratio, and confidence interval. Equivalently, the stratified analysis can be carried out using McNemar's test and estimation procedure (see Kleinbaum and Klein 2002, Chapter 8, or Kleinbaum 2002, Lesson 15).

example of a model whose number of parameters (101) is "large" relative to the number of subjects (200), so use of a conditional likelihood function is appropriate.

If an unconditional likelihood is used to fit model (22.15), the resulting (biased) estimated odds ratio for the exposure effect is the squared value of the estimated odds ratio obtained from using a conditional likelihood; that is, if $\hat{\beta}_{1U}$ and $\hat{\beta}_{1C}$ denote the estimates of β_1 using unconditional and conditional likelihoods, respectively, then

$$\widehat{OR}_U = e^{\hat{\beta}_{1U}} \equiv (\widehat{OR}_C)^2 = e^{2\hat{\beta}_{1C}}$$

Thus, for example, if a pair-matched analysis using a conditional likelihood to fit model (22.15) produces an estimated odds ratio of, say, 3, a corresponding analysis (involving the same model on the same data) using an unconditional likelihood would yield a biased estimate of 3^2, or 9.

Consider the data in the following table, which comes from the "Agent Orange Study" (Donovan, MacLennan, and Adena, 1984), a pair-matched case–control study involving 8502 matched pairs (i.e., $n = 17{,}004$):

		$D = 0$	
		$E = 1$	$E = 0$
$D = 1$	$E = 1$	2	125
	$E = 0$	121	8254

In this table, the data layout separates the 8502 case–control pairs into four cells depending on whether both the case and the control were exposed (2 *concordant* pairs), the case was exposed and the control was unexposed (125 *discordant* pairs), the case was unexposed and the control was exposed (121 *discordant* pairs), or both the case and the control are unexposed (8254 *concordant* pairs). The D and E variables are defined as follows: D = case-control status (1 = baby born with genetic anomaly, 0 = baby born without genetic anomaly), and E = father's status (1 = Vietnam vet, 0 = non-Vietnam vet). The matching variables are M_1 = time period of birth, M_2 = mother's age, M_3 = health insurance status, and M_4 = hospital. Since only the matching variables are being controlled, the analysis can be carried out using Mantel-Haenszel (or McNemar) statistics for a stratified analysis (see Kleinbaum and Klein 2002, Chapter 8, or Kleinbaum 2002, Lesson 15) without the need to perform a logistic regression analysis.

For these data, the Mantel-Haenszel (i.e., McNemar) test of H_0: "No (E, D) association" gives a 1-d.f. chi-square statistic of 0.0650 ($P = 0.80$) and an estimated Mantel-Haenszel odds ratio of 1.033.[8] An equivalent analysis (yielding the same results) based on logistic regression with a conditional likelihood uses the following logistic model:

$$\text{logit}[\text{pr}(D = 1)] = \beta_0 + \beta_1 E + \sum_{i=1}^{8501} \gamma_i V_i \qquad (22.16)$$

---

[8] McNemar's test is computed from the data in the preceding table using only the information from the $(125 + 121)$ discordant pairs. The test statistic is computed as $(125 - 121)^2/(125 + 121) = 0.0650$. The (Mantel-Haenszel) odds ratio is computed as the ratio of discordant pairs: $125/121 = 1.033$.

where the V_i denote dummy variables that distinguish the 8502 matched sets. Model (22.16) contains 8503 parameters, which is a large number (just over 50%) relative to the number of subjects in the study ($n = 17{,}004$).

Summarized output from fitting model (22.16) using a conditional likelihood is presented in Table 22.3. The output indicates that the odds ratio estimate is $\exp(0.0325) = 1.033$ and that the Wald test statistic is 0.0650 with a P-value of 0.80. These are the same results we obtained from the stratified analysis. The output does not show estimated coefficients for β_0 and for the $\{\gamma_i\}$; this is because these parameters, which distinguish the different matching strata, drop out of the conditional likelihood function (given by equation 22.17) for this model, and therefore cannot be estimated.

The output in Table 22.3 was obtained by using SAS's (version 9.1) *LOGISTIC* procedure.[9] The conditional likelihood function used by the LOGISTIC procedure for the general logistic model (written in logit form)

$$\text{logit}[\text{pr}(Y = 1)] = \beta_0 + \sum_{j=1}^{k} \beta_j X_j$$

takes the form

$$L_C(\mathbf{Y}; \boldsymbol{\beta}) = \frac{\displaystyle\prod_{i=1}^{n_1} \exp\left(\beta_0 + \sum_{j=1}^{k} \beta_j X_{ij}\right)}{\displaystyle\sum_{u}\left(\prod_{l=1}^{n_1} \exp\left(\beta_0 + \sum_{j=1}^{k} \beta_j X_{ulj}\right)\right)} \qquad (22.17)$$

where the sum in the denominator is over all partitions of the set $\{1, 2, \ldots, n\}$ into two subsets, the first of which contains n_1 elements. Expression (22.17) assumes, without loss of generality, that the first n_1 of the n subjects actually develop the disease; so X_{ij} denotes the value of variable X_j for the ith of these first n_1 subjects. The X_{ulj} term in the denominator, on the other hand, denotes

TABLE 22.3   **Edited computer output from SAS's PROC LOGISTIC for logistic regression of the "Agent Orange Data," using conditional maximum likelihood estimation for model (22.16)**

Variable	d.f.	Parameter Estimate	Standard Error	Wald Chi-Square	Pr > Chi-Square	Odds Ratio
E	1	0.0325	0.1275	0.0650	0.7987	1.033

[9] Earlier versions of the LOGISTIC procedure in SAS (e.g., Version 8) did not allow the use of conditional ML estimation. Previously, SAS's PHREG procedure for fitting a Cox Proportional Hazards (PH) model, a survival analysis procedure, was required for conditional ML estimation with SAS. STATA's *clogit* procedure is similar to SAS's (Version 9) LOGISTIC procedure, and carries out conditional ML estimation directly without using a *PH* program. However, SPSS requires a program for fitting the Cox PH model to perform conditional ML estimation. See Kleinbaum and Klein (2005) for details about the Cox PH model.

From Table 22.6, the odds ratio for the effect of EST, controlling for GALL and the matching variables (not shown in the table), is estimated as exp (2.209) = 9.107. The Wald statistic for testing whether this odds ratio differs significantly from one is given by

$$\chi^2_{1 \text{ d.f.}} = \left(\frac{2.209}{0.610}\right)^2 = 13.127 \qquad (P = 0.000)$$

which is highly significant. The estimated odds ratio and corresponding test statistic previously obtained from a stratified analysis that ignored the variable GALL were 9.67 and 21.125, respectively, versus 9.107 and 13.127 when controlling for GALL. Nevertheless, both analyses lead to the same conclusion: these data strongly suggest that estrogen use has a strong effect on the development of endometrial cancer. We can also compute a confidence interval for the (adjusted) odds ratio, using the output of Table 22.6. To obtain a 95% confidence interval for the adjusted odds ratio, we calculate

$$\exp [2.209 \pm 1.96(0.610)]$$

which yields lower and upper limits of 2.755 and 30.102, respectively.

Table 22.6 does not provide estimates of the coefficients of the V_i variables that control for the matching. This is because these coefficients drop out of the conditional likelihood (22.17) and therefore cannot be estimated. Although the V_i variables are not explicitly specified in the data listing given by Table 22.4, the matched pairs are identified in the listing by the STRATUM variable, which is used by the conditional ML program (e.g., SAS's LOGISTIC) program to control for the matching.

22.7 Summary

This chapter described the key features of logistic regression analysis, the most popular regression technique available for modeling dichotomous dependent variables. The *logistic model* is defined as a probability of the occurrence of one of two possible outcomes (say, 0 and 1), using the following formula:

$$\text{pr}(Y = 1) = \frac{1}{1 + \exp\left[-\left(\beta_0 + \sum_{j=1}^{k}\beta_j X_j\right)\right]}$$

The most important reason for the popularity of the logistic model is that the right-hand side of the preceding expression ensures that the predicted value of Y will always lie between 0 and 1. Using the logit form of the logistic model defined by the expression

$$\text{logit}[\text{pr}(Y = 1)] = \beta_0 + \sum_{j=1}^{k}\beta_j X_j$$

we can estimate an odds ratio; the general formula for an odds ratio comparing two specifications $\mathbf{X_A}$ and $\mathbf{X_B}$ of the set of predictors $\mathbf{X} = (X_1, X_2, \ldots, X_k)$ is

$$\text{OR}_{\mathbf{X_A} \text{ vs. } \mathbf{X_B}} = e^{\sum_{j=1}^{k}(X_{Aj} - X_{Bj})\beta_j}$$

Maximum likelihood estimates of the regression coefficients and associated estimated standard errors of the regression coefficients in a logistic model are typically obtained by using computer packages for logistic regression. These statistics can then be used to obtain numerical values for estimated adjusted odds ratios, to test hypotheses, and to obtain confidence intervals for population odds ratios based on standard maximum likelihood techniques.

When performing a logistic regression analysis, we must decide between two alternative ML procedures, called *unconditional* and *conditional*. The key distinction between the two involves whether the number of parameters in the model constitutes a "large" proportion of the total study size. If so, the situation is typical of matched data, requiring the use of the conditional approach to ensure validity of the odds ratio estimates.

This chapter included examples using both the unconditional approach and the conditional approach. SAS's LOGISTIC procedure (Version 9) can be used to implement either approach. Earlier versions of SAS required the use of SAS's PHREG program to carry out conditional ML estimation. PHREG is a survival analysis program for fitting a Cox Proportional Hazards model. Since the conditional likelihood for logistic regression has the same structure as the partial likelihood used for the Cox model, the survival analysis program (PHREG) can be "tricked" into carrying out the conditional estimation procedure by adding appropriate *STRATUM* and *SURVT* variables to the dataset.

Problems

1. A researcher was interested in determining risk factors for high blood pressure (hypertension) among women. Data from a sample group of 680 women were collected. The following table gives the observed relationship between hypertension and smoking:

	Hypertension	
	Yes	No
Smokers	$a = 28$	$b = 271$
Nonsmokers	$c = 13$	$d = 368$
	41	639

Let π be the probability of having hypertension, and suppose that the researcher used logistic regression to model the relationship between smoking and hypertension.

One model the researcher considered is

$$\log_e \frac{\pi}{1 - \pi} = \beta_0 + \beta_1 X_1 \qquad \text{(Model 0)}$$

where X_1 = smoking status (1 = smoker, 0 = nonsmoker). For this model, the estimated odds ratio for the effect of smoking on hypertension status can be computed by using the simple formula for an odds ratio in a 2×2 table—namely, ad/bc, where a, b, c, and d are the cell frequencies in the preceding table.

a. Based on this information, compute the point estimate $\hat{\beta}_1$ of β_1.

The researcher ultimately decided to use the following logistic regression model:

$$\log_e \frac{\pi}{1 - \pi} = \beta_0 + \beta_1 X_1 + \beta_2 X_2 + \beta_3 X_1 X_2 \qquad \text{(Model 1)}$$

where X_1 = smoking status (1 = smoker, 0 = nonsmoker) and X_2 = age. The following information was obtained:

Parameter	Estimate	SE
β_0	−2.8	1.2
β_1	0.706	0.311
β_2	0.0004	0.0001
β_3	0.0006	0.0003

$$-2 \ln \hat{L} = 303.84$$

The estimated covariance matrix for $\hat{\beta}_0$, $\hat{\beta}_1$, $\hat{\beta}_2$, and $\hat{\beta}_3$ is as follows:

$$
\begin{array}{cccc}
\hat{\beta}_0 & \hat{\beta}_1 & \hat{\beta}_2 & \hat{\beta}_3
\end{array}
$$

$$
\begin{bmatrix}
1.44 & 0.0001 & 0.0001 & 0.0001 \\
 & 0.0967 & 0.1 \times 10^{-8} & 2.0 \times 10^{-8} \\
 & & 1.0 \times 10^{-8} & 3.0 \times 10^{-8} \\
 & & & 9.0 \times 10^{-8}
\end{bmatrix}
$$

1. **b.** What is the estimated logistic regression model for the relationship between age and hypertension for nonsmokers?

   c. What is a 20-year-old smoker's predicted probability of having hypertension?

   d. Estimate the odds ratio comparing a 20-year-old smoker to a 21-year-old smoker. Interpret this estimated odds ratio.

   e. Find a 95% confidence interval for the population odds ratio being estimated in part (d).

   f. The log likelihood $(-2 \ln \hat{L})$ for the model consisting of the intercept, age, and smoking status was 308.00. Use this information, plus other information provided earlier, to perform a likelihood ratio test of the null hypothesis that $\beta_3 = 0$ in model 1.

2. A five-year follow-up study on 600 disease-free subjects was carried out to assess the effect of a (0–1) exposure variable E on the development or not of a certain disease. The variables AGE (continuous) and obesity status (OBS), the latter a (0–1) variable, were determined at the start of follow-up and were to be considered as control variables in analyzing the data. For this study, answer the following questions.

   a. State the logit form of a logistic model that assesses the effect of the (0, 1) exposure variable E, controlling for the confounding effects of AGE and OBS and for the interaction effects of AGE with E and OBS with E.

   b. Given the model described in part (a), give a formula for the odds ratio for the exposure-disease relationship that controls for the confounding and interactive effects of AGE and OBS.

   c. Use the formula described in part (b) to derive an expression for the estimated odds ratio for the exposure–disease relationship that considers both confounding and interaction when AGE = 40 and OBS = 1.

   d. Give a formula for a 95% confidence interval for the population adjusted odds ratio being estimated in part (c). In stating this confidence interval formula, write out the formula for the estimated variance of the log of the estimated adjusted odds ratio in terms of estimated variances and covariances of appropriate regression coefficients.

   e. State the null hypothesis (in terms of model parameters) for simultaneously testing for no interaction of either AGE with E or OBS with E.

   f. State the formula for the likelihood ratio statistic that tests the null hypothesis described in part (e). What is the large-sample distribution of this statistic, including its degrees of freedom, under the null hypothesis?

2. g. Give the formula for the Wald statistic for testing the null hypothesis described in part (e). What is the large-sample distribution of this statistic under the null hypothesis?

h. Assuming that both the likelihood ratio statistic described in part (f) and the Wald statistic described in part (g) are nonsignificant, state the logit form of the no-interaction model that is appropriate to consider at this point in the analysis.

i. For the no-interaction model given in part (h), give the formula for the (adjusted) odds ratio for the effect of exposure, controlling for AGE and OBS.

j. For the no-interaction model given in part (h), give the formula for a 95% confidence interval for the adjusted odds ratio for the effect of exposure, controlling for AGE and OBS.

k. Describe the Wald test for the effect of exposure, controlling for AGE and OBS, in the no-interaction model given in part (h). Include a description of the null hypothesis being tested, the form of the test statistic, and its distribution under the null hypothesis.

3. A study was conducted on a sample of 53 patients presenting with prostate cancer who had also undergone a laparotomy to ascertain the extent of nodal involvement (Collett 1991). The result of the laparotomy is a binary response variable, where 0 signifies the absence of, and 1 the presence of, nodal involvement. The purpose of the study was to determine variables that could be used to forecast whether the cancer has spread to the lymph nodes. Five predictor variables were considered, each measurable without surgery. The following printout provides information for fitting two logistic models based on these data. The five predictor variables were: age of patient at diagnosis, level of serum acid phosphatase, result of an X-ray examination (0 = negative, 1 = positive), size of the tumor as determined by a rectal exam (0 = small, 1 = large), and summary of the pathological grade of the tumor as determined from a biopsy (0 = less serious, 1 = more serious).

a. Which method of estimation do you think was used to obtain estimates of parameters for both models—conditional or unconditional ML estimation? Explain briefly.

b. For model I, test the null hypothesis of no effect of X-ray status on response. State the null hypothesis in terms of an odds ratio parameter; give the formula for the test statistic; state the distribution of the test statistic under the null hypothesis; and finally, carry out the test, using the computer output information for Model I. Is the test significant?

Model I

Variable Name	Coeff	StErr	P-value	$\widehat{OR}$	95% CI	
0 constant	0.057	3.460	0.987			
1 age	−0.069	0.058	0.232	0.933	0.833	1.045
2 acid	2.434	1.316	0.064	11.409	0.865	150.411
3 xray	2.045	0.807	0.011	7.731	1.589	37.611
4 tsize	1.564	0.774	0.043	4.778	1.048	21.783
5 tgrad	0.761	0.771	0.323	2.141	0.473	9.700

d.f.: 47   Dev: 48.126[*]

[*] The deviance statistic is a likelihood ratio statistic that compares a current model of interest to the baseline model containing as many parameters as there are data points. The difference in deviance statistics obtained for two (hierarchically ordered) models being compared is equivalent to the difference in log-likelihood statistics for each model. Thus a likelihood ratio test can equivalently be carried out by using differences in deviance statistics. See Section 24.5 for further details about deviances.

Model II

Variable Name	Coeff	StErr	P-value	$\widehat{OR}$		95% CI
0 constant	2.928	4.044	0.469			
1 age	−0.101	0.067	0.132	0.904	0.792	1.031
2 acid	1.462	1.559	0.349	4.314	0.203	91.678
3 xray	−19.171	20.027	0.338	0.000	0.000	5.259×10^8
4 tsize	1.285	0.894	0.151	3.613	0.626	20.845
5 tgrad	0.421	0.923	0.648	1.523	0.250	9.291
7 age*xray	0.257	0.278	0.356	1.292	0.750	2.228
8 acid*xray	7.987	9.197	0.385	2943.008	0.000	1.982×10^{11}
9 tsiz*xray	2.236	2.930	0.445	9.356	0.030	2920.383
10 tgrd*xray	−0.624	2.376	0.793	0.536	0.005	56.403

d.f.: 43   Dev: 44.474

c. Using the printout data for Model I, compute the point estimate and 95% confidence interval for the odds ratio for the effect of X-ray status on response for a person of age 50, with phosphatase acid level of .50, tsize equal to 0, and tgrad equal to 0.

d. State the logit form of Model II given in the accompanying computer printout.

e. Using the results for Model II, give an expression for the estimated odds ratio that describes the effect of X-ray status on the response, controlling for age, phosphatase acid level, tsize, and tgrad. Using this expression, compute the estimated odds ratio of the effect of X-ray status on the response for a person of age 50, with phosphatase acid level of .50, tsize equal to 0, and tgrad equal to 0.

f. For Model II, give an expression for the estimated variance of the estimated adjusted odds ratio relating X-ray status to response for a person of age 50, with phosphatase acid level of .50, tsize equal to 0, and tgrad equal to 0. Write this expression for the estimated variance in terms of estimated variances and covariances obtained from the variance-covariance matrix.

g. Using your answer to part (f), give an expression for a 95% confidence interval for the odds ratio relating X-ray status to response, for a person of age 50, with phosphatase acid level of .50, tsize equal to 0, and tgrad equal to 0.

h. For Model II, carry out a "chunk" test for the combined interaction of X-ray status with each of the variables age, phosphatase acid level, tsize, and tgrad. State the null hypothesis in terms of one or more model coefficients; give the formula for the test statistic and its distribution and degrees of freedom under the null hypothesis; and report the P-value. Is the test significant?

i. If you had to choose between Model I and Model II, which would you pick as the "better" model? Explain.

Assume that the following model has been defined as the initial model to be considered in a backward-elimination strategy to obtain a "best" model:

$$\text{logit}[\text{pr}(Y = 1)] = \alpha + \beta_1(\text{xray}) + \beta_2(\text{age}) + \beta_3(\text{acid}) + \beta_4(\text{tsize}) + \beta_5(\text{tgrad})$$
$$+ \beta_6(\text{age} \times \text{acid}) + \beta_7(\text{xray} \times \text{age}) + \beta_8(\text{xray} \times \text{acid})$$
$$+ \beta_9(\text{xray} \times \text{tsize}) + \beta_{10}(\text{xray} \times \text{tgrad}) + \beta_{11}(\text{xray} \times \text{age} \times \text{acid})$$

    j. For this model—and considering the variable xray to be the only "exposure" variable of interest, with the variables age, acid, tsize, and tgrad considered for control—which β's in the above model are coefficients of (potential) effect modifiers? Explain briefly.

    k. Assume that the only interaction term found significant is the product term (xray $\times$ age). What variables are left in the model at the end of the interaction assessment stage?

    l. Based on the (reduced) model described in part (k) (where the only significant interaction term is xray $\times$ age), what expression for the odds ratio describes the effect of xray on nodal involvement status?

    m. Suppose that, as a result of confounding and precision assessment, the variables (age $\times$ acid), tgrad, and tsize are dropped from the model described in part (k). What is your final model, and what expression for the odds ratio describes the effect of xray on nodal involvement status?

References

Breslow, N., and Day, N. 1980. *Statistical Methods in Cancer Research. Vol. I: The Analysis of Case-Control Studies.* Lyon, France: IIARC Scientific Publications, No. 32.

Collett, D. 1991. *Modeling Binary Data.* London: Chapman & Hall.

Cox, D. R. 1975. "Partial Likelihood." *Biometrika* 62: 269–76.

Dantes, H. G., Koopman, J. S., et al. 1988. "Dengue Epidemics on the Pacific Coast of Mexico." *International Journal of Epidemiology* 17(1): 178–86.

Donovan, J. W., MacLennan, R., and Adena, M. 1984. "Vietnam Service and the Risk of Congenital Anomalies; a Case-Control Study," *Medical Journal of Australia* 140: 394–97.

Hosmer, D. W., and Lemeshow, S. 1989. *Applied Logistic Regression.* New York: John Wiley & Sons.

Kleinbaum, D. G. 2002. *ActivEpi- A CD ROM Course on Fundamentals of Epidemiologic Research.* New York and Berlin: Springer Publishers.

Kleinbaum, D. G., and Klein, M. 2002. *Logistic Regression—A Self-Learning Text: Second Edition.* New York and Berlin: Springer Publishers.

Kleinbaum, D. G., and Klein, M. 2005. *Survival Analysis—A Self Learning Text: Second Edition.* New York and Berlin: Springer Publishers.

Kleinbaum, D. G., Kupper, L. L., and Morgenstern, H. 1982. *Epidemiologic Research—Principles and Quantitative Methods.* New York: Van Nostrand Reinhold.

Polytomous and Ordinal Logistic Regression

23.1 Preview

In this chapter, the standard logistic model is extended to consider outcome variables that have more than two categories. We first describe polytomous logistic regression, which is used when the categories of the outcome variable are nominal, that is, they do not have any natural order. We then describe ordinal logistic regression, which is appropriate when the categories of the outcome variable do have a natural order.

Examples of outcome variables with more than two levels might include (1) endometrial cancer subtypes: adenosquamous, adenocarcinoma, or other; (2) patients' preferred treatment regimen, selected from among three or more options; (3) tumor grade: well differentiated, moderately differentiated, or poorly differentiated; (4) disease symptoms that have been classified by subjects as being absent, mild, moderate, or severe; (5) invasiveness of a tumor classified as in situ, locally invasive, or metastatic. For the first two of these examples, the categories are essentially nominal (i.e., polytomous, without ordinality), whereas examples (3)–(5) consider ordinal categories.

When modeling an outcome variable, whether polytomous or ordinal, with more than two categories, the typical research question remains the same as when there are two outcome categories: What is the relationship of one or more predictor variables $(X_1, X_2, \ldots, X_k)$ to an outcome (Y) of interest? Also, since the model is an extension of binary logistic regression, the measure of effect of interest will be an odds ratio, whose formula involves the exponential of linear functions of the regression coefficients in the model. Furthermore, since the method of estimation used to fit both polytomous and ordinal logistic regression models is ML estimation, the procedures for carrying out hypothesis testing and confidence interval estimation are analogous to ML techniques used for binary logistic regression. The specific forms that the odds ratio takes, as well as examples of statistical inference procedures for polytomous and ordinal logistic regression, will be described and illustrated in the sections to follow.

23.2 Why Not Use Binary Regression?

A simple approach for the analysis of data with an outcome containing more than two categories is to choose an appropriate referent category out of the total collection of categories or combine categories into a single referent group, pool the remaining categories to form a second group, and then simply utilize the logistic modeling techniques for dichotomous outcomes. For example, if endometrial cancer subtypes is the outcome variable, one could pool "adenocarcinoma" and "other" into a referent category and then use "adenosquamous" as a second category. This can also be done for ordinal outcomes; for example, if the outcome symptom severity has four categories of severity, one might compare subjects with none or only mild symptoms to those with either moderate or severe symptoms.

One major disadvantage of dichotomizing a polytomous or ordinal outcome is loss of meaningful detail in describing the outcome of interest. For example, in the last scenario given above, we can no longer compare mild versus none or moderate versus mild. This loss of detail may, in turn, affect the conclusions made about the covariate-outcome relationships.

Another possible approach to the analysis of polytomous (or ordinal) outcomes would be to choose a referent category and then fit several separate dichotomous logistic models comparing the remaining categories individually with the chosen referent category. Such an approach may be criticized because the likelihood function for any separate dichotomous logistic model utilizes the data involving only the two categories of the outcome variable being considered. In contrast, the likelihood function for a polytomous or ordinal logistic regression model utilizes the data involving all categories of the outcome variable in a single model structure. In other words, different likelihood functions are used when fitting each dichotomous model separately than when fitting a polytomous model that considers all levels simultaneously. Consequently, both the estimation of the parameters and the estimation of the variances of the parameter estimates may differ when comparing the results from fitting separate dichotomous models to the results from the polytomous or ordinal model. In fact, in general, there is a power loss when doing separate dichotomous logistic models versus doing polytomous (or ordinal) logistic regression. (Note, however, for the special case of a polytomous model with one dichotomous predictor, fitting separate logistic models yields the same parameter estimates and variance estimates as fitting the polytomous model.)

23.3 An Example of Polytomous Logistic Regression: One Predictor, Three Outcome Categories

In this section, we illustrate polytomous logistic regression with one dichotomous predictor variable and an outcome (Y) that has three categories. This is the simplest case of a polytomous logistic regression model. In a later section, we discuss extending the polytomous model to more than one predictor variable and then to outcomes with more than three categories.

End Stage Renal Disease (ESRD) is a condition in which there has been an irreversible loss of renal function making the patient permanently dependent on renal replacement therapy (RRT), typically involving hemodialysis, in order to sustain life. ESRD patients suffer from increased morbidity and their life expectancy is about one third to one sixth of the general population. We consider data obtained from an ESRD surveillance system on 3,049 ESRD patients who initiated dialysis in (the U.S. state of) Georgia in 2002 (Volkova et al., 2006). For illustrative purposes, the outcome variable that we consider is "cause (of ESRD)," which has been categorized into the following three nominal categories: 2 = hypertension, 1 = diabetes, or 0 = other disease. There is no inherent order in the outcome variable, i.e., the 0, 1, and 2 coding of the categories of the variable cause is arbitrary.

The predictor (i.e., exposure) variable that we consider here is "race," which has been coded as 1 = black, 0 = white. Thus, the research question concerns whether there is an association between race and cause of ESRD. Other variables that we consider as control variables include "age at dialysis initiation," and "gender" (1 = female, 0 = male). The dataset is called *esrddata*, and this dataset can be obtained from the website.

With polytomous logistic regression, one of the categories of the outcome variable is designated as the reference category and each of the other categories is compared with this referent. The choice of reference category can be arbitrary and is at the discretion of the researcher. Changing the reference category does not change the general structure of the polytomous logistic regression model but does change the interpretation of the parameter estimates in the model.

In our three-outcome example, the Other Disease group has been designated as the reference category. We are therefore interested in modeling two main comparisons. We want to compare subjects whose ESRD was caused by Hypertension (category 2) to those subjects whose ESRD was caused by Other Disease (category 0), and we also want to compare subjects whose ESRD was caused by Diabetes (category 1) to those subjects whose ESRD was caused by Other Disease (category 0).

If we consider these two comparisons separately, the crude odds ratios can be calculated using data from Table 23.1. The odds ratio comparing Hypertension (category 2) to Other Disease (category 0) is 2.105, and the odds ratio comparing Diabetes (category 1) to Other Disease (category 0) is 1.484. Thus, without considering statistical significance, blacks appear to be twice as likely as whites to have their ESRD being caused by Hypertension relative to the ESRD cause being Other Disease. Similarly, blacks appear to be 1.5 times as likely as whites to have their ESRD being caused by Diabetes relative to the ESRD cause being Other Disease.

As previously described for a dichotomous outcome variable coded 0 and 1, the odds for developing an outcome of interest (e.g., Diabetes) equals

$$\text{odds} = \frac{P(D = 1)}{1 - P(D = 1)} = \frac{P(D = 1)}{P(D = 0)}$$

namely, the probability that the outcome equals one divided by the probability that the outcome equals zero. Also, for a dichotomous outcome variable coded as 0 or 1, recall that the logit form of the logistic model, logit $P(X)$, is defined as the natural log of the odds for developing a disease for a person with a set of independent variables specified by X, i.e.,

$$\text{logit } P(X) = \ln\left[\frac{P(D = 1 \mid X)}{P(D = 0 \mid X)}\right] = \alpha + \sum_{j=1}^{k} \beta_j X_j$$

TABLE 23.1   Crude data layout and odds ratios relating race to cause of End Stage Renal Disease (Georgia, 2002)

ESRD Cause	Black	White	Total	$\widehat{\text{OR}}$
Hypertension (2)	630	323	953	2.105
Diabetes (1)	773	562	1335	1.484
Other (0)	366	395	761	1.000
Total	1769	1280	3049	

However, when there are three categories of the outcome, there are three probabilities to consider:

$$P(D = 0|X), \qquad P(D = 1|X), \qquad \text{and} \qquad P(D = 2|X)$$

and the sum of the probabilities for the three outcome categories must be equal to one. Because each pairwise ratio comparison considers only two probabilities, the two probabilities in such a ratio do not sum to one. With three outcome categories, there are now two "odds-like" expressions to consider, one for each of the two pairwise ratio comparisons we are making:

$$\frac{P(D = 2|X)}{P(D = 0|X)}, \qquad \frac{P(D = 1|X)}{P(D = 0|X)}$$

Because each comparison considers only two probabilities, the probabilities in each ratio do not sum to one. Thus, the two "odds-like" expressions are not true odds. However, if we restrict our interest to just the two categories being considered in a given ratio, we may still roughly interpret the ratio as an odds; for example,

$$\frac{P(D = 2|X)}{P(D = 0|X)}$$

is a measure of how likely it is for a person with covariate values X to be in outcome category 2 rather than in outcome category 0. For ease of our subsequent discussion, we will use the term "odds" rather than "odds-like" for these expressions.

In polytomous logistic regression with three levels, we therefore define our model using two expressions for the natural log of these "odds-like" quantities. The first is the natural log of the probability that the outcome is in category 1 divided by the probability that the outcome is in category 0; the second is the natural log of the probability that the outcome is in category 2 divided by the probability that the outcome is in category 0, i.e.,

$$\ln\left[\frac{P(D = 1|X)}{P(D = 0|X)}\right], \qquad \ln\left[\frac{P(D = 2|X)}{P(D = 0|X)}\right]$$

Because our example has three outcome categories and one predictor (i.e., X is the scalar variable "race"), our polytomous model requires two regression expressions:

$$\text{Diabetes vs. Other:} \qquad \ln\left[\frac{P(D = 1|\text{race})}{P(D = 0|\text{race})}\right] = \alpha_1 + \beta_{11}(\text{race})$$

$$\text{Hypertension vs. Other: } \ln\left[\frac{P(D = 2|\text{race})}{P(D = 0|\text{race})}\right] = \alpha_2 + \beta_{21}(\text{race})$$

(23.1)

In these two models, corresponding parameters for the intercept (α) and the slope (β) are allowed to be different, i.e., when comparing outcome categories 1 versus 0, the parameters are denoted α_1 and β_{11}, whereas the parameters are denoted as α_2 and β_{21} when comparing categories 2 versus 0.

The results, based on fitting the polytomous model examining the association between "race" and "cause of ESRD," are presented in Table 23.2. These results were obtained using the LOGISTIC procedure in SAS (Version 9.1). There are two sets of parameter estimates. The output is listed in descending order, with α_2 labeled as intercept 2 and α_1 labeled as intercept 1. If cause = 2 had been designated as the reference category, the output would have been in ascending order.

TABLE 23.2 **Edited polytomous logistic regression output for ESRD data evaluating the relationship between race (X) and cause of ESRD (Y)**

```
                        Model Fit Statistics

                                          Intercept and
              Criterion     Intercept Only    Covariates
              -2 Log L         6534.134        6091.865

                        Type 3 Analysis of Effects
              Effect    DF     Wald ChiSq    Pr > ChiSq
              race       2      55.6927        <.0001
```

Parameter	cause	DF	Estimate	Std. Err.	Wald ChiSq	Pr > ChiSq	Symbol
Intercept	2	1	-0.2013	0.0750	7.1976	0.0073	α_2
Intercept	1	1	0.3526	0.0657	28.8456	<.0001	α_1
race	2	1	0.7443	0.0997	55.6924	<.0001	β_{21}
race	1	1	0.3950	0.0913	18.7135	<.0001	β_{11}

```
                        Odds Ratio Estimates
                                Point           95% Wald
              Effect    cause   Estimate   Confidence Limits
              race        2      2.105      1.731     2.559
              race        1      1.484      1.241     1.775
```

Computer packages vary in the presentation of output; in particular, the coding of the polytomous outcome variable must be understood to correctly read and interpret the computer output for a given package. For example, when the reference category for the outcome variable (i.e., Cause) is 0 (i.e., Other Disease), the SAS output shown in Table 23.2 will list the (intercept and race) parameters in descending order. Thus, the output corresponding to α_2, which corresponds to category 2 (i.e., Hypertension), precedes the output corresponding to α_1, corresponding to category 1 (i.e., Diabetes); similarly, the output for β_{21} precedes the output for β_{11}.

From Table 23.2, the two logit-like expressions for the fitted polytomous model are given by

$$\text{Hypertension versus Other: } \ln\left[\frac{\hat{P}(D=2|\text{race})}{\hat{P}(D=0|\text{race})}\right] = -0.2013 + (0.7443)(\text{race})$$

and

$$\text{Diabetes versus Other: } \ln\left[\frac{\hat{P}(D=1|\text{race})}{\hat{P}(D=0|\text{race})}\right] = 0.3526 + (0.3950)(\text{race})$$

Once a polytomous logistic regression model has been fit and the parameters (intercepts and beta coefficients) have been estimated, the estimates of the exposure-disease association can be calculated in a similar manner (using e^β-type formulae) as used for standard logistic regression (with a dichotomous outcome). For example, since the model considered in Table 23.2 contains a single (0,1) predictor variable (i.e., race), the two odds ratio estimates shown in this table are calculated by exponentiating the corresponding estimated regression coefficients $\hat{\beta}_{21}$ and $\hat{\beta}_{11}$, i.e.,

$$\widehat{\text{OR}}_{\text{race}}(\text{Hypertension versus Other}) = e^{\hat{\beta}_{21}} = e^{0.7443} = 2.105$$

$$\widehat{\text{OR}}_{\text{race}}(\text{Diabetes versus Other}) = e^{\hat{\beta}_{11}} = e^{0.3950} = 1.484$$

Notice that these odds ratio estimates are identical to the estimates shown in Table 23.1 for the crude data in which logistic modeling was not performed. In general, when a polytomous logistic regression model contains only one dichotomous predictor variable, the estimated odds ratios comparing separate outcome categories to a referent category are identical to corresponding crude odds ratios calculated directly from appropriate 2×2 frequency tables.[1]

Procedures for testing hypotheses and computing confidence intervals are straightforward generalizations of techniques that apply to logistic regression modeling with a dichotomous outcome variable. As with a standard logistic regression, we can use a likelihood ratio test to assess the significance of the independent variable X = race in our model. However, rather than testing one beta coefficient for each independent variable, we are now testing two at the same time in our example; i.e., the null hypothesis for the effect of race is given by H_0: $\beta_{11} = \beta_{21} = 0$. The corresponding likelihood ratio statistic is therefore

$$\text{LR}_{\text{race}} = -2 \ln L_R - (-2 \ln L_F) \tag{23.2}$$

which has approximately a chi-square distribution with 2 degrees of freedom ($d.f.$) under H_0. (The test has 2 $d.f.$ because H_0 involves setting two parameters, β_{11} and β_{21}, equal to 0.) The *full* model (F) identified in (23.2) refers to the polytomous logistic regression model defined by (23.1) and involving the single independent variable race. The *reduced* model (R) refers to the polytomous logistic regression model that (23.1) reduces to under the null hypothesis H_0: $\beta_{11} = \beta_{21} = 0$, namely, a model containing only intercept parameters:

$$\text{Reduced model: } \ln\left[\frac{P(D = 1|\text{race})}{P(D = 0|\text{race})}\right] = \alpha_1 \qquad \ln\left[\frac{P(D = 2|\text{race})}{P(D = 0|\text{race})}\right] = \alpha_2 \tag{23.3}$$

The *Model Fit Statistics* in Table 23.2 provide the values of log-likelihood statistics required to calculate the LR statistic (23.2). The LR statistic is therefore calculated as

$$\text{LR}_{\text{race}} = -2 \ln L_R - (-2 \ln L_F) = (6534.134) - (6091.865) = 442.269$$

which is highly significant (*P*-value < 0.0001) based on the chi-square distribution with 2 $d.f.$, thus indicating a significant effect of race in model (23.1).

While the likelihood ratio test allows for the assessment of the effect of an independent variable across all categories of the outcome simultaneously, it is possible that one might be interested in evaluating the effect of the independent variable on a single (non-referent) outcome category, e.g., Diabetes or Hypertension compared individually to Other Disease in our example. A Wald test can be performed in this situation. Continuing with our example, the null hypothesis for comparing Diabetes versus Other (i.e., category 1 vs. 0) is that β_{11} equals zero. From Table 23.2, the Wald statistic for testing H_0: $\beta_{11} = 0$ is equal to 18.7135, with a *P*-value less than 0.0001. The null hypothesis for comparing Hypertension versus Other (i.e., category 2 vs. 0) is

---

[1] The special case of a dichotomous predictor can be generalized to include multi-category or continuous predictors. For a polytomous outcome with three categories (as in our example), the odds ratio formula that compares any two levels ($X = X_A$ vs. $X = X_B$) of a predictor (X) is given by

$$\text{OR}_g = \exp[\beta_{g1}(X_A - X_B)] \quad \text{where } g = 1, 2$$

In our example, X = race with $X_A = 1$ and $X_B = 0$, so the above formula simplifies to

$$\text{OR}_g = \exp[\beta_{g1}(1 - 0)] = \exp[\beta_{g1}] = e^{\beta_{g1}}, \quad g = 1, 2$$

If race had been coded as $X_A = 1$ and $X_B = -1$, however, the odds ratio formula would change to

$$\text{OR}_g = \exp[\beta_{g1}(1 - (-1))] = \exp[2\beta_{g1}] = e^{2\beta_{g1}}, \quad g = 1, 2.$$

that β_{21} equals zero. The Wald statistic for testing H_0: $\beta_{21} = 0$ is equal to 55.6924, with a P-value less than 0.0001. The reader should not be surprised to see such small P-values since the sample size for this study ($n = 3049$) is quite large.

From the above significance testing, we conclude that race is statistically significant for both the Diabetes versus Other comparison (category 1 vs. 0), as well as for the Hypertension versus Other comparison (category 2 vs. 0). When using a polytomous logistic regression model for these data, we must either keep both betas (β_{11} and β_{21}) for the independent variable "race" or drop both betas. Even if only one estimated beta is significantly different from zero, both betas must be retained if the independent variable is to remain in the model.

Large-sample confidence interval estimation using a polytomous logistic regression model is also analogous to the standard (dichotomous) logistic regression situation. However, since two odds ratios are involved for the variable *race* (i.e., β_{11} and β_{21}), we must calculate two confidence interval estimates[2]. Using the estimated standard errors 0.0997 and 0.0913 in Table 23.2 for the estimated coefficients $\hat{\beta}_{21}$ and $\hat{\beta}_{11}$, respectively, the two large-sample 95% confidence intervals for the race variable are calculated as follows:

95% CI for $OR_2 = e^{\beta_{21}}$(Hypertension vs. Other):

$$\exp[\hat{\beta}_{21} \pm 1.96\, S_{\hat{\beta}_{21}}] = \exp[0.7443 \pm 1.96(0.0997)] = (1.731, 2.559)$$

95% CI for $OR_1 = e^{\beta_{11}}$(Diabetes vs. Other):

$$\exp[\hat{\beta}_{11} \pm 1.96\, S_{\hat{\beta}_{11}}] = \exp[0.3950 \pm 1.96(0.0913)] = (1.241, 1.775)$$

The above two 95% confidence intervals calculated above for $e^{\beta_{21}}$ and $e^{\beta_{11}}$ are shown in the output in Table 23.2. In general, SAS's LOGISTIC procedure automatically provides output for confidence intervals for e^{β} corresponding to each β regression coefficient in a polytomous logistic regression model.[3]

23.4 An Example: Extending the Polytomous Logistic Model to Several Predictors

Expanding the polytomous logistic regression model to add more independent variables is straightforward. If the model contains k predictors, we simply add k independent variables for each of the outcome comparisons. The procedures for calculation of odds ratios and confidence intervals, and for hypothesis testing, remain the same.

---

[2] As with point estimates of odds ratios, confidence interval estimation can be generalized from the special case of a single dichotomous predictor to include multi-category or continuous predictors. For a polytomous outcome with three categories (as in our example), the 95% confidence interval formula that compares any two levels ($X = X_A$ vs. $X = X_B$) of a predictor (X) is given by

$$\exp[\hat{\beta}_{g1}(X_A - X_B) \pm 1.96\,(X_A - X_B)S_{\hat{\beta}_{g1}}]\,, \; g = 1, 2$$

[3] As with dichotomous logistic regression models, the expression $e^{\hat{\beta}_{g1}}$ is a correct formula for an odds ratio estimate for a predictor that is coded as (0, 1). However, for multi-category or continuous predictors, $e^{\hat{\beta}_{g1}}$ gives the odds ratio for a one-unit-change in the predictor, which might not reflect an estimated odds ratio of interest (e.g., if the predictor is systolic blood pressure). Similarly, the large-sample confidence interval formula $\exp[\hat{\beta}_{g1} \pm 1.96 S_{\hat{\beta}_{g1}}]$ will not reflect the appropriate confidence interval for an odds ratio for a continuous predictor.

To illustrate, we return to our example that considers the relationship of race to cause of ESRD in 3,049 ESRD patients who initiated dialysis in Georgia in 2002. Suppose we now want to consider the effect of race controlling for age and gender. The age variable has been defined as a categorical variable called "ageg" containing 7 categories, using the following SAS code:

if age < 23 then ageg = 1;	(age < 23)
if age >= 23 and age < 40 then ageg = 2;	(23 ≤ age < 40)
if age >= 40 and age < 50 then ageg = 3;	(40 ≤ age < 50)
if age >= 50 and age < 60 then ageg = 4;	(50 ≤ age < 60)
if age >= 60 and age < 70 then ageg = 5;	(60 ≤ age < 70)
if age >= 70 and age < 80 then ageg = 6;	(70 ≤ age < 80)
if age >= 80 then ageg = 7;	(age ≥ 80)

The gender variable has been defined as gender = 1 if female and gender = 0 if male. The model now contains 8 predictor variables:

X_1 = race,
X_j = ageg_j, $j = 2, 3, \ldots, 7$, denote six (0, 1) dummy variables for seven age groups, with ageg = 1 as the referent group.
X_8 = gender.

Since there are three outcome categories, the polytomous logistic regression model is defined in terms of the following two logit-like functions:[4]

$$\begin{cases} \ln\left[\dfrac{P(D = 1|\text{race, ageg, gender})}{P(D = 0|\text{race, ageg, gender})}\right] = \alpha_1 + \beta_{11}(\text{race}) + \sum_{j=2}^{7}\beta_{1j}(\text{ageg}_j) + \beta_{18}(\text{gender}) \\[4mm] \ln\left[\dfrac{P(D = 2|\text{race, ageg, gender})}{P(D = 0|\text{race, ageg, gender})}\right] = \alpha_2 + \beta_{21}(\text{race}) + \sum_{j=2}^{7}\beta_{2j}(\text{ageg}_j) + \beta_{28}(\text{gender}) \end{cases} \quad (23.4)$$

The results from fitting the polytomous model defined by (23.4) are presented in Table 23.3. These results were obtained using the LOGISTIC procedure in SAS (version 9.1). Again, there

---

[4] The general form of the polytomous logistic regression model that allows G (≥ 2) categories of the outcome variable and k predictors is given as follows:

$$\ln\left[\frac{P(D = g|X)}{P(D = 0|X)}\right] = \alpha_g + \sum_{j=1}^{k}\beta_{gj}X_j, \quad g = 1, 2, \ldots, (G - 1)$$

where $X = (X_1, X_2, \ldots, X_k)$ denotes the collection of predictors in the model. Note that the number of logit-like functions is $(G - 1)$; e.g., if $G = 3$ (as in our example), then $(G - 1) = 2$; and, if $G = 5$, then $(G - 1) = 4$.

The general formula for the collection of $(G - 1)$ odds ratios that compares two different specifications of X is

$$\text{OR}_g = \exp\left[\sum_{j=1}^{k}\beta_{gj}(X_{Aj} - X_{Bj})\right] \quad g = 1, 2, \ldots, (G - 1)$$

and where $X_A = (X_{A1}, X_{A2}, \ldots, X_{Ak})$ and $X_B = (X_{B1}, X_{B2}, \ldots, X_{Bk})$.

are two sets of parameter estimates corresponding to the intercept and each predictor in each of the two models in (23.4). There are a total number of 18 parameters (including intercepts) that are estimated. The output is listed in descending order, with α_2 corresponding to the first intercept listed and α_1 corresponding to the second intercept; similarly, the estimates for each predictor are listed in descending order, with the estimate of β_{2j} preceding the estimate of β_{1j} for the jth predictor, $j = 1, 2, \ldots, k$.

Since the primary predictor (i.e., exposure) variable of interest is race, the odds ratio values of interest in Table 23.3 are 2.901 and 1.776, which are the two point estimates listed for race in the portion of the output labeled "Odds Ratio Estimates." The value 2.901 gives the estimated odds ratio for the effect of race comparing subjects whose ESRD cause is Hypertension (cause = 2) versus Other Disease (cause = 0) controlling for age and gender. The value 1.776 gives the odds ratio for the effect of race comparing subjects whose ESRD cause is Diabetes (cause = 1) versus Other Disease (cause = 0) controlling for age and gender. Since race has a (0, 1) coding, these "adjusted" odds ratios can be computed alternatively by exponentiating the appropriate regression coefficients for race listed in the "Analysis of Maximum Likelihood Estimates" portion of the output, i.e.,

$$\widehat{OR}_{race} \text{ (Hypertension versus Other}|\text{age, gender)} = e^{\hat{\beta}_{21}} = e^{1.0652} = 2.901$$

$$\widehat{OR}_{race} \text{ (Diabetes versus Other}|\text{age, gender)} = e^{\hat{\beta}_{11}} = e^{0.5745} = 1.776$$

The 95% confidence intervals in Table 23.3 were calculated using the standard large-sample formula for $e^{\beta_{g1}}$ given by

$$\exp[\hat{\beta}_{g1} \pm 1.96 \, S_{\hat{\beta}_{g1}}], \qquad g = 1, 2$$

The Wald chi-square statistics for testing H_0: $\beta_{g1} = 0$, $g = 1, 2$, were calculated using the standard format described in Chapter 21; i.e., the Wald test statistic

$$\left(\hat{\beta}_{g1}/S_{\hat{\beta}_{g1}}\right)^2, \qquad g = 1, 2$$

has an approximate chi-square distribution with 1 $d.f.$ for large samples under H_0: $\beta_{g1} = 0$.

The portion of Table 23.3 that is labeled "Type 3 Analysis of Effects" gives "Generalized Wald Tests" for testing each of the three predictors in the model (race, ageg, and gender, with the six dummy variables for ageg being considered collectively as a single predictor). These test statistics are not equivalent to the LR statistics given by (23.2). In large samples, values obtained for the generalized Wald statistics will typically be close to corresponding LR values; nevertheless, LR tests are considered by statisticians to have better statistical properties than Wald tests, so that LR tests are typically preferred.

To carry out a LR test for the significance of the race variable in model (23.4), we would need to compare the log-likelihood statistic $(-2 \ln L_F)$ for model (23.4) with the corresponding log-likelihood statistic $(-2 \ln L_R)$ for the reduced model that would be obtained under the null hypothesis H_0: $\beta_{11} = \beta_{21} = 0$ in model (23.4). The reduced model will not contain the race variable and can be written as follows:

$$\ln\left[\frac{P(D = 1|\text{ageg, gender})}{P(D = 0|\text{ageg, gender})}\right] = \alpha_1 + \sum_{j=2}^{7} \beta_{1j} \, (\text{ageg}_j) + \beta_{18}(\text{gender})$$

$$\ln\left[\frac{P(D = 2|\text{ageg, gender})}{P(D = 0|\text{ageg, gender})}\right] = \alpha_2 + \sum_{j=2}^{7} \beta_{2j} \, (\text{ageg}_j) + \beta_{28}(\text{gender})$$

$$(23.5)$$

TABLE 23.3 **Edited polytomous logistic regression output for ESRD data for evaluating the relationship between race (X_1) and cause of ESRD (Y) controlling for age and gender**

```
                        Model Fit Statistics

                                              Intercept and
              Criterion        Intercept Only     Covariates
              -2 Log L            6534.134         6136.928

                     Type 3 Analysis of Effects
              Effect      DF      Wald ChiSq      Pr > ChiSq
              race        2        95.8805          <.0001
              ageg       12       276.2004          <.0001
              gender      2        23.2241          <.0001
```

Analysis of Maximum Likelihood Estimates

Parameter	cause	DF	Estimate	Std. Err	Wald ChiSq	Pr > ChiSq	Symbol
Intercept	2	1	-0.7552	0.1235	37.3888	<.0001	α_2
Intercept	1	1	-0.4820	0.1315	13.4437	0.0002	α_1
race	2	1	1.0652	0.1088	95.8639	<.0001	β_{21}
race	1	1	0.5745	0.0993	33.4818	<.0001	β_{11}
ageg1	2	1	-2.5520	0.5188	24.1958	<.0001	β_{22}
ageg1	1	1	-3.0331	0.6241	23.6216	<.0001	β_{12}
ageg2	2	1	-0.4855	0.1533	10.0333	0.0015	β_{23}
ageg2	1	1	-0.6062	0.1644	13.6034	0.0002	β_{13}
ageg3	2	1	-0.2032	0.1511	1.8092	0.1786	β_{24}
ageg3	1	1	0.1263	0.1531	0.6806	0.4094	β_{14}
ageg4	2	1	0.4442	0.1397	10.1153	0.0015	β_{25}
ageg4	1	1	0.9505	0.1439	43.6399	<.0001	β_{15}
ageg5	2	1	0.5459	0.1367	15.9360	<.0001	β_{26}
ageg5	1	1	1.0288	0.1413	53.0194	<.0001	β_{16}
ageg6	2	1	0.8655	0.1382	39.2336	<.0001	β_{27}
ageg6	1	1	0.9632	0.1453	43.9475	<.0001	β_{17}
gender	2	1	-0.0642	0.1030	0.3891	0.5328	β_{28}
gender	1	1	0.3247	0.0958	11.4997	0.0007	β_{18}

Odds Ratio Estimates

Effect	cause	Point Estimate	95% Wald Confidence Limits	
race	2	2.901	2.344	3.591
race	1	1.776	1.462	2.158
ageg 1 vs 7	2	0.020	0.006	0.066
ageg 1 vs 7	1	0.027	0.006	0.117
ageg 2 vs 7	2	0.154	0.102	0.232
ageg 2 vs 7	1	0.308	0.201	0.472
ageg 3 vs 7	2	0.204	0.136	0.306
ageg 3 vs 7	1	0.641	0.429	0.959
ageg 4 vs 7	2	0.390	0.267	0.570
ageg 4 vs 7	1	1.462	0.999	2.141
ageg 5 vs 7	2	0.432	0.299	0.625
ageg 5 vs 7	1	1.581	1.089	2.297
ageg 6 vs 7	2	0.595	0.412	0.859
ageg 6 vs 7	1	1.481	1.015	2.162
gender	2	0.938	0.766	1.148
gender	1	1.384	1.147	1.669

TABLE 23.4   Odds ratios, 95% confidence intervals, and Wald *P*-values for race comparing models that control and do not control for ageg and gender

Variables in Model

Comparison	race, ageg, gender			race only		
	Odds Ratio	95% Conf. Int.	Wald *P*-value	Odds Ratio	95% Conf. Int.	Wald *P*-value
2 vs. 0 Hypertension vs. Other Disease	2.901	(2.344, 3.591)	< 0.0001	2.105	(1.731, 2.559)	< 0.0001
1 vs. 0 Diabetes vs. Other Disease	1.776	(1.462, 2.158)	< 0.0001	1.484	(1.241, 1.775)	< 0.0001

From Table 23.3, we find that $-2 \ln L_F = 6136.938$. The log-likelihood statistic obtained from the reduced model (output not provided) is $-2 \ln L_R = 6236.806$. The corresponding LR statistic is then computed as

$$\text{LR}_{\text{race}} = -2 \ln L_R - (-2 \ln L_R) = (6236.806) - (6136.938) = 99.868$$

which is highly significant (*P*-value < 0.0001) based on the chi-square distribution with 2 *d.f.*, thus indicating a significant effect of race in model (23.4) after adjusting for age and gender. Notice that the generalized Wald statistic for race provided in Table 23.3 is 95.8805, which is close but not identical to the LR statistic. Since race is the primary predictor of interest, the generalized Wald tests provided for ageg and gender in Table 23.3, as well as corresponding LR tests for ageg and gender, are not discussed further. (Note, however, that the *d.f.* = 12 for the generalized Wald test for the significance of ageg, because there are two sets of six dummy variables that define the ageg variable in model (23.4).)

Table 23.4 compares the estimated odds ratios and corresponding 95% confidence intervals for race in model (23.4), which controls for ageg and gender, with those for model (23.1), which does not control for ageg and gender. If we compare the model with three predictor variables (race, ageg, and gender) with the model with only race included, the effect of race in the reduced model (with race only) is weaker both for the comparison of Hypertension to Other Disease (odds ratio estimates: 2.901 vs. 2.105) and for the comparison of Diabetes to Other Disease (odds ratio estimates: 1.776 vs. 1.484). These results suggest that ageg and/or gender act as confounders of the relationship between race and cause of ESRD. The numerical results based on the model containing only race suggest a bias toward the null value of one when both ageg and gender are not included as covariates.

The 95% confidence intervals in the table are wider when ageg and gender are included in the models than when they are not (2 vs. 0 widths: 1.427 vs. 0.828; 1 vs. 0 widths: 0.696 vs. 0.534). This indicates a slight loss of precision when ageg and gender are included along with race in the model. Nevertheless, the model that contains all three predictors is preferred because the estimated odds ratios are meaningfully different when ageg and gender are included than when ageg and gender are ignored (i.e., ageg and gender are apparently confounders).

23.5 Ordinal Logistic Regression: Overview

In this section, the standard logistic model is extended to consider ordinal outcome variables, i.e., the categories of the outcome variable have a natural ordering. We restrict attention to the most popular form of ordinal logistic regression model called the *proportional odds (or cumulative logit) model* (see Ananth and Kleinbaum, 1997, for a description of other types of ordinal logistic regression models).

Ordinal variables have a natural ordering among the outcome levels. An example is cancer tumor grade, ranging from well-differentiated to moderately differentiated to poorly differentiated tumors. Another example is length of inpatient hospital stay (in days), ranging from short, to intermediate, to long. A third example is time since previous mammography, ranging from within one year, to greater than one year, to never having had a mammography.

An ordinal outcome variable can always be modeled with a polytomous logistic model, as discussed in the previous section, but it can also be modeled with more precision using ordinal logistic regression, provided that certain assumptions are met. Ordinal logistic regression, unlike polytomous regression, takes into account any inherent ordering of the levels in the outcome variable, thus making better use of the ordinal information.

23.6 A "Simple" Hypothetical Example: Three Ordinal Categories and One Dichotomous Exposure Variable

Table 23.5 provides a 3×2 table of hypothetical data on 275 hospital patients collected to assess the relationship between an ordinal outcome variable for length of hospital stay (D) and a $(0, 1)$ exposure variable (E) that distinguishes between two types of patients:

TABLE 23.5 **Crude data relating patient type (E) to hospital length of stay (D)**

	$E = 1$	$E = 0$	Total
$D = 2$ (long)	20	30	50
$D = 1$(medium)	30	70	100
$D = 0$ (short)	25	100	125
Total	75	200	275

If we treat the outcome variable D as a polytomous outcome by ignoring the natural ordering of the categories of D, the analysis (without controlling for other variables) would consider two odds-ratio estimates obtained from two 2×2 sub-tables that compare separate outcome categories (say, $D = 1$ and $D = 2$) to a referent category (say $D = 0$), as shown in Table 23.6.

However, if the natural ordering of the outcome (D) variable is to be considered, the use of the ordinal logistic regression procedure that involves the proportional odds model requires a different partitioning of the overall data into two sub-tables. In particular, we must consider the possible ways to collapse the ordinal categories into 2×2 tables that preserve the natural ordering of the categories of D. Using the data in Table 23.5, there are two such tables that can be formed, which are shown in Table 23.7. These tables compare category 0 to categories 1 and 2 combined, and also compare categories 0 and 1 combined to category 2. We have not combined categories 0 and 2 for comparison with category 1, since that would disrupt the natural ordering from 0 through 2.

The proportional odds model makes an important assumption. Under this model, the odds ratio for the effect of an exposure variable for any 2×2 table that is obtained by collapsing one or more of the rows of the 3×2 table given in Table 23.5 will be the same regardless of where

TABLE 23.6 Tables and estimated odds ratios treating D as a polytomous outcome

	$E = 1$	$E = 0$	Total
$D = 1$ (medium)	30	70	100
$D = 0$ (short)	25	100	125
Total	55	170	225

$$\widehat{OR} = \frac{30 \times 100}{70 \times 25} = 1.71$$

	$E = 1$	$E = 0$	Total
$D = 2$ (long)	20	30	50
$D = 0$ (short)	25	100	125
Total	45	130	175

$$\widehat{OR} = \frac{20 \times 100}{30 \times 25} = 2.67$$

TABLE 23.7 Collapsed 2×2 tables obtained from Table 23.5 for a proportional odds model

	$E = 1$	$E = 0$	Total
$D = 1$ or 2	50	100	150
$D = 0$	25	100	125
Total	75	200	275

$$\widehat{OR} = \frac{50 \times 100}{100 \times 25} = 2.00$$

	$E = 1$	$E = 0$	Total
$D = 2$	20	30	50
$D = 0$ or 1	55	170	225
Total	75	200	275

$$\widehat{OR} = \frac{20 \times 170}{30 \times 55} = 2.06$$

the cut-point is made, provided the natural ordering of the outcome variable is not disrupted. In other words, the odds ratio is *invariant* to how the outcome categories are dichotomized.

In Table 23.7, the odds ratios from the two collapsed tables are almost identical (2.00 versus 2.06) and thus provide evidence that the proportional odds assumption is not violated. It would be unusual for the collapsed-table odds ratios to match perfectly. The odds ratios do not have to be exactly equal; as long as they are "close," the proportional odds assumption may be considered reasonable.

There is also a statistical test called the *score test* designed to evaluate whether a model constrained by the proportional odds assumption (i.e., an ordinal model) is significantly different from the corresponding model in which the odds ratio parameters are not constrained by the proportional odds assumption (i.e., a polytomous model). The test statistic is distributed approximately chi-square, with degrees of freedom (*d.f.*) equal to $k(G-2)$, where k is the number of predictor variables in the model and G is the number of ordinal categories of the outcome variable.

For the ordinal 3×2 data given in Table 23.5, the score test is equivalent to testing the null hypothesis that the two odds ratios being estimated by collapsing one or more of the rows of the table are equal. In other words, for these data, the null hypothesis that the proportional odds assumption is satisfied is equivalent to the equality of the two odds ratios estimated in Table 23.7. Since we have $G = 3$ ordinal outcome categories and $k = 1$ predictor (E), the *d.f.* for the Score test is $k(G-2) = 1$. The score test result for these data (chisq = 0.0082, *P*-value = .9281) are nonsignificant, so that we can conclude that the proportional odds assumption is satisfied.[5]

The proportional odds model for the analysis of the data in Table 23.5 is given by the following expression:

$$P(D \geq g|E) = \frac{1}{1 + \exp[-(\alpha_g + \beta E)]}, \qquad g = 1, 2 \qquad (23.6)$$

where $\alpha_1 > \alpha_2$.

The above formula for the proportional odds model differs from the binary logistic model in that model (23.6) is formulated as the probability of an inequality, i.e., $P(D \geq g|E)$, whereas the formula for a binary outcome is the probability of an equality, i.e., $P(D = 1|E)$. Moreover, the proportional odds model differs from the polytomous logistic regression model (defined generally in footnote 4 of the previous section) in that the regression coefficient β in model (23.6) is not subscripted by g, as in the polytomous model.

An equivalent definition of the proportional odds model can be written in terms of the odds of an inequality as follows:

$$\text{Odds}_g(E) = \frac{P(D \geq g|E)}{1 - P(D \geq g|E)} = \frac{P(D \geq g|E)}{P(D < g|E)} \qquad (23.7)$$

If we substitute the formula (23.6) for $P(D \geq g|E)$ into the expression (23.7), we find that the formula for the Odds simplifies to:

$$\text{Odds}_g(E) = \frac{P(D \geq g|E)}{P(D < g|E)} = \exp[\alpha_g + \beta E] \qquad (23.8)$$

Using (23.8), we can then derive the formula for the odds ratio comparing an exposed person ($E = 1$) to an unexposed person ($E = 0$), as follows:

$$OR_{E=1 \text{vs.} E=0} = \frac{\text{Odds}_g(E=1)}{\text{Odds}_g(E=0)} = \frac{\exp[\alpha_g + \beta(1)]}{\exp[\alpha_g + \beta(0)]} = \exp[(\alpha_g - \alpha_g) + \beta(1-0)] = e^{\beta}$$

$$(23.9)$$

From (23.9), even though the intercept terms of the proportional odds model (α_g) are subscripted by g, a single odds ratio is obtained for the comparison of exposed and unexposed groups. This single odds ratio (given by e^{β}) is the common odds ratio that is assumed by the proportional odds model when the ordinal data (e.g., Table 23.5) are partitioned into the two 2×2 tables given in Table 23.7.

---

[5] If the proportional odds assumption is inappropriate, there are other ordinal logistic models available that make alternative assumptions about the ordinal nature of the outcome. Examples include a continuation ratio model, a partial proportional odds model, and a stereotype regression model. A discussion of these models is beyond the scope of the current presentation. (See the review by Ananth and Kleinbaum 1997.)

Using simple algebra, we can rewrite (23.9) to express the odds for exposed subjects in terms of the odds for unexposed subjects as follows:

$$\text{Odds}_g(E = 1) = e^{\beta}\text{Odds}_g(E = 0) \qquad (23.10)$$

Equation (23.10) essentially says that the odds for exposed persons for any value of g can be obtained from the odds for unexposed persons for that same value of g by multiplying the latter by the proportionality constant e^{β} that does not depend on g. In other words, the odds for exposed persons is *proportional* to the odds for unexposed persons no matter which of the 2×2 collapsed tables is considered (i.e., no matter what value of g is used). This is the reason why model (23.6), and its more general form (given in footnote 11 later in this section), is called the *proportional odds* model.

The results from fitting model (23.6) to the data in Table 23.5 are presented in Table 23.8. The results were obtained using PROC LOGISTIC in SAS.[6]

**TABLE 23.8   Edited computer output for analysis of data from Table 24.5
using the proportional odds model**

```
       Score Test for the Proportional Odds Assumption
     Chi-Square          DF        Pr > ChiSq
     0.0082              1          0.9281

          Model Fit Statistics
     -2 Log L            562.218

     Analysis of Maximum Likelihood Estimates
   Parameter    DF   Estimate   bbb StdErr   Wald Chi-Sq   Pr > ChiSq

   Intercept 2  1    -1.0228      0.2301       19.7665      <.0001
   Intercept 1  1     0.7021      0.2241        9.8139       0.0017
   E            1     0.7044      0.2543        7.6702       0.0056

               Odds Ratio Estimates
               Point           95% Wald
   Effect     Estimate    Confidence Limits
   E           2.023       1.229       3.333
```

---

[6] The SAS program code used to produce the output in Table 23.8 is given as follows:

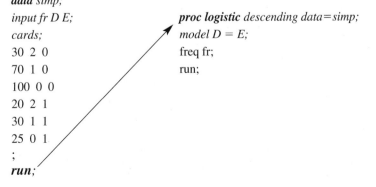

data simp;

input fr D E;   **proc logistic** *descending data=simp;*

cards;   *model D = E;*

30 2 0   *freq fr;*

70 1 0   *run;*

100 0 0

20 2 1

30 1 1

25 0 1

;

run;

The output provides the estimates for two intercepts (α_2 and α_1, respectively), but there is only one estimated β for the effect of E. The estimated odds ratio for E is $e^{0.7044} = 2.023$. Note that this estimated odds ratio lies between 2.00 and 2.06, the two estimated odds ratios obtained from partitioning the data into two 2×2 tables (Table 23.7) that preserve the ordering of the outcome variable.

The Wald chi-square test for the significance of E (i.e., the test of H_0: $\beta = 0$ versus H_A: $\beta \neq 0$) is highly significant (two-tailed P-value = .0056). The 95% confidence limits for a confidence interval for e^{β} are 1.239 and 3.33.

The results for this hypothetical dataset indicate that exposed patients are about twice (i.e., 2.023) as likely as unexposed patients to have a "longer" rather than a "shorter" hospital stay, regardless of whether "longer" is categorized strictly or broadly (i.e., just long or medium and/or long).

23.7 Ordinal Logistic Regression Example Using Real Data with Four Ordinal Categories and Three Predictor Variables

Acinetobacter species are aerobic gram-negative bacilli that can cause healthcare-associated infections and can survive for prolonged periods of time in the environment and on healthcare workers' hands. *Acinetobacter* infections have become increasingly difficult to treat due to the emergence of strains resistant to almost all commonly prescribed antimicrobial agents. Outbreaks caused by multidrug-resistant (MDR) *Acinetobacter* have been reported in hospitals all over the world; more recently, they have become a significant problem in military medical facilities.

We consider an example of the use of the proportional odds model for the Multidrug-Resistant (MDR) *Acinetobacter* Infection Study (Sunenshine et al., 2006), a retrospective matched-pair cohort study[7] that investigated the impact of MDR *Acinetobacter* infection status on length of hospitalization stay using patient records from January 2003 through August 2004 from two tertiary care hospitals in Baltimore City. The data considered here compare the outcomes of 56 patients with MDR *Acinetobacter* infection (exposed patients) to 56 pair-matched patients with susceptible *Acinetobacter* infection (unexposed susceptible referents).

The predictor (exposure) variable of primary interest is *Acinetobacter* infection status (acstatus), coded as 0 for susceptible patients and 1 for infected patients. The outcome variable is the ordinal variable ordhospdays, defined from terciles of hospital length of stay (hospdays) using the following 3 categories (SAS code follows):

if hospdays <= 9 then ordhospdays = 0;
if hospdays > 9 and hospdays <= 23 then ordhospdays = 1;
if hospdays > 23 then ordhospdays = 2;

(The SAS datafile used here is called acineto.dat and can be obtained from the publisher's website.)

The matching variable is exposure time prior to infection (*exptime*); in particular, a susceptible referent patient had to have a pre-infection hospital length of stay (LOS) within 5% of

---

[7] In their paper, Sunenshine et al. (2006) did not use a proportional odds model to analyze their data, but rather treated the outcome variable as a binary outcome and used conditional logistic regression to account for the matching.

the matched MDR *Acinetobacter* patient's pre-infection hospital length of stay. The variable exptime was redefined[8] as a categorical variable called newextime with the following 8 categories (SAS code follows):

if exptime = 0 then newexptime = 0;
if exptime >= 1 and exptime < 5 then newexptime = 1;
if exptime >= 5 and exptime < 10 then newexptime = 2;
if exptime >= 10 and exptime < 15 then newexptime = 3;
if exptime >= 15 and exptime < 20 then newexptime = 4;
if exptime >= 20 and exptime < 25 then newexptime = 5;
if exptime >= 25 and exptime < 30 then newexptime = 6;
if exptime >= 30 then newexptime = 7.

There were two other control variables (not involved in the matching): a measure of severity-of-illness prior to *Acinetobacter* infection called "apache" score; and, a summary measure of co-morbidities, called "charlson" score, calculated using data from the past medical history recorded in the medical chart of the patient.

Table 23.9 presents the 3×2 table of the crude data that describes the association between the exposure variable *Acinetobacter* infection status (acstatus) and the ordinal outcome variable hospital length of stay (ordhospdays). Here, the coding of ordhospdays as 0, 1, or 2 reflects the ordinal nature of the outcome. For example, it is necessary that the interval "$9 < \text{hospdays} \leq 23$" be coded (as ordhospdays = 1) between the interval "$0 \leq \text{hospdays} \leq 9$" (coded as ordhospdays = 0) and the interval "hospdays > 23" (coded as ordhospdays = 2). This contrasts with polytomous logistic regression, in which the coding is not necessarily reflective of an underlying order in the outcome variable.

Under the proportional odds assumption, the odds ratio obtained for any 2×2 table obtained by collapsing one of the rows of the 3×2 table given in Table 23.9 should be the same

TABLE 23.9   Crude data relating acinetobacter infection status (acstatus) and hospital length of stay (ordhospdays)

	E = acstatus		
D = ordhospdays	1 = infected	0 = susceptibles	Total
2 = (hospdays > 23)	24	13	37
1 = (9 < hospdays ≤ 23)	16	23	39
0 = (0 ≤ hospdays ≤ 9)	16	20	36
Total	56	56	112

---

[8] SAS's LOGISTIC procedure does not allow the use of conditional ML estimation (typically used for matched data when the number of matching strata is large relative to the study size) when fitting an ordinal logistic regression model. Consequently, the matching variable *exptime* was redefined as an 8-category variable (*newexptime*) in order to justify the use of unconditional ML estimation that may be used when the number of matching strata (e.g., 8) is small relative to the total number of observations ($n = 112$). Such pooling of the original matched pairs strata is reasonable whenever a referent patient from one pair would have been eligible to be chosen as the referent in another pair because of similar exposure times (i.e., two such matched pairs are said to be *exchangeable*).

TABLE 23.10   **Collapsed 2 × 2 tables obtained from Table 23.9 with the natural ordering of the "ordhospdays" variable preserved**

	E = acstatus	
D = ordhospdays	1 = infected	0 = susceptibles
1–2	40	36
0	16	20

$$\widehat{OR} = \frac{40 \times 20}{36 \times 16} = 1.39$$

	E = acstatus	
D = ordhospdays	1 = infected	0 = susceptibles
2	24	13
0–1	32	43

$$\widehat{OR} = \frac{24 \times 43}{13 \times 32} = 2.48$$

regardless of where the cut-point is made, provided the natural ordering of the outcome variable is not disrupted. There are two such 2×2 tables, as shown in Table 23.10. These tables compare category 0 to categories 1 and 2 combined, and categories 0 and 1 combined with category 2. Again, we cannot combine categories 0 and 2 for comparison with category 1, since that would disrupt the natural ordering from 0 through 2.

The two odds ratios shown in Table 23.10 are 1.39 and 2.48, which are different enough from one another to suggest that the proportional odds assumption may be violated. However, because three control variables (apache, charlson, and the categorical variable newexptime) are being considered in the analysis in addition to the exposure variable (acstatus), a more appropriate comparison should compare two estimated *adjusted* odds ratios for the effect of acstatus on ordhospdays controlling for apache, charlson, and newextime.[9]

Moreover, the *score test* (described earlier) for the proportional odds assumption that considers all four predictors in the same model (23.11) provides a comparison of such estimated adjusted odds ratios.[10] The *score test* result for this model [chisq = 8.7459, d.f. = $k(G - 2) = 10(3 - 2) = 10$, P-value = .5564] is non-significant. Thus, even though the two unadjusted odds ratios shown in Table 23.10 are somewhat different, we can nevertheless argue from the score test result that the proportional odds assumption seems to be satisfied for these data.

---

[9] Estimated adjusted odds ratios for the effect of acstatus, controlling for apache, charlson, and newexptime, can be computed by fitting two separate binary logistic models using the data in the 2×2 tables shown in Table 23.10, where each model would contain all four predictor variables. The adjusted odds ratios are $e^{\hat{\beta}_1}$, where $\hat{\beta}_1$ is the coefficient of acstatus in each model.

[10] The *score test* for the proportional odds model (23.11) containing the four predictor variables (*acstatus, apache, charlson, and newexptime*) simultaneously tests the proportional odds assumption for all four variables in the model, rather than one variable at a time adjusted for the other three variables. The latter (single variable) test is not automatically provided in SAS's LOGISTIC procedure.

The proportional odds model[11] for the analysis of the *Acinetobacter* infection data described above is:

$$P(D \geq g|X)$$

$$= \frac{1}{1 + \exp\left[-\left(\alpha_g + \beta_1(\text{acstatus}) + \beta_2(\text{apache}) + \beta_3(\text{charlson}) + \sum_{j=0}^{6}\gamma_j(\text{newexptime}_j)\right)\right]}$$

$$(23.11)$$

where $g = 1, 2$; $\alpha_1 > \alpha_2$; $\text{newexptime}_0, \ldots, \text{newexptime}_6$ are 7 dummy variables for the 8 categories of the categorical variable newexptime; and $X = (\text{acstatus}, \text{apache}, \text{charlson}, \text{newexptime}_0, \text{newexptime}_1, \ldots, \text{newexptime}_6)$.

As with the simpler version of the proportional odds model defined earlier by (23.6), the regression coefficients $\beta_1, \beta_2, \beta_3, \gamma_0, \gamma_1, \ldots,$ and γ_6 in model (23.11) are not subscripted by g.[12] Also, we can equivalently write model (23.11) in Odds form as:

$$\text{Odds}_g(X) = \frac{P(D \geq g|X)}{P(D < g|X)}$$

$$= \exp\left[\alpha_g + \beta_1(\text{acstatus}) + \beta_2(\text{apache}) + \beta_3(\text{charlson})\right.$$

$$\left. + \sum_{j=0}^{6}\gamma_j(\text{newexptime}_j)\right], \qquad g = 1, 2 \qquad (23.12)$$

---

[11] The general form of the proportional odds model (used by SAS) that allows $G(\geq 3)$ categories of the outcome variable and k predictors is given as follows:

$$P(D \geq g|X) = \frac{1}{1 + \exp\left[-\left(\alpha_g + \sum_{j=1}^{k}\beta_j X_j\right)\right]}, \qquad g = 1, 2, \ldots, G - 1$$

where $X = (X_1, X_2, \ldots, X_k)$ denotes the collection of predictors in the model, and $\alpha_1 > \alpha_2 > \cdots > \alpha_{G-1}$. Note that the number of intercept terms is $(G - 1)$, e.g., if $G = 4$ (as in the *Acinetobacter* example), then $(G - 1) = 3$.

[12] An alternate formulation of the proportional odds model (used by STATA) is given by:

$$P(D^* \leq g|X) = \frac{1}{1 + \exp\left[-\left(\alpha_g^* - \sum_{j=1}^{k}\beta_j^* X_j\right)\right]}, \qquad g = 1, 2, \ldots, G - 1$$

where $D^* = 1, 2, \ldots, G$. Two important features of this formula that differ from the general formula in footnote 11 are the direction of the inequality ($D^* \leq g$) and the negative sign before $\sum_{j=1}^{k}\beta_j^* X_j$. In terms of the beta coefficients, these two key differences "cancel out" so that $\beta_j = \beta_j^*$. Consequently, if the same data are fit for each formulation of the model, the same parameter estimates of the β_j's would be obtained for each model. However, the intercepts for the two formulations differ as follows: $\alpha_g = -\alpha_g^*$.

Using (23.12), the formula for the odds ratio[13] comparing an exposed person (acstatus = 1) to an unexposed person (acstatus = 0) controlling for apache, charlson, and newexptime is given as follows:

$$OR_{E=1 \text{ vs. } E=0|\text{apache, charlson, newexptime}} = \frac{\text{Odds}_g(E = 1|\text{apache, charlson, newexptime})}{\text{Odds}_g(E = 0|\text{apache, charlson, newexptime})}$$

$$= \frac{\exp\left[\alpha_g + \beta_1(1) + \beta_2(\text{apache}) + \beta_3(\text{charlson}) + \sum_{j=0}^{6}\gamma_j(\text{newexptime}_j)\right]}{\exp\left[\alpha_g + \beta_1(0) + \beta_2(\text{apache}) + \beta_3(\text{charlson}) + \sum_{j=0}^{6}\gamma_j(\text{newexptime}_j)\right]}$$

$$= e^{\beta_1} \qquad\qquad (23.13)$$

From (23.13), even though the intercept terms of the proportional odds model (α_g) are subscripted by g, a single odds ratio, e^{β_1}, is obtained for the comparison of exposed and unexposed groups.

The results from fitting model (23.11) to the *Acinetobacter* data are presented in Table 23.11. The results were obtained using PROC LOGISTIC in SAS.[14]

As mentioned earlier, the *score test* shown in the output is non-significant (P-value = 0.5564), which supports the use of the proportional odds model for this analysis. The degrees of freedom of 10 for this test are obtained from the formula $k(G - 2)$, where $k = 10$ is the number of predictors in the model and $G = 3$ is the number of ordinal categories.

The output shows estimates for two intercepts (α_2 and α_1, respectively), but only one estimated coefficient for the effect of each predictor in the model. The estimated odds ratio for the *adjusted* effect of acstatus is $e^{0.6865} = 1.99$. This odds ratio value lies between 1.39 and 2.48, the two estimated odds ratios obtained by partitioning the data into two 2 × 2 tables (in Table 23.10) that preserve the ordering of the outcome variable.

---

[13] The general formula for the odds ratio that compares two different specifications of X is:

$$OR_{\mathbf{X}_A, \mathbf{X}_B} = \exp\left[\sum_{j=1}^{k}\beta_j(X_{Aj} - X_{Bj})\right]$$

where $\mathbf{X}_A = (X_{A1}, X_{A2}, \ldots, X_{Ak})$ and $\mathbf{X}_B = (X_{B1}, X_{B2}, \ldots, X_{Bk})$. Notice that the subscript g does not appear in the formula.

[14] The SAS program code used to produce the output in Table 23.11 is given as follows:

> **proc logistic** *descending data = acineto;*
> *class newexptime;*
> *model ordhospdays = acstatus apache charlson newexptime;*
> *run;*

Each of the 112 persons considered in the analysis was listed on a separate line of data in the datafile used by SAS; hence, no "freq" statement was used in the program code as previously described in footnote 6.

TABLE 23.11 **Edited computer output for analysis of *Acinetobacter* data using the proportional odds model (23.11)**

```
Score Test for the Proportional Odds Assumption
        Chi-Square          DF          Pr > ChiSq
          7.4396            10            0.5564

      Model Fit Statistics
      -2 Log L      206.617
```

Analysis of Maximum Likelihood Estimates

Parameter	DF	Estimate	Standard Error	Wald Chi-Square	Pr > ChiSq
Intercept 2	1	-2.0965	0.5248	15.9583	<.0001
Intercept 1	1	-0.1671	0.4802	0.1211	0.7278
acstatus	1	0.6865	0.3876	3.1380	0.0765
apache	1	0.0392	0.0130	9.0900	0.0026
charlson	1	-0.1755	0.0650	7.2931	0.0069
newexptime 0	1	-1.9109	0.5169	13.6648	0.0002
newexptime 1	1	-0.6888	0.4972	1.9192	0.1659
newexptime 2	1	0.2996	0.4174	0.5153	0.4729
newexptime 3	1	-0.7062	0.5713	1.5277	0.2165
newexptime 4	1	-0.0715	0.5443	0.0173	0.8954
newexptime 5	1	0.5357	0.5388	0.9885	0.3201
newexptime 6	1	0.7573	0.5394	1.9715	0.1603

Odds Ratio Estimates

Effect	Point Estimate	95% Wald Confidence Limits	
acstatus	1.987	0.930	4.247
apache	1.040	1.014	1.067
charlson	0.839	0.739	0.953
newexptime 0 vs 7	0.025	0.005	0.121
newexptime 1 vs 7	0.084	0.018	0.388
newexptime 2 vs 7	0.227	0.057	0.901
newexptime 3 vs 7	0.083	0.016	0.441
newexptime 4 vs 7	0.156	0.031	0.784
newexptime 5 vs 7	0.287	0.060	1.367
newexptime 6 vs 7	0.358	0.075	1.717

The Wald chi-square test for the significance of acstatus (i.e., H_0: $\beta_1 = 0$) is of borderline significance (the two-tailed P-value is .0765, whereas the one-tailed P-value is .0383). The 95% confidence limits for e^{β_1} are 0.930 and 4.247, which indicates a moderate lack of precision.

The results for this hypothetical dataset indicate that patients with *Acinetobacter* infection were about twice (i.e., 1.99) as likely as susceptible patients to have a "long" rather than a "short" hospital stay, regardless of whether "long" is categorized strictly or broadly (e.g., just long or medium and/or long). Note that these results are similar to those obtained for our hypothetical example considered in Section 23.6.

23.8 Summary

In this chapter, the standard "binary" logistic regression model has been extended to handle an outcome variable that has more than two categories. When the categories of the outcome have no natural order, polytomous logistic regression is appropriate, whereas ordinal logistic regression is preferred when the categories have a natural order.

We illustrated the use of polytomous logistic regression to assess the relationship of "race" to "cause" of End Stage Renal Disease (ESRD) in 3,049 patients who initiated dialysis in Georgia in 2002 (Volkova et al., 2006). The three categories of ESRD cause (2 = Hypertension, 1 = Diabetes, or 0 = Other Disease) had no natural ordering, so polytomous logistic regression was appropriate. The polytomous logistic regression model provides separate odds ratio estimates that compare different categories of the outcome with a referent category. The regression coefficients in this model are different for different outcome categories.

The proportional odds model is the most popular model used for carrying out ordinal logistic regression, but requires a "proportional odds assumption" to be satisfied. A test procedure for assessing the proportional odds assumption is the *score test*, which is part of the output in most computer packages (e.g., SAS, STATA, SPSS) that fit the proportional odds model. We have referenced alternative models to consider if the proportional odds assumption is not satisfied (Ananth and Kleinbaum, 1997).

In contrast to the polytomous logistic regression model, the regression coefficients for each predictor in the proportional odds model do not vary by outcome category, and a single odds ratio estimate is obtained for each (binary) predictor. Such a single odds ratio estimate represents a weighted average of estimated odds ratios obtained from different ways of combining the outcome categories into strata that preserve the natural ordering of the outcome variable.

We illustrated ordinal logistic regression using data from a Multidrug-Resistant (MDR) *Acinetobacter* Infection Study (Sunenshine et al., 2006); this is a retrospective matched-pair cohort study that investigated the impact of MDR *Acinetobacter* infection status on length of hospitalization stay using patient records from January 2003 through August 2004 from two tertiary care hospitals in Baltimore City. The outcome variable, days of hospital stay, was treated as an ordinal variable, so that ordinal logistic regression was an appropriate data analytic procedure.

Problems

1. A cross-sectional study was carried out to assess the relationship of alcohol and smoking to blood pressure in 2,500 men ages 20 years or older in 4 North American population groups, each group utilizing a different clinic. The outcome variable was blood pressure status (BP): normotensive = 0, moderate hypertensive = 1, and severe hypertensive = 2. The primary exposure variables were alcohol status (ALC: non-drinker = 0, light drinker = 1, and heavy drinker = 2, with 0 being the referent category) and smoking status (SMK: none = 0, less than 2 packs per day = 1, and 2+ packs per day = 2, with 0 being the referent category). Variables considered for control included AGE (under 40 = 0, at least 40 = 1), PA (physical activity: not regular = 0, regular = 1), and CLINIC (4 categories, with clinic 4 being the referent category).

   a. State the logit form of a no-interaction ordinal (proportional odds) logistic model that would describe the relationship of the outcome BP (treated ordinally) with the predictors ALC, SMK, AGE, PA, and CLINIC. In stating this model, make sure to treat all predictors as categorical variables.

   (*Note:* In questions to follow, we will refer to the exposure variables ALC and SMK as *E* variables; the control variables AGE, PA, and CLINIC will be referred to as *V* variables.)

   b. Consider the following (crude) odds ratio estimates obtained from the crude 3×3 table relating ALC to BP:

   $\widehat{OR}_1 = \widehat{OR}_{BP=2 \text{ vs. } 0-1}(ALC = 2 \text{ vs. } 0),$ $\widehat{OR}_2 = \widehat{OR}_{BP=2 \text{ vs. } 0-1}(ALC = 1 \text{ vs. } 0)$

   $\widehat{OR}_3 = \widehat{OR}_{BP=1-2 \text{ vs. } 0}(ALC = 2 \text{ vs. } 0),$ $\widehat{OR}_4 = \widehat{OR}_{BP=1-2 \text{ vs. } 0}(ALC = 1 \text{ vs. } 0)$

What relationships among the above odds ratios would one expect to see if the proportional odds model was appropriate for the variable ALC?

1. c. What does the score test for the proportional odds model allow one to evaluate and how does it work (i.e., state the null hypothesis and how to proceed depending on whether you rejects or does not reject the null hypothesis)?

d. For the model described in part (a), give an expression for the ODDS of being a moderate or severe hypertensive, i.e., BP $\geq$ 1, who is a heavy-drinking, 45-year-old, 2+ pack-a-day smoker, who does regular physical activity, and comes from clinic 4.

e. For the model described in part (a), give an expression for the ODDS RATIO for being a moderate or severe hypertensive, i.e., BP $\geq$ 1, that compares a heavy-drinking, 2+ pack-a-day smoker to a light-drinking, non-smoker, controlling for AGE, PA, and CLINIC.

f. For the model described in part (a), give an expression for the ODDS RATIO for being a normotensive, i.e., BP $=$ 0, that compares a heavy-drinking, 2+ pack-a-day smoker to a light-drinking, non-smoker, controlling for AGE, PA, and CLINIC.

g. Suppose that the model in part (a) was extended to allow for two-way interactions (i.e., products involving two variables) between alcohol and age, alcohol and physical activity, alcohol and clinic, smoking and age, smoking and physical activity, smoking and clinic, and alcohol and smoking. Assuming that alcohol status and smoking status are exposure (E) variables of interest, and that AGE, PA, and CLINIC are control (V) variables, state which variables in the interaction model are E_iV_j variables and which are $E_iE_{i'}$ variables. (Make sure to define the variables explicitly, keeping in mind that all the variables in the model are being treated as categorical variables).

h. Starting with the model described in part (g), suppose that it was decided to determine a "best" model by first testing for significant interaction among the E_iV_j variables. Describe how one would *simultaneously test* for the significance of at least one of the E_iV_j product terms in this model. Make sure to state the null hypothesis in terms of model parameters, describe the formula for the test statistic, and give the distribution and degrees of freedom of the test statistic under the null hypothesis.

i. Assume that the only significant interaction terms among the E_iV_j variables were product terms involving alcohol status with age and smoking status with physical activity.
State the model that remains for further assessment after this interaction assessment. Make sure to write the model in terms of the variables alcohol status, smoking status, age, physical activity, and appropriate product terms.

j. Considering the answer to part (i), what is the formula for the odds ratio that compares heavy-drinking, 2+ pack-a-day smokers to light-drinking non-smokers, controlling for AGE, PA, and CLINIC?

k. Using the odds ratio formula described in part (j), give a formula for a 95% confidence interval for the odds ratio for a 30-year old man who does not do regular physical activity.
    Note: The general 95% CI formula for an odds ratio of the form **exp**[L], where $\hat{L} = c_1\hat{\beta}_1 + \cdots + c_k\hat{\beta}_k$ is given by:

$$\exp[\hat{L} \pm 1.96\sqrt{\hat{Var}(\hat{L})}]$$

where

$$\mathrm{Var}(\hat{L}) = \sum_{j=1}^{k} c_j^2 \mathrm{Var}(\hat{\beta}_j) + 2\sum_{j<j'}^{k}\sum c_j c_{j'} \mathrm{Cov}(\hat{\beta}_j, \hat{\beta}_{j'})$$

1. l. Suppose that one decides that there is no interaction between alcohol and smoking, that CLINIC is not a confounder but needs to be considered for reasons of precision, and that the only two significant interaction terms are between ALC and AGE and between SMK and PA. What is your "final" odds ratio expression for a moderate or severe hypertensive (i.e., BP ≥ 1) that compares a heavy-drinking, 2+ pack-a-day smoker to a light-drinking non-smoker, controlling for AGE, PA, and CLINIC?

2. A 2005 study was conducted to evaluate the influence of fear avoidance beliefs (FAB), chronicity of low back pain (CHR), and severity of low back pain (SEV) on disability (DIS) in 209 patients from 7 different regions in Spain who were treated for low back pain (LBP) in health-care centers belonging to the Spanish National Health Service. The sample was balanced for acute, sub-acute, and chronic LBP (i.e., the three categories of the CHR variable). Validated scales and questionnaires were used to assess DIS and FAB during the first visit (day 1) and also 14 days later (day 15). The analyses performed in the study were linear regression analyses using disability measurement (DIS) treated as a continuous outcome variable. However, in the questions to follow, we will treat the disability outcome (DIS) as either a polytomous or ordinal outcome variable in a logistic regression model. The variables being considered are listed as follows:

DIS = disability level at day 15 (0 = none, 1 = mild, 2 = moderate, 3 = severe)
FAB = fear avoidance measure at day 1 (low = 0, high = 1)
CHR = chronicity of LBP at day 1 (1 = acute, 2 = subacute, 3 = chronic)
SEV = severity of LBP at day 1 (1 = low, 2 = medium, 3 = high)
SL = sick leave status at day 1 (0 = not on sick leave, 1 = on sick leave)
SEX = male (= 0) or female (= 1)
AGE (continuous variable)

a. State the logit form of a no-interaction proportional odds logistic model for the relationship of disability outcome DIS to FAB, CHR, SEV, SL, SEX, and AGE. Make sure to treat the variables CHR and SEV as nominal (rather than ordinal) variables and to treat AGE as a continuous variable. In stating this model, let "acute" denote the referent category for CHR, and let "low" denote the referent category for SEV.

b. Based on the answer to part (a) above,

i. What is the formula for the ODDS for moderate or severe disability (DIS = 2 or 3) to none or mild disability (DIS = 0 or 1) for a subject with high fear avoidance behavior (FAB = 1), chronic low back pain (CHR = 3), severe low back pain (SEV = 3), who is on sick leave, is female, and is 40 years old?

ii. What is the formula for the ODDS RATIO that compares the odds for moderate or severe disability (DIS = 2 or 3) to none or mild disability (DIS = 0 or 1) for a subject with high fear avoidance behavior (FAB = 1), chronic low back pain (CHR = 3), and high severity of low back pain (SEV = 3) to the corresponding odds for a subject with low fear avoidance behavior (FAB = 0), acute low back pain (CHR = 1), and low severity of low back pain (SEV = 1), controlling for SL, SEX, and AGE?

c. Consider the following 4 × 2 table that describes the crude relationship between fear avoidance behavior (FAB) and disability outcome (DIS):

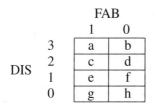

i. If the proportional odds assumption is satisfied for these data, describe the 2 × 2 sub-tables whose corresponding odds ratios are assumed to be equal (i.e., use the table above to assign letters to the appropriate cells and corresponding odds ratios in the sub-tables below):

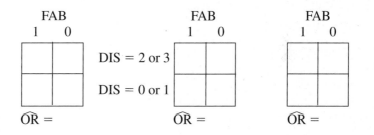

ii. Give two reasons why one might want to consider using a polytomous logistic regression model instead of a proportional odds logistic regression model for analyzing these data.

Suppose that one decided to use polytomous logistic regression, instead of ordinal logistic regression, to analyze these data. Also, suppose that the variables CHR and SEV are treated as ordinal variables. Suppose also that the variables FAB, CHR, and SEV are considered as *exposure* (i.e., *E*) *variables*, with the variables SL, SEX, and AGE treated as *control* (i.e., *V*) *variables*.

Suppose further that this revised model is expanded to allow for two-way interactions (i.e., products of two variables) between FAB and each of the control variables SL, SEX, and AGE, two-way interactions between CHR and each of the control variables SL, SEX, and AGE, two-way interactions between SEV and each of the control variables SL, SEX, and AGE, and two-way interactions among FAB, CHR, and SEV. This "initial" model can thus be written as follows:

$$\ln\left[P(\text{DIS} = g|X)/P(\text{DIS} = 0|X)\right] = \alpha_g + \beta_{1g}(\text{FAB} + \beta_{2g}\text{CHR}) + \beta_{3g}(\text{SEV} + \gamma_{1g}\text{SL})$$
$$+ \gamma_{2g}(\text{SEX} + \gamma_{3g}\text{AGE}) + \delta_{F1g}(\text{FAB} \times \text{SL})$$
$$+ \delta_{F2g}(\text{FAB} \times \text{SEX}) + \delta_{F3g}(\text{FAB} \times \text{AGE})$$
$$+ \delta_{C1g}(\text{CHR} \times \text{SL}) + \delta_{C2g}(\text{CHR} \times \text{SEX})$$
$$+ \delta_{C3g}(\text{CHR} \times \text{AGE}) + \delta_{S1g}(\text{SEV} \times \text{SL})$$
$$+ \delta_{S2g}(\text{SEV} \times \text{SEX}) + \delta_{S3g}(\text{SEV} \times \text{AGE})$$
$$+ \delta_{FC1g}(\text{FAB} \times \text{CHR}) + \delta_{FS2g}(\text{FAB} \times \text{SEV})$$
$$+ \delta_{CS3g}(\text{CHR} \times \text{SEV})$$

$$g = 1, 2, 3$$

2. d. Using the above polytomous logistic regression model, describe how one would *simultaneously test* for the significance of all $E_i V_j$ product terms in this model. Make sure to state the null hypothesis in terms of model parameters, describe the formula for the test statistic, and give the distribution and degrees of freedom of the test statistic under the null hypothesis.

e. At the end of the interaction assessment stage, suppose that it was determined that the variables (CHR $\times$ SEV) and (CHR $\times$ SL) are the only product terms remaining in the model as significant interaction effects. For the **reduced model** obtained from the interaction results, give a formula for the **ODDS RATIO** that compares the **odds** for severe disability (DIS = 3) to mild disability (DIS = 1) for a subject with high fear avoidance behavior (FAB = 1), chronic low back pain (CHR = 3), and high severity low back pain (SEV = 3) to the corresponding **odds** for a subject with low fear avoidance behavior (FAB = 0), acute low back pain (CHR = 1), and low severity low back pain (SEV = 1), controlling for SL, SEX, and AGE.

References

Ananth, C. V., Kleinbaum, D. G. 1977. Regression models for ordinal responses: A review of methods and applications. *International Journal of Epidemiology*, 26:1323–1333.

Sunenshine, R. H., Wright, M. O., Maragakis, L. L., Harris, A. D., Song, X., Hebden, J., Cosgrove, S. E., Anderson, A., Carnell, J., Jernigan, D. B., Kleinbaum, D. G., Perl, T. M., Standiford, H. C., and Srinivasan, A. 2006 "Impact of Multidrug-Resistant *Acinetobacter* Infection on Mortality and Length of Hospitalization." Unpublished.

Volkova, N., McClellan, W., Kleinbaum, D. G., Klein, M., and Flanders, W. D. 2006. "Neighborhood economic deprivation and racial differences in ESRD incidence." Unpublished.

Poisson Regression Analysis

24.1 Preview

Poisson regression analysis is a regression technique available for modeling dependent variables that describe *count* (i.e., *discrete*) *data*. The purpose of this chapter is to discuss the Poisson regression model and several key features of the model, particularly how a rate ratio can be estimated using Poisson regression. We also use real-life data to demonstrate how Poisson regression may be applied. Maximum likelihood procedures are used to estimate the parameters in a Poisson regression model. Therefore, the general principles and inference-making procedures described in Chapter 22 on ML estimation carry over directly to the likelihood functions appropriate for Poisson regression analysis

24.2 The Poisson Distribution

The methodology of Poisson regression analysis assumes that the underlying distribution of the response variable Y under consideration is Poisson. The Poisson probability distribution with parameter μ is given by the formula

$$\text{pr}(Y; \mu) = \frac{\mu^Y e^{-\mu}}{Y!} \quad , \quad Y = 0, 1, \ldots, \infty \tag{24.1}$$

Theoretically, a Poisson random variable can take any nonnegative integer value. From (24.1), for example, the probability that Y takes the value 10 is

$$\text{pr}(Y = 10; \mu) = \frac{\mu^{10} e^{-\mu}}{10!} = \frac{\mu^{10} e^{-\mu}}{3,628,800}$$

This value of this probability changes as a function of the value of μ.

The Poisson distribution is often used to model the occurrence of rare events, such as the number of new cases of lung cancer developing in some population over a certain period of time or the number of automobile accidents occurring at a certain location per year. An interesting statistical attribute of the Poisson distribution is that $E(Y) = \text{Var}(Y) = \mu$ when Y has the Poisson distribution (24.1). Also, from statistical theory, the Poisson distribution provides a good approximation to the binomial distribution for rare events.

To illustrate an application of the Poisson distribution, consider the results of a study to investigate whether exposure to magnetic fields is associated with the development of breast cancer in men (a notably rare disease). In a 1976–1980 study of 50,582 male telecom workers in New York (N.Y.), two cases of breast cancer were found out of 206,667 person-years of follow-up (Matanoski et al., 1991). The question being addressed, therefore, is whether observing two breast cancers in men working in the telecom industry is (statistically) excessive when compared with the breast cancer rate in the general adult male population.

To answer this question, we first must recognize that the parameter that we are estimating for the N.Y. male telecom workers is a *rate*, not a proportion. By *rate*, we mean the frequency of occurrence (i.e., *count*) of an event (i.e., *developing breast cancer*) over an accumulation of person-time follow-up information (i.e., *person-years*). Mathematically, we define a rate, which we denote by λ, as

$$\lambda = \frac{E(Y)}{PT}$$

where $E(Y)$ denotes the expected value (or true average value) of the frequency Y in the population under study, and PT denotes the accumulated person-time of follow-up. Typically, PT is obtained by summing up the follow-up time contributions of all persons studied, recognizing that not all study subjects may have been followed for the same amount of time. The estimated rate obtained from a sample is defined as

$$\hat{\lambda} = \frac{Y}{PT}$$

Because the numerator (Y) of an estimated rate is not a subset of the denominator (PT), a rate is not a proportion. Thus, the range of possible values for a rate is $0 \leq \lambda < \infty$, whereas the range for a proportion π is $0 \leq \pi \leq 1$. (See Kleinbaum's *ActivEpi*, 2003 for further discussion about the terms *rate* and *risk*, where only the latter is a proportion.)

In order to answer the question at hand, let us assume that published literature about breast cancer in men indicated that, in the general adult male population in N.Y., the expected breast cancer rate for men was about 15 new cases for every 10,000,000 person-years of follow-up. We may therefore address our question by carrying out a statistical test of hypothesis, where our null hypothesis can be stated as follows:

$$H_0: \lambda = \frac{15}{10,000,000} \quad (= \text{expected rate for N.Y. male telecom workers})$$

Using the above expected rate, we can alternatively state the null hypothesis in terms of the expected number of male breast cancer cases:

$$H_0: \mu = E(Y) = 206,667 \times \frac{15}{10,000,000} = 0.31 \text{ expected cases}$$

If we now assume that Y, the observed number of cases, has the Poisson distribution with expected value $\mu = 0.31$ under H_0, we can compute the probability of observing 2 or more male breast cancer cases using the Poisson distribution formula (24.1). This probability will therefore give us the P-value for a one-sided alternative to H_0, namely, $H_A: \mu > 0.31$.

Since $\mu = 0.31$ under H_0, the probability of obtaining any value of Y under H_0 is computed using the formula:

$$\text{pr}(Y; 0.31) = \frac{0.31^Y e^{-0.31}}{Y!}$$

We can then compute the desired P-value as follows:

$$
\begin{aligned}
P\text{-value} &= \text{pr}(Y \geq 2 \,|\, \mu = 0.31) \\
&= 1 - \text{pr}(Y = 0 \,|\, \mu = 0.31) - \text{pr}(Y = 1 \,|\, \mu = 0.31) \\
&= 1 - \frac{0.31^0 e^{-0.31}}{0!} - \frac{0.31^1 e^{-0.31}}{1!} \\
&= 1 - 0.7334 - 0.2274 \\
&= 0.0392
\end{aligned}
$$

Thus, we would reject H_0 at the 5% level, and conclude that the breast cancer rate for male telecom workers in N.Y. is higher than that for the general adult male population in N.Y.

24.3 An Example of Poisson Regression

To illustrate the utility of Poisson regression analysis, let us consider a data analysis situation where Poisson regression has been used quite successfully. Table 24.1 gives nonmelanoma skin cancer data for women stratified by age in two metropolitan areas: Dallas–Ft. Worth and Minneapolis–St. Paul (Scotto, Kopf, and Urbach 1974). In this example, the dependent variable Y is a count, the number of cases of skin cancer. Since eight age strata and two metropolitan areas are involved, we let Y_{ij} denote the count for the ith age stratum in the jth area, where i ranges from 1 to 8 for the eight age groups and $j = 0$ (Minneapolis–St. Paul) or $j = 1$ (Dallas–Ft. Worth). We also let ℓ_{ij} denote the population size for the ith age stratum in the jth area. For these data, one analysis goal is to determine whether the risk for skin cancer adjusted for age is higher in one metropolitan area than in the other. The term *risk* in this context essentially means the probability associated with an event of interest—for example, the probability of developing skin cancer. We will let λ_{ij} denote the true (i.e., population) risk in the (i, j)th group. The ratio

$$RR_i = \frac{\lambda_{i1}}{\lambda_{i0}}$$

is commonly referred to as the *relative risk* or *risk ratio,* which in this case is the population risk for Dallas–Ft. Worth in the ith age group *divided by* the population risk for Minneapolis–St. Paul in the ith age group. If $RR_i = 1$, then the population risks are the same in the ith age group; if $RR_i > 1$, however, then the risk for Dallas–Ft. Worth is higher than the risk for Minneapolis–St. Paul in this age group.

As indicated in the last column of Table 24.1, the estimated risk ratios in all age groups are greater than 1, which clearly suggests that the Dallas–Ft. Worth area has a higher overall incidence of skin cancer than Minneapolis–St. Paul. Our objective here is to use Poisson regression analysis to determine whether such a data pattern is statistically significant and to obtain an estimate of the overall risk ratio that adjusts for the effect of age.

TABLE 24.1 **Comparison of incidence of nonmelanoma skin cancer among women in Minneapolis–St. Paul and Dallas–Ft. Worth**[†]

Age Group (yr)	Minneapolis–St. Paul		Dallas–Ft. Worth		Estimated Risk Ratio[*]
	No. of Cases	Population Size	No. of Cases	Population Size	
15–24	1	172,675	4	181,343	3.81
25–34	16	123,065	38	146,207	2.00
35–44	30	96,216	119	121,374	3.14
45–54	71	92,051	221	111,353	2.57
55–64	102	72,159	259	83,004	2.21
65–74	130	54,722	310	55,932	2.33
75–84	133	32,185	226	29,007	1.89
85+	40	8,328	65	7,538	1.80

Source: Adapted from Scotto, Kopf, and Urbach (1974)

[*]With Minneapolis–St. Paul as the reference group
[†]The SAS data file for these data is called "melanoma.dat" and can be obtained from the publisher's website http://www.thomsonedu.com/statistics/kleinbaum.

Where does the Poisson distribution enter into this problem? Notice, first, that the count Y_{ij} is, in theory, a binomial random variable with mean $\mu_{ij} = \ell_{ij}\lambda_{ij}$. We know from statistical theory that the binomial distribution can be approximated by a Poisson distribution with the same mean, provided that the population size is large and the binomial probability parameter is small, so that the expected binomial count (i.e., the mean μ) is small relative to the population size. In other words, the Poisson distribution provides a good approximation to the binomial distribution for rare events. The data in Table 24.1 satisfy this requirement reasonably well, since all stratum-specific counts are quite small relative to the corresponding population sizes.

To develop a Poisson regression model for the above situation, we need to define a model for the expected number of skin cancer cases, $E(Y_{ij})$, in terms of the predictor variables of interest. Here, two underlying predictor variables are of interest: "age" and "area." Since "age" has been categorized into eight groups, we will use seven dummy variables to index them.[1] The variable "area," which contains two categories, requires only one dummy variable. Thus, one possible model for the expected number of skin cancer cases in the (i, j)th group can be written as

$$E(Y_{ij}) = \mu_{ij} = \ell_{ij}\lambda_{ij}, \qquad i = 1, 2, \dots, 8; \quad j = 0, 1$$

where

$$\ln \lambda_{ij} = \alpha + \sum_{k=1}^{7} \alpha_k U_k + \beta E$$

with

$$U_k = \begin{cases} 1 & \text{if } k = i, \\ 0 & \text{otherwise} \end{cases} \qquad k = 1, 2, \dots, 7$$

$$E = \begin{cases} 1 & \text{if } j = 1 \qquad \text{(Dallas–Ft. Worth)} \\ 0 & \text{if } j = 0 \qquad \text{(Minneapolis–St. Paul)} \end{cases}$$

---

[1] Alternatively, the model can be defined by using eight dummy variables for "age" and one dummy variable for "area"; when eight dummy variables are used for "age," using an intercept term is redundant. The eight-dummy-variable alternative was used in the published analysis of this data set.

Using this model, we can write the risks λ_{ij} in terms of the parameters α_i and β to obtain

$$\ln \lambda_{i0} = \alpha + \alpha_i \qquad \text{and} \qquad \ln \lambda_{i1} = \alpha + \alpha_i + \beta, \qquad i = 1, 2, \ldots, 7;$$

and, for $i = 8$,

$$\ln \lambda_{80} = \alpha \qquad \text{and} \qquad \ln \lambda_{81} = \alpha + \beta$$

since $U_k = 0, k = 1, 2, \ldots, 7$ for $i = 8$. Hence,

$$\ln \lambda_{i1} - \ln \lambda_{i0} = (\alpha + \alpha_i + \beta - \alpha - \alpha_i) = \beta, \qquad i = 1, 2, \ldots, 7$$

Also,

$$\ln \lambda_{81} - \ln \lambda_{80} = (\alpha + \beta - \alpha) = \beta$$

In other words,

$$RR_i = \frac{\lambda_{i1}}{\lambda_{i0}} = \exp\left[\ln\left(\frac{\lambda_{i1}}{\lambda_{i0}} \right) \right] = \exp[\ln \lambda_{i1} - \ln \lambda_{i0}] = \exp[\beta] = e^{\beta}, \quad i = 1, 2, \ldots, 8$$

Thus, using this model, we can estimate the risk ratio for any age group by fitting the model, estimating the coefficient of the E-variable, and then exponentiating this estimate. Since the estimated risk ratio $e^{\hat{\beta}}$ is independent of i, we can interpret $e^{\hat{\beta}}$ as being an estimate of an overall risk ratio adjusted for age.

The example just described illustrates the type of model used in performing a Poisson regression analysis. In general, instead of having two variables (like age and area) to consider, we may have several (say k) predictor variables $X_1, X_2, \ldots, X_k$ to examine. Nevertheless, the general method of fitting a Poisson regression model is still to use the Poisson model formulation to derive a likelihood function that can then be maximized so that parameter estimates, estimated standard errors, maximized likelihood statistics, and other statistical information can be produced. Since packaged programs can now carry out such analyses, a user need only specify the model to be fit; the program then determines the likelihood function, maximizes it, and computes relevant statistics. We shall return later to the same example to illustrate methods of Poisson regression analysis numerically.

The preceding example (strictly speaking) involves a model for estimating the *risk* of developing a disease. A more general and popular application of Poisson regression involves modeling *rates of disease development* for different subgroups of interest. An estimated rate is generally defined as

$$\hat{\lambda} = \frac{Y}{\ell}$$

where Y is the observed count of adverse health outcomes (e.g., the number of cases of skin cancer or the number of new cases of heart disease) for a subgroup of interest, and ℓ denotes the accumulated length of (disease-free) follow-up time for all persons in the subgroup. Thus, $\hat{\lambda}$ measures the number of adverse health outcomes (i.e., health failures) relative to the total amount of follow-up time for all persons in a given subgroup. If, for example, the data in Table 24.1 were based on a one-year follow-up study of the Minneapolis–St. Paul and Dallas–Ft. Worth populations, then the numbers in the table giving population size might be considered as *person-years* of follow-up time for the age–area subgroups. The ratio of two rates (e.g., $\lambda_{i1}/\lambda_{i0}$) is commonly referred to as a *rate ratio*. Other terms used are *incidence density ratio* (abbreviated *IDR*) and *hazard ratio*.

24.4 Poisson Regression: General Considerations

We are now ready to describe the general Poisson regression analysis framework. The dependent variable Y is, as already mentioned, typically a count of health failures obtained for each of a number of subgroups that are described by a set of predictor variables $X_1, X_2, \ldots, X_k$. For subgroup i, $i = 1, 2, \ldots, n$, let Y_i denote the observed number of failures, and let ℓ_i denote the total length of follow-up time for all persons in that subgroup. Let $\mathbf{X}_i = (X_{i1}, X_{i2}, \ldots, X_{ik})$ denote the set of values of $X_1, X_2, \ldots, X_k$ specific to subgroup i, let $\boldsymbol{\beta} = (\beta_0, \beta_1, \ldots, \beta_k)$ be a set of unknown regression coefficient parameters, and let $\lambda(\mathbf{X}_i, \boldsymbol{\beta})$ denote some specific function of $\mathbf{X}_i$ and $\boldsymbol{\beta}$ (e.g., $\exp(\beta_0 + \sum_{j=1}^{k} \beta_j X_{ij})$) that represents the failure rate for subgroup i (i.e., $\lambda(\mathbf{X}_i, \boldsymbol{\beta})$ measures the rate at which failures occur per unit of follow-up time). Then the expected number of failures in the ith subgroup is

$$E(Y_i) = \mu_i = \ell_i \lambda(\mathbf{X}_i, \boldsymbol{\beta}), \qquad i = 1, 2, \ldots, n \qquad (24.2)$$

where Y_i is a Poisson random variable. It is required that $\lambda(\mathbf{X}_i, \boldsymbol{\beta}) > 0$.[2]

Under the assumption that Y_i is Poisson with mean μ_i,[3] so that

$$\mathrm{pr}(Y_i; \mu_i) = \frac{\mu_i^{Y_i} e^{-\mu_i}}{Y_i!}, \qquad i = 1, 2, \ldots, n \qquad (24.3)$$

it follows from (24.2) and (24.3) that

$$\mathrm{pr}(Y_i; \boldsymbol{\beta}) = \frac{[\ell_i \lambda(\mathbf{X}_i, \boldsymbol{\beta})]^{Y_i} e^{-\ell_i \lambda(\mathbf{X}_i, \boldsymbol{\beta})}}{Y_i!} \qquad (24.4)$$

where $Y_i = 0, 1, \ldots, \infty$ and $i = 1, 2, \ldots, n$.

Note that the only real conceptual difference between Poisson regression and standard multiple linear regression is that the former involves the Poisson distribution whereas the latter involves the normal distribution. In each instance, the analysis goal is the same—namely, to fit to the data a regression equation that will accurately model $E(Y)$ as a function of a set of predictor variables $X_1, X_2, \ldots, X_k$.[4]

In the most general sense, then, regression analysis pertains to modeling the mean of the dependent variable under consideration as a function of certain predictor variables. The form of likelihood function that is used to estimate the regression coefficient set $\boldsymbol{\beta}$ is determined by the assumptions made about the distribution of that dependent variable.

As we did earlier to obtain the likelihood function (21.12), let us assume that the set $\mathbf{Y} = (Y_1, Y_2, \ldots, Y_n)$ constitutes a mutually independent set of Poisson random variables, with Y_i

---

[2] Formula (24.2) is analogous (but not equivalent) to the formula for the mean $\mu = n\pi$ of a binomial random variable; here, ℓ_i is similar to n, and $\lambda(\mathbf{X}_i, \boldsymbol{\beta})$ is similar to π.

[3] That the Poisson distribution is useful for modeling certain types of health count data can be loosely argued on the basis of the well-known Poisson approximation to the binomial distribution (see, e.g., Remington and Schork (1985), chapter 5). If $Y \frown \mathrm{Bin}(n, \pi)$, and if n is large and π is very small, then $Y \frown \mathrm{Poi}(\mu = n\pi)$. For many health outcomes (e.g., the development of a rare disease), the length ℓ_i of follow-up time (analogous to n) is large, and the rate $\lambda(\mathbf{X}_i, \boldsymbol{\beta})$ of occurrence of the health outcome in question (analogous to π) is small, thus suggesting the Poisson model.

[4] If, in the likelihood (21.12) of Chapter 21 we replace $E(Y_i) = \beta_0 + \beta_1 X_i$ with, say, $E(Y_i) = \beta_0 + e^{\beta_1 X_i}$, then we change from a *linear* (in the coefficients) model to a *nonlinear* model, and hence from a *linear* regression analysis to a *nonlinear* one. The major effect of this change is that we have to solve a set of nonlinear (as opposed to linear) likelihood equations in the β's. This solution generally requires some sort of computer-assisted iteration procedure.

having the probability distribution (24.4). Then, the *likelihood function for Poisson regression analysis* is of the general form

$$L(\mathbf{Y}; \boldsymbol{\beta}) = \prod_{i=1}^{n} \text{pr}(Y_i; \boldsymbol{\beta}) = \prod_{i=1}^{n} \left\{ \frac{[\ell_i \lambda(\mathbf{X}_i, \boldsymbol{\beta})]^{Y_i} e^{-\ell_i \lambda(\mathbf{X}_i, \boldsymbol{\beta})}}{Y_i!} \right\}$$

$$= \frac{\left\{ \prod_{i=1}^{n} [\ell_i \lambda(\mathbf{X}_i, \boldsymbol{\beta})]^{Y_i} \right\} \exp\left[-\sum_{i=1}^{n} \ell_i \lambda(\mathbf{X}_i, \boldsymbol{\beta}) \right]}{\prod_{i=1}^{n} Y_i!} \qquad (24.5)$$

where $E(Y_i) = \mu_i = \ell_i \lambda(\mathbf{X}_i, \boldsymbol{\beta})$, $i = 1, 2, \ldots, n$.

To utilize the likelihood function (24.5) in practice, the investigator must specify a particular form for the rate function $\lambda(\mathbf{X}_i, \boldsymbol{\beta})$. Such a specification should be based on the process under study and on previous knowledge of, and experience with, the relationships among the variables under consideration. Examples of possible choices for $\lambda(\mathbf{X}_i, \boldsymbol{\beta})$ are $e^{\lambda_i^*}$ when $\lambda_i^* = \beta_0 + \sum_{j=1}^{k} \beta_j X_{ij}$, λ_i^* itself when $\lambda_i^* > 0$, and $\ln \lambda_i^*$ when $\lambda_i^* > 1$.

Recall that the maximum likelihood estimators $\hat{\beta}_0, \hat{\beta}_1, \ldots, \hat{\beta}_k$ of $\beta_0, \beta_1, \ldots, \beta_k$ are obtained from (24.5) as the solutions of the $(k + 1)$ equations

$$\frac{\partial}{\partial \beta_j} [\ln L(\mathbf{Y}; \boldsymbol{\beta})] = 0 \qquad j = 0, 1, \ldots, k \qquad (24.6)$$

The solution to the set of ML equations given by (24.6) must generally be obtained by a computer-based iteration procedure. Frome (1983) discusses the use of algorithms for solving the system of equations (24.6). In particular, he argues for the use of a computational algorithm referred to as *iteratively reweighted least squares* (IRLS).[5] Several statistical packages, such as SAS (using PROC GENMOD), can be utilized to find the ML estimator $\hat{\boldsymbol{\beta}}$ of $\boldsymbol{\beta}$ based on the likelihood (24.5). In addition, the estimated covariance matrix $\hat{\mathbf{V}}(\hat{\boldsymbol{\beta}})$ of $\boldsymbol{\beta}$, measures of goodness of fit of the model under consideration, and certain regression diagnostic statistics (i.e., indices useful for detecting influential observations and multicollinearity) can be obtained as part of the computer output.

For an application of the above procedures, we return to the data in Table 24.1 describing nonmelanoma skin cancer data for women stratified by age in Minneapolis–St. Paul and Dallas–Ft. Worth (adapted from Scotto, Kopf, and Urbach (1974), and reanalyzed by Frome and Checkoway (1985)). We previously considered the following Poisson regression model for the expected number of skin cancer cases in subgroup (i, j), $i = 1, 2, \ldots, 8$ and $j = 0, 1$:

$$E(Y_{ij}) = \mu_{ij} = \ell_{ij} \lambda_{ij}$$

where

$$\ln \lambda_{ij} = \alpha + \sum_{k=1}^{7} \alpha_k U_k + \beta E$$

---

[5] The fact that $E(Y_i) = \text{Var}(Y_i) = \mu_i = \ell_i \lambda(\mathbf{X}_i, \boldsymbol{\beta})$ means that the variance of the response variable is *not* constant (i.e., it varies as a function of ℓ_i and $\mathbf{X}_i$, thus requiring a weighted-least-squares regression analysis). And because this variance is a mathematical function of $\boldsymbol{\beta}$, the weights in such a weighted regression analysis necessarily change as a function of the change in the estimate $\hat{\boldsymbol{\beta}}$ at each step of the iteration process (i.e., a reweighting is required at each step). This is the reason for the terminology "iteratively reweighted least squares," or IRLS for short.

Here, the U_k's were 0–1 dummy variables indexing the age strata, and E was a 0–1 variable delineating metropolitan area (1 = Dallas–Ft. Worth, 0 = Minneapolis–St. Paul). For this model, the risk (or more generally, rate) ratio

$$RR_i = \frac{\lambda_{i1}}{\lambda_{i0}}$$

reduced to the expression

$$RR_i = e^{\beta}$$

where e^{β} is independent of i and represents an overall rate ratio parameter adjusted for age.

The likelihood function L for the preceding model, based on the assumption that the count Y_{ij} follows the Poisson distribution with mean $\mu_{ij} = \ell_{ij}\lambda_{ij}$ and that the $\{Y_{ij}\}$ are a set of mutually independent random variables, is given by the expression

$$L = \prod_{i=1}^{8}\left\{\left[\frac{(\ell_{i0}\lambda_{i0})^{Y_{i0}}e^{-\ell_{i0}\lambda_{i0}}}{Y_{i0}!}\right]\left[\frac{(\ell_{i1}\lambda_{i1})^{Y_{i1}}e^{-\ell_{i1}\lambda_{i1}}}{Y_{i1}!}\right]\right\}$$

where $\lambda_{i0} = \exp(\alpha + \alpha_i)$, $\lambda_{i1} = \exp(\alpha + \alpha_i + \beta)$ for $i = 1,\ldots, 7$, $\lambda_{80} = \exp(\alpha)$, and $\lambda_{81} = \exp(\alpha + \beta)$.

The use of a Poisson regression computer package would then maximize this likelihood function to produce the nine parameter estimates

$$\{\hat{\alpha}, \hat{\alpha}_1, \hat{\alpha}_2,\ldots, \hat{\alpha}_7, \hat{\beta}\}$$

along with a (9×9) estimated covariance matrix. The computer output for these data, using SAS's GENMOD procedure, is given next. From this output, we see that the estimates of β and its standard error are

$$\hat{\beta} = 0.806, \quad S_{\hat{\beta}} = 0.0522$$

Thus, the point estimate of the adjusted rate ratio is given by

$$e^{\hat{\beta}} = e^{0.806} = 2.2389$$

The GENMOD Procedure

Criteria For Assessing Goodness of Fit

Criterion	DF	Value	Value/DF
Deviance	7	8.3426	1.1918
Scaled Deviance	7	8.3426	1.1918
Pearson Chi-Square	7	8.2189	1.1741
Scaled Pearson X2	7	8.2189	1.1741
Log Likelihood	.	7201.7897	.

Analysis Of Parameter Estimates

Parameter	DF	Estimate	Std Err	Chi Square	Pr > Chi
INTERCEPT	1	-5.4851	0.1037	2798.4771	0.0000
CITY ← E	1	0.8064	0.0522	238.5764	0.0000
U1	1	-6.1743	0.4577	181.9416	0.0000
U2	1	-3.5441	0.1675	447.7882	0.0000
U3	1	-2.3270	0.1275	333.2712	0.0000
U4	1	-1.5791	0.1138	192.4421	0.0000
U5	1	-1.0870	0.1109	96.0623	0.0000
U6	1	-0.5221	0.1086	23.1044	0.0000
U7	1	-0.1156	0.1109	1.0866	0.2972

An approximate large-sample 95% confidence interval for e^β is calculated as

$$\exp[\hat{\beta} \pm 1.96\, S_{\hat{\beta}}] = \exp[0.806 \pm 1.96(0.0522)]$$
$$= \exp(0.806 \pm 0.1023)$$

which gives the 95% confidence limits

$$(e^{0.7037}, e^{0.9083}) = (2.0212, 2.4801)$$

A large-sample test of H_0: $\beta = 0$ versus H_A: $\beta \neq 0$ can be based on the Wald statistic

$$Z = \frac{\hat{\beta} - 0}{S_{\hat{\beta}}}$$

which is approximately $N(0, 1)$ under H_0: $\beta = 0$.

For our example,

$$Z = \frac{0.806 - 0}{0.0522} = 15.44 \qquad (P\text{-value} \approx 0)$$

Thus, this particular Poisson regression analysis indicates that there is a statistically significant effect due to area and that the overall (adjusted for age) rate of nonmelanoma skin cancer in women in Dallas–Ft. Worth is approximately 2.2 times the corresponding adjusted rate for women in Minneapolis–St. Paul; a 95% confidence interval for the (adjusted) rate ratio is (2.0212, 2.4801). We will return to this example to illustrate how to evaluate interaction, confounding, and goodness of fit.

24.5 Measures of Goodness of Fit

Measures of the goodness of fit of Poisson regression models are obtained via comparisons of maximized likelihood values. Suppose that Y_i has the Poisson distribution (24.3) and that $Y_1, Y_2, \ldots, Y_n$ are mutually independent; then, expressed as a general function of $\mu_1, \mu_2, \ldots, \mu_n$ (i.e., ignoring the predictors $X_1, X_2, \ldots, X_k$ completely), the likelihood function takes the form

$$L(\mathbf{Y}; \boldsymbol{\mu}) = \prod_{i=1}^{n} \frac{\mu_i^{Y_i} e^{-\mu_i}}{Y_i!} = \frac{\left(\prod_{i=1}^{n} \mu_i^{Y_i}\right) \exp\left(-\sum_{i=1}^{n} \mu_i\right)}{\prod_{i=1}^{n} Y_i!} \tag{24.7}$$

where $\boldsymbol{\mu} = (\mu_1, \mu_2, \ldots, \mu_n)$. The system of ML equations

$$\frac{\partial}{\partial \mu_i}[\ln L(\mathbf{Y}; \boldsymbol{\mu})] = 0 \qquad i = 1, 2, \ldots, n$$

leads to the solution $\hat{\mu}_i = Y_i$, $i = 1, 2, \ldots, n$. Thus, the maximized likelihood value for the likelihood function (24.7) is

$$L(\mathbf{Y}; \hat{\boldsymbol{\mu}}) = \frac{\left(\prod_{i=1}^{n} Y_i^{Y_i}\right) \exp\left(-\sum_{i=1}^{n} Y_i\right)}{\prod_{i=1}^{n} Y_i!} \tag{24.8}$$

where $\hat{\boldsymbol{\mu}} = (\hat{\mu}_1, \hat{\mu}_2, \ldots, \hat{\mu}_n) = (Y_1, Y_2, \ldots, Y_n)$.

The value of the maximized likelihood $L(\mathbf{Y}; \hat{\boldsymbol{\mu}})$ based on (24.7) will be larger (for any set of data) than that achieved by maximizing a likelihood such as (24.5) when $(k + 1) < n$. This is because (24.7) imposes no restrictions on the structure of μ_i, whereas (24.5) imposes the restriction $\mu_i = \ell_i \lambda(\mathbf{X}_i, \boldsymbol{\beta})$. In other words, (24.5) can be thought of as the likelihood function under H_0: $\mu_i = \ell_i \lambda(\mathbf{X}_i, \boldsymbol{\beta})$, $i = 1, 2, \ldots, n$, whereas (24.7) is the likelihood under H_A: "μ_i is unrestricted in structure, $i = 1, 2, \ldots, n$."

Thus, if $L(\mathbf{Y}; \hat{\boldsymbol{\beta}})$ is the maximized likelihood value under (24.5), where $\hat{\boldsymbol{\beta}}$ is the ML estimator of $\boldsymbol{\beta}$, then

$$-2 \ln \left[\frac{L(\mathbf{Y}; \hat{\boldsymbol{\beta}})}{L(\mathbf{Y}; \hat{\boldsymbol{\mu}})} \right] \tag{24.9}$$

is a likelihood-ratio-type statistic reflecting the goodness of fit of the model $\mu_i = \ell_i \lambda(\mathbf{X}_i, \boldsymbol{\beta})$ relative to the model where no structure has been imposed on μ_i. Since the objective of any regression analysis is to obtain a parsimonious description of the data, the model $\mu_i = \ell_i \lambda(\mathbf{X}_i, \boldsymbol{\beta})$ involving $(k + 1)$ parameters will (we hope) provide a maximized likelihood value almost as large as can be obtained by the baseline (and uninformative) model that involves as many parameters (namely, n) as data points. By "almost as large," we mean that $L(\mathbf{Y}; \hat{\boldsymbol{\beta}})$ will not be significantly smaller than $L(\mathbf{Y}; \hat{\boldsymbol{\mu}})$ based on a likelihood ratio test using (24.9).

The quantity

$$D(\hat{\boldsymbol{\beta}}) = -2 \ln \left[\frac{L(\mathbf{Y}; \hat{\boldsymbol{\beta}})}{L(\mathbf{Y}; \hat{\boldsymbol{\mu}})} \right] \tag{24.10}$$

is the goodness-of-fit statistic employed to assess whether $L(\mathbf{Y}; \hat{\boldsymbol{\beta}})$ is significantly less than $L(\mathbf{Y}; \hat{\boldsymbol{\mu}})$ and thus to suggest meaningful lack of fit to the data of the assumed regression model $\mu_i = \ell_i \lambda(\mathbf{X}_i, \boldsymbol{\beta})$. The quantity $D(\hat{\boldsymbol{\beta}})$ is also called the *deviance* for the Poisson regression model $\mu_i = \ell_i \lambda(\mathbf{X}_i, \boldsymbol{\beta})$, and it can be thought of as a measure of residual variation about (or deviation from) the fitted model. Under H_0: $\mu_i = \ell_i \lambda(\mathbf{X}_i, \boldsymbol{\beta})$, the deviance $D(\hat{\boldsymbol{\beta}})$ is typically (although not strictly legitimately) assumed to have (for large samples) an approximate chi-square distribution with $(n - k - 1)$ degrees of freedom, where n is the number of parameters (i.e., the number of subgroups, cells, or categories) specified in the likelihood (24.7), and $(k + 1)$ is the number of parameters (i.e., β_j's) in the likelihood (24.5). Thus, a very approximate test for goodness-of-fit of the model $\mu_i = \ell_i \lambda(\mathbf{X}_i, \boldsymbol{\beta})$ to a given data set can be performed by comparing the calculated value of $D(\hat{\boldsymbol{\beta}})$ to an appropriate upper-tail value of the chi-square distribution with $(n - k - 1)$ degrees of freedom.

With $\hat{Y}_i = \ell_i \lambda(\mathbf{X}_i, \hat{\boldsymbol{\beta}})$ denoting the predicted response in cell i under model (24.2), the quantity (24.10) can be written in the form

$$D(\hat{\boldsymbol{\beta}}) = 2 \sum_{i=1}^{n} \left[Y_i \ln \left(\frac{Y_i}{\hat{Y}_i} \right) - (Y_i - \hat{Y}_i) \right] \tag{24.11}$$

Hence, $D(\hat{\boldsymbol{\beta}})$ behaves like SSE $= \sum_{i=1}^{n} (Y_i - \hat{Y}_i)^2$ in standard multiple linear regression analysis. When the fitted model exactly predicts the observed data (i.e., $Y_i = \hat{Y}_i$, $i = 1, 2, \ldots, n$), then $D(\hat{\boldsymbol{\beta}}) = 0$; and the larger the discrepancy between observed and predicted responses, the larger is the value of $D(\hat{\boldsymbol{\beta}})$.

When the predicted values are all of reasonable size (i.e., $\hat{Y}_i > 3$, $i = 1, 2, \ldots, n$), then (24.11) can be reasonably approximated by the more familiar Pearson-type observed-versus-predicted chi-square statistic of the form

$$\chi^2 = \sum_{i=1}^{n} \frac{(Y_i - \hat{Y}_i)^2}{\hat{Y}_i} \tag{24.12}$$

As a word of caution, the statistic (24.12) can be misleadingly large when certain $\hat{Y}_i$-values are very small.

The deviances for various models in a hierarchical class can be used to produce likelihood ratio tests. In particular, consider again the likelihood (24.5) involving the parameter set $\boldsymbol{\beta} = (\beta_0, \beta_1, \ldots, \beta_k)$, with deviance $D(\hat{\boldsymbol{\beta}})$ given by (24.10). Now, for $0 < r < k$, suppose that we wish to test whether the last $(k - r)$ parameters in $\boldsymbol{\beta}$ are equal to 0; i.e., our null hypothesis is H_0: $\beta_{r+1} = \beta_{r+2} = \cdots = \beta_k = 0$. Under H_0, the (null hypothesis) likelihood can be obtained by replacing $\boldsymbol{\beta}$ in (24.5) with $\boldsymbol{\beta}_r$, where

$$\boldsymbol{\beta}_r = (\beta_0, \beta_1, \ldots, \beta_r, 0, 0, \ldots, 0)$$

If we denote this likelihood function by $L(\mathbf{Y}; \boldsymbol{\beta}_r)$, and if $\hat{\boldsymbol{\beta}}_r$ is the maximum likelihood estimator of $\boldsymbol{\beta}_r$ using $L(\mathbf{Y}; \boldsymbol{\beta}_r)$, then the likelihood ratio test of H_0 is performed using the test statistic

$$-2 \ln \left[\frac{L(\mathbf{Y}; \hat{\boldsymbol{\beta}}_r)}{L(\mathbf{Y}; \hat{\boldsymbol{\beta}})} \right] \qquad (24.13)$$

which has approximately a chi-square distribution with $(k - r)$ degrees of freedom for large samples when H_0 is true.

Furthermore, expression (24.13) is exactly equal to the deviance difference

$$D(\hat{\boldsymbol{\beta}}_r) - D(\hat{\boldsymbol{\beta}}) \qquad (24.14)$$

To see this, recall the general definition of $D(\hat{\boldsymbol{\beta}})$ given by expression (24.10). Using (24.10) and (24.14), we have

$$D(\hat{\boldsymbol{\beta}}_r) - D(\hat{\boldsymbol{\beta}}) = -2 \ln \left[\frac{L(\mathbf{Y}; \hat{\boldsymbol{\beta}}_r)}{L(\mathbf{Y}; \hat{\boldsymbol{\mu}})} \right] + 2 \ln \left[\frac{L(\mathbf{Y}; \hat{\boldsymbol{\beta}})}{L(\mathbf{Y}; \hat{\boldsymbol{\mu}})} \right]$$

$$= -2 \ln \left[\frac{L(\mathbf{Y}; \hat{\boldsymbol{\beta}}_r)}{L(\mathbf{Y}; \hat{\boldsymbol{\beta}})} \right]$$

which is exactly the likelihood ratio test statistic (24.13). Under H_0: $\beta_{r+1} = \beta_{r+2} = \cdots = \beta_k = 0$, the difference $D(\hat{\boldsymbol{\beta}}_r) - D(\hat{\boldsymbol{\beta}})$ has approximately a chi-square distribution with $(k - r)$ degrees of freedom in large samples.

Thus, when we use Poisson regression to analyze a set of data, members of a set of candidate models within a hierarchical class can be compared by considering differences between pairs for deviances for these models.

24.6 Continuation of Skin Cancer Data Example

We again consider the data in Table 24.1 giving skin cancer counts for women stratified by age in Minneapolis–St. Paul and Dallas–Ft. Worth. For these data, we used ML estimation to fit the following Poisson regression model for the expected number of skin cancer cases:

$$E(Y_{ij}) = \mu_{ij} = \ell_{ij} \lambda_{ij}$$

where

$$\ln \lambda_{ij} = \alpha + \sum_{k=1}^{7} \alpha_k U_k + \beta E$$

The set $\{U_k\}$ are 0–1 dummy variables indexing the age strata, and E contrasts metropolitan areas ($1 =$ Dallas–Ft. Worth, $0 =$ Minneapolis–St. Paul). We will refer to this model as **Model 1.** For this model, the estimated rate ratio adjusted for age was $e^{\hat{\beta}} = 2.2389$, and a 95% confidence interval for the true adjusted rate ratio was (2.0212, 2.4801). Also, a large-sample test for H_0: $\beta = 0$ versus H_A: $\beta \neq 0$ yielded a Z-statistic of 15.44 (P-value ≈ 0), which is highly significant.

Two additional questions of interest for these data are:

1. Is "age" an effect modifier? That is, does the "area" effect (as measured by a rate ratio parameter) differ for different age strata?

2. If the answer to question 1 is no, is "age" a confounder? That is, does "age" need to be in the model in some form in order to produce a valid estimate of the "area" effect?

At this point, readers may wish to review the definitions and properties of effect modifiers and confounders discussed in Chapter 11.

To answer question 1 directly, we could modify model 1 to include interaction terms, as follows:

$$\textbf{Model 2: } \ln \lambda_{ij} = \alpha + \sum_{k=1}^{7} \alpha_k U_k + \beta E + \sum_{k=1}^{7} \delta_k (EU_k)$$

Using the preceding interaction model, we can test for effect modification by testing H_0: $\delta_1 = \delta_2 = \cdots = \delta_7 = 0$, using a likelihood ratio χ^2 statistic with 7 degrees of freedom. This test involves comparing model 1 (without interaction terms) to model 2 (with seven EU_k terms). Earlier, we looked at the computer output based on fitting model 1. The computer output resulting from fitting model 2 (again using SAS's GENMOD procedure) is presented next. From this latter block of output, we see that the deviance for model 2 is exactly zero, because model 2 fits the data perfectly (i.e., we have fit a model with 16 parameters to a data set of size $n = 16$). Thus, the likelihood ratio test statistic for testing H_0: $\delta_1 = \delta_2 = \cdots = \delta_7 = 0$ is obtained by subtracting the deviance for model 2 (i.e., zero) from the deviance for model 1, which is just the deviance for model 1. Therefore, an equivalent way to carry out this particular interaction test is to use the deviance previously obtained for model 1, which has the value 8.34. When compared with upper-tail chi-square values for which

d.f. = [Number of Y_{ij}'s] − [Number of parameters in model 1]
$= 16 - 9 = 7,$

the value 8.34 provides no evidence for lack of fit of model 1 (i.e., there are no large deviations of observed Y_{ij} values from predicted $\hat{Y}_{ij}$ values). This indicates that adding more terms (e.g., of the form EU_k) to model 1 will not significantly improve the fit of that model.

SAS Output for Model 2

The GENMOD Procedure

Criteria For Assessing Goodness of Fit

Criterion	DF	Value	Value/DF
Deviance	0	0.0000	.
Scaled Deviance	0	0.0000	.
Pearson Chi-Square	0	0.0000	.
Scaled Pearson X2	0	0.0000	.
Log Likelihood	.	7205.9610	.

(continued)

SAS Output for Model 2 (continued)

Analysis of Parameter Estimates

Parameter	DF	Estimate	Std Err	Chi Square	Pr > Chi
INTERCEPT	1	-5.3385	0.1581	1139.9829	0.0000
CITY	1	0.5792	0.2010	8.3076	0.0039
U1	1	-6.7207	1.0124	44.0657	0.0000
U2	1	-3.6094	0.2958	148.8871	0.0000
U3	1	-2.7347	0.2415	128.2000	0.0000
U4	1	-1.8289	0.1977	85.5824	0.0000
U5	1	-1.2232	0.1866	42.9868	0.0000
U6	1	-0.7040	0.1808	15.1595	0.0001
U7	1	-0.1504	0.1803	0.6957	0.4042
CU1	1	0.7581	1.1360	0.4454	0.5045
CU2	1	0.1135	0.3594	0.0996	0.7523
CU3	1	0.5664	0.2866	3.9068	0.0481
CU4	1	0.3659	0.2429	2.2694	0.1320
CU5	1	0.2126	0.2325	0.8364	0.3604
CU6	1	0.2776	0.2265	1.5026	0.2203
CU7	1	0.0549	0.2288	0.0577	0.8102

(EU_k, $k = 1, 2, \ldots, 7$ for CU1–CU7)

To answer question 2 about confounding, we need to see whether $\hat{\beta}$, or $e^{\hat{\beta}}$, changes meaningfully if we ignore (i.e., don't control for) "age." In particular, we need to drop the age terms (i.e., the "$\sum_{k=1}^{7} \alpha_k U_k$" component) from Model 1 to see whether the estimated coefficient of E changes meaningfully from its value of 0.806 (or whether the estimate of the rate ratio changes meaningfully from its value of $e^{\hat{\beta}} = 2.2389$). If we fit

Model 3: $\ln \lambda_{ij} = \alpha + \beta_0 E$

to these data, we obtain $\hat{\beta}_0 = 0.743$ and a **crude** rate ratio estimate of

$$\widehat{RR}_c = e^{\hat{\beta}_0} = e^{0.743} = 2.1043$$

There is enough change here to suggest that "age" is a confounder and so should be controlled at the analysis stage.

To this point, we have treated age as a *categorical* variable, using seven terms of the form $\alpha_k U_k$ ($k = 1, 2, \ldots, 7$) plus a constant α in model 1 to reflect the eight age strata. An alternative analysis that treats age as an interval variable is suggested by plotting $\ln \hat{\lambda}_{ij}$ versus $\ln T_i$ where

$$T_i = \frac{[\text{Midpoint of } i\text{th age interval}] - 15}{35}, \qquad i = 1, 2, \ldots, 8$$

These plots are presented in Figure 24.1 and the values used for the plots are given in Table 24.2. Figure 24.1 shows that the plotted points for each city graph essentially as a straight line, with the two lines being essentially parallel. These results suggest using a parsimonious model involving only a single linear effect of "age," with no interaction terms involving "age" and E.

In particular, consider the following model:

Model 4: $\ln \lambda_{ij} = \alpha + \theta \ln T_i + \beta E$

Model 4 says that

$$\lambda_{i0} = e^{\alpha} T_i^{\theta} \qquad \text{and} \qquad \lambda_{i1} = e^{\alpha} T_i^{\theta} e^{\beta}$$

so that

$$RR_i = \frac{\lambda_{i1}}{\lambda_{i0}} = e^{\beta}$$

FIGURE 24.1   **Plots of ln $\hat{\lambda}_{ij}$ versus ln T_i, where $T_i = [(\text{midpoint of } i\text{th age interval}) - 15]/35$, using values in Table 24.2 from the study of nonmelanoma skin cancer in two cities (Scotto, Kopf, and Urbach 1974)**

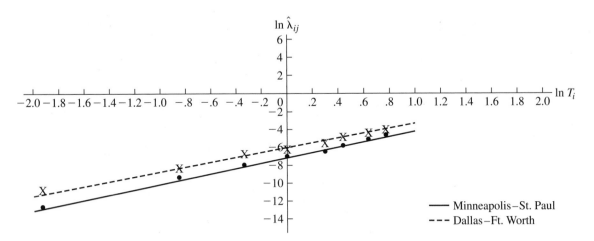

When $i = 4$ (so that $T_4 = 1$) and $j = 0$, we have $\lambda_{40} = e^{\alpha}T_4^{\theta} = e^{\alpha}(1)^{\theta} = e^{\alpha}$, so $\ln \lambda_{40} = \alpha$. Hence, α (the intercept term in Model 4) is the natural logarithm of the rate for the 45- to 54-year-old age group ($i = 4$) in Minneapolis–St. Paul ($j = 0$). The following block of computer output (using SAS's GENMOD procedure) is based on fitting Model 4.

SAS Output for Model 4

```
                          The GENMOD Procedure

                  Criteria For Assessing Goodness of Fit

Criterion                   DF              Value            Value/DF
Deviance                    13            14.8476             1.1421
Scaled Deviance             13            14.8476             1.1421
Pearson Chi-Square          13            14.6475             1.1267
Scaled Pearson X2           13            14.6475             1.1267
Log Likelihood               .          7198.5372

                  Analysis of Parameter Estimates

Parameter      DF      Estimate      Std Err     ChiSquare      Pr > Chi
INTERCEPT       1       -7.0764       0.0476    22071.7488        0.0000
CITY            1        0.8051       0.0522      238.0716        0.0000
LOGT            1        2.2894       0.0627     1334.0067        0.0000
```

From the above computer output, the following point estimates and standard errors can be obtained:

$$\hat{\alpha} = -7.076 \qquad S_{\hat{\alpha}} = 0.048$$

$$\hat{\beta} = 0.805 \qquad S_{\hat{\beta}} = 0.052$$

$$\hat{\theta} = 2.289 \qquad S_{\hat{\theta}} = 0.063$$

The deviance value for model 4 is 14.85, with $(16 - 3) = 13$ degrees of freedom, indicating a good fit to the data. The estimated adjusted rate ratio for Model 4 is

$$e^{\hat{\beta}} = e^{0.805} = 2.2367$$

TABLE 24.2   **Values of ln $\hat{\lambda}_{ij}$ and ln T_i, where $T_i = $ [(midpoint of ith age interval) $- 15]/35$, using the data in Table 24.1 from the study of nonmelanoma skin cancer in two cities (Scotto, Kopf, and Urbach 1974)**

Age Group	Midpoint	T_i	ln T_i	ln $\hat{\lambda}_{i0}$	ln $\hat{\lambda}_{i1}$
15–24	20	0.1429	−1.94591	−12.0592	−10.7219
25–34	30	0.4286	−0.84730	−8.9479	−8.2552
35–44	40	0.7143	−0.33647	−8.0732	−6.9275
45–54	50	1.0000	0.00000	−7.1674	−6.2223
55–64	60	1.2857	0.25131	−6.5617	−5.7698
65–74	70	1.5714	0.45199	−6.0425	−5.1953
75–84	80	1.8571	0.61904	−5.4889	−4.8548
85+	90	2.1429	0.76214	−5.3385	−4.7533

$\hat{\lambda}_{i0} = Y_{i0}/\ell_{i0} = $ Crude rate for Minneapolis–St. Paul
$\hat{\lambda}_{i1} = Y_{i1}/\ell_{i1} = $ Crude rate for Dallas–Ft. Worth

The 95% confidence interval for e^β is given by

$$\exp [0.805 \pm 1.96(0.052)] = \exp (0.805 \pm 0.1019)$$
$$= (e^{0.7031}, e^{0.9069})$$
$$= (2.020, 2.477)$$

Thus, using Model 4 leads to essentially the same conclusions obtained using Model 1, with a very small gain in precision (since the confidence interval for e^β using Model 4 is slightly narrower than the one using Model 1). Since Model 4 contains fewer parameters than Model 1, without compromising on validity or precision, and since Model 4 describes the linear effect of "age" in a clear-cut manner, it is the model of choice!

A summary of the results of fitting various models to the data set under consideration can be presented in a Poisson ANOVA table, as illustrated in Table 24.3. As this table indicates, Models 1 and 4 clearly have the best deviance values; but Model 4 is our choice for the "best" model. Model 1 has a deviance of 8.3, whereas Model 4 has a deviance of 14.8—but Model 1 *should* have a smaller deviance (i.e., should fit the data better) than Model 4, since it contains nine parameters, whereas Model 4 contains only three. The issue is really whether Model 1 fits the

TABLE 24.3   **Poisson ANOVA table for the skin cancer data of Table 24.1**

	Model for ln λ_{ij}	Number of Parameters	$D(\beta)$	d.f.
	α	1	2794.3	15
Model 3:	$\alpha + \beta E$	2	2573.3	14
	$\alpha + \theta \ln T_i$	2	272.7	14
Model 4:	$\alpha + \theta \ln T_i + \beta E$	3	14.8	13
	$\alpha + \sum_{k=1}^{7} \alpha_k U_k$	8	268.4	8
Model 1:	$\alpha + \sum_{k=1}^{7} \alpha_k U_k + \beta E$	9	8.3	7
	α_{ij}	16	0.0	0

data *significantly* better than Model 4. To address this issue, we can look at the difference in the two deviance values (namely, $14.8 - 8.3 = 6.5$), just as we would look at the difference in SSE values in standard multiple regression analysis. Let us assume that

(Deviance for model 4) − (Deviance for Model 1)

is very approximately a chi-square random variable for large n with $(13 - 7) = 6$ degrees of freedom. Since $\chi^2_{6, 0.60} = 6.211$, the P-value for testing H_0: "Model 1 and Model 4 fit the data equally well" is about .40. Hence, there is absolutely no evidence that Model 1 provides a better fit to the data than Model 4.

Finally, a pseudo-R^2 for Model 4 can be computed as

$$\text{Pseudo-}R^2 = \frac{2794.3 - 14.8}{2794.3} = .9947$$

which indicates a superb fit to the data.

24.7 A Second Illustration of Poisson Regression Analysis

We now consider an example given by Frome (1983). The data to be used appear in Table 24.4. The basic response variable Y is "number of lung cancer deaths," which is assumed to have a Poisson distribution. More specifically, Y_{ij} denotes the observed number of lung cancer deaths in row i and column j, $i = 1, 2, \ldots, 9$ and $j = 1, 2, \ldots, 7$; thus, there are $n = 9 \times 7 = 63$ subgroups. The rows represent "years of smoking" (defined as age minus 20 years) in five-year categories from 15 through 19 to 55 through 59; the columns represent "number of cigarettes smoked per day," starting at 0 for nonsmokers and going up to 35 or more for the heaviest smokers. The variable ℓ_{ij} denotes the number of man-years at risk for cell (i, j). The variable T_i, defined as the midpoint of the ith "years of smoking" category divided by 42.5, will be employed when we fit some dose–response models to the data in Table 24.4; the variable D_j will denote the dosage level variable for the jth "number of cigarettes smoked per day" category.

One particular form of failure rate model $\lambda(\mathbf{X}_{ij}, \boldsymbol{\beta})$ to be fit to these data is the standard two-way cross-classification model *in exponentiated form*. (Failure rates are always nonnegative, and using an exponential function ensures that this will be the case for all estimated rates.) In particular, consider modeling the rate $\lambda(\mathbf{X}_{ij}, \boldsymbol{\beta}) \equiv \lambda_{ij}$ for cell (i, j) as

$$\lambda_{ij} = e^{\mu + \alpha_i + \delta_j} \tag{24.15}$$

where μ is the overall mean, α_i is the fixed effect of the ith row, and δ_j is the fixed effect of the jth column. Here, as in standard two-way ANOVA, we impose the constraints $\sum_{i=1}^{9} \alpha_i = \sum_{j=1}^{7} \delta_j = 0$, so a total of $1 + (9 - 1) + (7 - 1) = 15$ parameters must be estimated using model (24.15).

As with standard regression model representations of ANOVA-type data (see elsewhere in this book), the "X" variables underlying model (24.15) are the usual dummy variables used to index the various rows and columns; their appearance has been suppressed for notational convenience. An equivalent way to write (24.15) in dummy variable regression notation is $\lambda_{ij} = \exp(\mu + \sum_{k=1}^{8} \alpha_k R_k + \sum_{\ell=1}^{6} \delta_\ell C_\ell)$, where the R_k's and C_ℓ's denote dummy variables indexing the rows and columns, respectively.

TABLE 24.4   **Man-years at risk (ℓ_{ij}) and observed number (Y_{ij}) of lung cancer deaths (in parentheses)**

Years of Smoking*	$42.5T_i$	Number of Cigarettes Smoked per Day						
		0 ($D_1 = 0$)	1–9 ($D_2 = 5.2$)	10–14 ($D_3 = 11.2$)	15–19 ($D_4 = 15.9$)	20–24 ($D_5 = 20.4$)	25–34 ($D_6 = 27.4$)	35+ ($D_7 = 40.8$)
15–19	17.5	10,366 (1)	3,121 (0)	3,577 (0)	4,317 (0)	5,683 (0)	3,042 (0)	670 (0)
20–24	22.5	8,162 (0)	2,937 (0)	3,286 (1)	4,214 (0)	6,385 (1)	4,050 (1)	1,166 (0)
25–29	27.5	5,969 (0)	2,288 (0)	2,546 (1)	3,185 (0)	5,483 (1)	4,290 (4)	1,482 (0)
30–34	32.5	4,496 (0)	2,015 (0)	2,219 (2)	2,560 (4)	4,687 (6)	4,268 (9)	1,580 (4)
35–39	37.5	3,512 (0)	1,648 (1)	1,826 (0)	1,893 (0)	3,646 (5)	3,529 (9)	1,336 (6)
40–44	42.5	2,201 (0)	1,310 (2)	1,386 (1)	1,334 (2)	2,411 (12)	2,424 (11)	924 (10)
45–49	47.5	1,421 (0)	927 (0)	988 (2)	849 (2)	1,567 (9)	1,409 (10)	556 (7)
50–54	52.5	1,121 (0)	710 (3)	684 (4)	470 (2)	857 (7)	663 (5)	255 (4)
55–59	57.5	826 (2)	606 (0)	449 (3)	280 (5)	416 (7)	284 (3)	104 (1)

Note: If ℓ_{ij} really equalled the number of people in cell (i, j) from which the observed number Y_{ij} of lung cancer cases developed, then Y_{ij} could be treated as a binomial random variable with sample size ℓ_{ij} and unknown probability (or risk) of lung cancer death π_{ij}, in which case categorical data analysis methods might be utilized (although having several cells with very few deaths is problematic).

The quantity T_i is the midpoint of the ith "years of smoking" category divided by 42.5; D_j is the dosage level variable for the jth "cigarettes per day" category.

*Age minus 20 years

The Poisson model-based likelihood function for the data in Table 24.4 and under model (24.15) is

$$\prod_{i=1}^{9}\prod_{j=1}^{7}\left[\frac{(\lambda_{ij})^{Y_{ij}}e^{-\lambda_{ij}}}{(Y_{ij})!}\right]$$

where $\lambda_{ij} = \exp(\mu + \alpha_i + \delta_j)$, $\alpha_9 = -\sum_{i=1}^{8}\alpha_i$, and $\delta_7 = -\sum_{j=1}^{6}\delta_j$. For the data in Table 24-4, using IRLS methods leads to the following estimates (in exponentiated form):

$$e^{\hat{\mu}} = 7.69 \times 10^{-5}$$

$$e^{\hat{\alpha}_1} = 0.039, \quad e^{\hat{\alpha}_2} = 0.117, \quad e^{\hat{\alpha}_3} = 0.247, \quad e^{\hat{\alpha}_4} = 1.105, \quad e^{\hat{\alpha}_5} = 1.144,$$

$$e^{\hat{\alpha}_6} = 3.017, \quad e^{\hat{\alpha}_7} = 3.823, \quad e^{\hat{\alpha}_8} = 6.047, \quad e^{\hat{\alpha}_9} = 10.052,$$

$$e^{\hat{\delta}_1} = 1.00, \quad e^{\hat{\delta}_2} = 3.39, \quad e^{\hat{\delta}_3} = 8.16, \quad e^{\hat{\delta}_4} = 10.10, \quad e^{\hat{\delta}_5} = 18.20,$$

$$e^{\hat{\delta}_6} = 22.60, \quad e^{\hat{\delta}_7} = 36.80$$

Given these estimates, the predicted number $\hat{Y}_{ij}$ of cancer deaths in cell (i, j) is $\hat{Y}_{ij} = \ell_{ij}e^{\hat{\mu}+\hat{\alpha}_i+\hat{\delta}_j}$. For example, when $i = 4$ and $j = 5$, then

$$\hat{Y}_{45} = \ell_{45}\,e^{\hat{\mu}}e^{\hat{\alpha}_4}\,e^{\hat{\delta}_5}$$

$$= (4,687)(7.69 \times 10^{-5})(1.105)(18.20)$$

$$= 7.25$$

which should be compared with the actual observed value $Y_{45} = 6$. It can be shown that the deviance for this fitted model, as calculated using (24.11), has the numerical value of 51.47.

(Formula (24.12) is not appropriate for these data, since several $\hat{Y}_{ij}$'s are close to 0 in value.) When compared with critical values of the chi-square distribution with $(63 - 15) = 48$ degrees of freedom, the value 51.47 does not suggest any significant lack of fit for the cross-classification model (24.15).

However, using model (24.15) involves estimating 15 parameters (whereas n is only 63); hence, we are not achieving a parsimonious description of the data. In addition, it is of considerable interest with these data to fit a model whose parameters can realistically be interpreted in terms of the mathematical theory of carcinogenesis.

In what follows, we will consider the *four-parameter nonlinear* model described by Frome (1983)—namely,

$$\lambda_{ij} = \lambda(T_i, D_j; \gamma, \alpha, \theta, \delta) = (\gamma + \alpha D_j^\theta)T_i^\delta \qquad (24.16)$$

where T_i and D_j are as defined earlier.[6]

By using the mathematical identity $e^{\ln a} = a$ for $a > 0$, we can write (24.16) in an equivalent form considered by Frome (1983):

$$\lambda_{ij} = [e^{\ln \gamma} + e^{(\ln \alpha + \theta \ln D_j)}]e^{\delta \ln T_i} = [e^{\beta_3} + e^{(\beta_1 + \beta_2 X_{2j})}]e^{\beta_0 X_{1i}} \equiv \lambda(\mathbf{X}_{ij}, \boldsymbol{\beta}) \qquad (24.17)$$

where $\mathbf{X}_{ij} = (X_{1i}, X_{2j}) = (\ln T_i, \ln D_j)$ and $\boldsymbol{\beta} = (\beta_0, \beta_1, \beta_2, \beta_3) = (\delta, \ln \alpha, \theta, \ln \gamma)$. Expressing (24.16) in the exponential form (24.17) ensures that predicted rates are always positive.

The fitting of model (24.17) by IRLS gives the estimates $\hat{\beta}_0 = 4.46$, $\hat{\beta}_1 = 1.82$, $\hat{\beta}_2 = 1.29$, and $\hat{\beta}_3 = 2.94$. The estimated standard errors for these four estimators (obtained as the square roots of the appropriate diagonal elements of the estimated covariance matrix) are, respectively, 0.33, 0.66, 0.20, and 0.58. For example, an approximate large-sample 95% confidence interval for $\alpha (= e^{\beta_1})$ would be

$$\exp\left(\hat{\beta}_1 \pm 1.96 \sqrt{\widehat{\mathrm{Var}}\,\hat{\beta}_1}\right)$$

giving $\exp[1.82 \pm 1.96(0.66)] = \exp(1.82 \pm 1.29)$, and thus the interval $(e^{0.53}, e^{3.11}) = (1.70, 22.42)$.

Finally, a summary of the results of fitting various subset models of (24.16) is provided in Table 24.5.

As discussed earlier, calculating the various differences between deviances in the ANOVA-type table presented in Table 24.5 will provide likelihood ratio tests for the importance of the parameters γ, δ, α, and θ. First of all, the difference $(445.10 - 180.82) = 264.28$, when compared with upper-tail values of the χ_1^2 distribution, argues strongly for rejecting H_0: $\delta = 0$ in favor of H_A: $\delta \neq 0$. This means that the (multiplicative) effect of time since first exposure is important. Second, the difference $(180.82 - 61.84) = 118.98$ demands rejection of H_0: $\alpha = 0$ in favor of H_A: $\alpha \neq 0$, suggesting that the amount smoked is an important variable. Finally, the difference $(61.84 - 59.58) = 2.26$ does not lead to rejection of H_0: $\theta = 1$, so the first power of dosage seems most appropriate. Since the deviance value of 61.84 with 60 degrees of freedom for the model $\lambda_{ij} = (\gamma + \alpha D_j)T_i^\delta$ does not suggest any lack of fit, this model seems preferable.[7]

---

[6] In model (24.16), γ represents the background (i.e., $D_j = 0$ for a nonsmoker) rate at age 62.5 (i.e., $T_i = [\text{age} - 20]/42.5 = 1$ at age 62.5), αD_j^θ describes the effect of dosage (i.e., amount smoked) on lung cancer death rates, and T_i^δ is the multiplicative effect (on $\gamma + \alpha D_j^\theta$) of the time elapsed since the smoking habit was started. Frome (1983) provides further discussion of scientific evidence in favor of using a model like (24.16) to describe the incidence of lung cancer.

[7] The reduction in the deviance due to adding a parameter (or a group of parameters) is *order-dependent*; consequently, a different order of parameter additions can lead to a different final model.

TABLE 24.5   Summary of analyses of data in Table 24.4

Model for λ_{ij}	Number of Parameters	$D(\beta)$	d.f.*
γ	1	445.10	62
γT_i^δ	2	180.82	61
$(\gamma + \alpha D_j)T_i^\delta$	3	61.84	60
$(\gamma + \alpha D_j^\theta)T_i^\delta$	4	59.58	59

*d.f. = 63 − (Number of parameters in fitted model).

24.8 Summary

This chapter described the general form and several key features of the Poisson regression model. The "typical" Poisson regression model expresses a log *rate* as a linear function of a set of predictors. The Poisson regression method, nevertheless, allows for more complicated nonlinear models as well. The dependent variable is a count of the number of occurrences of an event of interest, such as the number of cases of a disease that occur over a given follow-up time period. For the "typical" Poisson regression model, the natural measures of effect that are estimated are *rate ratios*.

   Maximum likelihood estimation is used to estimate the regression coefficients of a Poisson regression model. The likelihood function assumes that the underlying response (a count) has the Poisson distribution. A measure of goodness of fit of a Poisson regression model is obtained by using the deviance statistic, which is a likelihood-ratio-type statistic reflecting the fit of a current model of interest relative to the baseline (uninformative or "saturated") model (which involves as many parameters as there are data points). The difference in deviance statistics obtained for two (hierarchically ordered) models being compared is equivalent to the difference in log-likelihood statistics for each model, so a likelihood ratio test for Poisson regression can be carried out equivalently by using differences in deviance statistics. Finally, a tabular summary of the results of fitting various models to the same data set can be presented in a *Poisson ANOVA table,* which contains deviance and corresponding d.f. (degrees of freedom) information for each model being fit.

   It is important to emphasize again that the mean and variance of the Poisson distribution are equal. However, it is common to observe count data for which sample variances are either larger than sample means (called "over-dispersion") or smaller than sample means (called "under-dispersion"), with the former phenomenon being more prevalent. For most popular categorical data analysis procedures (e.g., logistic and Poisson regression), there are statistical methods available to account appropriately for over-dispersed or under-dispersed data [e.g., see the excellent book by McCullagh and Nelder (1989)].

Problems

1. Suppose that, for each of three age groups (25–34, 35–44, and 45–54), we have recorded yearly sex-specific lung cancer mortality rates for the five-year period 1990 through 1994. These data are to be analyzed by Poisson regression methods to see whether the change (if any) in log rate over time varies by age–sex group, and, if so, to quantify that variation. In what follows, we will index the six age–sex groups, as follows:

   group 1($i = 1$):   25–34-year-old females

   group 2($i = 2$):   35–44-year-old females

group 3($i = 3$):   45–54-year-old females
group 4($i = 4$):   25–34-year-old males
group 5($i = 5$):   35–44-year-old males
group 6($i = 6$):   45–54-year-old males

Consider the following model for the expected cell count $E(Y_{ik})$ for age–sex group i in year
$(1990 + k)$, where $i = 1, 2, 3, 4, 5, 6$ and $k = 0, 1, 2, 3, 4$:

$$E(Y_{ik}) = \ell_{ik}\lambda_{ik}$$

where

$$\ln \lambda_{ik} = \sum_{i=1}^{6} \alpha_i A_i + \beta k \qquad (1)$$

Here, ℓ_{ik} and λ_{ik} are, respectively, the person-years at risk and the (unknown) population lung
cancer mortality rate in cell (i, k). The independent variables in model (1) are defined as follows:

$$A_i = \begin{cases} 1 & \text{if age–sex group } i \\ 0 & \text{otherwise} \end{cases}$$

$$k = (\text{year} - 1990)$$

1. **a.** What is the total number n of data points, or pairs (ℓ_{ik}, Y_{ik}), for this data set?
  b. Based on model (1), what is the expected cell count for a 40-year-old male in 1992, writ-
     ten as a function of α_5 and β?
  c. For the ith age–sex group, how does model (1) describe the *change* in log rate over time?
  d. What does model (1) assume about the effect of age–sex group on the *change* in log rate
     over time?
  e. Find a general expression for

$$RR_{ik} = \frac{\lambda_{ik}}{\lambda_{10}}$$

   the rate ratio comparing the mortality rate for age group i and year $(1990 + k)$ to the mortality rate for the reference category "25–34-year-old females in 1990."
  f. Suppose that it is of interest to assess whether the *change* in log rate over time actually
     varies by age–sex group (i.e., whether there is a group-by-time interaction). By adding
     appropriate cross-product terms to model (1), construct a model that will permit such an
     assessment and then discuss how one would interpret this model.
     Consider the following Poisson regression ANOVA table, based on fitting various
     models to the data under study:

Model for $\ln \lambda_{ik}$	Number of Parameters	$D(\hat{\beta})$	d.f.
(1) α	1	300	29
(2) $\alpha + \beta k$	2	200	28
(3) $\sum_{i=1}^{6} \alpha_i A_i$	6	175	24
(4) $\sum_{i=1}^{6} \alpha_i A_i + \beta k$	7	60	23
(5) $\sum_{i=1}^{6} \alpha_i A_i + \beta_1 k + \beta_2 k^2$	8	59	22
(6) $\sum_{i=1}^{6} \alpha_i A_i + \beta k + \gamma(A_1 k)$	8	25	22
(7) $\sum_{i=1}^{6} \alpha_i A_i + \beta k + \sum_{i=1}^{5} \gamma_1(A_i k)$	12	20	18
(8) α_{ik}	30	0	0

1. Using this table, answer the following questions:

g. Ignoring (for now) the variable "time," is there evidence that average mortality rates differ among the six age–sex groups?

h. Is adding the linear time term (βk) to model (3) worthwhile?

i. Assuming that the change in log rate over time is the same for all age–sex groups, do the data argue for adding a quadratic time term to a model that already contains a linear component of time?

j. Is there evidence in the data that the change in log rate over time differs for different age–sex groups?

k. Carry out a test of H_0: "All six age–sex groups have the same slope" versus H_A: "All age–sex groups except group 1 have the same slope."

l. Of the models that fit the data well (use $\alpha = .1$ for any test of lack of fit that you perform), which one would you choose as your final model? Why?

m. For your final model chosen in part (l), calculate a pseudo-R^2-value.

n. Now, assume that model (6) has been fit to the data, resulting in the following regression coefficient estimates and standard errors:

Parameter	Point Estimate	Standard Error
$\hat{\alpha}_1$	0.50	0.25
$\hat{\alpha}_2$	1.00	0.40
$\hat{\alpha}_3$	1.50	0.30
$\hat{\alpha}_4$	1.25	0.30
$\hat{\alpha}_5$	1.50	0.50
$\hat{\alpha}_6$	1.75	0.40
$\hat{\beta}$	0.50	0.20
$\hat{\gamma}_1$	-3.00	0.50

o. Find and interpret approximate 95% confidence intervals for the following parameters: (1) the common slope for age–sex groups 2 through 6, and (2) the slope for age–sex group 1. [*Note:* The estimated covariance between $\hat{\beta}$ and $\hat{\gamma}_1$ is $\widehat{\text{Cov}}(\hat{\beta}, \hat{\gamma}_1) = -.10$.]

2. A five-year follow-up study was carried out to assess the relationship of diet and weight to the incidence of stomach cancer in 40- to 50-year-old males in a certain metropolitan area. Let Y_{ij} denote the number of cases of stomach cancer found in the ith weight category of diet type j, as indicated in the following table:

	Low-cholesterol Diet ($j = 1$)	High-cholesterol Diet ($j = 2$)
Low Weight ($i = 1$)	Y_{11}	Y_{12}
Medlow Weight ($i = 2$)	Y_{21}	Y_{22}
Medhigh Weight ($i = 3$)	Y_{31}	Y_{32}
High Weight ($i = 4$)	Y_{41}	Y_{42}

Consider the following Poisson regression model for these data:

Model 1: $E(Y_{ij}) = \ell_{ij}\lambda_{ij}$, $\qquad i = 1, 2, 3, 4$ and $j = 1, 2$

where ℓ_{ij} = man-years at risk in the (i, j)th cell, λ_{ij} = rate of development of stomach cancer in the (i, j)th cell, where λ_{ij} is modeled as follows:

$$\ln \lambda_{ij} = \sum_{i=1}^{4} \alpha_i V_i + \beta C \qquad \text{where} \qquad V_i = \begin{cases} 1 & \text{if Weight group } i \\ 0 & \text{otherwise} \end{cases}$$

$$C = \begin{cases} 1 & \text{if Low-cholesterol Diet } (j=1) \\ 0 & \text{if High-cholesterol Diet } (j=2) \end{cases}$$

2. a. Based on this model, give an expression for the rate ratio (RR) comparing low-cholesterol diet subjects to high-cholesterol diet subjects, adjusted for weight group.
   b. Give a large-sample formula for the 95% confidence interval for the RR described in part (a).
   c. How would one test whether there is a significant interaction between diet type and weight group?
   d. Assuming that there is no interaction between weight group and diet type, consider the following model as an alternative to Model 1:

   Model 2: $\ln \lambda_{ij} = \alpha + \gamma W_i + \beta C$

   where W_i denotes the midpoint (in pounds) of weight group i.

   Describe how to use a "difference between deviances" approach to evaluate whether Model 2 fits the data significantly better than Model 1. (State the null hypothesis, describe the test statistic in terms of deviances, and state the distribution of the test statistic, including degrees of freedom, under the null hypothesis.) How might one criticize the use of this approach?

3. The following set of questions relates to using Poisson regression methods to analyze data from an in vitro study of human chromosome damage. In this study, using Poisson regression is appropriate because we have "count" response data, where each count is the number of broken chromosomes in a sample of 100 cells taken from each of $n = 40$ individuals. (A Poisson distribution is appropriate for describing counts per unit of *space*—e.g., 100 cells—as well as for describing counts per unit of *time*.)

   The design of the study described here involved randomly assigning each of the cell samples for the 40 individuals to one of four treatment groups consisting of ten individuals each. The 10 sets of 100 cells in each treatment group were exposed to a particular combination of two drugs, A and B, as follows:

Treatment Group j	Number of Individuals n_j	Drug Combination
1	10	neither A nor B
2	10	A only
3	10	B only
4	10	both A and B
	40	

The exposed cells were then examined for chromosome breakage, and the number of broken chromosomes in the 100 cells for each individual was counted. *The investigator wishes to assess the effects of drug use, and of drug interaction, on the rate of chromosome damage.*

   Let Y_{ij} denote the random variable representing the number of chromosome breaks counted for each individual i ($i = 1, 2, \ldots, 10$) within treatment group j ($j = 1, 2, 3, 4$)

Also, let $\ell_{ij} = 100$, where ℓ_{ij} denotes the amount of "person-space" of observation. Assume that the drugs act in a multiplicative fashion on the rate of chromosome breakage (λ_{ij}). The Poisson regression model under consideration can be written as

$$\ln \lambda_{ij} = \beta_0 + \sum_{k=1}^{p} \beta_k X_{ijk}$$

where X_{ijk} denotes the value of the kth predictor for the ith individual in the jth treatment group.

3. a. Based on the preceding scenario, write out the form of the Poisson regression model for $\ln \lambda_{ij}$ as a function of the predictors of interest in this study. In doing so, *make sure to explicitly define a set of categorical predictors to include in this model; these predictors should reflect the primary objective of the study.*

b. Based on the model specified in part (a), describe *two* alternative ways one can test the null hypothesis of no interaction between the drugs. Include in the answer the null hypothesis (in terms of regression coefficients), the formula for the test statistic(s), the degrees of freedom for each test, and the distribution of the test statistic under the null hypothesis.

c. Fill in the blanks in the following Poisson regression ANOVA table, which is based on fitting various models to describe the rate of chromosome breakage:

	Model for $\ln \lambda_{ij}$	Number of Parameters	Deviance	Deviance d.f.
(I)	Baseline (i.e., β_0)	1	90	39
(II)	Main effect of drug A only	—	85	—
(III)	Main effect of drug B only	—	84	—
(IV)	Main effects of both drug A and drug B	—	40	—
(V)	All main effects and interaction	—	30	—

d. According to the ANOVA table, which models provide good fit to the data? Explain briefly.

e. According to the ANOVA table, does drug A have a significant main effect? Explain briefly.

f. According to the ANOVA table, does drug B have a significant effect over and above the effect of drug A? Explain briefly.

g. According to the ANOVA table, do drugs A and B have a significant interaction effect? Explain briefly.

h. According to the ANOVA table (and in light of your answers to parts (d) through (g)), provide an appropriate expression for the rate ratio for the effect of drug A given drug B status relative to baseline and an expression for the rate ratio for the effect of drug B given drug A status relative to baseline.

4. For the nonmelanoma skin cancer data considered in Sections 24.3 and 24.6 of the main body of this chapter, consider the following additional model that was fit to these data:

Model 5: $\ln \lambda_{ij} = \alpha + \theta \ln T_i + \beta E + \delta E(\ln T_i)$

where T_i is defined as

$$T_i = \frac{[\text{Midpoint of the } i\text{th age interval}] - 15}{35}, \qquad i = 1, 2, \ldots, 8,$$

and E denotes city (1 = Dallas–Ft. Worth, 0 = Minneapolis–St. Paul).

Edited Output for Model 5

```
              Criteria for Assessing Goodness of Fit

Criterion                        DF              Value          Value/DF
Deviance                         12            10.6302           0.8859
Scaled Deviance                  12            10.6302           0.8859
Pearson Chi-Square               12            10.6319           0.8860
Scaled Pearson X2                12            10.6319           0.8860
Log Likelihood                    .          7200.6459

              Analysis of Parameter Estimates

Parameter      DF     Estimate     Std Err     Chi Square    Pr > Chi
INTERCEPT       1      -7.1400      0.0597     14324.9448      0.0001
CITY            1       0.8869      0.0686       167.2316      0.0001
LNT             1       2.4916      0.1237       405.9034      0.0001
CITYLNT         1      -0.2811      0.1435         3.8386      0.0501
SCALE           0       1.0000      0.0000
```

Note: $\text{CITY} = E$, $\text{LNT} = \ln T_i$, and $\text{CITYLNT} = E(\ln T_i)$

4. a. Using the available edited computer output, calculate the estimated rate ratio comparing the two cities for the following two age-groups: **i)** 15–24 and **ii)** 45–54.
(*Note:* For the 15–24 age-group, $T_i = 1/7$; for the 45–54 age-group, $T_i = 1$.)

b. What do your results in part (a) suggest about interaction?

c. Carry out both a Wald test and a Likelihood Ratio test for the significance of the coefficient of the product term $E(\ln T_i)$. (*Hint:* To carry out the LR test, use the deviance statistic $D(\boldsymbol{\beta})$ given in Table 24.3 for Model 4.).

d. Do the results from part (c) support the conclusion that the product term $E(\ln T_i)$ needs to remain in the model?

e. Are the tests conducted in part (c) equivalent to testing whether either Model 4 or Model 5 fits the data?

f. Which model appears to be the best model among Models 3, 4, and 5? Briefly justify any answer.

(*Note:* The SAS datafile for these data is called "melanoma.dat" and can be obtained from the publisher's website.)

5. The following questions concern Poisson regression models fit to fictitious follow-up study data in which rates of disease are modeled as a function of age and smoking status. The SAS program codes and output for six models are provided on pages 688–692 following the last problem in this chapter. The dependent variable is called COUNT and the independent variables are AGE and SMOKE. There are three age groups (defined by age midpoint) and two levels of smoking status. The data are given in the following table:

	SMOKE = 0			SMOKE = 1		
	AGE = 25	AGE = 45	AGE = 75	AGE = 25	AGE = 45	AGE = 75
Person-time	1,000	3,500	2,000	100	1,500	1,000
Count	15	36	35	4	28	32

Note: AGE = 25 if $15 \leq$ age < 35
AGE = 45 if $35 \leq$ age < 55
AGE = 75 if $55 \leq$ age < 95

Output for Problem 5

```
                           MODEL 1

                     The GENMOD Procedure

                      Model Information

          Description                    Value
          Data Set                       WORK.ONE
          Distribution                   POISSON
          Link Function                  LOG
          Dependent Variable             COUNT
          Offset Variable                LOG_TIME
          Observations Used              6


             Criteria for Assessing Goodness of Fit

        Criterion              DF        Value       Value/DF
        Deviance                3       6.3636       2.1212
        Scaled Deviance         3       6.3636       2.1212
        Pearson Chi-Square       3      7.2783       2.4261
        Scaled Pearson X2        3      7.2783       2.4261
        Log Likelihood          .      350.6333


                 Analysis of Parameter Estimates

   Parameter      DF    Estimate     Std Err      ChiSquare     Pr > Chi
   INTERCEPT       1    -4.8391      0.2839       290.5207      0.0001
   AGE             1     0.0098      0.0049         4.0497      0.0442
   SMOKE           1     0.5784      0.1663        12.1051      0.0005
```

```
                           MODEL 2

                     The GENMOD Procedure

                      Model Information

          Description                    Value
          Data Set                       WORK.ONE
          Distribution                   POISSON
          Link Function                  LOG
          Dependent Variable             COUNT
          Observations Used              6


             Criteria for Assessing Goodness of Fit

        Criterion              DF        Value       Value/DF
        Deviance                3      18.7971       6.2657
        Scaled Deviance         3      18.7971       6.2657
        Pearson Chi-Square       3     17.3596       5.7865
        Scaled Pearson X2        3     17.3596       5.7865
        Log Likelihood          .     344.4165


                 Analysis of Parameter Estimates

   Parameter      DF    Estimate     Std Err      ChiSquare     Pr > Chi
   INTERCEPT       1     2.4336      0.2487        95.7434      0.0001
   AGE             1     0.0177      0.0040        19.4544      0.0001
   SMOKE           1    -0.2955      0.1651         3.2033      0.0735
```

(continued)

Output for Problem 5 (continued)

```
                              MODEL 3
                      The GENMOD Procedure
                      Model Information

           Description                       Value
           Data Set                          WORK.ONE
           Distribution                      POISSON
           Link Function                     LOG
           Dependent Variable                COUNT
           Offset Variable                   LOG_TIME
           Observations Used                 6

              Criteria for Assessing Goodness of Fit
       Criterion                DF          Value        Value/DF
       Deviance                  2          6.2371        3.1186
       Scaled Deviance           2          6.2371        3.1186
       Pearson Chi-Square        2          7.2916        3.6458
       Scaled Pearson X2         2          7.2916        3.6458
       Log Likelihood            .        350.6965          .

                 Analysis of Parameter Estimates

   Parameter     DF      Estimate     Std Err     ChiSquare    Pr > Chi
   INTERCEPT      1       -4.7668      0.3474      188.2664      0.0001
   AGE            1        0.0084      0.0061        1.8770      0.1707
   SMOKE          1        0.3763      0.5936        0.4017      0.5262
   AGE_SMK        1        0.0036      0.0100        0.1264      0.7222
```

```
                              MODEL 4
                      The GENMOD Procedure
                      Model Information

           Description                       Value
           Data Set                          WORK.ONE
           Distribution                      POISSON
           Link Function                     LOG
           Dependent Variable                COUNT
           Offset Variable                   LOG_TIME
           Observations Used                 6

              Criteria for Assessing Goodness of Fit
       Criterion                DF          Value        Value/DF
       Deviance                  4         10.3962        2.5991
       Scaled Deviance           4         10.3962        2.5991
       Pearson Chi-Square        4         10.6195        2.6549
       Scaled Pearson X2         4         10.6195        2.6549
       Log Likelihood            .        348.6170          .

                 Analysis of Parameter Estimates

   Parameter     DF      Estimate     Std Err     ChiSquare     Pr > Chi
   INTERCEPT      1       -4.3252      0.1078     1608.8402       0.0001
   SMOKE          1        0.6208      0.1651       14.1426       0.0002
```

(continued)

Output for Problem 5 (continued)

```
                                    MODEL 5

                            The GENMOD Procedure

                             Model Information

              Description                        Value
              Data Set                           WORK.TWO
              Distribution                       POISSON
              Link Function                      LOG
              Dependent Variable                 COUNT
              Offset Variable                    LOG_TIME
              Observations Used                  6

                  Criteria for Assessing Goodness of Fit

          Criterion                 DF          Value        Value/DF
          Deviance                   2         0.3925         0.1962
          Scaled Deviance            2         0.3925         0.1962
          Pearson Chi-Square         2         0.4225         0.2112
          Scaled Pearson X2          2         0.4225         0.2112
          Log Likelihood             .       353.6188            .

                    Analysis of Parameter Estimates

     Parameter     DF      Estimate     Std Err     ChiSquare    Pr > Chi
     INTERCEPT      1       -4.1355      0.2310      320.5959     0.0001
     AGE1           1       -0.4568      0.2657        2.9545     0.0858
     AGE2           1        0.0769      0.2657        0.0839     0.7721
     SMOKE          1        0.6305      0.1688       13.9433     0.0002

                    Estimated Covariance Matrix

                  Intercept          AGE1            AGE2           SMOKE
     Intercept     0.05334        -0.05133        -0.05116       -0.004509
     AGE1         -0.05133         0.07062         0.05531       -0.008207
     AGE2         -0.05116         0.05531         0.07059       -0.009300
     SMOKE        -0.004509       -0.008207       -0.009300       0.02851
```

```
                                    MODEL 6

                            The GENMOD Procedure

                             Model Information

              Description                        Value
              Data Set                           WORK.TWO
              Distribution                       POISSON
              Link Function                      LOG
              Dependent Variable                 COUNT
              Offset Variable                    LOG_TIME
              Observations Used                  6

                  Criteria for Assessing Goodness of Fit

          Criterion                 DF          Value        Value/DF
          Deviance                   0         0.0000            .
          Scaled Deviance            0         0.0000            .
          Pearson Chi-Square         0         0.0000            .
          Scaled Pearson X2          0         0.0000            .
          Log Likelihood             .       353.8151            .
```

(continued)

Output for Problem 5 (continued)

```
                         MODEL 6 (Continued)

                  Analysis of Parameter Estimates

    Parameter      DF      Estimate     Std Err    ChiSquare    Pr > Chi
    INTERCEPT      1       -4.1997      0.2582     264.5628     0.0001
    AGE1           1       -0.3773      0.3073       1.5072     0.2196
    AGE2           1        0.1542      0.3086       0.2495     0.6174
    SMOKE          1        0.9808      0.5627       3.0380     0.0813
    AGE1_SMK       1       -0.3848      0.6166       0.3896     0.5326
    AGE2_SMK       1       -0.3773      0.6136       0.3781     0.5386
```

References

Baker, R. J., and Nelder, J. A. 1978. *Generalized Linear Interactive Modeling (GLIM), Release 3,* Oxford: Numerical Algorithms Group.

Frome, E. L. 1983. "The Analysis of Rates Using Poisson Regression Models." *Biometrics* 39: 665–74.

Frome, E. L., and Checkoway, H. 1985. "Use of Poisson Regression Models in Estimating Incidence Rates and Ratios." *American Journal of Epidemiology* 121(2): 309–23.

Kleinbaum D. G. 2003. *ActivEpi.* New York and Berlin: Springer Publisher.

Kleinbaum, D. G., and Klein, M. 2005. *Survival Analysis—A Self-Learning Text.* 2nd edition. New York and Berlin: Springer Publishers.

Matanoski, G. M., Breysse, P. N., Elliott E. A. 1991. "Electromagnetic field exposure and male breast cancer." *Lancet* 337(8743): 737.

McCullagh, P. and Nelder, J. A. (1989). *Generalized Linear Models*, Second Edition. London and New York: Chapman and Hall.

Remington, R. D., and Schork, M. A. 1985. *Statistics with Applications to the Biological and Health Sciences.* Englewood Cliffs, N.J.: Prentice-Hall.

Scotto, J., Kopf, A. W., and Urbach, F. 1974. "Non-Melanoma Skin Cancer among Caucasians in Four Areas of the United States." *Cancer* 34: 1333–38.

Stokes, M. E., and Koch, G. G. 1983. "A Macro for Maximum Likelihood Fitting of Log-Linear Models to Poisson and Multinomial Counts with Contrast Matrix Capability for Hypothesis Testing." *Proceedings of the Eighth Annual SAS Users Group International Conference,* pp. 795–800.

Analysis of Correlated Data
Part 1: The General Linear
Mixed Model

25.1 Preview

Up to this point in the text, we have described regression-modeling techniques for which a *single response* is measured on each observational unit (e.g., subject), where the response may be either continuous (e.g., multiple linear regression), categorical (e.g., logistic regression), or a count variable (e.g., Poisson regression). For many studies, however, *two or more responses* are observed on each unit. The responses on each unit will generally be correlated, thereby requiring an analysis that accounts for such correlation.

As a simple example of *correlated data*, several blood pressure measurements may be taken over time on each of several subjects. Since the same response variable (i.e., blood pressure, treated as a continuous variable) is being measured at different times, we typically refer to such data as *repeated measures data*, and the collection of responses on the same subject is often called a *cluster* of responses. When such repeated measures are taken over time, the study is called a *longitudinal study*. In this chapter, we will only consider continuous responses, although more general methods are available (e.g., generalized linear mixed models; see Diggle, Heagerty, Liang, and Zeger, 2002; Zeger and Liang, 1986; or Kleinbaum and Klein, 2002).

Figure 25.1 illustrates repeated measures data on blood pressure (BP) for two different persons. Notice that person 1 has blood pressure measurements taken at 5 different times, whereas person 2 has blood pressure measurements taken at 4 times. Moreover, the times at which person 1 is observed are not all the same as the times at which person 2 is observed. This graph illustrates that, *in any repeated measures study, it is possible that different subjects can have different numbers of observations and that the observations on different subjects can occur at different times.*

FIGURE 25.1 **Longitudinal data on two subjects**

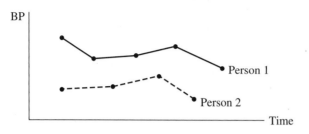

The above example also leads us to introduce the following general mathematical notation for a continuous outcome/response variable for the ith subject/cluster in any type of repeated measures/correlated data study:

$$\mathbf{Y}_i = \begin{bmatrix} Y_{i1} \\ Y_{i2} \\ \vdots \\ Y_{in_i} \end{bmatrix}, \qquad i = 1, 2, \ldots, K \tag{25.1}$$

where K denotes the number of subjects and n_i denotes the number of observations on the i-th subject. $\mathbf{Y}_i$ is often referred to as the **response vector** for the ith subject.[1]

For the $K = 2$ subjects in Figure 25.1, the two response vectors can then be denoted as

$$\mathbf{Y}_1 = \begin{bmatrix} Y_{11} \\ Y_{12} \\ Y_{13} \\ Y_{14} \\ Y_{15} \end{bmatrix} \quad \text{for } i = 1 \text{ and } \quad \text{and } \mathbf{Y}_2 = \begin{bmatrix} Y_{21} \\ Y_{22} \\ Y_{23} \\ Y_{24} \end{bmatrix} \quad \text{for } i = 2 \text{ and}$$
$$n_1 = 5 \text{ (person 1)} \qquad\qquad n_2 = 4 \text{ (person 2)}$$

The above notation, nevertheless, does not distinguish the different times at which these two subjects are measured. To make such a distinction, we would have to lay out the data as shown in Table 25.1.

In Table 25.1, note that there is a separate line of data for each response on each subject. Also, the columns of this table describe the different variables that may be used in the analysis. The last column shown gives the times at which each subject is observed, and indicates that different subjects can be observed at different times.

---

[1] The term **response vector** used to describe $\mathbf{Y}_i$ derives from "matrix" terminology (see Appendix B on Matrices and Their Relationship to Regression Analysis). The analysis of repeated measures data can be conveniently described using matrices; in fact, most published literature on this subject uses matrix notation almost exclusively. Although this chapter will not emphasize a matrix-based discussion of repeated measures analysis, we will provide alternative matrix formulae for the interested reader.

TABLE 25.1   Data layout for longitudinal data on two subjects

Subject ID	Y	Repeat Number	Time
1	Y_{11}	1	T_{11}
1	Y_{12}	2	T_{12}
1	Y_{13}	3	T_{13}
1	Y_{14}	4	T_{14}
1	Y_{15}	5	T_{15}
2	Y_{21}	1	T_{21}
2	Y_{22}	2	T_{22}
2	Y_{23}	3	T_{23}
2	Y_{24}	4	T_{24}

Table 25.1 provides an example of the general data layout for repeated measures data that we present in Table 25.2 below. Table 25.2 is organized so that observations are grouped by subjects/clusters. The first column denotes the subject ID. The second column indicates which repeated measure is observed. The third column indicates the time at which the repeated measure is observed; this column would be omitted if the study is not longitudinal (i.e., the data are not gathered over time). The fourth column describes the response variable. The remaining columns (X's) describe any variables to be considered as predictors (i.e., independent variables) of interest, i.e., X_{ijg} denotes the jth value of the gth predictor for subject i, where $i = 1, 2, \ldots, K$; $j = 1, 2, \ldots, n_i$; and $g = 1, 2, \ldots, p$.

It is possible for clustered data not to be longitudinal. For example, suppose the responses in each cluster are measurements of blood pressure taken within a few minutes of each other on

TABLE 25.2   General data layout for repeated measures data

Subject (i)	Repeat(j)	Time (T_{ij})	Y_{ij}	X_{ij1}	X_{ij2}	...	X_{ijp}
1	1	T_{11}	Y_{11}	X_{111}	X_{112}	...	X_{11p}
1	2	T_{12}	Y_{12}	X_{121}	X_{122}	...	X_{12p}
$\vdots$	$\vdots$	$\vdots$	$\vdots$	$\vdots$	$\vdots$		$\vdots$
1	n_1	T_{1n_1}	Y_{1n_1}	X_{1n_11}	X_{1n_12}	...	X_{1n_1p}
$\vdots$	$\vdots$	$\vdots$	$\vdots$	$\vdots$	$\vdots$		$\vdots$
i	1	T_{i1}	Y_{i1}	X_{i11}	X_{i12}	...	X_{i1p}
i	2	T_{i2}	Y_{i2}	X_{i21}	X_{i22}	...	X_{i2p}
$\vdots$	$\vdots$	$\vdots$	$\vdots$	$\vdots$	$\vdots$		$\vdots$
i	n_i	T_{in_i}	Y_{in_i}	X_{in_i1}	X_{in_i2}	...	X_{in_ip}
$\vdots$	$\vdots$	$\vdots$	$\vdots$	$\vdots$	$\vdots$		$\vdots$
K	1	T_{K1}	Y_{K1}	X_{K11}	X_{K12}	...	X_{K1p}
K	2	T_{K2}	Y_{K2}	X_{K21}	X_{K22}	...	X_{K2p}
$\vdots$	$\vdots$	$\vdots$	$\vdots$	$\vdots$	$\vdots$		$\vdots$
K	n_K	T_{Kn_K}	Y_{Kn_K}	X_{Kn_K1}	X_{Kn_K2}	...	X_{Kn_Kp}

members of the same family. Then, for all practical purposes, the data are gathered at one point in time (i.e., cross-sectionally), and so are not longitudinal in nature. Nevertheless, responses on persons in the same family are unlikely to be independent, since family members typically share both lifestyle and genetic factors. If the data are not gathered longitudinally, then the data layout (Table 25.2) will not contain a column for Time; moreover, the Repeat column in the table will not necessarily contain observations listed by the time order at which they were observed.

Prior to the development of high-speed computer programs that allow iterative solution of a complex system of equations required for maximum likelihood estimation, the classical approach to the analysis of repeated measures (i.e., clustered) data required the partitioning of sums of squares according to various sources of variation analogous to the classical ANOVA techniques described in Chapters 17 through 20. This ANOVA approach is equivalent to using a regression (ANOVA) model that contains both fixed and random effects (defined in Chapter 17), where the factor *Subjects* is considered a random factor. Inferences about main effects and interactions are carried out using *F* tests based on appropriate ratios of mean square terms.

The ANOVA approach to repeated measures analysis has a number of limitations. First, the classical approach considers only continuous outcomes assumed to be normally distributed and requires that all independent variables in the model must be categorized, including interval variables. Such categorization generally forfeits information on interval predictor variables that are validly measured, so researchers may be disinclined to categorize such interval variables. Also, the assumption of normality for the outcome variable may not be appropriate, particularly when the response of interest is a binary (0 or 1) variable or a count variable. Furthermore, the ANOVA approach works most efficiently when the study design is *balanced* in the sense that each subject is observed the same number of times and at the same times; however, repeated measures designs often involve unbalanced data.

In light of the above limitations, other approaches to the analysis of repeated measures data have been developed, including the use of a more general form of the multiple linear regression model with both fixed and random effects, called the *general linear mixed model*. The general linear mixed model allows for independent variables to be of any type (not just categorical) and uses the method of maximum likelihood (described in Chapter 21) to estimate parameters and to make statistical inferences. The SAS MIXED procedure can carry out the computations required to fit such a model. In this chapter, we focus on the general linear mixed model approach used by MIXED, and we illustrate this approach to model correlated data when only fixed effects are considered. In the next chapter (26), which is Part 2 of this topic, we focus on linear mixed models that contain random effects. We also briefly describe the ANOVA "partitioning" approach and, using an example, compare the results from using MIXED with those obtained from using the SAS GLM procedure (which performs the ANOVA approach). Also, in Chapter 26, we give a brief overview of how such analyses can be carried out for discrete-type (e.g., binary or count) outcomes using *generalized linear mixed models*.

25.2 Examples

In this section, we introduce three examples of repeated measures data. The analysis of each dataset will be considered in later sections of this chapter.

25.2.1 A Study of the Effect of an Air Pollution Episode on Pulmonary Function

Let's first consider a hypothetical (but realistic) study of 40 school children who were examined under normal conditions, then during the week of an air pollution alert, and then on three successive weeks following the alert. The study objective was to determine whether FEV1 (often referred to as "lung capacity"), the volume of air exhaled in the first second of a forced exhalation, was depressed during the alert. A secondary objective was the identification of sensitive subgroups or individuals most severely affected by the pollution episode. Table 25.3 lists the data for the first 15 subjects in this study for the two primary variables of interest: FEV1 (the outcome variable) and Week, the primary predictor variable indicating the five weeks during which measurements were taken.[2] The first week's measurements were taken prior to the pollution alert, the second week's measurements were taken during the pollution alert, and the remaining measurements were taken during the next three weeks.

TABLE 25.3 Pulmonary function scores (FEV1) on the first 15 of 40 school children over five weeks with a pollution episode during week two

Subj	Week	FEV1	Subj	Week	FEV1	Subj	Week	FEV1
1	1	9.43	6	1	9.12	11	1	12.44
1	2	5.71	6	2	7.71	11	2	8.68
1	3	5.86	6	3	6.75	11	3	7.48
1	4	7.70	6	4	9.55	11	4	6.40
1	5	5.89	6	5	8.20	11	5	7.61
2	1	11.15	7	1	10.30	12	1	7.88
2	2	9.48	7	2	7.24	12	2	4.88
2	3	10.11	7	3	6.99	12	3	6.43
2	4	8.89	7	4	7.66	12	4	3.47
2	5	9.19	7	5	5.67	12	5	5.77
3	1	8.40	8	1	8.67	13	1	11.81
3	2	4.42	8	2	4.10	13	2	6.03
3	3	4.89	8	3	6.51	13	3	4.45
3	4	2.80	8	4	5.80	13	4	5.85
3	5	4.34	8	5	5.46	13	5	4.43
4	1	9.20	9	1	9.75	14	1	9.17
4	2	4.51	9	2	9.41	14	2	5.59
4	3	6.20	9	3	9.15	14	3	5.45
4	4	4.01	9	4	9.42	14	4	5.93
4	5	5.16	9	5	7.99	14	5	7.84
5	1	10.40	10	1	9.05	15	1	8.30
5	2	9.17	10	2	6.45	15	2	8.87
5	3	9.22	10	3	5.33	15	3	8.75
5	4	7.48	10	4	5.94	15	4	8.68
5	5	8.66	10	5	6.46	15	5	6.46

---

[2] A file containing the complete dataset can be obtained from the publisher's website using the URL http://www.thomsonedu.com/statistics/kleinbaum. The data in Table 25.3, and in the complete dataset, are hypothetical, although real study data from an analogous study were described and analyzed by Laird and Ware (1982).

This study is an example of a ***repeated measures (longitudinal) study***, in which each child (i.e., subject) has an FEV1 measurement taken at each of five consecutive weeks. The study design is ***balanced*** in that each subject has the *same number of repeated measurements* (i.e., 5) that are *measured at the same times*.

As mentioned above, the primary study objective was to determine whether FEV1 levels were significantly lowered during and possibly after the week of the pollution alert, i.e., whether the FEV1 levels measured during the first week (prior to the pollution alert) were lowered during the week of the pollution episode and possibly in the three weeks following the episode. A direct way to address this analysis objective would be to compare the mean FEV1 scores for all 40 children for each of the five weeks (see Table 25.4).

TABLE 25.4 **Mean FEV1 scores on 40 school children over five weeks with a pollution episode during week two**

Week j	K	Mean FEV1
1	40	9.81375
2	40	6.84825
3	40	7.00175
4	40	6.95150
5	40	6.99850
Average	40	7.52275

From Table 25.4, we can see that the mean FEV1 value of 9.81 for Week 1, prior to the pollution alert, is higher than each of the mean FEV1 values for Weeks 2 through 5. These summary data suggest that FEV1 levels were depressed (i.e., lowered) during and after the pollution alert and that, moreover, average FEV1 levels didn't seem to change during the three weeks subsequent to the week of the pollution alert. Statistically, we might hope to show that:

a. There is a statistically significant difference among the average FEV1 scores over all five weeks.

b. There is a statistically significant difference between average FEV1 at week 1 and the average of the average FEV1 values over Weeks 2 through 5, i.e., the "contrast"

$$\overline{FEV1}_1 - \frac{\overline{FEV1}_2 + \overline{FEV1}_3 + \overline{FEV1}_4 + \overline{FEV1}_5}{4}$$

is statistically different from zero.

Using the data from Table 25.4, we obtain the following calculation:

Contrast (Week 1 vs. Weeks 2–5)

$$= 9.81375 - \frac{6.84825 + 7.00175 + 6.95150 + 6.99850}{4} = 2.8637$$

As we now discuss, evaluating statistical significance is a little more complicated than we might think.

The comparison being considered in item (a) above essentially is a One-Way ANOVA-type comparison: the single factor (i.e., nominal independent variable) of interest is "Week," the outcome Y is FEV1, and the null hypothesis of interest can be stated as

$$H_0: \mu_1 = \mu_2 = \mu_3 = \mu_4 = \mu_5$$

where μ_j denotes the population mean FEV1 score for Week j that is being estimated by $\overline{\text{FEV1}}_j$. The classical One-Way ANOVA multiple linear regression model for this situation (see Chapter 17) can be written as follows:

$$Y = \beta_0 + \beta_1 W_1 + \beta_2 W_2 + \beta_3 W_3 + \beta_4 W_4 + E \qquad (25.2)$$

where Y denotes FEV1, W_1 through W_4 represent four dummy variables for distinguishing among the five weeks, and E denotes the error term in the model.[3]

We can alternatively write the model in classical One-Way ANOVA format as follows:

$$Y_{ij} = \mu + \tau_j + E_{ij}, \qquad i = 1,\ldots,40 \text{ and } j = 1,\ldots,5 \qquad (25.3)$$

where Y_{ij} denotes the FEV1 measurement at the jth week for subject i, μ denotes the overall mean FEV1 over all five weeks, τ_j denotes the effect of the jth week, E_{ij} denotes the error at the jth week for subject i, and $\sum_{j=1}^{5} \tau_j = 0$. This ANOVA model assumes that the variable Week is a fixed factor.

Nevertheless, there is a problem with using a classical multiple linear regression analysis program (e.g., PROC REG in SAS), or an ANOVA program, to analyze these data: the classical regression methodology assumes that the responses for subjects in the study are mutually independent. Such an assumption is clearly questionable in this particular study because the five FEV1 measurements on each subject are likely to be correlated with one another. In other words, for the ANOVA model (25.3), we may reasonably assume that $\text{corr}(E_{ij}, E_{i'j}) = 0$ whenever $i \neq i'$, but we must allow $\text{corr}(E_{ij}, E_{ij'}) \neq 0$ whenever $j \neq j'$. Moreover, the five sample means (see Table 25.4) are likely correlated with each other because each of these means involves responses from the same 40 subjects in the study; consequently, any test about a contrast of such sample means must also account for such correlations. Thus, we need to use a data analysis approach that takes into account the correlations among the observations on the same subject.

The approach that we describe in this chapter generalizes the classical linear model approach by allowing for such correlations among observations on the same subject. Note, however, that the goal of both the uncorrelated response approach and the correlated response approach is essentially the same in this setting: to describe the effect of one or more predictors (i.e., independent variables) on an outcome (i.e., dependent variable) of interest.

So far, regarding our FEV1 example involving a pollution alert, we have only focused on the primary question of interest. Recall that a *secondary objective* concerned identifying sensitive subgroups or individuals most severely affected by the pollution episode. To address this secondary question, we must carry out an analysis that provides us with information about the effect of the pollution alert on each of the 40 schoolchildren individually, rather than about the effect of the pollution alert on the entire sample of schoolchildren. As we will see in Chapter 26, when we return to this question, such an individual-specific analysis requires us to consider a

---

[3] If SAS's MIXED procedure is used to analyze these data, one way to define the dummy variables is to use a CLASS statement with the variable Week; another way is to define the dummy variables directly. If a CLASS variable is used and Week is coded as $W_i = i$ for week i in the original dataset, then SAS will (by default) make the referent group be Week 5, the larger of the values for Week. By defining dummy variables directly, one might alternatively make the referent group Week 1, especially since Week 1 is the week prior to the pollution alert (i.e., the week prior to when lowered FEV1 levels might be expected).

regression model involving one or more random effects for subjects (i.e., school children) in addition to fixed effects for the variable Week. Without specifying such a model at this point, we can nevertheless provide some insight into the structure of such a subject-specific analysis. For example, let's compare the data on the first two of the 40 schoolchildren in Table 25.3:

Subj	Week	FEV1
1	1	9.43
1	2	5.71
1	3	5.86
1	4	7.70
1	5	5.89
2	1	11.15
2	2	9.48
2	3	10.11
2	4	8.89
2	5	9.19

The mean FEV1 value over the five weeks for subject 1 is 6.92, whereas the corresponding mean FEV1 value for subject 2 is 9.76. Thus, over all five weeks, subject 1 had an average that was 2.84 units lower than the average for subject 2. However, simply comparing 5-week averages does not directly address how the pollution alert comparably affected each subject because the pollution alert did not begin until week two. A more appropriate way to compare these two subjects, therefore, would be to consider the estimated linear contrast

$$L_i = \text{FEV1}_{i1} - \frac{(\text{FEV1}_{i2} + \text{FEV1}_{i3} + \text{FEV1}_{i4} + \text{FEV1}_{i5})}{4}$$

for each subject, where FEV1_{ij} is the FEV1 measurement for the ith subject ($i = 1, 2$) at week j ($j = 1, 2, 3, 4, 5$). It is easy to see that

$$L_1 = 9.43 - 6.29 = 3.14 \text{ for subject 1 and } L_2 = 11.15 - 9.42 = 1.73 \text{ for subject 2.}$$

Thus, when comparing week 1 with the average of weeks 2 through 5 separately for each subject, we see that the FEV1 value for subject 1 was depressed 1.41 units more than the FEV1 value for subject 2. In other words, subject 1 was somewhat more adversely affected by the pollution alert than was subject 2.

We have just illustrated a type of subject-specific comparison that could be done to answer the secondary question. In Section 25.4, we will illustrate the analysis of the primary question involving these data by fitting a linear mixed model containing only fixed effects using SAS's MIXED procedure. In Chapter 26, we will return to these data to consider the analysis of the secondary objective involving a mixed model containing random effects.

25.2.2 A Study of the Posture of Computer Operators

Over the past few decades, the use of computers for all kinds of professional and personal activities has grown exponentially. Accompanying this growth, public health specialists have become increasingly concerned about the possible health consequences of spending long hours

W_4 are defined so that the referent group is week 5, then for measurements taken during the third week,

$$W_{i33} = 1 \text{ and } W_{i31} = W_{i32} = W_{i34} = 0;$$

and, for measurements taken during the fifth (referent) week,

$$W_{i51} = W_{i52} = W_{i53} = W_{i54} = 0$$

The subject-specific matrix form of this model is

$$\mathbf{Y}_i = \mathbf{X}_i\boldsymbol{\beta} + \mathbf{E}_i, \quad i = 1, \dots, 40 \tag{25.8}$$

where $\mathbf{Y}_i$ denotes the collection of 5 FEV1 measurements on the ith child, $\mathbf{X}_i$ denotes the collection of dummy variable values for each of the five weeks for subject i, $\boldsymbol{\beta}$ denotes the parameter vector for the set of β's in this model, and $\mathbf{E}_i$ denotes the set of error terms for the ith subject. Assuming again that the referent group is week 5, then $\mathbf{X}_i$, $\boldsymbol{\beta}$, and $\mathbf{E}_i$ can be written as follows:

$$\mathbf{X}_i = \begin{bmatrix} 1 & W_{i11} & W_{i12} & W_{i13} & W_{i14} \\ 1 & W_{i21} & W_{i22} & W_{i23} & W_{i24} \\ 1 & W_{i31} & W_{i32} & W_{i33} & W_{i34} \\ 1 & W_{i41} & W_{i42} & W_{i43} & W_{i44} \\ 1 & W_{i51} & W_{i52} & W_{i53} & W_{i54} \end{bmatrix} = \begin{bmatrix} 1 & 1 & 0 & 0 & 0 \\ 1 & 0 & 1 & 0 & 0 \\ 1 & 0 & 0 & 1 & 0 \\ 1 & 0 & 0 & 0 & 1 \\ 1 & 0 & 0 & 0 & 0 \end{bmatrix}, \quad \boldsymbol{\beta} = \begin{bmatrix} \beta_0 \\ \beta_1 \\ \beta_2 \\ \beta_3 \\ \beta_4 \end{bmatrix}, \quad \mathbf{E}_i = \begin{bmatrix} E_{i1} \\ E_{i2} \\ E_{i3} \\ E_{i4} \\ E_{i5} \end{bmatrix}$$

$$\tag{25.9}$$

Note that the $\mathbf{X}_i$ matrix contains 5 rows (for the 5 weeks) and 5 columns (for the intercept variable, always taking the value 1, and the 4 dummy variables).

If the referent group is week 5, it also follows (as for any One-Way ANOVA model) that the estimated coefficients (i.e., the $\hat{\beta}$'s) in all three models (25.2, 25.7, 25.8) are determined as follows:

$$\hat{\beta}_0 = \overline{\text{FEV1}}_5$$
$$\hat{\beta}_1 = \overline{\text{FEV1}}_1 - \overline{\text{FEV1}}_5, \hat{\beta}_2 = \overline{\text{FEV1}}_2 - \overline{\text{FEV1}}_5, \hat{\beta}_3 = \overline{\text{FEV1}}_3 - \overline{\text{FEV1}}_5, \tag{25.10}$$
$$\hat{\beta}_4 = \overline{\text{FEV1}}_4 - \overline{\text{FEV1}}_5$$

In other words, if the referent group is week 5, then the estimated intercept $\hat{\beta}_0$ will be the mean FEV1 value at week 5, and the estimated regression coefficient $\hat{\beta}_g$, $g = 1, 2, 3, 4$, will be the difference between the mean FEV1 value at week g and the mean FEV1 value at week 5. These estimated coefficients are identical to the estimates that would result if we assumed that the observations on the same subject were mutually independent (i.e., the classical One-Way ANOVA assumption). Nevertheless, the variances of the estimates in (25.10) will differ depending on the manner and extent to which the observations on the same subject are correlated; consequently, statistical inference procedures about the regression coefficients will vary with the "correlation structure" of the observations on the same subject.

For the FEV1 data, there was a secondary study objective: to identify the extent to which the pollution episode severely affected some study subjects more than other subjects. This secondary objective requires a regression model different from the fixed effect model given by (25.2), or equivalently (25.7) and (25.8). This new (subject-specific) model now requires one or more random effects for subjects in addition to the fixed effects. The subject-specific scalar form of an appropriate linear mixed model for this situation is:

$$Y_{ij} = \beta_0 + \beta_1 W_{ij1} + \beta_2 W_{ij2} + \beta_3 W_{ij3} + \beta_4 W_{ij4} + b_{i0} + b_{i1} Z_{ij1} + E_{ij} \tag{25.11}$$

which contains two random effects b_{i0} and b_{i1} and the random coefficient factor Z_{ij1} in addition to the fixed factor dummy variables W_{ij1} through W_{ij4} and their corresponding fixed effects (i.e., β_1 through β_4). The Z_{ij1} variable is defined as follows:

$$Z_{ij1} = 0 \text{ if week } 1 \ (j = 1) \text{ and } Z_{ij1} = 1 \text{ if weeks } 2, 3, 4 \text{ or } 5 \ (j \neq 1)$$

(We can equivalently restate b_{i0} in model (25.11) as $b_{i0} Z_{ij0}$, where Z_{ij0} is identically equal to 1 for each subject at each week, i.e., model (25.11) effectively considers two random coefficient factors Z_{ij0} and Z_{ij1}).

In Chapter 26, where we consider random effects for these data, we will describe and illustrate with computer results why this model is appropriate. Here, we show the subject-specific matrix form of this model:

$$\mathbf{Y}_i = \mathbf{X}_i \boldsymbol{\beta} + \mathbf{Z}_i \mathbf{b}_i + \mathbf{E}_i, \qquad i = 1, 2, \ldots, 40 \tag{25.12}$$

where $\mathbf{Y}_i$, $\mathbf{X}_i$, $\boldsymbol{\beta}$, and $\mathbf{E}_i$ are the same as in model (25.8) and matrix expression (25.9), and where $\mathbf{b}_i$ and $\mathbf{Z}_i$ are defined as follows:

$$\mathbf{b}_i = \begin{bmatrix} b_{i0} \\ b_{i1} \end{bmatrix}, \qquad \mathbf{Z}_i = \begin{bmatrix} 1 & 0 \\ 1 & 1 \\ 1 & 1 \\ 1 & 1 \\ 1 & 1 \end{bmatrix} \tag{25.13}$$

The first column (i.e., the column of 1's) of the $\mathbf{Z}_i$ matrix denotes the values of the variable Z_{i0} corresponding to the random effect b_{i0} for each of the 5 observations on subject i, and the second column denotes the values of the variable Z_{i1} corresponding to the random effect b_{i1} for the 5 observations on subject i.

In Chapter 26, we show that use of the above random effects model (25.11) will lead to identical estimated regression coefficients given by (25.10) as when using the fixed effects model (25.7). Nevertheless, the estimated variances of these estimated coefficients based on (25.11) will differ from corresponding estimated variances for the fixed effects model (25.7) even when correlation structures are appropriately taken into account. As we shall discuss in the next section, corresponding regression coefficient variances in the two models are different because the two models have different correlation structures.

25.3.2 Correlation Structures

The correlation structure involving a set of observations

$$\mathbf{Y}_i = \begin{bmatrix} Y_{i1} \\ Y_{i2} \\ \vdots \\ Y_{in_i} \end{bmatrix}$$

on the same subject (i) is defined by the set of all pairwise correlations of the form

$$\rho_{ijj'} = \text{Corr}(Y_{ij}, Y_{ij'})$$

where j and j' denote two possibly different observations of Y for the ith subject. Recall that, in general, $|\rho_{ijj'}| \leq 1$, for any i, j and j', and that $\rho_{ijj'} \equiv 1$ whenever $j = j'$.

For example, in the FEV1 study, the correlation structure for the ith child can be generally described by the following (5 rows $\times$ 5 columns) correlation matrix:

$$\mathbf{C}_i = \begin{bmatrix} 1 & \rho_{i12} & \rho_{i13} & \rho_{i14} & \rho_{i15} \\ \rho_{i12} & 1 & \rho_{i23} & \rho_{i24} & \rho_{i25} \\ \rho_{i13} & \rho_{i23} & 1 & \rho_{i34} & \rho_{i35} \\ \rho_{i14} & \rho_{i24} & \rho_{i34} & 1 & \rho_{i45} \\ \rho_{i15} & \rho_{i25} & \rho_{i35} & \rho_{i45} & 1 \end{bmatrix} \qquad (25.14)$$

Any such correlation matrix is always "symmetric" since $\rho_{ijj'} \equiv \rho_{ij'j}$; in other words, the set of correlations above the diagonal of 1's are the identical mirror image of the set of correlations below this diagonal.

The simplest form that a subject-specific correlation structure ($\mathbf{C}_i$) can take is the *independent correlation structure* (or matrix), i.e.,

$$\mathbf{C}_i^{(\mathbf{IND})} = \begin{bmatrix} 1 & 0 & 0 & 0 & 0 \\ 0 & 1 & 0 & 0 & 0 \\ 0 & 0 & 1 & 0 & 0 \\ 0 & 0 & 0 & 1 & 0 \\ 0 & 0 & 0 & 0 & 1 \end{bmatrix} \qquad (25.15)$$

This is the structure that is assumed in classical ANOVA, when it is assumed that all observations on all subjects are mutually independent. The above matrix, with ones on the diagonal and zeros elsewhere, is typically referred to as an *identity matrix* and is denoted as $\mathbf{I}$. In particular, the above (5 $\times$ 5) identity matrix would be written as $\mathbf{I}_5$; and, in general, an ($n \times n$) identity matrix would be written as $\mathbf{I}_n$.

When one wants to consider non-zero correlations among observations on the same subject, there is a large variety of choices for the correlation structure. For example, *an exchangeable correlation structure* assumes that all the correlations between pairs of observations on a given subject are identical, i.e.,

$$\mathbf{C}_i^{(\mathbf{EXCH})} = \begin{bmatrix} 1 & \rho & \rho & \rho & \rho \\ \rho & 1 & \rho & \rho & \rho \\ \rho & \rho & 1 & \rho & \rho \\ \rho & \rho & \rho & 1 & \rho \\ \rho & \rho & \rho & \rho & 1 \end{bmatrix} \qquad (25.16)$$

Notice that the above exchangeable correlation structure (or matrix) also assumes that the correlation between any pair of observations on any subject *does not vary from subject to subject*. This assumption, stated mathematically, says that

$$\rho_{ijj'} = \rho_{jj'} \equiv \rho \quad \text{for all } i \qquad (25.17)$$

An alternative version of an exchangeable correlation structure would allow the pairwise correlations to be identical for a given subject, but different for different subjects, i.e.,

$$\rho_{ijj'} \equiv \rho_i \qquad (25.18)$$

Another correlation structure called *autoregressive* 1 (*AR*1) assumes that observations "closer to each other" (e.g., in time or space) are more highly correlated than observations that are "farther apart." The pairwise correlations decrease according to the following formula:

$$\rho_{jj'} = \rho^{|T_{ij} - T_{ij'}|} \qquad \text{when } j \neq j' \tag{25.19}$$

where T_{ij} and $T_{ij'}$ denote the times at which the jth and j'th observations, respectively, are measured on the ith subject. If subjects are measured at equally-spaced time intervals (i.e., $|T_{ij} - T_{ij'}| = c|j - j'|$ for some positive constant c, the corresponding (5 × 5) AR1 correlation matrix is given by:

$$\mathbf{C}_i^{(\mathrm{AR1})} = \begin{bmatrix} 1 & \rho & \rho^2 & \rho^3 & \rho^4 \\ \rho & 1 & \rho & \rho^2 & \rho^3 \\ \rho^2 & \rho & 1 & \rho & \rho^2 \\ \rho^3 & \rho^2 & \rho & 1 & \rho \\ \rho^4 & \rho^3 & \rho^2 & \rho & 1 \end{bmatrix} \tag{25.20}$$

For this matrix, the correlation ρ between the 1st and 2nd observations is the same as the correlations between the 2nd and 3rd, 3rd and 4th, and 4th and 5th observations; this is because the two responses in each of these pairs are exactly one equally-spaced time occasion apart. In contrast, the correlation ρ^2 between the 1st and 3rd observations, which are two equally-spaced time occasions apart, is the same as the correlations between the 2nd and 4th, and 3rd and 5th, observations. Also, notice that, as with our example of an exchangeable correlation structure, we have assumed that the above AR1 correlation structure is the same for all subjects.

There are many other correlation structures that one might consider. Another structure often considered, called the *unstructured correlation structure*, takes the following form, again assuming that such a structure does not vary by subject:

$$\mathbf{C}_i^{(\mathrm{UN})} = \begin{bmatrix} 1 & \rho_{12} & \rho_{13} & \rho_{14} & \rho_{15} \\ \rho_{12} & 1 & \rho_{23} & \rho_{24} & \rho_{25} \\ \rho_{13} & \rho_{23} & 1 & \rho_{34} & \rho_{35} \\ \rho_{14} & \rho_{24} & \rho_{34} & 1 & \rho_{45} \\ \rho_{15} & \rho_{25} & \rho_{35} & \rho_{45} & 1 \end{bmatrix} \tag{25.21}$$

An unstructured correlation matrix has the property that all possible correlations are allowed to be different. Such a correlation matrix is often considered as a starting point for assessing whether a simpler correlation structure (e.g., exchangeable) might suffice, or when it is unclear what simpler structure might be specified.

The *assumption* in the above examples that the *underlying correlation structure is the same for all subjects* (i.e., $\rho_{ijj'} = \rho_{jj'}$) is typically made to avoid considering a model that has more parameters than can be estimated validly and precisely with the available data (i.e., the model is *over-parameterized* and there is not enough information to yield reliable parameter estimates). In contrast, assuming that each subject has the same set of correlation parameters reduces the number of correlation parameters substantially for a study containing K subjects. In our FEV1 pollution study, for example, with $K = 40$, if we assume the same AR1 correlation structure for each subject, then the number of possible correlation parameters to be estimated is reduced from 40 to 1.

We have so far considered a correlation matrix $\mathbf{C}_i$ at the subject-specific level. Alternatively, we might consider the larger correlation matrix for *all* study subjects. If, for example, there are K subjects, each of whom has n Y observations, then the correlation matrix, say $\mathbf{C}$, for all K subjects will have Kn rows and Kn columns (e.g., for $K = 40$ and $n = 5$, the dimensions of $\mathbf{C}$ will be 200×200). When considering the overall correlation matrix $\mathbf{C}$, an important characteristic of repeated measures/clustered data is the assumption that *different responses within the same cluster* (e.g., Y_{ij} and $Y_{ij'}$, $j \neq j'$) *are correlated, whereas responses from different clusters* (e.g., Y_{ij} and $Y_{i'j'}$, $i \neq i'$) *are uncorrelated.* Thus, in the FEV1 study involving $n = 5$ responses on each of $K = 40$ subjects, there are $Kn = 200$ response observations in total, and so the corresponding correlation matrix for all study subjects will be of the following form:

$$\mathbf{C}(200 \times 200) = \begin{bmatrix} \mathbf{C}_1 & \mathbf{0} & \mathbf{0} & \cdots & \mathbf{0} \\ \mathbf{0} & \mathbf{C}_2 & \mathbf{0} & \cdots & \mathbf{0} \\ \mathbf{0} & \mathbf{0} & \mathbf{C}_3 & \cdots & \mathbf{0} \\ \vdots & \vdots & \vdots & \vdots & \vdots \\ \mathbf{0} & \mathbf{0} & \mathbf{0} & \cdots & \mathbf{C}_{40} \end{bmatrix} \tag{25.22}$$

Each of the 40 components (written in bold) within $\mathbf{C}$ is a (5×5) matrix. The 40 $\mathbf{C}_i$ matrices within $\mathbf{C}$ are the correlation matrices for all $K = 40$ subjects. The $\mathbf{0}$ matrices, each of which is (5×5) and has all entries equal to zero, reflect the assumption that the observations for any one subject are independent of the observations for any other subject. This large $\mathbf{C}$ matrix is called a *block diagonal matrix*, since the $\mathbf{C}_i$'s on the diagonal of the matrix are the only non-zero matrices.

If we assume that all subjects have the same set of correlation parameters and that $n_i \equiv n$ for all i, the above block diagonal matrix will simplify to the following block diagonal matrix:

$$\mathbf{C}(200 \times 200) = \begin{bmatrix} \mathbf{C} & \mathbf{0} & \mathbf{0} & \cdots & \mathbf{0} \\ \mathbf{0} & \mathbf{C} & \mathbf{0} & \cdots & \mathbf{0} \\ \mathbf{0} & \mathbf{0} & \mathbf{C} & \cdots & \mathbf{0} \\ \vdots & \vdots & \vdots & \vdots & \vdots \\ \mathbf{0} & \mathbf{0} & \mathbf{0} & \cdots & \mathbf{C} \end{bmatrix} \tag{25.23}$$

A crucial question about the above discussion of correlation structures is **"How do we choose the correlation structure appropriate for the dataset we are analyzing?"**

The answer to this question is not straightforward, but rather requires consideration of several approaches from which the investigator(s) must make a reasoned decision. Typically, but not always, the investigator makes an "educated guess" about the appropriate correlation structure (e.g., exchangeable), fits a model using the guessed or postulated structure, tries other "guesses" that might also be reasonable, and compares numerical results based on the use of different structures. A postulated correlation structure is typically referred to as a *working correlation structure*. Choices among different correlation structures might be based on which correlation structures are most biologically/clinically reasonable, on which models have better "goodness of fit" to the data, or on the extent to which the modeling results vary for different structures. In particular, if the conclusions about the study questions turn out to be essentially the same regardless of which correlation structure (among reasonable guesses) is chosen, then any of these correlation structures could suffice. (Note that conclusions can still be invalid if the wrong regression model is used.) In general, it is more important to make reasonably valid statistical conclusions about the study questions than to determine exactly the perfectly correct correlation structure.

One other frequently used method related to the choice of the correlation structure involves accounting for the possibility that a chosen working correlation structure is actually not the correct correlation structure. The method used is generally referred to as *empirical (or robust) variance estimation*. A so-called "robust/empirical standard error estimator" is a modified standard error estimator for an estimated regression coefficient that combines an estimator of the chosen working correlation structure with an estimator of the correct (but unknown) correlation structure to produce an asymptotically unbiased estimator of the standard error.

When testing for the significance of a specific regression coefficient, say $\hat{\beta}$, the test statistic used is of the form

$$T = \frac{\hat{\beta}}{S_{\hat{\beta},\text{emp}}}$$

where $S_{\hat{\beta},\text{emp}}$ denotes the empirical standard error estimate for the estimated regression coefficient $\hat{\beta}$. If the assumed linear mixed model is correctly specified, then T will have an approximate t-distribution[7] (and, for large samples, an approximate Z-distribution) under the null hypothesis H_0: $\beta = 0$. Because this "empirical standard error estimate" involves in its computation the choice of the working correlation structure, the numerical value of this estimate can be somewhat different for different choices of the working correlation structure, especially for small and intermediate size samples. In other words, it is possible that the use of, say, an exchangeable working correlation structure may yield a significant test result, whereas the use of, say, an AR1 working correlation structure may yield a non-significant test result, or vice-versa. So, even though the use of an empirical standard error estimate helps to correct for possible misspecification of the correlation structure, the choice of the working correlation structure can still be an important concern when sample sizes are not large.

25.3.3 Covariance Structures and Their Relationship to Correlation Structures

As we have previously described (in Chapter 6), the correlation parameter is a measure of the linear association between two random variables, X and Y, that generally ranges between -1 and 1. It is a dimensionless, scale-invariant measure. In a repeated measures study, these two random variables can be two different responses, Y_{ij} and $Y_{ij'}$, for the same outcome variable Y measured on the same subject i. In this chapter, we have used the notation $\rho_{ijj'} = \text{corr}(Y_{ij}, Y_{ij'})$ for such a correlation.

An alternative measure of association between two variables that has a specific mathematical relationship with the correlation is the **covariance**. The covariance between X and Y is defined as the expected value, or population average, of the product of X minus its mean (μ_X) and Y minus its mean (μ_Y), i.e.,

$$\nu_{X,Y} = \text{Cov}(X, Y) = E[(X - \mu_X)(Y - \mu_Y)]$$

Note that $\nu_{X,Y}$ has dimensions and is not scale-invariant. Also, the covariance of X with itself is called the variance of X, i.e.,

$$\nu_{X,X} = \text{Cov}(X, X) \equiv \text{Var}(X)$$

---

[7] SAS's MIXED procedure provides several options for determining the denominator degrees of freedom (DDFM) associated with t and F statistics for the predictors in one's model. See Chapter 26, Section 26.4.7, for further details.

Analogously, if we are considering repeated measures data

$$
\mathbf{Y}_i = \begin{bmatrix} Y_{i1} \\ Y_{i2} \\ \vdots \\ Y_{in_i} \end{bmatrix} \quad \text{for subject } i,
$$

we can describe the covariance between any two observations for subject i, and the corresponding variance of any one of these observations, as

$$
\nu_{ijj'} = \text{Cov}(Y_{ij}, Y_{ij'}) = E[(Y_{ij} - \mu_{Yij})(Y_{ij'} - \mu_{Y_{ij'}})]
$$

and

$$
\sigma_{ij}^2 = \nu_{ijj} = \text{Cov}(Y_{ij}, Y_{ij}) = \text{Var}(Y_{ij})
$$

The mathematical relationship between the correlation and the covariance can be stated as follows:

$$
\text{Corr}(X, Y) = \frac{\text{Cov}(X, Y)}{\sqrt{\text{Var}(X)\,\text{Var}(Y)}}, \quad \text{i.e., } \rho_{XY} = \frac{\nu_{XY}}{\sigma_X \sigma_Y} \tag{25.24}
$$

Analogously, for repeated measures data, we can write this relationship as

$$
\text{Corr}(Y_{ij}, Y_{ij'}) = \frac{\text{Cov}(Y_{ij}, Y_{ij'})}{\sqrt{\text{Var}(Y_{ij})\,\text{Var}(Y_{ij'})}}, \quad \text{i.e., } \rho_{ijj'} = \frac{\nu_{ijj'}}{\sigma_{ij} \sigma_{ij'}} \tag{25.25}
$$

We can alternatively, using simple algebra, write the covariance in terms of the correlation as follows:

$$
\text{Cov}(Y_{ij}, Y_{ij'}) = \text{Corr}(Y_{ij}, Y_{ij'}) \sqrt{\text{Var}(Y_{ij})\,\text{Var}(Y_{ij'})} \tag{25.26}
$$

The above scalar relationships between covariance and correlation can also be described in matrix terms. In particular, corresponding to a given covariance matrix (also called a variance-covariance matrix) $\mathbf{V}$, there is a correlation matrix $\mathbf{C}$.

The simplest form of *covariance* matrix corresponds to an independent *correlation* matrix $\mathbf{C}_i^{(\text{IND})}$, as illustrated by (25.15) for the identity matrix $\mathbf{I}_5$. The corresponding independent covariance matrix $\mathbf{V}_i^{(\text{IND})}$ will have all response variable covariances (located off the diagonal) equal to zero, with the (response) variances on the diagonal. If all the variances are assumed to be equal to σ^2, say, for all ($n_i = 5$) responses, then we can write $\mathbf{V}_i^{(\text{IND})} = \sigma^2 \mathbf{I}_5$. If, however, the variances for different responses are not assumed to be equal, i.e., $\text{Var}(Y_{ij}) = \sigma_j^2$ for $j = 1, \ldots, n_i$, then the covariance matrix corresponding to a (5×5) independent correlation structure ($n_i = 5$) will take the following form:

$$
\mathbf{V}_i^{(\text{IND})} = \begin{bmatrix} \sigma_1^2 & 0 & 0 & 0 & 0 \\ 0 & \sigma_2^2 & 0 & 0 & 0 \\ 0 & 0 & \sigma_3^2 & 0 & 0 \\ 0 & 0 & 0 & \sigma_4^2 & 0 \\ 0 & 0 & 0 & 0 & \sigma_5^2 \end{bmatrix} \tag{25.27}
$$

As another example, for the FEV1 study, an exchangeable correlation matrix for the ith subject (assuming $\rho_{ijj'} = \rho_{jj'} \equiv \rho$ for all i) would be given by the following (5×5) correlation matrix:

$$\mathbf{C}_i^{(\text{EXCH})} = \begin{bmatrix} 1 & \rho & \rho & \rho & \rho \\ \rho & 1 & \rho & \rho & \rho \\ \rho & \rho & 1 & \rho & \rho \\ \rho & \rho & \rho & 1 & \rho \\ \rho & \rho & \rho & \rho & 1 \end{bmatrix} \qquad (25.28)$$

A covariance matrix that corresponds to the above correlation matrix is given by

$$\mathbf{V}_i^{(\text{Het EXCH})} = \begin{bmatrix} \sigma_1^2 & v_{12} & v_{13} & v_{14} & v_{15} \\ v_{12} & \sigma_2^2 & v_{23} & v_{24} & v_{25} \\ v_{13} & v_{23} & \sigma_3^2 & v_{34} & v_{35} \\ v_{14} & v_{24} & v_{34} & \sigma_4^2 & v_{45} \\ v_{15} & v_{25} & v_{35} & v_{45} & \sigma_5^2 \end{bmatrix} \qquad (25.29)$$

where $v_{ijj'} = v_{jj'} = \rho\sigma_j\sigma_{j'}$ for $j \neq j'$. The above covariance matrix in (25.29) allows for different covariances for all $j \neq j'$, even though the corresponding exchangeable correlation matrix $\mathbf{C}_i^{(\text{EXCH})}$ in (25.28) assumes that all pairwise correlations are identical. Unequal covariances are obtained here because there are different variances σ_j^2 for different observations on subject i. We call such a covariance matrix a *heterogeneous exchangeable covariance matrix* (hence the notation **Het EXCH**).

If all the variances are assumed to be the same, i.e., $\sigma_j^2 \equiv \sigma^2$, the covariance structure would simplify as follows:

$$\mathbf{V}_i^{(\text{Hom EXCH})} = \begin{bmatrix} \sigma^2 & v & v & v & v \\ v & \sigma^2 & v & v & v \\ v & v & \sigma^2 & v & v \\ v & v & v & \sigma^2 & v \\ v & v & v & v & \sigma^2 \end{bmatrix} = \sigma^2 \begin{bmatrix} 1 & \rho & \rho & \rho & \rho \\ \rho & 1 & \rho & \rho & \rho \\ \rho & \rho & 1 & \rho & \rho \\ \rho & \rho & \rho & 1 & \rho \\ \rho & \rho & \rho & \rho & 1 \end{bmatrix} = \sigma^2 \mathbf{C}_i^{(\text{EXCH})} \qquad (25.30)$$

where $v_{jj'} \equiv \rho\sigma^2 \equiv v$, i.e., the covariance is the same between any two observations j and j' for subject i. The above matrix is a *homogeneous exchangeable covariance matrix*. (A popular special case of this matrix is a *compound symmetric (CS) matrix*, which requires ρ to be non-negative.) Notice that the two covariance structures $\mathbf{V}_i^{(\text{Het EXCH})}$ and $\mathbf{V}_i^{(\text{Hom EXCH})}$ yield the same exchangeable correlation structure $\mathbf{C}_i^{(\text{EXCH})}$.

The scalar relationship between covariance and correlation given in equation (25.26) can be expressed generally in matrix form as follows:

$$\mathbf{V}_i = \mathbf{D}_i^{1/2}\mathbf{C}_i\mathbf{D}_i^{1/2} \qquad (25.31)$$

where $\mathbf{V}_i$ and $\mathbf{C}_i$ denote the covariance matrix and its corresponding correlation matrix, respectively, and $\mathbf{D}_i^{1/2}$ denotes a diagonal matrix whose diagonal elements are the square roots of corresponding variances obtained from the $\mathbf{V}_i$ matrix. For example, for the (5×5) heterogeneous covariance matrix (25.29), the corresponding (5×5) diagonal matrix $\mathbf{D}_i^{1/2}$ is given by:

$$\mathbf{D}_i^{1/2} = \begin{bmatrix} \sigma_1 & 0 & 0 & 0 & 0 \\ 0 & \sigma_2 & 0 & 0 & 0 \\ 0 & 0 & \sigma_3 & 0 & 0 \\ 0 & 0 & 0 & \sigma_4 & 0 \\ 0 & 0 & 0 & 0 & \sigma_5 \end{bmatrix} = \mathrm{Diag}(\sigma_1, \sigma_2, \sigma_3, \sigma_4, \sigma_5) \qquad (25.32)$$

For the (5×5) exchangeable covariance matrix given by (25.30), in which all the variances are assumed equal (to σ^2), the corresponding (5×5) diagonal matrix $\mathbf{D}_i^{1/2}$ is given by:

$$\mathbf{D}_i^{1/2} = \begin{bmatrix} \sigma & 0 & 0 & 0 & 0 \\ 0 & \sigma & 0 & 0 & 0 \\ 0 & 0 & \sigma & 0 & 0 \\ 0 & 0 & 0 & \sigma & 0 \\ 0 & 0 & 0 & 0 & \sigma \end{bmatrix} = \sigma \mathbf{I}_5$$

Thus, we have seen that the pairwise associations among repeated continuous responses on the same subject can alternatively be described by correlation or covariance structures, which can be described easily in matrix form. Further, the mathematical relationship between correlation and covariance matrices can be expressed succinctly using (25.31).

25.3.4 Covariance Structure for the General Linear Mixed Model

We have previously defined the general linear mixed model in both scalar and matrix forms, as follows:

Regression Scalar Form:

$$Y = (\beta_0 + \beta_1 X_1 + \beta_2 X_2 + \cdots + \beta_p X_p) + (b_0 + b_1 Z_1 + b_2 Z_2 + \cdots + b_q Z_q) + E$$

or, equivalently, using summation notation:

$$\boxed{Y = \beta_0 + \sum_{g=1}^{p} \beta_g X_g + b_0 + \sum_{h=1}^{q} b_h Z_h + E} \qquad (25.33)$$

Subject-specific Scalar Form:

$$Y_{ij} = (\beta_0 + \beta_1 X_{ij1} + \beta_2 X_{ij2} + \cdots + \beta_p X_{ijp}) + (b_{i0} + b_{i1} Z_{ij1} + b_{i2} Z_{ij2} + \cdots + b_{iq} Z_{ijq}) + E_{ij}$$

where $i = 1, 2, \ldots, K$ and $j = 1, 2, \ldots, n_i$, or, equivalently, using summation notation:

$$\boxed{Y_{ij} = \beta_0 + \sum_{g=1}^{p} \beta_g X_{ijg} + b_{i0} + \sum_{h=1}^{q} b_{ih} Z_{ijh} + E_{ij}} \qquad (25.34)$$

Subject-specific Matrix Form:

$$\boxed{\mathbf{Y}_i = \mathbf{X}_i \boldsymbol{\beta} + \mathbf{Z}_i \mathbf{b}_i + \mathbf{E}_i} \quad i = 1, 2, \ldots, K \qquad (25.35)$$

In each of the above (equivalent) models, there are two random components: the collection of random effects (denoted by $\mathbf{b}_i$ in the matrix version) and the collection of random errors (denoted by $\mathbf{E}_i$ in the matrix version).

Since $\mathbf{b}_i$ contains $(q + 1)$ scalar random components, i.e.,

$$\mathbf{b}_i = \begin{bmatrix} b_{i0} \\ b_{i1} \\ b_{i2} \\ \vdots \\ b_{iq} \end{bmatrix}$$

its associated covariance and correlation matrices are $(q + 1) \times (q + 1)$ matrices that we denote as $\mathbf{G}$ and $\mathbf{C}^{\mathbf{G}}$, respectively.

Similarly, $\mathbf{E}_i$ contains n_i scalar random components, i.e.,

$$\mathbf{E}_i = \begin{bmatrix} E_{i1} \\ E_{i2} \\ E_{i3} \\ \vdots \\ E_{in_i} \end{bmatrix},$$

and its associated covariance and correlation matrices are $(n_i \times n_i)$ matrices that we denote as $\mathbf{R}_i$ and $\mathbf{C}_i^{\mathbf{R}_i}$, respectively. We typically assume that each subject has the same number $(q + 1)$ of random components and that the matrices $\mathbf{G}$ and $\mathbf{C}^{\mathbf{G}}$ do not vary with i. However, since the number of observations per subject (n_i) may vary with i, we need to retain the subscript i when denoting the $(n_i \times n_i)$ $\mathbf{R}_i$ matrix.

Since the subject-specific response vector $\mathbf{Y}_i$ can be written as the linear sum of three matrices, one of which $(\mathbf{X}_i\boldsymbol{\beta})$ involves fixed (i.e., non-random) parameters and the other two of which $(\mathbf{Z}_i\mathbf{b}_i$ and $\mathbf{E}_i)$ involve random variables, it is possible to express the covariance structure $\mathbf{V}_i$ of $\mathbf{Y}_i$ (as well as the corresponding correlation structure $\mathbf{C}_i$) in terms of the covariance structures $\mathbf{G}$ and $\mathbf{R}_i$.

The matrix version for this relationship is:

$$\boxed{\mathbf{V}_i = \mathbf{Z}_i\mathbf{G}\mathbf{Z}_i' + \mathbf{R}_i} \qquad (25.36)$$

Equation (25.36) tells us that we can specify the covariance structure $\mathbf{V}_i$ of $\mathbf{Y}_i$ by choosing the $\mathbf{G}$ and $\mathbf{R}_i$ matrices that correspond to the random effects and the error terms in our model. When using SAS's MIXED procedure to fit a general linear mixed model, the user is required to do exactly this, i.e., specify the $\mathbf{G}$ and $\mathbf{R}_i$ matrices that will identify the working covariance structure (or matrix) $\mathbf{V}_i$ to be used for model fitting.

From equation (25.36), if the user does not want to consider any random effects in the model, then the user specifies that $\mathbf{G} = \mathbf{0}$; the covariance structure is then determined entirely by the $\mathbf{R}_i$ matrix, which involves variances and covariances for the n_i error components in $\mathbf{E}_i$.

Such a model (without random effects) is typically referred to as a *marginal model*.[8] As mentioned previously, the choice of covariance/correlation structure, in this case (when $\mathbf{G} = \mathbf{0}$) only for $\mathbf{R}_i$, represents a "guess" from consideration of several reasonable choices. We have already mentioned several possible choices, including exchangeable (**EXCH**), autoregressive (**AR1**), and unstructured (**UN**). In fact, the SAS MIXED procedure offers a huge array of over 50 choices for a correlation/covariance structure for either $\mathbf{R}_i$ and/or $\mathbf{G}$. We have also mentioned that an *empirical* (or *robust*) estimator of the standard error of any estimated regression coefficient can be computed to account for a possibly incorrect choice of one's "guessed"covariance/correlation structure.

Whether or not the model contains random effects (i.e., whether or not $\mathbf{G} = \mathbf{0}$), the simplest choice for the $\mathbf{R}_i$ matrix is $\sigma_e^2 \mathbf{I}_{n_i}$, where σ_e^2 denotes the "common variance" for each observation (Y_{ij}) on the ith subject. If, in fact, the researcher specifies a non-zero $\mathbf{G}$ matrix when using $\sigma_e^2 \mathbf{I}_{n_i}$ for the $\mathbf{R}_i$ matrix, we then say that the covariance matrix $\mathbf{V}_i$ for $\mathbf{Y}_i$ has a "conditionally independent" structure; by this, we mean that once we "condition on" (i.e., fix) the random effects, the observations $Y_{i1}, Y_{i2}, \ldots, Y_{in_i}$ are (conditionally) mutually independent.

In general, since $\mathbf{G}$ may be chosen not equal to $\mathbf{0}$ (and $\mathbf{R}_i$ is always non-zero), the form that the correlation/covariance structure for $\mathbf{Y}_i$ takes will depend on the combined choices for both the $\mathbf{G}$ and the $\mathbf{R}_i$ matrices. The fact that both matrices may be specified for a given model therefore allows the investigator great flexibility in choosing the working covariance matrix $\mathbf{V}_i$.

There is at least one situation[9] in which two different choices for $\mathbf{G}$ and $\mathbf{R}_i$ will result in the same overall covariance structure $\mathbf{V}_i$. This situation occurs when either

i) $\mathbf{G} = \mathbf{0}$ and $\mathbf{R}_i = \sigma^2 \mathbf{C}_i^{(\mathbf{EXCH})} = \mathbf{CS}, \rho \geq 0$

ii) $\mathbf{G} = \sigma_0^2$, a scalar, and $\mathbf{R}_i = \sigma_e^2 \mathbf{I}_{n_i}$

For a given dataset, the investigator then must use a computer program (e.g., SAS's MIXED procedure) to carry out the analysis, specifying the following:

1.   the outcome vector $\mathbf{Y}_i$;

2.   the predictors $\mathbf{X}_i$ (covariates for fixed effects $\boldsymbol{\beta}$) and $\mathbf{Z}_i$ (covariates for random effects $\mathbf{b}_i$);

---

[8] A *marginal model* can be distinguished from a *conditional/transitional model* in that one or more of the predictors (i.e., X's) in the latter type model may be previous values of the outcome variable, e.g., $Y_{i(j-1)}$, which denotes the Y response on the ith subject that immediately precedes Y_{ij}. (See Diggle, Heagerty, Liang, and Zeger, 2002 for further discussion on conditional/transitional models.)

Technically, a marginal model can be defined using the general linear mixed model formula $\mathbf{Y}_i = \mathbf{X}_i \boldsymbol{\beta} + \mathbf{Z}_i \mathbf{b}_i + \mathbf{E}_i$ by re-expressing this formula as $\mathbf{Y}_i = \mathbf{X}_i \boldsymbol{\beta} + \mathbf{E}_i^*$, where $\mathbf{E}_i^* = \mathbf{Z}_i \mathbf{b}_i + \mathbf{E}_i$. If one is not interested in differentiating $\mathbf{Z}_i \mathbf{b}_i$ from $\mathbf{E}_i$, then the variances and covariances among the elements of $\mathbf{E}_i^*$ are summarized in a $\mathbf{V}_i$ matrix (which is still a function of $\mathbf{G}$ and $\mathbf{R}_i$, but where now the specific forms for $\mathbf{G}$ and $\mathbf{R}_i$ are not of interest).

[9] Using (i) together with the variance-covariance formula (25.36), it follows that

$$\mathbf{V}_i = \mathbf{Z}_i \mathbf{G} \mathbf{Z}_i' + \mathbf{R}_i = \mathbf{0} + \mathbf{CS} = \mathbf{CS}.$$

Using (ii) and matrix algebra (details omitted, but see Appendix B), where $\mathbf{Z}_i$ is a ($n_i \times 1$) column matrix of ones, it follows that

$$\mathbf{V}_i = \mathbf{Z}_i \mathbf{G} \mathbf{Z}_i' + \mathbf{R}_i = \sigma_0^2 \mathbf{Z}_i \mathbf{Z}_i' + \sigma_e^2 \mathbf{I}_{n_i} = \sigma^2 \mathbf{C}_i^{(\mathbf{EXCH})} = \mathbf{CS}$$

since $\sigma^2 = (\sigma_0^2 + \sigma_e^2)$ and $\rho = \sigma_0^2/(\sigma_0^2 + \sigma_e^2) \geq 0$.

3. the correlation/covariance structure of the model in terms of the matrices $\mathbf{G}$ and $\mathbf{R}_i$; and,

4. whether or not robust/empirical standard error estimates one to be used to correct for the possibility of incorrectly specifying the $\mathbf{G}$ and $\mathbf{R}_i$ matrices.

Also, the data layout for the computer should agree with the general format described in Table 25.2. In the next section, we illustrate such an analysis (using MIXED) for the Air Pollution Study previously described.

25.3.5 ML Estimation in the General Linear Mixed Model

By exploiting the assumption that the random components in $\mathbf{b}_i$ and $\mathbf{E}_i$ for the general linear mixed model (25.35) are jointly normally distributed, estimation of effects (the β's and b's) in the general linear mixed model is typically carried out using likelihood-based methods. Two such likelihood-based methods implemented in SAS's MIXED procedure are **maximum likelihood** (ML) and **restricted maximum likelihood** (REML), both of which involve maximizing (log-) likelihood functions (Littell et. al., 1996). REML estimation differs from ML estimation; the former method adjusts certain parameter estimates for the number ($p + 1$) of fixed effect parameters being estimated, so that REML estimators generally are equivalent to unbiased estimators obtained from ANOVA methods (via partitioning sums of squares) for balanced data sets. In contrast, ML estimators are always slightly biased, and, when the data are unbalanced, are typically more biased than REML estimators. This is why REML is the default choice for SAS's MIXED procedure, although both REML and ML estimates will differ negligibly for large samples.

Regardless of whether REML or ML is used, tests of hypotheses about fixed effects parameters can be carried out using Wald and/or Likelihood Ratio tests, as described previously in Chapter 21. If, as is typically assumed, the response variable (Y_{ij}) is normally distributed, then (approximate) t tests and F tests can be used instead of Wald tests and Likelihood Ratio tests, respectively. Also, confidence interval estimation for fixed effects parameters is carried out using previously described large sample methods. Statistical inference methods for random effects, however, are somewhat more complicated, and are described in Chapter 26, Section 26.4.6.

25.4 Example: Study of Effects of an Air Pollution Episode on FEV1 Levels

In this section, we illustrate the use of a linear mixed model using the previously described data involving FEV1 measurements on 40 schoolchildren before, during, and after an air pollution episode. We begin by illustrating the type of computer statements required to implement SAS's MIXED procedure. We then describe and compare the results from fitting marginal models to the data, considering both model-based (non-empirical) standard errors and empirical standard errors. We also illustrate results for unbalanced data obtained by arbitrarily deleting several observations from the original FEV1 dataset.

25.4.1 Computer Statements for Fitting a Linear Mixed Model

As a simple example of computer program code using SAS's MIXED procedure to analyze the FEV1 air pollution study data described earlier, we provide the code for using a compound symmetric covariance structure when fitting a linear mixed model to these data on 40 subjects, where we set $\mathbf{G} = \mathbf{0}$ and $\mathbf{R}_i = \sigma^2 \mathbf{C}_i^{(\text{EXCH})}$:

```
proc mixed data=fev1;
   class subj week;
   model fev1=week/ s;
   repeated/ type=cs subject=subj r;
run;
```

These statements tell the MIXED procedure to use the FEV1 dataset and to fit a linear mixed model in which the outcome variable is "fev1", the independent variables are four fixed factor dummy variables (X's) for the 5 weeks (based on the *class* statement), and the covariance structure for the $\mathbf{R}_i$ matrix (from the repeated statement) is compound symmetric ("*cs*"). The *subject* option allows the user to specify the variable "subj" that identifies the cluster (a cluster is a child, so there are 40 clusters). This model does not contain any random effects, which would require a *random* statement. Also, the "*s*" tells the program to provide output that gives the "solutions" for the estimated regression coefficients, and the "*r*" tells the program to print out the estimated $\mathbf{R}_i$ matrix for the first person ($i = 1$) in the data set; note that $\mathbf{R}_i$ will be identical for all 40 subjects since $n_i \equiv 5$ for all i.

An alternative set of statements that will yield the same (**CS**) covariance structure using a random intercept model (where $\mathbf{G} = \sigma_0^2$, a scalar, and $\mathbf{R}_i = \sigma_e^2 \mathbf{I}_5$) is given as follows:

```
proc mixed data=fev1;
   class subj week;
   model fev1=week/ s;
   random intercept/ subject=subj s g;
run;
```

The latter statements tell the MIXED procedure[10] to use the FEV1 dataset and to fit a linear mixed model in which the outcome variable is FEV1, the independent variables are four fixed factor dummy variables (X's) for the 5 weeks, and one random factor covariate[11] (Z_0) for a random intercept (b_{i0}). The "*random intercept*" statement is the code required to specify that the model contains one random effect b_{i0} (for the intercept), from which it follows (by default) that the $\mathbf{G}$ matrix is a (1×1) scalar value giving the variance (σ_0^2) for this random effect.

---

[10] SAS's MIXED procedure is being illustrated here primarily because the authors of this text are more familiar with SAS than with other computer packages. Nevertheless, other packages such as STATA and SPSS have their own software for fitting a general linear mixed model. For complete details about such software for any of these packages, the reader should refer to the User's Guide provided by the computer package of choice.

[11] Using the matrix version (25.35) of the general linear mixed model, the single random factor Z_0 will correspond to a $\mathbf{Z}_i$ matrix consisting of a (5×1) column vector of 1's.

Since the above code contains no "*repeated*" statement, the MIXED procedure uses (by default) an $\mathbf{R}_i$ matrix of the form $\sigma_e^2 \mathbf{I}_5$, which assumes conditional mutual independence among the responses for each subject, where σ_e^2 is assumed to be a common variance for all error terms (E_{ij}). As in the previous set of statements, the variable "subj" identifies the cluster (one cluster for each of the 40 children). The "*g*" tells the program to print out the estimated $\mathbf{G}$ matrix, which here is the estimate of the scalar variance σ_0^2.[12]

25.4.2 Model-based Results for Marginal Models: FEV1 Data

To illustrate the results from fitting a linear mixed model, we once again consider the analysis of the FEV1 data. Table 25.7 provides edited computer output for several choices of correlation/covariance structures for marginal models (i.e., $\mathbf{G} = \mathbf{0}$) considered for these data. The "marginal" linear mixed model used here takes the form:

$$\mathbf{Y}_i = \mathbf{X}_i\boldsymbol{\beta} + \mathbf{E}_i, \qquad i = 1, \ldots, 40 \qquad\qquad (25.8\ repeated)$$

where $\mathbf{Y}_i$ denotes the collection of 5 FEV1 measurements on the ith child, and where $\mathbf{X}_i$, $\boldsymbol{\beta}$, and $\mathbf{E}_i$ are defined in (25.9). Note that the dummy variables (X_1, X_2, X_3, X_4) are defined so that the referent group is week 5. Also, **model-based** (non-empirical) standard errors are shown here; empirical standard errors will be shown separately later. The (marginal model) covariance structure used for each computer run is of the form $\mathbf{V}_i = \mathbf{Z}_i\mathbf{G}\mathbf{Z}_i' + \mathbf{R}_i$ where $\mathbf{G} = \mathbf{0}$ always and where $\mathbf{R}_i$ is chosen to be either independent (**IND**), compound symmetric (**CS**), auto-regressive (**AR1**), or unstructured (**UN**).

Column 1 in Table 25.7 presents the estimated regression coefficients and associated standard errors for the intercept and for each of the four dummy variables. The estimated regression coefficients are identical for all four choices of the correlation structure (i.e., $\mathbf{R}$ matrix), yet the standard errors are different for each different choice of $\mathbf{R}$. In general, the estimated regression coefficients call also differ with the choice of $\mathbf{R}$; but, for these balanced data, all choices for $\mathbf{R}$ yield the same estimated coefficients. We will later show that, when the data set is unbalanced, the estimated regression coefficients vary as a function of $\mathbf{R}$. The fact that the standard errors differ with the choice of $\mathbf{R}$ should not be surprising, since the primary reason for carrying out a correlated data analysis is to account for the possible effects that different correlation structures can have on statistical inferences.

Since the estimated regression coefficients are identical for each choice of correlation structure, let's focus on the interpretation of the estimated model for a given correlation structure, say compound symmetric (CS). Since the referent group for the "Week" variable is week 5, the intercept β_0 in the model represents the true FEV1 mean for week 5, whereas β_j is the difference in true FEV1 mean levels between week j and week 5 ($j = 1, 2, 3, 4$).

Thus, from the output (Table 25.7), the estimated intercept is $\hat{\beta}_0 = \overline{\text{FEV1}}_5 = 6.9985$, and the estimated differences between mean FEV1 levels for week j and week 5, $j = 1, \ldots, 4$, are, respectively,

$$\hat{\beta}_1 = \overline{\text{FEV1}}_1 - \overline{\text{FEV1}}_5 = 2.8152, \qquad \hat{\beta}_2 = \overline{\text{FEV1}}_2 - \overline{\text{FEV1}}_5 = -0.1502$$

$$\hat{\beta}_3 = \overline{\text{FEV1}}_3 - \overline{\text{FEV1}}_5 = 0.00325, \qquad \hat{\beta}_4 = \overline{\text{FEV1}}_4 - \overline{\text{FEV1}}_5 = -0.0470$$

---

[12] To obtain a robust/empirical standard error for either of the above runs, the first line of code would be replaced by: *proc mixed data=fev1 empirical;*

TABLE 25.7   **Edited output based on FEV data for different marginal models (G = 0) without empirical standard error option, using REML in the SAS mixed procedure**

Column 1	Column 2[†]	Column 3

Model-based IND

Effect	Week	Estimate	Standard Error
Intercept		6.9985	0.2590
week	1	2.8152	0.3663
week	2	−0.1502	0.3663
week	3	0.003250	0.3663
week	4	−0.04700	0.3663

Type 3 Tests of Fixed Effects

Effect	Num DF	Den DF	F Value	Pr > F
week	4	195	24.50	<.0001

	Num DF	Den DF	F Value	Pr > F
contrast	1	195	97.79	<.0001

$\mathbf{V} \equiv \mathbf{R} = \sigma_e^2 \mathbf{I}_5$, where

$$\hat{\sigma}_e^2 = 2.6837$$

Model-based CS

Effect	Week	Estimate	Standard Error
Intercept		6.9985	0.2590
week	1	2.8152	0.2439
week	2	−0.1502	0.2439
week	3	0.003250	0.2439
week	4	−0.04700	0.2439

Type 3 Tests of Fixed Effects

Effect	Num DF	Den DF	F Value	Pr > F
week	4	195	55.26	<.0001

	Num DF	Den DF	F Value	Pr > F
contrast	1	195	220.51	<.0001

$\mathbf{V} \equiv \mathbf{R} = \sigma^2 \mathbf{C}^{(\text{EXCH})}$, where $\hat{\mathbf{C}}^{(\text{EXCH})}$ is

1.0000	0.5565	0.5565	0.5565	0.5565
0.5565	1.0000	0.5565	0.5565	0.5565
0.5565	0.5565	1.0000	0.5565	0.5565
0.5565	0.5565	0.5565	1.0000	0.5565
0.5565	0.5565	0.5565	0.5565	1.0000

$\hat{\sigma}^2 = \hat{\sigma}_0^2 + \hat{\sigma}_e^2 = 2.6837$, where $\hat{\sigma}_0^2 = 1.4936$, $\hat{\sigma}_e^2 = 1.1901$

Model-based AR1

Effect	Week	Estimate	Standard Error
Intercept		6.9985	0.2526
week	1	2.8152	0.3359
week	2	−0.1502	0.3198
week	3	0.003250	0.2902
week	4	−0.04700	0.2306

Type 3 Tests of Fixed Effects

Effect	Num DF	Den DF	F Value	Pr > F
week	4	195	44.51	<.0001

	Num DF	Den DF	F Value	Pr > F
contrast	1	195	133.91	<.0001

$\mathbf{V} \equiv \mathbf{R} = \mathbf{AR1}$, where $\hat{\mathbf{C}}^{(\text{AR1})}$ is

1.0000	0.5835	0.3405	0.1987	0.1159
0.5835	1.0000	0.5835	0.3405	0.1987
0.3405	0.5835	1.0000	0.5835	0.3405
0.1987	0.3405	0.5835	1.0000	0.5835
0.1159	0.1987	0.3405	0.5835	1.0000

Model-based UN

Effect	Week	Estimate	Standard Error
Intercept		6.9985	0.2449
week	1	2.8153	0.2546
week	2	−0.1502	0.2078
week	3	0.003250	0.2102
week	4	−0.04700	0.2419

Type 3 Tests of Fixed Effects

Effect	Num DF	Den DF	F Value	Pr > F
week	4	195	37.64	<.0001

	Num DF	Den DF	F Value	Pr > F
contrast	1	195	143.78	<.0001

$\mathbf{V} \equiv \mathbf{R} = \mathbf{UN}$, where $\hat{\mathbf{C}}^{(\text{UN})}$ is

1.0000	0.3774	0.3027	0.3542	0.3350
0.3774	1.0000	0.7636	0.6713	0.6767
0.3027	0.7636	1.0000	0.6576	0.6763
0.3542	0.6713	0.6576	1.0000	0.6299
0.3350	0.6767	0.6763	0.6299	1.0000

[†] The F statistic shown for Type 3 Tests of Fixed Effects tests H_0: $\mu_1 = \mu_2 = \mu_3 = \mu_4 = \mu_5$. The F statistic shown for the "contrast" tests H_0: $\mu_1 - \dfrac{\mu_2 + \mu_3 + \mu_4 + \mu_5}{4} = 0$. Although the estimation procedure used to fit the models is REML (a likelihood procedure), the resulting test statistics are (either exact or approximate) F statistics given the normality assumptions. Similarly, such F statistics would be obtained if ML is used instead of REML, although the numerical values of the F statistics may differ slightly from REML values. Correspondingly, test statistics for individual regression coefficients can be described as (exact or appropriate) t statistics. Even if random effects are added to the model, such F and t statistics will be obtained if both the random components and the error terms are assumed to be normally distributed. For large samples, these F and t statistics essentially become χ^2 and Z statistics, respectively.

Also, regardless of the correlation structure (i.e., **R** matrix) shown in Table 25.7, the predicted FEV1 mean values for each of the five weeks are simply the average FEV1 levels for these five weeks (as previously shown in Table 25.4).[13]

$$\overline{\text{FEV1}}_1 = 9.81375, \ \overline{\text{FEV1}}_2 = 6.84825, \ \overline{\text{FEV1}}_3 = 7.00175,$$
$$\overline{\text{FEV1}}_4 = 6.96150, \ \overline{\text{FEV1}}_5 = 6.99850$$

These results indicate that the average FEV1 level prior to the pollution alert (namely, $\overline{\text{FEV1}}_1$) was almost 3 units higher than during the week of the pollution alert (namely, $\overline{\text{FEV1}}_2$); further, for weeks 3, 4, and 5, mean FEV1 levels remained depressed to about the same level as for week 2.

Column 2 in Table 25.7 gives hypothesis testing results about the overall effect of "week", as well as about the contrast that compares the FEV1 mean for week 1 with the average of the FEV means for the other four weeks. The null hypothesis for the first of these tests can be stated as:

$$H_0\!: \mu_1 = \mu_2 = \mu_3 = \mu_4 = \mu_5, \text{ or equivalently, } H_0\!: \beta_1 = \beta_2 = \beta_3 = \beta_4 = 0$$

where μ_j denotes the true (i.e., population) mean FEV1 value at week j and β_j denotes the coefficient of the jth dummy variable in the (marginal) model (25.8). The corresponding F statistics for different choices of **R** are all different,[14] but are nevertheless all highly significant. Thus, we can conclude that there are significant differences among the estimated mean FEV1 levels over the five weeks.

The estimated average FEV1 scores for each of the five weeks suggest that the pollution episode may have caused a significant overall *reduction* in lung capacity (i.e., FEV1 level) *that remained essentially the same* over the three weeks following the episode. A significance test about this assertion can be carried out by comparing the observed FEV1 mean value for the first week with the average of the observed FEV1 mean values for the other four weeks. We previously showed in Section 25.2 that this "contrast" is estimated from the data to be:

$$\overline{\text{FEV1}}_1 - \frac{\overline{\text{FEV1}}_2 + \overline{\text{FEV1}}_3 + \overline{\text{FEV1}}_4 + \overline{\text{FEV1}}_5}{4}$$

$$= 9.81375 - \frac{6.84825 + 7.00175 + 6.95150 + 6.99850}{4} = 2.8637$$

---

[13] The estimated FEV1 mean level for each week is identical to the corresponding predicted mean FEV1 level for that week because the only predictors in the marginal model are the 4 dummy variables for the 5 weeks, with no other covariates being included in the model.

[14] The denominator degrees of freedom (Den DF) for each test in column 2 is 195, which is determined from the typical formula $(n - p^*)$, where $n \ (= 200)$ is the total number of observations in the study and p^* is the number of fixed effect parameters in the model (in our example, $p = 4$ for the four dummy variables X_1, X_2, X_3, and X_4, so that $p^* = p + 1 = 5$). This choice for Den DF (also called DDFM in SAS's MIXED procedure) is called the *Residual option*. MIXED also allows the user to make other choices for DDFM. For example, if the model contains random effects, it might be argued that the DDFM should involve subtracting from $(n - p^*)$ the number (q^*) of random effects in the model times one less than the number of subjects in the study, i.e., DDFM $= (n - p^*) - q^*(K - 1)$. Thus, for the FEV1 data (where $K = 40$), the DDFM for a model containing a single random effect $(q^* = 1)$ for the intercept would be DDFM $= (200 - 5) - (1 \times 39) = 156$: for two random effects $(q^* = 2)$, DDFM $= (200 - 5) - (2 \times 39) = 117$. See Chapter 26, Section 26.4.7, for further discussion on choices for DDFM.

which is meaningfully different from zero. The null hypothesis for testing for the statistical significance of this contrast is given by either of the following equivalent statements[15]:

$$H_0: \mu_1 - \frac{\mu_2 + \mu_3 + \mu_4 + \mu_5}{4} = 0, \text{ or equivalently, } H_0: \beta_1 - \frac{\beta_2 + \beta_3 + \beta_4}{4} = 0$$

Using SAS's MIXED procedure, the following program code for $\mathbf{R} = \mathbf{CS}$ (i.e., compound symmetric) that uses a CLASS statement for the variable "week" would perform this test:[16]

```
proc mixed data=fev1;
   class subj week;
   model fev1=week/ s;
   repeated/ type=cs subject=subj r;
   contrast 'week'
      week 1−.25−.25−.25−.25;
   run;
```

(The contrast statement in the above code effectively uses the first of the two null hypotheses given above, since the weight for the first week is 1 and the weight for each of the other 4 weeks is $-\frac{1}{4} = -.25$.)

The following output is obtained using the above code:

		Contrasts		
	Num	Den		
Label	DF	DF	F Value	Pr > F
week	1	195	220.51	<.0001

From the above output, we conclude that the estimated contrast is highly significantly different from zero, so that the estimated mean FEV1 level for week 1 is statistically different from the average of the other four estimated means. Column 2 of Table 25.7 provides corresponding F statistics for this contrast for 4 different choices of the $\mathbf{R}$ matrix. All four F statistics are numerically different from one another because different $\mathbf{R}$ matrices are used. Nevertheless, for each of these choices for $\mathbf{R}$, the contrast comparing the estimated mean FEV1 value at week 1 and the average of the estimated mean FEV1 values over the other 4 weeks is highly significant.

---

[15] The two null hypotheses are equivalent since it follows that

$$\beta_1 - \frac{\beta_2 + \beta_3 + \beta_4}{4} = (\mu_1 - \mu_5) - \frac{(\mu_2 - \mu_5) + (\mu_3 - \mu_5) + (\mu_4 - \mu_5)}{4}$$
$$= \mu_1 - \frac{\mu_2 + \mu_3 + \mu_4 + \mu_5}{4}$$

[16] If, instead of using a CLASS statement for "Week", four dummy variables $w1$, $w2$, $w3$, and $w4$ are defined by the user, the model and contrast statements would be modified as follows:

```
model fev1=w1 w2 w3 w4/ s;
contrast 'week' w1 1 w2−.25 w3 −.25 w4−.25;
```

25.4.3 Results Using Empirical Standard Errors: FEV1 Data

From the model-based output given in Table 25.7, we concluded that, for any of four stated choices for the $\mathbf{R}$ matrix, there was a statistically significant difference among the estimated mean FEV1 values for the 5 weeks; moreover, an estimated contrast of interest was also statistically significantly different from zero. These analyses do not clearly identify which of the four correlation structures considered is most appropriate for these data, nor do they consider the wide variety of other correlation structures that could be used (particularly in SAS's MIXED procedure).

As mentioned previously, the choice of appropriate correlation/covariance structure requires consideration of several approaches from which the investigator(s) must make a "reasoned" decision. Such approaches include possibly identifying a particular correlation structure that seems the most biologically or clinically reasonable, deciding which model and associated correlation structure has the best "goodness of fit" to the data, and assessing the extent to which numerical results (and attendant statistical inferences and study conclusions) differ among different correlation structures. The fact that we reached the same general conclusions about the overall effects of the pollution alert on FEV1 mean levels for the four $\mathbf{R}$ matrices that we considered is possibly reason enough not to investigate other correlation structures. We note, however, that a logical choice of correlation structure for these data is autoregressive ($\mathbf{AR1}$), the logic being that the farther apart in time are observations on the same subject, the smaller should be the pairwise correlations.

Nevertheless, the availability of empirical standard errors suggests that it would be useful to assess the effects on numerical results of possibly making an incorrect choice of correlation/covariance structure. For this purpose, Table 25.8 has been produced, which provides corresponding output for the same four marginal models ($\mathbf{G} = \mathbf{0}$) considered in Table 25.7. The only difference between the computer codes used for Tables 25.7 and 25.8 is that the first line of code for producing Table 25.8 adds the word "empirical" as follows:

proc mixed data=fev1 empirical;

From column 1 of Table 25.8, we can see that corresponding estimated regression coefficients are identical for the four choices of $\mathbf{R}$, and also that corresponding empirical standard errors are identical (although not shown, corresponding empirical estimated regression coefficient covariance estimates are also identical for all four choices of $\mathbf{R}$). Thus, for these data, once we account for a possibly incorrect choice for the correlation/covariance structure, any statistical inferences about model parameters are identical regardless of the choice of working correlation structure. Such a finding is not generally true when using empirical standard errors, as we will illustrate in the next section for unbalanced data.

From column 2 of Table 25.8, we see that all F statistics for testing the null hypothesis of equal mean FEV1 scores over the five weeks, i.e.,

$$H_0: \ \mu_1 = \mu_2 = \mu_3 = \mu_4 = \mu_5 \ \text{(or equivalently, } H_0: \ \beta_1 = \beta_2 = \beta_3 = \beta_4 = 0\text{)},$$

as well as the F statistics for testing

$$H_0: \ \mu_1 - \frac{\mu_2 + \mu_3 + \mu_4 + \mu_5}{4} = 0 \ \text{(or equivalently, } H_0: \ \beta_1 - \frac{\beta_2 + \beta_3 + \beta_4}{4} = 0\text{)},$$

are identical for all the different choices of $\mathbf{R}$. These results are obtained because the empirical standard errors (as well as empirical covariance estimates) for the estimated regression coefficients do not vary with the choice for $\mathbf{R}$.

TABLE 25.8   **Edited output based on FEV data for different marginal models (G = 0) with empirical standard error option, using REML in the SAS mixed procedure**

| | Column 1 | | Column 2 | | Column 3 |

Empirical IND

Column 1

Effect	Week	Estimate	Standard Error
Intercept		6.9985	0.2418
week	1	2.8152	0.2514
week	2	−0.1502	0.2052
week	3	0.003250	0.2076
week	4	−0.04700	0.2389

Column 2 — Type 3 Tests of Fixed Effects

Effect	Num DF	Den DF	F Value	Pr > F
week	4	195	38.61	<.0001
contrast	1	195	147.46	<.0001

Column 3

$$\mathbf{V} \equiv \mathbf{R} = \sigma_e^2 \mathbf{I}_5, \text{ where}$$

$$\hat{\sigma}_e^2 = 2.6837$$

Empirical CS

Column 1

Effect	Week	Estimate	Standard Error
Intercept		6.9985	0.2418
week	1	2.8152	0.2514
week	2	−0.1503	0.2052
week	3	0.003250	0.2076
week	4	−0.04700	0.2389

Column 2 — Type 3 Tests of Fixed Effects

Effect	Num DF	Den DF	F Value	Pr > F
week	4	195	38.61	<.0001
contrast	1	195	147.46	<.0001

Column 3

$$\mathbf{V} \equiv \mathbf{R} = \sigma^2 \mathbf{C}^{(\text{EXCH})} \text{ where } \hat{\mathbf{C}}^{(\text{EXCH})} \text{ is}$$

1.0000	0.5565	0.5565	0.5565	0.5565
0.5565	1.0000	0.5565	0.5565	0.5565
0.5565	0.5565	1.0000	0.5565	0.5565
0.5565	0.5565	0.5565	1.0000	0.5565
0.5565	0.5565	0.5565	0.5565	1.0000

$$\hat{\sigma}^2 = \hat{\sigma}_0^2 + \hat{\sigma}_e^2 = 2.6837, \text{ where}$$
$$\hat{\sigma}_0^2 = 1.4936, \hat{\sigma}_e^2 = 1.1901$$

Empirical AR1

Column 1

Effect	Week	Estimate	Standard Error
Intercept		6.9985	0.2418
week	1	2.8152	0.2514
week	2	−0.1503	0.2052
week	3	0.003250	0.2076
week	4	−0.04700	0.2389

Column 2 — Type 3 Tests of Fixed Effects

Effect	Num DF	Den DF	F Value	Pr > F
week	4	195	38.61	<.0001
contrast	1	195	147.46	<.0001

Column 3

$$\mathbf{V} \equiv \mathbf{R} = \mathbf{AR1} \text{ where } \hat{\mathbf{C}}^{(\text{AR1})} \text{ is}$$

1.0000	0.5835	0.3405	0.1987	0.1159
0.5835	1.0000	0.5835	0.3405	0.1987
0.3405	0.5835	1.0000	0.5835	0.3405
0.1987	0.3405	0.5835	1.0000	0.5835
0.1159	0.1987	0.3405	0.5835	1.0000

Empirical UN

Column 1

Effect	Week	Estimate	Standard Error
Intercept		6.9985	0.2418
week	1	2.8153	0.2514
week	2	−0.1502	0.2052
week	3	0.003250	0.2076
week	4	−0.04700	0.2389

Column 2 — Type 3 Tests of Fixed Effects

Effect	Num DF	Den DF	F Value	Pr > F
week	4	195	38.61	<.0001
contrast	1	195	147.46	<.0001

Column 3

$$\mathbf{V} \equiv \mathbf{R} = \mathbf{UN} \text{ where } \hat{\mathbf{C}}^{(\text{UN})} \text{ is}$$

1.0000	0.3774	0.3027	0.3542	0.3350
0.3774	1.0000	0.7636	0.6713	0.6767
0.3027	0.7636	1.0000	0.6576	0.6763
0.3542	0.6713	0.6576	1.0000	0.6299
0.3350	0.6767	0.6763	0.6299	1.0000

The column 3 numerical results in Table 25.8 are exactly the same as those in Table 25.7; this is because these computations of estimated working correlation matrices are done independently of empirical standard error calculations.

25.4.4 Results for Unbalanced Data: FEV1 Study

So far, we have provided numerical illustrations based on fitting a particular linear mixed model to a balanced dataset, i.e., each subject has the same number of repeated observations that are measured at the same times. In this section, we show that the same linear mixed model can also be applied to an unbalanced dataset. To illustrate this, we arbitrarily removed 24 observations from 15 of the 40 subjects in the FEV1 study. Table 25.9 gives the data for the 15 subjects that now each have less than 5 observations; the remaining 25 subjects still have FEV1 data for all 5 weeks. Notice, for example, that subjects 1, 2, and 24 are each missing one FEV1 measurement at week 5, subject 3 is missing FEV1 data at weeks 3 and 4, and subject 4 is missing FEV1 measurements at weeks 4 and 5. The use of a "." to denote a missing observation allows the SAS MIXED procedure to keep track of which week has a missing observation.[17]

TABLE 25.9   Unbalanced data for 24 subjects from FEV1 dataset

Subj	Week	FEV1	Subj	Week	FEV1	Subj	Week	FEV1
1	1	9.43	7	1	10.30	24	1	10.63
1	2	5.71	7	2	.	24	2	5.15
1	3	5.86	7	3	6.99	24	3	6.07
1	4	7.70	7	4	.	24	4	7.44
1	5	.	7	5	5.67	24	5	.
2	1	11.15	8	1	8.67	26	1	.
2	2	9.48	8	2	4.10	26	2	7.51
2	3	10.11	8	3	.	26	3	7.56
2	4	8.89	8	4	.	26	4	4.88
2	5	.	8	5	5.46	26	5	7.54
3	1	8.40	10	1	9.05	28	1	.
3	2	4.42	10	2	.	28	2	6.87
3	3	.	10	3	5.33	28	3	5.79
3	4	.	10	4	5.94	28	4	8.30
3	5	4.34	10	5	.	28	5	8.19
4	1	9.20	11	1	12.44	30	1	10.24
4	2	4.51	11	2	8.68	30	2	8.48
4	3	6.20	11	3	.	30	3	.
4	4	.	11	4	6.40	30	4	.
4	5	.	11	5	7.61	30	5	.
6	1	9.12	12	1	7.88	32	1	11.50
6	2	7.71	12	2	4.88	32	2	.
6	3	6.75	12	3	6.43	32	3	6.69
6	4	.	12	4	.	32	4	.
6	5	8.20	12	5	.	32	5	.

---

[17] If the data are listed without a "." line, the SAS MIXED procedure can keep track of the missing week by adding "week" (bolded for emphasis) in the "*repeated*" statement as follows:

repeated **week**/ *type=cs subject=subj r;*

Table 25.10 presents results obtained from fitting marginal linear mixed models to the unbalanced FEV1 dataset using the same four **R** matrices previously considered for the balanced dataset.

TABLE 25.10   **Edited output based on unbalanced FEV data for different marginal models (G = 0), using REML in the SAS mixed procedure**

IND

Effect	Week	Estimate	Model-based SE	Empirical SE
Intercept		7.0347	0.2856	0.2660
week	1	2.7748	0.3876	0.2838
week	2	−0.1633	0.3899	0.2630
week	3	0.00559	0.3925	0.2656
week	4	0.2075	0.4038	0.2874

Type 3 Tests of Fixed Effects

Effect	Num DF	Den DF	Model-based F Value	Empirical F Value
week	4	170	22.08	34.37

	Num DF	Den DF	Model-based F Value	Empirical F Value
contrast	1	195	86.91	123.04

$\mathbf{V} \equiv \mathbf{R} = \sigma_e^2 \mathbf{I}_5$ where $\hat{\sigma}_e^2 = 2.6093$

CS

Effect	Week	Estimate	Model-based SE	Empirical SE
Intercept		7.0104	0.2727	0.2533
week	1	2.8027	0.2704	0.2729
week	2	−0.1572	0.2706	0.2453
week	3	−0.01109	0.2739	0.2421
week	4	0.1070	0.2809	0.2710

Type 3 Tests of Fixed Effects

Effect	Num DF	Den DF	Model-based F Value	Empirical F Value
week	4	170	47.42	35.53

	Num DF	Den DF	Model-based F Value	Empirical F Value
contrast	1	195	187.33	130.02

$\mathbf{V} \equiv \mathbf{R} = \sigma^2 \mathbf{C}^{(\mathbf{EXCH})}$ where $\hat{\mathbf{C}}^{(\mathbf{EXCH})}$ is

1.0000	0.5269	0.5269	0.5269
0.5269	1.0000	0.5269	0.5269
0.5269	0.5269	1.0000	0.5269
0.5269	0.5269	0.5269	1.0000

$\hat{\sigma}^2 = \hat{\sigma}_0^2 + \hat{\sigma}_e^2 = 2.5913$, where $\hat{\sigma}_0^2 = 1.3653$, $\hat{\sigma}_e^2 = 1.2260$

AR1

Effect	Week	Estimate	Model-based SE	Empirical SE
Intercept		7.0552	0.2715	0.2521
week	1	2.7642	0.3514	0.2718
week	2	−0.1900	0.3370	0.2580
week	3	−0.08110	0.3119	0.2424
week	4	0.05857	0.2609	0.2749

Type 3 Tests of Fixed Effects

Effect	Num DF	Den DF	Model-based F Value	Empirical F Value
week	4	170	41.57	36.30

	Num DF	Den DF	Model-based F Value	Empirical F Value
contrast	1	195	124.38	135.03

$\mathbf{V} \equiv \mathbf{R} = \mathbf{AR1}$ where $\hat{\mathbf{C}}^{(\mathbf{AR1})}$ is

1.0000	0.5825	0.3393	0.1976
0.5825	1.0000	0.5825	0.3393
0.3393	0.5825	1.0000	0.5825
0.1976	0.3393	0.5825	1.0000

UN

Effect	Week	Estimate	Model-based SE	Empirical SE
Intercept		7.0145	0.2542	0.2506
week	1	2.8020	0.2716	0.2677
week	2	−0.1762	0.2488	0.2451
week	3	−0.04812	0.2427	0.2390
week	4	0.09776	0.2737	0.2693

Type 3 Tests of Fixed Effects

Effect	Num DF	Den DF	Model-based F Value	Empirical F Value
week	4	171	36.89	37.92

	Num DF	Den DF	Model-based F Value	Empirical F Value
contrast	1	195	136.94	140.69

$\mathbf{V} \equiv \mathbf{R} = \mathbf{UN}$ where $\hat{\mathbf{C}}^{(\mathbf{UN})}$ is

1.0000	0.3886	0.3193	0.3268
0.3886	1.0000	0.7821	0.5896
0.3193	0.7821	1.0000	0.6852
0.3268	0.5896	0.6852	1.0000

[†] The estimated working covariance matrices given in column 3 are all (4 × 4) matrices because the first subject in the unbalanced dataset has only 4 repeated observations.

From column 1 of Table 25.10, it can be seen that the sets of estimated regression coefficients differ for each choice of **R**; in contrast, for the balanced dataset, the sets of estimated coefficients were identical (see Tables 25.7 and 25.8). Also, as with the balanced dataset, the sets of model-based standard errors vary, as expected, with varying choices for **R**. Note, however, that the sets of empirical standard errors also vary as **R** varies; in contrast, empirical standard errors did not change as a function of **R** when the dataset was balanced (see Table 25.8).

From column 2 of Table 25.10, we can also see that the model-based F statistics for the overall comparison among the five weeks, and for the contrast of interest, differ, as expected, as **R** differs. However, in contrast to the balanced dataset results, the empirical-based F statistics also differ as **R** differs. Although the overall conclusion for the unbalanced dataset is the same (i.e., a significant effect of "week" and a significant "contrast"), the fact that the empirical standard errors and corresponding F statistics differ with **R** indicates that, in general, numerical results using the empirical standard error option can vary when analyzing data sets of small to intermediate size that are unbalanced (e.g., there are missing data).

Finally, since Table 25.10 is based on the use of the unbalanced data set, the estimated working correlation matrices in column 3 of Table 25.10 differ (as expected) from corresponding numerical results in Tables 25.7 and 25.8.

25.4.5 Summary of Analyses for the FEV1 Data

From the analyses of the FEV1 dataset that we have carried out in this section, we summarize the results as follows:

1. With regard to the primary question of whether or not lung capacity as measured by FEV1 was depressed during the pollution episode:
   a. The observed FEV1 mean level prior to the pollution alert (week 1) was almost 3 units higher than during week 2 of the pollution alert; further, the observed mean FEV1 levels for weeks 3, 4, and 5 remained depressed at about the same level as for week 2.
   b. There is a significant difference among the estimated FEV1 mean levels over the five weeks.
   c. There is also a significant difference between the FEV1 sample mean for week 1 and the average of the FEV1 sample means over the other four weeks.
   d. The statistical conclusions in (b) and (c) above are essentially the same regardless of the choice of working correlation structure.

2. The following results were obtained based on using marginal models (varying **R**, **G** = **0**) to analyze the complete (*balanced*) data set:
   a. The estimated regression coefficients for the effect of "week" were identical regardless of the choice of **R** and, as expected, regardless of whether model-based or empirical standard errors were used.
   b. The set of empirical standard errors was identical regardless of the choice of **R**. For a given **R**, however, the estimated empirical standard errors were numerically different from corresponding model-based estimated standard errors. (The finding of identical estimated empirical standard errors for different choices of **R** for these balanced data, however, is not true in general for unbalanced data.)
   c. Hypothesis testing (using F statistics) resulted in finding a significant difference among estimated mean FEV1 levels over the five weeks, regardless of the choice of **R**, and regardless of whether model-based or empirical standard error options

were used. Nevertheless, using model-based standard errors, F statistic values differed numerically as $\mathbf{R}$ differed. In contrast, using empirical standard errors, F statistic values were identical regardless of the choice for $\mathbf{R}$.

d. The contrast that compared the estimated mean FEV1 level for week 1 with the estimated average mean FEV1 level over the other four weeks was found to be statistically (as well as meaningfully) different from zero when either model-based or empirical standard errors options were used.

e. The estimated $\mathbf{R}$ matrix differed, as expected, as $\mathbf{R}$ differed. Nevertheless, for a given choice of $\mathbf{R}$, the estimated $\mathbf{R}$ matrix was identical when either model-based or empirical standard error options were used; this is because the $\mathbf{R}$ matrix, being a "working-covariance structure," is estimated independently of empirical standard error calculations.

f. A reasonable choice of correlation structure for these data is autoregressive (**AR1**); because of the longitudinal nature of the study, the farther apart in time subject-specific observations are, the smaller should be the pairwise correlations. Nevertheless, the global statistical results essentially do not differ for the four $\mathbf{R}$ matrices considered, which suggests that the choice of correlation structure does not affect the general conclusions to be drawn from the study data.

3. The following results were obtained based on using marginal models (varying $\mathbf{R}$, $\mathbf{G} = \mathbf{0}$) to analyze an unbalanced subset of the original data:

a. The set of estimated regression coefficients for the effect of "week" differed (slightly) as the choice of $\mathbf{R}$ differed when either model-based or empirical standard error options were used. This illustrates the general principle that the estimated regression coefficients may differ in value for different choices of the working covariance structure, especially for unbalanced data sets.

b. The set of estimated empirical standard errors differed (slightly) with the choice of $\mathbf{R}$. This illustrates the general principle that the empirical standard errors may differ in value for different choices of the working covariance structure, especially for small to intermediate size studies with unbalanced data.

c. The general linear mixed model defined by (25.4) can be used to analyze unbalanced data in which different subjects can have different numbers of responses measured at different times.

25.5 Summary—Analysis of Correlated Data: Part 1

In this chapter, we have extended the multiple linear regression model to consider outcomes (i.e., responses) that are observed two or more times on each unit of analysis (e.g., a subject, or more generally, a *cluster*). The responses on each unit will generally be correlated, thereby requiring an analysis that accounts for such correlation. We illustrated this type of data using three examples:

a. A Study of the Effect of an Air Pollution Episode on Pulmonary Function
b. A Study of the Posture of Computer Operators
c. A Study of Treatments for Heartburn

We provided general notation for the "response vector" $\mathbf{Y}_i$ that characterizes the collection of n_i responses within the ith cluster, and we described the general data layout for such clustered data.

We then described in detail the general linear mixed model approach assuming normally distributed outcome variables and, correspondingly, normally distributed random effects and error terms. We described this model in both general scalar form and general matrix form as follows:

Subject-specific scalar form: $Y_{ij} = (\beta_0 + \beta_1 X_{ij1} + \beta_2 X_{ij2} + \cdots + \beta_p X_{ijp})$
$$+ (b_{i0} + b_{i1}Z_{ij1} + b_{i2}Z_{ij2} + \cdots + b_{iq}Z_{ijq}) + E_{ij}$$

Subject-specific matrix form: $\mathbf{Y}_i = \mathbf{X}_i\boldsymbol{\beta} + \mathbf{Z}_i\mathbf{b}_i + \mathbf{E}_i$

The X_{ij}'s (or $\mathbf{X}_i$) in the above models denote covariates for fixed effects and the Z_{ij}'s (or $\mathbf{Z}_i$) denote covariates for random effects. Correspondingly, the β's (or $\boldsymbol{\beta}$) denote fixed effects and the b_i's (or $\mathbf{b}_i$) denote random effects.

We distinguished among different *working correlation structures* that could be considered for a given model (including independent, exchangeable, autoregressive, and unstructured matrices), and we described the relationship between a covariance structure and its corresponding correlation structure. We also showed that the covariance structure (or matrix) $\mathbf{V}_i$ for $\mathbf{Y}_i$ corresponding to the general linear mixed model had the general matrix form $\mathbf{V}_i = \mathbf{Z}_i\mathbf{G}\mathbf{Z}_i' + \mathbf{R}_i$, where $\mathbf{G}$ denotes the covariance structure (or matrix) for the random component ($\mathbf{b}_i$) and $\mathbf{R}_i$ denotes the covariance structure (or matrix) for the error component ($\mathbf{E}_i$). And we described the use of a robust or empirical standard error option to correct for the possibility of misspecifying the correlation structure in a given analysis.

Finally, we described the analysis of the Air Pollution Study using SAS's MIXED procedure for marginal models ($\mathbf{G} = \mathbf{0}$) with different choices for the working correlation structure; also, we considered the entire (balanced) data set as well as an unbalanced data set in which not all subjects had the same number of FEV1 observations.

In the next chapter, which is the second part of our discussion on the analysis of correlated data, we focus on the use of random effects when fitting a linear mixed model.

Problems

1. Consider the following four correlation matrices that apply to clusters having five responses per cluster:

$$\mathbf{A} = \begin{bmatrix} 1 & 0.61 & 0.45 & 0.39 & 0.25 \\ 0.61 & 1 & 0.43 & 0.51 & 0.35 \\ 0.45 & 0.43 & 1 & 0.29 & 0.22 \\ 0.39 & 0.51 & 0.29 & 1 & 0.53 \\ 0.25 & 0.35 & 0.22 & 0.53 & 1 \end{bmatrix} \quad \mathbf{B} = \begin{bmatrix} 1 & 0 & 0 & 0 & 0 \\ 0 & 1 & 0 & 0 & 0 \\ 0 & 0 & 1 & 0 & 0 \\ 0 & 0 & 0 & 1 & 0 \\ 0 & 0 & 0 & 0 & 1 \end{bmatrix}$$

$$\mathbf{C} = \begin{bmatrix} 1 & 0.50 & 0.25 & 0.125 & 0.0625 \\ 0.50 & 1 & 0.50 & 0.25 & 0.125 \\ 0.25 & 0.50 & 1 & 0.50 & 0.25 \\ 0.125 & 0.25 & 0.50 & 1 & 0.50 \\ 0.0625 & 0.125 & 0.25 & 0.50 & 1 \end{bmatrix} \quad \mathbf{D} = \begin{bmatrix} 1 & 0.46 & 0.46 & 0.46 & 0.46 \\ 0.46 & 1 & 0.46 & 0.46 & 0.46 \\ 0.46 & 0.46 & 1 & 0.46 & 0.46 \\ 0.46 & 0.46 & 0.46 & 1 & 0.46 \\ 0.46 & 0.46 & 0.46 & 0.46 & 1 \end{bmatrix}$$

1. **a.** Matrix **A** is an example of what kind of correlation structure?
   b. Matrix **B** is an example of what kind of correlation structure?
   c. Matrix **C** is an example of what kind of correlation structure?
   d. Matrix **D** is an example of what kind of correlation structure?

2. Consider the following four covariance matrices that apply to clusters having four responses per cluster:

$$\mathbf{P} = \begin{bmatrix} 2.50 & 1.25 & 0.625 & 0.3125 \\ 1.25 & 2.50 & 1.25 & 0.625 \\ 0.625 & 1.25 & 2.50 & 1.25 \\ 0.3125 & 0.625 & 1.25 & 2.50 \end{bmatrix} \quad \mathbf{Q} = \begin{bmatrix} 4.00 & 1.952 & 1.721 & 1.593 \\ 1.952 & 4.50 & 1.826 & 1.690 \\ 1.721 & 1.826 & 3.50 & 1.491 \\ 1.593 & 1.690 & 1.491 & 3.00 \end{bmatrix}$$

$$\mathbf{R} = \begin{bmatrix} 4.00 & 1.84 & 1.84 & 1.84 \\ 1.84 & 4.00 & 1.84 & 1.84 \\ 1.84 & 1.84 & 4.00 & 1.84 \\ 1.84 & 1.84 & 1.84 & 4.00 \end{bmatrix} \quad \mathbf{S} = \begin{bmatrix} 4.00 & 1.732 & 1.061 & 0.234 \\ 1.732 & 3.00 & 1.837 & 0.810 \\ 1.061 & 1.837 & 4.50 & 1.984 \\ 0.234 & 0.810 & 1.984 & 3.50 \end{bmatrix}$$

   a. Which of the above matrices are heterogeneous (as compared to homogeneous) covariance matrices?
   b. What is the correlation structure defined by each of the above matrices?

3. Consider the data layout for a correlated data analysis provided in Table 25.2. Using this data layout as a guiding framework, describe a study involving either longitudinal data (and/or other type of correlated data) that relates to a study in which you are already involved or to a study that you might wish to undertake in an area of your own interest. Assume that your outcome (i.e., response) variable is continuous.

   In your description, make sure to:
   a. state whether the study design is longitudinal or of some other type involving correlated data;
   b. state the primary research objective of the study;
   c. give a brief outline of the study design;
   d. describe the variables being measured or observed, and summarize such information using the same format as the data layout in Table 25.2;
   e. specify at least one "marginal" regression model that you would like to fit to the data under consideration. Also, describe any random effects model or conditional/transitional model (see footnote 8 in the main body of the chapter) that you might wish to consider; and,
   f. describe what you think is the appropriate correlation structure for each cluster in your analysis (e.g., exchangeable, auto-regressive, other).

4. Consider again the study described in the main body of this chapter concerning the effect of an air pollution episode on pulmonary function measurements (FEV1) taken on each of $K = 40$ schoolchildren over five weeks. The subject-specific scalar and matrix versions of the model defined by expressions (25.7) and (25.8) use week 5 as the reference level. In this exercise, you are to consider how the model and analysis would be modified if week 1 is used as the reference level.

4. a. State the subject-specific scalar form of the model [i.e., analogous to expression (25.7)] in which week 1 is used as the reference level.

(*Note:* Denote the coefficients in this model by β'_g, $g = 0, 1, 2, 3, 4$.)

b. State the subject-specific matrix form of the model [i.e., analogous to expression (25.8)] in which week 1 is used as the reference level, making sure to specify the components and corresponding dimensions of the matrices involved.

c. Using the subject-specific scalar model defined in part (a), express each of the estimated regression coefficients ($\hat{\beta}'_g$, $g = 0, 1, 2, 3, 4$) in terms of sample mean FEV1 levels

$$\overline{FEV1}_j, \quad j = 1, 2, 3, 4, 5$$

d. Express the null hypothesis

$$H_0: \ \mu_1 - \frac{\mu_2 + \mu_3 + \mu_4 + \mu_5}{4} = 0$$

in terms of the regression coefficients β'_g ($g = 0, 1, 2, 3, 4$) for the model specified in part (a).

e. Will the computed test statistic resulting from testing H_0 using the model in part (a) be identical to the computed test statistic resulting from testing H_0 using model (25.2), or equivalently model (25.7), in which the reference level is week 5? Explain briefly.

f. In order to carry out the computations needed to fit the model described in part (a) using SAS's MIXED procedure, four dummy variables $d1$ through $d4$ (with the referent category being week 1) are defined to incorporate the effects of the five weeks into the model. Complete the computer code below (i.e., fill in the blanks in the code) by adding the appropriate statements to the "contrast" statement:

```
proc mixed data=fev1;
class subj;
model fev1=d1 d2 d3 d4/ s;
repeated/ type=cs subject=subj rcorr;
contrast 'week' d1 ___ d2 ___ d3 ___ d4 ___ ;
run;
```

(*Hint:* See footnote 16 in the main text for this chapter, which gives the code when the referent group is week 5. Here, you must modify the contrast statement so that the referent group is week 1.) The edited computer output shown below was obtained based on the code specified in part (f) in which the referent group is week 1. The questions that follow are based on this computer output.

Estimated R Correlation Matrix for subj 1

Row	Col1	Col2	Col3	Col4	Col5
1	1.0000	0.5565	0.5565	0.5565	0.5565
2	0.5565	1.0000	0.5565	0.5565	0.5565
3	0.5565	0.5565	1.0000	0.5565	0.5565
4	0.5565	0.5565	0.5565	1.0000	0.5565
5	0.5565	0.5565	0.5565	0.5565	1.0000

Covariance Cov Parm	Parameter Subject	Estimates Estimate
CS	subj	1.4936
Residual		1.1901

(continued)

Output for Problem 4 (continued)

```
                      Solution for Fixed Effects

                          Standard
 Effect          Estimate      Error      DF     t Value      Pr > |t|
 Intercept         9.8137     0.1873      195       52.39       <.0001
 d1               -2.9655     0.2610      195      -11.36       <.0001
 d2               -2.8120     0.2782      195      -10.11       <.0001
 d3               -2.8623     0.2920      195       -9.80       <.0001
 d4               -2.8153     0.2514      195      -11.20       <.0001

             Type 3 Tests of Fixed Effects
                     Num        Den
         Effect       DF         DF      F Value      Pr > F
           d1          1        195       129.14      <.0001
           d2          1        195       102.14      <.0001
           d3          1        195        96.08      <.0001
           d4          1        195       125.36      <.0001

 Contrasts
                         Num        Den
            Label         DF         DF      F Value      Pr > F
            week          1        195       147.46      <.0001
```

4. g. What are the predicted FEV1 values for weeks 1 through 5, respectively?

   h. What is the estimate of the contrast that compares week 1 with the average of the other four weeks? Is this estimated contrast statistically significantly different from zero?

   i. The output shown above does not provide the results for testing whether or not there is a significant difference among the estimated mean FEV1 levels for all 5 weeks.

     i. Specify a "contrast" statement (SAS code) that would perform such a test.

    ii. Will the results obtained from the contrast statement specified in the preceding question be the same as the results previously obtained in Table 25.7, for which the referent group is week 5? Explain briefly.

  j. i. What correlation structure has been assumed to generate the above output?

    ii. What is the estimated correlation between any two responses on the same subject?

   iii. How can the correlation in the preceding question (**j. ii.**) be computed using the output information provided for "*Covariance Parameter Estimates*"?

   iv. Does the output shown above provide test results that allow for the possibility that the assumed correlation structure may be incorrectly specified? Explain briefly.

    v. Based on the output summarized in Table 25.7, would you expect the numerical test results obtained for different working correlation structures to be identical when the referent group is week 1? Explain briefly.

5. This problem considers the (fictitious) study designed to compare two treatments for the relief of heartburn, described by the dataset of Table 25.6. One of two treatments (A = active = 1, P = placebo = 0) has been randomly allocated to two groups of 15 subjects each, and each subject receives the same treatment for both meals. The outcome variable is a continuous measure of physical discomfort measured by a questionnaire administered to each subject two hours after receiving the treatment. The computer code and corresponding edited computer output from using SAS's MIXED procedure are given next:

```
proc mixed data = hrtbrn;
class subj;
model discom = trt/s;
repeated/ type = cs subject = subj r;
run;

                Solution for Fixed Effects Empirical REML
   Effect          Estimate      Std Error      DF     t Value      Pr > |t|
   Intercept         62.17                      28
   trt              -11.00         4.4168        28     2.4905        .0190

                    Type 3 Tests of Fixed Effects
                        Num        Den
           Effect        DF         DF      F Value       Pr > F
           trt           1          28      6.2026         .0190
```

5. a. Why is it appropriate to do a correlated data analysis here?

  b. Why doesn't it matter what correlation structure is chosen (other than "independence") for carrying out a correlated data analysis of these data?

  c. Specify the subject-specific scalar model form used in this analysis.

  d. Specify the subject-specific matrix model form used in this analysis.

  e. What do you conclude about whether or not there is a significant difference between the effects of the two treatments on the amount of discomfort experienced by study subjects? Explain briefly.

6. Consider a different (fictitious) study to compare two treatments for the relief of heartburn. This new study involves a two-period "cross-over" design in which each subject is given two symptom-provoking meals and receives (with random assignment) an active treatment (A) after one meal and a placebo (P) after the other meal. In other words, each subject receives both treatments in this study; in contrast, for the previously described study (see question 5), each subject received either the active treatment or the placebo after both meals. The dataset is given below:

A:P				P:A		
Subject Number	A	P		Subject Number	P	A
1	65	75		16	85	55
2	60	60		17	60	70
3	70	75		18	80	70
4	35	50		19	55	30
5	50	45		20	50	50
6	40	65		21	70	40
7	50	80		22	70	65
8	55	50		23	65	35
9	20	40		24	60	50
10	30	35		25	90	70
11	65	50		26	85	80
12	45	55		27	65	70
13	30	50		28	60	45
14	55	70		29	75	65
15	25	40		30	55	45
Mean:	46.33	56.00		Mean:	68.33	56.00

Overall Mean A: 51.17   Overall Mean P: 62.17

The computer code and corresponding edited computer output from using SAS's MIXED procedure are given below:

```
proc mixed data = hrtbrn2;
class subj;
model discom = trt/s;
repeated/type = cs subject = subj r;
run;
```

```
              Solution for Fixed Effects l REML
Effect          Estimate      Std Error      DF      t Value      Pr > |t|
Intercept       62.1667        2.7822        29       22.34        <.0001
trt            -11.0000        2.2540        29       -4.88        <.0001

                Type 3 Tests of Fixed Effects
                   Num        Den
        Effect     DF          DF      F Value      Pr > F
         trt        1          29       23.82       <.0001
```

6. a. Specify the subject-specific scalar model form used for this analysis.

  b. What do you conclude about whether or not there is a significant difference between the effects of the two treatments on the amount of discomfort experienced by study subjects? Explain briefly.

  c. Why are the Den DF equal to 29 in this analysis, whereas the Den DF were equal to 28 in the analysis used for the data of question 5? Suppose that the investigator wished to consider the possibility that the sequence used, i.e., (A:P) versus (P:A), had an effect on the response in addition to a possible treatment effect. To carry out such an analysis, he/she modified the model by adding a dichotomous variable, denoted as "seq," to the model. The computer code and output for this latter model are provided below:

```
proc mixed data = hrtbr2n;
class subj;
model discom = trt/seq/s;
repeated/type = cs subject = subj r;
run;
```

```
              Solution for Fixed Effects Empirical REML
Effect          Estimate      Std Error      DF      t Value      Pr > |t|
Intercept       67.1667        3.5373        28       19.13        <.0001
trt            -11.0000        2.2540        29       -4.88        <.0001
seq            -11.0000        4.7417        28       -2.32        0.0279

                Type 3 Tests of Fixed Effects
                   Num        Den
        Effect     DF          DF      F Value      Pr > F
         trt        1          29       23.82       <.0001
         seq        1          28        5.38       0.0279
```

  d. Specify the subject-specific scalar model form used that includes both "trt" and "seq."

  e. What do you conclude about whether or not there is a significant difference between the effects of the two treatments on the amount of discomfort experienced by study subjects controlling for the variable "seq"? Explain briefly.

  f. How would you evaluate whether or not it was necessary to control for the "seq" variable in order to determine whether or not there was a significant difference between the two treatments?

References

Diggle, P. J., Heagerty, P., Liang, K. Y., and Zeger, S. L., 2002. *Analysis of Longitudinal Data,* 2nd Edition. Oxford: Oxford University Press.

Kleinbaum, D. G. and Klein, M. 2002. *Logistic Regression– A Self-Learning Text,* 2nd Edition. New York and Berlin: Springer Publishers.

Laird, N. M. and Ware, J. H. 1982. "Random Effects Models for Longitudinal Data, *Biometrics*, 38 (4): 963–974.

Littell, R. C., Milliken, G. A., Stroup, W. W., and Wolfinger, R. D. 1996. *SAS System for Mixed Models.* Cary NC: SAS Institute.

Ortiz, D. J., Marcus, M., Gerr, F., Jones, W., and Cohen, S. 1997. "Estimation of Within Subject Variability in Upper Extremity Posture Measured With Manual Goniometry Among Video Display Terminal Users." *Applied Ergonomics* 28(2): 139–143.

Zeger, S. L. and Liang, K. Y. 1986. "Longitudinal Data Analysis for Discrete and Continuous Outcomes." *Biometrics* 42: 121–30.

Analysis of Correlated Data
Part 2: Random Effects
and Other Issues

26.1 Preview

In this chapter, we continue our discussion of the general linear mixed model for analyzing correlated data by focusing on the use of random effects in such models. In Chapter 17 on Analysis of Variance methods, recall that we introduced and distinguished between *fixed effects* and *random effects*. Here, we revisit this distinction in the context of correlated data. We then return to the analysis of the data from the Air Pollution Study described in the previous chapter, and we consider a model containing random effects to address the secondary objective of the analysis, the identification of sensitive subgroups or individuals most severely affected by the pollution episode. We then consider the use of random effects to analyze the data from the Study of the Posture of Computer Operators previously introduced in Chapter 25. We also use these data to describe the ANOVA "partitioning" approach for analyzing repeated measures data. Finally, we give a brief overview of how such analyses can be carried out for discrete-type (e.g., binary or count) outcomes using *generalized linear mixed models*.

26.2 Random Effects Revisited

We begin by reviewing the "classical" definitions of a *fixed factor* and a *random factor*, as described previously in Chapter 17:

Fixed Factor: A variable in a regression model whose possible values (i.e., levels) are the only ones of interest.

Random Factor: A variable in a regression model whose levels are regarded as a random sample from some large population of levels.

The above definitions can then be used to distinguish a fixed effect from a random effect:

Fixed Effect: A coefficient in a regression model corresponding to a fixed factor. Such an effect is considered a population parameter and is typically denoted by a Greek symbol (e.g., α, β, γ).

Random Effect (Definition 1): A term in a regression model corresponding to a random factor. Such an effect is considered a random variable and is typically denoted by a Latin letter (e.g., a, b, g).

When applying the above definitions to epidemiologic studies, we typically postulate that:

a. Subjects, litters, observers, families, households are *random factors*

b. Gender, age, marital status, day of the week, education are *fixed factors*

c. Locations, treatments, clinics, exposures, time may be considered as *either random or fixed factors,* depending on the context of the study.

When in doubt, one approach for deciding how to classify a particular study variable is to consider the following question: "If I was able to replicate the study, would I want a given factor to have the exact same categories as observed in the current study?" Equivalently, would I want a replicate study to use the same treatments, days of the week, or subjects as used in the current study? If your answer is:

yes: treat the factor as fixed;

no: treat the factor as random

An alternative way to define a random effect, which gives another perspective on how to interpret a random effect, is given by the following second definition.

Random Effect (Definition 2): A random effect is a random variable that is included in a mixed model *to account for the effect of the natural heterogeneity among subjects* on the prediction of a response variable (Y) of interest, e.g.,

i. One random effect for the intercept: $Y_{ij} = (\beta_0 + \beta_1 X_{ij}) + b_{i0} + E_{ij}$

ii. Two random effects (intercept and slope): $Y_{ij} = (\beta_0 + \beta_1 X_{ij}) + (b_{i0} + b_{i1} X_{ij}) + E_{ij}$

The idea of heterogeneity among subjects when there is a single random effect for the intercept is illustrated in Figures 26.1 and 26.2. Figure 26.1 considers three subjects (Harry, Barry, and Gary), each of whom has his own straight-line model relating $X = $ AGE to $Y = $ SBP. These subject-specific straight-line models differ in their subject-specific intercepts because of the random effects b_{H0}, b_{B0}, and b_{G0}, for which $E(b_{H0}) = E(b_{B0}) = E(b_{G0}) = 0$. Furthermore, each subject-specific model differs from the population model identified by the equation $E(Y \mid X) = \beta_0 + \beta_1 X$.

Figure 26.2 plots four repeated (X, Y) measurements on Harry, where again Harry's line differs from the population line because of Harry's subject-specific intercept. Each of these four measurements are sample observations of SBP taken at different ages. Because of random variation, each repeated (X, Y) measurement does not lie exactly on Harry's line, and the differences between the measurements and Harry's line are denoted by the error terms $e_{Hj} = Y_{Hj} - E(Y_H \mid X_{Hj}, b_{H0})$, $j = 1, 2, 3, 4$.

FIGURE 26.1   Heterogeneity among subjects with random intercepts

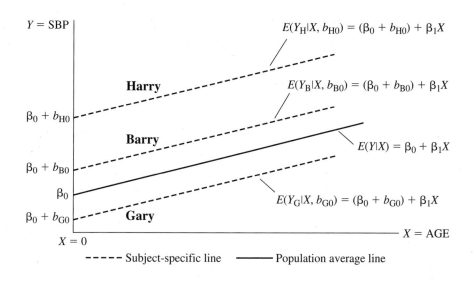

FIGURE 26.2   Heterogeneity with a random intercept

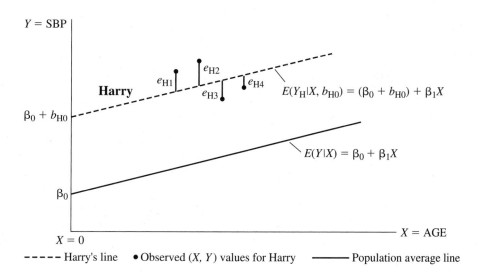

Figure 26.3 considers three different subjects Holly, Dolly, and Molly, each of whom has her own straight-line model relating $X = $ AGE to $Y = $ SBP. These subject-specific straight-line models differ in both their intercepts (because of the random effects b_{H0}, b_{D0}, and b_{M0}) and slopes (because of the random effects b_{H1}, b_{D1}, and b_{M1}); all these random effects are assumed to have expected (or true average values) of zero. Furthermore, each subject-specific model differs from the population model identified by the equation $E(Y|X) = \beta_0 + \beta_1 X$.

FIGURE 26.3 **Heterogeneity among subjects with random intercepts and slopes**

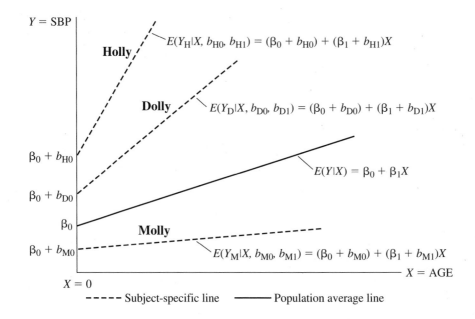

---- Subject-specific line ——— Population average line

Figure 26.4 plots four repeated (X, Y) measurements on Dolly, where again Dolly's line differs from the population line because of Dolly's subject specific-intercept and subject-specific slope. As with Harry in Figure 26.2, each of the four measurements on Dolly are sample observations of SBP taken at different ages. Because of random variation, each repeated measurement does not lie exactly on Dolly's line, and the differences between the measurements and Dolly's line are denoted by the error terms $e_{Dj} = Y_{Dj} - E(Y_D|X_{Dj}, b_{D0}, b_{D1}), j = 1, 2, 3, 4$.

FIGURE 26.4 **Heterogeneity with a random intercept and a random slope**

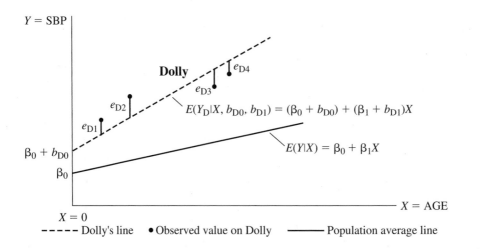

---- Dolly's line • Observed value on Dolly ——— Population average line

From the above discussion and Figures 26.1 through 26.4, we see that "heterogeneity of effects among subjects" indicates that each subject in the population under study has his or her own regression model, e.g.,

$$E(Y|X, b_{i0}, b_{i1}) = (\beta_0 + b_{i0}) + (\beta_1 + b_{i1})X \quad \text{for subject } i,$$

whose general form will depend on the number and type of random effects. Each subject-specific model will differ from a population average model, e.g.,

$$E(Y|X) = \beta_0 + \beta_1 X,$$

whose form contains population (average) parameters only since $E(b_{i0}) = E(b_{i1}) = 0$.

Thus, it may be argued that *another reason for considering random effects in a mixed model*, other than to identify individual subject effects as previously mentioned for the FEV1 data, *is to determine whether or not the appropriate model needs to account for such "heterogeneity of effects among subjects."* That is, we may want to answer the question: "Is a model that incorporates subject-specific effects more appropriate than a model that ignores subject-specific effects?" A decision to include specific types of random effects (e.g., random slopes) in a mixed model should always involve subject-matter considerations. A related question is: "Assuming that random effects are needed, how many random effects need to be used and how should they be defined in the model?" These questions suggest that tests about the overall significance of all random effects considered together, as well as tests of significance about individual random effects, are of interest. We describe such tests in the context of a second example in the next section.

26.3 Results for Models with Random Effects Applied to Air Pollution Study Data

The mixed linear model analyses previously described in Chapter 25, Section 25.4 for the FEV1 dataset have addressed the following primary objective: determine whether or not true average FEV1 levels were depressed during and after the pollution alert when considering the 40 school-children as a sample from a larger population of schoolchildren. In this section, we address the secondary objective: identify sensitive subgroups or individuals most severely affected by the pollution episode. This second objective focuses on drawing conclusions about individual subjects in the study, rather than on making statistical inferences about the population mean FEV1 levels over the five weeks of the study. When individuals are the focus of the study, the appropriate approach to the analysis is to consider a model with one or more random effects, since such random effects are defined to represent subject-specific effects.

26.3.1 A Subject-Specific Model with a Random Intercept

The simplest form of subject-specific model for the FEV1 data involves adding to model (25.7) of Chapter 25 a single subject-specific random effect b_{i0} to the population-average intercept β_0, an addition that may be stated in scalar and matrix terms, respectively, as follows:

$$Y_{ij} = \beta_0 + \beta_1 W_{ij1} + \beta_2 W_{ij2} + \beta_3 W_{ij3} + \beta_4 W_{ij4} + b_{i0} + E_{ij} \tag{26.1}$$

which contains one random effect b_{i0} for the ith subject in addition to the fixed factors W_1 through W_4 and their corresponding fixed effects (i.e., β_1, β_2, β_3, β_4). The subject-specific matrix form of this model is given as:

$$\mathbf{Y}_i = \mathbf{X}_i\boldsymbol{\beta} + \mathbf{Z}_i\mathbf{b}_i + \mathbf{E}_i, \qquad i = 1, 2, \ldots, 40 \tag{26.2}$$

where $\mathbf{Y}_i$ denotes the set of 5 FEV1 measurements on the ith child, $\mathbf{X}_i$ denotes the set of inter-cept and dummy variable values for each of the five weeks for subject i, $\boldsymbol{\beta}$ denotes the set of five β's in this model, and $\mathbf{E}_i$ denotes the set of five error terms for the ith subject. [The $\mathbf{X}_i$ matrix and the $\boldsymbol{\beta}$ and $\mathbf{E}_i$ vectors were defined earlier in (25.9).] The $\mathbf{Z}_i$ matrix in matrix model (26.2) is the (5×1) column vector containing all ones, i.e.,

$$\mathbf{Z}_i = \begin{bmatrix} 1 \\ 1 \\ 1 \\ 1 \\ 1 \end{bmatrix}$$

and the $\mathbf{b}_i$ vector is simply the (1×1) scalar b_{i0}. (We will see shortly that the above simple subject-specific model is not appropriate for answering the secondary objective.)

Since model (26.2) contains both fixed and random effects, we need to specify both an $\mathbf{R}_i$ matrix and a $\mathbf{G}$ matrix in order to define and fit this model. Here, the $\mathbf{G}$ matrix is a scalar [i.e., a (1×1) matrix] that specifies the variance (say, σ_0^2) of the random intercept effect b_{i0}. The sim-plest form of $\mathbf{R}_i$ matrix that can be specified is an independence structure, i.e., $\sigma_e^2 \mathbf{I}_5$, where σ_e^2 denotes the variance of any error component (E_{ij}). The assumption of "conditional" indepen-dence (i.e., $\mathbf{R}_i = \sigma_e^2 \mathbf{I}_5$) is a typical choice when the model contains random effects; this assump-tion basically states that, once the random effects are accounted for (i.e., conditional on b_{i0} being fixed), the remaining error terms (E_{ij}'s) are (conditionally) independent of one another.

Table 26.1 shows the estimated fixed effects, the overall F test concerning the equality of the five FEV1 population means, the F test for the significance of the contrast of interest, and the estimated correlation structures for model (26.2). This table also shows the previous output (Table 25.7) for a marginal model in which $\mathbf{R}$ is CS and $\mathbf{G} = \mathbf{0}$.

From Column 1, we can see that the estimated regression coefficients, as well as both model-based and empirical standard errors, are identical for both sets of $\mathbf{R}$ and $\mathbf{G}$ matrices. From Column 2, we also see that the F statistic values of 55.26 (model-based) and 38.61 (empir-ical) are also identical for the two choices of covariance structures. These results are as expected since we have previously stated that the general covariance matrix formula

$$\mathbf{V}_i = \mathbf{Z}_i \mathbf{G} \mathbf{Z}_i' + \mathbf{R}_i$$

yields the same overall covariance structure for $\mathbf{V}_i$ when the $\mathbf{G}$ and $\mathbf{R}$ matrices are either

$$\{\mathbf{G} = \sigma_0^2, \ \mathbf{R} = \sigma_e^2 \mathbf{I}_5\} \quad \text{or} \quad \{\mathbf{G} = \mathbf{0}, \ \mathbf{R} = CS\}.$$

The only difference in the output for the two fitted models is the 156 denominator DF (Den DF) for the random intercept model versus the 195 Den DF for the marginal model in Col-umn 2. We have previously mentioned that SAS's MIXED procedure allows the user to choose among several options for the Den DF, so we could have used 195 for both models. However, when a random effects model is used (see Footnote 14 in Chapter 25), the popular choice for the Den DF involves subtracting from the residual DF of 195 the value $q^*(K - 1)$ to account for the number (q^*) of random effects being estimated. Since $q^* = 1$ and there are $K = 40$ values of b_{i0} to be estimated (one for each subject), $q^*(K - 1) = 39$, so that the resulting Den DF value is $(195 - 39) = 156$. Since the marginal model does not require estimation of random effects, one does not need to subtract additional DF from the residual DF of 195 for this model.

unstructured ($\mathbf{R} = \mathbf{UN}$). The program code for $\mathbf{R} = \mathbf{CS}$ using SAS's MIXED procedure is given as follows:

```
proc mixed data = sf;
  class subj day time;
  model sf = day time day*time/s;
  repeated/ type = cs subject = subj s rcorr;
  run;
```

(The program code for autoregressive and unstructured covariance matrices would involve replacing the *type=cs* statement by *type=ar(1)* and *type=un*, respectively.)

From the output (see Table 26.7), the same conclusions about the significance of the effects are made regardless of the choice of $\mathbf{R}$; that is, the main effects of Day and Time, as well as the interaction effects of Day with Time, are all non-significant ($P > .05$). As expected, model-based estimated standard errors differ as $\mathbf{R}$ differs. As with the FEV1 dataset previously analyzed, empirical estimated standard errors are identical regardless of the choices for $\mathbf{R}$. Similarly, model-based F statistic values vary as $\mathbf{R}$ varies, whereas empirical F statistic values are identical regardless of the choice for $\mathbf{R}$. Furthermore, as expected, the estimated correlations in the unstructured correlation matrix are all different. Nevertheless, since all such correlations are relatively high and roughly similar in value, a CS/exchangeable correlation structure seems reasonable to use.

The above analyses, involving three choices for $\mathbf{R}$, support the conclusion that there are no statistically significant main effects or interaction effects involving the predictors Day and Time. If one decided to postulate a correlation structure appropriate for these data before considering the numerical results in Table 26.7, one might initially consider an autoregressive (**AR1**) structure since the study is longitudinal (so that correlations decrease as the time between repeated observations increases). Nevertheless, we might expect repeated observations on the same day to be more correlated than observations on different days, which suggests a more complex correlation structure than either **AR1** or compound symmetric (**CS**). There are many other choices for the correlation structure still available, including structures that result from the use of random effects. In particular, we might want to account for the natural heterogeneity among "Subjects," a third predictor variable, when modelling of the response variable *SF*.

26.4.3   A Random Intercept Model for the SF Data

The simplest random effects model involving Subjects uses a single random effect associated with the intercept. The subject-specific scalar form for this model can be written as follows:

$$SF_{ij} = \beta_0 + b_{i0} + \beta_1 D_{ij1} + \beta_2 D_{ij2} + \beta_3 T_{ij} + \beta_4(D_{ij1}T_{ij}) + \beta_5(D_{ij2}T_{ij}) + E_{ij} \qquad (26.10)$$

where D_{ij1}, D_{ij2} and T_{ij} are the values of variables D_1, D_2, and T, respectively, for the *j*th observation on the *i*th subject as defined previously in the marginal model (26.8). Model (26.10) contains two random components, E_{ij} and b_{i0}. As for the marginal model, E_{ij} denotes a random error term. The new term, b_{i0}, denotes a random effect for the (random) Subjects factor, which gives a random intercept ($\beta_0 + b_{i0}$) for subject *i*. The typical distributional assumptions made about E_{ij} and b_{i0} are that they are each normally distributed as $N(0, \sigma_e^2)$ and $N(0, \sigma_0^2)$, respectively, and that b_{i0} and E_{ij} are mutually independent for all *i, j*.

TABLE 26.7 Edited output based on modeling shoulder flexion data using predictors day and time for different marginal models (G = 0), using REML in SAS's MIXED procedure

CS

Column 1

Effect	Estimate	Model-based SE	Empirical SE
Intercept	16.7895	2.8984	3.0368
day 1	1.6842	1.9257	1.6274
day 2	1.7368	1.9257	2.4243
time	−1.2105	1.9257	1.2990
day × time 1	1.6842	2.7234	1.7771
day × time 2	2.4211	2.7234	1.9879

Column 2[†,‡] — Type 3 Tests of Fixed Effects

Effect	Num DF	Den DF	Model-based F Value	Empirical F Value
day	2	36	2.74	1.33
time	1	18	0.02	0.02
day × time	2	36	0.42	0.84

Column 3[◊] — $R = \sigma^2 C^{(EXCH)}$ where $\hat{C}^{(EXCH)}$ is

1.0000	0.7792	0.7792	0.7792	0.7792	0.7792
0.7792	1.0000	0.7792	0.7792	0.7792	0.7792
0.7792	0.7792	1.0000	0.7792	0.7792	0.7792
0.7792	0.7792	0.7792	1.0000	0.7792	0.7792
0.7792	0.7792	0.7792	0.7792	1.0000	0.7792
0.7792	0.7792	0.7792	0.7792	0.7792	1.0000

$\sigma^2 = \hat{\sigma}_0^2 + \hat{\sigma}_1^2$, where $\hat{\sigma}_0^2 = 124.38$, $\hat{\sigma}_1^2 = 35.2304$

AR1

Column 1

Effect	Estimate	Model-based SE	Empirical SE
Intercept	16.7895	2.9017	3.0368
day 1	1.6842	3.1298	1.6274
day 2	1.7368	2.4390	2.4243
time	−1.2105	1.8158	1.2990
day × time 1	1.6842	2.6325	1.7771
day × time 2	1.6842	2.6671	1.9879

Column 2[†,‡] — Type 3 Tests of Fixed Effects

Effect	Num DF	Den DF	Model-based F Value	Empirical F Value
day	2	36	1.05	1.33
time	1	18	0.03	0.02
day × time	2	36	0.44	0.84

Column 3[◊] — $R = AR1$ where $\hat{C}^{(AR1)}$ is

1.0000	0.8042	0.6467	0.5201	0.4183	0.3364
0.8042	1.0000	0.8042	0.6467	0.5201	0.4183
0.6467	0.8042	1.0000	0.8042	0.6467	0.5201
0.5201	0.6467	0.8042	1.0000	0.8042	0.6467
0.4183	0.5201	0.6467	0.8042	1.0000	0.8042
0.3364	0.4183	0.5201	0.6467	0.8042	1.0000

UN

Column 1

Effect	Estimate	Model-based SE	Empirical SE
Intercept	16.7895	3.1200	3.0368
day 1	1.6842	1.6720	1.6274
day 2	1.7368	2.4907	2.4243
time 3	−1.2105	1.3346	1.2990
day × time 1	1.6842	1.8218	1.7771
day × time 2	2.4211	2.0424	1.9879

Column 2[†,‡] — Type 3 Tests of Fixed Effects

Effect	Num DF	Den DF	Model-based F Value	Empirical F Value
day	2	36	1.26	1.33
time	1	18	0.02	0.02
day × time	2	36	0.80	0.84

Column 3[◊] — $R = UN$ where $\hat{C}^{(UN)}$ is

1.0000	0.8525	0.8655	0.7820	0.7301	0.7953
0.8525	1.0000	0.8821	0.8660	0.7401	0.8466
0.8655	0.8821	1.0000	0.8333	0.7307	0.7442
0.7820	0.8660	0.8333	1.0000	0.6006	0.6802
0.7301	0.7401	0.7307	0.6006	1.0000	0.9041
0.7953	0.8466	0.7442	0.6802	0.9041	1.0000

[†] All F statistics are non-significant with $P > .05$.

[‡] Denominator DF's (i.e., DDFM) were determined using MIXED's "between-within" option, which is the default method used when the program code uses a repeated statement. See Section 26.4.7 for further discussion of DDFM.

[◊] All 3 correlation matrices are (6 × 6) because each subject is observed 6 times.

The subject-specific matrix version of (26.10) can be written as follows:

$$\mathbf{SF}_i = \mathbf{X}\boldsymbol{\beta} + \mathbf{Z}\mathbf{b}_i + \mathbf{E}_i \tag{26.11}$$

where $\mathbf{SF}_i(6 \times 1)$, $\mathbf{X}(6 \times 6)$, $\boldsymbol{\beta}(6 \times 1)$, and $\mathbf{E}_i(6 \times 1)$ are defined as for the marginal model (26.9), and where

$$\mathbf{Z}(6 \times 1) = \begin{bmatrix} 1 \\ 1 \\ 1 \\ 1 \\ 1 \\ 1 \end{bmatrix}, \qquad \mathbf{b}_i(1 \times 1) = b_{i0}$$

Table 26.8 provides edited computer output based on fitting the random intercept model (26.10), or equivalently (26.11), together with the previous output shown in Table 26.7 for a compound symmetric ($\mathbf{R} = \mathbf{CS}$) marginal model fit to these same data. The program code for fitting the random intercept model using SAS's MIXED procedure is given as follows:

```
proc mixed data = sf;
  class subj day time;
  model sf = day time day*time/s;
  random intercept/ subject = subj g;
run;
```

(An equivalent way to write the *random* statement in program code for the random intercept model for the sf data is*: random subj / g;)*

From column 1 of Table 26.8, we can see that the estimated regression coefficients, as well as both model-based and empirical estimated standard errors, are identical for both choices of covariance structures.

From column 2, we also see that corresponding model-based and empirical F statistics differ *within* each separate model, yet both models produce identical sets of F statistics. These results are as expected since we have previously shown that the general covariance formula

$$\mathbf{V}_i = \mathbf{Z}_i \mathbf{G} \mathbf{Z}_i' + \mathbf{R}_i$$

yields the same overall covariance structure for $\mathbf{V}_i$ when the $\mathbf{G}$ and $\mathbf{R}$ matrices are either

$$\{\mathbf{G} = \sigma_0^2, \ \mathbf{R} = \sigma_e^2 \mathbf{I}\} \qquad \text{or} \qquad \{\mathbf{G} = \mathbf{0}, \ \mathbf{R} = \mathbf{CS}\}$$

The estimated working correlation structures for the two models are seen to be identical in column 3 of Table 26.8.

The denominator degrees of freedom (i.e., den DF or DDFM) shown in column 2 are also identical for both models even though two different (default) methods were used: "contain" for the random intercept model versus "between-within" for the marginal model. We have previously mentioned that SAS's MIXED procedure allows the user to choose among several options for DDFM. We provide further discussion on these different DDFM options in Section 26.4.7.

TABLE 26.8 Edited output based on modeling shoulder flexion data using predictors day and time, comparing random intercept model (26.10) with marginal CS model (26.11) with marginal CS model using REML in SAS's MIXED procedure

Column 1

$\mathbf{G} = \sigma_0^2, \ \mathbf{R} = \sigma_e^2 \mathbf{I}_6$

Effect	Estimate	Model-based SE	Empirical SE
Intercept	16.7895	2.8984	3.0368
day 1	1.6842	1.9257	1.6274
day 2	1.7368	1.9257	2.4243
time	−1.2105	1.9257	1.2990
day × time 1	1.6842	2.7234	1.7771
day × time 2	2.4211	2.7234	1.9879

$\mathbf{G} = 0, \ \mathbf{R} = \mathbf{CS}$

Effect	Estimate	Model-based SE	Empirical SE
Intercept	16.7895	2.8984	3.0368
day 1	1.6842	1.9257	1.6274
day 2	1.7368	1.9257	2.4243
time	−1.2105	1.9257	1.2990
day × time 1	1.6842	2.7234	1.7771
day × time 2	2.4211	2.7234	1.9879

Column 2 [†,‡]

Type 3 Tests of Fixed Effects

Effect	Num DF	Den DF	Model-based F Value	Empirical F Value
day	2	36	2.74	1.33
time	1	18	0.02	0.02
day × time	2	36	0.42	0.84

Type 3 Tests of Fixed Effects

Effect	Num DF	Den DF	Model-based F Value	Empirical F Value
day	2	36	2.74	1.33
time	1	18	0.02	0.02
day × time	2	36	0.42	0.84

Column 3

$\mathbf{V} = \mathbf{ZGZ'} + \mathbf{R}$ where $\hat{\mathbf{C}}_V$ is

$$
\begin{bmatrix}
1.0000 & 0.7792 & 0.7792 & 0.7792 & 0.7792 & 0.7792 \\
0.7792 & 1.0000 & 0.7792 & 0.7792 & 0.7792 & 0.7792 \\
0.7792 & 0.7792 & 1.0000 & 0.7792 & 0.7792 & 0.7792 \\
0.7792 & 0.7792 & 0.7792 & 1.0000 & 0.7792 & 0.7792 \\
0.7792 & 0.7792 & 0.7792 & 0.7792 & 1.0000 & 0.7792 \\
0.7792 & 0.7792 & 0.7792 & 0.7792 & 0.7792 & 1.0000
\end{bmatrix}
$$

$\hat{\sigma}_0^2 = 124.38, \ \hat{\sigma}_e^2 = 35.2304$

$\mathbf{V} = \sigma^2 \mathbf{C}^{(\text{EXCH})}$ where $\hat{\mathbf{C}}^{(\text{EXCH})}$ is

$$
\begin{bmatrix}
1.0000 & 0.7792 & 0.7792 & 0.7792 & 0.7792 & 0.7792 \\
0.7792 & 1.0000 & 0.7792 & 0.7792 & 0.7792 & 0.7792 \\
0.7792 & 0.7792 & 1.0000 & 0.7792 & 0.7792 & 0.7792 \\
0.7792 & 0.7792 & 0.7792 & 1.0000 & 0.7792 & 0.7792 \\
0.7792 & 0.7792 & 0.7792 & 0.7792 & 1.0000 & 0.7792 \\
0.7792 & 0.7792 & 0.7792 & 0.7792 & 0.7792 & 1.0000
\end{bmatrix}
$$

$\hat{\sigma}^2 = \hat{\sigma}_0^2 + \hat{\sigma}_e^2$, where $\hat{\sigma}_0^2 = 124.38, \ \hat{\sigma}_e^2 = 35.2304$

[†] All F statistics are non-significant with $P > .05$.

[‡] Denominator DFs (i.e., DDFM) for random intercept model (26.11) were determined using MIXED's "contain" option, which is the default method used when the program code uses a random statement. In contrast, the DDFMs for the marginal CS model were determined using MIXED's "between-within" option, which is the default method used when the program code uses a repeated statement. See Section 26.4.7 for further discussion about DDFM.

We have shown during our previous analysis of the FEV1 pollution study data that any model containing random effects produces both estimated fixed effects and estimated random effects. The computer program code can specify such output via the "*s*" term (bolded for emphasis) in the "*random*" statement below:

```
model sf = day time day*time/s;
random intercept/ subject = subj  s  g;
```

Nevertheless, for the SF posture measurement data, the investigator was not specifically interested in identifying individual subject heterogeneity, but rather was focused on assessing the effects of the fixed factors Day and Time. Consequently, a subject-specific random effect estimate $(\hat{b}_{i0})$ for each of the 19 study subjects was not requested in the program code (i.e., the "*s*" term was not used).

The results from using the random intercept model indicate that neither the main effects of Day or Time, nor the interaction effects of Day with Time, were statistically significant. This finding supported the investigators' interest in using posture measurements that were not influenced by either the day of the week or the time of day when attempting to evaluate the extent to which the postures of computer operators are associated with muscular-skeletal disorders in a subsequent study.

26.4.4 Random Effects 3-Way ANOVA Model for the SF Data

A more complex random effects model involving Subjects allows for the random effect interactions with the two fixed effect predictor variables Day and Time. This model can be written in subject-specific scalar form as follows:

$$SF_{ij} = \beta_0 + b_{i0} + \beta_1 D_{ij1} + \beta_2 D_{ij2} + \beta_3 T_{ij} + \beta_4 (D_{ij1} T_{ij}) + \beta_5 (D_{ij2} T_{ij})$$
$$+ b_{i1} D_{ij1} + b_{i2} D_{ij2} + b_{i3} T_{ij} + E_{ij} \tag{26.12}$$

where D_{ij1}, D_{ij2} and T_{ij} are the values of variables D_1, D_2, and T, respectively, for the jth observation on the ith subject as defined in previous models (26.8) and (26.10). Model (26.12), however, now contains random effects b_{i0}, b_{i1}, b_{i2}, and b_{i3} in addition to the random error term E_{ij}. As in model (26.10), b_{i0} denotes a random intercept effect for the (random) Subjects factor. The random components b_{i1} and b_{i2} denote two random effects that reflect the interaction of Subjects with Day, and the random component b_{i3} denotes a fourth random effect that reflects the interaction of Subjects with Time. The typical distributional assumptions made about the E_{ij}, b_{i0}, b_{i1}, b_{i2}, and b_{i3} are that they are each normally distributed as $N(0, \sigma_e^2)$, $N(0, \sigma_0^2)$, $N(0, \sigma_{D1}^2)$, $N(0, \sigma_{D2}^2)$, and $N(0, \sigma_T^2)$, respectively, and that E_{ij}, b_{i0}, b_{i1}, b_{i2}, and b_{i3} are mutually independent for all i, j. Moreover, since the random effects b_{i1} and b_{i2} concern the two dummy variables (D_{ij1} and D_{ij2}) for the three-level nominal variable Day, it is typically assumed (as in SAS's MIXED procedure) that the corresponding variance components σ_{D1}^2 and σ_{D2}^2 are equal, i.e., $\sigma_{D1}^2 = \sigma_{D2}^2 \equiv \sigma_D^2$, say.

The subject-specific matrix version of (26.12) can be written as follows:

$$\mathbf{SF}_i = \mathbf{X}\boldsymbol{\beta} + \mathbf{Z}\mathbf{b}_i + \mathbf{E}_i \tag{26.13}$$

where $\mathbf{SF}_i(6 \times 1)$, $\mathbf{X}(6 \times 6)$, $\boldsymbol{\beta}(6 \times 1)$, and $\mathbf{E}_i(6 \times 1)$ are defined as in models (26.9) and (26.11), and where

$$\mathbf{Z}(6 \times 4) = \begin{bmatrix} 1 & 1 & 0 & 1 \\ 1 & 1 & 0 & 0 \\ 1 & 0 & 1 & 1 \\ 1 & 0 & 1 & 0 \\ 1 & 0 & 0 & 1 \\ 1 & 0 & 0 & 0 \end{bmatrix} \quad \text{and} \quad \mathbf{b}_i(4 \times 1) = \begin{bmatrix} b_{i0} \\ b_{i1} \\ b_{i2} \\ b_{i3} \end{bmatrix}$$

Table 26.9 provides edited computer output based on fitting model (26.12), or equivalently (26.13), together with the previous output shown in Table 26.8 for a random intercept model (26.10) involving these same data. The program code for fitting model (26.12) using SAS's MIXED procedure is given as follows:

```
proc mixed data = sf;
  class subj day time;
  model sf = day time day*time/s;
  random intercept day time/ subject = subj g;
run;
```

(An equivalent way to write the *random* statement in program code is: *random subj subj*day subj*time / g;*)

In Column 1 of Table 26.9, the set of model-based estimated standard errors for model (26.12) containing four random effects is seen to be different from the corresponding set of model-based estimated standard errors for the random intercept model (26.10). This result is as expected, since the addition of the three random effects in model (26.12) changes the $\mathbf{G}$ matrix and corresponding overall covariance structure $\mathbf{V}$.

The difference in correlation structures between model (26.12) and model (26.10) is also reflected in column 2 of Table 26.9 by the different sets of model-based F statistic values, i.e., $\{1.78, 0.02. 0.79\}$ versus $\{2.74, 0.02. 0.42\}$.

From Column 3 of Table 26.9, the estimated correlation structure for model (26.12) is not "exchangeable," as is the correlation structure for the random intercept model (26.10).

We may wonder why regression model (26.12), which contains 4 random effects, is being considered at all for this analysis. The answer is that such a model makes sense if we consider the analysis problem as involving the three predictors Day, Time, and Subjects, each of which explains to some extent the total variability involved in all 114 observations on the 19 subjects in the study. In other words, we may wish to consider model (26.12) as equivalent to a *Three-Way ANOVA model*, with the factor Subjects treated as random and the factors Day and Time treated as fixed.

By using such an ANOVA approach, we are effectively allowing for a correlation structure that reflects the partitioning of the total variability (as described in Chapters 17–20 for classical One- and Two-Way ANOVA data) into variance contributions for the three factors being considered. The ANOVA approach for the analysis of repeated measures data is described in the next section. The reader should realize, nevertheless, that the ML approach for fitting a linear mixed model is preferred, since it includes the ANOVA approach as a special case and also applies to a much larger variety of correlated data analysis problems.

TABLE 26.9 Edited output based on modeling shoulder flexion data using predictors day and time and comparing mixed model (26.12) with 4 random effects with random intercept model (26.10), using REML in SAS's MIXED procedure

Column 1

$G(4 \times 4)$, $R = \sigma_e^2 I_6$

Effect		Estimate	Model-based SE	Empirical SE
Intercept		16.7895	2.8984	3.0368
day	1	1.6842	1.9577	1.6274
day	2	1.7368	1.9577	2.4243
time		−1.2105	1.5374	1.2990
day × time	1	1.6842	1.9697	1.7771
day × time	2	2.4211	1.9697	1.9879

$G = \sigma_0^2$, $R = \sigma_e^2 I_6$

Effect		Estimate	Model-based SE	Empirical SE
Intercept		16.7895	2.8984	3.0368
day	1	1.6842	1.9257	1.6274
day	2	1.7368	1.9257	2.4243
time		−1.2105	1.9257	1.2990
day × time	1	1.6842	2.7234	1.7771
day × time	2	2.4211	2.7234	1.9879

Column 2[†,‡]

Type 3 Tests of Fixed Effects

Effect	Num DF	Den DF	Model-based F Value	Empirical F Value
day	2	36	1.78	1.33
time	1	18	0.02	0.02
day × time	2	36	0.79	0.84

Type 3 Tests of Fixed Effects

Effect	Num DF	Den DF	Model-based F Value	Empirical F Value
day	2	36	2.74	1.33
time	1	18	0.02	0.02
day × time	2	36	0.42	0.84

Column 3

$V = ZGZ' + R$ where $\hat{C}^V$ is

1.0000	0.8593	0.7719	0.7467	0.7719	0.7467
0.8593	1.0000	0.7467	0.7719	0.7467	0.7719
0.7719	0.7467	1.0000	0.8593	0.7719	0.7467
0.7467	0.7719	0.8593	1.0000	0.7467	0.7719
0.7719	0.7467	0.7719	0.7467	1.0000	0.8593
0.7467	0.7719	0.7467	0.7719	0.8593	1.0000

$V = ZGZ' + R$ where $\hat{C}^V$ is

1.0000	0.7792	0.7792	0.7792	0.7792	0.7792
0.7792	1.0000	0.7792	0.7792	0.7792	0.7792
0.7792	0.7792	1.0000	0.7792	0.7792	0.7792
0.7792	0.7792	0.7792	1.0000	0.7792	0.7792
0.7792	0.7792	0.7792	0.7792	1.0000	0.7792
0.7792	0.7792	0.7792	0.7792	0.7792	1.0000

$\hat{\sigma}_0^2 = 124.38$, $\hat{\sigma}_e^2 = 35.2304$

† All F statistics are non-significant with $P > .05$.

‡ Denominator DFs (i.e., DDFM) for random intercept model (26.10) were determined using MIXED's "contain" option, which is the default method used when the program code uses a random statement. See Section 26.4.7 for further discussion about DDFM.

26.4.5 Repeated Measures ANOVA

The ANOVA approach for the analysis of correlated/repeated measures data typically involves the following steps:

1. **Specify the ANOVA model**; make sure to distinguish fixed factors from random factors and to state the distributional (i.e., normality and independence) assumptions for the random factors (including defining the relevant variance components).

2. **Partition the total sums of squares** according to the various sources of variation, including subject-to-subject variation.

3. **Form the ANOVA table** that corresponds to the ANOVA model specified in Step 1 and that incorporates the partitioning of the total sums of squares delineated in Step 2.

4. **Determine the formulas for expected mean squares** in terms of fixed effects and variance component parameters.

5. **Determine the appropriate *F*-statistic** for each test of hypothesis of interest as the ratio of the mean square for the factor being evaluated to the mean square for the source of variation that yields the same expected mean square as the numerator expected mean square under the null hypothesis.

6. **Carry out *F* tests of interest using a convenient computer program for ANOVA**; then **summarize your results**, using, as appropriate, the hypothesis testing results, sample means, estimated standard errors for estimated effects of interest, and confidence intervals for contrasts of interest involving population means.

SAS's GLM procedure is particularly well-suited for carrying out balanced (and some unbalanced) repeated measures ANOVA procedures for continuous outcome data. Nevertheless, SAS's MIXED procedure uses a ML approach applied to a more general model than the (ANOVA) model used by GLM. Users of other packages should check the list of procedures available in such packages to determine whether programs similar to MIXED or GLM are available.

Returning to the posture measurement study involving Shoulder Flexion (**SF**) data (introduced in Section 25.2), we can write the classical ANOVA effects version of this model as follows:

$$Y_{ijk} = \mu + b_i + \delta_j + \tau_k + (\delta\tau)_{jk} + b_{ij} + b_{ik} + E_{ijk} \qquad (26.14)$$

where μ denotes the overall population mean, $\{\delta_j\}$ are the fixed effects of the Day factor, $\{\tau_k\}$ are the fixed effects of the Time factor, $\{(\delta\tau)_{jk}\}$ are the fixed interaction effects of Day with Time, and $\{b_i\}$, $\{b_{ij}\}$, and $\{b_{ik}\}$ are the appropriate random effects due to Subjects and their interactions with Day and Time, respectively. The error term E_{ijk} is a three-way interaction term that is assumed to represent only random error.

For the above model, it is assumed that b_i, b_{ij}, b_{ik}, and E_{ijk} are mutually independent, that b_i is distributed as $N(0, \sigma_0^2)$, that b_{ij} is distributed as $N(0, \sigma_D^2)$, that b_{ik} is distributed as $N(0, \sigma_T^2)$, and that E_{ijk} is distributed as $N(0, \sigma_e^2)$. The terms σ_0^2, σ_D^2, and σ_T^2 are called the *variance components* of the model, and σ_e^2 is called the *error component*.

The total variability, in terms of sums of squares, associated with model (26.14) can be partitioned as shown in Figure 26.5. We first partition the total variation into between-subjects variation and within-subjects variation. We conduct a further partitioning by dividing the within-subjects variation into variations contributed separately by Factor D (Day), by Factor T (Time), and by their

FIGURE 26.5   **Partitioning the total sums of squares for three-way ANOVA of the SF posture measurement data involving day (factor D), time (factor T), and subjects (factor S)**

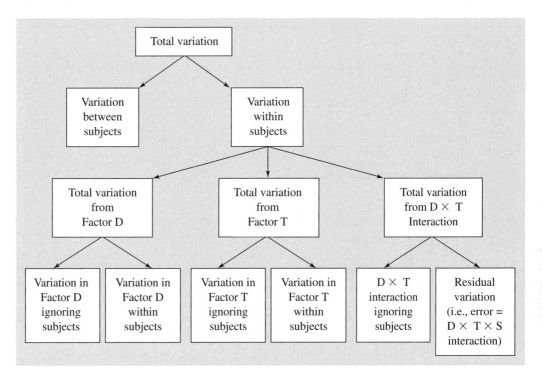

(D × T) interaction. This partitioning comes from the within-subjects box because the variation explained by the two factors and their interaction is derived from responses "within" subjects. A final partitioning is then carried out by separately dividing the total variation of each factor and associated interactions into two sources: the variation in the factor (or interaction) "ignoring" subjects, and the variation in the factor (or interaction) "within" subjects.

The boxes at the bottom of Figure 26.5 indicate that the "Variation in Factor D Ignoring Subjects" describes the main effect of Factor D (i.e., Day), whereas the "Variation in Factor D Within Subjects" describes the Subjects-by-Factor D Interaction; a similar interpretation holds for Factor T. The residual variation (σ_e^2) is actually the variation due to the interaction of Factor D with Factor T within Subjects (i.e., D × T × S interaction); but, because there is only one observation per subject at each combination of the levels of Factor D and Factor T, we must assume that these three-factor interaction effects represent random error E.

From the above partitioning, Table 26.10 can be produced to summarize the repeated measures ANOVA based on model (26.14). A pure estimate of the error for each subject is not possible because each subject is observed only once at each (day, time) combination, Therefore, we must assume that the mean square for this three-way interaction estimates σ_e^2.

From Table 26.10, we see that all three F-tests are non-significant. The F statistics in this table are correct because the expected mean squares of the numerator and denominator for each F statistic both estimate the same quantity under the null hypothesis being tested, i.e., the mean square terms in each F statistic represent two independent estimates of the same variance under the null hypothesis of interest.

TABLE 26.10 **ANOVA table for repeated measures ANOVA of shoulder flexion (SF) data in Table 25.5**

Source	DF	MS	F
Between Subjects	18	781.51	
Within Subjects	95		
Day	2	96.56	1.78 (P = .18)
Subjects × Day	36	54.39	
Time	1	0.71	0.02 (P =.88)
Subjects × Time	18	30.51	
Day × Time	2	14.63	0.79 (P = .50)
Subjects × Day × Time (i.e., Error)	36	18.43	
Total (corrected)	113		

For example, under the null hypothesis of no Day-by-Time interaction, MS(Day × Time) estimates σ_e^2 as does MS (Error). Similarly, under the null hypothesis of no main effect of Day, MS(Day) estimates the same quantity that is estimated by MS (Subjects × Day). We can therefore conclude that neither Day nor Time, nor the Day × Time interaction, are statistically significant factors in predicting Shoulder Flexion response for these data.

The repeated measures ANOVA results in Table 26.10 are identical to the model-based results in Table 26.9 previously obtained from SAS's MIXED procedure for the linear mixed model (26.12) containing random effects for Subjects, Subjects-by-Day, and Subjects-by-Time. Thus, the correlation structure chosen for model (26.12) is identical to the (unstated) correlation structure being assumed for the three-way repeated measures ANOVA analysis. Note, however, that the ANOVA analysis derived from partitioning sums of squares does not directly provide empirical standard errors, nor does it allow for the large variety of other choices for the correlation structure available with SAS's MIXED procedure.

In general ANOVA parlance, the study design that gave rise to the SF data is called a Balanced Repeated Measures Design with Two Crossover Factors. General ANOVA tables, expected mean square terms, and *F* test formulae for this type of design are provided in a section of the publisher's website for this text. These tables consider the cases when both factors are fixed and when either one or both factors are random. The website also considers three other commonly used repeated measures designs involving one or two factors, which may either be time-dependent (e.g., crossover) or time-independent (e.g., nest) variables.

26.4.6 Testing Hypotheses about Random Effects

In this section, we describe three methods for testing hypotheses about random effects, namely, assessing whether or not one or more variance components corresponding to random effects in a mixed linear model are equal to zero.

Method 1: Approximate LR test (Snijders and Bosker, 1999):

Method 1 involves a Likelihood Ratio (LR) test that compares two models (one with, and one without, the random effects of interest) using a (large sample) *chi square statistic with r d.f.*; *r* is the number of *variances* and *covariances* in the **G** matrix that are restricted to be equal to 0 under the null hypothesis that the variance components of the random effects in question are zero. The two-tailed approximate P-value for this test statistic must be halved to get the approximate one-tailed P-value since variance components cannot be negative.

■ **Example 26.1** We consider a simple random intercept linear mixed model for the SF data set involving the single predictor variable Time (T):

$$SF_{ij} = \beta_0 + b_{i0} + \beta_1 T_{ij} + E_{ij}$$

where b_{i0} is $N(0, \sigma_0^2)$ and E_{ij} is $N(0, \sigma_e^2)$, and where b_{i0} and E_{ij} are mutually independent. The null and alternative hypotheses of interest are

$$H_0: \sigma_0^2 = 0 \quad \text{and} \quad H_A: \sigma_0^2 > 0.$$

To carry out the approximate LR test, the above full (F) model would be compared to the following reduced (R) model:

$$Y_{ij} = \beta_0 + \beta_1 T_{ij} + E_{ij}$$

The LR statistic is then computed in the usual way as follows:

$$\text{LR} = -2 \ln L_R - (-2 \ln L_F)$$

which is approximately χ_1^2 (i.e., chi-square with 1 d.f.) under H_0.

To justify that d.f. = r = 1, we note that the **G** matrix for the full model equals the scalar σ_0^2, so that r = {number of variances and covariances equal to zero under H_0} = 1. Also, the two-tailed approximate P-value must be halved since only the one-sided alternative hypothesis $H_A: \sigma_0^2 > 0$ is relevant.   ■

■ **Example 26.2**   We now consider a linear mixed model containing a random intercept and a random slope. Here, we will test for the significance of the variance component for the random slope, i.e., the null hypothesis of interest is $H_0: \sigma_T^2 = 0$ and the alternative hypothesis is $H_A: \sigma_T^2 > 0$. The full (F) model is

$$Y_{ij} = (\beta_0 + b_{i0}) + (\beta_1 + b_{i1})T_{ij} + E_{ij}$$

where b_{i0} is $N(0, \sigma_0^2)$, b_{i1} is $N(0, \sigma_T^2)$, and E_{ij} is $N(0, \sigma_e^2)$, and b_{i0} and b_{i1} are independent of E_{ij} but not necessarily of each other.   ■

To carry out the approximate LR test for this model, we need to compare the above full model with the following reduced (R) model:

$$Y_{ij} = (\beta_0 + b_{i0}) + \beta_1 T_{ij} + E_{ij}$$

The LR statistic has the following structure:

$$\text{LR} = -2 \ln L_R - (-2 \ln L_F) \sim \chi_2^2$$

under $H_0: \sigma_T^2 = 0$, where $\text{Var}(b_{i1}) = \sigma_T^2$.

To justify that d.f. = r = 2, the **G** matrix is a (2 × 2) matrix

$$\mathbf{G} = \begin{bmatrix} \sigma_0^2 & \sigma_{0T} \\ \sigma_{0T} & \sigma_T^2 \end{bmatrix}$$

involving the 3 parameters σ_0^2, σ_T^2, and $\text{Cov}(b_{i0}, b_{i1}) = \sigma_{0T}$. Under the null hypothesis that $\sigma_T^2 = \text{Var}(b_{i1}) = 0$, it must also follow that $\sigma_{0T} = \text{Cov}(b_{i0}, b_{i1}) = 0$, so that there are actually two parameters, σ_T^2 and σ_{0T}, that must be set equal to zero under H_0. Thus, $r = \{$number of variances and covariances equal to zero under $H_0\} = 2$. Also the two-tailed approximate P-value must be halved.

Method 2: Approximate Wald test:

SAS's MIXED procedure provides point estimates, estimated standard errors, Wald test Z statistics, and two-tailed P-values for each variance component and the error variance for the mixed model under consideration. As with the approximate LR test described above, the two-tailed P-value for the Wald test statistic must be halved to get the approximate P-value for a one-tailed test.

In particular, for a variance component σ_R^2, say, the null and alternative hypotheses being tested are H_0: $\sigma_R^2 = 0$ and the one-sided alternative H_A: $\sigma_R^2 > 0$, respectively, and the corresponding test statistic is:

$$Z = \frac{\hat{\sigma}_R^2}{S_{\hat{\sigma}_R^2}}$$

which is approximately $N(0, 1)$ for large samples under H_0.

■ **Example 26.3** We now return to the random effects model (26.12) involving mutually independent random effects for Subjects, Subjects × Day, and Subjects × Time:

$$SF_{ij} = \beta_0 + b_{i0} + \beta_1 D_{ij1} + \beta_2 D_{ij2} + \beta_3 T_{ij} + \beta_4(D_{ij1}T_{ij}) + \beta_5(D_{ij2}T_{ij}) + b_{i1}D_{ij1}$$
$$+ b_{i2}D_{ij2} + b_{i3}T_{ij} + E_{ij}$$

where

b_{i0} is distributed $N(0, \sigma_0^2)$,
b_{i1} and b_{i2} are independently distributed $N(0, \sigma_D^2)$,
b_{i3} is distributed $N(0, \sigma_T^2)$, and
E_{ij} is distributed $N(0, \sigma_e^2)$,

and where σ_0^2, σ_D^2, and σ_T^2 are the *variance components* associated with these random effects.

Table 26.11 provides computer output for estimated variance component parameters not previously shown in the output of Table 26.9 for this model. (This extra output can be obtained using SAS's MIXED procedure using the *covtest* option with the PROC statement.) Using the approximate Wald test procedure of Method 2, this table indicates that the (Subjects × Time) random effect (i.e., Cov Parm = TIME in output) is not significant (one-tailed $P = 0.2747/2 = 0.1374$) and therefore may be dropped from the model. For a reduced model without this random component b_{i3}, we could then determine whether or not the variance component for

TABLE 26.11 Variance component information from SAS's MIXED procedure for model (26.12) fit to SF data

Cov Parm	Ratio	Estimate	Std Error	Z	Pr > |z|	Estimated Variance Components Wald Tests for	
Intercept (σ_0^2)	6.4667	119.1725	43.5088	2.74	0.0062	σ_0^2	S
Day (σ_D^2)	0.9758	17.9834	6.7684	2.66	0.0079	σ_D^2	S
Time (σ_T^2)	0.2185	4.0263	3.6859	1.09	0.2747	σ_T^2	NS
Residual (σ_e^2)	1.0000	18.4279	4.3435	4.24	0.0000	σ_e^2	

S = significant NS = non-significant

(Subjects $\times$ Day) is significant, and, if not, reduce the model further to evaluate whether the random intercept for Subjects is important. ∎

Method 3: Approximate Mixture Test (Verbeke and Molenberghs, 2001):

The test statistic here is a "mixture" of two chi-square statistics. The two-tailed P-value is not halved when using this Mixture Test. Generally, SAS's MIXED procedure does not directly compute this test statistic, but it does provide the information necessary to carry out the Mixture Test. However, when comparing a random intercept only model to a model without a random intercept, as illustrated in Example 26.1 above for the Approximate LR method (i.e., Method 1 above), the Mixture Test is identical to the Approximate LR test. (Correspondingly, the P-value for the Mixture test equals the halved P-value for the approximate LR test.) We therefore illustrate the Mixture Test procedure using the second of the two examples previously considered for Method 1 (the approximate LR test).

∎ **Example 26.2 revisited**   We again consider a linear mixed model for the SF data involving the single predictor Time and containing a random intercept and a random slope. We will test for the significance of the variance component for the random slope; i.e., we test $H_0: \sigma_T^2 = 0$ versus $H_A: \sigma_T^2 > 0$. The full (F) model is

$$Y_{ij} = (\beta_0 + b_{i0}) + (\beta_1 + b_{i1})T_{ij} + E_{ij}$$

where b_{i0} is $N(0, \sigma_0^2)$, b_{i1} is $N(0, \sigma_T^2)$, and E_{ij} is $N(0, \sigma_e^2)$, and it is assumed that the $\mathbf{R}_i$ matrix is $\sigma_e^2 \mathbf{I}_6$ and that the $\mathbf{G}$ matrix is

$$\mathbf{G} = \begin{bmatrix} \sigma_0^2 & \sigma_{0T} \\ \sigma_{0T} & \sigma_T^2 \end{bmatrix}$$

As with Method 1, there are two parameters, σ_T^2 and σ_{0T}, that must be set equal to zero under H_0. To test H_0 using the Mixture Test, we again, as with Method 1, compute the LR statistic that compares the above full (F) model with the following reduced (R) model:

$$Y_{ij} = (\beta_0 + b_{0i}) + \beta_1 T_{ij} + E_{ij}$$

The LR statistic thus has the following formula:

$$\text{LR} = -2 \ln L_R - (-2 \ln L_F)$$

where L_R and L_F denote the maximized likelihoods for the reduced and full models, respectively. The Mixture Test then assumes this LR statistic has approximately the following "mixture" distribution:

$$\text{LR} \sim 0.5\,\chi_2^2 + 0.5\,\chi_1^2$$

where χ_2^2 and χ_1^2 denote chi-square random variables with 2 and 1 d.f. respectively. Essentially this "mixture" distribution reflects the viewpoint that the appropriate d.f. for LR should be some average of 2 d.f. (corresponding to two parameters, σ_T^2 and σ_{0T}, that need to set equal to zero under H_0) and 1 d.f. (corresponding to the one parameter, σ_T^2, that is the primary parameter of interest).

The P-value for this "mixture" test is given by the expression

$$P = 0.5 \Pr(\chi_2^2 > \text{LR}) + 0.5 \Pr(\chi_1^2 > \text{LR})$$

For example, if LR is computed to be, say, 4.6, then

$$P = 0.5 \Pr(\chi_2^2 > 4.6) + 0.5 \Pr(\chi_1^2 > 4.6) \approx 0.5(0.10) + 0.5(0.032) = 0.066$$

which is non-significant at the 5% level of significance. ∎

26.4.7 Denominator Degrees of Freedom in SAS's MIXED Procedure

We previously indicated in the footnotes to Tables 26.8 and 26.9 that SAS's MIXED procedure provides several options for determining the denominator degrees of freedom (DDFM) associated with F statistics for testing hypotheses about the predictors in the linear mixed model. Four of these options are:

1. Residual: default for independence correlation structure
2. BetWithin: default for marginal model, i.e., $\mathbf{G} = \mathbf{0}$
3. Contain: default for variance components (VC) model, i.e., $\mathbf{G} \neq \mathbf{0}$
4. Satterthwaite: needed when denominator MS in F statistic is a linear function of MS terms

TABLE 26.12   **Denominator degrees of freedom (DDFM) and F test results for different DDFM options for the SF posture data for using the predictors day (D), time (T), and day × time interaction (DT)**

| | | | Correlation Structure-Specific DDFM Values | | | | | F statistics | | | | |
			IND	CS	AR1	VC1	VC3	IND	CS	AR1	VC1	VC3
DDFM Option with Model-based SE	Residual	D	108	108	108	108	108	0.60	2.74	1.05	2.74	1.78
		T	108	108	108	108	108	0.00	0.02	0.03	0.02	0.02
		DT	108	108	108	108	108	0.09	0.42	0.44	0.42	0.79
	BetWithin	D		36	36	36	36		2.74	1.05	2.74	1.78
		T	NA-RES	18	18	18	18	NA-RES	0.02	0.03	0.02	0.02
		DT		36	36	36	36		0.42	0.44	0.42	0.79
	Contain	D		108	108	90	36		2.74	1.05	2.74	1.78
		T	NA-RES	108	108	90	18	NA-RES	0.02	0.03	0.02	0.02
		DT		108	108	90	36		0.42	0.44	0.42	0.79
	Satterthwaite	D		90	97.5	90	36		2.74	1.05	2.74	1.78
		T	NA-RES	90	99.8	90	18	NA-RES	0.02	0.03	0.03	0.02
		DT		90	82.7	90	36		0.42	0.44	0.42	0.79
DDFM Option with Empirical SE	Residual	D		108	108	108	108		1.33	1.33	1.33	1.33
		T	NA	108	108	108	108	NA	0.02	0.02	0.02	0.02
		DT		108	108	108	108		0.84	0.84	0.84	0.84
	BetWithin	D		36	36	36	36		1.33	1.33	1.33	1.33
		T	NA	18	18	18	18	NA	0.02	0.02	0.02	0.02
		DT		36	36	36	36		0.84	0.84	0.84	0.84
	Contain	D		108	108	90	36		1.33	1.33	1.33	1.33
		T	NA	108	108	90	18	NA	0.02	0.02	0.02	0.02
		DT		108	108	90	36		0.84	0.84	0.84	0.02
	Satterthwaite	D							1.33	1.33	1.33	1.33
		T	NA	NA-BW	NA = BW	NA-CO	NA-CO	NA	0.02	0.02	0.02	0.02
		DT							0.84	0.84	0.84	0.84

IND: independent, CS: compound symmetric, AR1: autoregressive
VC1: random intercept model, VC3: 3-way repeated measures ANOVA model
NA: not applicable   NA-RES: uses Residual DDFM
NA-BW: uses BetWithin DDFM   NA-CO: uses Contain DDFM

Table 26.12 presents, for the SF posture measurement data, the DDFM values and corresponding F statistics that result from the different DDFM options when both model-based and empirical standard errors are used with different correlation structures. The following items attempt to summarize the results in Table 26.12:

1. When an independent (**IND**) correlation structure is chosen, SAS's MIXED procedure only allows the residual option to be used with model-based standard errors. The corresponding F statistics, {0.60, 0.00, 0.09}, are computed using a Two-Way Fixed Effects ANOVA (as described in Chapter 19), where the two factors are Day (D) and Time (T), and where there are 19 replications of the SF response for each of the six ($= 3 \times 2$) combinations of Day with Time (thus giving $18 \times 6 = 108$ DDFM).

2. Using the residual option, the DDFM values equal 108 for D, T, and DT, regardless of the choice of correlation structure; the value 108 can be determined using the standard formula for residual DDFM, namely, $(n - p^*)$, where $n = 114$ (total number of observations) and $p^* = 6$ (number of fixed-effect model parameters) for model (26.12).

3. For a given choice of (non-**IND**) correlation structure (CS, AR1, VC1, VC3) using model-based standard errors, the same set of F statistics for D, T, and DT are obtained regardless of which DDFM option is used. The set of F statistics obtained always corresponds to the three-way repeated measures ANOVA (**VC3**) model (26.12). However, as expected, the set of F statistics differs for different correlation structures, e.g., {2.74, 0.02, 0.42} for **CS**, {1.05, 0.03, 0.44} for **AR1**, and {1.78, 0.02, 0.79} for **VC3**.

4. When empirical standard errors are used, the same set of F statistics, namely {1.33, 0.02, 0.84}, is obtained regardless of the DDFM option and regardless of the (non-**IND**) correlation structure chosen.

5. The BetWithin DDFM option always yields the same set of DDFM values for D, T, and DT (namely 36, 18, and 36), regardless of the (non-**IND**) correlation structure chosen. This set of DDFM values corresponds to the DDFM values obtained for the three-way repeated measures ANOVA model (26.12).

6. For a random intercept (**VC1**) model, the Contain (default) option, using either model-based or empirical standard errors, yields DDFM values of 90, 90, and 90 for D, T, and DT, respectively, The value 90 is obtained using the formula $n - p^* - q^*(K - 1)$, where $n = 114$ observations, $p^* = 6$ fixed-effect parameters, $q^* = 1$ random effect, and $K = 19$ clusters.

7. For the three-way repeated measures ANOVA (**VC3**) model (26.12), the Contain (default) option, using either model-based or empirical standard errors, yields DDFM values of 36, 18, 36 for D, T, and DT, respectively.

8. When model-based standard errors are used, the Satterthwaite option yields DDFM values identical to those for the Contain option for the random effects models **VC1** ({90, 90, 90}) and **VC3** ({36, 18, 36}). When empirical standard errors are used, specifying the Satterthwaite option in the code allows only the default option to be used (BW for marginal models and CO for random effects models).

The above summary describes a complicated pattern of results corresponding to different DDFM options and different choices for the working correlation structure. One problematic feature of these results (from summary items 3 and 4 above) indicates that, for a given correlation structure, the set of F statistics does not change even when the DDFM values change, regardless of whether or not model-based or empirical standard errors are used. This is somewhat problematic because conclusions about statistical significance can vary depending on the choice of the DDFM value used.

The above discussion indicates that it is difficult to provide firm recommendations on how to choose the appropriate DDFM option when analyzing continuous correlated data. Perhaps the simplest recommendation is to let the user/investigator decide based on the study under consideration. Nevertheless, we offer the following suggestions, recognizing that the reader may wish to pursue this topic further before accepting these suggestions:

   I. If a marginal model is used (i.e., $\mathbf{G} = \mathbf{0}$), choose the Residual DDFM option regardless of the $\mathbf{R}$ matrix chosen.

  II. If a linear mixed model with random effects ($\mathbf{G} \neq \mathbf{0}$) is used and all the effects in the model other than the cluster effects are treated either as all fixed or as all random factors, choose the Contain DDFM option regardless of the number of random effects in the model and regardless of whether or not an $\mathbf{R}$ matrix other than $\sigma_e^2 \mathbf{I}$ is chosen.

 III. Use the Satterthwaite option only if the variables in the model other than cluster effects are a mixture of fixed and random effects (e.g., if Day is a fixed factor and Time is a random factor, or vice versa, for the SF example).

We recommend the Residual DDFM option for marginal models ($\mathbf{G} = \mathbf{0}$) because the DDFM reflects only the difference between the total number of observations and the number of fixed-effect model parameters when there are no random effects to consider.

The Contain DDFM option is recommended for linear mixed models with random effects to reflect the partitioning of sums of squares corresponding to the random effects chosen. Thus, for example, if only a random effect for the intercept is used to model the SF data, the DDFM values will be 90 each for testing each of the D, T, and DT effects; in contrast if the three-way repeated measures ANOVA model (26.12) is used, the DDFMs will be 36, 18, and 36 for assessing the D, T, and DT effects.

We recommend the Satterthwaite option only for situations in which the denominator mean square is a linear function of mean square terms for two or more predictor variables, at least one of which is a fixed factor and at least one of which is a random factor. In such cases, the DDFM is determined by a complex formula (see authors' website) that is readily computed by SAS's MIXED procedure. The use of this formula often, in fact, leads to DDFM values that are not integer values. For the balanced SF data set considered in this chapter, however, the two factors Days and Time are both fixed factors, so that the use of the Satterthwaite option is not necessary.

26.4.8 Summary of Analyses for the SF Data

From the analyses of the SF posture measurement dataset that we have carried out, we summarize the results with regard to the primary question of whether or not there are statistically significant effects of Day, Time, and/or (Day $\times$ Time) interaction:

   1. Based on the sample means provided in Chapter 25, Table 25.5, there appears to be a possible main effect of the factor Day, and some suggestion of a (Day $\times$ Time) interaction effect, but very little indication of a main effect of Time.

2. Regardless of the correlation structure chosen (for marginal or random effects models), and regardless of whether model-based or empirical standard errors were used, there were no significant effects of Day, Time, and (Day × Time) interaction in this sample of computer operators. These results support the investigators' contention that, in their subsequent study of the possible relationship of posture measurements to muscular-skeletal disorders, they will not need to control for the effects of day of the week or time of day.

3. Regardless of the DDFM option used and the correlation structure chosen, the set of F statistic values obtained for the effects of Day, Time and (Day × Time) interaction were the same as those obtained for a three-way repeated measures ANOVA, where Day and Time are fixed crossover factors and Subjects is a random factor.

4. Although the same conclusions were found regardless of correlation structure chosen, we suggest that the most appropriate correlation structure for these data derives from the three-way repeated measures ANOVA model given by (26.14), or equivalently by (26.12) and (26.13). This model allows for a correlation structure that cannot be specified for any choice of **R** matrix using a marginal model (**G = 0**). Moreover, the correlation structure for model (26.14) makes sense since it allows for random effects for Subjects, Subjects × Day, and Subjects × Time, in addition to fixed effects for Day, Time, and Day × Time, thus reflecting all possible main effects and two-way interaction effects for the three factors Day, Time, and Subjects.

26.5 Recommendations about Choice of Correlation Structure

The choice of a working correlation structure is a very important part of the process when analyzing continuous response correlated data, especially since an investigator has to make a hopefully educated guess about the correlation structure and thus needs to be concerned that an inappropriate choice might be made. As we discussed in Chapter 25, Section 25.3, the choice process can involve several approaches. Here is a list of several options that may be considered when attempting to specify a working correlation structure:

1. **Use what makes sense clinically/biologically.**
   This option should certainly be considered when analyzing any correlated data set, although the choice of correlation structure may not be obvious, especially when the study design has both longitudinal and cross-sectional features (e.g., when obtaining repeated measures on family members at several different times). Possible choices are *exchangeable* (e.g., when clusters are households studied cross-sectionally) and *autoregressive* (e.g., when clusters are individual subjects who are measured over time).

2. **Try several different correlation structures.**
   Even if one has a logical choice for the correlation structure based on clinical/biological features of the study, if the same overall study conclusions are obtained for a variety of different plausible correlation structures, one can be more confident that intra-cluster correlations have been taken into account appropriately in the data analysis.

3. **Use an unstructured correlation matrix.**

This choice makes sense if it is believed that the correlation structure is quite complicated and there is little support for a simpler (e.g., more patterned) structure. Nevertheless, choosing the **unstructured** option is actually "structured" to provide separate estimates for all possible pairwise correlations. As a result, the number of parameters to be estimated is often too large for the data to support a valid and stable numerical solution (e.g., the model fitting algorithm sometimes does not converge to a solution when the **unstructured** option is chosen).

4. **Start with an unstructured correlation matrix and simplify.**

The estimated correlations based on using an **unstructured** correlation matrix may suggest a more simplified (or patterned) structure (e.g., if all estimated correlations are numerically close in value, this might suggest an exchangeable correlation structure).

5. **Start with a Toeplitz (i.e., stationary) structure and simplify.**

A **Toeplitz (i.e., stationary) m-dependent** correlation structure, which we have not previously described, has the property that correlations k units apart are the same for $k = 1, 2, \ldots, m$, while correlations more than m units apart are zero. For example, a correlation structure that is **stationary 1-dependent** (i.e., $m = 1$) has the following property:

$$\text{corr}(Y_{ij}, Y_{i(j+1)}) = \rho_1 \quad \text{whereas} \quad \text{corr}(Y_{ij}, Y_{i(j+m')}) = 0 \quad \text{for } m' > 1$$

A subject with $n_i = 4$ repeated observations would therefore have a **stationary 1-dependent** structure of the following form:

$$\mathbf{C}(\rho_1) = \begin{bmatrix} 1 & \rho_1 & 0 & 0 \\ \rho_1 & 1 & \rho_1 & 0 \\ 0 & \rho_1 & 1 & \rho_1 \\ 0 & 0 & \rho_1 & 1 \end{bmatrix}$$

As another example, a correlation structure that is **stationary 2-dependent** (i.e., $m = 2$) has the following property: $\text{corr}(Y_{ij}, Y_{i(j+1)}) = \rho_1$, $\text{corr}(Y_{ij}, Y_{i(j+2)}) = \rho_2$, whereas $\text{corr}(Y_{ij}, Y_{i(j+m')}) = 0$ for $m' > 2$. A subject with $n_i = 5$ repeated observations would therefore have a **stationary 2-dependent** structure of the following form:

$$\mathbf{C}(\rho_1, \rho_2) = \begin{bmatrix} 1 & \rho_1 & \rho_2 & 0 & 0 \\ \rho_1 & 1 & \rho_1 & \rho_2 & 0 \\ \rho_2 & \rho_1 & 1 & \rho_1 & \rho_2 \\ 1 & \rho_2 & \rho_1 & 1 & \rho_1 \\ 1 & 1 & \rho_2 & \rho_1 & 1 \end{bmatrix}$$

Toeplitz (i.e., stationary) is an alternative to **AR1** that allows for correlations to be different as occasions are farther apart in time, yet does not follow the **AR1** structure of ρ, ρ^2, ρ^3, etc. If a mixed model analysis does not yield stable numerical results when an unstructured correlation matrix is used, then a **Toeplitz** structure using the maximum value of m possible for the number of repeated measures observed on a given subject is a reasonable substitute choice. For example, if a balanced design involves 6 repeated measures per subject, then a **Toeplitz 5-dependent** correlation structure can serve as a starting structure for possible simplification.

6.   **Use Goodness of Fit (GOF) procedures AIC and/or BIC.**

Akaike's Information Criterion (**AIC**) and Bayes' Information Criterion (**BIC**) are two goodness of fit (**GOF**) measures that are functions of the likelihood function adjusted for the number of fixed-effect parameters in the fitted linear mixed model. The smaller is the value of **AIC** and/or **BIC**, the better the model is said to fit. Consequently, either or both these criteria can be used to decide on the appropriate correlation structure using the following process: For each correlation structure chosen, rank-order the **AIC** and/or **BIC** values from smallest to largest. Choose the correlation structure that yields the smallest value of **AIC** and/or **BIC**, or choose a structure that makes etiologic sense and also yields **AIC** or **BIC** values close to the smallest values observed.

7.   **Use the estimated correlation matrix of the *Y*'s as a fixed (numerical) structure.**

The main idea here is to let the data suggest a fixed correlation structure to use in a marginal model. However, using a correlation structure derived from the *Y*'s does not account for the fixed effects in the model (i.e., a more appropriate approach would estimate the correlation structure using residual error values after accounting for fixed effects).

8.   **Start with an independence correlation structure, fit the model, obtain residuals, then use the correlation matrix of the residuals as a fixed (numerical) structure.**

This approach estimates correlations using residual error values as described in item 7. It effectively gives the starting correlation structure used for estimating an unstructured correlation matrix.

9.   **Use a random effects model.**

This approach is appropriate for accounting for heterogeneity of subject-specific effects, estimating subject-specific effects, defining a correlation structure not clearly specified by an **R** matrix in a marginal model, and modelling the effects of predictor variables that are considered to be random factors.

10.   **For any of the above options, decide between using model-based and empirical standard errors.**

A safe (i.e., conservative) choice is to always use empirical standard errors, assuming that one can never be certain as to what is the appropriate working correlation structure. However, if one can make a strong case for a specific correlation structure, then standard errors based on that specific correlation structure may provide more efficient statistical inferences (i.e., more powerful hypothesis tests and narrower confidence intervals).

26.6 Analysis of Data for Discrete Outcomes

To address situations where the (repeated measures) response is binary, a count variable, or continuous but not normally distributed, a method involving the use of *generalized estimating equations* (GEE) has been developed (Zeger and Liang, 1986). The GEE approach uses a procedure called *quasi-likelihood estimation*, which is a generalization of maximum likelihood estimation. In recent years, the GEE approach has become very popular among researchers who need to consider nonlinear regression models for correlated binary or count response data (e.g., the logistic model discussed in Chapters 22 and 23, and the Poisson regression model discussed in Chapter 24). Some currently available software can apply the GEE method, such as SAS's GENMOD and GLIMMIX procedures. The GENMOD procedure does not allow random effects, whereas the GLIMMIX procedure allows both fixed and random effects. SAS's NLMIXED procedure is an approximate likelihood-based procedure (i.e., it does not use GEE) that is also available. A thorough discussion of GEE methods is beyond the scope of this text. For further information, see Kleinbaum and Klein (2002) and/or Diggle, Heagerty, Liang, and Zeger (2002).

Problems

1. The dataset for this problem derives from the posture measurement study described in the main body of this chapter. Here, we consider the data on shoulder flexion (SF) for 19 subjects that were each observed by two different raters on each of the three days (Monday, Wednesday, and Friday) and at two time periods (AM and PM) during each day. Note that our previous version of this example considered only one rater, but now there are two raters who have independently taken the posture measurements. Thus, we now consider 12 observations to have been made on each subject, with each observation corresponding to one of the 12 combinations of 3 days, 2 times, and 2 raters. In this problem, *we assume that the investigator is interested only in assessing whether the SF measurements vary significantly by day of measurement*. For this purpose, we have provided in the table below the average of the four SF scores taken on each subject (at 2 times and by 2 raters) for each of the three days.

Average shoulder flexion score by day of week

Subject	Monday	Wednesday	Friday
1	10.50	7.78	6.00
2	4.00	2.50	16.75
3	25.75	28.00	22.75
4	20.00	22.50	20.00
5	4.75	8.25	6.50
6	17.25	22.00	18.00
7	24.00	9.25	12.50
8	45.50	50.00	41.75
9	10.00	10.00	5.50
10	28.25	31.25	41.25
11	21.75	22.25	15.50
12	23.00	29.00	26.00
13	10.50	7.00	7.00
14	22.00	19.50	22.50
15	7.25	5.00	0.00
16	19.50	9.00	9.00
17	23.25	26.50	17.50
18	28.50	35.25	8.75
19	3.00	1.75	2.50
Mean of Averages:	18.36	18.25	15.78

a. Assuming that the only important effect is that of Day (i.e., the factors Time and Rater are assumed *not* to have important effects), state the subject-specific scalar form of a random intercept model for analyzing the above data. In stating this model, make sure to describe the the assumptions made on the random effects (including the error term) in the model.

   1. b. State the null hypothesis that there is no significant effect of the factor Day in terms of a statement about parameters in your model given in part (a).

   c. Based on a comparison of averages at the bottom of the table, does there appear to be a meaningful effect of the factor Day? Explain.

   d. Describe how the data in the above table need to be reorganized in order for the MIXED procedure to carry out the analysis. (*Hint*: Use the format given in Table 25.2 of Chapter 25.)

   e. Based on the computer output provided below, is there a significant effect of the factor Day? Explain by specifying the F statistic, its degrees of freedom, and its P-value appropriate for these data.

   f. Based on the computer output, compute the estimate of the (exchangeable) correlation assumed by the model. (*Hint*: The output provides estimates of the variance of the subject-specific random intercept effect and the variance of the error term; you need to combine this information to compute the correlation coefficient estimate.)

   g. Use the output to test whether there is a significant random effect for Subjects. If such a test was non-significant, why might you be concerned regarding the model that has been fit, and how might you redo the analysis?

Program Statements for SAS's MIXED procedure for problem 1:

```
PROC MIXED data= mnsf covtest;
CLASS subj day;
MODEL mnsf = day/s;
RANDOM intercept/ subject=subj;
PROC MIXED data= mnsf covtest empirical;
CLASS subj day;
MODEL mnsf = day/s;
RANDOM intercept/ subject=subj;
PROC MIXED data= mnsf covtest empirical;
CLASS subj day;
MODEL mnsf = day/s;
REPEATED / subject=subj type=ar(1) rcorr;
```

Edited Output:

```
Covariance Parameter Estimates
Cov Parm          Subject        Estimate        StdErr        Z Value        Pr Z
Intercept          subj          113.90          40.8944       2.79           0.0027
Residual                          25.8839         6.1009        4.24           <.0001

Type 3 Tests of Fixed Effects (Model-based)
Effect             NDF            DDF           F Value        Pr > F
day                 2             36            1.5646         0.2231

Type 3 Tests of Fixed Effects (Empirical- Random Intercept)
Effect             NDF            DDF           F Value        Pr > F
day                 2             36            1.19           0.317

Type 3 Tests of Fixed Effects (Empirical- AR(1))
Effect             NDF            DDF           F Value        Pr > F
day                 2             36            1.19           0.317
```

2. The analysis described in Problem 1 for the posture measurement data on Shoulder Flexion (SF) may be criticized because information is lost when the 4 observations for a given subject on a given day are combined into an average score, rather than being treated individually in the analysis. The dataset shown below allows for an analysis that considers all 12 observations per subject. In analyzing this dataset, we assume that the 4 observations for a given subject on a given day are true *replicates* (i.e., we assume that neither the time of day observed nor the rater used, factors which combine to provide the 4 observations for a given subject, are important factors for predicting SF response).

   a. How should the subject-specific scalar model for problem 1 be modified to consider the data layout below? [*Hint*: You will need to add a second random effect to the model that reflects the interaction of Day with Subjects.]

   b. Use the computer information provided below to test whether there is a significant main effect of the factor Day. [*Hint*: See the ANOVA table provided below.] What do you conclude?

   c. Use the computer output provided below to test whether there is a significant interaction effect between Subjects and Day. If such a test is non-significant, why might you be concerned regarding the model that has been fit, and how might you redo the analysis?

Shoulder flexion scores by day of week

Subject	Sample Size	Monday				Wednesday				Friday			
1	12	15	5	17	5	11	9	10	1	8	10	5	1
2	12	0	8	1	7	−2	1	7	4	18	18	19	12
3	12	29	16	36	22	25	26	31	30	13	26	27	25
4	12	17	18	21	24	14	24	22	30	16	20	20	24
5	12	6	17	10	14	8	8	7	10	9	4	8	5
6	12	18	18	15	18	16	24	22	26	20	1	2	17
7	12	12	7	19	10	10	7	12	8	7	17	12	14
8	12	41	49	46	48	50	52	46	52	30	44	41	52
9	12	8	12	9	11	7	11	7	15	1	8	5	8
10	12	27	34	31	31	36	22	39	28	42	45	38	40
11	12	17	27	17	26	20	18	27	24	17	10	19	16
12	12	23	20	25	24	20	32	29	5	19	30	24	31
13	12	5	15	8	14	6	6	7	9	9	4	6	9
14	12	25	18	23	18	15	18	21	24	16	22	27	25
15	12	5	10	7	7	0	1	7	12	2	−2	0	0
16	12	20	17	22	19	5	11	7	13	3	13	6	14
17	12	28	21	24	20	27	20	32	27	16	15	22	17
18	12	33	25	28	28	38	27	42	34	5	14	9	7
19	12	2	4	1	5	2	5	0	0	2	0	6	2
Averages:			18.36				18.25				15.78		

Program Statements for SAS's GLM procedure for Problem 2:

```
PROC MIXED DATA=sf2 covtest;
CLASS subj day;
MODEL sf= day/s;
RANDOM intercept day/subject=subj g;
run;
PROC MIXED DATA=sf2 covtest empirical;
CLASS subj day;
MODEL sf= day/s;
RANDOM intercept day/subject=subj g;
run;
PROC MIXED DATA=sf2 covtest empirical;
CLASS subj day;
MODEL sf= day/s;
REPEATED intercept /subject=subj g;
run;
```

Edited Computer Output:

```
Covariance Parameter Estimates
Cov Parm          Subject       Estimate          StdErr        Z Value        Pr Z
Intercept           subj         113.75          40.5659           2.80       0.0025
day                 subj         17.9537          5.5630           3.23       0.0006
Residual                         22.0482          2.3845           9.25      <.0001
-2 Res Log Likelihood                             1475.1

Type 3 Tests of Fixed Effects (Model-based)
Effect           NDF             DDF          F Value        Pr > F
 day              2               36            1.95         0.1571

Type 3 Tests of Fixed Effects (Empirical)
Effect           NDF             DDF          F Value        Pr > F
 day              2               36            1.84         0.1730
```

3. A study by Holder, Plikaytis, and Carlone (1996) compared two laboratory protocols designed to measure antibody levels in an enzyme-linked immunosorbent assay (ELISA) for *Streptococcus pneumoniae*. One protocol incorporates a "blocking step" that is thought to increase the specificity of the assay, maximizing the yield of the specific antibody; the other protocol does not use the blocking step. The data shown below provide the ELISA results on six specimens (i.e., samples), each with *Streptococcus pneumoniae* Serogroup 4. This is a balanced repeated measures ANOVA design with six measurements on each sample, three of which use the protocol with the blocking step and three of which do not use the blocking step.

Antibody yields for two protocols measuring
***streptococcus pneumoniae* serogroup 4**
Specimen (or Sample) Number

		1	2	3	4	5	6	Sample Mean
Blocking Step	Yes	7.3	1.0	2.8	2.3	28.4	0.4	7.66
		9.3	1.0	3.1	3.2	29.2	0.5	
		6.4	1.0	2.8	2.8	35.9	0.4	
	No	9.3	1.5	3.6	3.7	27.2	0.6	7.67
		9.1	1.4	3.4	3.6	25.5	0.6	
		9.3	2.0	4.5	3.6	28.3	0.8	

3. a. Should the factor Blocking Step be considered a fixed or random factor? Explain.

   b. State the subject-specific scalar regression model for this analysis. (*Hint*: For these data, the subjects are the samples.)

   c. State the subject-specific matrix model for this analysis.

   d. State the null hypothesis that there is no significant effect of Blocking Step in terms of a statement about the parameters in the regression model given in part (b).

   e. Based on a comparison of sample means, does there appear to be a meaningful effect of the Blocking Step factor? Explain

   f. Below are given computer results for a repeated-measures analysis of these data. Based on these results, is there a significant effect of the Blocking Step factor? Explain by specifying the F statistic, its degrees of freedom, and its P-value appropriate for these data.

   g. Use the computer output to test whether there is a significant interaction effect between Specimen (or Samples) and Blocking Step. If such a test is non-significant, why might you be concerned about the model that has been fit, and how might you redo the analysis?

Program Statements for SAS's MIXED procedure for Problem 3:

```
proc mixed data=serogroup4 covtest;
class block sample;
model result= block/s;
random intercept block/subject=sample g;
run;
proc mixed data=serogroup4 covtest empirical;
class block sample;
model result= block/s;
random intercept block/subject=sample g;
run;
proc mixed data=serogroup4 covtest;
class block sample;
model result= block/s;
random intercept/subject=sample g;
run;
proc mixed data=serogroup4 covtest empirical;
class block sample;
model result= block/s;
random intercept/subject=sample g;
run;
proc mixed data=serogroup4 covtest empirical;
class block sample;
model result= block/s;
repeated /subject=sample type=ar(1);
run;
```

(continued)

Output for Problem 3 (continued)

```
Covariance Parameter Estimates (Random intercept block)
Cov Parm          Subject          Estimate          StdErr          Z Value          Pr Z
Intercept         sample           116.64            74.4682         1.57             0.0586
block             sample           1.5818            1.3957          1.13             0.1285
Residual                           1.8225            0.5261          3.46             0.0003

Type 3 Tests of Fixed Effects (Model-based)
Effect            NDF              DDF              F Value          Pr > F
block             1                5                0.00             0.9901

Type 3 Tests of Fixed Effects (Empirical)
Effect            NDF              DDF              F Value          Pr > F
block             1                29               0.00             0.9838

Covariance Parameter Estimates (Random intercept only)
Cov Parm          Subject          Estimate          StdErr          Z Value          Pr Z
Intercept         sample           117.30            74.4651         1.58             0.0576
Residual                           2.6407            0.6935          3.81             <.0001

Type 3 Tests of Fixed Effects (Model-based- Random Intercept only)
Effect            NDF              DDF              F Value          Pr > F
block             1                29               0.00             0.9838

Type 3 Tests of Fixed Effects (Empirical- Random Intercept only)
Effect            NDF              DDF              F Value          Pr > F
block             1                29               0.00             0.9887

Type 3 Tests of Fixed Effects (Empirical- Repeated AR1)
Effect            NDF              DDF              F Value          Pr > F
block             1                5                0.13             0.7321
```

4. A study by Heffner et al. (1974) investigated the effects of an amphetamine on the behavior of rats. Before the study began, 24 "thirsty" rats were trained to press a lever to obtain water. The rats were categorized into three groups (slow, medium, and fast) of equal size according to their initial press rates. Each rat received three doses of the drug (i.e., the amphetamine under study), as well as a placebo, on separate occasions and in random order. One hour after the drug injection, an experimental session began in which the rat received water after pressing the lever a pre-specified number of times. Half of the rats received water after two presses of the lever, whereas the other half received water after five presses.

   The response measured was the lever press rate (i.e., LPR, the total number of lever presses divided by elapsed time in seconds) used by a thirsty rat to press the lever and receive water. The primary research question was whether the drug affected the LPR. Also of interest was the question of whether there was an effect on the response (LPR) of the number of presses (PRS) required to obtain water (two versus five) and/or the initial press rate (IPR)(slow, medium, or fast), and whether there were any interaction effects. The data are given below:

   a. Which of the factors Drug, IPR and PRS should be treated as fixed factors and which as random factors? Explain.

   b. Consider an analysis of these data that focuses only on the effect of the factor Drug (i.e., ignore the factors IPR and PRS in the analysis). Thus, for each of the 24 rats, we have four responses corresponding to the administration of each of the four levels of the factor Drug. State a subject-specific scalar model for analyzing these data, making sure to state the assumptions typically made about the random effects (including the error term) in the model. (*Note*: The subjects are the rats in this example.)

4. c. State the subject-specific matrix model that corresponds to the subject-specific scalar model stated in part (b).

d. Based on a comparison of sample means, does there appear to be a meaningful effect of the factor Drug? Explain.

e. Use the computer output provided below to test for the significance of the factor Drug. Make sure to describe the null hypothesis being tested, the F statistic used, its degrees of freedom, and the resulting P-value for this test. What do you conclude?

f. Use the computer output below to test for the significance of the subject (i.e., Rat) factor. Why might you expect the results for this test to be significant?

g. Explain why it is not possible to test whether there is a significant interaction between Drug and Rat.

Lever press rate (LPR) for 24 thirsty rats

IPR	PRS	Rat Number	Drug 1 (Placebo)	2	3	4	Mean
		1	0.81	0.80	0.82	0.50	
	2	2	0.77	0.78	0.79	0.51	0.76
		3	0.80	0.82	0.83	0.52	
		4	0.95	0.95	0.91	0.60	
Slow		5	2.18	2.44	1.92	0.92	
	5	6	2.02	2.20	1.75	0.82	1.80
		7	2.06	2.28	1.86	0.80	
		8	2.28	2.46	1.90	0.90	
		9	1.03	1.13	1.04	0.82	
	2	10	0.96	0.93	1.02	0.63	0.98
		11	0.98	1.00	0.98	0.74	
		12	1.17	1.20	1.18	0.91	
Medium		13	2.62	2.58	2.21	1.03	
	5	14	2.60	2.60	2.34	1.14	2.09
		15	2.39	2.41	2.09	0.90	
		16	2.70	2.64	2.23	1.02	
		17	1.20	1.24	1.27	0.96	
	2	18	1.25	1.23	1.30	1.01	1.20
		19	1.23	1.20	1.18	0.95	
		20	1.31	1.42	1.41	1.08	
Fast		21	2.98	2.64	2.34	1.28	
	5	22	3.10	2.85	2.40	1.35	2.34
		23	2.80	2.48	2.16	1.01	
		24	3.21	2.92	2.56	1.40	
		Mean	1.81	1.80	1.60	0.91	

IPR = initial press rate; PRS = Number of presses

Program Statements for SAS's MIXED procedure for Problem 4:

```
proc mixed data=rats covtest;
class rat drug;
model lpr= drug/s;
random intercept/subject=rat g;
run;
proc mixed data=rats covtest empirical;
class rat drug;
model lpr= drug/s;
random intercept/subject=rat g;
run;
proc mixed data=rats covtest;
class rat drug;
model lpr= drug/s;
repeated /subject=rat type=ar(1);
run;
```

Edited Computer Output:

```
Covariance Parameter Estimates
Cov Parm          Subject          Estimate          StdErr          Z Value          Pr Z
Intercept           rat              0.3419          0.1079            3.17          0.0008
Residual                             0.09489         0.01616           5.87          <.0001

Type 3 Tests of Fixed Effects (Model-based)
Effect            NDF              DDF             F Value         Pr > F
drug               3                69             45.72           <.0001

Type 3 Tests of Fixed Effects (Empirical- Random intercept)
Effect            NDF              DDF             F Value         Pr > F
drug               3                69             47.63           <.0001

Type 3 Tests of Fixed Effects (Empirical- Repeated AR1)
Effect            NDF              DDF             F Value         Pr > F
drug               3                69             47.63           <.0001
```

5. Consider the same study by Heffner et al. and the dataset described in Problem 4, in which the response measured was the lever press rate (LPR) used by a thirsty rat to press a lever and receive water. The three factors identified as possible predictors were Drug (3 doses plus a placebo), IPR (initial press rate, categorized as slow, medium, or fast), and PRS (number of presses required to obtain water: two or five). Another factor of importance is the random factor Rat (with 24 levels). Furthermore, the factors IPR and PRS were combined into a single Factor A with 6 levels according to the following categorization:

IPR	PRS	Level of Factor A
slow	2	1
slow	5	2
medium	2	3
medium	5	4
fast	2	5
fast	2	6

In this problem, we consider both Factor A and Drug together in a repeated measures analysis to evaluate whether there is an effect of either (or both) factors and their interaction on the response.

5. **a.** Describe the data layout for this analysis using the format given by Table 25.2 of Chapter 25.

   b. Using the data provided in Problem 4 together with the above definition of Factor A, we obtained following table of sample means:

Factor A

Drug	1	2	3	4	5	6	Average
1 (Placebo)	0.8325	2.1350	1.0350	2.5775	1.2475	3.0225	1.8083
2	0.8375	2.3450	1.0650	2.5575	1.2725	2.7225	1.8000
3	0.8375	1.8575	1.0550	2.2175	1.2900	2.3650	1.6038
4	0.5325	0.8600	0.7750	1.0225	1.0000	1.2600	0.9083
Average	0.7600	1.7994	0.9825	2.0938	1.2025	2.3425	1.5301

Based on this table, describe whether there appears to be a meaningful main effect of Factor A, a main effect of Drug, and/or an interaction effect between Factor A and Drug. Explain.

   c State the formula for a subject-specific scalar model for the analysis of these data, and specify the assumptions typically made about the random effects (and error term) for this model.

   d. For the model stated in part (c), use a flow diagram to describe how the variation (sums of squares) is partitioned into contributions from various sources of predictors. [*Hint*: Consider Figure 26.5 in the text.]

   e. Using the computer output provided below, carry out tests of whether the main effect of Factor A, the main effect of Drug, and the Factor A-by-Drug interaction is significant. For each test, state the null hypothesis being tested, the form of the F statistic, its degrees of freedom under the null hypothesis, and the resulting P-value. What do you conclude?

Program Statements for SAS's MIXED procedure for Problem 5:

```
data rats;
input id prs rat ipr drug lpr;
if ipr=1 and prs=1 then factora=1;
if ipr=1 and prs=2 then factora=2;
if ipr=2 and prs=1 then factora=3;
if ipr=2 and prs=2 then factora=4;
if ipr=3 and prs=1 then factora=5;
if ipr=3 and prs=2 then factora=6;
cards;
(the data is inserted here)
;
run;
```

Program Code for Problem 5 *(continued)*

```
(continued)
proc mixed data=rats covtest;
class rat factora drug;
model lpr= factora drug factora*drug/s;
random intercept/subject=rat(factora) g;
run;
proc mixed data=rats covtest;
class rat factora drug;
model lpr= factora drug factora*drug/s;
random intercept/subject=rat g;
run;
proc mixed data=rats covtest empirical;
class rat factora drug;
model lpr= factora drug factora*drug/s;
random intercept/subject=rat(factora) g;
run;
proc mixed data=rats covtest empirical;
class rat factora drug;
model lpr= factora drug factora*drug/s;
repeated/subject=rat(factora) type=ar(1);
run;
```

Edited Computer Output *(same output obtained for first two program codes):*

```
Covariance Parameter Estimates
Cov Parm             Subject         Estimate          StdErr        Z Value        Pr Z
Intercept       rat(factora)         0.01108        0.003823           2.90      0.0019
Residual                            0.001539        0.000296           5.20     <.0001
```

| Type 3 Tests of Fixed Effects (Model-based- Random Intercept) | | | | |
Effect	NDF	DDF	F Value	Pr > F
factora	5	18	143.06	<.0001
drug	3	54	2818.00	<.0001
factora*drug	15	54	279.93	<.0001

| Type 3 Tests of Fixed Effects (Empirical- Random intercept) | | | | |
Effect	NDF	DDF	F Value	Pr > F
factora	5	18	227.13	<.0001
drug	3	54	3486.01	<.0001
factora*drug	15	54	10261.5	<.0001

| Type 3 Tests of Fixed Effects (Empirical- Repeated AR1) | | | | |
Effect	NDF	DDF	F Value	Pr > F
factora	5	18	227.13	<.0001
drug	3	54	3486.01	<.0001
factora*drug	15	54	10261.5	<.0001

6. Consider the same study by Heffner et al. and the dataset described in Problems 4 and 5, where the response measured was the lever press rate (LPR) taken by a thirsty rat to press a lever and receive water. The three factors identified as possible predictors were Drug (3 doses plus a placebo), IPR (initial press rate, categorized as slow, medium, or fast) and PRS (number of presses required to obtain water: two or five). Another factor of importance is the random factor Rat (with 24 levels). In this problem, we consider the effects of IPR and PRS individually, as well as the effects of Drug and Rat and various interactions among these factors in a repeated measures analysis.

The ANOVA model for this situation is a modification of the model used in Problem 5, except that the effect of Factor A is split into individual components. The structure for the subject-specific scalar version of this model involves the following fixed effect parameters and random effects:

μ = overall mean
α_j = jth fixed effect of IPR, $j = 1, 2, 3$
β_k = kth fixed effect of PRS, $k = 1, 2$
γ_l = lth fixed effect of Drug, $l = 1, 2, 3, 4$
$(\alpha\beta)_{jk}$ = fixed interaction effect of the jth level of IPR with the kth level of PRS
$(\alpha\gamma)_{jl}$ = fixed interaction effect of the jth level of IPR with the lth level of Drug
$(\beta\gamma)_{kl}$ = fixed interaction effect of the kth level of PRS and the lth level of Drug
$(\alpha\beta\gamma)_{jkl}$ = 3-way fixed interaction effect for the (j, k, l)-combination of IPR, PRS and Drug
$S_{i(jk)}$ = random effect of rat i within levels j and k of IPR and PRS, respectively
$E_{l(ijk)}$ = random error of the lth level of Drug within levels j and k of IPR and PRS, respectively, for rat i.

6. **a.** Using the computer output based on fitting the above model, carry out tests for main effects and interactions of each predictor in the model. (Assume that all such tests are orthogonal.) What do you conclude about whether or not the Drug factor has a significant effect on the response?

   b. Use the output to test whether there is a significant random effect of the factor Rat.

Program Statements for SAS's GLM procedure for Problem 6:

```
proc mixed data=rats covtest;
class ipr prs drug rat;
model lpr= ipr prs drug ipr*prs ipr*drug prs*drug ipr*prs*drug/s;
random intercept/subject=rat(ipr*prs) g;

run;
proc mixed data=rats covtest;
class ipr prs drug rat;
model lpr= ipr prs drug ipr*prs ipr*drug prs*drug ipr*prs*drug;
random intercept/subject=rat g;

run;
```

Edited Computer Output (same output obtained for first two program codes):

Covariance Parameter Estimates

Cov Parm	Subject	Estimate	StdErr	Z Value	Pr Z
Intercept	rat(ipr*prs)	0.01108	0.003823	2.90	0.0019
Residual		0.001539	0.000296	5.20	<.0001

Type 3 Tests of Fixed Effects (Model-based- Random intercept)

Effect	NDF	DDF	F Value	Pr > F
ipr	2	18	42.40	<.0001
prs	1	18	629.59	<.0001
drug	3	54	2818.00	<.0001
ipr*prs	2	18	0.47	0.6333
ipr*drug	6	54	17.02	<.0001
prs*drug	3	54	1319.73	<.0001
ipr*prs*drug	6	54	22.94	<.0001

(continued)

e. ii. State the subject-specific scalar model that is being fit by each of the two program statements. In what way do the program statements differ?

iii. Is the correlation structure assumed by the model stated in part (e, ii) an exchangeable correlation structure? Explain.

iv. Assuming (perhaps incorrectly) that the F tests for each effect are "orthogonal," what would you conclude about which effects are significant and which effects are not significant?

v. Based on the information provided by the "NOTE" in the Edited Computer Output and by the "Covariate Parameter Estimates," should you be concerned about the appropriateness of these analyses? Explain.

Another set of analyses that was carried out for these data is described by the following computer code and edited output using SAS's MIXED procedure.

Program statements for SAS's MIXED procedure for Problem 8:

```
    PROC MIXED DATA=sf2 covtest;
    CLASS subj day time rater;
③  MODEL sf= day time rater day*time day*rater time*rater day*time*rater/s;
    RANDOM intercept /subject=subj;
    run;
    PROC MIXED DATA=sf2 covtest empirical;
    CLASS subj day time rater;
④  MODEL sf= day time rater day*time day*rater time*rater day*time*rater/s;
    RANDOM intercept /subject=subj;
    run;
```

Edited Output:

```
    Covariance Parameter Estimates
    Cov Parm        Subject       Estimate        StdErr      Z Value         Pr Z
    Intercept          subj        118.68        40.5250         2.93       0.0017
    Residual                       34.6456         3.4820         9.95       <.0001
```

	Type 3 Tests of Fixed Effects (Model-Based)				
	Effect	NDF	DDF	F Value	Pr > F
	day	2	198	5.28	0.0058
③	time	1	198	0.47	0.4933
	rater	1	198	5.01	0.0263
	day*time	2	198	0.68	0.5057
	day*rater	2	198	0.32	0.7247
	time*rater	1	198	0.79	0.3752
	day*time*rater	2	198	0.03	0.9666

	Type 3 Tests of Fixed Effects (Empirical)				
	Effect	NDF	DDF	F Value	Pr > F
	day	2	198	1.84	0.0161
④	time	1	198	0.50	0.4818
	rater	1	198	12.85	0.0004
	day*time	2	198	0.80	0.4523
	day*rater	2	198	0.82	0.4414
	time*rater	1	198	1.89	0.1709
	day*time*rater	2	198	0.08	0.9240

vi. State the subject-specific scalar model that is being fit by each of the two program statements (③ and ④). In what way do these program statements differ?

e.   vii. Is the correlation structure assumed by the model stated in part (e, vi) an exchangeable correlation structure? Explain.

    viii. Assuming (perhaps incorrectly) that the F tests for each effect are "orthogonal," what would you conclude about which effects are significant and which effects are not significant?

    ix. Based on the information by the "Covariate Parameter Estimates," should you be concerned about the appropriateness of these analyses? (*Note*: There is no "NOTE" statement output for these analyses.) Explain.

f. We will now consider the analysis of the SF data in which we have treated the Day and Time factors as fixed and the Rater factor as random.

i. How can the investigators justify a decision to treat the Rater factor as random?

Consider the following computer program code:

⑤
```
PROC MIXED DATA=sf2 covtest;
CLASS subj day time rater;
MODEL sf= day time day*time/s;
RANDOM rater day*rater time*rater day*time*rater;
RANDOM intercept day time rater day*time day*rater time*rater/subject=subj;
run;
```

    ii. State the subject-specific scalar model that is being fit by the above program statements (⑤).

    iii. Is the correlation structure assumed by the model stated in part (f.ii) an exchangeable correlation structure? Explain.

The code described in ⑤ can be shown from the output to provide an inappropriate analysis. Another code and its corresponding output are given as follows:

⑥
```
PROC MIXED DATA=sf2 covtest;
CLASS subj day time rater;
MODEL sf= day time day*time;
RANDOM rater/g;
run;
```

Covariance Parameter Estimates

Cov Parm	Estimate	StdErr	Z Value	Pr Z
rater	0.2069	2.1583	0.10	0.4618
Residual	150.10	14.2788	10.51	<.0001

Type 3 Tests of Fixed Effects (Model-based; empirical not allowed)

Effect	NDF	DDF	F Value	Pr > F
day	2	221	1.22	0.2975
time	1	221	0.11	0.7419
day*time	2	221	0.16	0.8540

    iv. State the subject-specific scalar model that is being fit by the above program code ⑥.

    v. Is the correlation structure assumed by the model stated in part (f. iv.) an exchangeable correlation structure? Explain.

    vi. Assuming that the above analysis is appropriate, what can you conclude about whether there are significant effects of Day, Time, Day × Time, and/or Rater? Explain.

    vii. What other approaches might you take to carry out the analysis when considering Day and Time as fixed effects and Rater as a random effect?

References

Diggle, P. J., Heagerty, P., Liang, K. Y., and Zeger, S. L. 2002. *Analysis of Longitudinal Data,* 2nd Edtion, Oxford: Oxford University Press.

Heffner, T. G., Drawbaugh, R. B., Zigmond, M. J. 1974. "Amphetamine and operant behavior in rats: relationship between drug effect and control response rate." *J Comp Physiol Psychol.* 86(6):1031–43.

Holder. P., Plikaytis, B. D., and Carlone, G. 1996. "Need for Blocking to Reduce Non-specific Binding," Paper presented at WHO, Workshop on Pneumococcal ELISA Standardization, Atlanta, May 15–16, 1996.

Kleinbaum, D. G., and Klein, M. 2002. *Logistic Regression– A Self-Learning Text,* 2nd Edition. New York and Berlin: Springer Publishers.

Littell, R. C., Milliken, G. A., Stroup, W. W., and Wolfinger, R. D. 1996. *SAS System for Mixed Models.* Cary NC : SAS Institute.

Ortiz, D. J., Marcus, M., Gerr, F., Jones, W., and Cohen, S. 1997. "Estimation of Within Subject Variability in Upper Extremity Posture Measured With Manual Goniometry Among Video Display Terminal Users." *Applied Ergonomics* 28(2) : 139–143.

Rikkers, L. F., Rudman, D., Galambos, J. T., Fulenwidr, J. T., Millikan, W. J., Kutner, M. H., Smith, R. B., Salam, A. A., Sones, P. J., and Warren, W. D. 1978. "A Randomized Controlled Trial of Distal Spenoral Shunt." *Annals of Surgery* 188: 271–82.

Snijders, T. and Bosker, R. 1999. *Multilevel Analysis.* London: Sage Publishers.

Verbeke, G., and Molenberghs, G. 2001. *Linear Mixed Models for Longitudinal Data,* New York and Berlin: Springer Publishers.

Zeger, S. L., and Liang, K. Y. 1986. "Longitudinal Data Analysis for Discrete and Continuous Outcomes." *Biometrics* 42: 121–30.

27

Sample Size Planning for Linear and Logistic Regression and Analysis of Variance

27.1 Preview

Sample size planning is an essential component of study design. In this chapter, we present practical approaches for sample size determination for linear and logistic regression and for ANOVA models. Unfortunately, the mathematical theory for sample size calculations can be fairly complicated for some of these multivariable methods. As a result, in practice, researchers sometimes choose to depend on simple but error-prone approaches, such as approximate power and sample size charts (Pearson and Hartley, 1951; Neter, Wasserman and Kutner, 1990) and even rules of thumb (Tabachnik and Fidell, 2001). Such shortcuts can be costly. In an undersized study, effects which are large enough to be scientifically important may turn out not to be statistically significant. In both undersized and oversized studies, resources are wasted.

With recently developed methods and continual improvements in computer software, rigorous sample size planning for regression methods is now much more feasible in everyday practice. In this chapter, we begin with a review of basic concepts and formulae relevant to sample size planning. We then present two approaches for sample size calculation for linear and logistic regression. The first is an approximate, but simple, approach that has been shown to yield fairly accurate sample sizes and for which even manual computation is feasible. The second is based on more traditional theory for sample size determination, and is best implemented using computer software. For the latter approach, we present examples for linear regression and ANOVA using two popular programs, SAS®, version 9.1 (SAS Institute, 2002–2003), and PASS® 2005 (Hintz, 2004).

27.2.3 Sample Size Planning: The Relationship Among α, $(1-\beta)$, Δ, σ^2, and n

In real-world applications, sample size planning does not involve a single calculation. Instead, it is usually a process involving examination of the required sample sizes for various different combinations of the following quantities: α, the Type I error rate or significance level; $(1 - \beta)$ or power (β is the Type II error rate); Δ, the minimum effect size that would be of scientific significance and that, therefore, scientists would like to be able to detect; and σ^2, the population variance.

For a fixed sample size, α and β are inversely related in the following sense. If one tries to guard against making a Type I error by choosing a small rejection region, the non-rejection region (and hence β) will be large. Conversely, protecting against a Type II error necessitates using a large rejection region, leading to a large value for α.

Increasing the sample size generally decreases β, for any given value of α. For fixed α and β, the larger Δ is, the smaller the sample size required to detect that effect.

Typically, researchers will study tables and graphs of sample sizes for various combinations of β and Δ (α is usually fixed), and will select the sample sizes that result in high power for desirable combinations of these parameters, and that are also practically feasible given resource constraints.

Researchers can also observe the variation in sample sizes for a range of values of σ^2, the population variance, in order to understand the impact this parameter has on sample size; in practice, of course, the researcher generally does not have control over the population variance.

27.3 Sample Size Planning for Linear Regression

In this section, we present simple approaches to sample size calculations for linear regression where manual calculation is feasible and specialized software is not necessary. We begin with the simple linear regression model where the main independent variable of interest is binary and where the dependent variable is assumed to be normally distributed. The inference-making tool in this case is an equal population variances t test comparing two population means; for the large sample case, the sample size calculation will be approximately the same as discussed in Section 27.2.1. Next, we consider the simple linear regression model with a continuous covariate; in this setting, a test for zero slope of the straight line is mathematically equivalent to a test for zero correlation between the dependent and independent variables; from this relationship, another simple sample size formula arises. For multiple regression with a single predictor of primary interest, and p other predictors mainly for control of confounding, it has been shown that the simple linear regression sample size formulae can be easily adjusted to give approximate but fairly accurate results.

27.3.1 Sample Size Determination for Simple Linear Regression with a Binary Predictor

Consider the simple linear regression model (5.3) in Chapter 5:

$$Y = \beta_0 + \beta_1 X + E$$

Let X be a binary independent variable indexing two groups or treatments, coded as 1 for one group, 0 for the other. The main inference of interest in this regression, performed to investigate whether or not independent variable X is a statistically significant predictor of Y, is the test for zero slope (H_0: $\beta_1 = 0$ vs H_A: $\beta_1 \neq 0$), which was described in Section 5.7.

As described in Chapter 5, the simple linear regression model relates the mean value of Y to values of X. Therefore, the t test for the slope compares the average value of Y for two values of X, which are one unit apart. This is analogous to a two-sample t test (since Y is assumed to be normally distributed and to have constant variance σ^2 for all values of X). As such, the sample size formula (27.1) can be used for simple linear regression with a binary predictor. Although the formula is based on the standard normal distribution rather than the t-distribution, for large sample sizes the approaches converge and the sample size calculations are accurate (Kupper and Hafner, 1989). Also, Δ, the effect size of interest is replaced by $|\beta_1^*|$ ($\beta_1^* \neq 0$), the minimum absolute value of the slope for which it would be important to reject the null hypothesis of zero slope.

■ **Example 27.3** Example 27.1 can be modeled with the following simple linear regression:

$$Y = \beta_0 + \beta_1 X + E$$

where

Y = quantity of active ingredient, in mg

X = 1 if new formulation of drug, 0 if standard formulation

To test whether the true average quantities of the active ingredient are different, a test of H_0: $\beta_1 = 0$ vs H_A: $\beta_1 \neq 0$ will be conducted. For $\alpha = 0.05$, $\sigma = 2.0$ mg and a power of 0.8 to detect a slope value $|\beta_1|$ of at least 1.5 (corresponding to an absolute difference in average quantity of 1.5 mg), the sample size calculation is

$$n \geq 2\left[\frac{(2.0)(Z_{0.975} + Z_{0.8})}{1.5}\right]^2 = 2\left[\frac{(2.0)(1.96 + 0.842)}{1.5}\right]^2 = 27.9$$

At least 28 doses of each formulation of the drug need to be sampled. ■

27.3.2 Sample Size Determination for Simple Linear Regression with a Continuous Predictor

Consider again the simple linear regression model (5.3):

$$Y = \beta_0 + \beta_1 X + E$$

This time, let X be a normally distributed independent variable representing the continuous predictor of interest. The main inference of interest in this regression again involves a t test for zero slope. In Section 6.6.1, it was shown that this test is mathematically equivalent to the test of H_0: $\rho_{YX} = 0$ vs. H_A: $\rho_{YX} \neq 0$, the test for the correlation between Y and X. For the latter test, it can be shown that the minimum sample size, n_s, necessary to reject the null hypothesis H_0 in favor of the alternative H_A when $\rho_{XY} = \rho$ ($\rho \neq 0$) with significance level α and power $1 - \beta$ is

$$n_s \geq \left[\frac{Z_{1-\alpha/2} + Z_{1-\beta}}{C(\rho)}\right]^2 + 3 \tag{27.3}$$

where $C(\rho)$ is the Fisher's Z transformation discussed in Section 6.6.2, namely,

$$C(\rho) = \frac{1}{2} \ln\left(\frac{1 + \rho}{1 - \rho}\right)$$

■ **Example 27.4**   Researchers are planning to investigate the association between APGAR scores taken five minutes after birth and infant blood pressure five hours after birth. They will use $\alpha = 0.05$ and $\beta = 0.20$, and want to detect a correlation of at least $\rho = 0.30$. Using formula (27.3), we can determine that (X, Y) data should be collected from at least 85 newborns:

$$n_s \geq \left[\frac{Z_{1-\alpha/2} + Z_{1-\beta}}{C(\rho)}\right]^2 + 3 = \left[\frac{1.96 + 0.842}{\frac{1}{2} \ln\left(\frac{1 + 0.3}{1 - 0.3}\right)}\right]^2 + 3 = 85.0$$

■

27.3.3 Sample Size Determination for Multiple Linear Regression: A Simple Approach

Hsieh, Bloch, and Larsen (1998) have shown that sample sizes for multiple regression models where a single independent variable is primarily of interest and other predictors are included mainly for control of confounding can be easily approximated from the simple linear regression sample size formulae above. In particular, if n_s is the sample size for a simple linear regression of Y on X_1, the sample size, n_m, for a multiple regression of Y regressed on $X_1, \ldots, X_k$ is approximately

$$n_m = \frac{n_s}{1 - \rho^2_{X_1(X_2, \ldots, X_k)}} \qquad (27.4)$$

where $\rho^2_{X_1(X_2, \ldots, X_k)}$ is the chosen value of the population squared multiple correlation between the main independent variable of interest, X_1, and the control variables $X_2, \ldots, X_k$. The basic idea is that the simple linear regression sample size is adjusted by multiplying by the variance inflation factor for the primary independent variable of interest. Hsieh et al. show the approximate calculation in formula (27.4) to be reasonably accurate for normally and non-normally distributed independent variables.

■ **Example 27.5**   Researchers want to conduct a study evaluating two types of surgery for nearsightedness. The outcome will be the patient's refractive error, measured in diopters, one year after surgery. Researchers want to control for the patient's age and baseline (i.e., pre-surgical) refractive error. They want to use $\alpha = 0.05$ and $\beta = 0.20$ and want to detect an absolute difference in average error of at least 1 diopter (i.e., $|\beta_1^*| \geq 1$). The variance, σ^2, of the refractive error is believed to be at most 1, and the population squared multiple correlation between the dummy variable indexing surgery type and the control variables age and baseline refractive error is chosen to be no more than 0.25.

Assuming equal sample sizes for the two surgeries, the per-surgery type sample size for the simple linear regression using the dummy variable for surgery type is:

$$n \geq 2\left[\frac{\sigma(Z_{1-\alpha/2} + Z_{1-\beta})}{|\beta_1^*|}\right]^2 = 2\left[\frac{(1)(Z_{0.975} + Z_{0.8})}{1}\right]^2 = 2(1.96 + 0.842)^2 = 15.70$$

or 16 patients per type of surgery. Using the adjustment suggested by Hsieh et al., the total sample size for the multiple regression would be $32/(1 - 0.25)$ or 43 (i.e., 21 or 22 patients for each of the two surgery groups). ∎

Example 27.6   Consider again Example 27.4. Suppose that researchers want to control for mother's age and whether or not the birth was Cesarian. They believe that the squared multiple correlation between the main independent variable, APGAR score, and the two control variables is 0.5. Using the simple linear regression sample size determined in Example 27.4 and formula 27.4, the minimum required sample size is $85/(1 - 0.5) = 170$. ∎

27.4 Sample Size Planning for Logistic Regression

The theory behind power and sample size calculations is more difficult for logistic regression than for linear regression. We present sample size calculations for the simple logistic regression model, and then for the multiple logistic model using the approach of Hsieh et al., which has already been discussed in previous sections. Both sets of calculations are simple enough to do manually; alternatively, PASS software can be used to perform the same calculations.

27.4.1 Sample Size Determination for Simple Logistic Regression with a Binary Predictor

Consider the simple logistic regression model:

$$\text{logit}[\text{pr}(Y = 1)] = \log_e\left(\frac{\pi}{1 - \pi}\right) = \beta_0 + \beta_1 X$$

where Y is the binary outcome (coded $Y = 1$ for the event of interest, $Y = 0$ otherwise), π is the probability of the event of interest, and X is a binary covariate indexing two groups or treatments (coded $X = 1$ for one group, $X = 0$ for the other). The main inference of interest in this regression, performed to investigate whether or not covariate X is a statistically important predictor of Y, involves a hypothesis test for zero slope: H_0: $\beta_1 = 0$ vs H_A: $\beta_1 \neq 0$. The Wald chi-square test for the slope was described in Section 21.3.1. Also, it can easily be shown that this large sample test is mathematically equivalent to the test comparing two proportions described in Section 27.2.2. Therefore, sample size calculations for a logistic regression with a single binary covariate can be performed using formula (27.2).

Example 27.7   Consider again Example 27.2. The researchers want to conduct a logistic regression of Y (coded 1 if a dose is unacceptable, 0 if it is acceptable) on X (coded 1 if new formulation, 0 if standard formulation). They plan to perform a test of H_0: $\beta_1 = 0$ versus H_A: $\beta_1 \neq 0$, where $\beta_1 = $ slope of the logit model. Suppose that previous studies of the standard formulation suggest that π_1, the true proportion of acceptable doses for the standard formulation, is 0.90. If an absolute difference, Δ, in proportions of at least 0.05 would be important to detect, then, for $\alpha = 0.05$ and $\beta = 0.2$, the sample size for each group is given by equation (27.2)—at least 435 doses of each formulation would be required. ∎

27.4.2 Sample Size Determination for Simple Logistic Regression with a Continuous Predictor

Consider again the simple logistic regression model:

$$\text{logit}[\text{pr}(Y = 1)] = \log_e\left(\frac{\pi}{1 - \pi}\right) = \beta_0 + \beta_1 X$$

where Y is the response variable (coded 1 if the event of interest occurs, 0 if it does not), π is the probability of the event of interest, and X is a standardized, normally distributed covariate. The main inference of interest in this regression, performed to investigate whether or not covariate X is a statistically significant predictor, involves testing $H_0: \beta_1 = 0$ vs $H_A: \beta_1 \neq 0$. This test is equivalent to testing whether or not the true average value of X is the same at the two values of Y, and as such it can be considered to be a two-sample test comparing means. Hsieh et al. (1998) present a modified version of equation (27.1) for sample size calculations. The minimum sample size necessary, n_s, is given by:

$$n_s \geq \frac{(Z_{1-\alpha/2} + Z_{1-\beta})^2}{\pi_1(1 - \pi_1)\beta_1^{*2}} \tag{27.5}$$

where π_1 is the probability of the outcome at the mean value of covariate X and $|\beta_1^*|$ is the minimum absolute value of the slope that would be important to detect if the alternative hypothesis were true. Note that, since the variable X is entered in standardized form, β_1^* measures the effect of a one standard deviation change in X.

■ **Example 27.8** Consider again Examples 27.1 and 27.2. Suppose that the researchers want to conduct a logistic regression of Y (coded 1 if a dose is unacceptable, 0 if it is acceptable) on X, the weight, in mg, of doses of the new formulation of the drug. X is assumed to be normally distributed, and will be entered into the regression as a standardized variable. The researchers plan to perform a test of $H_0: \beta_1 = 0$ versus $H_A: \beta_1 \neq 0$, where β_1 = slope of the logit model. Suppose that the manufacturing process for each formulation is calibrated to provide an average dose weight of 15 mg with a standard deviation of 2 mg; and, at this average weight, experience has shown that 3 out of every 100 doses of the standard drug would be deemed unacceptable. If a minimum difference, β_1^*, in the slope of 1 unit (corresponding to a large odds ratio of 2.72) would be important to detect, then, for $\alpha = 0.05$ and $\beta = 0.2$, the approximate sample size for each group is given by:

$$n_s \geq \frac{(Z_{1-\alpha/2} + Z_{1-\beta})^2}{\pi_1(1 - \pi_1)\beta_1^{*2}} = \frac{(Z_{0.975} + Z_{0.8})^2}{(0.03)(1 - 0.03)(1)} = \frac{(1.96 + 0.842)^2}{0.0291} = 269.8$$

At least 270 doses of the new formulation need to be sampled. ■

27.4.3 Sample Size Determination for Multiple Logistic Regression

For multiple logistic regression, when one covariate is primarily of interest and one or more other covariates are included mainly for control, the sample size for the simple linear regression, n_s, may be calculated using whichever of (27.2) or (27.5) above is appropriate, and then the adjustment of Hsieh et al. shown in equation (27.4) may be applied to obtain the approximate sample size for the multiple logistic regression model.

■ **Example 27.9** Tearing of the anterior cruciate ligament (ACL) is a common knee injury suffered by athletes in many sports. Surgery to replace the torn ACL with a muscle graft or cadaver ligament is often performed to repair the injury. Suppose that an orthopedic surgeon wants to evaluate two forms of the surgery: the standard surgery in which a single replacement ligament is used, versus the "double-bundle" technique in which two replacement ligaments are used. The following logistic regression is planned:

$$\text{logit}[\text{pr}(Y = 1)] = \beta_0 + \beta_1 X_1 + \beta_2 X_2 + \beta_3 X_3$$

where $Y = 1$ if the replacement ligament tears within 2 years after surgery, $Y = 0$ if it does not tear; $X_1 = 1$ if standard surgery is performed, $X_1 = 0$ if the "double bundle" technique is used; X_2 = patient's age at the time of surgery; X_3 = patient's weight at the time of surgery. X_1 is the primary predictor of interest; X_2 and X_3 are control variables.

The researchers want to use $\alpha = 0.05$ and $\beta = 0.2$. They believe that the re-tear rate is 2% for the standard surgery and would like to have a power of 0.8 of detecting an absolute difference in re-tear rates of at least 1%. Since the patients will be randomized to the two types of surgery, there will be little correlation between X_1 and the two control variables; a population multiple squared correlation, $\rho^2_{X_1(X_2, X_3)}$, of 0.0001 will be assumed.

Using formula (27.2), the required minimum sample size, n_s, for a simple logistic regression of Y on X_1 would be:

$$n_s \geq 2 \left[\frac{(1.96)\sqrt{(2)(0.015)(1 - 0.015)} + (0.842)\sqrt{(0.02)(1 - 0.02) + (0.01)(1 - 0.01)}}{0.01} \right]^2$$

$$= 4{,}637.7$$

At least 4,638 patients will be needed (i.e., at least 2,319 patients will have to be evaluated for each type of surgery). For the multiple logistic regression, the approximate minimum sample size, n_m, is found using formula (27.3):

$$n_m = \frac{n_s}{1 - \rho^2_{X_1(X_2, X_3)}} = \frac{4638}{1 - 0.0001} = 4638.5$$

At least 4,639 patients will be needed (i.e., at least 2,320 patients will have to be evaluated for each type of surgery). ■

Note that the required sample size in this example would, practically speaking, be impossible to achieve. In such situations, as described in Section 27.3.2, researchers would need to study tables and graphs of sample sizes for various combinations of power and effect size, the purpose being to determine whether acceptable combinations of these parameters exist for which the required sample size is practically feasible.

■ **Example 27.10** Planning for emergency medical care needs in environments involving mass gatherings (e.g., large sports events and concerts) requires an understanding of the type of care that will most likely be needed: trauma (i.e., injury) care or care for non-injury medical conditions. The type of care needed will probably be associated with a patient's age (assumed to be normally distributed). Suppose that planning for the relevant study to investigate this association is being conducted. Other covariates that will need to be controlled are type of event (concert versus sports event), patient's gender, and the type of weather during the event.[3] The logistic regression model of interest is:

$$\text{logit}[\text{pr}(Y = 1)] = \beta_0 + \beta_1 X_1 + \beta_2 X_2 + \beta_3 X_3 + \beta_4 X_4$$

where $Y = 1$ if the patient requires trauma care, $Y = 0$ if care for a non-injury medical condition is required; $X_1 =$ age of the patient, in years (the primary covariate of interest); $X_2 = 1$ if the patient is male, $X_2 = 0$ if female, $X_3 = 1$ if the event is a concert, $X_3 = 0$ if it is a sporting event; $X_4 =$ amount of precipitation, in inches, on the day of the event.

How many patients will need to be studied? The researchers will use $\alpha = 0.05$ and $\beta = 0.2$. They believe that 20% of the care necessary will be for traumas, and, once again, believe that a minimum difference in the slope of 1 unit would be important to detect. The multiple correlation between X_1 and the control variables is assumed to be low; a multiple squared correlation, $\rho^2_{X_1(X_2, X_3, X_4)}$, of 0.01 will be assumed.

The minimum necessary sample size is given by formula (27.5) and by Hsieh, Bloch and Larsen's formula (27.4):

$$n_s \geq \frac{(Z_{1-\alpha/2} + Z_{1-\beta})^2}{\pi_1(1 - \pi_1)\beta_1^{*2}} = \frac{(Z_{0.975} + Z_{0.8})^2}{(0.2)(1 - 0.2)(1)} = \frac{(1.96 + 0.842)^2}{0.16} = 49.1$$

$$n_m = \frac{n_s}{1 - 0.01} = \frac{49.1}{0.99} = 49.6$$

At least 50 patients should be included in the study. ■

27.5 Power and Sample Size Determination for Linear Models: A General Approach

The sample size formulae presented in the preceding sections are conceptually simple and are also easy to implement, even without the use of specialized software. Indeed, these approaches are probably the simplest approaches available that yield reasonably accurate sample size calculations. However, these simple approaches are not based on the standard theory for power and sample size calculations for regression models. The standard theory provides a more general approach which allows for power (and sample size) determination for a variety of experimental designs and models and a variety of contrasts related to the models. For example, in a multiple linear regression, a *multiple* partial F test for the joint importance of several independent variables, controlling for several other variables, may be of interest, rather than a partial F test about just one particular independent variable. As another example, in ANOVA, it may be of more interest to determine the sample size necessary for powerful contrasts in a multiple comparisons analysis, rather than for the overall test for the significance of a single factor. The theory also generalizes to more complicated models such as ANOVA models involving two or more factors, models with repeated measures, and MANOVA models (Koele, 1982; Muller et al., 1992).

---

[3] A similar study has been conducted by Milsten et al. (2003).

This theory is well-documented in numerous books and papers (e.g., Scheffé, 1959; Cohen, 1977; Muller et al., 1992). We present just the technical details which are necessary for the implementation of the theory. Analysts typically will rely on the use of specialized software to implement these methods. PASS and, where possible, SAS examples will be presented.

27.5.1 Power and Sample Size Determination for Multiple Linear Regression

Consider the multiple linear regression model

$$Y = \beta_0 + \beta_1 X_1 + \cdots + \beta_k X_k + \beta_1^* X_1^* + \cdots + \beta_p^* X_p^* + E \qquad (27.6)$$

with the usual assumptions (see Section 8.4.1). Suppose that the hypothesis test of interest is a test for the importance of $X_1, \ldots, X_k$ given $X_1^*, \ldots, X_p^*$ (i.e., a test of H_0: $\beta_1 = \cdots = \beta_k = 0$ vs H_A: at least one of these $\beta_j \neq 0$, given $X_1^*, \ldots, X_p^*$ are in the model). It was shown in Section 10.7.1 that, for sample size n, the test statistic can be written as:

$$F = \frac{R^2_{Y(X_1, \ldots, X_k) \mid X_1^*, \ldots, X_p^*}/k}{(1 - R^2_{Y \mid X_1, \ldots, X_k, X_1^*, \ldots, X_p^*})/(n - p - k - 1)},$$

where $F \sim F_{k, n-p-k-1}$ under the null hypothesis. The steps necessary to determine the power of this size α test are as follows:

1. Determine the critical value $F_{k, n-p-k-1, 1-\alpha}$.

2. Estimate (or provide values for), using a reasoned approach, $\rho^2_{Y \mid X_1, \ldots, X_k, X_1^*, \ldots, X_p^*}$ (the population squared multiple correlation), and $\rho^2_{Y(X_1, \ldots, X_k) \mid X_1^*, \ldots, X_p^*}$, the population squared multiple-partial correlation for variables $X_1, \ldots, X_k$ given $X_1^*, \ldots, X_p^*$.

3. Calculate the power as $\Pr(F_\lambda > F_{k, n-p-k-1, 1-\alpha})$, where F_λ follows a non-central F distribution with non-centrality parameter $\lambda = \dfrac{n\rho^2_{Y(X_1, \ldots, X_k) \mid X_1^*, \ldots, X_p^*}}{1 - \rho^2_{Y \mid X_1, \ldots, X_k, X_1^*, \ldots, X_p^*}}$. This parameter λ captures the effect size that is to be detected. Typically, users rely on software to perform this probability calculation.

There is no straightforward way, in this general setting, to calculate sample sizes for multiple linear regression models. Instead, the power calculation above has be repeated with different values of n until a sample size that yields an acceptable power is achieved. Computer software makes this iterative approach feasible.

■ **Example 27.11**   This example illustrates the use of PASS software to perform a sample size calculation for multiple linear regression. The presentation here will not provide insight into all the capabilities of the software. Instead, it will focus simply on the solution to the current problem. It is left to the reader to explore the additional capabilities of PASS.

Consider again Example 27.5. Researchers want to conduct a study evaluating two surgeries for near-sightedness. The outcome will be the patient's refractive error, measured in diopters, one year after surgery. Researchers want to control for the patient's age and baseline refractive error. They want to use $\alpha = 0.05$ and $\beta = 0.20$ and want to detect a minimum difference in average refractive error of 1 diopter.

Suppose that the population squared multiple correlation between the dependent and all independent variables is projected to be between 0.25 and 0.70; the population squared multiple correlation between the dependent variable and the control variables, age and baseline refractive error, is projected to be 0.05. (Note that, therefore, the population squared partial correlation between the dependent variable and main independent variable of interest, surgery type, given that age and baseline refractive error are in the model, will therefore be between 0.20 and 0.65.)

After starting PASS, the user should select 'Multiple Regression (Y is continuous)' from the list of available procedures and then complete the subsequent dialog box as shown in Figure 27.1 below[4]. To perform the calculation, the play button (blue arrow near the top left of the dialog box) is clicked.

Figure 27.2 shows a screenshot of the output from PASS. Both a table and a graph are provided by the software by default. They show that a sample size of between 8 and 32 is required to achieve a power of at least 0.8, with a significance level of 0.05, assuming that the two control variables have a ρ^2 of R2(C) = 0.05 and that the main independent variable has a partial ρ^2 of R2(T|C) between 0.20 and 0.65 (power increases as the latter partial ρ^2 increases for fixed N).

PASS could be re-run to explore how the required sample size would vary for different values of power, significance level, and squared correlations.

FIGURE 27.1   PASS dialog box for the multiple linear regression sample size calculation for Example 27.11

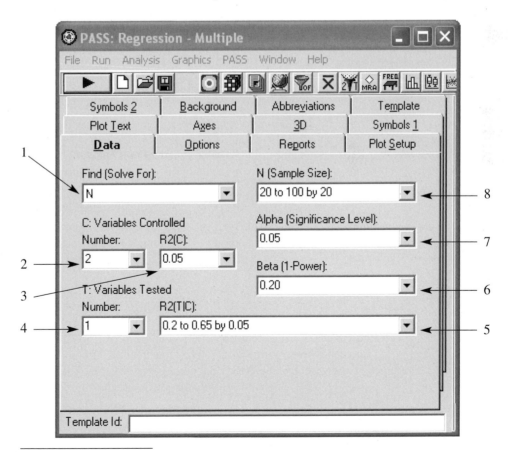

FIGURE 27.2   PASS output for Example 27.11

Numeric Results

Power	N	Alpha	Beta	Cnt	Ind. Variables Tested R2	Cnt	Ind. Variables Controlled R2
0.80504	32	0.05000	0.19496	1	0.20000	2	0.05000
0.81303	25	0.05000	0.18697	1	0.25000	2	0.05000
0.81376	20	0.05000	0.18624	1	0.30000	2	0.05000
0.80061	16	0.05000	0.19939	1	0.35000	2	0.05000
0.81959	14	0.05000	0.18041	1	0.40000	2	0.05000
0.81996	12	0.05000	0.18004	1	0.45000	2	0.05000
0.84882	11	0.05000	0.15118	1	0.50000	2	0.05000
0.80084	9	0.05000	0.19916	1	0.55000	2	0.05000
0.87644	9	0.05000	0.12356	1	0.60000	2	0.05000
0.86891	8	0.05000	0.13109	1	0.65000	2	0.05000

Report Definitions
Power is the probability of rejecting a false null hypothesis.
N is the number of observations on which the multiple regression is computed.
Alpha is the probability of rejecting a true null hypothesis. It should be small.
Beta is the probability of accepting a false null hypothesis. It should be small.
Cnt refers to the number of independent variables in that category.
R2 is the amount that is added to the overall R-Squared value by these variables.
Ind. Variables Tested are those variables whose regression coefficients are tested against zero.
Ind. Variables Controlled are those variables whose influence is removed from experimental error.

Summary Statements
A sample size of 32 achieves 81% power to detect an R-Squared of 0.20000 attributed to 1 independent variable(s) using an F-Test with a significance level (alpha) of 0.05000. The variables tested are adjusted for an additional 2 independent variable(s) with an R-Squared of 0.05000.

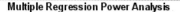

Multiple Regression Power Analysis

Chart Section

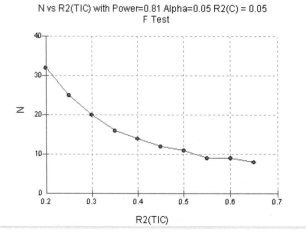

Explanation of values in the dialog box:

1. PASS allows one to calculate the power of a test, or the sample size necessary to achieve that power. We have selected sample size calculation.

2. The number of control variables, p, is specified here.

3. The population squared correlation between the dependent and control variables is specified here. This must be a single value. PASS uses the notation R2(C) for this population parameter.

4. The number of main independent variables of interest, k, is specified here.

5. The population squared (multiple) partial correlation between the dependent variable and the k main independent variable(s) of interest, controlling for the remaining p independent variables, is specified here as R2(T|C). The sum of R2(T|C) and R2(C) above must be less than 1. A range of R2(T|C) values can be specified, as shown here. [Note that the sum of R2(C) and R2(T|C) equals the assumed value of the population squared multiple correlation coefficient for Model (27.6).]

6. The desired Type II error rate (Beta), or equivalently the power, is specified here. A single value or a range of values may be specified.

7. The desired significance level is specified here. A single value or a range of values may be specified.

8. This box is inactive in the current calculation. If, in box 1, we had chosen to calculate power, a range of sample sizes could have been specified in box 8. ■

■ **Example 27.12** We re-do the previous example using the PROC POWER procedure in SAS software. The SAS program (Figure 27.3) and output are shown below. The SAS code is annotated with explanatory notes. The output shows the same results as seen in the PASS output.

FIGURE 27.3 SAS program for the multiple linear regression sample size calculation in Example 27.12

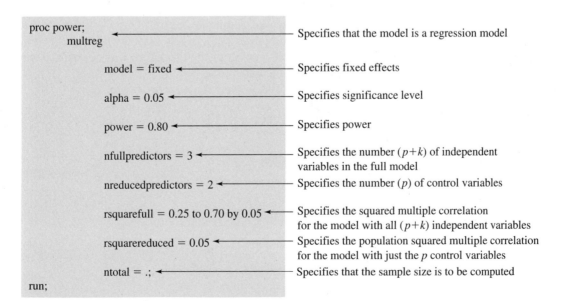

Edited SAS output (PROC POWER) for Example 27.12

```
                    The POWER Procedure
          Type III F Test in Multiple Regression

                 Fixed Scenario Elements
       Method                                          Exact
       Model                                           Fixed X
       Number of Predictors in Full Model              3
       Number of Predictors in Reduced Model           2
       Alpha                                           0.05
       R-square of Reduced Model                       0.05
       Nominal Power                                   0.8

                     Computed N Total
                     R-square    Actual       N
            Index      Full      Power      Total
              1        0.25      0.805        32
              2        0.30      0.813        25
              3        0.35      0.814        20
              4        0.40      0.801        16
              5        0.45      0.820        14
              6        0.50      0.820        12
              7        0.55      0.849        11
              8        0.60      0.801         9
              9        0.65      0.876         9
             10        0.70      0.869         8
```

∎

27.5.2 Power and Sample Size for One-Way Fixed Effects ANOVA

In Chapter 17, it was pointed out that ANOVA can be viewed as a special case of multiple linear regression. As such, the sample size determination methods already presented for multiple linear regression can be applied to ANOVA models, including the simple approaches in Sections 27.3.1 and 27.3.3. Indeed, in its simplest form, One-Way ANOVA with a two-level, fixed effects factor is the equivalent of a two-sample t test with equal population variances (see Sections 17.1 and 17.3.2); therefore, formula (27.1) could be used in sample size planning. For more general ANOVA models, the procedure for determining power (and sample size) described in Section 27.5.1 can be used. However, practical implementation of the procedure will be easier if we restate the procedure specifically for ANOVA models.

Consider the fixed effects One-Way ANOVA model presented in Chapter 17:

$$Y_{ij} = \mu + \alpha_i + E_{ij} \qquad \text{for } i = 1, 2, \ldots, k \qquad \text{and} \qquad j = 1, 2, \ldots, n_i \qquad (27.7)$$

where

Y_{ij} = jth observation from the ith population

μ_i = mean for population i

$\mu = \dfrac{\mu_1 + \mu_2 + \cdots + \mu_k}{k}$ = true overall mean response

$\alpha_i = \mu_i - \mu$ = differential effect of population i

E_{ij} = error term for the jth observation from the ith population and where the usual assumptions are made (see Section 17.2). A hypothesis test that is always of interest is the overall F test with hypotheses H_0: $\mu_1 = \mu_2 \ldots = \mu_k$ vs H_A: the k population means are not all equal. It was shown in Section 17.3 that the test statistic can be written as:

$$F = \frac{\text{MST}}{\text{MSE}} = \frac{\sum\limits_{i=1}^{k} n_i (\bar{Y}_{i.} - \bar{Y})^2 / (k - 1)}{\sum\limits_{i=1}^{k} \sum\limits_{j=1}^{n_i} (Y_{ij} - \bar{Y}_{i.})^2 / (n - k)}$$

where $F \sim F_{k-1, n-k}$ under the null hypothesis. The steps necessary to determine the power of this size α test are as follows:

1. Determine the critical value $F_{k-1, n-k, 1-\alpha}$.

2. Using a reasoned approach, provide a value for the non-centrality parameter $\lambda = \frac{n\sigma_m^2}{\sigma^2}$, where n is the total sample size ($n = \sum_{i=1}^{k} n_i$), σ^2 is the variance of the out-come that is common to each population (i.e., the within-population variance), and σ_m^2 is the between-population variance $\left(\sigma_m^2 = \sum_{i=1}^{k} \frac{n_i(\mu_i - \bar{\mu})^2}{n} \right.$, where $\bar{\mu}$ is the weighted average group mean, $\bar{\mu} = \sum_{i=1}^{k} \frac{n_i\mu_i}{n} \left. \right)$. Once again, λ captures the effect size that is to be detected. In practice, the value of σ^2 that is used is based on prior research or pilot studies, or is a sound guess based on expert knowledge. The value of σ_m^2 is obtained by specifying both the values of $\mu_1, \mu_2, \ldots, \mu_k$ under the alternative hypothesis that would be of interest to detect and the sample sizes $n_1, n_2, \ldots, n_k$.

3. Calculate the power as $\Pr(F_\lambda > F_{k-1, n-k, 1-\alpha})$, where F_λ follows a *non-central F distribution* with non-centrality parameter λ. Typically, users rely on software to perform this probability calculation.

As in the case of linear regression, iterative implementation of this procedure with different values of n allows the user to determine the overall sample size that yields an acceptable power. Note that this procedure is easily adapted to the case where the main hypothesis test of interest involves some other contrast of interest, as illustrated in the Example 27.13.

■ **Example 27.13**   Researchers in Bangladesh want to compare the average zooplankton counts per liter in surface waters (ponds) in three rural regions of Bangladesh. It is believed that zoo-plankton harbor cholera bacteria and, therefore, that elevated zooplankton counts in surface waters may put rural inhabitants at increased risk for cholera. How many ponds should be studied in each region? Previous studies show that the standard deviation of counts per liter is approximately 1,000 per liter of water sampled. If the average count per liter in one, well-studied area is 5,000 per liter, and, in actuality, in other regions it is four-fifths of that value, the researchers would like to be able to reject the null hypothesis with $\alpha = 0.05$ and power 0.8.

Again, we use PASS to perform the sample size determination. The user selects 'One Way ANOVA' from the list of available procedures and then completes the subsequent dialog box as

FIGURE 27.4 **PASS dialog box for the One-Way ANOVA sample size calculation in Example 27.13**

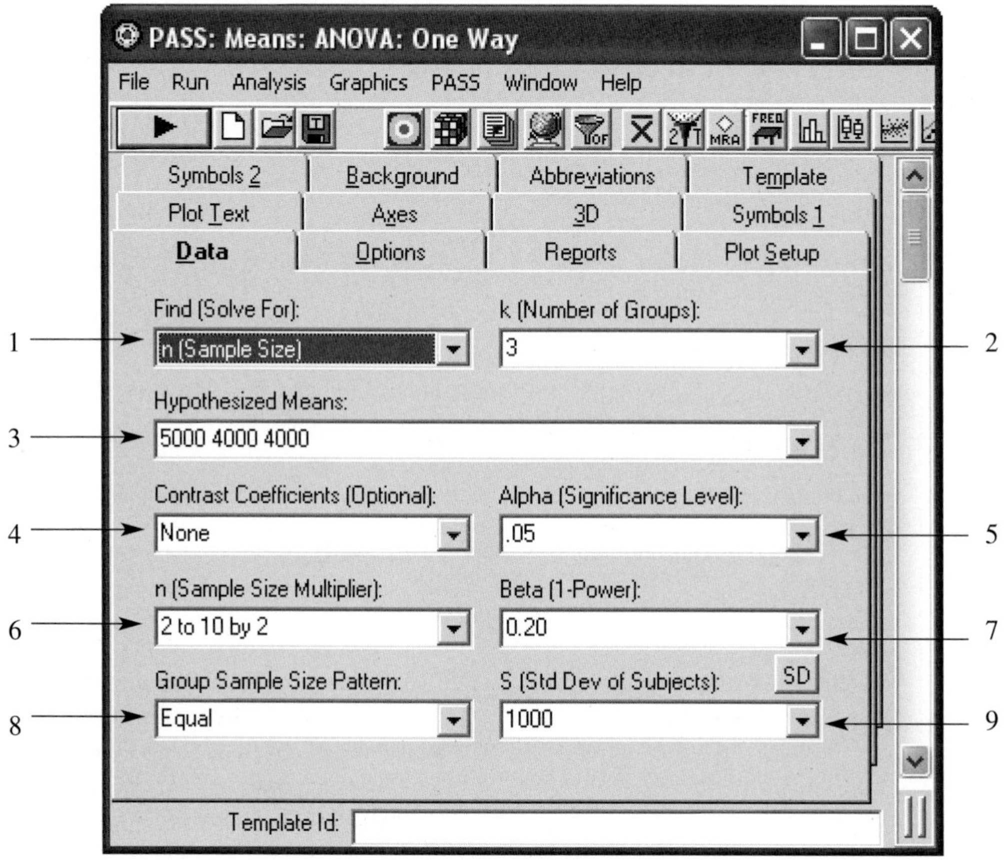

shown in Figure 27.4 above. The output from PASS, shown in Figure 27.5, indicates that 16 ponds need to be sampled in each of the three regions of the study.

Explanation of values in the dialog box:

1. PASS allows one either to calculate the power of a test or to determine the sample size necessary to achieve that power. We have selected sample size determination option here.

2. The number of populations or groups for which means are to be compared is specified here.

3. The values of the population means under the alternative hypothesis are specified here.

4. If the calculation concerns a user-defined contrast, rather than the standard null hypothesis that all population means are equal, the coefficients for the contrast are specified here.

argued that there is little to be learned from such calculations, since it can be assumed that, if the null hypothesis of no effect was not rejected, then presumably the power of the method was not high enough to conclude that the observed effect size was significantly different from zero (Lenth, 2000, 2001).

A situation in which retrospective power calculations may be useful is when pilot study results are used to conduct power calculations for a future, larger study. Of course, it could be argued that these calculations are actually prospective with regard to the future study. Regardless of how they are viewed, software packages (including PASS and the Analyst application in SAS) often do allow for such power calculations.

Problems

In the problems below, assume that a significance level, α, of 0.05 and a power, $(1 - \beta)$, of 0.8 are desired, unless otherwise stated.

1. A clinical trial is to be conducted to investigate and compare the effects of a new cholesterol-lowering drug with the effects of a standard drug. Determine the sample sizes necessary for the analysis scenarios in parts (a) and (b) below.
   a. A two-tailed, two-sample t test will be performed to compare the average cholesterol levels of hypercholesteremic patients treated with the new drug and hypercholesteremic patients treated with the standard drug. Under the homogeneous variance assumption, the population-specific standard deviation of cholesterol levels is assumed to be 25 mg/dl. It will be important to detect an absolute difference in average true cholesterol levels of 25 mg/dl or larger.
   b. A test comparing the proportions of new drug-treated patients and standard drug-treated patients who complain of side effects is to be performed. Previous studies suggest that 25% of patients treated with the standard drug complain of side effects. It is important to detect an absolute difference in population proportions of 0.1 or more.

2. For the scenario in Problem 1(a),
   a. Determine the sample size for power levels ranging from 0.6 to 0.9, in increments of 0.05. Plot the sample size versus power. Comment on the observed relationship.
   b. Determine the sample size for absolute differences in average cholesterol levels ranging from 20 mg/dl to 70 mg/dl, in increments of 10 mg/dl. Plot the sample size versus the absolute difference. Comment on the observed relationship.

3. For the scenario in Problem 1(b),
   a. Determine the required sample size for power levels ranging from 0.6 to 0.9, in increments of 0.10. Plot the sample size versus power. Comment on the observed relationship.
   b. Determine the required sample size for absolute differences in proportions of patients who report side effects, ranging between 0.01 and 0.10, in increments of 0.01. Comment on the observed relationship.

4. In Example 27.3, the manufacturer of the drugs would like to determine how much more powerful the hypothesis testing method would be if larger samples were collected. Plot the sample size versus the power for an appropriate range of levels of power. What, approximately, would the power of the testing method be if:
   a. 50 doses of each drug were sampled?
   b. 100 doses of each drug were sampled?

5. A study is being planned to investigate the association between systolic blood pressure (SBP) and independent variables age (AGE), smoking history (SMK = 0 if non-smoker, SMK = 1 if a current or previous smoker), and body size as measured by the Quetelet Index described in Problem 2 of Chapter 5. SBP, AGE and QUET are continuous variables. The statistical analysis will involve multiple linear regression with SBP as the dependent variable.

   a. Suppose that SMK is the only independent variable to be included in the regression model. Determine the approximate sample size necessary to detect a true slope with an absolute value of at least 1.5. From previous research, the population standard deviation of SBP is approximately 3 mm Hg.

   b. Suppose that SMK, AGE, and QUET are all included as predictors in the regression analysis, with SMK being the main predictor of interest and AGE and QUET being included to control for confounding. Using the method described in Section 27.3.3, determine the approximate sample size necessary to detect a true slope for SMK that has an absolute value of at least 1.5. From previous research, the population standard deviation of SBP is about 3 mm Hg, and the population squared multiple correlation between SMK and the other two independent variables is approximately 0.10.

   c. In part (b), re-calculate the required sample size for population squared multiple correlations ranging from 0.10 to 0.90, in increments of 0.1. Comment on the relationship between the required sample size and the population squared multiple correlation.

   d. Suppose that AGE is the only independent variable to be included in the regression model. Determine the sample size necessary to detect a true correlation between AGE and SBP of at least 0.40 in absolute value.

   e. Suppose that AGE, SMK, and QUET are all included as predictors in the regression analysis, with AGE being the main predictor of interest. Using the method described in Section 27.3.3, determine the sample size necessary to detect a correlation between SBP and AGE of at least 0.40 in absolute value, assuming that the population squared multiple correlation between AGE and the other two independent variables is 0.10.

6. Researchers in Example 27.9 feel that at most 200 patients can be enrolled in the study for each type of surgery. Redo the sample size calculations for that example and identify at least one combination of power and effect size (absolute difference in tear rates) for which the required sample size is 200 or less per surgery type.

For the remaining problems, it is recommended that a software package such as PASS or SAS be employed.

7–12. Redo Examples 27.11, 27.13, and 27.15 to determine the approximate sample size if:
   a. The desired power is 0.9 (all other parameters are as given in the examples).
   b. The desired significance level is 0.1 (all other parameters are as given in the examples).

References

Cohen, J. 1977. *Statistical Power Analysis for the Behavioral Sciences* (Rev. ed.). New York: Academic Press.

Dupont, W. 1988. "Power Calculations for Matched Case-Control Studies." *Biometrics*. 44: 1157–1168.

Hintz, J. 2004. NCSS and PASS. Number Cruncher Statistical Systems. Kaysville, UT.

Hsieh, F. Y., Bloch, D. A., and Larsen, M. D. 1998. "A Simple Method of Sample Size Calculation For Linear and Logistic Regression." *Statistics in Medicine* 17: 1623–1634.

Koele, P. 1982. "Calculating Power in Analysis of Variance." *Psychological Bulletin* 92(2): 513–516.

Kupper, L. K., and Hafner, K. B. 1989. "How Appropriate are Popular Sample Size Formulas?" *The American Statistician* 43: 101–105.

Lenth, R. V. 2001. "Some Practical Guidelines for Effective Sample Size Determination." *The American Statistician* 55(3): 187–193.

Lenth, R. V. 2003. "Two Sample-size Practices That I Don't Recommend." Available: http://www.stat.uiowa.edu/~rlenth/Power/2badHabits.pdf. Accessed Dec. 29, 2006.

Milsten, A. M., Seaman, K. G., Liu, P., Bissell, R. A., and Maguire, B. J. 2003. "Variables Influencing Medical Usage Rates, Injury Patterns, and Levels of Care for Mass Gatherings." *Prehospital and Disaster Medicine* 18(4): 334–345.

Muller, K. E., LaVange, L. M., Ramey, S. L., and Ramey, C. T. 1992. "Power Calculations for General Linear Multivariate Models Including Repeated Measures Applications." *Journal of the American Statistical Association* 87: 1209–1226.

Neter, J., Wasserman, W., and Kutner, M. H. 1990. *Applied Linear Statistical Models,* Third Edition. Boston: Irwin.

Pearson, E. S., and Hartley, H. O. 1951. "Charts of the Power Function for Analysis of Variance Tests, Derived from the Non-Central F Distribution." *Biometrika* 38: 112–130.

SAS Institute Inc., 2002–2003. SAS® 9.1. Cary, NC.

Scheffé, H. 1959. *The Analysis of Variance*. New York: Wiley.

Tabachnick, B. G., and Fidell, L. S. 2001. *Using Multivariate Statistics,* Fourth Edition. Boston: Allyn and Bacon.

Thomas, L., and Krebs C. J. 1997. (A Review of Statistical Power Analysis Software.) *Bulletin of the Ecological Society of America* 78(2): 126–139.

Appendix—Tables

TABLE A.1   Standard Normal Cumulative Probabilities

z	0.00	0.01	0.02	0.03	0.04	0.05	0.06	0.07	0.08	0.09
-3.8	0.0001	0.0001	0.0001	0.0001	0.0001	0.0001	0.0001	0.0001	0.0001	0.0001
-3.7	0.0001	0.0001	0.0001	0.0001	0.0001	0.0001	0.0001	0.0001	0.0001	0.0001
-3.6	0.0002	0.0002	0.0001	0.0001	0.0001	0.0001	0.0001	0.0001	0.0001	0.0001
-3.5	0.0002	0.0002	0.0002	0.0002	0.0002	0.0002	0.0002	0.0002	0.0002	0.0002
-3.4	0.0003	0.0003	0.0003	0.0003	0.0003	0.0003	0.0003	0.0003	0.0003	0.0002
-3.3	0.0005	0.0005	0.0005	0.0004	0.0004	0.0004	0.0004	0.0004	0.0004	0.0003
-3.2	0.0007	0.0007	0.0006	0.0006	0.0006	0.0006	0.0006	0.0005	0.0005	0.0005
-3.1	0.0010	0.0009	0.0009	0.0009	0.0008	0.0008	0.0008	0.0008	0.0007	0.0007
-3.0	0.0014	0.0013	0.0013	0.0012	0.0012	0.0011	0.0011	0.0011	0.0010	0.0010
-2.9	0.0019	0.0018	0.0018	0.0017	0.0016	0.0016	0.0015	0.0015	0.0014	0.0014
-2.8	0.0026	0.0025	0.0024	0.0023	0.0023	0.0022	0.0021	0.0021	0.0020	0.0019
-2.7	0.0035	0.0034	0.0033	0.0032	0.0031	0.0030	0.0029	0.0028	0.0027	0.0026
-2.6	0.0047	0.0045	0.0044	0.0043	0.0041	0.0040	0.0039	0.0038	0.0037	0.0036
-2.5	0.0062	0.0060	0.0059	0.0057	0.0055	0.0054	0.0052	0.0051	0.0049	0.0048
-2.4	0.0082	0.0080	0.0078	0.0076	0.0073	0.0071	0.0069	0.0068	0.0066	0.0064
-2.3	0.0107	0.0104	0.0102	0.0099	0.0096	0.0094	0.0091	0.0089	0.0087	0.0084
-2.2	0.0139	0.0136	0.0132	0.0129	0.0125	0.0122	0.0119	0.0116	0.0113	0.0110
-2.1	0.0179	0.0174	0.0170	0.0166	0.0162	0.0158	0.0154	0.0150	0.0146	0.0143
-2.0	0.0228	0.0222	0.0217	0.0212	0.0207	0.0202	0.0197	0.0192	0.0188	0.0183
-1.9	0.0287	0.0281	0.0274	0.0268	0.0262	0.0256	0.0250	0.0244	0.0239	0.0233
-1.8	0.0359	0.0351	0.0344	0.0336	0.0329	0.0322	0.0314	0.0307	0.0301	0.0294
-1.7	0.0446	0.0436	0.0427	0.0418	0.0409	0.0401	0.0392	0.0384	0.0375	0.0367
-1.6	0.0548	0.0537	0.0526	0.0516	0.0505	0.0495	0.0485	0.0475	0.0465	0.0455
-1.5	0.0668	0.0655	0.0643	0.0630	0.0618	0.0606	0.0594	0.0582	0.0571	0.0559
-1.4	0.0808	0.0793	0.0778	0.0764	0.0749	0.0735	0.0721	0.0708	0.0694	0.0681
-1.3	0.0968	0.0951	0.0934	0.0918	0.0901	0.0885	0.0869	0.0853	0.0838	0.0823
-1.2	0.1151	0.1131	0.1112	0.1093	0.1075	0.1057	0.1038	0.1020	0.1003	0.0985
-1.1	0.1357	0.1335	0.1314	0.1292	0.1271	0.1251	0.1230	0.1210	0.1190	0.1170
-1.0	0.1587	0.1562	0.1539	0.1515	0.1492	0.1469	0.1446	0.1423	0.1401	0.1379
-0.9	0.1841	0.1814	0.1788	0.1762	0.1736	0.1711	0.1685	0.1660	0.1635	0.1611
-0.8	0.2119	0.2090	0.2061	0.2033	0.2005	0.1977	0.1949	0.1922	0.1894	0.1867
-0.7	0.2420	0.2389	0.2358	0.2327	0.2297	0.2266	0.2236	0.2206	0.2177	0.2148
-0.6	0.2743	0.2709	0.2676	0.2643	0.2611	0.2578	0.2546	0.2514	0.2483	0.2451
-0.5	0.3085	0.3050	0.3015	0.2981	0.2946	0.2912	0.2877	0.2843	0.2810	0.2776
-0.4	0.3446	0.3409	0.3372	0.3336	0.3300	0.3264	0.3228	0.3192	0.3156	0.3121
-0.3	0.3821	0.3783	0.3745	0.3707	0.3669	0.3632	0.3594	0.3557	0.3520	0.3483
-0.2	0.4207	0.4168	0.4129	0.4090	0.4052	0.4013	0.3974	0.3936	0.3897	0.3859
-0.1	0.4602	0.4562	0.4522	0.4483	0.4443	0.4404	0.4364	0.4325	0.4286	0.4247
-0.0	0.5000	0.4960	0.4920	0.4880	0.4840	0.4801	0.4761	0.4721	0.4681	0.4641

Note: Table entry is the area under the standard normal curve to the left of the indicated *z*-value, thus giving $P(Z < z)$.

TABLE A.1 Standard Normal Cumulative Probabilities (*continued*)

z	0.00	0.01	0.02	0.03	0.04	0.05	0.06	0.07	0.08	0.09
0.0	0.5000	0.5040	0.5080	0.5120	0.5160	0.5199	0.5239	0.5279	0.5319	0.5359
0.1	0.5398	0.5438	0.5478	0.5517	0.5557	0.5596	0.5636	0.5675	0.5714	0.5753
0.2	0.5793	0.5832	0.5871	0.5910	0.5948	0.5987	0.6026	0.6064	0.6103	0.6141
0.3	0.6179	0.6217	0.6255	0.6293	0.6331	0.6368	0.6406	0.6443	0.6480	0.6517
0.4	0.6554	0.6591	0.6628	0.6664	0.6700	0.6736	0.6772	0.6808	0.6844	0.6879
0.5	0.6915	0.6950	0.6985	0.7019	0.7054	0.7088	0.7123	0.7157	0.7190	0.7224
0.6	0.7257	0.7291	0.7324	0.7357	0.7389	0.7422	0.7454	0.7486	0.7517	0.7549
0.7	0.7580	0.7611	0.7642	0.7673	0.7703	0.7734	0.7764	0.7794	0.7823	0.7852
0.8	0.7881	0.7910	0.7939	0.7967	0.7995	0.8023	0.8051	0.8078	0.8106	0.8133
0.9	0.8159	0.8186	0.8212	0.8238	0.8264	0.8289	0.8315	0.8340	0.8365	0.8389
1.0	0.8413	0.8438	0.8461	0.8485	0.8508	0.8531	0.8554	0.8577	0.8599	0.8621
1.1	0.8643	0.8665	0.8686	0.8708	0.8729	0.8749	0.8770	0.8790	0.8810	0.8830
1.2	0.8849	0.8869	0.8888	0.8907	0.8925	0.8943	0.8962	0.8980	0.8997	0.9015
1.3	0.9032	0.9049	0.9066	0.9082	0.9099	0.9115	0.9131	0.9147	0.9162	0.9177
1.4	0.9192	0.9207	0.9222	0.9236	0.9251	0.9265	0.9279	0.9292	0.9306	0.9319
1.5	0.9332	0.9345	0.9357	0.9370	0.9382	0.9394	0.9406	0.9418	0.9429	0.9441
1.6	0.9452	0.9463	0.9474	0.9484	0.9495	0.9505	0.9515	0.9525	0.9535	0.9545
1.7	0.9554	0.9564	0.9573	0.9582	0.9591	0.9599	0.9608	0.9616	0.9625	0.9633
1.8	0.9641	0.9649	0.9656	0.9664	0.9671	0.9678	0.9686	0.9693	0.9699	0.9706
1.9	0.9713	0.9719	0.9726	0.9732	0.9738	0.9744	0.9750	0.9756	0.9761	0.9767
2.0	0.9772	0.9778	0.9783	0.9788	0.9793	0.9798	0.9803	0.9808	0.9812	0.9817
2.1	0.9821	0.9826	0.9830	0.9834	0.9838	0.9842	0.9846	0.9850	0.9854	0.9857
2.2	0.9861	0.9864	0.9868	0.9871	0.9875	0.9878	0.9881	0.9884	0.9887	0.9890
2.3	0.9893	0.9896	0.9898	0.9901	0.9904	0.9906	0.9909	0.9911	0.9913	0.9916
2.4	0.9918	0.9920	0.9922	0.9924	0.9927	0.9929	0.9931	0.9932	0.9934	0.9936
2.5	0.9938	0.9940	0.9941	0.9943	0.9945	0.9946	0.9948	0.9949	0.9951	0.9952
2.6	0.9953	0.9955	0.9956	0.9957	0.9959	0.9960	0.9961	0.9962	0.9963	0.9964
2.7	0.9965	0.9966	0.9967	0.9968	0.9969	0.9970	0.9971	0.9972	0.9973	0.9974
2.8	0.9974	0.9975	0.9976	0.9977	0.9977	0.9978	0.9979	0.9979	0.9980	0.9981
2.9	0.9981	0.9982	0.9982	0.9983	0.9984	0.9984	0.9985	0.9985	0.9986	0.9986
3.0	0.9986	0.9987	0.9987	0.9988	0.9988	0.9989	0.9989	0.9989	0.9990	0.9990
3.1	0.9990	0.9991	0.9991	0.9991	0.9992	0.9992	0.9992	0.9992	0.9993	0.9993
3.2	0.9993	0.9993	0.9994	0.9994	0.9994	0.9994	0.9994	0.9995	0.9995	0.9995
3.3	0.9995	0.9995	0.9995	0.9996	0.9996	0.9996	0.9996	0.9996	0.9996	0.9997
3.4	0.9997	0.9997	0.9997	0.9997	0.9997	0.9997	0.9997	0.9997	0.9997	0.9998
3.5	0.9998	0.9998	0.9998	0.9998	0.9998	0.9998	0.9998	0.9998	0.9998	0.9998
3.6	0.9998	0.9998	0.9999	0.9999	0.9999	0.9999	0.9999	0.9999	0.9999	0.9999
3.7	0.9999	0.9999	0.9999	0.9999	0.9999	0.9999	0.9999	0.9999	0.9999	0.9999
3.8	0.9999	0.9999	0.9999	0.9999	0.9999	0.9999	0.9999	0.9999	0.9999	0.9999
3.9	1.0000									

TABLE A.1 Standard Normal Cumulative Probabilities (*continued*)

z	$P(Z < z)$
-4.265	0.00001
-3.891	0.00005
-3.719	0.0001
-3.291	0.0005
-3.090	0.001
-2.576	0.005
-2.326	0.01
-2.054	0.02
-1.960	0.025
-1.881	0.03
-1.751	0.04
-1.645	0.05
-1.555	0.06
-1.476	0.07
-1.405	0.08
-1.341	0.09
-1.282	0.10
-1.036	0.15
-0.842	0.20
-0.674	0.25
-0.524	0.30
-0.385	0.35
-0.253	0.40
-0.126	0.45
0	0.50

z	$P(Z < z)$
0	0.50
0.126	0.55
0.253	0.60
0.385	0.65
0.524	0.70
0.674	0.75
0.842	0.80
1.036	0.85
1.282	0.90
1.341	0.91
1.405	0.92
1.476	0.93
1.555	0.94
1.645	0.95
1.751	0.96
1.881	0.97
1.960	0.975
2.054	0.98
2.326	0.99
2.576	0.995
3.090	0.999
3.291	0.9995
3.719	0.9999
3.891	0.99995
4.265	0.99999

TABLE A.2   Percentiles of the *t* Distribution

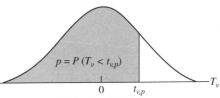

$$p = P\,(T_v < t_{v,p})$$

Student's *t* distribution

100p / df	55	65	75	85	90	95	97.5	99	99.5	99.95
1	0.158	0.510	1.000	1.963	3.078	6.314	12.706	31.821	63.657	636.619
2	0.142	0.445	0.816	1.386	1.886	2.920	4.303	6.965	9.925	31.599
3	0.137	0.424	0.765	1.250	1.638	2.353	3.182	4.541	5.841	12.924
4	0.134	0.414	0.741	1.190	1.533	2.132	2.776	3.747	4.604	8.610
5	0.132	0.408	0.727	1.156	1.476	2.015	2.571	3.365	4.032	6.869
6	0.131	0.404	0.718	1.134	1.440	1.943	2.447	3.143	3.707	5.959
7	0.130	0.402	0.711	1.119	1.415	1.895	2.365	2.998	3.499	5.408
8	0.130	0.399	0.706	1.108	1.397	1.860	2.306	2.896	3.355	5.041
9	0.129	0.398	0.703	1.100	1.383	1.833	2.262	2.821	3.250	4.781
10	0.129	0.397	0.700	1.093	1.372	1.812	2.228	2.764	3.169	4.587
11	0.129	0.396	0.697	1.088	1.363	1.796	2.201	2.718	3.106	4.437
12	0.128	0.395	0.695	1.083	1.356	1.782	2.179	2.681	3.055	4.318
13	0.128	0.394	0.694	1.079	1.350	1.771	2.160	2.650	3.012	4.221
14	0.128	0.393	0.692	1.076	1.345	1.761	2.145	2.624	2.977	4.140
15	0.128	0.393	0.691	1.074	1.341	1.753	2.131	2.602	2.947	4.073
16	0.128	0.392	0.690	1.071	1.337	1.746	2.120	2.583	2.921	4.015
17	0.128	0.392	0.689	1.069	1.333	1.740	2.110	2.567	2.898	3.965
18	0.127	0.392	0.688	1.067	1.330	1.734	2.101	2.552	2.878	3.922
19	0.127	0.391	0.688	1.066	1.328	1.729	2.093	2.539	2.861	3.883
20	0.127	0.391	0.687	1.064	1.325	1.725	2.086	2.528	2.845	3.850
21	0.127	0.391	0.686	1.063	1.323	1.721	2.080	2.518	2.831	3.819
22	0.127	0.390	0.686	1.061	1.321	1.717	2.074	2.508	2.819	3.792
23	0.127	0.390	0.685	1.060	1.319	1.714	2.069	2.500	2.807	3.768
24	0.127	0.390	0.685	1.059	1.318	1.711	2.064	2.492	2.797	3.745
25	0.127	0.390	0.684	1.058	1.316	1.708	2.060	2.485	2.787	3.725
26	0.127	0.390	0.684	1.058	1.315	1.706	2.056	2.479	2.779	3.707
27	0.127	0.389	0.684	1.057	1.314	1.703	2.052	2.473	2.771	3.690
28	0.127	0.389	0.683	1.056	1.313	1.701	2.048	2.467	2.763	3.674
29	0.127	0.389	0.683	1.055	1.311	1.699	2.045	2.462	2.756	3.659
30	0.127	0.389	0.683	1.055	1.310	1.697	2.042	2.457	2.750	3.646
35	0.127	0.388	0.682	1.052	1.306	1.690	2.030	2.438	2.724	3.591
40	0.126	0.388	0.681	1.050	1.303	1.684	2.021	2.423	2.704	3.551
45	0.126	0.388	0.680	1.049	1.301	1.679	2.014	2.412	2.690	3.520
50	0.126	0.388	0.679	1.047	1.299	1.676	2.009	2.403	2.678	3.496
60	0.126	0.387	0.679	1.045	1.296	1.671	2.000	2.390	2.660	3.460
70	0.126	0.387	0.678	1.044	1.294	1.667	1.994	2.381	2.648	3.435
80	0.126	0.387	0.678	1.043	1.292	1.664	1.990	2.374	2.639	3.416
90	0.126	0.387	0.677	1.042	1.291	1.662	1.987	2.368	2.632	3.402
100	0.126	0.386	0.677	1.042	1.290	1.660	1.984	2.364	2.626	3.390
120	0.126	0.386	0.677	1.041	1.289	1.658	1.980	2.358	2.617	3.373
140	0.126	0.386	0.676	1.040	1.288	1.656	1.977	2.353	2.611	3.361
160	0.126	0.386	0.676	1.040	1.287	1.654	1.975	2.350	2.607	3.352
180	0.126	0.386	0.676	1.039	1.286	1.653	1.973	2.547	2.603	3.345
200	0.126	0.386	0.676	1.039	1.286	1.653	1.972	2.345	2.601	3.340
∞	0.126	0.385	0.674	1.036	1.282	1.645	1.960	2.326	2.576	3.291

$$p = P(X_\nu^2 < X_{\nu,p}^2)$$

X^2 distribution

TABLE A.3 Percentiles of the Chi-square Distribution

100p df	0.5	1	2.5	5	10	20	30	40	50	60	70	80	90	95	97.5	99	99.5	99.95
1	0.0001	0.0002	0.001	0.004	0.016	0.064	0.148	0.275	0.455	0.708	1.074	1.642	2.706	3.841	5.024	6.635	7.879	12.116
2	0.010	0.020	0.051	0.103	0.211	0.446	0.713	1.022	1.386	1.833	2.408	3.219	4.605	5.991	7.378	9.210	10.597	15.202
3	0.072	0.115	0.216	0.352	0.584	1.005	1.424	1.869	2.366	2.946	3.665	4.642	6.251	7.815	9.348	11.345	12.838	17.730
4	0.207	0.297	0.484	0.711	1.064	1.649	2.195	2.753	3.357	4.045	4.878	5.989	7.779	9.488	11.143	13.277	14.860	19.997
5	0.412	0.554	0.831	1.145	1.610	2.343	3.000	3.655	4.351	5.132	6.064	7.289	9.236	11.070	12.833	15.086	16.750	22.105
6	0.676	0.872	1.237	1.635	2.204	3.070	3.828	4.570	5.348	6.211	7.231	8.558	10.645	12.592	14.449	16.812	18.548	24.103
7	0.989	1.239	1.690	2.167	2.833	3.822	4.671	5.493	6.346	7.283	8.383	9.803	12.017	14.067	16.013	18.475	20.278	26.018
8	1.344	1.646	2.180	2.733	3.490	4.594	5.527	6.423	7.344	8.351	9.524	11.030	13.362	15.507	17.535	20.090	21.955	27.868
9	1.735	2.088	2.700	3.325	4.168	5.380	6.393	7.357	8.343	9.414	10.656	12.242	14.684	16.919	19.023	21.666	23.589	29.666
10	2.156	2.558	3.247	3.940	4.865	6.179	7.267	8.295	9.342	10.473	11.781	13.442	15.987	18.307	20.483	23.209	25.188	31.420
11	2.603	3.053	3.816	4.575	5.578	6.989	8.148	9.237	10.341	11.530	12.899	14.631	17.275	19.675	21.920	24.725	26.757	33.137
12	3.074	3.571	4.404	5.226	6.304	7.807	9.034	10.182	11.340	12.584	14.011	15.812	18.549	21.026	23.337	26.217	28.300	34.821
13	3.565	4.107	5.009	5.892	7.042	8.634	9.926	11.129	12.340	13.636	15.119	16.985	19.812	22.362	24.736	27.688	29.819	36.478
14	4.075	4.660	5.629	6.571	7.790	9.467	10.821	12.078	13.339	14.685	16.222	18.151	21.064	23.685	26.119	29.141	31.319	38.109
15	4.601	5.229	6.262	7.261	8.547	10.307	11.721	13.030	14.339	15.733	17.322	19.311	22.307	24.996	27.488	30.578	32.801	39.719
16	5.142	5.812	6.908	7.962	9.312	11.152	12.624	13.983	15.338	16.780	18.418	20.465	23.542	26.296	28.845	32.000	34.267	41.308
17	5.697	6.408	7.564	8.672	10.085	12.002	13.531	14.937	16.338	17.824	19.511	21.615	24.769	27.587	30.191	33.409	35.718	42.879
18	6.265	7.015	8.231	9.390	10.865	12.857	14.440	15.893	17.338	18.868	20.601	22.760	25.989	28.869	31.526	34.805	37.156	44.434
19	6.844	7.633	8.907	10.117	11.651	13.716	15.352	16.850	18.338	19.910	21.689	23.900	27.204	30.144	32.852	36.191	38.582	45.973
20	7.434	8.260	9.591	10.851	12.443	14.578	16.266	17.809	19.337	20.951	22.775	25.038	28.412	31.410	34.170	37.566	39.997	47.498
21	8.034	8.897	10.283	11.591	13.240	15.445	17.182	18.768	20.337	21.991	23.858	26.171	29.615	32.671	35.479	38.932	41.401	49.011
22	8.643	9.542	10.982	12.338	14.041	16.314	18.101	19.729	21.337	23.031	24.939	27.301	30.813	33.924	36.781	40.289	42.796	50.511
23	9.260	10.196	11.689	13.091	14.848	17.187	19.021	20.690	22.337	24.069	26.018	28.429	32.007	35.172	38.076	41.638	44.181	52.000
24	9.886	10.856	12.401	13.848	15.659	18.062	19.943	21.752	23.337	25.106	27.096	29.553	33.196	36.415	39.364	42.980	45.559	53.479
25	10.520	11.524	13.120	14.611	16.473	18.940	20.867	22.616	24.337	26.143	28.172	30.675	34.382	37.652	40.646	44.314	46.928	54.947
26	11.160	12.198	13.844	15.379	17.292	19.820	21.792	23.579	25.336	27.179	29.246	31.795	35.563	38.885	41.923	45.642	48.290	56.407
27	11.808	12.879	14.573	16.151	18.114	20.703	22.719	24.544	26.336	28.214	30.319	32.912	36.741	40.113	43.195	46.963	49.645	57.858
28	12.461	13.565	15.308	16.928	18.939	21.588	23.647	25.509	27.336	29.249	31.391	34.027	37.916	41.337	44.461	48.278	50.993	59.300
29	13.121	14.256	16.047	17.708	19.768	22.475	24.577	26.475	28.336	30.283	32.461	35.139	39.087	42.557	45.722	49.588	52.336	60.735
30	13.787	14.953	16.791	18.493	20.599	23.364	25.508	27.442	29.336	31.316	33.530	36.250	40.256	43.773	46.979	50.892	53.672	62.162
35	17.192	18.509	20.569	22.465	24.797	27.836	30.178	32.282	34.336	36.475	38.859	41.778	46.059	49.802	53.203	57.342	60.275	69.199
40	20.707	22.164	24.433	26.509	29.051	32.345	34.872	37.134	39.335	41.622	44.165	47.269	51.805	55.758	59.342	63.691	66.766	76.095
45	24.311	25.901	28.366	30.612	33.350	36.884	39.585	41.995	44.335	46.761	49.452	52.729	57.505	61.656	65.410	69.957	73.166	82.876
50	27.991	29.707	32.357	34.764	37.689	41.449	44.313	46.864	49.335	51.892	54.723	58.164	63.167	67.505	71.420	76.154	79.490	89.561
60	35.534	37.485	40.482	43.188	46.459	50.641	53.809	56.620	59.335	62.135	65.227	68.972	74.397	79.082	83.298	88.379	91.952	102.695
70	43.275	45.442	48.758	51.739	55.329	59.898	63.346	66.396	69.334	72.358	75.689	79.715	85.527	90.531	95.023	100.425	104.215	115.578
80	51.172	53.540	57.153	60.391	64.278	69.207	72.915	76.188	79.334	82.566	86.120	90.405	96.578	101.879	106.629	112.329	116.321	128.261
90	59.196	61.754	65.647	69.126	73.291	78.558	82.511	85.993	89.334	92.761	96.524	101.054	107.565	113.145	118.136	124.116	128.299	140.782
100	67.328	70.065	74.222	77.929	82.358	87.945	92.129	95.808	99.334	102.946	106.906	111.667	118.498	124.342	129.561	135.807	140.169	153.167
120	83.852	86.923	91.573	95.705	100.624	106.806	111.419	115.465	119.334	123.289	127.616	132.806	140.233	146.567	152.211	158.950	163.648	177.603
140	100.655	104.034	109.137	113.659	119.029	125.758	130.766	135.149	139.334	143.604	148.269	153.854	161.827	168.613	174.648	181.840	186.847	201.683
160	117.679	121.346	126.870	131.756	137.546	144.783	150.158	154.856	159.334	163.898	168.876	174.828	183.311	190.516	196.915	204.530	209.824	225.481
180	134.884	138.820	144.741	149.969	156.153	163.868	169.588	174.580	179.334	184.173	189.446	195.743	204.704	212.304	219.044	227.056	232.620	249.048
200	152.241	156.432	162.728	168.279	174.835	183.003	189.049	194.319	199.334	204.434	209.985	216.609	226.021	233.994	241.058	249.445	255.264	272.423

TABLE A.4 Percentiles of the *F* Distribution

Upper 25% point of the *F* distribution (*p* = 0.75)

$$p = P(F_{\nu_1,\nu_2} < F_{\nu_1,\nu_2,p})$$

F distribution

DEGREES OF FREEDOM FOR NUMERATOR

ν₂＼ν₁	1	2	3	4	5	6	7	8	9	10	11	12	13	14	15	16	17	18	19	20	25	30	40	50	100	150	200
1	5.83	7.50	8.20	8.58	8.82	8.98	9.10	9.19	9.26	9.32	9.37	9.41	9.44	9.47	9.49	9.52	9.53	9.55	9.57	9.58	9.63	9.67	9.71	9.74	9.80	9.81	9.82
2	2.57	3.00	3.15	3.23	3.28	3.31	3.34	3.35	3.37	3.38	3.39	3.39	3.40	3.41	3.41	3.41	3.42	3.42	3.42	3.43	3.44	3.44	3.45	3.46	3.47	3.47	3.47
3	2.02	2.28	2.36	2.39	2.41	2.42	2.43	2.44	2.44	2.44	2.45	2.45	2.45	2.45	2.46	2.46	2.46	2.46	2.46	2.46	2.46	2.47	2.47	2.47	2.47	2.47	2.47
4	1.81	2.00	2.05	2.06	2.07	2.08	2.08	2.08	2.08	2.08	2.08	2.08	2.08	2.08	2.08	2.08	2.08	2.08	2.08	2.08	2.08	2.08	2.08	2.08	2.08	2.08	2.08
5	1.69	1.85	1.88	1.89	1.89	1.89	1.89	1.89	1.89	1.89	1.89	1.89	1.89	1.89	1.89	1.88	1.88	1.88	1.88	1.88	1.88	1.88	1.88	1.88	1.87	1.87	1.87
6	1.62	1.76	1.78	1.79	1.79	1.78	1.78	1.78	1.77	1.77	1.77	1.77	1.76	1.76	1.76	1.76	1.76	1.76	1.76	1.76	1.75	1.75	1.75	1.75	1.74	1.74	1.74
7	1.57	1.70	1.72	1.72	1.71	1.71	1.70	1.70	1.69	1.69	1.69	1.68	1.68	1.68	1.68	1.68	1.67	1.67	1.67	1.67	1.66	1.66	1.66	1.66	1.65	1.65	1.65
8	1.54	1.66	1.67	1.66	1.66	1.65	1.64	1.64	1.63	1.63	1.63	1.62	1.62	1.62	1.62	1.62	1.61	1.61	1.61	1.61	1.60	1.60	1.59	1.59	1.58	1.58	1.58
9	1.51	1.62	1.63	1.63	1.62	1.61	1.60	1.60	1.59	1.59	1.58	1.58	1.57	1.57	1.57	1.57	1.57	1.56	1.56	1.56	1.55	1.55	1.54	1.54	1.53	1.53	1.53
10	1.49	1.60	1.60	1.59	1.59	1.58	1.57	1.56	1.56	1.55	1.55	1.54	1.54	1.54	1.53	1.53	1.53	1.53	1.52	1.52	1.52	1.51	1.51	1.50	1.49	1.49	1.49
11	1.47	1.58	1.58	1.57	1.56	1.55	1.54	1.53	1.53	1.52	1.52	1.51	1.51	1.51	1.50	1.50	1.50	1.49	1.49	1.49	1.49	1.48	1.47	1.47	1.46	1.46	1.46
12	1.46	1.56	1.56	1.55	1.54	1.53	1.52	1.51	1.51	1.50	1.50	1.49	1.49	1.48	1.48	1.48	1.47	1.47	1.47	1.47	1.46	1.45	1.45	1.44	1.43	1.43	1.43
13	1.45	1.55	1.55	1.53	1.52	1.51	1.50	1.49	1.49	1.48	1.47	1.47	1.46	1.46	1.46	1.45	1.45	1.45	1.45	1.45	1.44	1.43	1.42	1.42	1.41	1.41	1.40
14	1.44	1.53	1.53	1.52	1.51	1.50	1.49	1.48	1.47	1.46	1.46	1.45	1.45	1.44	1.44	1.44	1.43	1.43	1.43	1.43	1.42	1.41	1.41	1.40	1.39	1.39	1.39
15	1.43	1.52	1.52	1.51	1.49	1.48	1.47	1.46	1.46	1.45	1.44	1.44	1.43	1.43	1.43	1.42	1.42	1.42	1.41	1.41	1.40	1.40	1.39	1.38	1.37	1.37	1.37
16	1.42	1.51	1.51	1.50	1.48	1.47	1.46	1.45	1.44	1.44	1.44	1.43	1.43	1.42	1.42	1.41	1.41	1.41	1.40	1.40	1.40	1.39	1.38	1.37	1.36	1.36	1.36
17	1.42	1.51	1.50	1.49	1.47	1.46	1.45	1.44	1.43	1.43	1.42	1.41	1.41	1.41	1.40	1.40	1.39	1.39	1.39	1.39	1.38	1.37	1.36	1.36	1.35	1.34	1.34
18	1.41	1.50	1.49	1.48	1.46	1.45	1.44	1.43	1.42	1.42	1.41	1.41	1.40	1.40	1.39	1.39	1.38	1.38	1.38	1.38	1.37	1.36	1.35	1.35	1.34	1.33	1.33
19	1.41	1.49	1.49	1.47	1.46	1.44	1.43	1.42	1.41	1.41	1.41	1.40	1.40	1.39	1.39	1.38	1.38	1.38	1.37	1.37	1.36	1.35	1.34	1.34	1.33	1.32	1.32
20	1.40	1.49	1.48	1.47	1.45	1.44	1.43	1.42	1.41	1.40	1.40	1.39	1.39	1.38	1.37	1.37	1.37	1.36	1.36	1.36	1.35	1.34	1.33	1.33	1.31	1.31	1.30
21	1.40	1.48	1.48	1.46	1.44	1.43	1.42	1.41	1.40	1.39	1.39	1.38	1.37	1.37	1.37	1.36	1.36	1.35	1.35	1.35	1.34	1.33	1.32	1.32	1.30	1.30	1.29
22	1.40	1.48	1.47	1.45	1.44	1.42	1.41	1.40	1.39	1.39	1.38	1.37	1.37	1.36	1.36	1.36	1.35	1.35	1.34	1.34	1.33	1.32	1.31	1.31	1.29	1.29	1.28
23	1.39	1.47	1.47	1.45	1.43	1.42	1.41	1.40	1.39	1.38	1.37	1.37	1.36	1.36	1.35	1.35	1.34	1.34	1.33	1.33	1.32	1.32	1.31	1.30	1.28	1.28	1.28
24	1.39	1.47	1.46	1.44	1.43	1.41	1.40	1.39	1.38	1.38	1.37	1.36	1.36	1.35	1.35	1.34	1.34	1.33	1.33	1.33	1.32	1.31	1.30	1.29	1.28	1.27	1.27
25	1.39	1.47	1.46	1.44	1.42	1.41	1.40	1.39	1.38	1.37	1.37	1.36	1.35	1.35	1.34	1.34	1.33	1.33	1.33	1.32	1.31	1.31	1.29	1.29	1.27	1.27	1.26
26	1.38	1.46	1.45	1.44	1.42	1.41	1.39	1.38	1.37	1.37	1.36	1.35	1.35	1.34	1.34	1.33	1.33	1.33	1.32	1.32	1.31	1.30	1.29	1.28	1.26	1.26	1.26
27	1.38	1.46	1.45	1.43	1.42	1.40	1.39	1.38	1.37	1.36	1.36	1.35	1.34	1.34	1.33	1.33	1.32	1.32	1.32	1.32	1.30	1.30	1.28	1.28	1.26	1.25	1.25
28	1.38	1.46	1.45	1.43	1.41	1.40	1.39	1.38	1.37	1.36	1.35	1.35	1.34	1.34	1.33	1.33	1.32	1.32	1.31	1.31	1.30	1.29	1.28	1.27	1.25	1.25	1.25
29	1.38	1.45	1.45	1.43	1.41	1.40	1.38	1.37	1.36	1.35	1.35	1.34	1.34	1.33	1.33	1.32	1.32	1.31	1.31	1.31	1.30	1.29	1.28	1.27	1.25	1.24	1.24
30	1.38	1.45	1.44	1.42	1.41	1.39	1.38	1.37	1.36	1.35	1.35	1.34	1.33	1.33	1.32	1.32	1.31	1.31	1.30	1.30	1.29	1.28	1.27	1.26	1.25	1.24	1.24
32	1.37	1.45	1.44	1.42	1.40	1.39	1.37	1.36	1.35	1.34	1.34	1.33	1.33	1.32	1.32	1.31	1.31	1.30	1.30	1.30	1.28	1.28	1.26	1.26	1.24	1.23	1.23
34	1.37	1.44	1.43	1.41	1.40	1.38	1.37	1.36	1.35	1.34	1.33	1.33	1.32	1.31	1.31	1.30	1.30	1.29	1.29	1.29	1.28	1.27	1.26	1.25	1.23	1.22	1.22
36	1.37	1.44	1.43	1.41	1.39	1.38	1.36	1.35	1.34	1.33	1.33	1.32	1.32	1.31	1.31	1.30	1.29	1.29	1.28	1.28	1.27	1.26	1.25	1.24	1.22	1.22	1.22
38	1.37	1.44	1.43	1.41	1.39	1.37	1.36	1.35	1.34	1.33	1.32	1.32	1.31	1.30	1.30	1.29	1.29	1.28	1.28	1.28	1.27	1.26	1.24	1.24	1.22	1.21	1.21
40	1.36	1.44	1.42	1.40	1.39	1.37	1.36	1.35	1.34	1.33	1.32	1.31	1.31	1.30	1.30	1.29	1.28	1.28	1.28	1.27	1.26	1.25	1.24	1.23	1.21	1.21	1.20
42	1.36	1.43	1.42	1.40	1.38	1.37	1.35	1.34	1.33	1.32	1.32	1.31	1.30	1.30	1.29	1.29	1.28	1.28	1.27	1.27	1.26	1.25	1.23	1.23	1.21	1.20	1.20
44	1.36	1.43	1.42	1.40	1.38	1.36	1.35	1.34	1.33	1.32	1.31	1.31	1.30	1.29	1.29	1.28	1.28	1.27	1.27	1.26	1.25	1.24	1.23	1.22	1.20	1.19	1.19
46	1.36	1.43	1.42	1.40	1.38	1.36	1.35	1.34	1.33	1.32	1.31	1.30	1.30	1.29	1.28	1.28	1.27	1.27	1.26	1.26	1.25	1.24	1.23	1.22	1.20	1.19	1.19
48	1.36	1.43	1.41	1.39	1.37	1.36	1.34	1.33	1.32	1.31	1.31	1.30	1.29	1.29	1.28	1.28	1.27	1.27	1.26	1.26	1.25	1.24	1.22	1.22	1.19	1.19	1.18
50	1.35	1.43	1.41	1.39	1.37	1.36	1.34	1.33	1.32	1.31	1.30	1.30	1.29	1.28	1.28	1.27	1.27	1.26	1.26	1.26	1.24	1.23	1.22	1.21	1.19	1.18	1.18
60	1.35	1.42	1.41	1.38	1.37	1.35	1.33	1.32	1.31	1.30	1.29	1.29	1.28	1.27	1.27	1.26	1.26	1.25	1.25	1.25	1.24	1.22	1.21	1.20	1.18	1.17	1.16
70	1.35	1.41	1.40	1.38	1.36	1.34	1.33	1.32	1.31	1.30	1.29	1.28	1.27	1.27	1.26	1.25	1.25	1.25	1.24	1.24	1.23	1.21	1.20	1.19	1.16	1.16	1.15
80	1.34	1.41	1.39	1.38	1.36	1.34	1.33	1.31	1.30	1.29	1.28	1.28	1.27	1.26	1.25	1.25	1.25	1.24	1.24	1.23	1.22	1.21	1.19	1.18	1.16	1.15	1.14
90	1.34	1.41	1.39	1.37	1.35	1.34	1.32	1.31	1.30	1.29	1.28	1.27	1.26	1.26	1.25	1.25	1.24	1.23	1.23	1.23	1.21	1.20	1.19	1.18	1.15	1.14	1.13
100	1.34	1.41	1.39	1.37	1.35	1.33	1.32	1.30	1.29	1.28	1.27	1.27	1.26	1.26	1.25	1.24	1.23	1.23	1.23	1.22	1.21	1.20	1.18	1.17	1.14	1.13	1.13
125	1.34	1.40	1.39	1.36	1.34	1.33	1.31	1.30	1.29	1.28	1.27	1.26	1.25	1.25	1.24	1.24	1.23	1.23	1.22	1.22	1.20	1.19	1.17	1.16	1.14	1.12	1.12
150	1.33	1.40	1.38	1.36	1.34	1.32	1.31	1.30	1.28	1.27	1.27	1.26	1.25	1.24	1.24	1.23	1.23	1.22	1.22	1.21	1.20	1.18	1.17	1.16	1.13	1.12	1.11
200	1.33	1.40	1.38	1.36	1.34	1.32	1.30	1.29	1.28	1.27	1.26	1.25	1.24	1.24	1.23	1.23	1.22	1.22	1.21	1.21	1.19	1.18	1.16	1.15	1.12	1.11	1.10
300	1.33	1.39	1.38	1.35	1.33	1.32	1.30	1.29	1.28	1.26	1.26	1.25	1.24	1.23	1.23	1.22	1.22	1.21	1.21	1.20	1.19	1.17	1.15	1.14	1.11	1.10	1.09
500	1.33	1.39	1.37	1.35	1.33	1.31	1.30	1.28	1.27	1.26	1.25	1.24	1.23	1.23	1.22	1.22	1.21	1.21	1.20	1.20	1.18	1.16	1.14	1.13	1.10	1.09	1.08
1000	1.32	1.39	1.37	1.35	1.33	1.31	1.29	1.28	1.27	1.26	1.25	1.24	1.23	1.23	1.22	1.21	1.21	1.20	1.20	1.20	1.18	1.16	1.14	1.13	1.10	1.08	1.07

DEGREES OF FREEDOM FOR DENOMINATOR

TABLE A.4 Percentiles of the F Distribution (*continued*)

Upper 10% point of the F distribution

DEGREES OF FREEDOM FOR NUMERATOR

den\num	1	2	3	4	5	6	7	8	9	10	11	12	13	14	15	16	17	18	19	20	25	30	40	50	100	150	200
1	39.9	49.5	53.6	55.8	57.2	58.2	58.9	59.4	59.9	60.2	60.5	60.7	60.9	61.1	61.2	61.3	61.5	61.6	61.7	61.7	62.1	62.3	62.5	62.7	63.0	63.1	63.2
2	8.53	9.00	9.16	9.24	9.29	9.33	9.35	9.37	9.38	9.39	9.40	9.41	9.41	9.42	9.42	9.43	9.43	9.44	9.44	9.44	9.45	9.46	9.47	9.47	9.48	9.48	9.49
3	5.54	5.46	5.39	5.34	5.31	5.28	5.27	5.25	5.24	5.23	5.22	5.22	5.21	5.20	5.20	5.20	5.19	5.19	5.19	5.18	5.17	5.17	5.16	5.15	5.14	5.14	5.14
4	4.54	4.32	4.19	4.11	4.05	4.01	3.98	3.95	3.94	3.92	3.91	3.90	3.89	3.88	3.87	3.86	3.86	3.85	3.85	3.84	3.83	3.82	3.80	3.80	3.78	3.77	3.77
5	4.06	3.78	3.62	3.52	3.45	3.40	3.37	3.34	3.32	3.30	3.28	3.27	3.26	3.25	3.24	3.23	3.22	3.22	3.21	3.21	3.19	3.17	3.16	3.15	3.13	3.12	3.12
6	3.78	3.46	3.29	3.18	3.11	3.05	3.01	2.98	2.96	2.94	2.92	2.90	2.89	2.88	2.87	2.86	2.85	2.85	2.84	2.84	2.81	2.80	2.78	2.77	2.75	2.74	2.73
7	3.59	3.26	3.07	2.96	2.88	2.83	2.78	2.75	2.72	2.70	2.68	2.67	2.65	2.64	2.63	2.62	2.61	2.61	2.60	2.59	2.57	2.56	2.54	2.52	2.50	2.49	2.48
8	3.46	3.11	2.92	2.81	2.73	2.67	2.62	2.59	2.56	2.54	2.52	2.50	2.49	2.48	2.46	2.45	2.45	2.44	2.43	2.42	2.40	2.38	2.36	2.35	2.32	2.31	2.31
9	3.36	3.01	2.81	2.69	2.61	2.55	2.51	2.47	2.44	2.42	2.40	2.38	2.36	2.35	2.34	2.33	2.32	2.32	2.30	2.30	2.27	2.25	2.23	2.22	2.19	2.18	2.17
10	3.29	2.92	2.73	2.61	2.52	2.46	2.41	2.38	2.35	2.32	2.30	2.28	2.27	2.26	2.24	2.23	2.22	2.22	2.21	2.20	2.17	2.16	2.13	2.12	2.09	2.08	2.07
11	3.23	2.86	2.66	2.54	2.45	2.39	2.34	2.30	2.27	2.25	2.23	2.21	2.19	2.18	2.17	2.16	2.15	2.14	2.13	2.12	2.10	2.08	2.05	2.04	2.01	1.99	1.99
12	3.18	2.81	2.61	2.48	2.39	2.33	2.28	2.24	2.21	2.19	2.17	2.15	2.13	2.12	2.10	2.09	2.08	2.08	2.07	2.06	2.03	2.01	1.99	1.97	1.94	1.93	1.92
13	3.14	2.76	2.56	2.43	2.35	2.28	2.23	2.20	2.16	2.14	2.12	2.10	2.08	2.07	2.05	2.04	2.03	2.02	2.01	2.01	1.98	1.96	1.93	1.92	1.88	1.87	1.86
14	3.10	2.73	2.52	2.39	2.31	2.24	2.19	2.15	2.12	2.10	2.07	2.05	2.04	2.02	2.01	2.00	1.98	1.98	1.97	1.96	1.93	1.91	1.89	1.87	1.83	1.82	1.82
15	3.07	2.70	2.49	2.36	2.27	2.21	2.16	2.12	2.09	2.06	2.04	2.02	2.00	1.99	1.97	1.96	1.95	1.94	1.93	1.92	1.89	1.87	1.85	1.83	1.79	1.78	1.77
16	3.05	2.67	2.46	2.33	2.24	2.18	2.13	2.09	2.06	2.03	2.01	1.99	1.97	1.95	1.94	1.93	1.92	1.91	1.90	1.89	1.86	1.84	1.81	1.79	1.76	1.74	1.74
17	3.03	2.64	2.44	2.31	2.22	2.15	2.10	2.06	2.03	2.00	1.98	1.96	1.94	1.93	1.91	1.90	1.89	1.88	1.87	1.86	1.83	1.81	1.78	1.76	1.73	1.71	1.71
18	3.01	2.62	2.42	2.29	2.20	2.13	2.08	2.04	2.00	1.98	1.95	1.93	1.92	1.90	1.89	1.87	1.86	1.85	1.84	1.84	1.80	1.78	1.75	1.74	1.70	1.68	1.68
19	2.99	2.61	2.40	2.27	2.18	2.11	2.06	2.02	1.98	1.96	1.93	1.91	1.89	1.88	1.86	1.85	1.84	1.83	1.82	1.81	1.78	1.76	1.73	1.71	1.69	1.66	1.65
20	2.97	2.59	2.38	2.25	2.16	2.09	2.04	2.00	1.96	1.94	1.91	1.89	1.87	1.86	1.84	1.83	1.82	1.81	1.80	1.79	1.76	1.74	1.71	1.69	1.65	1.64	1.63
21	2.96	2.57	2.36	2.23	2.14	2.08	2.02	1.98	1.95	1.92	1.90	1.87	1.86	1.84	1.83	1.81	1.80	1.79	1.78	1.78	1.74	1.72	1.69	1.67	1.63	1.62	1.61
22	2.95	2.56	2.35	2.22	2.13	2.06	2.01	1.97	1.93	1.90	1.88	1.86	1.84	1.83	1.81	1.80	1.79	1.78	1.77	1.76	1.73	1.70	1.67	1.65	1.61	1.60	1.59
23	2.94	2.55	2.34	2.21	2.11	2.05	1.99	1.95	1.92	1.89	1.87	1.84	1.83	1.81	1.80	1.78	1.77	1.76	1.75	1.74	1.71	1.69	1.66	1.64	1.59	1.58	1.57
24	2.93	2.54	2.33	2.19	2.10	2.04	1.98	1.94	1.91	1.88	1.85	1.83	1.81	1.80	1.78	1.77	1.76	1.75	1.74	1.73	1.70	1.67	1.64	1.62	1.58	1.56	1.56
25	2.92	2.53	2.32	2.18	2.09	2.02	1.97	1.93	1.89	1.87	1.84	1.82	1.80	1.79	1.77	1.76	1.75	1.74	1.73	1.72	1.68	1.66	1.63	1.61	1.56	1.55	1.54
26	2.91	2.52	2.31	2.17	2.08	2.01	1.96	1.92	1.88	1.86	1.83	1.81	1.79	1.77	1.76	1.75	1.73	1.72	1.71	1.71	1.67	1.65	1.61	1.59	1.55	1.54	1.53
27	2.90	2.51	2.30	2.17	2.07	2.00	1.95	1.91	1.87	1.85	1.82	1.80	1.78	1.76	1.75	1.74	1.72	1.71	1.70	1.70	1.66	1.64	1.60	1.58	1.54	1.52	1.52
28	2.89	2.50	2.29	2.16	2.06	2.00	1.94	1.90	1.87	1.84	1.81	1.79	1.77	1.75	1.74	1.73	1.71	1.70	1.69	1.69	1.65	1.63	1.59	1.57	1.53	1.51	1.50
29	2.89	2.50	2.28	2.15	2.06	1.99	1.93	1.89	1.86	1.83	1.80	1.78	1.76	1.75	1.73	1.72	1.71	1.69	1.68	1.68	1.64	1.62	1.58	1.56	1.52	1.50	1.49
30	2.88	2.49	2.28	2.14	2.05	1.98	1.93	1.88	1.85	1.82	1.79	1.77	1.75	1.74	1.72	1.71	1.70	1.69	1.68	1.67	1.63	1.61	1.57	1.55	1.51	1.49	1.48
32	2.87	2.48	2.26	2.13	2.04	1.97	1.91	1.87	1.83	1.81	1.78	1.76	1.74	1.72	1.71	1.69	1.68	1.67	1.66	1.65	1.62	1.59	1.56	1.53	1.49	1.47	1.46
34	2.86	2.47	2.25	2.12	2.02	1.96	1.90	1.86	1.82	1.79	1.77	1.75	1.73	1.71	1.70	1.68	1.67	1.66	1.65	1.64	1.61	1.58	1.54	1.52	1.47	1.46	1.45
36	2.85	2.46	2.24	2.11	2.01	1.94	1.89	1.85	1.81	1.78	1.76	1.73	1.72	1.70	1.68	1.67	1.66	1.65	1.64	1.63	1.59	1.56	1.53	1.51	1.46	1.44	1.43
38	2.84	2.45	2.23	2.10	2.01	1.94	1.88	1.84	1.80	1.77	1.75	1.72	1.70	1.69	1.67	1.66	1.65	1.63	1.62	1.61	1.58	1.55	1.52	1.49	1.45	1.43	1.42
40	2.84	2.44	2.23	2.09	2.00	1.93	1.87	1.83	1.79	1.76	1.74	1.71	1.70	1.68	1.66	1.65	1.64	1.62	1.61	1.61	1.57	1.54	1.51	1.48	1.43	1.42	1.41
42	2.83	2.43	2.22	2.08	1.99	1.92	1.86	1.82	1.78	1.75	1.73	1.71	1.69	1.67	1.65	1.64	1.62	1.61	1.60	1.60	1.56	1.53	1.50	1.47	1.42	1.41	1.40
44	2.82	2.43	2.21	2.08	1.98	1.91	1.86	1.81	1.77	1.75	1.72	1.70	1.68	1.66	1.65	1.63	1.62	1.60	1.59	1.59	1.55	1.52	1.49	1.46	1.41	1.39	1.39
46	2.82	2.42	2.21	2.07	1.98	1.91	1.85	1.81	1.77	1.74	1.71	1.69	1.67	1.65	1.64	1.63	1.60	1.60	1.59	1.58	1.54	1.51	1.48	1.46	1.40	1.39	1.38
48	2.81	2.42	2.20	2.07	1.97	1.90	1.85	1.80	1.77	1.74	1.71	1.69	1.66	1.65	1.63	1.62	1.59	1.59	1.58	1.57	1.53	1.51	1.47	1.45	1.40	1.38	1.37
50	2.81	2.41	2.20	2.06	1.97	1.90	1.84	1.80	1.76	1.73	1.70	1.68	1.66	1.64	1.63	1.61	1.60	1.58	1.58	1.57	1.53	1.50	1.46	1.44	1.39	1.37	1.36
60	2.79	2.39	2.18	2.04	1.95	1.87	1.82	1.77	1.74	1.71	1.68	1.66	1.64	1.62	1.60	1.59	1.58	1.56	1.55	1.54	1.50	1.48	1.44	1.41	1.36	1.34	1.33
70	2.78	2.38	2.16	2.03	1.93	1.86	1.80	1.76	1.72	1.69	1.66	1.64	1.62	1.60	1.59	1.57	1.55	1.54	1.53	1.53	1.49	1.46	1.42	1.39	1.34	1.31	1.30
80	2.77	2.37	2.15	2.02	1.92	1.85	1.79	1.75	1.71	1.68	1.65	1.63	1.61	1.59	1.57	1.56	1.55	1.52	1.52	1.51	1.47	1.44	1.40	1.38	1.32	1.30	1.28
90	2.76	2.36	2.15	2.01	1.91	1.84	1.78	1.74	1.70	1.67	1.65	1.62	1.60	1.58	1.56	1.55	1.54	1.52	1.51	1.50	1.46	1.43	1.39	1.36	1.30	1.28	1.27
100	2.76	2.36	2.14	2.00	1.91	1.83	1.78	1.73	1.69	1.66	1.64	1.61	1.59	1.57	1.56	1.54	1.53	1.52	1.50	1.49	1.45	1.42	1.38	1.35	1.29	1.27	1.26
125	2.75	2.35	2.13	1.99	1.89	1.82	1.77	1.72	1.68	1.65	1.62	1.60	1.58	1.56	1.54	1.53	1.51	1.50	1.49	1.48	1.44	1.41	1.36	1.34	1.27	1.25	1.23
150	2.74	2.34	2.12	1.98	1.89	1.81	1.76	1.71	1.67	1.64	1.61	1.59	1.57	1.55	1.53	1.52	1.50	1.48	1.48	1.47	1.43	1.40	1.35	1.33	1.26	1.23	1.22
200	2.73	2.33	2.11	1.97	1.88	1.80	1.75	1.70	1.66	1.63	1.60	1.58	1.55	1.54	1.52	1.51	1.49	1.48	1.47	1.46	1.41	1.38	1.34	1.31	1.24	1.22	1.20
300	2.72	2.32	2.10	1.96	1.87	1.79	1.74	1.69	1.65	1.62	1.59	1.57	1.55	1.53	1.51	1.50	1.48	1.47	1.46	1.45	1.40	1.37	1.32	1.29	1.21	1.19	1.18
500	2.72	2.31	2.09	1.96	1.86	1.79	1.73	1.68	1.64	1.61	1.58	1.56	1.54	1.52	1.50	1.49	1.47	1.46	1.45	1.44	1.39	1.36	1.31	1.28	1.21	1.18	1.16
1000	2.71	2.31	2.09	1.95	1.85	1.78	1.72	1.68	1.64	1.61	1.58	1.55	1.53	1.51	1.49	1.48	1.46	1.45	1.44	1.43	1.38	1.35	1.30	1.27	1.20	1.16	1.15

DEGREES OF FREEDOM FOR DENOMINATOR

Upper 5% point of the F distribution

DEGREES OF FREEDOM FOR NUMERATOR

den \ num	1	2	3	4	5	6	7	8	9	10	11	12	13	14	15	16	17	18	19	20	25	30	40	50	100	150	200
1	161	200	216	225	230	234	237	239	241	242	243	244	245	245	246	246	247	247	248	248	249	250	251	252	253	253	254
2	18.5	19.0	19.2	19.2	19.3	19.3	19.4	19.4	19.4	19.4	19.4	19.4	19.4	19.4	19.4	19.4	19.4	19.4	19.4	19.4	19.4	19.5	19.5	19.5	19.5	19.5	19.5
3	10.1	9.55	9.28	9.12	9.01	8.94	8.89	8.85	8.81	8.79	8.76	8.74	8.73	8.71	8.70	8.69	8.68	8.67	8.67	8.66	8.63	8.62	8.59	8.58	8.55	8.54	8.54
4	7.71	6.94	6.59	6.39	6.26	6.16	6.09	6.04	6.00	5.96	5.94	5.91	5.89	5.87	5.86	5.84	5.83	5.82	5.81	5.80	5.77	5.75	5.72	5.70	5.66	5.65	5.65
5	6.61	5.79	5.41	5.19	5.05	4.95	4.88	4.82	4.77	4.74	4.70	4.68	4.66	4.64	4.62	4.60	4.59	4.58	4.57	4.56	4.52	4.50	4.46	4.44	4.41	4.39	4.39
6	5.99	5.14	4.76	4.53	4.39	4.28	4.21	4.15	4.10	4.06	4.03	4.00	3.98	3.96	3.94	3.92	3.91	3.90	3.90	3.87	3.83	3.81	3.77	3.75	3.71	3.70	3.69
7	5.59	4.74	4.35	4.12	3.97	3.87	3.79	3.73	3.68	3.64	3.60	3.57	3.55	3.53	3.51	3.49	3.48	3.47	3.46	3.44	3.40	3.38	3.34	3.32	3.27	3.26	3.25
8	5.32	4.46	4.07	3.84	3.69	3.58	3.50	3.44	3.39	3.35	3.31	3.28	3.26	3.24	3.22	3.20	3.19	3.17	3.16	3.15	3.11	3.08	3.04	3.02	2.97	2.96	2.95
9	5.12	4.26	3.86	3.63	3.48	3.37	3.29	3.23	3.18	3.14	3.10	3.07	3.05	3.03	3.01	2.99	2.97	2.96	2.94	2.94	2.89	2.86	2.83	2.80	2.76	2.74	2.73
10	4.96	4.10	3.71	3.48	3.33	3.22	3.14	3.07	3.02	2.98	2.94	2.91	2.89	2.86	2.85	2.83	2.81	2.80	2.79	2.77	2.73	2.70	2.66	2.64	2.59	2.57	2.56
11	4.84	3.98	3.59	3.36	3.20	3.09	3.01	2.95	2.90	2.85	2.82	2.79	2.76	2.74	2.72	2.70	2.69	2.67	2.66	2.65	2.60	2.57	2.53	2.51	2.46	2.44	2.43
12	4.75	3.89	3.49	3.26	3.11	3.00	2.91	2.85	2.80	2.75	2.72	2.69	2.66	2.64	2.62	2.60	2.58	2.57	2.56	2.54	2.50	2.47	2.43	2.40	2.35	2.33	2.32
13	4.67	3.81	3.41	3.18	3.03	2.92	2.83	2.77	2.71	2.67	2.63	2.60	2.58	2.55	2.53	2.51	2.50	2.48	2.47	2.46	2.41	2.38	2.34	2.31	2.26	2.24	2.23
14	4.60	3.74	3.34	3.11	2.96	2.85	2.76	2.70	2.65	2.60	2.57	2.53	2.51	2.48	2.46	2.44	2.43	2.41	2.40	2.39	2.34	2.31	2.27	2.24	2.19	2.17	2.16
15	4.54	3.68	3.29	3.06	2.90	2.79	2.71	2.64	2.59	2.54	2.51	2.48	2.45	2.42	2.40	2.38	2.37	2.35	2.34	2.33	2.28	2.25	2.20	2.18	2.12	2.10	2.10
16	4.49	3.63	3.24	3.01	2.85	2.74	2.66	2.59	2.54	2.49	2.46	2.42	2.40	2.37	2.35	2.33	2.32	2.30	2.29	2.28	2.23	2.19	2.15	2.12	2.07	2.05	2.04
17	4.45	3.59	3.20	2.96	2.81	2.70	2.61	2.55	2.49	2.45	2.41	2.38	2.35	2.33	2.31	2.29	2.27	2.26	2.23	2.23	2.18	2.15	2.10	2.08	2.02	2.00	1.99
18	4.41	3.55	3.16	2.93	2.77	2.66	2.58	2.51	2.46	2.41	2.37	2.34	2.31	2.29	2.27	2.25	2.23	2.22	2.20	2.19	2.14	2.11	2.06	2.04	1.98	1.96	1.95
19	4.38	3.52	3.13	2.90	2.74	2.63	2.54	2.48	2.42	2.38	2.34	2.31	2.28	2.26	2.23	2.21	2.20	2.18	2.17	2.16	2.11	2.07	2.03	2.00	1.94	1.92	1.91
20	4.35	3.49	3.10	2.87	2.71	2.60	2.51	2.45	2.39	2.35	2.31	2.28	2.25	2.22	2.20	2.18	2.17	2.15	2.14	2.12	2.07	2.04	1.99	1.97	1.91	1.89	1.88
21	4.32	3.47	3.07	2.84	2.68	2.57	2.49	2.42	2.37	2.32	2.28	2.25	2.22	2.20	2.18	2.16	2.14	2.12	2.11	2.10	2.05	2.01	1.96	1.94	1.88	1.86	1.84
22	4.30	3.44	3.05	2.82	2.66	2.55	2.46	2.40	2.34	2.30	2.26	2.23	2.20	2.17	2.15	2.13	2.11	2.10	2.08	2.07	2.02	1.98	1.94	1.91	1.85	1.83	1.82
23	4.28	3.42	3.03	2.80	2.64	2.53	2.44	2.37	2.32	2.27	2.24	2.20	2.18	2.15	2.13	2.11	2.09	2.08	2.06	2.05	2.00	1.96	1.91	1.88	1.82	1.80	1.79
24	4.26	3.40	3.01	2.78	2.62	2.51	2.42	2.36	2.30	2.25	2.22	2.18	2.15	2.13	2.11	2.09	2.07	2.05	2.04	2.03	1.97	1.94	1.89	1.86	1.80	1.78	1.77
25	4.24	3.39	2.99	2.76	2.60	2.49	2.40	2.34	2.28	2.24	2.20	2.16	2.14	2.11	2.09	2.07	2.05	2.04	2.02	2.01	1.96	1.92	1.87	1.84	1.78	1.76	1.75
26	4.23	3.37	2.98	2.74	2.59	2.47	2.39	2.32	2.27	2.22	2.18	2.15	2.12	2.09	2.07	2.05	2.03	2.02	2.00	1.99	1.94	1.90	1.85	1.82	1.76	1.74	1.73
27	4.21	3.35	2.96	2.73	2.57	2.46	2.37	2.31	2.25	2.20	2.17	2.13	2.10	2.08	2.06	2.04	2.02	2.00	1.99	1.97	1.92	1.88	1.84	1.81	1.74	1.72	1.71
28	4.20	3.34	2.95	2.71	2.56	2.45	2.36	2.29	2.24	2.19	2.15	2.12	2.09	2.06	2.04	2.02	2.00	1.99	1.97	1.96	1.91	1.87	1.82	1.79	1.73	1.71	1.69
29	4.18	3.33	2.93	2.70	2.55	2.43	2.35	2.28	2.22	2.18	2.14	2.10	2.08	2.05	2.03	2.01	1.99	1.97	1.96	1.94	1.89	1.85	1.81	1.77	1.71	1.69	1.67
30	4.17	3.32	2.92	2.69	2.53	2.42	2.33	2.27	2.21	2.16	2.13	2.09	2.06	2.04	2.01	1.99	1.98	1.96	1.95	1.93	1.88	1.84	1.79	1.76	1.70	1.67	1.66
32	4.15	3.29	2.90	2.67	2.51	2.40	2.31	2.24	2.19	2.14	2.10	2.07	2.04	2.01	1.99	1.97	1.95	1.94	1.92	1.91	1.85	1.82	1.77	1.74	1.67	1.64	1.63
34	4.13	3.28	2.88	2.65	2.49	2.38	2.29	2.23	2.17	2.12	2.08	2.05	2.02	1.99	1.97	1.95	1.93	1.92	1.90	1.89	1.83	1.80	1.75	1.71	1.65	1.62	1.61
36	4.11	3.26	2.87	2.63	2.48	2.36	2.28	2.21	2.15	2.11	2.07	2.03	2.00	1.98	1.95	1.93	1.92	1.90	1.88	1.87	1.81	1.78	1.73	1.69	1.62	1.60	1.59
38	4.10	3.24	2.85	2.62	2.46	2.35	2.26	2.19	2.14	2.09	2.05	2.02	1.99	1.96	1.94	1.92	1.90	1.88	1.87	1.85	1.80	1.76	1.71	1.68	1.61	1.58	1.57
40	4.08	3.23	2.84	2.61	2.45	2.34	2.25	2.18	2.12	2.08	2.04	2.00	1.97	1.95	1.92	1.90	1.89	1.87	1.85	1.84	1.78	1.74	1.69	1.66	1.59	1.56	1.55
42	4.07	3.22	2.83	2.59	2.44	2.32	2.24	2.17	2.11	2.06	2.03	1.99	1.96	1.94	1.91	1.89	1.87	1.86	1.84	1.83	1.77	1.73	1.68	1.65	1.57	1.55	1.53
44	4.06	3.21	2.82	2.58	2.43	2.31	2.23	2.16	2.10	2.05	2.01	1.98	1.95	1.92	1.90	1.88	1.86	1.84	1.83	1.81	1.76	1.72	1.67	1.63	1.56	1.53	1.52
46	4.05	3.20	2.81	2.57	2.42	2.30	2.22	2.15	2.09	2.04	2.00	1.97	1.94	1.91	1.89	1.87	1.85	1.83	1.82	1.80	1.75	1.71	1.65	1.62	1.55	1.52	1.51
48	4.04	3.19	2.80	2.57	2.41	2.29	2.21	2.14	2.08	2.03	1.99	1.96	1.93	1.90	1.88	1.86	1.84	1.82	1.81	1.79	1.74	1.70	1.64	1.61	1.54	1.51	1.49
50	4.03	3.18	2.79	2.56	2.40	2.29	2.20	2.13	2.07	2.03	1.99	1.95	1.92	1.89	1.87	1.85	1.83	1.81	1.80	1.78	1.73	1.69	1.63	1.60	1.52	1.50	1.48
60	4.00	3.15	2.76	2.53	2.37	2.25	2.17	2.10	2.04	1.99	1.95	1.92	1.89	1.86	1.84	1.82	1.80	1.78	1.76	1.75	1.69	1.65	1.59	1.56	1.48	1.45	1.44
70	3.98	3.13	2.74	2.50	2.35	2.23	2.14	2.07	2.02	1.97	1.93	1.89	1.86	1.84	1.81	1.79	1.77	1.75	1.74	1.72	1.66	1.62	1.57	1.53	1.45	1.42	1.40
80	3.96	3.11	2.72	2.49	2.33	2.21	2.13	2.06	2.00	1.95	1.91	1.88	1.84	1.82	1.79	1.77	1.75	1.73	1.72	1.70	1.64	1.60	1.54	1.51	1.43	1.39	1.38
90	3.95	3.10	2.71	2.47	2.32	2.20	2.11	2.04	1.99	1.94	1.90	1.86	1.83	1.80	1.78	1.76	1.74	1.72	1.70	1.69	1.63	1.59	1.53	1.49	1.41	1.38	1.36
100	3.94	3.09	2.70	2.46	2.31	2.19	2.10	2.03	1.97	1.93	1.89	1.85	1.82	1.79	1.77	1.75	1.73	1.71	1.69	1.68	1.62	1.57	1.52	1.48	1.39	1.36	1.34
125	3.92	3.07	2.68	2.44	2.29	2.17	2.08	2.01	1.96	1.91	1.87	1.83	1.80	1.77	1.75	1.73	1.71	1.69	1.67	1.66	1.60	1.55	1.49	1.45	1.36	1.33	1.31
150	3.90	3.06	2.66	2.43	2.27	2.16	2.07	2.00	1.94	1.89	1.85	1.82	1.79	1.76	1.73	1.71	1.69	1.67	1.66	1.64	1.58	1.54	1.48	1.44	1.34	1.31	1.29
200	3.89	3.04	2.65	2.42	2.26	2.14	2.06	1.98	1.93	1.88	1.84	1.80	1.77	1.74	1.72	1.69	1.67	1.66	1.64	1.62	1.56	1.52	1.46	1.41	1.32	1.28	1.26
300	3.87	3.03	2.63	2.40	2.24	2.13	2.04	1.97	1.91	1.86	1.82	1.78	1.75	1.72	1.70	1.68	1.66	1.64	1.62	1.61	1.54	1.50	1.43	1.39	1.30	1.26	1.23
500	3.86	3.01	2.62	2.39	2.23	2.12	2.03	1.96	1.90	1.85	1.81	1.77	1.74	1.71	1.69	1.66	1.64	1.62	1.61	1.59	1.53	1.48	1.42	1.38	1.28	1.23	1.21
1000	3.85	3.00	2.61	2.38	2.22	2.11	2.02	1.95	1.89	1.84	1.80	1.76	1.73	1.70	1.68	1.65	1.63	1.61	1.60	1.58	1.52	1.47	1.41	1.36	1.26	1.22	1.19

DEGREES OF FREEDOM FOR DENOMINATOR

TABLE A.4 Percentiles of the *F* Distribution (*continued*)

Upper 2.5% point of the F distribution

DEGREES OF FREEDOM FOR NUMERATOR

den\num	1	2	3	4	5	6	7	8	9	10	11	12	13	14	15	16	17	18	19	20	25	30	40	50	100	150	200
1	548	800	864	900	922	937	948	957	963	969	973	977	980	983	985	987	989	990	992	993	998	1001	1006	1008	1013	1015	1016
2	38.5	39.0	39.2	39.2	39.3	39.3	39.4	39.4	39.4	39.4	39.4	39.4	39.4	39.4	39.4	39.4	39.4	39.4	39.4	39.4	39.5	39.5	39.5	39.5	39.5	39.5	39.5
3	17.4	16.0	15.4	15.1	14.9	14.7	14.6	14.5	14.5	14.4	14.4	14.3	14.3	14.3	14.3	14.2	14.2	14.2	14.2	14.2	14.1	14.1	14.0	14.0	14.0	13.9	13.9
4	12.2	10.6	9.98	9.60	9.36	9.20	9.07	8.98	8.90	8.84	8.79	8.75	8.71	8.68	8.66	8.63	8.61	8.59	8.58	8.56	8.50	8.46	8.41	8.38	8.32	8.30	8.29
5	10.0	8.43	7.76	7.39	7.15	6.98	6.85	6.76	6.68	6.62	6.57	6.52	6.49	6.46	6.43	6.40	6.38	6.36	6.34	6.33	6.27	6.23	6.18	6.14	6.08	6.06	6.05
6	8.81	7.26	6.60	6.23	5.99	5.82	5.70	5.60	5.52	5.46	5.41	5.37	5.33	5.30	5.27	5.24	5.22	5.20	5.18	5.17	5.11	5.07	5.01	4.98	4.92	4.89	4.88
7	8.07	6.54	5.89	5.52	5.29	5.12	4.99	4.90	4.82	4.76	4.71	4.67	4.63	4.60	4.57	4.54	4.52	4.50	4.48	4.47	4.40	4.36	4.31	4.28	4.21	4.19	4.18
8	7.57	6.06	5.42	5.05	4.82	4.65	4.53	4.43	4.36	4.30	4.24	4.20	4.16	4.13	4.10	4.08	4.05	4.03	4.02	4.00	3.94	3.89	3.84	3.81	3.74	3.72	3.70
9	7.21	5.71	5.08	4.72	4.48	4.32	4.20	4.10	4.03	3.96	3.91	3.87	3.83	3.80	3.77	3.74	3.72	3.70	3.68	3.67	3.60	3.56	3.51	3.47	3.40	3.38	3.37
10	6.94	5.46	4.83	4.47	4.24	4.07	3.95	3.85	3.78	3.72	3.66	3.62	3.58	3.55	3.52	3.50	3.47	3.45	3.44	3.42	3.35	3.31	3.26	3.22	3.15	3.13	3.12
11	6.72	5.26	4.63	4.28	4.04	3.88	3.76	3.66	3.59	3.53	3.47	3.43	3.39	3.36	3.33	3.30	3.28	3.26	3.24	3.23	3.16	3.12	3.06	3.03	2.96	2.93	2.92
12	6.55	5.10	4.47	4.12	3.89	3.73	3.61	3.51	3.44	3.37	3.32	3.28	3.24	3.21	3.18	3.15	3.13	3.11	3.09	3.07	3.01	2.96	2.91	2.87	2.80	2.78	2.76
13	6.41	4.97	4.35	4.00	3.77	3.60	3.48	3.39	3.31	3.25	3.20	3.15	3.12	3.08	3.05	3.03	3.00	2.98	2.96	2.95	2.88	2.84	2.78	2.74	2.67	2.65	2.63
14	6.30	4.86	4.24	3.89	3.66	3.50	3.38	3.29	3.21	3.15	3.09	3.05	3.01	2.98	2.95	2.92	2.90	2.88	2.86	2.84	2.78	2.73	2.67	2.64	2.56	2.54	2.53
15	6.20	4.77	4.15	3.80	3.58	3.41	3.29	3.20	3.12	3.06	3.01	2.96	2.92	2.89	2.86	2.84	2.81	2.79	2.77	2.76	2.69	2.64	2.59	2.55	2.47	2.45	2.44
16	6.12	4.69	4.08	3.73	3.50	3.34	3.22	3.12	3.05	2.99	2.93	2.89	2.85	2.82	2.79	2.76	2.74	2.72	2.70	2.68	2.61	2.57	2.51	2.47	2.40	2.37	2.36
17	6.04	4.62	4.01	3.66	3.44	3.28	3.16	3.06	2.98	2.92	2.87	2.82	2.79	2.75	2.72	2.70	2.67	2.65	2.63	2.62	2.55	2.50	2.44	2.41	2.33	2.30	2.29
18	5.98	4.56	3.95	3.61	3.38	3.22	3.10	3.01	2.93	2.87	2.81	2.77	2.73	2.70	2.67	2.64	2.62	2.60	2.58	2.56	2.49	2.44	2.38	2.35	2.27	2.24	2.23
19	5.92	4.51	3.90	3.56	3.33	3.17	3.05	2.96	2.88	2.82	2.76	2.72	2.68	2.65	2.62	2.59	2.57	2.55	2.53	2.51	2.44	2.39	2.33	2.30	2.22	2.19	2.18
20	5.87	4.46	3.86	3.51	3.29	3.13	3.01	2.91	2.84	2.77	2.72	2.68	2.64	2.60	2.57	2.55	2.52	2.50	2.48	2.46	2.40	2.35	2.29	2.25	2.17	2.14	2.13
21	5.83	4.42	3.82	3.48	3.25	3.09	2.97	2.87	2.80	2.73	2.68	2.64	2.60	2.56	2.53	2.51	2.48	2.46	2.44	2.42	2.36	2.31	2.25	2.21	2.13	2.10	2.09
22	5.79	4.38	3.78	3.44	3.22	3.05	2.93	2.84	2.76	2.70	2.65	2.60	2.56	2.53	2.50	2.47	2.45	2.43	2.41	2.39	2.32	2.27	2.21	2.17	2.09	2.06	2.05
23	5.75	4.35	3.75	3.41	3.18	3.02	2.90	2.81	2.73	2.67	2.62	2.57	2.53	2.50	2.47	2.44	2.42	2.39	2.37	2.36	2.29	2.24	2.18	2.14	2.06	2.03	2.01
24	5.72	4.32	3.72	3.38	3.15	2.99	2.87	2.78	2.70	2.64	2.59	2.54	2.50	2.47	2.44	2.41	2.39	2.36	2.35	2.33	2.26	2.21	2.15	2.11	2.02	2.00	1.98
25	5.69	4.29	3.69	3.35	3.13	2.97	2.85	2.75	2.68	2.61	2.56	2.51	2.48	2.44	2.41	2.38	2.36	2.34	2.32	2.30	2.23	2.18	2.12	2.08	2.00	1.97	1.95
26	5.66	4.27	3.67	3.33	3.10	2.94	2.82	2.73	2.65	2.59	2.54	2.49	2.45	2.42	2.39	2.36	2.34	2.31	2.29	2.28	2.21	2.16	2.09	2.05	1.97	1.94	1.92
27	5.63	4.24	3.65	3.31	3.08	2.92	2.80	2.71	2.63	2.57	2.51	2.47	2.43	2.39	2.36	2.34	2.31	2.29	2.27	2.25	2.18	2.13	2.07	2.03	1.94	1.91	1.90
28	5.61	4.22	3.63	3.29	3.06	2.90	2.78	2.69	2.61	2.55	2.49	2.45	2.41	2.37	2.34	2.32	2.29	2.27	2.25	2.23	2.16	2.11	2.05	2.01	1.92	1.89	1.88
29	5.59	4.20	3.61	3.27	3.04	2.88	2.76	2.67	2.59	2.53	2.48	2.43	2.39	2.36	2.32	2.30	2.27	2.25	2.23	2.21	2.14	2.09	2.03	1.99	1.90	1.87	1.86
30	5.57	4.18	3.59	3.25	3.03	2.87	2.75	2.65	2.57	2.51	2.46	2.41	2.37	2.34	2.31	2.28	2.26	2.23	2.21	2.20	2.12	2.07	2.01	1.97	1.88	1.85	1.84
32	5.53	4.15	3.56	3.22	3.00	2.84	2.71	2.62	2.54	2.48	2.43	2.38	2.34	2.31	2.28	2.25	2.22	2.20	2.18	2.16	2.09	2.04	1.98	1.93	1.85	1.82	1.80
34	5.50	4.12	3.53	3.19	2.97	2.81	2.69	2.59	2.52	2.45	2.40	2.35	2.31	2.28	2.25	2.22	2.20	2.17	2.15	2.13	2.06	2.01	1.95	1.90	1.82	1.78	1.77
36	5.47	4.09	3.50	3.17	2.94	2.78	2.66	2.57	2.49	2.43	2.37	2.33	2.29	2.25	2.22	2.20	2.17	2.15	2.13	2.11	2.04	1.99	1.92	1.88	1.79	1.76	1.74
38	5.45	4.07	3.48	3.15	2.92	2.76	2.64	2.55	2.47	2.41	2.35	2.31	2.27	2.23	2.20	2.17	2.15	2.13	2.11	2.09	2.01	1.96	1.90	1.85	1.76	1.73	1.71
40	5.42	4.05	3.46	3.13	2.90	2.74	2.62	2.53	2.45	2.39	2.33	2.29	2.25	2.21	2.18	2.15	2.13	2.11	2.09	2.07	1.99	1.94	1.88	1.83	1.74	1.71	1.69
42	5.40	4.03	3.45	3.11	2.89	2.73	2.61	2.51	2.43	2.37	2.32	2.27	2.23	2.20	2.16	2.14	2.11	2.09	2.07	2.05	1.98	1.92	1.86	1.81	1.72	1.69	1.67
44	5.39	4.02	3.43	3.09	2.87	2.71	2.59	2.50	2.42	2.36	2.30	2.26	2.22	2.18	2.15	2.12	2.10	2.07	2.05	2.03	1.96	1.91	1.84	1.80	1.70	1.67	1.65
46	5.37	4.00	3.42	3.08	2.86	2.70	2.58	2.48	2.41	2.34	2.29	2.24	2.20	2.17	2.13	2.11	2.08	2.06	2.04	2.02	1.94	1.89	1.82	1.78	1.69	1.65	1.63
48	5.35	3.99	3.40	3.07	2.84	2.69	2.56	2.47	2.39	2.33	2.27	2.23	2.19	2.15	2.12	2.09	2.07	2.04	2.02	2.01	1.93	1.88	1.81	1.77	1.67	1.64	1.62
50	5.34	3.97	3.39	3.05	2.83	2.67	2.55	2.46	2.38	2.32	2.26	2.22	2.18	2.14	2.11	2.08	2.06	2.03	2.01	1.99	1.92	1.87	1.80	1.75	1.66	1.62	1.60
60	5.29	3.93	3.34	3.01	2.79	2.63	2.51	2.41	2.33	2.27	2.22	2.17	2.13	2.09	2.06	2.03	2.01	1.98	1.96	1.94	1.87	1.82	1.74	1.70	1.60	1.56	1.54
70	5.25	3.89	3.31	2.97	2.75	2.59	2.47	2.38	2.30	2.24	2.18	2.14	2.10	2.06	2.03	2.00	1.97	1.95	1.93	1.91	1.83	1.78	1.71	1.66	1.56	1.52	1.50
80	5.22	3.86	3.28	2.95	2.73	2.57	2.45	2.35	2.28	2.21	2.16	2.11	2.07	2.03	2.00	1.97	1.95	1.92	1.90	1.88	1.81	1.75	1.68	1.63	1.53	1.49	1.47
90	5.20	3.84	3.26	2.93	2.71	2.55	2.43	2.34	2.26	2.19	2.14	2.09	2.05	2.02	1.98	1.95	1.93	1.91	1.88	1.86	1.79	1.73	1.66	1.61	1.50	1.46	1.44
100	5.18	3.83	3.25	2.92	2.70	2.54	2.42	2.32	2.24	2.18	2.12	2.08	2.04	2.00	1.97	1.94	1.91	1.89	1.87	1.85	1.77	1.71	1.64	1.59	1.48	1.44	1.42
125	5.15	3.80	3.22	2.89	2.67	2.51	2.39	2.30	2.22	2.15	2.10	2.05	2.01	1.97	1.94	1.91	1.89	1.86	1.84	1.82	1.74	1.68	1.61	1.56	1.45	1.40	1.38
150	5.13	3.78	3.20	2.87	2.65	2.49	2.37	2.28	2.20	2.13	2.08	2.03	1.99	1.95	1.92	1.89	1.87	1.84	1.82	1.80	1.72	1.67	1.59	1.54	1.42	1.38	1.35
200	5.10	3.76	3.18	2.85	2.63	2.47	2.35	2.26	2.18	2.11	2.06	2.01	1.97	1.93	1.90	1.87	1.84	1.82	1.80	1.78	1.70	1.64	1.56	1.51	1.39	1.35	1.32
300	5.07	3.73	3.16	2.83	2.61	2.45	2.33	2.23	2.16	2.09	2.04	1.99	1.95	1.91	1.88	1.85	1.82	1.80	1.77	1.75	1.67	1.62	1.54	1.48	1.36	1.31	1.28
500	5.05	3.72	3.14	2.81	2.59	2.43	2.31	2.22	2.14	2.07	2.02	1.97	1.93	1.89	1.86	1.83	1.80	1.78	1.76	1.74	1.65	1.60	1.52	1.46	1.34	1.28	1.25
1000	5.04	3.70	3.13	2.80	2.58	2.42	2.30	2.20	2.13	2.06	2.01	1.96	1.92	1.88	1.85	1.82	1.79	1.77	1.74	1.72	1.64	1.58	1.50	1.45	1.32	1.26	1.23

DEGREES OF FREEDOM FOR DENOMINATOR

TABLE A.4 Percentiles of the *F* Distribution (continued)

Upper 1% point of the F distribution

DEGREES OF FREEDOM FOR NUMERATOR

	1	2	3	4	5	6	7	8	9	10	11	12	13	14	15	16	17	18	19	20	25	30	40	50	100	150	200
1	4052	5000	5403	5625	5764	5859	5928	5981	6022	6056	6083	6106	6126	6143	6157	6170	6181	6192	6201	6209	6240	6261	6287	6303	6334	6345	6350
2	98.5	99.0	99.2	99.2	99.3	99.3	99.4	99.4	99.4	99.4	99.4	99.4	99.4	99.4	99.4	99.4	99.4	99.4	99.4	99.4	99.5	99.5	99.5	99.5	99.5	99.5	99.5
3	34.1	30.8	29.5	28.7	28.2	27.9	27.7	27.5	27.3	27.2	27.1	27.1	27.0	26.9	26.9	26.8	26.8	26.8	26.7	26.7	26.6	26.5	26.4	26.4	26.2	26.2	26.2
4	21.2	18.0	16.7	16.0	15.5	15.2	15.0	14.8	14.7	14.5	14.5	14.4	14.3	14.2	14.2	14.2	14.1	14.1	14.0	14.0	13.9	13.8	13.7	13.7	13.6	13.5	13.5
5	16.3	13.3	12.1	11.4	11.0	10.7	10.5	10.3	10.2	10.1	9.96	9.89	9.82	9.77	9.72	9.68	9.64	9.61	9.58	9.55	9.45	9.38	9.29	9.24	9.13	9.09	9.08
6	13.7	10.9	9.78	9.15	8.75	8.47	8.26	8.10	7.98	7.87	7.79	7.72	7.66	7.60	7.56	7.52	7.48	7.45	7.42	7.40	7.30	7.23	7.14	7.09	6.99	6.95	6.93
7	12.2	9.55	8.45	7.85	7.46	7.19	6.99	6.84	6.72	6.62	6.54	6.47	6.41	6.36	6.31	6.28	6.24	6.21	6.18	6.16	6.06	5.99	5.91	5.86	5.75	5.72	5.70
8	11.3	8.65	7.59	7.01	6.63	6.37	6.18	6.03	5.91	5.81	5.73	5.67	5.61	5.56	5.52	5.48	5.44	5.41	5.38	5.36	5.26	5.20	5.12	5.07	4.96	4.93	4.91
9	10.6	8.02	6.99	6.42	6.06	5.80	5.61	5.47	5.35	5.26	5.18	5.11	5.05	5.01	4.96	4.92	4.89	4.86	4.83	4.81	4.71	4.65	4.57	4.52	4.41	4.38	4.36
10	10.0	7.56	6.55	5.99	5.64	5.39	5.20	5.06	4.94	4.85	4.77	4.71	4.65	4.60	4.56	4.52	4.49	4.46	4.43	4.41	4.31	4.25	4.17	4.12	4.01	3.98	3.96
11	9.65	7.21	6.22	5.67	5.32	5.07	4.89	4.74	4.63	4.54	4.46	4.40	4.34	4.29	4.25	4.21	4.18	4.15	4.12	4.10	4.01	3.94	3.86	3.81	3.71	3.67	3.66
12	9.33	6.93	5.95	5.41	5.06	4.82	4.64	4.50	4.39	4.30	4.22	4.16	4.10	4.05	4.01	3.97	3.94	3.91	3.88	3.86	3.76	3.70	3.62	3.57	3.47	3.43	3.41
13	9.07	6.70	5.74	5.21	4.86	4.62	4.44	4.30	4.19	4.10	4.02	3.96	3.91	3.86	3.82	3.78	3.75	3.72	3.69	3.66	3.57	3.51	3.43	3.38	3.27	3.24	3.22
14	8.88	6.51	5.56	5.04	4.69	4.46	4.28	4.14	4.03	3.94	3.86	3.80	3.75	3.70	3.66	3.62	3.59	3.56	3.53	3.51	3.41	3.35	3.27	3.22	3.11	3.08	3.06
15	8.68	6.36	5.42	4.89	4.56	4.32	4.14	4.00	3.89	3.80	3.73	3.67	3.61	3.56	3.52	3.49	3.45	3.42	3.40	3.37	3.28	3.21	3.13	3.08	2.98	2.94	2.92
16	8.53	6.23	5.29	4.77	4.44	4.20	4.03	3.89	3.78	3.69	3.62	3.55	3.50	3.45	3.41	3.37	3.34	3.31	3.28	3.26	3.16	3.10	3.02	2.97	2.86	2.83	2.81
17	8.40	6.11	5.19	4.67	4.34	4.10	3.93	3.79	3.68	3.59	3.52	3.46	3.40	3.35	3.31	3.27	3.24	3.21	3.19	3.16	3.07	3.00	2.92	2.87	2.76	2.73	2.71
18	8.29	6.01	5.09	4.58	4.25	4.01	3.84	3.71	3.60	3.51	3.43	3.37	3.32	3.27	3.23	3.19	3.16	3.13	3.10	3.08	2.98	2.92	2.84	2.78	2.68	2.64	2.62
19	8.18	5.93	5.01	4.50	4.17	3.94	3.77	3.63	3.52	3.43	3.36	3.30	3.24	3.19	3.15	3.12	3.08	3.05	3.03	3.00	2.91	2.84	2.76	2.71	2.60	2.57	2.55
20	8.10	5.85	4.94	4.43	4.10	3.87	3.70	3.56	3.46	3.37	3.29	3.23	3.18	3.13	3.09	3.05	3.02	2.99	2.96	2.94	2.84	2.78	2.69	2.64	2.54	2.50	2.48
21	8.02	5.78	4.87	4.37	4.04	3.81	3.64	3.51	3.40	3.31	3.24	3.17	3.12	3.07	3.03	2.99	2.96	2.93	2.90	2.88	2.79	2.72	2.64	2.58	2.48	2.44	2.42
22	7.95	5.72	4.82	4.31	3.99	3.76	3.59	3.45	3.35	3.26	3.18	3.12	3.07	3.02	2.98	2.94	2.91	2.88	2.85	2.83	2.73	2.67	2.58	2.53	2.42	2.38	2.36
23	7.88	5.66	4.76	4.26	3.94	3.71	3.54	3.41	3.30	3.21	3.14	3.07	3.02	2.97	2.93	2.89	2.86	2.83	2.80	2.78	2.69	2.62	2.54	2.48	2.37	2.34	2.32
24	7.82	5.61	4.72	4.22	3.90	3.67	3.50	3.36	3.26	3.17	3.09	3.03	2.98	2.93	2.89	2.85	2.82	2.79	2.76	2.74	2.64	2.58	2.49	2.44	2.33	2.29	2.27
25	7.77	5.57	4.68	4.18	3.85	3.63	3.46	3.32	3.22	3.13	3.06	2.99	2.94	2.89	2.85	2.81	2.78	2.75	2.72	2.70	2.60	2.54	2.45	2.40	2.29	2.25	2.23
26	7.72	5.53	4.64	4.14	3.82	3.59	3.42	3.29	3.18	3.09	3.02	2.96	2.90	2.86	2.81	2.78	2.75	2.72	2.69	2.66	2.57	2.50	2.42	2.36	2.25	2.21	2.19
27	7.68	5.49	4.60	4.11	3.78	3.56	3.39	3.26	3.15	3.06	2.99	2.93	2.87	2.82	2.78	2.75	2.71	2.68	2.66	2.63	2.54	2.47	2.38	2.33	2.22	2.18	2.16
28	7.64	5.45	4.57	4.07	3.75	3.53	3.36	3.23	3.12	3.03	2.96	2.90	2.84	2.79	2.75	2.72	2.68	2.65	2.63	2.60	2.51	2.44	2.35	2.30	2.19	2.15	2.13
29	7.60	5.42	4.54	4.04	3.73	3.50	3.33	3.20	3.09	3.00	2.93	2.87	2.81	2.77	2.73	2.69	2.66	2.63	2.60	2.57	2.48	2.41	2.33	2.27	2.16	2.12	2.10
30	7.56	5.39	4.51	4.02	3.70	3.47	3.30	3.17	3.07	2.98	2.91	2.84	2.79	2.74	2.70	2.66	2.63	2.60	2.57	2.55	2.45	2.39	2.30	2.25	2.13	2.09	2.07
32	7.50	5.34	4.46	3.97	3.65	3.43	3.26	3.13	3.02	2.93	2.86	2.80	2.74	2.70	2.65	2.62	2.58	2.55	2.53	2.50	2.41	2.34	2.25	2.20	2.08	2.04	2.02
34	7.44	5.29	4.42	3.93	3.61	3.39	3.22	3.09	2.98	2.89	2.82	2.76	2.70	2.66	2.61	2.58	2.54	2.51	2.49	2.46	2.37	2.30	2.21	2.16	2.04	2.00	1.98
36	7.40	5.25	4.38	3.89	3.57	3.35	3.18	3.05	2.95	2.86	2.79	2.72	2.67	2.62	2.58	2.54	2.51	2.48	2.45	2.43	2.33	2.26	2.18	2.12	2.00	1.96	1.94
38	7.35	5.21	4.34	3.86	3.54	3.32	3.15	3.02	2.92	2.83	2.75	2.69	2.64	2.59	2.55	2.51	2.48	2.45	2.42	2.40	2.30	2.23	2.14	2.09	1.97	1.93	1.90
40	7.31	5.18	4.31	3.83	3.51	3.29	3.12	2.99	2.89	2.80	2.73	2.66	2.61	2.56	2.52	2.48	2.45	2.42	2.39	2.37	2.27	2.20	2.11	2.06	1.94	1.90	1.87
42	7.28	5.15	4.29	3.80	3.49	3.27	3.10	2.97	2.86	2.78	2.70	2.64	2.59	2.54	2.50	2.46	2.43	2.40	2.37	2.34	2.25	2.18	2.09	2.03	1.91	1.87	1.85
44	7.25	5.12	4.26	3.78	3.47	3.24	3.08	2.95	2.84	2.75	2.68	2.62	2.56	2.52	2.47	2.44	2.40	2.37	2.35	2.32	2.22	2.15	2.07	2.01	1.89	1.84	1.82
46	7.22	5.10	4.24	3.76	3.44	3.22	3.06	2.93	2.82	2.73	2.66	2.60	2.54	2.50	2.45	2.42	2.38	2.35	2.33	2.30	2.20	2.13	2.04	1.99	1.86	1.82	1.80
48	7.19	5.08	4.22	3.74	3.43	3.20	3.04	2.91	2.80	2.71	2.64	2.58	2.53	2.48	2.44	2.40	2.37	2.33	2.31	2.28	2.18	2.12	2.02	1.97	1.84	1.80	1.78
50	7.17	5.06	4.20	3.72	3.41	3.19	3.02	2.89	2.79	2.70	2.63	2.56	2.51	2.46	2.42	2.38	2.35	2.32	2.29	2.27	2.17	2.10	2.01	1.95	1.82	1.78	1.76
60	7.08	4.98	4.13	3.65	3.34	3.12	2.95	2.82	2.72	2.63	2.56	2.50	2.44	2.39	2.35	2.31	2.28	2.25	2.22	2.20	2.10	2.03	1.94	1.88	1.75	1.70	1.68
70	7.01	4.92	4.07	3.60	3.29	3.07	2.91	2.78	2.67	2.59	2.51	2.45	2.40	2.35	2.31	2.27	2.23	2.20	2.18	2.15	2.05	1.98	1.89	1.83	1.70	1.65	1.62
80	6.96	4.88	4.04	3.56	3.26	3.04	2.87	2.74	2.64	2.55	2.48	2.42	2.36	2.31	2.27	2.23	2.20	2.17	2.14	2.12	2.01	1.94	1.85	1.79	1.65	1.61	1.58
90	6.93	4.85	4.01	3.53	3.23	3.01	2.84	2.72	2.61	2.52	2.45	2.39	2.33	2.29	2.24	2.21	2.17	2.14	2.11	2.09	1.99	1.92	1.82	1.76	1.62	1.57	1.55
100	6.90	4.82	3.98	3.51	3.21	2.99	2.82	2.69	2.59	2.50	2.43	2.37	2.31	2.27	2.22	2.19	2.15	2.12	2.09	2.07	1.97	1.89	1.80	1.74	1.60	1.55	1.52
125	6.84	4.78	3.94	3.47	3.17	2.95	2.79	2.66	2.55	2.47	2.39	2.33	2.28	2.23	2.19	2.15	2.11	2.08	2.05	2.03	1.93	1.85	1.76	1.69	1.55	1.50	1.47
150	6.81	4.75	3.91	3.45	3.14	2.92	2.76	2.63	2.53	2.44	2.37	2.31	2.25	2.20	2.16	2.12	2.09	2.06	2.03	2.00	1.90	1.83	1.73	1.66	1.52	1.46	1.43
200	6.76	4.71	3.88	3.41	3.11	2.89	2.73	2.60	2.50	2.41	2.34	2.27	2.22	2.17	2.13	2.09	2.06	2.03	2.00	1.97	1.87	1.79	1.69	1.63	1.48	1.42	1.39
300	6.72	4.68	3.85	3.38	3.08	2.86	2.70	2.57	2.47	2.38	2.31	2.24	2.19	2.14	2.10	2.06	2.03	1.99	1.97	1.94	1.84	1.76	1.66	1.59	1.44	1.38	1.35
500	6.69	4.65	3.82	3.36	3.05	2.84	2.68	2.55	2.44	2.36	2.28	2.22	2.17	2.12	2.07	2.04	2.00	1.97	1.94	1.92	1.81	1.74	1.63	1.57	1.41	1.34	1.31
1000	6.66	4.63	3.80	3.34	3.04	2.82	2.66	2.53	2.43	2.34	2.27	2.20	2.15	2.10	2.06	2.02	1.98	1.95	1.92	1.90	1.79	1.72	1.61	1.54	1.38	1.32	1.28

DEGREES OF FREEDOM FOR DENOMINATOR

TABLE A.4 Percentiles of the *F* Distribution (continued)

Upper 0.5% point of the F distribution

DEGREES OF FREEDOM FOR NUMERATOR

Denom. df	1	2	3	4	5	6	7	8	9	10	11	12	13	14	15	16	17	18	19	20	25	30	40	50	100	150	200
1	****	****	****	****	****	****	****	****	****	****	****	****	****	****	****	****	****	****	****	****	****	****	****	****	****	****	****
2	199	199	199	199	199	199	199	199	199	199	199	199	199	199	199	199	199	199	199	199	199	199	199	199	199	199	199
3	55.6	49.8	47.5	46.2	45.4	44.8	44.4	44.1	43.9	43.7	43.5	43.4	43.3	43.2	43.1	43.0	42.9	42.9	42.8	42.8	42.6	42.5	42.3	42.2	42.0	42.0	41.9
4	31.3	26.3	24.3	23.2	22.5	22.0	21.6	21.4	21.1	21.0	20.8	20.7	20.6	20.5	20.4	20.4	20.3	20.3	20.2	20.2	20.0	19.9	19.8	19.7	19.5	19.4	19.4
5	22.8	18.3	16.5	15.6	14.9	14.5	14.2	14.0	13.8	13.6	13.5	13.4	13.3	13.2	13.1	13.1	13.0	13.0	12.9	12.9	12.8	12.7	12.5	12.5	12.3	12.2	12.2
6	18.6	14.5	12.9	12.0	11.5	11.1	10.8	10.6	10.4	10.3	10.1	10.0	9.95	9.88	9.81	9.76	9.71	9.66	9.62	9.59	9.45	9.36	9.24	9.17	9.03	8.98	8.95
7	16.2	12.4	10.9	10.1	9.52	9.16	8.89	8.68	8.51	8.38	8.27	8.18	8.10	8.03	7.97	7.91	7.87	7.83	7.79	7.75	7.62	7.53	7.42	7.35	7.22	7.17	7.15
8	14.7	11.0	9.60	8.81	8.30	7.95	7.69	7.50	7.34	7.21	7.10	7.01	6.94	6.87	6.81	6.76	6.72	6.68	6.64	6.61	6.48	6.40	6.29	6.22	6.09	6.04	6.02
9	13.6	10.1	8.72	7.96	7.47	7.13	6.88	6.69	6.54	6.42	6.31	6.23	6.15	6.09	6.03	5.98	5.94	5.90	5.86	5.83	5.71	5.62	5.52	5.45	5.32	5.28	5.26
10	12.8	9.43	8.08	7.34	6.87	6.54	6.30	6.12	5.97	5.85	5.75	5.66	5.59	5.53	5.47	5.42	5.38	5.34	5.31	5.27	5.15	5.07	4.97	4.90	4.77	4.73	4.71
11	12.2	8.91	7.60	6.88	6.42	6.10	5.86	5.68	5.54	5.42	5.32	5.24	5.16	5.10	5.05	5.00	4.96	4.92	4.89	4.86	4.74	4.65	4.55	4.49	4.36	4.31	4.29
12	11.8	8.51	7.23	6.52	6.07	5.76	5.52	5.35	5.20	5.09	4.99	4.91	4.84	4.77	4.72	4.67	4.63	4.59	4.56	4.53	4.41	4.33	4.23	4.17	4.04	3.99	3.97
13	11.4	8.19	6.93	6.23	5.79	5.48	5.25	5.08	4.94	4.82	4.72	4.64	4.57	4.51	4.46	4.41	4.37	4.33	4.30	4.27	4.15	4.07	3.97	3.91	3.78	3.74	3.71
14	11.1	7.92	6.68	6.00	5.56	5.26	5.03	4.86	4.72	4.60	4.51	4.43	4.36	4.30	4.25	4.20	4.16	4.12	4.09	4.06	3.94	3.86	3.76	3.70	3.57	3.53	3.50
15	10.8	7.70	6.48	5.80	5.37	5.07	4.85	4.67	4.54	4.42	4.33	4.25	4.18	4.12	4.07	4.02	3.98	3.95	3.91	3.88	3.77	3.69	3.58	3.52	3.39	3.35	3.33
16	10.6	7.51	6.30	5.64	5.21	4.91	4.69	4.52	4.38	4.27	4.18	4.10	4.03	3.97	3.92	3.87	3.83	3.80	3.76	3.73	3.62	3.54	3.44	3.37	3.25	3.20	3.18
17	10.4	7.35	6.16	5.50	5.07	4.78	4.56	4.39	4.25	4.14	4.05	3.97	3.90	3.84	3.79	3.75	3.71	3.67	3.64	3.61	3.49	3.41	3.31	3.25	3.12	3.07	3.05
18	10.2	7.21	6.03	5.37	4.96	4.66	4.44	4.28	4.14	4.03	3.94	3.86	3.79	3.73	3.68	3.64	3.60	3.56	3.53	3.50	3.38	3.30	3.20	3.14	3.01	2.96	2.94
19	10.1	7.09	5.92	5.27	4.85	4.56	4.34	4.18	4.04	3.93	3.84	3.76	3.70	3.64	3.59	3.54	3.50	3.46	3.43	3.40	3.29	3.21	3.11	3.04	2.91	2.87	2.85
20	9.94	6.99	5.82	5.17	4.76	4.47	4.26	4.09	3.96	3.85	3.76	3.68	3.61	3.55	3.50	3.46	3.42	3.38	3.35	3.32	3.20	3.12	3.02	2.96	2.83	2.78	2.76
21	9.83	6.89	5.73	5.09	4.68	4.39	4.18	4.01	3.88	3.77	3.68	3.60	3.54	3.48	3.43	3.38	3.34	3.31	3.27	3.24	3.13	3.05	2.95	2.88	2.75	2.71	2.68
22	9.73	6.81	5.65	5.02	4.61	4.32	4.11	3.94	3.81	3.70	3.61	3.54	3.47	3.41	3.36	3.31	3.27	3.24	3.21	3.18	3.06	2.98	2.88	2.82	2.69	2.64	2.62
23	9.63	6.73	5.58	4.95	4.54	4.26	4.05	3.88	3.75	3.64	3.55	3.47	3.41	3.35	3.30	3.25	3.21	3.18	3.15	3.12	3.00	2.92	2.82	2.76	2.62	2.58	2.56
24	9.55	6.66	5.52	4.89	4.49	4.20	3.99	3.83	3.69	3.59	3.50	3.42	3.35	3.30	3.25	3.20	3.16	3.12	3.09	3.06	2.95	2.87	2.77	2.70	2.57	2.52	2.50
25	9.48	6.60	5.46	4.84	4.43	4.15	3.94	3.78	3.64	3.54	3.45	3.37	3.30	3.25	3.20	3.15	3.11	3.08	3.04	3.01	2.90	2.82	2.72	2.65	2.52	2.47	2.45
26	9.41	6.54	5.41	4.79	4.38	4.10	3.89	3.73	3.60	3.49	3.40	3.33	3.26	3.20	3.15	3.11	3.07	3.03	3.00	2.97	2.85	2.77	2.67	2.61	2.47	2.43	2.40
27	9.34	6.49	5.36	4.74	4.34	4.06	3.85	3.69	3.56	3.45	3.36	3.28	3.22	3.16	3.11	3.07	3.03	2.99	2.96	2.93	2.81	2.73	2.63	2.57	2.43	2.38	2.36
28	9.28	6.44	5.32	4.70	4.30	4.02	3.81	3.65	3.52	3.41	3.32	3.25	3.18	3.12	3.07	3.03	2.99	2.95	2.92	2.89	2.77	2.69	2.59	2.53	2.39	2.35	2.32
29	9.23	6.40	5.28	4.66	4.26	3.98	3.77	3.61	3.48	3.38	3.29	3.21	3.15	3.09	3.04	2.99	2.95	2.92	2.88	2.86	2.74	2.66	2.56	2.49	2.36	2.31	2.29
30	9.18	6.35	5.24	4.62	4.23	3.95	3.74	3.58	3.45	3.34	3.25	3.18	3.11	3.06	3.01	2.96	2.92	2.89	2.85	2.82	2.71	2.63	2.52	2.46	2.32	2.28	2.25
32	9.09	6.28	5.17	4.56	4.17	3.89	3.68	3.52	3.39	3.29	3.20	3.12	3.06	3.00	2.95	2.90	2.86	2.83	2.80	2.77	2.65	2.57	2.47	2.40	2.26	2.22	2.19
34	9.01	6.22	5.11	4.50	4.11	3.84	3.63	3.47	3.34	3.24	3.15	3.07	3.01	2.95	2.90	2.85	2.81	2.78	2.75	2.72	2.60	2.52	2.42	2.35	2.21	2.16	2.14
36	8.94	6.16	5.06	4.46	4.06	3.79	3.58	3.42	3.30	3.19	3.10	3.03	2.96	2.90	2.85	2.81	2.77	2.73	2.70	2.67	2.56	2.48	2.37	2.30	2.17	2.12	2.09
38	8.88	6.11	5.02	4.41	4.02	3.75	3.54	3.39	3.26	3.15	3.06	2.99	2.92	2.87	2.82	2.77	2.73	2.70	2.66	2.63	2.52	2.44	2.33	2.27	2.12	2.08	2.05
40	8.83	6.07	4.98	4.37	3.99	3.71	3.51	3.35	3.22	3.12	3.03	2.95	2.89	2.83	2.78	2.74	2.70	2.66	2.63	2.60	2.48	2.40	2.30	2.23	2.09	2.04	2.01
42	8.78	6.03	4.94	4.34	3.95	3.68	3.48	3.32	3.19	3.09	3.00	2.92	2.86	2.80	2.75	2.71	2.67	2.63	2.60	2.57	2.45	2.37	2.26	2.20	2.06	2.00	1.98
44	8.74	5.99	4.91	4.31	3.92	3.65	3.45	3.29	3.16	3.06	2.97	2.89	2.83	2.77	2.72	2.68	2.64	2.60	2.57	2.54	2.42	2.34	2.24	2.17	2.03	1.97	1.95
46	8.70	5.96	4.88	4.28	3.90	3.62	3.42	3.26	3.14	3.03	2.94	2.87	2.80	2.75	2.70	2.65	2.61	2.58	2.54	2.51	2.40	2.32	2.21	2.14	2.00	1.95	1.92
48	8.66	5.93	4.85	4.25	3.87	3.60	3.40	3.24	3.11	3.01	2.92	2.85	2.78	2.72	2.67	2.63	2.59	2.55	2.52	2.49	2.37	2.29	2.19	2.12	1.97	1.92	1.90
50	8.63	5.90	4.83	4.23	3.85	3.58	3.38	3.22	3.09	2.99	2.90	2.82	2.76	2.70	2.65	2.61	2.57	2.53	2.50	2.47	2.35	2.27	2.16	2.10	1.95	1.90	1.87
60	8.49	5.79	4.73	4.14	3.76	3.49	3.29	3.13	3.01	2.90	2.82	2.74	2.68	2.62	2.57	2.53	2.49	2.45	2.42	2.39	2.27	2.19	2.08	2.01	1.86	1.81	1.78
70	8.40	5.72	4.66	4.08	3.70	3.43	3.23	3.08	2.95	2.85	2.76	2.68	2.62	2.56	2.51	2.47	2.43	2.39	2.36	2.33	2.21	2.13	2.02	1.95	1.80	1.74	1.71
80	8.33	5.67	4.61	4.03	3.65	3.39	3.19	3.03	2.91	2.80	2.72	2.64	2.58	2.52	2.47	2.43	2.39	2.35	2.32	2.29	2.17	2.08	1.97	1.90	1.75	1.69	1.66
90	8.28	5.62	4.57	3.99	3.62	3.35	3.15	3.00	2.87	2.77	2.68	2.61	2.54	2.49	2.44	2.39	2.35	2.32	2.28	2.25	2.13	2.05	1.94	1.87	1.71	1.65	1.62
100	8.24	5.59	4.54	3.96	3.59	3.33	3.13	2.97	2.85	2.74	2.66	2.58	2.52	2.46	2.41	2.37	2.33	2.29	2.26	2.23	2.11	2.02	1.91	1.84	1.68	1.62	1.59
125	8.17	5.53	4.49	3.91	3.54	3.28	3.08	2.93	2.80	2.70	2.61	2.54	2.47	2.42	2.37	2.32	2.28	2.24	2.21	2.18	2.06	1.98	1.86	1.79	1.63	1.56	1.53
150	8.12	5.49	4.45	3.88	3.51	3.25	3.05	2.89	2.77	2.67	2.58	2.51	2.44	2.38	2.33	2.29	2.25	2.21	2.18	2.15	2.03	1.94	1.83	1.76	1.59	1.53	1.49
200	8.06	5.44	4.41	3.84	3.47	3.21	3.01	2.86	2.73	2.63	2.54	2.47	2.41	2.35	2.30	2.26	2.21	2.18	2.14	2.11	1.99	1.91	1.79	1.71	1.54	1.48	1.44
300	8.00	5.39	4.36	3.80	3.43	3.17	2.97	2.82	2.69	2.59	2.51	2.43	2.37	2.31	2.26	2.21	2.17	2.14	2.10	2.07	1.95	1.87	1.75	1.67	1.50	1.43	1.39
500	7.95	5.35	4.33	3.76	3.40	3.14	2.94	2.79	2.66	2.56	2.48	2.40	2.34	2.28	2.23	2.19	2.14	2.11	2.07	2.04	1.92	1.84	1.72	1.64	1.46	1.39	1.35
1000	7.91	5.33	4.30	3.74	3.37	3.11	2.92	2.77	2.64	2.54	2.45	2.38	2.32	2.26	2.21	2.16	2.12	2.09	2.05	2.02	1.90	1.81	1.69	1.61	1.43	1.36	1.31

DEGREES OF FREEDOM FOR DENOMINATOR

TABLE A.4 Percentiles of the F Distribution (continued)

Upper 0.1% point of the F distribution

DEGREES OF FREEDOM FOR NUMERATOR

den \ num	1	2	3	4	5	6	7	8	9	10	11	12	13	14	15	16	17	18	19	20	25	30	40	50	100	150	200
1	····	····	····	····	····	····	····	····	····	····	····	····	····	····	····	····	····	····	····	····	····	····	····	····	····	····	····
2	999	999	999	999	999	999	999	999	999	999	999	999	999	999	999	999	999	999	999	999	999	999	999	999	999	999	999
3	167	148	141	137	135	133	132	131	130	129	129	128	128	128	127	127	127	127	127	126	126	126	125	125	124	124	124
4	74.1	61.2	56.2	53.4	51.7	50.5	49.7	49.0	48.5	48.1	47.7	47.4	47.2	46.9	46.8	46.6	46.5	46.3	46.2	46.1	45.7	45.4	45.1	44.9	44.5	44.3	44.3
5	47.2	37.1	33.2	31.1	29.8	28.8	28.2	27.6	27.2	26.9	26.6	26.4	26.2	26.1	25.9	25.8	25.7	25.6	25.5	25.4	25.1	24.9	24.6	24.4	24.1	24.0	24.0
6	35.5	27.0	23.7	21.9	20.8	20.0	19.5	19.0	18.7	18.4	18.2	18.0	17.8	17.7	17.6	17.4	17.4	17.3	17.1	17.1	16.9	16.7	16.4	16.3	16.0	15.9	15.9
7	29.2	21.7	18.8	17.2	16.2	15.5	15.0	14.6	14.3	14.1	13.9	13.7	13.6	13.4	13.3	13.2	13.1	13.1	13.0	12.9	12.7	12.5	12.3	12.2	12.0	11.9	11.8
8	25.4	18.5	15.8	14.4	13.5	12.9	12.4	12.0	11.8	11.5	11.2	11.1	11.1	10.9	10.8	10.8	10.7	10.6	10.5	10.5	10.3	10.1	9.92	9.80	9.57	9.49	9.45
9	22.9	16.4	13.9	12.6	11.7	11.1	10.7	10.4	10.1	9.89	9.72	9.57	9.44	9.33	9.24	9.15	9.08	9.01	8.95	8.90	8.69	8.55	8.37	8.26	8.04	7.96	7.93
10	21.0	14.9	12.6	11.3	10.5	9.93	9.52	9.20	8.96	8.75	8.59	8.45	8.32	8.22	8.13	8.05	7.98	7.91	7.86	7.80	7.60	7.47	7.30	7.19	6.98	6.91	6.87
11	19.7	13.8	11.6	10.3	9.58	9.05	8.66	8.35	8.12	7.92	7.76	7.63	7.51	7.41	7.32	7.24	7.17	7.11	7.06	7.01	6.81	6.68	6.52	6.42	6.21	6.14	6.10
12	18.6	13.0	10.8	9.63	8.89	8.38	8.00	7.71	7.48	7.29	7.14	7.00	6.89	6.79	6.71	6.63	6.57	6.51	6.45	6.40	6.22	6.09	5.93	5.83	5.63	5.56	5.52
13	17.8	12.3	10.2	9.07	8.35	7.86	7.49	7.21	6.98	6.80	6.65	6.52	6.41	6.31	6.23	6.16	6.09	6.03	5.98	5.93	5.75	5.63	5.47	5.37	5.17	5.10	5.07
14	17.1	11.8	9.73	8.62	7.92	7.44	7.08	6.80	6.58	6.40	6.26	6.13	6.02	5.93	5.85	5.78	5.71	5.66	5.59	5.56	5.38	5.25	5.10	5.00	4.81	4.74	4.71
15	16.6	11.3	9.34	8.25	7.57	7.09	6.74	6.47	6.26	6.08	5.94	5.81	5.71	5.62	5.54	5.46	5.40	5.35	5.29	5.25	5.07	4.95	4.80	4.70	4.51	4.44	4.41
16	16.1	11.0	9.01	7.94	7.27	6.80	6.46	6.19	5.98	5.81	5.67	5.55	5.44	5.35	5.27	5.20	5.14	5.09	5.04	4.99	4.82	4.70	4.54	4.45	4.26	4.19	4.16
17	15.7	10.7	8.73	7.68	7.02	6.56	6.22	5.96	5.75	5.58	5.44	5.32	5.22	5.13	5.05	4.99	4.92	4.87	4.82	4.78	4.60	4.48	4.33	4.24	4.05	3.98	3.95
18	15.4	10.4	8.49	7.46	6.81	6.35	6.02	5.76	5.56	5.39	5.25	5.13	5.03	4.94	4.87	4.80	4.74	4.68	4.64	4.59	4.42	4.30	4.15	4.06	3.87	3.80	3.77
19	15.1	10.2	8.28	7.27	6.62	6.18	5.85	5.59	5.39	5.22	5.08	4.97	4.87	4.78	4.70	4.64	4.58	4.52	4.47	4.43	4.26	4.14	3.99	3.90	3.71	3.65	3.61
20	14.8	9.95	8.10	7.10	6.46	6.02	5.69	5.44	5.24	5.08	4.94	4.82	4.72	4.64	4.56	4.49	4.44	4.38	4.33	4.29	4.12	4.00	3.86	3.77	3.58	3.51	3.48
21	14.6	9.77	7.94	6.95	6.32	5.88	5.56	5.31	5.11	4.95	4.81	4.70	4.60	4.51	4.44	4.37	4.31	4.26	4.21	4.17	4.00	3.88	3.74	3.64	3.46	3.39	3.36
22	14.4	9.61	7.80	6.81	6.19	5.76	5.44	5.19	4.99	4.83	4.70	4.58	4.49	4.40	4.33	4.26	4.20	4.15	4.10	4.06	3.89	3.78	3.63	3.54	3.35	3.28	3.25
23	14.2	9.47	7.67	6.70	6.08	5.65	5.33	5.09	4.89	4.73	4.60	4.48	4.39	4.30	4.23	4.16	4.10	4.05	4.00	3.96	3.79	3.68	3.53	3.44	3.25	3.19	3.16
24	14.0	9.34	7.55	6.59	5.98	5.55	5.23	4.99	4.80	4.64	4.51	4.39	4.30	4.21	4.14	4.07	4.02	3.96	3.92	3.87	3.71	3.59	3.45	3.36	3.17	3.10	3.07
25	13.9	9.22	7.45	6.49	5.89	5.46	5.15	4.91	4.71	4.56	4.42	4.31	4.22	4.13	4.06	3.99	3.94	3.88	3.84	3.79	3.63	3.52	3.37	3.28	3.09	3.03	2.99
26	13.7	9.12	7.36	6.41	5.80	5.38	5.07	4.83	4.64	4.48	4.35	4.24	4.14	4.06	3.99	3.92	3.86	3.81	3.77	3.72	3.56	3.44	3.30	3.21	3.02	2.95	2.92
27	13.6	9.02	7.27	6.33	5.73	5.31	5.00	4.76	4.57	4.41	4.28	4.17	4.08	3.99	3.92	3.86	3.80	3.75	3.70	3.66	3.49	3.38	3.23	3.14	2.96	2.89	2.86
28	13.5	8.93	7.19	6.25	5.66	5.24	4.93	4.69	4.50	4.35	4.22	4.11	4.01	3.93	3.86	3.80	3.74	3.69	3.64	3.60	3.43	3.32	3.18	3.09	2.90	2.83	2.80
29	13.4	8.85	7.12	6.19	5.59	5.18	4.87	4.64	4.45	4.29	4.16	4.05	3.96	3.88	3.80	3.74	3.68	3.63	3.59	3.54	3.38	3.27	3.12	3.03	2.84	2.78	2.74
30	13.3	8.77	7.05	6.12	5.53	5.12	4.82	4.58	4.39	4.24	4.11	4.00	3.91	3.82	3.75	3.69	3.63	3.58	3.53	3.49	3.33	3.22	3.07	2.98	2.79	2.73	2.69
32	13.1	8.64	6.94	6.01	5.43	5.02	4.72	4.48	4.30	4.14	4.02	3.91	3.81	3.73	3.66	3.60	3.54	3.49	3.44	3.40	3.24	3.13	2.98	2.89	2.70	2.64	2.60
34	13.0	8.52	6.83	5.92	5.34	4.93	4.63	4.40	4.22	4.06	3.94	3.83	3.74	3.65	3.56	3.52	3.46	3.41	3.37	3.33	3.16	3.05	2.91	2.82	2.63	2.56	2.52
36	12.8	8.42	6.74	5.84	5.26	4.86	4.56	4.33	4.14	3.99	3.87	3.76	3.67	3.59	3.51	3.45	3.40	3.34	3.30	3.26	3.10	2.98	2.84	2.75	2.56	2.49	2.46
38	12.7	8.33	6.66	5.76	5.19	4.79	4.49	4.26	4.08	3.93	3.80	3.70	3.60	3.52	3.45	3.39	3.34	3.28	3.24	3.20	3.04	2.92	2.78	2.69	2.50	2.43	2.40
40	12.6	8.25	6.59	5.70	5.13	4.73	4.44	4.21	4.02	3.87	3.75	3.64	3.55	3.47	3.40	3.34	3.28	3.23	3.19	3.14	2.98	2.87	2.73	2.64	2.44	2.38	2.34
42	12.5	8.18	6.53	5.64	5.07	4.68	4.38	4.16	3.97	3.83	3.70	3.59	3.50	3.42	3.35	3.29	3.23	3.18	3.14	3.10	2.94	2.83	2.68	2.59	2.40	2.33	2.29
44	12.4	8.12	6.48	5.59	5.02	4.63	4.34	4.11	3.93	3.78	3.66	3.55	3.46	3.38	3.31	3.25	3.19	3.14	3.10	3.06	2.89	2.78	2.64	2.55	2.35	2.28	2.25
46	12.3	8.06	6.42	5.54	4.98	4.59	4.30	4.07	3.89	3.74	3.62	3.51	3.42	3.34	3.27	3.21	3.15	3.10	3.06	3.02	2.86	2.74	2.60	2.51	2.31	2.24	2.21
48	12.3	8.00	6.38	5.50	4.94	4.55	4.26	4.03	3.85	3.70	3.58	3.48	3.38	3.31	3.24	3.18	3.12	3.07	3.03	2.98	2.82	2.71	2.56	2.47	2.28	2.21	2.17
50	12.2	7.96	6.34	5.46	4.90	4.51	4.22	4.00	3.82	3.67	3.55	3.44	3.35	3.27	3.20	3.14	3.09	3.04	2.99	2.95	2.79	2.68	2.53	2.44	2.25	2.18	2.14
60	12.0	7.77	6.17	5.31	4.76	4.37	4.09	3.86	3.69	3.54	3.42	3.32	3.23	3.15	3.08	3.02	2.96	2.91	2.87	2.83	2.67	2.55	2.41	2.32	2.12	2.05	2.01
70	11.8	7.64	6.06	5.20	4.66	4.28	3.99	3.77	3.60	3.45	3.33	3.23	3.14	3.06	2.99	2.93	2.88	2.83	2.78	2.74	2.58	2.47	2.32	2.23	2.03	1.95	1.92
80	11.7	7.54	5.97	5.12	4.58	4.20	3.92	3.70	3.53	3.39	3.27	3.16	3.07	3.00	2.93	2.87	2.81	2.76	2.72	2.68	2.52	2.41	2.26	2.16	1.96	1.89	1.85
90	11.6	7.47	5.91	5.06	4.53	4.15	3.87	3.65	3.48	3.34	3.22	3.11	3.02	2.95	2.88	2.82	2.76	2.71	2.67	2.63	2.47	2.36	2.21	2.11	1.91	1.83	1.79
100	11.5	7.41	5.86	5.02	4.48	4.11	3.83	3.61	3.44	3.30	3.18	3.07	2.99	2.91	2.84	2.78	2.73	2.68	2.63	2.59	2.43	2.32	2.17	2.08	1.87	1.79	1.75
125	11.4	7.30	5.77	4.93	4.40	4.03	3.75	3.54	3.37	3.23	3.11	3.00	2.92	2.84	2.77	2.71	2.66	2.61	2.56	2.52	2.36	2.25	2.10	2.01	1.79	1.71	1.67
150	11.3	7.24	5.71	4.88	4.35	3.98	3.71	3.49	3.32	3.18	3.06	2.96	2.87	2.80	2.73	2.67	2.61	2.56	2.52	2.48	2.32	2.21	2.06	1.96	1.74	1.66	1.62
200	11.2	7.15	5.63	4.81	4.29	3.92	3.65	3.43	3.26	3.12	3.00	2.90	2.82	2.74	2.67	2.61	2.56	2.51	2.46	2.42	2.26	2.15	2.00	1.90	1.68	1.60	1.55
300	11.0	7.07	5.56	4.75	4.22	3.86	3.59	3.38	3.21	3.07	2.95	2.85	2.76	2.69	2.62	2.56	2.50	2.46	2.41	2.37	2.21	2.10	1.94	1.85	1.62	1.53	1.48
500	11.0	7.00	5.51	4.69	4.18	3.81	3.54	3.33	3.16	3.02	2.91	2.81	2.72	2.64	2.58	2.52	2.46	2.41	2.37	2.33	2.17	2.05	1.90	1.80	1.57	1.48	1.43
1000	10.9	6.96	5.46	4.65	4.14	3.78	3.51	3.30	3.13	2.99	2.87	2.77	2.69	2.61	2.54	2.48	2.43	2.38	2.34	2.30	2.14	2.02	1.87	1.77	1.53	1.44	1.38

DEGREES OF FREEDOM FOR DENOMINATOR

TABLE A.5 Values of $\frac{1}{2} \ln \frac{1+r}{1-r}$

r	0.000	0.001	0.002	0.003	0.004	0.005	0.006	0.007	0.008	0.009
0.000	0.0000	0.0010	0.0020	0.0030	0.0040	0.0050	0.0060	0.0070	0.0080	0.0090
0.010	0.0100	0.0110	0.0120	0.0130	0.0140	0.0150	0.0160	0.0170	0.0180	0.0190
0.020	0.0200	0.0210	0.0220	0.0230	0.0240	0.0250	0.0260	0.0270	0.0280	0.0290
0.030	0.0300	0.0310	0.0320	0.0330	0.0340	0.0350	0.0360	0.0370	0.0380	0.0390
0.040	0.0400	0.0410	0.0420	0.0430	0.0440	0.0450	0.0460	0.0470	0.0480	0.0490
0.050	0.0501	0.0511	0.0521	0.0531	0.0541	0.0551	0.0561	0.0571	0.0581	0.0591
0.060	0.0601	0.0611	0.0621	0.0631	0.0641	0.0651	0.0661	0.0671	0.0681	0.0691
0.070	0.0701	0.0711	0.0721	0.0731	0.0741	0.0751	0.0761	0.0771	0.0782	0.0792
0.080	0.0802	0.0812	0.0822	0.0832	0.0842	0.0852	0.0862	0.0872	0.0882	0.0892
0.090	0.0902	0.0912	0.0922	0.0933	0.0943	0.0953	0.0963	0.0973	0.0983	0.0993
0.100	0.1003	0.1013	0.1024	0.1034	0.1044	0.1054	0.1064	0.1074	0.1084	0.1094
0.110	0.1105	0.1115	0.1125	0.1135	0.1145	0.1155	0.1165	0.1175	0.1185	0.1195
0.120	0.1206	0.1216	0.1226	0.1236	0.1246	0.1257	0.1267	0.1277	0.1287	0.1297
0.130	0.1308	0.1318	0.1328	0.1338	0.1348	0.1358	0.1368	0.1379	0.1389	0.1399
0.140	0.1409	0.1419	0.1430	0.1440	0.1450	0.1460	0.1470	0.1481	0.1491	0.1501
0.150	0.1511	0.1522	0.1532	0.1542	0.1552	0.1563	0.1573	0.1583	0.1593	0.1604
0.160	0.1614	0.1624	0.1634	0.1644	0.1655	0.1665	0.1676	0.1686	0.1696	0.1706
0.170	0.1717	0.1727	0.1737	0.1748	0.1758	0.1768	0.1779	0.1789	0.1799	0.1810
0.180	0.1820	0.1830	0.1841	0.1851	0.1861	0.1872	0.1882	0.1892	0.1903	0.1913
0.190	0.1923	0.1934	0.1944	0.1954	0.1965	0.1975	0.1986	0.1996	0.2007	0.2017
0.200	0.2027	0.2038	0.2048	0.2059	0.2069	0.2079	0.2090	0.2100	0.2111	0.2121
0.210	0.2132	0.2142	0.2153	0.2163	0.2174	0.2184	0.2194	0.2205	0.2215	0.2226
0.220	0.2237	0.2247	0.2258	0.2268	0.2279	0.2289	0.2300	0.2310	0.2321	0.2331
0.230	0.2342	0.2353	0.2363	0.2374	0.2384	0.2395	0.2405	0.2416	0.2427	0.2437
0.240	0.2448	0.2458	0.2469	0.2480	0.2490	0.2501	0.2511	0.2522	0.2533	0.2543
0.250	0.2554	0.2565	0.2575	0.2586	0.2597	0.2608	0.2618	0.2629	0.2640	0.2650
0.260	0.2661	0.2672	0.2682	0.2693	0.2704	0.2715	0.2726	0.2736	0.2747	0.2758
0.270	0.2769	0.2779	0.2790	0.2801	0.2812	0.2823	0.2833	0.2844	0.2855	0.2866
0.280	0.2877	0.2888	0.2898	0.2909	0.2920	0.2931	0.2942	0.2953	0.2964	0.2975
0.290	0.2986	0.2997	0.3008	0.3019	0.3029	0.3040	0.3051	0.3062	0.3073	0.3084
0.300	0.3095	0.3106	0.3117	0.3128	0.3139	0.3150	0.3161	0.3172	0.3183	0.3195
0.310	0.3206	0.3217	0.3228	0.3239	0.3250	0.3261	0.3272	0.3283	0.3294	0.3305
0.320	0.3317	0.3328	0.3339	0.3350	0.3361	0.3372	0.3384	0.3395	0.3406	0.3417
0.330	0.3428	0.3439	0.3451	0.3462	0.3473	0.3484	0.3496	0.3507	0.3518	0.3530
0.340	0.3541	0.3552	0.3564	0.3575	0.3586	0.3597	0.3609	0.3620	0.3632	0.3643
0.350	0.3654	0.3666	0.3677	0.3689	0.3700	0.3712	0.3723	0.3734	0.3746	0.3757
0.360	0.3769	0.3780	0.3792	0.3803	0.3815	0.3826	0.3838	0.3850	0.3861	0.3873
0.370	0.3884	0.3896	0.3907	0.3919	0.3931	0.3942	0.3954	0.3966	0.3977	0.3989
0.380	0.4001	0.4012	0.4024	0.4036	0.4047	0.4059	0.4071	0.4083	0.4094	0.4106
0.390	0.4118	0.4130	0.4142	0.4153	0.4165	0.4177	0.4189	0.4201	0.4213	0.4225
0.400	0.4236	0.4248	0.4260	0.4272	0.4284	0.4296	0.4308	0.4320	0.4332	0.4344
0.410	0.4356	0.4368	0.4380	0.4392	0.4404	0.4416	0.4429	0.4441	0.4453	0.4465
0.420	0.4477	0.4489	0.4501	0.4513	0.4526	0.4538	0.4550	0.4562	0.4574	0.4587
0.430	0.4599	0.4611	0.4623	0.4636	0.4648	0.4660	0.4673	0.4685	0.4697	0.4710
0.440	0.4722	0.4735	0.4747	0.4760	0.4772	0.4784	0.4797	0.4809	0.4822	0.4835
0.450	0.4847	0.4860	0.4872	0.4885	0.4897	0.4910	0.4923	0.4935	0.4948	0.4061
0.460	0.4973	0.4986	0.4999	0.5011	0.5024	0.5037	0.5049	0.5062	0.5075	0.5088
0.470	0.5101	0.5114	0.5126	0.5139	0.5152	0.5165	0.5178	0.5191	0.5204	0.5217
0.480	0.5230	0.5243	0.5256	0.5279	0.5282	0.5295	0.5308	0.5321	0.5334	0.5347
0.490	0.5361	0.5374	0.5387	0.5400	0.5413	0.5427	0.5440	0.5453	0.5466	0.5480

TABLE A.5 Values of $\frac{1}{2}\ln\dfrac{1+r}{1-r}$ (continued)

r	0.000	0.001	0.002	0.003	0.004	0.005	0.006	0.007	0.008	0.009
0.500	0.5493	0.5506	0.5520	0.5533	0.5547	0.5560	0.5573	0.5587	0.5600	0.5614
0.510	0.5627	0.5641	0.5654	0.5668	0.5681	0.5695	0.5709	0.5722	0.5736	0.5750
0.520	0.5763	0.5777	0.5791	0.5805	0.5818	0.5832	0.5846	0.5860	0.5874	0.5888
0.530	0.5901	0.5915	0.5929	0.5943	0.5957	0.5971	0.5985	0.5999	0.6013	0.6027
0.540	0.6042	0.6056	0.6070	0.6084	0.6098	0.6112	0.6127	0.6141	0.6155	0.6170
0.550	0.6184	0.6198	0.6213	0.6227	0.6241	0.6256	0.6270	0.6285	0.6299	0.6314
0.560	0.6328	0.6343	0.6358	0.6372	0.6387	0.6401	0.6416	0.6431	0.6446	0.6460
0.570	0.6475	0.6490	0.6505	0.6520	0.6535	0.6550	0.6565	0.6579	0.6594	0.6610
0.580	0.6625	0.6640	0.6655	0.6670	0.6685	0.6700	0.6715	0.6731	0.6746	0.6761
0.590	0.6777	0.6792	0.6807	0.6823	0.6838	0.6854	0.6869	0.6885	0.6900	0.6916
0.600	0.6931	0.6947	0.6963	0.6978	0.6994	0.7010	0.7026	0.7042	0.7057	0.7073
0.610	0.7089	0.7105	0.7121	0.7137	0.7153	0.7169	0.7185	0.7201	0.7218	0.7234
0.620	0.7250	0.7266	0.7283	0.7299	0.7315	0.7332	0.7348	0.7364	0.7381	0.7398
0.630	0.7414	0.7431	0.7447	0.7464	0.7481	0.7497	0.7514	0.7531	0.7548	0.7565
0.640	0.7582	0.7599	0.7616	0.7633	0.7650	0.7667	0.7684	0.7701	0.7718	0.7736
0.650	0.7753	0.7770	0.7788	0.7805	0.7823	0.7840	0.7858	0.7875	0.7893	0.7910
0.660	0.7928	0.7946	0.7964	0.7981	0.7999	0.8017	0.8035	0.8053	0.8071	0.8089
0.670	0.8107	0.8126	0.8144	0.8162	0.8180	0.8199	0.8217	0.8236	0.8254	0.8273
0.680	0.8291	0.8310	0.8328	0.8347	0.8366	0.8385	0.8404	0.8423	0.8442	0.8461
0.690	0.8480	0.8499	0.8518	0.8537	0.8556	0.8576	0.8595	0.8614	0.8634	0.8653
0.700	0.8673	0.8693	0.8712	0.8732	0.8752	0.8772	0.8792	0.8812	0.8832	0.8852
0.710	0.8872	0.8892	0.8912	0.8933	0.8953	0.8973	0.8994	0.9014	0.9035	0.9056
0.720	0.9076	0.9097	0.9118	0.9139	9.9160	0.9181	0.9202	0.9223	0.9245	0.9266
0.730	0.9287	0.9309	0.9330	0.9352	0.9373	0.9395	0.9417	0.9439	0.9461	0.9483
0.740	0.9505	0.9527	0.9549	0.9571	0.9594	0.9616	0.9639	0.9661	0.9684	0.9707
0.750	0.9730	0.9752	0.9775	0.9799	0.9822	0.9845	0.9868	0.9892	0.9915	0.9939
0.760	0.9962	0.9986	1.0010	1.0034	1.0058	1.0082	1.0106	1.0130	1.0154	1.0179
0.770	1.0203	1.0228	1.0253	1.0277	1.0302	1.0327	1.0352	1.0378	1.0403	1.0428
0.780	1.0454	1.0479	1.0505	1.0531	1.0557	1.0583	1.0609	1.0635	1.0661	1.0688
0.790	1.0714	1.0741	1.0768	1.0795	1.0822	1.0849	1.0876	1.0903	1.0931	1.0958
0.800	1.0986	1.1014	1.1041	1.1070	1.1098	1.1127	1.1155	1.1184	1.1212	1.1241
0.810	1.1270	1.1299	1.1329	1.1358	1.1388	1.1417	1.1447	1.1477	1.1507	1.1538
0.820	1.1568	1.1599	1.1630	1.1660	1.1692	1.1723	1.1754	1.1786	1.1817	1.1849
0.830	1.1870	1.1913	1.1946	1.1979	1.2011	1.2044	1.2077	1.2111	1.2144	1.2178
0.840	1.2212	1.2246	1.2280	1.2315	1.2349	1.2384	1.2419	1.2454	1.2490	1.2526
0.850	1.2561	1.2598	1.2634	1.2670	1.2708	1.2744	1.2782	1.2819	1.2857	1.2895
0.860	1.2934	1.2972	1.3011	1.3050	1.3089	1.3129	1.3168	1.3209	1.3249	1.3290
0.870	1.3331	1.3372	1.3414	1.3456	1.3498	1.3540	1.3583	1.3626	1.3670	1.3714
0.880	1.3758	1.3802	1.3847	1.3892	1.3938	1.3984	1.4030	1.4077	1.4124	1.4171
0.890	1.4219	1.4268	1.4316	1.4366	1.4415	1.4465	1.4516	1.4566	1.4618	1.4670
0.900	1.4722	1.4775	1.4828	1.4883	1.4937	1.4992	1.5047	1.5103	1.5160	1.5217
0.910	1.5275	1.5334	1.5393	1.5453	1.5513	1.5574	1.5636	1.5698	1.5762	1.5825
0.920	1.5890	1.5956	1.6022	1.6089	1.6157	1.6226	1.6296	1.6366	1.6438	1.6510
0.930	1.6584	1.6659	1.6734	1.6811	1.6888	1.6967	1.7047	1.7129	1.7211	1.7295
0.940	1.7380	1.7467	1.7555	1.7645	1.7736	1.7828	1.7923	1.8019	1.8117	1.8216
0.950	1.8318	1.8421	1.8527	1.8635	1.8745	1.8857	1.8972	1.9090	1.9210	1.9333
0.960	1.9459	1.9588	1.9721	1.9857	1.9996	2.0140	2.0287	2.0439	2.0595	2.0756
0.970	2.0923	2.1095	2.1273	2.1457	2.1649	2.1847	2.2054	2.2269	2.2494	2.2729
0.980	2.2976	2.3223	2.3507	2.3796	2.4101	2.4426	2.4774	2.5147	2.5550	2.5988
0.990	2.6467	2.6996	2.7587	2.8257	2.9031	2.9945	3.1063	3.2504	3.4534	3.8002

r	z
0.9999	4.95172
0.99999	6.10303

Source: Albert E. Waugh, *Statistical Tables and Problems,* McGraw-Hill Book Company, New York, 1952, Table A11, pp. 40–41, with the kind permission of the author and publisher.

Note: To obtain $\frac{1}{2}\ln(1+r)/(1-r)$ when r is negative, use the negative of the tabulated value corresponding to the absolute value of r; e.g., if $r = -0.242$, $\frac{1}{2}\ln(1 - 0.242)/(1 + 0.242) = -0.2469$.

833

TABLE A.6 **Upper α Point of Studentized Range,** $q_{k,v} = R/S$, k = sample size for range R, v = number of degrees of freedom for S

(entry = $q_{k,v,1-\alpha}$, where $P(q_{k,v} > q_{k,v,1-\alpha}) = \alpha$)

$\alpha = .05$

v \ k	2	3	4	5	6	7	8	9	10	11	12	13	14	15	16	17	18	19	20
1	18.0	27.0	32.8	37.1	40.4	43.1	45.4	47.4	49.1	50.6	52.0	53.2	54.3	55.4	56.3	57.2	58.0	58.8	59.6
2	6.08	8.33	9.80	10.9	11.7	12.4	13.0	13.5	14.0	14.4	14.7	15.1	15.4	15.7	15.9	16.1	16.4	16.6	16.8
3	4.50	5.91	6.82	7.50	8.04	8.48	8.85	9.18	9.46	9.72	9.95	10.2	10.3	10.5	10.7	10.8	11.0	11.1	11.2
4	3.93	5.04	5.76	6.29	6.71	7.05	7.35	7.60	7.83	8.03	8.21	8.37	8.52	8.66	8.79	8.91	9.03	9.13	9.23
5	3.64	4.60	5.22	5.67	6.03	6.33	6.58	6.80	6.99	7.17	7.32	7.47	7.60	7.72	7.83	7.93	8.03	8.12	8.21
6	3.46	4.34	4.90	5.30	5.63	5.90	6.12	6.32	6.49	6.65	6.79	6.92	7.03	7.14	7.24	7.34	7.43	7.51	7.59
7	3.34	4.16	4.68	5.06	5.36	5.61	5.82	6.00	6.16	6.30	6.43	6.55	6.66	6.76	6.85	6.94	7.02	7.10	7.17
8	3.26	4.04	4.53	4.89	5.17	5.40	5.60	5.77	5.92	6.05	6.18	6.29	6.39	6.48	6.57	6.65	6.73	6.80	6.87
9	3.20	3.95	4.41	4.76	5.02	5.24	5.43	5.59	5.74	5.87	5.98	6.09	6.19	6.28	6.36	6.44	6.51	6.58	6.64
10	3.15	3.88	4.33	4.65	4.91	5.12	5.30	5.46	5.60	5.72	5.83	5.93	6.03	6.11	6.19	6.27	6.34	6.40	6.47
11	3.11	3.82	4.26	4.57	4.82	5.03	5.20	5.35	5.49	5.61	5.71	5.81	5.90	5.98	6.06	6.13	6.20	6.27	6.33
12	3.08	3.77	4.20	4.51	4.75	4.95	5.12	5.27	5.39	5.51	5.61	5.71	5.80	5.88	5.95	6.02	6.09	6.15	6.21
13	3.06	3.73	4.15	4.45	4.69	4.88	5.05	5.19	5.32	5.43	5.53	5.63	5.71	5.79	5.86	5.93	5.99	6.05	6.11
14	3.03	3.70	4.11	4.41	4.64	4.83	4.99	5.13	5.25	5.36	5.46	5.55	5.64	5.71	5.79	5.85	5.91	5.97	6.03
15	3.01	3.67	4.08	4.37	4.59	4.78	4.94	5.08	5.20	5.31	5.40	5.49	5.57	5.65	5.72	5.78	5.85	5.90	5.96
16	3.00	3.65	4.05	4.33	4.56	4.74	4.90	5.03	5.15	5.26	5.35	5.44	5.52	5.59	5.66	5.73	5.79	5.84	5.90
17	2.98	3.63	4.02	4.30	4.52	4.70	4.86	4.99	5.11	5.21	5.31	5.39	5.47	5.54	5.61	5.67	5.73	5.79	5.84
18	2.97	3.61	4.00	4.28	4.49	4.67	4.82	4.96	5.07	5.17	5.27	5.35	5.43	5.50	5.57	5.63	5.69	5.74	5.79
19	2.96	3.59	3.98	4.25	4.47	4.65	4.79	4.92	5.04	5.14	5.23	5.31	5.39	5.46	5.53	5.59	5.65	5.70	5.75
20	2.95	3.58	3.96	4.23	4.45	4.62	4.77	4.90	5.01	5.11	5.20	5.28	5.36	5.43	5.49	5.55	5.61	5.66	5.71
24	2.92	3.53	3.90	4.17	4.37	4.54	4.68	4.81	4.92	5.01	5.10	5.18	5.26	5.32	5.38	5.44	5.49	5.55	5.59
30	2.89	3.49	3.85	4.10	4.30	4.46	4.60	4.72	4.82	4.92	5.00	5.08	5.15	5.21	5.27	5.33	5.38	5.43	5.47
40	2.86	3.44	3.79	4.04	4.23	4.39	4.52	4.63	4.73	4.82	4.90	4.98	5.04	5.11	5.16	5.22	5.27	5.31	5.36
60	2.83	3.40	3.74	3.98	4.16	4.31	4.44	4.55	4.65	4.73	4.81	4.88	4.94	5.00	5.06	5.11	5.15	5.20	5.24
120	2.80	3.36	3.68	3.92	4.10	4.24	4.36	4.47	4.56	4.64	4.71	4.78	4.84	4.90	4.95	5.00	5.04	5.09	5.13
∞	2.77	3.31	3.63	3.86	4.03	4.17	4.29	4.39	4.47	4.55	4.62	4.68	4.74	4.80	4.85	4.89	4.93	4.97	5.01

Source: From pp. 176–177 of *Biometrika Tables for Statisticians*, Vol. I, by E. S. Pearson and H. O. Hartley, published by the Biometrika Trustees, Cambridge University Press, Cambridge, 1954. Reproduced with the kind permission of the authors and the publisher. Corrections of ±1 in the last figure, supplied by James Pacheres, have been incorporated in 41 entries.

TABLE A.6 **Upper α Point of Studentized Range** (*continued*)

$\alpha = .01$

v \ k	2	3	4	5	6	7	8	9	10	11	12	13	14	15	16	17	18	19	20
1	90.0	135	164	186	202	216	227	237	246	253	260	266	272	277	282	286	290	294	298
2	14.0	19.0	22.3	24.7	26.6	28.2	29.5	30.7	31.7	32.6	33.4	34.1	34.8	35.4	36.0	36.5	37.0	37.5	37.9
3	8.26	10.6	12.2	13.3	14.2	15.0	15.6	16.2	16.7	17.1	17.5	17.9	18.2	18.5	18.8	19.1	19.3	19.5	19.8
4	6.51	8.12	9.17	9.96	10.6	11.1	11.5	11.9	12.3	12.6	12.8	13.1	13.3	13.5	13.7	13.9	14.1	14.2	14.4
5	5.70	6.97	7.80	8.42	8.91	9.32	9.67	9.97	10.2	10.5	10.7	10.9	11.1	11.2	11.4	11.6	11.7	11.8	11.9
6	5.24	6.33	7.03	7.56	7.97	8.32	8.61	8.87	9.10	9.30	9.49	9.65	9.81	9.95	10.1	10.2	10.3	10.4	10.5
7	4.95	5.92	6.54	7.01	7.37	7.68	7.94	8.17	8.37	8.55	8.71	8.86	9.00	9.12	9.24	9.35	9.46	9.55	9.65
8	4.74	5.63	6.20	6.63	6.96	7.24	7.47	7.68	7.87	8.03	8.18	8.31	8.44	8.55	8.66	8.76	8.85	8.94	9.03
9	4.60	5.43	5.96	6.35	6.66	6.91	7.13	7.32	7.49	7.65	7.78	7.91	8.03	8.13	8.23	8.32	8.41	8.49	8.57
10	4.48	5.27	5.77	6.14	6.43	6.67	6.87	7.05	7.21	7.36	7.48	7.60	7.71	7.81	7.91	7.99	8.07	8.15	8.22
11	4.39	5.14	5.62	5.97	6.25	6.48	6.67	6.84	6.99	7.13	7.25	7.36	7.46	7.56	7.65	7.73	7.81	7.88	7.95
12	4.32	5.04	5.50	5.84	6.10	6.32	6.51	6.67	6.81	6.94	7.06	7.17	7.26	7.36	7.44	7.52	7.59	7.66	7.73
13	4.26	4.96	5.40	5.73	5.98	6.19	6.37	6.53	6.67	6.79	6.90	7.01	7.10	7.19	7.27	7.34	7.42	7.48	7.55
14	4.21	4.89	5.32	5.63	5.88	6.08	6.26	6.41	6.54	6.66	6.77	6.87	6.96	7.05	7.12	7.20	7.27	7.33	7.39
15	4.17	4.83	5.25	5.56	5.80	5.99	6.16	6.31	6.44	6.55	6.66	6.76	6.84	6.93	7.00	7.07	7.14	7.20	7.26
16	4.13	4.78	5.19	5.49	5.72	5.92	6.08	6.22	6.35	6.46	6.56	6.66	6.74	6.82	6.90	6.97	7.03	7.09	7.15
17	4.10	4.74	5.14	5.43	5.66	5.85	6.01	6.15	6.27	6.38	6.48	6.57	6.66	6.73	6.80	6.87	6.94	7.00	7.05
18	4.07	4.70	5.09	5.38	5.60	5.79	5.94	6.08	6.20	6.31	6.41	6.50	6.58	6.65	6.72	6.79	6.85	6.91	6.96
19	4.05	4.67	5.05	5.33	5.55	5.73	5.89	6.02	6.14	6.25	6.34	6.43	6.51	6.58	6.65	6.72	6.78	6.84	6.89
20	4.02	4.64	5.02	5.29	5.51	5.69	5.84	5.97	6.09	6.19	6.29	6.37	6.45	6.52	6.59	6.65	6.71	6.76	6.82
24	3.96	4.54	4.91	5.17	5.37	5.54	5.69	5.81	5.92	6.02	6.11	6.19	6.26	6.33	6.39	6.45	6.51	6.56	6.61
30	3.89	4.45	4.80	5.05	5.24	5.40	5.54	5.65	5.76	5.85	5.93	6.01	6.08	6.14	6.20	6.26	6.31	6.36	6.41
40	3.82	4.37	4.70	4.93	5.11	5.27	5.39	5.50	5.60	5.69	5.77	5.84	5.90	5.96	6.02	6.07	6.12	6.17	6.21
60	3.76	4.28	4.60	4.82	4.99	5.13	5.25	5.36	5.45	5.53	5.60	5.67	5.73	5.79	5.84	5.89	5.93	5.98	6.02
120	3.70	4.20	4.50	4.71	4.87	5.01	5.12	5.21	5.30	5.38	5.44	5.51	5.56	5.61	5.66	5.71	5.75	5.79	5.83
∞	3.64	4.12	4.40	4.60	4.76	4.88	4.99	5.08	5.16	5.23	5.29	5.35	5.40	5.45	5.49	5.54	5.57	5.61	5.65

TABLE A.7 Orthogonal Polynomial Coefficients

k	POLYNOMIAL	1	2	3	4	5	6	7	8	9	10	(Σp_i^2)
3	Linear	−1	0	1								2
	Quadratic	1	−2	1								6
4	Linear	−3	−1	1	3							20
	Quadratic	1	−1	−1	1							4
	Cubic	−1	3	−3	1							20
5	Linear	−2	−1	0	1	2						10
	Quadratic	2	−1	−2	−1	2						14
	Cubic	−1	2	0	−2	1						10
	Quartic	1	−4	6	−4	1						70
6	Linear	−5	−3	−1	1	3	5					70
	Quadratic	5	−1	−4	−4	−1	5					84
	Cubic	−5	7	4	−4	−7	5					180
	Quartic	1	−3	2	2	−3	1					28
	Quintic	−1	5	−10	10	−5	1					252
7	Linear	−3	−2	−1	0	1	2	3				28
	Quadratic	5	0	−3	−4	−3	0	5				84
	Cubic	−1	1	1	0	−1	−1	1				6
	Quartic	3	−7	1	6	1	−7	3				154
	Quintic	−1	4	−5	0	5	−4	1				84
	Sextic	1	−6	15	−20	15	−6	1				924
8	Linear	−7	−5	−3	−1	1	3	5	7			168
	Quadratic	7	1	−3	−5	−5	−3	1	7			168
	Cubic	−7	5	7	3	−3	−7	−5	7			264
	Quartic	7	−13	−3	9	9	−3	−13	7			616
	Quintic	−7	23	−17	−15	15	17	−23	7			2184
	Sextic	1	−5	9	−5	−5	9	−5	1			264
	Septic	−1	7	−21	35	−35	21	−7	1			3,432
9	Linear	−4	−3	−2	−1	0	1	2	3	4		60
	Quadratic	28	7	−8	−17	−20	−17	−8	7	28		2,772
	Cubic	−14	7	13	9	0	−9	−13	−7	14		990
	Quartic	14	−21	−11	9	18	9	−11	−21	14		2,002
	Quintic	−4	11	−4	−9	0	9	4	−11	4		468
	Sextic	4	−17	22	1	−20	1	22	−17	4		1,980
	Septic	−1	6	−14	14	0	−14	14	−6	1		858
	Octic	1	−8	28	−56	70	−56	28	−8	1		12,870
10	Linear	−9	−7	−5	−3	−1	1	3	5	7	9	330
	Quadratic	6	2	−1	−3	−4	−4	−3	−1	2	6	132
	Cubic	−42	14	35	31	12	−12	−31	−35	−14	42	8,580
	Quartic	18	−22	−17	3	18	18	3	−17	−22	18	2,860
	Quintic	−6	14	−1	−11	−6	6	11	1	−14	6	780
	Sextic	3	−11	10	6	−8	−8	6	10	11	3	660
	Septic	−9	47	−86	92	56	−56	−42	86	−47	9	29,172
	Octic	1	−7	20	−28	14	14	−28	20	−7	1	2,860
	Novic	−1	9	−36	84	−126	126	−84	36	−9	1	48,620

TABLE A.8A Bonferroni Corrected Jackknife Residual Critical Values

$\alpha = 0.1$

k	n=5	10	15	20	25	50	100	200	400	800
1	6.96	3.50	3.27	3.22	3.21	3.27	3.39	3.54	3.70	3.86
2	31.82	3.71	3.33	3.25	3.23	3.28	3.39	3.54	3.70	3.86
3		4.03	3.41	3.29	3.25	3.28	3.40	3.54	3.70	3.86
4		4.60	3.51	3.33	3.27	3.29	3.40	3.54	3.70	3.86
5		5.84	3.63	3.37	3.30	3.29	3.40	3.54	3.70	3.86
6		9.92	3.81	3.43	3.33	3.30	3.40	3.54	3.70	3.86
7		63.66	4.06	3.50	3.36	3.30	3.40	3.54	3.70	3.86
8			4.46	3.58	3.39	3.31	3.40	3.54	3.70	3.86
9			5.17	3.69	3.44	3.31	3.40	3.54	3.70	3.86
10			6.74	3.83	3.49	3.32	3.40	3.54	3.70	3.86
15				7.45	3.99	3.36	3.41	3.54	3.70	3.86
20					8.05	3.41	3.42	3.55	3.70	3.86
40						4.50	3.47	3.55	3.70	3.86
80							3.92	3.58	3.70	3.86

$\alpha = 0.05$

k	5	10	15	20	25	50	100	200	400	800
1	9.92	4.03	3.65	3.54	3.50	3.51	3.60	3.73	3.87	4.02
2	63.66	4.32	3.73	3.58	3.53	3.51	3.60	3.73	3.87	4.02
3		4.77	3.83	3.62	3.55	3.52	3.60	3.73	3.87	4.02
4		5.60	3.95	3.67	3.58	3.53	3.61	3.73	3.87	4.02
5		7.45	4.12	3.73	3.61	3.53	3.61	3.73	3.87	4.02
6		14.09	4.36	3.81	3.65	3.54	3.61	3.73	3.87	4.02
7		127.32	4.70	3.89	3.69	3.55	3.61	3.73	3.87	4.02
8			5.25	4.00	3.73	3.56	3.61	3.73	3.88	4.02
9			6.25	4.15	3.79	3.57	3.61	3.73	3.88	4.02
10			8.58	4.33	3.85	3.61	3.61	3.73	3.88	4.02
15				9.46	4.50	3.67	3.62	3.73	3.88	4.02
20					10.21		3.63	3.74	3.88	4.02
40						5.04	3.69	3.75	3.88	4.03
80							4.23	3.78	3.88	4.03

$\alpha = 0.01$

k	5	10	15	20	25	50	100	200	400	800
1	22.33	5.41	4.55	4.29	4.17	4.03	4.06	4.15	4.27	4.40
2	318.31	5.96	4.68	4.35	4.20	4.04	4.06	4.15	4.27	4.40
3		6.87	4.85	4.42	4.24	4.05	4.06	4.15	4.27	4.40
4		8.61	5.08	4.50	4.28	4.06	4.06	4.15	4.27	4.40
5		12.92	5.37	4.60	4.33	4.07	4.07	4.15	4.27	4.40
6		31.60	5.80	4.72	4.39	4.07	4.07	4.15	4.27	4.40
7		636.62	6.43	4.86	4.45	4.08	4.07	4.15	4.27	4.40
8			7.50	5.05	4.53	4.09	4.07	4.15	4.27	4.40
9			9.57	5.29	4.62	4.10	4.07	4.15	4.27	4.40
10			14.82	5.62	4.72	4.12	4.08	4.15	4.27	4.40
15				16.33	5.81	4.18	4.09	4.15	4.27	4.40
20					17.60	4.28	4.10	4.16	4.27	4.40
40						6.44	4.18	4.17	4.27	4.40
80							4.97	4.21	4.28	4.40

TABLE A.8B Bonferroni Corrected Studentized Residual Critical Values

$\alpha = 0.1$

k	n=5	10	15	20	25	50	100	200	400	800
1	1.70	2.26	2.48	2.61	2.71	2.98	3.23	3.44	3.64	3.82
2	1.41	2.21	2.46	2.60	2.70	2.98	3.23	3.44	3.64	3.82
3		2.14	2.43	2.59	2.69	2.98	3.22	3.44	3.64	3.82
4		2.05	2.40	2.57	2.69	2.98	3.22	3.44	3.64	3.82
5		1.92	2.37	2.56	2.68	2.98	3.22	3.44	3.64	3.82
6		1.71	2.32	2.54	2.66	2.97	3.22	3.44	3.64	3.82
7		1.41	2.27	2.51	2.65	2.97	3.22	3.44	3.64	3.82
8			2.19	2.49	2.64	2.97	3.22	3.44	3.64	3.82
9			2.09	2.45	2.62	2.96	3.22	3.44	3.64	3.82
10			1.94	2.41	2.60	2.96	3.22	3.44	3.64	3.82
15				1.95	2.45	2.94	3.21	3.44	3.64	3.82
20					1.96	2.92	3.21	3.44	3.64	3.82
40						2.54	3.18	3.43	3.64	3.82
80							2.96	3.41	3.63	3.82

$\alpha = 0.05$

k	5	10	15	20	25	50	100	200	400	800
1	1.71	2.36	2.61	2.77	2.87	3.16	3.40	3.61	3.81	3.99
2	1.41	2.30	2.59	2.75	2.86	3.15	3.40	3.61	3.81	3.99
3		2.22	2.56	2.73	2.85	3.15	3.40	3.61	3.81	3.99
4		2.11	2.52	2.71	2.84	3.15	3.40	3.61	3.81	3.99
5		1.95	2.47	2.69	2.82	3.14	3.40	3.61	3.81	3.99
6		1.72	2.42	2.67	2.81	3.14	3.40	3.61	3.81	3.99
7		1.41	2.35	2.64	2.79	3.13	3.40	3.61	3.81	3.99
8			2.25	2.60	2.78	3.13	3.39	3.61	3.81	3.99
9			2.13	2.56	2.76	3.13	3.39	3.61	3.81	3.99
10			1.96	2.51	2.73	3.13	3.39	3.61	3.81	3.99
15				1.96	2.54	3.10	3.39	3.61	3.81	3.99
20					1.97	3.07	3.38	3.61	3.81	3.99
40						2.62	3.35	3.60	3.80	3.99
80							3.08	3.58	3.80	3.99

$\alpha = 0.01$

k	5	10	15	20	25	50	100	200	400	800
1	1.73	2.54	2.87	3.06	3.19	3.51	3.77	3.99	4.18	4.35
2	1.41	2.45	2.83	3.03	3.17	3.51	3.77	3.99	4.18	4.35
3		2.33	2.78	3.01	3.15	3.51	3.77	3.99	4.18	4.35
4		2.18	2.72	2.98	3.14	3.50	3.77	3.99	4.18	4.35
5		1.98	2.65	2.94	3.11	3.50	3.77	3.99	4.18	4.35
6		1.73	2.57	2.91	3.09	3.49	3.77	3.99	4.18	4.35
7		1.41	2.47	2.86	3.07	3.49	3.76	3.99	4.18	4.35
8			2.35	2.81	3.04	3.48	3.76	3.99	4.18	4.35
9			2.19	2.75	3.01	3.47	3.76	3.98	4.18	4.35
10			1.99	2.68	2.97	3.47	3.76	3.98	4.18	4.35
15				1.99	2.70	3.43	3.75	3.98	4.17	4.35
20					1.99	3.38	3.74	3.98	4.17	4.35
40						2.75	3.69	3.97	4.17	4.35
80							3.31	3.94	4.17	4.34

TABLE A.9 Critical Values for Leverages, n = sample size, k = number of predictors

$\alpha = .10$

n \ k	1	2	3	4	5	6	7	8	9	10	15	20	40	80
10	0.626	0.759	0.847	0.911	0.956	0.984	0.997	1.000						
15	0.481	0.595	0.679	0.748	0.806	0.855	0.897	0.932	0.959	0.980				
20	0.394	0.491	0.565	0.627	0.682	0.731	0.775	0.815	0.851	0.883	0.988			
25	0.335	0.419	0.484	0.540	0.589	0.635	0.676	0.715	0.751	0.784	0.918	0.992		
30	0.293	0.366	0.424	0.474	0.519	0.560	0.599	0.635	0.669	0.701	0.837	0.937		
40	0.236	0.295	0.342	0.383	0.420	0.455	0.487	0.518	0.547	0.576	0.701	0.806	0.888	
60	0.172	0.214	0.248	0.279	0.306	0.332	0.356	0.380	0.402	0.424	0.524	0.612	0.737	
80	0.137	0.170	0.197	0.221	0.242	0.263	0.283	0.301	0.319	0.337	0.418	0.491	0.625	
100	0.114	0.141	0.164	0.183	0.201	0.219	0.235	0.250	0.266	0.280	0.348	0.410	0.625	0.941
200	0.064	0.079	0.091	0.102	0.111	0.121	0.130	0.138	0.146	0.155	0.192	0.227	0.353	0.568
400	0.036	0.043	0.050	0.055	0.060	0.065	0.070	0.075	0.079	0.083	0.104	0.122	0.190	0.311
800	0.020	0.024	0.027	0.030	0.032	0.035	0.037	0.040	0.042	0.044	0.055	0.065	0.100	0.164

$\alpha = .05$

n \ k	1	2	3	4	5	6	7	8	9	10	15	20	40	80
10	0.683	0.802	0.879	0.933	0.969	0.990	0.999	1.000						
15	0.531	0.639	0.719	0.782	0.835	0.880	0.916	0.946	0.969	0.986				
20	0.436	0.531	0.602	0.662	0.714	0.761	0.802	0.839	0.872	0.901	0.991			
25	0.372	0.454	0.518	0.573	0.621	0.665	0.705	0.742	0.776	0.807	0.931	0.994		
30	0.325	0.398	0.455	0.505	0.549	0.589	0.627	0.662	0.695	0.726	0.855	0.947		
40	0.261	0.321	0.368	0.409	0.446	0.480	0.512	0.543	0.572	0.600	0.722	0.823	0.898	
60	0.190	0.233	0.268	0.298	0.326	0.352	0.376	0.400	0.422	0.444	0.543	0.630	0.751	
80	0.151	0.185	0.212	0.236	0.258	0.279	0.299	0.318	0.336	0.353	0.435	0.508	0.638	
100	0.126	0.154	0.176	0.196	0.215	0.232	0.248	0.264	0.279	0.294	0.363	0.425	0.638	0.946
200	0.070	0.085	0.098	0.108	0.119	0.128	0.137	0.146	0.154	0.162	0.201	0.236	0.362	0.570
400	0.039	0.047	0.053	0.059	0.064	0.069	0.074	0.079	0.083	0.088	0.108	0.127	0.196	0.317
800	0.021	0.025	0.029	0.032	0.034	0.037	0.039	0.042	0.044	0.046	0.057	0.067	0.103	0.168

$\alpha = .01$

n \ k	1	2	3	4	5	6	7	8	9	10	15	20	40	80
10	0.785	0.875	0.930	0.965	0.986	0.997	1.000	1.000						
15	0.629	0.724	0.792	0.844	0.887	0.921	0.948	0.969	0.984	0.994				
20	0.524	0.612	0.677	0.731	0.777	0.817	0.852	0.883	0.910	0.933	0.996			
25	0.450	0.529	0.589	0.640	0.685	0.724	0.761	0.794	0.824	0.851	0.953	0.997		
30	0.394	0.466	0.521	0.568	0.610	0.648	0.683	0.716	0.746	0.774	0.889	0.964		
40	0.318	0.377	0.424	0.464	0.501	0.534	0.565	0.595	0.622	0.649	0.763	0.855	0.917	
60	0.231	0.275	0.310	0.341	0.369	0.395	0.420	0.443	0.465	0.487	0.584	0.668	0.778	
80	0.183	0.218	0.246	0.271	0.293	0.314	0.334	0.353	0.372	0.389	0.471	0.543	0.666	
100	0.152	0.181	0.205	0.225	0.244	0.262	0.279	0.295	0.310	0.325	0.394	0.456	0.666	0.956
200	0.085	0.100	0.113	0.124	0.135	0.145	0.154	0.163	0.172	0.180	0.219	0.255	0.383	0.598
400	0.046	0.054	0.061	0.067	0.073	0.078	0.083	0.088	0.092	0.097	0.118	0.138	0.208	0.330
800	0.025	0.029	0.033	0.036	0.039	0.041	0.044	0.046	0.049	0.051	0.062	0.073	0.110	0.175

Notes for Table A.10

Cook and Weisberg (1982) suggested that any d_i value greater than 1 may deserve closer scrutiny. If the model is correct, d_i can be expected to be less than about 1.0. This simple approximation to a critical value for d_i keeps the regression coefficient estimates based on deleting the ith observation roughly within a 50% confidence region of the original estimates. The value of 1 is based on the fact that d_i is roughly like an F random variable with k and $n - k - 1$ degrees of freedom. The question being addressed is whether the set of regression coefficients differs with the ith subject deleted from what it is with all subjects included. The suggestion corresponds to comparing d_i values to the median (50th percentile) values of an F random variable with k and $n - k - 1$ degrees of freedom.

Muller and Chen Mok (1997) studied Cook's statistic, under the assumption that all predictors are Gaussian. They were motivated by their impression from various data analyses that comparing the statistic to the median of an F rarely highlights troublesome observations. They described computational forms for the exact distribution function, as well as a convenient F approximation, with accuracy increasing with sample size. They conducted computer simulations to verify the accuracy of their exact and approximate results, as well as the suggestion of Cook and Weisberg (1982). The simulation results were consistent with the impression that the Cook and Weisberg rule for evaluating Cook's statistic proves to be inaccurate. The approximation of Muller and Chen Mok worked well for evaluating a single value, but it performed inaccurately for evaluating the maximum value of a set, even with 200 observations. Hence, in Table A.10 we provide comparison values of $(n - k - 1)d_i$ for its maximum value based on n observations. The multiplication by the degrees of freedom simplifies the task of preserving numerical accuracy.

Table A.10 depends on assuming Gaussian predictors. Jensen and Ramirez (1996, 1997) discussed the contribution of Cook's statistic for fixed predictors. Muller and Chen Mok (1997) also briefly considered the issue. In considering fixed predictors, they recalled Obenchain's (1977) suggestion to ignore the statistic and concentrate on its two components, the residual and the leverage. Ultimately, the uncertainty surrounding the use of Cook's statistic reflects both the need for ongoing research in diagnostics and the impossibility of providing hard and fast rules for diagnostics.

References

Cook, R. D., and Weisberg, S. 1982. "Residuals and Influence in Regression." New York: Chapman & Hall.

Jensen, D. R., and Ramirez, D. E. 1996. "Computing the CDF of Cook's D(I) Statistic." In A. Prat and E. Ripoll, eds., *Proceedings of the 12th Symposium in Computational Statistics*, pp. 65–66. Barcelona, Spain: Institut d'Estadistica de Catalunya.

Muller, K. E., and Chen Mok, M. 1997. "The Distribution of Cook's Statistic." *Communications in Statistics: Theory and Methods* 26(3) pp. 525–546.

Obenchain, R. L. 1977. Letter to the Editor. *Technometrics* 19: 348–49.

TABLE A.10 Critical Values for the Maximum of n Values of Cook's $(n - k - 1)d_i$ (Bonferroni correction used), n observations and k predictors

$\alpha = 0.1$

k	n=5	10	15	20	25	50	100	200	400	800
1	14.96	11.13	11.84	12.68	13.46	16.39	19.97	23.94	28.70	33.80
2	40.53	12.21	12.09	12.63	13.22	15.65	18.64	22.09	25.96	30.12
3		13.30	12.09	12.35	12.79	14.84	17.48	20.52	23.86	27.50
4		15.21	12.18	12.14	12.45	14.23	16.62	19.36	22.30	25.97
5		19.33	12.44	12.03	12.21	13.76	15.95	18.49	21.39	24.51
6		31.06	12.94	12.01	12.04	13.39	15.43	17.81	20.36	23.51
7		96.01	13.79	12.08	11.94	13.10	15.02	17.27	19.75	22.42
8			15.26	12.26	11.90	12.85	14.70	16.83	19.20	21.73
9			18.00	12.55	11.91	12.66	14.40	16.52	18.62	21.45
10			23.93	13.02	11.97	12.50	14.16	16.16	18.43	20.55
15				27.66	13.60	12.01	13.39	15.16	17.00	19.34
20					30.94	11.83	12.92	14.53	16.31	18.35
40						15.95	12.26	13.56	15.10	16.83
80							13.49	13.05	14.39	15.85

$\alpha = 0.05$

k	n=5	10	15	20	25	50	100	200	400	800
1	24.97	15.24	15.55	16.37	17.18	20.41	24.31	28.83	33.88	40.15
2	82.06	16.56	15.63	16.01	16.56	19.08	22.33	26.05	30.20	33.96
3		18.16	15.50	15.49	15.85	17.93	20.72	24.14	27.57	32.06
4		21.28	15.59	15.14	15.33	17.06	19.63	22.49	25.83	29.31
5		28.40	15.94	14.95	14.96	16.41	18.70	21.39	24.42	28.24
6		50.22	16.70	14.91	14.70	15.91	17.97	20.54	23.48	26.68
7		192.90	17.99	15.00	14.55	15.50	17.49	20.00	22.35	25.67
8			20.32	15.25	14.48	15.19	17.05	19.31	22.06	24.44
9			24.78	15.69	14.49	14.92	16.69	18.85	21.34	24.29
10			34.72	16.38	14.58	14.70	16.38	18.42	20.49	23.33
15				39.98	16.94	14.03	15.36	17.16	19.39	21.75
20					44.63	13.79	14.81	16.52	18.46	20.32
40						19.50	13.92	15.22	16.83	18.76
80							15.55	14.58	15.99	17.52

$\alpha = 0.01$

p	n=5	10	15	20	25	50	100	200	400	800
1	77.29	28.72	26.88	27.24	27.92	31.46	36.10	41.22	49.42	68.39
2	415.27	30.97	26.13	25.65	25.81	28.12	32.61	37.34	44.99	57.70
3		35.12	25.66	24.22	24.33	26.17	29.15	34.23	37.55	52.58
4		44.09	25.82	23.58	23.20	24.56	27.31	31.26	35.28	40.60
5		66.83	26.66	23.20	22.49	23.39	25.84	29.44	34.14	36.91
6		150.47	28.48	23.12	22.00	22.55	24.35	28.42	31.04	36.91
7		964.09	31.80	23.34	21.71	21.79	24.19	26.87	31.04	33.55
8			37.84	23.93	21.59	21.26	23.28	25.83	29.31	33.55
9			50.10	24.93	21.64	20.76	22.23	25.62	28.21	30.50
10			80.67	26.54	21.83	20.37	22.11	24.53	28.21	30.50
15				92.09	27.02	19.16	20.22	22.40	25.64	27.73
20					102.32	18.82	19.18	21.32	23.31	25.21
40						29.95	18.04	19.32	21.17	22.91
80							20.67	18.57	20.12	22.90

Appendix—Matrices
and Their Relationship
to Regression Analysis

B.1 Preview

Statisticians have found matrix mathematics to be a very useful vehicle for compactly presenting the concepts, methods, and formulae of regression analysis and other multivariable methods. Moreover, matrix formulation of such topics has had the important practical implication of permitting extremely efficient and accurate use of the computer for carrying out multivariable analyses on large data sets.

  This appendix will summarize some of the more elementary but important notions and manipulations of matrix algebra and will use this tool to describe the general least-squares procedures of multiple regression. Admittedly, the material in this appendix is somewhat more mathematical than that in the main body of the text, but it is not absolutely necessary knowledge for the applied user of the multivariable methods we describe. Nevertheless, the reader who is comfortable with the mathematical level used here should find the matrix formulation of regression analysis a powerful and unifying supplement that may facilitate the learning of more advanced multivariable methods.

B.2 Definitions

A *matrix* may be simply defined as a rectangular array of numbers. For example,

$$A = \begin{bmatrix} 2 & 3 & 1 \\ 1 & 1 & 2 \end{bmatrix}, \quad B = \begin{bmatrix} 2 & 1 \\ 3 & 2 \end{bmatrix}, \quad C = \begin{bmatrix} 1 \\ 1 \\ 3 \end{bmatrix}$$

are all matrices.

The *dimensions* of a matrix are the number of rows and the number of columns that it has. For example, the matrix **A** above has two rows and three columns:

$$
\begin{array}{cccc}
 & \text{column} & \text{column} & \text{column} \\
 & 1 & 2 & 3 \\
 & \downarrow & \downarrow & \downarrow \\
\text{row 1} \rightarrow & \begin{bmatrix} 2 & 3 & 1 \\ \text{row 2} \rightarrow & 1 & 1 & 2 \end{bmatrix}
\end{array}
$$

It is customary to say that **A** is a 2 × 3 matrix or to write $\mathbf{A}_{2\times3}$. The dimensions of the matrices **B** and **C** above are 2 × 2 and 3 × 1, respectively. An example of a 1 × 4 matrix is any matrix with one row and four columns, such as $\mathbf{D}_{1\times4} = [-2 \quad 3 \quad 3 \quad 0]$. Matrices that contain only one row or only one column are often referred to as *vectors*; thus the matrices

$$
\mathbf{C} = \begin{bmatrix} 1 \\ 1 \\ 3 \end{bmatrix} \quad \text{and} \quad \mathbf{D} = [-2 \quad 3 \quad 3 \quad 0]
$$

are examples of a column vector and a row vector, respectively. Also, the matrix **B** is called a *square matrix* because it has the same number of rows and columns.

The numbers forming the rectangular array of a matrix are called the *elements* of the matrix. If we let a_{ij} denote the element in the ith row and jth column of the matrix **A** above,

$$a_{11} = 2, \qquad a_{12} = 3, \qquad a_{13} = 1$$
$$a_{21} = 1, \qquad a_{22} = 1, \qquad a_{23} = 2$$

It is often informative to write

$$\mathbf{A}_{2\times3} = ((a_{ij}))$$

indicating that the matrix **A** (represented by a capital letter) with two rows and three columns has typical element a_{ij} (represented by the corresponding lowercase letters). Thus, if you were given that

$$\mathbf{B}_{3\times2} = ((b_{ij}))$$

where $b_{21} = 3, b_{31} = 2, b_{11} = -2, b_{12} = 6, b_{22} = 0, b_{32} = 1$, you should construct **B** to be

$$
\mathbf{B} = \begin{bmatrix} -2 & 6 \\ 3 & 0 \\ 2 & 1 \end{bmatrix}
$$

B.3 Matrices in Regression Analysis

Given any set of multivariable data suitable for a regression analysis, a number of key matrices can be defined that directly correspond to the basic components of the regression model being postulated. For example, consider the following $n = 12$ pairs of observations on $Y = $ WGT and $X = $ HGT:

Variable	Child											
	1	2	3	4	5	6	7	8	9	10	11	12
Y (WGT)	64	71	53	67	55	58	77	57	56	51	76	68
X (HGT)	57	59	49	62	51	50	55	48	42	42	61	57

We can, in correspondence with the straight-line regression model $Y = \beta_0 + \beta_1 X + E$, define four matrices: $\mathbf{Y}$, the vector of observations on Y; $\mathbf{X}$, the matrix of independent variables; $\boldsymbol{\beta}$, the vector of parameters to be estimated; and $\mathbf{E}$, the vector of errors.

For the data given, these matrices are defined as follows:

$$\mathbf{Y}_{12\times1} = \begin{bmatrix} 64 \\ 71 \\ 53 \\ 67 \\ 55 \\ 58 \\ 77 \\ 57 \\ 56 \\ 51 \\ 76 \\ 68 \end{bmatrix}, \quad \mathbf{X}_{12\times2} = \begin{bmatrix} 1 & 57 \\ 1 & 59 \\ 1 & 49 \\ 1 & 62 \\ 1 & 51 \\ 1 & 50 \\ 1 & 55 \\ 1 & 48 \\ 1 & 42 \\ 1 & 42 \\ 1 & 61 \\ 1 & 57 \end{bmatrix}, \quad \boldsymbol{\beta}_{2\times1} = \begin{bmatrix} \beta_0 \\ \beta_1 \end{bmatrix}, \quad \mathbf{E}_{12\times1} = \begin{bmatrix} E_1 \\ E_2 \\ E_3 \\ E_4 \\ E_5 \\ E_6 \\ E_7 \\ E_8 \\ E_9 \\ E_{10} \\ E_{11} \\ E_{12} \end{bmatrix}$$

Notice that the first column of the $\mathbf{X}$ matrix of independent variables contains only 1's. This is the general convention used for any regression model containing a constant term (or intercept) β_0; motivation for adopting this convention follows by imagining the β_0 term to be of the form $\beta_0 X_0$, where X_0 is a dummy variable always taking the value 1. The vector of errors $\mathbf{E}$ contains random (and unobservable) error values, one for each pair of observations; these errors represent the differences between the observed Y-values and their (unknown) expected values under the given model.

B.4 Transpose of a Matrix

The transpose $\mathbf{A}'$ of a matrix $\mathbf{A}$ is defined to be that matrix whose (i, j)th element a'_{ij} is equal to the (j, i)th element of $\mathbf{A}$. For example, if

$$\mathbf{A} = \begin{bmatrix} 2 & 3 & 1 \\ 1 & 1 & 2 \end{bmatrix}$$

then

$$\mathbf{A}' = \begin{bmatrix} 2 & 1 \\ 3 & 1 \\ 1 & 2 \end{bmatrix}$$

since $a'_{11} = a_{11} = 2$, $a'_{12} = a_{21} = 1$, $a'_{21} = a_{12} = 3$, $a'_{22} = a_{22} = 1$, $a'_{31} = a_{13} = 1$, $a'_{32} = a_{23} = 2$. Another way of looking at this is that the first column of $\mathbf{A}$ becomes the first row of $\mathbf{A}'$, the second column of $\mathbf{A}$ becomes the second row of $\mathbf{A}'$, and so on. Thus, if $\mathbf{A}$ is $r \times c$, then $\mathbf{A}'$ is $c \times r$.

As examples, the transposes of

$$\mathbf{A}_{3\times 2} = \begin{bmatrix} 1 & 2 \\ 3 & 1 \\ 1 & 1 \end{bmatrix}, \quad \mathbf{B}_{3\times 3} = \begin{bmatrix} 1 & 0 & 2 \\ 0 & 4 & -5 \\ 2 & -5 & 3 \end{bmatrix}, \quad \mathbf{C}_{4\times 1} = \begin{bmatrix} 0 \\ 1 \\ 2 \\ -2 \end{bmatrix}$$

are

$$\mathbf{A}'_{2\times 3} = \begin{bmatrix} 1 & 3 & 1 \\ 2 & 1 & 1 \end{bmatrix}, \quad \mathbf{B}'_{3\times 3} = \begin{bmatrix} 1 & 0 & 2 \\ 0 & 4 & -5 \\ 2 & -5 & 3 \end{bmatrix}, \quad \mathbf{C}'_{1\times 4} = \begin{bmatrix} 0 & 1 & 2 & -2 \end{bmatrix}$$

Also, the transpose of the matrix $\mathbf{X}_{12\times 2}$ of the previous section is given by

$$\mathbf{X}'_{2\times 12} = \begin{bmatrix} 1 & 1 & 1 & 1 & 1 & 1 & 1 & 1 & 1 & 1 & 1 & 1 \\ 57 & 59 & 49 & 62 & 51 & 50 & 55 & 48 & 42 & 42 & 61 & 57 \end{bmatrix}$$

In the examples above, note that the matrix $\mathbf{B}$ is such that $\mathbf{B} = \mathbf{B}'$ (equality here means that corresponding elements are equal). A matrix satisfying this condition is said to be a *symmetric matrix*. Note that a symmetric matrix $\mathbf{A}$ must always be square, since otherwise $\mathbf{A}$ and $\mathbf{A}'$ would have different dimensions and so could not possibly be equal. A necessary and sufficient condition for the square matrix $\mathbf{A} = ((a_{ij}))$ to be symmetric is that $a_{ij} = a_{ji}$ for every $i \neq j$.

Correlation matrices such as

$$\mathbf{R}_1 = \begin{bmatrix} 1 & r_{XY} \\ r_{XY} & 1 \end{bmatrix} \quad \text{or} \quad \mathbf{R}_2 = \begin{bmatrix} 1 & r_{12} & r_{13} \\ r_{12} & 1 & r_{23} \\ r_{13} & r_{23} & 1 \end{bmatrix}$$

are always symmetric.

An important special case where the above condition for symmetry is satisfied is when $a_{ij} = a_{ji} = 0$ for every $i \neq j$. A square matrix having this property is said to be a *diagonal matrix*, the general form of which is given (for the 3×3 case) by

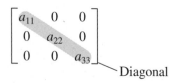

Diagonal

The most often used diagonal matrix is the *identity matrix* **I**, which has 1's on the diagonal; for example, the 3 × 3 identity matrix is

$$\mathbf{I} = \begin{bmatrix} 1 & 0 & 0 \\ 0 & 1 & 0 \\ 0 & 0 & 1 \end{bmatrix}$$

We will see shortly that an identity matrix serves the same algebraic function for matrix multiplication that the number 1 serves for ordinary scalar multiplication.

B.5 Matrix Addition

The sum of two matrices, say **A** and **B**, is obtained by adding together the corresponding elements of each matrix. Clearly, such addition can be performed only when the two matrices have the same dimensions. Thus, for example, we can add the matrices

$$\mathbf{A}_{2\times3} = \begin{bmatrix} 2 & 3 & 1 \\ 1 & 1 & 2 \end{bmatrix} \quad \text{and} \quad \mathbf{B}_{2\times3} = \begin{bmatrix} 4 & 1 & 5 \\ 1 & 3 & 1 \end{bmatrix}$$

(since they have the same dimensions) to get

$$\mathbf{A}_{2\times3} + \mathbf{B} = \begin{bmatrix} 2+4 & 3+1 & 1+5 \\ 1+1 & 1+3 & 2+1 \end{bmatrix} = \begin{bmatrix} 6 & 4 & 6 \\ 2 & 4 & 3 \end{bmatrix}$$

We could not, however, add the matrices **A** and **C**, where

$$\mathbf{C}_{3\times2} = \begin{bmatrix} 5 & 4 \\ 1 & 4 \\ 2 & 6 \end{bmatrix}$$

since the dimensions of these matrices are not the same.

An example of a more abstract use of matrix addition would be to sum the two 12 × 1 vectors

$$\mathbf{D}_{12\times1} = \begin{bmatrix} \beta_0 + 57\beta_1 \\ \beta_0 + 59\beta_1 \\ \vdots \\ \beta_0 + 57\beta_1 \end{bmatrix} \quad \text{and} \quad \mathbf{E}_{12\times1} = \begin{bmatrix} E_1 \\ E_2 \\ \vdots \\ E_{12} \end{bmatrix}$$

to obtain

$$\mathbf{D}_{12\times1} + \mathbf{E} = \begin{bmatrix} \beta_0 + 57\beta_1 + E_1 \\ \beta_0 + 59\beta_1 + E_2 \\ \vdots \\ \beta_0 + 57\beta_1 + E_{12} \end{bmatrix}$$

Actually, you may recognize that each element of the matrix $\mathbf{D} + \mathbf{E}$ is obtained by substituting for X in the right side of the straight-line regression equation

$$Y = \beta_0 + \beta_1 X + E$$

each of the 12 X (HGT) values given in the data set of Section B.3.

B.6 Matrix Multiplication

Multiplication of two matrices is somewhat more complicated than addition. The first rule to remember is that the product $\mathbf{AB}$ of two matrices $\mathbf{A}$ and $\mathbf{B}$ can exist if and only if the *number of columns* of $\mathbf{A}$ *is equal to the number of rows of* $\mathbf{B}$. Thus, if $\mathbf{A}$ is 2×3 and $\mathbf{B}$ is 3×4, the product $\mathbf{AB}$ exists since $\mathbf{A}$ has 3 columns and $\mathbf{B}$ has 3 rows. However, the product $\mathbf{BA}$ does not exist, since the number of columns of $\mathbf{B}$ (i.e., 4) is not equal to the number of rows of $\mathbf{A}$ (i.e., 2). Notationally, therefore, a matrix product can exist only if the dimensions of the matrices can be represented as follows:

Equal numbers

$$\mathbf{A}_{m\times n} \times \mathbf{B}_{n\times p} = \mathbf{AB}_{m\times p}$$

Product dimensions

Note also from the expression above that the dimensions $m \times p$ of the product matrix $\mathbf{AB}$ are given by the number of rows of the *pre*multiplier (i.e., $\mathbf{A}$) and the number of columns of the *post*multiplier (i.e., $\mathbf{B}$). For example, if $\mathbf{A}$ is 2×3 and $\mathbf{B}$ is 3×4, the dimensions of the product $\mathbf{AB}$ are 2×4.

Now, to carry out matrix multiplication, consider the two matrices

$$\mathbf{A}_{2\times3} = \begin{bmatrix} 2 & 1 & 0 \\ 0 & 3 & 1 \end{bmatrix} \quad \text{and} \quad \mathbf{B}_{3\times2} = \begin{bmatrix} 1 & -2 \\ 1 & 0 \\ 3 & 2 \end{bmatrix}$$

Since $\mathbf{A}$ is 2×3 and $\mathbf{B}$ is 3×2, the product $\mathbf{AB}$ will be a 2×2 matrix. If we let the elements of $\mathbf{AB}$ be denoted by

$$\mathbf{AB}_{2\times2} = ((c_{ij})) = \begin{bmatrix} c_{11} & c_{12} \\ c_{21} & c_{22} \end{bmatrix}$$

we can obtain the upper-left-hand corner element c_{11} by working with the first row of $\mathbf{A}$ and the first column of $\mathbf{B}$, as follows:

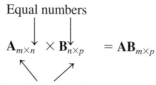

$$\begin{array}{ccc} \mathbf{A} & \mathbf{B} & \mathbf{AB} \\ \begin{bmatrix} 2 & 1 & 0 \\ 0 & 3 & 1 \end{bmatrix} & \begin{bmatrix} 1 & -2 \\ 1 & 0 \\ 3 & 2 \end{bmatrix} = & \begin{bmatrix} (2 \times 1) + (1 \times 1) \\ + (0 \times 3) = 3 & c_{12} \\ c_{21} & c_{22} \end{bmatrix} \end{array}$$

What we have done here is to calculate the product of each element in row 1 of **A** with the corresponding element in column 1 of **B**, and then add up these three products to obtain $c_{11} = 3$:

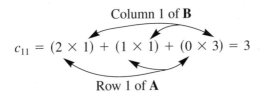

Column 1 of **B**

$$c_{11} = (2 \times 1) + (1 \times 1) + (0 \times 3) = 3$$

Row 1 of **A**

To find the element in the second row and first column of **AB** (i.e., c_{21}), we work with the second row of **A** and the first column of **B**, as follows:

$$
\begin{matrix}
\mathbf{A} & \mathbf{B} & & \mathbf{AB}
\end{matrix}
$$

$$
\begin{bmatrix} 2 & 1 & 0 \\ 0 & 3 & 1 \end{bmatrix}
\begin{bmatrix} 1 & -2 \\ 1 & 0 \\ 3 & 2 \end{bmatrix}
=
\begin{bmatrix} 3 & c_{12} \\ (0 \times 1) + (3 \times 1) & \\ + (1 \times 3) = 6 & c_{22} \end{bmatrix}
$$

Thus, for the element c_{21}, we find

Column 1 of **B**

$$c_{21} = (0 \times 1) + (3 \times 1) + (1 \times 3) = 6$$

Row 2 of **A**

Continuing this process, we find

$$c_{12} = (2 \times -2) + (1 \times 0) + (0 \times 2) = -4$$
$$c_{22} = (0 \times -2) + (3 \times 0) + (1 \times 2) = 2$$

Thus,

$$
\mathbf{AB}_{2\times2} =
\begin{bmatrix} 2 & 1 & 0 \\ 0 & 3 & 1 \end{bmatrix}
\begin{bmatrix} 1 & -2 \\ 1 & 0 \\ 3 & 2 \end{bmatrix}
=
\begin{bmatrix} 3 & -4 \\ 6 & 2 \end{bmatrix}
$$

In general, if $\mathbf{A}_{m\times n} = ((a_{ij}))$ and $\mathbf{B}_{n\times p} = ((b_{ij}))$, the (i, j)th element c_{ij} of the product

$$\mathbf{AB}_{m\times p} = ((c_{ij}))$$

is defined to be

$$c_{ij} = \sum_{l=1}^{n} a_{il}b_{lj} \qquad i = 1, 2, \ldots, m; \quad j = 1, 2, \ldots, p$$

Thus, as another example, if

$$
\mathbf{A}_{2\times2} =
\begin{bmatrix} -1 & 3 \\ 2 & 2 \end{bmatrix}
\quad \text{and} \quad
\mathbf{B}_{2\times1} =
\begin{bmatrix} 0 \\ 1 \end{bmatrix}
$$

then $m = 2, p = 1, n = 2$, and

$$c_{11} = \sum_{l=1}^{2} a_{1l}b_{l1} = (-1 \times 0) + (3 \times 1) = 3$$

$$c_{21} = \sum_{l=1}^{2} a_{2l}b_{l1} = (2 \times 0) + (2 \times 1) = 2$$

so that

$$\mathbf{AB}_{2\times2} = \begin{bmatrix} -1 & 3 \\ 2 & 2 \end{bmatrix} \begin{bmatrix} 0 \\ 1 \end{bmatrix} = \begin{bmatrix} 3 \\ 2 \end{bmatrix}$$

Here are a few other examples for additional practice:

1. Find **AI** and **IA**, where

$$\mathbf{A}_{2\times2} = \begin{bmatrix} -1 & 3 \\ 2 & 2 \end{bmatrix} \quad \text{and} \quad \mathbf{I}_{2\times2} = \begin{bmatrix} 1 & 0 \\ 0 & 1 \end{bmatrix}$$

2. Find **X'X**, where

$$\mathbf{X}_{3\times2} = \begin{bmatrix} 1 & 10 \\ 1 & 15 \\ 1 & 20 \end{bmatrix}$$

The answer to problem 1 is

$$\mathbf{AI} = \mathbf{IA} = \mathbf{A}$$

which indicates why **I** is generally referred to as the identity matrix, since, like the scalar identity 1, the product of any square matrix (e.g., **A**) with an appropriate identity matrix (of the right dimensions) will always yield the original matrix **A**, whether **I** premultiplies or postmultiplies **A**.

The answer to problem 2 is

$$\mathbf{X'X}_{2\times2} = \begin{bmatrix} 1 & 1 & 1 \\ 10 & 15 & 20 \end{bmatrix} \begin{bmatrix} 1 & 10 \\ 1 & 15 \\ 1 & 20 \end{bmatrix} = \begin{bmatrix} 3 & 45 \\ 45 & 725 \end{bmatrix}$$

which is a symmetric matrix, as will be the case whenever any matrix is multiplied (pre or post) by its own transpose.

B.7 Inverse of a Matrix

The definition of an *inverse matrix* parallels the basic property of the *reciprocal* of an ordinary (scalar) number. That is, for any nonzero number a, its reciprocal $1/a$ satisfies the equation

$$a \times \frac{1}{a} = \frac{1}{a} \times a = 1$$

In words, when a is either pre- or postmultiplied by its reciprocal, the result is the scalar 1. Analogously, we say that a square matrix $\mathbf{A}$ has an inverse $\mathbf{A}^{-1}$ if and only if

$$\mathbf{A}\mathbf{A}^{-1} = \mathbf{A}^{-1}\mathbf{A} = \mathbf{I}$$

That is, *the product of $\mathbf{A}$ by its inverse must be equal to an identity matrix*. We hasten to add at this point that *only the inverses of square matrices are being considered*. Thus, if $\mathbf{A}$ is $n \times n$, $\mathbf{A}^{-1}$ must also be $n \times n$.

We shall not describe here any of the many algorithms available for actually computing the inverse of a matrix.[1] For most practical applications, one can use standard computer packages for such computations. We wish to emphasize instead that one important attribute of inverses in statistical analyses is that their use permits the solution of matrix equations for a matrix of unknowns (e.g., the vector $\boldsymbol{\beta}$ of unknown regression parameters), in a manner similar to how division is used in ordinary algebra. Most specifically, in regression analysis, the use of an inverse is crucial because it provides a means for efficiently solving the least-squares equations, as well as providing compact formulae for additional components of the analysis (e.g., the variances of and covariances among the estimated regression coefficients).

Some examples of matrix inverses are given as follows:

$$\mathbf{A} = \begin{bmatrix} 2 & 0 & 1 \\ 1 & -1 & 2 \\ 1 & 0 & 0 \end{bmatrix} \quad \text{and} \quad \mathbf{A}^{-1} = \begin{bmatrix} 0 & 0 & 1 \\ 2 & -1 & -3 \\ 1 & 0 & -2 \end{bmatrix}$$

(The reader can check this out by multiplying $\mathbf{A}$ and $\mathbf{A}^{-1}$ together to obtain $\mathbf{I}_{3 \times 3}$.) The inverse of

$$\mathbf{X'X} = \begin{bmatrix} n & \sum_{i=1}^{n} X_i \\ \sum_{i=1}^{n} X_i & \sum_{i=1}^{n} X_i^2 \end{bmatrix}$$

is

$$(\mathbf{X'X})^{-1} = \begin{bmatrix} \dfrac{\sum_{i=1}^{n} X_i^2}{n \sum_{i=1}^{n} (X_i - \overline{X})^2} & \dfrac{-\overline{X}}{\sum_{i=1}^{n} (X_i - \overline{X})^2} \\ \dfrac{-\overline{X}}{\sum_{i=1}^{n} (X_i - \overline{X})^2} & \dfrac{1}{\sum_{i=1}^{n} (X_i - \overline{X})^2} \end{bmatrix}$$

---

[1] For further details, see N. R. Draper and H. Smith, *Applied Regression Analysis* (New York: Wiley, 1966); W. Mendenhall, *Introduction to Linear Models and the Design and Analysis of Experiments* (Belmont, Calif.: Wadsworth, 1968).

B.8 Matrix Formulation of Regression Analysis

We have previously seen in Section B.3 that when fitting a straight-line model

$$Y = \beta_0 + \beta_1 X + E$$

to a set of data consisting of n pairs of observations on the variables X and Y, we can define several matrices to characterize the regression problem under consideration: $\mathbf{Y}$, the $n \times 1$ vector of observations on Y; $\mathbf{X}$, the $n \times 2$ matrix of independent variables; $\boldsymbol{\beta}$, the 2×1 vector of parameters; and $\mathbf{E}$, the $n \times 1$ vector of random errors.

In general, whenever we are considering a regression problem involving p independent variables using a model such as

$$Y = \beta_0 + \beta_1 X_1 + \beta_2 X_2 + \cdots + \beta_p X_p + E$$

we can analogously construct appropriate matrices based on the multivariable data set being considered. Thus, if the data on the ith individual consist of the $p + 1$ values

$$Y_i, X_{i1}, X_{i2}, \ldots, X_{ip} \qquad i = 1, 2, \ldots, n$$

the following matrices can be constructed:

$$\mathbf{Y}_{n \times 1} = \begin{bmatrix} Y_1 \\ Y_2 \\ \vdots \\ Y_n \end{bmatrix} = \begin{array}{c} \text{vector of} \\ \text{observations on } Y \end{array}$$

$$\mathbf{X}_{n \times (p+1)} = \begin{bmatrix} 1 & X_{11} & X_{12} & \cdots & X_{1p} \\ 1 & X_{21} & X_{22} & \cdots & X_{2p} \\ \vdots & \vdots & \vdots & & \vdots \\ 1 & X_{n1} & X_{n2} & \cdots & X_{np} \end{bmatrix} = \begin{array}{c} \text{matrix of} \\ \text{independent} \\ \text{variables} \end{array}$$

$$\boldsymbol{\beta}_{(p+1) \times 1} = \begin{bmatrix} \beta_0 \\ \beta_1 \\ \vdots \\ \beta_p \end{bmatrix} = \text{vector of parameters}$$

$$\mathbf{E}_{n \times 1} = \begin{bmatrix} E_1 \\ E_2 \\ \vdots \\ E_n \end{bmatrix} = \text{vector of random errors}$$

Using the matrices above in conjunction with the notions of matrix addition and multiplication, we can formulate the general regression model in matrix terms as follows:

$$\mathbf{Y}_{n \times 1} = \mathbf{X}_{n \times (p+1)} \boldsymbol{\beta}_{(p+1) \times 1} + \mathbf{E}_{n \times 1} \tag{B.1}$$

This compact equation summarizes in a single statement the n equations

$$Y_i = \beta_0 + \beta_1 X_{i1} + \beta_2 X_{i2} + \cdots + \beta_p X_{ip} + E_i \qquad i = 1, 2, \ldots, n$$

Note that this equivalence follows from the following matrix calculations:

$$\mathbf{X}\boldsymbol{\beta} + \mathbf{E} = \begin{bmatrix} 1 & X_{11} & X_{12} & \cdots & X_{1p} \\ 1 & X_{21} & X_{22} & \cdots & X_{2p} \\ \vdots & \vdots & & & \vdots \\ 1 & X_{n1} & X_{n2} & \cdots & X_{np} \end{bmatrix} \begin{bmatrix} \beta_0 \\ \beta_1 \\ \vdots \\ \beta_p \end{bmatrix} + \begin{bmatrix} E_1 \\ E_2 \\ \vdots \\ E_n \end{bmatrix}$$

$$= \begin{bmatrix} \beta_0 + \beta_1 X_{11} + \beta_2 X_{12} + \cdots + \beta_p X_{1p} \\ \beta_0 + \beta_1 X_{21} + \beta_2 X_{22} + \cdots + \beta_p X_{2p} \\ \vdots \\ \beta_0 + \beta_1 X_{n1} + \beta_2 X_{n2} + \cdots + \beta_p X_{np} \end{bmatrix} + \begin{bmatrix} E_1 \\ E_2 \\ \vdots \\ E_n \end{bmatrix}$$

$$= \begin{bmatrix} \beta_0 + \beta_1 X_{11} + \beta_2 X_{12} + \cdots + \beta_p X_{1p} + E_1 \\ \beta_0 + \beta_1 X_{21} + \beta_2 X_{22} + \cdots + \beta_p X_{2p} + E_2 \\ \vdots \\ \beta_0 + \beta_1 X_{n1} + \beta_2 X_{n2} + \cdots + \beta_p X_{np} + E_n \end{bmatrix}$$

Based on the general matrix equation given by (B.1), a description of all the essential features of regression analysis can be expressed in matrix notation. In particular, the least-squares solution for the estimates of the regression coefficients in the parameter vector $\boldsymbol{\beta}$ can now be compactly written. This least-squares solution, in matrix terms, is that vector $\boldsymbol{\beta}$ which minimizes the error sum of squares (given in matrix terms)

$$(\mathbf{Y} - \mathbf{X}\hat{\boldsymbol{\beta}})'(\mathbf{Y} - \mathbf{X}\hat{\boldsymbol{\beta}})$$

The solution to this minimization problem, which is obtained via the use of matrix calculus, yields the following easy-to-remember matrix formula:

$$\hat{\boldsymbol{\beta}} = (\mathbf{X}'\mathbf{X})^{-1}\mathbf{X}'\mathbf{Y}$$

where $\hat{\boldsymbol{\beta}}'_{1 \times (p+1)} = [\hat{\beta}_0 \ \ \hat{\beta}_1 \ \ \cdots \ \ \hat{\beta}_p]$ denotes the vector of estimated regression coefficients.

Thus, although we have pointed out in the text the futility of *explicitly* giving the solutions to the least-squares equations for models of more complexity than a straight line, we can at least *implicitly* express these solutions in matrix notation and, because of modern computer technology, conveniently carry through with the computation of the least-squares solutions using this matrix representation.

At this point it is not our intention to carry through completely with the matrix formulation of every other aspect of a regression analysis. The reader is referred to Draper and Smith (1966; see footnote 1) for a fuller treatment of this matrix approach. However, some additional matrix results will be summarized below to give more of an indication of the utility of the matrix approach:

1. The *vector of predicted responses* is given by $\hat{\mathbf{Y}} = \mathbf{X}\hat{\boldsymbol{\beta}}$.

2. The *error sum of squares* SSE is given by SSE $= \mathbf{Y}'\mathbf{Y} - \hat{\boldsymbol{\beta}}'\mathbf{X}'\mathbf{Y}$.

3. The *regression sum of squares* is given by $\hat{\boldsymbol{\beta}}'\mathbf{X}'\mathbf{Y} - n\bar{Y}^2$.

4. The *variances of the regression coefficients*, that is, $\sigma_{\hat{\beta}_j}^2$, $j = 0, 1, 2, \ldots, p$, are given by the diagonal elements of the matrix $(\mathbf{X}'\mathbf{X})^{-1}\sigma^2$.

Finally, although we do not present the details here, any test of a (linear) statistical hypothesis concerning some subset of the regression coefficients can be formulated and carried out entirely by means of matrix operations.

Appendix—Answers
to Selected Problems

Chapter 3

2. nominal, ordinal, interval, ratio

3. a. 0.8413   **b.** −0.842

4. a. 18.475   **b.** 0.699

5. a. −1.350   **b.** 0.05

6. a. 2.51   **b.** 0.025

7. a. 0   **b.** 0   **c.** 0

8. standard normal

9. a. 3.0   **b.** 3   **c.** 2.8 or 3.11

10. e

11. a. 5.0   **b.** (187.44, 192.56)

12. $t_{0.975,27} = 2.052$

13. (24.66, 35.33)

14. 3.6858

15. a. significant difference   **b.** significant difference

16. nonsignificant difference

17. $t_{0.995,10} = 3.619$

18. b

19. a. Type I error   **b.** correct decision   **c.** correct decision   **d.** Type II error

20. a, b

21. c

22. b

23. a. $1 - \alpha$ **b.** α **c.** β **d.** $1 - \beta$

24. Accept H_0.

25. b

Chapter 5

1. **a.** Dry Weight (Y) does increase with increasing Age (X), but the relationship may not be linear. An exponential relationship between X and Y may better fit the data. Log Dry Weight (Z) increases linearly with increasing Age (X).

   b. $Y = \beta_0 + \beta_1 X + E \qquad Z = \beta_0' + \beta_1' X + E$

   c. $\hat{Y} = -1.885 + 0.235X \qquad \hat{Z} = -2.689 + 0.196X$

   d. The regression line for Log_{10} Dry Weight regressed on Age has a better fit. It is more appropriate to run a linear regression of Z on X.

   e. 95% confidence intervals: for β_1': (0.190, 0.202), for β_0': ($-2.759, -2.620$)

   f. ($-1.149, -1.096$)

3. **a.** The relationship between Time (Y) and Inc (X) does not appear to be linear.

   b. $\hat{\beta}_0 = 19.626 \qquad \hat{\beta}_1 = 0.0007$

   c. $\hat{Y} = 19.626 + 0.0007X$. The regression line fits the data poorly.

   d. The linearity assumption is not met.

   e. $T = 2.023$, $P = 0.0582$ (from SAS output). We do not reject H_0, since P-value > 0.05.

   f. The scatter plot suggests that a parabola would better fit the data.

5. **a.** $\hat{Y} = 2.174 + 1.177X$. The line fits the data well.

   b. No.

   c. $T = 13.5 \qquad P < 0.0001$ (from SAS output).
   Since P-value < 0.05, we reject H_0.

   d. $T = 0.954 \qquad 0.15 < P < 0.25$.
   Since P-value > 0.05, we do not reject H_0.

   e. (44.146, 47.276)

7. **a.** $\hat{Y}_1 = -122.345 + 6.227X \qquad \hat{Y}_2 = -1.697 + 0.299X$

   b. Y_2 regressed on X.

   c. $T = -57.934$. Critical value: $t_{17} \sim 2.898$ under H_0 at $\alpha = 0.01$. We see that $|T| = 57.934 > 2.898$, so we reject H_0 at $\alpha = 0.01$.

   d. (0.264, 0.334)

   e. (11.18, 12.32)

9. **a.** $\hat{\beta}_0 = 2.936 \qquad \hat{\beta}_1 = -1.785$

   b. $\hat{Y} = 2.936 - 1.785X$. The line fits the data well.

   d. $\hat{Y} = 862.979 \hat{X}^{-1.785}$

   e. (6.266, 9.311), (321.366, 528.445)

   f. One could plot the transformed data (X', Y') and then draw the estimated regression line on the plot. Then one could compare the fit of the estimated line of (X, Y) versus that of (X', Y'). If one did this comparison, the straight-line regression of (X, Y) gives a better fit.

11. **a.** $\hat{Y} = 3.707 - 0.012X$.

    b. $T = -8.684 \qquad P = 0.001$ (from SAS output).
    Since P-value < 0.05, we reject H_0.

    c. Including the data from the three experiments, rather than just using the average values, would provide more information and might improve the sensitivity of the analysis.

    d. (1.243, 3.737)

e. It is inappropriate, since the estimated model relies on data that do not include any information for average growth rate when exposed to a gas with a molecular weight of 200.

f. The chosen X values are not uniformly distributed in the experiment. There are large gaps between the X values of 39.9, 83.8, and 131.3. This may result in a fitted line subject to inaccuracies for predicting Y based on X.

13. b. From the SAS output: $\hat{\beta}_0 = 0.116$ $\hat{\beta}_1 = 0.005$

c. $T = 0.811$ $P = 0.433$ (from SAS output).
Since P-value > 0.10, we do not reject H_0.

d. $T = 0.046$ $P = 0.964$ (from SAS output).
Since P-value > 0.10, we do not reject H_0.

e. $\hat{Y} = 0.116 + 0.005X$

f. The line does not differ from the line plotted in part (e). The evidence suggests that there is no significant linear relationship. Determining a well-fitting line is difficult, given the dispersion of the data.

15. a. As X increases, the dispersion of Y increases.

b. $\hat{\beta}_0 = -2.546$ $\hat{\beta}_1 = 0.032$ $\hat{Y} = -2.546 + 0.032X$

c. This chart implies a better linear relationship between X and Y.

d. $\hat{\beta}_0 = -6.532$ $\hat{\beta}_1 = 1.430$ $\hat{Y} = -6.532 + 1.430X$

e. The natural log transformation provides the best representation. The natural log plot better illustrates the linear relationship and the dispersion of the data is more similar at each level of toluene exposure. The first plot indicates that there may be a violation of homoscedasticity for the untransformed data.

17. a. Yes.

b. $\hat{Y} = 1.643 + 1.057X$. The line appears to fit the data well.

c. 95% CI for β_1: (0.580, 1.534). The 95% CI does not include the null value of zero, indicating that there is a significant linear relationship at $\alpha = 0.05$.

d. No, it is not appropriate, since the data used to estimate the regression line do not inlude a $10 million advertising expenditure in its range.

19. a. There may be a slight negative linear relationship.

b. $Y = \beta_0 + \beta_1 X + E$ $\hat{\beta}_0 = 76.008$ $\hat{\beta}_1 = -0.015$.
The baseline OWNEROCC = 76%, and as OWNCOST increases by $1,000, the percentage of OWNEROCC decreases by $\sim 2\%$.

c. $\hat{Y} = 76.008 - 0.015X$. The line fits the data well.

d. $T = -2.607$ $P = 0.0155$ (from SAS output).
Since P-value < 0.05, we reject H_0.

e. $(-0.027, -0.003)$. We are 95% confident that the true slope is between -0.027 and -0.003. Since the interval does not contain zero, we conclude that the slope is not equal to zero (at $\alpha = 0.05$).

Chapter 6

1. a. (1) $r = 0.86$ **(2)** $r = 0.999$

b. (1) (0.546, 0.964) **(2)** (0.996, 1.000)

c. (1) $r^2 = 0.744$, so 74% of the variation in Y is explained with the help of X.
(2) $r^2 = 0.998$, so 99% of the variation in Z is explained with the help of X.

d. The regression using Log_{10} Dry Weight appears to fit better. This agrees with Chapter 5, Problem 1(d).

3. a. $r = 0.980$

b. Substitute $r(S_Y/S_X)$ and $\dfrac{(n-1)}{(n-2)}(S_Y^2 - \hat{\beta}_1^2 S_X^2)$ for $\hat{\beta}_1$ and $S_{Y|X}^2$ in formula (5.9).

c. $T' = 13.929$; $T \sim t_8$ under H_0: $\rho = 0$. The P-value for this test is less than 0.001; therefore, we reject H_0.

d. The graph of Y vs X does not illustrate a linear relationship.

5. a. $r^2 = 0.601$, $r = 0.775$. 60% of the variation in SBP (Y) is explained by AGE (X).

b. (0.504, 0.907). Since $\rho = 0$ is not included in the interval, we reject H_0: $\rho = 0$ at $\alpha = 0.01$.

7. a. $r^2 = 0.1853$, $r = 0.430$. 19% of the variation in SBP (Y) is explained by AGE (X).

b. $T = 2.02$; Critical value: $t_{18,0.975} = 2.101$. At $\alpha = 0.05$, since $|T| <$ critical value, we would not reject H_0.

c. $(-0.015, 0.733)$. Since $\rho = 0$ is included in the interval, we do not reject H_0: $\rho = 0$ at $\alpha = 0.05$.

9. a. $r^2 = 0.9101$, $r = 0.954$. 91% of the variation in Y is explained by X.

b. $T = 13.49$; Critical value: $t_{18,0.975} = 2.101$. At $\alpha = 0.05$, since $|T| >$ critical value, we would reject H_0.

c. (0.885, 0.982). Since $\rho = 0$ is not included in the interval, we reject H_0: $\rho = 0$ at $\alpha = 0.05$.

11. a. $r^2 = 0.9728$, $r = 0.986$, so 98% of the variation in Y_2 is explained by X.

b. $T = 24.640$; Critical value: $t_{17,0.975} = 2.110$. At $\alpha = 0.05$, since $|T| >$ critical value, we would reject H_0.

c. (0.963, 0.995). Since $\rho = 0$ is not included in the interval, we reject H_0: $\rho = 0$ at $\alpha = 0.05$.

13. a. $r = 0.971$

b. (0.813, 0.996). Since $\rho = 0$ is not included in the interval, we reject H_0: $\rho = 0$ at $\alpha = 0.05$.

15. $Z = 3.62$; Critical value $= 1.96$. Since $|Z|$ is $>$ critical value, we reject the null hypothesis.

17. a. $r^2 = 0.9558$, $r = 0.978$. 96% of the variation in Y is explained by X. *includes 16th obs.

b. $T = 17.424$; Critical value: $t_{14,0.975} = 2.145$. At $\alpha = 0.05$, we would reject H_0.

c. (0.936, 0.993). Since $\rho = 0$ is not included in the interval, we reject H_0: $\rho = 0$ at $\alpha = 0.05$.

19. a. $r^2 = 0.0310$, $r = -0.176$. 3% of the variation in Y is explained by X.

b. $T = 0.759$; Critical value: $t_{18,0.975} = 2.101$. At $\alpha = 0.05$, we would not reject H_0.

c. $(-0.574, 0.289)$. Since $\rho = 0$ is included in the interval, we do not reject H_0: $\rho = 0$ at $\alpha = 0.05$.

21. a. $r^2 = 0.9813$, $r = 0.991$. 98% of the variation in Y is explained by X.

b. $T = 55.088$; Critical value: $t_{38,0.975} = 2.0$. At $\alpha = 0.05$, we would reject H_0.

c. (0.985, 0.995). Since $\rho = 0$ is not included in the interval, we reject H_0: $\rho = 0$ at $\alpha = 0.05$.

23. a. $r^2 = 0.9044$, $r = 0.951$. 90% of the variation in Y is explained by X.

b. $T = 6.155$; Critical value: $t_{4,0.975} = 2.776$. At $\alpha = 0.05$, we would reject H_0.

c. (0.611, 0.995). Since $\rho = 0$ is not included in the interval, we reject H_0: $\rho = 0$ at $\alpha = 0.05$.

25. a. $r^2 = 0.2207$, $r = 0.470$. 22% of the variation in Y is explained by X.
  b. $T = 2.608$; Critical value: $t_{24,0.975} = 2.064$. At $\alpha = 0.05$, we would reject H_0.
  c. (0.101, 0.725). Since $\rho = 0$ is not included in the interval, we reject H_0: $\rho = 0$ at $\alpha = 0.05$.

Chapter 7

1. a. (1)

	d.f.	Sum of Squares	Mean Sum of Squares	F
Regression	1	6.0785	6.0785	26.18
Residual	9	2.0896	0.2322	
	10	8.1681		

a. (2)

	d.f.	Sum of Squares	Mean Sum of Squares	F
Regression	1	4.2211	4.2211	5355.59
Residual	9	0.0071	0.0008	
	10	4.2282		

  b. (1) H_0: $\beta_1 = 0$ H_A: $\beta_1 \neq 0$; Critical value: $F_{1,9,0.95} = 5.12$. Since $26.18 > 5.12$, we reject H_0 at $\alpha = 0.05$.
  b. (2) H_0: $\beta_1 = 0$ H_A: $\beta_1 \neq 0$; Critical value: $F_{1,9,0.95} = 5.12$. Since $5355.59 > 5.12$, we reject H_0 at $\alpha = 0.05$.

5. a.

	d.f.	Sum of Squares	Mean Sum of Squares	F
Regression	1	450.8673	450.8673	4.09
Residual	18	1982.9141	110.1619	
	19	2433.7814		

  b. H_0: $\beta_1 = 0$ H_A: $\beta_1 \neq 0$; Critical value: $F_{1,18,0.95} = 4.41$. $4.09 < 4.41$, so we do not reject H_0 at $\alpha = 0.05$.
  c. $T^2 = (2.023)^2 = 4.09$. The values are the same.
  d. The hypotheses for each test are equivalent. As mentioned in the text, the F statistic and T statistic are equivalent after squaring T. Using the information that the tests of hypotheses are equivalent (as are the test statistics), one may infer that the resulting P-values for each test should also be equivalent.

9. a.

	d.f.	Sum of Squares	Mean Sum of Squares	F
Regression	1	76858486.06	76858486.06	60.76
Residual	28	35419546.54	1264983.00	
	29	112278032.60		

  b. H_0: $\beta_1 = 0$ H_A: $\beta_1 \neq 0$; Critical value: $F_{1,28,0.95} = 4.20$.
  Since $60.76 > 4.20$, we reject H_0 at $\alpha = 0.05$.

13. a.

	d.f.	Sum of Squares	Mean Sum of Squares	F
Regression	1	12846354	12846354	302.99
Residual	14	593585	42399	
	15	134399939		

  b. H_0: $\beta_1 = 0$ H_A: $\beta_1 \neq 0$; Critical value: $F_{1,14,0.95} = 4.60$

Since $302.99 > 4.60$, we would reject H_0 and conclude that there is a significant linear relationship of Y on X at $\alpha = 0.05$.

17. a.

	d.f.	Sum of Squares	Mean Sum of Squares	F
Regression	1	177.5297	177.5297	3045.56
Residual	58	3.3809	0.0583	
	59	180.9106		

b. $H_0: \beta_1 = 0$ $H_A: \beta_1 \neq 0$; Critical value: $F_{1,58,0.95} = 4.00$. Since $3045.56 > 4.00$, we reject H_0 at $\alpha = 0.05$.

21. a.

	d.f.	Sum of Squares	Mean Sum of Squares	F
Regression	1	132.6203	132.6203	6.80
Residual	24	468.3412	19.5142	
	25	600.9615		

b. $H_0: \beta_1 = 0$ $H_A: \beta_1 \neq 0$; Critical value: $F_{1,24,0.95} = 4.26$. Since $6.80 > 4.26$, we reject H_0 at $\alpha = 0.05$.

Chapter 8

1. a. i $\hat{Y} = 145.771$ **ii** $\hat{Y} = 135.825$ **iii** $\hat{Y} = 141.475$

As QUET increases from 3.0 to 3.5, average SBP increases by an estimated 4.296 points, from 141.475 to 145.771.

b. SBP on AGE: $R^2 = 0.601$; SBP on AGE and SMK: $R^2 = 0.730$; SBP on AGE, SMK, and QUET: $R^2 = 0.761$.

The model using AGE and SMK to predict SBP appears to be the best choice.

5. a.

	d.f.	Sum of Squares	Mean Sum of Squares	F
Regression	3	25974.40	8658.00	80.87
Residual	21	2248.23	107.06	
Total	24	28222.63		

b. $R^2 = 0.920$. There is a strong positive linear relationship between education resources and student performance.

9. a. As temperature increases from 20 to 25, the average oxygen consumption increases by an estimated 0.197 units.

b. As weight increases from 0.25 to 0.5, the average oxygen consumption increases by an estimated 0.148 units.

c. i $R^2 = 0.019$ **ii** $R^2 = 0.814$ **iii** $R^2 = 0.943$

13. a. $\hat{Y} = 6.874 - 0.004X_2 - 0.234X_3$ **b.** $\hat{Y} = 3.734$

c. $R^2 = 0.553$. The model explains about half of the variation in Yield (Y). The model has a limited ability to predict the yield for a company using 1989 ranking and P-E ratio as predictors.

15. a. $\hat{Y} = 130.468 + 0.089$ zooplankton $- 0.025$ phytoplankton

b. $R^2 = 0.1332$. The model fit is poor.

Chapter 9

1. a. i $H_0: \beta_1 = 0$ $H_A: \beta_1 \neq 0$ (Full model: $Y = \beta_0 + \beta_1 X_1 + E$)
$F(X_1)_{1,30} = 45.18$, $P = 0.0001$. At $\alpha = 0.05$, we reject H_0.
ii $H_0: \beta_1 = \beta_2 = 0$ H_A: at least one $\beta_i \neq 0$ ($i = 1, 2$)
(Full model: $Y = \beta_0 + \beta_1 X_1 + \beta_2 X_2 + E$)
$F(X_1, X_2)_{2,29} = 39.16$, $P < 0.0001$. At $\alpha = 0.05$, we reject H_0.
iii $H_0: \beta_1 = \beta_2 = \beta_3 = 0$ H_A: at least one $\beta_i \neq 0$ ($i = 1, 2, 3$)
(Full model $Y = \beta_0 + \beta_1 X_1 + \beta_2 X_2 + \beta_3 X_3 + E$).
$F(X_1, X_2, X_3)_{3,28} = 29.71$, $P < 0.0001$. At $\alpha = 0.05$, we reject H_0.
b. One would choose the parsimonious model tested in (a(i)).

5. a. i $H_0: \beta_1 = 0$ $H_A: \beta_1 \neq 0$ (Full model: $Y = \beta_0 + \beta_1 X_1 + E$)
$F(X_1)_{1,40} = 0.40$, $P > 0.25$. At $\alpha = 0.05$, we do not reject H_0.
ii $H_0: \beta_2 = 0$ $H_A: \beta_2 \neq 0$ (Full model: $Y = \beta_0 + \beta_2 X_2 + E$)
$F(X_2)_{1,40} = 0.19$, $P > 0.25$. At $\alpha = 0.05$, we do not reject H_0.
iii $H_0: \beta_3 = 0$ $H_A: \beta_3 \neq 0$ (Full model: $Y = \beta_0 + \beta_3 X_3 + E$)
$F(X_3)_{1,40} = 7.58$, $P = 0.009$. At $\alpha = 0.05$, we reject H_0.
b. $H_0: \beta_1 = \beta_2 = \beta_3 = 0$ H_A: at least one $\beta_i \neq 0$ ($i = 1, 2, 3$)
(Full model: $Y = \beta_0 + \beta_1 X_1 + \beta_2 X_2 + \beta_3 X_3 + E$)
$F(X_1, X_2, X_3)_{3,38} = 2.74$, $0.05 < P < 0.10$. At $\alpha = 0.05$, we do not reject H_0.
c. $H_0: \beta_4 = \beta_5 = 0$ H_A: at least one $\beta_i \neq 0$ ($i = 4, 5$)
(Full model: $Y = \beta_0 + \beta_1 X_1 + \beta_2 X_2 + \beta_3 X_3 + \beta_4 X_1 X_3 + \beta_5 X_2 X_3 + E$)
$F(X_1 X_3, X_2 X_3 \mid X_1, X_2, X_3)_{2,36} = 0.362$, $P > 0.25$. At $\alpha = 0.05$, we do not reject H_0.
d. $H_0: \beta_3 = 0$ $H_A: \beta_3 \neq 0$ (Full model $Y = \beta_0 + \beta_1 X_1 + \beta_2 X_2 + \beta_3 X_3 + E$).
$F(X_3 \mid X_1, X_2)_{1,38} = 6.85$, $0.01 < P < 0.025$. At $\alpha = 0.05$, we reject H_0.
e. X_3 is associated with Y, but the other two independent variables are not.

9. a. $H_0: \beta_1 = \beta_2 = 0$ H_A: at least one $\beta_i \neq 0$ ($i = 1, 2$)
(Full model: $Y = \beta_0 + \beta_1 X_1 + \beta_2 X_2 + E$)
$F(X_1, X_2)_{2,4} = 20.03$, $P = 0.0082$. At $\alpha = 0.05$, we reject H_0.
b. i $H_0: \beta_1 = 0$ $H_A: \beta_1 \neq 0$ in the model $Y = \beta_0 + \beta_1 X_1 + E$.
$F(X_1)_{1,5} = 40.06$, $P = 0.0032$. At $\alpha = 0.05$, we reject H_0. Note: We use the residual MS from the largest model. See Section 9.5.2.
ii $H_0: \beta_2 = 0$ $H_A: \beta_2 \neq 0$ in the model $Y = \beta_0 + \beta_1 X_1 + \beta_2 X_2 + E$.
$F(X_2 \mid X_1)_{1,4} = 0.01$, $P = 0.9344$. At $\alpha = 0.05$, we do not reject H_0.
c. i $H_0: \beta_2 = 0$ $H_A: \beta_2 \neq 0$ in the model $Y = \beta_0 + \beta_2 X_2 + E$.
$F(X_2)_{1,5} = 37.83$, $0.001 < P < 0.005$. At $\alpha = 0.05$, we reject H_0.
ii $H_0: \beta_1 = 0$ $H_A: \beta_1 \neq 0$ (Full model: $Y = \beta_0 + \beta_1 X_1 + \beta_2 X_2 + E$)
$F(X_1 \mid X_2)_{1,4} = 9.80$, $P = 0.0352$. At $\alpha = 0.05$, we reject H_0.

d.

Source	d.f.	SS	MS	F	R^2
$X_1 \mid X_2$	1	1402.315	1402.315	9.8	0.91
$X_2 \mid X_1$	1	1.098	1.098	0.01	
Residual	4	572.393	143.098		

e. X_1 is the only necessary predictor.

13. a. $H_0: \beta_1 = \beta_2 = 0$ H_A: at least one $\beta_i \neq 0$ $(i = 1, 2)$
(X_1 = OWNCOST, X_2 = URBAN)
(Full model: $Y = \beta_0 + \beta_1 X_1 + \beta_2 X_2 + E$)
$F(X_1, X_2)_{2,23} = 8.52$, $P = 0.017$. At $\alpha = 0.05$, we reject H_0.
b. i $H_0: \beta_1 = 0$ $H_A: \beta_1 \neq 0$ (Full model: $Y = \beta_0 + \beta_1 X_1 + E$)
$F(X_1)_{1,24} = 8.84$, $P = 0.0068$. At $\alpha = 0.05$, we reject H_0.
ii $H_0: \beta_2 = 0$ $H_A: \beta_2 \neq 0$ (Full model: $Y = \beta_0 + \beta_1 X_1 + \beta_2 X_2 + E$)
$F(X_2 \mid X_1)_{1,23} = 8.21$, $P = 0.0088$. At $\alpha = 0.05$, we reject H_0.
c. i $H_0: \beta_2 = 0$ $H_A: \beta_2 \neq 0$ (Full model: $Y = \beta_0 + \beta_2 X_2 + E$)
$F(X_1)_{1,24} = 13.17$, $0.001 < P < 0.005$. At $\alpha = 0.05$, we reject H_0.
ii $H_0: \beta_1 = 0$ $H_A: \beta_1 \neq 0$ (Full model: $Y = \beta_0 + \beta_1 X_1 + \beta_2 X_2 + E$)
$F(X_1 \mid X_2)_{1,23} = 4.42$, $P = 0.0467$. At $\alpha = 0.05$, we reject H_0.
d.

Source	d.f.	SS	MS	F	R^2
$X_1 \mid X_2$	1	66.334	66.334	4.42	0.43
$X_3 \mid X_2$	1	123.158	123.158	8.21	
Residual	23	345.183	15.008		

e. Both predictors are necessary.

Chapter 10

1. a. Age with an r value of 0.7752.
b. i $r_{SBP,SMK \mid AGE} = 0.568$ **ii** $r_{SBP,QUET \mid AGE} = 0.318$
c. $H_0: \rho_{SBP,SMK \mid AGE} = 0$ $H_A: \rho_{SBP,SMK \mid AGE} \neq 0$
$F(SMK \mid AGE)_{1,29} = 13.83$, $P < 0.001$. At $\alpha = 0.05$, we reject H_0.
d. $H_0: \rho_{SBP,QUET \mid AGE,SMK} = 0$ $H_A: \rho_{SBP,QUET \mid AGE,SMK} \neq 0$
$T_{28} = 1.91$, $P = 0.066$. At $\alpha = 0.05$, we do not reject H_0.
e. Based on the results for a–d, we find that the following variables (ranked in order of their significance) helped explain the variation in SBP: (1) AGE, (2) SMK, (3) QUET.
f. $r^2_{SBP(QUET,SMK) \mid AGE} = 0.401$
$H_0: \rho_{SBP(QUET,SMK) \mid AGE} = 0$ $H_A: \rho_{SBP(QUET,SMK) \mid AGE} \neq 0$
$F(QUET, SMK \mid AGE)_{2,28} = 9.371$, $P < 0.001$.
The highly significant P-value suggests that both SMK and QUET are important variables, but there is room for debate since the increase in r^2 going from model 1, with only AGE (0.601), to model 3, with all 3 variables (0.761), is small (0.160).
5. a. i $r^2_{YX_1 \mid X_3} = 0.076$.   **ii** $r^2_{YX_2 \mid X_3} = 0.214$.
iii $r^2_{YX_2 \mid X_3} = 0.215$. The computations are nearly equivalent with the difference being due to round-off error.
b. X_2 should be considered next for entry into the model because X_2 has a higher partial correlation than does X_1.
c. $H_0: \rho_{YX_2 \mid X_3} = 0$ $H_A: \rho_{YX_2 \mid X_3} \neq 0$
$T_{17} = 2.154$, $0.02 < P < 0.05$. At $\alpha = 0.05$, we reject H_0.
d. $r^2_{YX_1 \mid X_2,X_3} = 0.082$
$H_0: \rho_{YX_1 \mid X_2,X_3} = 0$ $H_A: \rho_{YX_1 \mid X_2,X_3} \neq 0$
$T_{16} = 1.199$, $P = 0.248$. At $\alpha = 0.05$, we do not reject H_0.

e. $r^2_{Y(X_1,X_2)|X_3} = 0.279$
   $H_0: \rho_{Y(X_1,X_2)|X_3} = 0 \qquad H_A: r_{Y(X_1,X_2)|X_3} \neq 0$
   $F(X_1, X_2X_3)_{2,16} = 3.098, 0.05 < P < 0.10$. At $\alpha = 0.05$, we do not reject H_0.

f. Based on the above results, only X_3 should be included in the model at $\alpha = 0.05$.

9. a. i $H_0: \rho_{YX_1} = 0 \qquad H_A: \rho_{YX_1} \neq 0$
   $F(X_1)_{1,45} = 0.89, P = 0.35$. At $\alpha = 0.05$, we do not reject H_0.

   ii $H_0: \rho_{YX_2} = 0 \qquad H_A: \rho_{YX_2} \neq 0$
   $F(X_2)_{1,45} = 197.58, P < 0.0001$. At $\alpha = 0.05$, we reject H_0.

b. i $H_0: \rho_{YX_1|X_2} = 0 \qquad H_A: \rho_{YX_1|X_2} \neq 0$
   $F(X_1, X_2)_{1,44} = 98.605, P < 0.001$. At $\alpha = 0.05$, we reject H_0.

   ii $H_0: \rho_{YX_2|X_1} = 0 \qquad H_A: \rho_{YX_2|X_1} \neq 0$
   $F(X_2|X_1)_{1,44} = 709.641, P < 0.001$. At $\alpha = 0.05$, we reject H_0.

c. Both X_1 and X_2 should be included in the model with X_2 being more important than X_1.

13. a. $r^2_{Y|X_1, X_2} = 0.909$ **b.** $r_{YX_2|X_1} = -\sqrt{0.002} = -0.045$

c. $r_{YX_1|X_2} = \sqrt{0.710} = 0.843$

d. $H_0: \rho_{YX_2|X_1} = 0 \qquad H_A: \rho_{YX_2|X_1} \neq 0$
   $T_4 = -0.09, P > 0.90$. At $\alpha = 0.05$, we do not reject H_0.

e. $H_0: \rho_{YX_1|X_2} = 0 \qquad H_A: \rho_{YX_1|X_2} \neq 0$
   $T_4 = 3.131, 0.02 < P < 0.05$. At $\alpha = 0.05$, we reject H_0.

f. X_1 should be included in the model, while X_2 should not be included.

17. a. $r^2_{Y|X_1, X_2} = 0.426$ **b.** $r_{YX_2|X_1} = -\sqrt{0.263} = -0.513$

c. $r_{YX_1|X_2} = -\sqrt{0.161} = -0.401$

d. $H_0: \rho_{YX_2|X_1} = 0 \qquad H_A: \rho_{YX_2|X_1} \neq 0$
   $T_{23} = -2.865, 0.001 < P < 0.01$. At $\alpha = 0.05$, we reject H_0.

e. $H_0: \rho_{YX_1|X_2} = 0 \qquad H_A: \rho_{YX_1|X_2} \neq 0$
   $T_{23} = -2.1, 0.02 < P < 0.05$. At $\alpha = 0.05$, we reject H_0.

f. Both variables should be included in the model with X_2 being the more important predictor of Y.

Chapter 11

1. a. $WGT = \beta_0 + \beta_1 HGT + \beta_2 AGE + \beta_3 AGE^2 + E$

b. $\hat{\beta}_1$ does not change when either AGE or AGE^2 are removed from the model. However, $\hat{\beta}_1$ changes "significantly" when both AGE and AGE^2 are removed from the model. Thus, there is confounding due to AGE and AGE^2.

c. AGE^2 can be dropped from the model because $\hat{\beta}_1$ does not change significantly.

d. AGE^2 should not be retained in the model, because the 95% CI for β_1 is narrower when AGE^2 is absent from the model.

e. Considering the change in $\hat{\beta}_1$ and the width of the 95% CI, the final model should be $WGT = \beta_0 + \beta_1 HGT + \beta_2 AGE + E$.

f. Revise the initial model as
   $WGT = \beta_0 + \beta_1 HGT + \beta_2 AGE + \beta_3 AGE^2 + \beta_4 HGT*AGE + \beta_5 HGT*AGE^2 + E$.

g. We would test for interaction by performing a multiple-partial F test for $H_0: \beta_4 = \beta_5 = 0$. If this test proved significant, we would perform separate partial F tests to assess $H_0: \beta_4 = 0$ and $H_0: \beta_5 = 0$.

3. a. There is no confounding due to X_2, because $\hat{\beta}_1$ does not change when X_2 is removed from the model.

b. $r_{YX_1} = 0.265$ $r_{YX_1|X_2} = \sqrt{0.5} = 0.707$
Since the two correlation coefficients are significantly different, we conclude that confounding exists.

c. The conclusions for confounding depend on the definition of confounding.

d. Since $H_0: \beta_2 = 0$ is rejected ($P = 0.0005$), we conclude that confounding exists, which is contradictory to part (a).

5. a. i $H_0: \beta_1 = 0$ $H_A: \beta_1 \neq 0$ $F(X_1)_{1,40} = 0.4, P = 0.528$. At $\alpha = 0.05$, we do not reject H_0.

ii $H_0: \beta_2 = 0$ $H_A: \beta_2 \neq 0$ $F(X_2)_{1,40} = 0.19, P = 0.669$. At $\alpha = 0.05$, we do not reject H_0.

iii $H_0: \beta_3 = 0$ $H_A: \beta_3 \neq 0$ $F(X_3)_{1,40} = 7.58, P = 0.009$. At $\alpha = 0.05$, we reject H_0.

b. $H_0: \beta_1 = \beta_2 = \beta_3 = 0$ H_A: At least one $\beta_i \neq 0$ $(i = 1, 2, 3)$
$F(X_1, X_2, X_3)_{3,38} = 2.737, 0.05 < P < 0.1$. At $\alpha = 0.05$, we do not reject H_0.

c. $H_0: \rho_{Y(X_1X_3,X_2X_3)\,|\,X_1,X_2,X_3} = 0$ $H_A: \rho_{Y(X_1X_3,X_2X_3)\,|\,X_1,X_2,X_3} \neq 0$
$F(X_1X_3, X_2X_3|X_1, X_2, X_3)_{2,36} = 0.362, P > 0.25$. At $\alpha = 0.05$, we do not reject H_0.
The two regression lines are parallel.

d. $H_0: \rho_{YX_3\,|\,X_1,X_2} = 0$ $H_A: \rho_{YX_3\,|\,X_1,X_2} \neq 0$
$F(X_3|X_1, X_2)_{1,38} = 6.855, P = 0.014$. At $\alpha = 0.05$, we do not reject H_0.

e. First fit the full model
$$Y = \beta_0 + \beta_1 X_1 + \beta_2 X_2 + \beta_3 X_3 + E \quad (1)$$
Next, fit the reduced models
$$Y = \beta_0 + \beta_1 X_1 + \beta_3 X_3 + E \quad (2)$$
$$Y = \beta_0 + \beta_2 X_2 + \beta_3 X_3 + E \quad (3)$$
$$Y = \beta_0 + \beta_3 X_3 + E \quad (4)$$
We assess confounding by noting how $\hat{\beta}_3$ changes for the different models. In particular, if $\hat{\beta}_3$ from model (2), (3), or (4) differs from $\hat{\beta}_3$ from model (1), then X_1, X_2, or X_1 and X_2, respectively, are confounders. To assess precision, we note how the $100(1 - \alpha)\%$ CI's for $\hat{\beta}_3$ change. We only eliminate potential confounders from the model if the width of the CI for $\hat{\beta}_3$ does not widen significantly.

f. From the information provided, we can assess the confounding effects of X_1 or X_2 alone with respect to X_3 but not for X_1 and X_2 taken together.

7. a. No, there is no meaningful change in the estimate for $\hat{\beta}_1$ when X_2 is added to the model.

b. No, the confidence interval for $\hat{\beta}_1$ is narrower when only X_1 is in the model.

c. No, there is not evidence suggesting that including X_2 to the model improves the precision and/or the validity of the estimated relationship between X_1 and Y.

9. a. Let OWNCOST $= X_1$ and INCOME $= X_2$
$H_0: \beta_1 = \beta_2 = 0$ H_A: at least one β_i does not equal 0.
(Full model: $Y = \beta_0 + \beta_1 X_1 + \beta_2 X_2 + E$)
$F(X_1, X_2)_{2,23} = 6.38, P = 0.006$. At $\alpha = 0.05$, we reject H_0.

b. $H_0: \beta_1 = 0$ $H_A: \beta_1 \neq 0$ (Full model: $Y = \beta_0 + \beta_1 X_1 + \beta_2 X_2 + E$)
$F(X_1|X_2)_{1,23} = 11.47, P = 0.003$. At $\alpha = 0.05$, we reject H_0.

c. $H_0: \beta_2 = 0$ $H_A: \beta_2 \neq 0$ (Full model $Y = \beta_0 + \beta_1 X_1 + \beta_2 X_2 + E$)
$F(X_2|X_1)_{1,23} = 4.87, P = 0.038$. At $\alpha = 0.05$, we reject H_0.

d. Including X_2 does meaningfully change $\hat{\beta}_1$, and it should therefore be included in the model as a confounder, assuming there is no interaction between X_1 and X_2.

Chapter 12

1. a. For smokers: $\hat{Y} = 79.225 + 20.118X$. For nonsmokers: $\hat{Y} = 49.312 + 26.303X$.

b. $H_0: \beta_{1SMK} = \beta_{1\overline{SMK}}$ $H_A: \beta_{1SMK} < \beta_{1\overline{SMK}}$
$T_{28} = 0.892, 0.15 < P < 0.25$. At $\alpha = 0.05$, we do not reject H_0 that the slopes for smokers and nonsmokers are the same.

c. $H_0: \beta_{0SMK} = \beta_{0\overline{SMK}}$ $H_A: \beta_{0SMK} \neq \beta_{0\overline{SMK}}$
$T_{28} = -1.24, 0.20 < P < 0.30$. At $\alpha = 0.05$, we do not reject H_0, and we conclude that the two intercepts are equal.

d. The straight lines for smokers and nonsmokers are coincident since both tests failed to reject H_0.

5. a. For NY: $\hat{Y} = 2.174 + 1.177X$.   For CA: $\hat{Y} = 8.030 + 1.036X$.

b. $H_0: \beta_{1NY} = \beta_{1CA}$ $H_A: \beta_{1NY} > \beta_{1CA}$
$T_{33} = 1.115, 0.10 < P < 0.15$. At $\alpha = 0.05$, we do not reject H_0, and we conclude that the slopes are the same for NY and CA.

c. $H_0: \beta_{0NY} = \beta_{0CA}$ $H_A: \beta_{0NY} > \beta_{0CA}$
$T_{33} = 1.219, 0.85 < P < 0.90$. At $\alpha = 0.05$, we do not reject H_0 that the two intercepts are equal for NY and CA.

d. Since the tests for equal slopes and equal intercepts did not lead to rejection, we can conclude that the lines are coincident.

e. $H_0: \rho_{NY} = \rho_{CA}$ $H_A: \rho_{NY} \neq \rho_{CA}$
$Z = 0.252, P > 0.80$. At $\alpha = 0.05$, we do not reject H_0, and we conclude that the correlation coefficients for each straight line regression are not significantly different.

9. a. SBP $= \beta_0 + \beta_1$AGE $+ \beta_2$QUET $+ \beta_3$SMK $+ \beta_4$AGE*SMK $+ \beta_5$QUET*SMK $+ E$
Smokers: SBP $= (\beta_0 + \beta_3) + (\beta_1 + \beta_4)$AGE $+ (\beta_2 + \beta_5)$QUET $+ E$
Nonsmokers: SBP $= \beta_0 + \beta_1$AGE $+ \beta_2$QUET $+ E$

b. Smokers: $\widehat{\text{SBP}} = 48.076 + 1.466(\text{AGE}) + 6.744(\text{QUET})$
Nonsmokers: $\widehat{\text{SBP}} = 48.613 + 1.029(\text{AGE}) + 10.451(\text{QUET})$

c. $H_0: \beta_4 = \beta_5 = 0$ H_A: at least one $\beta_i \neq 0$ (Full Model given in part (a).)
$F(\text{QUET*SMK, AGE*SMK}|\text{AGE, QUET, SMK})_{2,26} = 0.222, P > 0.25$
At $\alpha = 0.05$, we do not reject H_0 and conclude the two lines are coincident.

d. $H_0: \beta_3 = \beta_4 = \beta_5 = 0$ H_A: At least one $\beta_i \neq 0$ (Full Model given in part (a).)
$F(\text{SMK, QUET*SMK, AGE*SMK}|\text{AGE, QUET})_{3,26} = 4.562, 0.01 < P < 0.025$
At $\alpha = 0.05$, we reject H_0 and conclude the two lines are not coincident.

13. a. $R = 1$ and TD $= 1$: $\widehat{\text{SBPL}} = (\hat{\beta}_0 + \hat{\beta}_2 + \hat{\beta}_4 + \hat{\beta}_9) + \hat{\beta}_1(\text{SBP1}) + (\hat{\beta}_6 + \hat{\beta}_{11})\text{RW}$
$R = 0$ and TD $= 1$: $\widehat{\text{SBPL}} = (\hat{\beta}_0 + \hat{\beta}_3 + \hat{\beta}_4 + \hat{\beta}_7) + \hat{\beta}_1(\text{SBP1}) + (\hat{\beta}_6 + \hat{\beta}_{13})\text{RW}$

b. $H_0: \beta_{11} = \beta_{12} = \beta_{13} = \beta_{14} = 0$ H_A: At least one $\beta_i \neq 0$.
$F(X_{11}, X_{12}, X_{13}, X_{14}|X_1, \ldots, X_{10})_{4,89} = 1.410, 0.1 < P < 0.25$. At $\alpha = 0.05$, we do not reject H_0, and we conclude that the lines are parallel.

c. H_0: The three regression lines corresponding to rural, town, and urban background are parallel (i.e., $H_0: \beta_7 = \beta_8 = \beta_9 = \beta_{10} = \beta_{11} = \beta_{12} = \beta_{13} = \beta_{14} = 0$).

H_A: At least one $\beta_i \neq 0$.

$F(X_7, \ldots, X_{14}|X_1, \ldots, X_6)_{8,89} = 1.103$ with $P > 0.25$. At $\alpha = 0.05$, we do not reject H_0, and we conclude that the three regression lines are parallel.

d. H_0: $\beta_2 = \beta_3 = \beta_4 = \beta_5 = \beta_7 = \beta_8 = \beta_9 = \beta_{10} = \beta_{11} = \beta_{12} = \beta_{13} = \beta_{14} = 0$

H_A: At least one $\beta_i \neq 0$.

$$F(X_2, X_3, X_4, X_5, X_7 \ldots, X_{14}|X_1, X_6)_{12,89}$$

$$= \frac{\dfrac{\text{SS Regression (full mode)} - \text{SS Regression }(X_1, X_6)}{12}}{\text{MS Residual (full model)}}$$

17. a. $Y = \beta_0 + \beta_1 X + \beta_2 Z + \beta_3 XZ + E$ where $Z = 1$ if cool, 0 if warm.

b. For cool: $\hat{Y} = 104.003 + 2.465X$. For warm: $\hat{Y} = 96.830 + 3.485X$.

c. H_0: $\beta_2 = \beta_3 = 0$ H_A: At least one $\beta_i \neq 0$ (Full model given in part (a).)

$F(Z, XZ|X)_{2,14} = 41.875$, $P = 0.0076$. At $\alpha = 0.05$, we reject H_0 and conclude that the lines do not coincide.

d. H_0: $\beta_3 = 0$ H_A: $\beta_3 \neq 0$ (Full model given in part (a).)

$F(XZ|X, Z)_{1,14} = 9.699$, $P < 0.01$. At $\alpha = 0.05$, we reject H_0 and conclude that the lines are not parallel.

e. Baseline sales are higher during the warm season than in the cool season. Advertising expenditures are higher in the cool season than in the warm season. By spending more money in advertising during the cool season, retailers are able to surpass the sales revenue of the warm season.

21. a. $Y = \beta_0 + \beta_1 X + \beta_2 Z + \beta_3 XZ + E$

b. H_0: $\beta_2 = \beta_3 = 0$ H_A: At least one $\beta_i \neq 0$ (Full model given in part (a).)

$F(Z, X_1 Z|X_1)_{2,50} = 2.704$, $0.05 < P < 0.10$. At $\alpha = 0.05$, we do not reject H_0, and we conclude that the lines coincide.

d. H_0: $\beta_3 = 0$ H_A: $\beta_3 \neq 0$ (Full model given in part (a).)

$F(X_1 Z|X_1, Z)_{1,50} = 3.587$, $P = 0.064$. At $\alpha = 0.05$, we do not reject H_0, and we conclude that the lines are parallel.

e. The change in refraction-baseline refractive relationship is the same for males and females.

Chapter 13

1. a. SBP $= \beta_0 + \beta_1(\text{QUET}) + \beta_2 \text{SMK} + E$

b. For smokers: SBP $= (\beta_0 + \beta_2) + \beta_1(\text{QUET}) + E$; $\overline{\text{SBP}}_{(adj)} = 148.548$

For nonsmokers: SBP $= \beta_0 + \beta_1(\text{QUET}) + E$; $\overline{\text{SBP}}_{(adj)} = 139.977$

c. H_0: $\beta_2 = 0$ H_A: $\beta_2 \neq 0$ in the model SBP $= \beta_0 + \beta_1(\text{QUET}) + \beta_2 \text{SMK} + E$.

$T_{29} = 2.707$, $P = 0.011$. At $\alpha = 0.05$, we reject H_0 and conclude that mean SBP differs for smokers and for nonsmokers, after adjusting for QUET.

d. Finding the 95% confidence interval for the true difference in adjusted mean SBP is equivalent to finding the 95% confidence interval for $\hat{\beta}_2$. The 95% confidence interval for $\hat{\beta}_2$ is (2.094, 15.048).

5. a. VIAD $= \beta_0 + \beta_1 \text{IQM} + \beta_2 \text{IQF} + \beta_3 Z + E$ where $Z_1 = 1$ if female, 0 if male.

b. For males: VIAD (adj) $= -3.307$ vs -3.00 unadjusted.

For females: VIAD (adj) $= 1.889$ vs 1.60 unadjusted.

c. $H_0: \beta_3 = 0$ $H_A: \beta_3 \neq 0$ in the model VIAD $= \beta_0 + \beta_1 IQM + \beta_2 IQF + \beta_3 Z + E$
$T_{16} = 1.659, P = 0.117$. At $\alpha = 0.05$, we do not reject H_0, and we conclude that the mean scores do not significantly differ by gender, after adjusting for IQM and IQF.

d. 95% CI: $(-1.446, 11.838)$

9. a. LN_BRNTL $= \beta_0 + \beta_1 WGT + \beta_2 Z_1 + \beta_3 Z_2 + \beta_4 Z_3 + E$, where $Z_1 = 1$ if 100 ppm, 0 otherwise; $Z_2 = 1$ if 500 ppm, 0 otherwise; $Z_3 = 1$ if 1000 ppm, 0 otherwise.

b. $\hat{\beta}_0 = -0.764$ $\hat{\beta}_1 = 0.0006$ $\hat{\beta}_2 = 0.828$ $\hat{\beta}_3 = 3.571$ $\hat{\beta}_4 = 4.214$

c.

PPM_TOLU	Adjusted Means	Unadjusted Means
50	−0.537	−0.548
100	0.291	0.282
500	3.034	3.019
1000	3.677	3.668

d. $H_0: \beta_2 = \beta_3 = \beta_4 = 0$ H_A: At least one $\beta_i \neq 0$ $(i = 2, 3, 4)$
(Full model: LN_BRNTL $= \beta_0 + \beta_1 WGT + \beta_2 Z_1 + \beta_3 Z_2 + \beta_4 Z_3 + E$)
$F(Z_1, Z_2, Z_3 \mid WGT)_{3,55} = 1662.526, P < 0.001$. At $\alpha = 0.05$, we reject H_0 and conclude that the adjusted means significantly differ.

11. a. Using PPM_TOLU as the reference group, we define the cross-product terms as
$XZ_1 = WGT$ if PPM_TOLU $= 100$,   0 otherwise
$XZ_2 = WGT$ if PPM_TOLU $= 500$,   0 otherwise
$XZ_3 = WGT$ if PPM_TOLU $= 1000$,   0 otherwise
in which X stands for WGT.

b. The appropriate regression model is
LN_BRNTL $= \beta_0 + \beta_1 X + \beta_2 Z_1 + \beta_3 Z_2 + \beta_4 Z_3 + \beta_5 XZ_1 + \beta_6 XZ_2 + \beta_7 XZ_3 + E$
in which X stands for WGT.

c. The hypothesis of no interaction is tested in terms of model parameters as $H_0: \beta_5 = \beta_6 = \beta_7 = 0$.

13. a. $Y = \beta_0 + \beta_1 AGE + \beta_2 Z + E$

b. $Y = \beta_0 + \beta_1 AGE + \beta_2 Z + \beta_3 AGE*Z + E$
$H_0: \beta_3 = 0$ $H_A: \beta_3 \neq 0$ (Full model: $Y = \beta_0 + \beta_1 AGE + \beta_2 Z + \beta_3 AGE*Z + E$)
$T_{26} = -1.677, P = 0.106$. At $\alpha = 0.05$, we do not reject H_0, and we conclude that the ANACOVA model in part (a) is appropriate.

c.

Location	Adjusted Means	Unadjusted Means
Intown/inner	81.526	82.187
Outer	85.46	84.807

$H_0: \beta_2 = 0$ $H_A: \beta_2 \neq 0$
$T_{27} = 1.212, P = 0.236$. At $\alpha = 0.05$, we do not reject H_0, and we conclude that the adjusted means do not significantly differ.

15. a. Let X denote OWNCOST and Y denote OWNEROCC.
$Y = \beta_0 + \beta_1 X + \beta_2 Z + E$

b. $Y = \beta_0 + \beta_1 X + \beta_2 Z + \beta_3 XZ + E$
$H_0: \beta_3 = 0$ $H_A: \beta_3 \neq 0$
$F(XZ \mid X, Z)_{1,22} = \dfrac{0.696}{17.580} = 0.04$, with $P \gg 0.25$.

At $\alpha = 0.05$, we do not reject H_0 and conclude the ANACOVA model in part (a) is appropriate.

c.

Location	Adjusted Means	Unadjusted Means
Urban $\geq 75\%$	56.15%	64.5%
Urban $\leq 75\%$	60.06%	69.25%

H_0: $\beta_2 = 0$ H_A: $\beta_2 \neq 0$

$T_{23} = -2.191$, $P = 0.039$.

At $\alpha = 0.05$, we reject H_0 and conclude the adjusted means significantly differ.

Chapter 14

1. a. The plot of jacknife residuals versus the predicted values shows a distinct pattern, indicating that an assumption has been violated. In this case, the obvious curvilinear pattern indicates a violation of the linearity assumption (much as the simple plot of Y vs X did in Chapter 5, Problem 1). A linearizing transformation should be applied, e.g., $\log_{10}$(dry weight) can be used as the dependent variable.

b. The skewness statistic ($=1.66$) and kurtosis statistic ($=3.21$) suggest that at least a moderate violation of the normality assumption has occurred. The normal probability plot also looks fairly non-linear. With such few data values, it is difficult to conclude that a *gross* violation of the linearity assumption has occurred. Since a violation of the linearity assumption has occurred (see part (a)), no attempt should be made to correct for the possible violation of normality until the linearity issue is addressed.

c. Observation 11 has a Cook's distance value greater than 1. No observations have leverage values greater than $2(k + 1)/n = 0.36$. The data for observation 11 should be double-checked and corrected, if necessary; if the recorded value is correct, it should be judged as to plausibility. If judged to be implausible, the observation can be removed from the analysis. If plausible, no corrective action is taken; instead, the analysis can be run with and without the observation included and the regression results compared to judge the impact of the outlier.

3. a. The plot of jackknife residuals versus the predicted values shows a distinct pattern, indicating that an assumption has been violated. In this case, the obvious curvilinear pattern indicates a violation of the linearity assumption (much as the simple plot of Y vs X did in Chapter 5, Problem 3). In this problem, it may help to add an X^2 term to the model.

b. The skewness and kurtosis statistics are both less than 1 in magnitude; the normal probability plot is not grossly non-linear. This suggests that there is no gross violation of the normality assumpton.

c. No Cook's distance values are greater than 1. Four observations have leverage values greater than $2(k + 1)/n = 0.20$. The data for these observations should be double-checked and corrected, if necessary; if the recorded values are correct, they should be judged as to plausibility. If judged to be implausible, an observation can be removed from the analysis. If plausible, no corrective action is taken; instead, the analysis can be run with and without the observation included and the regression results compared to judge the impact of the outlier.

19. a. The plot of jackknife residuals vs. redicted values looks like a random scatter of points; since no pattern is evident, no assumptions appear to be violated based on this plot.

b. The skewness and kurtosis statistics are both less than 1 in magnitude; the normal probability plot is fairly linear. This suggests that there is no gross violation of the normality assumption.

c. No Cook's distance values are greater than 1. The leverage value for observation 18 is greater than $2(k + 1)/n = 0.42$. The data for this observation should be double-checked and corrected, if necessary.

d. None of the variance inflation factors are larger than 10 and none of the condition indexes is greater than 30. Therefore, no collinearity problem exists.

25. a. $Y = \beta_0 + \beta_1 (\text{ADVERTISING}) + E$, where Y denotes sales.

    b. The largest studentized residual (absolute value) = 1.527, which is no cause for alarm.

    c. The plot of the jackknife residuals versus the predictor does not suggest any troublesome problems.

Chapter 15

1. b. From the computer output, we find:

    (1) Degree 1: $\hat{Y} = -1.932 + 0.246X$

    (2) Degree 2: $\hat{Y} = 3.172 - 0.781X + 0.047X^2$

    (3) ln Y on X: ln $\hat{Y} = -6.21 + 0.451X$

    (4) The above fitted equations are plotted on the graphs presented for 1(a).

c.

Source	d.f.	SS	MS	F
Regression	1	12.705	12.705	43.69
Lack of fit	4	4.419	1.105	57.03
Residual	16	4.651	0.2908	
Pure error	12	0.232	0.0194	
Total	17	17.357		

d.

Source	d.f.	SS	MS	F
Degree 1(X)	1	12.705	12.705	43.69
Regression	2	16.61	8.305	
Degree 2 ($X^2 \mid X$)	1	3.905	3.905	78.46
Lack of fit	3	0.514	0.171	8.85
Residual	15	0.746	0.0497	
Pure error	12	0.232	0.0194	
Total	17	17.357		

e. $r_{XY}^2 = 0.732$; $r^2(quadratic) = 0.957$

f. Test for significance of straight-line regression of Y on X

H_0: The straight-line regression is not significant.

$F_{1,16} = 43.69$, $P < 0.001$. At $\alpha = 0.05$, we reject H_0 and conclude that the straight-line regression is significant.

Test for adequacy of straight-line model

H_0: The straight-line model is adequate.

$F_{4,12} = 56.959$, $P < 0.001$. At $\alpha = 0.05$, we reject H_0 and conclude that the straight-line model is not adequate.

g. Test for significance of quadratic regression

H_0: The quadratic regression is not significant.

$F_{2,15} = 166.77$, $P = 0.0001$. At $\alpha = 0.05$, we reject H_0 and conclude that the quadratic regression is significant.

Test for addition of X^2 term

H_0: The addition of X^2 to a model already containing X is not significant.

Partial $F(X^2 | X)_{1,15} = 78.41$, $P = 0.0001$. At $\alpha = 0.05$, we reject H_0 and conclude that the addition of X^2 is significant.

Test for adequacy of quadratic model

H_0: The quadratic model is adequate.

$F_{3,12} = 8.84$, $P = 0.002$. At $\alpha = 0.05$, we reject H_0 and conclude that the quadratic model is not adequate.

h. Test for significance of straight-line regression of ln Y on X

H_0: The straight-line regression is not significant.

$F_{1,16} = 4277.167$, $P = 0.0001$. At $\alpha = 0.05$, we reject H_0 and conclude that the straight-line regression is significant.

Test for adequacy of straight-line model of ln Y on X

H_0: The straight-line model is adequate.

$F_{4,12} = 0.896$, $P = 0.471$. At $\alpha = 0.05$, we do not reject H_0 and conclude that the straight- line model is adequate.

i. R^2 (straight-line regression of ln Y on X) $= 0.9965$

R^2 (quadratic regression of Y on X) $= 0.957$

A comparison of the above two R^2 shows that the straight-line fit of ln Y on X provides a better fit.

j. (1) Homoscedasticity assumption appears to be much more reasonable when using ln Y on X than when using Y on X.

(2) The straight-line regression of ln Y on X is preferred.

k. The independence assumption is violated.

5. b. Test for significance of straight-line regression

H_0: The straight-line regression is not significant.

$F_{1,24} = 978.04$, $P < 0.001$. At $\alpha = 0.05$, we reject H_0 and conclude that the straight-line regression is significant.

Test for adequacy of straight-line model

H_0: The straight-line model is adequate.

$F_{16,8} = 1.57$, $P > 0.25$. At $\alpha = 0.05$, we do not reject H_0 and conclude that the straight-line model is adequate.

c. Test for addition of X^2 to the model

H_0: The addition of X^2 is not significant.

Partial $F(X^2 | X)_{1,23} = 0.55$, $P > 0.25$. At $\alpha = 0.05$, we do not reject H_0.

d. The straight-line model is most appropriate.

9. a. Fit the model VOC_SIZE $= \beta_0 + \beta_1(\text{AGE}^*) + \beta_2(\text{AGE}^*)^2 + \beta_3(\text{AGE}^*)^3 + E$, where AGE* $=$ AGE $- 2.867$.

$\hat{\beta}_0 = 741.84 \quad \hat{\beta}_1 = 645.60 \quad \hat{\beta}_2 = 70.43 \quad \hat{\beta}_3 = -31.18$

b. Using variables-added-in-order tests, the best model includes AGE*, $(\text{AGE}^*)^2$, and $(\text{AGE}^*)^3$.

c. Using variables-added-last tests, the best model includes AGE*, $(\text{AGE}^*)^2$, and $(\text{AGE}^*)^3$.

d. The only large predictor correlation is between AGE* and $(\text{AGE}^*)^3$. The largest condition index ($\text{CI}_3 = 6.05$) suggests that the centered data do not have any serious collinearity problems.

f. The estimated regression coefficients differ from those in Problem 8, but the best model includes the linear, quadratic, and cubic terms as in Problem 8. Also, the sums of squares for the variables-added-in-order test are the same as in Problem 8. The centering of AGE greatly reduced the previous collinearity problems. Centering does not affect the residual diagnostics.

13. a. The estimated equation is

$\hat{Y} = 10.64 + 94.42(\text{LIN\_TOL}) + 14.59(\text{QUAD\_TOL}) - 0.73(\text{CUB\_TOL})$,

where LN_TOL, QUAD_TOL, and COL_TOL denote the centered PPM_TOLU orthogonal polynomials.

b. Using the variable-added-in-order tests, the best model includes LIN_TOL and QUAD_TOL .

c. Using the variable-added-last tests, the best model includes LIN_TOL and QUAD_TOL, which is the same model as in part (b).

d. The orthogonal polynomials are uncorrelated with each other, which implies that any collinearities are eliminated as shown by the condition indices.

e. The residual plots suggest that the variances increase as the predicted values increase.

Chapter 16

1. a. The final model by forward selection is $\widehat{\text{WGT}} = 37.600 + 0.053(\text{AGE*HGT})$

b. The final model by backward elimination is $\widehat{\text{WGT}} = 37.600 + 0.053(\text{AGE*HGT})$

c. The best model resulting from this approach is:

$\widehat{\text{WGT}} = 6.553 + 0.722(\text{HGT}) + 2.050(\text{AGE})$

d. The first two approaches result in the model including only the AGE*HGT interaction. It is difficult to interpret results for such a model, since the variables that make up the interaction term are not included. The third approach results in a model that is easier to interpret. It is possible to conduct modified forward and backward stepwise strategies in which interaction terms are only considered for inclusion if the terms that make up the interaction term are already in the model.

3. For smokers: the final estimated model is $\widehat{\text{SBP}} = 102.200 + 0.252(\text{AGE*QUET})$

For nonsmokers: the final estimated model is $\widehat{\text{SBP}} = 93.073 + 0.250(\text{AGE*QUET})$

These two models are different from those obtained from 2(d) by putting SMK = 1 (for smokers) and SMK = 0 (for nonsmokers).

5. The best regression models using the sequential procedure of adding AGE first for Females and Males are

Females: $\widehat{\text{DEP}} = 190.012 - 1.099\text{AGE} - 1.217\text{MC}$

Males: $\widehat{\text{DEP}} = 270.056 - 2.514\text{AGE} - 1.065\text{MC}$

For females, the above model was selected as best because of its high r^2 (0.401), satisfactory $C(P)$(2.238), and low MSE (3274.364). The next best model was the full model with $r^2 = 0.413$, $C(P) = 4.0$, MSE = 3478.318. Emphasizing parsimony, we find that the above model is more favorable than the full model.

The reasoning is similar for selecting the model shown above for males.

7. a. Using the $C(P)$ criterion exclusively, we find that the three models with the most favorable $C(P)$ values contain AGE alone (1.23), AGE and WGT (2.63), or all three variables (4). None of these models is particularly impressive, and since it would be hard to argue that either of the multiple-variable models is better than the model with AGE alone, we could take AGE alone as the best model. Upon further investigation, the difficulty is seen to be partly due to none of these models having a significant overall F test.

b. Since there seems to be no rationale for grouping the variables (age, weight, and height) in any way, chunks are taken to be the three variable specific pairs of linear/quadratic terms (i.e., AGE_C and AGE_CSQ; HGT_C and HGT_CSQ; WGT_C and WGT_CSQ, where the _C terms are the centered variables, and the _CSQ terms are the squared centered terms). A plausible forward chunkwise strategy is to treat each chunk/pair as a distinct entity that cannot be split and then proceed in the usual forward manner (use $\alpha = 0.10$):

Step 1: WGT_C, WGT_CSQ added to the model. F test $= 2.78$
Step 2: AGE_C, AGE_CSQ added to the model. F test $= 3.49$
Step 3: Stop HGT_C, HGT_CSQ not significant. F test $= 0.16$

c. The all possible regressions method yields the following "best" models for each of the model sizes:

Number in Model	Variables	$C(P)$	R^2	MSE
1	WGT_CSQ	8.026	0.078	0.771
2	WGT_C, WGT_CSQ	5.213	0.300	0.631
3	WGT_C, WGT_CSQ, AGE_CSQ	4.337	0.432	0.554
4	WGT_C, WGT_CSQ, AGE_C, AGE_CSQ	3.312	0.571	0.456
5	WGT_C, WGT_CSQ, AGE_C, AGE_CSQ, HGT_CSQ	5.225	0.576	0.497
6	Full Model	7.00	0.586	0.539

The best model is most likely the four-variable model including WGT_C, WGT_CSQ, AGE_C, and AGE_CSQ. The R^2, $C(P)$, and MSE for this model are better than for any of the smaller models; and they are similar to, if not better than, the statistics for the larger models, which of course are less parsimonious.

d. Any model containing only first-order terms (part (a)) is seriously deficient. The model in parts (b) and (c) is the best one.

9. a. A model containing X_1, X_3, and X_4 is the best model by this method. The model has a relatively high R^2, a satisfactory $C(P)$, and one of the lowest MSEs. The model also has the benefit of parsimony compared to the larger models.

b. The selected model contains X_1, X_3, and X_4.

c. The selected model contains X_1, X_3, and X_4.

d. All three methods selected the same model, and it appears to be the best model for the reasons cited in (a).

11. a. The model containing X_1 and X_3 would be the recommended model. Its R^2, $C(P)$, and MSE are clearly superior to the best one-variable model statistics, and they are similar to the statistics for the less parsimonious full model.

b. The results of the stepwise regression show that the model containing X_1 and X_3 is the best model.

c. The model: $Y = \beta_0 + \beta_1 X_1 + \beta_3 X_3 + E$ appears to be best. This model is chosen, given the results in parts (a) and (b).

Chapter 17

1. a.

Treatment	Mean	S
1	7.5	1.643
2	5	1.265
3	4.333	1.033
4	5.167	1.472
5	6.167	2.041

b. ANOVA table:

Source	d.f.	SS	MS	F
Treatment	4	36.467	9.117	3.90
Error	25	58.50	2.34	
Total	29	94.967		

c. H_0: $\mu_1 = \mu_2 = \mu_3 = \mu_4 = \mu_5 = 0$

H_A: At least two treatments have different population means.

$$F_{4,25} = \frac{9.117}{2.34} = 3.896, \text{ with } P = 0.0136.$$

At $\alpha = 0.05$, we reject H_0 and conclude that at least two treatments have different population means.

d. Estimates of true effects $(\mu_i - \mu)$ where μ is the overall mean:

Treatment

i	$(\bar{Y}_i - \bar{Y})$
1	1.8667
2	−0.6333
3	−1.3000
4	−0.4667
5	0.5333
Total $\sum_{i=1}^{5}(\bar{Y}_i - \bar{Y})$	0.000

e. We define X_i such that

$X_i = 1$, for treatment i; 0, otherwise

where $i = 1, 2, 3, 4$. Then the appropriate regression model is

$Y = \beta_0 + \alpha_1 X_1 + \alpha_2 X_2 + \alpha_3 X_3 + \alpha_4 X_4 + E$

where the regression coefficients are as follows:

$\beta_0 = \mu_5$

$\alpha_1 = \mu_2 - \mu_5, \alpha_2 = \mu_2 - \mu_5, \alpha_3 = \mu_3 - \mu_5, \alpha_4 = \mu_4 - \mu_5$

$X_i = 1$ for treatment $i(i = 1, 2, 3, 6)$; $X_5 = -1$ for treatment 5; $X_i = 0$ otherwise.

The regression coefficients are:

$\beta_0 = \mu, \alpha_1 = \mu_1 - \mu, \alpha_2 = \mu_2 - \mu, \alpha_3 = \mu_3 - \mu, \alpha_4 = \mu_4 - \mu$, and also

$-(\alpha_1 + \alpha_2 + \alpha_3 + \alpha_4) = \mu_5 - \mu$

f. Hand calculated 95% confidence intervals for $(\mu_1 - \mu_3)$ and $(\mu_1 - \mu_2)$ are as follows:

Comparison	Scheffé	Tukey-Kramer	Bonferroni
$(\mu_1 - \mu_3)$	(0.238, 6.096)	(0.57, 5.764)	(0.34, 5.994)
$(\mu_1 - \mu_2)$	(−0.429, 5.429)	(−0.097, 5.097)	(−0.327, 5.327)

Conclusions: Treatments 1 and 3 significantly differ; treatments 1 and 2 do not significantly differ. We conclude that all other remaining comparisons, which involve smaller (in absolute value) pairwise sample mean differences, are not significant using $\alpha = 0.05$. Of the three methods, Scheffé's method gives the widest interval, the Bonferroni method gives the next widest, and the Tukey-Kramer method gives the narrowest interval.

5. a.

Source	d.f.	SS	MS	F
Parties	3	5625	1875.00	37.31
Error	16	804	50.25	
Total	19	6429		

b. $H_0: \mu_1 = \mu_2 = \mu_3 = \mu_4$

H_A: at least two parties have different mean authoritarianism scores.

$F_{(3,16)} = 37.31$, with $P < 0.001$. At $\alpha = 0.05$, we reject H_0 and conclude that the authoritarianism scores of the members of different political parties significantly differ.

c. Regression model:

$Y = \beta_0 + \alpha_1 X_1 + \alpha_2 X_2 + \alpha_3 X_3 + E$ where $X_i = 1$ if party #i, 0 otherwise; $i = 1, 2, 3$

d. The Tukey-Kramer method confidence intervals for different comparisons are then given as follows:

Comparison	Confidence Interval	Remark
P_3 vs P_4:	$(\bar{Y}_3 - \bar{Y}_4) \pm 12.93 = (32.07, 57.93)$	significant
P_3 vs P_2:	$(\bar{Y}_3 - \bar{Y}_2) \pm 12.93 = (2.07, 27.93)$	significant
P_3 vs P_1:	$(\bar{Y}_3 - \bar{Y}_1) \pm 12.93 = (-2.93, 22.93)$	not significant, since 0 in interval
P_1 vs P_4:	$(\bar{Y}_1 - \bar{Y}_4) \pm 12.93 = (22.07, 47.93)$	significant
P_1 vs P_2:	$(\bar{Y}_1 - \bar{Y}_2) \pm 12.93 = (-7.93, 17.93)$	not significant
P_2 vs P_4:	$(\bar{Y}_2 - \bar{Y}_4) \pm 12.93 = (17.07, 42.93)$	significant

At $\alpha = 0.05$, all differences except for $(\bar{Y}_1 - \bar{Y}_2)$ and $(\bar{Y}_3 - \bar{Y}_1)$ are significant.

9. a. $H_0: \mu_1 = \mu_2 = \mu_3$ H_A: At least two mean generation times differ.

$F_{(3,20)} = 46.99$, with $P < 0.001$. At $\alpha = 0.05$, we reject H_0 and conclude that between strains significantly differ.

b.

Source	d.f.	SS	MS	F
Strains	3	134713.00	44904.33	46.99
Error	20	19113.244	955.662	
Total	23	153826.244		

c. Using the Tukey-Kramer method confidence intervals for pairwise comparisons, we conclude that

$$\mu_C > (\mu_D = \mu_A) > \mu_B$$

where μ_A, μ_B, μ_C, and μ_D are population means from Strains A, B, C, and D respectively.

13. a.

Source	d.f.	SS	MS	F
Dosage	3	566.628	188.876	14.71
w/in Dosage	44	564.96	12.84	
Total	47	1131.588		

17. a. $Y_{ij} = \mu + \alpha_i + E_{ij}$, $i = 1, 2, 3; j = 1, \ldots, 6$.

b. MST = 4.667 $F = 5.53$ MSE = 0.844

c. $H_0: \mu_1 = \mu_2 = \mu_3$

H_A: There is a significant difference in attitude toward advertising by practice type.

$F_{2,15} = 5.53$, with $P = 0.0159$. At $\alpha = 0.05$, we reject H_0 and conclude that there are significant differences in attitude toward advertising by practice type.

d. GP vs IM: $3.6667 - 2.00 \pm 1.4398$ $(0.2269, 3.1065)$ significant
GP vs FP: $3.6667 - 2.333 \pm 1.4398$ $(-0.1061, 2.7735)$ not significant
FP vs IM: $2.333 - 2.00 \pm 1.4398$ $(-1.1068, 1.7728)$ not significant

e. Same as 17(a).

f.

Source	d.f.	SS	MS	F
Model	2	26.333	13.167	20.43
Error	15	9.667	0.6444	
Total	17	35.999		

g. $H_0: \mu_1 = \mu_2 = \mu_3$ H_A: At least two means differ by practice type.
$F_{2,15} = 20.43$, $P = 0.0001$. At $\alpha = 0.05$, we reject H_0 and conclude that there is a significant difference in influence on prescription writing habits by practice type.

h. GP vs IM: $4.3333 - 1.5000 \pm 1.2578$ $(1.5755, 4.0911)$ significant
GP vs FP: $4.3333 - 2.1667 \pm 1.2578$ $(0.9088, 3.4244)$ significant
FP vs IM: $2.1667 - 1.500 \pm 1.2578$ $(-0.5911, 1.9245)$ not significant

21. a. $Y_{ij} = \mu + \alpha_i + E_{ij}$, $i = 1, \ldots, 3; j = 1, \ldots, n_i$. Clear zone sizes are fixed-effect factors.

b.

Source	d.f.	SS	MS	F	P-value
Model	2	14.704	7.352	5.462	0.0073
Error	48	64.592	1.346		
Total	50	79.296			

c. $H_0: \mu_1 = \mu_2 = \mu_3$ H_A: At least two mean five-year changes differ by clear zone.
$F_{2,48} = 5.462$, with $P = 0.0073$. At $\alpha = 0.05$, we reject H_0 and conclude that mean five-year changes significantly differ by clear zone size.

d. μ_1 vs μ_3: $(0.3368, 2.2188)$ significant
μ_1 vs μ_2: $(-0.2562, 1.6603)$ not significant
μ_2 vs μ_3: $(-1.6603, 0.2562)$ not significant
The mean five-year refractive changes significantly differ between 3.0 mm and 4.0 mm.

Chapter 18

1. a. The rats are the blocks, and the three chemicals are the treatments.

b. SSE = 25.0   MSE = 1.786   Type I SS (chem) = 25.0   MS (chem) = 12.5

c. $H_0: \mu_1 = \mu_2 = \mu_3$ H_A: The mean irritation scores differ by chemical type.
$F_{2, 14} = 7.00$ with $P = 0.0078$. At $\alpha = 0.05$, we reject H_0 and conclude the toxic effects of the three chemicals significantly differ.

d. 98% CI on $\mu_I - \mu_{II}$: $(-3.0, 0.5)$ **e.** $R^2 = 0.635$

f. Fixed-effects ANOVA model: $Y_{ij} = \mu + \tau_i + \beta_j + E_{ij}$, $i = 1, 2, 3; j = 1, 2, \ldots, 8$
where
Y_{ij} = observation on the jth rat for the ith chemical effect
μ = overall mean
τ_i = ith chemical effect
β_j = jth rat effect
E_{ij} = error due to observation on the jth rat for the ith chemical effect (E_{ij}'s are independent and assumed to be normally distributed).

Regression model: $Y = \beta_0 + \alpha_1 X_1 + \alpha_2 X_2 + \sum_{j=1}^{7} \beta_j Z_j + E$

$$\text{where } X_1 = \begin{cases} 1 & \text{if chemical I} \\ 0 & \text{if chemical II,} \\ -1 & \text{if chemical III} \end{cases} \quad X_2 = \begin{cases} 0 & \text{if chemical I} \\ 1 & \text{if chemical II,} \\ -1 & \text{if chemical III} \end{cases} \quad Z_j = \begin{cases} -1 & \text{if rat 8} \\ 1 & \text{if rat } (j = 1, 2, \ldots, 7) \\ 0 & \text{otherwise} \end{cases}$$

g. Assumptions underlying the model: (i) additivity of the model (no interaction), (ii) homogeneity of variance, (iii) normality of the errors, (iv) independence of the errors.

5. $H_0: \mu_1 = \mu_2 = \mu_3$ H_A: At least two mean ESP scores differ by person.
$F_{2,8} = 15.12$ with $P = 0.0019$. At $\alpha = 0.05$, we reject H_0 and conclude that there are significant differences in ESP ability by person. Note that, since the F test for blocking on days is not significant, one may conclude that blocking on days is not necessary.

9. a.–f.

Source	d.f.	SS	MS	F
Treatments	4	160	40	5.00
Blocks	5	240	48	6.00
Error	20	160	8	

g. Test of treatments: $F_{4,20} = 5.00$ with $0.005 < P < 0.01$. At $\alpha = 0.05$, we reject H_0 and conclude that there is a significant main effect of treatments.
Test of blocks: $F_{5,20} = 6.00$ with $0.001 < P < 0.005$. At $\alpha = 0.05$, we reject H_0 and conclude that there is a significant main effect of blocks.

Chapter 19

1. a. The two factors are levorphanol and epinephrine.
 b. Both factors should be considered fixed.
 c. Rearrangement of the data into a two-way ANOVA layout:

Levels of Epinephrine

	Absence −	Presence +
Absence −	(Control) 1.90, 1.80, 1.54, 4.10, 1.89	(Epinephrine only) 5.33, 4.85, 5.26, 4.92, 6.07
Presence +	(Levorphanol only) 0.82, 3.36, 1.64, 1.74, 1.21	(Levorphanol and Epinephrine) 3.08, 1.42, 4.54, 1.25, 2.57

Levels of Levorphanol

d. Table of sample means:

Epinephrine

Levorphanol	−	+	Total (Row Mean)
−	2.25	5.28	3.77
+	1.75	2.57	2.16
Total (Col. Mean)	2.00	3.93	2.96

The presence of levorphanol appears to reduce stress, whereas the presence of epinephrine appears to increase stress. In the presence of epinephrine, levorphanol reduces stress by an average of 2.71 units, whereas in the absence of epinephrine, levorphanol reduces stress by only 0.50 units, suggesting the possibility of an interaction.

e.

Source	d.f.	SS	MS	F
Levorphanol	1	12.832	12.932	12.60
Epinephrine	1	18.586	18.586	18.25
Interaction	1	6.161	6.161	6.05
Error	16	16.298	1.019	
Total	19	53.877		

f. Main effect for levorphanol

H_0: Levorphanol has no significant main effect on stress.

H_A: Levorphanol has a significant main effect on stress.

$F_{1,16} = 12.60, P = 0.0027$

At $\alpha = 0.05$, we reject H_0 and conclude that levorphanol has a significant main effect on stress.

Main effect for epinephrine

H_0: Epinephrine has no significant main effect on stress.

H_A: Epinephrine has a significant main effect on stress.

$F_{1,16} = 18.25, P = 0.0066$

At $\alpha = 0.05$, we reject H_0 and conclude that epinephrine has a significant main effect on stress.

Interaction

H_0: There is no significant interaction between levorphanol and epinephrine on stress.

H_A: There is a significant interaction between levorphanol and epinephrine on stress.

$F_{1,16} = 6.05, P = 0.0267$

At $\alpha = 0.05$, we reject H_0 and conclude that there is significant interaction between levorphanol and epinephrine on stress.

5. a. Yes. There is an apparent same-direction interaction, which is reflected in the fact that the difference in mean waiting times between suburban and rural court locations is larger for State 1 than for State 2.

b. Main effect of State: $F_{1,594} = 217.45, P < 0.001$

Main effect of Court Location: $F_{2,594} = 184.86, P < 0.001$

Interaction effect: $F_{2,594} = 10.96, P < 0.001$

All effects are highly statistically significant.

c. Regression model:

$$Y = \beta_0 + \beta_1 S + \beta_2 C_1 + \beta_3 C_2 + \beta_4 SC_1 + \beta_5 SC_2 + E$$

where

$S = 1$ if State 1, -1 if State 2 $C_1 = -1$ if Urban, 1 if Rural, 0 if Suburban

$C_2 = -1$ if Urban, 1 if Suburban, 0 if Rural

d. We might consider a model of the form

$$Y = \beta_0' + \beta_1' S + \beta_2' C + \beta_3' SC + E$$

where S is as defined in (c) and where C is a variable taking on three values (i.e., 0, 1, and 2) that increase directly with the degree of urbanization. One difficulty here is how to determine the appropriate values for C.

9. a. "Species": fixed factor; "Locations": random factor, unless only the four locations chosen are of interest.

b.

Source	F (Mixed)	P-value	F (Both fixed)	P-value
Species	9.109 (d.f. = 2,6)	$0.01 < P < 0.025$	9.455 (d.f. = 2,48)	$P < 0.001$
Locations	0.811 (d.f. = 3,6)	$P > 0.25$	0.841 (d.f. = 3,48)	$P > 0.25$
Interaction	1.038 (d.f. = 6,48)	$P > 0.25$	1.038 (d.f. = 6,48)	$P > 0.25$

13. a.

Source	d.f.	SS	F	P
AIRTEMP	3	0.03	0.32	0.813
DOSE	2	0.06	1.04	0.370
A*D	6	0.05	0.29	0.936
Error	24	0.70		
TOTAL	35			

b. From the ANOVA table, we see that the main effects are not significant, and neither is the interaction term.

c. An appropriate multiple regression model is:

$$Y = \mu + \sum_{i=2}^{4} \alpha_i X_i + \sum_{j=2}^{3} \beta_j Z_j + \sum_{i=2}^{4}\sum_{j=2}^{3} X_i Z_j \gamma_{ij} + E$$

in which

$$X_i \begin{cases} -1 & \text{for AIRTEMP} = 21 \\ 1 & \text{for level } i \text{ of AIRTEMP}, i = 2, 3, 4 \\ 0 & \text{otherwise} \end{cases}$$

$$Z_j \begin{cases} -1 & \text{for DOSE} = 0 \\ 1 & \text{for level } j \text{ of DOSE}, j = 2, 3 \\ 0 & \text{otherwise} \end{cases}$$

$\alpha_i = \mu_{i.} - \mu_{..}, i = 2, 3, 4$

$\beta_j = \mu_{j.} - \mu_{..}, j = 2, 3$

$\gamma_{ij} = \mu_{ij} - \mu_i + \mu_{..}, i = 2, 3, 4; j = 2, 3$

AIRTEMP = 21 and DOSE = 0 correspond to control levels and hence are of least interest. The larger AIRTEMPS and DOSES are of primary interest, so they are directly parameterized in the model.

d. A natural polynomial model is

$Y = \beta_0 + \beta_1(\text{AIRTEMP}) + \beta_2(\text{DOSE}) + \beta_3(\text{AIRTEMP} * \text{DOSE}) + E.$

Note that higher-order terms can be added to the model if deemed reasonable.

17. a. The ANOVA model:

$Y_{ijk} = \mu + \alpha_i + \beta_j + \gamma_{ij} + E_{ijk}, i = 1, 2; j = 1, 2; k = 1, \ldots, 6$

α_i = effect for the ith school type

β_j = effect for the jth reputation category

γ_{ij} = interaction effect for the ith school type and jth reputation category

E_{ijk} = error term for the observation ijk

The factors are fixed.

b. d.f. = 3 $F_{3,20} = 2.572$, with $0.05 < P < 0.10$.

c. Main effect of school type:

H_0: There is no significant main effect of school type on starting salary.

H_A: There is a significant main effect of school type on starting salary.

$F_{1,20} = 0.52, P = 0.4786$. At $\alpha = 0.05$, we do not reject H_0, and we conclude that school type does not have a significant main effect on starting salary.

Main effect of reputation rank:

H_0: There is no significant main effect of reputation rank on starting salary.

H_A: There is a significant main effect of reputation rank on starting salary.

$F_{1,20} = 6.21, P = 0.0216$. At $\alpha = 0.05$, we reject H_0 and we conclude that reputation rank has a significant main effect on starting salary.

Test of interaction:

H_0: There is no significant interaction between school type and reputation rank with respect to starting salary.

H_A: There is significant interaction between school type and reputation rank with respect to starting salary.

$F_{1,20} = 0.99, P = 0.3323$. At $\alpha = 0.05$, we do not reject H_0, and we conclude that there is no significant interaction between school type and reputation rank.

Chapter 20

1. a. Table of sample means:

		HI	MED	LO	Total
	HI	135	147.5	165	152.5
Modern	MED	145	155	163	155.91
Rank	LO	161.67	143.33	121.67	142.22
	Total	147.22	148.75	154.23	

Traditional Rank (column header spanning HI, MED, LO)

The preceding table indicates than the sample mean blood pressure for males with low modern rank is lower than for other modern rank categories. It also illustrates that the sample mean blood pressure for males with low traditional rank is higher than the other traditional rank categories. Finally, the table hints at an interaction effect: for persons with high modern rank, mean blood pressure increases with decreasing traditional rank, whereas mean blood pressure for persons with low modern rank decreases with decreasing traditional rank. In other words, this interaction suggests that persons with incongruous cultural roles (i.e., HI-LO, LO-HI) tend to have higher blood pressures than persons with more congruent cultural roles.

b. Regression model:

$$Y = \beta_0 + \alpha_1 X_1 + \alpha_2 X_2 + \beta_1 Z_1 + \beta_2 Z_2 + \gamma_{11} X_1 Z_1 + \gamma_{12} X_1 Z_2 + \gamma_{21} X_2 Z_1 + \gamma_{22} X_2 Z_2 + E$$

where

$$X_i = \begin{cases} -1 & \text{if LO modern rank} \\ 1 & \text{if modern rank } i \ (i = 1 \text{ for HI}, i = 2 \text{ for MED}) \\ 0 & \text{otherwise} \end{cases}$$

and

$$Z_i = \begin{cases} -1 \cdot \text{if LO traditional rank} \\ 1 \ \ \text{if traditional rank } i \ (i = 1 \text{ for HI}, i = 2 \text{ for MED}) \\ 0 \ \ \text{otherwise} \end{cases}$$

c. Modern main effect:

(i) $F(X_1, X_2)_{2,27} = 2.076, 0.10 < P < 0.25$, which is not significant.

(ii) $F(X_1, X_2|Z_1, Z_2)_{2,25} = 1.768, 0.10 < P < 0.25$, which is not significant.

Traditional main effect:

(i) $F(Z_1, Z_2)_{2,27} = 0.578, P > 0.25$, which is not significant.

(ii) $F(Z_1, Z_2|X_1, X_2)_{2,25} = 0.397, P > 0.25$, which is not significant.

Interaction:

$F(X_1Z_1, X_1Z_2, X_2Z_1, X_2Z_2 | X_1, X_2, Z_1, Z_2)_{4,21} = 15.069, P < 0.001$, which is highly significant.

d. $Y = \beta_0 + \beta_1X_1 + \beta_2X_2 + \beta_3X_1X_2 + E$, where

$$X_1 = \begin{cases} 0 & \text{if LO modern rank} \\ 1 & \text{if MED modern rank} \\ 2 & \text{if HI modern rank} \end{cases} \quad X_2 = \begin{cases} 0 & \text{if LO traditional rank} \\ 1 & \text{if MED traditional rank} \\ 2 & \text{if HI traditional rank} \end{cases}$$

The difficulty arises with regard to assigning numerical values to the categories of each factor. The coding scheme for X_1 and X_2 given here assumes that the categories are "equally spaced," which may not really be the case.

5. a. Table of means:

Social Class

		Lo	Med	Hi	Row means
Number	0	14.571	12.00	13.50	13.5
of times	1	9.00	12.25	7.40	9.4
victimized	2+	8.00	2.33	5.75	5.8
Col means		11.3	9.7	8.8	

The preceding table suggests a downward trend in confidence with increasing number of victimizations and a slight downward trend with increasing social class status score.

b. From the ANOVA table given in the question, we obtain

$F(\text{Victim})_{2,31} = 8.81, P < 0.001$, which is significant.

$F(\text{SCLS})_{2,31} = 0.501, P > 0.25$, which is not significant.

$F(\text{Interaction})_{4,31} = 1.21, P > 0.25$, which is not significant.

c. We can use the following regression model:

$Y = \beta_0 + \alpha_1X_1 + \alpha_2X_2 + \beta_1Z_1 + \beta_2Z_2 + \gamma_{11}X_1Z_1 + \gamma_{12}X_1Z_2 + \gamma_{21}X_2Z_1 + \gamma_{22}X_2Z_2 + E$

where

$$X_1 = \begin{cases} -1 & \text{if no. of times victimized} = 0 \\ 1 & \text{if no. of times victimized} = 1 \\ 0 & \text{otherwise} \end{cases}$$

$$X_2 = \begin{cases} -1 & \text{if no. of times victimized} = 0 \\ 1 & \text{if no. of times victimized} = 2+ \\ 0 & \text{otherwise} \end{cases}$$

$$Z_j = \begin{cases} -1 & \text{if social class status} = \text{LO} \\ 1 & \text{if social class status} = j \ (j = 1 \text{ for MED}, j = 2 \text{ for HI}) \\ 0 & \text{otherwise} \end{cases}$$

We then obtain the following F values from the regression results given in the question:

Main effect of VICTIM:

$F(X_1, X_2)_{2,37} = 9.03$, $P < 0.001$, which is significant.

$F(X_1, X_2 | Z_1, Z_2)_{2,35} = 8.61$, $P < 0.001$, which is significant.

Main effect of SCLS:

$F(Z_1, Z_2)_{2,37} = 0.695$, $P > 0.25$, which is not significant.

$F(Z_1, Z_2 | X_1, X_2)_{2,35} = 0.707$, $P > 0.25$, which is not significant.

Interaction:

$F(X_1Z_1, X_1Z_2, X_2Z_1, X_2Z_2 | X_1, X_2, Z_1, Z_2)_{4,31} = 1.10$,

$P > 0.25$, which is not significant.

d. The error term may have a nonconstant variance.

9. a. A tabulation of DOSAGE *SEX sample sizes reveals that all combinations of these factors have 10 subjects except for 9 in the DOSAGE $= 0$, SEX $=$ Male combination. We could call this a nearly orthogonal design since we would only need one more observation in the cell with 9 to make a perfectly orthogonal design (equal cell sizes).

b.

Source	TYPE I		TYPE III	
	F	P-value	F	P-value
DOSAGE	2.65	$0.05 < P < 0.10$	2.64	$0.05 < P < 0.10$
SEX	0.22	$P > 0.25$	0.20	$P > 0.25$
DOSAGE*SEX	0.14	$P > 0.25$	0.14	$P > 0.25$

c. At $\alpha = 0.05$, no effects are significant.

d. Since no effects are significant, no multiple-comparison tests are justified.

13. a. $Y_{ijk} = \mu + \alpha_i + \beta_j + \gamma_{ij} + E_{ijk}$, $i = 1, 2, 3$; $j = 1, 2, 3$; $k = 1, \ldots, n_{ij}$;

$n_{11} = 3$, $n_{12} = 6$, $n_{13} = 11$, $n_{21} = 4$, $n_{22} = 5$, $n_{23} = 6$, $n_{31} = 4$, $n_{32} = 4$, $n_{33} = 8$

α_i = effect for the ith clear zone ($i = 1$: 3.0 mm, $i = 2$: 3.5 mm, $i = 3$: 4.0 mm)

β_j = effect for the jth baseline curvature category

γ_{ij} = interaction effect for the ith clear zone and jth curvature

E_{ijk} = error term for the observation ijk

b. d.f. $= 8$ $F_{8,42} = 3.595$ $0.001 < P < 0.005$

c. Using the SAS output: At $\alpha = 0.10$, we would conclude that there is a significant interaction between CLRZONE and BASECURV. We would also conclude that CLRZONE and BASECURV have significant main effects on refractive error.

Chapter 21

1. a. Yes, they are identical if we assume that the linear regression model is fit to normally distributed data.

b. $H_0: \beta_1 = 0$ $H_A: \beta_1 \neq 0$

(Full Model: $Y = \beta_0 + \beta_1 DRUG + \beta_2 SBP + \beta_3 QUET + E$ where DRUG $= 1$ if drug, 0 if placebo)

Test Statistic: $Z = \dfrac{\hat{\beta}_1}{\sqrt{\widehat{\text{Var}}_{\hat{\beta}_1}}}$, which is approximately standard normal under H_0.

Critical Value: If $Z > 1.96$, then reject H_0 at $\alpha = 0.05$.

c. No; however, as N increases, they approach one another.

  d. $H_0: \beta_1 = 0$ $H_A: \beta_1 \neq 0$
  Test statistic: $-2 \log L \text{ (reduced)} - (-2 \log L \text{ (full)}) = \chi^2$ with 1 d.f. where
  Full model equals: $Y = \beta_0 + \beta_1 DRUG + \beta_2 SBP + \beta_3 QUET + E$
  Reduced model equals: $Y = \beta_0 + \beta_2 SBP + \beta_3 QUET + E$
  Under H_0 the test statistic is approximately chi-square with 1 d.f.
  Critical value: if $\chi^2 > 3.841$, then reject H_0 at $\alpha = 0.05$.
  e. Yes, since $\chi^2 = Z^2$ when Z is normally distributed.
  f. 95% confidence interval for β_1: $\hat{\beta}_1 \pm 1.96 * \sqrt{\widehat{\text{Var}}_{\hat{\beta}_1}}$

Chapter 22

  1. a. $\hat{OR} = 2.925 \Rightarrow \hat{\beta}_1 = \ln(2.925) = 1.073$
  b. $\text{logit}[\Pr(Y = 1)] = -2.8 + 0.706X_1 + 0.0004X_2 + 0.0006X_3$
  c. $\Pr(Y = 1) = 0.112$
  d. $\hat{OR}_{20 \text{ vs } 21} = 0.999$
  The odds for hypertension for a 20-year-old smoker are essentially equal to those for a 21-year-old smoker.
  e. 95% CI for $\hat{OR}_{20 \text{ yr old smoker vs } 21 \text{ yr old smoker}}$: $(0.9982, 0.9998)$
  f. $H_0: \beta_3 = 0$ $H_A: \beta_3 \neq 0$
  Test Statistic: $-2 \log L \text{ (reduced)} - (-2 \log L \text{ (full)}) = \chi^2$ with 1 d.f.
  $\chi^2 = 308.00 - 303.84 = 4.16$, with $0.025 < P < 0.05$.
  At $\alpha = 0.05$, we reject H_0 and conclude that there is significant interaction between age and smoking.

Chapter 23

  1. a. $\ln \dfrac{P(D \geq g|X)}{P(D < g|X)} = \alpha_g + \beta_1 ALC1 + \beta_2 ALC2 + \beta_3 SMK1 + \beta_4 SMK2 + \beta_5 AGE$
  $$+ \beta_6 PA + \beta_7 CL1 + \beta_8 CL2 + \beta_9 CL3$$
  where $g = 1$ (moderate hypertensive), $g = 2$ (severe hypertensive)
  ALC1 $= 1$ if light drinker, 0 otherwise
  ALC2 $= 1$ if heavy drinker, 0 otherwise
  SMK1 $= 1$ if <2 packs per day, 0 otherwise
  SMK2 $= 1$ if $2+$ packs per day, 0 otherwise
  CL1 $= 1$ if Clinic 1, 0 otherwise
  CL2 $= 1$ if Clinic 2, 0 otherwise
  CL3 $= 1$ if Clinic 3, 0 otherwise
  b. $\hat{OR}_1 = \hat{OR}_3$ and $\hat{OR}_2 = \hat{OR}_4$
  c. The Score test allows you to evaluate the proportional odds (p.o.) assumption collectively for all predictor variables in a p.o. model. H_0: the p.o. assumption is satisfied. If the Score test does not reject H_0, then you can conclude that the p.o. model is appropriate. If the Score test rejects H_0, then you have to use some other approach for carrying out ordinal logistic regression or you can use polytomous logistic regression instead.
  d. $\exp[\alpha_1 + \beta_2 + \beta_4 + \beta_5 + \beta_6]$
  e. $OR_{BP \geq 1} = \exp[-\beta_1 + \beta_2 + \beta_4]$
  f. $OR_{BP=0} = OR_{BP<1} = 1/OR_{BP \geq 1} = \exp[\beta_1 - \beta_2 - \beta_4]$

g. $E_i E_j$ variables: ALC1 × SMK1, ALC1 × SMK2, ALC2 × SMK1, ALC1 × SMK2
$E_i V_j$ variables: ALC1 × AGE, ALC2 × AGE, ALC1 × PA, ALC2 × PA,
ALC1 × CL1, ALC2 × CL1, ALC1 × CL2, ALC2 × CL2, ALC1 × CL3, ALC2 × CL3,
SMK1 × AGE, SMK2 × AGE, SMK1 × PA, SMK2 × PA,
SMK1 × CL1, SMK2 × CL1, SMK1 × CL2, SMK2 × CL2, SMK1 × CL3, SMK2 × CL3

h. H_0: the coefficients of the 20 $E_i V_j$ variables listed above are all equal to 0.
LR statistic = $(-2 \ln L_R) - (-2 \ln L_F) \sim \chi^2$ with 20 df under H_0,
where R denotes the reduced model that contains the following predictor variables:
ALC1, ALC2, SMK1, SMK2, AGE, PA, CL1, CL2, CL3, and
ALC1 × SMK1, ALC1 × SMK2, ALC2 × SMK1, ALC1 × SMK2

i. For $g = 1, 2$, $\ln \dfrac{P(D \geq g|X)}{P(D < g|X)}$

$$= \alpha_g + \beta_1 ALC1 + \beta_2 ALC2 + \beta_3 SMK1 + \beta_4 SMK2 + \beta_5 AGE + \beta_6 PA$$
$$+ \beta_7 CL1 + \beta_8 CL2 + \beta_9 CL3 + \beta_{10}(ALC1 \times AGE) + \beta_{11}(ALC2 \times AGE)$$
$$+ \beta_{12}(SMK1 \times PA) + \beta_{13}(SMK2 \times PA) + \beta_{14}(ALC1 \times SMK1)$$
$$+ \beta_{15}(ALC1 \times SMK2) + \beta_{16}(ALC2 \times SMK1) + \beta_{17}(ALC2 \times SMK2)$$

j. $OR_{BP \geq 1} = \exp[-\beta_1 + \beta_2 + \beta_4 + \beta_{17} - \beta_{10} AGE + \beta_{11} AGE + \beta_{13} PA]$

k. 30 year old: AGE = 0; no regular physical activity: PA = 0
95% CI: $\exp[\hat{L} \pm 1.96\sqrt{V\hat{a}r(\hat{L})}]$ where

$$Var\,\hat{L} = Var\hat{\beta}_1 + Var\hat{\beta}_2 + Var\hat{\beta}_4 + Var\hat{\beta}_{17} - 2\,Cov(\hat{\beta}_1, \hat{\beta}_2) - 2\,Cov(\hat{\beta}_1, \hat{\beta}_4)$$
$$- 2\,Cov(\hat{\beta}_1, \hat{\beta}_{17}) + 2\,Cov(\hat{\beta}_2, \hat{\beta}_4) + 2\,Cov(\hat{\beta}_2, \hat{\beta}_{17}) + 2\,Cov(\hat{\beta}_4, \hat{\beta}_{17}).$$

l. $OR_{BP \geq 1} = \exp[-\beta_1 + \beta_2 + \beta_4 - \beta_{10} AGE + \beta_{11} AGE + \beta_{13} PA]$

2. a. $\ln Pr(DIS \geq g|X)/Pr(DIS < g|X) = \alpha_g + \beta_1 FAB + \beta_2 CHR1 + \beta_3 CHR2$
$$+ \beta_4 SEV1 + \beta_5 SEV2 + \gamma_1 SL + \gamma_2 SEX + \gamma_3 AGE$$

where $g = 1, 2, 3$ and $\alpha_1 > \alpha_2 > \alpha_3$
and CHR1 = 1 if CHR = 2 and 0 otherwise
CHR2 = 1 if CHR = 3 and 0 otherwise
SEV1 = 1 if SEV = 2 and 0 otherwise
SEV2 = 1 if SEV = 3 and 0 otherwise

b. i. $ODDS(X) = \exp[\alpha_2 + \beta_1 + \beta_3 + \beta_5 + \gamma_1 + \gamma_2 + 40\gamma_3]$
where

ii. $OR(X^*, X) = \exp[\beta_1 + \beta_3 + \beta_5]$

d. H_0: $\delta_{F1g} = \delta_{F2g} = \delta_{F3g} = \delta_{C1g} = \delta_{C2g} = \delta_{C3g} = \delta_{S1g} = \delta_{S2g} = \delta_{S3g} = 0$, for $g = 1, 2, 3$
LR statistic = $(-2\ln L_R) - (-2\ln L_F) \sim \chi^2$ with $9 \times 3 = 27$ d.f. under H_0

e. $OR_{X^*, X}(DIS = 3 \text{ vs } DIS = 1)$
$$= OR_{X^*, X}(DIS = 3 \text{ vs } DIS = 0)/\, OR_{X^*, X}(DIS = 1 \text{ vs } DIS = 0)$$
$$= \exp\,[\beta_{13} + 2\beta_{23} + 2\beta_{33} + 2\delta_{C13}SL + 8\delta_{CS33}]/\exp[\beta_{11} + 2\beta_{21} + 2\beta_{31}$$
$$+ 2\delta_{C11}SL + 8\delta_{CS31}]$$
$$= \exp\,[(\beta_{13} - \beta_{11}) + 2(\beta_{23} - \beta_{21}) + 2(\beta_{33} - \beta_{31}) + 2(\delta_{C13} - \delta_{C11})SL$$
$$+ 8(\delta_{CS33} - \delta_{CS31})]$$

Chapter 24

1. **a.** $n = 30$ (6 age–sex groups $\times$ 5 years)

   b. Here, $i = 5$ and $k = 1992 - 1990 = 2$, so $E(Y_{52}) = l_{52}\lambda_{52} = l_{52}e^{(\alpha_5 + 2\beta)}$

   c. Log rate changes linearly with time. In particular, for the ith group,
   $$\ln \lambda_{ik} = \alpha_i + \beta k$$
   so α_i is the intercept and β is the slope of the straight line relating the response $\ln \lambda_{ik}$ to the time variable $k = [\text{year}] - 1960$.

   d. Model (1) assumes no interaction between age–sex group and time in the sense that the change in log rate over time (as measured by β) does not depend on i. Since $\ln \lambda_{ik} = \alpha_i + \beta k$, a graph of $\ln \lambda_{ik}$ versus k for each i would plot as a series of parallel straight lines—that is, lines all with the same slope (β) but possibly different intercepts (the α_i's). A lack of parallelism would reflect interaction between age–sex groups and time because the change in log rate over time would differ for different age–sex groups.

   e. $\ln \text{IDR}_{ik} = \ln \lambda_{ik} - \ln \lambda_{10} = (\alpha_i + \beta k) - (\alpha_1 + \beta * 0) = (\alpha_i - \alpha_1) + \beta k$, so that
   $$\text{IDR}_{ik} = e^{\alpha_i - \alpha_1}e^{\beta k}$$
   Note that this is a function of both age–sex group (i) and time (k).

   f. An appropriate model is
   $$E(Y_{ik}) = l_{ik}\lambda_{ik}$$
   where
   $$\ln \lambda_{ik} = \sum_{i=1}^{6}\alpha_i A_i + \beta k + \sum_{i=1}^{5}\gamma_i(A_i k)$$
   For age–sex group i, then,
   $$\ln \lambda_{ik} = \alpha_i + \beta k + \gamma_i k$$
   $$= \alpha_i + (\beta + \gamma_i)k$$
   $$= \alpha_i + \delta_i k$$
   where $\delta_i = \beta + \gamma_i$. Hence the slope for group i (namely, δ_i) is now a function of i. Only when all six δ_i's (or equivalently, all six γ_i's) are equal will the straight lines be parallel.

   g. Yes, since $D(\hat{\beta})_{(1)} - D(\hat{\beta})_{(3)} = 300 - 175 = 125$, which is highly significant when compared to appropriate upper-tail χ^2-values with $29 - 24 = 5$ d.f.

   h. Yes, since $D(\hat{\beta})_{(3)} - D(\hat{\beta})_{(4)} = 175 - 60 = 115$, which is highly significant when compared to appropriate upper-tail χ^2-values with $24 - 23 = 1$ d.f.

   i. No, since $D(\hat{\beta})_{(4)} - D(\hat{\beta})_{(5)} = 60 - 59 = 1$, which is clearly not significant when compared to appropriate upper-tail χ^2-values with $23 - 22 = 1$ d.f.

   j. Yes, since $D(\hat{\beta})_{(4)} - D(\hat{\beta})_{(7)} = 60 - 20 = 40$, which is highly significant when compared to appropriate upper-tail χ^2-values with $23 - 18 = 5$ d.f.

   k. H_0 is rejected, since $D(\hat{\beta})_{(4)} - D(\hat{\beta})_{(6)} = 60 - 25 = 35$, which is highly significant when compared to appropriate upper-tail χ^2-values with $23 - 22 = 1$ d.f.

   l. Only models (6) and (7) are candidates to be the final model. Model (6) has a deviance of 25 based on 22 d.f., indicating a good fit to the data. Model (7) has a deviance of 20 based on 18 d.f., also indicating a good fit. All other candidate models have significant lack of fit. Note that $D(\hat{\beta})_{(6)} - D(\hat{\beta})_{(7)} = 25 - 20 = 5$, which is clearly not significant when compared to appropriate upper-tail χ^2-values with $22 - 18 = 4$ d.f. Hence, model (6) certainly fits the data as well as model (7), and it also characterizes very specifically the type of interaction present in the data (namely, that the group 1 slope differs from the slope common to the other five groups). Model (6) is our choice as the final model.

m. For model (6),

$$\text{pseudo } R^2 = \frac{300 - 25}{300} = 0.917$$

which is indicative of a good model.

n. $\hat{\beta} \pm 1.96 * SE_{\hat{\beta}} = 0.50 \pm 1.96(0.20) = (0.108, 0.892)$

o. The point estimate of δ_1 is $\hat{\delta}_1 = \hat{\beta} + \hat{\gamma}_1 = 0.50 - 3.00 = -2.50$. The variance of the estimator $\hat{\delta}_1$ is $\text{Var}(\hat{\delta}_1) = \text{Var}(\hat{\beta}) + \text{Var}(\hat{\gamma}_1) + 2 \text{Cov}(\hat{\beta}_1, \hat{\gamma}_1)$, which equals

$$\text{Var}(\hat{\delta}_1) = (0.20)^2 + (0.50)^2 + 2(-0.10) = 0.09$$

Finally, an approximate 95% CI for δ_1 is

$$\hat{\delta}_1 \pm 1.96 \sqrt{\widehat{\text{Var}}(\hat{\delta}_1)} = -2.50 \pm 1.96\sqrt{0.90} = (-3.088, -1.912).$$

These two confidence intervals suggest that log rate increases linearly ($\beta > 0$) for groups 2 through 6, but decreases ($\delta_1 < 0$) for group 1.

4. a. $RR = \exp[0.8869 + (-.2811) \ln T]$

 i. 15–24: $T = 1/7$, so $\widehat{RR}_{15-24} = \exp[0.8869 + (-0.2811) \ln(1/7)] = 4.20$

 ii. 45–54: $T = 1$, so $\widehat{RR}_{45-54} = \exp[0.8869 + (-0.2811)(\ln 1)] = 2.43$

b. The results in part **a** suggest that there is interaction of City with the variable $\ln T$, which is the natural log of Age Group centered around it's midpoint.

c. $H_0: \beta = 0$, where β is the coefficient of the product term $E \times \ln T_i$ in Model 5.
Wald Test: Chi-square statistic (1 d.f. under H_0) = 3.8386; 2-tailed P-value = 0.0501;
LR Test: Chi-square statistic (1 d.f. under H_0)
C = Dev(Model 4) − Dev(Model 5) = 14.8 − 10.63 = 4.17 (.025 < P < .05).

d. Yes, the results suggest that the product term is significant at the .05 level.

e. No. A test for fit uses the Dev statistic for one model only.

f. Model 4 is likely the best model choice, since the graph shown by Figure 24.1 shows excellent fit using the linear model involving $\ln T$, i.e., Model 4.

5. b. The person-time information for each subgroup would have to be the same, which is what is being assumed by Model 2. Otherwise, different effects would result from Model 1 and Model 2.

c. The model is $\ln[\lambda] = \beta_0 + \beta_1 AGE + \beta_2 SMOKE + \beta_3 AGE \times SMOKE$ where λ is the rate for a given subgroup. The null hypothesis is $H_0: \beta_3 = 0$.
The Wald statistic is given by $Z^2 = [.0036/.0100]^2 = .1264$
The two-tailed P-value for this test is .7222, which is > .05.
We conclude from the Wald test that the interaction term in Model 3 is not significant.

d. The deviance for Model 1 is of borderline significance (.05 < P < .10), whereas the deviance for ModeL 5 is not significant. This means that Model 1 questionably fits the data, whereas Model 5 fits the data very well. Thus, there should be a strong preference for Model 5.

e. LR = Deviance Model 4 − Deviance Model 5 = 10.3962 − 0.3925 = 10.0037.
The LR statistic has a chi-square distribution with 2 degrees of freedom under H_0. The P-value is less than .01. The two age variables should not be dropped from Model 5.

g. Model 6 is a saturated model.

h. $X^* = (75, \text{SMOKE})$, $X = (25, \text{SMOKE})$, so the rate ratio is $\exp[.0098(75 - 25) + .5784 (\text{SMOKE} - \text{SMOKE}] = 1.632$, 95% CI for $\exp[(50)\beta_1]$ is $\exp[0.49 \pm 1.96 (0.0049)(50)] = (1.01, 2.64)$

l. The best model is Model 5. It fits the data well. Its only competitor of the models provided is Model 1, which does not fit nearly as well. There is no significant interaction. Model 2 is out because it does not use an offset, and, as a result, does not fit the data well.

6. a. The sample size is 72.

c. Model 2: $\ln(E(C)) = \alpha + \beta_1' \ln(\text{DT}) + \beta_2' \ln(\text{WTGP}) + \gamma_1'(\text{GEN})$

$$+ \sum_{k=1}^{2} \gamma_{k+1}' (\text{AGEGP})_k + \delta_1' \ln(\text{DT}) \times (\text{GEN})$$

$$+ \delta_2' \ln(\text{WTGP}) \times (\text{GEN}) + \sum_{k=1}^{2} \delta_k^* \ln(\text{DT}) \times (\text{AGEGP})_k$$

$$+ \sum_{k=1}^{2} \delta_k^{**} \ln(\text{WTGP}) \times (\text{AGEGP})_k + \ln(\text{PT})$$

d. SAS code:
model C = LDT LWTGP GEN AGEGP1 AGEGP2 LDTGEN LWTGPGEN
              SDTAGEGP1 LDTAGEGP2 LWTGPAGEGP1 LWTGPAGEGP2/
              dist=poisson link=log offset=LPT;
where
LDT = $\ln$(DT), LWTGP = $\ln$(WTGP), AGEGPK = AGEGP$_k$ for $k = 1, 2$
LDTGEN = $\ln$(DT) $\times$ (GEN), LWTGPGEN = $\ln$(WTGP) $\times$ (GEN)
LDTAGEGK = $\ln$(DT) $\times$ (AGEGP)$_k$, LWTGPAGEGPK = $\ln$(WTGP) $\times$ (AGEGP)$_k$, for $k = 1, 2$

e. $X^* = (\ln 3, \ln 4, \text{GEN}, \text{AGEGP1}, \text{AGEGP2})$, $X = (\ln 1, \ln 1, \text{GEN}, \text{AGEGP1}, \text{AGEGP2})$
Using the model defined in part **c**:
    RR $= \exp[L]$ where
  $L = \beta_1'(\ln 3) + \beta_2'(\ln 4) + \delta_1'(\ln 3) \times (\text{GEN}) + \delta_2'(\ln 4) \times (\text{GEN})$

$$+ \sum_{k=1}^{2} \delta_k^*(\ln 3) \times (\text{AGEGP})_k + \sum_{k=1}^{2} \delta_k^{**}(\ln 4) \times (\text{AGEGP})_k$$

f. 95% CI: $\exp[\hat{L} \pm 1.96\sqrt{\widehat{\text{Var}}(\hat{L})}]$ where L is given in the answer to part **e**.

g. LR = Dev(Reduced Model) − Dev(Full Model) $\sim \chi^2_{(66-60)} = \chi^2_6$ under H_0 where
Reduced Model = Model 2 without 6 interaction terms, Full Model = Model 2
H_0: Coefficients of the six interaction terms in Model 2 are all 0

h. No. Model 2 is not the saturated model since Model 2 contains 12 parameters rather than the 72 parameters required for a saturated model in this example.

i. No. Model 1 is not the saturated model since Model 1 contains 24 parameters rather than the 72 parameters required for a saturated model.

Chapter 25

1. a. Matrix **A** is unstructured.
   b. Matrix **B** is independence.
   c. Matrix **C** is autoregressive.
   d. Matrix **D** is exchangeable.

2. a. **Q** and **S** are heterogeneous wereas **P** and **R** are homogeneous covariance matrices.

   b. **P:** autoregressive ($\rho = .5$); **Q:** exchangeable ($\rho = .46$); **R:** exchangeable ($\rho = .46$); **S:** autoregressive ($\rho = .5$).

3. The answer will vary with the reader.

4. a. Either of two ways to write the subject-specific scalar model:

   i. $Y = \beta_0 + \beta_1 R_1 + \beta_2 R_2 + \beta_3 R_3 + \beta_4 R_4 + E$,

   where $R_g = 1$ if week $g + 1$, otherwise $= 0$, $g = 1, \ldots, 4$

   ii. $Y_{ij} = \beta_0 + \beta_1 R_{ij1} + \beta_2 R_{ij2} + \beta_3 R_{ij3} + \beta_4 R_{ij4} + E_{ij}$, $i = 1, \ldots, 40, j = 1, \ldots, 5$

   If the (scalar) dummy variables R_1, R_2, R_3, and R_4 are defined so that the referent group is week 1, then $R_{ijg} = 1$ if $g = j - 1$ whereas $R_{ijg} = 0$ if $g \neq j - 1$.

   b. The subject-specific matrix form of the model in which week 1 is used as the reference level is given by

   $\mathbf{Y}_i = \mathbf{X}_i \boldsymbol{\beta} + \mathbf{E}_i, i = 1, \ldots, 40$,

   where $\mathbf{Y}_i$ denotes the collection of 5 FEV1 measurements on the ith child, $\mathbf{X}_i$ denotes the intercept variable (1.0, 1) together with the collection of dummy variable values for each of the five weeks for subject i, $\boldsymbol{\beta}$ denotes the parameter vector for the collection of β's in this model, and $\mathbf{E}_i$ denotes the collection of error terms on the ith subject. Assuming again that the referent group is week 5, then $\mathbf{X}_i$, $\boldsymbol{\beta}$, and $\mathbf{E}_i$ can be written as follows:

   $$\mathbf{X}_i = \begin{bmatrix} 1 & R_{i11} & R_{i12} & R_{i13} & R_{i14} \\ 1 & R_{i21} & R_{i22} & R_{i23} & R_{i24} \\ 1 & R_{i31} & R_{i32} & R_{i33} & R_{i34} \\ 1 & R_{i41} & R_{i42} & R_{i43} & R_{i44} \\ 1 & R_{i51} & R_{i52} & R_{i53} & R_{i54} \end{bmatrix} = \begin{bmatrix} 1 & 0 & 0 & 0 & 0 \\ 1 & 1 & 0 & 0 & 0 \\ 1 & 0 & 1 & 0 & 0 \\ 1 & 0 & 0 & 1 & 0 \\ 1 & 0 & 0 & 0 & 1 \end{bmatrix}, \boldsymbol{\beta}' = \begin{bmatrix} \beta_0' \\ \beta_1' \\ \beta_2' \\ \beta_3' \\ \beta_4' \end{bmatrix}, \mathbf{E}_i = \begin{bmatrix} E_{i1} \\ E_{i2} \\ E_{i3} \\ E_{i4} \\ E_{i5} \end{bmatrix}$$

   where $\mathbf{Y}_i$ is (5×1), $\mathbf{X}_i$ is (5×5), $\boldsymbol{\beta}^*$ is (5×1), and $\mathbf{E}_i$ is (5×1).

   c. $\hat{\beta}_0 = \overline{\text{FEV}}1_1, \hat{\beta}_1 = \overline{\text{FEV}}1_2 - \overline{\text{FEV}}1_1, \hat{\beta}_2 = \overline{\text{FEV}}1_3 - \overline{\text{FEV}}1_1,$

   $\hat{\beta}_3 = \overline{\text{FEV}}1_4 - \overline{\text{FEV}}1_1, \quad \hat{\beta}_4 = \overline{\text{FEV}}1_5 - \overline{\text{FEV}}1_1$

   d. $\beta_0' = \mu_1, \beta_1' = \mu_2 - \mu_1, \beta_2' = \mu_3 - \mu_1, \beta_3' = \mu_4 - \mu_1, \beta_4' = \mu_5 - \mu_1,$ so

   $$\mu_1 - \frac{\mu_2 + \mu_3 + \mu_4 + \mu_5}{4} = \beta_0' - \frac{(\beta_1' + \beta_0') + (\beta_2' + \beta_0') + (\beta_3' + \beta_0') + (\beta_4' + \beta_0')}{4}$$

   $$= -\frac{\beta_1' + \beta_2' + \beta_3' + \beta_4'}{4}$$

   Thus, the null hypothesis can alternatively be stated as H_0: $-\dfrac{\beta_1' + \beta_2' + \beta_3' + \beta_4'}{4} = 0.$

   e. Yes. The coding of the dummy variables will not affect the overall test statistic, even though the estimates of individual regression coefficients will change with the coding.

   f. *contrast 'week' d*1 $- .25$ *d*2 $- .25$ *d*3 $- .25$ *d*4 $- .25$;

   g. $\widehat{\text{FEV}}1_1 = \hat{\beta}_0' = 9.8137, \widehat{\text{FEV}}1_2 = \hat{\beta}_0' + \hat{\beta}_1' = 9.8137 + (-2.9655) = 6.8482,$

   $\widehat{\text{FEV}}1_3 = \hat{\beta}_0' + \hat{\beta}_2' = 9.8137 + (-2.8120) = 7.0017, \widehat{\text{FEV}}1_4 = \hat{\beta}_0' + \hat{\beta}_3' = 9.8137$

   $+ (-2.8623) = 6.9514,$

   $\widehat{\text{FEV}}1_5 = \hat{\beta}_0' + \hat{\beta}_4' = 9.8137 + (-2.8153) = 6.9984$

   h. $\hat{\mu}_1 - \dfrac{\hat{\mu}_2 + \hat{\mu}_3 + \hat{\mu}_4 + \hat{\mu}_5}{4}$

   $= 9.8137 - \dfrac{6.8482 + 7.0017 + 6.9514 + 6.9984}{4} = 2.8637$

i. The output does not provide the results for testing whether or not there is a significant difference among the mean FEV1 scores for all 5 weeks.
  i. contrast 'week'   d1 1, d2 1, d3 1, d4 1;
  ii. Yes, the results obtained from the contrast statement specified in the preceding question will be the same as the results previously obtained in Table 25.8.

j. **i.** exchangeable
   ii. $\hat{\rho} = .5565$
   iii. $\hat{\rho} = \dfrac{1.4936}{1.1901 + 1.4936} = .5565$
   iv. No, the output uses a model-based estimate of the standard error rather than an empirical or robust standard error; the latter corrects for possible mis-specification of the correlation structure.
   v. Yes, the estimated correlation structure should not depend on how the predictor variables are coded for the computer.

5. **c.** $Y_{ij} = \beta_0 + \beta_1(\text{trt})_{ij} + E_{ij}, i = 1, \ldots, 15, j = 1, 2,$
   where $(\text{trt})_{ij} = 1$ if the jth treatment on the ith subject is the active treatment, $= 0$ if the jth treatment on the ith subject is the placebo.

   d. $Y_i = X_i\beta + E_i, i = 1, \ldots, 15,$ where
   $$X_i(2 \times 2) = \begin{bmatrix} 1 & \text{trt}_{i1} \\ 1 & \text{trt}_{i2} \end{bmatrix}, \quad \beta(2 \times 1) = \begin{bmatrix} \beta_0 \\ \beta_1 \end{bmatrix}, \quad E_i(2 \times 1) = \begin{bmatrix} E_{i0} \\ E_{i1} \end{bmatrix}$$

6. **a.** $Y_{ij} = \beta_0 + \beta_1(\text{trt})_{ij} + E_{ij}, i = 1, \ldots, 15, j = 1, 2,$
   where $(\text{trt})_{ij} = 1$ if the jth treatment on the ith subject is the active treatment, $= 0$ if the jth treatment on the ith subject is the placebo.

   c. In this analysis, each subject gets both treatment and placebo, so there are 30 independent difference scores contributing to the variance due to the effect of treatment, which results in 29 df for error. In the analysis for the data of question 5, each subject gets either the treatment or placebo for both meals, so the 15 subjects receiving the active treatment contribute $15 - 1 = 14$ d.f. to error and the 15 subjects receiving the placebo contribute another 14 d.f. to error, so the error d.f. $= 14 + 14 = 28$ for all subjects together. (*Note*: The error d.f. of 4.4168 for treatment from the data of Problem 5 is considerably higher than the error d.f. of 2.2540 for treatment from the data of Problem 6; the latter error is smaller because each subject is serving as its own control, which is not the case for the data of Problem 5.)

   d. $Y_{ij} = \beta_0 + \beta_1(\text{trt})_{ij} + \beta_1(\text{seq})_{ij} + E_{ij}, i = 1, \ldots, 15, j = 1, 2,$
   where $(\text{seq})_{ij} = \begin{cases} 1 \text{ for subjects receiving the active treatment first and the placebo second,} \\ 0 \text{ for subject receiving the placebo first and the active treatment second.} \end{cases}$

Chapter 26

1. **a.** $\text{SF}_{ij} = \beta_0 + b_{i0} + \beta_1 D_{ij1} + \beta_2 D_{ij2} + E_{ij}, i = 1, \ldots, 19; j = 1, 2$
   where E_{ij} and b_{i0} are each assumed to be normally distributed as $N(0, \sigma^2)$ and $N(0, \sigma_0^2)$, respectively, and b_{i0} and E_{ij} are mutually independent for all i, j. Also, D_{ij1} and D_{ij2} are dummy variables that distinguish the three days, and SF_{ij} denotes the shoulder flexion measurement for the ith subject on the jth day.

   b. $H_0: \beta_1 = \beta_2 = 0$

   c. Yes, there appears to be a larger difference between Friday (15.78) and the other days (18.36 and 18.25).

d. To use SAS's MIXED procedure, the table needs to be reorganized so that there are $19 \times 3 = 57$ lines of data, with three lines per subject. There should be three columns of data per subject, containing the data for the following variables, subject id, day, and sf, where day denotes the day on which measurements are taken (coded, say, as 1 = Monday, 2 = Wednesday, 3 = Friday), and sf denotes the shoulder flexion measurement on a subject on a given day.

e. The factor Day is not significant, regardless of whether model-based standard errors with a random intercept model, empirical standard errors with a random intercept model, or empirical standard errors with a marginal model and an AR1 correlation structure are used. The F statistic obtained when an empirical standard error is used is $F = 1.19$ ($P = 0.317$) for both the random intercept model and the AR1 marginal model. The F statistic obtained for the model-based random intercept model is $F = 1.5646$ ($P = 2231$). The use of an empirical standard error makes most sense here, since such use considers the possibility of mis-specification of the correlation structure.

f. $\hat{\rho} = \dfrac{113.90}{113.90 + 25.8839} = 0.8148$

g. A test of significance for the random effect of Subjects can be performed using an approximate Wald test based on the output provided for "Covariance Parameter Estimates." The null hypothesis is $H_0: \sigma_0^2 = 0$, where σ_0^2 is the variance component associated with the random intercept b_{i0} in the random intercept only model stated in the answer to part (a). For this test, the Z value is $Z = 2.79$, which has a two-tailed P-value of .0027. However, because the alternative hypothesis for this test is $H_A: \sigma_0^2 > 0$, the two-tailed P-value must be halved, so that the correct P-value is .0014, which is significant well-below the .01 level. Consequently, using the approximate Wald test, we would conclude that the random effect of Subjects is significant, which justifies using a random effects model with at least a random intercept. If this test had been non-significant, there would be several sources of concern:

i. Perhaps a different conclusion would have been obtained from the more appropriate mixture test (in this case, also equivalent to the approximate LR test).

ii. Perhaps the use of either a marginal model with an AR1 correlation structure or simply an independence correlation structure would be more appropriate. (Note, however, that the computer output suggests that the results of the test for the effect of the Day variable would not change from non-significance no matter what other model or correlation structure is used.)

2. a. $SF_{ij} = \beta_0 + b_{i0} + (\beta_1 + b_{i1})D_{ij1} + (\beta_2 + b_{i2})D_{ij2} + E_{ij}, i = 1, \ldots, 19; j = 1, 2, \ldots, 12$, where E_{ij}, b_{i0}, b_{i1}, and b_{i2} are each assumed to be normally distributed as $N(0, \sigma^2)$, $N(0, \sigma_0^2)$, $N(0, \sigma_1^2)$, and $N(0, \sigma_2^2)$, respectively, noting that the variance components corresponding to b_{i1}, and b_{i2} are typically assumed to be equal. It is also assumed that b_{i0}, b_{i1}, b_{i2}, and E_{ij} are mutually independent for all i, j. Also, the variables D_{ij1} and D_{ij2} are dummy variables that distinguish the three days, and SF_{ij} denotes the jth shoulder flexion measurement for the ith subject.

b. The factor Day is not significant. The F statistics are 1.95 ($P = .1571$) and 1.84 ($P = .1730$) for model-based and empirical standard errors, respectively.

c. A test of significance for the interaction random effect of Subjects $\times$ Day can be performed using an approximate Wald test based on the output provided for "Covariance Parameter Estimates." The null hypothesis is $H_0: \sigma_1^2 = \sigma_2^2 = 0$, where σ_1^2 and σ_2^2 are the variance components associated with the random slopes b_{i1} and b_{i2}, respectively, in the random mixed model stated in the answer to part (a). For this test, the Z value is $Z = 3.23$, which has a two-tailed P-value of .0006. However, because the alternative hypothesis for this test is $H_A: \sigma_1^2 = \sigma_2^2 > 0$, the two-tailed P-value must be halved, so that the correct P-value is .0003, which is significant well-below the .01 level. Consequently, using the approximate Wald test, we would conclude that the interaction random effect of Subjects × Day is significant, which justifies using random slopes in the model. If this test had been non-significant, then:

   i. Perhaps a different conclusion would have been obtained from the more appropriate mixture test (in this case, also equivalent to the approximate LR test).

   ii. The use of a random intercept only model would be considered next.

3. a. The factor Blocking Step should be considered a fixed factor because there are only two categories (i.e., levels) of this factor of interest.

  b. $Y_{ij} = \beta_0 + b_{i0} + (\beta_1 + b_{i1})B_{ij} + E_{ij}, i = 1, \ldots, 6$ (sample); $j = 1, \ldots, 6$(replicate), where E_{ij}, b_{i0}, and b_{i1} are each assumed to be normally distributed as $N(0, \sigma^2)$, $N(0, \sigma_0^2)$, and $N(0, \sigma_1^2)$, respectively. It is also assumed that b_{i0}, b_{i1}, and E_{ij} are mutually independent for all i, j. [*Note*: this assumes that the random effects b_{i0} and b_{i1} are uncorrelated whereas a more general assumption (i.e., an unstructured **G** matrix) would allow the random effect of b_{i0} to be correlated with b_{i1}.] Also, the variable B_{ij} is a (0,1) dummy variable that distinguishes the two Blocking Step categories, and Y_{ij} denotes the jth antibody measurement for the ith sample.

  c. The subject-specific matrix form of the model is given by
$$\mathbf{Y}_i = \mathbf{X}_i\boldsymbol{\beta} + \mathbf{Z}_i\mathbf{b}_i + \mathbf{E}_i, i = 1, \ldots, 40$$
where $\mathbf{Y}_i$ (6×1) denotes the collection of 6 antibody measurements from the ith sample, $\mathbf{X}_i$ (6×2) denotes the values corresponding to the constant (always 1) together with values (0 or 1) for the Blocking Step variable, $\mathbf{Z}_i$ (6×2) denotes the values corresponding to the two random effects (for intercept and slope), $\boldsymbol{\beta}$ (2×1) denotes the parameter vector for the collection for β's in this model, $\mathbf{b}_i$ (2×1) is the vector of random effects (b_{i0} and b_{i1}), and $\mathbf{E}_i$ (6×1) denotes the collection of error terms on the ith subject.

  d. $H_0: \beta_1 = 0$

  e. No, the sample means for each Blocking Step category (7.66, 7.67) are almost identical.

5. a. To use SAS's MIXED procedure, the data needs to be reorganized so that there are $24 \times 4 = 96$ lines of data, with four lines per rat. There should be five columns of data per rat, containing the data for the following variables: rat id, IPR, PRS, Drug, and LPR. The variable Factor A can be defined as shown in the programming statements provided.

  c. $LPR_{ij} = \beta_0 + b_{i0} + \beta_1 D_{ij1} + \beta_2 D_{ij2} + \beta_3 D_{ij3} + \beta_4 A_{ij1} + \beta_5 A_{ij2} + \beta_6 A_{ij3} + \beta_7 A_{ij4}$
$+ \beta_8 A_{ij5} + \delta_{11} D_{ij1} \times A_{ij1} + \delta_{21} D_{ij2} \times A_{ij1} + \delta_{31} D_{ij3} \times A_{ij1} + \delta_{12} D_{ij1} \times A_{ij2}$
$+ \delta_{22} D_{ij2} \times A_{ij2} + \delta_{32} D_{ij3} \times A_{ij2} + \delta_{13} D_{ij1} \times A_{ij3} + \delta_{23} D_{ij2} \times A_{ij3} + \delta_{33} D_{ij3}$
$\times A_{ij3} + \delta_{14} D_{ij1} \times A_{ij4} + \delta_{24} D_{ij2} \times A_{ij4} + \delta_{34} D_{ij3} \times A_{ij4} + \delta_{15} D_{ij1} \times A_{ij5}$
$+ \delta_{25} D_{ij2} \times A_{ij5} + \delta_{35} D_{ij3} \times A_{ij5} + E_{ij}, i = 1, \ldots, 24; j = 1, 2, 3, 4,$

where E_{ij} and b_{i0} are each assumed to be normally distributed as $N(0, \sigma^2)$ and $N(0, \sigma_0^2)$, respectively, and b_{i0} and E_{ij} are mutually independent for all i, j. Also, the variables D_{ij1}, D_{ij2}, and D_{ij3} are dummy variables that distinguish the four treatments (i.e., three drugs and a placebo) that are administered to each rat, the variables A_{ij1}, A_{ij2}, A_{ij3}, A_{ij4}, and A_{ij5} are dummy variables that distinguish the six categories of Factor A, and LPR_{ij} denotes the lever press rate for the ith rat on the jth treatment.

d. The null hypotheses, F statistics and P-values can be stated as follows:

Main effect of Drug: H_0: $\beta_1 = \beta_2 = \beta_3 = 0$, $F_{3,54}$(empirical) $= 3486.01$ ($P < .0001$)

Main effect of Factor A: H_0: $\beta_4 = \beta_5 = \beta_6 = \beta_7 = \beta_8 = 0$, $F_{5,18}$(empirical) $= 227.13$ ($P < .0001$)

Interaction of Drug with Factor A: H_0: $\delta_{gh} = 0$ for $g = 1, 2, 3$; $h = 1, 2, 3, 4, 5$ $F_{15,54}$(empirical) $= 279.93$ ($P < .0001$)

Conclusion: The interaction of Factor A with Drug, and both main effects of Factor A and Drug are all highly significant.

7. a. Both Group and Time should be considered as fixed factors because the 2 levels of each of these variables are the only levels of interest.

b. MRUS$_{ij} = \beta_0 + b_{i0} + \beta_1 G_{ij} + \beta_2 T_{ij} + \beta_3 G_{ij} \times T_{ij} + E_{ij}$, where

c. $i = 1, \ldots, 21$ (subject); $j = 1, 2$ (Time).

d. Main effect of Group H_0: $\beta_1 = 0$. However, should first test for Group-by-Time interaction, i.e., H_0: $\beta_3 = 0$, since this test considers whether the change over time differs between the two groups.

g. The approximate Wald test for the random intercept is highly significant ($P = .0050/2 = .0025$) well-below the 5% level, which suggests that a random intercept only model is appropriate.

8. a. Sample means:

	Rater 1					**Rater 2**			
	Monday	Wednesday	Friday			Monday	Wednesday	Friday	
AM	17.4	16.2	13.3	15.6		19.0	19.7	15.6	18.1
PM	18.0	17.0	15.7	16.9		18.5	18.5	16.8	17.9
	17.7	16.6	14.5	16.3		18.8	19.1	16.2	18.0

c. Day-by-Time Interaction: Small amount [AM: (M $= 18.2$, W $= 18.0$, F $= 14.5$), PM: (M $= 18.3$, W $= 17.8$, F $= 16.3$)] since bigger drop from W to F for AM than for PM.

Day-by-Rater Interaction: Small amount [Rater1: (M $= 17.7$, W $= 16.6$, F $= 14.5$), Rater 2: (M $= 18.8$, W $= 19.1$, F $= 16.2$)] since decrease from M to W for Rater 1 but increase from M to W for Rater 2.

Time-by-Rater Interaction: Yes [Rater 1: (AM $= 15.6$, PM $= 16.9$), Rater 2: (AM $= 18.1$, PM $= 17.9$)] since increase from AM to PM for Rater 1 but slight decrease from AM to PM for Rater 1.

d. Day-by-Time-by-Rater Interaction: Small amount since Day-by-Time interaction for Rater 1 is slightly different than Day-by-Time interaction for Rater 2.

e. i. The Rater factor should be considered fixed if the two raters being considered are the only two raters about which the investigators want to draw conclusions.

ii. $SF_{ij} = (\beta_0 + b_{i0}) + (\beta_1 + b_{i1})D_{ij1} + (\beta_2 + b_{i2})D_{ij2} + (\beta_3 + b_{i3})T_{ij} + (\beta_4 + b_{i4})R_{ij}$
$+ (\beta_5 + b_{i5})(D_{ij1} \times T_{ij}) + (\beta_6 + b_{i6})(D_{ij2} \times T_{ij}) + (\beta_7 + b_{i7})(T_{ij} \times R_{ij})$
$+ (\beta_8 + b_{i8})(D_{ij1} \times R_{ij}) + (\beta_9 + b_{i9})(D_{ij2} \times R_{ij}) + \beta_{10}(D_{ij1} \times T_{ij} \times R_{ij})$
$+ \beta_{11}(D_{ij2} \times T_{ij} \times R_{ij}) + E_{ij}$, $i = 1, \ldots, 19$ (subject); $j = 1, 2, \ldots, 12$, and

where D_{ij1} and D_{ij2} denote the values of two dummy variables D_1 and D_2 for the 3 days (Monday, Wednesday, Friday) and T_{ij} denotes the values of a binary variable T for time of day (typically coded as $0/1$), and R_{ij} denotes the values of a binary variable T for rater (typically coded as $0/1$) for the jth observation on the ith subject. It is also assumed that $E_{ij}, b_{i0}, b_{i1}, \ldots, b_{i9}$ are each assumed to be normally distributed as $N(0, \sigma^2), N(0, \sigma_0^2), N(0, \sigma_1^2), \ldots, N(0, \sigma_9^2)$, respectively, noting that the variance components for Subjects-by-Day, i.e., σ_1^2 and σ_2^2, corresponding to b_{i1}, and b_{i2}, are typically assumed to be equal, and the variance components for Subjects-by-Day-by-Time, i.e., σ_5^2 and σ_6^2, corresponding to b_{i5}, and b_{i6}, are typically assumed to be equal, and the variance components for Subjects-by-Day-by-Rater, i.e., σ_8^2 and σ_9^2, corresponding to b_{i8}, and b_{i9}, are typically assumed to be equal.

iii. No, the correlation structure is much more complicated than exchangeable. The correlation structure would be exchangeable only if the model contained a single random effect for Subjects (i.e., random intercept only).

iv. The test for the main effect of Rater is the only test statistic that is significant, i.e., $F_{1,18}$ (Empirical) $= 12.85$ $(P = .0021)$.

v. Yes, the "NOTE" indicates that the iteration process used to obtain ML estimates was faulty, particularly because one or more of the matrices involved in the calculations did not have an inverse (see Appendix B on matrices). Thus, the results for this model are questionable, and one or more alternative models should be considered instead.

vi. $\mathrm{SF}_{ij} = (\beta_0 + b_{i0}) + \beta_1 D_{ij1} + \beta_2 D_{ij2} + \beta_3 T_{ij} + \beta_4 R_{ij} + \beta_5 (D_{ij1} \times T_{ij}) + \beta_6 (D_{ij2} \times T_{ij})$
$+ \beta_7 (T_{ij} \times R_{ij}) + \beta_8 (D_{ij1} \times R_{ij}) + \beta_9 (D_{ij2} \times R_{ij}) + \beta_{10} (D_{ij1} \times T_{ij} \times R_{ij})$
$+ \beta_{11} (D_{ij2} \times T_{ij} \times R_{ij}) + E_{ij},$

where $i = 1, \ldots, 19$ (subject); $j = 1, 2, \ldots, 12$. The two programs differ in that the first program computes model-based standard errors whereas the second program computes empirical standard errors.

f. i. The Rater factor could be considered a random factor if the investigators were interested in drawing conclusions to a larger population of raters instead of just the two raters that were used in the study.

ii. $\mathrm{SF}_{ij} = (\beta_0 + b_{i0}) + (\beta_1 + b_{i1})D_{ij1} + (\beta_2 + b_{i2})D_{ij2} + (\beta_3 + b_{i3})T_{ij} + r_{i1}$
$+ (\beta_4 + b_{i4})(D_{ij1} \times T_{ij}) + (\beta_5 + b_{i5})(D_{ij2} \times T_{ij}) + r_{i2}T_{ij} + r_{i3}D_{ij1} + r_{i4}D_{ij2}$
$+ r_{i5}(D_{ij1} \times T_{ij}) + r_{i6}(D_{ij2} \times T_{ij}) + r_{i01} + r_{i02}T_{ij} + r_{i03}D_{ij1} + r_{i04}D_{ij2} + E_{ij},$

where $E_{ij}, b_{i0}, b_{i1}, b_{i2}, b_{i3}, b_{i4}, b_{i5}$, are each assumed to be normally distributed as $N(0, \sigma^2), N(0, \sigma_0^2), N(0, \sigma_1^2), \ldots, N(0, \sigma_5^2)$ and $r_{i1}, r_{i2}, r_{i3}, r_{i4}, r_{i5}, r_{i6}, r_{i01}, r_{i02}, r_{i03}, r_{i04}$ are assumed to be normally distributed as $N(0, \sigma_{1r}^2), \ldots, N(0, \sigma_{10,r}^2)$, respectively.

iv. $\mathrm{SF}_{ij} = \beta_0 + \beta_1 D_{ij1} + \beta_2 D_{ij2} + \beta_3 T_{ij} + r_{i1} + \beta_4 (D_{ij1} \times T_{ij}) + \beta_5 (D_{ij2} \times T_{ij}) + E_{ij}.$

Chapter 27

1. a. $n \geq 2\left[\dfrac{\sigma(Z_{1-\alpha/2} + Z_{1-\beta})}{\Delta}\right]^2 = 2\left[\dfrac{(25)(1.96 + 0.842)}{25}\right]^2 = 15.7.$ At least 16 patients will be needed in each treatment group.

b. $n \geq \left[\dfrac{(1.96)\sqrt{(2)(0.3)(1 - 0.3)} + (0.842)\sqrt{(0.25)(1 - 0.25) + (0.35)(1 - 0.35)}}{0.10}\right]^2$

$= 328.6.$ At least 329 patients will be needed in each treatment group.

3. a.

Power	Sample Size Required in Each Group
0.6	206
0.7	259
0.8	329
0.9	440

The required sample size increases as the power increases.

b.

Differences in Proportions	Sample Size Required in Each Group
0.01	29281
0.02	7550
0.03	3397
0.04	1933
0.05	1251
0.06	878
0.07	652
0.08	504
0.09	402
0.10	329

The required sample size increases dramatically as the difference in proportions decreases.

5. a. $n \geq 2\left[\dfrac{(3.0)(Z_{0.975} + Z_{0.8})}{1.5}\right]^2 = 2\left[\dfrac{(3.0)(1.96 + 0.842)}{1.5}\right]^2 = 62.8.$

At least 63 smokers and 63 non-smokers are required.

b. Using the adjustment suggested by Hsieh et al., the total sample size for the multiple regression would be $126/(1 - 0.10)$ or 140 (i.e., 70 smokers and 70 non-smokers).

c.

Squared Multiple Correlation	Total Sample Size Required
0.1	140
0.2	158
0.3	180
0.4	210
0.5	252
0.6	315
0.7	420
0.8	630
0.9	1260

The required sample size increases as the squared multiple correlation between the main predictor and the other predictors increases.

d. $n_s \geq \left[\dfrac{Z_{1-\alpha/2} + Z_{1-\beta}}{C(\rho)}\right]^2 + 3 = \left[\dfrac{1.96 + 0.842}{\frac{1}{2}\ln\left(\dfrac{1 + 0.4}{1 - 0.4}\right)}\right]^2 + 3 = 46.7.$

At least 47 subjects should be sampled.

e. Using the adjustment suggested by Hsieh et al, the total sample size for the multiple regression would be $47/(1 - 0.10)$ or 53.

Index